The Elements and Their Atomic Weights

Element	*Symbol*	*Atomic Number*	*Atomic Weight*[a] *(Relative to* ^{12}C *= 12.00000)*
Actinium	**Ac**	89	(227)
Aluminum	**Al**	13	26.98154
Americium	**Am**	95	(243)
Antimony	**Sb**	51	121.75
Argon	**Ar**	18	39.948
Arsenic	**As**	33	74.9216
Astatine	**At**	85	(210)
Barium	**Ba**	56	137.33
Berkelium	**Bk**	97	(247)
Beryllium	**Be**	4	9.01218
Bismuth	**Bi**	83	208.9806
Boron	**B**	5	10.811
Bromine	**Br**	35	79.904
Cadmium	**Cd**	48	112.40
Calcium	**Ca**	20	40.08
Californium	**Cf**	98	(251)
Carbon	**C**	6	12.011
Cerium	**Ce**	58	140.12
Cesium	**Cs**	55	132.9055
Chlorine	**Cl**	17	35.453
Chromium	**Cr**	24	51.996
Cobalt	**Co**	27	58.9332
Copper	**Cu**	29	63.546
Curium	**Cm**	96	(247)
Dysprosium	**Dy**	66	162.50
Einsteinium	**Es**	99	(254)
Erbium	**Er**	68	167.26
Europium	**Eu**	63	151.96
Fermium	**Fm**	100	(253)
Fluorine	**F**	9	18.9984
Francium	**Fr**	87	(223)
Gadolinium	**Gd**	64	157.25
Gallium	**Ga**	31	69.72
Germanium	**Ge**	32	72.59
Gold	**Au**	79	196.9665
Hafnium	**Hf**	72	178.49
Hahnium	**Ha**	105	(260)
Helium	**He**	2	4.00260
Holmium	**Ho**	67	164.9303
Hydrogen	**H**	1	1.0080
Indium	**In**	49	114.82
Iodine	**I**	53	126.9045
Iridium	**Ir**	77	192.22
Iron	**Fe**	26	55.847
Krypton	**Kr**	36	83.80
Lanthanum	**La**	57	138.9055
Lawrencium	**Lr**	103	(257)
Lead	**Pb**	82	207.2
Lithium	**Li**	3	6.941
Lutetium	**Lu**	71	174.97
Magnesium	**Mg**	12	24.305
Manganese	**Mn**	25	54.9380
Mendelevium	**Md**	101	(256)
Mercury	**Hg**	80	200.59
Molybdenum	**Mo**	42	95.94
Neodymium	**Nd**	60	144.24
Neon	**Ne**	10	20.179
Neptunium	**Np**	93	237.0482
Nickel	**Ni**	28	58.71
Niobium	**Nb**	41	92.9064
Nitrogen	**N**	7	14.0067
Nobelium	**No**	102	(254)
Osmium	**Os**	76	190.2
Oxygen	**O**	8	15.9994
Palladium	**Pd**	46	106.4
Phosphorus	**P**	15	30.9738
Platinum	**Pt**	78	195.09
Plutonium	**Pu**	94	(242)
Polonium	**Po**	84	(210)
Potassium	**K**	19	39.102
Praseodymium	**Pr**	59	140.9077
Promethium	**Pm**	61	(147)
Protactinium	**Pa**	91	231.0359
Radium	**Ra**	88	226.0254
Radon	**Rn**	86	(222)
Rhenium	**Re**	75	186.2
Rhodium	**Rh**	45	102.9055
Rubidium	**Rb**	37	85.467
Ruthenium	**Ru**	44	101.07
Rutherfordium	**Rf**	104	(261)
Samarium	**Sm**	62	150.35
Scandium	**Sc**	21	44.9559
Selenium	**Se**	34	78.96
Silicon	**Si**	14	28.086
Silver	**Ag**	47	107.868
Sodium	**Na**	11	22.9898
Strontium	**Sr**	38	87.62
Sulfur	**S**	16	32.06
Tantalum	**Ta**	73	180.9479
Technetium	**Tc**	43	98.9062
Tellurium	**Te**	52	127.60
Terbium	**Tb**	65	158.9254
Thallium	**Tl**	81	204.37
Thorium	**Th**	90	232.0381
Thulium	**Tm**	69	168.9342
Tin	**Sn**	50	118.69
Titanium	**Ti**	22	47.90
Tungsten	**W**	74	183.85
Uranium	**U**	92	238.029
Vanadium	**V**	23	50.941
Xenon	**Xe**	54	131.30
Ytterbium	**Yb**	70	173.04
Yttrium	**Y**	39	88.9059
Zinc	**Zn**	30	65.37
Zirconium	**Zr**	40	91.22

[a]Atomic weight values in parentheses are the mass numbers of the most stable radioactive isotope.

Chemistry

Chemistry

Gary E. Maciel
COLORADO STATE UNIVERSITY

Daniel D. Traficante
MASSACHUSETTS INSTITUTE OF TECHNOLOGY

David Lavallee
COLORADO STATE UNIVERSITY

D. C. HEATH AND COMPANY

LEXINGTON, MASSACHUSETTS / TORONTO

Published simultaneously in Canada.

Printed in the United States of America.

International Standard Book Number: 0–669–84830–1

Library of Congress Catalog Card Number: 77–088504

Preface

This textbook has been written for students enrolled in general chemistry courses. *Chemistry* begins at an elementary level but advances to a fair degree of sophistication by the second half of the book. A background of high school chemistry and simple algebra is assumed, although the former is not really necessary for students with some aptitude for science and mathematics. Although the material covered is fairly standard for first-year college chemistry courses, this book is distinctive on two counts: (1) the unusual emphasis on certain specific points of view and (2) the order of presentation of topics.

1. One concept that occurs early in the book and is amplified periodically is the ever-present competition between ordering and randomizing influences in matter. Special emphasis is also given to the concept of a functional group—a concept that is not usually treated in depth before the sophomore course in organic chemistry. We find it extremely useful in teaching some of the descriptive chemistry and basic spectroscopy, and therefore we have introduced it in Chapter 3.

2. Chemical reactions, stoichiometry, and chemical equilibrium are also treated in considerable depth at a much earlier stage than one usually finds in freshman chemistry books; this is especially true of Chapters 4 and 5 on chemical equilibrium. Topics such as historical developments in atomic physics, kinetic theory, atomic structure, and molecular orbital theory often precede chemical equilibrium in first-year chemistry texts; but we see no loss in postponing detailed discussions of such material until later in the text.

The early treatment of equilibrium allows for repetition and reinforcement of a topic that presents difficulties for many students. Furthermore, the concept of equilibrium follows the material on chemical reactions and stoichiometry quite naturally. The mathematical exercises in chemical equilibrium often provide the first real challenge for students with a high school background. We have found these exercises invaluable in maintaining student interest. Early presentation of chemical equilibrium also emphasizes fundamental topics that are important in the laboratory and provides maximum opportunity for continual review as more sophisticated aspects of chemical equilibrium are presented.

To prepare the student for chemical equilibrium in Chapter 4, we introduce many basic ideas in Chapters 2 and 3—ideas that are amplified and reinforced later in separate chapters. These ideas include the concepts of atoms, molecules, ions, and reactions; simple ideas of thermochemistry, reaction rates, order and randomizing motions; and even the gas laws. The gas laws are explored briefly in Chapters 2 and

3, but are treated again in Chapter 15, which also contains material on the kinetic theory of gases and on deviations from ideal gas behavior.

The book is organized so that, after completing Chapters 1 through 5, the student has a working knowledge of the concepts of atoms, molecules, and ions (while not yet understanding the electronic basis of their existences), the general properties of solids, liquids, and gases (while not yet knowing the various physico-chemical aspects of these states, covered in later chapters), as well as a good basic understanding of reactions, stoichiometry, chemical equilibrium, acids and bases, pH, buffers, solubility products, and so on. This opens up new opportunities for meaningful laboratory experiments based upon these concepts. The companion laboratory manual by M. Bacon and C. Josefson takes advantage of this opportunity. Most other laboratory programs would also benefit from this type of text coverage.

Since the chapters of this text are largely self-sufficient, a more conventional order of presentation can be readily adopted.

Further advantageous features of this text are the large number of exercises and worked-out examples. In addition, a set of study problems is included immediately after the worked-out examples to provide the student with a self-testing device. If the student has fully understood the textual material and examples immediately preceding a study problem, he or she should be able to work out that problem. Answers to all study problems are given at the end of the chapters.

Although the descriptive chemistry is integrated throughout the text, much of it is concentrated in two chapters dealing with molecular structure (Chapters 7 and 8), in two chapters concerned with periodic trends (Chapters 9 and 17), and in a chapter on transition metal compounds (Chapter 18). This last chapter also provides a format for summarizing, reviewing, and integrating many of the more physical concepts covered in earlier chapters. A considerable portion of descriptive material is also included in Spotlights, which appear parallel to appropriate textual material and serve as a vehicle for selected historical notes and pertinent biographies. The intention is to make the book a "friendlier" source of knowledge for students and to provide information of a more practical nature.

Since the descriptive chemistry has been interspersed, there are no separate chapters devoted to organic chemistry, nuclear chemistry, or biochemistry. However, considerable coverage of these particular topics occurs in the sections on descriptive chemistry. We believe that this approach is pedagogically preferable and it increases the likelihood that these topics will actually be covered in the course. Separate chapters on organic chemistry, biochemistry, nuclear chemistry, the environment, and so on, are often left completely out of a one-year course due to time limitations.

A final feature of the book is that much of the more sophisticated material toward the ends of chapters is set off as "optional." This material can be used at the discretion of the instructor, depending upon time constraints and personal interests. An example of the optional material is a novel statistical approach for understanding chemical equilibrium presented in Chapter 11.

In addition to the Laboratory Manual, the textbook is supplemented by a Study Guide, which covers the same material as the text but with different points of view, examples, and exercises.

In a project of this magnitude, involving three authors, there are many people we should thank. There are many students, colleagues, and other people who have made useful suggestions and who have proofread or checked portions of the manuscript; special mention is due Professors Gary Schnuelle and Karl Lindfors, Dr. Carolyn Lavallee, and Ms. Mary Severson. Of course, thanks are due to typists, especially Maxine Maciel and Doris Traficante. And finally we wish to thank the many first-year chemistry students who asked the "hows" and "whys" that led us to consider concepts in greater depth than is required for more sophisticated questions in more advanced courses.

GARY E. MACIEL
DANIEL D. TRAFICANTE
DAVID LAVALLEE

Contents

3

Atoms, Molecules, and Reactions

4

Chemical Equilibrium

5

Acid-Base Equilibria in Aqueous Solutions

6
Atomic Structure

7
Bonding and Molecular Structure

8
Additional Aspects of Covalent Bonding and Spectroscopy

9
Descriptive Chemistry: Trends and Patterns

10
Energy Changes in Chemistry

11
The Driving Force of Chemical Equilibrium

12
Electrochemistry

13
Kinetics and Mechanisms

14
Solids

15

Gases

16

The Liquid State

17

Descriptive Chemistry. Vertical Trends for the Heavier Main Group Elements

18

Coordination Chemistry of the Transition Elements

To the Student

There are many reasons for enrolling in a first-year college chemistry course. It probably is a required part of the curriculum you have chosen, for chemistry is at the very core of many technical fields. Premedical, preveterinary, and prenursing programs require a substantial amount of course work in chemistry, as do nearly all major programs in the basic sciences (physics, biology, and chemistry) and the applied sciences (such as agriculture and engineering). First-year chemistry, therefore, is not just another academic hurdle that must be cleared for advancement in one's major. Rather, it can provide a foundation of essential concepts and information for many fields of endeavor.

Another reason for taking a first-year college chemistry course is that many important social, economic, and political issues of our time are related to scientific matters. The current, and sometimes conflicting, issues of energy and environment are obvious examples. Satisfactory resolution of such issues will depend to an important degree on the application of chemical principles. Thus, to be properly informed, today's citizen should understand the fundamentals of chemistry.

Knowledge of basic science, in fact, will soon be as essential to enlightened living as knowledge of arithmetic has been for centuries. In order to make intelligent decisions on such diverse matters as clothing fabrics, insecticides, fertilizers, water treatment and pollution, household fire alarms, transportation, and energy, each person should be able to judge opposing views on the basis of an educational foundation in science.

This book will help you learn the important basic concepts of chemistry. If you are conscientious, you can achieve a knowledge of chemistry surpassing that of the brightest scientists of former centuries. You can learn not only basic information about atoms and molecules but also the more subtle structural concepts that determine the architecture of substances as diverse as ice and hemoglobin. You can learn how to correlate chemical behavior in terms of the periodic table of the elements and come to understand the stabilities of substances in terms of the concepts of thermodynamics and kinetics. You can learn why some substances are colored while others are not, how plastics are made, how batteries and semiconductors function. You can explore the nature of forces that bind matter together into the substances with which we are familiar, and you can learn how to predict what types of molecules might exist and what types can't. You can learn the basic strategies that determine how substances are analyzed and how they are synthesized. You will develop all of these abilities through mastery of a relatively small number of chemical principles.

More than one reading will usually be necessary to learn the concepts in a chapter of this text. We strongly urge you to read the chapters with a questioning, even challenging, attitude. Continually ask yourself *why* a certain statement is made or *what* a specific equation really means. Try not to leave a paragraph without understanding its essence and the principle involved. Do not pass one of the worked-out examples without following it closely and being convinced either that you understand it or that you know what specific questions you should ask your instructor about the solution. We also urge you to work through all the study problems. These problems are usually placed after a worked-out example to enable you to test yourself as you progress through a chapter. Answers are provided at the end of each chapter.

You may test your learning further by doing the exercises at the end of each chapter. We encourage you to work out as many exercises as you can; answers to about half of them are provided at the end of each chapter. Do not underestimate the importance of being able to work out numerical problems. It is often possible to think that you understand textual material when, in fact, you are only superficially familiar with it; numerical problems provide an excellent test for determining whether you truly have a working knowledge of important concepts. Chapter summaries and a list of student checkpoints, which appear at the end of each cahpter, will also help you evaluate your comprehension of the material presented in that chapter. Do not overlook the figures. These are designed either to summarize relationships that are worth remembering or to provide some aid for understanding the essence of a concept. You should develop the ability to examine a plot (graph) and comprehend the relationship it is meant to convey.

Some of you may wish to supplement your study of this text with a related source that covers the same material, but with different wording, points of view, examples or exercises. This kind of supplement can be very helpful, and for this purpose we highly recommend the Study Guide by Professor Karl Lindfors, written to accompany this text.

Most of you are enrolled in a course that includes a laboratory section. The laboratory experience should strengthen your grasp of the concepts you encounter in this textbook. We urge you to continually relate your laboratory experience to these concepts. This will not only help you enjoy the laboratory section of the course, but it will also aid your understanding of this text and your lectures.

Finally, a few words on your background and preparation. The mathematics in the text is arithmetic and simple algebra. A brief review of the algebraic principles and methods used is given in Appendix E. While we assume a high school chemistry background, the textual material starts from "ground zero."

Chemistry

1

The Central Role of Chemistry in Science

1.1 Science, Chemistry, and Man

Chemistry is unquestionably one of the most important subdivisions of science. The concepts of chemistry bridge physics, geology, biology, and medicine. Applications of chemistry range throughout most manufacturing activities and have repeatedly led to important advances in medicine. When the modern world is increasingly concerned with such technical problems as the energy crisis, pollution, the relative worth of nuclear power generation, and supersonic transports—their effect on the ozone layer and the relation to skin cancer—it is practically impossible to be a well-informed citizen and voter without some knowledge of chemistry.

Science is often blamed for many of the ills of the modern world—the weapons with potentially huge destructive powers and the polluted air, oceans, lakes, and rivers. Likewise science is also often credited with having brought civilized man to a standard of living that even kings did not enjoy in earlier times. Science really deserves neither the blame nor the credit. Science has developed because of human ingenuity and perseverance and is applied only as people choose to apply it. Science and the technology based on it have vastly increased our capacity for accomplishing things, good or bad. But human beings are responsible for the decisions about how science is applied. As more people are educated in science, a larger fraction of the population will be able to contribute enlightened voices on important matters relating to how science will be used in the future.

1.2 The Scientific Method

A suitable prelude to the study of chemistry is a consideration of the nature and scope of the subject. An interesting parallel is the relation between chemistry and natural philosophy, an important field of intellectual activity that later became known as science. We can view **science** as the systematic discovery and organization of the correlated bodies of knowledge, based on observation, experimentation, and logical conclusions. **Chemistry** is one branch of science; it is the study of substances, their compositions, properties, and interactions, and their relations with external forces.

It is worth emphasizing that chemistry, like science in general, changes and expands. It is not merely a fixed body of knowledge that is to be studied, learned, and religiously respected in a formal way, but an evolving activity.

To ensure a firm foundation for this study of chemistry, let us attach concrete meanings to some of the key words in the above paragraphs. Science, as we know it today, is a relatively recent activity in human history. It owes its origin substantially to the emphasis that Sir Francis Bacon put on inductive reasoning and on experimental observations for studying natural phenomena. The importance of such procedure was first realized in the seventeenth century, when it became apparent that the only reliable criteria for judging theories of natural phenomena were experimental results.

Science can be thought of as the development and the results of a type of thinking that is commonly called the scientific method. This method can be described in discrete steps, or rules. These are arbitrary, and one can construct several equally valid sets of rules. The eminent scientist Percy Bridgeman has defined the **scientific method** as simply "doing one's utmost with his mind, no holds barred." It is Bridgeman's expression and not an arbitrary set of guidelines that really characterizes a

Spotlight

Sir Francis Bacon (1561–1626) was an important influence in England in the late 1500s and early 1600s. An essayist and philosopher, he also held important state offices under King James I and was a member of Parliament from 1584 to 1614. In spite of attention required by his legal and political activities, Bacon's main interest was philosophy, especially as an approach to scientific study. Finding the prevailing Aristotelian deduction unsatisfactory, he devoted himself to developing a methodology of scientific activity based on empirical data. Although he did not complete the detailed description of the grand method he had outlined, he provided a great stimulus to scientific inquiry carried out in ways now known as the scientific method. (Francis Bacon is sometimes confused with an earlier English scholar, and friar, Roger Bacon (1214–1294), who also argued eloquently for an experimental basis in scientific study, although he was not a great experimental scientist.)

Sir Francis Bacon's insistence that scientific inquiry should be founded on observable facts and not just by theoretical deduction made an important contribution to the evolution of scientific thought. He did not make any significant advances in scientific fact or theory himself. Indeed, his only documented scientific experiment may have led to his demise. He died from an illness that apparently followed a chill suffered while he was stuffing a chicken with snow to see whether cold would retard decay.

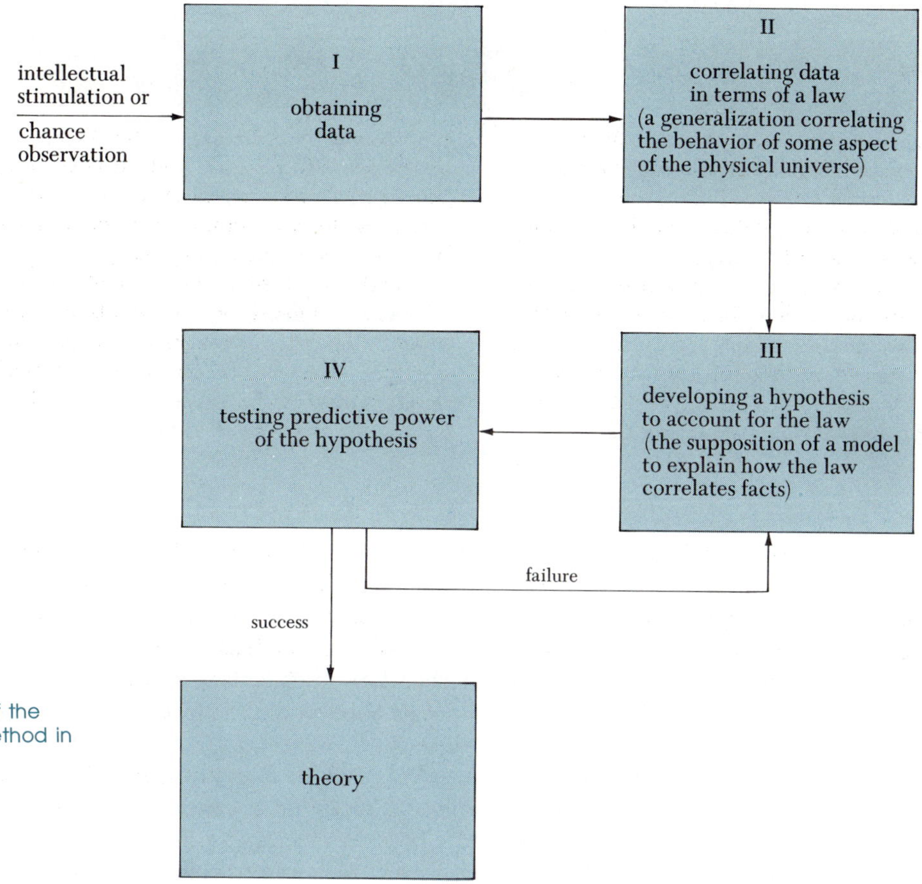
intellectual stimulation or
chance observation
I
obtaining data
II
correlating data in terms of a law (a generalization correlating the behavior of some aspect of the physical universe)
III
developing a hypothesis to account for the law (the supposition of a model to explain how the law correlates facts)
IV
testing predictive power of the hypothesis
failure
success
theory

Figure 1.1
One view of the scientific method in action.

scientist and the scientific method of thought. Nevertheless, it is worth while to review the usual description of the scientific method for specific guidelines.

As shown in Fig. 1.1, the scientific method can be pictured as a four-step process. While the basic concepts on which the scientific method is founded are clear, there is considerable diversity in the formulation of the method. For instance, what we refer to here as a *theory* is sometimes called a *hypothesis,* and the term *model* is sometimes used in place of either of these words.

The impetus for initiating a scientific inquiry can take numerous forms. Often a scientist makes an unexpected observation in an experiment designed for some presumably unrelated purpose. Or the scientist may read about a scientific development, experimental or conceptual, that provides some new insight on a subject of special interest or that appears to be inconsistent with some view of nature or prevailing theory. In any case, the scientist feels that a significant gap in knowledge and understanding of nature exists and accepts the challenge of filling that gap. As the first step, the scientist carries out an appropriate set of experiments and obtains a set of facts, called data. The second step calls for examining these data and finding relations that permit drafting a generalization on the basis of inductive reasoning. Any generalization that relates or correlates the data can be thought of as an empirical relation, or a **law.** The law may apply strictly only to phenomena of the specific kind that led to its development. The third step is initiated as the scientist reflects

Spotlight

A hypothetical example of how the scientific method might work can be imagined for the study of gravity. According to legend, Newton's attention was first drawn to the phenomenon when an apple fell on his head. A scientist might then design a set of experiments (for example, dropping various objects different distances) to explore such a phenomenon. From examining the data from these experiments, the scientist might then formulate a mathematical relation, a law of gravity, a law that is found to predict correctly not only the behavior of a falling apple, but the motions of planets in the solar system also. Faced with a successful law, the scientist then seeks for insight into how or why nature manifests this law. The scientist is probing for a hypothesis that will account for the law in the most fundamental terms. If such a hypothesis survives the tests to which it is subjected to explain related phenomena, or other fundamental forces, it achieves the status of a successful theory. In the case of gravity, Einstein's theory of relativity is a brilliant step towards a satisfactory theory, although its success has not yet been tested fully.

on the law, probably in the light of others related to it. On the basis of experience, intuition, or unpredictable "flashes of insight," the scientist takes the important step of imagining a model that can account for the law. There becomes established in the scientist's mind a definite picture of one feature of nature's design that would lead to phenomena of the kind the law describes. Such a view, or model, is referred to as a **hypothesis.** By the very nature of its conception, the hypothesis is consistent with the specific law that gave birth to it.

The real test of the model's validity lies in its ability to predict behavior in related phenomena that were not involved directly in its conception. It is the habit and duty of the scientist to subject the hypothesis scrupulously to exacting sets of experiments that will test its ability to predict such related behavior. If the deductions do not always lead to correct predictions of results in the experimental tests, then the hypothesis is modified, or a new one is conceived that will allow all predictions to be successful. Such success constitutes entry into the fourth step, the formulation of a theory. A **theory** is simply a hypothesis that has been found to make correct predictions time and time again and has graduated to the higher level. The validity of the theory is as far-reaching as the range of successfully predicted experimental results. Some theories are general; they can be applied to any case in nature. Others may be valid only over a limited range of cases, a limitation that the scientist must recognize.

Actual scientific experience may not fit neatly into the specific route outlined above, and some scientists use different labels to categorize the steps. The specific labels are not very important, but the overall strategy of the scientific method is. In some cases "the scientist" may be an entire research team, or possibly a succession of scientists covering a span of years, decades, or even centuries. In such cases each individual contributes a particular step in the overall progress.

The day-to-day conduct of science is like most other human endeavors in that the characteristics or personality of an individual scientist may well dictate the detailed style for pursuing the scientific method. Some scientists may be adventuresome or "flashy" in their style; some may be conservative and require far more experimental data before constructing a hypothesis. In all cases the ultimate test of the hypothesis must be its ability to account for experimental results, whether it has resulted from a "giant leap forward" or from a routine, day-by-day evolution.

1.3 Specialization in Science

A common view of science is that it has evolved from what was previously called natural philosophy. As this evolution has progressed, science has developed along more or less scientific routes in certain specialized areas. Accordingly, certain characterizing terms have grown up with science; for example, science is frequently divided arbitrarily into the categories *life science, physical science,* and *mathematics.* Mathematics is concerned with the language and formulation of logical and computational operations. Some observers view this as distinct from science, which they consider to be concerned primarily with understanding natural phenomena. In any case, no one disputes the effect mathematics has had on life science and physical science, or the importance mathematics will have in their continued development. Life science is specifically concerned with the nature of life itself, whereas physical science deals with all aspects of natural phenomena not covered by the first category. Of course, these areas are not mutually exclusive.

Physical science has itself tended to evolve into more specialized areas, such as geology, astronomy, physics, and chemistry. Geology is concerned with understanding the earth, its formation, and its makeup. Astronomy is similarly concerned with the physical universe beyond the earth. Physics has been concerned with the underlying physical laws that govern the behavior of all known natural phenomena. And chemistry has traditionally been subject to further classification. One such division recognizes the following: **physical chemistry,** the techniques of physics applied to understanding and correlating the properties of substances; **organic chemistry,** the study of compounds in which the element carbon is dominant; **inorganic chemistry,** the study of all compounds outside the realm of organic chemistry; **biochemistry,** the chemistry of life systems; and **analytical chemistry,** concerned with methods of determining the composition and identity of materials. While it is useful for a student to know these terms, the intrinsic value of such distinctions has been diminishing in recent years. Historically the trend in science has been towards greater specialization; however, important modern advances in science have been made by researchers whose work precludes classification into one specialized area. Some areas of chemistry have become so interwoven that new names have arisen to accommodate the trend. Examples are **physical organic chemistry** (physical and organic), **biophysical chemistry** (biochemistry and physical chemistry), and **organometallic chemistry** (organic and the metallic aspects of inorganic).

The interdisciplinary nature of science, and the importance of applying a unifying set of principles to the entire body of science, has become increasingly apparent in recent years. An examination of the research subjects of recent Nobel laureates in physics, chemistry, and medicine and physiology confirms this view. Table 1.1 summarizes some recent Nobel awards in the sciences.

The student who is starting to study chemistry should be aware of the central position of chemistry in science. Chemistry is involved in all other branches of science. To fully appreciate this, let us examine Fig. 1.2. The figure is based on the assumption that fundamental science can logically be divided into four areas—physics, chemistry, geology, and biology. The diagram also shows explicitly five "bridging fields"—biochemistry, geophysics, geochemistry, chemical physics, and paleontology—in which two of the four main divisions shown as circles are roughly of equal importance. The diagram also indicates definite ties between these bridging fields and chemistry. Also shown in Fig. 1.2 are a few of the applied sciences that are closely related to biology, physics, and geology. These are medicine, agriculture, engineering, electronics, mining, and oceanography. The central position of chemistry in this diagram parallels the position that chemistry occupies in science.

TABLE 1.1 Some Recent Nobel Prize Winners in Chemistry, Physics, and Medicine and Physiology

Chemistry

1955	V. du Vigneaud. Studies of sulfur compounds and synthesis of a polypeptide hormone[a]
1956	C. N. Hinshelwood and N. N. Semenov. Investigations into mechanisms of chemical reactions[b]
1957	A. R. Todd. Studies on nucleotides and nucleotide enzymes[a]
1958	F. Sanger. Studies of protein structure, especially of insulin[a]
1959	J. Heyrovsky. Discoveries of methods of electrochemical analysis
1960	W. F. Libby. Development of techniques of radiocarbon dating[b]
1961	M. Calvin. Studies on photosynthesis[a]
1962	J. C. Kendrew and M. F. Perutz. Studies on globular proteins[a]
1963	K. Ziegler and G. Natta. Developments in synthetic polymer chemistry
1964	D. C. Hodgkin. Structure determination on medicinally important compounds[a]
1965	R. B. Woodward. Synthesis of complicated organic compounds, including chlorophyll[a]
1966	R. S. Mulliken. Developments in the theory of the electronic structure of molecules[b,c]
1967	M. Eigen, R. G. W. Norrish and G. Porter. Investigations of very fast chemical reactions[b]
1968	L. Onsager. Developments in irreversible thermodynamics[c]
1969	O. Hassel and D. H. R. Barton. Studies of molecular conformations
1970	L. F. Leloir. Discovery of sugar nucleotides[a,b]
1971	G. Herzberg. Studies in electronic structure[b,c] and geometry of molecules
1972	C. B. Anfinsen, S. Moore, and W. H. Stein. Research relating to chemical structure and biological reactions of the protein ribonuclease[a,b]
1973	E. O. Fischer and G. Wilkinson. Research on the merging of organic and metallic compounds
1974	P. J. Flory. Study of polymers (e.g., plastics)[b,c,]
1975	J. W. Cornforth and V. Prelog. Studies in stereochemistry (geometric arrangements of atoms in molecules)[b]
1976	W. Lipscomb. Studies on the structures of boron compounds

Physics

1952	F. Bloch and E. M. Purcell. Discovery of nuclear magnetic resonance[b,d]
1953	F. Zernike. Invention of the phase-contrast microscope[b]
1954	M. Born and W. Bothe. Developments in quantum mechanics[d]
1955	W. E. Lamb, Jr., and P. Kusch. Developments in magnetic properties of electrons and their relation to spectra[d]
1961	R. Hofstadter and R. L. Mossbauer. Studies of the nucleus, including resonance absorption of gamma radiation in the solid state[d]
1964	C. H. Townes, N. G. Basov, and A. M. Prochorov. Discovery of the maser-laser principle[d]
1970	L. E. F. Néel and H. Allvén. Discovery and work in ferromagnetism and antiferromagnetism with applications in solid state physics[d]
1972	J. Bardeen, L. N. Cooper and J. R. Schrieffer. Development of superconductivity theory of certain metals at very low temperatures[d]

Medicine and Physiology

1953	H. A. Krebs and F. A. Lipmann. Discoveries of citric acid cycle and coenzyme A
1955	A. H. T. Theorell. Studies of oxidation enzymes

[a]Substances of biological importance. [b]Applications in biological studies. [c]Applications in physics. [d]Applications in chemistry.

TABLE 1.1 **(continued)**

Medicine and Physiology

1957	D. Bovet. Investigation of the effects of synthetic compounds on body substances
1958	G. W. Beadle, E. L. Tatum and J. Lederberg. Studies on the chemical basis of genetics
1962	J. D. Watson, M. H. F. Wilkins, and F. H. C. Crick. Discovery of the molecular structure of DNA
1963	A. L. Hodgkin, A. F. Huxley, and J. C. Eccles. Studies of nerve impulses, including electrochemical aspects
1967	H. K. Hartline, G. Wald, and R. Granit. Discoveries of primary visual processes in the eye, including photochemical aspects
1968	R. W. Holley, H. G. Khorana, and M. Nirenberg. Studies on the molecular basis of genetics
1969	A. D. Hershey, M. Delbrück, and S. Luria. Studies of viral genetics, including molecular aspects of the genetic code
1970	J. Axelrod, U. von Euler, and B. Katz. Basic research in the chemistry of nerve transmission
1971	E. W. Sutherland, Jr. Discovery concerning the mechanisms of the action of hormones
1972	G. M. Edelman and R. Porter. Determination of an antibody's exact chemical structure

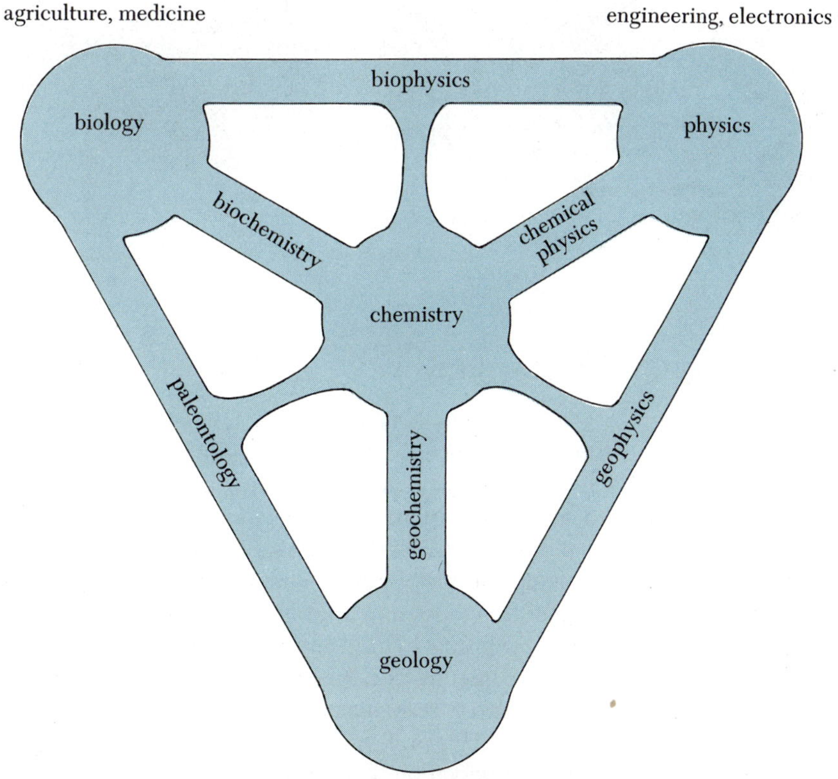

Figure 1.2 The central role of chemistry in science.

The Alchemist, David Teniers the Younger. John G. Johnson Collection, Philadelphia.

1.4 The Evolution of Chemistry

During the whole technical development of civilization, chemistry has been closely involved. Since it is so centrally important in many fields today—such as medicine, biology, geology, industrial technology, and agriculture—many students in diverse academic disciplines must study it. Even during ancient times, when chemistry was not recognizable as such, the subject was central because of the importance of materials such as metals, glass, and medicinal remedies. Of necessity, most of the techniques developed and most of the materials produced during this time arose largely from an approach based on trial and error. From about the fifteenth century through the seventeenth, the accumulated mass of information and techniques began to evolve into a recognizable human activity, and even into a profession known as alchemy. This profession was characterized by a shroud of secrecy, a reluctance to record observations carefully (to avoid use by others), and an unfortunate measure of dishonesty. It was concerned primarily with producing valuable materials from less valuable ones. While it was only partially successful in achieving that goal, it had the valuable purpose of giving birth to useful experimental procedures and apparatus. Hence, the stage was set for a period that is sometimes recognized as the "medical-chemical period," covering the seventeenth century and part of the eighteenth. During this time of greater openness and honesty, experiments and theories were being joined largely for the first time, a trend nurtured by the influence of Bacon. The primary obvious result was the development of useful medicines and remedies, but equally important was the rise of scientific thought.

The time from 1702 to 1777 is frequently referred to as the phlogiston period. During that time, what appeared to be a reasonable theory of combustion was proposed. It was based on a set of *qualitative* observations (that is, they had no numerical characterization) on the nature of burning; experiments seemed to suggest the existence of a substance known as phlogiston. This substance was supposed to escape when a material burned, leaving a lighter material. Because combusted materials frequently are fluffier and appear lighter than the materials from which they were derived, it was commonly assumed that they were lighter; actually, they were only less dense. For seventy-five years this assumption was not subjected to any critical experimentation, and the phlogiston theory was used to correlate a large body of chemical information. As early as 1630, the scientist Jean Rey in France had shown that tin gains weight when burned. The influence of Bacon, however, was not felt early enough to rescue such observations from the fate of being either ignored or rationalized away during the phlogiston period by those who were anxious for simple and neat solutions. Such "solutions" seemed to work. Not until about 1777 was the scientific world presented with the quantitative researches of Antoine Lavoisier on weight relations in combustion. This work immediately led to the downfall of the phlogiston theory and its replacement by a more modern theory of combustion, which is largely consistent with the present view.

Spotlight

A very plausible case could be made for the chemical behavior of charcoal and metals by the phlogiston theory. The observations and "plausible" conclusions can be summarized symbolically as follows, where the symbols "phl" and "me" represent phlogiston and a metal:

1. When charcoal burns, it disappears completely and must therefore be pure phlogiston.

$$\text{phl (solid)} \xrightarrow{\text{burning}} \text{phl (in some invisible form)}$$

2. When metals (me) burn, so-called calxes are obtained, which appear to be lighter. (It was assumed that the burning metals lost phlogiston.)

$$\text{me (phl)} \xrightarrow{\text{burning}} \text{me (calx)} + \text{phl (in some invisible form)}$$

3. When calxes are heated with charcoal, metals are obtained. (This was assumed to be a combination of metal with phlogiston.)

$$\text{me (calx)} + \text{phl (some invisible form)} \longrightarrow \text{me (phl)}$$

This picture was self-consistent so long as critical experiments, with weighing, were avoided.

Spotlight

Science Service

Antoine Laurent Lavoisier (1743–1794) was born in Paris to a family that had considerable commercial wealth. After studying both law and the sciences, and after being admitted to the French bar in 1765, Lavoisier chose a career in chemistry, which led him into a wide range of basic and applied studies. His definitive work on weight relations in combustion led to the downfall of the phlogiston theory and a reclassification of the elements. His many successes in this and other fields (for instance, agricultural and biological chemistry) also led ultimately to his own downfall. The fame generated by his scientific contributions brought recognition from the "establishment" and led to some affiliations with agencies that became targets of the Revolution. Lavoisier, who had been generous of his time, talents, and money in public service, was arrested, tried, and guillotined. The trial judge refused his pleas for a few days' respite for summarizing the results of unreported experiments, remarking, "The Republic has no need of Genius."

The study of the subject from 1777 to the present can properly be considered modern chemistry. Essentially, the quantitative aspects introduced during this time, through weight relations, distinguish this period from earlier ones.

1.5 Quantitative Science

The Concept of Units

Chemistry made dramatic progress after the work of Lavoisier. It is clear that an important change took place in the growing science during Lavoisier's time. This change was the introduction of quantitative observations, or numerical measurements, as the criteria for determining changes in chemical phenomena. The continuing transformation of the character of chemistry from the descriptive and qualitative to the more quantitative was gradual and evolutionary. A cornerstone of this quantitative development was the systematic employment of specific measures of quantities associated with particular observations and phenomena. These specific quantities have become known as **units,** and a self-consistent set of such quantities is known as a system of units, or a set of units. Units are indispensable for communicating quantitative information effectively and efficiently. For example, if you want to buy some fruit, it is uninformative to tell a merchant that you want to buy "seven," unless you indicate to what the "seven" refers. It might be seven bananas, seven apples, seven dozen bananas, or seven pounds of grapes. The *seven* has no meaning for a transaction unless it is associated with a particular unit. In this example the unit could be one banana, one apple, one dozen bananas, or one pound of grapes. It is similarly true in science that to report events, properties, and transformations with meaning, one must specify them with the appropriate units. We shall become familiar with various types of units as the corresponding concepts are introduced.

It is worthwhile noting that no properly constituted set of units is any more fundamental than any other. Each set can be converted to another by the appropriate set of conversion factors. For example, there is nothing more fundamental about referring to a person's height in terms of inches than there is in referring to it in terms of feet, yards, or eighths of an inch. One approach may be more convenient than another, however. Even in yards, a person's height can be presented with an arbitrary accuracy; that is, a young woman's height could be reported as 1.75 yd, which is just as detailed as 5 ft 3 in., or 63 in. This is also the case in scientific work; any self-consistent set of units is acceptable, but some sets of units are simply more convenient than others.

The Metric System

For most scientific work, the metric system of units is more convenient than the English system of units, which is in popular use in the United States. The metric system is especially convenient, because different units of a given type, of length or of mass, for example, are related by factors of ten. This makes it simple to manipulate such quantities in computation. These factor-of-ten relations are easily noted by specific prefixes. For example, the prefix *centi* means $\frac{1}{100}$; thus, *centimeter* (abbreviated cm) stands for $\frac{1}{100}$ meter (abbreviated m) and *centigram* (abbreviated cg) stands for $\frac{1}{100}$ gram (abbreviated g). Some of the nomenclature of the metric system is summarized in Table 1.2. In the English system, one finds conversion factors such as 12 (between inches and feet), 3 (between feet and yards), and 16 (between ounces and pounds). It is for convenience that the metric system is usually used in science. The United States is one of the few countries in which the English system still survives.

As an example of how simple certain types of mathematical manipulations are in the metric system, consider a length of 1000 m (about 1094 yd). This specific length can be expressed readily in other metric units by simply taking account of the meaning of the prefixes defined in Table 1.2.

$$\begin{aligned}
1000\text{ m} &= (\tfrac{1}{1000})(1{,}000{,}000)\text{ m} = (\tfrac{1}{1000})(1{,}000{,}000\text{ m}) = \tfrac{1}{1000}\text{ Mm}\\
1000\text{ m} &= 1\text{ km}\\
1000\text{ m} &= (10)(100)\text{ m} = (10)(100\text{ m}) = 10\text{ hm}\\
1000\text{ m} &= (100)(10)\text{ m} = (100)(10\text{ m}) = 100\text{ dam}\\
1000\text{ m} &= (10{,}000)(\tfrac{1}{10})\text{ m} = (10{,}000)(\tfrac{1}{10}\text{ m}) = 10{,}000\text{ dm}\\
1000\text{ m} &= (100{,}000)(\tfrac{1}{100})\text{ m} = (100{,}000)(\tfrac{1}{100}\text{ m}) = 100{,}000\text{ cm}\\
1000\text{ m} &= (1{,}000{,}000)(\tfrac{1}{1000})\text{ m} = (1{,}000{,}000)(\tfrac{1}{1000}\text{ m}) = 1{,}000{,}000\text{ mm}
\end{aligned}$$

In each of these mathematical manipulations, we have simply reexpressed the number 1000 as the product of two factors. The second factor in each case is recognized as being one of the several defined units of length in the metric system (Table 1.2). In the third example, the quantity 1000 m is reexpressed as (10)(100 m), and the

TABLE 1.2 **Common Prefixes in the Metric System**

Prefix	*Symbol*	*Meaning*	*Examples*
nano	n	one-billionth (10^{-9})	nanosecond (ns) nanometer (nm)
micro	μ	one-millionth (10^{-6})	microvolt (μV) microsecond (μsec)
milli	m	one-thousandth (10^{-3})	millimeter (mm) milligram (mg)
centi	c	one-hundredth (10^{-2})	centimeter (cm) centigram (cg)
deci	d	one-tenth (10^{-1})	decimeter (dm) deciliter (dl)
deka	da	ten (10)	dekagram (dag) dekameter (dam)
hecto	h	one hundred (10^2)	hectogram (hg) hectosecond (hs)
kilo	k	one thousand (10^3)	kilogram (kg) kilometer (km) kilovolt (kV)
mega	M	one million (10^6)	megaton (Mton) megahertz (MHz)

factor 100 m is recognized as the definition of one hectometer (hm), giving 10 hectometers as the result. Let us contrast this with the various common ways we might represent a 1000-yd length in English units.

$$1000 \text{ yd} = (1000 \text{ yd})\left(3\,\frac{\text{ft}}{\text{yd}}\right)\left(\frac{1}{5280}\,\frac{\text{mi}}{\text{ft}}\right) = 0.5682 \text{ mi}$$

$$1000 \text{ yd} = (1000 \text{ yd})\left(3\,\frac{\text{ft}}{\text{yd}}\right) = 3000 \text{ ft}$$

$$1000 \text{ yd} = (1000 \text{ yd})\left(3\,\frac{\text{ft}}{\text{yd}}\right)\left(12\,\frac{\text{in.}}{\text{ft}}\right) = 36{,}000 \text{ in.}$$

Certainly the relations shown in the first grouping are mathematically simpler, because they involve only numbers like 10, 100, and 1000.

Even within the metric system, there are a few different systems of units that have been adopted by different groups of scientists. Until recently, the most common such sets in use were the **cgs** (centimeter-gram-second) and **mks** (meter-kilogram-second) systems. Recently, by international agreement, the **SI** (Standard International) metric system has been adopted for future use in science. This system, summarized in Table 1.3 along with the cgs and English system, is conveniently compatible with what has largely been used in chemistry for many years, with the exception of the unit for energy. Chemists have typically used the calorie as the energy unit (see Table 1.3), whereas the joule is the preferred unit in the SI system.

TABLE 1.3 **Units**

Physical Quantity	*Distance*	*Volume*	*Mass*	*Force*	*Energy (Work)*	*Electric Charge*	*Temperature*[a]
Usual symbol	l	V	m	F	E (W)	q	T
Standard International (SI) unit	meter,[b] m	cubic decimeter, dm^3; or cm^3 or ml	kilogram,[b] kg	newton,[b] N (force that accelerates a mass of 1 kg by 1 m/s^2)	joule, J[b] (work performed by a force of 1 N acting for 1 m)	coulomb,[b] C (charge transferred by a current of 1 ampere (A) flowing for 1 s)	degree Kelvin, °K
cgs unit	centimeter, cm	cubic centimeter, cm^3 or ml	gram, g	dyne (force that accelerates a mass of 1 g by 1 cm/s^2)	erg (work performed by a force of 1 dyne acting for 1 cm) $= 10^{-7}$ J	statcoulomb (charge that repels an equal charge of the same sign 1 cm away with a force of 1 dyne)	degree centigrade (or Celsius) °C
English unit	foot, ft	cubic foot, ft^3	slug (mass that is accelerated 1 ft/s^2 by a force of 1 lb)	pound, lb	foot-pound, ft-lb (work performed by a force of 1 lb acting for 1 ft)		degree Fahrenheit, °F
Conversion factors	1 ft = 30.480 cm	1 ft^3 = 2.832×10^4 cm^3	1 slug = 14.59×10^3 g	1 dyne = 2.248×10^{-6} lb = 1.00×10^{-5} N	1 ft-lb = 1.356 J	1 coulomb = 2.998×10^9 statcoulombs	°C = °K − 273.15 = $(\frac{5}{9})$(°F − 32)
Other common units	inch, in. (2.540 cm); millimeter, mm ($\frac{1}{10}$ cm); angstrom, Å (10^{-8} cm)	liter, l[b] (1000 ml); pint 473.2 ml); gallon, gal (231 in.^{3}, 3.785 l)	milligram, mg (10^{-3} g)		calorie, cal (4.184×10^7 erg; 4.184 J); liter atmosphere, l-atm (101.3 J); BTU (252.0 cal)	electron charge (4.803×10^{-10} statcoulomb); faraday (96,487 C)	

[a]Temperature units are not strictly grouped into the cgs and English systems as they are represented here. Here, the Fahrenheit scale is frequently used in applications that use English units, and the Celsius (or Kelvin) scale is usually used in scientific work, which uses cgs (or mks) units.
[b]Units of the mks (meter-kilogram-second) system.

In most examples and exercises in this text we shall use units that are current, like the calorie, but some examples and exercises will involve SI units for those quantities, like the joule, in which there is a big difference from current popular usage in chemistry. The student should develop a facility with SI units, as the day will likely come when they are used universally.

Throughout this book we shall meet many examples in which the methods of manipulating units and conversion factors will be shown. For our present purposes, it is appropriate to introduce a general approach for converting from one set of units to another. This approach, called the **unit factor method**, can be summarized in the following steps and example.

Step 1. Algebraically state the equation that relates the units in which the problem is stated initially and the units into which you wish to convert it.

Step 2. By algebraic transposition, manipulate the equation obtained in step 1 to give a fractional expression for unity in which the "new units" appear in the numerator and the original units appear in the denominator.

Step 3. Multiply the physical quantity of the original units by the form of unity obtained in step 2. Since multiplying a quantity by unity cannot change its actual value, the resulting number gives the same physical quantity in the new units.

EXAMPLE 1.1 Express 1.34 lb in terms of grams.

SOLUTION *Step 1.* One pound is equivalent to 453.6 g. Algebraically, this is stated

$$1 \text{ lb} = 453.6 \text{ g}$$

Step 2. Dividing both sides of the above equation by 1 lb yields

$$\frac{1 \text{ lb}}{1 \text{ lb}} = 1 = \frac{453.6 \text{ g}}{1 \text{ lb}} = 453.6 \frac{\text{g}}{\text{lb}}$$

Step 3. $1.34 \text{ lb} = (1.34 \cancel{\text{lb}})(1) = (1.34 \text{ lb})\left(453.6 \frac{\text{g}}{\cancel{\text{lb}}}\right) = 608 \text{ g}$

When one becomes used to this approach in such problems, then there is no need to proceed in these discrete steps.

Note that the units cancel in the last expression of Example 1.1: lb × g/lb = g. This fact provides us with a check on our use of the conversion factors.

EXAMPLE 1.2 Convert 5.424 kilojoules (kJ) into calories.

SOLUTION From Table 1.3 we see that 1 cal is 4.184 J. Thus,

$$1 \text{ cal} = 4.184 \text{ J} = 4.184\left(\frac{1}{1000}\right)(1000 \text{ J}) = \frac{4.184}{1000} \text{ kJ}$$

$$\frac{1}{4.184/1000} \frac{\text{cal}}{\text{kJ}} = 1 = \frac{1000}{4.184} \frac{\text{cal}}{\text{kJ}}$$

Then $$5.424\ \text{kJ} = (5.424\ \cancel{\text{kJ}})\left(\frac{1000\ \text{cal}}{4.184\ \cancel{\text{kJ}}}\right) = 1296\ \text{cal}$$

EXAMPLE 1.3 Convert 1.201 mi/h into SI units.

SOLUTION The appropriate SI unit is meters per second (m/s). What is needed is a conversion factor that will convert mi/h into m/s. We can think of the needed conversion as made up of the product of two conversion factors, one taking miles to meters and one taking 1/h to 1/s.

$$\left(1.201\frac{\cancel{\text{mi}}}{\cancel{\text{h}}}\right)\left(?\ \frac{\text{m}}{\cancel{\text{mi}}}\right)\left(??\frac{\cancel{\text{h}}}{\text{s}}\right) = ???\ \frac{\text{m}}{\text{s}} \tag{1}$$

In this expression, "???" represents the final answer and "?" and "??" represent the unknown conversion factors. These factors can be obtained by rules 1 and 2 above as follows:

$$1\ \text{in.} = 0.02540\ \text{m}; \qquad 1 = 0.02540\frac{\text{m}}{\text{in.}}$$

$$1\ \text{mi} = 5280\ \text{ft} = 5280 \times 12\ \text{in.} = (5280 \times 12\ \text{in.})\left(0.02540\frac{\text{m}}{\text{in.}}\right)$$

$$1\ \text{mi} = 5280 \times 12 \times 0.02540\ \text{m}$$

$$1 = 5280 \times 12 \times 0.02540\frac{\text{m}}{\text{mi}} \tag{2}$$

Similarly,

$$1\ \text{h} = (1\ \cancel{\text{h}})\left(60\frac{\text{s}}{\cancel{\text{min}}}\right)\left(60\frac{\cancel{\text{min}}}{\cancel{\text{h}}}\right) = 60 \times 60\ \text{s}$$

$$1 = \frac{1}{60 \times 60}\frac{\text{h}}{\text{s}} \tag{3}$$

Then, using (2) and (3) in (1), we get

$$1.201\frac{\text{mi}}{\text{h}} = \left(1.201\frac{\cancel{\text{mi}}}{\cancel{\text{h}}}\right)\left(5280 \times 12 \times 0.02540\frac{\text{m}}{\cancel{\text{mi}}}\right)\left(\frac{1}{60 \times 60}\frac{\cancel{\text{h}}}{\text{s}}\right) = 0.5369\frac{\text{m}}{\text{s}}$$

In problems like those shown in Examples 1.2 and 1.3, the "unwanted" units of the conversion factors cancel. Thus, in Example 1.3 we see that we are left with m/s = m/s. If improper conversion factors had been used, if one had been omitted, if an additional one had been used, or if one of them had been inverted, we should not have obtained the check m/s = m/s. Using units in calculations of this type frequently relieves the student of deciding whether to multiply or divide by a given conversion factor to obtain the proper result. Figure 1.3 summarizes some useful conversion factors.

STUDY PROBLEM 1(a)

How many kilojoules are 13,564 calories?

Figure 1.3
Some useful conversion factors in chemistry.

Type of Measurement	*SI Unit*	*Conversion Factor*	*Other Units*
length	meter, m	$\times 3.281$ ft/m → ; ← $\times 0.3048$ m/ft	foot
		$\times 39.37$ in./m → ; ← $\times 0.02540$ m/in.	inch
volume	cubic centimeter, cm^3 or ml[a]	$\times 3.531 \times 10^{-5}$ ft^3/cm^3 → ; ← $\times 2.832 \times 10^4$ cm^3/ft^3	cubic foot
		$\times 2.642 \times 10^{-4}$ gal/cm^3 → ; ← $\times 3785$ cm^3/gal	gallon
mass	kilogram, kg	$\times 2.205$ lb/kg → ; ← $\times 0.4536$ kg/lb	pound[b]
energy	joule, J	$\times 0.2390$ cal/J → ; ← $\times 4.184$ J/cal	calorie
		$\times 0.009871$ liter atm/J → ; ← $\times 101.3$ J/liter atm	liter atmosphere

[a]The cubic centimeter may also be abbreviated cc. In this book, however, we shall use cm^3 or ml.
[b]The pound is a unit of force and not mass. In most laboratory operations, mass and weight concepts are used interchangeably. A 0.4536-kg mass exerts a weight of 1 lb at sea level.

1.6 Reporting Numbers, Significant Figures

As science in general, and chemistry in particular, evolved into a quantitative subject, scientists were faced with new responsibilities. In reporting a measured number or one computed from experimental data by a mathematical equation, the scientist wants to present a number that represents reality as it can best be judged. This includes two distinct yet related aspects. First, the scientist wants to give the best estimate for determining the number. At the same time the scientist wants to present it in a way that will not be misleading with regard to how well the number has been determined. For example, if one has measured the length of an object with a ruler having centimeters as the smallest division, as shown in Fig. 1.4, it would be unrealistic and misleading, and perhaps even intellectually dishonest, to report the result as if it were known to the nearest 0.01 mm. A report like "15.325 cm" would simply be unreasonable, since it is not even certain that the measurement could definitely distinguish between 15.3 cm and 15.4 cm. Similarly, caution must be exercised in reporting the results of calculations. For a simple example: Suppose you want to compute the area of a rectangle from the measured length and width. The length has been measured very accurately with a precise meter stick having

0.1 mm divisions. In that case you might report a length of, say, 3.02156 m. Let us suppose that when you measured the width, this precise instrument was no longer available, and you had to use a meter stick that was graduated in centimeters. In that case, you could not distinguish between a length of 1.3281 m and 1.3284 m; but you could probably distinguish 1.328 m from 1.329 m. Hence, a value such as 1.328 m would be perhaps the best estimate. Now, if the calculation of area were carried out by the usual rules of elementary arithmetic, you might be tempted to report the result

$$\text{area} = (3.02156\ \text{m})(1.328\ \text{m}) = 4.01263168\ \text{m}^2$$

Would such a result be honest? It implies knowledge of the area to within one part in 401263168, or within a percentage error of about 0.00000025%. However, the length of one side is known only to within one part in 1328, or about 0.07%. The answer to this question is then, No! To handle problems of this sort properly, one wishes to report not only accurate value numbers, but also a meaningful indication of their reliability.

For this dual purpose the scientist relies on statistical analysis and the theory of errors. These methods make up a sophisticated and well-studied subject in itself. More simply, however, the concept of significant figures will help you go a long way towards achieving the two ends mentioned.

Here we will consider how to represent one's precision properly, where we understand **precision** as the degree of exactness with which a measurement can be reproduced. Note that a precise measurement is not necessarily an accurate one. By **accuracy** we mean how close our measurement corresponds to the true value. A reported weight of 2.38654 g is considered precise, especially if the same number is obtained in repeated weighings of the same object. But if the balance used was not calibrated or adjusted properly, it could give a constant and reproducible error, say 0.00003 g, in all weighings, so that none of the weighings would be truly accurate.

Within the context of reporting measured or calculated results, a **significant figure** is defined as a numerical symbol that has some quantitative significance in a reported number. In other words, the significant figure has its origins in a measurement, or at least an estimate, and is not a blind guess. For example, if the mass of an object is reported at 1.364 g, and if we consider each of the numerical symbols significant, the following is implied. We are positive that the first three symbols (counting from

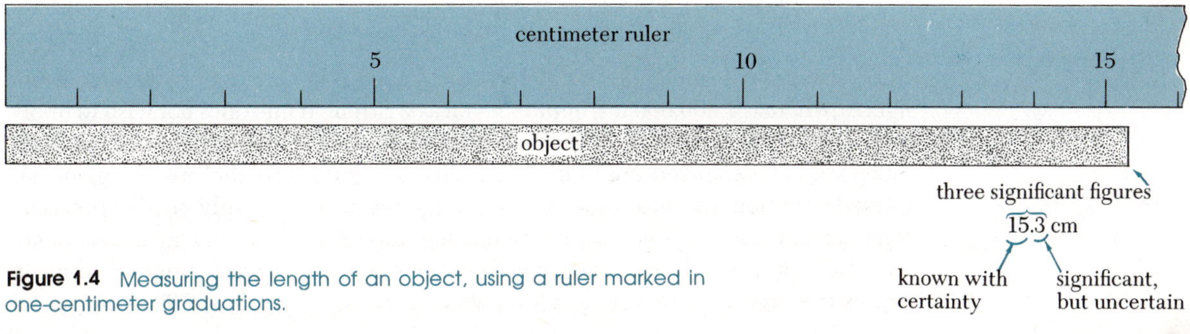

Figure 1.4 Measuring the length of an object, using a ruler marked in one-centimeter graduations.

left to right—1, 3, and 6—are correct; and we believe the true mass lies near 1.364. The true mass might be 1.362 g or 1.363 g or 1.366 g or 1.365 g. It is probably not very far from 1.364 g; it is probably not 1.369 g. A result such as 1.364 g is the kind that could come from measurements on a balance that weighs "to the nearest mg." In such a case, we believe, the actual mass lies closer to 1.364 g than to 1.363 g or 1.365 g; we are not really sure. There is uncertainty about the fourth digit, that is, the 4. We conclude that the measured result 1.364 g carries four "significant figures." We should not be tempted to report the numbers 1.3640 g or 1.36400 g from measurements with a balance weighing to the nearest mg. The former number carries five significant figures and implies unequivocal knowledge of 1.364, and uncertainty only in the fifth figure, 0. The latter carries six significant figures and implies unequivocal knowledge of the first five figures—1, 3, 6, 4, and 0—and uncertainty only for the sixth; clearly any such implications or claims would be unfounded for measured numbers originating from the balance described above.

If the measurement had been made on a balance that read only to the nearest tenth of a gram, then the reported result "1.3 g," with only two significant figures, would be appropriate. This implies knowledge that the actual mass is near 1.3 g; it might be 1.2 g or 1.4 g, but probably is not 0.9 g or 1.9 g. A reported result of 1.30 g or 1.300 g would be misleading, since each of these carries several significant figures, and implies levels of precision not attainable with a balance that weighs a 1.3-g sample to the nearest 0.1 g. As a general guideline, avoid placing unwarranted figures, including zeros, on the tail end (right side) of numerical results. One should not, for example, write 2.3010 J if the number is known only to two significant figures (2.3 J).

In following this guideline one is faced with what may appear to be occasional dilemmas. Suppose one wants to express in milligrams a certain mass given in kilograms, say 1.3 kg. The conversion factor is given by

$$1\ \text{kg} = (1000\ \text{g})\left(1000\,\frac{\text{mg}}{\text{g}}\right) = 1{,}000{,}000\ \text{mg}$$

Both sides are then divided by 1 kg.

$$1 = 1{,}000{,}000\,\frac{\text{mg}}{\text{kg}}$$

Hence, one is tempted to write

$$1.3\ \text{kg} = (1.3\ \text{kg})(1) = (1.3\ \cancel{\text{kg}})\left(1{,}000{,}000\,\frac{\text{mg}}{\cancel{\text{kg}}}\right) = 1{,}300{,}000\ \text{mg}$$

But our statement cannot be completely true, since 1.3 kg carries only two significant figures, whereas 1,300,000 mg implies seven as written. If one does not wish to imply seven significant figures, then some other notation besides 1,300,000 should be used. Since we have changed only units, and have not improved the measurement nor altered precision, the mass expressed in milligrams must not imply greater precision than the same mass as it was expressed in kilograms. A convenient way to accommodate these apparently opposing considerations is to use **exponential notation** in expressing very large or very small numbers. The number 1.3×10^6 (which is 1.3

times one million) has the same numerical value as 1,300,000; but 1.3×10^6 carries only two significant figures, the number of significant figures before the powers of ten. The exponential form then permits us to write the desired relation

$$1.3 \text{ kg} = 1.3 \times 10^6 \text{ mg}$$

without any loss or change of meaning.

The number 6 in the mathematical expression 10^6 is called an **exponent**, and the number 10 is called the **base.** The expression means that the base is multiplied by itself a number of times equal to the exponent. Thus, $10^6 = 10 \times 10 \times 10 \times 10 \times 10 \times 10 = 1{,}000{,}000$. The number represented by the symbols 10^6 or 1,000,000 is exact; it is an **integer**, which can be thought of as resulting from a tally, or a counting of distinct objects or symbols. When multiplied by a factor, such as 1.3 above, the exponent does not alter the precision, but merely changes the scale.

Exponential notation has another advantage; it permits writing very large or very small numbers, either of which would otherwise include several zeros, in a compact manner. Small numbers can be represented by negative exponents. Thus, the symbol 8.2×10^{-3} has the same meaning as $(8.2)(\frac{1}{10}) \times (\frac{1}{10}) \times (\frac{1}{10})$, or 0.0082. In general, the symbol "10^{-n}" has the meaning $(\frac{1}{10})^n$ or one-tenth multiplied by itself n times.

EXAMPLE 1.4 How many significant figures are there in each of the following quantities: 4.7×10^{20} μg; 1.27×10^{-6} km; 0.3100×10^3 g?

SOLUTION Recognizing that only the integers preceding the powers of ten need be counted, by inspection we have two, three, and four.

STUDY PROBLEM 1(b)

How many significant figures are there in the following quantities: 11.2×10^7 cal; 0.4100×10^3 g; 7.07×10^{-3} kg?

EXAMPLE 1.5 Reexpress 13.6 cm, 11.2 km, 1 mm, and 0.11 mm as meters.

SOLUTION

$$13.6 \text{ cm} = 13.6 \cancel{\text{cm}} \times \left(10^{-2} \frac{\text{m}}{\cancel{\text{cm}}}\right) = 13.6 \times 10^{-2} \text{ m}$$

$$11.2 \text{ km} = 11.2 \cancel{\text{km}} \times \left(10^{3} \frac{\text{m}}{\cancel{\text{km}}}\right) = 11.2 \times 10^{3} \text{ m}$$

$$1 \text{ mm} = 1 \cancel{\text{mm}} \times \left(10^{-3} \frac{\text{m}}{\cancel{\text{mm}}}\right) = 1 \times 10^{-3} \text{ m}$$

$$0.111 \text{ mm} = 0.111 \cancel{\text{mm}} \times \left(10^{-3} \frac{\text{m}}{\cancel{\text{mm}}}\right)$$

$$= 0.111 \times 10^{-3} \text{ m} = 1.11 \times 10^{-4} \text{ m}$$

STUDY PROBLEM 1(c)

Reexpress 0.56 mg, 21 kg and 15×10^{-3} μg (micrograms) as grams.

EXAMPLE 1.6 How many significant figures are there in each of the following: 7.324; 0.0003; 1.000; 0.09030; 0.025×10^4; 0.0017×10^{-16}; 2.00×10^8?

SOLUTION By inspection, counting right from the leftmost nonzero integer, we obtain four, one, four, four, two, two, three.

STUDY PROBLEM 1(d)

How many significant figures are there in each of the following quantities: 2.112 g; 0.004707 kg; 0.03100 km; 100 cal; 0.042×10^{-2} kcal; 0.4200×10^{6} g?

So far, we have explored only accounting for the precision of numerical values that result from individual experiments and unit conversions. We have only briefly mentioned the problems associated with properly representing the results of calculations. Such considerations can be seen in the following examples.

Suppose we are assigned the task of determining the **density**, or mass per unit volume, of a liquid by measuring the mass of a sample of it contained in a cylindrical vessel of measured dimensions, and we are concerned with properly representing the precision of the determination. In determining the density, the basic relation we have to work with is density = mass/volume. In representing the precision, it is convenient to treat this problem in the following three parts: (1) determine the precision with which the volume is known; (2) determine the precision with which the mass is known; and (3) properly combine the information from the first two parts.

1. For the first part of this problem, we need the formula for the volume of a cylinder,

$$\text{volume} = \pi r^2 l$$

where r is the radius, l is the length of the cylinder, and π is the universal mathematical constant, 3.14159265 . . . . Suppose that l is measured to be 7.4 cm, and the radius of the cylinder is found to be 4.75 mm. If we convert both lengths to units of centimeters, the volume in cm^3 can be expressed

$$V = \pi(0.475\ \text{cm})^2(7.4\ \text{cm}) \qquad (1.1)$$

In this expression, the figure π is basically different in kind from the numbers 0.475 and 7.4; it is a precisely defined mathematical constant, which is known to more significant figures (π = 3.14159265 . . .) than we shall be concerned with. This number cannot contribute to any uncertainty in the value of V, and hence is in no way involved in determining its precision. The least precisely known factor in Eq. (1.1) is the 7.4, which has two significant figures and an uncertainty of roughly one or two parts in seventy-four; that is, the true value might be 7.2, 7.3, 7.4, 7.5, or 7.6. The general sense of the precision of this value should be conveyed approximately into the value of V. Hence, the calculated V is also known to two significant figures:

$$V = (3.14)(0.226)(7.4)\ \text{cm}^3 = 5.2_5 \qquad (1.2)$$

The nonsignificant 5 in this result is shown lowered, and is retained for the next step to avoid unnecessarily degrading the overall, multistep calculation. In writing the left side of Eq. (1.2), we have written π as 3.14 and $(0.475)^2 = 0.225625$ as 0.226, each showing only three significant figures. We have rounded off 3.141592 . . . and 0.225625 to three significant figures each, one more than the number of significant figures we expect in the result (two). As customary procedure, for any factor that is known to more significant figures than the final answer of a many-step calculation, generally "carry" one more digit than you would retain in the final result (for example, retain three digits in 3.14 and 0.226 instead of two). This is to ensure that unnecessary round-off errors do not accumulate in a computation involving more than one step. In rounding off 3.141592 to 3.14 and 0.225625 to 0.226 we have used the customary round-off rule. This rule tells one to *leave a final significant digit alone* (4 in 3.14, for instance) *if the nonsignificant digit that follows it* (1 in 3.14159) *is less than* 5, or to *raise the last significant digit* (5 in 0.22562) *by one if the nonsignificant digit that follows it* (6 in 0.22562) *is larger than* 5. *If the first nonsignificant digit is 5, then the last significant digit is raised by one if it is odd, or left alone if it is even;* for example, 3.953_5 to 3.954, and 6.14_5 to 6.14, where the nonsignificant 5 is shown lowered.

2. The second part of this problem involves determining the mass of the sample. The procedure for determining mass in the density experiment is usually based on the method of **difference weighing.** According to this method, the mass of a liquid or other kind of sample is determined as the difference between the measured mass of a vessel containing the sample and the measured mass of the empty vessel. Suppose the mass of the vessel itself is reliably known to be 37.15026 g from a previous experiment in which an extremely fine balance was available. Also, suppose that with the balance available at the time of the density determination, the mass of the vessel completely filled with the liquid sample is found to be 49.188 g. What is the measured mass of the liquid sample? We might be tempted to obtain it simply as

$$\begin{array}{r} 49.18800\ \text{g} \\ -\underline{37.15026\ \text{g}} \\ 12.03774\ \text{g} \end{array}$$

However, the total mass of sample plus vessel is known only to the nearest 0.001 g (which is 1 mg), and the difference cannot be known more precisely than that. Thus, in the final result, we round off the net sample mass to the nearest milligram, giving 12.038 g.

3. Finally, in calculating the density, we write

$$d = \frac{m}{V} = \frac{m}{\pi r^2 l} = \frac{12.038\ \text{g}}{5.2_5\ \text{cm}^3} = 2.2930\frac{\text{g}}{\text{cm}^3},\ \text{or}\ 2.3\frac{\text{g}}{\text{cm}^3}$$

The result cannot be known with less uncertainty than about one or two parts in fifty-two, which is the actual precision of the denominator, 5.2_5. This information is qualitatively imparted to the final answer by rounding it off to two significant figures, giving 2.3 g/cm³.

The plausibility arguments given in the above example can be summarized in the following set of rules, which provide a useful qualitative framework for indicating precision in terms of the concept of significant figures. The student is encouraged to remember these rules in working subsequent numerical problems.

I. *In the arithmetic operations addition or subtraction, the answer should not carry an absolute precision that is greater than that of the number carrying the poorest absolute precision.* To accomplish this, first arrange the numbers to be added or subtracted in a column with their decimal points placed directly one above the other. Then, identify the number that has its last (rightmost) significant figure furthest to the

TABLE 1.4 **Significant Figures**

Expression	*Result*	*Rules Applied*	*Comments*
2.0162 + 37.81	39.83	I	2.0162 first rounded off to 2.02
16.92 kg + 1.2 mg	16.92 kg	I	First convert to one choice of units
(11.34)(18)	2.0×10^2	II	2.04×10^2 rounded off to two significant figures
$\dfrac{44.2 + 0.921}{33}$	1.4	I, then II	1.37 rounded off to two significant figures
$(2.061 \times 10^{-3}) \times (4.100 \times 10^{-4})$	8.450×10^{-7}	II	8.4501×10^{-7} rounded off to four significant figures
(0.00021)(0.0300)	6.3×10^{-6}	II	

left; this is the precision-limiting number. Second, round off the remaining numbers so that each has its last significant figure in the same place relative to the decimal point as the precision-limiting number does. Finally, perform the addition or subtraction.

II. *In multiplication or division, the answer should carry the same number of significant figures as the smallest number of significant figures carried by any of the quantities being multiplied or divided.*

Some simple examples of applying these rules are given in Table 1.4 and in the following examples.

EXAMPLE 1.7 Round off each of the following numbers to three significant figures: (a) 0.1462; (b) 1.9991×10^{-4}; (c) 15.25×10^3; (d) 0.088752.

SOLUTION Using the rules given above in italics for rounding off gives:

(a) 0.1462 to 0.146 (2 is less than 5)
(b) 1.991×10^{-4} to 2.00×10^{-4} (9 is greater than 5)
(c) 15.25×10^3 to 15.2×10^3 (2 is even)
(d) 0.088752 to 0.0888 (7 is odd)

STUDY PROBLEM 1(e)

Round off each of the following numbers to two significant figures: 10.66; 0.0421×10^2; 5.45; 3.952×10^{-7}.

EXAMPLE 1.8 Compute the following sum: 9.41 + 13.235 + 3.42261 + 118.315.

SOLUTION First arrange the sum as indicated in the first step above.

Step 1:

```
   9.41
  13.235
   3.42261
 118.315
```

Numbers in the shaded area are "excess" digits, to be rounded off as indicated in the second step.

Step 2:

$$\begin{array}{r} 9.41 \\ 13.24 \\ 3.42 \\ \underline{118.32} \\ 144.39 \end{array} \quad \text{(required sum)}$$

EXAMPLE 1.9 Evaluate the following sum, expressing the result in grams:

$$13.6\text{ g} + 14.21\text{ mg} + 151.56\text{ kg}$$

SOLUTION Before adding, we must first transform all the values into the same units; we choose the largest unit, kilograms in this case.

$$14.21\text{ mg} = 14.21\text{ mg} \times \left(10^{-6}\,\frac{\text{kg}}{\text{mg}}\right) = 0.00001421\text{ kg}$$

$$13.6\text{ g} = 13.6\text{ g} \times \left(10^{-3}\,\frac{\text{kg}}{\text{g}}\right) = 0.0136\text{ kg}$$

Then, placing them in a column, and using rule I,

$$\begin{array}{lr} 0.0136 & \text{kg} \\ 0.00001421 & \text{kg} \\ \underline{151.56} & \underline{\text{kg}} \\ 151.57 & \text{kg} \end{array}$$

Finally, converting to grams, we get

$$151.57\text{ kg} = 151.57 \times 10^3\text{ g}$$

EXAMPLE 1.10 Evaluate the expression $\dfrac{11.25 + 0.02114}{3.1}$.

SOLUTION By rule I, the numerator is $11.25 + 0.02114 = 11.27$. Then, by rule II,

$$\frac{11.27}{3.1} = 3.6$$

STUDY PROBLEM 1(f)

What is the volume of liquid in a container, filled to its 200-cm^3 capacity, and also containing a steel ball of volume 3.311 cm^3?

STUDY PROBLEM 1(g)

What is the area of a rectangle with length 11.3 cm and width 1.466 cm?

1.7 The Value of Mathematical and Instrumental Methods

The development of chemistry in a quantitative direction has been an accelerating process. As in most areas of science, much more has been accomplished within the last few years than in the preceding few decades; these in turn were more productive in many ways than the preceding few centuries. It can be argued that furthermore, the past few centuries were more productive in the sense of specific scientific achievements than the entire previous history of mankind. This acceleration towards quantitative relations in chemistry has gone hand in hand with developments in other fields, especially physics and mathematics, on which many developments of chemistry have relied. In parallel with these developments has been the enormous progress that has been realized in the technology of instrumentation—in the design and manufacture of devices for making specific measurements. Over the years chemists have accumulated a variety of measuring devices, and these have given rise to an increasing number of mathematically founded hypotheses and theories. Consequently, a body of theoretical chemistry has emerged, and physical chemistry has assumed an important place alongside the more descriptive aspects of chemistry.

For many years, some of the types of mathematical relations and calculations that chemists envisioned did not realize their potential since the required computations were simply not practical due to the length of time involved. This situation gave rise to methods of approximation, many of which were quite successful. These methods tried to keep the physical concepts intact while sacrificing mathematical rigor and precision in an effort to simplify the computations. Some approximations were so drastic that they tended to invalidate the significance of computed results. Some problems were so difficult that they were not amenable to solutions with any sort of reasonable approximation. For many years such problems remained unsolved and largely dormant.

The development of electronic computers has strongly affected the trend towards a more quantitative character in chemistry. This effect is felt at every level of chemistry, from the pocket-size calculators many students used for solving first-year chemistry problems to the huge, sophisticated computer centers engaged in the complicated calculations required in certain areas of chemical research, or the computer control of processes in the chemical industry. It appears safe to predict that computers will be even more important to chemistry in the future.

SUMMARY

The scientific method has evolved as a powerful intellectual method for answering questions about the nature of our world. Chemistry occupies a central position in the overall scheme of science. As science has become more highly developed, there has been a trend towards quantitative relations. Units are important in science, and the metric system, especially the Standard International system, is the system of choice. Another important concept in quantitative science is precision, and significant figures provide a convenient way of expressing precision properly.

STUDENT CHECKPOINTS

After studying this chapter, the student should be able to:

1. Explain the essence of the scientific method.
2. Show how chemistry is centrally important to other sciences.
3. Describe the need for units and their meaning.
4. Relate the prefixes used in the metric system to their values.
5. Convert familiar units to Standard International units.
6. Convert from one unit system to another.

7. Use the unit factor method.
8. Define *significant figures.*
9. Round off numbers.
10. Use significant figures in working problems.

EXERCISES

1.1. Distinguish between the members of each of the following pairs: (a) hypothesis and theory; (b) data and empirical relation; (c) experiment and data.

1.2. State a concept from each of the following areas: physics, biology, geology, chemistry.

1.3. Give a qualitative statement about the length of an object and a quantitative statement about the length of an object.

1.4.* Convert exactly 3 mi into inches.

1.5. Restate each of the following as some number from one to ten times the appropriate power of ten: (a) 0.00114; (b) 778,642; (c) 0.613; (d) 25.461.

1.6.* Express each of the following as some number between one and ten times the appropriate power of ten: (a) 3264.51; (b) 0.0006020; (c) 46,192; (d) $\frac{1}{20}$.

1.7. Convert each of the following into kilograms: (a) 3.4 mg; (b) 16 g; (c) 64.432 g; (d) 11.4 lb.; (e) 87 hg; (f) 3 dag; (g) 12.2 ng; (h) 0.0039 Mg.

1.8.* Convert each of the following into meters: (a) 0.466 km; (b) 43.621 cm; (c) 11 mm; (d) 55.624 hm; (e) 27.4 nm.

1.9. Restate each of the following in terms of Standard International (SI) units: (a) 16.65 in.; (b) 22.77×10^6 cal; (c) 26.9 lb; (d) 0.064 gal; (e) 42.8 liter atm.

1.10.* Restate each of the following in terms of SI units: (a) 67.4 kcal; (b) 42.6 yd; (c) 83.7 in.^{3}; (d) 46.4 lb/in.^{2}.

1.11.* A length of roughly 5 m is to be measured by a "meter stick" (the analog of a yardstick). Three significant figures are desired in the measurement. What is the size of the smallest divisions that must be marked on the meter stick to provide this level of precision?

1.12. Describe clearly the difference between precision and accuracy.

1.13. How does one try to make sure that precise measurements are also accurate?

1.14. What is the largest number of significant figures that will result from measurements of the lengths of the following objects, using a meter stick in which the smallest marked divisions are centimeters: (a) a mouse; (b) a professional basketball player; (c) a cocker spaniel; (d) a locomotive; (e) an Olympic-size swimming pool; (f) a dime; (g) a kernel of corn?

1.15.* How many significant figures are there in each of the following: (a) 11.22; (b) 0.0600; (c) 2×10^{32}; (d) 0.7×10^{-7}?

1.16.* Round off each of the following numbers to three significant figures: (a) 0.55555; (b) 0.66666; (c) 0.4445; (d) 0.44466; (e) 0.6665.

1.17. Round off each of the following numbers to two significant figures: (a) 0.00200; (b) 7.004×10^3; (c) 4.7×10^{-16}; (d) 100; (e) 3001.

1.18.* Compute the following sum, paying strict attention to significant figures: $16.2 + 372 + 1.677 + 32.0009$.

NOTE: Exercises with asterisks after the number have answers furnished at the end of the chapter.

1.19.* Evaluate the expression

$$\frac{42 + 0.211 + 1.6}{1.391}$$

1.20. How many significant figures are there in the results of the following arithmetic manipulations: (a) 15.0041 + 15.0040; (b) 15.0041 − 15.0040?

1.21.* Evaluate the following expressions, paying strict attention to significant figures:
(a) $13.47 + 0.144 + 3 =$
(b) $4.16 \times 10^3 + 2 + 55 =$

1.22. Express the following in centimeters: 32 m + 642 mm.

1.23.* Express the following in grams: 32.666 kg + 31 hg.

1.24. Evaluate the following expressions, paying strict attention to significant figures:
(a) $414.3 - 2 \times 10^2 =$
(b) $76.5 \times 10^{-3} - 0.005 =$

1.25.* Evaluate the following expressions, paying strict attention to significant figures:
(a) $9.2645 \times 17 \times 15.113 \times 10^{-4} =$
(b) $46.1 \times 0.000012 =$

1.26. Calculate the density of a substance, a 0.68-cm^3 sample of which is found to have a mass of 4341.1 mg. Express the result in g/cm^3.

1.27.* A cylindrical sample of a metal is found to have a radius of 3.4 cm and a length of 0.0962 m. Its mass is measured to be 2.4651 kg. What is the density of this sample in g/cm^3?

1.28. Describe the difference between "percentage error" and "absolute error" (which is the actual size of an error, say, 3 ft, or 2 g, or 11 cal).

1.29.* Evaluate the following expression:

$$\frac{(13.64 + 12 + 1.003) - (4.0021 + 2.1977)}{(0.118 - 0.0010) \times 4.22}$$

1.30. One is told that in an experiment a chemical sample was heated for 80 min. Express this time in seconds.

ANSWERS TO SELECTED EXERCISES

1.4 190,080 in. **1.6** (a) 3.26451×10^3; (b) 6.020×10^{-4}; (c) 4.6192×10^4; (d) 5.0×10^{-2}. **1.8** (a) 466 m; (b) 0.43621 m; (c) 0.011 m; (d) 5562.4 m; (e) 2.74×10^{-8} m. **1.10** (a) 282 kJ; (b) 39.0 m; (c) 1.37×10^{-3} m^3; (d) 3.26×10^3 g/cm^2. **1.11** 0.1 m. **1.15** (a) four; (b) three; (c) one; (d) one. **1.16** (a) 0.556; (b) 0.667; (c) 0.444; (d) 0.445; (e) 0.666. **1.18** 422. **1.19** 32. **1.21** (a) 17; (b) 4.22×10^3. **1.23** 3.58×10^4 g. **1.25** (a) 0.24; (b) 5.5×10^{-4}. **1.27** 7.1 g/cm^3. **1.29** 41.

ANSWERS TO STUDY PROBLEMS

1(a) 56.75 kJ **1(b)** three; four; three. **1(c)** 5.6×10^{-4} g; 2.1×10^4 g; 1.5×10^{-8} g. **1(d)** four; four; four; three; two; four. **1(e)** 11; 4.2; 5.4; 4.0×10^{-7}. **1(f)** 197 cm^3. **1(g)** 16.6 cm^2.

2

Substances and Their Classification

2.1 Overview

The three states of matter are the solid state, the liquid state and the gaseous state. The gross features of these three states can be described in terms of the concepts of order and random motion in the submicroscopic character of matter; this motion is related to the concept of temperature. Many of the physical characteristics of the gaseous state, including temperature, can be related mathematically by "gas laws," some aspects of which are presented in this chapter and in the next. Matter can be classified as substances, compounds, elements and various types of mixtures and all of these categories can be defined in terms of processes and properties.

Note that some of the many ideas introduced in this chapter will be dealt with in greater depth in later chapters. The examples, study problems, exercises, and student checkpoints indicate which subjects should be emphasized and mastered in this chapter.

2.2 Matter

Definitions

Chemistry is concerned with the forms of matter, and how they interact under various conditions, including situations in which energy changes are involved. **Energy** is the ability to perform work, an important consideration in chemistry. **Matter** is anything that has mass and occupies space. **Space** is simply a portion of the universe or some volume within it. **Mass** is the fundamental property of matter that is associated with the concepts of weight and inertia. We are familiar with **weight** as the force exerted on an object by gravitational attraction, a force that is proportional to the mass of the object. The object is said to have **inertia**, the tendency to remain in motion if initially in motion or to stay at rest if initially at rest. We can think of this tendency as a direct measure of what we call the mass of the object. We can distinguish further between weight and mass by recognizing that the mass of an object is independent of its location, whereas the weight of an object depends on its proximity to another body—the one that determines the gravitational force, like the earth. An astronaut weighs less on the moon than he does on earth, but his mass is the same in both places. The relation between weight w and mass m is summarized in the equation $w = mg$, where g is the acceleration constant of gravity; this constant is larger for the earth than for the moon.

Solids, Liquids, and Gases

Matter is the stuff of which the universe seems to be made. It is what one can touch, smell, weigh, mold, shape, or dissect. In the scientific view, there are several kinds of classifications into which it is convenient to divide matter. One useful classification of a sample of matter is based on its behavior when it is transferred into a new container under the influence of gravity. For some materials, such as a diamond, a piece of steel, a cube of ice, or a gold ring, the object is completely unaltered in the process. Its volume remains unchanged; even its shape remains intact. Such materials are solids. A second type of matter is exemplified by a pint of water, a cup of tea, or a pound of mercury; for this kind of matter, the liquids, transferral into a new container does not result in a different amount of space occupied by the sample—its volume, like the volume of the solid, remains unchanged. For liquids, however, the sample assumes the shape of whatever portion of the new container it occupies. Because of gravity, the occupied portion will include the bottom of the container, and continue upward only to a point that accounts for the volume of the sample. For a third type of matter, the gases, transferral to a new container results in not only a change of shape but also a change of volume. A sample of air, of oxygen, or of nitrogen at room temperature, for example, very quickly occupies the entire volume open to it when it is transferred from a small container to a large container. The gas is soon distributed uniformly in every part of the container, whether the container is a pint jar, or an entire irregularly shaped room.

These three different behaviors, shown in Fig. 2.1, can be stated in terms of fluidity and tendency to expand. They lead us to the classification of matter into the categories solid, liquid, and gas. In this way, we can define a **solid** as a sample of matter that retains both its shape and volume when transferred into a new container under the stress of gravity. Similarly, we define a **liquid** as a sample that retains its volume but not its shape when transferred in this way. Finally, we define a **gas** as a material that retains neither shape nor volume when it is subjected to this operation. Most samples of matter under conditions that are readily attainable in a laboratory fall into one of these three categories. Solids and liquids are often referred to as **condensed phases.**

What appear to be exceptions to the simple definitions given here for solids,

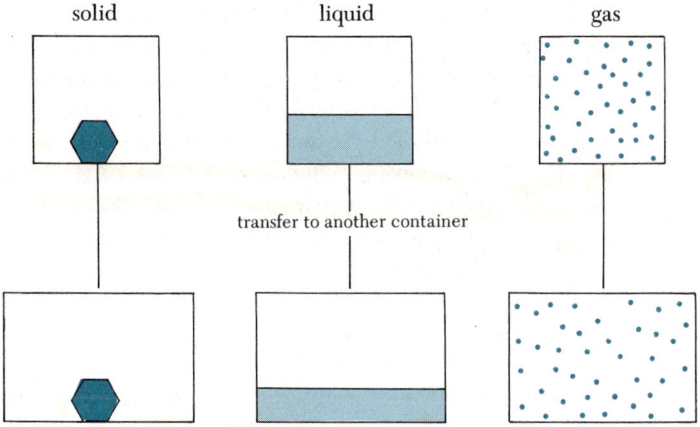

Figure 2.1
Functional definition of solids, liquids and gases.

liquids, and gases can readily be found. Thus, a finely granulated solid material (powder) appears to flow when poured from one vessel to another, superficially showing the fluidlike property of a liquid. Tar and window glass, which are really liquids, do not appear to be fluid unless they are subjected to intense stress; but if sufficient force is applied, they slowly flow.

To a scientist the word *solid* usually implies more than just the behavior described in terms of retaining its shape and volume. That is a useful and popular definition, but to many scientists the term *solid* implies a highly regular submicroscopic structure. A more precise term to impart this meaning of regularity is *crystalline solid*. Although this ambivalent interpretation of the word *solid* exists, it is safe to assume that if this word is used in the context of specifying a state of matter, as in "solids, liquids, and gases," the more precise meaning of crystalline solid is actually intended.

A Preliminary Submicroscopic View

Serious students of chemistry are interested in discovering the submicroscopic nature of matter, that is, the fundamental structural building blocks of matter. This almost incomprehensibly minute level of structure involves the atoms and molecules, which are currently viewed as the submicroscopic structural units. Some reasonable guesses about the submicroscopic nature of matter can be made from common-sense observation of the three states of matter.

Whatever the submicroscopic structural units are, for solids they appear to have a substantial tendency to stick together. In this tendency, we find a ready explanation for the fact that a solid retains both its shape and its volume when it is placed in a new container. Gases show a totally different kind of behavior; and it seems reasonable to interpret that behavior as based on a negligible tendency for the units to stick together. Liquids occupy a somewhat intermediate position.

One idea views matter in a constant state of war between two competing influences. On one side, tending to hold the submicroscopic structural units together, condensed in some sort of orderly way, are the attractive interactions between these units. These attractive interactions are essentially electrical in origin. On the other side is an influence associated with the constant motion of matter on the submicroscopic level. Thus, the submicroscopic structural units, the molecules, of a cold, inanimate piece of ice, sitting motionless in an ice tray, are in a frenzied state of motion. This submicroscopic motion is what tends to destroy the ordering tendencies we have just mentioned. In these terms we can view solid materials as those in which the attractive, ordering influences win out in this submicroscopic competition, whereas in gases the disordering influences of the randomizing motion win the competition. Liquids fall into some intermediate category, and they are therefore more complicated to describe.

Vapor Pressure

A readily demonstrated manifestation of submicroscopic motion is the phenomenon vapor pressure.* **Vapor pressure is the pressure exerted by a pure vapor in contact with a solid or a liquid sample** (the condensed phase) **of the same substance.** This pressure results from submicroscopic particles (molecules) that have escaped into the gaseous region from the solid or the liquid sample. These particles exist in constant motion, colliding randomly with each other, the walls of the container, and the surface of the solid or liquid. A so-called steady state comes about, in which some particles are continually escaping from the condensed phase and some particles are continually falling back into the condensed phase and being trapped there. Plausibly, a situation is reached in which the number of particles that exist in the enclosed volume of the vapor is constant in time; this occurs when the number of particles escaping from the condensed phase in some small interval of time equals the number being trapped in the condensed phase during that time. An experimental setup that could be used to demonstrate this phenomenon is shown in Fig. 2.2.

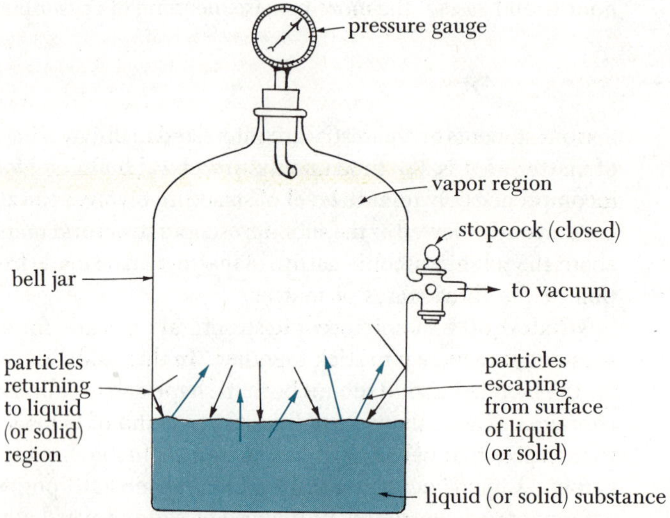

Figure 2.2
A simple apparatus for observing the vapor pressure of a liquid (or solid) substance. The vapor region contains only particles that have escaped from the liquid (or solid) region after the vessel has been evacuated.

*__Pressure__ is defined as the force exerted upon a unit area of a surface. The internal pressure in a gaseous region can be defined in terms of a force exerted on the surface of the container that defines the region. These points will be discussed more fully in Chap. 15.

Later, we shall learn how one can calculate what the vapor pressure would be by knowing the volume of the enclosure and the number of particles present in the vapor region in the steady state. For the present, we can accept the plausible statement that this pressure is proportional to the number of particles in the gaseous region. (This statement will be justified on the basis of a theoretical model in Chap. 15.) Several factors determine the number of particles in the vapor phase. If the sample is of a kind that we later learn to classify as a "pure substance," there is essentially only one determining factor. It is the fraction of the particles in the condensed phase that have enough energy in their submicroscopic motions to overcome the attractive forces that hold them in the condensed phase. In an experimental setup of the type shown in Fig. 2.2, the vapor pressure that is measured by an appropriate guage is directly proportional to the number of particles in the vapor region. This number in turn indicates the fraction of the submicroscopic particles in the condensed phase that have enough energy to escape. We find experimentally that if we increase the temperature of the solid or liquid sample, the measured pressure increases; hence, we conclude that the fraction of particles having the required amount of energy increases as the temperature increases. This is an indication that the submicroscopic motions tend to increase in intensity with an increase in temperature. We can thus appreciate the relation of temperature to submicroscopic motion.

Order in Matter

The submicroscopic ordering that the attractive influences bring about can be viewed as of two kinds—short-range order and long-range order. Short-range order deals with specifying structural details at close range, generally distances about the size of a few atoms. Long-range order specifies details at longer distances. More precisely, short-range order is the degree to which the nature of some reference point determines the structural details of matter in the immediate vicinity; short-range order is important if species at close range greatly influence each other's structures. Long-range order is associated with how well one can predict the structure at large distances—several atoms or molecules away—by knowing the structure at a particular reference point. As an analogy of short-range and long-range order, let us consider a warehouse in which checkerboards are stored. Within each checkerboard there is obviously a short-range order; by knowing that a given square is red, one can be absolutely confident of the color of the square separated from it by a fixed number of squares. Now suppose that all the checkerboards are arranged in a stack on top of each other in a prescribed fashion, so that each pattern is placed directly above the corresponding one below it. In this case, knowledge of the color and position of one square in a single checkerboard provides knowledge of the positions and colors not only of all squares in the same checkerboard, but also of all the squares in all the checkerboards in the entire stack, even in a checkerboard separated from the main one by a hundred intervening ones in a stack. This constitutes a case of profound long-range order.

Long-range order and short-range order are shown for collections of black and white marbles in various arrangements in Fig. 2.3. Remembering these concepts will help us in thinking of crystalline solids as having both short-range order and long-range order. Liquids will be seen to show short-range order and no long-range

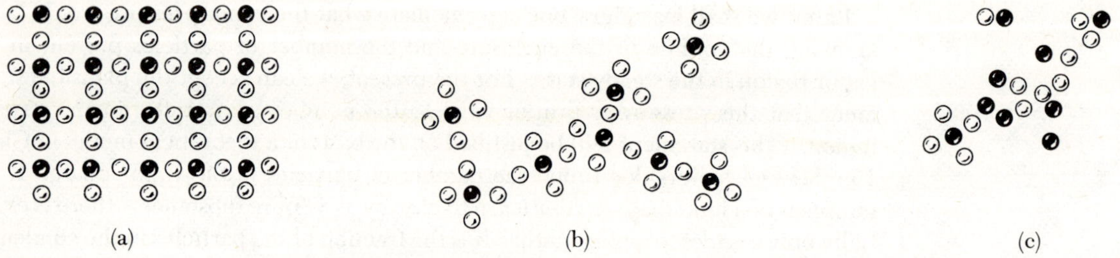

Figure 2.3 Ordering in three arrangements of a set of black and white marbles. (a) Arrangement with short- and long-range order. (b) Arrangement with only short-range order. (c) Arrangement with no recognizable order.

order. And gases will be found to have neither; gases are entirely random in both the long-range and short-range senses.

STUDY PROBLEM 2(a)

Classify each of the following in terms of whether it has short-range or long-range order, or neither: (a) lawn clippings dumped in a pile; (b) the cells of a honeycomb; (c) the cans of beer in cases, 24 per case, and the cases randomly stored in a warehouse.

2.3 The States of Matter

Liquids

Liquids represent an intermediate stage in the competition between submicroscopic order and disorder, and describing them in submicroscopic detail is comparatively the most difficult. Nevertheless, the liquid state in chemistry has long been of interest for important reasons. Many substances ordinarily occur in a liquid state. Also, chemical transformations are often carried out in the liquid state, since it is frequently the most convenient form in which to conduct the desired process. In addition, the liquid state is very important in industrial processes, because matter can be transferred on a large scale conveniently in the liquid form—by pipelines, tank cars, and related equipment. Understanding liquid state chemistry in living systems has been of continuing interest in the last several decades; most of the fundamental life processes, it is believed, occur in a liquid medium.

Solids

The solid state has long fascinated chemists and their predecessors in history. This state of matter has always occupied an important practical status because of the great value civilizations have placed on metals and gems. Also, most of the craft and construction activities of a civilization rely on solid materials of reliable strength, reasonable cost, and workable nature. In the scientific view, solids have been of interest because they are highly ordered and thus have unique properties. The solid state has lately taken on new importance because of the development in the electronics industry of "solid state" devices, for example, transitors in radio and television sets and integrated circuits in pocket calculators. Research on this subject has provided new levels of understanding, a situation that in turn has given rise to useful new applications.

Gases, Temperature

From several points of view, the gaseous state was the first to be well characterized. During the seventeenth and eighteenth centuries various sets of coordinated experiments led to the discovery of empirical laws that relate certain characteristics of a sample of gas. These laws relate the pressure, volume, and temperature of a sample of gas containing a fixed amount of matter.

The **pressure** of a gas is represented by the symbol P and is defined as the force exerted on a unit area of the walls of the vessel in which the sample of gas is contained. The volume of a sample of gas is designated by the symbol V, and simply shows the volume of the container, since any gas will completely occupy the entire volume of a vessel in which it is placed. To understand the temperature of a gas, or of any sample of matter, one needs to understand concepts that are considerably more subtle than pressure and volume. Temperature is related to the state of motion of submicroscopic particles, and the energy of such motions.

From our everyday experience, we have a general idea of temperature as a measure of "hotness or coldness." As children, we understood temperature usually from the sensations of hot or cold conveyed from the nerve endings in the skin. **Temperature** is associated with the direction of spontaneous heat flow between objects of different temperatures that are brought into contact with each other. We know that heat flows spontaneously from the object of higher temperature to the object of lower temperature until a common, intermediate temperature is attained by both. This flow is irrespective of the size of the objects. The direction of spontaneous heat flow from hot to cold is a manifestation of the second law of thermodynamics. We can use this characteristic as a functional definition of temperature (see Fig. 2.4). From a submicroscopic view, temperature is a measure of the vigor, or intensity, of the submicroscopic motions that cause the randomizing tendencies in matter. While this is a useful connection to remember, one can precisely define temperature without the submicroscopic interpretation. Indeed, the origin and general application

Figure 2.4 A functional representation of temperature. (a) Objects of different temperatures are brought into contact; heat flows from the object of higher temperature (T_A) to the object of lower temperature (T_B). (b) Heat flow stops when a common temperature that is intermediate between the two initial temperatures is reached. Imagine that object A is an initially hot piece of steel and object B an initially cold block of steel. After they are placed in contact, heat flows until both pieces of steel achieve the same warm temperature, T_{AB}.

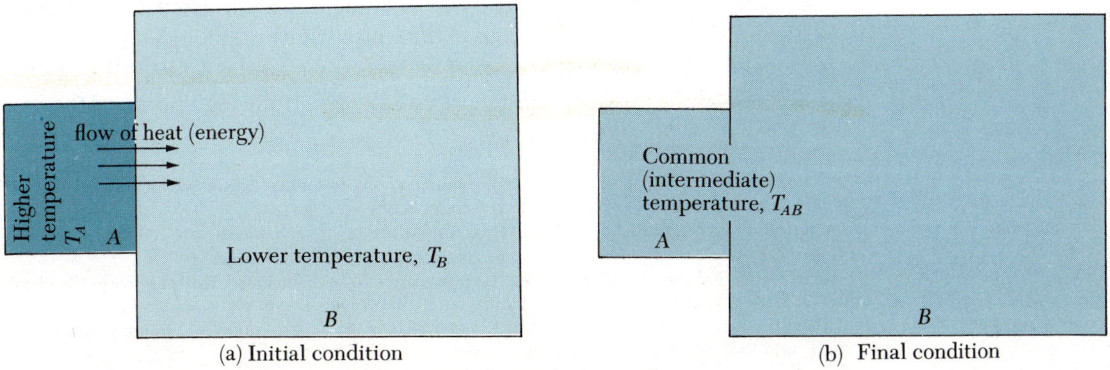

of the concept of temperature are concerned only with matter on a macroscopic level. A scale of temperature is usually defined entirely on the basis of experiments on macroscopic samples.

The conventional approach in making these definitions is to choose two precisely characterized conditions of "hotness" and to assign specific numbers to them arbitrarily, so that the idea of temperature is put on a quantitative basis. Historically this approach has developed by assigning a specific number for the temperature of freezing water (or melting ice) in an ice-water mixture, and assigning another number for the temperature of boiling water (or condensing steam) in a steam-water mixture. Both of these reference conditions are specified further by defining the pressure under which they are set up. This pressure is taken to be one **standard atmosphere,** the pressure (10.13 Newton/cm^2) that a column of mercury 76 cm high would exert at its bottom. This is the atmospheric pressure on some arbitrarily chosen "average day" at sea level; the unit is denoted by the symbol "atm." The temperature scale that is used in scientific work, as well as in nonscientific considerations in most countries except the United States, is called the **centigrade** (C), or **Celsius,** scale. In this scale the ice point is defined as exactly 0 °C and the boiling point exactly 100 °C.* From these two defined limits, the centigrade temperature scale can be constructed on the basis of a phenomenon characteristic of matter and known as thermal expansion. For a sample of liquid mercury, **thermal expansion** means simply that as the sample is made hotter, its volume increases. If the mercury is contained in a glass bulb connected to a capillary tube, then a variation in its volume is indicated with considerable sensitivity by a change in the length of the mercury column in the capillary. Hence, the length of the mercury column depends on how hot the mercury is—on its temperature, that is—so that the column of mercury is substantially shorter when the glass bulb is contained in an ice bath than when it is in a bath of boiling water. If the positions of the mercury meniscus (curved upper border) for these two conditions are marked on a scale as 0° and 100°, then the distance between them can be divided into one hundred equal parts, each representing one centigrade degree. This is shown schematically in Fig. 2.5. The device so defined is called a **thermometer.** For any intermediate "hotness," or temperature, between the ice point and the boiling point of water, the column of mercury is found to extend somewhere between the two reference points. One simply finds where the column extends, reads the number on the scale that corresponds to it, and thereby defines the temperature of the sample in which the thermometer has been inserted.†

The same considerations are used in defining the Fahrenheit (F) scale, which is used in nonscientific considerations in the United States, although it probably won't be for much longer. On the **Fahrenheit scale,** the ice point is arbitrarily assigned the value 32 °F and the boiling point the value 212 °F.‡ If the high points of the column

*Anders Celsius (1701–1744), a Swedish astronomer, originally set his scale at 100 for melting ice and 0 for boiling water, the reverse of the modern Celsius scale.

†For various practical reasons, especially for temperature ranges outside the 0°–100° span, other liquids besides mercury are sometimes chosen for measuring lower temperatures.

‡Gabriel Daniel Fahrenheit (1686–1736), a German scientist, was the first to use mercury as the expanding liquid. He set his scale at 0 for a particular freezing mixture of ice, salt, and water and at 96 for the "blood of a healthy man." Later measurements showed 98.6 °F to be the "normal" human temperature, and 32 °F and 212 °F became the reference points.

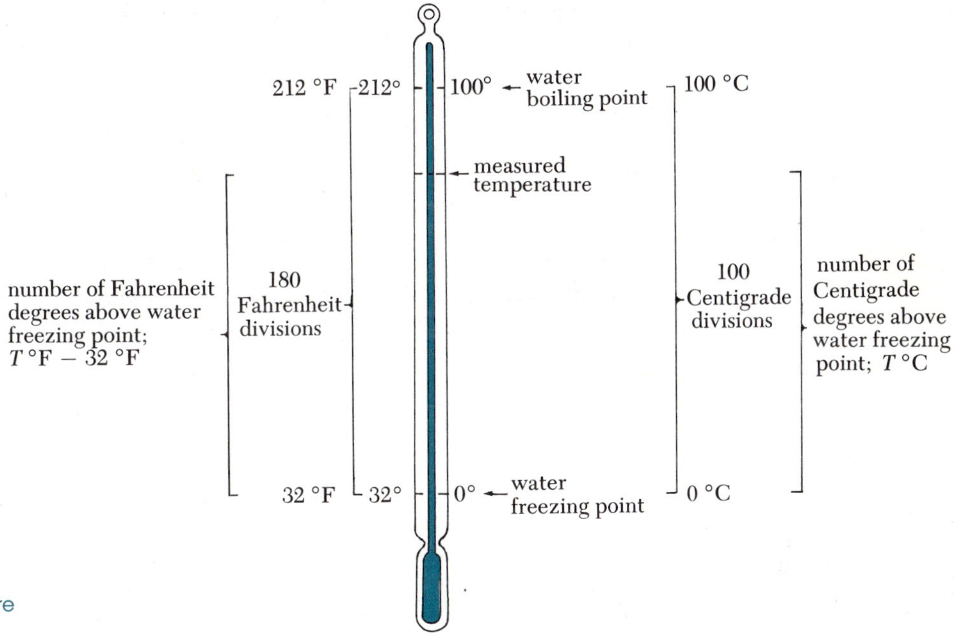

Figure 2.5 Simple mercury thermometer and the definition of temperature scales.

of mercury under these two specified conditions are marked with these values, then a Fahrenheit scale is defined. The distance between these two points on the thermometer scale is divided into 180 equal spaces and each of these is given the significance of one Fahrenheit degree. Thus we see that 0 °C and 32 °F specify the same sets of conditions, that is, the same temperature; and accordingly, 100 °C and 212 °F are also equivalent temperatures. Furthermore, since only 100 centigrade degrees span the distance between the two chosen reference points, whereas 180 Fahrenheit degrees are required to span the same temperature difference, it follows that one centigrade degree must be equivalent to $\frac{180}{100}$, or nine-fifths, of a Fahrenheit degree. By these relations, one can algebraically compute the temperature specified in °C from a temperature specified in °F, or vice versa. This relation is shown algebraically in the following two equations and diagramatically in Fig. 2.6:

$$\text{degrees F} = \tfrac{9}{5}(\text{degrees C}) + 32 \tag{2.1a}$$

$$\text{degrees C} = \tfrac{5}{9}(\text{degrees F} - 32) \tag{2.1b}$$

With these concepts defined we can reasonably characterize the behavior of most gases by the empirical laws discovered during the seventeenth through the nineteenth centuries. **Boyle's law** states that *a sample of gas of fixed weight at a given temperature has a volume that is inversely proportional to its pressure.*

This law can be stated mathematically as follows:

$$V = (\text{constant})\left(\frac{1}{P}\right) \quad \text{for fixed temperature and weight of the sample} \tag{2.2}$$

For example, if a 0.28-g sample of N_2 gas has a volume of 270 cm^3 at a pressure of

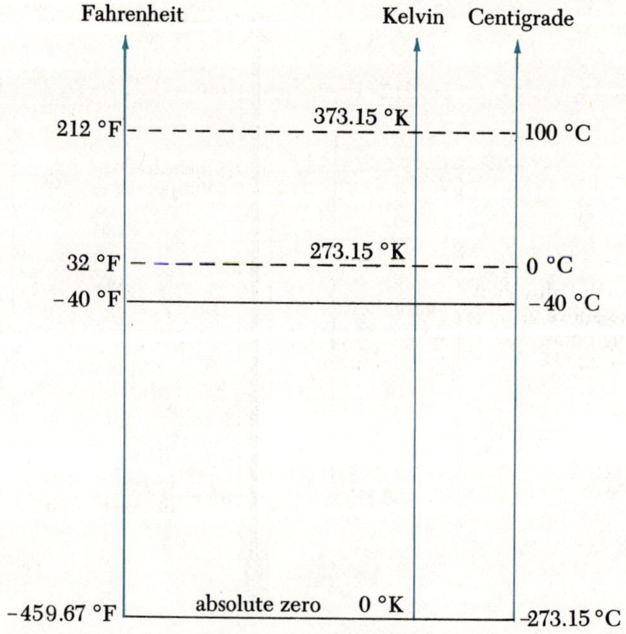

Figure 2.6 Showing the relation of Fahrenheit, Centigrade, and Kelvin temperature scales.

1.0 atm and 50 °C, then at a pressure of 3.0 atm at the same temperature, this same sample would have a volume of

$$(270\ \text{cm}^3)\left(\frac{1.0\ \text{atm}}{3.0\ \text{atm}}\right) = 90\ \text{cm}^3$$

That is, increasing the pressure by a factor of 3 reduces the volume to one-third its initial value.

Charles's law states that a gaseous sample of fixed weight has a volume that varies linearly with temperature if the pressure is constant. A plot of V vs T would be a straight line (see Fig. 15.2). Charles's law can be stated more conveniently if a new temperature scale is defined, the so-called **absolute,** or **Kelvin** (K), scale. This scale is defined by the equation

$$\text{degrees K} = \text{degrees C} + 273.15 \tag{2.3}$$

The size of the absolute, or Kelvin, degree is the same as the size of the Celsius degree; however, one scale is shifted from the other in the amount 273.15 degrees. This relation is shown in Fig. 2.6.

EXAMPLE 2.1 (a) Express 39.5 °F in the Celsius and in the Kelvin scales. (b) Express 92 °K as a Fahrenheit measurement.

SOLUTION (a) By Eq. (2.1b),

$$\text{deg C} = \tfrac{5}{9}(39.5 - 32) = 4.2 \quad (4.2\ °\text{C})$$

By Eq. (2.3),

$$\text{deg K} = 4.2 + 273.15 = 277.4 \quad (277.4\ °\text{K})$$

(b) By Eq. (2.1a) and (2.3),

$$\begin{aligned}\text{deg F} &= \tfrac{9}{5}(92 - 273.15) + 32 \\ &= -326 + 32 = -294 \quad (-294\ ^\circ\text{F})\end{aligned}$$

STUDY PROBLEM 2(b)

Express 192 °F and −13 °F in Celsius degrees and in Kelvin degrees.

In terms of the Kelvin temperature scale, Charles's law can be restated as a simple proportionality: *The volume of a gaseous sample of fixed weight is directly proportional to the absolute temperature if pressure is constant.*

$$V = (\text{constant})(T) \qquad \text{for fixed pressure and weight of sample} \tag{2.4}$$

These so-called gas laws are often combined in a single expression, frequently called the *ideal gas law* (Equation 2.5):

$$P = (\text{const})\left(\frac{nT}{V}\right) \tag{2.5}$$

In this equation, V and P are the volume and pressure in a specified choice of units; T is the absolute temperature; n is a quantity that represents the amount of sample by the number of some basic structural unit, for example, molecules; and the value of the proportionality constant (const) depends on the units chosen for V, n, and T. Equation (2.5) tells us that the pressure of a gas is directly proportional to the temperature if n and V are held constant; directly proportional to the amount (that is, mass) of sample if V and T are held constant; and inversely proportional to the volume of the container if n and T are held constant.

EXAMPLE 2.2 What is the pressure of a sample of gas of fixed mass and volume at a temperature of 90 °C if its pressure was 1.00 atm at 25 °C?

SOLUTION Using Eq. (2.5) and recognizing that n and V are held constant, we see that P is directly proportional to T (in °K). Then, using Eq. (2.3) to convert from °C to °K, and where the subscript "i" indicates "initial" and the subscript "f" indicates "final," we get

$$\frac{P_f}{P_i} = \frac{T_f}{T_i}$$

where $T_f = 90 + 273 = 363$ °K and $T_i = 25 + 273 = 298$ °K.

$$\begin{aligned}P_f &= P_i\frac{T_f}{T_i} \\ &= (1.00\ \text{atm})\left(\frac{363\ ^\circ\text{K}}{298\ ^\circ\text{K}}\right) \\ &= \frac{363}{298}\ \text{atm} = 1.22\ \text{atm}\end{aligned}$$

EXAMPLE 2.3 What is the final pressure of a sample of gas of fixed mass if its pressure is initially 0.90 atm at a volume of 250 cm³, and it is then compressed to a volume of 100 cm³ while its temperature is held constant?

SOLUTION Using Eq. (2.5), and recognizing that n and T are held constant, we see that P is inversely proportional to V:

$$\frac{P_f}{P_i} = \frac{V_i}{V_f}$$

$$P_f = (P_i)\frac{V_i}{V_f}$$

$$= (0.90 \text{ atm})\left(\frac{250 \text{ cm}^3}{100 \text{ cm}^3}\right) = 2.2 \text{ atm}$$

EXAMPLE 2.4 What is the volume of a sample of gas of fixed mass if its pressure is maintained constant while the temperature is lowered from 100 °C to 10 °C? At 100 °C, the volume was 135 ml.

SOLUTION From Charles's law, or from a rearranged form of Eq. (2.5) ($V = \text{const} \cdot nT/P$), we see that with n and P held constant, V is directly proportional to T (°K):

$$\frac{V_f}{V_i} = \frac{T_f}{T_i}$$

$$V_f = (V_i)\frac{T_f}{T_i}$$

$$= (135 \text{ ml})\left[\frac{(10 + 273) \text{ °K}}{(100 + 273) \text{ °K}}\right]$$

$$= (135 \text{ ml})\left(\frac{283}{373}\right) = 102 \text{ ml}$$

EXAMPLE 2.5 What is the volume of a sample of gas of fixed mass if temperature is maintained constant while pressure is increased from 1.20 atm to 2.22 atm? The volume was 255 ml at 1.20 atm.

SOLUTION From Boyle's law, or from the rearranged form of Eq. (2.5) given in the preceding example, we see that with n and T held constant, V is inversely proportional to P:

$$\frac{V_f}{V_i} = \frac{P_i}{P_f}$$

$$V_f = (V_i)\frac{P_i}{P_f}$$

$$= (255 \text{ ml})\left(\frac{1.20 \text{ atm}}{2.22 \text{ atm}}\right) = 138 \text{ ml}$$

STUDY PROBLEM 2(c)

What is the pressure of a sample of gas of fixed mass and volume at a temperature of 40 °C if pressure was 0.95 atm at 0 °C?

STUDY PROBLEM 2(d)

What is the final pressure of a sample of gas of fixed mass if its pressure is initially 1.20 atm at a volume of 200 cm^3, and it is then compressed to a volume of 110 cm^3 while temperature is held constant?

STUDY PROBLEM 2(e)

What is the final volume of a sample of gas of fixed mass with an initial volume of 140 ml at 0.90 atm if the pressure is increased to 1.70 atm while the temperature is maintained constant?

The word *ideal* as used in the term *ideal gas law* indicates that the law is strictly valid only for a limiting, or ideal, case. This limiting case is perhaps never quite achieved exactly, but in so many different situations it is approached closely enough that it can be used for numerical calculations. The ideal gas law is reasonably well obeyed for most gases in which the pressure is not excessively high and the temperature is not too low. (See Chap. 15 for more detailed discussion.)

2.4 Classification of Matter According to Complexity

Composition, Properties, and Processes

Ideas on the complexity of specific samples of matter can be developed relative to a pure substance. A **pure substance,** frequently referred to by just the word *substance,* can be defined as a specific form of matter with a specific set of characteristics, including a precisely defined composition. By *composition,* we mean the relative amounts of various fundamental types of matter, the elements. A specific substance has a specific composition; that is, it is composed of one or more elements in definite proportions by weight or mass. Furthermore, it has a fixed set of properties under a given set of conditions, that is, pressure and temperature.

The concept of a property is critical here, especially in defining precisely the distinctions between various kinds of substances and in defining more complicated materials. A property of a sample is a characteristic of the sample. It is often a measured response to a specific stimulus. A property can be a noted qualitative result or a measured quantitative result from the interaction of the sample and something else. We do something to the sample (for example, heat it, treat it with oxygen, or shine light on it) and we note what happens. This stimulus, or interaction, can often be viewed as some sort of process. If it is a process that changes the composition of the substance, then the associated property is a **chemical property** and the process itself a **chemical process.** If the process does not involve a change in composition, then it is referred to as a **physical process**, and the property associated with this physical process is a **physical property.** In summary, properties and the processes leading to their definition are called chemical if compositional changes are involved, and physical otherwise.

Another classification for properties distinguishes intensive and extensive types. **Extensive properties** are those that depend on the amount of a sample; examples are volume and weight. **Intensive properties** are independent of size—they are the same for a small quantity of a given substance as for a large quantity of the same substance under the same conditions. Examples of intensive properties are density, color, odor, hardness, corrosion resistance, and temperature. A small piece of copper metal has the same density, color, odor, hardness, resistance to corrosion, and temperature as a larger piece of copper under the same conditions.

Elements and Compounds

Having thus characterized materials, we can now formulate precise definitions to distinguish explicitly among various kinds. A substance that cannot undergo any transformation in which it is decomposed into simpler substances must itself consist of only one type of matter; this is referred to as an elemental substance, or simply, an **element.** A substance that can be decomposed into simpler substances in a chemical process or that can be formed in a chemical process from more than one element obviously contains more than one element; it is referred to as a compound substance, or simply, a **compound.** Since compounds are substances, they must have precisely defined compositions. The elements they comprise must occur in precisely defined proportions that are always the same for any sample of a given compound. Elemental composition is one of the properties characteristic of a particular compound. The compound called sodium chloride, which we know as common table salt, consists of 39.3% (by weight) of the element sodium and 60.7% (by weight) of the element chlorine; and this is always true for any pure sample of sodium chloride regardless of its origin. When a sample of this compound is melted at very high temperatures, it can be decomposed into the elements of which it is made by a chemical process known as **electrolysis** (the direct use of electrical energy to bring about a chemical change). This decomposition is depicted in Fig. 2.7. The materials produced in the electrolysis of sodium chloride are always obtained in the sodium-to-chlorine mass

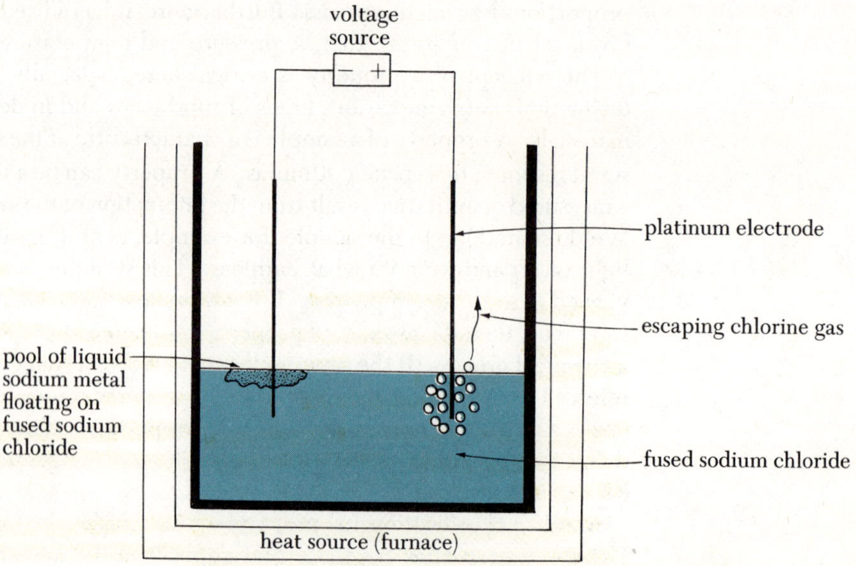

Figure 2.7
Electrolysis of fused sodium chloride at 1000 °C, showing the chlorine gas and liquid sodium metal produced.

ratio 39.3:60.7. The task of determining the mass percentage of each element in a material, that is, the elemental composition, is called **analysis.**

Looking at sodium chloride in another way, we realize that it can be prepared from the elements sodium and chlorine just as it can be decomposed into them. Constructing compounds is called **synthesis.** It is found in synthesizing sodium chloride from elemental sodium and chlorine that the starting materials combine in relative amounts that correspond to 39.3% sodium and 60.7% chlorine. These percentages, the same as the percentages found by analysis, are always the same for this specific compound, irrespective of when, where, why, or by whom the compound is produced; it is independent of the phase of the moon, the tide, or any other coincidental details associated with preparation. More important, the same combination ratio holds regardless of the relative amounts of sodium and chlorine initially used. If the elements had been mixed in a ratio greater than 39.3:60.7 then sodium chloride would still have been produced, but some sodium would have been left over. A ratio less than 39.3:60.7, on the contrary, would have left some chlorine unreacted. This ratio is a clearly and precisely defined characteristic of sodium chloride. Every compound has its own fixed, characteristic composition in terms of the elements of which it is composed.

There is an interesting and important relation between a compound and the elements of which it is composed. This is the overall relation between the properties of a compound and the properties of its constituent elements; and it is dramatically exemplified by sodium chloride. As a general rule, we find these properties to be widely different. Sodium chloride is a crystalline material that appears colorless if viewed at close range or with a microscope, and it has the appearance of being white when granulated.* It dissolves substantially in water to produce a new liquid material with the characteristic salty taste, the familiar "salt water." In contrast, the element chlorine is a greenish yellow gas under typical conditions (say, atmospheric pressure and room temperature). It is poisonous, and a very harsh chemical that must be treated with caution. It dissolves to some extent in water, producing a liquid material that has a lethal effect on bacteria; thus it can be used in purifying water in swimming pools and so on. The very reactive element sodium occurs at room temperature as a solid metal. It has the characteristic shiny luster of metals, and in almost no way resembles the solid salt that is produced from it by the action of chlorine. Metallic sodium reacts explosively on contact with water, producing a caustic solution (sometimes called lye) with properties that are known as basic; the properties of lye include the ability to transform fatty substances into soap. Such observations clearly show that the elements composing a compound are combined not only in fixed proportions but in such a way that they no longer have their former chemical properties. The compound has a new set of intensive properties entirely its own!

*Individual crystals of sodium chloride have a colorless and translucent appearance because sodium chloride does not strongly absorb light (radiation) with frequencies in the region to which the human eye is sensitive. In later chapters some detailed attention will be given to *spectroscopy,* the study of how matter absorbs radiation. The white appearance of granulated sodium chloride when the substance is viewed at a distance and without a microscope is due to the efficient reflection of visible light of all frequencies by the small surfaces of the crystals.

A total of 85 elements have been found to occur naturally on the earth, either as pure elemental substances or in combination with other elements, that is, as compounds. Many elements are familiar to the nonscientist as well as to the scientist. Such elements as iron, gold, oxygen, uranium, platinum, silver, nickel, copper, aluminum, and iodine have familiar names, which are common words in day-to-day nonscientific usage. They are materials with which everyone is familiar. Many other examples could be quoted. In addition to the 85 naturally occurring elements, there are another 19 that have been "artificially produced." These have been discovered as a result of the high-energy processes that characterize the scientific studies nuclear physics and nuclear chemistry. Such elements as fermium, californium, einsteinium, and berkelium are in this category. These elements are radioactive.

A set of symbols based on the abbreviations of the names of elements or their Latin equivalents has been devised to simplify the representation of elements. These are given in alphabetical order in Table 2.1. In Tables 2.2 and 2.3 are listed

TABLE 2.1 **The Elements: Abbreviations and Discoveries**

Element	*Symbol*	*Discovery*	*Element*	*Symbol*	*Discovery*
Actinium	Ac	Debierne (1899); Giesel (1902)	Erbium	Er	Mosander (1843)
Aluminum	Al	Oersted (1825); Wöhler (1827)	Europium	Eu	Demarcay (1896)
Americium	Am	Seaborg, Ghiorso et al. (1944)	Fermium	Fm	Ghiorso et al. (1953)
Antimony	Sb[a]	Ancients	Fluorine	F	Moisson (1886)
Argon	Ar	Rayleigh, Ramsay (1894)	Francium	Fr	Perey (1939)
Arsenic	As	Magnus (1250)	Gadolinium	Gd	Marignac (1880); Boisbaudran (1886)
Astatine	At	Corson, Segré et al. (1940)			
Barium	Ba	Crawford, (1800); Davy (1808)	Gallium	Ga	Boisbaudran (1875)
Berkelium	Bk	Thompson, Ghiorso, Seaborg (1949)	Germanium	Ge	Winkler (1886)
			Gold	Au[c]	Ancients
Beryllium	Be	Vauquelin (1798)	Hafnium	Hf	Coster, Von Hevesey (1923)
Bismuth	Bi	Pott (1739); Geoffroy (1753)	Helium	He	Janssen (1868)
Boron	B	Homberg (1702); Gay-Lussac, Thenard (1808)	Holmium	Ho	Delafontaine et al. (1878); Cleve (1879)
Bromine	Br	Balard (1826)	Hydrogen	H	Cavendish (1766)
Cadmium	Cd	Stromeyer (1817)	Indium	In	Reich, Richter (1863)
Calcium	Ca	Davy (1808)	Iodine	I	Courtois (1811); Gay-Lussac (1813)
Californium	Cf	Thompson et al. (1950)			
Carbon	C	Ancients	Iridium	Ir	Tennant (1803)
Cerium	Ce	Berzelius, Klaproth (1803)	Iron	Fe[d]	Ancients
Cesium	Cs	Bunsen, Kirchhoff (1860)	Krypton	Kr	Ramsay, Travers (1898)
Chlorine	Cl	Scheele, Lavoisier (1774)	Lanthanum	La	Mosander (1839)
Chromium	Cr	Vauquelin (1797)	Lawrencium	Lw	Ghiorso et al. (1961)
Cobalt	Co	Brandt (1735); Cronstedt (1750)	Lead	Pb[e]	Ancients
Copper	Cu[b]	Ancients	Lithium	Li	Arfvedson (1817)
Curium	Cm	Seaborg, James, Ghiorso (1944)	Lutetium	Lu	Urbain (1907); Von Welsbach (1908); James (1911)
Dysprosium	Dy	Boisbaudran (1886)			
Einsteinium	Es	Ghiorso et al. (1952)	Magnesium	Mg	Grew (1695); Black (1755)

The Latin names from which certain elemental symbols have been derived are given herewith. [a]*Stibium* for antimony. [b]*Cuprum* for copper. [c]*Aurum* for gold. [d]*Ferrum* for iron. [e]*Plumbum* for lead.

Element	*Symbol*	*Discovery*	*Element*	*Symbol*	*Discovery*
Manganese	Mn	Gahn (1774)	Ruthenium	Ru	Klaus (1844)
Mendelevium	Md	Ghiorso et al. (1955)	Samarium	Sm	Boisbaudran, (1879); von Welsbach (1885)
Mercury	Hg[f]	Ancients	Scandium	Sc	Nilson (1879)
Molybdenum	Mo	Scheele (1778); Hjelm (1782)	Selenium	Se	Berzelius (1817)
Neodymium	Nd	Mosander (1841)	Silicon	Si	Berzelius (1824)
Neon	Ne	Ramsay, Travers (1898)	Silver	Ag[h]	Ancients
Neptunium	Np	McMillan, Abelson (1940)	Sodium	Na[i]	Davy (1807)
Nickel	Ni	Cronstedt (1751)	Strontium	Sr	Scheele (1779); Davy (1808)
Niobium	Nb	Hatchett (1801)	Sulfur	S	Ancients
Nitrogen	N	D. Rutherford (1772)	Tantalum	Ta	Ekeberg (1802)
Nobelium	No	Ghiorso et al. (1958)	Technetium	Tc	Perrier, Segré (1937)
Osmium	Os	Tennant (1803)	Tellurium	Te	Von Reichenstein (1782); Klaproth (1798)
Oxygen	O	Scheele (1774); Priestley (1774)	Terbium	Tb	Mosander (1843)
Palladium	Pd	Wollaston (1803)	Thallium	Tl	Crookes, Lamy (1861)
Phosphorus	P	Brand (1669)	Thorium	Th	Berzelius (1828)
Platinum	Pt	Ancients	Thulium	Tm	Cleve (1879)
Plutonium	Pu	Seaborg et al. (1940)	Tin	Sn[j]	Ancients
Polonium	Po	Curie (1898)	Titanium	Ti	Gregor (1791)
Potassium	K[g]	Davy (1807)	Tungsten	W[k]	de Elhuyar brothers (1783)
Praseodymium	Pr	Mosander (1841); Boisbaudran (1879); Von Welsbach (1885)	Uranium	U	Klaproth (1789)
Promethium	Pm	Marinsky, Glendenin (1945)	Vanadium	V	Sefström (1830); Wohler (1831)
Protactinium	Pa	Hahn, Meitner (1917)	Xenon	Xe	Ramsay, Travers (1898)
Radium	Ra	M. and Mme. Curie (1898)	Ytterbium	Yb	Marignac et al. (1878)
Radon	Rn	Dorn (1900)	Yttrium	Y	Gadolin (1794); Mosander (1843)
Rhenium	Re	Noddack et al. (1925)	Zinc	Zn	Ancients
Rhodium	Rh	Wollaston (1803)	Zirconium	Zr	Klaproth (1789); Berzelius (1824)
Rubidium	Rb	Bunsen, Kirchhoff (1861)			

[f]*Hydrargyrum* for mercury. [g]*Kalium* for potassium. [h]*Argentum* for silver. [i]*Natrium* for sodium. [j]*Stannum* for tin. [k]The German word *Wolfram* is the origin of the symbol "W."

TABLE 2.2 Estimated Elemental Composition of the Earth (Percentage by Weight)

Solid Crust		*Ocean*		*Atmosphere*	
Oxygen	46.5	Oxygen	85.8	Nitrogen	75.2
Silicon	28.0	Hydrogen	10.7	Oxygen	23.4
Aluminum	8.1	Chlorine	1.9	Argon	1.3
Iron	5.1	Sodium	1.1	Carbon	0.12
Calcium	3.5	Magnesium	0.13	Neon	0.000014
Sodium	3.0	Sulfur	0.09	Krypton	0.000003
Potassium	2.5	Calcium	0.04	Helium	0.0000007
Magnesium	2.2	Potassium	0.04	All others	0.0000003
Titanium	0.5	All others	0.20		
Hydrogen	0.2				
All others	0.4				

TABLE 2.3 **Approximate Elemental Composition of the Human Body (Percentage by Weight)**

Oxygen	65.0	Nitrogen	3.0
Carbon	18.0	Calcium	2.0
Hydrogen	10.0	Phosphorus	1.0

some of the most important elements that occur in the earth's atmosphere, crust, and ocean, and in the human body. You can see that oxygen is by far the most abundant element on earth.

STUDY PROBLEM 2(f)

Identify each of the following with the most appropriate boldface word or words in Sec. 2.4 to this point: (a) hardness; (b) freezing; (c) gold; (d) burning; (e) water or copper; (f) weight; (g) color.

Elements of Biological Importance

Elements vary in their relative importance to life on Earth. From Table 2.3, which is concerned only with the human body, it is apparent that oxygen, carbon, hydrogen, nitrogen, calcium, and phosphorus are important. These same elements, together with potassium, magnesium, manganese, iron, copper, zinc, and sulfur, are essential to all known plants and animals. In addition, sodium, molybdenum, cobalt, silicon, chlorine, and iodine are essential to several classes of plants and animals. Vanadium, cadmium, boron, selenium, bromine, and fluorine are required by a wide variety of species of at least one class. The essentiality of barium, strontium, chromium, and tin is more limited, but well demonstrated.

Deficiencies in essential elements can lead to serious illness. And excesses can in some cases be equally dangerous. The human body, for example, requires low levels of chromium and selenium. There are narrow "tolerance ranges," however; if they are exceeded, illness and even death can result.

Spotlight

In an organism, balancing mechanisms exist to prevent overaccumulation of certain elements. These controls, called **homeostatic** (same state) **mechanisms,** ordinarily function very well—ridding the body of excess elements through screening action of the intestinal tract and kidneys. These mechanisms fail, however, if intake is less than the minimum excretion level, if severe overloading of the excretory system occurs, or if diseases, especially hereditary ones, cause a deficiency in chemical species needed for the homeostatic action. For certain elements, including nonessential ones, the homeostatic mechanisms are not very efficient and the elements can accumulate in the body. Examples are lead, mercury, and cadmium. This is very serious, since exposure to low levels of such elements over a long period, as well as the more obvious short-term exposure to high levels, can be very toxic. The hazard of gradual accumulation has led to great concern about the release of elements such as lead, mercury, and cadmium into the atmosphere and into water supplies.

Figure 2.8
Elements, compounds, and mixtures.

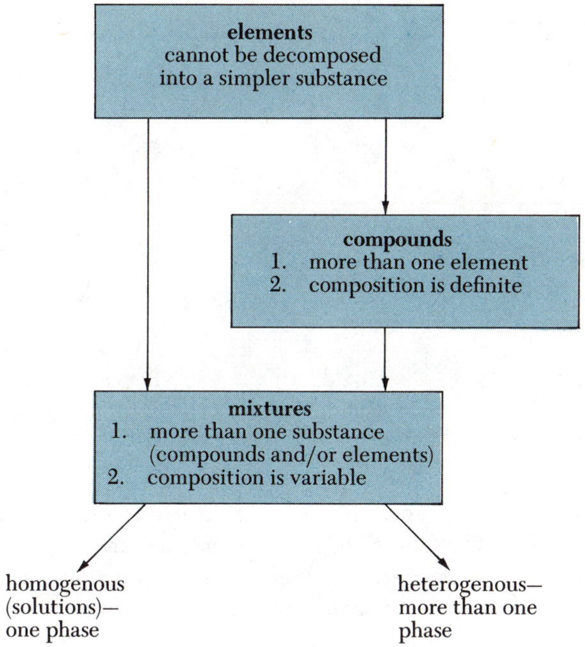
elements
cannot be decomposed
into a simpler substance
compounds
1. more than one element
2. composition is definite
mixtures
1. more than one substance (compounds and/or elements)
2. composition is variable
homogenous (solutions)— one phase
heterogenous— more than one phase

Mixtures, Solutions, and Dispersions

Not only the chemical composition but any intensive property of a given substance is characteristic of that substance. Hence, it is clear that such properties are uniform throughout any portion of the substance. Such a sample of matter, with uniform intensive properties, is described as **homogeneous**, and as possessing a single phase. Accordingly, a **phase** is defined as a bounded region of matter having uniform intensive properties throughout. If the entire sample consists of a single phase, it is therefore homogeneous.

A sample of matter containing more than one substance is a **mixture.** Thus, a mixture is not a single, pure substance, and the relative proportions of its components may vary widely. That a substance has a definite composition is perhaps the most important single criterion differentiating mixtures from substances. The composition of a mixture can be varied a little to give a slightly different mixture. In contrast, the composition of a pure substance can not be varied at all; if it were varied, the substance would no longer be pure. Some of these definitions are emphasized schematically in Fig. 2.8.

If the mixture consists of only one phase (that is, is uniform throughout), then it is referred to as a homogeneous mixture. If more than one phase is present in a mixture, then it is referred to as **heterogeneous.** Homogeneous mixtures are called **solutions,** and "nearly uniform" mixtures are called heterogeneous **dispersions.**

Solutions

For solutions, it is common to refer to the component present in the largest amount as the **solvent**, and to any component present in smaller amount as a **solute.** Before being mixed, the solvent and the solute may be gaseous, liquid, or solid. Hence there

TABLE 2.4 **Types of Solutions**

Solvent[a]	*Solute*	*Example*
Gas	Gas	Air (oxygen in nitrogen)
	Liquid	Moist air (water in air)
	Solid	Mothball odor (P-dichlorobenzene in air)
Liquid	Gas	Carbonated water (carbon dioxide in water)
	Liquid	Alcohol in water
	Solid	Salt in water
Solid	Gas	Hydrogen in platinum
	Liquid	Benzene in iodine
	Solid	Metallic alloys

[a]Note that the state of the solution is the state of the solvent.

are nine possible "types" of solutions, referred to the physical states of the separate components. These nine possibilities, and an example of each, are given in Table 2.4. As the table shows, there are three general types of solutions relative to the physical state of the solution: gaseous, solid, and liquid solutions.

GASEOUS SOLUTIONS. All gases mix to form solutions. In some cases, when a gaseous solution is formed from a mixture of two gas samples, a chemical reaction can occur that will produce solid or liquid materials. An example of this kind of behavior is seen if gaseous ammonia and gaseous hydrogen chloride are mixed; solid ammonium chloride is formed, which appears as a dense white smoke. This process is depicted in Fig. 2.9. Air is a homogeneous mixture of elemental substances such as nitrogen,

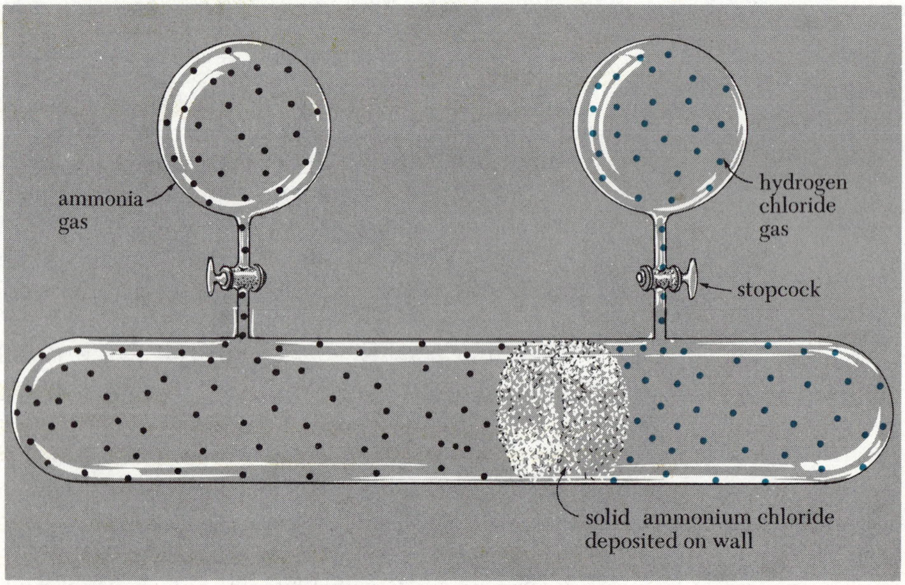

Figure 2.9 Formation of a solid ammonium chloride from a gaseous solution of ammonia and hydrogen chloride prepared by introducing both components into a common chamber. (The presence of ammonia gas and hydrogen chloride gas is represented symbolically by dots.)

oxygen, and argon, and compounds such as carbon dioxide, and unfortunately, various pollutants; air can be viewed as a gaseous solution. Some pollutants also occur as dispersions of tiny droplets or solid particles in air; these are the aerosols.

SOLID SOLUTIONS. Some materials combine to form homogeneous solid solutions. Many of these have had great importance in the technologies of both modern and earlier ages. Some solid solutions of metals, referred to as **alloys**, prove to be structurally important materials. Some of these have been used throughout much of recorded time; examples are brass, an alloy of copper and zinc, and bronze, an alloy of copper, tin, and zinc. Other alloys have been developed and used more recently. In this category are alloys of iron with nickel and with other metals, which constitute some of the "miracle" stainless steels that are now in widespread use. Another solid solution that is very important today is the type known as a semiconductor; semiconductors are largely responsible for the so-called solid state revolution in electronics (see Chap. 14).

LIQUID SOLUTIONS. Some substances, when mixed together, form homogeneous liquid mixtures, that is, liquid solutions. Examples of this classification are alcohol in water, sugar in water, gasoline (which is a mixture of liquid "hydrocarbons"), and many medicinal preparations such as "tincture of iodine" (a solution of iodine in alcohol). Liquid solutions frequently provide a convenient method of bringing substances together for chemical reactions; usually the reacting substances are present as solutes. Furthermore, liquid solutions provide a vehicle for introducing solute quantities of known amounts. The "bookkeeping" is accomplished by following **concentration,** the amount of solute material contained in a given amount of solvent (or alternatively, in a given amount of solution).

Some examples of concentration units are grams of solute per 100 g of solvent, milligrams of solute per milliliter of solution, and percentage of solute by weight. There is a much more convenient convention for defining concentrations in chemistry that will be introduced later. An extremely important phenomenon that occurs in some solutions is the appearance of electrically charged particles that result when certain solute substances are added to the solvent water. This process, called ionization, occupies an important place in our understanding of many aspects of solution chemistry (see Chap. 3 and subsequent chapters).

Dispersions

The heterogeneity of most heterogeneous mixtures is obvious because of nonuniformity in appearance; thus a sample of concrete, a mixture of sand and water, a bowl of soup, or a sample of granite are each heterogeneous, since each clearly consists of more than one phase. Some mixtures contain nonuniformities that are not so readily obvious, because the individual macroscopic particles are too small for straightforward visual detection; in such mixtures one phase is more or less evenly distributed throughout the other. The whole *appears* homogeneous. Such a mixture is referred to as a **dispersion,** and the major and minor components are referred to as the **dispersing medium** and the **dispersed phase.** Like solutions, dispersions exist in many types. The categories are the same as the nine shown for solutions in Table 2.4, except for one; for mixtures of gases, only solutions have been observed.

TABLE 2.5 Types of Dispersions

Dispersing Medium	*Dispersed Phase*	*Name*	*Example*
Gas	Gas	——	Unknown
	Liquid	Aerosol or fog	San Francisco fog
	Solid	Aerosol or smoke	Cigarette smoke
Liquid	Gas	Foam	Whipped cream
	Liquid	Emulsion	Homogenized milk (an oil in water)
	Solid	Sol	Precipitated sulfur in water
Solid	Gas	——	Air in pumice stone
	Liquid	Gel	"Jello" (water in gelatin)
	Solid	——	"Ruby glass" (glass colored with dispersed metal)

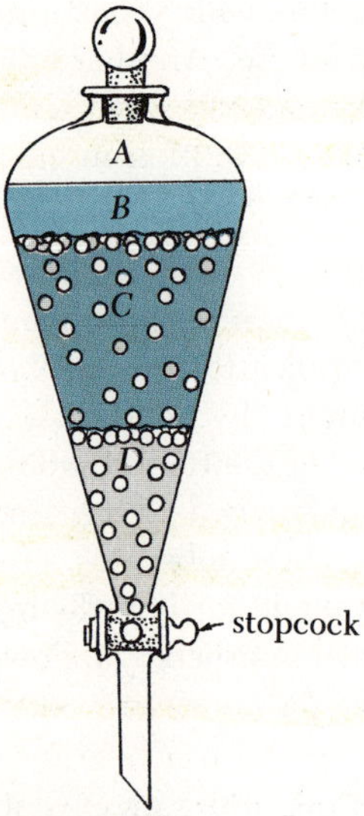

Figure 2.10
Sketch of an emulsion in a separatory funnel containing: Region *A*, air; Region *B*, oil; Region *C*, emulsion of soap solution and oil; Region *D*, soap solution. (White circles in Regions *C* and *D* represent soap bubbles.)

The particle size of the dispersed phase in a dispersion determines whether the mixture is referred to as a colloid or a suspension. The limits are arbitrary, but a particle size less than 1 nm (10^{-9} m) is considered in a true solution; between 1 nm and 1 μm (1 micrometer, 10^{-6} m), in a **colloid;** and greater than 1 μm, in a **suspension.** Special names are used for various types of colloids and these are given, along with an example of each in Table 2.5. An example of an emulsion, which is a liquid dispersed in a liquid, is shown in Fig. 2.10. This case results from vigorously shaking a soap solution with mineral oil; these are liquids that don't mutually dissolve.

Whereas suspensions usually "settle out" after some time, colloids do not. An example of this is cigarette smoke, in which the solid particles are so small that they are prevented from settling by the random submicroscopic motion of the gas, which is the air. The formation of colloids often interferes with laboratory separation procedures. An example of the difficulty experienced in trying to filter a colloidal solid from a liquid is represented in Fig. 2.11.

STUDY PROBLEM 2(g)

Identify each of the following with the most appropriate boldface word or words in the section on mixtures, solutions, and dispersions: (a) water in the ocean; (b) salt in the ocean; (c) soil; (d) pure water; (e) gasoline; (f) vegetable soup; (g) Coca Cola; (h) smoke; (i) a leaf.

Separation Processes

Any mixture, whether it is homogeneous or heterogeneous, can be thought of as a combination of substances that largely retain their identities—the identities that they had in their pure states. When sugar is dissolved in water, many of the properties of the solution are directly attributable to properties of water, the appearance, for instance; whereas others reflect the characteristics of pure sugar, like the taste. Fur-

Figure 2.11
The importance of particle size in separations by filtration. (a) Large, crystalline (solid) particles separated from a clear (supernatant) solution. (b) Fine, solid particles, a colloid, suspended in the fluid, and not separated from it.

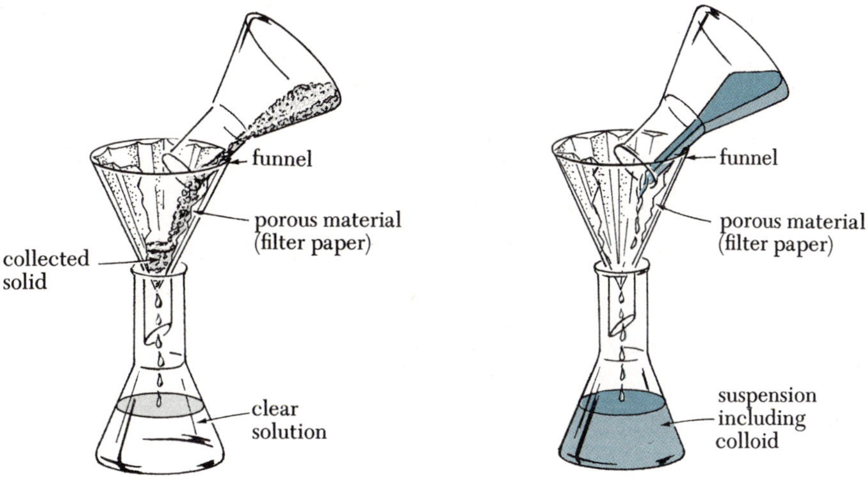

(a) Separation of crystalline (solid) substance

(b) No separation

thermore, a simple physical process can be used to separate the components of the mixture into the substances of which it is composed. This is in direct contrast to the nature of a compound. In a **compound** the constituents, or elements, tend to lose their identities, since the compound has a set of properties peculiarly its own and vastly different from properties of the component elements. Moreover, it is not possible to separate a compound into its component elements in any other way besides a chemical process. Hence, we can readily distinguish between a mixture and a compound by the kinds of processes that can be used to separate them into their components. A physical process suffices to separate the components of a mixture; only a chemical process can be used to separate the components of a compound.

It is easy to visualize physical processes of separation for heterogeneous mixtures. For example, a mixture of gravel and water can be separated by simply pouring off the water, whereupon the gravel is left behind; or perhaps more elegantly, by filtering the gravel away from the water. Separating substances of a homogeneous mixture is often not so trivial; nevertheless, it can be accomplished by different physical processes. These include such processes as crystallization, distillation, extraction, and adsorption. **Crystallization** involves separating materials by lowering the temperature of a mixture until one component freezes out as a crystalline material. **Distillation** depends on raising the temperature until one component evaporates or boils out as a vapor that can be recondensed. **Extraction** involves removing a solute from a solvent by allowing it to dissolve in another solvent that is immiscible with the original solution. **Adsorption** is the process by which certain substances in a liquid or gaseous solution become attached to the surface of a solid or a liquid; in this way such substances can be separated from other substances in the solution. The three last-mentioned processes are shown pictorially in Figs. 2.12, 2.13, and 2.14.

During a physical process in which the components of a mixture are being separated, the composition of the mixture can vary. For example, in separating a solution

Figure 2.12
Distillation apparatus.

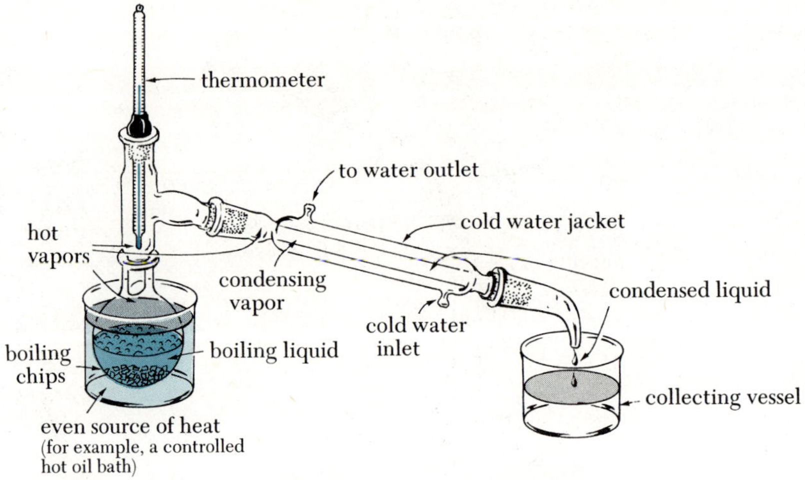

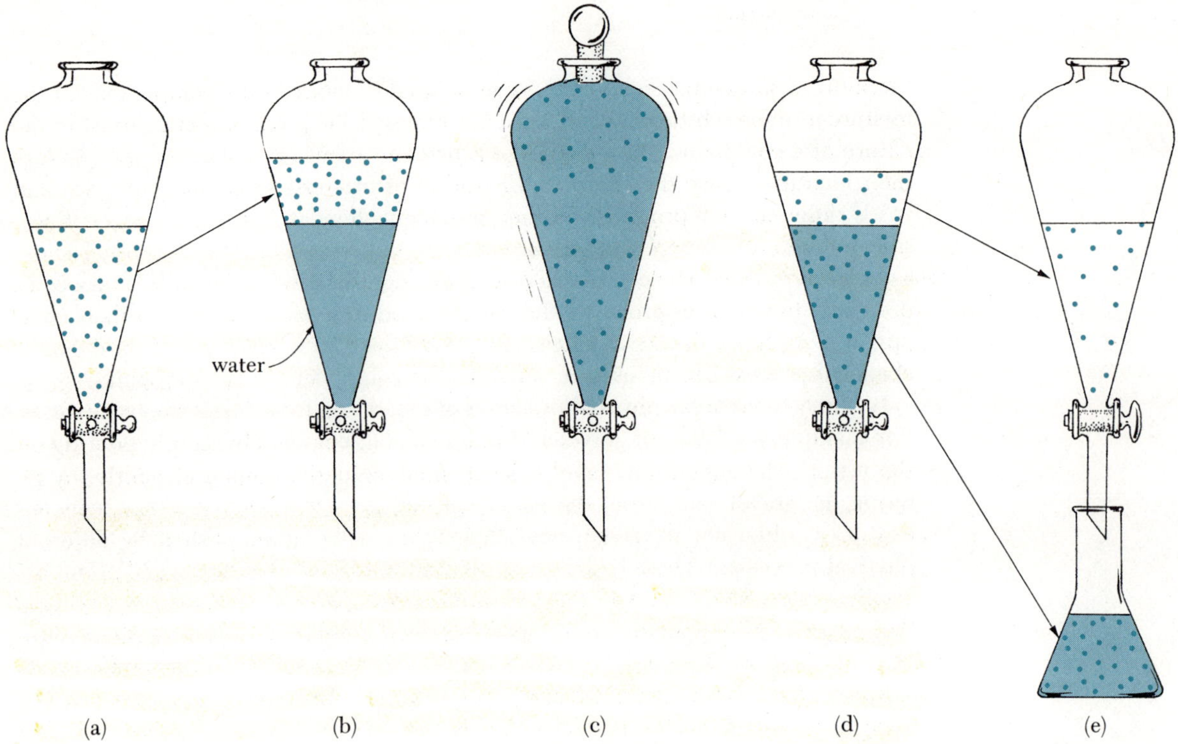

Figure 2.13 The extraction of acetic acid from a solution in ether by water. (The presence of acetic acid is represented symbolically by dots. (a) Original solution of acetic acid in ether. (b) Water added to the original solution (no agitation). The denser water becomes the lower layer. (c) The mixture is shaken vigorously. All components are mixed thoroughly. (d) The agitated mixture is allowed to settle and separate. The denser water layer (now containing some acetic acid) again goes to the bottom. (e) A solution of acetic acid in water is drawn off into a flask. The procedure is repeated until all the acetic acid has been removed (extracted) from the ether.

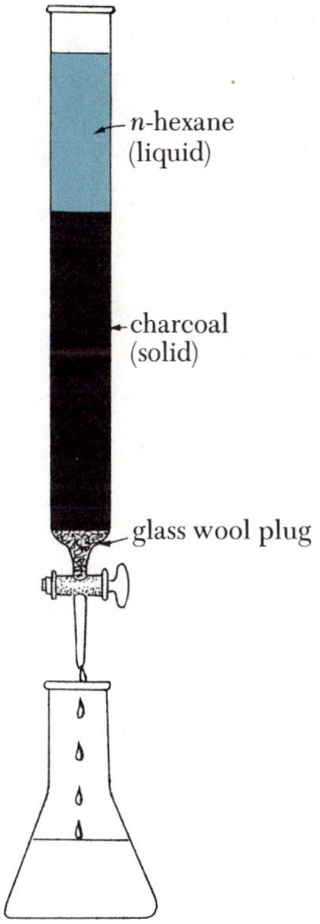

Figure 2.14
Removal of impurities from a sample of *n*-Hexane, C_6H_{14}, by passing it through a column of charcoal (carbon), to which the impurities become adsorbed.

of water and sugar by crystallization and removal of crystalline sugar, one obtains a final solution that is richer in water and more dilute in sugar than the original solution. On the other hand, the composition of the compounds is not altered during a physical process. The elemental compositions of the water and the sugar are unchanged during the crystallization.

Figure 2.15 on page 52 represents the separation relations between chemical and physical processes, elements, compounds, and homogeneous and heterogeneous mixtures.

STUDY PROBLEM 2(h)

Using the terms introduced in Sec. 2.4, characterize the materials that go into the making of English tea flavored by milk and sugar; and characterize the whole result of the brewing.

STUDY PROBLEM 2(i)

(a) List three specific cases in which physical processes are used in purifying some common products of industry. (b) List three specific examples in which chemical processes are used in the industrial manufacture of some common product.

SUMMARY

Matter can be classified in three states—solid, liquid, and gas—by operational definitions based on fluidity and change of sample volume in a new container. These three states also differ in long-range and short-range submicroscopic order. A fundamental competition in matter exists between the randomizing influence of motion at a submicroscopic level and the interaction among the submicroscopic particles that tends to cause order. Vapor pressure is a simple manifestation of this submicroscopic motion. Temperature is a measure of the intensity of this motion, and is measured in scientific work on the Celsius (centigrade) and Kelvin (absolute) scales. The physical behavior of gases can be described by a set of simple proportionalities (Charles's law and Boyle's law) or by the ideal gas law; the laws mathematically relate the pressure, volume, absolute temperature, and amount (that is, mass) of a gaseous sample.

Matter can also be classified into elements, compounds, and mixtures (homogenous or heterogeneous). Substances (elements and compounds) are characterized by fixed sets of physical properties and chemical properties, including chemical composition. Intensive properties can be related to chemical and physical processes. Physical processes can separate mixtures into component substances, but only chemical processes can alter a substance. Each compound has a definite set of properties, which differ from the properties of the component elements.

Figure 2.15
(a) Schematic diagram of the relation between categories of composition, their complexity, and the separation process.
(b) A specific example: successive separations performed on a heterogeneous mixture of solid mercuric oxide and a solution of mercuric chloride in water.

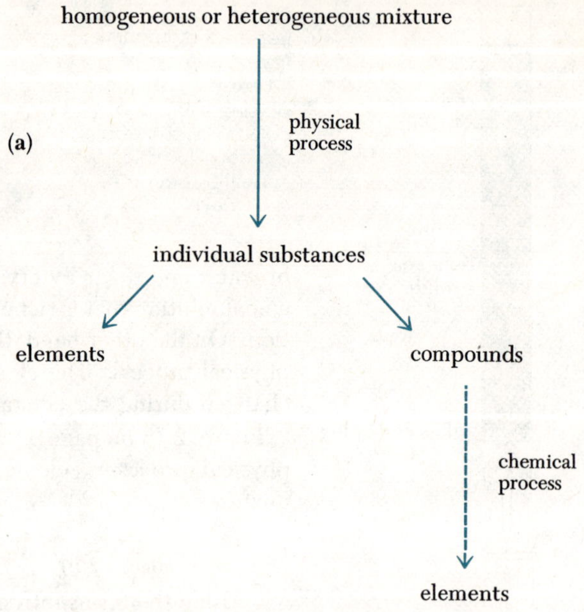

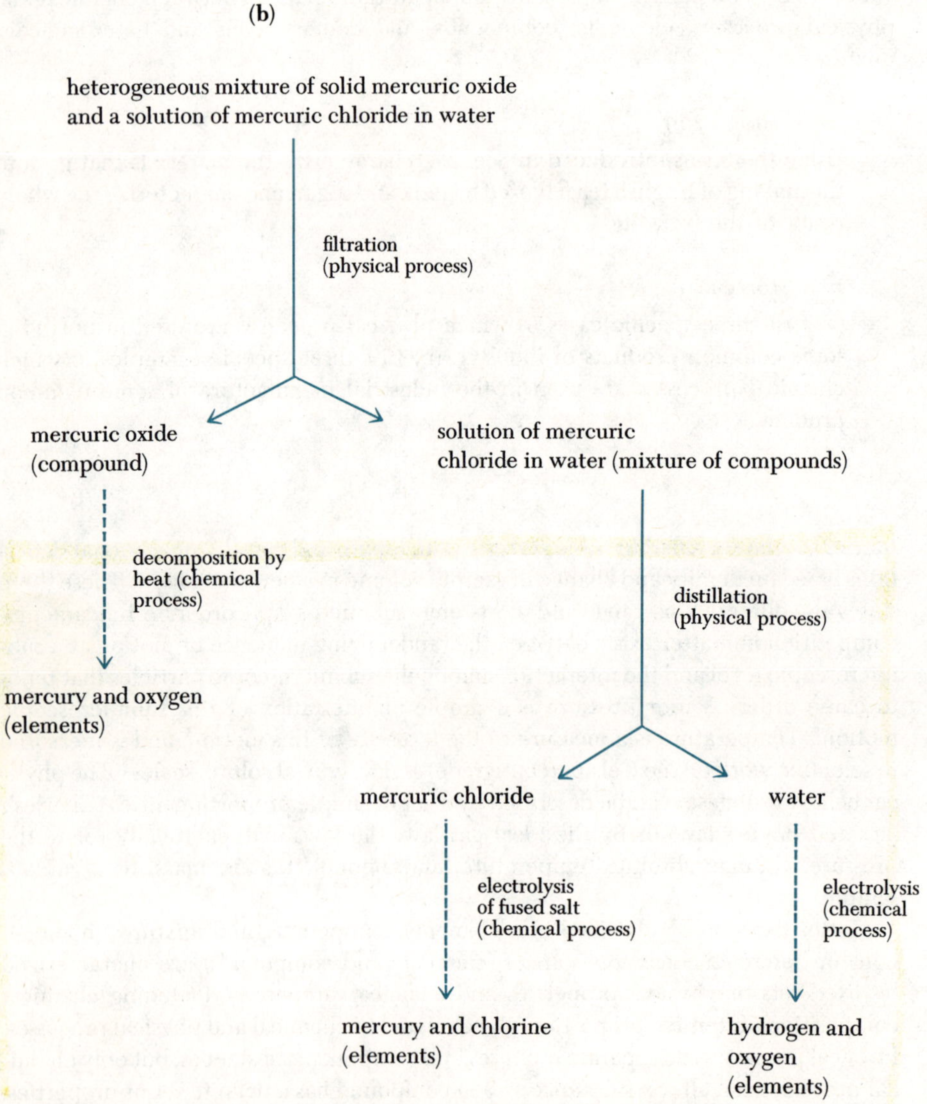

STUDENT CHECKPOINTS

After studying this chapter, the student should be able to:

1. Classify familiar, everyday matter as solids, liquids, or gases, according to the functional definitions of these states.
2. Explain the difference between short-range and long-range order.
3. Interpret the competition between order and the randomizing submicroscopic motions.
4. Convert from one temperature scale to another among the Fahrenheit, Celsius, and absolute scales.
5. Apply Boyle's law and Charles's law.
6. Relate processes (physical or chemical) and intensive properties (physical or chemical).
7. Differentiate among elements, compounds, and mixtures and explain the kinds of processes that can change them.
8. Define the terms *phase, homogeneous,* and *heterogeneous,* and explain the relation among the conditions they describe.

EXERCISES

2.1.* Which of the following qualify as matter: mountain, human body, glass of water, thought, flame, sky, theory?

2.2. Under what conditions, if any, is it possible for an object to have (a) mass, but no weight? (b) weight, but no mass?

2.3. Which substance, do you expect, would evaporate more rapidly under a given set of circumstances (say temperature or wind velocity): a substance with a high vapor pressure, or one with a low vapor pressure? Why?

2.4. Categorize each of the following cases as an example of short-range order or long-range order, or both: (a) ice cubes in a tray in a refrigerator; (b) ice cubes in an ice bucket, (c) golf balls in a bucket; (d) golf balls spread out on a driving range; (e) the cells of a honeycomb.

2.5.* Convert each of the following temperatures in degrees Fahrenheit into temperature on the Celsius scale: (a) -100 °F; (b) -30 °F; (c) 0 °F; (d) 100 °F.

2.6. Convert each of the following temperatures in degrees Celsius into temperature on the Kelvin scale: (a) -17 °C; (b) 79 °C; (c) 340 °C.

2.7.* Convert each of the following temperatures in degrees Fahrenheit into temperature on the Kelvin scale: (a) -117 °F; (b) 6 °F; (c) 59 °F; (d) 1703 °F.

2.8. Convert each of the following temperatures in degrees Kelvin into temperature on the Fahrenheit scale: (a) 671 °K; (b) 57 °K; (c) 2 °K.

2.9.* Calculate the final volume of a 2.80-g sample of N_2 gas whose temperature has been maintained at 0 °C while the pressure has been increased from 1.00 atm to 2.50 atm. The volume at 1.00 atm was 2.24 liters.

2.10. Calculate the final volume of a 0.700-g sample of Cl_2 gas whose pressure has been maintained at 0.50 atm and whose temperature has been increased from 0 °C to 100 °C. The volume at 0 °C was 448 ml.

2.11.* What is the final pressure of a sample of ammonia gas, NH_3, in a 224-ml flask at 0 °C if its pressure was 2.00 atm at 273 °C?

2.12. What is the final pressure of a 0.729-g sample of HCl gas that has been held at a temperature of 80 °C, if its pressure is 700 torr (mmHg) when the volume is 629 ml, and the volume is then reduced to 250 ml?

2.13.* If a certain sample of gas containing a fixed mass has a volume of 150 cm^3 at 20 °C and a pressure of 1.00 atm, what would the pressure be at 100 °C for a volume of 110 cm^3?

2.14. Give five examples of extensive properties and five examples of intensive properties.

2.15. Give three examples of physical properties and three examples of chemical properties.

2.16.* Categorize the following processes as physical or chemical: (a) distilling whiskey; (b) separating cream from milk; (c) burning paper; (d) developing a photograph; (e) flashing a photographic flash bulb; (f) pouring coffee through a filter; (g) freezing water.

2.17. List two elements in which the human body is richer (has a higher percentage by weight) than (a) the earth's crust; (b) the ocean; (c) the atmosphere.

2.18. Considering the low abundance of nitrogen in the earth's crust and in the ocean, how do animals, including man, obtain the nitrogen required for their bodies?

2.19.* Considering the much greater abundance of the element oxygen in the ocean than in air, why can we not breathe under water in the ocean?

2.20. Name six elemental substances you have seen or otherwise had experience with.

2.21.* Characterize each of the following as compound, element, or mixture: (a) uranium; (b) brass; (c) pea soup; (d) snow; (e) diamond.

2.22. Characterize each of the following as compound, element, or mixture: (a) wood; (b) tea; (c) stainless steel; (d) platinum; (e) DDT; (f) helium.

2.23.* Describe how one can determine whether a colorless liquid is a pure substance or a mixture.

2.24. Give two examples of each of the following: (a) a solid element; (b) a liquid element; (c) a gaseous element.

2.25. Give two examples of each of the following: (a) a solid compound; (b) a liquid compound; (c) a gaseous compound.

2.26. Give two common examples of each of the following: (a) a solid mixture; (b) a liquid mixture; (c) a gaseous mixture.

2.27. How many phases are there in each of the following: (a) a glass of Coke as usually served in a cafe; (b) clean air; (c) a salad; (d) a glass of water into which a penny and a dime have been dropped?

2.28.* Which of the following is homogeneous: (a) a jar of steel ball bearings; (b) a small plastic container; (c) clean seawater; (d) a small piece of pine (wood); (e) Italian salad dressing; (f) a shiny new copper penny?

2.29. Give an additional example of each of the nine types of solutions shown in Table 2.4.

2.30. Which of the following solutions has the highest concentration of salt: (a) 1.0 g salt in 190 g water; (b) 0.011 lb salt in 20 ml water; (c) a solution of salt in water that is 10% salt by weight?

2.31.* How many grams of an 8.00% (by weight) solution of sugar in water must be measured out to provide 13.1 g of sugar?

2.32. How many milliliters of a 12% (by volume) solution of alcohol in water must be measured out to provide an amount of alcohol equivalent to 2.2 ml of pure alcohol?

2.33.* How many milliters of a 7.20% (by weight) solution of chloroform in alcohol must be measured out to provide a total of 12.3 g of chloroform if the density of the solution is 0.842 g/ml?

2.34. Suppose one has a mixture consisting of (a) copper granules, each of dimensions about 2 mm by 2 mm by 2 mm, (b) iron balls of 5-mm diameter, and (c) a solution of paraffin (wax) and ether in benzene. Describe a set of physical processes that can be used to separate this mixture into pure substances.

2.35. Describe the physical properties on which the separation processes of Example 2.34 depend. Are these extensive or intensive properties?

2.36. Suppose you have just discovered a way to synthesize a fabulous new anticancer drug, but your production method provides this drug with a 0.1% impurity that is a deadly poison. Both the drug and the impurity are solids at room temperature. Discuss what steps you would take to try to eliminate the deadly impurity from the beneficial drug, and indicate which properties govern the success or failure of your approaches.

ANSWERS TO SELECTED EXERCISES

2.1 mountain, human body, glass of water, flame, sky. 2.5 (a) −73 °C; (b) −34 °C; (c) −18 °C; (d) 38 °C. 2.7 (a) 190 °K; (b) 259 °K; (c) 288 °K; (d) 1201 °K. 2.9 0.896 liters. 2.11 1.00 atm. 2.13 1.74 atm. 2.16 (a) physical; (b) physical; (c) chemical; (d) chemical; (e) chemical; (f) physical; (g) physical. 2.19 Our bodies have no suitable mechanism for using the oxygen of water for the purpose for which we use the oxygen of the air. 2.21 (a) element; (b) mixture; (c) mixture; (d) compound; (e) element (carbon). 2.23 If its composition can be changed by physical processes, it's a mixture. 2.28 (a) no (air, steel); (b) yes; (c) yes; (d) no; (e) no; (f) yes. 2.31 164 g. 2.33 203 ml.

ANSWERS TO STUDY PROBLEMS

2(a) no order; long-range and short-range order; short-range order. 2(b) 89 °C = 362 °K; −25 °C = 248 °K. 2(c) 1.09 atm. 2(d) 2.18 atm. 2(e) 74 ml. 2(f) (a) (intensive) physical property; (b) physical process; (c) element; (d) chemical process; (e) pure substance; (f) (extensive) physical property; (g) (intensive) physical property. 2(g) (a) solvent; (b) solute; (c) heterogeneous mixture; (d) homogeneous; (e) solution; (f) heterogeneous mixture; (g) solution; (h) colloid; (i) heterogeneous mixture. 2(h) water:solvent; tea leaves—heterogeneous mixture; milk—suspension; sugar—compound; the final brew–a suspension. 2(i) many possible answers.

3

Atoms, Molecules, and Reactions

3.1 The Experimental Origins

Early Ideas

From the time of the early Greek philosophers, the idea that matter consists of discrete tiny particles has been one of the principal foundations in natural philosophy. During the early 1800s, this idea achieved a firm status because it was confirmed by experimental observations and laws. In the preceding century two very crucial patterns of experimental behavior had been summarized and described in terms of two so-called natural laws. These are the law of conservation of mass and the law of definite composition. They were based largely on the experiments of the pioneering chemists Antoine Lavoisier and Joseph Proust, and their contemporaries.

Law of Conservation of Mass

The law of conservation of mass seems almost trivially obvious today because it is so basic to the philosophical outlook of our time. When it was first demonstrated, however, it was by no means a trivial matter, and indeed, it was not universally accepted at once. The law states that *in chemical reactions, mass is neither created nor destroyed.* Irrespective of what other properties may be altered during a chemical transformation, the total mass of materials at the end is exactly the same as the total

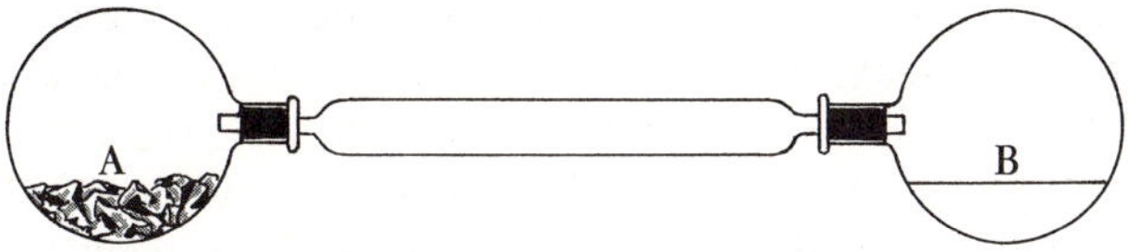

Figure 3.1
Simple glass apparatus for conducting experiments to demonstrate the law of conservation of mass. Reactant *A* (aluminum) and reactant *B* (liquid bromine) can be mixed by tilting the apparatus.

at the beginning.* The kind of experiment that was used to demonstrate this law is shown in Fig. 3.1. The apparatus represented there shows separate containers for two reactants. **Reactants are substances that are combined in a chemical reaction.** The apparatus is designed so that the reactants can be weighed, together with the glass container, before the reaction begins. They can be mixed in a completely closed system; the reaction is thus permitted to proceed. No materials are allowed to enter or escape the reaction vessel. At the end of the reaction the total mass is again determined, and it is always found to be the same as the initial mass within the limits that can be determined by even the best available balances. As an example, suppose that the left chamber contains a sample of aluminum metal (for example, strips of aluminum foil) and the right chamber contains liquid bromine, and suppose that the entire apparatus, including the glass, the bromine, and the aluminum, has a total mass of 234.1563 g before any reaction is permitted to occur.† If the apparatus is

*During the twentieth century, it has been learned that the law of conservation of mass is not strictly true. In ordinary chemical reactions, however, it is true to the extent that it can be tested experimentally; that is, it is true within the ordinary limits of experimental precision. Only in what are called nuclear transformations, in which extremely large amounts of energy are involved, and the identities of elements are no longer retained, must the "chemical" law of conservation of mass be modified.

†In mass measurements, the kind of experimental accuracy represented by figures such as 234.1563 g is readily available in most modern chemical laboratories.

Spotlight

Joseph Proust (1754–1826) was a French chemist. A student of apothecary chemistry and a balloonist in his youth, he avoided the turmoil of the French Revolution because he was working in Spain at the time. His careful work on weight relations in compounds and in chemical reactions had a profound influence on chemistry.

Picture Collection, The Branch Libraries, The New York Public Library

then tilted so that the liquid bromine pours onto the aluminum metal, a vigorous reaction ensues that involves the liberation of light and also considerable quantities of heat. (The liberation of heat and the resulting rise in temperature would prompt a prudent experimentalist to remove heat from the reacting system by applying a cooling bath.) At the end of a period of vigorous reaction it would be obvious that the appearance of the materials had changed, and that a new, white solid had formed —clear evidence that a chemical reaction had taken place. If the apparatus is weighed again, it is found to have exactly the same mass as before the reactants were allowed to come into contact with each other. Experiments of this kind, with any combination of materials that can undergo chemical reaction, are always found to lead to this same result: the total mass does not change!

The Law of Definite Composition

The **law of definite composition,** or the **law of definite proportions,** states that *the quantity of one element needed to combine with a fixed weight of another element in the formation of a particular compound is always the same.* That is, the elemental composition of a particular compound is always the same; it is a fixed property. This is simply another statement of one of the criteria that we specified in our earlier definition of a compound (Chap. 2). Suppose, for example, that we want to form the compound known as aluminum bromide. If we use 2.70 g of aluminum, we find that 24.0 g of bromine will always be required to consume just exactly all the aluminum. If one-tenth of this amount of aluminum (0.270 g) is used, then one-tenth of the amount of bromine (2.40 g) will be needed. If more bromine than that dictated by the 24.0:2.70 ratio is made available, then some of it will be left over when the reaction is complete, and it will be found that the aluminum has combined with only the amount of bromine corresponding to that ratio. Thus, any sample of the compound known as aluminum bromide will always contain aluminum and bromine in the weight ratio 2.70:24.0, and this can be considered a fundamental and identifying property of that compound. Similar statements can be made about any compound, using the ratio or ratios characteristic of each.

3.2 Dalton's Atomic Theory

The Postulates

These basic laws of chemistry and the general pattern of available experimental data were digested about 1808 by an English schoolteacher named John Dalton. Using that basis, he developed what became known as **Dalton's atomic theory,** a simple explanation for the known patterns of behavior. A similar approach was presented a few years earlier by the English scientist William Higgins. Dalton's theory stated that all matter consisted of tiny particles called **atoms.** He ascribed to these atoms certain properties, most of which still hold true today in a liberal sense. Indeed, Dalton's main postulates constitute the basic foundations of modern atomic theory. These main features that have survived are summarized as follows.

Spotlight

John Dalton was born in Eaglesfield, England, ten years before the American Revolution began. After an early education received from his father, he attended the village school, and assumed teaching duties there for two years beginning at the age of twelve. Then, after a brief period of farm work, he assumed another teaching position, where he assisted his brother Jonathan and was able to study science, literature, and languages under John Gough, a gifted blind scientist. During his period with Gough, the young Dalton developed a keen interest in meteorology, a subject on which he worked for many years. At age 26, the year he became a tutor at New College, Manchester, Dalton published *Meteorological Observations and Essays,* which is considered one of the main foundations of meteorology as a science.

The Bettmann Archive

In 1793 Dalton studied red-green color blindness, an affliction from which he suffered. Because of his extensive studies, this problem is still widely known as Daltonism.

During the 1790s, Dalton's interests turned largely to the study of gases and weight relations in chemistry. His studies led him to Dalton's law of partial pressures (Chap. 15) and to the hypothesis now known as Dalton's atomic theory, which has been one of the most significant developments in the history of chemistry.

Dalton's later years were rewarding in many ways, including well-earned financial security and numerous honors. He remained active in teaching nearly until his death in 1844.

1. Each elemental substance is composed of basic particles of that substance, called atoms; and these atoms have a particular set of properties—size, mass, and so on. Thus, the elemental substance sodium consists of what we call sodium atoms, each of which is identical to every other sodium atom.* The basic atomic view is that a sample of metal is more like a collection of extremely tiny ball bearings than like a continuously solid block.
2. The atoms of different elements have different chemical and physical properties. For example, the atoms that compose a sample of sodium have different chemical reactivity, size, shape, mass, and the like from the atoms that compose a sample of oxygen.
3. The atoms of one element are never transformed into atoms of another element in a chemical reaction, nor are atoms created or destroyed in chemical reactions.†

*The idea that all atoms of a given element are identical is modified slightly by the idea of isotopes (Sec. 6.2); however, this detail does not alter the basic conceptual approach.

†Only in high-energy physics (for example, with high-energy particle accelerators and cyclotrons) are the elemental identities of atoms ever changed. In chemistry, statement 3 of the Dalton theory is inviolate.

On the basis of this postulate it is easy to understand why the law of conservation of mass is obeyed. Thus, in the simple experiment pictured in Fig. 3.1, if a fixed number of aluminum atoms and a fixed number of bromine atoms (corresponding to the fixed weights of the materials used) are present at the beginning of the reaction, then the same atoms of the same identities, having the same masses, will be present at the end of the reaction, in the same numbers. Consequently, the total mass of the apparatus must remain unchanged.

4. Compounds are composed of collections of atoms of more than one element in definite ratios. These ratios can generally be expressed in terms of integers or simple fractions. As we shall see, the aluminum bromide produced in the reaction of aluminum and bromine consists of a collection of aluminum atoms and bromine atoms in the number ratio 1:3. Thus, in any sample of aluminum bromide, one can find three times as many bromine atoms as aluminum atoms, and this ratio is a characteristic of that particular substance.

5. Chemical changes, also referred to as chemical reactions or chemical processes, involve a rearrangement of atoms, in one step or several. Rearrangement results from the union or separation of atoms, in which the atomic combinations that exist before the change are replaced with new ones. Let us consider the above example of the reaction of aluminum metal with liquid bromine to form solid aluminum bromide. Before the reaction, the aluminum atoms are combined in a manner characteristic of the shiny solid metal, and the bromine atoms are combined in a manner that gives rise to a red liquid. After these two samples of matter are combined, and have reacted, a new solid substance is formed in which bromine atoms and aluminum atoms are combined in a ratio that turns out to be three bromine atoms per aluminum atom. It is this new combination that gives rise to the appearance of the new white solid.

The Nature of Atoms

We think of atoms as the fundamental building blocks of both elemental and compound substances. In elemental substances only one type of atom is present, whereas in compound substances, more than one type of atom is present and the various types are present in distinct ratios. Furthermore, chemical reactions change these ratios and the geometrical arrangements that give rise to these ratios. The reaction of aluminum and bromine at the submicroscopic scale can be thought of as the combination of aluminum atoms and bromine atoms in some geometrical arrangement that is compatible with the ratio 1:3.

To appreciate how small atoms are, let us consider a typical copper penny, which after a few years of circulation might weigh about 3.1 g. Current scientific knowledge lets us reach the astonishing conclusion that such a penny contains about 3×10^{22} copper atoms (that is, 30,000,000,000,000,000,000,000 atoms). This is a number of such astronomical size that it is difficult for our minds to picture its meaning. To give some idea of how small each copper atom must be, we can calculate that to count enough copper atoms just to cover the head of a pin, one would have to count them for about 2 million years at a rate of three per second.

Molecules, Chemical Bonds, and Ions

Some other useful concepts come directly or indirectly from Dalton's original theory. One of these is the concept of a molecule. Dalton's theory leads one to the view that the fundamental submicroscopic units of which compounds are made are discrete clusters or aggregates of atoms that are arranged in definite ways. These atoms are pictured as being so tightly held together that they can be recognized as units, referred to as **molecules.** The forces that are responsible for holding the atoms together in these molecules are referred to as **chemical bonds.**

Since Dalton's time much has been learned about molecules and the bonds that hold them together. It has also been found that not all compounds can be described in terms of molecules. We shall learn that for some compounds, it is not accurate to define individual clusters of atoms as molecules; some compounds are arranged as extremely large aggregates that amount to entire macroscopic particles, for instance, a crystal of table salt. Such large aggregates are constructed of regular repetitive arrays of atoms. These arrays are called **crystal lattices.** In these lattices, positively and negatively charged atoms or clusters of atoms are arranged in an ordered manner at precisely determined positions. In some simple cases these positions are analogous to the points of intersection in a three-dimensional array of tic-tac-toe; most other arrangements are more complicated to describe. The array of positively and negatively charged sodium and chlorine atoms in the crystal lattice of table salt is shown in Fig. 3.2. The electrically positive and negative atoms or clusters of atoms are known as **ions.** Logical arguments and considerations have been developed that allow the chemist to predict which atoms or groupings of atoms are likely to become ions of a positive nature and which negative. At this point it is sufficient for us to be aware of the fact that ions exist and are important structural building blocks in many substances.

Those compounds that can be thought of as consisting of collection of molecules of a specific kind are called **molecular compounds.** We shall refer to those that do not fall into this category as nonmolecular compounds. Compounds in which ions are not present are molecular compounds; compounds formed from aggregates of ions are nonmolecular.

collection gathered as a whole!

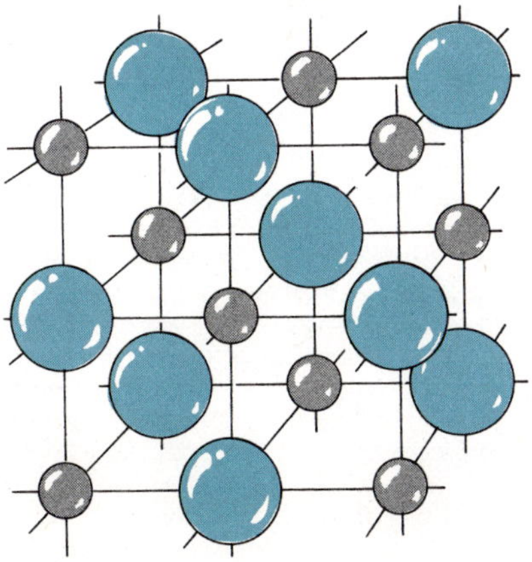

Figure 3.2
Crystal lattices of sodium chloride. Small gray circles represent Na^+ and larger color circles, Cl^-.

3.3 Symbolism

Formulas

For representing the presence of molecules, atoms, and ions, and also other concepts, conveniently and efficiently, it is useful to call on the abbreviations that we have given earlier for the elements. We now assign to these symbols an additional meaning; we use them not only as abbreviations for the names of elements, but also as symbols for identifying individual atoms of the corresponding elements. For example, the symbol "Cl" takes on at least two meanings for us, depending on the use and context; it immediately identifies the element under discussion, namely chlorine, and in some contexts represents one atom of chlorine. With this convention in mind, it is now a straightforward matter to represent symbolically compounds of both the molecular and nonmolecular types.

In Fig. 3.3, we show three representations of one molecule of a molecular compound called methane. The molecular formula of methane is CH_4; this formula tells us that each molecule of methane consists of one carbon atom and four hydrogen atoms. The **molecular formula** in general denotes the atomic composition of a molecule by means of atomic symbols (C and H in this instance) and subscripts (like the 4 in CH_4). The subscript indicates how many atoms of a given element occur in the molecule (for example, four hydrogen atoms in a CH_4 molecule). The absence of a subscript denotes the occurrence of one atom; there is one carbon atom in a CH_4

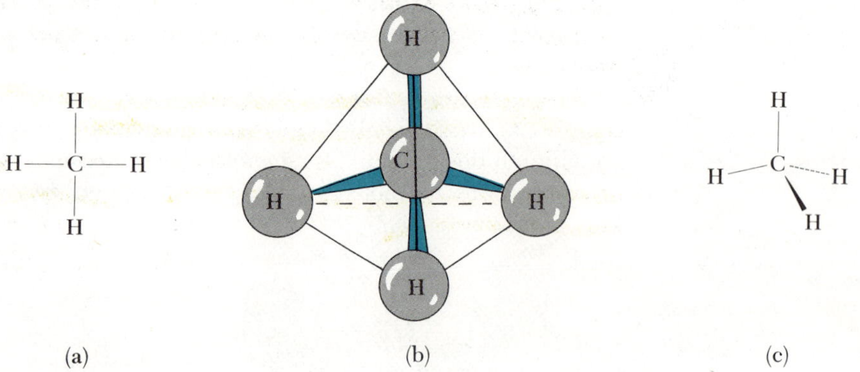

Figure 3.3 A molecule of methane, CH_4. (a) The structural formula, which does not imply the geometrical arrangement of atoms but only the pattern of attachments. (b) One type of geometrical representation of the molecule. The outline of a regular tetrahedron is given only to indicate that the geometry of methane resembles a formation in which the hydrogen atoms are at the corners of a regular tetrahedron with the carbon atom at its center. (c) Another attempt at representing the geometry of methane. A line in regular type, for instance, H—C, represents a hypothetical "attachment line" in the plane of the page. A dashed line represents an attachment line that begins at the carbon atom in the plane of the page and recedes from the plane as it goes back to the hydrogen atom. A wedged line represents an attachment line that begins at the carbon atom in the plane of the page, and comes out to the hydrogen atom above the plane.

molecule. The molecular formula provides no direct information about the geometry of the molecule or about connectivity among the atoms, that is, which atoms are bound to which. Part (a) of Fig. 3.3 shows a type of formula that does provide information on connectivity. With straight lines representing bonds, this formula clearly shows that the carbon atom is bonded to each of the four hydrogen atoms, which in turn are bonded only to the carbon atom and not to each other. Such formulas are called **structural formulas,** because they delineate the gross features of bonding structure in the molecule—which atoms are attached to which. Even a structural formula, however, provides no information about the shape of the molecule.

The shape of a methane molecule is shown is parts (b) and (c) of Fig. 3.3. As shown in part (b), a pictorial representation of the shape of CH_4, the hydrogen atoms are arranged as if they were at the corners of a regular tetrahedron (a three-dimensional shape having four equivalent triangular faces), with the carbon atom at the center. Part (c) is, in a sense, a compromise notation between parts (a) and (b). The connectivity is clearly shown, and a symbolism is adopted with attempts to portray the three-dimensional shape of the CH_4 molecule. The three-dimensional character is represented on a two-dimensional page by letting normal straight lines represent attachment lines (bonds) directed within a plane (say, the plane of the page), and letting a dashed line represent a bond directed behind the plane and a "wedge" a bond directed above the plane. Such formulas are common for representing geometrical aspects of chemistry, that is, chemical behavior and properties that depend on the geometrical arrangements of bonds. Such aspects are referred to as **stereochemistry.**

A molecular formula can be used to describe the atomic composition of any molecule. Besides telling us the composition of a single molecule, the molecular formula also tells us the atomic composition of any sample of the corresponding compound. Thus, the molecular formula CH_4 tells us that in any sample of methane containing any number of molecules, there are four times as many hydrogen atoms as carbon atoms. This is because the entire sample of methane consists of molecular units in which this 4:1 ratio prevails. The formula CH_4O represents a molecule of a substance that contains one carbon atom, one oxygen atom, and four hydrogen atoms per molecule. It also gives the ratio of the number of carbon, hydrogen, and oxygen atoms in any macroscopic sample of the substance.

EXAMPLE 3.1 Write the molecular formula of a substance, a 100-g sample of which contains carbon atoms, hydrogen atoms, and nitrogen atoms in the proportions 5:5:1. This substance has one nitrogen atom per molecule.

SOLUTION The 5:5:1 proportion would be characteristic of a sample of any size, and also holds for individual molecules. Hence, the molecular formula is C_5H_5N.

STUDY PROBLEM 3(a)

Write the molecular formula of a substance of which any sample has half as many fluorine atoms as hydrogen atoms, two-thirds as many hydrogen atoms as carbon atoms, and four hydrogen atoms per molecule.

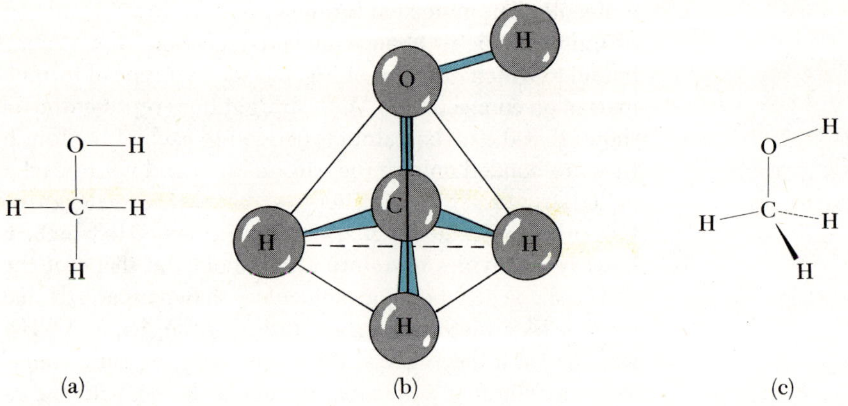

Figure 3.4
A molecule of methanol (methyl alcohol), CH_3OH. (a) The structural formula. (b) and (c) Attempts at three-dimensional representation of geometrical arrangements of atoms.

Functional Groups

Usually, in representing the substance called methanol (methyl alcohol), chemists use the formula CH_3OH instead of the formula CH_4O. The reason for the preferred notation, in which all H's are not grouped together as H_4 in the formula, is associated with the concept of a **functional group.** This concept constitutes a very important simplifying approach that chemists often use. It takes into account that certain recognizable groups of atoms occur over and over again in a variety of different types of molecules, just as certain patterns of notes or rhythms occur over and over again in specific types of music. These groups or patterns are recognized in different situations. That is, certain collections of atoms seem to occur in nearly the same arrangements in a variety of molecules. We see a simple example of this in methane and methanol. As Figs. 3.3 and 3.4 show, both types of molecules contain the CH_3 aggregate, in which an "attachment line" connects the carbon atom to each of three hydrogen atoms in the structural formula. In methane, the CH_3 aggregate is attached to H, whereas in methanol it is attached to OH. This CH_3 aggregate is given the name *methyl group.* The symbol OH in the formula CH_3OH represents a *hydroxyl group,* and is represented in this way to impart the meaning that the OH group is an entity that is recognizable not only in this type of molecule but in others also.

Compounds that contain the hydroxyl group usually have a set of chemical properties associated with acidity. This condition depends on the involvement of the OH group in reactions that remove a hydrogen atom in the form of a positively charged species, H^+, called a hydrogen ion (positively charged hydrogen atom). In CH_3OH, this acidity is very weak. In some cases, the acidity is very strong, for example, sulfuric acid, which can be represented by the formula $SO_2(OH)_2$. (Many analogous generalizations based on functional groups are explored in later chapters, especially Chap. 8.)

Molecular and Nonmolecular Compounds. The Compound Formula

Table 3.1 contains the molecular formulas and brief descriptions of some typical molecular substances. For some of the molecules included in the table, a formula notation that emphasizes functional groups is given. If a given functional group occurs more than once in the molecular formula, the group is enclosed in parentheses

TABLE 3.1 Typical Molecular and Nonmolecular Compounds

Name	Formula	*Formula in Terms of Functional Groups (indicated)*	*Melting Point, °C*	*General Description (atmospheric pressure and room temperature)*
Molecular				
Ammonia	NH_3		−77.7	Reactive gas with pungent odor, turns pink litmus blue.
Carbon dioxide	CO_2		−56.6[a]	Gas given off by respiration of animals, taken up by plants; "dry ice" in the solid state.
Iodine monochloride	ICl		27.2	Corrosive red solid.
Benzene	C_6H_6		5.5	Volatile colorless liquid, important industrial chemical.
Nitrobenzene	$C_6H_5NO_2$	$\underline{C_6H_5}\overline{NO_2}$	5.7	Volatile liquid, odor familiar in shoe polish.
Phenol	C_6H_6O	$\underline{C_6H_5}\overline{OH}$	43	Aromatic solid often referred to as carbolic acid in medicinal applications.
Boron trifluoride	BF_3		−126.7	Extremely reactive gas; reacts rapidly with moist air.
Isooctane	C_8H_{18}	$\underline{(CH_3)_3C}CH_2CH(\overline{CH_3})_2$	−107.4	Volatile liquid, important component of gasoline.
Nickel tetracarbonyl	C_4O_4Ni	$Ni(\underline{CO})_4$	−25	Reactive, poisonous gas.
Hydrogen chloride	HCl		−114.8	Reactive gas, turns blue litmus red.
Ferrocene	$C_{10}H_{10}Fe$	$Fe(\underline{C_5H_5})_2$	174	Stable solid, a "pioneering organometallic substance."
Acetic acid	$C_2H_4O_2$	$\underline{CH_3}\overline{\underline{COOH}}$	16.6	Clear, colorless liquid; the sour, acidic constituent of vinegar.
Nonmolecular				
Sodium chloride	NaCl		801	Stable colorless solid, common table salt.
Calcium chloride	$CaCl_2$		772	White solid, collects moisture rapidly, used as drying agent and for melting snow on highways.
Hexammine cobalt chloride	$N_6H_{18}Cl_3Co$	$Co(NH_3)_6Cl_3$	Above 215[b]	Red crystals.
Magnesium sulfate	$MgSO_4$	$(Mg\underline{SO_4})$[c]	Above 1100[b]	Colorless solid, useful as a drying agent.
Calcium carbonate	$CaCO_3$	$(Ca\underline{CO_3})$[c]	1339	Colorless solid; in various solid forms it occurs as coral, sea shells, cement, stalagmites, etc.
Potassium nitrate	KNO_3	$(K\underline{NO_3})$[c]	334	Colorless solid, mined as saltpeter for an industrial source of nitrogen.

[a] Solid CO_2 melts only at high pressure. At one atmosphere, solid CO_2 sublimes at −78 °C; that is, it vaporizes directly without passing through the liquid state.

[b] Substance decomposes at the temperature given.

[c] Depending on one's point of view, one may or may not wish to consider ions such as SO_4^{2-}, CO_3^{2-} and NO_3^- in ionic substances as functional groups.

and the number of times it occurs is represented by a subscript at the right of the parentheses.

For compound substances that are not molecular compounds, a different interpretation applies. We have mentioned as an example of the structural arrangements typical of nonmolecular compounds the crystal lattice of sodium chloride, as shown in Fig. 3.2. In this case ions are involved, and this is indicated explicitly by the electrical charge on each ion. For ionic crystals, typified by this example, it makes no sense to speak of molecules; throughout the entire crystal there is just a continuous lattice of **ions** (atoms or groups of atoms that are electrically charged). Since there are no molecules for such compounds, a molecular formula would have no meaning. In place of molecules, we introduce the more general concept of the **formula unit,** a hypothetical entity that represents the atomic composition of the substance; in place of the molecular formula, we use the more general term *compound formula.* The **compound formula** is constructed from atomic symbols in the same way that the molecular formula is constructed for a molecular compound; it is merely a shorthand notation that gives the atomic composition of the nonmolecular compound in the briefest possible way. Although sodium chloride does not exist in the solid state as discrete molecules, we use the symbol NaCl for the formula unit to show that sodium chloride is composed of equal numbers of chlorine and sodium atoms (actually sodium ions and chloride ions).* Now that the symbol used to represent the formula unit has been defined, the meaning of the formula unit itself becomes clear; it is simply a collection of atoms (or ions), the identity and number of which correspond to the compound formula. The formula unit for magnesium bromide is represented by the compound formula $MgBr_2$, which does not imply the existence of molecules having the molecular formula $MgBr_2$ but simply indicates the relative numbers of atoms present in the compound. We shall often refer to symbols such as CH_4, NaCl, and $MgBr_2$ as compound formulas, realizing that in some cases they stand for molecules, while in other cases they represent formula units that do not qualify as molecules.

Many compound formulas are written with some atomic symbols placed within parentheses followed by a subscript; an example is $Ba(NO_3)_2$. This notation simply means that the collection of atoms within the parentheses occurs in the formula unit the number of times indicated by the subscript. Thus, $Ba(NO_3)_2$ signifies that the group of four atoms NO_3 occurs twice in the formula unit—each formula unit contains one barium atom, two nitrogen atoms, and six oxygen atoms.

3.4 Chemical Composition

Relative Masses

If the compound formula is available, that provides the first step in being able to calculate the chemical composition by weight of any substance. The other requirement for such calculation is a knowledge of the masses of the atoms of the elements

*Although nonmolecular solids do not exist as a collection of molecules, it is possible to produce discrete molecules of the same molecular formula (for example, NaCl) in the gas phase at very low pressures and very high temperatures. While such gaseous molecular species are of theoretical interest, substances like NaCl are usually encountered in chemistry as the nonmolecular solids, or as solutes in liquid solutions (for example, in water).

concerned. We might initially expect that the actual atomic masses would be needed; however, a little thought may convince you that only the relative masses of the various kinds of atoms are needed for computing the composition (percentage by weight of each element) of a substance from the corresponding compound formula. If we know, for example, that a certain substance called sulfur dioxide has a molecular formula SO_2, and if we know that each sulfur atom weighs about twice as much as each oxygen atom, then we can conclude that sulfur dioxide is about 50% sulfur by weight and 50% oxygen by weight. We conclude this without knowing the actual masses of either sulfur or oxygen atoms; only knowledge of their relative atomic masses is necessary! In reality, the historical order of developments was largely the reverse of this logical sequence; studies of chemical composition led to information on the relative atomic masses of the elements. The interplay between knowledge of the relative atomic masses, compound formulas, and chemical composition was complicated, and constitutes one of the interesting and important stories in the history of chemistry. Our presentation of the overall picture will include the sense of these developments without necessarily preserving the correct chronology.

Quantitative mass measurements in chemical research sparked much activity among chemists. They sought to determine the chemical composition of compounds and the masses of individual substances that were consumed or produced in chemical reactions. The logic used to lead from this experimental work to determining relative atomic masses was not always perfect or unequivocal. In many cases deductions began with educated guesses, which were based on intuition and experience. In some of those cases the early chemist was correct; in others not. Recent work has erased the doubts of those earlier controversies.

The Periodic Table

One of the important milestones that helped to clear up many uncertainties about compound formulas and relative atomic masses was the development of a periodic table of the elements. Mendeleev and Meyer independently established such an arrangement about 1869. A modern form of this table is shown on the inside cover of this textbook. To construct their periodic tables, both Mendeleev and Meyer arranged elements with similar chemical and physical properties vertically in a "family." In each family, the relative atomic masses were greater in descending order within the column. The families of similar elements were then arranged next to each other in order of increasing relative atomic masses; this order was deduced from the relative masses of elements involved in specific compounds and reactions. In this way, a regular array of the known elements was formed in which the elements were arranged in order of increasing relative atomic mass across a row until a particular set of recognizable properties recurred in another element; then a new row was begun. It was noted that the alkali metal elements (lithium, sodium, potassium, rubidium, and cesium), which come substantially separated in the list of elements as arranged in increasing order of relative atomic masses, have chemical and physical properties that are in many ways similar. Furthermore, immediately preceding the positions of these elements in this ordering come the elements helium, neon, argon, krypton, and xenon, known as the rare, or noble gases. This group of elements is characterized by showing almost no chemical properties, seeming essentially

Spotlight

An important approach in early work on relative atomic masses was to determine the weights of individual elements that would combine with a fixed weight of some arbitrary "reference element." These amounts were called combining weights. Originally hydrogen was chosen as a reference element, but later oxygen became more popular. Because of different choices of the reference element, and also the different amounts of the reference element specified for comparison, there was a lot of confusion in the development of the concept of combining weights. Also, it is often found that the amount of one element that combines with a fixed weight of another element in a chemical reaction depends on the conditions; different conditions sometimes lead to different products. As an illustration, carbon and oxygen combine to form either carbon monoxide, CO, or carbon dioxide, CO_2, depending on the temperature and proportion of carbon and oxygen available. Another difficulty with the combining weight concept is that not all elements combine in reactions with any single reference element.

For these reasons the definition became extended to a more general form—the combining weight of an element is the weight that is related to 16 g of oxygen in one or more of the following ways: (1) the mass of the element that will just react with 16 g of oxygen; (2) the mass of the element that will just replace 16 g of oxygen, or be replaced by that amount, in a chemical reaction; or (3) the mass of the element that is "equivalent" to 16 g of oxygen in analogous reactions. By these criteria, the relative masses of numerous elements could be related quantitatively to a single reference element. The values obtained, however, could depend on the specific reaction conditions, and many uncertainties and controversies arose because of this. Nevertheless, this approach was a useful stepping-stone in the development of quantitative chemistry, and provided an interesting and important chapter in the history of chemistry. The accompanying table gives an idealized example of the derivation of combining weights for a few elements. In the table only one ambiguity appears; it is related to hydrogen.

Determination of Combining Weights for Some Elements

	Defining Reaction	*Criterion*[a]	*Element*	*Combining Weight, g (referred to 16 g oxygen)*
(a)	Burning of 2.0 g hydrogen with 16.0 g oxygen to form 18.0 g water	1	H	2.0
(b)	Oxidation of 1.0 g hydrogen with 16.0 g oxygen to form 17.0 g hydrogen peroxide	1	H	1.0
(c)	Burning of 24.3 g magnesium with 16.0 g oxygen to form 40.3 g magnesium oxide	1	Mg	24.3
(d)	Oxidation of 63.5 g copper with 16.0 g oxygen to form 79.5 g copper oxide	1	Cu	63.5

	Defining Reaction	*Criterion*[a]	*Element*	*Combining Weight, g (referred to 16 g oxygen)*
(e)	Burning of 12.0 g carbon in a restricted oxygen supply (consuming 16.0 g), to form 28.0 g carbon monoxide	1	C	12.0
(f)	Replacement of 32.0 g sulfur in 95.5 g copper sulfide by 16.0 g oxygen to form 79.5 g copper oxide ("roasting" copper sulfide in oxygen atmosphere)	2	S	32.0
(g)	Formation of 95.5 g copper sulfide from 32.0 g sulfur and 63.5 g copper in analogy to case (d)	3	S	32.0

[a]Refer to the text of this Spotlight for numbered criteria.

inert.* Immediately preceding the rare gases in the atomic mass order are the elements fluorine, chlorine, bromine, and iodine, which as a group have similar chemical and physical properties; these similarities led to the classification of these elements as a family called the halogens. When such similarities were taken into account, a pattern, or table, of the elements emerged. Both Mendeleev and Meyer found gaps in their tables, spaces that corresponded to elements not yet discovered at that time. Later developments corroborated the periodic arrangement by filling these gaps.

Early chemical evidence often led to ambiguities in the knowledge about relative atomic masses of certain elements; more than one value was compatible with the information available at the time. Recognizing patterns within the periodic table was useful in resolving these apparent ambiguities. The correct relative atomic masses were chosen by placing the element in a position in the table that corresponded to its known chemical and physical properties, and noting which of the alternative relative atomic masses corresponded most closely to this position.

Suppose that a primitive periodic table had been available to an early chemist who just discovered a new element, which he called germanium. The table available at the time would have had several gaps in it, corresponding to elements not yet discovered. Suppose further that different experiments to determine the relative atomic mass led to two different values, 36.3 and 72.6, on a scale in which the mass of an

*The discovery of compounds formed from xenon and fluorine was made in 1960 by Professor Neil Bartlett, now of the University of California, Berkeley. Until then the rare gas "inertness" seemed to be inviolate. Such interesting chemical discoveries do not, however, undermine the concept of a periodic table.

oxygen atom was arbitrarily placed at 16.0. The chemist had to decide which of these two values was the correct relative atomic mass for germanium. The choice between 36.3 and 72.6 could easily have been made simply by considering how well each of these two possible values fitted into the established scheme of the periodic table. The value 72.6 would have placed germanium intermediate between gallium and arsenic in the order of increasing relative atomic masses of the elements.

This position places germanium in the same column as carbon, silicon, tin, and lead in the Mendeleev-Meyer scheme.* The chemical and physical properties of germanium are in many ways analogous to the properties of carbon, silicon, tin, and lead. All form tetrahalides—CCl_4, $SiCl_4$, $GeCl_4$, $SnCl_4$ and $PbCl_4$; and all form hydrides—CH_4, SiH_4, GeH_4, SnH_4 and PbH_4. And with germanium placed between gallium and arsenic in their *row*, germanium falls between silicon and tin in the *column* containing carbon. This position reflects known trends of properties for elements in this column; for instance, the boiling points of the hydrides are in the increasing order CH_4, SiH_4, GeH_4, SnH_4, PbH_4. Therefore, by the known patterns of chemical and physical properties, such a position for germanium in the table would be consistent. In contrast, the value 36.3 for germanium places it between chlorine and potassium, a region inconsistent with germanium's known chemical properties and also inconsistent with the lack of vacant spaces between chlorine and potassium in the table. Hence, the value 72.6 is correct for germanium, on the chosen scale. Familiarity with the periodic table is one of the most important aspects of learning chemistry. We will discuss the construction and interpretation of this table in considerable detail in Chaps. 6 and 7.

To account for the formulas of most compounds in terms of the periodic table, one needs to learn several theories (in Chaps. 6 and 7). We can, however, make some simple generalizations about one small but important class of compounds without knowing the theoretical basis for the periodic table. That class comprises the nonmolecular **binary compounds**, which are compounds containing two elements. Of these elements, one comes from the left side of the table (say, the first two or three columns) and one from the right side (say, the six and seventh columns; we omit from consideration the eighth column, the noble gases). A few examples of nonmolecular binary compounds are NaCl, KBr, LiF, $BaCl_2$, $AlCl_3$, $MgCl_2$, $CaCl_2$, $BaBr_2$, KI, Na_2S, and CsF. When formulas for substances of this kind are put together, the symbol corresponding to the element from the left side of the periodic table is written on the left, and the symbol corresponding to the element on the right side of the table is written on the right. Another significant generalization is that the compound formulas of these compounds fit in with the concept of *ionic valence*. This concept takes account of the following facts. In nonmolecular binary compounds, elements from the first column of the periodic table exist as electrically positive particles—as positive ions. Each of these ions has one unit of positive charge,

*Another useful aid for choosing between possible values of the atomic masses comes from the work of DuLong and Petit. From their own experiments and from organizing the results of others, these investigators concluded in 1819 that the amount of energy needed to bring about a one-degree increase in temperature in an elemental solid was approximately the same for a given number of atoms of any element. Since this relation prevails independent of the identity of the element, experiments on a sample of known mass could give a measure of the relative number of atoms that corresponds to that mass. Hence, relative atomic masses could be determined within certain broad limits of uncertainty.

where the unit of negative electrical charge is the charge of the fundamental particle of electricity, called the **electron** (see Chap. 6); elements in the second column of the table exist as +2 ions; elements of the third column exist as +3 ions; ions of the sixth column tend to exist as −2 ions; and elements of the seventh column exist as −1 ions. Then, if we think of forming electrically neutral compounds by combining positively and negatively charged ions, we account for the generalization that the two elements in such a compound come from opposite sides of the periodic table.

Furthermore, obtaining neutral compounds requires combining the positively and negatively charged ions in specific ratios; for example,

Reactant Ions and Electrical Charges		*Neutral Product*
Li^+ and F^- +1 −1	$\longrightarrow$	LiF
Ba^{2+} and two Cl^- +2 −2	$\longrightarrow$	$BaCl_2$
Al^{3+} and three Cl^- +3 −3	$\longrightarrow$	$AlCl_3$
two Na^+ and S^{2-} +2 −2	$\longrightarrow$	Na_2S

As a bookkeeping method (not as a valid theoretical model), this view is also useful in accounting for the existence of some molecular compounds formed between hydrogen and an element on the right side of the periodic table. Some examples are HCl (viewed for bookkeeping purposes as H^+ and Cl^-), H_2O (viewed as H^+ and O^{2-}), and H_2S (viewed as H^+ and S^{2-}). To go beyond generalizing for these simple cases, one must explore detailed theoretical models of chemical bonding (Chap. 7).

EXAMPLE 3.2 Give formulas for the compounds consisting of (a) cesium and fluorine, (b) gallium and chlorine, (c) oxygen and aluminum.

SOLUTION Using the scheme outlined above for nonmolecular binary compounds, we have

(a) Cs^{+1} and $F^{-1} \rightarrow$ CsF
(b) Ga^{3+} and three $Cl^- \rightarrow GaCl_3$
(c) two Al^{3+} and three $O^{2-} \rightarrow Al_2O_3$

STUDY PROBLEM 3(b)

Give formulas for the compounds consisting of aluminum and bromine; selenium and potassium; sulfur and barium.

Since chemistry has been increasingly important to the fundamental understanding of biology, it is interesting to consider the periodic table from the point of view of biology. Table 3.2, which we can call the "biochemical periodic table," summarizes the importance of the elements in biological systems.

TABLE 3.2 **The Biochemical Periodic Chart**

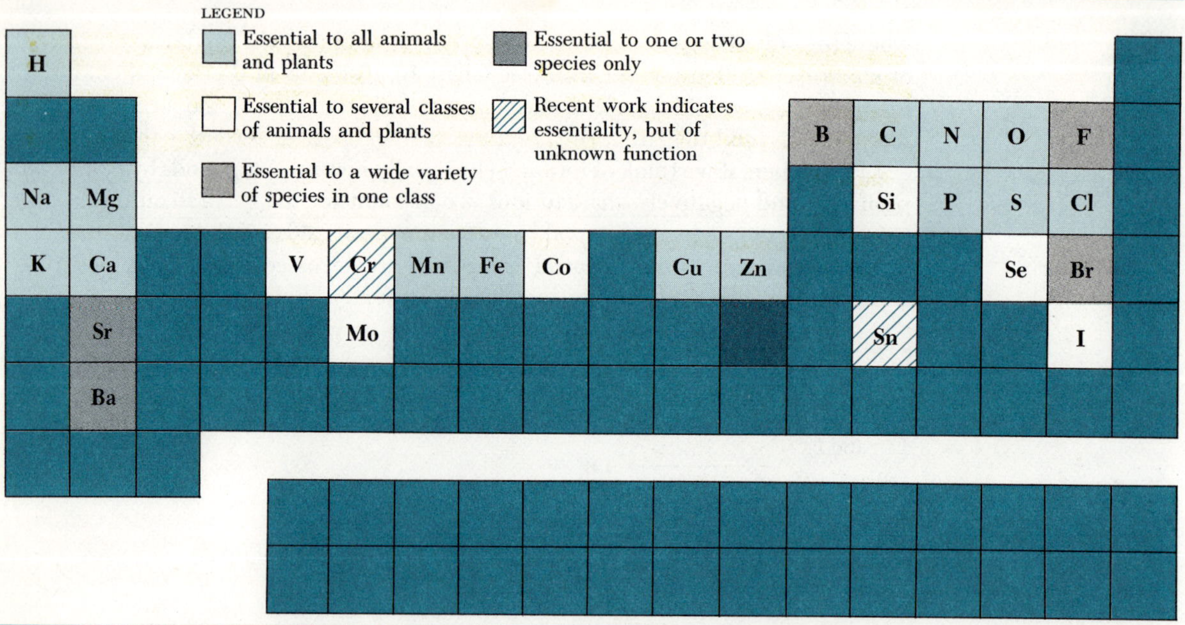

Atomic Weights

On the basis of concepts so far outlined, a set of relative atomic masses has evolved, historically known as **atomic weights.** This set is based on the choice of some particular, arbitrary standard. At least three standards have been in common use during the past hundred years; but recently an agreement was reached on a reference that is unlikely to be challenged henceforth. This choice is based on specifying a particular value for the atomic weight of a particular type of atom. The atom chosen is carbon and the value chosen for carbon is exactly 12 in a set of units known as **atomic mass units** (abbreviated amu) and sometimes called **daltons.*** The specification of this value was entirely arbitrary; but now it has been made, all other atomic masses are defined precisely in terms of this relative scale. Specifying 12 amu as the atomic weight of carbon is no more or less arbitrary than specifying that a particular piece of metal in a vault in Paris defines the length one meter. Once a scale is established, other values are determined relative to it. The masses of the atoms of all other elements have been related to this number, and are popularly referred to as atomic weights. The resulting values of all the elements are given in a table of elements and in the periodic table on the inside cover of this book.

The concept of an atomic weight is extremely powerful and establishes for the chemist a quantitative basis for calculations in which chemical composition is essential. Some examples of the types of problem solving that are based on this concept follow.

EXAMPLE 3.3 Calculate the percentage composition of a compound called ammonia, which has the molecular formula NH_3.

*The atomic weight reference atom is the carbon 12 isotope. Isotopes are discussed in Chap. 6.

SOLUTION Lacking some contrary specification, one can assume that the term *percentage composition* means "elemental composition in percentage by weight." Since we know from the molecular formula that there is a 1:3 ratio in the number of nitrogen and hydrogen atoms, all we need to do is take into account the relative masses of each type of atom to solve the problem. Each hydrogen atom weighs 1.008 amu and each nitrogen atom weighs 14.007 amu. Hence, one molecule of ammonia contains 14.007 amu of nitrogen and 3(1.008) amu of hydrogen. Thus the total mass of NH_3 is

$$14.007 \text{ amu} + 3(1.008)\text{amu}$$

Then, the elemental percentages are computed directly:

$$\% \text{ nitrogen} = (100)\frac{14.007 \text{ amu}}{14.007 \text{ amu} + 3.024 \text{ amu}} = 82.24\%$$

$$\% \text{ hydrogen} = (100)\frac{3.024 \text{ amu}}{14.007 \text{ amu} + 3.024 \text{ amu}} = 17.76\%$$

Since each molecule can be viewed as having the same percentage composition and since an entire sample is merely a collection of such molecules, the calculated percentages are valid for any given weight or sample of ammonia.

EXAMPLE 3.4 Calculate the percentage composition of sodium phosphate, for which the compound formula is Na_3PO_4.

SOLUTION From the compound formula and from reasoning similar to that used in Example 3.3, we see that each formula unit of Na_3PO_4 contains 3(22.99) amu of sodium, 30.97 amu of phosphorus, and 4(16.00) amu of oxygen. Accordingly,

$$\% \text{ sodium} = (100)\frac{68.97 \text{ amu}}{68.97 \text{ amu} + 30.97 \text{ amu} + 64.00 \text{ amu}} = 42.07\%$$

$$\% \text{ phosphorus} = (100)\frac{30.97 \text{ amu}}{68.97 \text{ amu} + 30.97 \text{ amu} + 64.00 \text{ amu}} = 18.89\%$$

$$\% \text{ oxygen} = (100)\frac{64.00 \text{ amu}}{68.97 \text{ amu} + 30.97 \text{ amu} + 64.00 \text{ amu}} = 39.04\%$$

Since this percentage composition describes a formula unit, and since this unit is representative of the composition of any sample of the compound, these percentages constitute the desired result. These results can be used to compute the number of grams of each element in any macroscopic sample of known total mass. For example, in a 213-g sample of Na_3PO_4, there would be

$$\text{no. g of sodium} = \left(\frac{42.07}{100}\right)(213 \text{ g}) = 89.6 \text{ g Na}$$

$$\text{no. g of phosphorus} = \left(\frac{18.89}{100}\right)(213 \text{ g}) = 40.2 \text{ g P}$$

$$\text{no. g of oxygen} = \left(\frac{39.04}{100}\right)(213 \text{ g}) = 83.2 \text{ g O}$$

EXAMPLE 3.5 Derive the molecular formula of a molecular compound whose composition is 25.0% hydrogen and 75.0% carbon.

SOLUTION

$$\frac{\text{mass of H}}{\text{mass of C}} = \frac{(\text{mass per H atom})(\text{no. H atoms})}{(\text{mass per C atom})(\text{no. C atoms})}$$

$$= \frac{(1.008 \text{ amu})(\text{no. H atoms})}{(12.01 \text{ amu})(\text{no. C atoms})}$$

From the weight percentages given, the ratio of masses is 25.0/75.0. Thus, we can write

$$\frac{25.0}{75.0} = \left(\frac{1.008}{12.01}\right)\left(\frac{\text{no. H atoms}}{\text{no. C atoms}}\right)$$

Rearranging gives

$$\frac{\text{no. H atoms}}{\text{no. C atoms}} = \left(\frac{12.01}{1.008}\right)\left(\frac{25.0}{75.0}\right) = 3.97$$

This result, which is approximately 4, leads us to the formula CH_4.

On the basis of the above result, we have been led to conclude that the molecular formula is CH_4. But we must recognize that the formulas C_2H_8, C_3H_{12}, C_4H_{16}, . . . also give this ratio, and the composition data cannot distinguish among them. Later we shall see how to make such choices. In the meantime, we note that the formula CH_4 is called the **simplest formula, or the empirical formula**. The simplest formula of any compound has the smallest set of integer subscripts; for CH_4, this means 1 for C and 4 for H.

EXAMPLE 3.6 Calculate the number of grams of copper needed to produce a substance with compound formula CuO by direct combination of 10.0 g of oxygen.

SOLUTION A direct combination will, according to the compound formula, consume copper and oxygen atoms in the number ratio 1:1. From the atomic weights, we see that each copper atom weighs (63.55/16.00) times as much as each oxygen atom. Hence the mass of copper required is given by

$$\text{g copper} = \frac{\text{at.wt of Cu}}{\text{at.wt of O}}(\text{g oxygen})$$

$$= \frac{63.55 \text{ amu}}{16.00 \text{ amu}} \times 10.0 \text{ g} = 39.7 \text{ g}$$

STUDY PROBLEM 3(c)

Calculate the percentage composition of hydrazine, N_2H_4.

STUDY PROBLEM 3(d)

What is the simplest formula of a compound whose composition is 7.7% hydrogen and 92.3% carbon?

STUDY PROBLEM 3(e)

How many grams of iron are required to produce a substance with compound formula FeO by direct combination of iron with 12.2 g of oxygen?

3.5 Gram-Atom

As the above examples show, the concept of atomic weight opens some important kinds of quantitative problems to solution. However, there are other, more convenient ways in which atomic weight can be used for making calculations. These more efficient approaches rely on some additional definitions and concepts. Of these, the gram-atomic weight or simply gram-atom, abbreviated "g-atom," is especially important. The **gram-atom** of any element is defined as that quantity of the element that has a mass in grams numerically equal to the atomic weight (in amu). Since the atomic weight of barium is 137.34 amu, one g-atom of barium is defined as 137.34 g of barium. Since hydrogen has an atomic weight of only 1.008 amu, one g-atom of the element hydrogen is a sample of only 1.008 g of hydrogen.

Significance

From these definitions, we see that the mass of one g-atom of one element bears a simple relation to the mass of one g-atom of another element; this relation is determined simply by the ratio of the corresponding atomic weights. One g-atom of uranium weighs 238.03/1.008 times as much as one g-atom of hydrogen, simply because the atomic weight of uranium is 238.03/1.008 times the atomic weight of hydrogen. Also, we know that the ratio of atomic weights of two elements is the same as the corresponding ratio of the masses of an individual atom of each element.* One g-atom of uranium weighs 238.03/1.008 times as much as one g-atom of hydrogen because a uranium atom has a mass that is 238.03/1.008 times the mass of a single hydrogen atom. By the same logic, the only reason that one g-atom of bromine (a mass of 79.90 g) weighs 79.90/1.008 times as much as one g-atom of hydrogen is that each individual bromine atom has a mass 79.90/1.008 times the mass of a hydrogen atom.

It is enlightening to reorder some of these relations algebraically as follows:

$$\begin{aligned}\begin{matrix}\text{no. U atoms}\\ \text{in one g-atom U}\end{matrix} &= \frac{\text{mass of one g-atom of U}}{\text{mass of one U atom}}\\ &= \frac{(238.03/1.008)(\text{mass of one g-atom of H})}{(238.03/1.008)(\text{mass of one H atom})}\\ &= \frac{\text{mass of one g-atom of H}}{\text{mass of one H atom}}\\ &= \text{no. of H atoms in one g-atom of H}\end{aligned}$$

Similarly

$$\begin{aligned}\begin{matrix}\text{no. Br atoms}\\ \text{in one g-atom Br}\end{matrix} &= \frac{\text{mass of one g-atom of Br}}{\text{mass of one Br atom}}\\ &= \frac{(79.90/1.008)(\text{mass of one g-atom of H})}{(79.90/1.008)(\text{mass of one H atom})}\\ &= \frac{\text{mass of one g-atom of H}}{\text{mass of one H atom}}\\ &= \text{no. of H atoms in one g-atom of H}\end{aligned}$$

*As will be seen in Chap. 6, this ratio of masses for individual atoms is an average for all isotopes.

Summarizing yields

$$\text{no. U atoms per g-atom U} = \text{no. of H atoms per g-atom H}$$
$$= \text{no. of Br atoms per g-atom Br}$$

This kind of reasoning can be applied to any element, and we are led to conclude that whatever number of atoms constitutes a g-atom of a given element applies to any element. That is, the number of atoms contained in a g-atom of an element is the same for all elements. This number is known as **Avogadro's number,** denoted here by the symbol N_A. Its value was not known when chemists were developing concepts of atomic weight and gram-atoms. But the early chemists realized that for most applications such knowledge was not prerequisite. Modern experimental work has provided a reliable value for Avogadro's number; it is, to six significant figures, 6.02217×10^{23}. This is a huge number and its magnitude is outside the usual references of human experience. It clearly indicates the incredibly tiny size and mass of an individual atom. The value of Avogadro's number should be committed to memory for four significant figures, 6.022×10^{23}.

EXAMPLE 3.7 How many mercury atoms are there in a very tiny droplet of liquid mercury, weighing 1.0 mg, that is barely visible to the naked eye?

SOLUTION Since we now know the number of atoms in a g-atom of any element, we simply need to determine the number of g-atoms in 1.0 mg of mercury. The atomic weight of Hg is 200.6 amu. By the unit factor method (Sec. 1.5),

$$1 \text{ g-atom Hg} = 200.6 \text{ g Hg}$$
$$1 = \frac{1}{200.6}\,\frac{\text{g-atom Hg}}{\text{g Hg}}$$
$$1.0 \text{ mg Hg} = (1.0 \text{ mg Hg})(1)(1)$$
$$= (1.0 \text{ mg Hg})\left(\frac{1}{1000}\,\frac{\text{g Hg}}{\text{mg Hg}}\right)\left(\frac{1}{200.6}\,\frac{\text{g-atom Hg}}{\text{g Hg}}\right)$$
$$= 0.50 \times 10^{-5} \text{ g-atom Hg}$$

Then,

$$(0.50 \times 10^{-5} \text{ g-atom Hg})\left(6.022 \times 10^{23}\,\frac{\text{Hg atoms}}{\text{g-atom Hg}}\right) = 3.0 \times 10^{18} \text{ Hg atoms}$$

STUDY PROBLEM 3(f)

How many magnesium atoms are there in a 1.0-mg chip of magnesium?

EXAMPLE 3.8 How much would a sample of pure iron containing 4.0×10^{15} atoms weigh? Or more precisely expressed, what would the mass be?

SOLUTION From the atomic weight of iron, 55.8 amu,

$$1 \text{ g-atom Fe} = 55.8 \text{ g Fe}$$
$$1 = 55.8\,\frac{\text{g Fe}}{\text{g-atom Fe}}$$

Also,

$$6.022 \times 10^{23} \text{ atoms Fe} = 1 \text{ g-atom Fe}$$

$$1 = \frac{1}{6.022 \times 10^{23}} \frac{\text{g-atom Fe}}{\text{Fe atom}}$$

$$4.0 \times 10^{15} \text{ Fe atoms} = (4.0 \times 10^{15} \cancel{\text{Fe atom}})\left(\frac{1}{6.023 \times 10^{23}} \frac{\cancel{\text{g-atom Fe}}}{\cancel{\text{Fe atom}}}\right)$$

$$\times \left(55.8 \frac{\text{g Fe}}{\cancel{\text{g-atom Fe}}}\right)$$

$$= \frac{4.0 \times 10^{15} \times 55.8}{6.022 \times 10^{23}} \text{ g Fe} = 3.7 \times 10^{-7} \text{ g Fe}$$

This mass is about a thousand times smaller than what could be weighed on balances available in most well-equipped laboratories.

STUDY PROBLEM 3(g)

What is the mass of a small piece of solid sulfur containing 12.5×10^{17} atoms?

STUDY PROBLEM 3(h)

How much would a sample containing 1,000,000,000,000,000 iron atoms weigh? More precisely, what would the mass be?

3.6. The Mole

Molecular Weight and the Gram-Mole

We can extend the approach used in defining the atomic weight and the gram-atom to even more useful concepts—the molecular weight and gram-mole. **The molecular weight** (mol wt) of a substance is defined as the sum of all the atomic weights of the atoms occurring in the formula of the substance, where each atomic weight is counted as many times as the corresponding atomic symbol appears in the formula. This definition applies even to nonmolecular compounds (*formula weight* would perhaps be a better term). Since the molecular weight is defined as the sum of atomic weights, it must be expressed in the same units, namely amu's, or daltons. For example, the molecular weight of the substance HCl is by definition simply the sum of atomic weights of hydrogen and chlorine: 1.008 amu + 35.453, or 36.461 amu. Similarly, the molecular weight of the substance sodium sulfate, for which the compound formula is Na_2SO_4, is simply twice the atomic weight of sodium plus the atomic weight of sulfur plus four times the atomic weight of oxygen, or 142.04 amu. The molecular weight of any substance can be calculated in this way.

Using the definition of molecular weight and the concept of the gram-atom, we can draw an important conclusion about the relation between molecular weight and the number of gram-atoms of each element in a compound. If we have a sample

Spotlight

Molecular or atomic weights are now most accurately and most conveniently determined by an instrument called a mass spectrometer, or a mass spectrograph. This instrument functions on the basis of the effects that an electric field has on charged particles and a magnetic field has on moving charged particles. The charged particles are ions made from individual atoms or molecules, usually by electron bombardment in an ionization chamber. The electric field causes the ions to be accelerated in a beam that has its geometrical characteristics determined by "slits." These moving charges are then "bent" by a magnetic field, of a direction perpendicular to the electric field. Mathematical analysis shows that the amount by which the beam is bent is determined by the strengths of the applied electric and magnetic fields and by the ratio of the electric charge of the ion to its mass. Hence, it is straightforward to determine the mass of the ion (which is very nearly equal to the mass of the corresponding neutral species—atom or molecule). A typical mass spectrometer system is shown in the accompanying sketch.

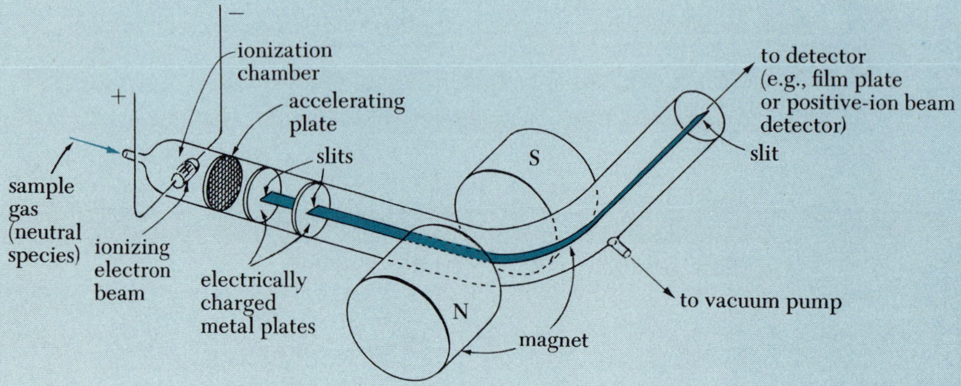

of a substance with a mass in grams numerically equal to the molecular weight, then the number of gram-atoms of each element present equals the number of atoms of that element in the compound formula. For example, a sample of 142.04 g of sodium sulfate, Na_2SO_4, contains two g-atoms of sodium, one g-atom of sulfur, and four g-atoms of oxygen.

This relation makes it convenient for us to define an additional quantity of material known as the gram-molecular weight, or simply gram-mole. The **gram-mole** is defined as an amount of a substance with a mass in grams that is numerically equal to the molecular weight in atomic mass units. Thus one gram-mole of HCl is defined as a sample of HCl with a mass of 36.461 g, and this quantity of material contains 1.008 g of hydrogen and 35.453 g of chlorine. Similarly, one gram-mole of sodium sulfate consists of a sample with mass 142.04 g, which comprises 45.98 g of sodium, 32.06 g of sulfur, and 64.00 g of oxygen. The relations embodied in these definitions provide a convenient framework for relating not only elemental compositions to molecular formulas but also quantities of materials produced to quantities of materials consumed in chemical reactions. These definitions can be summarized as follows:

1. The gram-mole is the amount of substance with mass numerically equal to the molecular weight in grams.
2. The molecular weight is the sum of the atomic weights represented in the formula, in other words, mass (in amu's) corresponding to the formula unit of the substance.
3. The molecular weight is the mass (in amu's) of the formula unit; that is, it is numerically equal to the mass of the gram-molecule (g-mol) expressed in grams.

EXAMPLE 3.9 What are the molecular weights of acetic acid, CH_3CO_2H, and potassium perchlorate, $KClO_4$?

SOLUTION From our definitions and the given formulas, we have for CH_3CO_2H,

$$\text{mol wt} = 2(12.01 \text{ amu}) + 2(16.00 \text{ amu}) + 4(1.008 \text{ amu})$$
$$= 60.05 \text{ amu}$$

Similarly, for $KClO_4$, we have

$$\text{mol wt} = 39.10 \text{ amu} + 35.45 \text{ amu} + 4(16.00 \text{ amu})$$
$$= 138.55 \text{ amu}$$

EXAMPLE 3.10 How many gram-moles are there in a 4.523-g sample of CH_3CO_2H?

SOLUTION Since the molecular weight of CH_3CO_2H is given in the preceding example as 60.05 amu, we can write

$$1 \text{ g-mol } CH_3CO_2H = 60.05 \text{ g } CH_3CO_2H$$
$$1 = \frac{1}{60.05} \frac{\text{g-mol } CH_3CO_2H}{\text{g } CH_3CO_2H}$$
$$4.523 \text{ g } CH_3CO_2H = (4.523 \text{ g } CH_3CO_2H)(1)$$
$$= (4.523 \text{ g } CH_3CO_2H)\left(\frac{1}{60.05} \frac{\text{g-mol } CH_3CO_2H}{\text{g } CH_3CO_2H}\right)$$
$$= 0.07532 \text{ g-mol } CH_3CO_2H$$

STUDY PROBLEM 3(i)

What are the molecular weights of benzene, C_6H_6, and sodium hydrogen phosphate, Na_2HPO_4?

STUDY PROBLEM 3(j)

How many g-mol are there in 41.62 g of C_6H_6?

It is instructive to consider the relations between gram-atoms, gram-moles, atoms, molecules, and Avogadro's number. Let us consider 1 g-mol of methyl alcohol (CH_3OH, mol wt 32.04 amu). This sample weighs 32.04 g and contains 1 g-atom carbon, 4 g-atoms hydrogen, and 1 g-atom oxygen. In the atomic view, the sample contains 6.022×10^{23} carbon atoms, $4 \times 6.022 \times 10^{23}$ hydrogen atoms, and 6.022

$\times$ 10^{23} oxygen atoms. This collection of atoms adds up to 6.022×10^{23} CH_3OH molecules. The relations are summarized as follows:

One gram-mole of methyl alcohol, CH_3OH, weighs 32.04 g, and contains

$$\begin{array}{lcl}
12.01\ \text{g C} & = \text{one g-atom of C} & = 6.022 \times 10^{23}\ \text{C atoms} \\
4(1.008)\ \text{g H} & = \text{four g-atom of H} & = 4 \times 6.022 \times 10^{23}\ \text{H atoms} \\
16.00\ \text{g O} & = \text{one g-atom of O} & = 6.022 \times 10^{23}\ \text{O atoms} \\
\hline
32.04\ \text{g } CH_4O & & 6.022 \times 10^{23} \times (\text{one C atom, four H atoms, and one O atom})
\end{array}$$

$$\underbrace{\qquad\qquad\qquad\qquad\qquad\qquad}_{6.022 \times 10^{23}\ CH_3OH\ \text{molecules}}$$

This kind of analysis can be carried out for any molecular substance, and leads to the following general statement: *One gram-mole of any molecular substance contains Avogadro's number of molecules of that substance.* The same kind of analysis holds for nonmolecular compounds also; however, in such cases the word *molecule* must be replaced by the more general term *formula unit.* The more general statement reads: *One gram-mole of any substance contains Avogadro's number of formula units of that substance.* This statement applies even to atomic systems for which the formula unit is merely one atom.

A Mole Is a Number

From the last statement in italics, and the corresponding statement earlier about the gram-atom (1 g-atom of any element contains N_A atoms of that element), chemists have found it convenient to define a closely related and more general concept, the mole. **A mole** of any species is Avogadro's number of that species. Thus, a mole of methane, CH_4, is 6.022×10^{23} CH_4 molecules. A mole of Na_2SO_4 is 6.022×10^{23} Na_2SO_4 formula units. A mole of sodium atoms is 6.022×10^{23} Na atoms. A mole of doughnuts is 6.022×10^{23} doughnuts. A mole is a number, like one dozen. This analogy is graphically emphasized in Fig. 3.5. The Standard International (SI) unit for the mole is **mol**; we will use this abbreviation in the solutions of computational problems.

We now know that 6.022×10^{23} atoms can be called either a mole of atoms or one gram-atom; hence, we see that "1 g-atom" of any element functionally means the same thing as "1 mol of atoms" of that element. Also, since we now know that

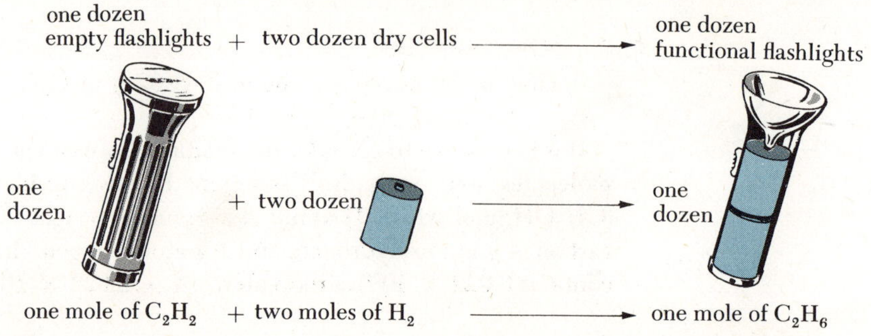

Figure 3.5
A mole is a number.

6.022×10^{23} molecules of a given substance can be called 1 g-mol of the substance, or alternatively, 1 mol of molecules of the substance, we see that the terms *g-mol* and *mole* functionally refer to the same sample. Only the definitions are different, one referring to molecular weight and the other to Avogadro's number.

What has developed is a picture in which gram-atoms and gram-moles bear a relation to each other that is analogous to that between atoms and formula units. Avogadro's number is the bridge that connects these two relations (Fig. 3.6). The

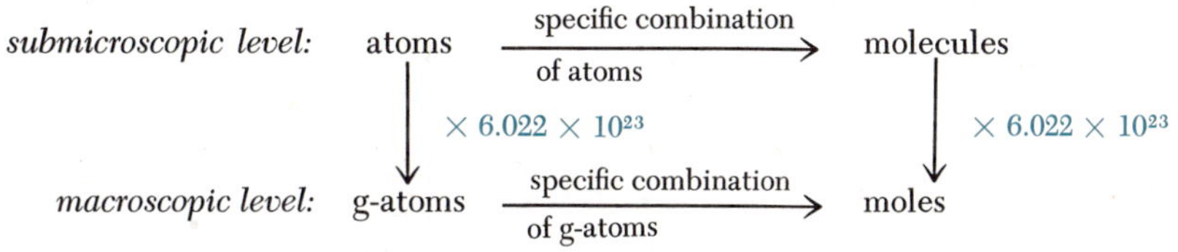

Figure 3.6 Parallel relations among related submicroscopic and macroscopic chemical units.

concept of the mole unifies the macroscopic side of this analogy. The function of Avogadro's number as a bridge between submicroscopic units and macroscopic samples is emphasized in Table 3.3. Note that in this table the term *mol wt* is used as an abbreviation for *molecular weight.* This abbreviation is used frequently in this text. It should be noted that use of the term *mole* can cause confusion with some elemental substances. For example, under normal conditions, elemental hydrogen exists as a diatomic (containing two atoms) molecule, H_2. The phrase *a mole of hydrogen* does not distinguish between Avogadro's number of hydrogen atoms (1.008 g) and Avogadro's number of hydrogen molecules (2.016 g). For this reason, we shall often use the term *gram-atom,* abbreviation *g-atom,* to mean "N_A atoms" and *mole* to mean "N_A molecules." When we use *mole* in reference to atoms, we shall note this explicitly with statements such as "one mole of H."

Applications

The use of the mole concept in chemical calculations is amplified in the following examples.

EXAMPLE 3.11 How many moles of sulfur atoms are contained in 1 mol of molecular sulfur in its common form, S_8.

SOLUTION The phrase *mole of sulfur atoms* is functionally synonymous with *g-atom of sulfur.* Hence, 1 mol of S_8 contains 8 g-atoms of sulfur, or 8 mol of S atoms.

EXAMPLE 3.12 How many fluorine atoms are contained in a 10.0-g coating of Teflon (simplest formula CF_2) in a frying-pan?

SOLUTION Using the unit factor method to convert g CF_2 to mol CF_2, we get

$$10.0 \text{ g} = (10.0 \text{ g}) \frac{1}{(12.01 + 2 \times 19.0)} \frac{\text{g}}{\text{mol}} = \frac{10.0}{50.0} \text{ mol} = 0.200 \text{ mol}$$

One mole of CF_2 contains 2 g-atoms F. Thus, 0.200 mol of CF_2 contains 2 × (0.200 g-atom F), or 0.400 mol F atoms.

$$\text{no. F atoms} = (0.400 \text{ g-atom}) \left(6.022 \times 10^{23} \frac{\text{F atoms}}{\text{g-atom}} \right) = 2.41 \times 10^{23} \text{ F atoms}$$

TABLE 3.3 **Avogadro's Number as the Bridge between Submicroscopic Units and Macroscopic Samples**

Submicroscopic Structural Unit			$N_A = 6.022 \times 10^{23}$ *(the bridge)*	*Macroscopic Sample Unit*		
Formula Unit	*Formula*	*Mass*		*N_A Submicroscopic Structural Units (Formula Units)*	*Formula*	*Mass*
I. Atom	Elemental symbol e.g., Cu, H	at. wt (amu)		**I.** One g-atom (or one mole of atoms)	Elemental symbol, e.g., Cu, H	at. wt in g
Example: One hydrogen atom	H	1.008 amu		Example: One g-atom of hydrogen 6.022×10^{23} H atoms	H	1.008 g H
II. Molecule	Molecular formula, e.g., NH_3, CH_4	mol wt (amu)		II. One g-mol or one mole	Molecular formula, e.g., NH_3, CH_4	mol wt in g
Example: One methane molecule	CH_4	16.042 amu		Example: One mole of methane, 6.022×10^{23} CH_4 molecules	CH_4	16.042 g CH_4
III. Nonmolecular formula unit	Simplest formula, e.g., $CaSO_4$, NaCl	mol wt (amu)		**III.** One g-mol or one mole	Simplest formula	mol wt in g
Example: One formula unit of sodium chloride	NaCl	58.44 amu		Example: One mole of sodium chloride, 6.022×10^{23} NaCl formula units	NaCl	58.44 g NaCl

The unit factor method can be used to give the final result in one expression.

$$10.0\ g\ CF_2 \times \frac{1\ mol\ CF_2}{50.0\ g\ CF_2} \times \frac{2\ g\text{-atom F}}{1\ mol\ CF_2} \times \frac{6.022 \times 10^{23}\ atoms\ F}{1\ g\text{-atom F}} = 2.41 \times 10^{23}\ F\ atoms$$

g CF_2 | mole CF_2 | g-atom CF_2 | atoms F

The units shown at the vertical lines show the progression from left to right through the expression, starting with g CF_2 and ending with F atoms.

EXAMPLE 3.13 How many atoms are contained in a 3.1-g copper penny?

SOLUTION The atomic weight of Cu is 63.55 amu. Hence, 63.55 g Cu constitutes 1 mol Cu atoms, that is, 6.022×10^{23} Cu atoms. Then,

$$3.1\ g\ Cu = (3.1\ g\ Cu)\left(\frac{1}{63.55}\frac{mol\ Cu}{g\ Cu}\right)\left(6.02 \times 10^{23}\frac{Cu\ atoms}{mol\ Cu}\right)$$

$$= 2.9 \times 10^{22}\ Cu\ atoms$$

EXAMPLE 3.14 What is the mole ratio of ethyl alcohol, CH_3CH_2OH, to water in a liquor that is 49% alcohol (98 proof), 49% water, and 2% miscellaneous flavors, by weight?

SOLUTION Let us choose a specific size of sample for discussion, say 100 g. This will contain 49 g H_2O and 49 g CH_3CH_2OH. The water has mol wt 18.02 amu, and the alcohol has mol wt 2(12.01) + 16.00 + 6(1.01) = 46.07 amu. Then,

$$\frac{no.\ mol\ H_2O}{no.\ mol\ CH_3CH_2OH} = \frac{(49\ g\ H_2O)(1\ mol\ H_2O/18.02\ g\ H_2O)}{(49\ g\ CH_3CH_2OH)(1\ mol\ CH_3CH_2OH/46.07\ g\ CH_3CH_2OH)}$$

$$= \left(\frac{46.07}{18.06}\right)\frac{mol\ H_2O}{mol\ CH_3CH_2OH}$$

$$= 2.6\frac{mol\ H_2O}{mol\ CH_3CH_2OH}$$

Spotlight

The element fluorine (*fluere*, "to flow") was isolated in 1886 by Moissan, who obtained the gas by the electrolysis of a solution of potassium fluoride, KF, in anhydrous hydrofluoric acid, HF. The high reactivity of the element is largely responsible for its relatively late isolation. Fluorine is a widely distributed element, found to some extent in almost all rocks, natural waters, and plant and animal matter. Despite the large amounts of sodium chloride found in the sea and in salt beds, fluorine is more abundant than chlorine; on earth, fluorine constitutes 0.065% of the earth's crust, while chlorine constitutes 0.055%. The chief fluorine-containing minerals are cryolite, $3NaF \cdot AlF_3$, fluorite, CaF_2, and fluorapatite, $Ca_5(PO_4)_3F$.

Elemental fluorine, F_2, is a canary-yellow gas under standard conditions. It is used in the manufacture of refrigerants (for example, Freon), uranium hexafluoride, UF_6, for the separation of uranium isotopes, and fluorine-containing polymers (for example, Teflon), and as a rocket fuel when used with hydrogen or lithium.

STUDY PROBLEM 3(k)

How many lead atoms are present in a 10-g sample of polluted water from a river that is found to contain $6 \times 10^{-4}\%$ $PbCO_3$?

STUDY PROBLEM 3(l)

What is the mole ratio in a sample that is 20% KCl and 80% NaCl?

EXAMPLE 3.15 Work the problem stated in Example 3.3 using the concepts of the mole and the gram-atom.

SOLUTION Consider a sample of ammonia that contains 1 mol nitrogen atoms, that is N_A nitrogen atoms. According to the molecular formula NH_3, this sample must contain three times as many H atoms as N atoms. This would be $3 \times N_A$ H atoms, or 3.00 g-atom H. Thus, the hypothetical sample under consideration would consist of 14.01 g of nitrogen and 3×1.008 g of hydrogen. Hence,

$$\%\,\text{N} = \frac{14.01\text{ g}}{14.01\text{ g} + 3.024\text{ g}} \times 100 = 82.2\%$$

$$\%\,\text{H} = \frac{3.024\text{ g}}{14.01\text{ g} + 3.024\text{ g}} \times 100 = 17.8\%$$

EXAMPLE 3.16 Work the problem stated in Example 3.4 by using the concepts of the mole and the gram-atom.

SOLUTION We choose to consider a sample that consists of 1 mol Na_3PO_4. Such a sample contains 3 g-atoms sodium (3×22.99 g sodium), 1 g-atom phosphorus (30.97 g phosphorus), and 4 g-atoms oxygen (4×16.00 g oxygen). Thus,

$$\%\,\text{P} = \frac{30.97\text{ g}}{30.97\text{ g} + 68.97\text{ g} + 64.00\text{ g}} \times 100 = 18.89\%$$

$$\%\,\text{Na} = \frac{68.97\text{ g}}{30.97\text{ g} + 68.97\text{ g} + 64.00\text{ g}} \times 100 = 42.07\%$$

$$\%\,\text{O} = \frac{64.00\text{ g}}{30.97\text{ g} + 68.97\text{ g} + 64.00\text{ g}} \times 100 = 39.04\%$$

EXAMPLE 3.17 Work the problem stated in Example 3.5 by the concepts of the mole and the gram-atom.

SOLUTION For convenience, let us assume a sample size of 100.0 g. (Clearly our answer must be independent of the size of sample we choose to consider!) In this case, the sample will contain 25.0 g hydrogen and 75.0 g carbon. In terms of gram-atoms, these are

$$25.0\text{ g H} = (25.0\text{ g H})\left(\frac{1}{1.008}\,\frac{\text{g-atom H}}{\text{g H}}\right) = \frac{25.0}{1.008}\text{ g-atom H} = 24.8\text{ g-atom H}$$

$$75.0\text{ g C} = (75.0\text{ g C})\left(\frac{1}{12.01}\,\frac{\text{g-atom C}}{\text{g C}}\right) = \frac{75.0}{12.01}\text{ g-atom C} = 6.24\text{ g-atom C}$$

$$\frac{24.8 \text{ g-atom H}}{6.24 \text{ g-atom C}} = 3.97 \frac{\text{g-atom H}}{\text{g-atom C}}$$

Then, simply multiplying both numerator and denominator by Avogadro's number, we obtain

$$\left(3.97 \frac{\cancel{\text{g-atom}} \text{ H}}{\cancel{\text{g-atom}} \text{ C}}\right) \left(\frac{6.022 \times 10^{23} \dfrac{\text{H atoms}}{\cancel{\text{g-atom}} \text{ H}}}{6.022 \times 10^{23} \dfrac{\text{C atoms}}{\cancel{\text{g-atom}} \text{ C}}} \right) = 3.97 \frac{\text{H atoms}}{\text{C atoms}}$$

Then, rounding off reveals four hydrogen atoms for each carbon atom. As discussed in Example 3.5, this leads to a simplest formula, CH_4.

In Example 3.17, it is seen that the number of gram-atoms of hydrogen is numerically equal to the number of grams of hydrogen divided by the atomic weight of hydrogen. An analogous relation is also seen for carbon. Such a relation is true of any element: *The number of gram-atoms of an element is numerically equal to the number of grams of the element divided by the atomic weight of the element. Similarly, the number of moles of a compound is numerically equal to the number of grams of the compound divided by the molecular weight of the compound.*

STUDY PROBLEM 3(m)

What is the simplest formula of a compound containing 11.2% hydrogen and 88.8% oxygen?

STUDY PROBLEM 3(n)

What is the percentage elemental composition of sodium formate, $NaHCO_2$?

EXAMPLE 3.18 How much CuO will be produced in the reaction considered in Example 3.6?

SOLUTION From the relevant atomic weights, we compute the mol wt of CuO to be 16.00 + 63.55 = 79.55 amu. Then,

$$79.55 \text{ g CuO} = 1 \text{ mole CuO}$$

$$1 = \frac{79.55}{1} \frac{\text{g CuO}}{\text{mol CuO}}$$

Also,

$$16.00 \text{ g O} = 1 \text{ mol O}$$

$$1 = \frac{1}{16.00} \frac{\text{mol O}}{\text{g O}}$$

Thus,

$$10.0 \text{ g O} = (10.0 \text{ g O})(1) = (10.0 \cancel{\text{g O}})\left(\frac{1}{16.00} \frac{\text{mol O}}{\cancel{\text{g O}}}\right) = \frac{10.0}{16.00} \text{mol O}$$

As discussed in Example 3.6, we know that one oxygen atom (in combination with one copper atom) leads to one formula unit of CuO. Thus, recognizing the correspondences identified in Table 3.3 and Fig. 3.6, we see that 1 mol O atoms (or 1 g-atom O) will lead to 1 mol CuO. Then, 10.0 g O, which is (10.0/16.00) mol O, will give (10.0/16.00) mol CuO. The amount of CuO produced is

$$\begin{aligned}\frac{10.0}{16.00}\text{ mol CuO} &= \left(\frac{10.0}{16.00}\text{ mol CuO}\right)(1)\\ &= \left(\frac{10.0}{16.00}\text{mol CuO}\right)\left(\frac{79.55}{1}\ \frac{\text{g CuO}}{\text{mol CuO}}\right)\\ &= \frac{10.0 \times 79.55}{16.00}\text{ g CuO}\\ &= 49.7\text{ g CuO}\end{aligned}$$

Alternatively, the unit factor method can be used directly in one step:

$$10.0\text{ g O} = (10.0\text{ g O})\left(\frac{1}{16.00}\ \frac{\text{mol O}}{\text{g O}}\right)\left(1\,\frac{\text{mol CuO}}{\text{mol O}}\right)\left(79.55\,\frac{\text{g CuO}}{\text{mol CuO}}\right) = 49.7\text{ g CuO}$$

g O | mole O | mole CuO | g CuO

Note that the sequence of units at the vertical lines corresponds to the train of thought given in the solution of the problem, that is, from grams of oxygen to moles O, to moles CuO, to grams of CuO. (Note also that 10.0 g O is not really equal to 49.7 g CuO; the equal sign is used here to express only a chemical equivalence in the sense of the indicated reaction.)

In the last example, we have again seen the following very useful numerical relations, which apply to any species:

$$\text{no. grams} = \text{no. moles} \times \text{mol wt} \tag{3.1}$$

$$\text{no. moles} = \frac{\text{no. grams}}{\text{mol wt}} \tag{3.2}$$

In Example 3.18, the number of moles in a 10.0-g sample of O was found to be

$$(10.0\text{ g O})\left(\frac{1}{16.00}\ \frac{\text{mol O}}{\text{g O}}\right) = \frac{10.0\text{ g O}}{16.00\text{ (g O/mol O)}} = \left(\frac{\text{no. grams}}{\text{mol wt}}\right)\text{ mol O}$$

And the number of grams in a $\frac{10.0}{16.00}$ mol sample of CuO was found to be

$$\left(\frac{10.0}{16.00}\text{mol CuO}\right)\left(79.55\,\frac{\text{g CuO}}{\text{mol CuO}}\right) = \Big((\text{no. moles}) \times (\text{mol wt})\Big)\text{ g CuO}$$

EXAMPLE 3.19 Calculate the amount of CO_2 produced in the complete combustion of 1 kg of a gasoline-type fuel with formula C_8H_{18} in a "clean" engine system that produces CO_2 but no other carbon-containing products (that could contribute to smog).

SOLUTION The molecular weight of C_8H_{18} is $8 \times 12.01 + 18 \times 1.008 = 114.22$ amu. According to the information given, one molecule of the fuel, which contains 8 carbon atoms, will produce 8 molecules of CO_2 (with molecular weight $12.01 + 2 \times 16.00 = 44.01$ amu). Hence, 1 mol C_8H_{18} will produce 8 mol CO_2. The available fuel consists of

$$(1000\ \text{g})\left(\frac{1}{114.22}\frac{\text{mol}}{\text{g}}\right) = \frac{1000}{114.22}\ \text{mol}$$

Thus, the number of moles of CO_2 is $8 \times \frac{1000}{114.22}$ mole, and the amount of CO_2 in grams is

$$\left(\frac{8000}{114.22}\ \cancel{\text{mol CO}_2}\right)\left(44.01\ \frac{\text{g CO}_2}{\cancel{\text{mol CO}_2}}\right) = 3082\ \text{g CO}_2$$

Using just one equation in the unit factor method gives

$$(1000\ \cancel{\text{g C}_8\text{H}_{18}})\left(\frac{1}{114.22}\ \frac{\cancel{\text{mol C}_8\text{H}_{18}}}{\cancel{\text{g C}_8\text{H}_{18}}}\right)\left(8\ \frac{\cancel{\text{mol CO}_2}}{\cancel{\text{mol C}_8\text{H}_{18}}}\right)\left(44.01\ \frac{\text{g CO}_2}{\cancel{\text{mol CO}_2}}\right) = 3082\ \text{g CO}_2$$

STUDY PROBLEM 3(o)

How many grams of water will be produced in a reaction with 540 mg of hydrogen with oxygen?

The Mole Concept in the Ideal Gas Equation

In terms of the concept of the mole, the ideal gas equation, Eq. (2.5), can now be written

$$P = \frac{nRT}{V} \qquad \text{or} \qquad PV = nRT \tag{3.3}$$

In this equation n represents the number of moles of gas in the sample. If P is expressed in atmospheres, V in liters, and T in degrees K, the constant R is found to be 0.08205 liter atm mol^{-1} deg^{-1}. R is called the gas constant.

EXAMPLE 3.20 Calculate the volume of one mole of any ideal gas at standard temperature and pressure (0 °C and 1 atm).

SOLUTION From Eq. (3.3), we write for one mole of gas

$$\begin{aligned} V &= \frac{nRT}{P} \\ &= \frac{(1\ \text{mol})(0.08205\ \text{liter atm deg}^{-1}\ \text{mol}^{-1})(273.15\ \text{deg})}{1\ \text{atm}} \\ &= 22.4\ \text{liters} \end{aligned}$$

EXAMPLE 3.21 Calculate the volume of 4.65 g of ammonia at standard temperature and pressure.

SOLUTION From Eq. (3.2) and (3.3), we write (recognizing the mol wt of NH_3 as 17.03 amu)

$$V = \frac{nRT}{P}$$

$$= \frac{[4.65\text{ g}/(17.03\text{ g/mol})](0.08205\text{ liter atm deg}^{-1}\text{ mol}^{-1})(273.15°)}{1.00\text{ atm}}$$

$$= 6.12\text{ liter}$$

EXAMPLE 3.22 Calculate the pressure of a 1.92-g sample of gaseous methane, CH_4, contained in a 250-ml vessel at 25 °C.

SOLUTION From Eq. (3.2) and (3.3), we write

$$P = \frac{nRT}{V}$$

$$= \frac{[1.92\text{ g}/(16.04\text{ g/mol})](0.08205\text{ liter atm mol}^{-1}\text{ deg}^{-1})(298°)}{0.250\text{ liter}}$$

$$= 11.7\text{ atm}$$

EXAMPLE 3.23 Calculate the number of moles of nitrous oxide, N_2O, in a gaseous sample that occupies 3.88 liters at 20 °C and 0.951 atm.

SOLUTION By Eq. (3.3),

$$n = \frac{PV}{RT}$$

$$= \frac{(0.951\text{ atm})(3.88\text{ liter})}{(0.08205\text{ liter atm deg}^{-1}\text{ mol}^{-1})(293\text{ deg})}$$

$$= 0.153\text{ mol}$$

EXAMPLE 3.24 Calculate the mass of a gaseous nitric oxide (NO) sample that occupies 1680 ml at 30 °C and 0.905 atm.

SOLUTION Using Eq. (3.2) and (3.3), we get

$$n = \frac{PV}{RT}$$

$$= \frac{(0.905\text{ atm})(1680\text{ ml}\cdot\frac{1}{1000}\text{ liter/ml})}{(0.08205\text{ liter atm deg}^{-1}\text{ mol}^{-1})(303\text{ deg})}$$

$$= 0.0612\text{ mol}$$

Then, using Eq. (3.1), we get

$$\text{no. g NO} = (\text{no. moles NO})(\text{mole wt NO})$$

$$= (0.0612\text{ mol})\left[(14.01 + 16.00)\left(\frac{\text{g}}{\text{mol}}\right)\right]$$

$$= 1.84\text{ g}$$

STUDY PROBLEM 3(p)

Calculate the volume of 0.246 mol of any ideal gas at 40 °C and 1.22 atm.

STUDY PROBLEM 3(q)

Calculate the pressure of a 0.888-g gaseous sample of ethylene, C_2H_4, contained in a 800-ml vessel at 20 °C.

STUDY PROBLEM 3(r)

Calculate the mass of a gaseous CF_4 sample that occupies 1.22 liters at 25 °C and 0.890 atm.

3.7 Chemical Equations

Meaning

The subject of weight relations in chemistry is called **chemical stoichiometry**. In our consideration of stoichiometry thus far, we have concerned ourselves primarily with one substance at a time; in considering chemical reactions, we have largely taken for granted the pertinent relations between the substances consumed and the substances produced. Now, the Dalton atomic theory, the atomic-molecular view, and their related concepts lead us to a powerful concept that allows us to handle these relations concisely and precisely. The conceptual vehicle that makes this possible is the chemical equation.

The **chemical equation** is a shorthand notation for presenting both qualitative and quantitative facts about the chemical reaction. The chemical equation provides qualitative information by identifying the substances consumed, i.e., the **reactants**, and the substances produced in the reaction, i.e., the **products**. The quantitative information obtained from chemical equations refers to relative amounts of reactants and of products involved. The reactants and products in a chemical reaction are identified symbolically in a chemical equation by their chemical formulas. The reactant and product symbols are separated in a chemical equation by the chemical "equals" sign, the opposing arrows $\rightleftharpoons$; the reactants are placed to the left of this symbol and the products to the right. For example, the reaction of sodium metal with bromine liquid to form sodium bromide is represented by Eq. (3.4):

$$2Na + Br_2 \rightleftharpoons 2NaBr \tag{3.4}$$

In this equation, the subscript 2 for the symbol Br_2 indicates explicitly that liquid bromine exists as molecules that consist of two bromine atoms.

The symbol "$\rightleftharpoons$" in any chemical equation takes on the meaning "is transformed into" or "are transformed into," signifying that any substances represented on the left side are transformed into substances on the right side. Equation (3.4) thus states: "Each Br_2 molecule reacts with two Na atoms to form two NaBr units."

Balancing

Quantitative information can be derived from proper chemical equations, since such equations are balanced. By **balancing a chemical equation,** we simply mean ensuring that the same amounts of each type of matter (elements) are represented on each side

Spotlight

Lithium (*litheos*, "stony") was discovered in 1917 by Arfvedson, and isolated in 1855 by Bunsen and Matthiessen by electrolysis of the fused chloride. It is not an abundant element, but is widely distributed and occurs in the minerals lepidolite, $K(Li, Al)(AlSi_3O_{10})(O, OH, F)_2$, spodumene, $LiAl \cdot (Si_2O_6)$, and amblygonite, $Li(AlF)PO_4$. Lithium has been detected in meteorites, mineral waters, tea, coffee, blood, milk, and the ashes of many plants, notably tobacco. The metal can be prepared by the electrolysis of the fused chloride (not the aqueous solution):

$$2Li^+ + 2Cl^- \rightleftharpoons 2Li(s) + Cl_2(s)$$

Lithium is a soft, silvery metal that is the least dense of all the solid elements (0.534 g/cc). It is more reactive than any of the other second period elements except fluorine, and vigorously reacts with water to give lithium hydroxide, LiOH, and hydrogen:

$$2Li(s) + 2H_2O(l) \rightleftharpoons 2Li^+ + 2OH^- + H_2(g)$$

Lithium salts give a splendid red color to flames and are used in fireworks. An alloy with an elasticity and tensile strength similar to steel is formed from zinc and aluminum containing 0.1% lithium.

Lithium is too metallic to form a negative ion even in the gas phase, where it can exist as a dimer, Li_2. Nor does it form any negative complex ions, and its chemistry in aqueous solutions is limited to that of Li^+.

of the equation. In other words, chemical equations are balanced so that they will conform to both the law of conservation of mass and Dalton's atomic theory. If the chemical equation is to conform precisely, the symbols used must adhere to the requirement that the same number of atoms of any given type must appear in both the reactant side (left) and the product side (right). This preserves the idea that atoms are merely reshuffled in chemical reactions—not created, nor destroyed, nor altered in identity. Thus, the symbols in a chemical equation represent the formula units (that is, atoms, molecules) in a chemical reaction, which must be balanced in numbers of atoms of each element involved.

This requirement introduces the idea of coefficients in chemical equations. The number 2 that precedes the symbol NaBr in Eq. (3.4) is an example of a coefficient in a chemical equation; it has a meaning similar to the meaning of coefficients in ordinary algebraic equations. In algebra, a coefficient determines how many times the immediately following symbol is to be counted. In a chemical equation, a **coefficient** shows that the chemical species (or formula unit) represented by the formula written to the right of the coefficient is counted a number of times equal to the coefficient. Thus that number 2 in Eq. (3.4) indicates two NaBr formula units. Determining what coefficients must occur in a chemical equation is popularly referred to then as balancing the equation. In Eq. (3.4), this process can be carried out by inspection, and a brief procedure of trial and error. In this particular instance, an alternative equation that would be equally valid is

$$Na + \tfrac{1}{2}Br_2 \rightleftharpoons NaBr$$

We shall generally avoid fractional coefficients and present chemical equations with the smallest integral coefficients consistent with balancing; thus fractions will be avoided. For many chemical reactions it is possible to write the balanced equations in a straightforward manner from knowledge of the identities of the reactants and products. This is especially true if advantage is taken, whenever appropriate, of the concept of a functional group. Some simple examples of chemical equations that are straightforward to balance follow:

$$C + O_2 \rightleftharpoons CO_2$$
$$2C + O_2 \rightleftharpoons 2CO$$
$$I_2 + Cl_2 \rightleftharpoons 2ICl$$
$$2H_2O_2 \rightleftharpoons 2H_2O + O_2$$
$$CoCl_2 + 6H_2O \rightleftharpoons Co(H_2O)_6Cl_2{}^{*}$$
$$\underline{CH_3}H + \underline{HO}NO_2 \rightleftharpoons \underline{CH_3}NO_2 + \underline{HO}H^{\dagger}$$

It is often useful in chemical equations to indicate the physical state of each solid, liquid, or gaseous species by means of symbols—s, l, and g. Examples of this notation are

$$C(s) + O_2(g) \rightleftharpoons CO_2(g)$$
$$2H_2O_2(l) \rightleftharpoons 2H_2O(l) + O_2(g)$$
$$CoCl_2(s) + 6H_2O(l) \rightleftharpoons Co(H_2O)_6Cl_2(s)$$

Many of the reactions we shall consider in this textbook, and most that are met in the laboratory, are carried out in liquid solutions. In representing solutes of liquid solutions in chemical equations, we shall adopt the convention of using no specific symbol, for instance no "l" to represent the physical state. Thus, the equation

$$C_6H_{12} + Br_2 \rightleftharpoons C_6H_{12}Br_2$$

represents the reaction between hexene, C_6H_{12}, and bromine in a liquid solution (for example, with CCl_4 as the solvent). Balancing chemical equations for some types of reactions, especially oxidation-reduction reactions, can involve special difficulties and require special procedures. (See Chaps. 8 and 12.)

EXAMPLE 3.25 Write a balanced chemical equation for the reaction of solid aluminum with gaseous oxygen, O_2, to form the binary solid compound aluminum oxide (alumina).

SOLUTION From the guidelines in Sec. 3.4, we can predict that the formula for aluminum oxide is Al_2O_3 (thinking of aluminum as a $+3$ ion and oxygen as a -2 ion). Then, since two atoms of aluminum are required for each three oxygen atoms, we can write

$$2Al(s) + \tfrac{3}{2}O_2(g) \rightleftharpoons Al_2O_3(s),$$

or alternatively

$$4Al(s) + 3O_2(g) \rightleftharpoons 2Al_2O_3(s)$$

This reaction occurs vigorously if a high-temperature torch is used to heat the aluminum and a good oxygen source is provided.

EXAMPLE 3.26 Write a balanced chemical equation for the complete combustion (burning with O_2) of octane, C_8H_{18}, a reaction that produces only water and carbon dioxide.

SOLUTION Knowing the reactants and products, we can immediately write an unbalanced representation of the reaction:

$$C_8H_{18}(g) + O_2(g) \rightarrow CO_2(g) + H_2O(g)$$

---

*The formula unit H_2O can be handled here as a functional group for balancing.
†Functional groups have been identified by underscoring.

To begin balancing the equation, we note that there are 8 carbon and 18 hydrogen atoms on the left side; thus, the coefficients on the right side for CO_2 and H_2O are 8 and 9, respectively.

$$C_8H_{18}(g) + O_2(g) \rightarrow 8CO_2(g) + 9H_2O(g)$$

This leaves only oxygen to be balanced. Since 25 oxygen atoms now appear on the right side, the coefficient for O_2 must be $\frac{25}{2}$:

$$C_8H_{18}(g) + \tfrac{25}{2}O_2(g) \rightarrow 8CO_2(g) + 9H_2O(g)$$

or when multiplied by two,

$$2C_8H_{18}(g) + 25O_2(g) \rightleftharpoons 16CO_2(g) + 18H_2O(g)$$

STUDY PROBLEM 3(s)

Write a balanced chemical equation for the reaction of cesium metal (a solid), with gaseous chlorine, Cl_2, to produce the binary solid cesium chloride.

Interpretation in Moles

A straightforward way of extracting quantitative information from balanced chemical equations is to "scale up" our interpretation by a factor of Avogadro's number. In other words, if we alter our interpretation of chemical symbols and formulas from a submicroscopic view (with formula units) to a macroscopic view (with moles), then a convenient scheme is at hand for treating problems in chemical stoichiometry. Thus, instead of viewing atomic symbols in terms of atoms, we view them in terms of moles of atoms (gram-atoms); instead of viewing molecular formulas in terms of

Spotlight

The oxide of beryllium was discovered by Vauquelin in 1798 in the mineral beryl, $Be_3Al_2(SiO_3)_6$; hence its name. The element was not isolated, however, until 1828, independently by Wohler and Bussey, through the action of potassium on beryllium chloride. The principal source of the element is beryl, which includes emerald and aquamarine gem stones. Beryllium is obtained by electrolysis of the fused chloride or by the reaction of magnesium with beryllium fluoride, BeF_2:

$$Mg(s) + BeF_2(s) \rightleftharpoons Be(s) + MgF_2(s)$$

Beryllium carbide is formed by heating BeO with carbon:

$$2BeO(s) + 3C(s) \rightleftharpoons Be_2C(s) + 2CO(g)$$

It is an extremely hard material that can easily scratch quartz, but silicon carbide (SiC, "carborundum") is the most widely used industrial abrasive. Beryllium carbide slowly hydrolyzes in water to produce methane:

$$Be_2C(s) + 12H_2O \rightleftharpoons CH_4(g) + 2Be(H_2O)_4^{2+} + 4OH^-$$

Thus it can be called beryllium methanide.

Similarly, boron carbides are formed from boron or its oxide, and although several carbides of varying composition have been reported, B_4C(or $B_{12}C_3$) seems to be the best choice for a true compound. With the possible exception of boron nitride, BN, the carbide B_4C is the only known substance with a hardness equal to that of diamond. It is very inert and is not attacked even by strong oxidizing agents.

molecules, we view them in terms of moles. In general, instead of picturing a formula unit (which is submicroscopic) when we view a formula in an equation, we can imagine an entire mole of the indicated substance. This alteration in points of view is shown symbolically below for the reaction represented in Eq. (3.4):

$$2\text{Na (atoms)} \quad + \text{Br}_2 \text{ (molecule)} \quad \rightleftharpoons 2\text{NaBr (formula units)} \tag{3.4a}$$

$$\downarrow \times N_A \qquad\qquad \downarrow \times N_A \qquad\qquad \downarrow \times N_A$$

$$2\,N_A\,\text{Na (atoms)} \quad + N_A\,\text{Br}_2 \text{ (molecules)} \quad \rightleftharpoons 2\,N_A\,\text{NaBr (formula units)}$$

which is the same as

$$2 \text{ moles Na (atoms)} + 1 \text{ mole Br}_2 \text{ (molecules)} \rightleftharpoons 2 \text{ moles NaBr (formula units)} \tag{3.4b}$$

Interpreting chemical equations in moles, while leading to the same quantitative weight relations as an interpretation based on formula units, has a superior pedagogical advantage. The formula for any compound is well defined, and so is the meaning of a mole of substance corresponding to that formula. Thus, the meaning of Eq. (3.4b) is clear, as the mole quantities are readily visualized. For nonmolecular compounds, however, the formula unit (for example, NaBr in Eq. 3.4a) may have meaning only in atomic bookkeeping. We need not bother further with distinctions between these views, since they give rise to equivalent calculations. Nevertheless, for calculations in chemical stoichiometry, we shall generally develop a habit of thought based on the mole approach.

Applying Chemical Equations

A balanced chemical equation quantitatively relates the number of moles of each substance consumed or produced in a chemical reaction. Since the concept of molecular weight and its relation to the definition of a mole allows one to correlate numbers of moles with sample masses, the chemical equation thus provides a way of computing the amounts (masses) of all substances consumed or produced in chemical reactions. This makes it possible to solve a wide variety of problems of practical importance in pure and applied chemistry. Some examples of these are given below.

EXAMPLE 3.27 How many moles of CO_2 gas molecules are produced by the thermal decomposition of 0.30 mol of calcium carbonate, $CaCO_3$, to form calcium oxide, CaO?

SOLUTION By inspection, the balanced equation is

$$CaCO_3(s) \rightleftharpoons CaO(s) + CO_2(g)$$

According to this equation, there is a one-to-one correspondence between the number of moles of CO_2 gas produced and the number of moles of $CaCO_3(s)$ consumed. Thus, 0.30 mol $CO_2(g)$ will be produced in the decomposition.

EXAMPLE 3.28 How many grams of calcium oxide will be produced by the thermal decomposition of 2.34 g of calcium carbonate?

Spotlight

Nitrogen ("niter-forming," *niter* meaning sodium nitrate, $NaNO_3$) was discovered in 1772 by Daniel Rutherford, who found that animals confined in a closed airspace removed one of the components of air (oxygen) and the remaining gas, primarily N_2, couldn't support life. Nitrogen occurs chemically combined in the proteins, amino acids, and so on contained in all forms of life, and as nitrate deposits in Stassfurt, Germany, and in Chile. Elemental nitrogen in the N_2 form makes up 78% of the atmosphere and is commercially produced from this source by first liquefying air, and then boiling away the lower-boiling nitrogen from oxygen. It is conveniently prepared in the laboratory by heating a solution of ammonium nitrite (or a mixture of ammonium chloride and sodium nitrite):

$$NH_4^+ + NO_2^- \rightleftharpoons N_2(g) + 2H_2O$$

Nitrogen is a colorless, odorless, and tasteless gas under ordinary conditions. It is relatively inert, usually requiring a high temperature before it reacts with other elements. "Active nitrogen" is very reactive and was first prepared by Strutt in 1910. If an electric discharge is passed through nitrogen at a low pressure, a golden yellow afterglow persists after the discharge has been stopped, indicating the presence of active nitrogen. It is very reactive and combines directly with many elements, including metals, sulfur, phosphorus, and iodine. For many years the nature of active nitrogen was unknown and many species were suggested as possibilities. Not until 1958 was it shown that the main species is represented by individual nitrogen atoms; the high chemical reactivity is attributable to this species. The recombination of nitrogen atoms produces nitrogen molecules in high-energy states, and these molecules emit the radiation as light when they drop to a lower energy state.

Nitrogen can form negative ions, for example the nitride ion, N^{3-}. This ion is not stable in water and reacts (hydrolyzes) to form ammonia:

$$N^{3-} + 3H_2O \rightleftharpoons NH_3(g) + 3OH^-$$

Nitrogen also forms the azide ion, N_3^-, which is stable in water. In combination with other elements, nitrogen forms a variety of important positive and negative ions, for example NH_4^+ and NO_3^-.

SOLUTION Using the same reasoning as that of the preceding example, we note that the number of moles of CaO produced is the same as the number of moles of $CaCO_3$ consumed. Hence, to determine the amount of CaO produced, one simply needs to determine the number of moles of $CaCO_3$ decomposed and then convert that number to a mass of CaO. We first compute the molecular weight of $CaCO_3$ to be 100.1 amu. Then, using Eqs. (3.2) and (3.1) gives

$$\text{no. moles } CaCO_3 = \frac{\text{mass of } CaCO_3}{\text{mol wt } CaCO_3} = \frac{2.34 \text{ g}}{100.1 \text{ g/mol}} = 0.0234 \text{ mol}$$

$$\begin{aligned}\text{mass of CaO} &= (\text{mol wt of CaO}) \times (\text{no. moles CaO}) \\ &= (56.1 \text{ g/mol})(0.0234 \text{ mol}) \\ &= 1.31 \text{ g}\end{aligned}$$

Alternatively,

$$(2.34 \text{ g } \cancel{CaCO_3})\left(\frac{1}{100.1}\frac{\cancel{\text{mol } CaCO_3}}{\text{g } \cancel{CaCO_3}}\right)\left(1\frac{\cancel{\text{mol CaO}}}{\cancel{\text{mol } CaCO_3}}\right)\left(56.1\frac{\text{g CaO}}{\cancel{\text{mol CaO}}}\right) = 1.31 \text{ g CaO}$$

EXAMPLE 3.29 The complete combustion of octane, C_8H_{18}, produces only water and carbon dioxide. How much carbon dioxide is produced in the complete combustion of one ton of octane?

SOLUTION From Example 3.26, we take the balanced equation for this process, $2C_8H_{18}(g) + 25O_2(g) \rightleftharpoons 16CO_2(g) + 18H_2O(g)$. From this equation, we see an 8:1 ratio between the number of moles of CO_2 produced and the number of moles of C_8H_{18} burned. Then, using the unit factor method, and using the molecular weights of C_8H_{18} and CO_2 (114.22 amu and 44.01 amu), we find that from 1.00 g C_8H_{18},

$$(1.00 \text{ g } C_8H_{18}) \stackrel{\circ}{=} (1.00 \text{ } \cancel{\text{g } C_8H_{18}})\left(\frac{1}{114.22}\frac{\cancel{\text{mol } C_8H_{18}}}{\cancel{\text{g } C_8H_{18}}}\right)\left(8\frac{\cancel{\text{mol } CO_2}}{\cancel{\text{mol } C_8H_{18}}}\right)\left(44.01\frac{\text{g } CO_2}{\cancel{\text{mol } CO_2}}\right)$$

$$= \frac{8 \times 44.01}{114.22} \text{ g } CO_2 = 3.08 \text{ g } CO_2 \text{ are produced.}^*$$

This ratio holds whether we think of masses in grams or of weights in pounds or tons. Hence, one ton of octane produces 3.08 tons of carbon dioxide.

Example 3.29 indicates the seriousness of relying on combustion engines for motive power, even if exhausts can be "cleaned up" by promoting complete combustion. We must still be concerned about the increasing amounts of CO_2 being introduced into the atmosphere. If the atmosphere becomes laden with CO_2, it may act as a "blanket" around the earth and increase its average temperature. The mechanism by which this warming may occur is called the **greenhouse effect**, and is analogous to the heating that occurs inside a greenhouse. It has been estimated that if our CO_2 production from all sources continues to increase at the present rate, it will take only 50 to 100 years for the earth's temperature to increase to a point such that the polar ice caps will melt, and cause the oceans to rise enough to partially submerge all coastal cities! And this is in spite of all the green foliage on the earth that continuously absorbs CO_2 and releases oxygen.

EXAMPLE 3.30 How many grams of bromine, Br_2, must be consumed by reaction with aluminum to produce 10.0 g of $AlBr_3$?

SOLUTION The balanced equation is: $2Al(s) + 3Br_2(l) \rightleftharpoons 2AlBr_3(s)$. There is a 3:2 ratio of the number of moles of Br_2 consumed to the number of moles of $AlBr_3$ produced. We need only compute the number of moles of $AlBr_3$ in a 10.0-g sample, apply this ratio, and obtain the number of moles of Br_2 required. Since the mol wt of $AlBr_3$ is 266.7 amu, we have

$$\text{no. moles } AlBr_3 = (10.0 \text{ g } AlBr_3)\left(\frac{1}{266.7}\frac{\text{mol } AlBr_3}{\text{g } AlBr_3}\right)$$

$$= \frac{10.0}{266.7} \text{ mol } AlBr_3$$

$$\text{mass } Br_2 \text{ consumed} = (\text{no. } \cancel{\text{moles } Br_2})\left(159.8\frac{\text{g } Br_2}{\cancel{\text{mol } Br_2}}\right)$$

$$= \left[\left(\frac{3}{2}\right)\left(\frac{10.0}{266.7}\right)\right](159.8) \text{ g } Br_2$$

$$= 8.99 \text{ g } Br_2$$

*The symbol $\stackrel{\circ}{=}$ indicates an "equivalence" in the sense of a chemical transformation; this symbol does not stand for an algebraic equality. For example, 1.00 g C_8H_{18} does not equal 3.08 g CO_2; but 1.00 g C_8H_{18} is transformed in a chemical reaction into 3.08 g CO_2.

Or using the unit factor method in one step, we get

$$10.0\text{ g AlBr}_3 \stackrel{*}{=}$$

$$(10.0\text{ g AlBr}_3)\left(\frac{1}{266.7}\,\frac{\text{mol AlBr}_3}{\text{g AlBr}_3}\right)\left(\frac{3}{2}\,\frac{\text{mol Br}_2}{\text{mol AlBr}_3}\right)\left(159.8\,\frac{\text{g Br}_2}{\text{mol Br}_2}\right) = \left(\frac{3}{2}\right)\left(\frac{10.0}{266.7}\right)(159.8)\text{ g Br}_2$$

g $AlBr_3$ | mol $AlBr_3$ | mol Br_2 | g Br_2

$$= 8.99\text{ g Br}_2$$

EXAMPLE 3.31 Calculate the number of grams of potassium chloride, KCl, produced from the reaction occurring in a mixture of 39.1 g of potassium and 30.1 g of chlorine, according to the equation

$$2K + Cl_2 \rightleftharpoons 2KCl$$

SOLUTION We first note from the balanced chemical equation that for each 2 g-atom potassium, 1 mol Cl_2 is required for complete reaction. We then calculate, in these terms, the amounts of each reactant available:

$$\text{no. moles K} = (39.1\text{ g K})\left(\frac{1}{39.1}\,\frac{\text{mol K}}{\text{g K}}\right) = 1.00\text{ mol K}$$

$$\text{no. moles Cl}_2 = (30.1\text{ g Cl})\left(\frac{1}{70.9}\,\frac{\text{mol Cl}_2}{\text{g Cl}_2}\right) = 0.425\text{ mol Cl}_2$$

$$\frac{\text{no. moles K}}{\text{no. moles Cl}_2} = \frac{1.00}{0.425} = 2.35$$

which is more than the value 2 required for exactly the correct proportions of K and Cl_2. Hence, more potassium is available than that required by the amount of chlorine present; we say that chlorine is the *limiting reagent.* No more KCl can be produced than that dictated by the 0.425 mol Cl_2 available; the number of moles of KCl produced will be twice the number of moles of Cl_2 available, or

$$0.425\text{ mol Cl}_2 \stackrel{*}{=} (0.425\text{ mol Cl}_2)\left(2\frac{\text{mol KCl}}{\text{mol Cl}_2}\right)\left(74.6\frac{\text{g KCl}}{\text{mol KCl}}\right) = 63.4\text{ g KCl}$$

The amount of unreacted potassium is given by subtracting the amount caused to react (2 × 0.425 mol) from the amount available (1.00 mol),

$$(1.00 - 0.850)\text{ mol K} = 0.150\text{ mol K}$$

$$= (0.150\text{ mol K})\left(39.1\frac{\text{g K}}{\text{mol K}}\right)$$

$$= 5.86\text{ g K}$$

STUDY PROBLEM 3(t)

How many grams of CO_2 are produced in the combustion of 100 g of butane, C_4H_{10}, by a reaction analogous to that shown in Example 3.29?

STUDY PROBLEM 3(u)

If a mixture of 12.0 g of hydrogen gas, H_2, and 15.0 g of oxygen gas, O_2, is ignited, how much water, H_2O, is produced?

3.8 Net Equations; Ionic Solutions

Cancellation

When one is writing balanced chemical equations, it is customary to represent only those species that do not appear on both sides of the equation. To include a particular formula on both sides of a chemical equation implies that the species in question was present before the reaction and after the reaction. This tends to draw attention away from the species of interest—those involved in change; for the chemical equation is designed to represent transformation. Formulas that occur on both sides of an equation represent species that experience no net change during the reaction and hence are of no direct interest when one is concerned with the net reaction. For

Spotlight

Oxygen (*oxys,* "acid," and *genes,* "forming"; "acid former") was discovered in 1774 by Joseph Priestley, who obtained the element by heating mercuric oxide using the sun's rays through a magnifying glass. Karl Scheele liberated oxygen from several substances in 1771–1772, but did not publish his results until 1777. Oxygen is an abundant and widely distributed element, making up about one-half of the earth's crust. By weight, it forms about one-fifth of the air, about two-thirds of the human body, about eight-ninths of the water, and a high percentage of the minerals. The bulk of the oxygen in air comes from green foliage, which consumes CO_2 and liberates O_2 through the process called photosynthesis.

Thus, the plant and animal kingdoms are linked by a cycle; a well-kept lawn 50 ft × 50 ft liberates enough oxygen for a family of four. The element is commercially obtained, along with nitrogen, from the distillation of liquid air. The most common laboratory preparation consists in heating potassium chlorate, $KClO_3$; manganese dioxide, MnO_2, acts as a catalyst:

$$2KClO_3(s) \rightleftharpoons 3O_2(g) + 2KCl(s)$$

The most convenient method of preparation is from the controlled dropping of water onto sodium peroxide:

$$2Na_2O_2(s) + 2H_2O(l) \rightleftharpoons O_2(g) + 4Na^+ + 4OH^-$$

Oxygen, in its stable O_2 form, is a colorless, odorless, and tasteless gas at room temperature; but liquid oxygen is pale blue. Oxygen is very reactive and can combine directly with all the elements except the halogens and a few of the noble metals and noble gases; oxygen occurs in compounds with all the elements except He, Ne, and possibly Ar. The variety of compounds in which oxygen exists is virtually limitless.

Oxygen supports combustion and is essential for respiration in the animal kingdom; marine life depend on its slight solubility in water. The chief commercial uses of oxygen gas are (1) in oxyacetylene and oxyhydrogen torches for welding and cutting metals; (2) in the jet-piercing process of drilling rock; (3) in open-hearth furnaces for melting and refining iron and steel; (4) in medicine for hospital patients, aviators flying in high altitude aircraft, and breathing-gas mixtures for submariners; and (5) in rockets using liquid oxygen as the oxidant. It is also useful in purifying water supplies and in treating sewage.

example, in describing the reaction of hydrogen gas with oxygen gas to form water—burning H_2 in air (which contains oxygen)—it would be algebraically correct in a formal sense to include the formula of nitrogen molecules, N_2 (which are also present in air), or to include any other coincidental substances that happen to be present during the transformation:

$$2H_2 + O_2 + N_2 \rightleftharpoons 2H_2O + N_2 \tag{3.5}$$

However, since these coincidental substances do not themselves undergo any net change, they would have to be represented to the same extent on each side of the equation; it is therefore algebraically equally correct to cancel them by subtracting them from each side. Practically, and for emphasizing net change, it is customary to make such cancellations. Thus, the only formulas that should appear in a balanced equation for a net reaction are those that represent species that actually undergo change. This point is particularly important in describing the reactions that occur between the species present in ionic solutions.

Electrolytes and Ionic Equations

When some substances are dissolved in water or in certain other solvents, they break up into charged particles called ions. In many cases these ionic species in solution can be traced directly to such ions in a crystal lattice of the pure solute. An example

Spotlight

Joseph Priestley was born in Yorkshire, England, in 1733. He was an excellent student and the product of a classical education, which did not include science. As a nonconformist, he espoused a variety of unpopular causes; for instance, he was against slavery and supported religious tolerance, the French Revolution, and the American colonies in their dispute with the British king. His pro-American sympathies led him to become acquainted with Benjamin Franklin, whose influence led to a career in science for Priestley.

Priestley carried out important research on gases. Although he is usually credited with the discovery of oxygen, Scheele had isolated it a couple of years earlier, but had not reported it before Priestley's announcement.

After considerable trouble in England because of his unpopular political stands, he moved to the United States in 1794. He died in Pennsylvania in 1804.

Culver Pictures

TABLE 3.4 **Some Typical Ions and Electrolytes**

Formula	*Name*	*Electrolytes*[a]
F^-	fluoride ion	NaF (sodium fluoride), LiF (lithium fluoride)
Cl^-	chloride ion	NaCl (sodium chloride), $CaCl_2$ (calcium chloride)
Br^-	bromide ion	KBr (potassium bromide), $MgBr_2$ (magnesium bromide)
I^-	iodide ion	NaI (sodium iodide), LiI (lithium iodide)
NO_3^-	nitrate ion	HNO_3 (nitric acid), $Mn(NO_3)_2$ (manganese nitrate)
SO_4^{2-}	sulfate ion	H_2SO_4 (sulfuric acid), $MgSO_4$ (magnesium sulfate)
CO_3^{2-}	carbonate ion	Na_2CO_3 (sodium carbonate), $CaCO_3$ (calcium carbonate)
HCO_3^-	bicarbonate ion	NH_4HCO_3 (ammonium bicarbonate), $NaHCO_3$ (sodium bicarbonate)
MnO_4^-	permanganate ion	$KMnO_4$ (potassium permanganate), $Ca(MnO_4)_2$ (calcium permanganate)
PO_4^{3-}	phosphate ion	Na_3PO_4 (sodium phosphate), $Ca_3(PO_4)_2$ (calcium phosphate)
ClO_4^-	perchlorate ion	$AgClO_4$ (sodium perchlorate), $Ba(ClO_4)_2$ (barium perchlorate)
S^{2-}	sulfide ion	Na_2S (sodium sulfide), BaS (barium sulfide)
HS^-	bisulfide ion	LiHS (lithium bisulfide), NH_4HS (ammonium bisulfide)
$CH_3CO_2^-$	acetate ion	$NaCH_3CO_2$ (sodium acetate), $Mg(CH_3CO_2)_2$ (magnesium acetate)
OH^-	hydroxide ion	KOH (potassium hydroxide), NaOH (sodium hydroxide)
NH_4^+	ammonium ion	$(NH_4)_2\ CO_3$ (ammonium carbonate), NH_4I (ammonium iodide)
Na^+	sodium ion	NaBr (sodium bromide), NaHS (sodium bisulfide)
K^+	potassium ion	$KHCO_3$ (potassium bicarbonate), K_2S (potassium sulfide)
Li^+	lithium ion	Li_2SO_4 (lithium sulfate), Li_3PO_4 (lithium phosphate)
Ca^{2+}	calcium ion	$CaSO_4$ (calcium sulfate), $Ca(NO_3)_2$ (calcium nitrate)
Mg^{2+}	magnesium ion	$MgCO_3$ (magnesium carbonate), $Mg(ClO_4)_2$ (magnesium perchlorate)
Mn^{2+}	manganese ion	$MnCl_2$ (manganese chloride), $Mn(ClO_4)_2$ (manganese perchlorate)
Ba^{2+}	barium ion	BaF_2 (barium fluoride), BaS (barium sulfide)
Ag^+	silver ion	$AgNO_3$ (silver nitrate), Ag_2S (silver sulfide)
Zn^{2+}	zinc ion	$ZnBr_2$ (zinc bromide), $Zn(ClO_4)_2$ (zinc perchlorate)

[a]Either ion is derived from the electrolyte (by dissolving in water) or it exists within the electrolyte.

is table salt with its corresponding salt solution, both of which contain positive sodium ions (sodium atoms with positive electrical charge) and negative chloride ions (chlorine atoms with negative electrical charge). In some cases, the formation of ions, a process called **ionization,** occurs only when the substance is dissolved. In the undissolved, pure state, the solute substance may exist as a molecular, nonionic substance. Hydrogen chloride, a gas at room temperature and 1 atmosphere, is an example of such a substance. Table 3.4 lists some ions commonly encountered in chemistry, together with typical substances in which they exist or from which they are generated on dissolving in water.

The presence of ions in solution can be measured directly through experiments with electrical conductivity, the flow of electrical current. An electrical current consists simply of the movement of electrical charges. Since ions have electrical charges, their motion in a solution can give rise to an electrical current. A typical and useful procedure is to apply an electric field across two plates immersed in the solution of interest, thereby making one plate electrically positive and the other electrically negative. Then one measures the current of ions that flows between the plates. The current is measured by a current meter, an ammeter, placed in the external part of what then constitutes a complete circuit. A simplified version of this apparatus is shown in Fig. 3.7. The amounts and mobilities of ions in the solution are then indicated by the corresponding electrical current that is found to flow.

Spotlight

Several simple ions—Na^+, K^+, Mg^{2+}, Ca^{2+}, Cl^-, F^-, Br^- and I^-—have important functions in biological systems. For example, in animals sodium ions are primarily responsible for maintaining a proper balance of the pressure inside and outside cells (osmotic pressure; see Chap. 16). One cause of high blood pressure in humans is an elevated concentration of Na^+ within the cells. The high Na^+ concentration tends to cause a lowering of the osmotic pressure within the cell, which is balanced by an infusion of water. As the water enters the cell, the cell expands against the surrounding fluids and vessels, which can compress and constrict capillaries, requiring greater pressure for the blood to be forced through the narrowed capillaries. Hence, high blood pressure is the result.

The concentrations of Na^+ and K^+ are also important in regulating the activities of a number of enzymes involved in the energy-releasing conversions of ATP to ADP and PO_4^{3-} (see Chap. 8). The detailed nature of these regulating influences is not yet understood, and is one of the many biochemical puzzles being studied in biochemical research laboratories.

Calcium is an element that is very important in the "structural engineering" of animals. Bones and teeth are formed of mineral-like materials based on Ca^{2+}, for example, calcium phosphate, $Ca_3(PO_4)_2$. The ion Ca^{2+} is also involved in bringing about chemical processes in muscle contraction.

The properties of Sr^{2+} and Ca^{2+} are so similar that Sr^{2+} readily replaces Ca^{2+} in bones and teeth. An isotope of strontium, ^{90}Sr, accompanies the fallout from nuclear detonations. Danger exists in that the ^{90}Sr accumulates in bones, where the full effects of its long-term residency in the body as part of the bone material is felt over a long period. This is one of the reasons that scientists overwhelmingly favor bans on atmospheric testing of nuclear bombs.

Large currents results from large concentrations of ions in solution, and the substances that produce such solutions are referred to as **strong electrolytes.** Weak currents are evidence of small concentrations of ions in solution, and substances that give rise to such solutions are **weak electrolytes.** The absence of a measurable conductivity demonstrates an absence of ions in solution, and the solutes that produce such solutions are **nonelectrolytes.** Examples of these three categories are given

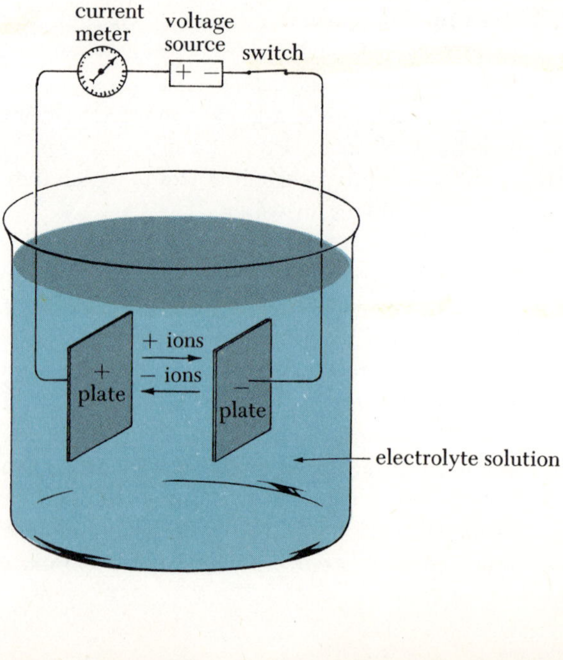

Figure 3.7
Apparatus for measuring the electrical conductivity of a solution.

in Table 3.5. Comparing this table with Table 3.1 reveals a strong correlation between the substances we have identified here as strong electrolytes and the substance that we earlier cited as examples of nonmolecular compounds. A strong correlation also exists between what we describe here as nonelectrolytes and what we have earlier classified as molecular compounds. These correlations are not surprising in view of the structural role of ions in both the electrolyte-nonelectrolyte and the molecular-nonmolecular classifications. A common solvent for electrolytes is water.

The ions that result from a strong electrolyte in a sense act largely as if they were independent of one another. For example, the chloride ions that come from dissolving a sample of sodium chloride in a large excess of water essentially react independently of the sodium ions that come from the same solute substance. The chloride ions behave essentially the same whether the other ions present are Na^+ or K^+, or some other **cation,** a positive ion. Hence, we frequently find that one type of ion or others resulting from a given electrolyte may not be involved in a chemical reaction that consumes another ion resulting from the same electrolyte. An example of this is a reaction that occurs when a solution of sodium chloride in water is mixed with a solution of silver nitrate in water. When these two clear, colorless solutions are added to each other, a white **precipitate**, a solid formed in the reaction, immediately forms. Analysis of this solid substance shows it to have the formula AgCl. We can represent this reaction in terms of the ions present as follows:

$$Ag^+ + NO_3^- + Na^+ + Cl^- \rightleftharpoons AgCl(s) + NO_3^- + Na^+$$

We note that in this equation each of the ions Na^+ and NO_3^- occurs on both sides of the equation, and can be subtracted from each side. In other words, sodium ions and nitrate ions play no part in the transformation, except that they must be present in the original solutions as companion ions to make it possible for the chloride ions

TABLE 3.5 Conductivity Categories of Some Typical Substances in Solutions

Strong Electrolytes		*Weak Electrolytes*		
Positive Ions	*Negative Ions*	*Positive Ions*	*Negative Ions*	*Nonelectrolytes*
Sodium chloride, NaCl		Acetic acid, CH_3CO_2H		Chlorobenzene, C_6H_5Cl
Na^+	Cl^-	H^+	$CH_3CO_2^-$	Nitrobenzene, $C_6H_5NO_2$
Calcium chloride, $CaCl_2$		Ammonia, NH_3 (in water)		Ferrocene, $Fe(C_5H_5)_2$
Ca^{2+}	Cl^-	NH_4^+	OH^-	Nickel tetracarbonyl, $Ni(CO)_4$
Potassium nitrate, KNO_3		Phenol, C_6H_5OH		
K^+	NO_3^-	H^+	$C_6H_5O^-$	Carbon tetrachloride, CCl_4
Hexamminecobalt chloride, $Co(NH_3)_6Cl_3$		Hydrogen fluoride, HF (hydrofluoric acid)		Isooctane, C_8H_{18}
$Co(NH_3)_6^{+3}$	Cl^-	H^+	F^-	Iodine monochloride, ICl
Hydrogen chloride, HCl (hydrochloric acid)		Mercuric chloride, $HgCl_2$		Benzene, C_6H_6
H^+	Cl^-	Hg^{2+}	Cl^-	
Ammonium acetate, $NH_4O_2CCH_3$				
NH_4^+	$CH_3CO_2^-$			

and silver ions to exist in solution.* Such ions are often called **spectator ions**—they just "look on" without taking part in the reaction. But this involvement as companion ions is perhaps no more pertinent than that of the glass container in which the reaction takes place, so it is customary to exclude these symbols from the equation:

$$Ag^+ + Cl^- \rightleftharpoons AgCl(s) \tag{3.6}$$

Solutions in which water is the solvent are typically called **aqueous solutions**. To be explicit, one can add the notation "aq" to the right of the usual symbol for species in dilute, aqueous solution. According to this convention, the above reaction is represented

$$Ag^+(aq) + Cl^-(aq) \rightleftharpoons AgCl(s) \tag{3.7}$$

If the context in which an equation is presented makes it obvious that the reaction occurs in aqueous solution, then one often omits the designation "aq." The reaction shown in Eq. (3.7) forms the basis for an analytical technique for determining the amount of chloride ion present in a sample, as indicated in Exercise 3.61.† Additional examples of what are called net ionic chemical equations follow:

$$Ba^{2+}(aq) + SO_4{}^{2-}(aq) \rightleftharpoons BaSO_4(s) \quad \text{Forms basis of analysis for } SO_4{}^{2-}. \tag{3.8}$$

$$H^+(aq) + OH^-(aq) \rightleftharpoons H_2O \tag{3.9}$$

$$HClO_4(aq) \rightleftharpoons H^+(aq) + ClO_4{}^- \tag{3.10}$$

$$Pb(s) + PbO_2(s) + 4H^+ + 2SO_4{}^{2-} \rightleftharpoons 2PbSO_4(s) + 2H_2O \quad \text{Fundamental reaction of lead (automobile) storage battery.} \tag{3.11}$$

$$Zn(s) + 2MnO_2(s) + 8NH_4{}^+ \rightleftharpoons Zn^{2+} + 2Mn^{3+} + 8NH_3 + 4H_2O \quad \text{Primary reaction of common "dry cell."} \tag{3.12}$$

The designation "aq" has been omitted from the symbols for ions in Eq. (3.11) and (3.12) because the systems referred to are most definitely not dilute.

The concept of a net chemical equation derives from a resolve to use the symbolism of chemical equations in the most effective way, providing as much information as possible with a concise notation. For the same reason, we always try in a chemical equation to represent substances in solution by symbols that imply the form in which the substances actually exist. Hence, it would be misleading to represent the reaction that occurs between aqueous sodium chloride solutions and aqueous silver nitrate solutions as

$$NaCl + AgNO_3 \rightleftharpoons AgCl + NaNO_3$$

*The high repulsive forces and potential energy associated with a collection of ions with the same charge make it impossible to prepare a "bottle of Ag^+" or a "flask of Cl^-." Companion ions (or counterions) must be present.

†The analytical technique of Exercise 3.61 generally involves converting any chlorine present in a sample of the Cl^- form. After suitable procedures that remove any "interfering ions" (ions that would also form a precipitate when brought into contact with Ag^+), the sample is treated with aqueous silver nitrate solution. The resulting AgCl(s) precipitate is separated by filtration and weighed. The weight of the AgCl(s) gives the information necessary to specify the amount of chlorine present in the original sample. This technique, based on weighing a precipitate, is called a *gravimetric* method.

because the substances $AgNO_3$, NaCl, and $NaNO_3$ simply do not exist as such in aqueous solution. Equations (3.6) and (3.7) get much closer to the truth. Using chemical equations to tell the true story assumes special importance in some of the subsequent examples.

EXAMPLE 3.32 Write a balanced net equation for the reaction that occurs when an aqueous solution of potassium carbonate, K_2CO_3 (a strong electrolyte), is mixed with a solution of strontium nitrate, $Sr(NO_3)_2$ (a strong electrolyte), and a precipitate of strontium carbonate is formed.

SOLUTION The ions that are involved in this system are K^+, $CO_3{}^{2-}$, Sr^{2+}, and $NO_3{}^-$. Of these, only Sr^{2+} and $CO_3{}^{2-}$ are combined to form a product, and K^+ and $NO_3{}^-$ function only as companion ions. Hence, the net, balanced equation is

$$Sr^{2+} + CO_3{}^{2-} \rightleftharpoons SrCO_3(s)$$

STUDY PROBLEM 3(v)

Write a balanced net equation for the reaction that occurs when an aqueous solution of ferrous nitrate, $Fe(NO_3)_2$ (a strong electrolyte), is mixed with a solution of sodium sulfide (a strong electrolyte) and a precipitate of ferrous sulfide is formed.

Acids and Bases

Two important classes of electrolytes are acids and bases. A simple and useful definition of **acids** is that they are substances that produce the ions H^+ when they dissolve in water; examples are hydrochloric acid, HCl, nitric acid, HNO_3, and sulfuric acid, H_2SO_4. A compatible definition of **bases** classifies them as substances that produce hydroxide ions, OH^-, when dissolved in water; examples are sodium hydroxide, NaOH, potassium hydroxide, KOH, and barium hydroxide, $Ba(OH)_2$. Thus, when acids or bases are dissolved in water, we can visualize ionization as follows:

$$HCl(g) \rightleftharpoons H^+ + Cl^- \qquad (3.13)$$

$$HNO_3(l) \rightleftharpoons H^+ + NO_3{}^- \qquad (3.14)$$

$$NaOH(s) \rightleftharpoons Na^+ + OH^- \qquad (3.15)$$

$$Ba(OH)_2(s) \rightleftharpoons Ba^{2+} + 2OH^- \qquad (3.16)$$

According to this view, a strong acid is an acid that is a strong electrolyte and a strong base is a base that is a strong electrolyte. **Neutralization** is the reaction of a strong acid with a strong base. In such a reaction, hydrogen ions combine with hydroxide ions to form water according to Eq. (3.9). The other ions in solution do not take part in such a neutralization. These would be the chloride ions and sodium ions if hydrochloric acid and sodium hydroxide were the acid and base. Hence, we represent the neutralization of a strong acid by a strong base by the equation

$$H^+ + OH^- \rightleftharpoons H_2O$$

Spotlight

Chloride ion, Cl^-, and phosphate ion, PO_4^{3-}, are the companion ions primarily responsible for "electrically neutralizing" the essential positive ions in biological systems. The ion Br^- can often replace Cl^- in this function. Fluoride and iodide ions have very specific action in our biochemistry. Fluoride ions strengthen the mineral-like structure of teeth and bone. The best method for obtaining enough F^- is through having a fluorinated water supply available during the formation of the teeth (at levels about the same as fluoride in seawater, 1.3×10^{-4} weight percent). Use of toothpaste containing SnF_2 is good for the surface of the teeth (thus useful in preventing cavities) but does little to help in strong dental formation.

Iodide ion is necessary for the proper functioning of the thyroid gland. Its deficiency can cause lethargy and goiter (a massive enlargement of the gland). Most inland soil is deficient to some extent in I^-, which can be adequately obtained by using iodized salt—salt to which NaI has been added.

Heat and Chemical Reactions

Energy changes are important in chemistry. Not only do chemical reactions provide energy in many important processes, such as burning gasoline or metabolism in organisms, but energy changes are important characteristics of chemical reactions. (These topics are dealt with in detail in Chaps. 10, 11, and 12.)

Some reactions require an input of energy (heat); such reactions are called **endothermic.** Other reactions give off excess energy (heat) when they occur; these are called **exothermic.** These two types are represented pictorially in Fig. 3.8.

Anyone who has mixed a solution of a strong acid with a solution of a strong base has noted a sharp increase in temperature of the resulting solution, indicating that such reactions are exothermic. Similarly, anyone who has mixed a highly concentrated solution of sulfuric acid with water, or has dissolved solid calcium chloride in water, knows that such processes are exothermic (and can be dangerous!). In the

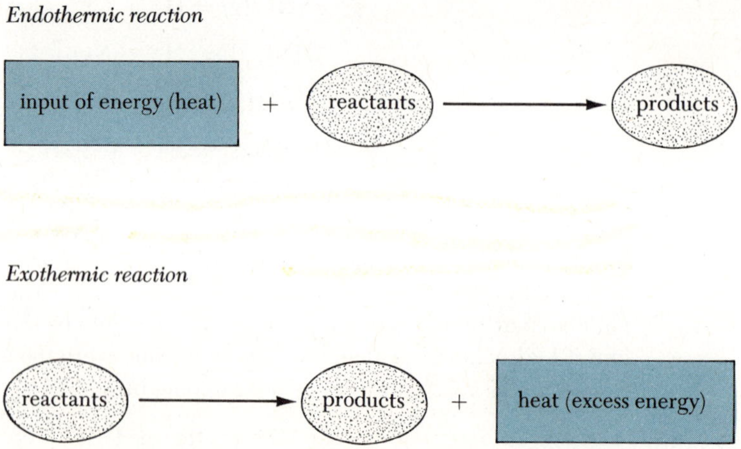

Figure 3.8 Endothermic and exothermic reactions.

biological realm, a complex array of chemical reactions that are exothermic overall is responsible for generating the heat that maintains the ambient temperature of an organism. For humans, this is 37 °C.

Also, numerous examples of endothermic chemical processes exist—processes that require energy to be added to the chemical system. In such processes the addition of heat to the chemical system facilitates the reaction.

3.9 Concentration

Molarity

An important property of a solution is the concentration of any of the substances present. We are now ready to define what has turned out to be the most convenient unit of concentration for most chemical considerations. This unit, or measure, of concentration is called **molarity**, and is defined as the number of moles of solute substance per liter of solution. A solution prepared by adding to one mole of NaOH enough water to produce one liter of solution is by definition a one-molar solution of NaOH, abbreviated 1 *M* NaOH. Similarly, a solution prepared by adding to 0.30 mole of HCl enough water to give 500 ml of a solution would be characterized as 0.60 *M* HCl; in concentration, 0.30 mol in 500 ml is equivalent to 0.60 mol in 1000 ml, even though there is a factor of 2 relating the total amounts. Concentration is an intensive property!

This method of specifying concentration is particularly convenient when considering what amounts of various substances in solution one wants to measure for chemical experiments or processes. This convenience results directly from the definition of molarity, by which the number of moles of a solute present in a given volume of solution simply equals the product of the volume (in liters) and the molarity:

$$\text{(volume in liters)(molarity)} = (V\,\cancel{\text{liters}})\left(M\frac{\text{mol}}{\cancel{\text{liter}}}\right) = VM\text{ mol} \tag{3.17}$$

Thus, 3 liters of a 2 *M* NaCl solution contains 6 moles of NaCl:

$$(3\,\cancel{\text{liters}})\left(2\,\frac{\text{mol NaCl}}{\cancel{\text{liter}}}\right) = 6\text{ mol NaCl}$$

This kind of relation makes possible some important simplifications in laboratory manipulations. If a particular substance is needed repeatedly in the form of a solution in a series of related experiments, then a "stock solution" can be prepared and used for all the experiments. A stock solution need be made up only once, with only one set of weighings necessary. From that point on, a specific number of moles of solute can be measured out simply by controlling the number of milliliters of solution dispensed. This is a convenient way of measuring desired amounts of materials—by using **volumetric** equipment rather than gravimetric equipment; this refers to volume-measuring devices like graduated cylinders, pipets, and burets, as shown in Fig. 5.1, as opposed to weight-measuring devices, or balances.

Calculations

The following examples illustrate how useful molarity can be.

EXAMPLE 3.33 How many chlorine atoms, which would be present as chloride ions, are contained in a 300-ml sample of a 1.00 *M* HCl solution?

SOLUTION To determine the number of chlorine atoms present, we simply need to find the number of moles of chlorine atoms (g-atoms Cl), which in this case is the same as the number of moles of HCl.

$$\text{no. Cl atoms} = \left(6.022 \times 10^{23} \frac{\text{Cl atoms}}{\text{mol HCl}}\right) \times \left(1.00 \frac{\text{mol HCl}}{\text{liter}}\right)(300 \text{ ml})\left(\frac{1}{1000}\frac{\text{liter}}{\text{ml}}\right)$$

$$= 1.81 \times 10^{23} \text{ Cl atoms}$$

EXAMPLE 3.34 How many grams of sodium hydroxide, NaOH, are required to react completely with 300 ml of 1.00 *M* aqueous HCl solution according to the equation

$$H^+(aq) + OH^-(aq) \rightleftharpoons H_2O$$

This equation results from realizing that both NaOH and HCl are strong electrolytes, and completely ionized in solution into Na^+ and OH^-, and H^+ and Cl^-.

SOLUTION According to the equation, there is a one-to-one ratio between the number of moles of H^+ and OH^-, and hence between the number of moles of HCl and NaOH required in the reaction. Furthermore, we know that if y represents the unknown mass in grams, then

$$\text{no. moles NaOH} = (y \text{ g NaOH})\left(\frac{1}{40.0}\frac{\text{mol NaOH}}{\text{g NaOH}}\right)$$

Also,

$$\text{no. moles HCl} = \left(1.00 \frac{\text{mol HCl}}{\text{liter}}\right)\left(\frac{300}{1000} \text{liter}\right) = 0.300 \text{ mol HCl}$$

Setting the number of moles of HCl equal to the number of moles of NaOH, we get

$$(y \text{ g NaOH})\left(\frac{1}{40.0}\frac{\text{mol NaOH}}{\text{g NaOH}}\right) = 0.300 \text{ mol NaOH}$$

$$y \text{ g NaOH} = (40.0 \text{ g NaOH})(0.300) = 12.0 \text{ g NaOH}$$

EXAMPLE 3.35 How many milliliters of 1.50 *M* acetic acid, CH_3CO_2H, are required to react completely with 200 ml of 0.500 *M* ammonia according to the following equation?

$$CH_3CO_2H(aq) + NH_3(aq) \rightleftharpoons CH_3CO_2^-(aq) + NH_4^+(aq)$$

SOLUTION We see a one-to-one mole ratio of the required CH_3CO_2H and NH_3:

$$\text{no. moles } CH_3CO_2H = \text{no. moles } NH_3$$

Then, if y represents the unknown volume of acetic acid solution in milliliters,

$$(y\ \text{ml})\left(\frac{1}{1000}\frac{\text{liter}}{\text{ml}}\right)\left(1.50\frac{\text{mol}}{\text{liter}}\right) = (200\ \text{ml})\left(\frac{1}{1000}\frac{\text{liter}}{\text{ml}}\right)\left(0.500\frac{\text{mol}}{\text{liter}}\right)$$

$$(y\ \text{ml})\left(\frac{1.50\ \text{mol}}{\text{liter}}\right) = (200\ \text{ml})\left(\frac{0.500\ \text{mol}}{\text{liter}}\right)$$

$$y\ \text{ml} = \left(\frac{200 \times 0.500}{1.50}\right)\text{ml} = 66.7\ \text{ml}$$

EXAMPLE 3.36 How many milliliters of 0.100 M $Na_2S_2O_3$ are needed to react completely with 60.0 ml of 0.300 M KI_3 in aqueous solution, according to the following equation?

$$2S_2O_3^{2-} + I_3^- \rightleftharpoons S_4O_6^{2-} + 3I^-$$

As this equation implies, both the sodium thiosulfate, $Na_2S_2O_3$, and the potassium triiodide, KI_3, are strong electrolytes.

SOLUTION The balanced equation shows that twice as many moles of thiosulfate ion as triiodide ion are required.

$$\text{no. moles } S_2O_3^{2-} = 2(\text{no. moles } I_3^-)$$

Then, if y is the unknown volume of thiosulfate solution,

$$(y\ \text{ml})\left(\frac{1}{1000}\frac{\text{liter}}{\text{ml}}\right)\left(0.100\frac{\text{mol}}{\text{liter}}\right) = 2(60\ \text{ml})\left(\frac{1}{1000}\frac{\text{liter}}{\text{ml}}\right)\left(0.300\frac{\text{mol}}{\text{liter}}\right)$$

$$y\ \text{ml} = (120\ \text{ml}) \cdot \frac{0.300}{0.100} = 360\ \text{ml}$$

The unit factor method gives in one step

$$(60.0\ \text{ml } I_3^-\ \text{soln}) \times \left(\frac{0.300\ \text{mol } I_3^-}{1000\ \text{ml } I_3^-\ \text{soln}}\right) \times \left(\frac{2\ \text{mol } S_2O_3^{2-}}{1\ \text{mol } I_3^-}\right) \times \left(\frac{1000\ \text{ml } S_2O_3^{2-}\ \text{soln}}{0.100\ \text{mol } S_2O_3^{2-}}\right) = 360\ \text{ml } S_2O_3^{2-}\ \text{soln}$$

ml I_3^- | moles I_3^- | moles $S_2O_3^{2-}$ | ml $S_2O_3^{2-}$

STUDY PROBLEM 3(w)

How many benzene molecules, C_6H_6, are there in 100 ml of a 0.300 M solution of benzene in carbon tetrachloride, CCl_4?

STUDY PROBLEM 3(x)

How many milliliters of a 0.50 M aqueous NaOH solution will react with 0.360 liter of a 0.250 M aqueous HCl (both strong electrolytes) according to Eq. (3.9)?

3.10 Nomenclature

Like any science, or indeed any subject, chemistry needs a satisfactory terminology so that concepts and relations can be communicated efficiently. Much of the terminology in chemistry is concerned with naming compounds, a subject called **chemical nomenclature.** Chemical nomenclature is a subject with many complexities and several conventions. Exploring the technicalities calls for more detailed exposure to the richly diverse structural features of chemistry than what we shall find in a beginning study. Nevertheless, some aspects of chemical nomenclature are straightforward.

For most electrolytes, the nomenclature is simple, since it is based on the names of the ions involved. The name of an electrolyte is formed by placing the name of the positive ion first and the name of the negative ion second—for example, *calcium nitrate* for $Ca(NO_3)_2$. The ion derived from a metal, say Ca^{2+} from calcium, simply takes its name from the parent metal—*calcium ion* for Ca^{2+}, *potassium ion* for K^+, *sodium ion* for Na^+, *magnesium ion* for Mg^{2+}, and so on. There are several exceptions, for example, *ferrous ion* for Fe^{2+}, *ferric ion* for Fe^{3+}, and *plumbous ion* for Pb^{2+}. For several positive ions and most negative ions, naming the ions depends on conventions that can be explored only in conjunction with the detailed chemistry associated with these ions and the species from which they are derived; some of these conventions will be considered later. For a few simple negative ions, the naming is achieved by attaching the suffix *ide* to the root word for the element. Examples are *chloride, fluoride, bromide, iodide, oxide, sulfide,* and *selenide* for Cl^-, F^-, Br^-, I^-, O^{2-}, S^{2-}, and Se^{2-}. Several examples of combining the names of the positive and negative ions are given in Table 3.4.

For nonelectrolytes, the nomenclature is generally more complex. Nevertheless, it is generally useful to remember the functional groups, since the names of functional groups can often be combined (or added to the name of some skeletal structure) to provide an acceptable name for a compound. If one recognizes, for example, that $C_6H_5NO_2$ is a compound in which a nitro (NO_2) group replaces a hydrogen atom of a benzene molecule, C_6H_6, then one is led to the name *nitrobenzene.* (The naming of a broad class of compounds based on what are called coordination complexes is considered in detail in Chap. 18.)

Some additional nomenclature is relevant to the periodic table. As indicated in Sec. 3.4, the elements are arranged in the periodic table according to vertical families. These families of elements, which have certain similarities in their properties, are often referred to as **groups,** and the members of a group, that is, of a given column, are called **congeners.**

Each of the groups of the periodic table is assigned a symbol consisting of a number and a letter, as seen in the table on the inside cover. In addition, some of the columns have acquired common names. The first column, on the left, is referred to as Group IA, and is called the alkali metals group. It consists of lithium, sodium, potassium, rubidium, and cesium; hydrogen is left out, since it is a special case. The second column is designated Group IIA, and is called the alkaline earth metals group. It comprises beryllium, magnesium, calcium, strontium, barium, and radium. The last column of the periodic table (helium, neon, argon, krypton, xenon, and radon) is sometimes designated Group VIIIA (sometimes Group 0); the elements of this group are referred to as the noble gases, or rare gases (earlier known as inert gases).

The column preceding the noble gases is Group VIIA, called the halogens (fluorine, chlorine, bromine, iodine, and astatine). Preceding the halogens is the column designated Group VIA, the chalcogens; they are oxygen, sulfur, selenium, tellurium, and polonium. While all the other groups, or columns, in the table are assigned specific symbols (IIIA, IVA, VA, IIIB, and so forth), names for the other groups are not common. Besides groups such as IA, IIA, and IIIA, there are also groups IB, IIB, IIIB, and the like. The "B" columns are distinct from the corresponding "A" columns; one finds certain chemical similarities, however, between groups with the same number designation but different letters (Groups IIA and IIB, for instance). The groups with the suffix A are often referred to as **main group elements**, or **representative elements.**

A given row of the periodic table is referred to as a **period.** The elements lithium, beryllium, boron, carbon, nitrogen, oxygen, fluorine, and neon constitute what is often called the first complete row, or the first complete period.

In Chap. 7, we shall explore a theoretical model that accounts for the form of the periodic table. In Chap. 9, we shall discuss certain trends in chemical properties relative to the periodic table.

3.11 Viewing Chemical Equations from Both Ends

Synthesis and Exploration

A chemical equation, by its very nature, provides information on what substances are consumed in the reaction—what substances react with each other—and what substances are produced. Depending on your requirements, you can view a chemical equation as answering many kinds of questions. The following two are noteworthy: (1) What happens when a particular set of reactants are mixed; in other words, what substances are produced? (2) What substances would one have to bring together to bring about a reaction that would produce some specific product? These two outlooks tend to guide the thinking in two somewhat different types of activity in chemical research. The first outlook, which we might consider the exploratory approach, is concerned with exploring what kinds of reactions are possible, what kinds of reactions certain types of substances or combinations of substances can undergo. One uses this approach in characterizing the chemical properties of a substance. The second outlook, which we can call the synthetic approach, is concerned with finding what reactions or combination of reactions can synthesize some specific desired substance, like a new plastic or a new anticancer drug. These two points of view are by no means mutually exclusive. Indeed, knowledge obtained from one of these approaches often provides valuable new leads in research activities in the other.

Percentage Completion and Percentage Yield

A false impression to be avoided is that all chemical reactions occur "completely," or "quantitatively." By these two words we mean that all the reactant materials if present in the proper proportions are transformed completely into products. We shall come to view the complete reaction as a special case rather than the general rule. It is true that a balanced, net chemical equation represents a process in which the substances on the left side react to produce the substances on the right side; but the mere writing of a chemical equation does not imply that the indicated reaction will actually occur quantitatively when the reactants are mixed in the appropriate proportions. Some reactions do indeed occur quantitatively, at least as far as we can tell using measurements we can make conveniently. This is not true, however, in the most general case. We can imagine millions of reactions for which chemical equations can be written but which do not occur quantitatively, or in some cases to any measurable extent, when tried in the laboratory.

The calculations that we have made in the above examples have been based on the assumption that the reactants are transformed quantitatively into products. Clearly, the computed results and the balanced equations on which they were based represent reality only to the extent that the chemical reactions occur. The above examples were chosen specifically because the indicated reactions are essentially complete, and difficulties with this assumption could be avoided. But in some cases, when one mixes reactants in the appropriate amounts, perhaps only 10% of them react according to the chemical equation of interest—in other cases, perhaps 70%; in another, perhaps 0.1%. The percentage of completion depends on many things, including the identity of the chemical reaction under consideration. Other relevant factors are the temperature, concentrations, and pressure, and other conditions under which the reaction is carried out.

We find it convenient to use chemical equations to represent both processes that occur quantitatively and processes that occur only to a slight extent.* In the latter case we must keep in mind that the written equation represents reality only for the portion of the reactants that undergo the indicated change. The reactants may undergo some other reaction, a side reaction, or perhaps no reaction at all. Thus, the stoichiometric relations that are implied by the balanced equation correspond to only that fraction of the reactants that undergo the specific reaction indicated. For example, when sodium acetate, which is a strong electrolyte, generating Na^+ and acetate ions, is dissolved in water, a process called hydrolysis occurs; it occurs only to the extent that a small fraction of 1% of the acetate ion is transformed. More than 99% of the acetate ions, $CH_3CO_2^-$, remain unchanged. The reaction that does occur to a small extent is represented by the equation

$$CH_3CO_2^-(aq) + H_2O \rightleftharpoons CH_3CO_2H(aq) + OH^-(aq)$$

Writing this equation does not imply that all the acetate ions present undergo a transformation; it describes only what happens to that fraction of the acetate ions (less than $\frac{1}{100}$) that undergo the transformation.

If one knows the extent to which incomplete chemical reactions occur, then one can still make certain types of stoichiometric calculations for actual systems of practical interest. The following two examples illustrate this point.

*Merely writing a chemical equation does not imply that the indicated reaction would actually occur. It is not uncommon to write a chemical equation for a reaction that essentially does not occur at all. The written equation can provide a definite framework for discussion of why the reaction does not occur, that is, why nature seems to prefer the species on the left side of the equation over the species on the right. A football coach may diagram a specific play that turns out not to proceed as planned. Nevertheless, the diagram of the unsuccessful play is a basis for determining why it is not successful, and for analyzing the relevant football strategy. Similarly, the chemical equation corresponding to an unsuccessful reaction indicates specifically what reaction is being discussed and provides a useful framework for considering the relevant chemistry.

EXAMPLE 3.37 When equal volumes of 0.020 *M* solutions of aqueous $Ca(NO_3)_2$ and aqueous sulfuric acid are mixed, it is found that the production of solid calcium sulfate occurs to the extent of about 45% according to the equation

$$Ca^{2+}(aq) + SO_4^{2-}(aq) \rightleftharpoons CaSO_4(s)$$

(This equation reflects that both solutes are strong electrolytes, the first providing the ions Ca^{2+} and NO_3^-, and the second giving the ions H^+ and SO_4^{2-}.) Compute the mass in grams of the calcium sulfate precipitated when 100-ml samples of each of these solutions are mixed together.

SOLUTION If the reaction were "complete," the number of moles of $CaSO_4(s)$ produced would equal the number of moles of calcium ion or sulfate ion available, namely ($\frac{100}{1000}$ liter) · (0.020 mole/liter) = (0.100)(0.020) mole. However, since the reaction is only 45% complete, the actual number of moles of $CaSO_4(s)$ precipitated will be (0.45) · (0.100)(0.020) mole. Then, the number of grams of $CaSO_4(s)$ prepared is given according to Eq. (3.1) as

$$\begin{aligned} \text{no. g } CaSO_4 &= (\text{no. moles } CaSO_4(s)) \cdot (\text{mol wt } CaSO_4) \\ &= [(0.45)(0.100)(0.020) \text{ mole } CaSO_4]\left(136.1 \frac{\text{g } CaSO_4}{\text{mole } CaSO_4}\right) \\ &= 0.12 \text{ g } CaSO_4 \end{aligned}$$

A related concept, which is pertinent mainly in synthetic chemistry, is the **percentage yield.** This is simply the fraction of a reaction product actually obtained relative to what would be obtained in a quantitative reaction; this fraction is expressed in percentage. Thus, in mathematical form,

$$\text{percentage yield} = \frac{\text{mass of product actually obtained}}{\text{mass of product for a quantitative reaction}} \times 100 \qquad (3.18)$$

In other words, for a reaction that goes to completion, the percentage yield is 100%. For a reaction in which only half as much product is obtained as would be obtained if the reaction were quantitative, the percentage yield would be 50%.

STUDY PROBLEM 3(y)

What is the percentage yield in a synthetic process in which 36 g of product are obtained from enough reactants to have produced 162 g of product if the reaction were quantitative?

EXAMPLE 3.38 If 200 ml of an 0.18 *M* solution of lead nitrate, $Pb(NO_3)_2$, and 300 ml of a 0.080 *M* solution of aluminum chloride, $AlCl_3$, are mixed, 5.0 g of lead chloride, $PbCl_2$, precipitate. What is the percentage yield of $PbCl_2$ for this reaction? Both $Pb(NO_3)_2$ and $AlCl_3$ are strong electrolytes.

SOLUTION We first write a balanced, net equation.

$$Pb^{2+}(aq) + 2Cl^-(aq) \rightleftharpoons PbCl_2(s)$$

Then we compute the number of gram-atoms of Pb^{2+} and Cl^- used, noting that there are three moles of Cl^- per mole of $AlCl_3$.

$$\text{no. moles Pb}^{2+} = \left(0.18\frac{\text{mol Pb(NO}_3)_2}{\text{liter}}\right)(0.20\text{ liter})\left(1\frac{\text{mol Pb}^{2+}}{\text{mol Pb(NO}_3)_2}\right)$$
$$= 3.6 \times 10^{-2}\text{ mol Pb}^{2+}$$

$$\text{no. moles Cl}^- = \left(0.080\frac{\text{mol AlCl}_3}{\text{liter}}\right)(0.30\text{ liter})\left(3\frac{\text{mol Cl}^-}{\text{mol AlCl}_3}\right)$$
$$= 7.2 \times 10^{-2}\text{ mol Cl}^-$$

The balanced chemical equation shows that Pb^{2+} and Cl^- are required in the ratio 1:2, and the above calculations show that this requirement is fulfilled in this particular reaction mixture. If the reaction were 100% complete, we should expect 3.6×10^{-2} mol of $PbCl_2$ to precipitate, because the balanced chemical equation shows a 1:1 correspondence between the number of moles of Pb^{2+} consumed and the number of moles of $PbCl_2$ produced. In grams, this is

$$(3.6 \times 10^{-2}\text{ mol PbCl}_2)\left(278.1\frac{\text{g PbCl}_2}{\text{mol PbCl}_2}\right) = 10\text{ g PbCl}_2$$

But since only 5.0 g of $PbCl_2$ precipitated, the percentage yield is

$$\frac{5.0\text{ g}}{10\text{ g}} \times 100\% = 50\%$$

To carry out calculations like those in Example 3.37, one must know the extent of completion of the relevant chemical reactions. This depends on a variety of factors, including the amounts of substances present. For the reaction in Example 3.37 the extent of completion strongly depends on the amount of water used to dissolve the given amount of starting material. If more water were present in the final mixture, then more $PbCl_2$ would remain dissolved in the final solution; hence, the amount of product as written, $PbCl_2(s)$, would be accordingly less. For each specific chemical reaction, there is a characteristic percentage completion for each set of conditions. After the reactants are mixed and enough time allowed for the reaction to proceed as far as it ever will, a point is reached that corresponds to this characteristic percentege completion. When that point is reached, the chemical system no longer changes. In this stable condition, the chemical system is said to exist in **chemical equilibrium.** Knowing what factors determine the position of this equilibrium permits one to predict the percentage completion of chemical reactions under a wide variety of conditions; consequently, it allows one to predict reliably the amounts of relevant substances present in a system at equilibrium.

SUMMARY

An understanding of the basis for mass relations in chemistry rests on the concepts of atoms and molecules, as articulated in Dalton's atomic theory. All compounds can be characterized as molecular or nonmolecular. A compound formula, which defines the atomic makeup of the formula unit, can be written for every compound. The atomic weight scale, which can be used to calculate the elemental compositions of compounds from their formulas, is the cornerstone of calculations in chemical stoichiometry, and is used to define molecular weights. In terms of atomic weights and molecular weights, the concepts of a gram-atom and a gram-mole are defined; these concepts are generalized through Avogadro's number to the concept of a mole. A mole of any species is Avogadro's number of the species, and corresponds to a gram-atom for elements and a gram-mole for compounds. The behavior of ideal gases can be summarized conveniently in terms of the mole by the ideal gas equation.

A chemical equation is a shorthand notation for representing a chemical reaction. The balanced equation is an expression of the law of conservation of mass. Using balanced chemical equations and the concept of the mole, one can compute the mass relations among reactants and products for any reaction. Net ionic chemical equations emphasize the importance of the ionization of electrolytes in aqueous solution. Two important classes of electrolytes are acids and bases. For solutions, mass relations are conveniently handled in terms of concentration, especially molarity.

Only some reactions proceed to completion. For most reactions, the percentage of completion is much less than 100%. In synthesis, this leads to the concept of percentage yield.

STUDENT CHECKPOINTS

After studying this chapter, the student should be able to:

1. Define and give examples of atoms, molecules, formulas, ions, and formula units.
2. Explain the concept of a functional group.
3. Differentiate between molecular and nonmolecular compounds.
4. Explain atomic weight and molecular weight and their relation to each other.
5. Calculate the elemental composition of any compound from its formula.
6. Predict the formulas of nonmolecular binary compounds.
7. Derive the simplest formula of a compound from its elemental composition.
8. Explain the difference between the gram-atom, the gram-mole, and the mole.
9. Relate Avogadro's number to the concepts of gram-atom, gram-mole, and mole.
10. Balance chemical equations.
11. Calculate weight relations in chemical reactions by using the concept of the mole.
12. Use the concept of the mole in calculations based on the ideal gas equation.
13. Define the term *electrolyte*, and its relation to net balanced chemical equations.
14. Define *molarity* and use this concept in stoichiometric calculations.
15. Name simple compounds from their molecular formulas.
16. Define *percentage completion* and *percentage yield* and give examples.

EXERCISES

3.1.* Write the molecular formula and the simplest formula for each of the following substances:

(a) Hydrazine 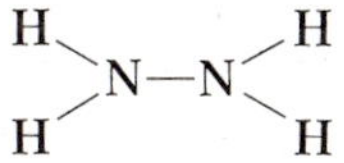

(b) Hydrogen peroxide 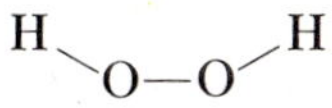

(c) Ammonia 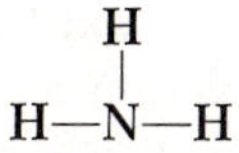

(d) Dimethylamine 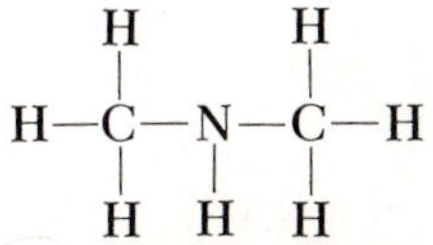

3.2.° What are the functional groups in each of the following substances?

(a) Ethylene glycol, the most common automobile antifreeze.

$$\text{HO}-\underset{\underset{H}{|}}{\overset{\overset{H}{|}}{C}}-\underset{\underset{H}{|}}{\overset{\overset{H}{|}}{C}}-\text{OH}$$

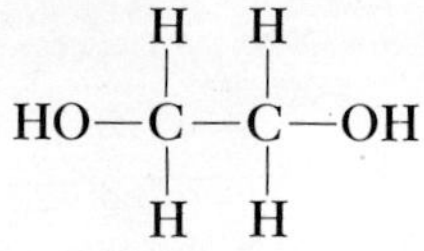

(b) Nitromethane, sometimes added to the fuel of high-speed engines.

$$\text{H}-\underset{\underset{H}{|}}{\overset{\overset{H}{|}}{C}}-\text{NO}_2$$

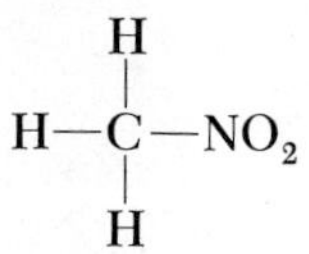

(c) Butyric acid, contributor to the odor of rancid fats.

$$\text{H}-\underset{\underset{H}{|}}{\overset{\overset{H}{|}}{C}}-\underset{\underset{H}{|}}{\overset{\overset{H}{|}}{C}}-\underset{\underset{H}{|}}{\overset{\overset{H}{|}}{C}}-\overset{\overset{O}{\parallel}}{C}-\text{OH}$$

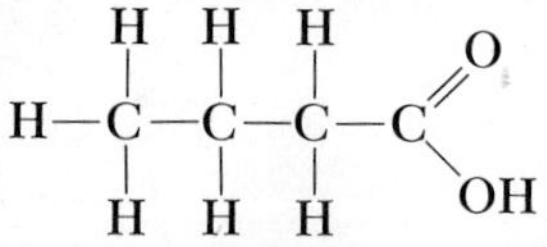

(d) Glycine, the simplest example of a biologically important class of compounds called amino acids.

$$\text{H}-\underset{\underset{H}{|}}{\overset{\overset{H}{|}}{N}}-\underset{\underset{H}{|}}{\overset{\overset{H}{|}}{C}}-\overset{\overset{O}{\parallel}}{C}-\text{OH}$$

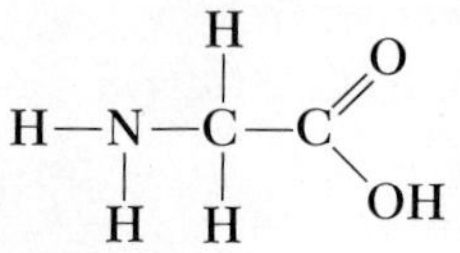

3.3. Write the simplest formula and the molecular formula of each of the compounds given in Example 3.2.

3.4.° By analogy with the cases given in Table 3.1, which of the following species are molecular compounds and which are nonmolecular compounds: (a) Na_2SO_4 (sodium sulfate); (b) $MgCl_2$ (magnesium chloride); (c) BCl_3 (boron trichloride); (d) $C_{10}H_{22}$ (decane); (e) $Ca(NO_3)_2$ (calcium nitrate); (f) KCl (potassium chloride); (g) $Fe(CO)_5$ (iron pentacarbonyl)?

3.5. Write the chemical formulas for binary nonmolecular compounds between (a) potassium and oxygen; (b) aluminum and sulfur; (c) cesium and bromine.

3.6.° On the basis of the periodic table of the elements, what pairs of the following elements would be expected to have somewhat similar chemical and physical properties: chlorine, Cl; potassium, K; antimony, Sb; tin, Sn; bismuth, Bi; bromine, Br; lead, Pb; lithium, Li?

3.7. Calculate the percentage composition by weight of a compound called naphthalene (sometimes used as mothballs), which has the molecular formula, $C_{10}H_8$.

3.8.° Calculate the percentage composition by weight of aluminum oxide (alumina), Al_2O_3.

3.9. Calculate the percentage composition by weight of a 50:50 mixture (in terms of mass) of NaCl and $NaClO_4$.

3.10. Derive the simplest formula for a compound whose percentage composition is 14.4% hydrogen and 85.6% carbon.

3.11.° Derive the simplest formula of a compound whose percentage composition is 29.1% sodium, 40.6% sulfur, and 30.3% oxygen.

3.12. Derive the simplest formula of a compound whose percentage composition is 42.1% sodium, 18.9% phosphorus, and 39.0% oxygen.

3.13. Calculate the number of grams of oxygen required to produce CuO by direct combination with 27.7 g of copper.

3.14.* How many grams of FeO are produced by direct combination of 24.4 g of oxygen with iron?

3.15. How many atoms are there in a 3.60-g sample of chlorine gas?

3.16. How many atoms are there in one ounce of gold? What is the current price of gold per atom?

3.17.* How many copper atoms are there in a 40.0-g sample of gold that has a 0.030% copper impurity?

3.18.* How much would a speck of iron containing 3.7×10^{19} atoms weigh?

3.19. How many gram-atoms of hydrogen are contained in each of the following: (a) 8.64 g of HCl; (b) 3.2 mol H_2SO_4; (c) 0.45 mol CH_3CO_2H; (d) 16.4 g CH_3CH_2OH?

3.20.* Calculate the molecular weight of each of the following: (a) $MgSO_4$; (b) Na_3PO_4; (c) $K_2S_2O_3$; (d) C_7H_8.

3.21. How many moles are contained in each of the following: (a) 32.4 g C_6H_6; (b) 12.7 g BF_3; (c) 0.642 g HBr; (d) 0.448 g ICl?

3.22.* How many molecules are contained in each of the following: (a) 0.339 mol H_2; (b) 7.09 mol CCl_4; (c) 4.36 g H_2O; (d) 1.11 g HCN?

3.23. How many molecules are contained in each of the following: (a) 6.06 g I_2; (b) a sample of NH_3 containing 0.542 g-atom of hydrogen; (c) a sample of C_6H_6 containing 4.2×10^{24} hydrogen atoms; (d) a sample of C_4H_{10} containing 2.40×10^{22} carbon atoms?

3.24.* How many oxygen atoms are contained in each of the following: (a) 2.30 moles $HClO_4$; (b) 0.451 mole O_2; (c) 11.4×10^{23} CO_2 molecules; (d) 4.77 g CO; (e) 0.813 g BaO?

3.25. How many "moles of sulfur atoms" are contained in (a) 3.66 g H_2SO_4, and (b) a sample of $Na_2S_2O_3$ containing 4.11×10^{-3} g-atom of oxygen?

3.26.* How many grams of chlorine are required to produce 3.24 g HCl according to the following reaction?

$$H_2(g) + Cl_2(g) \rightleftharpoons 2HCl(g)$$

3.27. What is the mass (or weight) percentage of chlorine in a sample that consists of 1.05 mole NaCl and 2.49 mole $NaClO_4$?

3.28.* What is the mass (or weight) percentage of oxygen in a sample that consists of 3.24 g $BaCO_3$ and 14.10 g $MgCO_3$?

3.29. Calculate the volume of a 3.87-g sample of gaseous hydrazine, H_2NNH_2, at 200 °C and 0.980 atm.

3.30.* What is the pressure of a 0.784-g sample of Cl_2 gas enclosed in a 200-ml flask at 25 °C?

3.31. How many moles of ethane gas, CH_3CH_3, are contained in a 300-ml vessel at pressure 0.800 atm and temperature 10 °C?

3.32.* What is the mass of a gaseous sample of CO_2 contained in a 500-ml flask at pressure 0.850 atm and temperature 25 °C?

3.33. What is the temperature of a 0.347-g sample of gaseous NO (nitric oxide) in a 200-ml vessel at pressure 1.37 atm?

3.34.* What is the molecular weight of a gas, a 0.505-g sample of which occupies a volume of 250 ml at 0 °C and 0.850 atm?

3.35. Derive the simplest formula and the molecular formula of a gaseous substance consisting of 92.3% carbon and 7.73% hydrogen; a sample of the gas of mass 727 mg at 100 °C and 0.950 atm occupies 300 ml.

3.36.* Write a chemical equation for the reaction of gaseous ammonia, NH_3, and gaseous HCl to produce solid ammonium chloride, NH_4Cl.

3.37. Write a balanced chemical equation for the decomposition of solid barium carbonate, $BaCO_3$, at high temperature to produce solid barium oxide and carbon dioxide gas.

3.38.* Write a balanced chemical equation for the reaction of solid germanium with gaseous chlorine, Cl_2, to form liquid germanium tetrachloride, $GeCl_4$.

3.39. Write a balanced chemical equation for the reaction of an aqueous solution of the strong electrolyte Na_2SO_4 with an aqueous solution of the strong electrolyte $Ba(NO_3)_2$.

3.40.* How many grams of silver oxide, Ag_2O, can be prepared from 0.161 mole of $AgNO_3(s)$ and sufficient NaOH(s) when both solids are dissolved in water and the following reaction occurs?

$$2Ag^+ + 2OH^- \rightleftharpoons Ag_2O(s) + H_2O$$

3.41. How many grams of water are produced when 11.4 g of nonane, $C_{10}H_{22}$, are burned in an oxygen, O_2, atmosphere, producing only H_2O and CO_2 as products?

3.42.* Write a balanced chemical equation for the reaction that occurs when an aqueous solution of the strong electrolyte $Pb(NO_3)_2$ is mixed with an aqueous solution of NaOH, producing solid PbO.

3.43. How many grams of NaOH, when dissolved in water, will be exactly neutralized by 0.250 mol of hydrochloric acid, HCl?

3.44. A mixture of 3.24 g NaCl, 1.63 g KNO_3, and 6.25 g $AgNO_3$ is dissolved in water. How much solid AgCl will be formed in the reaction that occurs?

3.45.* A mixture of 0.031 mol $Zn(NO_3)_2$, 4.34 g $NaNO_3$, 1.24 g Na_2S, and 0.055 g KCl is dissolved in water. How much solid ZnS will be formed in the reaction that occurs?

3.46. How many liters of O_2 gas at 25 °C and 1.00 atm pressure must be used to completely combust 3.46 g of propane, C_3H_8 (commonly used as a fuel in rural areas), with only CO_2 and water as the products?

3.47. How many moles of KCl are contained in 640 ml of a 0.688 *M* KCl solution?

3.48.* How many grams of NaCl are contained in 10 liters of a 0.50 *M* NaCl solution (this is approximately the NaCl concentration in seawater)?

3.49. How many sodium atoms (present as Na^+) are present in 400 ml of a 0.850 *M* solution of Na_2S in water?

3.50.* How many milliliters of a 0.150 *M* NaCl solution must be added to 0.560 g of $AgNO_3$ dissolved in water in order to convert all the silver to solid AgCl?

3.51. How many grams of KOH must be added to 250 ml of a 2.00 *M* aqueous HNO_3 solution to have a complete acid-base neutralization?

3.52.* How many milliliters of a 0.440 *M* NaOH solution will exactly neutralize 64.5 ml of a 0.395 *M* HNO_3 solution?

3.53. A mixture of 3.25 g KOH, 2.90 g NaOH, 1.92 g K_2SO_4, and 36.4 ml of a 1.54 *M* NaOH solution is added to 1000 ml of water. How many milliliters of a 1.37 *M* HNO_3 solution will be required to neutralize this mixture?

3.54.* How many milliliters of a 1.27 *M* acetic acid solution, CH_3CO_2H, will be exactly consumed by 0.499 g KOH in an acid-base neutralization?

3.55. A mixture of 100 ml of a 0.500 *M* aqueous ammonia solution, NH_3, and 0.404 g NaOH is added to 200 ml of water. How many milliliters of an 0.800 *M* HCl solution is required for complete neutralization?

3.56.* How many grams of metallic silver (elemental) can be produced according to the reaction $2Ag^+ + Cu(s) \rightleftharpoons 2Ag(s) + Cu^{2+}$ if 300 ml of a 0.350 *M* aqueous $AgNO_3$ solution is treated with excess copper metal?

3.57. How many grams of lead metal are consumed in the discharging of an automobile lead storage battery according to the equation

$$Pb(s) + PbO_2(s) + 4H^+ + 2SO_4{}^{2-} \rightleftharpoons 2PbSO_4(s) + 2H_2O$$

if 1.48 g of solid lead sulfate is produced?

3.58. A student is trying to prepare a pure sample of ethyl acetate by the reaction of acetic acid, CH_3CO_2H (the essence of vinegar), and ethyl alcohol, CH_3CH_2OH (the "spirits" of alcoholic beverages).

$$CH_3CO_2H + CH_3CH_2OH \rightleftharpoons \underset{\text{ethyl acetate}}{CH_3CO_2CH_2CH_3} + H_2O$$

Starting with a mixture of 4.65 g of acetic acid and 10.2 g ethyl alcohol, the student ultimately obtains 6.80 g ethyl acetate. What was the percentage yield in this preparation?

3.59.* Name the following compounds: MnS, $Pb(NO_3)_2$, $Zn(HCO_3)_2$, $Mg(ClO_4)_2$, $NaMnO_4$, Li_2CO_3, Ag_2S, BaI_2, CaF_2, $Al_2(SO_4)_3$, $Mn_3(PO_4)_2$, $CsNO_3$, $SrBr_2$.

3.60. Name the following compounds: C_6H_5F, CBr_4, IBr, $Fe(CO)_5$, CH_3NO_2.

3.61.* A common method of analyzing for chlorine in a sample is to first convert all chlorine present to the Cl^- form (by suitable chemical processes), and then determine the amount of Cl^- present by treatment with an aqueous $AgNO_3$ solution. Solid AgCl(s), which can be filtered and weighed, forms according to

$$Ag^+(aq) + Cl^-(aq) \rightleftharpoons AgCl(s)$$

A 3.721-g sample of a solid containing an unknown amount of chlorine in some form or forms is subjected to this analytical procedure. In the last step, 43.2 mg of AgCl(s) is obtained. What was the percentage of chlorine in the original sample?

ANSWERS TO SELECTED EXERCISES

3.1 (a) N_2H_4 and NH_2; (b) H_2O_2 and HO; (c) NH_3 and NH_3; (d) C_2H_7N and C_2H_7N. 3.2 (a) OH and CH_2; (b) CH_3 and NO_2; (c) CH_3, CH_2, and CO_2H (or CO and OH); (d) NH_2, CH_2, and CO_2H (or CO and OH). 3.4 (a) nonmolecular; (b) nonmolecular; (c) molecular; (d) molecular; (e) nonmolecular; (f) nonmolecular;

(g) molecular. **3.6** Cl and Br, K and Li, Sb and Bi, Sn and Pb. **3.8** 52.9% Al, 47.1% O. **3.11** $Na_2S_2O_3$. **3.14** 110 g. **3.17** 1.1×10^{20} Cu atoms. **3.18** 3.4 mg. **3.20** (a) 120.37 amu; (b) 163.94 amu; (c) 190.33 amu; (d) 92.14 amu. **3.22** (a) 2.04×10^{23}; (b) 4.27×10^{24}; (c) 1.46×10^{23}; (d) 2.47×10^{22}. **3.24** (a) 5.54×10^{24}; (b) 5.43×10^{23}; (c) 2.28×10^{24}; (d) 1.02×10^{23}; (e) 3.19×10^{21}. **3.26** 3.15 g Cl_2. **3.28** 50.8% O. **3.30** 1.35 atm. **3.32** 0.765 g. **3.34** 53.2 amu. **3.36** $NH_3(g) + HCl(g) \rightleftharpoons NH_4Cl(s)$. **3.38** $Ge(s) + 2Cl_2(g) \rightleftharpoons GeCl_4(l)$. **3.40** 18.6 g. **3.42** $Pb^{2+} + 2OH^- \rightleftharpoons PbO(s) + H_2O$. **3.45** 1.55 g ZnS(s). **3.48** 292 g. **3.50** 22.0 ml. **3.52** 57.9 ml. **3.54** 7.00 ml. **3.56** 11.3 g. **3.59** manganese sulfide, lead nitrate, zinc bicarbonate, magnesium perchlorate, sodium permanganate, lithium carbonate, silver sulfide, barium iodide, calcium fluoride, aluminum sulfate, manganese phosphate, cesium nitrate, strontium bromide. **3.61** 0.287% Cl.

ANSWERS TO STUDY PROBLEMS

3(a) $C_6H_4F_2$. **3(b)** $AlBr_3$; K_2Se; BaS. **3(c)** 12.58% H, 87.42% N. **3(d)** CH. **3(e)** 42.6 g Fe. **3(f)** 2.5×10^{19} Mg atoms. **3(g)** 6.66×10^{-5} g. **3(h)** 9.274×10^{-8} g. **3(i)** 78.11 amu, 141.96 amu. **3(j)** 0.5328 g-mol. **3(k)** 10^{17} Pb atoms. **3(l)** 5.10 mole NaCl/mole KCl. **3(m)** H_2O. **3(n)** 33.8% Na, 1.5% H, 17.7% C, 47.0% O. **3(o)** 4.83 g. **3(p)** 5.18 liters. **3(q)** 0.952 atm. **3(r)** 3.91 g. **3(s)** $2Cs(s) + Cl_2(g) \rightleftharpoons 2CsCl(s)$. **3(t)** 303 g. **3(u)** 16.9 g. **3(v)** $Fe^{2+} + S^{2-} \rightleftharpoons FeS(s)$. **3(w)** 1.81×10^{22} benzene molecules. **3(x)** 180 ml. **3(y)** 22.2%.

4

Chemical Equilibrium: An Introduction

4.1 The Condition of Equilibrium

The idea that not all chemical reactions proceed to completion has led to the concept of percentage completion. This concept must be considered in calculating amounts of substances converted during chemical reactions. It is extremely useful to be able to predict quantitatively what the percentage completion will be for any given chemical reaction in any particular set of circumstances. The ability to make such predictions for any chemical synthesis comes from understanding chemical equilibrium.

The Steady State

Chemical equilibrium is a special aspect of the **steady state**, a condition occurring in a variety of natural phenomena. In the steady state, two opposing processes occur at the same rate, so that the net result is no apparent change. The word *apparent* has significance here, because at some level of observation of a steady state (the molecular level in chemistry) processes could be perceived although the net result would be mutual cancellation. As a result of the cancellation, no change is observed on the gross, macroscopic level. An illustration is the populations of isolated twin cities separated by a bridge on which the traffic flow is the same in both directions. At the detailed level of individual cars, it is seen that changes are indeed occurring; but the populations of the cities are at a steady state since the same number of people

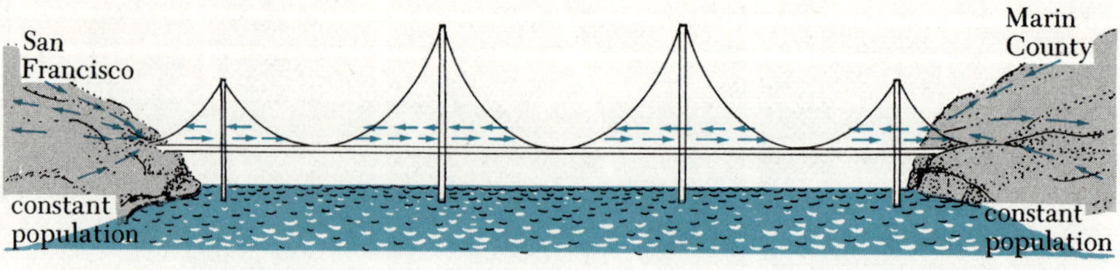

Figure 4.1 A steady state. Equal flow of cars (arrows) in both directions on the Golden Gate Bridge. The net result is no *apparent* change. The population on each side maintains a constant number. At the gross level, no change is obvious; but at the level of each car, crossings are taking place.

on the average are entering and leaving the city in any particular span of time. This situation is represented in Fig. 4.1.

Physical Equilibrium

An example of the steady state has been mentioned in connection with vapor pressure (Fig. 2.1.). That phenomenon, represented in Fig. 4.2, is associated with the fact that a closed system containing a solid or liquid substance, the condensed phase, and a free space above it will, if left alone, eventually come to a state in which a constant pressure exists in the space above the condensed phase. The pressure that is exerted by the molecules in the gas is called the **vapor pressure**. This pressure leads to a rate of reentry of molecules into the condensed phase that is precisely the same as the rate of escape of molecules from the condensed phase. Hence, the vapor pressure results from the steady state that exists. If we could see individual molecules, we should observe the many transfers between the gas phase and the condensed phase. However, at the macroscopic level, a steady state has the appearance of being static, since opposing submicroscopic processes occur at the same rate.

The steady state for vaporization, the physical process responsible for vapor pressure, is an example of **physical equilibrium**, the steady state for a physical process.

Figure 4.2 The buildup of the equilibrium vapor pressure by a condensed phase in an evacuated chamber. Individual molecular paths are shown by arrows. (a) At the instant the system is constructed, before any molecules have escaped from the condensed phase. (b) After some molecules have escaped from the condensed phase, generating a small vapor pressure, so small that the number of reentries into the condensed phase is neglected. (c) After more molecules have escaped, giving a higher vapor pressure, with an appreciable fraction of the gas-phase molecules reentering the condensed phase. (d) After the vapor pressure has reached its equilibrium value; rate of reentry into condensed phase equals rate of escape from it.

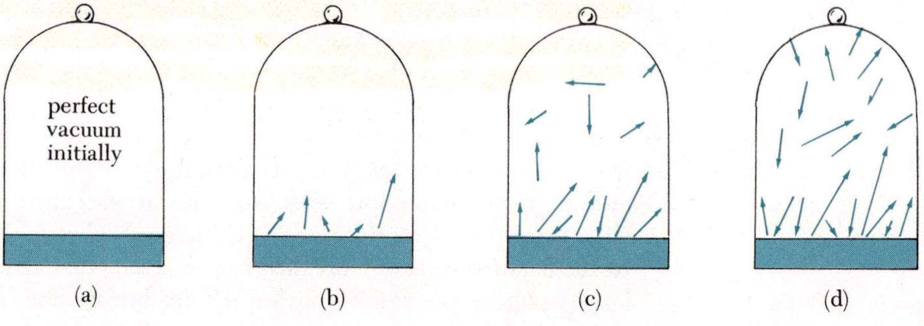

One qualitative guideline in physical equilibrium that will give us a preliminary understanding of the material to be currently emphasized is Le Chatelier's principle.

Le Chatelier's Principle

Le Chatelier's principle can be stated as follows: *When any system in a state of equilibrium is subjected to a stress or stimulus that upsets the equilibrium, the system responds by minimizing the stress if possible.*

One simple example of a physical equilibrium to which this principle can be applied is the equilibrium between a condensed phase and its equilibrium vapor pressure in a closed vessel. For such a system at equilibrium we can ask ourselves what the effects would be of adding heat to the condensed phase. This addition leads to an increase in temperature. By Le Chatelier's principle, we can consider the added heat (or the increase in temperature) a stress applied to a system initially at equilibrium. The way in which this added heat can be minimized as a stress is for some of the material in the condensed phase to vaporize, since this vaporization requires heat.* We thus expect that applying heat will result in a net increase in the number of molecules in the vapor phase—leading to a new condition of equilibrium with a higher vapor pressure. The net result can be summarized in the following statement: Le Chatelier's principle predicts that an increase in the temperature of a condensed phase will lead to an increase in equilibrium vapor pressure.

Another example of how Le Chatelier's principle can explain well-known physical phenomena is found in ice skating. Imagine a mixture of ice and liquid water in physical equilibrium in the piston-cylinder arrangement shown in Fig. 4.3; no net change of solid to liquid or liquid to solid is occurring. If the total pressure the piston exerts is 1 atm, then the temperature at which this system can be in equilibrium is 0 °C. Now, suppose that the pressure the piston exerts is suddenly increased to several atmospheres, a stress taking the system away from equilibrium. According to Le Chatelier's principle, the system will respond so as to reduce the effect of this sudden squeeze. Since ice cubes float in water, we know that ice is less dense than

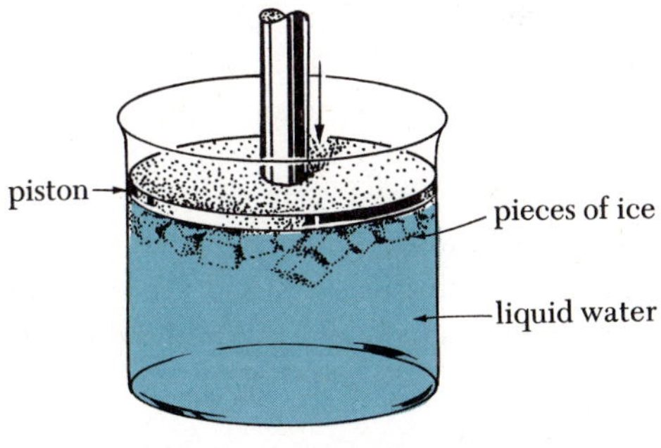

Figure 4.3 The effect of pressure on the melting point of ice in a piston-cylinder arrangement containing pieces of ice and liquid water.

*That vaporization requires heat is one reason why swimming on a hot dry day is so refreshing. The vaporization of water from your skin requires heat, which is removed from your body, giving rise to a marked cooling effect. Of course, if the water is cool, it provides direct cooling also.

liquid water. Thus, the volume of a given mass of liquid water is less than the volume of the same mass of ice; the liquid "takes up less space" than the solid. The response to the "squeeze" is for some of the ice to melt. The melting of ice requires heat, and this can come only at the expense of the energy due to random motions of the molecules. Hence, the average intensity of these random motions decreases; the temperature of the system drops so that the new equilibrium state can be achieved. The net result is that an increase in the pressure, having caused some of the ice to melt, has lowered the melting point of the ice. In this new steady state, as in the initial one, some individual water molecules are continually being transformed from the solid state to the liquid state and vice versa, but at exactly the same rate for both directions.

As pressure can be exerted on ice by the force pressing down on the blade of an ice skate, the freezing point of the ice below the blade drops, and some of the ice melts. The resulting thin film of water provides a lubricant that allows the skate to slide smoothly over the surface of what appears to be only ice.

Chemical Equilibrium

A phenomenon that is analogous to physical equilibrium but associated with chemical rather than physical processes is **chemical equilibrium.** This is a condition associated with the simultaneous occurrence in each direction, forward and reverse, of any chemical reaction, just as the cars represented in Fig. 4.1 are moving in both directions on the bridge. In this text, the symbol $\rightleftharpoons$ is used for chemical equations. This notation also reminds us that when the reacting system is in chemical equilibrium, the reactions occur in both directions and the two reactions occurring in opposite directions proceed at precisely the same rate. Thus, the symbolism

$$N_2(g) + 3H_2(g) \rightleftharpoons 2NH_3(g)$$

tells us not only that one mole of N_2 combines with three moles of H_2 to produce two moles of NH_3, but also that the reverse reaction, or the conversion of two moles of NH_3 to one mole of N_2 and three moles of hydrogen, also occurs at equilibrium, and that the two processes occur at the same rate at equilibrium. This behavior is directly analogous to the physical equilibrium associated with vapor pressure. When this kind of situation exists, there is no net change occurring in the bulk structural makeup or in the bulk chemical and physical properties of the system.

The analogy between chemical and physical equilibrium can be pursued further, and the approach to equilibrium from a nonequilibrium condition can be viewed similarly for the two cases. For equilibrium vapor pressure, the experiment depicted in Fig. 4.2 began with a substance in a condensed phase in contact with a vacuum; initially there were no molecules of the substance in the gaseous phase (Fig. 4.2a). However, as the escape into the gaseous phase occurred (Fig. 4.2b), a sufficient concentration of gas-phase molecules built up (Fig. 4.2c) so that the rate of reentry into the condensed phase became equal to the rate of escape from the condensed phase (Fig. 4.2d). An analogous view can be applied to chemical equilibrium.

Isomerization

Let us consider as an example of chemical equilibrium, a very simple type of chemical process, one in which a substance with a particular molecular formula and a particular molecular structure is transformed into another substance with the same

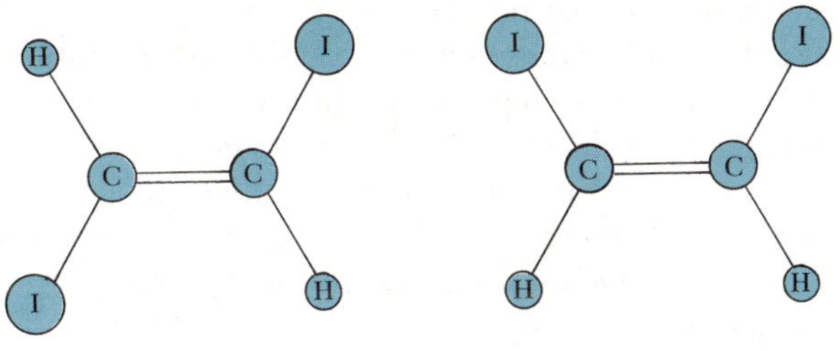

trans-diiodoethylene *cis*-diiodoethylene

Figure 4.4
Simple representation of the geometrical arrangements of the atoms (carbon, hydrogen, and iodine) in *trans*-diiodoethylene and *cis*-diiodoethylene. The centers of all atoms lie in a plane.

molecular formula but a different molecular structure; this new molecular structure brings about a different geometrical arrangement of a given set of atoms. Such a chemical reaction is known as **isomerization**. Compounds that have the same molecular formula but different structures are known as **isomers**. The occurrence of isomers is referred to as **isomerism**.[*] A specific example of a simple isomerization is the isomerization reaction that transforms the compound known as *cis*-diiodoethylene, with formula $C_2H_2I_2$, into what is known as *trans*-diiodoethylene, which has the same molecular formula. The prefix *cis* means that like atoms are on the "same side of the molecule," and *trans* means that they are on "opposite sides," as seen in Fig. 4.4. The isomerization reaction can be represented by the chemical equation (4.1):[†]

$$cis\text{-}C_2H_2I_2 \rightleftharpoons trans\text{-}C_2H_2I_2 \tag{4.1}$$

At 130 °C this reaction occurs at a rate that is convenient for observation, neither too fast nor too slow. If it were too fast, it would be over before one had time to study it; if too slow, one might not have the patience to wait for a measurable change.

Let us consider an experiment in which a 1 *M* solution of *cis*-diiodoethylene, stored initially in a refrigerator, is heated abruptly to about 130 °C.[‡] At the beginning of the experiment (time $t = 0$), the concentration of the trans isomer is zero. Accord-

[*]There are many examples of isomers in chemistry. A simple example of isomerism is CH_3CN and CH_3NC, in which the carbon and nitrogen atoms are arranged in a different order in the two isomers. The properties of these two isomers are much different, and CH_3NC is highly toxic. Isomerism is an extremely important concept in biological systems. There are thousands of instances in which a living system uses one isomer, and the other isomer is totally useless to it.

[†]A convenient experimental approach for carrying out isomerization reaction (4.1) would be to use a solution of diiodethylene(s) in cyclooctane solvent. Cyclooctane is a compound of molecular formula C_8H_{16}, with a structure shown here in shorthand form. Its boiling point is 149 °C, so that it does not boil off at the temperature of the experiment.

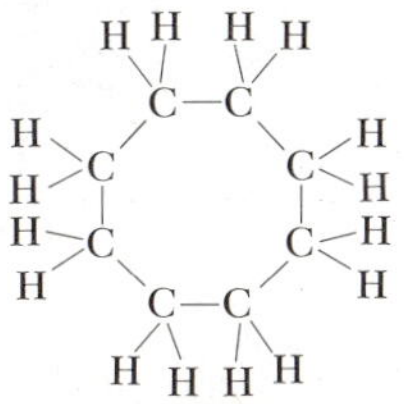

[‡]As we shall see in detail in Chap. 13, the rates of chemical reactions are profoundly influenced by temperature. Generally, lower temperatures give slower reactions and higher temperatures give faster reactions. For the isomerization reaction depicted in Eq. (4.1), storing the solution of *cis*-diiodoethylene at low temperature virtually prevents the occurrence of the reaction at an appreciable rate.

ingly, it is initially not possible for any cis isomer to be formed from trans isomer, because there is no trans material from which it could be produced. As time progresses and the reaction proceeds, the amount of cis isomer decreases as it is transformed into *trans*-diiodoethylene. At the same time, the concentration of the trans isomer increases continuously from its initial value of zero. As soon as this buildup begins, a second process makes its presence felt in the system. This additional process is the reverse of the isomerization depicted by Eq. (4.1); it corresponds to the equation's being read from right to left. Concerning the speeds of the forward and reverse isomerization reactions, let us make the following plausible assumption (see Chap. 13 for elaboration): The rate of the transformation of one isomer into another decreases if the concentration of the species being transformed decreases. That is, the rate of formation of trans isomer from cis isomer decreases continuously as the concentration of the cis isomer decreases. As the conversion of the cis isomer to the trans isomer proceeds, the concentration of trans isomer increases, thus increasing the rate of the trans-to-cis reaction. These relations are summarized in Figs. 4.5 and 4.6.

In Fig. 4.6 we see that what we expect to happen indeed occurs; namely, the rate of the reverse reaction continues to increase and the rate of the forward reaction continues to decrease, until ultimately the rates become equal. At that point, while the system is still in a dynamic condition—while changes are continually occurring with individual molecules—the overall result of the equal rates is that there is no longer any tendency for a net change in the concentration of *cis*- or *trans*-diiodoethylene. There is consequently no longer any tendency for the rates of the

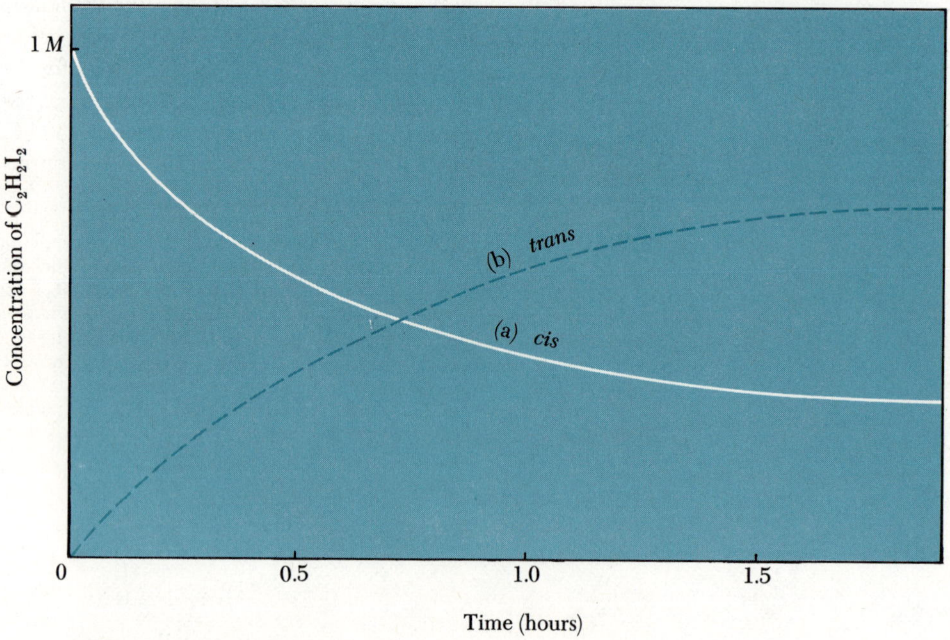

Figure 4.5 (a) Time dependence of the concentration of the cis isomer, starting with a 1 *M* solution of *cis*-diiodoethylene. (b) Time dependence of the concentration of trans isomer, starting with a 1 *M* solution of *cis*-diiodoethylene.

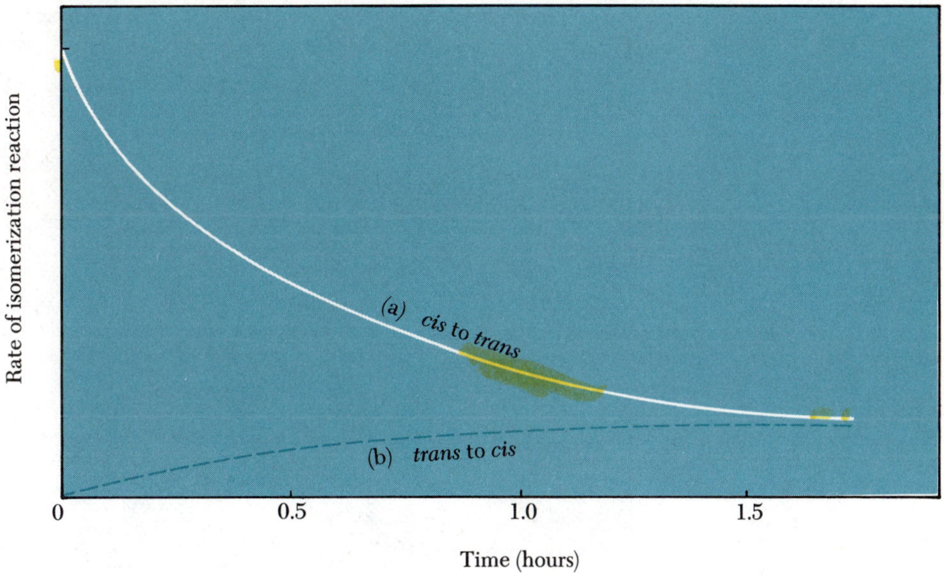

Figure 4.6 (a) Time dependence of the rate of *cis* to *trans* conversion of diiodoethylene, starting with only the *cis* isomer. (b) Time dependence of the rate of *trans* to *cis* conversion of diiodoethylene, starting with only the *cis* isomer.

forward or reverse processes to change, so the system has achieved a steady state. From this point on, so long as the system is left undisturbed, the forward and reverse reaction rates remain fixed at some constant value, with the rate of the net reaction equal to zero, and with the concentrations of reactants and products no longer changing. There is no longer any driving force for net change. This steady state constitutes chemical equilibrium. While the concentrations remain constant at equilibrium, any measured chemical or physical property of the system also remains constant.

With this picture in mind, we can visualize chemical equilibrium as a steady state in which opposing chemical reactions occur at the same rate, and in which the properties of the system, including the concentrations of all species, remain constant. This picture applies to any chemical reaction, no matter how simple or how complicated. For any arbitrarily chosen reaction, graphs analogous to Figs. 4.5 and 4.6 will generally differ in detail. They will, however, convey the common feature that as the reaction proceeds, the concentrations and the rates of forward and reverse processes change until those rates become equal, at which point the system is in chemical equilibrium.

The details of plots such as Figs. 4.5 and 4.6 will also depend, even for a specific reaction, on the initial conditions, such as the initial concentrations of the starting materials. For the cis and trans isomerization reaction of diiodoethylene, one set of initial conditions (initially only cis isomer present) was described above; three other sets are shown in Figs. 4.7 to 4.12. One set of initial conditions in the figures involves starting with the trans isomer, or starting with the right side of Eq. (4.1). If a solution

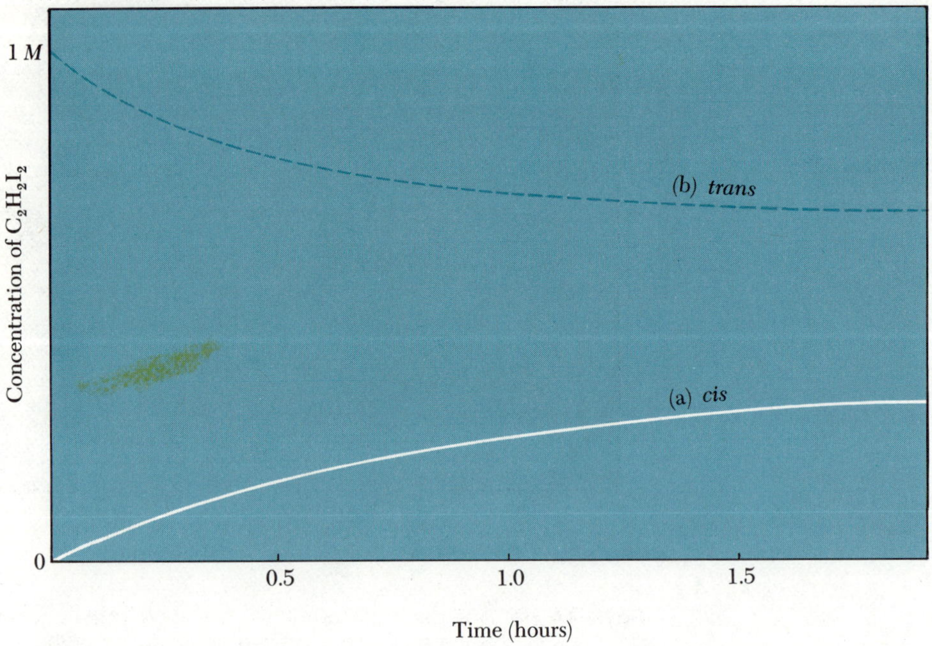

Figure 4.7 (a) Time dependence of the concentration of the cis isomer, starting with a 1 *M* solution of *trans*-diiodoethylene. (b) Time dependence of the concentration of the trans isomer, starting with a 1 *M* solution of *trans*-diiodoethylene.

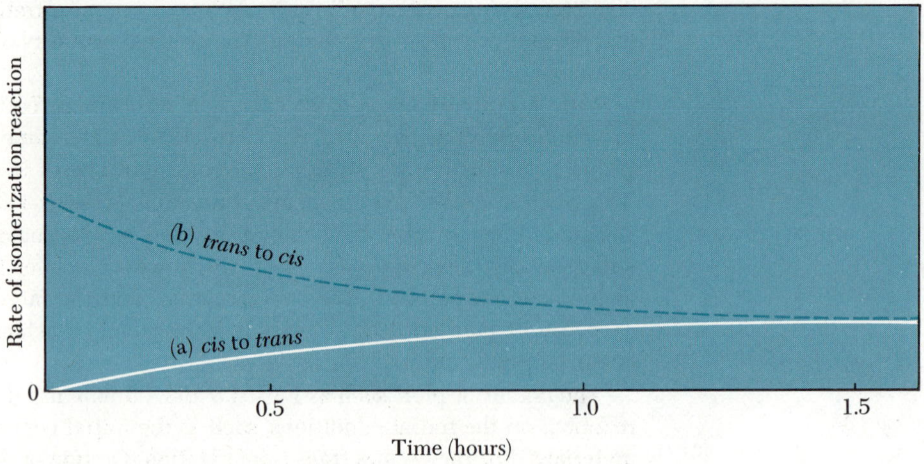

Figure 4.8 (a) Time dependence of the rate of cis to trans conversion of diiodoethylene, starting with only the trans isomer. (b) Time dependence of the rate of trans to cis conversion of diiodoethylene, starting with only the trans isomer.

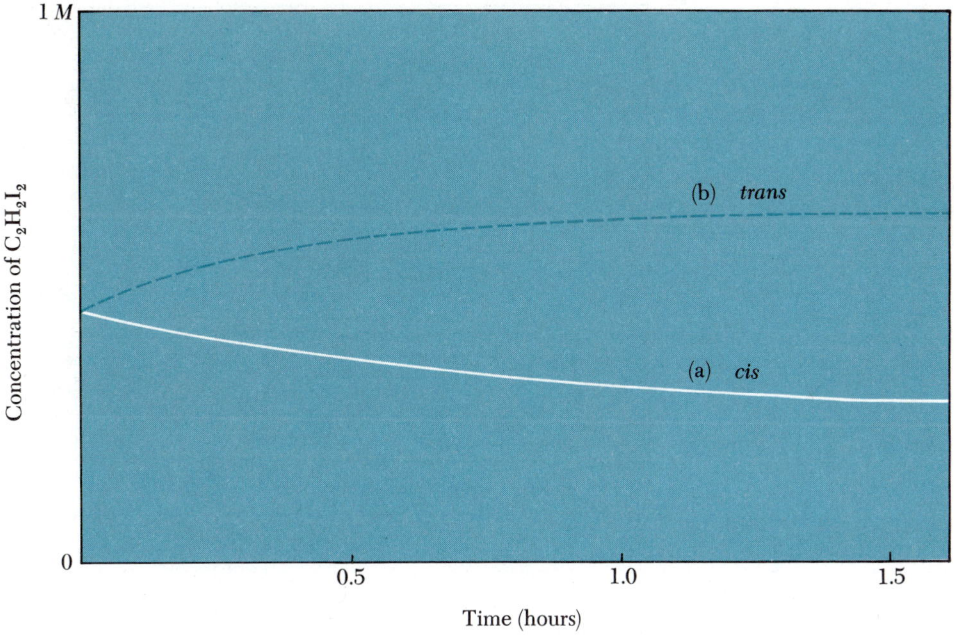

Figure 4.9 (a) Time dependence of the concentration of the cis isomer, starting with a solution 0.5 *M* in *cis*-$C_2H_2I_2$ and 0.5 *M* in *trans*-$C_2H_2I_2$. (b) Time dependence of the concentration of the trans isomer, starting with a solution 0.5 *M* in *cis*-$C_2H_2I_2$ and 0.5 *M* in *trans*-$C_2H_2I_2$.

of the trans isomer is heated at 130 °C, a partial conversion to the cis isomer is observed. Initially, there is in this case no *cis*-diiodoethylene present, so initially there is no reaction occurring in the direction from cis to trans. As the trans-to-cis conversion takes place, the concentration of trans isomer, and hence the rate of this conversion, decreases. At the same time a concentration of cis isomer is built up from zero, so the "reverse" process (cis to trans) makes its presence felt, with an importance that initially increases with time. These trends continue until ultimately equilibrium is achieved, when the rates of cis-to-trans and trans-to-cis processes become the same. Figures 4.7 and 4.8 illustrate this case.

Two important recurring features can be found in Figs. 4.5 through 4.12. One is the demonstration of achievement of a steady state, i.e., equal forward and reverse rates, as seen in Figs. 4.6, 4.8, 4.10, and 4.12. The other may be less obvious from the graphs but is equally important. Looking at Figs. 4.5, 4.7, 4.9, and 4.11, you can see that in each case, when equilibrium is reached, the ratio of the concentration of *trans*-diiodoethylene to the concentration of the cis isomer is $\frac{68.8}{31.2} = \frac{2.2}{1}$. Thus, when *cis*-diiodoethylene is the starting material, 68.8% of it is converted to the trans isomer at equilibrium (Fig. 4.5), whereas when *trans*-diiodoethylene is the starting material, only 31.2% of it is converted to cis isomer at equilibrium (Fig. 4.7), both cases corresponding to the 2.2:1 trans-to-cis ratio at equilibrium.

The unifying feature of the experimental behavior just described is that when equilibrium has been achieved, there is in each case a 2.2:1 ratio of the concentrations

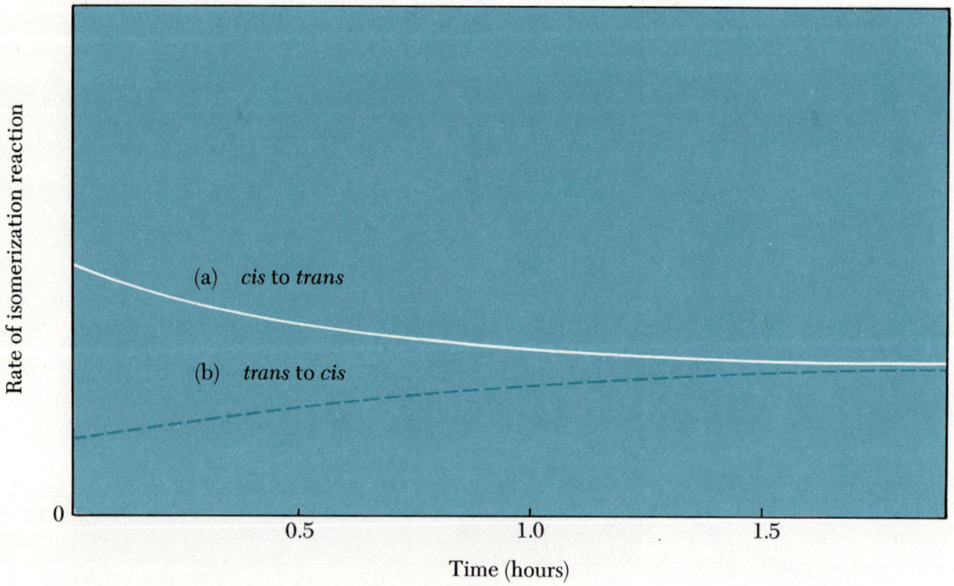

Figure 4.10 (a) Time dependence of the rate of cis to trans conversion of diiodoethylene, starting with a solution 0.5 *M* in *cis*-$C_2H_2I_2$ and 0.5 *M* in *trans*-$C_2H_2I_2$. (b) Time dependence of the rate of trans to cis conversion of diiodoethylene, starting with a solution 0.5 *M* in *cis*-$C_2H_2I_2$ and 0.5 *M* in *trans*-$C_2H_2I_2$.

of trans and cis isomers. This ratio is apparently independent of the direction from which equilibrium is approached—whether from pure cis starting material or from pure trans starting material. Indeed, one can show by several experiments of this kind at the same temperature that regardless of the initial amounts or concentrations of starting material and of whether the starting material corresponds to the right or the left side of Eq. (4.1) (or even mixtures), one always obtains the same ratio of concentrations at equilibrium. Figures 4.11 and 4.12 describe graphically the special case in which the starting material is a solution of a 2.2:1 trans-to-cis mixture in cyclooctane. In this last case, there are no changes in concentrations or rates as times passes, because the starting solution already has the 2.2:1 concentration ratio that corresponds to equilibrium; thus it has no tendency to experience change.

A convenient way of characterizing the equilibrium for the diiodoethylene isomerization is to define a quotient Q as

$$Q = \frac{[trans\text{-}C_2H_2I_2]}{[cis\text{-}C_2H_2I_2]} \tag{4.2}$$

For equations of this type, we shall adopt the convention that for any chemical species, if the chemical symbol is enclosed by brackets, it is understood to mean the molarity of the designated species. Thus, [*trans*-$C_2H_2I_2$] means molarity of *trans*-diiodoethylene. The equilibrium condition for the cis-to-trans isomerization of diiodoethylene at 130 °C is then $Q = 2.2$.

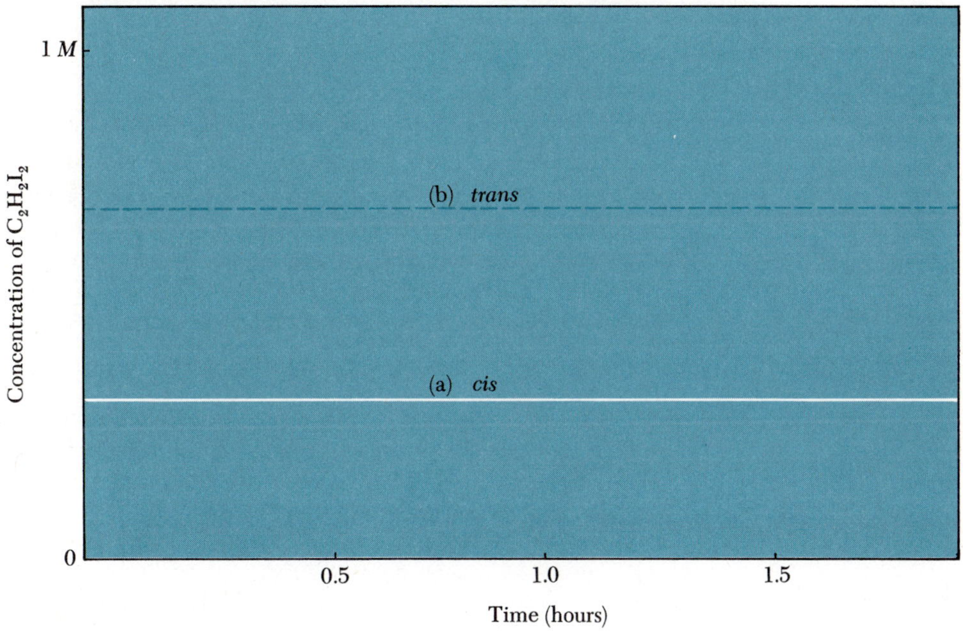

Figure 4.11 (a) Time dependence of the concentration of the cis isomer, starting with a solution 0.688 *M* in *trans*-$C_2H_2I_2$ and 0.312 *M* in *cis*-$C_2H_2I_2$. (b) Time dependence of the concentration of the trans isomer, starting with a solution 0.688 *M* in *trans*-$C_2H_2I_2$ and 0.312 *M* in *cis*-$C_2H_2I_2$.

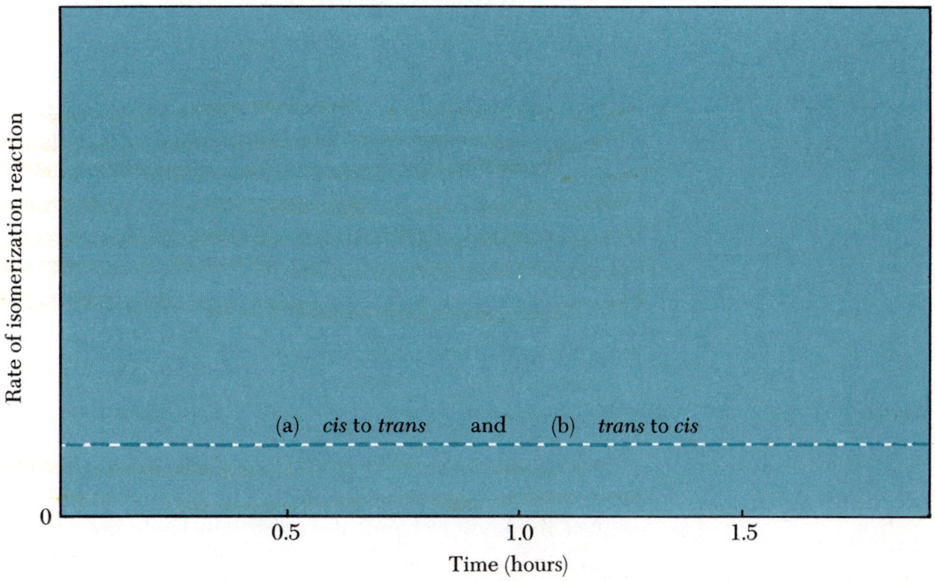

Figure 4.12 (a) Time dependence of the rate of cis to trans conversion of diiodoethylene, starting with a solution 0.688 *M* in *trans*-$C_2H_2I_2$ and 0.312 *M* in *cis*-$C_2H_2I_2$. (b) Time dependence of the rate of trans to cis conversion of diiodoethylene, starting with a solution 0.688 *M* in *trans*-$C_2H_2I_2$ and 0.312 *M* in *cis*-$C_2H_2I_2$.

4.2 Law of Chemical Equilibrium

An Experimental Fact

For any chemical reaction, detailed examination of concentration of reactants and products under differing initial conditions, as explored above for $C_2H_2I_2$, reveals that a quotient of concentrations analogous to Q in Eq. (4.2) can always be found; the constancy of this Q turns out to be the criterion for the equilibrium. However, in general this quotient is not a simple ratio of the concentrations of two species, as in Eq. (4.2). The general case can be described in terms of the following symbolic representation of a generalized chemical reaction:

$$aA + bB + \cdots \rightleftharpoons mM + nN + \cdots \tag{4.3}$$

In Eq. (4.3), the symbols A and B represent reactant species and N and M represent product species. The corresponding lower-case letters are the coefficients in the balanced chemical equation; in other words, a moles of species A plus b moles of species B, and so on react to produce m moles of species M plus n moles of species N, and so on. In these terms, the quotient we are seeking can be expressed as

$$Q = \frac{[M]^m[N]^n \ldots}{[A]^a[B]^b \ldots} \tag{4.4}$$

where the symbol $[M]^m$ means the mth power of the molarity of species M, with similar meanings for the other symbols. This quotient Q is called the **mass action quotient,** or the **mass action expression.** An experimental fact, based on the study of a wide variety of reactions and discovered about 1866 by the Norwegian scientists Guldberg and Waage, is that the mass action quotient Q is a constant at equilibrium. The value of the constant is characteristic of each reaction at a specific temperature. Stated in more formal terms, as follows, this set of observations constitutes the **law of chemical equilibrium:** *The state of equilibrium for a given chemical reaction at some specific temperature is characterized by the constancy of the mass action expression Q.* The quotient Q has a numerator expressed as the multiplied molarities of "product" species, each one raised to an exponent that equals the coefficient for that species in the balanced chemical equation, and a denominator expressed as the multiplied molarities of the "reactants," each one raised to a power that equals the coefficient of that species in the balanced chemical equation. For example, for the reaction $2H_2O + SO_3 \rightleftharpoons H_3O^+ + HSO_4^-$, we write

$$Q = \frac{[H_3O^+][HSO_4^-]}{[H_2O]^2[SO_3]}$$

The constant value for Q for a specific reaction at equilibrium is called the **equilibrium constant** and is often denoted by the symbol K. Thus, a mathematical statement of the law of chemical equilibrium is obtained by combining Eq. (4.4) with the equation $Q = K$ at equilibrium. Thus, at equilibrium

$$K = \frac{[M]^m[N]^n \ldots}{[A]^a[B]^b \ldots} \tag{4.5}$$

When the actual chemical symbols for a specific chemical reaction are used in place of the generalized notation A, B, C, D, . . . , Eq. (4.5) becomes a statement of the condition of chemical equilibrium for that particular reaction. This mathematical statement is often referred to as the equilibrium expression. For example, the equilibrium expression for the synthesis of ammonia from H_2 and N_2 is

$$3H_2(g) + N_2(g) \rightleftharpoons 2NH_3(g)$$

$$K = \frac{[NH_3(g)]^2}{[H_2(g)]^3[N_2(g)]}$$

The approach of a reacting system to the condition of chemical equilibrium can be represented as the approach of the value of Q to its value at equilibrium, K. In this light it is important to emphasize that the mass action expression Q equals K only at equilibrium; in any other state besides the equilibrium state, Q will not equal K. The extent to which Q differs from K is a quantitative measure of how far the system is from equilibrium.

In the isomerization reaction previously discussed, the form of Q is the simple one shown in Eq. (4.2), as there is only one reactant (*cis*-diiodoethylene) and only one product (*trans*-diiodoethylene). Also, we have seen that K for this reaction at 130 °C is 2.2. The approach to equilibrium in the experiments summarized in Figs. 4.5 to 4.12 can now be expressed as the time dependence of Q; this is summarized in Fig. 4.13, where it is seen that the value $Q = K = 2.2$ is achieved in each case when equilibrium is attained.

It is worth emphasizing that as in all of the cases shown in Fig. 4.13, the value of Q at time zero depends simply on the initial conditions, the initial concentrations of the pertinent species. For any given reaction, however, a particular value of Q is obtained when equilibrium has been achieved ($Q = K$), irrespective of the initial conditions. The value Q may initially be larger or smaller than the equilibrium constant, depending on which way the initial condition is displaced from the equilibrium. For example, in Fig. 4.13, we see a case (b) where the value of Q is initially larger than K, and consequently decreases as the reaction proceeds until equilibrium is attained. In the same figure, we see examples (a and c) in which the value of Q is initially less than the equilibrium value K, and hence increases as the reaction proceeds until K is reached. When the system is not in equilibrium, Q can a priori have any value; but any value besides the equilibrium constant represents an unstable situation, which in time will revert to an equilibrium if the opportunity is available. The length of time it takes for this equilibrium to be achieved is a subject pertinent to chemical kinetics (Chap. 13).

EXAMPLE 4.1 State the equilibrium condition (equilibrium expression) for each of the following reactions: (a) $H_2(g) + C_2H_4(g) \rightleftharpoons C_2H_6(g)$; (b) $Ag^+ + 2NH_3 \rightleftharpoons Ag(NH_3)_2^+$.

SOLUTION By the form of Eq. (4.5), which must be valid at equilibrium,

$$\text{(a)}\quad K = \frac{[C_2H_6(g)]}{[H_2(g)][C_2H_4(g)]} \qquad \text{(b)}\quad K = \frac{[Ag(NH_3)_2^+]}{[Ag^+][NH_3]^2}$$

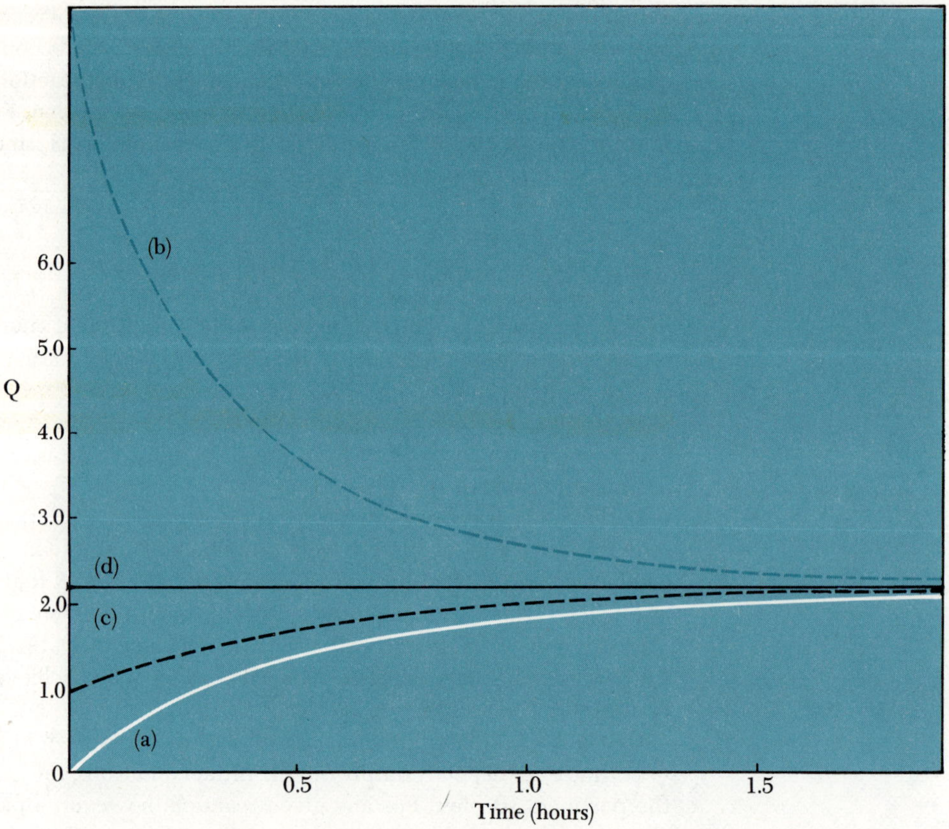

Figure 4.13 Time dependence of the Q values, [trans]/[cis], for the isomerization reaction. (a) Starting with only the cis isomer. (b) Starting with only the trans isomer. (c) Starting with a solution 0.5 M in *cis*-$C_2H_2I_2$ and 0.5 M in *trans*-$C_2H_2I_2$. (d) Starting with a solution 0.688 M in *trans*-$C_2H_2I_2$ and 0.312 M in *cis*-$C_2H_2I_2$.

STUDY PROBLEM 4(a)

Using Eq. (4.5), write the equation expressing the equilibrium condition of each of the following reactions:

$$2NO(g) + O_2(g) \rightleftharpoons 2NO_2(g)$$

$$2NO_2(g) \rightleftharpoons 2NO(g) + O_2(g)$$

$$NO(g) + \tfrac{1}{2}O_2(g) \rightleftharpoons NO_2(g)$$

Equilibrium Constants and Stability

Since the equilibrium constant is characteristic of the equilibrium for a particular reaction, we can call on it to give both qualitative and quantitative meaning to considerations of equilibrium phenomena. One way of doing this is to note that the value of an equilibrium constant, when viewed properly, is a direct measure of the tendencies for the species on the right or the left side of a chemical equation to be favored at equilibrium. Since concentrations of the species represented on the right side occur in the numerator of Q, and since concentrations of the species represented on the left side occur in the denominator of Q, equilibrium conditions that favor the species of the right-hand side, or the products, quite clearly correspond to equilibrium constants greater than 1. Similarly, chemical reactions in which the species of the left-hand side, the reactants, are favored at equilibrium have equilibrium constants less than 1. Chemical reactions with equilibrium constants in the neighborhood of unity can therefore be considered qualitatively as reactions in which reactants and products are roughly equally favored at equilibrium.

The idea that certain species are "favored" at equilibrium over other species is akin to the concept of stability. This qualitative concept can be thought of as a measure of the tendency of a species or a set of species to exist in the face of the possibility that other species could be formed from it by chemical reaction. If one species (for example, *trans*-diiodoethylene) is favored over another species (say, *cis*-diiodoethylene) at equilibrium, as measured by their relative equilibrium concentrations, then the former species, it is understood, is more stable than the latter. Similar statements can be made about sets of species—the reactant set or the product set. In light of the discussion of the preceding paragraph, then, we can make the following qualitative statements about a reaction at a given temperature:

1. If K for a given reaction is greater than unity, then the right-hand species are more stable than the left-hand species (products are favored).
2. If K for a given reaction is less than unity, then the left-hand species are more stable than the right-hand species (reactants are favored).
3. If K for a given reaction is near unity, then reactants and products are nearly the same in stability.

These general statements relate the equilibrium constant of a reaction, the equilibrium proper, and relative stabilities. The statements are qualitative, but they can be based on mathematical developments. They prove to be extremely useful guidelines, and we shall draw on them frequently.

The stability previously referred to is an important kind of stability, but not the only kind. The kind of stability about which the equilibrium constant tells us refers to which species are favored at equilibrium—if the system is allowed to reach equilibrium, no matter how long that might take. Depending on the identity of the reaction in question, it might take a tiny fraction of a second, as in a simple acid-base reaction in an aqueous solution, or it might require a very long time, even years, for equilibrium to be reached. The type of relative stability concerned with whether reactants or products are favored at equilibrium is called equilibrium stability, or **thermodynamic stability** (for reasons that will become clear when we study thermodynamics). Thus, in the reaction of $H_2(g)$ with $O_2(g)$ to produce $H_2O(g)$:

$$2H_2(g) + O_2(g) \rightleftharpoons 2H_2O(g)$$

the product is overwhelmingly favored at equilibrium ($K = 10^{57}$), so water, we say, is thermodynamically more stable than a mixture of H_2 and O_2 (in a 2:1 mole ratio).

The other important kind of stability is related to how long it takes to attain equilibrium—in other words, how fast the reaction proceeds. Thus, even though a species or mixture of species is thermodynamically unstable, the rate of the reaction that would transform these species may be so low that equilibrium is essentially never achieved. This kind of stability is called **kinetic stability**. A substance is kinetically stable if the reactions by which it would be transformed into some other species are too slow to be significant, even when the other species, the products, are thermodynamically more stable—they would be more stable if they would ever be formed.

Many laboratory and industrial situations arise that call for a chemical reaction in which the chemical equilibrium would provide a good percentage yield, with the products thermodynamically more stable than the reactants, but which takes too long to reach equilibrium. One tries in such cases to find a way of causing the reaction to proceed faster. One way that can often be used is to raise the temperature of the reacting system, for an increase in temperature generally increases the rate of a reaction. Another approach is to use a catalyst. A **catalyst** is a substance that participates in the reaction in the sense that it causes the reaction to proceed faster although it is never itself either consumed or produced in the overall reaction. The symbol for a catalyst does not appear in the net balanced equation for the overall net reaction. The catalyst acts as an intermediary. Most chemical reactions occur in a series of steps; and the catalyst participates in some of the steps in such a way that it does not undergo any overall change. The situation is analogous to that represented in Fig. 4.14, in which the change consists of the transfer of a large stone from

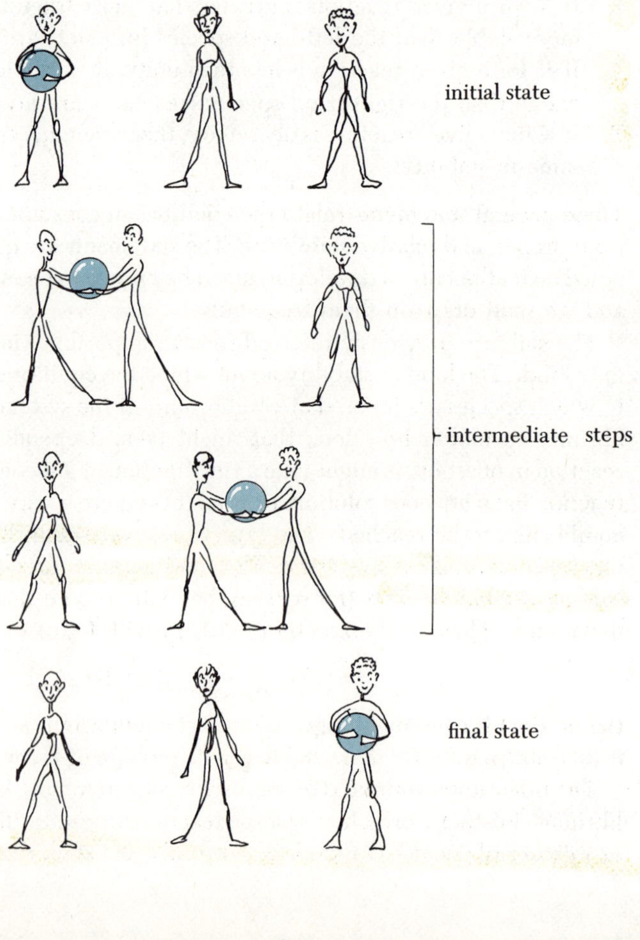

Figure 4.14
Simple analogy showing the transfer of a heavy stone from the man on the left (initial state) to the man on the right (final state) with the intermediary involvement of the man in the middle (the analog of a catalyst), who shows no change throughout.

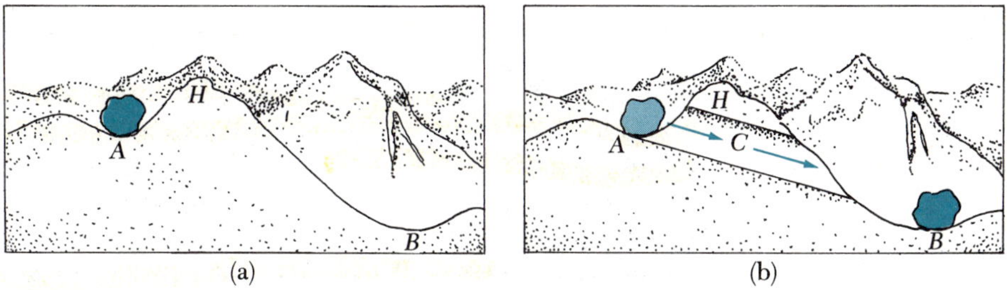

Figure 4.15 A simple physical analogy to thermodynamic and kinetic stabilities. (a) The boulder is thermodynamically more stable in state *B* than in state *A*, but it is kinetically stable in state *A* because it cannot spontaneously roll over the hump *H*. (b) The tunnel *C* is analogous to a catalyst, providing a favorable "pathway" for the thermodynamically more stable state to be achieved.

the man on the left to the man on the right. The man in the middle is an intermediary, representing intermediate steps, but he is the same at the end of the transfer as he was at the beginning. Like the man in the middle, the catalyst provides a convenient means, or pathway, for the total process to occur.

We can summarize the ideas of thermodynamic stability, kinetic stability, and the catalyst by the simple analogy shown in Fig. 4.15. In part (a) of the figure, we see a large stationary stone at position *A* on the hill. Clearly, position *B* in the valley would be thermodynamically more stable. But kinetically, the stone is stable at *A* because it cannot roll spontaneously over the hump *H*. In part (b) of the figure we see the analogy with the catalyst. If a tunnel were dug under the hump, as shown in position *C*, then a pathway would be available for the stone to roll down to the thermodynamically more stable position at *B*. The availability of the pathway through tunnel *C* is analogous to the availability of the pathway a catalyst provides in a chemical system, permitting a process to take place so that a more stable state can be attained. If the stone were provided with enough energy, it could roll up over hump *H*, and then down to *B*. This would be analogous to enhancing the speed of a chemical reaction by raising the temperature.

4.3 Le Chatelier's Principle and Chemical Equilibrium

Heat Transfers and Temperature Effects on Equilibrium

By knowing whether or not a given chemical reaction is exothermic or endothermic, one can use Le Chatelier's principle to predict qualitatively the effect that temperature variation will have on the value of the equilibrium constant, and hence on the position of equilibrium. Let us imagine a system undergoing a chemical reaction that is endothermic, considering this system in its equilibrium condition. Suppose that the equilibrium of such a system is disturbed by added heat, which raises the temperature. This temperature change can be considered a stress, and according to Le Chatelier's principle the system will respond in a way that minimizes that stress. The way in which the system can accomplish this is a shift in the equilibrium position in a direction that requires the absorption of some of the added heat. In the endothermic case under discussion this corresponds to transforming some more of the reactant to product, corresponding to the chemical equation's proceeding from left to right. Thus, the increase in temperature will increase the concentrations of species occurring in the numerator of Q and decrease the concentrations of species occurring in the denominator. Since the new value of Q must equal the value of K for the new equilibrium, we conclude that the new equilibrium is described by a larger value of K.

Thus, we can state that for an endothermic process an increase in temperature causes an increase in the value of K. A decrease in temperature, accompanying the removal of heat from the system at equilibrium, will bring about a decrease in K for the endothermic case. Following the same reasoning, it is clear that for an exothermic reaction an increase in temperature will cause a decrease in the value of K; and a decrease in temperature brings about an increase in K for the exothermic case.

For example, the decomposition of N_2O_4 gas into NO_2 is an endothermic reaction:

$$N_2O_4(g) \rightleftharpoons 2NO_2(g)$$

Therefore, an increase in the temperature increases the equilibrium constant, or favors NO_2 more at equilibrium, and a decrease in temperature decreases the equilibrium constant, or favors N_2O_4 at equilibrium. By contrast, the neutralization of a strong acid by a strong base

$$H^+ + OH^- \rightleftharpoons H_2O$$

is a very exothermic reaction, so increasing the temperature decreases the value of K (from a very large value to a smaller, but still very large, value). How sensitive the K value of a particular reaction is to a change in temperature depends on how much heat is absorbed or released as the reaction proceeds, that is, how endothermic or exothermic the reaction is.

EXAMPLE 4.2 The reaction of pyridine, C_5H_5N, with hydrochloric acid (in a solution with benzene, C_6H_6, as solvent) proceeds with evolution of heat. Predict the effect of decreasing the temperature on the equilibrium constant of this reaction: $C_5H_5N + HCl \rightleftharpoons C_5H_6NCl(s)$.

SOLUTION According to Le Chatelier's principle, the equilibrium constant for this exothermic reaction would increase if the temperature were decreased.

STUDY PROBLEM **4(b)**

A certain reaction is found to proceed faster the higher the temperature, but with a lower percentage yield. Is this reaction exothermic or endothermic?

Sudden Concentration Changes

Temperature changes, then, can affect the position of an equilibrium by altering the value of the equilibrium *constant*. Sudden changes of solute concentrations in a gaseous or liquid solution at equilibrium can also bring about a change in the position of an equilibrium, in other words, the equilibrium *concentrations*. Solute changes are not accompanied by alterations in the value of K. The equilibrium constant is indeed a constant for a given reaction, and is altered only if the temperature is changed. This case, like the case in which the temperature is varied, can be explained qualitatively by means of Le Chatelier's principle; it is also understandable mathematically in terms of the law of chemical equilibrium.

Let us consider the "bromination" of hexene, C_6H_{12}, in a solution of CCl_4:

$$C_6H_{12} + Br_2 \rightleftharpoons C_6H_{12}Br_2 \tag{4.6}$$

For this reaction, the equilibrium expression is

$$K = \frac{[C_6H_{12}Br_2]}{[C_6H_{12}][Br_2]} \tag{4.7}$$

Now, suppose that we have an equilibrium mixture of C_6H_{12}, Br_2, and $C_6H_{12}Br_2$ in solution at a particular temperature, and we suddenly add some Br_2. This sudden increase in the Br_2 concentration will upset the equilibrium. According to Le Chatelier's principle, this perturbation can be removed by an appropriate response of the system in a direction to undo this departure from equilibrium. The appropriate system response is for some portion of the Br_2 in solution to react according to Eq. (4.6), thereby reducing $[Br_2]$ from its instantaneous, nonequilibrium value, at the same time decreasing $[C_6H_{12}]$ and increasing $[C_6H_{12}Br_2]$ by a corresponding amount; this response persists until a new equilibrium is attained. The same conclusion would be drawn by considering Eq. (4.7). Thus, the sudden increase in $[Br_2]$ would increase the denominator, and make Q instantaneously smaller than its equilibrium value K. If the reaction shown in Eq. (4.6) then occurs to the appropriate extent, both $[C_6H_{12}]$ and $[Br_2]$ will decrease and $[C_6H_{12}Br_2]$ will increase by a corresponding amount. These concentration changes will increase the numerator and decrease the denominator of Q, i.e., restore it to its equilibrium value K. In this new equilibrium, $[Br_2]$ and $[C_6H_{12}Br_2]$ will be larger and $[C_6H_{12}]$ smaller than before the initial disruption of the equilibrium. This case is an example of how an increase in the amount of one reactant (in this case, Br_2) can increase the fraction of another reactant (C_6H_{12}) that is converted to product ($C_6H_{12}Br_2$). This can be thought of as "driving a reaction towards completion" by providing an excess of one of the reactants (presumably the less expensive one). This is an important concept in chemical synthesis and is frequently used to improve the percentage yield of a desired product.

EXAMPLE 4.3 Suppose you want to carry out the following reaction:

$$Pt(NH_3)_2Cl_2 + 2H_2O \rightleftharpoons Pt(H_2O)_2(NH_3)_2{}^{2+} + 2Cl^-$$

What reagent would you use in excess to drive the reaction towards completion?

SOLUTION Clearly, water is far less expensive than the platinum compound is, so water would be used in excess to increase the percentage of $Pt(NH_3)_2Cl_2$ converted to products.

STUDY PROBLEM 4(c)

Suppose one is removing water by carrying out the following dehydration reaction in a liquid solution $C_{10}H_{13}OH \rightleftharpoons C_{10}H_{12} + H_2O$. If one had some way to remove the water that is formed (say by boiling it off), how would this affect the percentage yield of $C_{10}H_{12}$?

Another way in which concentrations can be changed to alter an equilibrium is by changing the volume of the container in which a reaction is being carried out in the gas phase. Let us consider, for example, the following reaction occurring at about 400 °C in the enclosed piston-cylinder arrangement shown in Fig. 4.16(a):

$$N_2(g) + 3H_2(g) \rightleftharpoons 2NH_3(g) \quad (4.8)$$

At this temperature and a total pressure of 50 atm, if the starting mixture is a 1:3 mole ratio of N_2 to H_2, or alternatively pure ammonia, at equilibrium one-fourth of the nitrogen is present as ammonia, and the other three-fourths is N_2. We represent the equilibrium mixture in part (1), in which the symbol • stands for one mole of N_2, ∗ stands for one mole of H_2, and ∗ stands for one mole of NH_3. Now let us imagine that with the temperature kept constant at 400 °C, the volume is suddenly decreased to one-tenth its initial value when the piston is pushed down with ten times the initial force. At the instant this occurs, the pressure is suddenly increased from 50

Figure 4.16 (a) Le Chatelier's principle applied to the H_2–N_2–NH_3 system in a piston, cylinder arrangement at constant temperature. The initial condition shows 4 moles NH_3, 18 moles H_2, and 6 moles N_2 at 50 atm and 400 °C. The volume is suddenly decreased by a factor of 10 at the same temperature—the hypothetical instantaneous condition. The new equilibrium is arrived at from (2) by the combination of 3 moles N_2 and 9 moles H_2 to give 6 additional moles NH_3. (b) The effect of adding 28 moles Ne via the valve at the bottom of the cylinder to the initial system shown in part (a): no change takes place in the concentrations of H_2, N_2, or NH_3 even though the total pressure increases.

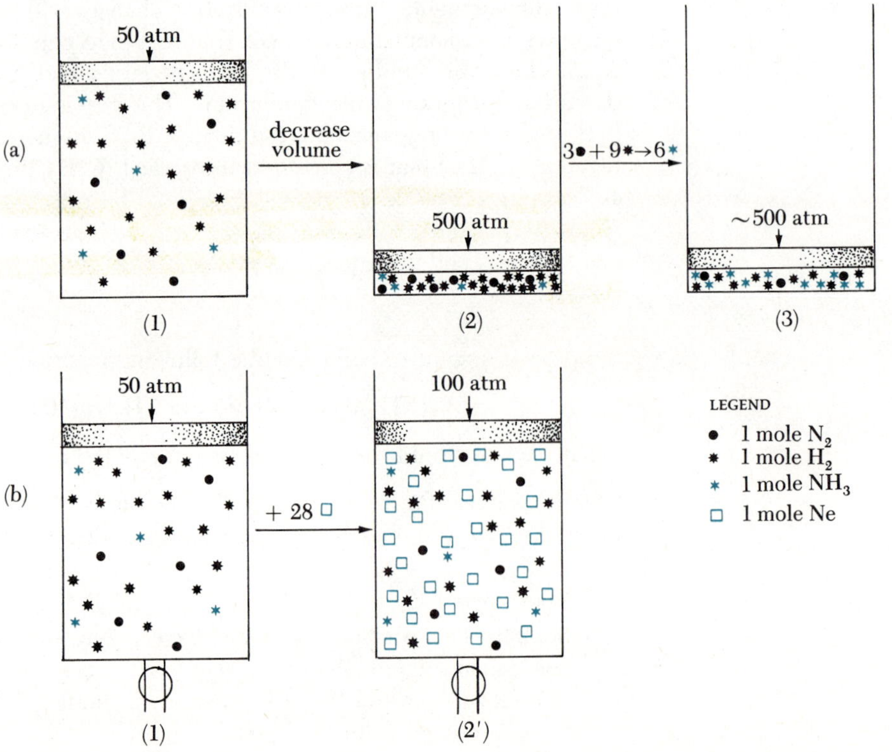

atm to 500 atm; this instantaneous situation is depicted in part (2). According to Le Chatelier's principle, the system will respond so as to decrease the stress of this sudden increase of pressure. As the right side of Eq. (4.8) represents two moles of gas, compared with four moles of gas on the left side, the way the system can relieve the sudden perturbation of equilibrium is by converting some N_2 and H_2 to NH_3. In this particular case, 3 mol N_2 and 9 mol H_2 are converted to 6 mol NH_3 to achieve the new equilibrium, represented in part (3). The same conclusion would be drawn by considering the equilibrium expression that corresponds to Eq. (4.8):

$$K = \frac{[NH_3]^2}{[N_2][H_2]^3} \qquad (4.9)$$

At the instant the volume is decreased by a factor of 10 (part (2)), each of the values $[NH_3]$, $[N_2]$, and $[H_2]$ suddenly increases by a factor of 10 (same number of moles, one-tenth of the initial volume), so the numerator increases by a factor of 10^2 and the denominator by 10^4. The net result is that Q suddenly decreases to $\frac{1}{100}$ of its equilibrium value K. To restore equilibrium, Q must increase by a factor of 100 to become equal to K again. This comes about by an increase in the numerator of Q and a decrease in the denominator, which occurs when 3 mol N_2 combines with 9 mol H_2 to produce 6 mol NH_3.

Let us also consider the effect on the same N_2–H_2–NH_3 equilibrium position of suddenly increasing the total pressure in the vessel shown in part (1) of Fig. 4.16b by introducing a nonreactive gas. Suppose the system is initially at equilibrium, and 28 mol neon gas is suddenly introduced through the valve shown at the bottom of the cylinder while the temperature and volume of the arrangement are held constant. This has the effect of doubling the total pressure in the vessel. The concentrations of N_2, H_2, and NH_3 are all unchanged, however, by this maneuver. Hence the equilibrium has really not been disturbed at all, so Le Chatelier's principle does not apply, and the system remains at the same equilibrium position as it had before the addition of neon.

An important distinction exists between the two ways of altering the equilibrium that we have discussed here relative to Le Chatelier's principle. Temperature change alters the position of equilibrium primarily by altering the equilibrium constant itself. In the case of sudden concentration changes—like the addition of some more reactant to a liquid solution, or a change in volume of a gaseous reaction—the equilibrium position changes, but the equilibrium constant remains fixed so long as the temperature is unchanged.

EXAMPLE 4.4 How would a sudden increase in the volume of the reaction vessel, at constant temperature, affect the equilibrium position and the equilibrium constant for the following reaction?

$$2CO(g) + O_2(g) \rightleftharpoons 2CO_2(g)$$

SOLUTION Using the reasoning just noted, we find that three moles of gas on the left side are converted to two moles of gas on the right side. Hence an increase in volume will shift the equilibrium towards the side of the equation with the larger number of moles, in this case the left side. However, since the temperature is held fixed, K remains constant.

STUDY PROBLEM 4(d)

Indicate how a decrease in volume at constant temperature will affect each of the following equilibria:

(a) $H_2(g) + Cl_2(g) \rightleftharpoons 2HCl(g)$
(b) $2NO(g) + Cl_2(g) \rightleftharpoons 2NOCl(g)$

4.4 Calculations on Chemical Equilibrium

Algebraic Manipulations of Equilibrium Constants

Such equations as (4.2), (4.4), and (4.5) are simply algebraic equations, and the symbols that occur in them are simply algebraic quantities. These equations and symbols can be manipulated according to the same mathematical rules that apply to algebraic equations in general. We shall see that these algebraic manipulations bear a one-to-one relation with corresponding manipulations of chemical equations. One such case would be to raise each side of Eq. (4.2) or (4.4) or (4.5) to any specified power p. This gives the equilibrium expression that would prevail for the chemical equation we should obtain simply by multiplying the corresponding chemical equation by the factor p. For the general case, let us multiply both sides of Eq. (4.3) by a constant p:

$$paA + pbB + \cdots \rightleftharpoons pmM + pnN + \cdots \tag{4.10}$$

The corresponding equilibrium expression is given, according to our mathematical form of the law of chemical equilibrium (Eq. 4.5), by

$$K' = \frac{[M]^{pm}[N]^{pn}\ldots}{[A]^{pa}[B]^{pb}\ldots} = \left(\frac{[M]^{m}[N]^{n}\ldots}{[A]^{a}[B]^{b}\ldots}\right)^{p} \tag{4.11}$$

where the prime refers to the case of chemical Eq. (4.10). Hence,

$$K' = K^{p} \tag{4.12}$$

where K without a prime corresponds to Eq. (4.3).

For example, the first and third chemical equations in Study Problem 4a differ only by a factor 2 ($p = 2$ converts the third equation to the first equation). Hence, the equilibrium constant for the first is the square of the equilibrium constant for the third.

$$K_3 = \frac{[NO_2(g)]}{[NO(g)]\,[O_2(g)]^{1/2}}$$

$$K_1 = \frac{[NO_2(g)]^2}{[NO(g)]^2[O_2(g)]} = \left(\frac{[NO_2(g)]}{[NO(g)]\,[O_2(g)]^{1/2}}\right)^2$$

$$K_1 = K_3{}^2$$

Suppose we choose a value -1 for p. Then Eq. (4.10) becomes

$$-aA - bB - \cdots \rightleftharpoons -mM - nN \cdots \tag{4.13a}$$

which can be written

$$mM + nN + \cdots \rightleftharpoons aA + bB + \cdots \tag{4.13b}$$

This is simply the reverse of the reaction expressed in Eq. (4.3). For Eq. (4.13a) and (4.13b),

$$K' = \frac{[M]^{-m}[N]^{-n}}{[A]^{-a}[B]^{-b}} = \frac{[A]^a[B]^b}{[M]^m[N]^n} = \frac{1}{K} \tag{4.14}$$

By comparing Eq. (4.3), (4.4), and (4.5) with Eq. (4.13) and (4.14), we see the following important relation. If in one's mind the direction of a reaction is reversed, the corresponding equilibrium constant of the reversed reaction is simply the inverse of the K of the original reaction. Therefore, in working with equilibrium constants, one must clearly state either in words or with an equation the direction in which the reaction should be viewed. For example, for the reaction *cis*-$C_2H_2I_2 \rightleftharpoons$ *trans*-$C_2H_2I_2$, using the rule embodied in Eq. (4.5), we write $K = [trans]/[cis]$. For the reverse process, *trans*-$C_2H_2I_2 \rightleftharpoons$ *cis*-$C_2H_2I_2$, by the same rule, we write $K' = [cis]/[trans]$. These two K's are simply reciprocals.

In another example, compare the first and second chemical equations of Study Problem 4(a). The second equation is the reverse of the first (that is, $p = -1$ converts the first equation to the second). Thus

$$K_2 = \frac{[NO(g)]^2[O_2(g)]}{[NO_2(g)]^2} = \frac{1}{K_1}$$

Another important manifestation of algebraic manipulations of chemical equations and equilibrium expressions is the case of consecutive reactions. This is the case in which some overall chemical reaction can be thought of as the result of two or more steps, each of which is a chemical reaction. In such a case, we readily see that the chemical equation for the overall reaction is the sum of the chemical equations for the individual steps, and the equilibrium constant for the overall reaction is the product of the K's for the individual steps. Consider, for example, the reaction of methane, CH_4, with oxygen to yield CO_2 and water (the "combustion" of methane),

$$CH_4(g) + 2O_2(g) \rightleftharpoons CO_2(g) + 2H_2O(g)$$

This process can be viewed in steps—the two steps, (a) and (b)—

(a) $$CH_4(g) + O_2(g) \rightleftharpoons CH_2O(g) + H_2O(g)$$

(b) $$CH_2O(g) + O_2(g) \rightleftharpoons CO_2(g) + H_2O(g)$$

Sum $$CH_4(g) + 2O_2(g) + CH_2O(g) \rightleftharpoons CH_2O(g) + CO_2(g) + 2H_2O(g)$$

After cancellation of $CH_2O(g)$ from both sides,

(c) $$CH_4(g) + 2O_2(g) \rightleftharpoons CO_2(g) + 2H_2O(g)$$

Also, for step (a)

$$K_a = \frac{[CH_2O][H_2O]}{[CH_4][O_2]}$$

and for step (b)

$$K_b = \frac{[CO_2][H_2O]}{[CH_2O][O_2]}$$

Then,

$$K_a \cdot K_b = \frac{[CH_2O][H_2O]}{[CH_4][O_2]} \frac{[CO_2][H_2O]}{[CH_2O][O_2]} = \frac{[CO_2][H_2O]^2}{[CH_4][O_2]^2}$$

Inspection of Eq. (c) shows that $K_aK_b = K_c$. Expressed in words, the equilibrium constant for the overall reaction (c) is the product of the K's for the steps (a) and (b). If reaction (c) = reaction (a) + reaction (b), then

$$K_c = K_a \cdot K_b \tag{4.15}$$

STUDY PROBLEM 4(e)

Show the relation between the following three chemical equations and the corresponding equilibrium constants.

$$Co(H_2O)_6{}^{3+} + 2NH_3 \rightleftharpoons Co(H_2O)_4(NH_3)_2{}^{3+} + 2H_2O \qquad K_1$$

$$Co(H_2O)_4(NH_3)_2{}^{3+} + 4NH_3 \rightleftharpoons Co(NH_3)_6{}^{3+} + 4H_2O \qquad K_2$$

$$Co(H_2O)_6{}^{3+} + 6NH_3 \rightleftharpoons Co(NH_3)_6{}^{3+} + 6H_2O \qquad K_3$$

A related case is the situation in which the chemical equation for an overall reaction of interest can be expressed as the difference between the chemical equations of two other chemical reactions. In this case, the equilibrium constant of the overall reaction is the ratio of the K's of the other two reactions. For example, suppose we wish to derive the equilibrium constant for the process

$$\text{(III)} \quad C_2H_4(g) + H_2(g) \rightleftharpoons C_2H_6(g) \qquad K_{III} = \frac{[C_2H_6]}{[C_2H_4][H_2]} = ?$$

and we know the values of the K's for the following two reactions:

$$\text{(II)} \quad C_2H_2(g) + 2H_2(g) \rightleftharpoons C_2H_6(g) \qquad K_{II} = \frac{[C_2H_6]}{[C_2H_2][H_2]^2}$$

$$\text{(I)} \quad C_2H_2(g) + H_2(g) \rightleftharpoons C_2H_4(g) \qquad K_I = \frac{[C_2H_4]}{[C_2H_2][H_2]}$$

Clearly, if we consider the reverse of reaction (I), we have

$$(-\text{I}) \quad C_2H_4(g) \rightleftharpoons C_2H_2(g) + H_2(g) \qquad \frac{1}{K_I} = \frac{[C_2H_2][H_2]}{[C_2H_4]}$$

Then, if we add chemical equation (II) to chemical equation (−I), effectively subtracting (I) from (II), we obtain

$$C_2H_2(g) + 2H_2(g) + C_2H_4(g) \rightleftharpoons C_2H_6(g) + C_2H_2(g) + H_2(g)$$

$$C_2H_4(g) + H_2(g) \rightleftharpoons C_2H_6(g)$$

which is the process of interest.

$$(K_{\mathrm{II}})\left(\frac{1}{K_{\mathrm{I}}}\right) = \left(\frac{[C_2H_6]}{[C_2H_2][H_2]^2}\right)\left(\frac{[C_2H_2][H_2]}{[C_2H_4]}\right) = \frac{[C_2H_6]}{[H_2][C_2H_4]} = K_{\mathrm{III}}$$

That is, $K_{\mathrm{III}} = (K_{\mathrm{II}}/K_{\mathrm{I}})$, the result expected from the general statement above. Although these combination relations were illustrated for specific cases, they are quite general.

If reaction (III) = reaction (II) − reaction (I), then

$$K_{\mathrm{III}} = \frac{K_{\mathrm{II}}}{K_{\mathrm{I}}} \tag{4.16}$$

STUDY PROBLEM 4(f)

Given the following equilibrium constants,

$$Ag^+ + 2NH_3 \rightleftharpoons Ag(NH_3)_2^+ \qquad K = 1.6 \times 10^7$$
$$Ag^+ + NH_3 \rightleftharpoons Ag(NH_3)^+ \qquad K = 2.3 \times 10^3$$

calculate the equilibrium constant for the process

$$Ag(NH_3)^+ + NH_3 \rightleftharpoons Ag(NH_3)_2^+$$

Some Typical Chemical Equilibria and Equilibrium Expressions

Gas Phase Reactions

Gas phase reactions are very important in pure and applied chemistry. One reason that such reactions are important is that the interpretations of these reactions in terms of molecular structures are not clouded by the conceptual uncertainties associated with the strong intermolecular forces characteristic of liquids and solids. Hence, many theoretical models can be tested most directly by experimental data from the gas phase. A more practical present-day reason for studying gas phase equilibria is that many important processes involved in current or projected methods of energy production make use of gas phase processes. Also, characterizing gas phase reactions is very important in understanding the causes of air pollution and in designing methods to combat it.

Below are three examples of chemical equations of gas phase reactions, with the corresponding equilibrium expressions.

$$H_2(g) + I_2(g) \rightleftharpoons 2HI \tag{4.17a}$$

$$K = \frac{[HI]^2}{[H_2][I_2]} \tag{4.17b}$$

$$N_2(g) + 3H_2(g) \rightleftharpoons 2NH_3(g) \tag{4.18a}$$

$$K = \frac{[NH_3]^2}{[N_2][H_2]^3} \tag{4.18b}$$

$$N_2(g) + 2O_2(g) \rightleftharpoons 2NO_2(g) \tag{4.19a}$$

$$K = \frac{[NO_2]^2}{[N_2][O_2]^2} \tag{4.19b}$$

Spotlight

Hydrogen, since it is the element of lowest atomic weight, the first element of the periodic table, is in many ways in a class by itself. Hydrogen and helium constitute the entire first row of the table; and as helium is chemically very inert, the chemical properties of hydrogen are in many ways set apart from any of the trends that one commonly finds in the table. The reasons for this become clear when one considers the relations between the positions of elements in the table and the electronic structures of atoms (see Chap. 6). Hydrogen comes in the same column in the periodic table as the alkali metals, Li, Na, K, Rb, and Co. Yet, hydrogen is nonmetallic. Its most common types of involvements in compounds are almost never parallel to the alkali metals. Indeed, nearly the only chemical similarity is that hydrogen and the alkali metals all form stable +1 ions.

The stable elemental form of hydrogen is diatomic, H_2. Only traces of H_2 are found in the atmosphere; more substantial quantities are sometimes found in volcanic and other kinds of natural gases, however. In mass, hydrogen accounts for only about 1% of the earth's crust; but hydrogen furnishes about 16% of the total number of atoms on the earth's surface (second only to oxygen). The largest portion of this hydrogen is combined with oxygen in water. Hydrogen appears to be by far the most abundant element in the universe, accounting for about 30% of the mass of the sun.

The element hydrogen occurs in all acids (as commonly defined; Sec. 3.5) and in all animal and plant tissue. Most elements form compounds with hydrogen. Combinations of hydrogen with elements on the left side of the periodic table are often called hydrides—lithium hydride for LiH, and calcium hydride for CaH_2. Combinations with elements on the right side of the table call for other kinds of nomenclature—hydrogen chloride (or hydrochloric acid) for HCl, germane for GeH_4, hydrogen sulfide for H_2S, silane for SiH_4. Combinations of carbon and hydrogen are compounds called **hydrocarbons**—CH_4, C_2H_6, C_8H_{18}, $C_{10}H_{20}$, $C_{12}H_8$, and so on. Such compounds are typically the main constituents of fossil fuels, including natural gas, petroleum, coal, and oil shale (arranged here in increasing order of average molecular weight).

The gas H_2 can readily, sometimes explosively, be generated by reaction of a reactive metal with water:

$$2Na(s) + 2H_2O \rightleftharpoons 2Na^+ + 2OH^- + H_2(g)$$

or by the electrolysis of water (see Sec. 12.5):

$$2H_2O \rightleftharpoons 2H_2(g) + O_2(g)$$

Hydrogen is used in a variety of important laboratory and industrial processes, including the addition of H_2 to a molecule, a process called **hydrogenation.** An example of hydrogenation is adding H_2 to diiodoethylene:

$$trans\text{-}C_2H_2I_2 + H_2(g) \rightleftharpoons C_2H_4I_2$$

Currently there is much interest in the possible use of $H_2(g)$ as a fuel. Hydrogen gas has the advantage of being a clean fuel; it burns producing only water and heat:

$$2H_2(g) + O_2(g) \rightleftharpoons 2H_2O(g)$$

If inexpensive ways could be found to produce $H_2(g)$, the results could have a profound positive effect on our energy problem. Current chemical research is exploring possible ways of producing $H_2(g)$ directly or indirectly from water, using energy from the sun.

The reaction summarized in Eq. (4.18a) is commercially important since it constitutes the basis of the Haber synthesis of ammonia. Chemical equation (4.19a) represents a process that concerns air pollution; NO_2 is a species often responsible for the reddish brown haze associated with air polluted by internal combustion engine exhausts or by steel mills. There are many methods by which NO_2 can be produced, both in air and in an engine or factory; and some of these methods are not well understood. Nevertheless, to the extent that the tendency for systems to approach equilibrium has an opportunity for manifesting itself in the dynamic and complex case of the atmosphere, the relations among O_2, N_2, and NO_2 concentrations given by Eq. (4.19b) are of interest.*

*It is often convenient to express equilibrium expressions, and the value K, for gas phase processes in terms of pressure instead of concentration. This will be discussed in Chap. 15.

EXAMPLE 4.5 At 520 °C, the equilibrium constant for the gas phase reaction of H_2 with I_2 to form 2HI is 63. Calculate the equilibrium mole ratio of HI to H_2 in an equilibrium mixture resulting from pure HI.

SOLUTION The chemical equation and the equilibrium expression for this process are

$$H_2(g) + I_2(g) \rightleftharpoons 2HI(g)$$

$$K = \frac{[HI]^2}{[H_2][I_2]} = 63$$

Clearly, the number of moles of H_2 produced by dissociation of HI will be exactly the same as the number of moles of I_2 produced. Hence, for this system, we can write $[I_2] = [H_2]$. Substituting $[H_2]$ for $[I_2]$ in the above equilibrium expression, we obtain

$$\frac{[HI]^2}{[H_2]^2} = 63$$

$$\frac{[HI]}{[H_2]} = \sqrt{63} = 7.9$$

$$\frac{\text{moles HI}}{\text{moles } H_2} = 7.9$$

Spotlight

Fritz Haber was born in Breslau, Germany, in 1868. After his education in Berlin, Heidelberg, Charlottenburg, and Karlsruhe, he taught chemistry at the technical high school in Karlsruhe until 1911. He then moved to Berlin as professor of physical chemistry at the University of Berlin and later as director of the Kaiser Wilhelm Institute. His research encompassed a range of studies in physical chemistry, including thermodynamic studies of gas reactions, flame chemistry, and electrochemistry. His thorough investigations of the system here represented by Eq. (4.18) led him to discover conditions under which ammonia could be produced directly from N_2 by this reaction. A key feature of these conditions was the use of very high pressures, over 200 atm. This can be appreciated by considering Le Chatelier's principle in light of the transformation of four moles of gas into two in the process shown in Eq. (4.18).

Haber was awarded the Nobel Prize in 1918 for his work on ammonia synthesis. His developments were of substantial practical importance, since they made possible for the first time the synthetic "fixation of nitrogen," the conversion of N_2 into some other, more chemically useful form of nitrogen. Nitrogen compounds, like ammonia and its derivative, are needed as fertilizers. Ammonia could be converted to nitric acid by reaction with oxygen ($2NH_3 + 4O_2 \rightleftharpoons 2HNO_3 + 2H_2O$), using the already known Ostwald process. Nitric acid is useful for making nitrate fertilizers and also explosives. As Germany had previously depended on Chilean saltpeter deposits ($NaNO_3$) as a nitrate source, a source conveniently blockaded by England, the industrialization by Carl Bosch of the Haber process gave Germany a new independence in manufacturing explosives. This encouraged Germany to launch World War I.

Haber's Jewish ancestry made necessary his exile from Germany during the Nazi era, and he died in Switzerland in 1934.

The fixation of nitrogen, which in nature is carried out by legumes, has remained an interesting and important subject in chemical research. Recent work has focused on processes that can function at low N_2 pressures, like 1 atm, and has centered about the development of new catalysts, or "chemical expediters."

Spotlight

Iodine is a member of the family of elements called halogens: fluorine, chlorine, bromine, iodine, and astatine. Astatine is highly radioactive, and has been isolated in only very small amounts, so its chemistry has not been explored nearly so completely as the chemistry of the other halogens. Iodine, like the other halogens, is diatomic in its stable elemental form. It is interesting to note the trend towards higher melting points in the series of elemental halogens: F_2, −223 °C; Cl_2, −102 °C; Br_2, −7.3 °C; I_2, 113 °C. Similarly for boiling points, we find F_2, −188 °C; Cl_2, −35 °C; Br_2, 58 °C; I_2, 183 °C. This tendency towards higher melting points and boiling points for a series of related molecular species that systematically range down a particular column of the periodic table is common. (Compare CH_4, SiH_4, GeH_4, SnH_4, and PbH_4.) Inspection of these melting points and boiling points shows that at room temperature, 25 °C, F_2 and Cl_2 are gases, Br_2 is a liquid, and I_2 is a solid at atmospheric pressure.

The elemental halogens have many features of chemical reactivities that vary in the order F_2, Cl_2, Br_2, I_2. In agreement with this order, F_2 can convert $2Cl^-$ to Cl_2, $2Br^-$ to Br_2, and $2I^-$ to I_2, Cl_2 can convert $2Br^-$ to Br_2 and $2I^-$ to I_2, and Br_2 can convert $2I^-$ to I_2, but none of the reverse processes occur very much. Thus, the reactions

$$F_2 + 2Cl^- \rightleftharpoons 2F^- + Cl_2$$
$$Cl_2 + 2Br^- \rightleftharpoons 2Cl^- + Br_2$$
$$Br_2 + 2I^- \rightleftharpoons 2Br^- + I_2$$

occur as written from left to right (large K values); and the reverse reactions have very small equilibrium constants.

The elemental halogens are generally isolated by electrolysis reactions, of which the following is typical:

$$2H_2O + 2Cl^- \rightleftharpoons 2OH^- + H_2(g) + Cl_2(g)$$

In the United States, iodine is often obtained by reaction of the iodides in kelp with chlorine:

$$Cl_2 + 2I^- \rightleftharpoons 2Cl^- + I_2$$

The elemental halogens find many uses in the laboratory, in industry, and in medicine. Chlorine is used in bleaching and as a germicide in water purification. Bromine is useful in producing chemical reagents important in synthetic organic chemistry. Iodine is used as an antiseptic in an alcohol solution, tincture of iodine. Iodides find use in photography and in treating disorders of the thyroid gland.

Hydrogen iodide, HI, like HCl and HBr, is a strong acid. A variety of oxygen-containing anions of iodine are known and well characterized: hypoiodite, IO^-; iodate, IO_3^-; periodate, IO_4^-. Analogous ions are also known for Br and Cl. In aqueous solution, I_2 reacts with an excess of iodide to form triiodide:

$$I_2 + I^- \rightleftharpoons I_3^-$$

STUDY PROBLEM 4(g)

What is the concentration of HI in an equilibrium gas phase mixture at 520 °C in which the concentrations of H_2 is 0.011 M and of I_2 is 0.0056 M?

Nonaqueous Solutions

The study of substances and chemical reactions in liquid solutions constitutes a very large proportion of chemical research and chemical applications. Reactions are typically carried out in liquid solution because relatively few substances are gases under conveniently achieved conditions, and reactions in the solid state are typically slow. Also, the liquid solution provides a convenient way of regulating and dispensing the amounts of reagents by volumetric techniques.

Most chemical reactions that involve organic compounds, which are the compounds of carbon and hydrogen and their derivatives, use some other solvent besides water, a nonaqueous solvent. This often happens because of solubility limitations—

most organic compounds are not very soluble in water. One of the many important kinds of reactions of organic compounds, including many of biochemical importance, is the esterification reaction between a carboxylic acid, RCO_2H, and an alcohol, R′OH.

$$RCO_2H + R'OH \rightleftharpoons RCO_2R' + H_2O \tag{4.20a}$$

Here R and R′ represent specific groups of carbon and hydrogen atoms, for example, CH_3, C_6H_5, and C_2H_5. A specific example of this is

$$C_6H_5CO_2H + CH_3OH \rightleftharpoons C_6H_5CO_2CH_3 + H_2O \tag{4.20b}$$

The product, RCO_2R' is called an ester, a class of typically sweet-smelling compounds that occur in a variety of natural and synthetic materials.

EXAMPLE 4.6 For this particular esterification (Eq. 4.20a) in which R is a CH_3 group and R′ is a C_2H_5 group, the carboxylic acid is called acetic acid and the alcohol is called ethyl alcohol; the ester produced in this case is ethyl acetate, and its formation at 25 °C is characterized by an equilibrium constant 4.0. Calculate the absolute yield in number of grams of ethyl acetate ($CH_3CO_2C_2H_5$) that would be produced from a reaction in which 0.10 mole of acetic acid and 0.10 mole of ethyl alcohol are allowed to react in a 1-liter solution in a suitable solvent, say dioxane, $C_4H_8O_2$.

SOLUTION From Eq. (4.5) and Eq. (4.20a), the chemical equation and the equilibrium expression for this reaction are

$$\underset{\text{acetic acid}}{CH_3CO_2H} + \underset{\text{ethyl alcohol}}{C_2H_5OH} \rightleftharpoons \underset{\text{ethyl acetate}}{CH_3CO_2C_2H_5} + H_2O \tag{4.20c}$$

$$\frac{[CH_3CO_2C_2H_5][H_2O]}{[CH_3CO_2H][C_2H_5OH]} = 4.0 \tag{4.21}$$

We recognize from Eq. (4.20a) that the two reactants are initially present in this example in the same amounts, and they react together on a 1:1 mole basis; hence, their concentrations remain the same as the reaction proceeds. This gives for the products

$$[CH_3CO_2C_2H_5] = [H_2O]$$

Similarly, for the reactants

$$[CH_3CO_2H] = [C_2H_5OH]$$

Then, Eq. (4.21) rewritten becomes

$$\frac{[CH_3CO_2C_2H_5]^2}{[CH_3CO_2H]^2} = 4.0$$

$$\frac{[CH_3CO_2C_2H_5]}{[CH_3CO_2H]} = 2.0$$

$$[CH_3CO_2H] = \tfrac{1}{2}[CH_3CO_2C_2H_5] \tag{4.22}$$

But assuming that there are no side reactions, from Eq. (4.20) the total number of moles of acetic acid, CH_3CO_2H, and ethyl acetate, $CH_3CO_2C_2H_5$, at equilibrium must equal the number of moles of acetic acid present initially, namely 0.10 mole. Furthermore, since the reaction takes place in one liter of solution, and since in such a condition the number of moles of any species is numerically equal to its molarity,

$$[CH_3CO_2C_2H_5] + [CH_3CO_2H] = 0.10$$

Substituting the expression for $[CH_3CO_2H]$ of Eq. (4.22) into this equation, we obtain

$$[CH_3CO_2C_2H_5] + \tfrac{1}{2}[CH_3CO_2C_2H_5] = 0.10$$

$$\tfrac{3}{2}[CH_3CO_2C_2H_5] = 0.10$$

$$[CH_3CO_2C_2H_5] = 0.067\ \text{(mol/liter) present at equilibrium}$$

$$\text{no. g } CH_3CO_2C_2H_5 = \left(0.067\frac{\text{mol } CH_3CO_2C_2H_5}{\text{liter}}\right)(1.0\ \text{liter}) \times \left(88.1\frac{\text{g } CH_3CO_2C_2H_5}{\text{mol } CH_3CO_2C_2H_5}\right) = 5.9\ \text{g } CH_3CO_2C_2H_5$$

Aqueous Solutions

Many natural chemical reactions take place in solution. Water is commonly the solvent, whereupon the solution is referred to as aqueous. Such are the reactions that take place in oceans, lakes, and rivers and internally within plants and animals. Many interesting geological deposits can be explained by reactions that occur in aqueous solutions. Also, the chemical basis of many biological functions involves aqueous solutions.

In some reactions carried out in aqueous solution, water can participate directly as a reactant or a product in the reaction, and in some cases it merely provides the medium for reaction. Examples of both types of reaction that occur in water are given in Eqs. (4.23a) and (4.24a).

$$H_2O + CO_2 \rightleftharpoons H_2CO_3 \tag{4.23a}$$

$$NH_3 + CH_3CO_2H \rightleftharpoons NH_4^+ + CH_3CO_2^- \tag{4.24a}$$

where CH_3CO_2H represents acetic acid and $CH_3CO_2^-$ the acetate ion. Chemical equation (4.23a) represents a reaction that has some importance in a wide range of applications, ranging from the acid-base properties of blood to the soft drink industry and its carbonated beverages. Equation (4.24a) represents a broad class of acid-base chemistry. The conditions for equilibrium in the cases of Eqs. (4.23a) and (4.24a) are

$$K = \frac{[H_2CO_3]}{[H_2O][CO_2]} \tag{4.23b}$$

$$K = \frac{[NH_4^+][CH_3CO_2^-]}{[NH_3][CH_3CO_2H]} \tag{4.24b}$$

While there is no fundamental difference in the characteristics of these two equilibrium expressions, there is a practical feature of Eq. (4.23b) that does not occur in Eq. (4.24b). That feature is the appearance of the concentration of the solvent, i.e.,

H_2O, in the equilibrium expression. If the solution in which the chemical reaction occurs is very dilute, consisting mainly of water, then $[H_2O]$ is very nearly equal to its value in pure water, which is 55.5 *M*:

$$1000\frac{\text{g}}{\text{liter}} \times \frac{1}{18.08}\frac{\text{mol}}{\text{g}} = 55.5\frac{\text{mol}}{\text{liter}}$$

Furthermore, if the solution is sufficiently dilute, say 1 *M*, in CO_2 and H_2CO_3, then the position of equilibrium does not appreciably influence the value of $[H_2O]$, at least to two significant figures. In such cases it is often convenient to treat $[H_2O]$ as a constant, 55.5.

$$K = \frac{[H_2CO_3]}{55.5[CO_2]} \tag{4.25}$$

$$55.5K = \frac{[H_2CO_3]}{[CO_2]} \tag{4.26}$$

Then, knowing that K is a constant, one can write

$$K' = 55.5K = \text{a constant} = \frac{[H_2CO_3]}{[CO_2]} \tag{4.27}$$

where K' can be viewed as a type of equilibrium constant, the value of which equals $[H_2CO_3]/[CO_2]$ at equilibrium. It is this equilibrium constant K', having the value 5.5×10^{-2}, and not K, that one would find tabulated for this system. For chemical reactions that involve water as a reactant or a product and that are carried out in aqueous solution, the number usually reported for an equilibrium constant corresponds to a value that has had 55.5 for $[H_2O]$ incorporated, as in Eq. (4.27). In such cases, the equilibrium expressions appear to have had the quantity $[H_2O]$ replaced by unity wherever it occurs. (Another view that justifies this procedure is given in Sec. 4.6.)

EXAMPLE 4.7 Write the equilibrium expressions for the following reactions carried out in aqueous solutions:

$$SO_3 + 3H_2O \rightleftharpoons SO_4^{2-} + 2H_3O^+$$
$$H_2O + Cl_2 \rightleftharpoons HClO + H^+ + Cl^-$$

SOLUTION

$$K = \frac{[SO_4^{2-}][H_3O^+]^2}{[SO_3]}$$

$$K = \frac{[HClO][H^+][Cl^-]}{[Cl_2]}$$

STUDY PROBLEM 4(h)

Write the equilibrium expressions for the following reactions carried out in aqueous solutions:

(a) $H_2C_2O_3 + H_2O \rightleftharpoons 2HCO_2H$
(b) $CH_3CO_2H + OH^- \rightleftharpoons CH_3CO_2^- + H_2O$

The electrolytes are an important class of compounds that often undergo reactions in aqueous solutions. These substances, when dissolved in water, ionize into submicroscopic electrically charged particles, called **ions**; these ions are positively or negatively charged subunits of the original formula units of the undissolved electrolytes. Some electrolytes, like sodium nitrate, $NaNO_3$, sulfuric acid, H_2SO_4, potassium bromide, KBr, ammonium sulfate, $(NH_4)_2SO_4$, and sodium perchlorate, $NaClO_4$, are extremely soluble, dissolving to an extent exceeding 30% by weight of the water. Some other electrolytes, such as sodium chloride, NaCl, and magnesium sulfate, $MgSO_4$, have moderate solubility in water. Some others have very low solubility in water, less than a few percent; examples of these are silver chloride, AgCl, calcium sulfate, $CaSO_4$, and zinc sulfide, ZnS.

There is essentially an infinite variety of chemical reactions involving ions or combinations of ions. In some cases *only* ions are involved; in others ions are involved in combination with neutral species. Some examples of ionic reactions and ionic equilibria in dilute aqueous solution are given in Eqs. (4.28) to (4.34).

$$Co(NH_3)_6{}^{3+} + 6CN^- \rightleftharpoons Co(CN)_6{}^{3-} + 6NH_3 \tag{4.28a}$$

$$K = \frac{[Co(CN)_6{}^{3-}][NH_3]^6}{[Co(NH_3)_6{}^{3+}][CN^-]^6} \tag{4.28b}$$

$$CH_3Cl + Br^- \rightleftharpoons CH_3Br + Cl^- \tag{4.29a}$$

$$K = \frac{[CH_3Br][Cl^-]}{[CH_3Cl][Br^-]} \tag{4.29b}$$

$$CO_3{}^{2-} + H_2O \rightleftharpoons HCO_3{}^- + OH^- \tag{4.30a}$$

$$K = \frac{[HCO_3{}^-][OH^-]}{[CO_3{}^{2-}]} \tag{4.30b}$$

$$2Fe^{3+} + Hg_2{}^{2+} \rightleftharpoons 2Fe^{2+} + 2Hg^{2+} \tag{4.31a}$$

$$K = \frac{[Fe^{2+}]^2[Hg^{2+}]^2}{[Fe^{3+}]^2[Hg_2{}^{2+}]} \tag{4.31b}$$

$$Co(H_2O)_6{}^{3+} + 6CN^- \rightleftharpoons Co(CN)_6{}^{3-} + 6H_2O \tag{4.32a}$$

$$K = \frac{[Co(CN)_6{}^{3-}]}{[Co(H_2O)_6{}^{3+}][CN^-]^6} \tag{4.32b}$$

$$H^+ + OH^- \rightleftharpoons H_2O \tag{4.33a}$$

$$K = \frac{1}{[H^+][OH^-]} \tag{4.33b}$$

$$Zn^{2+} + 4Cl^- \rightleftharpoons ZnCl_4{}^{2-} \tag{4.34a}$$

$$K = \frac{[ZnCl_4{}^{2-}]}{[Zn^{2+}][Cl^-]^4} \tag{4.34b}$$

In Eqs. (4.30), (4.32), and (4.33), one of the reactants or products is water. Since it was specified that the reactions were carried out in dilute aqueous solutions, the symbol H_2O does not appear explicitly in the final equilibrium expressions [explained in connection with Eq. (4.23)].

Chemical equation (4.33a) represents the neutralization reaction of a strong acid, which provides H^+, and a strong base, which provides OH^-. The equilibrium expression (4.33b) represents the equilibrium for the formation of water from the ions. It is known that at 25 °C

$$K = \frac{1}{[H^+][OH^-]} = 1.0 \times 10^{14} \tag{4.35}$$

This is a very large number, corresponding to a very favorable reaction, that is, one that proceeds to an equilibrium position very far to the right side of the chemical equation.

Solids in Chemical Equilibria

You may notice that so far, the examples of chemical equilibria for which we have written expressions have all been concerned with just one phase, either a liquid solution or a gaseous solution. The concept of chemical equilibrium is not limited to such cases, however. **Heterogeneous reactions**, which are reactions involving more than one phase, can also be treated by the principles of chemical equilibrium. In a reaction involving both the liquid and the gaseous phases, the equilibrium expression can be written in terms of the concentrations of gaseous and liquid-solution species, just as in homogeneous reactions. For example, for the reaction

$$H_2(g) + C_8H_{16} \rightleftharpoons C_8H_{18}$$

in which hydrogen gas is bubbled into a liquid solution containing octene, C_8H_{16}, and the product octane, C_8H_{18}, is present in solution, we write

$$K = \frac{[C_8H_{18}]}{[H_2(g)][C_8H_{16}]}$$

For a solid involved in a heterogeneous reaction, the situation is different, although it resembles the involvement of water in an equilibrium in aqueous solution. For the aqueous solution, it is as though the concentration of water were replaced by 1, since the symbol for water does not appear in the final equilibrium expression (Eq. 4.27). For solids, the same kind of simplification is found; the symbol for a pure solid substance does not appear in the final equilibrium expression. Suppose we were interested in writing an equilibrium expression for the following heterogeneous reaction, in which a small amount of solid barium hydroxide is added to a large volume of an aqueous hydrochloric acid solution:

$$Ba(OH)_2(s) + 2H^+ \rightleftharpoons Ba^{2+} + 2H_2O \tag{4.36a}$$

In this case, neither the symbol of $Ba(OH)_2(s)$ nor of H_2O appears in the equilibrium expression:

$$K = \frac{[Ba^{2+}]}{[H^+]^2} \tag{4.36b}$$

The absence of a symbol for $Ba(OH)_2(s)$ in the equilibrium expression (4.36b) is often rationalized in a variety of ways; one reason is that the concentration of the solid substance is constant in its pure solid state. A really satisfactory explanation requires the concept of activity (Sec. 4.6).

EXAMPLE 4.8 Write the equilibrium expression for each of the following reactions:

$$\text{(a)}\quad 3Ca^{2+}(aq) + 2PO_4^{3-}(aq) \rightleftharpoons Ca_3(PO_4)_2(s)$$
$$\text{(b)}\quad 2Na(s) + 2H^+(aq) \rightleftharpoons 2Na^+(aq) + H_2(g)$$
$$\text{(c)}\quad Zn(s) + Cu^{2+}(aq) \rightleftharpoons Zn^{2+}(aq) + Cu(s)$$

Note that in these cases, since it was not stated in the introductory sentences that the liquid medium for the ions is aqueous, the symbol "aq" is indispensable in each equation if one wishes to convey such information.

SOLUTION Recognizing that the symbol for solids will not be included in the equilibrium expression, and that the appropriate quantities for gases and the solutes of liquid solutions are their molarities, we write

$$\text{(a)}\quad K = \frac{1}{[Ca^{2+}(aq)]^3[PO_4^{3-}(aq)]^2}$$

$$\text{(b)}\quad K = \frac{[Na^+(aq)]^2[H_2(g)]}{[H^+(aq)]^2}$$

$$\text{(c)}\quad K = \frac{[Zn^{2+}(aq)]}{[Cu^{2+}(aq)]}$$

STUDY PROBLEM 4(i)

Write the equilibrium expression for each of the following reactions, in which the liquid medium is aqueous:

$$\text{(a)}\quad Ba^{2+} + SO_4^{2-} \rightleftharpoons BaSO_4(s)$$
$$\text{(b)}\quad PbO(s) + 2H^+ \rightleftharpoons Pb^{2+} + H_2O$$

The absence of an explicit symbol for a pure solid substance included in an equilibrium expression is noticeable in a variety of equilibrium problems. One example is the chemical reaction that occurs when solid calcium carbonate is heated:*

$$CaCO_3(s) \rightleftharpoons CaO(s) + CO_2(g) \tag{4.37a}$$

Since the symbols representing $CaCO_3(s)$ and $CaO(s)$ do not appear in the equilibrium expression, we write

$$K = [CO_2(g)] \tag{4.37b}$$

We now use the ideal gas relation, $PV = nRT$, from which an expression for the concentration n/V is readily obtained:

$$\frac{n}{V} = \frac{P}{RT}$$

*Equation (4.37a) represents the chemical reaction that occurs when marble, which is largely $CaCO_3$, experiences elevated temperatures, as in a fire. The calcium oxide that forms has a granular, "crumbly" consistency. Hence, there have been many irreplaceable losses of marvelous marble statues carved by ancient Greek and Roman sculptors. In Italy, the current rapid deterioration of some marble statues is due to air pollution.

Spotlight

Lime mortar is one of many examples of how insoluble calcium salts are used as structural materials. Another, chemically more complex case is Portland cement. Lime functions as mortar in the following way. Calcium oxide is produced by driving CO_2 from limestone, $CaCO_3$, at high temperature, giving "quicklime," CaO(s), according to Eq. (4.37a). On the construction job, the calcium oxide is slaked with water, producing "slaked lime," $Ca(OH)_2$:

$$CaO(s) + H_2O \rightleftharpoons Ca(OH)_2$$

Because of excess water, the slaked lime is present as a paste. As the paste sets, the excess water is eliminated by evaporation. At the same time, CO_2 is absorbed from the air and reacts with the $Ca(OH)_2$ as follows:

$$Ca(OH)_2(s) + CO_2 \rightleftharpoons CaCO_3(s) + H_2O$$

This produces $CaCO_3(s)$, which constitutes the hard binding substance of the mortar. The mortar also usually contains two to three times as much sand as lime.

Thus,

$$[CO_2(g)] = \frac{n}{V} = \frac{P_{CO_2}}{RT}$$

where P_{CO_2} is the pressure due to CO_2, called the CO_2 partial pressure. Then, substituting this expression for P into Eq. (4.37b), we obtain

$$K = \frac{P_{CO_2}}{RT}$$

$$P_{CO_2} = KRT \tag{4.38}$$

Thus, for a given temperature, the CO_2 partial pressure above a $CaCO_3$–CaO system at equilibrium has a fixed value, KRT.

Another example of a heterogeneous equilibrium involving both a solid phase and an aqueous solution phase is the formation of solid lead carbonate from lead ions and carbonate ions in aqueous solution.

$$Pb^{2+} + CO_3^{2-} \rightleftharpoons PbCO_3(s) \tag{4.39}$$

This reaction can be carried out in the laboratory simply by mixing aqueous solutions of lead nitrate, $Pb(NO_3)_2$, and sodium carbonate, Na_2CO_3. The equilibrium favors the formation of $PbCO_3(s)$ overwhelmingly. This equilibrium is important in determining the chemical form and fate of polluting lead in streams; carbonate ion is always present to some extent in natural waters because CO_2 from the atmosphere is soluble therein. With the chemical equation written in the manner of Eq. (4.39), the view is towards the formation of a lead carbonate precipitate from ions in solution. In the opposite sense,

$$PbCO_3(s) \rightleftharpoons Pb^{2+} + CO_3^{2-} \tag{4.40a}$$

the equation is concerned with the formation of the ions from the solid, the process that occurs when part or all of the solid dissolves. For the equilibrium expression, we write

$$K = [Pb^{2+}][CO_3^{2-}]$$

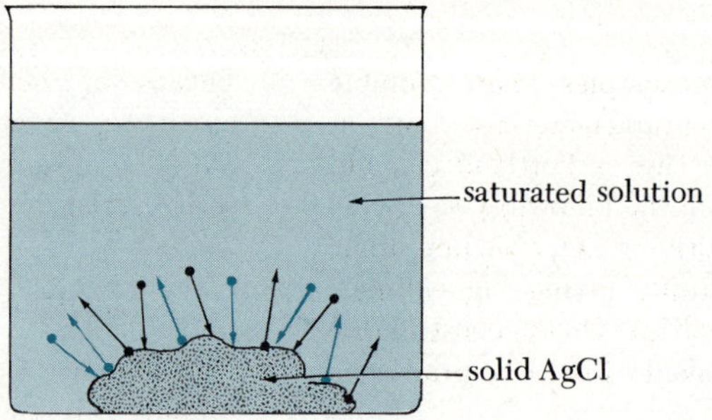

Figure 4.17 The equilibrium between solid AgCl and the ions Ag^+ and Cl^- in an aqueous solution that is *saturated* with AgCl. The Ag^+ and Cl^- ions are either going into solution from the solid or coming out of the solution to contribute to the solid. They are represented by black circles and color circles.

This equilibrium expression, for which it has been determined that K is 1.0×10^{-13} at 25 °C, gives the product of the concentrations of Pb^{2+} and CO_3^{2-} in equilibrium with solid $PbCO_3$. This product is always equal to that number if the ions are in equilibrium with solid at this temperature. No more $PbCO_3(s)$ can dissolve beyond what brings the product $[Pb^{2+}][CO_3^{2-}]$ to the value 1.0×10^{-13}. In other words, this K value determines the solubility limit of $PbCO_3(s)$. We know that the equation $[Pb^{2+}][CO_3^{2-}] = 1.0 \times 10^{-13}$ holds in a saturated solution of $PbCO_3(s)$—a solution in equilibrium with $PbCO_3(s)$ in which no more $PbCO_3(s)$ can dissolve. Hence, the value of the product $[Pb^{2+}][CO_3^{2-}]$ at equilibrium is popularly referred to as a solubility product, and given the symbol K_{sp}:

$$K_{sp} = 1.0 \times 10^{-13} = [Pb^{2+}][CO_3^{2-}] \tag{4.40b}$$

The **solubility product** K_{sp} is simply a special kind of equilibrium constant, one that characterizes solubility behavior quantitatively. Every solid that ionizes in this general way when it dissolves has its own characteristic K_{sp} value. From a knowledge of the K_{sp} value, one can compute directly the concentration of one of the constituent ions in a solution in contact with the solid if one knows the concentration of the other constituent ion. For example, if it is known that the concentration of the ion CO_3^{2-} in a certain stream is 2.4×10^{-10} *M*, then the maximum amount of Pb^{2+} that can coexist in solution in equilibrium with it in the stream can be calculated immediately:

$$1.0 \times 10^{-13} = [Pb^{2+}][2.5 \times 10^{-10}]$$

$$[Pb^{2+}] = \frac{1.0 \times 10^{-13}}{2.5 \times 10^{-10}} = 4.0 \times 10^{-4}\ M$$

This kind of information can be valuable in determining the form in which lead pollutants are transported through the aquatic environment.*

*The "effective solubility" of Pb^{2+}, or other metal ions, in water can be greatly increased by the formation of chemical complexes (see Chap. 18); formation of these complexes with species sometimes present in aquatic environments permits higher concentrations of "total lead" in solution, while the concentration of "free" Pb^{2+} is maintained at the relatively low value dictated by K_{sp} values. Lead salts and other pollutants can also exist as suspensions, or colloids, in a stream. Actual solubility in the water is therefore not an absolute requirement for the pollutant to exist in the stream and be transported by it.

4.5 Solubility Products

The Silver Chloride Solubility Equilibrium

Because of the complex acid-base chemistry in which the ion CO_3^{2-} participates in aqueous solutions (Sec. 5.5), applying Eq. (4.40b) is not always so straightforward as implied in the simple case just cited. Therefore, to explore the kinds of information one can derive from solubility products, we shall consider in more detail a less complicated example, the silver chloride solubility equilibrium:

$$AgCl(s) \rightleftharpoons Ag^+ + Cl^- \tag{4.41a}$$

$$K_{sp} = [Ag^+][Cl^-] \tag{4.41b}$$

A straightforward way to establish the equilibrium characterized by Eqs. (4.41a) and (4.41b) is to add an excess of solid silver chloride to water in a beaker and stir; one finds that a very small amount of AgCl(s) dissolves, generating the ions Ag^+ and Cl^-. When the AgCl(s) and the water are mixed, generation of these ions continues until an equilibrium is established between the ions and the solid substance from which they are derived. This means that ions are generated until their rate of formation from dissolving solid is exactly the same as the rate of formation of new crystals of AgCl(s) from the ions Ag^+ and Cl^- in solution. The resulting steady state is the equilibrium embodied in Eq. (4.41). This situation is pictured in Fig. 4.17, and the equation is explored in the following material.

EXAMPLE 4.9 The solubility of AgCl(s) in water at 25 °C is 1.1×10^{-5} mol per liter. Calculate K_{sp} for AgCl(s) at this temperature.

SOLUTION From the form of Eq. (4.41b), we see that the only data required for computing K_{sp} are the values of $[Ag^+]$ and $[Cl^-]$ at equilibrium. The meaning of the term *solubility* in this context is simply "the maximum amount of substance that can be dissolved in a specified volume of solvent." Hence, 1.1×10^{-5} mol is the maximum amount of AgCl(s) that can be dissolved in 1 liter of water at 25 °C. If one tries to dissolve a larger amount of solid, the material simply remains solid, settling to the bottom of the saturated dilute solution of the ions Ag^+ and Cl^-. The excess solid simply remains in equilibrium with the ions in solution, and it is this equilibrium that is governed by Eq. (4.41b). According to Eq. (4.41a), each mole of AgCl(s) that dissolves produces one mole of Ag^+ and one mole of Cl^- in solution. Thus, in a saturated solution of AgCl, the concentrations of these two ions are each 1.1×10^{-5} *M*. We can now substitute these values into Eq. (4.41b):

$$[Ag^+][Cl^-] = (1.1 \times 10^{-5})(1.1 \times 10^{-5})$$
$$K_{sp} = 1.2 \times 10^{-10}$$

STUDY PROBLEM 4(j)

The solubility of $BaSO_4$(s) in water at 25 °C is 0.00010 mol in 100 ml. Calculate K_{sp} for $BaSO_4$(s) at 25 °C.

Spotlight

The **gravimetric technique** for analyzing chloride ion, based on the reverse of Eq. (4.41a), is carried out by adding an excess of an aqueous solution of silver nitrate to a solution containing the unknown amount of chloride ion. The precipitate that forms is collected by filtration, and after being dried, is weighed accurately. The weight of AgCl(s) collected indicates directly the amount of chloride ion in the unknown.

In the **volumetric technique**, the unknown amount of chloride ion contained in a solution is determined by adding, dropwise and carefully, a solution of silver nitrate of precisely known concentration. By determining the precise volume of silver nitrate solution needed to react just completely with all the chloride ion, one can calculate directly the amount of chloride ion in the sample. When this volume of silver nitrate is reached, the next drop of solution added to the unknown solution does not react to form AgCl(s), since all the chloride ion has already been consumed. An indicator detects the occurrence of the excess Ag^+, and at this point the slow, dropwise addition of $AgNO_3$ solution is terminated. This point is the **endpoint**, and the entire dropwise addition is termed a **titration.**

Now, suppose that we have a saturated solution of silver chloride in water, i.e., one that is 1.1×10^{-5} *M* in Ag^+ and Cl^-, and suppose that to this solution we add a 0.5 *M* solution of sodium chloride, one that is concentrated enough so that a small amount added will alter the chloride ion concentration considerably without appreciably increasing the total volume of the solution. What, if anything, will happen when the saturated equilibrium solution is disturbed in this way? According to Le Chatelier's principle, an increase in chloride ion concentration is a stress on the equilibrium, a stress that will be relieved by a shift to a new equilibrium. The shift that can give this relief is the reaction of some silver ions with chloride ions to produce solid silver chloride, which reduces the temporary stress of an excess of chloride ions. At the same time, the concentration of Ag^+ is also reduced from its equilibrium value in the initial saturated solution. The solubility of silver chloride in water has thus been reduced by adding chloride ions from an external source because a portion of the initial number of silver ions in solution has now been removed from solution. In this phenomenon, called the **common ion effect**, one kind of ion is common to both the sparingly soluble electrolyte and the added electrolyte solution. In this case the common ion is Cl^-, the sparingly soluble electrolyte is AgCl, and the added electrolyte solution is aqueous NaCl.

Now let us suppose that we choose to perform a related experiment in which we start with a 0.5 *M* solution of sodium chloride and try to dissolve silver chloride in it. We are using a sodium chloride solution as the solvent, instead of pure water. We ask ourselves whether AgCl(s) will have as high a solubility in the aqueous sodium chloride as it does in pure water. In answering this question, we use the important principle that the final equilibrium is independent of the direction or manner of approach. So long as the total number of each type of ion (sodium ions, chloride ions, silver ions) and the total water molecules are the same for any two experiments, the same equilibrium will ultimately be achieved. In other words, the answer to our second experiment can be based on our answer to the first experiment. It doesn't matter whether the excess sodium chloride is present when we initially dissolve the silver chloride, or whether it is added afterwards; the net result will be the same. The solubility of silver chloride will be lower in the presence of an external source of the common ion, Cl^-. Le Chatelier's principle leads us to the conclusion that the presence or addition of a common ion reduces the solubility of an electrolyte.

These same conclusions could have been obtained through a more quantitative approach by the law of chemical equilibrium. Consider the experiment in which we add sodium chloride to a saturated silver chloride solution. In the saturated solution before it is disturbed, the concentrations of silver ions and chloride ions are each 1.1×10^{-5} *M*, according to Example 4.9. With the addition of sodium chloride, however, the chloride ion concentration necessarily increases; hence, the product of the silver ion concentration and chloride ion concentration no longer equals the equilibrium value, K_{sp}, and the equilibrium is destroyed. The only way equilibrium can be achieved again spontaneously is for this suddenly augmented product, $[Ag^+][Cl^-]$, to be decreased to the equilibrium value. This can be accomplished by a combination reaction between some silver ions and an equal number of chloride ions to form solid silver chloride, until the values of $[Ag^+]$ and $[Cl^-]$ are reduced to the point at which their product is again equal to K_{sp}. At this point the new equilibrium is reached, with the result that the final silver ion concentration is lower than it was before sodium chloride was added.

An analysis similar to the above can be developed for the case in which solid silver chloride is dissolved in a solution that already contains a substantial concentration of chloride ion. In that case, the silver ion and the chloride ion produced by the dissolution of silver chloride clearly cannot attain the levels that they would when solid AgCl was added to pure water since the solution already contains a substantial contribution of chloride ions; yet the product of $[Ag^+]$ and $[Cl^-]$ cannot exceed the K_{sp} value. These ideas are explored quantitatively in the following examples.

EXAMPLE 4.10 Compute the solubility of AgCl(s) in moles/liter in a 1.0 *M* aqueous NaCl solution and calculate the concentrations of Ag^+ and Cl^- in the saturated solution.

SOLUTION Before any AgCl(s) dissolves, the concentration of Cl^- is 1.0 *M*. Let us choose a convenient volume to consider in our calculation. One liter is convenient, because there is then a direct one-to-one numerical equivalence between the number of moles of a species present and its molarity. If into this 1 liter of solvent, y mol of AgCl(s) dissolves to give a saturated solution, then y equals the no. of mol/liter AgCl that dissolves, and the concentrations of Ag^+ and Cl^- increase each by y mol/liter. Then, when equilibrium is reached, the concentrations of Ag^+ and Cl^- are given by

$[Ag^+] = y$ (dissolved AgCl is the only source)

$[Cl^-] = 1.0 + y$ (dissolved AgCl plus a contribution from dissolved NaCl)

Now, substituting the K_{sp} value and these expressions for $[Ag^+]$ and $[Cl^-]$ into Eq. (4.41b) gives

$$[Ag^+][Cl^-] = 1.2 \times 10^{-10} \qquad (4.41c)$$

$$(y)(1.0 + y) = 1.2 \times 10^{-10} \qquad (4.41d)$$

There are two approaches to solving equations of this type, and both are considered here.

1. Expanding and rearranging the equation, we obtain

$$y^2 + 1.0y - 1.2 \times 10^{-10} = 0$$

This equation is a type well known in algebra, the quadratic equation,

$$ay^2 + by + c = 0$$

For the quadratic equation there are two solutions, given by the quadratic formula:

$$y = \frac{-b + \sqrt{b^2 - 4ac}}{2a} \quad \text{and} \quad y = \frac{-b - \sqrt{b^2 - 4ac}}{2a}$$

For the case in which we are interested, we make the following identifications: $a = 1$, $b = 1.0$, and $c = 1.2 \times 10^{-10}$. Substituting these values into the quadratic formula, we obtain

$$y = \frac{-1.0 + \sqrt{1.0 + 4.8 \times 10^{-10}}}{2} \quad \text{and} \quad y = \frac{-1.0 - \sqrt{1.0 + 4.8 \times 10^{-10}}}{2}$$

The second solution is negative, and while algebraically valid, is scientifically unacceptable; not less than zero mole/liter of AgCl(s) will dissolve, so y cannot be negative. Then, taking the first solution, we find

$$y = -0.50 + \frac{\sqrt{1.00000000048}}{2}$$

There are mathematical techniques for evaluating the above square root, giving 1.00000000024, from which we could obtain the value 1.2×10^{-10} for y. However, there is a more convenient mathematical approach that can be applied to many problems like this, in which equilibrium constants that have very large or very small values are involved; in this case, we have a very small value, 1.2×10^{-10}. This alternative approach is the method of successive approximations. It is shown here for this example.

2. The **method of successive approximations** starts with an estimate of the value of the unknown quantity in an expression in the equation of interest; this estimate is based on some qualitative chemical guideline. The next step is to use this estimated value in solving the equation, providing a new value of the unknown. This new value is then used in place of the original estimate in obtaining a second solution to the equation. This sequence is repeated until the value obtained in solving the equation is as close as one desires to the value obtained in the preceding step. Let us see how the method can be applied to the present example.

In the equation above that is to be solved, we look for any simple relation that is obvious without calculation, and that would simplify the problem by allowing us to make a guess. Such a relation is found in the relative magnitudes of 1.0 and y in the factor $(1.0 + y)$. From Example 4.9 we know that the solubility of AgCl(s) in pure water is only 1.1×10^{-5} *M*. From Le Chatelier's principle, we know that the solubility of AgCl(s) in 1 *M* NaCl must be less than 1.1×10^{-5} *M* due to the common ion effect. Thus, we can state that y is less than 1.1×10^{-5}. Using the mathematical symbol $<$ for "is less than," we write

$$y < 1.1 \times 10^{-5}$$

On this basis we can "guess" that y is *negligible* in comparison with 1.0 in the expression $(1.0 + y)$. That is, the value of y is so small that it is beyond the limit of significant figures in 1.0. Thus, we write the approximate relation

$$1.0 + y \cong 1.0$$

where the symbol $\cong$ means "is approximately equal to." Using this relation, we write

$$y(1.0 + y) \cong 1.0y$$

which when substituted into Eq. (4.41d) leads to

$$(y)(1.0 + y) \cong 1.0y \cong 1.2 \times 10^{-10}$$

Thus, we obtain our first solution,

$$y \cong 1.2 \times 10^{-10}$$

We now use this value of y in place of our initial guess of zero, and seek a second solution to Eq. (4.42d), by substituting the following approximation for $1.0 + y$:

$$1.0 + y \cong 1.0 + 1.2 \times 10^{-10}$$

into Eq. (4.42d),

$$y(1.0 + y) \cong y(1.0 + 1.2 \times 10^{-10}) \cong 1.2 \times 10^{-10}$$

$$y \cong \frac{1.2 \times 10^{-10}}{1.0 + 1.2 \times 10^{-10}}$$

Clearly 1.2×10^{-10} is negligible compared with 1.0 in the denominator of the expression on the right side (so our original approximation was valid), and again we obtain 1.2×10^{-10} for the value of y. Thus, the results of the two successive approximations, that is, the first and second solutions, are equal within the accuracy of two significant figures. The analysis is then complete. We note that the successive approximation method gave the same result to two significant figures, as the direct calculation by the quadratic formula. If a problem warrants greater accuracy, any desired level of precision can be obtained by a sufficient number of iterations of the cycle of successive approximations.

Returning to the meaning of y, we recall that its value was defined as equal to the number of moles of AgCl(s) that dissolve in 1 liter of the aqueous NaCl solution. This solubility is indeed much smaller than the solubility computed in Example 4.9 with pure water as the solvent. This gives us a quantitative measure of the common ion effect, due to Cl^- in this case. According to the equations above relating y to the Ag^+ and the Cl^- concentrations, these concentrations are given as

$$[Ag^+] = y = 1.2 \times 10^{-10} \text{ (mol/liter)}$$

$$[Cl^-] = 1.0 + y = 1.0 + 1.2 \times 10^{-10} = 1.0 \text{ (mol/liter)}$$

Suppose that we were faced with a problem of the kind treated in Example 4.10, but were concerned with predicting the solubility of AgCl(s) in a water supply that contains 1.0×10^{-6} *M* Cl^-. We should set up the problem as in Example 4.10, but in place of Eq. (4.41d), we should have

$$y(1.0 \times 10^{-6} + y) = 1.2 \times 10^{-10} \tag{4.41e}$$

Here is where we see a big difference in solving this problem compared with the solution given for Example 4.10. In Eq. (4.41e) it is *not* reasonable to approximate the expression $(1.0 \times 10^{-6} + y)$ by 1.0×10^{-6}, that is to neglect y in comparison with 1.0×10^{-6}. If we did, we should obtain from Eq. (4.41e)

$$y(1.0 \times 10^{-6}) = 1.2 \times 10^{-10}$$

$$y = \frac{1.2 \times 10^{-10}}{1.0 \times 10^{-6}} = 1.2 \times 10^{-4}$$

Clearly, 1.2×10^{-4} is not negligible compared with 1.0×10^{-6}; indeed, the former is 100 times bigger than the latter! Thus $1.0 \times 10^{-6} + y$ should not be approximated by 1.0×10^{-6}; in other words, y should not be neglected in comparison with 1.0×10^{-6}. Applying the quadratic formula to Eq. (4.41e) as in Example 4.10 gives $y = 1.1 \times 10^{-5}$, which is the solubility of AgCl(s) (in moles/liter) in this particular water source.

STUDY PROBLEM 4(k)

Knowing that K_{sp} for AgBr(s) is 3.3×10^{-13} at 25 °C, calculate the solubility of AgBr(s) at this temperature in (a) pure water; (b) a 0.60 *M* aqueous KBr solution.

EXAMPLE 4.11 Calculate the amount of precipitate formed and the concentration of ions present in solution when equilibrium is achieved after mixing 300 ml of 1.0×10^{-4} *M* $AgNO_3$ and 200 ml of 1.0×10^{-3} *M* NaCl.

SOLUTION First we consider the situation that would exist immediately after mixing, assuming that we could view the system before any solid AgCl is formed. At this point we should have a 500-ml solution containing the ions Ag^+ and NO_3^- at concentration (300 ml/500 ml) (1.0×10^{-4} *M*) and the ions Na^+ and Cl^- at concentration (200 ml/500 ml) (1.0×10^{-3} *M*). If no reaction occurred, the pertinent ions would have the concentrations:

$$[Ag^+] = 0.60 \times 10^{-4} \tag{4.42a}$$

$$[Cl^-] = 0.40 \times 10^{-3} \tag{4.42b}$$

The product $[Ag^+][Cl^-]$ for such a solution would have the value 2.4×10^{-8}, much larger than the K_{sp} value 1.2×10^{-10}. Since $[Ag^+][Cl^-] \neq K_{sp}$, this is clearly not an equilibrium, and a spontaneous process is initiated to bring the system to equilibrium by reducing the concentrations of Ag^+ and Cl^- in solution. This process is the precipitation reaction, which is the reverse of chemical equation (4.41a). The crucial question now is, How much precipitate forms? And the answer is, Just the amount that leaves sufficient Ag^+ and Cl^- in solution so that $[Ag^+][Cl^-] = 1.2 \times 10^{-10}$. A convenient way of determining this amount is to pretend that all the Ag^+ precipitates as AgCl(s), and then calculate the amount of this solid that dissolves according to Eq. (4.41a). Now, if the complete precipitation of Ag^+ occurred, the relevant concentrations would be:

$$[Ag^+] = 0 \tag{4.43a}$$

$$[Cl^-] = 4.0 \times 10^{-4} - 0.60 \times 10^{-4} = 3.4 \times 10^{-4} \tag{4.43b}$$

All the Cl^- cannot be used up in the precipitation since there is more Cl^- than Ag^+ in the system.

We can now consider the solubility of the AgCl(s) in this medium, which will give us the final concentrations of Ag^+ and Cl^- and the final amount of AgCl(s). Let us assume that to achieve equilibrium, y mol of the precipitated AgCl dissolves in the 500-ml solution that is characterized by Eq. (4.43). Then, at equilibrium, the number of moles/liter of AgCl that dissolves is

$$y \frac{\text{mol}}{0.5 \text{ liter}} = 2y \frac{\text{mol}}{\text{liter}}$$

$$[Ag^+] = 2y \tag{4.44a}$$

$$[Cl^-] = 3.4 \times 10^{-4} + 2y \tag{4.44b}$$

The factor 2 before the y in Eq. (4.44) indicates that y mol AgCl(s) dissolve in half a liter of solution. Then, using Eqs. (4.44a) and (4.44b) in Eq. (4.41b), we get

$$(2y)(3.4 \times 10^{-4} + 2y) = 1.2 \times 10^{-10} \tag{4.44c}$$

Now, we can initiate the method of successive approximations by neglecting, as a first guess, the quantity $2y$ compared with 3.4×10^{-4} in Eq. (4.44c):

$$(2y)(3.4 \times 10^{-4}) \cong 1.2 \times 10^{-10} \tag{4.44d}$$

$$y = \frac{1.2 \times 10^{-10}}{2(3.4 \times 10^{-4})} = 1.8 \times 10^{-7}$$

which is the end of the first cycle. Let us now use this value of y to make a second estimate of the approximated quantity, $3.4 \times 10^{-4} + 2y$ in Eq. (4.44c):

$$\begin{aligned}(3.4 \times 10^{-4} + 2y) &\cong 3.4 \times 10^{-4} + 2(1.8 \times 10^{-7}) \\ &= 0.00034 + 0.00000036 \\ &\cong 3.4 \times 10^{-4}\end{aligned}$$

If we now substituted this result for $3.4 \times 10^{-4} + 2y$ back into Eq. (4.44c), we should again obtain (4.44d), just as we did in the first cycle.

Hence, to two significant figures, which is all that is carried in this problem, we see that a second cycle of the method would give the same resultant value of y as the first cycle did. Thus, $y = 1.8 \times 10^{-7}$, and the final concentrations are

$$[Ag^+] = 2y = 3.6 \times 10^{-7} \tag{4.45a}$$

$$[Cl^-] = 3.4 \times 10^{-4} + 2y = 3.4 \times 10^{-4} \tag{4.45b}$$

Now, to determine the amount of AgCl(s) precipitate present at equilibrium, we note that the maximum number of moles of AgCl(s) that could possibly have formed (neglecting equilibrium considerations) is given by the information in Eq. (4.42a); this is the maximum concentration of Ag^+ if no precipitation occurred.

$$\begin{aligned} \text{maximum no. moles AgCl(s)} &= (0.60 \times 10^{-4}\ \text{mol/liter})(0.50\ \text{liter}) \\ &= 0.30 \times 10^{-4}\ \text{mol} \end{aligned}$$

The actual number of moles of Ag^+ in solution at equilibrium can be calculated from the information given by Eq. (4.45a):

$$(3.6 \times 10^{-7}\ \text{mol/liter})(0.50\ \text{liter}) = 1.8 \times 10^{-7}\ \text{mol}$$

Hence,

$$\begin{aligned} \text{no. moles AgCl(s)} &= 300 \times 10^{-7} - 1.8 \times 10^{-7} \\ &= 298 \times 10^{-7} \end{aligned}$$

or rounded off to two significant figures,

$$= 3.0 \times 10^{-5}\ \text{mol}$$

Therefore,

$$\begin{aligned} \text{no. g AgCl(s) precipitated} &= (3.0 \times 10^{-5}\ \text{mol})(143.3\ \text{g/mol}) \\ &= 4.3 \times 10^{-3}\ \text{g} \end{aligned}$$

STUDY PROBLEM 4(l)

Calculate the amount of precipitate formed and the concentration of ions remaining in solution after 150 ml of 2.0×10^{-4} M $AgNO_3$ and 300 ml of 1.2×10^{-3} M KCl are mixed together.

The AgCl(s) solubility example can be summarized in terms of the solubility product through the following representative situations:

1. In a saturated solution of AgCl in equilibrium with the solid, the product $[Ag^+][Cl^-]$ must equal the K_{sp} value, 1.2×10^{-10}.
2. If the product $[Ag^+][Cl^-]$ for a particular solution exceeds the value 1.2×10^{-10}, then the solution cannot be in equilibrium, since it contains more of the ions Ag^+ and Cl^- than a saturated solution can accommodate. Such a solution is said to be supersaturated; this means an unstable condition that must revert to equilibrium by the appropriate spontaneous process. That process is simply the combination of some Ag^+ and Cl^- ions to form solid AgCl, and this precipitation occurs spontaneously until the concentrations of Ag^+ and Cl^- are reduced to levels for which $[Ag^+][Cl^-] = 1.2 \times 10^{-10}$.
3. If the product $[Ag^+][Cl^-]$ for a particular solution is less than the K_{sp} value, there is no equilibrium between ions and solid, and it cannot be achieved by the formation of solid, which would decrease the product even more. Hence, no AgCl precipitate will form from such a solution.
4. If solid AgCl is mixed with pure water, equilibrium cannot exist until a small amount of AgCl(s) dissolves, just enough to bring the concentrations of Ag^+ and Cl^- in solution to the level at which the product $[Ag^+][Cl^-]$ equals K_{sp}.

Spotlight

One of the areas in chemistry in which solubility equilibria are important is **qualitative analysis**, identifying the constituents of a sample. The basic strategy can be seen in the scheme that is common for identifying individual metal ions in a solution that may contain many different ions. This scheme depends on adding a specific reagent under conditions that will cause only selected metal ions to combine with an anion of the added reagent, forming insoluble salts. For example, from a mixture containing the ions Ag^+ and Ca^{2+}, if an aqueous NaCl solution is added, AgCl(s) will precipitate

$$Ag^+ + Cl^- \rightleftharpoons AgCl(s)$$

Calcium chloride, $CaCl_2(s)$, will not precipitate, however, since AgCl(s) has a much smaller solubility product than $CaCl_2(s)$.

The accompanying diagram depicts a "qualitative analysis scheme" for a solution containing some combination of 18 different metal ions. The 18 possibilities are listed on the left side of the diagram. As indicated symbolically in the diagram, the first step in the scheme is treating the "unknown" solution with 0.1 *M* HCl. Three solid chlorides precipitate, Hg_2Cl_2, AgCl, and $PbCl_2$, because their solubility products, shown approximately in powers of ten, are exceeded (see rule 2 for the AgCl case). For all the other metal ions, the solubility products of the chloride salts (not shown) are much larger. After the solid Hg_2Cl_2, AgCl, and $PbCl_2$ are removed, by filtration or centrifugation (very rapid spinning, in which the denser phase settles to the bottom), the remaining solution still contains some unknown combination of the other 15 metal ions. This solution is then treated with a solution containing sulfide ion, S^{2-}, at a concentration of about 10^{-22} *M* (obtained in a solution containing 0.3 *M* HCl and H_2S; see Sec. 5.5). At this point, all the metal sulfides with solubility products substantially lower than about 10^{-24} (Bi_2S_3, CuS, CdS, HgS, SnS_2, and PbS from any Pb^{2+} ion that wasn't removed in the first step) precipitate, because their small solubility products are exceeded even with the $[S^{2-}]$ value of only 10^{-22} *M* (assuming metal ion concentrations of about 10^{-2} *M*). After these 6 metal sulfides are removed, the remaining solution is then treated with a solution containing 10^{-5} *M* S^{2-} (a solution containing $(NH_4)_2S$ and NH_3).

(Continued on page 164.)

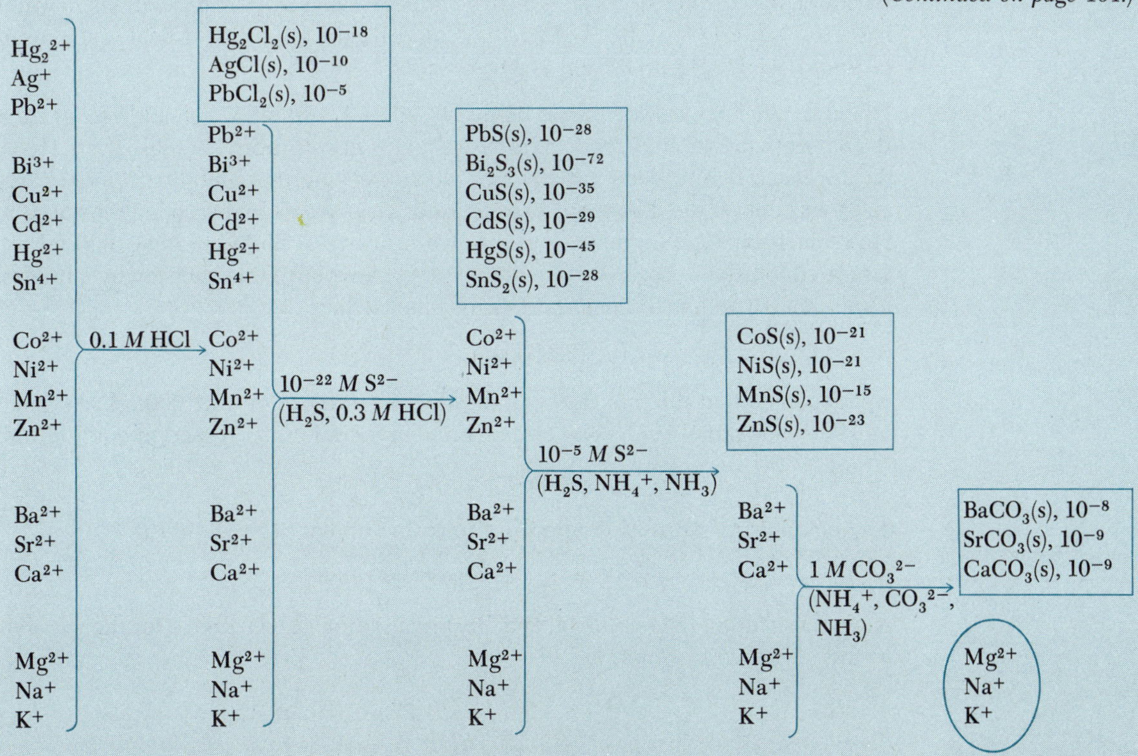

At this point, 4 additional metal sulfides, CoS, NiS, MnS, and ZnS precipitate; all have solubility products in the range 10^{-23} to 10^{-15}, values that are greatly exceeded in solutions containing 10^{-5} *M* S^{2-} and appreciable concentrations of Co^{2+}, Ni^{2+}, Mn^{2+}, or Zn^{2+}. (These four metal ions did not form precipitates earlier when treated with the 10^{-22} *M* S^{2-} solution, because such a low value of $[S^{2-}]$ is not enough to cause products like $[Zn^{2+}][S^{2-}]$ to exceed K_{sp} if $[Zn^{2+}] \leq 0.1$ *M*.) The sulfides of the remaining 6 metal ions have much larger solubility products, and remain in solution after the CoS(s), NiS(s), MnS(s), and ZnS(s) are removed by filtering or centrifuging. This solution is then treated with a solution containing 1 *M* CO_3^{2-}, and 3 solid metal carbonates ($BaCO_3$, $SrCO_3$, $CaCO_3$) precipitate, as their solubility products (10^{-8}, 10^{-9}, and 10^{-9}) are exceeded. The 3 metal ions remaining in solution, Mg^{2+}, Na^{+}, and K^{+}, do not form carbonate salts with small solubility products.

The net result of this scheme, based on solubility products, is that the 18 metal ions have been separated into 5 classes, corresponding to the 4 blocks and the oval in the diagram. Other experiments would be required to distinguish the individual metal ions within each of these classes.

Many modern chemical laboratories use a variety of instrumental techniques to perform qualitative analysis as described here. Especially important are spectroscopic approaches (see Chap. 6 and 8). The "sulfide scheme" is often used in teaching laboratories, because of the many chemical principles that can be learned from this scheme.

Other Solubility Cases

Solubility considerations have all been discussed here for one substance, AgCl, and one temperature, implicitly 25 °C. Similar considerations apply, however, to other substances, and to other temperatures.

EXAMPLE 4.12 Barium sulfate, which dissociates into the ions Ba^{2+} and SO_4^{2-} on dissolving in water, has a K_{sp} value 1.08×10^{-10} at 25 °C. Calculate the solubility of $BaSO_4$(s) in terms of grams $BaSO_4$(s) per 100 ml H_2O.

SOLUTION Whereas our final answer should be expressed in g/100 ml, we should recognize at the outset that solubility products are based on concentrations in moles/liter. Hence we approach the problem initially as if we are seeking the solubility in moles/liter, and then convert the final result to g/100 ml. Thus, we begin by asking the question, How much $BaSO_4$(s) in moles will dissolve in water at 25 °C to give 1000 ml of a saturated solution? Let us denote the unknown answer to this question by y (moles). Then, realizing that the net reaction of interest here is

$$BaSO_4(s) \rightleftharpoons Ba^{2+} + SO_4^{2-}$$

we recognize that if y mol $BaSO_4$(s) dissolves in water to give a 1000-ml solution, y mol Ba^{2+} and y mol SO_4^{2-} will be generated in the solution. Hence, at equilibrium

$$[Ba^{2+}] = [SO_4^{2-}] = y$$

The equilibrium expression appropriate to the above net reaction is

$$K_{sp} = [Ba^{2+}][SO_4^{2-}]$$

Now, substituting the numerical and algebraic symbols given above for the variables in this expression, we obtain

$$1.08 \times 10^{-10} = (y)(y)$$

$$y = \sqrt{1.08 \times 10^{-10}} = 1.04 \times 10^{-5}$$

As defined above, this is the value of the solubility of $BaSO_4(s)$ in moles/liter. What was requested in the statement of the problem is the solubility in g/100 ml. Clearly the solubility in moles/100 ml is one-tenth that number, that is, 1.0×10^{-6} mol/100 ml. Then,

$$1.04 \times 10^{-6}\frac{\text{mol}}{100\text{ ml}} = \left(1.04 \times 10^{-6}\frac{\text{mol}}{100\text{ ml}}\right)\left(233.4\frac{\text{g}}{\text{mol}}\right)$$

$$= 2.43 \times 10^{-4}\frac{\text{g}}{100\text{ ml}}$$

The concept of solubility product has important applications in a variety of scientific areas, from biology to geology. Examples 4.13, 4.14, and 4.15 indicate some specific situations in which these concepts apply.

EXAMPLE 4.13 Gout is a disease in man in which solid sodium acid urate, $NaC_5H_3N_4O_3$, is deposited in cartilage. The disease is associated with the production of excess uric acid, $C_5H_4N_4O_3$, in metabolism. The concentration of sodium ion in typical extracellular fluids in man is about 0.15 *M*. The solubility of solid sodium acid urate, a strong electrolyte, is about 0.085 g/100 ml in pure water at 25 °C. Calculate the minimum concentration of the acid urate ion, $C_5H_3N_4O_3^-$, that can cause formation of the solid deposits.

SOLUTION The overall approach here involves the equilibrium between the solid, denoted here by the symbol NaHU(s), and its ions, Na^+ and HU^-:

$$\text{NaHU(s)} \rightleftharpoons \text{Na}^+ + \text{HU}^-$$

As in the first three examples, the details of this solubility equilibrium are available from the corresponding K_{sp} value, defined by the equation

$$K_{sp} = [\text{Na}^+][\text{HU}^-]$$

Since the value of $[Na^+]$ is given in the statement of the problem, the value of $[HU^-]$ that would be in equilibrium with NaHU(s) can be calculated from a knowledge of K_{sp}. The value K_{sp} can be calculated from the statement of solubility in pure water in g/100 ml:

$$0.085\frac{\text{g}}{100\text{ ml}} = 0.85\frac{\text{g}}{\text{liter}} = \left(0.85\frac{\text{g}}{\text{liter}}\right)\left(\frac{1}{190.1}\frac{\text{mol}}{\text{g}}\right)$$

$$= 4.5 \times 10^{-3}\text{ mol/liter}$$

Thus, if one liter of pure water is saturated with sodium acid urate, each of the Na^+ and HU^- concentrations equals 4.5×10^{-3} *M*.* Using these values to compute K_{sp}, we obtain

$$K_{sp} = (4.5 \times 10^{-3})(4.5 \times 10^{-3}) = 2.0 \times 10^{-5}$$

*Because lithium acid urate is about three times as soluble as the sodium salt, an early remedy for gout was the partial replacement of Na^+ by Li^+ in the blood by ingestion of lithium salts. However, for this to be effective, such large amounts of lithium were required that lithium poisoning resulted.

Applying this K_{sp} value to the biological problem posed above, for which $[Na^+] = 0.15$, we obtain

$$2.0 \times 10^{-5} = (0.15)[HU^-]$$

$$[HU^-] = \frac{2.0 \times 10^{-5}}{0.15} = 1.3 \times 10^{-4}$$

Thus, 1.3×10^{-4} *M* is the minimum concentration of HU^- for which solid will precipitate when $[Na^+] = 0.15$. If $[HU^-]$ is less than 1.3×10^{-4}, then there will be no equilibrium with solid; solid will not form. Thus, 1.3×10^{-4} *M* corresponds to the saturation point, at which gout occurs.

EXAMPLE 4.14 Sea shells and the hard skeleton of coral are made up mainly of calcium carbonate, $CaCO_3(s)$. This substance is a strong electrolyte, ionizing into Ca^{2+} and CO_3^{2-} (the carbonate ion) on dissolution; it has a K_{sp} value 4.8×10^{-9}. For a region of the sea in which the calcium ion concentration is 0.015 *M*, calculate what the carbonate ion concentration would have to be if the shells and coral were to be in chemical equilibrium relative to the dissolution and precipitation of $CaCO_3(s)$.

SOLUTION The pertinent net reaction and solubility product are

$$CaCO_3(s) \rightleftharpoons Ca^{2+} + CO_3^{2-}$$

$$K_{sp} = 4.8 \times 10^{-9} = [Ca^{2+}][CO_3^{2-}]$$

In this case we do not have to solve for $[Ca^{2+}]$; it was given in the statement of the problem, ostensibly based on some direct determination or independent information on the seawater. (In this sense, this problem differs markedly from one in which the solubility of $CaCO_3(s)$ is to be computed in some medium; in such problems both concentrations are usually unknowns.) Using the given value of $[Ca^{2+}]$, we obtain

$$4.8 \times 10^{-9} = (1.5 \times 10^{-2}) \times [CO_3^{2-}]$$

$$[CO_3^{2-}] = 3.2 \times 10^{-7}$$

Thus, the equilibrium concentration of CO_3^{2-} would be 3.2×10^{-7} *M*.

In the cases so far considered, we have dealt only with electrolytes that give a single positive ion and a single negative ion from the ionization of each formula unit. Many electrolytes do not fit this category. An important example is calcium phosphate, $Ca_3(PO_4)_2$, the main constituent of bone. When solid samples of this substance dissolve sparingly in water, the following chemical equation applies:

$$Ca_3(PO_4)_2(s) \rightleftharpoons 3Ca^{2+} + 2PO_4^{3-} \qquad (4.46a)$$

When the solid is in equilibrium with the ions, and we have a saturated solution in contact with the solid, the law of chemical equilibrium (Eq. 4.4) applies. For this we write

$$K_{sp} = [Ca^{2+}]^3[PO_4^{3-}]^2 \qquad (4.46b)$$

where the coefficients 2 and 3 in chemical equation 4.46a appear as exponents in the mass action expression of Eq. (4.46b). A K_{sp} value about 1.0×10^{-25} has been determined for $Ca_3(PO_4)_2$.

EXAMPLE 4.15 Typical calcium ion concentrations in extracellular fluids in man are about 4×10^{-3} *M*. How far would the phosphate ion concentration have to drop, say in some body malady, before the decomposition of bone due to its dissolution according to Eq. (4.46a) would be expected?

SOLUTION Since the values of both K_{sp} and $[Ca^{2+}]$ are given, the only unknown in Eq. (4.46b) is the value $[PO_4^{3-}]$ requested here. Hence,

$$1.0 \times 10^{-25} = (4 \times 10^{-3})^3[PO_4^{3-}]^2$$

$$[PO_4^{3-}] = \sqrt{\frac{(1.0 \times 10^{-25})}{(4 \times 10^{-3})^3}}$$

$$= \sqrt{1.56 \times 10^{-18}}$$

$$= 1 \times 10^{-9}$$

At a phosphate ion concentration of 1×10^{-9} *M* and a calcium ion concentration of 4×10^{-3} *M*, the system Ca^{2+}–PO_4^{3-}–bone would be in chemical equilibrium. However, if the phosphate ion concentration should drop to a lower value (with the $[Ca^{2+}]$ value maintained constant), there would no longer be an equilibrium, and the $Ca_3(PO_4)_2(s)$ of the bone would begin to dissolve—to increase the value of $[Ca^{2+}]$ and $[PO_4^{3-}]$ until equilibrium is reestablished.

EXAMPLE 4.16 Calculate the solubility of $Ca_3(PO_4)_2(s)$ in pure water in g/100 ml, and the concentrations of the calcium and phosphate ions in solution at equilibrium.

SOLUTION Differing from the preceding example, this problem has as the only source of Ca^{2+} and PO_4^{3-} ions the dissolution of $Ca_3(PO_4)_3(s)$, according to Eq. (4.46a). It is convenient to consider one liter of water. Then, if y is the number of moles of $Ca_3(PO_4)_3(s)$ dissolving, we write the following algebraic expressions for the concentrations of ions present in the saturated solution:

$$[Ca^{2+}] = 3y \qquad \text{and} \qquad [PO_4^{3-}] = 2y$$

Substituting these expressions into Eq. (4.46b), we obtain

$$1.0 \times 10^{-25} = (3y)^3(2y)^2$$

The student should note that this expression is not $(3Ca^{2+})^3(2PO_4^{3-})^2$, which has no meaning! Nor is it $(3[Ca^{2+}])^3(2[PO_4^{3-}])^2$, which has a meaning but is incorrect. The correct expression above simply results from paying strict attention to the meanings of the symbols in Eq. (4.46b). The designation $[Ca^{2+}]$ means "the Ca^{2+} molarity," and the symbol $3y$ stands for it algebraically, according to our definition of y. An analogous statement applies to our using the symbol $2y$ for "the PO_4^{3-} molarity."

$$(27y^3)(4y^2) = 1.0 \times 10^{-25}$$

$$y^5 = \frac{1.0 \times 10^{-25}}{108} = 9.2 \times 10^{28}$$

$$y = (9.2 \times 10^{-28})^{1/5}$$

This equation can be solved by logarithms, which are defined as follows for **log *n*** (the logarithm of some number n)

$$10^{\log n} = n \tag{4.47}$$

Thus, ten raised to the power $\log n$ gives the number n. This applied to our example gives

$$\begin{aligned}\log y &= \log[(9.2 \times 10^{-28})^{1/5}] \\ &= \tfrac{1}{5}\log(9.2 \times 10^{-28}) \\ &= \tfrac{1}{5}(\log 9.2 + \log 10^{-28})\end{aligned}$$

Looking up the logarithm of 9.2 in Appendix D we obtain 0.9638, and substituting gives

$$\log y = \tfrac{1}{5}\{0.9638 - 28\} = \tfrac{1}{5}(-27.0362) = -5.407 = -6 + 0.593$$

Then again using Appendix D to find the number for which 0.593 is the logarithm, we obtain

$$\begin{aligned}y &= \text{antilog}(\log y) \\ &= \text{antilog}(-6 + 0.593) \\ &= [\text{antilog}(-6)][\text{antilog}(0.593)] \\ &= (10^{-6})(3.92) = 3.9 \times 10^{-6}\end{aligned}$$

Then,

$$[Ca^{2+}] = 3y = 12 \times 10^{-6}$$

and

$$[PO_4^{3-}] = 2y = 7.8 \times 10^{-6}$$

Hence, the Ca^{2+} concentration is 1.2×10^{-5} *M* and the PO_4^{3-} concentration is 7.8×10^{-6} *M*. Since y is numerically equal to the number of moles of $Ca_3(PO_4)_2(s)$ that dissolve per liter of water, we need only the mol wt of $Ca_3(PO_4)_2$ to express the result in the requested solubility units.

$$\begin{aligned}y\frac{\text{mol}}{\text{liter}} &= 3.9 \times 10^{-6}\frac{\text{mol}}{\text{liter}} \\ &= 3.9 \times 10^{-7}\frac{\text{mol}}{100\text{ ml}} \\ &= \left(3.9 \times 10^{-7}\frac{\text{mol}}{100\text{ ml}}\right)\left(310.2\frac{\text{g}}{\text{mol}}\right) \\ &= 1.2 \times 10^{-4}\frac{\text{g}}{100\text{ ml}}\end{aligned}$$

STUDY PROBLEM 4(m)

Calculate the solubility (in moles/liter) of $Ca_3(PO_4)_2(s)$ in a solution that is initially 0.010 *M* aqueous Na_3PO_4.

OPTIONAL

4.6 The Law of Chemical Equilibrium in Terms of Activities

Perspectives

As science progresses, the logical turn of events is to try to explain experimental facts or laws by a rigorous theory. The concept of chemical equilibrium and the law that describes it were discovered as generalizations of a large body of experimental data. Our development of these ideas in this chapter has followed this basically empirical approach. However, the law of chemical equilibrium can be derived directly from the principles of thermodynamics. **Thermodynamics** is the science of energy changes and related properties (see Chaps. 10, 11, and 12). Obtained by that approach, the law is a rigorous, generalized form that looks very similar to the form of Eq. (4.5). The difference is that instead of being a mathematical function of concentrations, it is an expression involving abstract thermodynamic quantities called activities. While it is not essential to use the concept of activity in carrying out the sort of equilibrium calculations typically experienced in a first-year chemistry course, this concept provides a theoretically more satisfactory basis for understanding equilibrium relations. Furthermore, it leads to a more satisfying justification of some of the algebraic simplifications we have used.

Activities

If we denote the activity of a species by the symbol A, then the rigorous, thermodynamic analog of Eq. (4.5) is

$$K = \frac{A_M{}^m A_N{}^n \ldots}{A_A{}^a A_B{}^b \ldots} \tag{4.48}$$

The **activity** of a species is a quantity that is developed in thermodynamic formalism to describe the tendency of a system in a specific state to undergo a transformation to a different state. It is sometimes referred to as a measure of the "escaping tendency" of a species, or the tendency of a species in a specific state to "escape" into another state, say from liquid to gas, or from one chemical form to another. While this is an abstract concept, it can be related quantitatively to experimental measurements. We shall not be concerned here with the details of these thermodynamic relations, but certain aspects of activity are easily summarized and are important to our understanding of the law of chemical equilibrium and its applications.

Standard States

One important fact is that the activity of a species represents its escaping tendency relative to some point of reference—the activity of a **standard state**. For example, it can be shown that the activity of water in a particular state is given by

$$A_{H_2O} = \frac{\text{escaping tendency of water in a particular state}}{\text{escaping tendency of water in its standard state}}$$

An analogous equation applies to any species in any state. Thus, we see that the activity of any species in its standard state is by definition unity. The choice of a

OPTIONAL

standard state is arbitrary but must be consistent with the standard states involved in other branches of chemical thermodynamics (see Chaps. 10, 11, and 12).

The accepted convention is to take the pure solid as the standard state for a solid material and the pure liquid for a liquid material. Alloys and concentrated liquid solutions are judged also by these criteria. This means that a pure solid or a pure liquid (one component) has an activity of unity, and wherever the activities of such substances occur in expressions such as K in Eq. (4.48), they are replaced by unity; hence, they do not appear explicitly. This gives formal justification for the fact that the symbol for the concentration of a pure liquid or pure solid does not appear explicitly in equilibrium expressions. Furthermore, since the concentration and properties of water in a dilute aqueous solution are nearly the same as for pure water, the activity of water in such solutions is nearly the same as the activity in its standard state, pure water. Hence, A_{H_2O} is often taken to be unity for such cases, and then does not appear explicitly in equilibrium expressions. For example, the equilibrium expression corresponding to Eq. (4.32a) would be written in terms of activities as follows:

$$K = \frac{A_{Co(NH_3)_6{}^{3+}} \cdot A^6_{H_2O}}{A_{Co(H_2O)_6{}^{3+}} \cdot A^6_{CN^-}} \cong \frac{A_{Co(NH_3)_6{}^{3+}}}{A_{Co(H_2O)^{3+}} \cdot A^6_{CN^-}} \tag{4.49}$$

Similarly, for a heterogeneous equilibrium involving a solid, for example the system represented by Eq. (4.41a), we set the activity of the pure solid to unity in the equilibrium expression written in terms of activities:

$$K = \frac{A_{Ag^+}A_{Cl^-}}{A_{AgCl(s)}} = A_{Ag^+}A_{Cl^-} \tag{4.50}$$

For species in the gas phase, the standard state of a substance is taken to be the substance with a partial pressure of 1 atm. The **partial pressure** is the contribution that a given gaseous species makes to the total pressure of a gas mixture, where the total pressure of the gas is the sum of all the partial pressures. Since it can be shown thermodynamically that the activity of a gaseous species is essentially proportional to its partial pressure, this choice of a standard state is equivalent to designating the activity of any gaseous species as the numerical value of its partial pressure in atmospheres (which is proportional to its concentration).

If the species of interest is present as a solute in a dilute solution, the standard state is taken to be a 1 *M* solution of that species.* Then, using the thermodynamic result that the activity of a solute species is approximately porportional to the concentration of the species, we find that the activity of any solute species is approximately the numerical value of the molarity of the species. For example, recalling that A_{H_2O} is approximately unity for dilute aqueous solutions, for the reaction represented in Eq. (4.30a), we should write

*The standard state chosen for solute species is a so-called hypothetical 1 *M* solution. This is an abstract thermodynamic concept with the following meaning. The hypothetical 1 *M* solution of a given solute has properties that a real 1 *M* solution would have if intermolecular interactions in the 1 *M* solution were the same (per molecule) as in an infinitely dilute solution; the hypothetical 1 *M* solution is essentially a solution for which the vapor pressure of the solute, for example, is 1 million times the vapor pressure of a solute in a 10^{-6} *M* solution.

$$K = \frac{A_{HCO_3^-}A_{OH^-}}{A_{CO_3^{2-}}A_{H_2O}} \cong \frac{[HCO_3^-][OH^-]}{[CO_3^{2-}]} \tag{4.51}$$

which is the same expression that was given in Eq. (4.30b). Similarly, Eqs. (4.49) and (4.50) become

$$K = \frac{[Co(NH_3)_6^{3+}]}{[Co(H_2O)_6^{3+}][CN^-]^6}$$

$$K = [Ag^+][Cl^-]$$

which are the same as Eqs. (4.32b) and (4.41b).

In all these cases, the activities are dimensionless numbers. They are numerically equal to partial pressures, concentrations, and so on for which specific units are assigned; but the activities themselves have no units. Hence, the equilibrium constant, which at equilibrium satisfies Eq. (4.48), is also dimensionless.

Clearly, the equilibrium expressions obtained from activities and the directions for evaluating activities give the same numerical formulas as those that one obtains by the approaches outlined in terms of molarity in Section 4.2; partial pressures are used instead of gas concentrations. These techniques let one apply the law of chemical equilibrium in a straightforward way to all kinds of chemical equilibria.

SUMMARY

Chemical equilibrium is a steady state in which the rates of the forward and reverse reactions are the same. Le Chatelier's principle tells us that when a system that is initially at equilibrium is suddenly perturbed so that the equilibrium is disturbed, the system responds if possible to minimize the effect of the disturbance. This principle is useful in accounting for changes in physical equilibrium, and for the common ion effect, the effect of sudden volume changes in gaseous reaction systems, and the effect of temperature changes on equilibria, and in predicting ways of improving the percentage yield of a reaction. The law of chemical equilibrium states that at equilibrium, the mass action quotient is a constant, the equilibrium constant K. The value of K is a characteristic of each reaction at a specific temperature and is a measure of the relative thermodynamic stabilities of reactants and products. For endothermic reactions, an increase in temperature causes an increase in the value of K, while for exothermic reactions, increasing temperature decreases K.

Equilibrium expressions are algebraic expressions and can be manipulated according to the rules of algebra. These manipulations bear a one-to-one correspondence with manipulations of balanced chemical equations. The algebraic character of equilibrium expressions also makes possible a wide variety of useful calculations that characterize the position of an equilibrium. Such calculations permit one to compute the concentrations of all pertinent solute species when equilibrium has been reached. One important type of equilibrium on which such calculations can be carried out is the solubility equilibrium, for which the equilibrium constant is called the solubility product. These calculations permit one to predict the solubilities of electrolytes and to account quantitatively for the common ion effect. One useful numerical technique for solving equilibrium problems is the method of successive approximations.

STUDENT CHECKPOINTS

After studying this chapter, the student should be able to:

1. Explain the concept of a steady state in both physical and chemical processes.
2. State Le Chatelier's principle and exemplify its effect on both physical and chemical equilibria.
3. Predict the effect of any sudden change on a system initially in a state of chemical equilibrium on the position of that equilibrium.
4. Write the equilibrium expression for any chemical reaction.
5. Describe qualitatively the relations between thermodynamic and kinetic stabilities and their relations with the equilibrium constant.
6. Define *catalyst,* and use a chemical reaction to illustrate a catalyst.
7. Manipulate equilibrium constants in direct correspondence with combining balanced chemical equations.
8. Calculate the position of equilibrium from equilibrium expressions.
9. Define solubility products and elaborate their use in predicting solubility behavior.
10. Explain the "common ion effect" at both a qualitative and a quantitative level.
11. Use the quadratic formula in equilibrium calculations.
12. Use the "method of successive approximations" in equilibrium calculations.

EXERCISES

4.1.* Indicate whether each of the following statements is true or false.
- (a) In a steady state, nothing is occurring in the system.
- (b) When a steady state is reached in a vapor pressure experiment, molecules are no longer escaping into the vapor phase.
- (c) The properties of a system do not change after a steady state is reached.

4.2. Give an example of the application of Le Chatelier's principle in which the stress applied to a system initially at physical equilibrium is caused by a sudden increase in volume. How does the system react?

4.3.* The melting of any solid substance requires energy added as heat. How is the solid-liquid equilibrium (for instance, ice cubes in water) affected by (a) the addition of heat? (b) the removal of heat?

4.4. Indicate whether each of the following statements is true or false.
- (a) Chemical equilibrium does not require a steady state.
- (b) The concentrations of all products must be equal at equilibrium.
- (c) All reactions require an addition of heat for chemical equilibrium to be achieved.
- (d) The odor of a solution does not change if the solution is in chemical equilibrium.

4.5.* When equilibrium is reached in a certain experiment in which a solution of *trans*-diiodoethylene is heated to 130 °C, it is found that the concentration of the cis isomer is 0.85 *M*. What is the equilibrium concentration of the trans isomer?

4.6. When equilibrium is reached in a certain experiment in which a solution of *cis*-diiodoethylene is heated to 130 °C, it is found that the amount of cis isomer present in solution is 0.54 g. How much of the trans isomer is present in solution?

4.7.* When equilibrium is reached in a certain experiment in which a solution containing 3.6 g of *trans*-diiodoethylene in 300 ml of cyclooctane, the solvent, is heated to 130 °C, what are the final concentrations of the cis and the trans isomers?

4.8. One heats a 500-ml solution of *trans*-diiodoethylene at 130 °C and wishes to produce 3.0 g of the cis isomer. How many grams of the trans isomer were required in the initial solution?

4.9.* One heats a solution containing 5.0 g *cis*-diiodoethylene for several hours at 130 °C and then quickly isolates all the trans isomer produced. What is the percentage yield in this synthesis of *trans*-diiodoethylene?

4.10. One heats a solution containing 7.0 g *trans*-diiodoethylene for several hours at 130 °C and then quickly isolates all the cis isomer produced. What is the percentage yield in this synthesis of *cis*-diiodoethylene?

4.11.* How many isomers are there for triiodoethylene, C_2HI_3?

4.12. Is there any starting combination of *cis*-diiodoethylene and *trans*-diiodoethylene that could lead to a trans-to-cis ratio different from 2.2:1 at equilibrium at 130 °C? Explain.

4.13.* Indicate whether each of the following statements relative to the *cis-trans* isomerization of diiodoethylene is true or false.

(a) If twice as much *cis*-diiodoethylene is used as a reactant, twice as much *cis*-diiodoethylene will be present when equilibrium is reached.

(b) If twice as much *cis*-diiodoethylene is used as a reactant, at equilibrium the ratio of trans-to-cis present will be half as large as it was initially.

(c) When equilibrium is reached in the *cis*- and *trans*-diiodoethylene system, the concentrations of both species are equal.

(d) When equilibrium is reached in the *cis*- and *trans*-diiodoethylene system, the rate of conversion of cis to trans is exactly the same as the rate of the trans-to-cis transformation.

(e) When equilibrium is reached in the *cis*- and *trans*-diiodoethylene system, no further chemical transformations occur.

4.14. Write the equilibrium expression for each of the following chemical reactions (occurring in liquid solution unless otherwise indicated):

(a) $CH_2O + H_2NOH \rightleftharpoons CH_2NOH + H_2O$

(b) $2H_2(g) + C_2H_2(g) \rightleftharpoons C_2H_6(g)$

(c) $2C_{19}H_{15} \rightleftharpoons C_{38}H_{30}$

(d) $Fe^{2+} + 6CN^- \rightleftharpoons Fe(CN)_6{}^{4-}$

(e) $Zn^{2+} + 4NH_3 \rightleftharpoons Zn(NH_3)_4{}^{2+}$

(f) $P_4(g) + 5O_2(g) \rightleftharpoons P_4O_{10}(s)$

(g) $4NH_3(g) + 5O_2(g) \rightleftharpoons 4NO(g) + 6H_2O(g)$

4.15.* Suppose that a solution of *cis*-difluoroethylene, $C_2H_2F_2$, in cyclooctane is heated for several hours at 130 °C and no *trans*-difluoroethylene is detected. All the starting material remains, unchanged. When the same type of experiment is carried out with *trans*-difluoroethylene initially present no cis isomer is detected after several hours. What can you conclude about the stabilities of the cis and trans isomers of difluoroethylene?

4.16. Suppose that two isomers, A and B, corresponding to a given molecular formula are known to exist. When a 1 *M* solution of A in CCl_4 solvent is heated to 55 °C, analysis shows that the concentrations of A remains at 1 *M*, and no B is detected. If an appropriate catalyst is added to the solution, however, the concentration of A is found to decrease and B is detected. The concentration of A continues to decrease and the concentration of B increases until after several hours, the concentrations no longer change, and the ratio of concentrations remains at $[A]/[B] = 11.4$. What can you conclude about the stabilities of the two isomers?

4.17.* How will the equilibria of the following gas phase reactions be affected by a sudden decrease in volume at constant temperature?

(a) $N_2O_4(g) \rightleftharpoons 2NO_2(g)$
(b) $2NO(g) + O_2(g) \rightleftharpoons 2NO_2(g)$
(c) $H_2O(g) + CO(g) \rightleftharpoons CO_2(g) + H_2(g)$
(d) $2H_2(g) + O_2(g) \rightleftharpoons 2H_2O(g)$
(e) $Cl_2(g) + PCl_3(g) \rightleftharpoons PCl_5(g)$

4.18. Indicate whether the following statements are true or false.

(a) If heat is required in a chemical reaction, a decrease in temperature will shift the equilibrium from right to left for the chemical equation as written.
(b) If heat is required in a chemical reaction, an increase in temperature causes an increase in the concentrations, or amounts, of products present at equilibrium.
(c) When the temperature is increased in an experiment in which a particular reaction is being studied, it is found that products begin appearing. From this we can conclude that the reaction proceeds with the absorption of heat.

4.19.* Indicate which of the following statements are true and which false.

(a) If the reaction indicated in Exercise 4.17a is initially at equilibrium, and the volume of the system is suddenly increased at the same temperature, the system responds with an increase in the amount of products.
(b) If the reaction indicated in Exercise 4.17a is initially at equilibrium and the volume of the system is suddenly increased at the same temperature, the system responds with an increase in the value of the equilibrium constant.
(c) If heat is released in a chemical reaction, a decrease in temperature causes an increase in the concentrations of products present at equilibrium.
(d) If heat is released in a chemical reaction, an increase in temperature causes a decrease in the value of the equilibrium constant.

4.20. The combustion of hydrogen [$2H_2(g) + O_2(g) \rightleftharpoons 2H_2O(g)$] occurs with the release of much heat—indeed explosively if a mixture of H_2 and O_2 is simply ignited. (a) How will the equilibrium constant for this reaction, which is likely to be important in future commercial energy schemes, be affected by an increase in temperature? (b) Is this likely to be an important consideration in the design of devices aimed at using the energy available? Why?

4.21.* Bleaches and swimming pool algacides often contain chlorine in some form. One form is sodium hypochlorite, NaClO, which is a strong electrolyte, from which the ion ClO^- can react with small concentrations of Cl^- present in the water supply, according to the reaction

$$H_2O + Cl^- + ClO^- \rightleftharpoons Cl_2 + 2OH^-$$

If a sudden breeze causes Cl_2 to evaporate from the surface of a swimming pool, thereby abruptly lowering the concentration of Cl_2 dissolved in the water, how will equilibrium be reestablished?

4.22. The following reaction proceeds in solution in benzene, C_6H_6, with the absorption of heat:

$$C_{38}H_{30} \rightleftharpoons 2C_{19}H_{15}$$

Indicate which of the following statements are true and which false.

(a) If this system is at equilibrium and the temperature is suddenly decreased, the concentration of $C_{38}H_{30}$ will decrease.

(b) If this system is at equilibrium and the temperature is suddenly increased, the system will respond with an increase in the equilibrium constant.

(c) If this system is at equilibrium and the concentration of $C_{19}H_{15}$ is suddenly decreased, say by adding a reagent that reacts with some of the $C_{19}H_{15}$ and removes it, the system will respond with a decrease in the concentration of $C_{38}H_{30}$ due to the occurrence of the reaction written above, from left to right.

(d) If this system is at equilibrium and the concentration of $C_{38}H_{30}$ is suddenly increased, as by merely adding more $C_{38}H_{30}$ to the solution, the system will respond with an increase in value of the equilibrium constant.

4.23.* Sulfur dioxide, SO_2, and sulfur trioxide, SO_3, are both commercially important gases. The dioxide is used as a preservative and the trioxide is important in the production of sulfuric acid ($SO_3 + H_2O \rightleftharpoons H_2SO_4$). The following chemical equations, with equilibrium constants K_1 and K_2, relate these two species:

$$O_2(g) + 2SO_2(g) \rightleftharpoons 2SO_3(g) \qquad K_1$$

$$SO_3(g) \rightleftharpoons SO_2(g) + \tfrac{1}{2}O_2(g) \qquad K_2$$

What is the mathematical relation between K_1 and K_2?

4.24. If K_1 is the equilibrium constant for the reaction (carried out in liquid solution at some particular temperature)

$$C_2H_2 + 2Br_2 \rightleftharpoons C_2H_2Br_4$$

and if K_2 is the equilibrium constant for the solution reaction

$$C_2H_2Br_2 + Br_2 \rightleftharpoons C_2H_2Br_4$$

carried out at the same temperature, then what is the value (in terms of K_1 and K_2) of the equilibrium constant for the reaction

$$C_2H_2 + Br_2 \rightleftharpoons C_2H_2Br_2$$

which is carried out in solution at the same temperature?

4.25.* At 450 °C, the equilibrium constant for the reaction

$$N_2(g) + 3H_2(g) \rightleftharpoons 2NH_3(g)$$

is 0.20 (referred to mol/liter). What is the value of K for the reaction

$$NH_3(g) \rightleftharpoons \tfrac{3}{2}H_2(g) + \tfrac{1}{2}N_2(g)$$

4.26. In a gaseous sample of HI at 520 °C, it is found that at equilibrium 3.1% of the iodine exists as I_2.

(a) What is the equilibrium constant at this temperature for the reaction

$$2HI(g) \rightleftharpoons H_2(g) + I_2(g)$$

(b) What is the equilibrium constant for the reaction

$$HI(g) \rightleftharpoons \tfrac{1}{2}H_2(g) + \tfrac{1}{2}I_2(g)$$

(c) What is the equilibrium constant for the reaction

$$H_2(g) + I_2(g) \rightleftharpoons 2HI(g)$$

(d) What is the concentration of H_2 in an equilibrium gas phase mixture at 520 °C in which the concentration of HI is 0.0044 M and of I_2 is 0.0022 M?

4.27.* Imagine that the gas phase reaction in Exercise 4.17b is at equilibrium. Helium gas, considered inert, is added to this system while its temperature and volume are held constant. How does the concentration of $O_2(g)$ change (if it does) to reestablish equilibrium?

4.28. The equilibrium constant for the reaction between $H_2(g)$ and $N_2(g)$ to form $NH_3(g)$ is about three times larger at 450 °C than it is at 500 °C. Is this reaction endothermic or exothermic?

4.29.* At a particular temperature, a 1.0-liter flask in which the system shown in Exercise 4.17e is in equilibrium is found to contain 0.050 mol PCl_3, 0.10 mol Cl_2, and 0.10 mol PCl_5.

(a) What is the equilibrium constant for this reaction?

(b) If 0.050 mol Cl_2 were added, what would the new concentration of $PCl_3(g)$ be when equilibrium is reestablished at the same temperature?

4.30. At 1273 °K, the equilibrium constant for the reaction

$$H_2O(g) + CO(g) \rightleftharpoons H_2(g) + CO_2(g)$$

is about 0.62.

(a) What is the equilibrium concentration of hydrogen if the concentrations of H_2O, CO, and CO_2 are all equal at 0.030 M?

(b) What is the final concentration of each component in an equilibrium mixture at 1000 °C formed from an initial mixture containing 0.060 mol H_2O and 0.060 mol CO in a 4.0-liter vessel?

4.31.* At 673 °K, the equilibrium constant for the reaction

$$N_2(g) + 3H_2(g) \rightleftharpoons 2NH_3(g)$$

is 0.51 (referred to moles/liter).

(a) Calculate the concentration of NH_3 present at 673 °K in an equilibrium system in which $[N_2] = 3.2 \times 10^{-2}$ *M* and $[H_2] = 6.0 \times 10^{-2}$ *M*.

(b) What would the equilibrium concentration of NH_3 be at 673 °K if one starts with 0.200 mol H_2 and 2.2×10^{-2} mol N_2 in a 3.0-liter vessel?

4.32. (a) In the reaction shown in Eq. (4.20c), what would be the effect on the yield of ethyl acetate (based on amount of acetic acid initially used) if a large excess of ethyl alcohol were used?

(b) If one starts with 10 mol ethyl alcohol and 0.10 mol acetic acid, what would be the absolute yield and the percentage yield of ethyl acetate when equilibrium is achieved?

4.33. If one dissolves 0.10 mol ethyl acetate and 10.0 mol water in a suitable solvent and waits for equilibrium to be achieved at 25 °C, what will the equilibrium concentrations of all four pertinent species be? (See Eq. 4.21.)

4.34.* Given the values of the equilibrium constants K_1 and K_2 for the first and second "complexation" reactions below, calculate the equilibrium constant K_3 for the third reaction.

$$Hg^{2+} + NH_3 \rightleftharpoons Hg(NH_3)^{2+} \qquad K_1 = 6.3 \times 10^8$$

$$Hg^{2+} + 2NH_3 \rightleftharpoons Hg(NH_3)_2{}^{2+} \qquad K_2 = 3.2 \times 10^{17}$$

$$Hg(NH_3)^{2+} + NH_3 \rightleftharpoons Hg(NH_3)_2{}^{2+} \qquad K_3 = ?$$

4.35. To a 0.020 *M* aqueous solution of mercuric nitrate, $Hg(NO_3)_2$, a strong electrolyte, a chemist adds enough ammonia for the resulting ammonia concentration to be 1.0 *M*. Using information from Exercise 4.34,

(a) Calculate the ratio $[Hg(NH_3)^{2+}]/[Hg^{2+}]$.

(b) Calculate the ratio $[Hg(NH_3)_2{}^{2+}]/[Hg^{2+}]$.

(c) What fraction of the mercury in this solution remains in the Hg^{2+} form?

4.36.* While Hg^{2+} is itself a toxic species, the neutral species $(CH_3)_2Hg$ is especially dangerous in animals; it is generated microbially in sludges and muddy lake bottoms that have been contaminated somehow with mercury. Structurally intermediate between Hg^{2+} and $(CH_3)_2Hg$ is the ion CH_3Hg^+, which can undergo a very favorable complexation reaction with ammonia—incidentally rendering the mercury less available for transformation into the highly dangerous $(CH_3)_2Hg$:

$$CH_3Hg^+ + NH_3 \rightleftharpoons CH_3Hg(NH_3)^+ \qquad K = 2.5 \times 10^8$$

If a laboratory sample of water obtained from the mud of a lake bottom is treated with a reagent so that at equilibrium the ratio $[CH_3Hg(NH_3)^+]/[CH_3Hg^+]$ is 1 million, what is the concentration of ammonia in the resulting solution?

4.37. How many grams of AgCl will dissolve in 1700 ml of water at 25 °C?

4.38.* The solubility of AgBr in water at 15° is 8.4×10^{-6} g per 100 cc of H_2O. Calculate K_{sp} for silver bromide at this temperature.

4.39. From the known solubility product of $BaSO_4$ (1.5×10^{-9}), what is the maximum concentration of Ba^{2+} that can exist at equilibrium in a solution that has a SO_4^{2-} concentration 0.50 *M*?

4.40.* From the known solubility product of silver iodide (1×10^{-16}), calculate the solubility (in moles/liter) of AgI in water.

4.41. A standard method for analyzing for sulfate ion in an aqueous solution is to treat the sample with a barium nitrate solution, and collect and weigh the resulting $BaSO_4$ precipitate. If the addition of the barium nitrate solution brings the value of $[Ba^{2+}]$ to 1.0 *M*, what is the error in this analysis due to SO_4^{2-} that is not precipitated and remains in solution (and hence is not accounted for in the weighing)?

4.42.* From the known K_{sp} value of lead chromate (2.0×10^{-14}), calculate the solubility (in moles/liter) of $PbCrO_4$ in each of the following solvents: (a) pure water; (b) a solution in which $[CrO_4^{2-}] = 0.00010\ M$; (c) a solution in which the chromate ion concentration is 0.50 *M*; (d) a 0.20 *M* $Pb(NO_3)_2$ solution.

4.43. Using information given in Exercise 4.42, compute the solubility (in grams) of lead nitrate in a 100-ml 0.30 *M* Na_2CrO_4 solution.

4.44.* Calculate the final concentrations of all ions present at equilibrium in a solution prepared by adding 0.100 mol NaCl, 0.300 mol KNO_3, 0.400 mol $NaNO_3$, 0.052 mol $AgNO_3$, and 0.020 mol KCl, and enough water to yield a 1-liter solution. How many grams of AgCl(s) are formed?

4.45. Using information given in Exercise 4.42, calculate the final concentrations of all ions present at equilibrium in a solution prepared by mixing 100 ml of a 0.14 *M* $Pb(NO_3)_2$ solution and 100 ml of a 0.27 *M* Na_2CrO_4 solution. How much $PbCrO_4$(s) is formed?

4.46.* From the known solubility product of $PbCl_2$ (1.7×10^{-5}), how many moles of $Pb(NO_3)_2$ will dissolve in 500 ml of a 0.60 *M* NaCl solution? What are the final concentrations of all ions?

4.47. How many grams of a 530-mg sample of solid $AgNO_3$ will remain undissolved when mixed with 1 liter of a 0.40 *M* KCl solution?

4.48. One wishes to prepare a solid sample of $PbCrO_4$ ($K_{sp} = 2.0 \times 10^{-14}$) from a 1.25-g sample of solid $Pb(NO_3)_2$. (a) The lead nitrate is dissolved in 10 ml water, and to this solution 10 ml of a 2 *M* Na_2CrO_4 solution is added. What is the percentage yield of the $PbCrO_4$ precipitate obtained (based on $Pb(NO_3)_2$)? (b) What would the percentage yield have been if 100 ml of the 2 *M* Na_2CrO_4 solution had been added?

4.49. Fluorite is a mineral with the formula CaF_2. Its solubility product is 2.3×10^{-10}. What is the solubility, in g per liter, of this mineral in water?

4.50.* A stagnant pond was once constructed for catching waste material from a factory. This pond, which no longer is used and has no outlet, contains 40,000 gal of very impure water. Among other impurities are Pb^{2+}, at a concentration of $2.5 \times 10^{-6}\ M$, and SO_4^{2-}, at a concentration of 4.0×10^{-4} *M*. Knowing that K_{sp} for lead sulfate is 1.2×10^{-8}, determine the percentage of evaporation of the pond that must take place before one might expect to find crystals of $PbSO_4$.

4.51. Explain what serious situation could develop in the skeletal system if through some combination of body dysfunction and dietary deficiencies, the Ca^{2+} concentration in body fluids dropped well below normal values. On what chemical principles do you base your conclusion?

ANSWERS TO SELECTED EXERCISES

4.1 (a) false; (b) false; (c) true. 4.3 (a) more solid melts; (b) more liquid freezes. 4.5 1.9 *M*. 4.7 [*cis*-$C_2H_2I_2$] = 0.013 *M*; [*trans*-$C_2H_2I_2$] = 0.030 *M*. 4.9 69%. 4.11 one. 4.13 (a) true; (b) false; (c) false; (d) true; (e) false. 4.15 no information on thermodynamic stabilities; both appear to be kinetically stable. 4.17 (a) shift to left; (b) shift to right; (c) no effect; (d) shift to right; (e) shift to right. 4.19 (a) true; (b) false; (c) true; (d) true. 4.21 by a shift to the right. 4.23 $K_1 = (1/K_2)^2$. 4.25 2.2. 4.27 no change. 4.29 (a) 20; (b) 0.040 *M*. 4.31 (a) 1.9×10^{-3} *M*; (b) 1.0×10^{-3} *M*. 4.34 5.1×10^8. 4.36 4.0×10^{-3} *M*. 4.38 2.0×10^{-13}. 4.40 1×10^{-8} mol/liter. 4.42 (a) 1.4×10^{-7} *M*; (b) 2.0×10^{-10} *M*; (c) 4.0×10^{-14} *M*; (d) 1.0×10^{-13} *M*. 4.44 $[Na^+] = 0.500$ *M*, $[K^+] = 0.320$ *M*, $[Ag^+] = 1.8 \times 10^{-9}$ *M*, $[Cl^-] = 0.068$ *M*, $[NO_3^-] = 0.752$ *M*; 7.45 g AgCl(s). 4.46 2.4×10^{-5} mol; $[Na^+] = [Cl^-] = 0.60$ *M*, $[Pb^{2+}] = 4.7 \times 10^{-5}$ *M*, $[NO_3^-] = 9.4 \times 10^{-5}$ *M*. 4.50 71% of the water must evaporate.

ANSWERS TO STUDY PROBLEMS

4(a) $K = \dfrac{[NO_2(g)]^2}{[NO(g)]^2[O_2(g)]}$; $K = \dfrac{[NO(g)]^2[O_2(g)]}{[NO_2(g)]^2}$; $K = \dfrac{[NO_2(g)]}{[NO(g)][O_2(g)]^{1/2}}$.

4(b) exothermic. 4(c) increase it. 4(d) (a) no change; (b) shift towards right. 4(e) Eq. (3) is the sum of Eqs. (1) and (2); $K_3 = K_1 \cdot K_2$. 4(f) 7.0×10^3. 4(g) 0.062 *M*. 4(h) $K = [HCO_2H(aq)]^2/[H_2C_2O_3(aq)]$; $K = [CH_3CO_2^-(aq)]/[CH_3CO_2H(aq)][OH^-(aq)]$. 4(i) $K = 1/([Ba^{2+}][SO_4^{2-}])$; $K = [Pb^{2+}]/[H^+]^2$. 4(j) 1.0×10^{-6}. 4(k) 5.7×10^{-7} mol/liter; 5.5×10^{-13} mol/liter. 4(l) $[Ag^+] = 1.6 \times 10^{-7}$ *M*, $[Cl^-] = 7.3 \times 10^{-4}$ *M*, 3.0×10^{-5} mol AgCl precipitates. 4(m) 3.3×10^{-8} mol/liter.

5

Acid-Base Equilibria in Aqueous Solutions

5.1 Scope and Importance

The law of chemical equilibrium and the computational methods associated with it have a scope that is as wide as the variety of chemical reactions. Within that broad scope, there is one chemical reaction that stands out as being of special interest to a wide range of chemical systems. This is the reaction of acids and bases in aqueous solutions.

Chemical equilibrium in the aqueous solution is of particular interest because water is such an important solvent in chemistry. Aqueous equilibria are involved in such diverse activities of practical importance as water softening, the formation of natural minerals like crystalline gems, stalagmites, and stalactites, and the transport and balance of many of the constituents in blood. The study of acids and bases in aqueous solution is of great interest because it can be applied in so many fields, ranging from industrial chemical production to subtle biological processes. From the submicroscopic point of view, it is difficult to consider any biological phenomenon in reasonable detail without becoming intimately involved with the concepts of acids and bases in aqueous solution. Many biological processes can occur only if a very precise acid-base balance is maintained in the organism. The nature of this relation and the way in which the balance is maintained are important aspects of modern biochemistry.

5.2 Concepts of Acids and Bases

Historical Overview

The qualitative concept of acids and bases seems to be nearly as old as recorded history. The concept has ranged in sophistication from the pictorial view harbored by the Greeks, who imagined acids as microscopic swords and bases as microscopic shields, to the more sophisticated theories of so-called hard and soft acids and bases, which are considered later, in Chapter 18. The early ideas were based on the accepted concept of an acid as something that produced a material with a sour taste when it was dissolved in water. Examples are vinegar and various mineral acids, such as "muriatic acid" (hydrochloric acid). Likewise, the idea of a base was associated with materials that gave rise to bitter solutions when they were dissolved in water. Historically, potash and lye (common terms for K_2O, K_2CO_3, KOH, and NaOH) were early models for this concept. As these ideas took hold, it also became apparent that substances classified as acids would undergo a transformation when mixed with materials classified as bases. In this transformation, both the acidic and the basic properties disappeared; that is, the acid and the base seemed to neutralize each other. This idea of neutralization of acids and bases is very old, and it was common to such old views as the sword-and-shield model of the Greeks.

The Arrhenius View

The Special Case of Electrolytes

The first modern theory of acids and bases is credited to the famous Swedish chemist Svante Arrhenius (1859–1927). He recognized on the basis of a large body of accumulated evidence that **acids could reasonably be classified as substances that give rise to hydrogen ions when dissolved in water.** Similarly, he recognized that common **bases could be classified as substances that give rise to hydroxide ions when dissolved in water.** According to the Arrhenius view, we can consider any acid, symbolized for our purposes by the formula HA, as a substance that provides, when it is dissolved in water, detectable concentrations of H^+. The H^+ ions arise because of an **ionization** reaction that can be summarized as

$$HA \rightleftharpoons H^+ + A^- \tag{5.1}$$

Some examples are

$$HCl \rightleftharpoons H^+ + Cl^-$$

$$HNO_3 \rightleftharpoons H^+ + NO_3{}^-$$

$$H_3PO_4 \rightleftharpoons H^+ + H_2PO_4{}^-$$

Similarly, bases were characterized by this view as substances whose behavior on dissolution in water can be represented by the symbolic equation

$$MOH \rightleftharpoons M^+ + OH^- \tag{5.2}$$

Some examples are

$$NaOH \rightleftharpoons Na^+ + OH^-$$

$$KOH \rightleftharpoons K^+ + OH^-$$

$$Ba(OH)_2 \rightleftharpoons Ba^{2+} + 2OH^-$$

According to this view, acids and bases are classified as **electrolytes**—particular types of electrolytes that give rise to the ions H^+ and OH^-, respectively, when ionization takes place. The ionization of an electrically neutral acid must give rise not only to H^+ but also to negative ions, since electrical neutrality of the resulting solution must be maintained. Similarly, the ionization of a basic electrolyte, which gives rise to hydroxide ions, must also maintain the appropriate electrical balance, as represented by the chemical equations given in the examples above.

The Strengths of Acids and Bases

One of the properties of acids or bases of both practical and theoretical interest is the strength of the acid or base. According to the Arrhenius point of view, strong acids are those that produce high concentrations of H^+ on dissolution in water and weak acids are those that produce small concentrations of H^+. Strong bases are those bases that give rise, when dissolved in water, to large concentrations of OH^-, small concentrations being generated by weak bases. In other words, one can view strong acids and bases as strong electrolytes of the acidic or basic type; whereas weak acids and bases can be viewed as weak electrolytes that produce the ions H^+ or OH^- on ionization. We are not mainly concerned here about the amount of an acid or base that will dissolve in a given amount of water. That is a matter of solubility, which is a related but different matter. What we are specifically concerned with here is how much of an acid or base that has dissolved actually ionizes once it is in solution—how much the substance acts like an electrolyte. From this point of view, the concepts of acid and base strength are perhaps more properly viewed as the percentage of ionization of an acid or a base when it is dissolved in water.

The term *strong acid* in common usage is usually reserved for substances that ionize to an extent closely approaching 100% when they are dissolved in water. Examples of typical strong acids are nitric acid, HNO_3; hydrochloric acid, HCl; hydrobromic acid, HBr; perchloric acid, $HClO_4$; and sulfuric acid, H_2SO_4. The ionization of each of these strong acids is represented by the corresponding chemical equations, which are given either in the above examples or below:

$$HBr \rightleftharpoons H^+ + Br^-$$

$$HClO_4 \rightleftharpoons H^+ + ClO_4^-$$

$$H_2SO_4 \rightleftharpoons H^+ + HSO_4^- \quad (5.3)$$

In the last of these chemical equations we note that only one hydrogen ion per H_2SO_4 formula unit is shown to ionize, producing the bisulfate ion, HSO_4^-. The bisulfate ion that is thus produced is itself an acid, which can dissociate according to the chemical equation

$$HSO_4^- \rightleftharpoons H^+ + SO_4^{2-} \quad (5.4)$$

This process occurs to an extent that is substantially less than 100%, however, and that strongly depends on the concentration of the sulfuric acid in the solution in question. The sum of the steps represented by Eq. (5.3) and (5.4) is the overall ionization of sulfuric acid, yielding two ions H^+:

$$H_2SO_4 \rightleftharpoons 2H^+ + SO_4^{2-} \quad (5.5)$$

Spotlight

Nitric acid, HNO_3, is formed naturally in small amounts in the atmosphere. The process is initiated by lightning, which forms oxides of nitrogen from O_2 and N_2. These oxides then dissolve in atmospheric water to form HNO_3. Nitric acid was prepared in 1650 by distilling a mixture of sulfuric acid and potassium nitrate, KNO_3, also called niter and saltpeter:

$$KNO_3(s) + H_2SO_4(l) \rightleftharpoons HNO_3(l) + KHSO_4(s)$$

This type of reaction is still one of the two commercial routes to HNO_3, except that sodium nitrate, $NaNO_3$, Chile saltpeter, is used because of the enormous deposits of this substance in Chile and Peru. At higher temperatures, the hydrogen sulfate salt formed in the above reaction reacts further with the nitrate:

$$KHSO_4(s) + KNO_3(s) \rightleftharpoons HNO_3(l) + K_2SO_4(s)$$

A more common commercial preparation is given below, and is referred to as the **Ostwald process,** the first step of which is the production of ammonia by the Haber process (Sec. 4.4).

(a) $N_2(g) + 3H_2(g) \rightleftharpoons 2NH_3(g)$
(b) $2NH_3(g) + \frac{5}{2}O_2(g) \rightleftharpoons 2NO(g) + 3H_2O$
(c) $2NO(g) + O_2(g) \rightleftharpoons 2NO_2(g)$
(d) $3NO_2(g) + H_2O(l) \rightleftharpoons 2H^+ + 2NO_3^- + NO(g)$

The nitric acid that is formed in the last step is drawn off and concentrated, and the NO is recycled to the third step (c). The sum of all these steps (3a + 3b + 3c +3d) is

$$3N_2(g) + 9H_2(g) + \tfrac{21}{2}O_2(g) \rightleftharpoons 4H^+ + 4NO_3^- + 2NO(g) + 7H_2O$$

Pure liquid HNO_3 undergoes autoionization; that is, two molecules of HNO_3 ionize each other:

$$2HNO_3(l) \rightleftharpoons NO_2^+ + NO_3^- + H_2O$$

In some respects this is analogous to the ionization of water (Eq. 5.32). Pure HNO_3 is a colorless liquid, and the slight yellow color sometimes present in actual samples is due to decomposition to NO_2 by light:

$$4HNO_3(l) \rightleftharpoons 4NO_2(g) + 2H_2O + O_2(g)$$

This decomposition also takes place when HNO_3 is heated, and it is quite marked above 68 °C.

Because the second step [Eq. (5.4)] does not occur completely in aqueous solution, the overall process [Eq. (5.5)] is not complete when sulfuric acid is dissolved in water, although virtually all of the H_2SO_4 ionizes to at least the HSO_4^- stage.

Some acids ionize much less than the bisulfate ion. Such acids are called **weak acids.** Examples of these are acetic acid, which we shall represent by the formula CH_3CO_2H; nicotinic acid, with the formula $C_5H_4NCO_2H$; carbolic acid (phenol), with a formula C_6H_5OH; bicarbonate ion, HCO_3^-; and water itself. The action of these substances as acids is represented by the chemical equations

$$CH_3CO_2H \rightleftharpoons H^+ + \underset{\text{acetate ion}}{CH_3CO_2^-}$$

$$C_5H_4NCO_2H \rightleftharpoons H^+ + C_5H_4NCO_2^-$$

$$C_6H_5OH \rightleftharpoons H^+ + C_6H_5O^-$$

$$HCO_3^- \rightleftharpoons H^+ + CO_3^{2-}$$

$$H_2O \rightleftharpoons H^+ + OH^- \qquad (5.6)$$

These processes occur to extents that correspond to less than a small percentage of ionization in 1 *M* solutions of the acids (or in pure water for the last case). The first three examples of these weak acids are typical examples of thousands of weak acids that occur naturally in living systems. The first of these, acetic acid, is the main acidic constituent in vinegar and results from the fermentation of sugars past the alcohol stage. The second of these acids, nicotinic acid, occurs in a variety of living systems, and is popularly known for its existence in tobacco. Phenol, or carbolic acid as it is referred to frequently in pharmacies, is used in a variety of popular remedies. Each of these weak acids is a representative of many thousands of analogous substances of various complexities that occur in biological systems.

The manifestation of weak acid characteristics by water itself is an interesting and important phenomenon. While HCl in a 1 *M* solution is approximately 100% ionized, and HF in a 1 *M* solution is about 3% ionized, and acetic acid in a 1 *M* solution is about 1% ionized, pure water is only about 2×10^{-7}% ionized. Although this is a very small value compared with the others, we shall see repeated examples of the importance of the small concentrations of H^+, and OH^- also, due to the ionization of water.

Most examples of strong Arrhenius-type bases are compounds of the metal hydroxide type. Examples of strong bases that dissociate to approximately 100% in 1 *M* solution are sodium hydroxide, potassium hydroxide, and the other alkali metal hydroxides; we recall that the alkali metals are those elements of the first column of the periodic table. Examples of weaker bases are silver hydroxide, AgOH; lead hydroxide, $Pb(OH)_2$; and zinc hydroxide, $Zn(OH)_2$. The dissociation reactions responsible for the basic character of these substances are represented in the corresponding chemical equations.

$$NaOH \rightleftharpoons Na^+ + OH^-$$

$$AgOH \rightleftharpoons Ag^+ + OH^-$$

$$\left.\begin{aligned} Pb(OH)_2 &\rightleftharpoons PbOH^+ + OH^- \\ PbOH^+ &\rightleftharpoons Pb^{2+} + OH^- \end{aligned}\right\} \text{sum } Pb(OH)_2 \rightleftharpoons Pb^{2+} + 2OH^-$$

$$\left.\begin{aligned} Zn(OH)_2 &\rightleftharpoons ZnOH^+ + OH^- \\ ZnOH^+ &\rightleftharpoons Zn^{2+} + OH^- \end{aligned}\right\} \text{sum } Zn(OH)_2 \rightleftharpoons Zn^{2+} + 2OH^-$$

The ionization of lead hydroxide and zinc hydroxide can each be thought of as two discrete steps, the sum of which is the overall dissociation that can give two hydroxide ions. These two steps have different tendencies to occur (different *K* values).

Representative of many weak bases is ammonia, NH_3. Ammonia qualifies as an Arrhenius base by the following reaction that it undergoes with water:

$$NH_3 + H_2O \rightleftharpoons NH_4^+ + OH^- \tag{5.7}$$

This reaction occurs to only a small extent, and hence gives rise to only a small concentration of OH^-.*

*In some contexts, especially in older publications, the weakly basic character of aqueous ammonia was attributed to a hypothetical species called ammonium hydroxide, NH_4OH. This was viewed as a species that produced small concentrations of hydroxide ion by the following ionization reaction:

$$NH_4OH \rightleftharpoons NH_4^+ + OH^-$$

Modern techniques have clearly shown that aqueous ammonia solutions do not contain NH_4OH as a primary species. Hence, we represent aqueous ammonia solutions as simply $NH_3(aq)$, usually deleting "aq" for convenience.

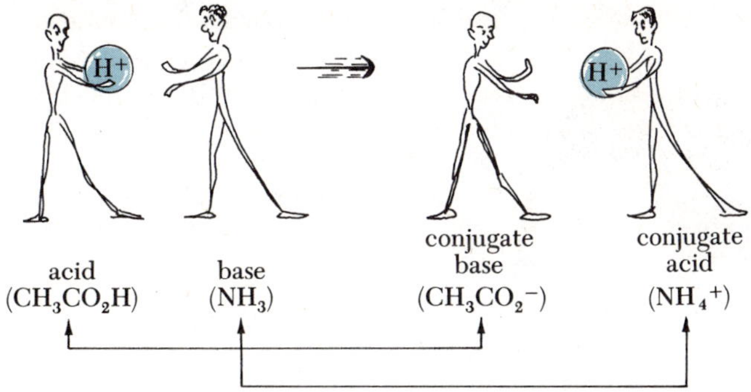

Figure 5.1
Acid-base chemistry as a "transaction".

The Brönsted-Lowry Approach

A Commerce in Protons. Proton Donors and Acceptors

A more general approach to acid-base phenomena than the Arrhenius view is called the Brönsted-Lowry approach. The Brönsted-Lowry approach, developed toward the end of the last century, focuses attention on the ion H^+, or a **proton**, and does not explicitly involve OH^- in its basic formulation. According to the **Brönsted-Lowry theory, an acid is a proton donor, that is, a source of H^+. A base is any proton acceptor, that is, any species that can attach itself to an H^+.** According to this view, acid-base phenomena are couched in a language similar to that of commerce, where the currency of exchange in acid-base chemistry is the proton. Any acid-base process simply involves a transaction in which the proton donor transfers a proton to the proton acceptor, as indicated pictorially in Fig. 5.1. To take a specific example, we can consider the reaction between the weak acid, acetic acid, and the weak base, ammonia:

$$CH_3CO_2H + NH_3 \rightleftharpoons CH_3CO^- + NH_4^+ \tag{5.8}$$

Conjugate Acids and Bases

We note that on the left side of Eq. (5.8) we have both the proton donor, CH_3CO_2H, and the proton acceptor, NH_3. On the right side we have the products of this proton transfer, $CH_3CO_2^-$ and NH_4^+. The ammonium ion, NH_4^+, could be transformed into ammonia merely by donating a proton to some other species; in other words, as a result of the acid-base process shown in Eq. (5.8), we have transformed the proton acceptor, NH_3, into a species that can be considered a proton donor, NH_4^+. We also recognize that the acetate ion, $CH_3CO_2^-$, can again become acetic acid by merely accepting an ion H^+ from some source; that is, the acetate ion is potentially a proton acceptor, a base. Hence, the acid-base process represented in Eq. (5.8) has transformed the proton donor, CH_3CO_2H, into a proton acceptor, $CH_3CO_2^-$, and the proton acceptor, NH_3, into the proton donor, NH_4^+. This is characteristic of all acid-base reactions viewed by the Brönsted-Lowry approach, and leads to a specific terminology that emphasizes this relation. In that terminology, we refer to the proton donor produced by the transfer of a proton to the reactant base as the **conjugate acid of the base.** Thus, in Eq. (5.8), the base we start with on the left side is NH_3; in the proton transfer process the NH_3 is transformed into the proton donor NH_4^+, and this is referred to as the conjugate acid of the NH_3 base. Similarly, the

Spotlight

The famous Danish physical chemist, Johannes Brönsted (1879–1947), is best known for his work on acid-base theory. Much of his work in this area concerned the effects of acids and bases on the rates of reactions (chemical kinetics, which is discussed extensively in Chap. 13). In addition, he made many contributions to other areas which are discussed in this text, including thermodynamics (Chaps. 10 and 11), electrochemistry (Chap. 12) and the properties of ions in solution (Chap. 15). Brönsted was known as a careful and accomplished experimentalist whose results provided a severe test for the theories being developed by other chemists in the first decades of the twentieth century. It is interesting to note that one of his schoolmates and closest friends, Niels Bjerrum, also became a leading Danish physical chemist.

United Press International Photo

proton acceptor on the right side of the equation, $CH_3CO_2^-$, is referred to as the conjugate base of acetic acid. We summarize these conjugate relations as follows:

Conjugate Pairs

Acid	*Base*
CH_3CO_2H	$CH_3CO_2^-$
NH_4^+	NH_3

This relation between the species on the left side and the species on the right side of the balanced net equation can be written for any acid-base process viewed by the Brönsted-Lowry approach. If we represent any Brönsted-Lowry acid by the symbol "HA" and any Brönsted-Lowry base by the symbol "B," then we can generalize these concepts by the following equation:

$$\underset{\text{acid}}{HA} + \underset{\text{base}}{B} \rightleftharpoons \underset{\text{base}}{A^-} + \underset{\text{acid}}{HB^+} \tag{5.9}$$

In this balanced equation, HA is referred to as the acid conjugate to the base A^-, and we say that the base A^- is conjugate to the acid HA. A similar relation exists between the base B and its conjugate acid HB^+. These relations are emphasized by the arrows shown in Eq. (5.9) and in Fig. 5.1.

In the generalized acid-base equation (5.9), the symbol "HA" is arbitrarily chosen to appear neutral, which means that its conjugate base A^- appears to have a negative charge. This does not mean to imply that all conjugate acid-base systems bear this electrical charge relation. Indeed, the symbol chosen arbitrarily for the base B gives it the appearance of being electrically neutral, a choice that gives the conjugate acid BH^+ a symbol that conveys the appearance of a positive charge. This particular collection of symbols emphasizes that Brönsted-Lowry acids and bases can be charged or uncharged. Of course, the acid form of a conjugate acid-base pair always has an electrical charge greater (more positive) than the base form by +1.

The Brönsted-Lowry approach and the chemical equations used to represent it (Eqs. 5.8 and 5.9) give a more graphic view of the essence of acidity and basicity than the simpler, Arrhenius approach. In the latter, acidity is viewed as the tendency for protons to be *made available* from acids through the ionization represented in Eq. (5.1), and basicity is viewed in terms of the availability of OH^-. In the Brönsted-Lowry approach, acidity is viewed as the tendency of a proton to be *transferred* from the acid to a base, which is not necessarily a source of hydroxide ions; the fate of the proton is then represented in the right side of the chemical equation in that it is shown to be a part of the new acid that has been formed (HB^+ in Eq. 5.9). This latter approach is generally closer to reality than the Arrhenius ionization approach in describing an actual net acid-base process, and need not be restricted to aqueous solutions. Actually, for aqueous solutions both approaches can be taken along in parallel to give the same general conclusions in equilibrium calculations; but the Brönsted-Lowry approach is more useful in maintaining proper perspective about the relation between chemical equations and the reality of chemical systems.

5.3 Acids and Bases in Water

Ionization in the Brönsted-Lowry View

In the Arrhenius view, acid character is manifested in water by ionization, following equations of the form shown in Eq. (5.1). The Brönsted-Lowry approach views the same process as a donation of a proton from the acid in question (symbolized by HA) to the base, H_2O; for example,

$$\underset{\text{acid}}{HA} + \underset{\text{base}}{H_2O} \rightleftharpoons \underset{\text{acid}}{H_3O^+} + \underset{\text{base}}{A^-} \qquad \text{(generalized)} \qquad (5.10)$$

$$\underset{\text{acid}}{HCl} + \underset{\text{base}}{H_2O} \rightleftharpoons \underset{\text{acid}}{H_3O^+} + \underset{\text{base}}{Cl^-} \qquad (5.11)$$

$$\underset{\text{acid}}{HSO_4^-} + \underset{\text{base}}{H_2O} \rightleftharpoons \underset{\text{acid}}{H_3O^+} + \underset{\text{base}}{SO_4^{2-}} \qquad (5.12)$$

The acid character of any acid HA in water is viewed in terms of its tendency to transfer a proton to water, thereby producing the conjugate base A^- and the conjugate acid of water, H_3O^+, called the **hydronium ion.** Protonated water is given the symbol H_3O^+, more or less for convenience. The actual structure of protonated water is at least as complicated to describe as the structure of water itself (see Chap. 8); it is highly mobile, each positive center changing very rapidly from one structure to another. On the average, a reasonable representation is $H^+(H_2O)_4$, or $H_9O_4^+$. Nevertheless, the symbol H_3O^+ conveys the same general meaning, protonated water, and is less cumbersome; so it is generally preferred in modern presentations. We recognize that just as Eq. (5.1) of the Arrhenius approach means the same as Eq. (5.10) of the Brönsted-Lowry approach, the symbol H^+ in the former view has the same meaning as the symbol H_3O^+ in the latter view. In either view, a strong acid is seen essentially as completely ionized, generating large concentrations of H^+ or H_3O^+. The same species that qualify as acids by the Arrhenius approach are also Brönsted-Lowry acids, and vice versa.

The Brönsted-Lowry (BL) view of a base, simply a molecule or ion capable of accepting H^+, is considerably more general than the Arrhenius view, which relies on the ability to produce OH^- in solution. Clearly, the BL view of bases readily includes the ion OH^-, since it is an excellent proton acceptor, hence a strong base.

$$\underset{\text{acid}}{HA} + \underset{\text{base}}{OH^-} \rightleftharpoons \underset{\text{acid}}{H_2O} + \underset{\text{base}}{A^-} \tag{5.13}$$

The BL approach also readily incorporates species like NH_3 as weak bases, because of their ability to accept protons.

$$\underset{\text{acid}}{HA} + \underset{\text{base}}{NH_3} \rightleftharpoons \underset{\text{acid}}{NH_4^+} + \underset{\text{base}}{A^-} \tag{5.14}$$

Thus, all Arrhenius bases are also Brönsted-Lowry bases, but the reverse is true only for compounds containing ionizable hydroxides or for aqueous solutions (where reactions like the one shown in Eq. (5.7) can generate OH^- in solution).

Spotlight

Several metal salts, it is well known, provide acidic solutions in water. This property is due to reactions of the metal ions that can formally be written as follows for some representative cases:

$$BeF_2(s) + 6H_2O \rightleftharpoons Be(OH)_3^- + 3H_3O^+ + 2F^-$$

$$AlCl_3(s) + 2H_2O \rightleftharpoons AlOH^{2+} + H_3O^+ + 3Cl^-$$

$$ZnCl_2(s) + 2H_2O \rightleftharpoons ZnOH^+ + H_3O^+ + 2Cl^-$$

Some metal oxides and hydroxides exhibit a property called **amphoterism**; such species are able to react with both bases and acids, since they have both acidic and basic properties. Examples are $ZnO(s)$ and $Al(OH)_3(s)$:

$$\underset{\text{base}}{ZnO(s)} + 2H_3O^+ \rightleftharpoons Zn^{2+} + 3H_2O$$

$$\underset{\text{acid}}{ZnO(s)} + 2OH^- + H_2O \rightleftharpoons Zn(OH)_4^{2-}$$

$$\underset{\text{acid}}{Al(OH)_3(s)} + OH^- \rightleftharpoons Al(OH)_4^-$$

$$\underset{\text{base}}{Al(OH)_3(s)} + H_3O^+ \rightleftharpoons [Al(OH)_2(H_2O)_2]^+$$

In Eqs. (5.10) and (5.13), we see water represented in the first case as a BL base, and in the other case as a BL acid. Water is interesting because, not only can it function as either an acid or a base in different reactions, but it can actually function as both in the same reaction. Eq. (5.15) represents the net reaction that occurs between any strong acid (which ionizes completely to form H_3O^+) and any strong base (which ionizes completely to form OH^-):

$$\underset{\text{acid}}{H_3O^+} + \underset{\text{base}}{OH^-} \rightleftharpoons \underset{\text{acid}}{H_2O} + \underset{\text{base}}{H_2O} \tag{5.15}$$

Such reactions are referred to as **neutralization.** This process for strong acids and strong bases is an extremely favorable reaction; that is, it has a very large equilibrium constant, and can be considered quantitative.

In the reaction between a strong acid and strong base, water is the only product that is represented in the balanced net ionic equation. The spectator ions are present, as we have said, both at the beginning and at the end of the reaction; but they do not appear explicitly in this shorthand representation. In the neutralization of hydrochloric acid with aqueous sodium hydroxide, sodium ions that were initially present in a sodium hydroxide solution and chloride ions that were present in a hydrochloric solution both become identified with an aqueous sodium chloride solution as the product; that is, the solution that results from the neutralization is equivalent to what one would get by dissolving the appropriate amount of sodium chloride in the same total amount of water.

For any neutralization reaction we could imagine the effect of driving off all the water from the solution produced by the neutralization, for example, simply by vaporizing it. The substances that are left behind by such removal of water result from the combination of positive ions, **cations**, and negative ions, **anions.** These substances are referred to as salts, the most common example of which is ordinary table salt, NaCl. The most common definition of a **salt** is "a substance that results from the neutralization of an acid with a base." However, we must realize that in most cases, unless some effort is made to remove the water, what is actually produced is a salt solution. In some cases, the solubility of the salt produced will be sufficiently low that solid material will form even without the removal of water. A typical example is the salt barium sulfate, which results from the neutralization reaction between barium hydroxide and sulfuric acid.

$$2H_3O^+ + SO_4^{2-} + Ba^{2+} + 2OH^- \rightleftharpoons 4H_2O + BaSO_4(s) \tag{5.16}$$

In writing Eq. (5.16) for the neutralization of a particular strong acid with a particular strong base, we have paid strict attention to the precept that a chemical equation should represent the actual chemical transformation that occurs as neatly and as precisely as possible in the concise notation of chemical formulas. Correspondingly, it is not correct to represent an acid-base reaction involving a weak acid or a weak base or both by Eq. (5.15). The ion H_3O^+ is simply not the main reactant species in the reaction of a weak acid and OH^- is not the main reactant species in the reaction of a weak base. For example, in representing the neutralization of a 1.0 *M* solution of acetic acid by aqueous sodium hydroxide solution, we do not write Eq. (5.15), since only about 1% of the acetic acid initially present is in the form of acetate

Spotlight

In aqueous dilute solutions, HF acts as a weak acid,

$$HF + H_2O \rightleftharpoons H_3O^+ + F^-$$

In concentrated solutions or when pure, however, HF becomes a strong acid because of the formation of stable complex species such as $H_2F_3^-$ and $H_3F_4^-$. For example,

$$\text{base} + 3HF(l) \rightleftharpoons H_2F_3^- + (\text{base})H^+$$

In fact, pure HF is almost as strong an acid as 100% H_2SO_4, and even concentrated HNO_3 acts as a base by accepting a proton from HF:

$$HNO_3(l) + 3HF(l) \rightleftharpoons H_2NO_3^+ + H_2F_3^-$$

The acidity of liquid HF can be increased by adding any of several inorganic fluorides that complex with F^-, for example, boron trifluoride, BF_3, and antimony pentafluoride, SbF_5:

$$BF_3(g) + 2HF(l) \rightleftharpoons BF_4^- + H_2F^+$$

$$SbF_5(l) + 2HF(l) \rightleftharpoons SbF_6^- + H_2F^+$$

In these reactions, we see formation of the very strong acid, H_2F^+, which can react according to

$$\text{base} + H_2F^+ \rightleftharpoons HF + (\text{base})H^+$$

ions and H_3O^+. We realize that instead, the predominant form of acetic acid before the chemical reaction occurs is to be represented by the chemical formula of acetic acid itself. Hence, we represent the neutralization as

$$\underset{\text{acid}}{CH_3CO_2H} + \underset{\text{base}}{OH^-} \rightleftharpoons \underset{\text{acid}}{CH_3CO_2^-} + \underset{\text{base}}{H_2O} \qquad (5.17)$$

Similarly, in discussing the neutralization of a weak base with an acid, we recognize that the weak base is predominantly in its un-ionized form before the neutralization takes place. Consequently, it is predominantly the un-ionized form that accounts for the net transformation, and it is this form that is represented on the left side of the balanced equation. For example, we write the chemical equation for the neutralization of aqueous ammonia by aqueous hydrochloric acid as

$$\underset{\text{acid}}{H_3O^+} + \underset{\text{base}}{NH_3} \rightleftharpoons \underset{\text{acid}}{NH_4^+} + \underset{\text{base}}{H_2O} \qquad (5.18)$$

Following this view one step further, the neutralization of a weak acid by a weak base does not involve the explicit inclusion of either H_3O^+ or OH^- in the balanced net equation, as neither of these species is initially present in high concentrations, and neither experiences an overall net transformation to the extent that the more predominant weak acid and weak base do. Hence, the neutralization of acetic acid with ammonia can be represented by the equation

$$CH_3CO_2H + NH_3 \rightleftharpoons CH_3CO_2^- + NH_4^+ \qquad (5.19)$$

One thing that should be remembered for reactions involving weak acids and weak bases is that we can no longer assume that such a reaction goes essentially 100% to completion. Indeed, we shall see that the extent to which such acid-base reactions occur depends on just how weak or how strong the acids and bases are.

Nevertheless, we shall see that for stoichiometry it still makes sense to speak of neutralizing acids with bases. Here, however, we are neutralizing strong acids with weak bases and weak acids with strong bases, unlike the strong acid–strong base neutralization.

EXAMPLE 5.1 How many milliliters of a 0.20 *M* NaOH solution must be added to exactly neutralize a 50-ml solution that is 0.10 *M* in $HClO_4$?

SOLUTION Both the solutes in this problem are strong electrolytes (NaOH a strong base, and $HClO_4$ a strong acid). The 50-ml solution contains

$$(50\ \cancel{\text{ml}})\left(\frac{1}{1000}\frac{\cancel{\text{liter}}}{\cancel{\text{ml}}}\right)\left(0.10\frac{\text{mol } H_3O^+}{\cancel{\text{liter}}}\right) = 0.0050 \text{ mol } H_3O^+$$

When neutralization is exact, 0.0050 mol OH^- must be provided. Let *V* represent the volume (in ml) of 0.20 *M* NaOH that is required.

$$0.0050 \text{ mol } OH^- = (V\ \cancel{\text{ml}})\left(\frac{1}{1000}\frac{\cancel{\text{liter}}}{\cancel{\text{ml}}}\right)\left(0.20\ \frac{\text{mol } OH^-}{\cancel{\text{liter}}}\right)$$

$$V = \left(\frac{1000}{0.20}\right)(0.0050) = 25$$

Thus, 25 ml of 0.20 *M* NaOH is required.

EXAMPLE 5.2 A solution is prepared by dissolving 0.050 mol $NaClO_4$, 0.060 mol KCl, and 0.010 mol HNO_3 in sufficient water to bring the total volume to 500 ml. Calculate the molarity of H_3O^+ in the resulting solution.

Spotlight

Often the most convenient way in which to measure amounts of acids and bases for chemical experiments is the volumetric technique. By knowing the concentration of an acid or a base in a sample, one can conveniently measure the number of moles of that acid or base simply by measuring the volume of the sample. Then, from a knowledge of the stoichiometry of the ionization of the given acid or base, one can equally well calculate the number of moles of H^+ or OH^-, or the "available" ions, present in a known volume of acid or base solution of known concentration. This amount is referred to as the **number of equivalents** of acid or base. The number of equivalents will equal the number of moles of the acidic or basic species only in the case where ionization of one mole of acid or base yields one mole of the ions H_3O^+ or OH^-. In other cases, the appropriate stoichiometric ratios must be taken into account. Thus, one mole of H_2SO_4 in aqueous solution provides two equivalents of acid according to Eq. (5.5). This concept of equivalents leads us to an alternative definition of concentrations for acids and bases, a convention that finds wide application in many laboratories. This alternative measure of concentration is known as the **normality**, which is defined simply as the number of equivalents of acid or base per liter of solution. The relation between the molarity of an acid or a base solution and the normality of the solution is simply determined by the stoichiometry of ionization of the acid or base. For example, a 0.1 *M* H_2SO_4 solution is also described as being 0.2 normal; the notation is 0.2 *N* H_2SO_4.

SOLUTION Since we are concerned only with acid-base chemistry in this example, and since only strong electrolytes are involved as solutes, the only substance of interest is the nitric acid. The number of moles of H_3O^+ from 0.010 mol HNO_3 is 0.010. Since the total volume is 500 ml,

$$\text{molarity} = \text{no. moles } H_3O^+ \text{ per liter} = \frac{0.010 \text{ mol } H_3O^+}{(500/1000) \text{ liter}} = 0.020\, M\, H_3O^+$$

EXAMPLE 5.3 A solution is prepared by dissolving 0.030 mol KOH, 0.30 mol $LiClO_4$, and 0.010 mol H_2SO_4 in sufficient water to bring the total volume to 200 ml. Characterize the acid-base properties of the resulting solution.

SOLUTION From 0.030 mol KOH, there is available 0.030 mol OH^-; and from 0.010 mol H_2SO_4, there is available 0.020 mol H_3O^+. The available OH^- will completely consume the 0.020 mol H_3O^+, leaving an excess of 0.010 mol OH^- in a total volume of 200 ml. Then,

$$\text{molarity } OH^- = \frac{0.010 \text{ mol } OH^-}{(200/1000) \text{ liter}} = 0.050\, M\, OH^-$$

EXAMPLE 5.4 How many milliters of a 0.20 M solution of NaOH will be required to exactly neutralize 20 ml of a 0.15 M H_2SO_4 solution?

SOLUTION Recognizing that for each mole of H_2SO_4 available, two moles of H_3O^+ are generated (Eq. 5.5), we know that a 20-ml 0.15 M H_2SO_4 solution provides

$$\left(\frac{20}{1000} \text{ liter}\right)\left(0.15 \frac{\text{mol } H_2SO_4}{\text{liter}}\right)\left(2 \frac{\text{mol } H_3O^+}{\text{mol } H_2SO_4}\right) = 0.0060 \text{ mol } H_3O^+$$

If V represents the volume (in milliliters) of 0.20 M NaOH to be added, then

$$(V\, \cancel{\text{ml}})\left(\frac{1}{1000} \frac{\cancel{\text{liter}}}{\cancel{\text{ml}}}\right)\left(0.20 \frac{\text{mol } OH^-}{\cancel{\text{liter}}}\right) = \text{no. moles of } OH^-$$

When exact neutralization is achieved, the number of moles of H_3O^+ equals the number of moles of OH^-, or 0.0060 mol OH^-:

$$(V\, \cancel{\text{ml}})\left(\frac{1}{1000} \frac{\cancel{\text{liter}}}{\cancel{\text{ml}}}\right)\left(0.20 \frac{\text{mol } OH^-}{\cancel{\text{liter}}}\right) = 0.0060 \text{ mol } OH^-$$

$$V = (0.0060)\left(\frac{1000}{0.20}\right) = 30$$

Thus, 30 ml of 0.20 M NaOH is the required amount.

In neutralizing an acid and a base, as in Examples 5.1 and 5.4, the point at which just enough acid has been added to react with all the base, or just enough base has been added to react with all the acid, is called the **equivalence point.** Those two examples were solved on the basis that in any neutralization between a strong acid and a strong base, when the equivalence point is reached,

$$\text{no. moles } H_3O^+ = \text{no. moles } OH^- \qquad (5.20)$$

This statement can be generalized to include weak acids and bases; in general, at the equivalence point,

$$\text{no. moles available H}^+ = \text{no. moles base} \tag{5.21}$$

EXAMPLE 5.5 How many milliliters of a 0.30 *M* H_2SO_4 solution will be required to react exactly with a 50-ml sample of a 0.20 *M* NH_3 solution (that is, to cause the equivalence point to be reached)?

SOLUTION Using the knowledge that each mole of H_2SO_4 provides two moles of H^+, we can write, using V_a to represent the number of milliliters of acid solution required,

$$(V_a \cancel{\text{ml}})\left(\frac{1}{1000}\frac{\cancel{\text{liter}}}{\cancel{\text{ml}}}\right)\cdot\left(0.30\frac{\cancel{\text{mol H}_2\text{SO}_4}}{\cancel{\text{liter}}}\right)\left(2\frac{\cancel{\text{mol H}^+}}{\cancel{\text{mol H}_2\text{SO}_4}}\right)\left(1\frac{\text{mol base}}{\cancel{\text{mol H}^+}}\right)$$

$$= (50\ \cancel{\text{ml}})\left(\frac{1}{1000}\frac{\cancel{\text{liter}}}{\cancel{\text{ml}}}\right)\left(0.20\frac{\text{mol base}}{\cancel{\text{liter}}}\right)$$

$$V_a = \frac{1}{2}(50)\left(\frac{0.20}{0.30}\right) = 16.7$$

Therefore, 16.7 ml of 0.30 *M* H_2SO_4 are required.

STUDY PROBLEM 5(a)

How many milliliters of a 0.18 *M* NaOH solution must be added to exactly neutralize 50 ml of a 0.25 *M* HCl solution?

STUDY PROBLEM 5(b)

How many liters of a 0.0011 *M* $Ca(OH)_2$ solution must be added to a solution containing 1.8×10^{-4} mol H_2SO_4 and 1.5×10^{-4} mol HCl to reach the equivalence point?

Volumetric Techniques; Titration

The convenience of using volumetric techniques for measuring and manipulating specific amounts of aqueous solutions of acids and bases has resulted in the development of an efficient set of volumetric techniques and associated apparatus. Many of these techniques have been extant over a hundred years. The most common types of volumetric apparatus in use during this period, and still in use, are the buret, the pipet (both graduated and volumetric), and the volumetric flask, which are shown in Fig. 5.2.

The most common type of acid-base experiment in aqueous solution chemistry is the **titration,** the dropwise addition of an aqueous solution of either an acid or a base to neutralize a solution of base or acid to which it is being added. The dropwise addition is made from a buret so that one can precisely determine the equivalence point; this is referred to as the **endpoint** of a titration. The endpoint can be recognized by using an "indicator," a chemical that imparts to the solution a clearly defined color that depends on whether a small excess of H_3O^+ or of OH^- exists. A

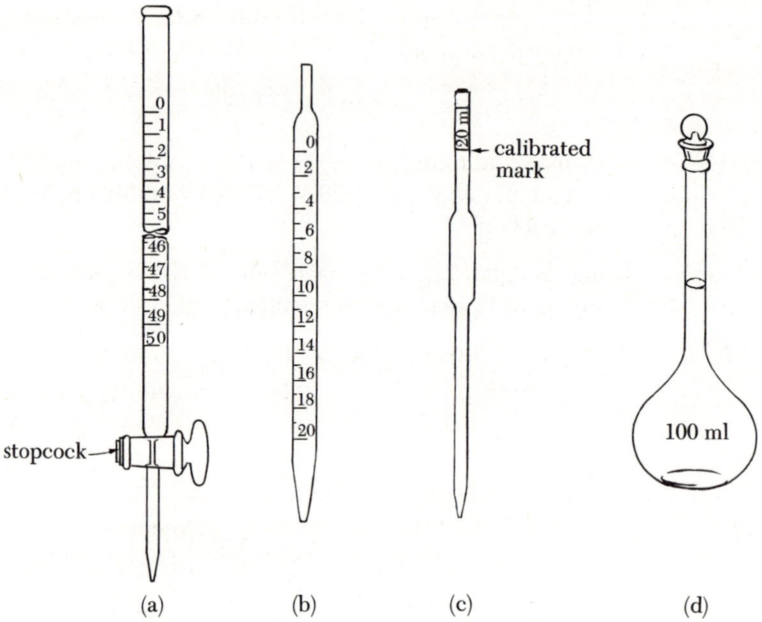

Figure 5.2 Standard volumetric equipment. (a) A 50-ml buret; (b) a 20-ml graduated pipet; (c) a 20-ml volumetric pipet; and (d) a 100-ml volumetric flask.

common indicator used for titrations involving strong acids and strong bases is **phenolphthalein,** a substance that is colorless if a slight excess of H^+ is present or that is reddish pink in a slight excess of OH^-. Hence, in the titration of an acid solution with a basic solution dispensed from a buret, the solution being titrated would be colorless until the endpoint, at which it would turn faintly pink. If another drop of the basic solution is added, taking the system past the endpoint, the solution would turn deep reddish pink. The following two examples show what kinds of experiments are typically carried out by this method.

EXAMPLE 5.6 Industrial wastes are being dumped into a small stream of low flow rate, and environmentalists suspect that substantial concentrations of acids are present downstream from the dumping site. A 100-ml sample of the stream water is titrated with a 0.0012 *M* NaOH solution in a 50-ml buret. At the beginning of the titration, the level in the buret is determined to be 1.67 ml; at the phenolphthalein endpoint, the buret level is 24.37 ml. Calculate the molarity of H^+ available in the stream sample.

SOLUTION The total volume of 0.0012 *M* NaOH solution dispensed in the titration was 24.37 − 1.67 = 22.70 ml. Then, using Eq. (5.21), and denoting the molarity of available H^+ in the sample as M_a, we write

$$(22.70\,\cancel{\text{ml}})\left(\frac{1}{1000}\frac{\cancel{\text{liter}}}{\cancel{\text{ml}}}\right)\left(0.0012\frac{\text{mol OH}^-}{\cancel{\text{liter}}}\right) = (100\,\cancel{\text{ml}})\left(\frac{1}{1000}\frac{\cancel{\text{liter}}}{\cancel{\text{ml}}}\right)\left(M_a\frac{\cancel{\text{mol H}^+}}{\cancel{\text{liter}}}\right)\left(1\frac{\text{mol OH}^-}{\cancel{\text{mol H}^+}}\right)$$

$$M_a = \left(\frac{22.70}{100}\right)(0.0012) = 2.7 \times 10^{-4}\left(\frac{\text{mol H}^+}{\text{liter}}\right)$$

EXAMPLE 5.7 A 0.152-g sample of a weak acid of unknown identity is found to require 23.14 ml of 0.103 *M* NaOH in a titration to the equivalence point. Calculate the equivalent weight of the unknown weak acid, the mass of the acid that provides one mole of H^+ in a titration with a strong base.

SOLUTION The number of moles of OH^- used in the titration was (23.14/1000)(0.103). Hence, 0.152 g of the acid corresponds to (23.14/1000)(0.103) mol H^+, or 2.38×10^{-3} mol acid.

$$0.152 \text{ g} = \frac{23.14}{1000}(0.103) \text{ mol H}^+ = 2.38 \times 10^{-3} \text{ mol acid}$$

$$1 \text{ mol acid} = \frac{0.152}{2.38 \times 10^{-3}} \text{ g} = 0.0639 \times 10^3 \text{ g} = 63.9 \text{ g}$$

Thus, the equivalent weight of the unknown weak acid is 63.9 g.*

STUDY PROBLEM 5(c)

What is the equivalent weight of an unknown acid, a 484-mg sample of which required 31.2 ml of 0.150 *M* NaOH in a titration to the equivalence point?

EXAMPLE 5.8 One wishes to standardize a 5-liter aqueous solution of NaOH, or determine its molarity, so that it can be used in titrations with unknown acids. A 25.0-ml sample of the NaOH solution is transferred into a flask with a 25.0-ml volumetric pipet, a few drops of phenolphthalein solution are added, and the solution is titrated with a 0.168 *M* HCl solution; 15.13 ml of titrant are required to reach the equivalence point. What is the molarity M_b of the NaOH solution?

SOLUTION Using Eq. (5.20), we write

$$(15.13 \text{ ml})\left(\frac{1 \text{ liter}}{1000 \text{ ml}}\right)\left(0.168\frac{\text{mol H}_3\text{O}^+}{\text{liter}}\right) = (25.0 \text{ ml})\left(\frac{1 \text{ liter}}{1000 \text{ ml}}\right)\left(M_b\frac{\text{mol OH}^-}{\text{liter}}\right)\left(1\frac{\text{mol H}_3\text{O}^+}{\text{mol OH}^-}\right)$$

$$M_b = \frac{15.13}{25.0}(0.168) = 0.102$$

Thus, the molarity of the NaOH solution is 0.102 *M*.

Acid-Base Equilibrium Constants

From the Brönsted-Lowry point of view, the ionization of any acid (represented by the general symbol "HA") in water is given by Eq. (5.10):

$$\underset{\text{acid}}{\text{HA}} + \underset{\text{base}}{\text{H}_2\text{O}} \rightleftharpoons \underset{\text{acid}}{\text{H}_3\text{O}^+} + \underset{\text{base}}{\text{A}^-} \tag{5.10}$$

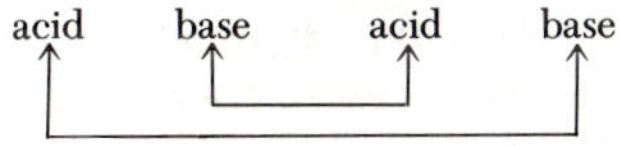

*If the unknown compound in Example 5.7 is a **monoprotic acid**—providing one mole of H^+ per mole of compound—then 63.9 is also its molecular weight (63.9 amu). However, if the unknown compound is a **diprotic**—providing two moles of H^+ per mole of compound—then its molecular weight is $2 \times 63.9 = 127.8$ amu.

The corresponding equilibrium expression, with the equilibrium constant designated K_a (*a* for acid), is

$$K_a = \frac{[H_3O^+][A^-]}{[HA]} \tag{5.22}$$

For the specific case of the weak acid CH_3CO_2H, acetic acid,

$$\underset{\text{acid}}{CH_3CO_2H} + \underset{\text{base}}{H_2O} \rightleftharpoons \underset{\text{acid}}{H_3O^+} + \underset{\text{base}}{CH_3CO_2^-} \tag{5.23a}$$

$$K_a = \frac{[H_3O^+][CH_3CO_2^-]}{[CH_3CO_2H]} \tag{5.23b}$$

In writing Eqs. (5.22) and (5.23b), we have used the convention described in Chap. 4 of omitting the symbol $[H_2O]$ in equilibrium expressions for reactions carried out in water as the medium.

In the Arrhenius approach, the analog of Eq. (5.10) is simply Eq. (5.1). One would write the following equilibrium expression corresponding to Eq. (5.1),

$$K = \frac{[H^+][A^-]}{[HA]} \tag{5.24}$$

Recalling that $[H^+]$ in the Arrhenius view means the same thing as $[H_3O^+]$ in the BL view, it is clear that this equilibrium expression (5.24) is equivalent to Eq. (5.22). We emphasize the equivalence of these expressions by writing

$$K_a = \frac{[H_3O^+][A^-]}{[HA]} = \frac{[H^+][A^-]}{[HA]} \tag{5.25}$$

Chemists tend to use the symbols “$[H^+]$” and ”$[H_3O^+]$” interchangeably as well as the two corresponding expressions for K_a.

Equations (5.22), (5.23), and (5.25) cover only the important special cases in which water is the reactant base. More generally, any acid-base reaction is represented in the BL approach by Eq. (5.9). A corresponding equilibrium expression can be written (Eq. 5.26):

$$HA + B \rightleftharpoons BH^+ + A^- \tag{5.9}$$

$$K = \frac{[HB^+][A^-]}{[HA][B]} \tag{5.26}$$

The strength of a particular acid HA in the presence of a certain base B depends on the position of the equilibrium for Eq. (5.9); that is, it depends on the value of the equilibrium constant K in Eq. (5.26). If the acid HA is stronger than the acid HB^+ that would be formed if the proton transfer occurred, then HA has a greater tendency to donate its proton to a base than HB^+ does; in other words, the tendency of HA to give up a proton and become A^- is greater than the tendency of HB^+ to donate a proton and become B. In such a case, the equilibrium would lie on the side favoring HB^+ and A^-, so the equilibrium constant would be greater than 1. On the other hand, if the acid HB^+ should happen to be stronger than the acid HA, then HB^+ would have a greater tendency to rid itself of its proton than HA does; and consequently, one would not expect that HB^+ could be formed by donating a proton from HA to B. Hence, the equilibrium in such a case would lie on the left side of Eq. (5.26), and the equilibrium constant would be less than 1.

We can summarize these considerations by stating that acid-base reactions that have equilibrium positions that lie preferentially on the right (have K values greater than 1) are those that transform a stronger acid into a weaker acid. On the contrary, those acid-base processes with equilibrium constants less than 1 are those in which a weaker acid is transformed into a stronger acid. Similarly, if the base B is stronger than the base A^-, the tendency of B to accept and attach a proton (donated from HA) will be greater than the tendency of A^- to accept a proton (donated from HB^+). If B is a stronger base than A^-, then the equilibrium should lie to the right, with an equilibrium constant greater than 1. However, if A^- is the stronger base, then the equilibrium constant for the reaction transforming B and HA into HB^+ and A^- should be less than 1. To expand our summary, we can state that acid-base reactions with equilibrium constants larger than 1 are those that transform a stronger acid and stronger base into a weaker acid and a weaker base. Reactions that transform weaker acids and weaker bases into stronger acids and stronger bases have equilibrium constants less than 1.

Relation between Arrhenius and Brönsted-Lowry Approaches

The Arrhenius view is largely limited to acid-base phenomena in aqueous solutions. Although the Brönsted-Lowry approach is by no means limited to aqueous media and is general for proton transfer reactions in any medium, its most frequent applications are also concerned with acid-base reactions in aqueous solutions. In that context, the symbolism of the Brönsted-Lowry approach is somewhat different from the Arrhenius representation. In the Arrhenius approach we have viewed the ionization of an acid in dissociation equations (5.1) and (5.24). The ionization is viewed in the BL approach as a proton transfer reaction between the acid and water as a base, as in Eqs. (5.10) and (5.22). Thus, the acidic character of any acid HA in water is generally viewed as its ability to transfer a proton to water, thereby producing the conjugate base A^- and the conjugate acid of water, H_3O^+, the hydronium ion. The relative acid strengths of various acids in water are measured as their relative tendencies for this type of proton transfer to water. According to the summary above, those acids that are largely ionized in water will then be the ones that are stronger acids than the acid formed in this transformation, namely H_3O^+. For such strong acids, the corresponding equilibrium constants will be larger than 1. For these same strong acids, it is clear that water is a stronger base than the conjugate base A^- of the acid; the tendency of HA to be rid of its proton is stronger than the similar tendency of the acid H_3O^+; this is equivalent to saying that H_2O has a greater tendency to attach a proton than A^- does.

Those acids that are only slightly ionized in water are those that have equilibrium constants for the ionization reaction less than 1. According to the summary above, these weak acids will all be weaker acids than the hydronium ion, which is equivalent to saying that their conjugate bases are stronger bases than water.

Spotlight

Ammonia reacts directly with acids to form ammonium salts. Gaseous ammonia and gaseous HCl are sometimes used to produce amonium chloride, NH_4Cl, smoke screens:

$$NH_3(g) + HCl(g) \rightleftharpoons NH_4Cl(s)$$

Ammonium salts liberate NH_3 when warmed with a base, according to the reaction

$$NH_4^+ + OH^- \rightleftharpoons NH_3(g) + H_2O(l)$$

This reaction constitutes the most convenient laboratory preparation of ammonia and is often used as the basis for the qualitative analysis of the ammonium ion; the liberated $NH_3(g)$ is detected by its characteristic choking odor or by its effect on moist red litmus paper. The litmus paper turns blue in the presence of a basic condition, characterized by excess OH^- ion; this condition is produced in the moisture of the litmus paper by the reaction

$$NH_3 + H_2O \rightleftharpoons NH_4^+ + OH^-$$

Ammonium salts also generally liberate ammonia when they are heated to about 300 °C:

$$NH_4Cl(s) \rightleftharpoons NH_3(g) + HCl(g) \qquad (300\ ^\circ C)$$

For this reason, NH_4Cl is used as a soldering flux. The HCl liberated when the NH_4Cl is heated cleans the metal surfaces by dissolving the oxide coating; in a specific example,

$$CuO(s) + 2HCl(g) \rightleftharpoons Cu^{2+} + 2Cl^- + H_2O(g)$$

On the basis of the reasoning of the preceding two paragraphs, we can conclude that the strongest acid that can exist in appreciable concentrations in aqueous solutions is the hydronium ion. If we should introduce an acid that is stronger than H_3O^+, it would largely transfer its proton to water, forming the hydronium ion. Hence, when a very strong acid, such as hydrochloric acid, is dissolved in water, a reaction of the type shown in Eq. (5.11) occurs essentially completely:

$$HCl + H_2O \rightleftharpoons H_3O^+ + Cl^- \tag{5.11}$$

This corresponds to an equilibrium constant much greater than 1 and an equilibrium far to the right for strong acids.

Base Strengths in Terms of the Conjugate Acids

As we have presented the Brönsted-Lowry approach for ammonia, it allows one to represent the base characteristics of weak bases in water in a straightforward and convenient manner. Equation (5.9) is equally general for representing basic properties in water as it is for representing acidic properties. In this case, one is concerned with the ability of a base B to abstract a proton from the water, which acts as the acid HA. The use of the Brönsted-Lowry approach for describing basic properties in water can be neatly summarized by Eq. (5.27), of which the ammonia case (Eq. 5.7) is a specific example.

$$H_2O + B \rightleftharpoons BH^+ + OH^- \tag{5.27a}$$

$$K_b = \frac{[BH^+]\,[OH^-]}{[B]} \tag{5.27b}$$

A strong base B will have a greater tendency to abstract a proton from the acid H_2O

than the tendency of the base OH^- to abstract a proton from the acid BH^+; consequently, the equilibrium will lie to the right and the equilibrium constant will be greater than 1. A weak base, B, will have a lower tendency to abstract a proton from water than the tendency of OH^- to abstract a proton from BH^+; the equilibrium will then lie to the left, and the equilibrium constant will be less than 1. Examples of weak bases are provided by ammonia and a class of compounds called **amines**; the amines are structurally related to ammonia in that one or more of the hydrogens in NH_3 is replaced by a group of carbon and hydrogen atoms—say CH_3. Thus, if R stands for such a group of carbon and hydrogen atoms, then the formula RNH_2 symbolizes one type of amine.* The basic properties of aqueous solutions of ammonia and amines can conveniently be described by the equations

$$H_2O + NH_3 \rightleftharpoons NH_4^+ + OH^- \tag{5.7}$$

$$H_2O + RNH_2 \rightleftharpoons RNH_3^+ + OH^- \tag{5.28}$$

From Eq. (5.7) one can define the "ionization constant" of ammonia, symbolized for bases as K_b, in terms of the analog of Eq. (5.27b),

$$K_b = \frac{[NH_4^+][OH^-]}{[NH_3]} \tag{5.29}$$

Examples of very strong bases are the anions (negative ions) CH_3^- and NH_2^-, for which the following reactions have equilibria extremely far to the right:

$$H_2O + CH_3^- \rightleftharpoons CH_4(g) + OH^- \tag{5.30}$$

$$H_2O + NH_2^- \rightleftharpoons NH_3 + OH^- \tag{5.31}$$

By considering such cases, one realizes that the strongest base that can be present in appreciable concentrations in water is the ion OH^-; any stronger base that might be added to water would undergo a reaction of the type represented in Eq. (5.27), for example, Eqs. (5.30) and (5.31).

An important feature of acid-base chemistry in aqueous solution is what is called the ionization of water. In the Brönsted-Lowry framework, this chemical process is cast into the form of Eq. (5.9) by identifying both HA and B with water.† We visualize water functioning as both a Brönsted-Lowry acid and a Brönsted-Lowry base, the former donating a proton to the latter. This is represented by the equation

$$H_2O + H_2O \rightleftharpoons H_3O^+ + OH^- \tag{5.32}$$

acid base acid base

Corresponding to Eq. (5.32) is the equilibrium expression

$$K = [H_3O^+][OH^-]$$

*Compounds of the type CH_3NH_2, $CH_3CH_2NH_2$, $(CH_3)_2NH$, and so on are called *amines*. Species in which ammonia is bound to certain metal ions, as in $Co(NH_3)_6^{3+}$, are called *ammine* complexes, or ammines (see Sec. 18.2).

†In the Arrhenius approach, the ionization of water is represented by the equation $H_2O \rightleftharpoons H^+ + OH^-$, with a corresponding equilibrium expression

$$K_w = [H^+][OH^-] = 1.0 \times 10^{-14}$$

(Here again $[H_2O]$ does not appear in the expression of a system for which water is the solvent.) We refer to this constant K by the symbol K_w, which is known to equal 1.0×10^{-14} at 25 °C. Hence, we have the equation

$$[H_3O^+][OH^-] = K_w = 1.0 \times 10^{-14} \qquad (5.33)$$

Equation (5.33) tells us that any aqueous solution in a state of equilibrium has a hydronium ion (or proton) concentration, the value of which multiplied by the hydroxide ion concentration is 1.0×10^{-14}. This is an extremely important consideration to keep in mind, as all aqueous solutions at equilibrium at 25 °C must conform to it. For a sample of pure water, the only source of H_3O^+ and of OH^- is the ionization (Eq. 5.32), and the concentrations of H_3O^+ and OH^- produced by the ionization are equal. Then, we can write for pure water, using Eq. (5.33):

$$[H_3O^+] = [OH^-]$$
$$[H_3O^+]^2 = 1.0 \times 10^{-14}$$
$$[H_3O^+] = \sqrt{1.0 \times 10^{-14}}$$
$$= 1.0 \times 10^{-7}$$

Thus, in pure water, the concentrations of both H_3O^+ and OH^- are 1.0×10^{-7} *M*.

EXAMPLE 5.9 Calculate the H_3O^+ concentration in a solution for which $[OH^-] = 0.015$ *M*.

SOLUTION Knowing that the equilibrium represented by Eq. (5.33) is essentially always valid for aqueous solutions, we write

$$K_w = 1.0 \times 10^{-14} = [H_3O^+][OH^-] = [H_3O^+][0.015]$$

$$[H_3O^+] = \frac{1.0 \times 10^{-14}}{0.015} = 6.7 \times 10^{-13}\,M$$

EXAMPLE 5.10 Calculate the value of $[OH^-]$ for a solution in which $[H^+]$ is 2.5×10^{-4} *M*.

SOLUTION From Eq. (5.33),

$$1.0 \times 10^{-14} = [H_3O^+][OH^-] = [H^+][OH^-] = (2.5 \times 10^{-4})[OH^-]$$

$$[OH^-] = \frac{1.0 \times 10^{-14}}{2.5 \times 10^{-4}} = 4.0 \times 10^{-11}\,M$$

STUDY PROBLEM 5(d)

Calculate the value of $[H^+]$ for a solution in which $[OH^-] = 9.2 \times 10^{-10}$ *M*.

STUDY PROBLEM 5(e)

What is the OH^- concentration if $[H_3O^+]$ is 6.6×10^{-3} *M*?

The very small value of K_w shows that in pure water only a very small fraction of the water is ionized (about 2×10^{-7}%). Noting that the equation for the ionization of water (5.32) is simply the reverse of the equation for the neutralization of a strong acid with a strong base (5.15), we can find it reasonable that if the neutralization of a strong acid with a strong base is overwhelmingly favorable, the ionization of water occurs to only a tiny extent (in pure water).

The Concept of pH

Before progressing further in our discussion of aqueous acid-base chemistry, it is advantageous to introduce some additional notation that has been developed specifically for acid-base problems. This notation, which is in widespread use in fields ranging far from basic chemistry, makes use of logarithms (see Eq. 4.47). Because equilibrium constants are often very large or very small numbers (say 1.5×10^{32} or 4.2×10^{-11}), it is often convenient to convert them to a logarithmic scale, say for drawing plots. One of the most useful defined quantities in chemistry is the pH. It is defined as

$$\text{pH} = -\log [H_3O^+] = -\log [H^+] \tag{5.34}$$

The definition of pH embodied in Eq. (5.34) is valid under any circumstances for aqueous systems. Thus, pH always means "the negative logarithm of the H_3O^+ concentration," irrespective of the source of the ions H_3O^+ in the system. The pH variable is very commonly used in basic and applied chemistry and in allied fields. It is straightforward to compute the pH, if $[H^+]$ is known, or vice versa. For example, if $[H^+] = 1.0$, then $\text{pH} = -\log(1.0) = 0$. Or if $[H^+] = 1.0 \times 10^{-6}$, then $\text{pH} = -\log (1.0 \times 10^{-6}) = -(-6) = 6$. Similarly, if $[H^+] = 3.5 \times 10^{-11}$, then

$$\begin{aligned} \text{pH} &= -\log (3.5 \times 10^{-11}) \\ &= -\{\log (3.5) + \log (10^{-11})\} \\ &= -(0.54 - 11) \\ &= -(-10.46) \\ &= 10.46 \end{aligned}$$

We see from these relations that large values of pH correspond to very low concentrations of H_3O^+ and small values of pH correspond to higher H_3O^+ concentrations. We know from Eq. (5.33) that in any aqueous solution at equilibrium

$$[H^+] = \frac{1.0 \times 10^{-14}}{[OH^-]}$$

Taking the logarithms of both sides yields

$$\begin{aligned} \log [H^+] &= \log \left\{ \frac{1.0 \times 10^{-14}}{[OH^-]} \right\} \\ &= \log (1.0 \times 10^{-14}) - \log [OH^-] \\ &= -14 - \log [OH^-] \end{aligned}$$

$$\text{pH} = -\log [H^+] = 14 + \log [OH^-] \tag{5.35}$$

Then, if pOH is defined as $-\log [OH^-]$, we have, for the condition 25 °C,

$$\text{pH} = 14 - \text{pOH} \tag{5.36}$$

TABLE 5.1 **Relations between $[H^+]$, $[OH^-]$, pH, and pOH**

H^+ *Molarity*	OH^- *Molarity*	*pH*	*pOH*
2.5	0.4×10^{-14}	−0.40	14.40
1.0	1.0×10^{-14}	0	14
1.0×10^{-3}	1.0×10^{-11}	3	11
1.7×10^{-5}	6.0×10^{-10}	4.77	9.23
1.0×10^{-7}	1.0×10^{-7}	7	7
1.0×10^{-10}	1.0×10^{-4}	10	4
1.0×10^{-14}	1.0	14	0
3.3×10^{-15}	3.0	14.48	−0.48

The variable pOH is used only seldom, since either $[OH^-]$ or pOH can always be calculated from a knowledge of $[H^+]$ or pH by using Eq. (5.33). Some of these relations are summarized in Table 5.1. For any row in the table, the product of the entries in the first two columns always equals 1.0×10^{-14} and the sum of the entries of the last two columns is in each case 14. For laboratory work in which the concept of pH is typically used, the concentration of H^+ or of OH^- is seldom much larger than about 1 *M*. Therefore the pH scale more commonly spans the range 0 to 14, corresponding to a range of $[H^+] = 1$ (with $[OH^-] = 10^{-14}$) to $[H^+] = 10^{-14}$ (with $[OH^-] = 1$).

STUDY PROBLEM 5(f)

Calculate the pH corresponding to each of the following concentrations: (a) $[H^+] = 3.4 \times 10^{-6}$; (b) $[H^+] = 7.7 \times 10^{-3}$; (c) $[OH^-] = 5.5 \times 10^{-4}$.

Unified Ranking of Acid and Base Strengths

We have emphasized that the Brönsted-Lowry approach considers acid-base chemistry within the framework of proton transfer, i.e., a proton exchange. The "currency" of this exchange is the ion H^+. The source of the currency is the acid and the acceptor of the currency is the base; the "transaction" is the transfer of a proton.

Clearly an acid cannot donate a proton if a base does not accept the proton; these two operations are simply two manifestations of the same overall "transaction." Nevertheless, it is for some purposes convenient to visualize an overall proton transfer as though it could be factored into two "half-reactions," one of which would be concerned only with the tendency of an acid to get rid of its proton, and the other with the acceptance of a proton by a base. Thus, for the generalized acid-base equation

$$HA + B \rightleftharpoons BH^+ + A^- \tag{5.9}$$

we can visualize the process as if it occurred by the two steps, or half-reactions:

$$HA \rightleftharpoons H^+ + A^- \tag{5.37}$$

$$H^+ + B \rightleftharpoons BH^+ \tag{5.38}$$

The sum of (5.37) and (5.38) gives Eq. (5.9). This artificial separation into steps is symbolized in Fig. 5.3. We recognize, according to the Brönsted-Lowry view, that both half-reactions (5.37) and (5.38) must occur simultaneously for any acid-base reaction. We choose to let the symbolism of Eq. (5.37) represent the behavior of a species in functioning as an acid. Then, the "strengths" of various acids are to be represented in terms of the tendencies for these half-reactions to occur, or the tendency of each acid to get rid of its proton. On this basis we find it convenient to rank qualitatively the relative strengths of various acids by listing their corresponding half-reactions, with the strongest acid at the top, the progressively weaker acids lower on the list, and the weakest acid at the bottom. Thus, at the top we shall have an entry such as $HClO_4 \rightleftharpoons H^+ + ClO_4^-$, and towards the bottom of such a list we might expect an entry such as $H_2O \rightleftharpoons H^+ + OH^-$, representing the weakly acidic character of H_2O. Similarly, we find it convenient to prepare a list of bases, with the tendency of each species to function as a base represented by a half-reaction of the form of Eq. (5.38). In preparing this list, we arbitrarily place the half-reactions for the strongest bases on the bottom, and represent decreasing base strength higher on the list. Thus, at the top we might find an entry such as $H^+ + H_2O \rightleftharpoons H_3O^+$, representing the basic character of the weak base H_2O. At the bottom, we should find entries such as $H^+ + CH_3^- \rightleftharpoons CH_4$, representing the strongly basic character of the CH_3^- species.

These convenient lists can be even more useful if we notice that the base half-reaction is exactly the same as an acid half-reaction, except that it is written in the reverse direction. Accordingly, for every entry of the form $HX \rightleftharpoons H^+ + X^-$ in the acid table, there can be a corresponding entry $H^+ + X^- \rightleftharpoons HX$ in the base table. This duplication is simply a manifestation of the fact that a half-reaction representing the acidic property of an acid also represents, viewed in reverse, the basic property of the conjugate base. One table is sufficient for both purposes if we arbitrarily choose one form or the other. Choosing the acid form for the half-reactions, we then

Figure 5.3
Artificially separating the steps of Fig. 5.1.

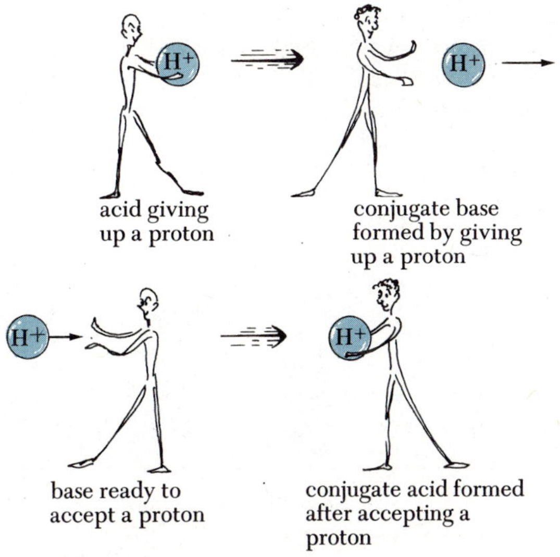

TABLE 5.2 **Relative Strengths of Acids and Bases**

increasing acid strength ↑

Acids		*Bases*	K_a
CF_3SO_3H	$\rightleftharpoons$	H^+ $CF_3SO_3^-$	$\gg 1$
$HClO_4$	$\rightleftharpoons$	$H^+ + ClO_4^-$	$\gg 1$
HCl	$\rightleftharpoons$	$H^+ + Cl^-$	$\gg 1$
HNO_3	$\rightleftharpoons$	$H^+ + NO_3^-$	$\gg 1$
H_2SO_4	$\rightleftharpoons$	$H^+ + HSO_4^-$	$\gg 1$
H_3O^+	$\rightleftharpoons$	$H^+ + H_2O$	1
Cl_3CCO_2H	$\rightleftharpoons$	$H^+ + Cl_3CCO_2^-$	1.0×10^{-1}
H_2SO_3	$\rightleftharpoons$	$H^+ + HSO_3^-$	1.5×10^{-2}
HSO_4^-	$\rightleftharpoons$	$H^+ + SO_4^{2-}$	1.2×10^{-2}
H_3PO_4	$\rightleftharpoons$	$H^+ + H_2PO_4^{2-}$	7.5×10^{-3}
HF	$\rightleftharpoons$	$H^+ + F^-$	6.7×10^{-4}
HCO_2H	$\rightleftharpoons$	$H^+ + HCO_2^-$	1.8×10^{-4}
CH_3CO_2H	$\rightleftharpoons$	$H^+ + CH_3CO_2^-$	1.8×10^{-5}
$CO_2(aq) + H_2O$	$\rightleftharpoons$	$H^+ + HCO_3^-$	4.4×10^{-7}
HSO_3^-	$\rightleftharpoons$	$H^+ + SO_3^{2-}$	1.0×10^{-7}
H_2S	$\rightleftharpoons$	$H^+ + HS^-$	1.0×10^{-7}
$H_2PO_4^-$	$\rightleftharpoons$	$H^+ + HPO_4^{2-}$	6.2×10^{-8}
NH_4^+	$\rightleftharpoons$	$H^+ + NH_3$	5.5×10^{-10}
HCN	$\rightleftharpoons$	$H^+ + CN^-$	4.9×10^{-10}
HCO_3^-	$\rightleftharpoons$	$H^+ + CO_3^{2-}$	4.8×10^{-11}
$CH_3NH_3^+$	$\rightleftharpoons$	$H^+ + CH_3NH_2$	2.7×10^{-11}
$C_7H_5NO_2S$[a]	$\rightleftharpoons$	$H^+ + C_7H_4NO_2S^-$	2.1×10^{-12}
HPO_4^{2-}	$\rightleftharpoons$	$H^+ + PO_4^{3-}$	2.2×10^{-13}
H_2O	$\rightleftharpoons$	$H^+ + OH^-$	1.0×10^{-14}
HS^-	$\rightleftharpoons$	$H^+ + S^{2-}$	1×10^{-15}
OH^-	$\rightleftharpoons$	$H^+ + O^{2-}$	$< 10^{-15}$
NH_3	$\rightleftharpoons$	$H^+ + NH_2^-$	$\ll 10^{-15}$
CH_4	$\rightleftharpoons$	$H^+ + CH_3^-$	$\ll 10^{-15}$

increasing base strength ↓

[a] Saccharin.

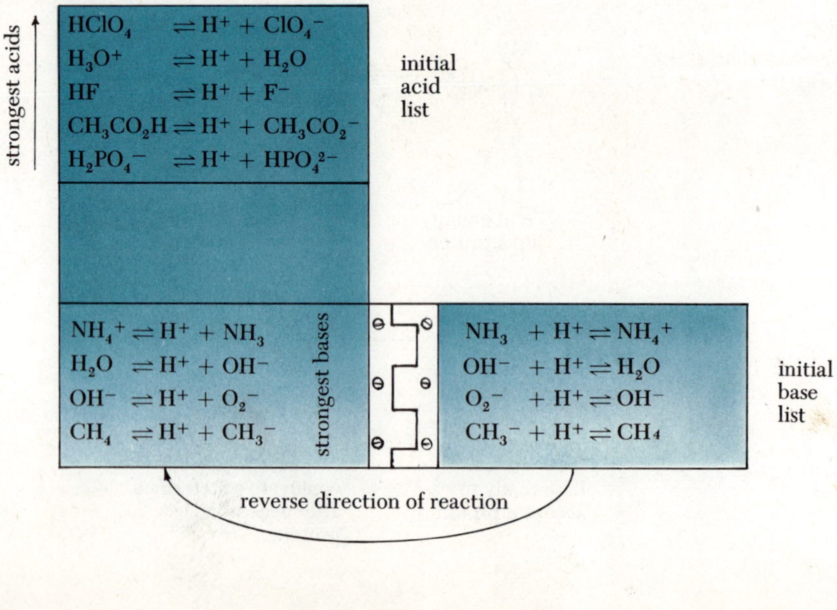

Figure 5.4
Evolution of Table 5.2. Reversing the direction of the reactions in the bottom portion is represented by swinging a hinged gate.

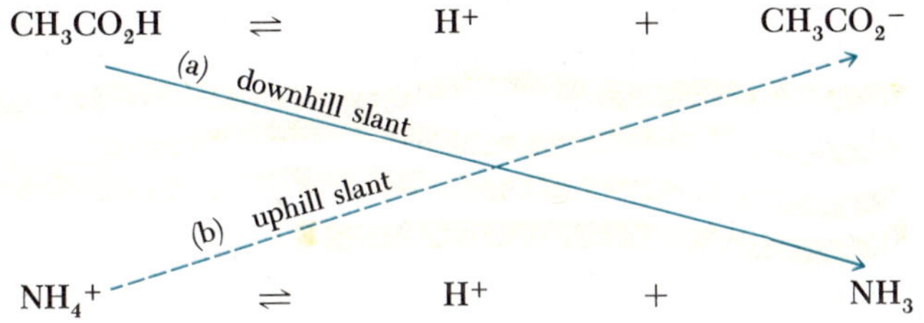

Figure 5.5 Using Table 5.2 for the reaction between CH_3CO_2H and NH_3 to form $CH_3CO_2^-$ and NH_4^+. (a) The downhill line that corresponds to the reaction in the forward direction, for which $K > 1$. (b) The uphill line that corresponds to the reverse direction, for which $K < 1$.

create a unified table of the form of Table 5.2. The evolution of Table 5.2 is outlined schematically in Fig. 5.4.

By the design of Table 5.2, it is clear that all the acid species are to be found on the left side of the arrows and all the basic species are to be found to the right of the arrows. Furthermore, the strongest acids are to be found at the top of the table along with the weakest bases, and the strongest bases and the weakest acids are to be found at the bottom. This grouping of strong acid with weak conjugate base and weak acid with strong conjugate base is to be expected. It simply shows that if a given acid tends readily to get rid of a proton, its conjugate base must not have a very great tendency to attach to a proton.

Using Table 5.2, it is easy to devise a simple procedure by which to judge quickly whether or not a specific acid-base reaction will occur in a given direction—whether the corresponding equilibrium constant is greater than 1. To accomplish this we draw a line from left to right connecting the reactant acid and the reactant base; these occur on the left side of the *overall* equation of interest. For example, if this reaction is between acetic acid and ammonia, described by Eq. (5.8), we then draw a line from left to right connecting CH_3CO_2H and NH_3 in the table. We find that this line, seen as the solid line in Fig. 5.5, slants downward as we go from left to right across the page. This downhill slant indicates that acetic acid is a stronger acid than NH_4^+, and consequently occurs higher on the table than the position corresponding to NH_4^+. From these relative acidities, we know that the reaction in question should have an equilibrium constant greater than 1, and we shall find for *any* acid-base reaction with K greater than 1, that the line connecting the reactant acid with the reactant base will always give a downhill slope, as we look from left to right on the table. We can think of these as downhill reactions, tending to occur spontaneously in the direction written. (This is analogous to the tendency of a ball placed on an incline to roll downhill spontaneously.)

Suppose we want to consider the reverse reaction, namely the reaction between ammonium ion and acetate ion to form ammonia and acetic acid. If we again draw a line between the reactant acid and reactant base, we obtain a line that slants uphill as we view the table from left to right (the dashed line in Fig. 5.5). This uphill slant corresponds to a reaction between a weak base ($CH_3CO_2^-$) and a weak acid (NH_4^+) to produce a stronger base (NH_3) and a stronger acid (CH_3CO_2H); such a reaction must have an equilibrium constant less than 1. This uphill slant is analogous to the tendency of a ball to roll uphill; of course, there is no tendency for such a process to occur spontaneously. From considerations of this type it is clear that any acid-base reaction for which the slant is uphill from left to right on Table 5.2 has an equilibrium constant less than 1.

By drawing the types of lines described above or by imagining them, we can immediately determine the direction in which acid-base reactions occur. The steepest downhill lines will occur for combinations of the strongest reactant acids and the strongest reactant bases; for such reactions the equilibrium constant is much greater than 1; i.e., we have $K \gg 1$. The steepest uphill lines correspond to the most unfavorable reactions, for which $K \ll 1$.

STUDY PROBLEM 5(g)

Predict the general magnitude of the equilibrium constant for the reaction between acetic acid and sulfate ion. Will this reaction tend to occur spontaneously?

Generalized Acid-Base Equilibrium Constants

Everything that has been said so far about the usefulness of the relations represented in Table 5.2 has been qualitative. There are various ways in which quantitative methods can be incorporated into our approaches. One obvious way is to tabulate equilibrium constants and work with them for all possible acid-base reactions. But even if we limited our scope to the acids and bases represented in Table 5.2, we should have to tabulate 729 equilibrium constants, one for each possible combination of a reactant acid and a reactant base. Fortunately that type of tabulation is not necessary.

As we have seen, the tendency of an acid to donate a proton is inextricably related to the tendency of its conjugate base to accept a proton—this relation being of an inverse type. Hence, if we tabulate an equilibrium constant that is characteristic of acid strength, that same constant provides information on the base strength of the conjugate base; so a tabulation of just acid strengths is sufficient for characterizing all acid-base processes. Then, by properly combining our information on the strength of a reactant acid and the strength of a reactant base (derived from the strength of the conjugate acid), we should, using a tabulation like Table 5.2, be able to compute the equilibrium constant for any acid-base process.

The equilibrium constant chosen to represent the strength of an acid is the ionization constant, K_a, defined in Eq. (5.25) for both the Arrhenius and Brönsted-Lowry views. Values of K_a, or estimates, for the acids represented in Table 5.2 are listed on the right side of that table. Although the table is designed in a way that seems to emphasize the Arrhenius approach, it is advantageous to remember that each K_a is a measure of the tendency of the corresponding acid to donate a proton to the base, water. For example, the entry for the hyposulfite ion, HSO_3^-, means that for the reaction

$$HSO_3^- + H_2O \rightleftharpoons H_3O^+ + SO_3^{2-}$$

we have

$$K_a = \frac{[H_3O^+][SO_3^{2-}]}{[HSO_3^-]} = 1.0 \times 10^{-7}$$

Note that for H_3O^+ as an acid the K_a value is unity. This is to be expected, as the implied proton transfer reaction is

$$H_3O^+ + H_2O \rightleftharpoons H_3O^+ + H_2O$$

for which

$$K_a = \frac{[H_3O^+]}{[H_3O^+]} = 1$$

To see how the K_a values tabulated in Table 5.2 can be used for computing acid-base equilibrium constants in general, let us consider the reaction of acetic acid with ammonia in aqueous solution:

$$CH_3CO_2H + NH_3 \rightleftharpoons NH_4^+ + CH_3CO_2^-$$

$$K = \frac{[NH_4^+][CH_3CO_2^-]}{[CH_3CO_2H][NH_3]} = ?$$

In determining the value of K for this case, let us recall that the symbols in brackets are algebraic variables that stand for the equilibrium concentration of species, and can be manipulated by the usual rules of algebra. With this in mind, let us multiply both the numerator and the denominator of the right side of the last equation by $[H_3O^+]$, and then rearrange the result a little:

$$K = \frac{[NH_4^+][CH_3CO_2^-]\,[H_3O^+]}{[CH_3CO_2H]\,[NH_3]\,[H_3O^+]} = \underbrace{\frac{[CH_3CO_2^-][H_3O^+]}{[CH_3CO_2H]}}_{(a)} \cdot \underbrace{\frac{[NH_4^+]}{[NH_3]\,[H_3O^+]}}_{(b)} \qquad (5.39)$$

The first factor on the right (a) clearly equals the ionization constant of acetic acid (1.8×10^{-5}), according to Eq. (5.23b). Similarly, the second factor on the right (b) is simply the inverse of K_a for the acid NH_4^+; this K_a value is also given in Table 5.2.

$$NH_4^+ + H_2O \rightleftharpoons H_3O^+ + NH_3 \qquad (5.40)$$

$$K_a = \frac{[H_3O^+]\,[NH_3]}{[NH_4^+]} = 5.5 \times 10^{-10} \qquad (5.41)$$

Hence, the value of K in Eq. (5.39) is obtained directly by substitution:

$$K = (a)(b) = (K_{a,CH_3CO_2H})\left(\frac{1}{K_{a,NH_4^+}}\right) = \frac{1.8 \times 10^{-5}}{5.5 \times 10^{-10}} = 3.3 \times 10^{4} \qquad (5.42)$$

Clearly, as we could tell qualitatively from the downhill slope in Table 5.2, this reaction proceeds strongly in the direction written.

That the equilibrium constant obtained in Eq. (5.42) equals the ratio of the K_a values of CH_3CO_2H and NH_4^+ should not surprise us. In Sec. 4.4, we learned that the equilibrium constant of a reaction that can be expressed as the difference of two reactions (one minus the other) is the ratio of the equilibrium constants of those two reactions. Clearly the net reaction

(I) $$CH_3CO_2H + NH_3 \rightleftharpoons NH_4^+ + CH_3CO_2^-$$

can be expressed as the difference between reactions (II) and (III).*

(II) $$CH_3CO_2H + H_2O \rightleftharpoons H_3O^+ + CH_3CO_2^-$$

(III) $$NH_4^+ + H_2O \rightleftharpoons H_3O^+ + NH_3$$

That is, (I) = (II) − (III). Hence, from Sec. 4.4, we expect to find the following relation among the equilibrium constants:

$$K_I = \frac{K_{II}}{K_{III}}$$

Since $K_{II} = K_{a,CH_3CO_2H}$, and $K_{III} = K_{a,NH_4^+}$ the result embodied in Eq. (5.42) follows directly.

STUDY PROBLEM 5(h)

Using information given in Table 5.2, derive the value of K for the reaction $HSO_3^- + CN^- \rightleftharpoons HCN + SO_3^{2-}$.

The kinds of manipulations represented in the preceding cases allow one to carry out a variety of useful computations. The following representative examples cover some of the common kinds of calculation. In each case, the first step in solving the problem is to determine *the dominant acid-base reaction;* this *will be the reaction between the strongest available acid and the strongest available base.* This statement can be justified on the basis of the above arguments if one notes that this particular acid-base combination gives the largest possible equilibrium constant for the overall acid-base process.

EXAMPLE 5.11 Calculate the pH of an aqueous solution that is formally 1.2 M in formic acid, HCO_2H.

SOLUTION This problem is concerned with the reaction between the strongest available acid, HCO_2H, and the strongest available base, H_2O. This is a simple ionization reaction, $HCO_2H + H_2O \rightleftharpoons H_3O^+ + HCO_2^-$ for which, according to Table 5.2,

$$K_a = \frac{[H_3O^+][HCO_2^-]}{[HCO_2H]} = 1.8 \times 10^{-4}$$

We solve the problem by first listing what the pertinent concentrations of the species would be if no ionization took place, and then listing what the concentrations would be if y mol/liter of the acid ionizes in achieving equilibrium. This amount of ioniza-

*If one wishes to stay closer to the notation used in Table 5.2, the reactions (I) and (II) could be represented as

(II) $$CH_3CO_2H \rightleftharpoons H^+ + CH_3CO_2^-$$

(III) $$NH_4^+ \rightleftharpoons H^+ + NH_3$$

Clearly, the relation (I) = (II) − (III) still holds, from which Eq. (5.42) still follows directly, according to Sec. 4.2 or to the type of development used to obtain Eq. (5.39).

tion would produce y mol/liter of H_3O^+ and y mol/liter of HCO_2^-. It is convenient to summarize these two situations.

Hypothetical Initial Concentrations	*Concentrations at Equilibrium*
$[HCO_2H] = 1.2$	$[HCO_2H] = 1.2 - y$
$[HCO_2^-] = 0$	$[HCO_2^-] = y$
$[H_3O^+] = 0$ (actually 1.0×10^{-7} for pure water)	$[H_3O^+] = y$*

*If we want to be truly rigorous, we should write $[H_3O^+] = y + 1.0 \times 10^{-7} - z$ for equilibrium. The number 1.0×10^{-7} corresponds to the value of $[H_3O^+]$ before HCO_2H ionization occurred. The symbol z represents the amount of H_3O^+ that combines with an equal amount of OH^- (initially available at 1.0×10^{-7} M before ionization of HCO_2H), so that $[OH^-]$ can be reduced to a value consistent with the constraint $[H^+][OH^-] = 1.0 \times 10^{-14}$. In other words, as a significant increase in $[H_3O^+]$ accompanies ionization of the acid, $[OH^-]$ must decrease accordingly for Eq. (5.33) to remain satisfied; and this occurs by a combination of a portion z of the "initial" 1.0×10^{-7} M H^+ with an equal amount of the initial 1.0×10^{-7} M OH^-. As the value of $1.0 \times 10^{-7} - z$ is very small, the statement that $[H_3O^+] = y$ at equilibrium is a very good approximation.

Substituting the equilibrium concentrations into the above equilibrium expression gives

$$\frac{y \cdot y}{1.2 - y} = 1.8 \times 10^{-4}$$

Now, using the method of successive approximations outlined in Chap. 4, we make an initial guess that y can be neglected in comparison with 1.2. Then,

$$\frac{y^2}{1.2} = 1.8 \times 10^{-4}$$

$$y^2 = 1.2 \times 1.8 \times 10^{-4} = 2.2 \times 10^{-4}$$

$$y = \sqrt{2.2 \times 10^{-4}}$$

$$= 1.5 \times 10^{-2}$$

From this we see that y is only 1% of 1.2 and that the approximation $1.2 - y \cong 1.2$ was valid to the precision of two significant figures. Hence, the result is

$$[H_3O^+] = y = 1.5 \times 10^{-2}$$

$$\begin{aligned} pH &= -\log [H^+] \\ &= -\log [H_3O^+] \\ &= -\log (1.5 \times 10^{-2}) \\ &= -\log 1.5 - \log 10^{-2} \\ &= -0.18 - (-2) \\ &= 1.82 \end{aligned}$$

STUDY PROBLEM 5(i)

Derive the concentrations of all principal species in a 0.80 M solution of HCO_2H in water.

EXAMPLE 5.12 What is the pH of an 0.80 *M* ammonia solution?

SOLUTION When ammonia is dissolved in water, the most important reaction that occurs is the reaction of H_2O (the strongest available acid) with NH_3 (the strongest available base):

$$H_2O + NH_3 \rightleftharpoons NH_4^+ + OH^- \tag{5.7}$$

The equilibrium for this reaction, as expressed by Eq. (5.29), determines the equilibrium concentrations of the pertinent species. We again write down the hypothetical "initial" concentrations and the concentrations of species at equilibrium, assuming that y mol/liter of NH_3 undergoes ionization reaction (5.7) in attaining equilibrium.

Hypothetical Initial Concentrations	*Concentrations at Equilibrium*
$[NH_3] = 0.80$	$[NH_3] = 0.80 - y$
$[NH_4^+] = 0$	$[NH_4^+] = y$
$[OH^-] = 0$ (actually 1.0×10^{-7} for pure water)	$[OH^-] = y$*

*An argument analogous to what was described in the footnote to Example 5.11 also applies here.

Substituting these expressions for the equilibrium concentrations into Eq. (5.29), we have

$$\frac{y \cdot y}{0.80 - y} = K_b$$

Now, either we can look up the value of K_b (1.8×10^{-5}) in the literature, or we can compute it from the information in Table 5.2. From the entry for NH_4^+ ionization, we know that K_a corresponding to Eq. (5.40) is 5.5×10^{-10}. Then, multiplying both the numerator and the denominator of the middle expression of Eq. (5.41) by $[OH^-]$, we obtain

$$K_a = K_a\frac{[OH^-]}{[OH^-]} = \frac{[H_3O^+][NH_3]}{[NH_4^+]} \cdot \frac{[OH^-]}{[OH^-]} = 5.5 \times 10^{-10}$$

This equation can be rearranged, and simplified by recalling that $[H_3O^+][OH^-] = 1.0 \times 10^{-14} = K_w$:

$$K_a = \frac{[NH_3]}{[NH_4^+][OH^-]} \cdot \frac{[H_3O^+][OH^-]}{1}$$

$$= \frac{[NH_3]}{[NH_4^+][OH^-]} \cdot K_w = \frac{1}{K_b} \cdot K_w$$

With additional algebraic rearrangement, we obtain the relation that we need:

$$K_b = \frac{[NH_4^+][OH^-]}{[NH_3]} = \frac{K_w}{K_a}$$

Now, substituting the values of K_w (1.0×10^{-14}) and K_a (5.5×10^{-10}), we get

$$K_b = \frac{[NH_4{}^+][OH^-]}{[NH_3]} = \frac{1.0 \times 10^{-14}}{5.5 \times 10^{-10}} = 1.8 \times 10^{-5}$$

Hence, for 0.80 M NH_3, we have from above

$$\frac{y^2}{0.80 - y} = 1.8 \times 10^{-5}$$

Using the method of successive approximations, we try neglecting y in comparison with 0.80; that is, we make the trial approximation $0.80 - y \cong 0.80$:

$$\frac{y^2}{0.8} = 1.8 \times 10^{-5}$$

$$y^2 = 0.8 \times 1.8 \times 10^{-5}$$

$$y = \sqrt{14.4 \times 10^{-6}}$$

$$= 3.8 \times 10^{-3}$$

This value of y is indeed negligible in comparison with 0.80; hence, the initial approximation was valid within the limits of precision of this calculation (two significant figures), and $[OH^-] = 3.8 \times 10^{-3}$ M. Then, by Eq. (5.35),

$$\begin{aligned} pH &= 14 + \log [OH^-] \\ &= 14 + \log (3.8 \times 10^{-3}) \\ &= 14 + 0.58 - 3 \\ &= 11.58 \end{aligned}$$

An important general result contained in this last example is the equation, for 25 °C,

$$K_b = \frac{K_w}{K_a} \tag{5.43a}$$

$$= \frac{1.0 \times 10^{-14}}{K_a} \tag{5.43b}$$

Equation (5.43) relates K_b for any base to the K_a value of its conjugate acid, where both K_a and K_b are ionization constants in water. This relation makes it convenient to use Table 5.2 to obtain both K_a values (directly) and K_b values (indirectly).

STUDY PROBLEM 5(j)

Compute K_b for F^- in water at 25 °C.

STUDY PROBLEM 5(k)

Calculate the pH of a 0.20 M ammonia solution.

STUDY PROBLEM 5(l)

Derive the relation shown in Eq. (5.43) for the specific case of the $NH_3/NH_4{}^+$ system.

EXAMPLE 5.13 What is the pH of an aqueous solution that is prepared as 0.73 *M* in sodium acetate?

SOLUTION When the salt of a strong base (NaOH in this case) and a weak acid (CH_3CO_2H in this case) is dissolved in water, a somewhat basic solution results. This phenomenon is sometimes referred to as **hydrolysis**; the hydrolysis reaction is easily understood as just another example of an acid-base reaction that can be described in terms of Table 5.2. When sodium acetate is added to water, the strongest reactant base available is $CH_3CO_2^-$ and the strongest available acid is H_2O. Hence, the main reaction is

$$H_2O + CH_3CO_2^- \rightleftharpoons CH_3CO_2H + OH^-$$

The uphill slant of the line that one visualizes on Table 5.2 for this reaction tells us immediately that the equilibrium constant is less than unity. The actual value of K, given by

$$K = \frac{[CH_3CO_2H][OH^-]}{[CH_3CO_2^-]}$$

can be obtained quite simply by recognizing that the acid-base reaction of interest is of the general type represented by Eq. (5.27a) and (5.27b). For this case, the equilibrium constant K is actually what we have denoted K_b in Eq. (5.27b). Furthermore, Eq. (5.43) provides a way of computing this K_b from the known value of K_a for acetic acid (1.8×10^{-5}, given in Table 5.2).

$$\frac{[CH_3CO_2H][OH^-]}{[CH_3CO_2^-]} = K_b = \frac{K_w}{K_a} = \frac{1.0 \times 10^{-14}}{1.8 \times 10^{-5}} = 0.56 \times 10^{-9}$$

Letting y represent the number of moles/liter of $CH_3CO_2^-$ that undergo hydrolysis in attaining equilibrium, we have

Hypothetical Initial Concentrations	*Concentrations at Equilibrium*
$[CH_3CO_2H] = 0$	$[CH_3CO_2H] = y$
$[CH_3CO_2^-] = 0.73$	$[CH_3CO_2^-] = 0.73 - y$
$[OH^-] = 0$	$[OH^-] = y$

Substituting these equilibrium concentrations into the above equilibrium expression gives

$$K_b = \frac{y^2}{0.73 - y} = 5.6 \times 10^{-10}$$

Then, if we neglect y in comparison with 0.73,

$$y^2 = 0.73 \times 5.6 \times 10^{-10}$$
$$y = \sqrt{4.1 \times 10^{-10}}$$
$$= 2.0 \times 10^{-5}$$

The approximation $0.73 - y = 0.73$ was valid.

Since $y = [OH^-]$, the solution is slightly basic (10^{-7} *M* OH^- would be neutral).

Spotlight

The cyanide ion, CN^-, hydrolyzes in water to form the weak hydrocyanic acid (prussic acid):

$$CN^- + H_2O \rightleftharpoons HCN + OH^- \qquad K = 2 \times 10^{-5}$$

Thus, solutions of metallic cyanides that ionize are basic. When these solutions are acidified, they liberate hydrogen cyanide, HCN, a poisonous gas with an odor of bitter almonds. Commercially, HCN is prepared by the oxidation of a mixture of methane and ammonia:

$$2CH_4(g) + 2NH_3(g) + 3O_2(g) \rightleftharpoons 2HCN(g) + 6H_2O$$

The cyanide ion forms a wide variety of complexes with metal ions. Typical examples are $Pt(CN)_2Cl_2$ and $Co(CN)_6{}^{3-}$. The cyanide ion functions in many reactions as if it were a halide ion, say Cl^- or I^-. Many features of the chemistry of halogens are also found for the CN group. For this reason the compound cyanogen, $(CN)_2$, is referred to as a pseudohalogen. Two examples of this similarity follow.

$$4HCN(g) + O_2(g) \rightleftharpoons 2(CN)_2(g) + 2H_2O$$
$$4HI(aq) + O_2(g) \rightleftharpoons 2I_2(s) + 2H_2O$$

$$AuCN(s) + CN^- \rightleftharpoons Au(CN)_2{}^- \text{ (cyanoaurate ion)}$$
$$AuCl(s) + Cl^- \rightleftharpoons AuCl_2{}^- \text{ (chloraurate ion)}$$

The formation of the cyanoaurate ion provides the basis of the commercially important cyanide process for extracting gold from its ore using an aqueous NaCN solution:

$$4Au(s) + 8CN^- + 2H_2O + O_2(g) \rightleftharpoons 4Au(CN)_2{}^- + 4OH^-$$

Then, from Eq. (5.35)

$$\begin{aligned} pH &= 14 + \log [OH^-] \\ &= 14 + \log (2.0 \times 10^{-5}) \\ &= 14 + 0.30 - 5 \\ &= 9.30 \end{aligned}$$

STUDY PROBLEM 5(m)

Derive the concentrations of all principal species in a 0.10 *M* aqueous solution of KCN.

EXAMPLE 5.14 What are the concentrations of the important species in an aqueous solution prepared as 0.40 *M* in methyl ammonium chloride, CH_3NH_3Cl?

SOLUTION CH_3NH_3Cl is a strong electrolyte that will completely ionize to the ions $CH_3NH_3{}^+$ and Cl^- on being dissolved, or prepared, in water. When the salt of a weak base (such as CH_3NH_2) and a strong acid (HCl in this case) is dissolved in water, a somewhat acidic solution results. According to Table 5.2, the strongest acid available when CH_3NH_3Cl is added to water is $CH_3NH_3{}^+$ and the strongest available base is H_2O. Hence, the main reaction is

$$CH_3NH_3{}^+ + H_2O \rightleftharpoons H_3O^+ + CH_3NH_2$$

for which an uphill line, and an equilibrium constant less than unity, is predicted from Table 5.2. The pertinent equilibrium expression is

$$K = \frac{[H_3O^+][CH_3NH_2]}{[CH_3NH_3{}^+]}$$

Since the main acid-base reaction is simply the "dissociation" reaction for the acid

$CH_3NH_3^+$, we can read the value of K for this reaction directly from Table 5.2:

$$K = K_{a,CH_3NH_3^+} = 2.7 \times 10^{-11}$$

Letting y represent the number of moles/liter of $CH_3NH_3^+$ that dissociate in achieving equilibrium, we write

Hypothetical Initial Concentrations	*Concentrations at Equilibrium*
$[CH_3NH_3^+] = 0.40$	$[CH_3NH_3^+] = 0.40 - y$
$[CH_3NH_2] = 0$	$[CH_3NH_2] = y$
$[H_3O^+] = 0$	$[H_3O^+] = y$

Substituting these equilibrium concentrations into the above equilibrium expression gives

$$\frac{y^2}{0.40 - y} = 2.7 \times 10^{-11}$$

Neglecting y in comparison with 0.40, we have

$$y^2 = 0.40 \times 2.7 \times 10^{-11}$$

$$y = \sqrt{10.8 \times 10^{-12}}$$

$$= 3.3 \times 10^{-6}$$

Hence, the approximation $0.40 - y = 0.40$ was valid. Therefore,

$$[CH_3NH_2] = [H_3O^+] = y = 3.3 \times 10^{-6}\ M$$

$$[CH_3NH_3^+] = 0.40 - y = 0.40\ M$$

to two significant figures. And

$$[OH^-] = \frac{1.0 \times 10^{-14}}{[H_3O^+]}$$

$$= \frac{1.0 \times 10^{-14}}{3.3 \times 10^{-6}}$$

$$= 3.0 \times 10^{-9}\ (\text{mol/liter})$$

EXAMPLE 5.15 What is the pH of a solution prepared by dissolving 0.64 mol sodium acetate and 0.46 mol HCl gas in enough water to provide 1 liter of solution?

SOLUTION The strongest acid available is H_3O^+ (from the complete ionization of HCl) and the strongest base available is the ion $CH_3CO_2^-$. We take a hypothetical starting point for our calculation of assuming that the following reaction occurs completely to the extent of 0.46 mole (the amount of the limiting reagent):

$$H_3O^+ + CH_3CO_2^- \rightleftharpoons CH_3CO_2H + H_2O$$

This hypothetical starting point would be a 1-liter solution that is 0.46 M in CH_3CO_2H and 0.18 M in $CH_3CO_2^-$ (the excess remaining after the reaction above), and that is neutral (that is, $[H_3O^+] = 1.0 \times 10^{-7}$). We can then search for the true equilibrium by asking what acid-base reaction will occur from that hypothetical starting point and the extent to which it will occur. The strongest acid available in our hypothetical initial solution is CH_3CO_2H and the strongest base is $CH_3CO_2^-$. However, the reaction

$$CH_3CO_2H + CH_3CO_2^- \rightleftharpoons CH_3CO_2H + CH_3CO_2^-$$

is really no reaction at all, so this is not involved in establishing the true equilibrium. The next strongest base after $CH_3CO_2^-$ is water, so the reaction of interest is the reaction of CH_3CO_2H with water, that is, the ionization of CH_3CO_2H, as represented by Eqs. (5.23a) and (5.23b). Setting up the initial and equilibrium conditions as in the preceding examples, and letting y represent the number of moles (per liter) of CH_3CO_2H that ionize in the transformation from the hypothetical initial conditions to the equilibrium condition, we have

Hypothetical Initial Concentrations	*Concentrations at Equilibrium*
$[CH_3CO_2H] = 0.46$	$[CH_3CO_2H] = 0.46 - y$
$[CH_3CO_2^-] = 0.18$	$[CH_3CO_2^-] = 0.18 + y$
$[H_3O^+] = 0$	$[H_3O^+] = y$

Substituting these expressions for the equilibrium concentration into equation (5.23b) and using K_a from Table 5.2,[*] we get

$$\frac{[H_3O^+][CH_3CO_2^-]}{[CH_3CO_2H]} = \frac{(y)(0.18 + y)}{(0.46 - y)} = 1.8 \times 10^{-5}$$

Now, making the initial approximation that the value of y can be neglected in comparison with either 0.18 or 0.46 (that is, $0.18 - y \cong 0.18$; $0.46 - y \cong 0.46$), we obtain

$$\frac{(y)(0.18)}{(0.46)} = 1.8 \times 10^{-5}$$

$$y = \frac{0.46}{0.18} \times 1.8 \times 10^{-5}$$

$$= 4.6 \times 10^{-5}$$

This is certainly negligible compared with either 0.18 or 0.46, so the initial approximation was valid. Thus, $[H_3O^+] = [H^+] = 4.6 \times 10^{-5}$.

$$pH = -\log [H^+] = -\log (4.6 \times 10^{-5}) = -(0.66 - 5) = 4.34$$

[*]We could equally well have approached Example 5.15 by recognizing that water is the second strongest acid available in the hypothetical initial condition; we should then have proceeded to find the equilibrium state by considering the reaction

$$H_2O + CH_3CO_2^- \rightleftharpoons CH_3CO_2H + OH^-$$

The results obtained by this approach would be the same as in the example worked out.

5.4 Buffers, Indicators, and Titrations

The Buffer Concept

pK_a

Some additional useful relations are obtained by logarithms. Let us take the logarithm of both sides of Eq. (5.22):

$$K_a = \frac{[H_3O^+][A^-]}{[HA]} \tag{5.22}$$

$$\log K_a = \log\left\{\frac{[H_3O^+][A^-]}{[HA]}\right\} = \log [H_3O^+] + \log \frac{[A^-]}{[HA]}$$

Rearranging this, we obtain

$$-\log [H_3O^+] = -\log K_a + \log\left(\frac{[A^-]}{[HA]}\right) \tag{5.44}$$

Now, we define the following new variable:

$$pK_a = -\log K_a \tag{5.45}$$

In terms of this new variable and the definition of pH (Eq. 5.34), Eq. (5.44) takes the form

$$pH = pK_a + \log \frac{[A^-]}{[HA]} \tag{5.46}$$

Among biochemists, Eq. (5.46) is often referred to as the **Henderson-Hasselbach equation.**

Values of pK_a are often tabulated, rather than K_a values. For a given acid HA, the value of pK_a is fixed for a given temperature, and the pH of a solution containing HA and A^- is determined by the ratio $[A^-]/[HA]$. If the concentrations of HA and A^- are equal, irrespective of their magnitude, then according to Eq. (5.46) the pH is precisely equal to the pK_a of HA; that is,

$$pH = pK_a + \log \frac{[A^-]}{[HA]} = pK_a + \log 1$$

$$= pK_a$$

$$pH = pK_a \qquad \text{for } [HA] = [A^-] \tag{5.47}$$

Buffers

If a solution contains nearly equal concentrations of an acid and its conjugate base, and if the magnitude of this concentration is relatively large, then it is said to be a **buffer solution.** Equation (5.47) tells us that the pH of such a buffer solution is equal to the pK_a of HA. For example, if a solution is 1.0 *M* in CH_3CO_2H ($pK_a = 4.7$) and is 1.0 *M* in the conjugate base, $CH_3CO_2^-$, then the pH of the solution is 4.7. For a buffer solution in which HA is $H_2PO_4^-$ ($pK_a = 7.2$) and A^- is HPO_4^{2-}, the equilibrium of interest is

$$H_2PO_4^- + H_2O \rightleftharpoons H_3O^+ + HPO_4^{2-}$$

$$K_a = \frac{[H_3O^+][HPO_4^{2-}]}{[H_2PO_4^-]}$$

Then, if $[H_2PO_4^-] = [HPO_4^{2-}]$, the pH would be 7.2.

The buffer relation implicit in Eq. (5.47) constitutes the basis of a convenient way of maintaining a desired value for the pH of a solution. One may wish to use such a solution for an experiment in which it is important to maintain a known and fixed pH. For example, if one wants to carry out a biologically significant experiment on a blood extract, maintaining the pH at 7.2 throughout the experiment, then one needs merely to dissolve an equal number of moles of NaH_2PO_4 and Na_2HPO_4 in the extract. In applying this buffer concept, one must be sure that the acid-base combination that is chosen for the buffer system will not react with the components to be studied, and that enough buffer is present to ensure maintenance of the desired pH even though small amounts of other acids or bases are added or generated during the experiment. This ability to maintain a preselected pH even when small amounts of acids or bases are introduced into the system is called the **buffer capacity.**

The manner in which a buffer manifests a capacity can be seen as follows. Suppose we have a solution made up by adding one mole of NaH_2PO_4 and one mole of Na_2HPO_4 with enought water to yield 1 liter of solution. Then, $[H_2PO_4^-] = [HPO_4^{2-}] = 1.0\ M$, and $pH = pK_a = 7.2$. Now, suppose that to this solution we add (a) 0.010 mole of HCl, or (b) 0.010 mole of NaOH; in each case, what would the resulting pH be? First, consider the addition of HCl. We note that "initially," the strongest acid available is H_3O^+ (0.010 mol of it), and the strongest base available is HPO_4^{2-} (1.0 mol of it). Thus, the main acid-base reaction that occurs in reattaining equilibrium is

$$H_3O^+ + HPO_4^{2-} \rightleftharpoons H_2PO_4^- + H_2O$$

This reaction consumes essentially all the added H_3O^+, since it has a very large equilibrium constant, approximately 10^7. At equilibrium, $[HPO_4^{2-}] = 1.0 - 0.010 = 0.99$, and $[H_2PO_4^-] = 1.0 + 0.010 = 1.01$. Then, by Eq. (5.46),

$$\begin{aligned} pH = pK_a + \log\frac{[HPO_4^{2-}]}{[H_2PO_4^-]} &= 7.2 + \log\frac{(0.99)}{(1.01)} \\ &= 7.2 + \log(0.98) = 7.2 + \log(9.8 \times 10^{-1}) \\ &= 7.2 + (0.99 - 1) = 7.2 - 0.01 \cong 7.2 \end{aligned}$$

Thus, to the precision of two significant figures, when 0.010 mol HCl is added to the buffer, the pH doesn't change at all! Now, consider the effect of adding 0.010 mol NaOH to the $H_2PO_4^-/HPO_4^{2-}$ buffer system. Initially, the strongest acid present is $H_2PO_4^-$ (1.0 mol of it) and the strongest base present is OH^- (0.010 mol of it), so the main acid-base reaction (also with a very large equilibrium constant, approximately 10^7) is

$$H_2PO_4^- + OH^- \rightleftharpoons H_2O + HPO_4^{2-}$$

Then, at equilibrium, $[HPO_4^{2-}] = 1.0 + 0.010 = 1.01$ and $[H_2PO_4^-] = 1.0 - 0.010 = 0.99$. Again, substituting into Eq. (5.46),

$$\begin{aligned} pH = pK_a + \log\frac{[HPO_4^{2-}]}{[H_2PO_4^-]} &= 7.2 + \log\frac{1.01}{0.99} \\ &= 7.2 + \log 1.02 = 7.2 + 0.01 \cong 7.2. \end{aligned}$$

Therefore, to a rough level of precision, when 0.010 mol NaOH is added to the $H_2PO_4^-/HPO_4^{2-}$ buffer system, the pH remains unchanged.

It is instructive to compare the additions of HCl and NaOH to the buffer system with corresponding additions to an aqueous solution without a buffer. Suppose we have 1 liter of such a solution, with pH = 7.2.* If 0.010 mol HCl is added to one liter of an unbuffered solution, the resulting pH is $-\log(0.01) = 2$; thus, we have a change from 7.2 to 2.0! Similarly, if 0.010 mol NaOH is added, we should have $[OH^-] = 0.010$, $[H^+] = 1.0 \times 10^{-14}/0.01 = 10^{-12}$, and pH = 12. This is a pH change from 7.2 to 12! Thus, the capacity of a buffer to maintain a fixed pH is of considerable consequence, and useful to remember.

STUDY PROBLEM 5(n)

Calculate the pH of each of the following two solutions before and after 0.0010 mol NaOH is added. (a) A 1000-ml solution that is made up as 0.50 *M* CH_3CO_2H and 0.50 *M* CH_3CO_2Na (acetic acid and sodium acetate). (b) A 1000-ml sample of pure water.

Indicators

Another important manifestation of Eq. (5.46) is the ability to use specific acid-base systems to find out what the pH is. Such acid-base systems are called **indicators.** The basis of an indicator is that the acid and its conjugate base interact differently with light, and therefore have different colors. For example, the common indicator, phenolphthalein, is colorless in its acid form HA, and pinkish in its conjugate base form A^-. As the pH of a solution is varied, say by addition of strong acids or bases, the ratio $[A^-]/[HA]$ for the phenolphthalein system changes in accordance with Eq. (5.46); and since A^- and HA have different colors, the color of the solution changes as the relative amounts of these two species are varied. By rewriting Eq. (5.46) for the specific case of an indicator system in an aqueous solution, we obtain

$$\text{pH} = \text{p}K_{\text{a(ind)}} + \log\left(\frac{[A^-]}{[HA]}\right)_{\text{ind.}} \tag{5.46a}$$

From this equation we see that if the pH of the solution happens to equal the pK_a of the indicator acid, then $[HA] = [A^-]$ for the indicator acid-base system. If the pH is somewhat higher or lower than the $pK_{a(ind)}$ value, then the ratio $[A^-]/[HA]$ is correspondingly larger or smaller than unity—and the color of the solution reflects the predominant species. The pH region for which a given indicator is best able to detect pH changes is the region about the $pK_{a(ind)}$ value. If the pH is varied to just above or below this value, accordingly the A^- or the HA form of the indicator grows at the expense of the other form and determines the color of the solution. Thus, if one wants to detect pH changes over a particular pH range, as in a titration (Sec. 5.3), then an indicator with a pK_a value near the pH range of interest must be chosen. Table 5.3 lists the pK_a values of some common indicators.

For a given solution, using only one indicator to determine the pH can often provide only limited information. The indicator can show whether the pH is greater

*Pure water has a pH = 7.0. A pH = 7.2 solution, which is nearly neutral, might result if the glassware in which the solution is contained has not been thoroughly rinsed free of traces of soap after it was last washed.

TABLE 5.3 Characteristics of Some Common Indicators

Indicator	$pK_{a(ind)}$	*Color for* $\{[A^-]/[HA]\} \ll 1$ *(Acid Side)*	*Useful pH Range*	*Color for* $\{[A^-]/[HA]\} \gg 1$ *(Basic Side)*
Methyl violet	0.8	Yellow	0 to 2	Violet
Thymol blue	2.0	Pink	1.2 to 2.8	Yellow
Bromphenol blue	3.8	Yellow	3.0 to 4.7	Violet
Methyl orange	3.8	Pink	3.1 to 4.4	Yellow
Bromcresol green	4.7	Yellow	4.0 to 5.6	Blue
Bromcresol purple	6.0	Yellow	5.2 to 6.8	Purple
Litmus (a mixture)	5.5–7.5	Red	4.7 to 8.2	Blue
Phenolphthalein	9.1	Colorless	8.3 to 10.0	Pink
Thymolphthalein	9.9	Colorless	9.3 to 10.5	Blue
Alizarin yellow G	11	Colorless	10.1 to 12.1	Yellow
Trinitrobenzene	13	Colorless	12.0 to 14.1	Orange

or smaller than the pH value of the color change of that specific indicator. For example, when one adds phenolphthalein, of $pK_a = 8.8$, to a solution of unknown pH, if the solution remains colorless, then it is clear that pH < 8.6; on the other hand, if the solution turns pink, then one knows that pH > 9.0. If one makes several tests on several samples of unknown solution, using different indicators, then the combination of such results can bracket the pH fairly well, especially if careful comparisons are made with solutions of known pH. Furthermore, the shade of coloring can often be used to pinpoint a pH rather closely. In modern chemical laboratories, electrochemical methods, represented by the pH meter, are often used to determine the pH of a solution. (Electrochemical methods are treated in some detail in Chap. 12.)

EXAMPLE 5.16 What is the approximate pH of a solution that turns yellow when a few drops of thymol blue indicator or a few drops of bromphenol blue indicator are added?

SOLUTION From Table 5.3, we can see that the pH must be between about 2.0 and 3.8. To estimate the pH with better precision one would need to make careful color comparisons with these indicators in solutions of known pH. A better method would be to use a pH meter, if one is available.

Titrations

The Equivalence Point

One of the most common applications of indicators in the laboratory is in detecting the equivalence point in an acid-base titration. This technique consists of the dropwise addition of an acid solution to a base solution, or vice versa. The **equivalence point** is the instant in the stepwise addition at which the number of moles of available H^+ (or base) added is exactly the same as the number of equivalents of base (or available H^+) initially present in the solution being titrated; it is the point at which Eq. (5.21) is satisfied. For the titration of strong acids with strong bases, the pH at the equivalence point will be essentially that of a neutral salt solution, approximately 7.

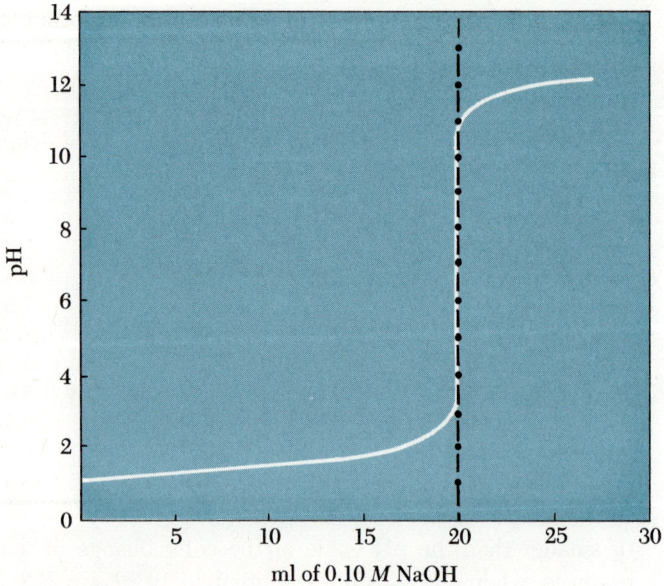

Figure 5.6 Titration curve for the titration of 20.0 ml of 0.10 *M* HCl with 0.10 *M* NaOH. The broken vertical line shows the equivalence point.

For example, suppose one is titrating 20.00 ml of 0.10 *M* HCl contained in a flask with 0.10 *M* NaOH in a buret, 20.00 ml being required to reach the equivalence point. If an excess of 0.04 ml, or about half a drop, of base is added, the equivalence point is passed by 0.04 ml, and the resulting $[OH^-]$ value would be (0.04/40.00)(0.10), or 0.00010 *M*, and the pH would be 10. Similarly, if we stopped short of the endpoint by 0.04 ml, there would be an excess of H_3O^+ in the amount 0.0001 *M*, so the pH would be 4. Hence, the pH changes from 10 to 7 to 4 as the titration proceeds from 19.96 ml to 20.00 ml to 20.04 ml of added base. Such a high sensitivity of pH to added titrant in the region of the equivalence point is referred to as a **sharp endpoint.**

A plot of pH vs the added volume of 0.10 *M* NaOH for this titration is shown in Fig. 5.6. It is readily seen from this kind of reasoning and from Fig. 5.6 that any indicator that gives a color change from about pH 4 to about pH 10 would be a good indicator for this kind of titration. For example, bromcresol purple, litmus, and phenolphthalein would be fine for the titration of a strong acid with a strong base. An entirely analogous situation and the same sort of titration curve would be observed if any strong base were being titrated with any strong acid.

STUDY PROBLEM 5(o)

What indicator would you choose for an acid-base titration with an equivalence point at pH 10.7? Justify your choice.

Titrations of Weak Acids

If weak acids or bases are involved in a titration, the endpoints are not so sharp as those obtained in titrations of strong acids with strong bases. For example, in the titration of a weak acid (in a flask) with a strong base (in a buret), the increase in pH in the region of the endpoint is not so steep for so large a pH range as what is seen in Fig. 5.6 for the titration of a 0.10 *M* solution of strong acid. A weak acid titration is shown graphically in Fig. 5.7. This plot covers a case identical to that of Fig. 5.6 except that the acid being titrated has a pK_a value 5, near the value of acetic acid.

Figure 5.8 is a superposition of separate titration curves of a series of four weak acids (each present initially in 20-ml solutions at 0.10 *M*) with different pK_a values. It is seen that the "best" endpoint (the one of the longest steep portion), corresponding to the lowest curve, is associated with the strongest weak acid in this set, with pK_a = 3. The "poorest" endpoint (shallowest pH change and the most

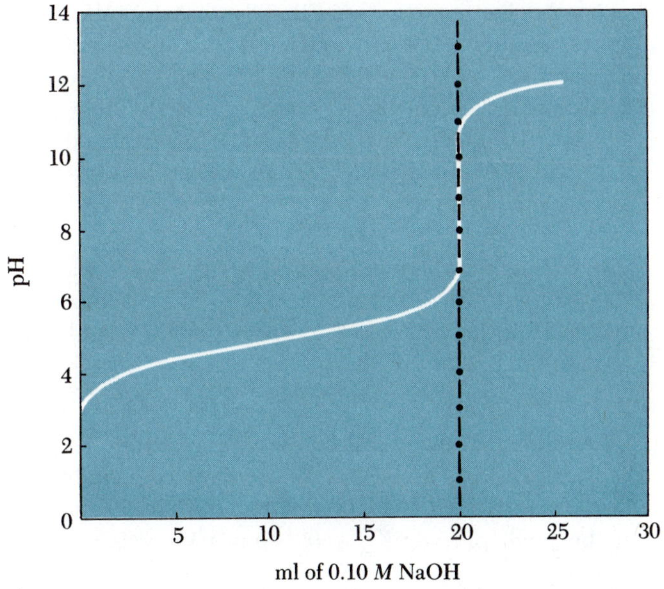

Figure 5.7
Titration curve for the titration of 20.0 ml of a 0.10 *M* solution of a weak acid having $K_a = 1.0 \times 10^{-5}$. Titrant is 0.10 *M* NaOH. The broken vertical line shows the equivalence point.

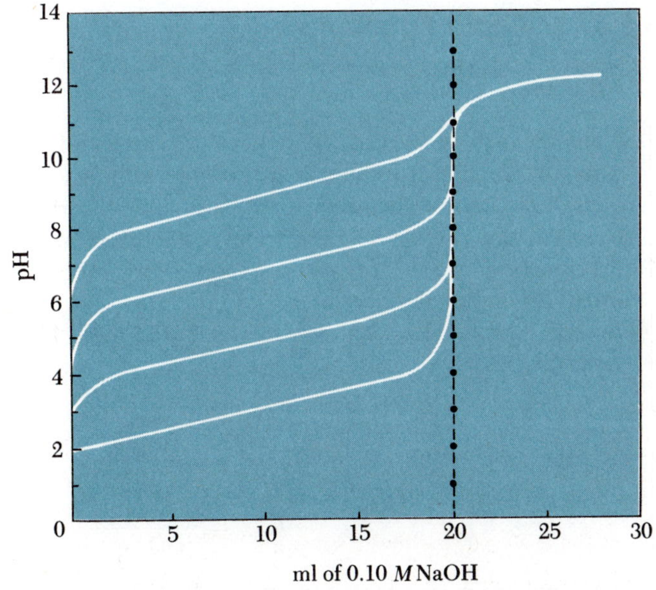

Figure 5.8
Titration curves for the titrations of 20.0 ml samples of 0.10 *M* solutions of four separate acids, with pK_a values of 3 (bottom curve), 5, 7, and 9 (top curve). Titrant in each case is 0.10 *M* NaOH. The broken vertical line shows the equivalence points.

difficult to detect) corresponds to the upper curve, which is associated with the weakest acid in the set, of $pK_a = 9$. The curves for the two acids with intermediate pK_a values (5 and 7) fall between the highest and lowest curves. From Fig. 5.8 one sees that the initial pH is lowest for the strongest weak acid. The portions of the curves corresponding to the addition of excess NaOH, beyond the equivalence point, are all the same; since the acid has already been neutralized before this region is reached, the identity of the acid would no longer be of any consequence. In an actual titration, one would ordinarily have no reason to go beyond the endpoint.

Another feature to note in Fig. 5.8 is that since the quality, or steepness and length, of the endpoint decreases as the strength of the acid decreases, the selection of an indicator for observing the endpoint becomes more critical and more difficult for weaker acids. It can also be seen from Fig. 5.8 that the midpoint of each titration, which is 10.0 ml of added 0.10 M NaOH, corresponds to a pH value equal to pK_a of the weak acid. Equation (5.47) tells us that this must be the case; at the midpoint of the titration, half the acid has been neutralized, so $[A^-] = [HA]$, and $pH = pK_a$.

EXAMPLE 5.17 When a sample of a weak acid of unknown pK_a was titrated to a phenolphthalein endpoint with 0.12 M NaOH, 22.16 ml of titrant was consumed. Then 11.08 ml of 0.12 M HCl was added to the titrated solution and the pH of the resulting solution was determined as 5.6—with a pH meter or a judicious combination of indicators with suitable control experiments. What is the pK_a of the weak acid?

SOLUTION When the neutralized solution was "back-titrated" with 11.08 ml of 0.12 M HCl, half of the conjugate base A^- was converted back to the acid form HA. Hence, at that point, $[HA] = [A^-]$, the midpoint of the titration; and according to Eq. (5.47), at this point $pH = pK_a + 0$. Thus, $pK_a = 5.6$.

OPTIONAL

5.5 Multiple Equilibria

The Carbon Dioxide System

Some chemical species can function in a variety of ways in acid-base reactions. Extremely important examples of this type of behavior are provided by the components of acid-base chemistry in the carbonate–bicarbonate–carbonic acid–carbon dioxide system. This system is extremely important in the respiratory processes and other functions of living organisms; it is also very important in the areas of geological formations and of the oceans.

When gaseous CO_2 dissolves in water, the main transformation that occurs is simply the dissolution of CO_2:

$$CO_2(g) \rightleftharpoons CO_2(aq) \tag{5.48a}$$

for which the equilibrium is defined by

$$K_{abs} = \frac{[CO_2(aq)]}{[CO_2(g)]} \tag{5.48b}$$

Such equilibrium constants are often called **absorption constants** (K_{abs}), referring to the absorption of a gas into a liquid. A small fraction of the dissolved CO_2 "hydrates" to carbonic acid, H_2CO_3:

$$CO_2(aq) + H_2O \rightleftharpoons H_2CO_3 \quad (5.49a)$$

$$K = \frac{[H_2CO_3]}{[CO_2(aq)]} = 2.58 \times 10^{-3} \quad (5.49b)$$

The carbonic acid functions as a diprotic acid. The first ionization produces bicarbonate ion, HCO_3^-.

$$H_2CO_3 + H_2O \rightleftharpoons H_3O^+ + HCO_3^- \quad (5.50a)$$

$$K_1' = \frac{[H_3O^+][HCO_3^-]}{[H_2CO_3]} = 1.7 \times 10^{-4} \quad (5.50b)$$

The second ionization produces carbonate ion, CO_3^{2-}.

$$HCO_3^- + H_2O \rightleftharpoons H_3O^+ + CO_3^{2-} \quad (5.51a)$$

$$K_2' = \frac{[H_3O^+][CO_3^{2-}]}{[HCO_3^-]} = 4.8 \times 10^{-11} \quad (5.51b)$$

The overall ionization of $CO_2(aq)$ that produces bicarbonate ions can be represented by combining Eqs. (5.49) and (5.50), and is characterized by the corresponding ionization constant K_1.

$$CO_2(aq) + 2H_2O \rightleftharpoons H_3O^+ + HCO_3^- \quad (5.52a)$$

$$K_1 = \frac{[H_3O^+][HCO_3^-]}{[CO_2(aq)]} = KK_1' = 4.4 \times 10^{-7} \quad (5.52b)$$

Similarly, the overall ionization of $CO_2(aq)$ that produces carbonate ions can be represented by combining Eqs. (5.51) and (5.52) and is characterized by the corresponding ionization constant K_2.

$$CO_2(aq) + 3H_2O \rightleftharpoons 2H_3O^+ + CO_3^{2-} \quad (5.53a)$$

$$K_2 = \frac{[H_3O^+]^2[CO_3^{2-}]}{[CO_2(aq)]} = K_1K_2' = 2.1 \times 10^{-17} \quad (5.53b)$$

All these equilibrium expressions—(5.48b), (5.49b), (5.50b), (5.51b), (5.52b), and (5.53b)—must hold true for the system to be in a state of chemical equilibrium. In solving problems with this system, we simply have to decide which equations are most useful for our purposes and how to manipulate and use them properly and to our best advantage. The first step in making such decisions is to determine which acid-base reaction is the main reaction, the reaction that dominates the acid-base properties of the system. This dominant acid-base reaction will be the one involving the strongest acid available and the strongest base available. The following examples show the kinds of reasoning used in analyzing multiple acid-base equilibria.

EXAMPLE 5.18 Calculate the concentrations of CO_2, H_2CO_3, HCO_3^-, CO_3^{2-}, H_3O^+, and OH^- in an aqueous solution into which $CO_2(g)$ is being bubbled at a pressure of one atmo-

Spotlight

The CO_2–H_2CO_2–HCO_3^- system is extremely important in maintaining the proper pH of blood. In mammals the tolerable range is about 7.40 ± 0.05. This narrow range must be maintained so that a wide variety of essential life processes can continue. Although other buffering systems are also important in body fluids, the CO_2–HCO_3^- equilibrium (Eq. 5.52a) is especially effective in blood because of the relatively large amounts of these species present in blood and because the dissolved CO_2 concentration is rapidly controllable by respiration.

From Eq. (5.52b) we can write, taking logarithms of both sides,

$$\log K_1 = \log [H_3O^+] + \log \frac{[HCO_3^-]}{[CO_2(aq)]}$$

$$-\log [H_3O^+] = pH = -\log K_1 + \log \frac{[HCO_3^-]}{[CO_2(aq)]}$$

Then inserting the value of pK_1 for 37 °C, the body temperature, one obtains

$$pH = 6.1 + \log \frac{[HCO_3^-]}{[CO_2(aq)]}$$

Thus, the ratio $[HCO_3^-]/[CO_2(aq)]$ can control the pH of blood.

Excessive accumulation or loss of CO_2 gives rise to a disturbance of the acid-base balance and pH control. Hyperventilation, for example, depletes CO_2 from the lungs and causes $[CO_2(aq)]$ to drop; this, according to the above equation, causes the pH to increase, a condition called respiratory alkalosis. The opposite situation, in which expulsion of CO_2 from the lungs is impeded, as in emphysema, results in an increase in CO_2 concentration in the blood. According to the above equation, this gives rise to a decrease in pH, a condition called respiratory acidosis.

OPTIONAL

sphere at a temperature of 30 °C; at this temperature and CO_2 pressure, the value of $[CO_2(g)]K_{abs}$ is found to be 0.033.*

SOLUTION According to Eq. (5.48b), we write

$$K_{abs} = \frac{[CO_2(aq)]}{[CO_2(g)]}$$

$$[CO_2(g)]K_{abs} = [CO_2(aq)]$$

Hence, from the statement of the problem

$$[CO_2(aq)] = 0.033 \text{ (mol/liter)}$$

Then, from Eq. (5.49b),

$$K = \frac{[H_2CO_3]}{[CO_2(aq)]} = 2.6 \times 10^{-3}$$

Then, rearranging gives

$$[H_2CO_3] = (2.6 \times 10^{-3})(3.3 \times 10^{-2}) = 8.6 \times 10^{-5}$$

As $CO_2(aq)$ is the main species in solution, not H_2CO_3, the main acid-base equilibrium to consider is given by Eq. (5.52). For this reaction

$$K_1 = 4.4 \times 10^{-7} = \frac{[H_3O^+][HCO_3^-]}{[CO_2(aq)]} = \frac{[HCO_3^-]^2}{0.033}$$

*The student may wonder if it is really legitimate to use the value 0.033 for $[CO_2(aq)]$. Shouldn't we have some sort of expression like $0.033 - y = [CO_2(aq)]$, since some of the $CO_2(aq)$ is transformed into H_2CO_3? The answer is no, because the equilibrium value of $[CO_2(aq)]$ is maintained at 0.033; $CO_2(g)$ is being constantly bubbled through the solution, maintaining the equilibrium of Eq. (5.48).

Here we have set the concentrations of H_3O^+ and HCO_3^- equal to each other considering Eq. (5.52a) the dominant reaction. Then,

$$[HCO_3^-]^2 = 3.3 \times 10^{-2} \times 4.4 \times 10^{-7} = 1.45 \times 10^{-8}$$

$$[HCO_3^-] = [H_3O^+] = \sqrt{1.45 \times 10^{-8}} = 1.2 \times 10^{-4} \text{ (mol/liter)}$$

From Eq. (5.33), we can obtain the value of $[OH^-]$:

$$[OH^-] = \frac{1.0 \times 10^{-14}}{[H_3O^+]} = \frac{1.0 \times 10^{-14}}{1.2 \times 10^{-4}} = 0.83 \times 10^{-10} \text{ (mol/liter)}$$

For calculating the value of $[CO_3^{2-}]$, we can use either Eq. (5.51b), with the values we have computed for $[HCO_3^-]$ and $[H_3O^+]$, or Eq. (5.53b). Choosing the latter, we write

$$K_2 = 2.1 \times 10^{-17} = \frac{[H_3O^+]^2[CO_3^{2-}]}{[CO_2(aq)]} = \frac{(1.2 \times 10^{-4})^2[CO_3^{2-}]}{(0.033)}$$

$$[CO_3^{2-}] = \frac{2.1 \times 10^{-17} \times 3.3 \times 10^{-2}}{(1.2 \times 10^{-4})^2} = 4.8 \times 10^{-11} \text{ (mol/liter)}$$

EXAMPLE 5.19 Compute the concentrations of CO_2, H_2CO_3, HCO_3^-, CO_3^{2-}, H_3O^+, and OH^- in a solution prepared as 0.20 *M* Na_2CO_3.

SOLUTION The strongest base initially present in an appreciable amount at the instant the solution is prepared is CO_3^{2-}, and the strongest acid available is H_2O (see Table 5.2). Hence, the main acid-base reaction is

(a) $$H_2O + CO_3^{2-} \rightleftharpoons HCO_3^- + OH^-$$

$$K = \frac{[HCO_3^-][OH^-]}{[CO_3^{2-}]}$$

This reaction is a simple acid-base reaction of the general type of Eq. (5.27a). The value of K can be obtained readily through Eq. (5.43) and the known value of K_a for HCO_3^-:

$$K = K_b = \frac{K_w}{K_a} = \frac{1.0 \times 10^{-14}}{4.8 \times 10^{-11}} = 2.1 \times 10^{-4}$$

Letting y stand for the number of moles/liter of CO_3^{2-} that react according to equation (a), we write

Hypothetical Initial Concentrations	*Concentrations at Equilibrium*
$[CO_3^{2-}] = 0.20$	$[CO_3^{2-}] = 0.20 - y$
$[HCO_3^-] = 0$	$[HCO_3^-] = y$
$[OH^-] = 0$	$[OH^-] = y$

$$K_b = \frac{y^2}{0.20 - y} = 2.1 \times 10^{-4}$$

OPTIONAL

Approximating $0.20 - y$ as 0.20, we obtain

$$y^2 \cong (0.20)(2.1 \times 10^{-4})$$

$$y \cong \sqrt{0.42 \times 10^{-4}} = 0.65 \times 10^{-2}$$

To see how good our first approximation was, we note that $0.20 - 0.0065 \cong 0.19$; in other words, even limiting our precision to two significant figures, we see that 0.65×10^{-2} is not completely negligible in comparison with 0.20. Hence, instead of approximating $0.20 - y$ by the value 0.20, we use our derived result, 0.19.

$$2.1 \times 10^{-4} = \frac{y^2}{0.20 - y} \cong \frac{y^2}{0.19}$$

$$y^2 \cong (0.19)(2.1 \times 10^{-4}) = 0.40 \times 10^{-4}$$

$$y = [OH^-] = [HCO_3^-] \cong \sqrt{0.40 \times 10^{-4}} \cong 0.63 \times 10^{-2} \text{ (mol/liter)}$$

Then, $0.20 - y = 0.19$, so our second approximation is valid to two significant figures, and we need not proceed to additional iterations.

$$[CO_3^{2-}] = 0.20 - y \cong 0.19 \text{ (mol/liter)}$$

Using Eq. (5.33),

$$[H_3O^+] = \frac{1.0 \times 10^{-14}}{[OH^-]} = \frac{1.0 \times 10^{-14}}{0.63 \times 10^{-2}} = 1.6 \times 10^{-12}$$

All that remains to be computed are the values of $[H_2CO_3]$ and $[CO_2(aq)]$, either of which could be obtained from a knowledge of the other through Eq. (5.49b). Having computed $[H_3O^+]$ and $[HCO_3^-]$ above, we can obtain $[H_2CO_3]$ from Eq. (5.50b) or $[CO_2(aq)]$ from Eq. (5.52b). (Alternatively, we could just as well calculate $[CO_2(aq)]$ from our knowledge of $[H_3O^+]$ and $[CO_3^{2-}]$ through Eq. 5.53b.)

$$K_1' = 1.7 \times 10^{-4} = \frac{[H_3O^+][HCO_3^-]}{[H_2CO_3]} = \frac{(1.6 \times 10^{-12})(0.63 \times 10^{-2})}{[H_2CO_3]}$$

$$[H_2CO_3] = \frac{(1.6 \times 10^{-12})(0.63 \times 10^{-2})}{1.7 \times 10^{-4}} = 5.9 \times 10^{-11} \text{ (mol/liter)}$$

$$K_1 = 4.4 \times 10^{-7} = \frac{[H_3O^+][HCO_3^-]}{[CO_2(aq)]} = \frac{(1.6 \times 10^{-12})(0.63 \times 10^{-2})}{[CO_2(aq)]}$$

$$[CO_2(aq)] = \frac{(1.6 \times 10^{-12})(0.63 \times 10^{-2})}{4.4 \times 10^{-7}} = 2.3 \times 10^{-8} \text{ (mol/liter)}$$

Note that by these values of $[H_2CO_3]$ and $[CO_2(aq)]$,

$$\frac{[H_2CO_3]}{[CO_2(aq)]} = \frac{5.9 \times 10^{-11}}{2.3 \times 10^{-8}} = 2.6 \times 10^{-3}$$

which, as it should be, is in agreement with Eq. (5.49b).

STUDY PROBLEM 5(p)

What is the concentration of $CO_2(aq)$ in a solution prepared as 0.10 *M* Na_2CO_3?

EXAMPLE 5.20 Compute the concentrations of CO_2, H_2CO_3, HCO_3^-, CO_3^{2-}, H_3O^+, and OH^- in a solution labeled 0.50 *M* $NaHCO_3$.

SOLUTION Looking at Table 5.2, we see that HCO_3^- is both the strongest available acid and the strongest available base in the initial solution. Hence, the main acid-base reaction is

(b) $$HCO_3^- + HCO_3^- \rightleftharpoons H_2O + CO_2(aq) + CO_3^{2-}$$

(c) $$K = \frac{[CO_2(aq)][CO_3^{2-}]}{[HCO_3^-]^2}$$

We chose Eq. (b) over the alternative (d),

(d) $$HCO_3^- + HCO_3^- \rightleftharpoons H_2CO_3 + CO_3^{2-}$$

because we can avoid falling into traps in the logic if we strictly adhere to the convention of representing the principal reactants and the principal products in the chemical equations chosen to represent the system. We know from Eq. (5.49b) that there is much more $CO_2(aq)$ than H_2CO_3 present in solution at equilibrium, so the former is the logical product to include in the equation for the most important acid-base reaction in this system. The desired *K* can be determined by simple algebraic manipulations.

$$K = \frac{[CO_2(aq)][CO_3^{2-}]}{[HCO_3^-]^2} \cdot \frac{[H_3O^+]}{[H_3O^+]} = \frac{[CO_2(aq)]}{[H_3O^+][HCO_3^-]} \cdot \frac{[H_3O^+][CO_3^{2-}]}{[HCO_3^-]}$$

$$= \frac{1}{K_1} \cdot \frac{K_2'}{1} = \frac{4.8 \times 10^{-11}}{4.4 \times 10^{-7}} = 1.09 \times 10^{-4}$$

Note that Eq. (b) is the *difference* between Eqs. (5.51a) and (5.52a). Now, letting $2y$ represent the number of moles/liter of HCO_3^- that react according to Eq. (b) to attain equilibrium, we write

Hypothetical Initial Concentrations	*Concentrations at Equilibrium*
$[HCO_3^-] = 0.50$	$[HCO_3^-] = 0.50 - 2y$
$[CO_2(aq)] = 0$	$[CO_2(aq)] = y$
$[CO_3^{2-}] = 0$	$[CO_3^{2-}] = y$

$$K = 1.09 \times 10^{-4} = \frac{[CO_2(aq)][CO_3^{2-}]}{[HCO_3^-]^2} = \frac{y^2}{(0.50 - 2y)^2}$$

As a first approximation, we neglect $2y$ in comparison with 0.50, and obtain

$$\frac{y^2}{(0.50)^2} = 1.09 \times 10^{-4}$$

$$y = \sqrt{0.25 \times 1.09 \times 10^{-4}} = 0.52 \times 10^{-2}$$

OPTIONAL

Then, a second approximation, $0.50 - 2y = 0.50 - 0.0104 = 0.49$, gives

$$\frac{y^2}{(0.49)^2} = 1.09 \times 10^{-4}$$

$$y = \sqrt{(0.49)^2 \times 1.09 \times 10^{-4}} = 0.51 \times 10^{-2}$$

Since $0.50 - 2y = 0.49$, as in the second approximation, there is no need to carry out any additional iterations in this scheme of successive approximations. Hence,

$$y = [CO_2(aq)] = [CO_3^{2-}] = 5.1 \times 10^{-3} \text{ (mol/liter)}$$

$$[HCO_3^-] = 0.50 - 2y = 0.49 \text{ (mol/liter)}$$

From these results we can compute $[H_2CO_3]$ directly from Eq. (5.49b):*

$$[H_2CO_3] = K[CO_2(aq)] = (2.58 \times 10^{-3})(5.1 \times 10^{-3}) = 1.3 \times 10^{-5} \text{ (mol/liter)}$$

Knowing $[H_2CO_3]$, $[HCO_3^-]$, $[CO_2(aq)]$, and $[CO_3^{2-}]$, we find it simple to compute $[OH^-]$ and $[H^+]$ using Eq. (5.50b), Eq. (5.51b), Eq. (5.52b), or Eq. (5.53b). For example, using Eq. (5.51b), we have

$$4.8 \times 10^{-11} = \frac{[H_3O^+][CO_3^{2-}]}{[HCO_3^-]} = \frac{[H_3O^+](5.1 \times 10^{-3})}{(0.49)}$$

$$[H_3O^+] = \frac{(0.49)(4.8 \times 10^{-11})}{(5.1 \times 10^{-3})} = 4.6 \times 10^{-9}$$

From Eq. (5.33),

$$[OH^-] = \frac{1.0 \times 10^{-14}}{4.6 \times 10^{-9}} = 2.2 \times 10^{-6} \text{ (mol/liter)}$$

STUDY PROBLEM 5(q)

What is the pH of a solution labeled 0.20 *M* $NaHCO_3$?

EXAMPLE 5.21 Calculate the ratio of concentrations of CO_2 and HCO_3^- in a solution for which the pH is 6.6. What would the pH be if this ratio could be increased by (a) a factor of 2? (b) a factor of 10? (c) a factor of 100? decreased by (d) a factor of 2? (e) a factor of 10? or (f) a factor of 100?

SOLUTION

$$pH = -\log[H_3O^+] = 6.6; \quad -6.6 = \log[H_3O^+]$$

$$\begin{aligned}[H_3O^+] &= \text{antilog}\,(-6.6) = \text{antilog}\,(-7 + 0.4) \\ &= \text{antilog}\,(-7) \times \text{antilog}\, 0.4 \\ &= 10^{-7} \times 2.5\end{aligned}$$

Then, from Eq. (5.52b)

*Strictly speaking, these results of Example 5.20 cannot be entirely correct. From the stoichiometry implied in Eq. (b) and since $CO_2(aq)$ and H_2CO_3 are interconvertible, we can see that the equation $[CO_3^{2-}] = [CO_2(aq)]$, which is implicit in the above treatment, is not strictly valid. This equation should be replaced by the equation $[CO_3^{2-}] = [CO_2(aq)] + [H_2CO_3]$. However, as $[H_2CO_3]$ is much less than $[CO_2(aq)]$, the error incurred in our approach is rather small.

$$K_1 = 4.4 \times 10^{-7} = \frac{[H_3O^+][HCO_3^-]}{[CO_2(aq)]} = \frac{(2.5 \times 10^{-7})[HCO_3^-]}{[CO_2(aq)]}$$

$$\frac{[HCO_3^-]}{[CO_2(aq)]} = \frac{4.4 \times 10^{-7}}{2.5 \times 10^{-7}} = 1.76; \qquad \frac{[CO_2(aq)]}{[HCO_3^-]} = \frac{1}{1.76} = 0.57$$

For case (a), if this ratio were $2 \times 0.57 = 1.14$, then from Eq. (5.52b),

$$[H_3O^+] = 4.4 \times 10^{-7} \frac{[CO_2(aq)]}{[HCO_3^-]}$$

$$= (4.4 \times 10^{-7})(1.14) = 5.0 \times 10^{-7} \text{ (mol/liter)}$$

$$\text{pH} = 6.3$$

Similarly, for case (b),

$$[H_3O^+] = (4.4 \times 10^{-7})(5.70) = 2.5 \times 10^{-6}, \quad \text{pH} = 5.6;$$

for (c), $[H_3O^+] = (4.4 \times 10^{-7})(57.0) = 2.5 \times 10^{-5}$, pH = 4.6;
for (d), $[H_3O^+] = (4.4 \times 10^{-7})(0.285) = 1.25 \times 10^{-7}$, pH = 6.9;
for (e), $[H_3O^+] = (4.4 \times 10^{-7})(0.0570) = 2.5 \times 10^{-8}$, pH = 7.6;
for (f), $[H_3O^+] = (4.4 \times 10^{-7})(5.7 \times 10^{-3}) = 2.5 \times 10^{-9}$, pH = 8.6.

This example shows how the ratio of concentrations of a conjugate acid-base pair determines the pH of the solution, a factor-of-ten change in the ratio corresponding to a change of one pH unit.

The Sulfide System

Another example of multiple acid-base equilibria is the system consisting of H_2S, HS^-, and S^{2-}. The relevant equilibria and equilibrium constants are

$$H_2S + H_2O \rightleftharpoons H_3O^+ + HS^- \tag{5.54a}$$

$$K_I = \frac{[H_3O^+][HS^-]}{[H_2S(aq)]} = 1 \times 10^{-7} \tag{5.54b}$$

$$HS^- + H_2O \rightleftharpoons H_3O^+ + S^{2-} \tag{5.55a}$$

$$K_{II} = \frac{[H_3O^+][S^{2-}]}{[HS^-]} = 1 \times 10^{-15} \tag{5.55b}$$

$$H_2S + 2H_2O \rightleftharpoons 2H_3O^+ + S^{2-} \tag{5.56a}$$

$$K_{III} = \frac{[H_3O^+]^2[S^{2-}]}{[H_2S(aq)]} = K_I K_{II} = 1 \times 10^{-22} \tag{5.56b}$$

The relation of these equations to Eqs. (5.49) to (5.53) should be obvious: $H_2S(aq)$ is analogous to $CO_2(aq)$; HS^- acts like HCO_3^-; and S^{2-} is the counterpart of CO_3^{2-}. As in the CO_2–HCO_3^-–CO_2^{2-} system, the pH determines ratios like

$$\frac{[HS^-]}{[H_2S(aq)]}, \quad \frac{[S^{2-}]}{[HS^-]}, \quad \text{and} \quad \frac{[S^{2-}]}{[H_2S]}$$

The quantitative relations are given by Eqs. (5.54b), and (5.55b), and (5.56b).

OPTIONAL

One of the main areas of importance in which these equilibrium relations apply is in the qualitative analysis of metal ions in solution. The analytical approach (see Sec. 4.5) depends on the differing solubility products of the sulfide salts of different metals. Thus, for a given S^{2-} concentration, some metal ions form a sulfide precipitate and others do not. Consider, for example, the Zn^{2+} situation. The ZnS(s) solubility product is 1×10^{-23}:

$$[Zn^{2+}][S^{2-}] = 1 \times 10^{-23}$$

Thus, even if the Zn^{2+} concentration is only about 10^{-3} *M*, if $[S^{2-}]$ is larger than about 10^{-20} *M*, a precipitate will form, since $[Zn^{2+}][S^{2-}]$ would otherwise exceed 10^{-23}. If $[S^{2-}]$ is smaller, however, Zn^{2+} will remain in solution. For some metal ions, like Ag^{+}, an even smaller S^{2-} concentration is enough to bring about precipitation (Ag_2S), while for other metal ions, like Na^{+}, even large S^{2-} concentrations (say, 0.1 *M*) do not lead to precipitates. Thus, the various metals can be distinguished. One can control the value of $[S^{2-}]$ by controlling the value of $[H_3O^{+}]$.

A classical method is to bubble H_2S through an aqueous solution containing the metal ions to be analyzed. In this way, a value about 0.1 *M* is easily established for $[H_2S(aq)]$, and the value of $[S^{2-}]$ is then determined according to Eq. (5.56b) by the value of $[H_3O^{+}]$. One common procedure uses metal-ion solutions that are 0.3 *M* in HCl. Then, with $[H_3O^{+}] = 0.3$ and $[H_2S(aq)] = 0.1$, one calculates from Eq. (5.56b)

$$\frac{(0.3)^2[S^{2-}]}{(0.1)} = 1 \times 10^{-22}$$

$$[S^{2-}] = 1 \times 10^{-22} \text{ (mol/liter)}$$

Under these conditions, if Zn^{2+} and Cd^{2+} were in the solution at moderately small concentrations (say, 10^{-2} *M*), ZnS(s) would not precipitate but CdS(s) would.

Another common procedure is to use an NH_4^{+}/NH_3 buffer in the solution containing the metal ions through which the H_2S is bubbled. For such a solution (see Sec. 5.4),

$$[NH_4^{+}] = [NH_3]$$

and

$$5.5 \times 10^{-10} = K_{a,NH_4^+} = \frac{[NH_3][H_3O^{+}]}{[NH_4^{+}]} = [H_3O^{+}]$$

Then, with $[S^{2-}] = 0.1$, Eq. (5.56b) gives

$$\frac{(5.5 \times 10^{-10})^2[S^{2-}]}{(0.1)} = 1 \times 10^{-22}$$

$$[S^{2-}] = \frac{(0.1)}{(30 \times 10^{-20})} \times 1 \times 10^{-22} = 3 \times 10^{-5} \text{ (mol/liter)}$$

Under these conditions, even Zn^{2+} would precipitate (as ZnS), as the product $[Zn^{2+}][S^{2-}]$ would otherwise exceed $K_{sp} = 10^{-22}$.

STUDY PROBLEM 5(r)

Using the data in Appendix C, to what value would you adjust the pH of a 0.1 *M* H_2S solution to precipitate Cu^{2+} as CuS(s), while leaving Co^{2+} in solution, assuming that both Cu^{2+} and Co^{2+} are present at concentrations of about 10^{-3} *M*?

SUMMARY

Acid-base chemistry is a very important segment of chemistry, especially as it relates to aqueous solutions. The Arrhenius view, focusing on electrolytes that give rise to the ions H^+ and OH^- when ionized in water, is adequate for most purposes in treating acid-base problems for aqueous solutions. The Brönsted-Lowry approach, which views acids as proton donors and bases as proton acceptors, is more general, and views acid-base chemistry as proton transfers. An important concept in this latter approach is the conjugate acid–conjugate base concept. The ionization of both acids and bases (strong or weak) is viewed within this context in the Brönsted-Lowry approach. Based on the concept of concentration (molarity), volumetric techniques provide a convenient way of carrying out acid-base neutralization. Titration is an important technique by which the amounts of acids or bases in a sample can be determined.

The law of chemical equilibrium provides a powerful framework for quantitatively accounting for, or predicting, the extent to which any specific acid-base process will occur. The pertinent equilibrium constants are, or can be related to, acid or base ionization constants (or dissociation constants), which are quantitative measures of the strengths of acids and bases. A table based on the acid ionization constants can be constructed that ranks the strengths of acids and the strengths of their conjugate bases. Many of the familiar algebraic techniques apply to solving acid-base equilibrium problems, with the pertinent equilibrium expressions. Using either qualitative arguments or the results of equilibrium calculations, one finds that the acid-base reactions that tend to proceed spontaneously, having equilibrium constants greater than 1, are reactions in which a stronger acid and a stronger base are converted to a weaker base and a weaker acid, respectively. In a solution containing appreciable concentrations of several acids and bases, the dominant acid-base reaction is between the strongest acid and the strongest base.

Because of the frequency of occurrence of very large and very small numbers in acid-base equilibrium considerations, it is often convenient to express some quantities with logarithms. For this reason, the pH scale and pK_a's have been defined. It is readily shown that pH, that is, $-\log [H_3O^+]$, is directly tied to the ratio of the concentrations of a conjugate acid-base pair. This leads to the idea of a buffer, which has a value of this ratio near 1 and has a predictable capacity for maintaining the pH of an experimental solution at or near some desired value. Also related to this conjugate acid-base ratio is the indicator, which shows a color change at the pH value equal to the pK_a of the acid form of the indicator. This is useful in determining the equivalence point of a titration.

STUDENT CHECKPOINTS

After studying this chapter, the student should be able to:

1. Describe the Arrhenius and Brönsted-Lowry models of acids and bases and the relations between these two models.
2. Explain the concept of conjugate acids and bases.
3. Differentiate between the Brönsted-Lowry and Arrhenius views of the ionization of acids and bases.
4. Define *neutralization* and *equivalence point.*
5. Compute the stoichiometric relations of acid-base reactions, including titrations.
6. Relate ionization constants (dissociation constants) and acid or base strength.
7. Explain the relations between the strengths of acids and the strengths of their conjugate bases.
8. Compute K_b from K_a and vice versa.
9. Compute $[H^+]$ from $[OH^-]$ and vice versa.
10. Compute pH from $[H^+]$ and vice versa.
11. Use Table 5.2 qualitatively and quantitatively.
12. Carry out equilibrium calculations on the ionization (dissociation) of weak acids and weak bases.
13. Compute the equilibrium constant for the reaction between a weak acid and a weak base from the two pertinent dissociation (ionization) constants.
14. Describe the basis and use of buffers and indicators and how they relate to the Henderson-Hasselbach equation.
15. Demonstrate mathematically the concept of buffer capacity.
16. Recognize the most important acid-base equilibrium in a complex system.

EXERCISES

5.1.* Indicate which of the following are strong acids, weak acids, strong bases, or weak bases: H_2SO_4, H_2O, NH_3, OH^-, $CH_3CO_2^-$, HCl, HS^-, HCO_3^-, S^{2-}.

5.2. Rank the following acids in increasing strength: HNO_3, H_2O, CH_3CO_2H, HSO_4^-, HF, H_3O^+, CH_4.

5.3.* Rank the following bases in increasing strength: SO_4^{2-}, CN^-, OH^-, H_2O, Cl^-, PO_4^{3-}, NH_2^-.

5.4. Write down all possible acid-base reactions (favorable or unfavorable) among the following species: NH_3, H_2O, HS^-, HCl. Identify the Brönsted-Lowry acids and bases in each equation.

5.5.* Write down all possible acid-base reactions (favorable or not) among the following species: CH_3CO_2H, HSO_4^-, H_2O, H_3O^+. Identify the Brönsted-Lowry acids and bases in each equation.

5.6. Write down all possible acid-base reactions (favorable or unfavorable) among the following species: PO_4^{3-}, NH_2^-, HF, CH_3CO_2H, H_2O. Identify the Brönsted-Lowry acids and bases in each equation.

5.7. Write down all possible acid-base reactions with equilibrium constants greater than 1 for any combination of species listed in Exercise 5.4. Identify the Brönsted-Lowry acids and bases in each equation.

5.8.* Write down all possible acid-base reactions with equilibrium constants greater than 1 for any combination of species listed in Exercise 5.5. Identify the Brönsted-Lowry acids and bases in each equation.

5.9. Write down all possible acid-base reactions with equilibrium constants greater than 1 for any combination of species listed in Exercise 5.6. Identify the Brönsted-Lowry acids and bases in each equation.

5.10. How many grams of KOH would, in solution, be neutralized by 32.0 ml of 0.550 *M* HCl?

5.11.* How many milliliters of a 0.102 *M* H_2SO_4 solution would be required to neutralize 100 ml of a 0.0350 *M* NaOH solution?

5.12.* What are the concentrations of all principal species present in a solution prepared by adding 0.100 mol HCl, 0.250 mol KOH, and 0.200 mol HNO_3 in enough water to make a 1-liter solution?

5.13. Write a chemical equation for the main net reaction that occurs when 0.1 mol of each of the following substances is dissolved (simultaneously) in 1000 ml of water; Na_2SO_4, NH_3, KOH, CH_3CO_2H, KNO_3.

5.14.* Write a chemical equation for the main net reaction that occurs when 0.30 mol ammonium acetate, $CH_3CO_2NH_4$, is added to 1 liter of 0.20 *M* HCl. Identify the Brönsted-Lowry acids and bases.

5.15. An unknown weak acid is isolated in pure form from a soil sample. It is found that a 0.120-g sample of this acid requires 11.73 ml of 0.0844 *M* NaOH in a titration to reach the equivalence point. What is the equivalent weight of the unknown acid?

5.16.* A sample of a weak acid isolated from a microbiological organism is found to have an equivalent weight of 236 g. A 166-mg sample of this substance is dissolved in 50 ml of a 0.0444 *M* NaOH solution. How many milliliters of a 0.122 *M* HCl solution must be added to reach the equivalence point of the resulting solution?

5.17.* Calculate the pH of each of the following solutions: (a) 0.22 *M* HNO_3; (b) 4.7×10^{-2} *M* HCl; (c) 7.1×10^{-9} *M* $HClO_4$.

5.18. Calculate the H_3O^+ concentration of each of the following solutions: (a) 1.6 *M* NaOH; (b) 9.2×10^{-3} *M* KOH; (c) 4.7×10^{-4} *M* NaOH.

5.19.* Calculate the pH of solutions with the following values of $[OH^-]$: (a) 3.2×10^{-13} *M* OH^-; (b) 6.2×10^{-3} *M* OH^-.

5.20.* Calculate the percentage ionization and the pH of each of the following solutions: (a) 0.50 *M* HF; (b) 0.020 *M* HCN; (c) 0.90 *M* KCl.

5.21. Calculate the pH of a 0.50 *M* NaF solution.

5.22. Calculate the pH of each of the following solutions: (a) 0.30 *M* NH_3; (b) 0.30 *M* NH_4Cl; (c) 0.30 *M* NH_4Cl and 0.30 *M* NH_3.

5.23. Calculate the K_a and pK_a values of a weak acid for which a 0.35 *M* solution is 2.0% ionized.

5.24.* Calculate the pH of a 0.080 *M* solution of a weak acid that is 1.3% ionized.

5.25. Show how the *K* value for the following reaction can be derived from Table 5.2: $HSO_4^- + NH_3 \rightleftharpoons NH_4^+ + SO_4^{2-}$. What would be the percentage completion of this reaction if 500 ml of a 0.20 *M* $NaHSO_4$ solution and 500 ml of a 0.20 *M* NH_3 solution were mixed?

5.26.* Determine the *K* value of the reaction $NH_4^+ + F^- \rightleftharpoons HF + NH_3$. What would be the percentage of completion of this reaction if 500 ml of a 0.20 *M* NH_4Cl solution and 500 ml of a 0.20 *M* NaF solution were mixed?

5.27.* In a titration of a 0.39 M HCO_2H solution with a 0.40 M NaOH, what would be the pH at the halfway point in the titration, that is, when half the 0.40 M NaOH required for the equivalence point has been added?

5.28. What is the pH of a solution that is 0.10 M in formic acid, HCO_2H, and 0.10 M in sodium formate, HCO_2Na?

5.29.* Calculate the pH of solutions resulting from the addition of 0.0010 mol NaOH to (a) 1000 ml pure water, and (b) 1000 ml of the solution described in Exercise 5.28.

5.30. Calculate the pH of a solution prepared by adding 0.050 mol methylamine, CH_3NH_2, and 0.025 mol HCl to enough water to give a 500-ml solution.

5.31. Calculate the pH of a solution resulting from the addition of 0.0010 mol HCl to (a) 500 ml of pure water; (b) the solution described in Exercise 5.30.

5.32.* What indicator or indicators would you choose for each of the titrations shown in Figure 5.8?

5.33. Plot a titration curve, of the general type shown in Figs. 5.6 through 5.8, for the titration of a 30-ml sample of 0.10 M NH_3 with a 0.10 M HCl solution. What is the pH halfway to the equivalence point?

5.34.* What is the main reaction that would occur when 10 ml of each of the following solutions is added to 10 ml of a 0.50 M $NaHCO_3$ solution: (a) 0.50 M NaOH; (b) 0.50 M HCl; (c) 0.50 M CH_3CO_2H; (d) 0.50 M Na_3PO_4; (e) 0.50 M NH_4Cl; (f) 0.50 M NH_3; (g) 0.50 M KCl?

5.35. Determine the value of the equilibrium constant for each of the reactions of Exercise 5.34.

5.36.* What is the main reaction that occurs when 50 ml of a 0.020 M CO_2 solution is mixed with 50 ml of a 0.020 M Na_2CO_3 solution?

5.37.* Calculate the equilibrium constant for the process referred to in Example 5.36. What are the concentrations of all species present in the resulting solution?

5.38. Calculate the concentrations of all pertinent species present in a solution prepared as 0.50 M NaHS.

5.39. What are the concentrations of all species present at equilibrium in a solution prepared by mixing 500 ml of a 0.10 M H_2S solution and 500 ml of a 0.10 M Na_2S solution?

5.40. Calculate the pH of a 0.70 M ammonium acetate, $CH_3CO_2NH_4$, solution.

5.41. Amino acids are biologically important molecules that are the building blocks of proteins. The simplest amino acid is glycine, $\underline{H_2N}—CH_2—\underline{CO_2H}$. The common feature of amino acids is that they all contain the underlined functional groups (amino, $\sim NH_2$, and carboxyl, $\sim CO_2H$). Let us assume that for a given amino acid, symbolized here as $H_2N\sim CO_2H$, the acidic character of the carboxyl group and of acetic acid can be considered comparable and that in basic property, the amino group can be considered comparable with CH_3NH_2. Calculate the pH and the concentrations of all pertinent species of a solution in which 0.0010 mol of the amino acid is dissolved in (a) 1000 ml of pure water; (b) 1000 ml of a pH 3 buffer; and (c) 1000 ml of a pH 11 buffer.

ANSWERS TO SELECTED EXERCISES

5.1 strong acids: H_2SO_4 and HCl; strong bases: OH^- and S^{2-}; weak acids: H_2O, OH^-, NH_3, HS^-, HCO_3^-; weak bases: H_2O, NH_3, $CH_3CO_2^-$, HS^-, HCO_3^-. **5.3** $NH_2^- > OH^- > PO_4^{3-} > CN^- > SO_4^{2-} > H_2O > Cl^-$. **5.5** $\underset{\text{acid}}{CH_3CO_2H} + \underset{\text{base}}{HSO_4^-} \rightleftharpoons \underset{\text{acid}}{H_2SO_4} + \underset{\text{base}}{CH_3CO_2^-}$; $\underset{\text{acid}}{CH_3CO_2H} + \underset{\text{base}}{H_2O} \rightleftharpoons \underset{\text{acid}}{H_3O^+} + \underset{\text{base}}{CH_3CO_2^-}$; $\underset{\text{acid}}{HSO_4^-} + \underset{\text{base}}{H_2O} \rightleftharpoons \underset{\text{acid}}{H_3O^+} + \underset{\text{base}}{SO_4^{2-}}$; $\underset{\text{acid}}{H_2O} + \underset{\text{base}}{HSO_4^-} \rightleftharpoons \underset{\text{acid}}{H_2SO_4} + \underset{\text{base}}{OH^-}$; $\underset{\text{base}}{HSO_4^-} + \underset{\text{acid}}{H_3O^+} \rightleftharpoons \underset{\text{base}}{H_2O} + \underset{\text{acid}}{H_2SO_4}$. **5.8** none. **5.11** 17.2 ml. **5.12** 0.050 M H_3O^+, 0.100 M Cl^-, 0.250 M K^+, 0.200 M NO_3^-. **5.14** $\underset{\text{base}}{CH_3CO_2^-} + \underset{\text{acid}}{H_3O^+} \rightleftharpoons \underset{\text{base}}{H_2O} + \underset{\text{acid}}{CH_3CO_2H}$. **5.16** 12.5 ml. **5.17** (a) 0.7; (b) 1.3; (c) 7.0. **5.19** (a) 1.5; (b) 11.8. **5.20** (a) 3.6%, pH = 1.7; (b) 1.6×10^{-2}%, pH = 5.5; (c) 100%, pH = 7. **5.24** 3.0. **5.26** 8.2×10^{-7}; 0.091%. **5.27** 3.7. **5.29** (a) 11; (b) 3.7. **5.32** bromcresol green or bromcresol purple or phenolphthalein (or litmus) for the $pK_a = 3$ case; phenolphthalein (or litmus) for the $pK_a = 5$ case; thymolphthalein or phenolphthalein or alizarin yellow for the $pK_a = 7$ case; alizarin yellow seems to be the best choice for the $pK_a = 9$ case (very difficult endpoint). **5.34** (a) $OH^- + HCO_3^- \rightleftharpoons CO_3^{2-} + H_2O$; (b) $H_3O^+ + HCO_3^- \rightleftharpoons CO_2(g) + 2H_2O$; (c) $CH_3CO_2H + HCO_3^- \rightleftharpoons CO_2(g) + H_2O + CH_3CO_2^-$; (d) $PO_4^{3-} + HCO_3^- \rightleftharpoons CO_3^{2-} + HPO_4^{2-}$; (e) $NH_4^+ + HCO_3^- \rightleftharpoons CO_2(aq) + H_2O + NH_3$; (f) $NH_3 + HCO_3^- \rightleftharpoons CO_3^{2-} + NH_4^+$; (g) no reaction. **5.36** $CO_2(aq) + CO_3^{2-} + H_2O \rightleftharpoons 2HCO_3^-$. **5.37** 9.2×10^3; $[CO_3^{2-}] = [CO_2(aq)] = 2.1 \times 10^{-4}$ M; $[HCO_3^-] = 0.020$ M; $[H_3O^+] = 4.6 \times 10^{-9}$ M; $[OH^-] = 2.2 \times 10^{-6}$ M.

ANSWERS TO STUDY PROBLEMS

5a 69 ml. **5b** 0.23 liter. **5c** 103 g. **5d** 1.1×10^{-5} M. **5e** 1.5×10^{-12} M. **5f** (a) 5.5; (b) 2.1; (c) 10.7. **5g** < 1; no. **5h** 2.0×10^2. **5i** $[H^+] = [HCO_2^-] = 1.2 \times 10^{-2}$ M; $[HCO_2H] = 0.79$ M; $[OH^-] = 8.3 \times 10^{-13}$ M. **5j** 1.5×10^{-11}. **5k** 11.3. **5l** Derivation leads to $K_{b,NH_3} = K_w/K_{a,NH_4^+}$. **5m** $[HCN] = [OH^-] = 1.4 \times 10^{-3}$ M; $[CN^-] = 0.099$ M; $[H_3O^+] = 7.1 \times 10^{-12}$ M. **5n** (a) 4.7 before; 4.7 after; (b) 7.0 before; 11 after. **5o** alizarin yellow. **5p** 2.2×10^{-8} M. **5q** 8.3. **5r** less than 1.97.

6

Atomic Structure

The constitution of matter is of ultimate importance to chemists. To consider the way in which substances interact, we must understand the submicroscopic nature of substances in a way that can explain their physical and chemical properties. Since we cannot actually see the smallest particles of matter, chemists and physicists have tried to describe these particles by using models and concepts that are drawn from their experience with large, observable objects and processes.

We have seen that a cornerstone of chemistry is the concept of the atom, the smallest particle of matter (uncharged) that retains elemental properties. A variety of models for atomic structure have been based on familiar objects and phenomena; these include geometric figures (Plato), billiard balls (Dalton), electrical charges (electron theory), the motion of planets (motion of electrons about a nucleus), and the vibrations of piano strings (wave nature of electrons). The behavior of these familiar objects and phenomena are all explainable in terms of classical physics, that is, physics as it was understood before the twentieth century. We shall discover that much of our sophisticated modern theory is actually based on such classical concepts as the attraction of unlike charges (the nucleus and electron) and repulsion of like charges (one electron and another), centripetal force of a particle in a circular orbit (electrons circling about the nucleus without being "pulled" into it, as planets circle the sun without succumbing to its gravitational pull), and the generation of a magnetic field by a moving charge (the magnetic moment caused by the "spinning" electron). Nevertheless, we shall learn that a strictly classical physics is incapable of describing matter at the submicroscopic level, and that some revolutionary ideas had to be introduced during the twentieth century in order to approach an understanding of the submicroscopic world.

These new ideas can be meshed with helpful classical concepts to form the basis of a relatively simple model that can explain or correlate much chemistry. The model for atomic structure can be used to explain properties of atoms and ions and also to explain how atoms combine to form compounds.

Models that we use to explain the behavior of atoms are not literal descriptions. They constitute the bases of theories, which are retained only so long as their predictions agree with careful observations.

6.1 Structure of Matter: Experiments and Models

Early Concepts

The idea of a smallest unit of matter was one of the first features of the natural philosophy of the Greeks. One of their early theories to explain the properties of substances was based on geometric figures. Plato's notion was that the five "elements"—earth, air, fire, water, and "the indefinite"—could best be represented by regular polyhedra. Fire was conceived as the "sharp-pointed" pyramid, and earth was imagined as a stable cube. The octahedron, dodecahedron, and icosahedron represented water, the indefinite, and air. Plato, to explain the diversity of properties shown by the several substances known at the time, postulated that substances could be formed from combinations of the faces from regular polyhedra (Fig. 6.1). This idea persisted for several centuries.

Dalton's Symbols for the Elements

	wt.
Hydrogen	1
Azote	5
Carbon	5.4
Oxygen	7
Phosphorus	9
Sulfur	13
Magnesia	20
Lime	24
Soda	28
Potash	42
Strontian	46
Barytes	68
Iron	50
Zinc	56
Copper	56
Lead	90
Silver	190
Gold	190
Platina	190
Mercury	167

Another early theory extrapolated bulk properties (macroscopic properties) of substances down to the submicroscopic, or atomic, level. This theory envisioned atoms (from *atomos*, Greek word for "small") of gaseous substances to be very round and smooth, while atoms of solids were thought to have several hooks to hold them together. The atoms of liquids were thought to be intermediate in nature, having fewer hooks than atoms of solids (Fig. 6.2).

Much experimental information was obtained during the next two thousand years, although not very rapidly nor systematically. By the end of the eighteenth century, the properties of numerous substances were known, and better theories were needed to explain them. Dalton proposed a model based on a billiard-ball type of atom—smooth, hard atoms that underwent elastic collisions with one another. This hypothesis gave better explanations for some experimental observations than the Greeks' models, but provided no basis for the combining of atoms of some elements and not of others. This atomic model could certainly not predict that atoms of different elements would combine to form molecules in specific atomic ratios, nor more specifically, what makes one ratio occur rather than another.

Dividing the Atom

An answer to the puzzle of specific atomic ratios in molecules came only when it was discovered that atoms are divisible. We shall find that atoms are the smallest electrically neutral particles that show the properties of an element, but this does not mean that an atom does not have "parts." We now know that the number and arrangement of particles that make up an atom determine its chemical properties.

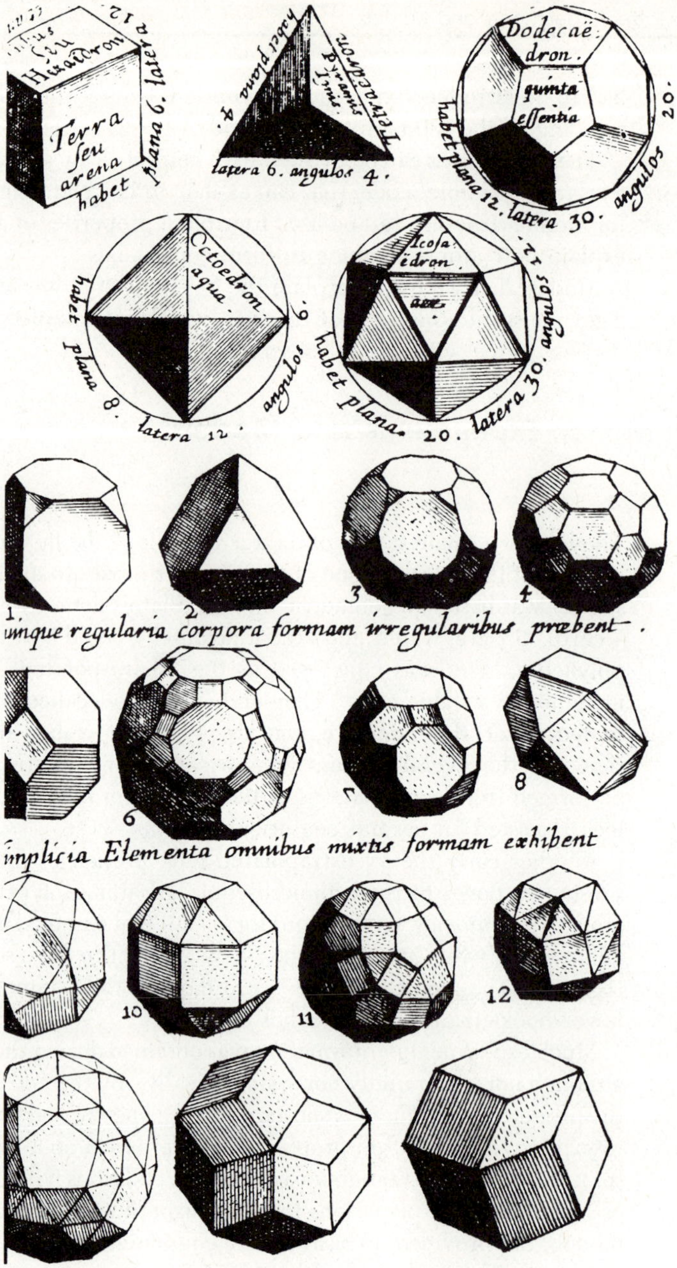

Figure 6.1
Plato's Polyhedral Theory. The persistence of Plato's Polyhedral Theory is shown in this figure, which corresponds to William Davisson's scheme for analyzing substances in terms of the five "perfect" solid forms (*Philosophia Pyrotechnia,* Part III, Paris, 1642!).

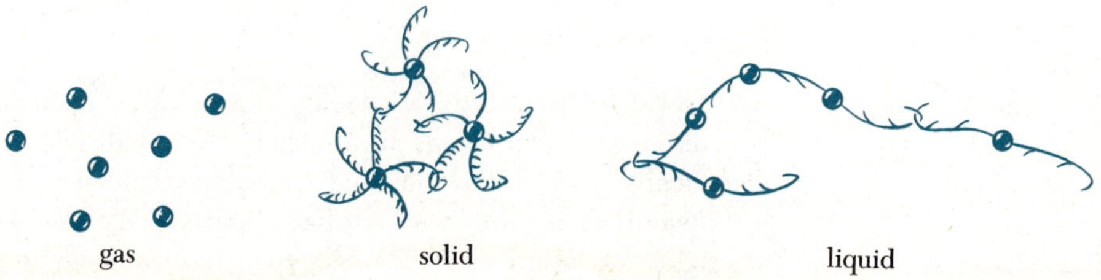

Figure 6.2
The structure of atoms according to Lucretius, about the first century A.D.

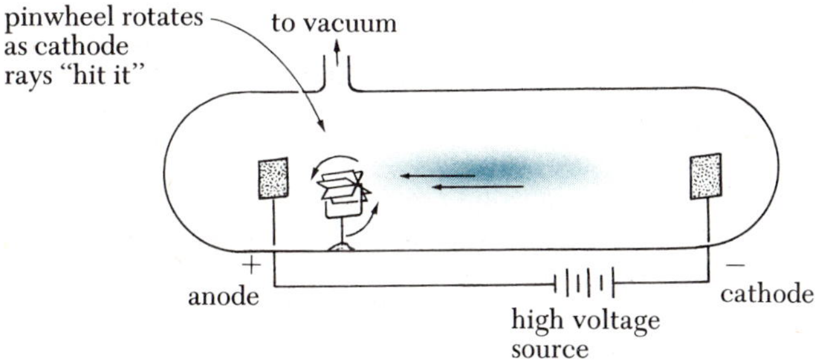

Figure 6.3
A Crookes tube experiment, in which the rays that emanate from the cathode are shown to act as particles.

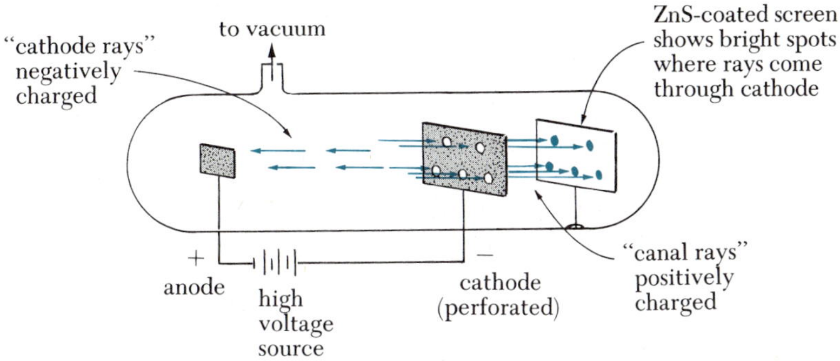

Figure 6.4
A Goldstein experiment, showing that rays also emanate from the anode.

"Static" electricity was a phenomenon known by the Greeks, who rubbed amber (*elektron,* in Greek) with cloth or fur to produce a "charge" capable of picking up bits of straw or paper. This frictional, or static, electricity was an object of curiosity for many centuries, and was considered fundamentally different from the flowing electricity of a circuit, which Luigi Galvani and Alessandro Volta studied during the eighteenth and nineteenth centuries. The later demonstration that these two forms of electricity, static and flowing, are really the same and involve a fundamental unit of electricity, the **electron,** gave the first solid evidence for "elementary particles," the subunits of atoms.

During the 1850s, Sir William Crookes (1832–1919) experimented with the conduction of electricity through gases. Air, or a high concentration of most gases (their pressure), has a high resistance to the passage of electric current. Crookes found that a gas at low pressure, however, passes current. As a result of passing current, the gas glows. He found that a pinwheel placed between the **anode** (the plate connected to the voltage source of positive polarity) and the **cathode** (the plate connected to the voltage source of negative polarity) of his evacuated gas discharge tube would turn when current was passed. Crookes's apparatus is shown in Fig. 6.3. The turning of the pinwheel showed that the current had imparted the momentum associated with it to the object that it struck; the conclusion was therefore that the basis of the current was particulate.

Soon after Crookes's experiments, Eugen Goldstein demonstrated that a luminous glow could also be seen behind a perforated cathode (Fig. 6.4). From this experimental fact and a knowledge of the direction of rotation of the pinwheel in Crookes's experiment, it was evident that there were two types of current—one flowing from the cathode, comprising **cathode rays,** and one flowing towards the cathode, comprising **canal rays.**

Studies of Radioactivity Provide an Experimental Method

Scientists soon undertook determining the nature of these cathode rays and canal rays. This task was made easier by a fortuitous discovery by physicists who were interested in what superficially appeared to be a completely different area of study; these physicists found that zinc sulfide, ZnS, would "glow" when various sorts of rays impinged on it. Radioactive substances had been discovered by Becquerel in 1895, and now three kinds of rays emanating from radioactive materials could be distinguished by monitoring them on a screen coated with ZnS. When a magnet or a set of electrically charged plates* was used, one type of ray, called the beta (β) ray, would be deflected by the magnet in one direction, towards the electrically positive plate, and was therefore determined to be made of negatively charged entities. A second type, called the alpha (α) ray, was deflected by the magnet in the opposite direction and was attracted towards the negatively charged plate; it was therefore identified as positively charged.† The third type, the gamma (γ) ray, was not deflected at all, and thus it was concluded that it did not consist of electrically charged entities. Figure 6.5 diagrams these experiments. By the use of the ZnS-coated screen and deflection by magnets, the rays in the gas discharge tubes were investigated.

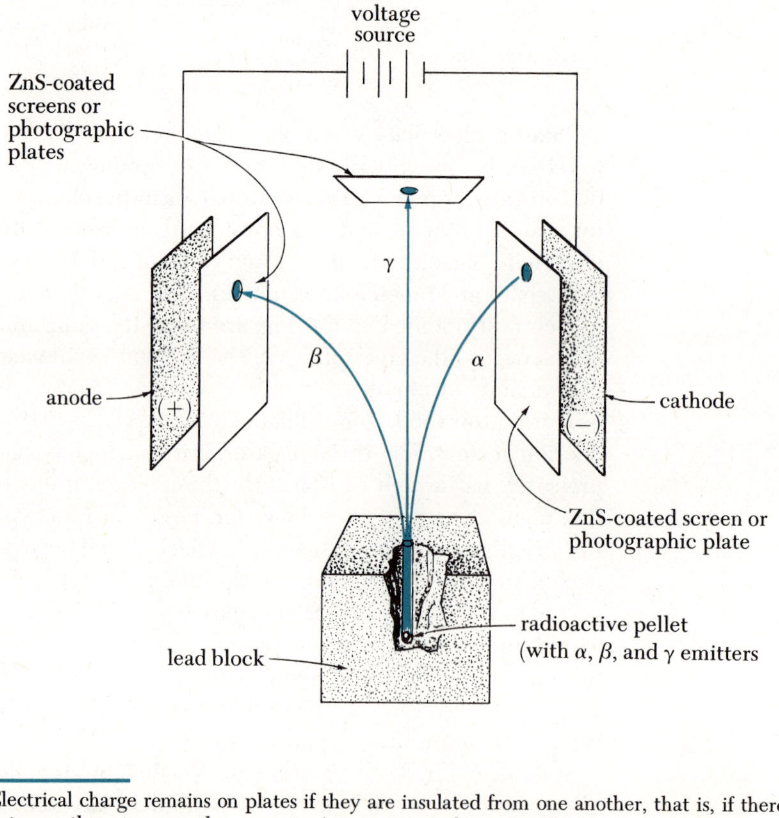

Figure 6.5 Determination of the sign of charge on "rays" from radioactive material. (The entire apparatus would be contained in an evacuated chamber.)

*Electrical charge remains on plates if they are insulated from one another, that is, if there is nothing between them to carry the current. A vacuum can be an insulator.

†The alpha ray consists of the ions $({}^{4}_{2}He)^{2+}$. These ions very readily pick up electrons to form neutral helium atoms. That heavy elements such as radium could disintegrate to form the light element helium was a convincing demonstration to scientists around 1900 that atoms were divisible.

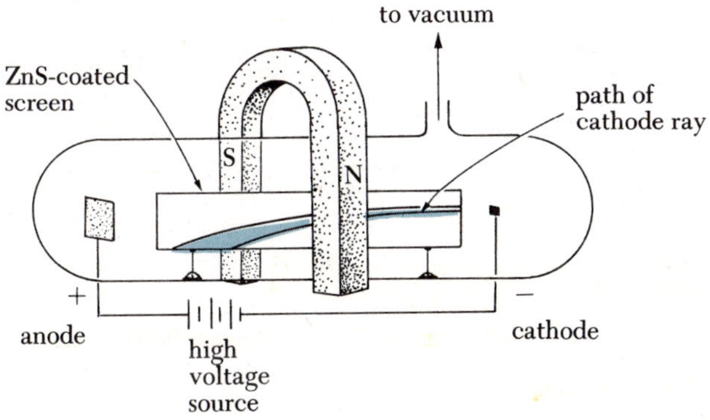

Figure 6.6
A Thomson experiment, in which a magnetic field is shown to deflect the cathode rays.

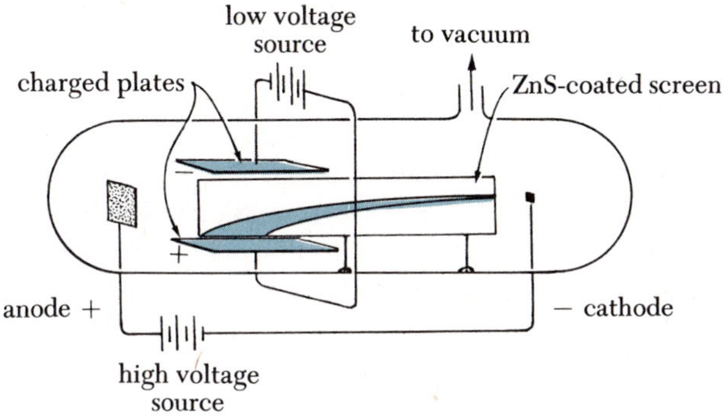

Figure 6.7
A demonstration in which charged plates "bend" cathode rays, showing that the rays are negatively charged. The path of cathode rays is found by a fluorescent glow of the ZnS-coated screen.

Thomson Finds Electrons

In 1897, J. J. Thomson showed that the cathode ray was deflected in a magnetic field in the same manner as the beta ray (Fig. 6.6). The cathode ray was also attracted towards a positively charged plate and therefore was regarded as made up of negatively charged species (Fig. 6.7). From the strength of the magnetic field and the angle of the deflection, Thomson calculated the mass-to-charge ratio of the particles in the beam. The mass-to-charge ratio did not change no matter which gas Thomson used in the tube or what metal was used in making the cathode. The cathode ray appeared to consist of some "fundamental particle" common to all gases. This particle is what is now known as the electron.

In a similar manner, the mass-to-charge ratio was calculated for the "canal rays" that emanated from the back of Goldstein's perforated cathode. The charge was found to be positive, and the mass-to-charge ratio differed when different gases were used. The smallest value was found, as we might expect, when hydrogen gas, H_2, was used in the tube.* The mass-to-charge ratio for the lightest positive ion, H^+, formed from hydrogen gas, is nearly 2000 times the mass-to-charge ratio of negatively charged particles found in cathode rays.

*Hoping that nature would tend to the simplest arrangement, scientists postulated that there would be a unit positive charge, equivalent in size to the unit negative charge. Since the lightest ion formed from hydrogen gas, H^+, was the lightest of the canal rays, they hoped the other rays would have masses that were multiples of the H^+ mass. This turned out to be approximately the case, even though the nucleus contains more than just protons.

At this point, scientists theorized that the atom consisted of negatively charged particles that were both relatively light and common to all gases, and positively charged particles having a total mass that was much greater than the mass of the electrons and that depended on the gas used. To calculate the actual masses of these particles instead of just relative masses, one had to determine the amount of charge on the electron.

Millikan Determines the Charge on the Electron

In 1909, R. A. Millikan determined the charge on the electron by the famous oil-drop experiment, represented in Fig. 6.8. Millikan found that the tiny droplets of oil produced in air by an "atomizer" would be ionized—caused to pick up an electric charge—when they were bombarded by x rays or β rays, the electrons emitted from certain radioactive materials. Not all the oil droplets were ionized, but a substantial, observable portion were. Under the gravitational force exerted by Earth, all the droplets had a tendency to fall from the top to the bottom of the apparatus. However, the motion of ionized droplets was greatly affected by applying an electric field to the plates, as shown in Fig. 6.8; they could be accelerated upward or downward, or even suspended in an essentially motionless state. By observing the motion of the particles in the absence of an electric field and under the influence of electric fields of various strengths, Millikan was able to determine the amount of electrical charge carried by each ionized droplet observed. He found that there was a minimum size of charge manifested by these droplets, presumably the smallest unit of charge available—which is the charge of one electron. The different charges borne by various droplets were shown to be integral multiples of this value. The value obtained by Millikan is 1.602×10^{-19} coulomb (C).

From Thomson's mass-to-charge ratio of the electron and Millikan's determination of the charge of an electron, the mass of the electron is readily computed at 9.109×10^{-28} g.

$$\left[\frac{\text{mass (g)}}{\text{charge (C)}}\right][\text{charge (C)}] = \text{mass of electron (g)}$$

$$5.686 \times 10^{-9}\,\frac{\text{g}}{\text{C}} \times 1.602 \times 10^{-19}\,\frac{\text{C}}{\text{electron}} = 9.109 \times 10^{-28}\,\frac{\text{g}}{\text{electron}}$$

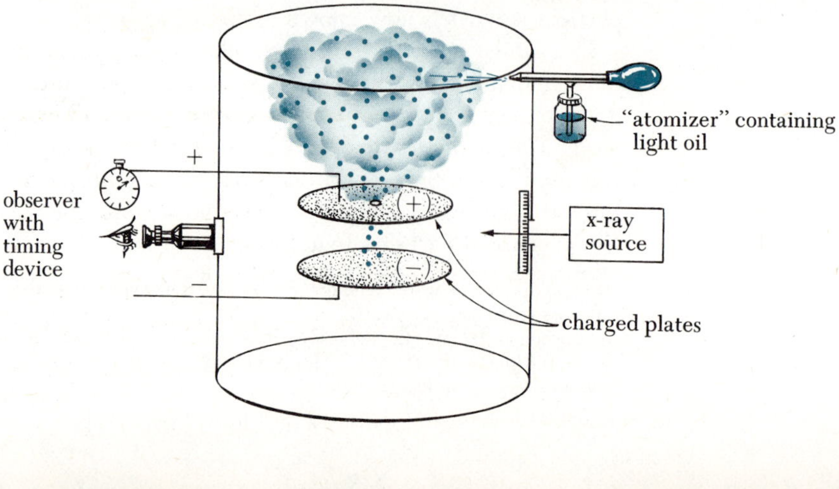

Figure 6.8
A schematic drawing of the Millikan oil drop experiment.

Similarly, the mass of the proton H^+ was determined to be 1.673×10^{-24} g. Expressed in atomic mass units, the proton mass is 1.0080 amu and the electron mass is 5.4876×10^{-4} amu, $\frac{1}{1837}$ the mass of the proton. Another elementary particle, called the **neutron**, has also been found to reside in the nucleus. The neutron has very nearly the same mass as the proton, but the neutron is electrically uncharged.*

Thomson suggested as a model for the atom a spherical shell in which the protons and electrons were embedded in a symmetric arrangement. His experiments had determined relative masses of elementary particles, but gave no clues about how these particles might be arranged. Using a model of charged particles, Thomson expected that the protons would repel one another and would be insulated from one another by the electrons. This seemingly "reasonable" model soon ran into trouble because of Rutherford's experiments.

Rutherford Proposes the "Nuclear" Atom

In 1911, Ernest Rutherford proposed a new atomic model based on the data of an experiment Geiger and Marsden had reported two years earlier. A "beam" of alpha particles (helium ions, He^{2+}) from a radioactive source was aimed at a thin metal foil, and the positions of the particles after they collided with the foil were detected on a ZnS screen. The experiments, represented in Fig. 6.9, showed to the surprise of the experimenters that some of the scattered particles (about one in 8000) were reflected from the foil. If the mass (the protons) of the atoms in the foil had been distributed throughout each atom, as Thomson had proposed, the heavier alpha particles (having about four times the mass of the proton) would have been only slightly deflected in their paths, and would not have been reflected. Thus, Rutherford concluded that the alpha particles were evidently hitting particles of greater mass. From the scattering data, Rutherford calculated that nearly all the mass of each atom had to be contained in the particles hit by the alpha particles. The other conclusion that could be drawn from the experiment was that the heavy part of the atom was very small compared with the whole atomic volume, since most alpha particles passed through the foil without being deflected at all. This heavy part of the atom, which had been shown to contain the protons, was called the **nucleus**. Rutherford concluded from his calculations that the nucleus of the atom occupied only about 10^{-15} of the total atomic volume; the radius of an atom is on the order of angstroms, 1×10^{-8} cm, whereas Rutherford determined the radius of the nucleus to be about 10^{-13} cm. Instead of protons and electrons evenly distributed in a spherical ball, a very different picture had emerged. By the new model, the atom still appeared spherical, but almost the entire volume of the sphere seemed to be composed of the nearly weightless, negatively charged electrons. Only a small spot in the center of the sphere contained the relatively heavy, positively charged proton. The idea that all the positive charge and nearly all the mass were concentrated at a mere pinpoint, while nearly all atomic volume was occupied by all the negatively charged

*The discovery of the neutron was an evolutionary development, involving several prominent scientists over about twelve years, beginning with a suggestion by Rutherford. The experimental information that led to the identification of the neutron came from experiments in which these particles were produced by bombardment of various types of targets with α particles. After a series of understandably incorrect interpretations of these kinds of experiments by eminent scientists, Chadwick concluded in 1932 that any particles with zero electrical charge and a mass nearly the same as that of a proton could account for the type of radiation observed after bombardment.

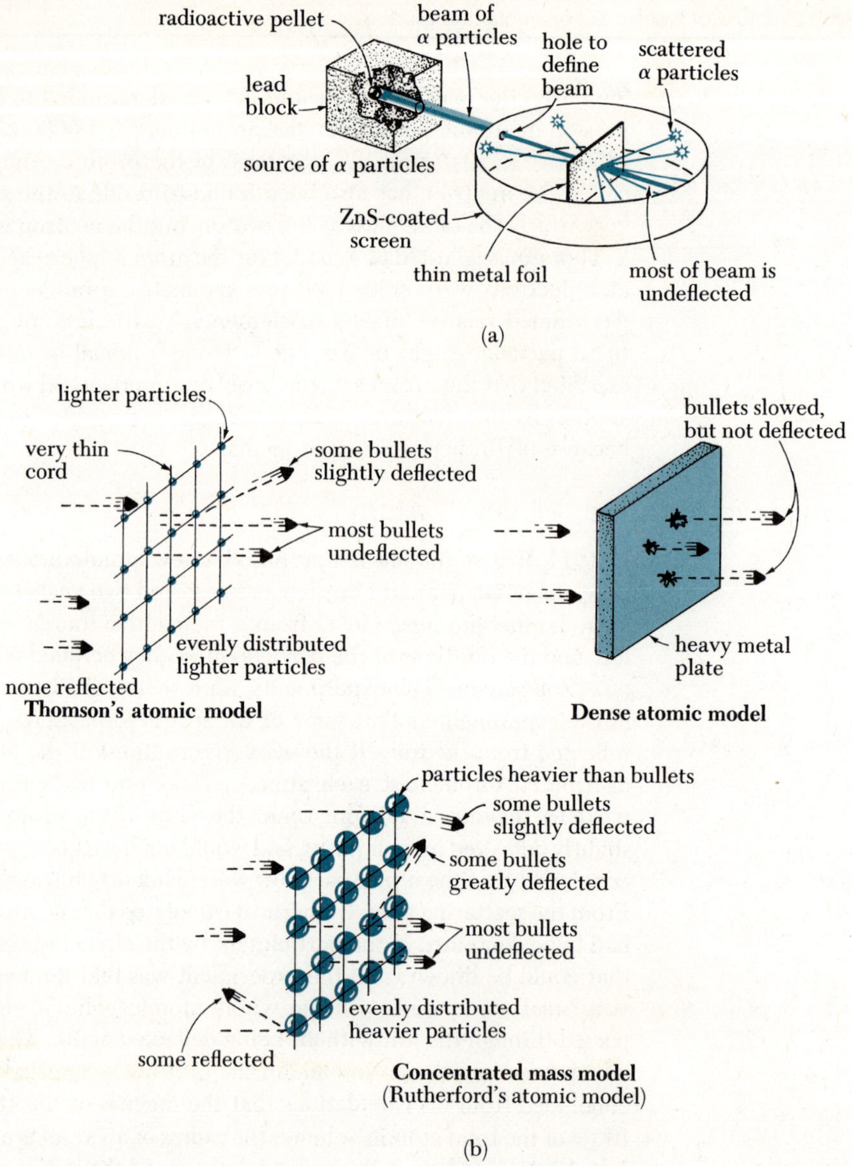

Figure 6.9
(a) A schematic drawing of the "alpha scattering experiment." (The entire region traversed by the beam is evacuated.)
(b) Some analogies of the atomic models in the "alpha scattering experiment."

particles, was intuitively "ridiculous"; but it was the only model consistent with the experiments.

Rutherford's model of the atom posed dilemmas not only with intuition but also with accepted theory. By classical theories of physics, including the assumption that the energy of a system could be varied continuously, the following two behavioral characteristics were expected: (1) If electrons and protons were stationary, they would merely rush together by electrostatic attraction. (2) If the electrons used centrifugal force to remain separated from the nucleus by orbiting very rapidly (as planets orbit the sun without being pulled into it by gravitation force), then the moving charge would be expected to radiate energy continuously. The electrons would then continuously lose energy and move in smaller and smaller orbits, eventually collapsing into the nucleus. This kind of behavior is, of course, not found—all the atoms of the physical universe have not simply "collapsed."

Niels Bohr Proposes a "Planetary" Atomic Model

A new view of the nature of physical interaction in atoms was necessary to solve the dilemmas just outlined. The new model for atomic structure was more fundamental, and required an even greater break with accepted concepts than the large advances already taken in the form of the concepts of subatomic particles or the Rutherford model of the nuclear atom.

Classical physics had as one of its implied philosophical postulates the assumption that energy could be distributed continuously among the various entities of the physical universe. The new physics, quantum mechanics, is in contrast based on postulates embracing the idea that specific physical systems can have only certain "allowed" values of energy.* In 1900, Max Planck proposed that the energy of light (or other forms of electromagnetic radiation) was contained in discrete amounts or packets, called **photons.** This idea was generalized to other forms of energy, believed to be packaged in discrete amounts, called **quanta.** The amount of energy in each photon was shown to be directly proportional to the frequency of the light, where the frequency is the number of full wavelengths (cycles) of light that pass a point per unit time (sec). This relation is expressed simply as

$$E = h\nu \tag{6.1}$$

where E is the energy in a quantum of light, or photon (expressed in ergs), ν is the frequency of the light (in Hz, which are cycles/second), and h is the proportionality constant, called **Planck's constant** (6.6256×10^{-27} erg sec). The Planck relation has far-reaching consequences in physical science.

Niels Bohr used Planck's quantum concept to resolve the dilemmas associated with viewing Rutherford's model of the atom in terms of classical physics. The way in which Bohr incorporated the idea of a quantum into atomic theory was to make the following two assumptions: (1) The electrons in an atom revolve around the nucleus only in certain allowed orbits, the allowed orbits being those for which the angular momentum mvr of the electron is an integral multiple of $h/2\pi$. (2) The only changes in energy that the atom can experience correspond to jumps between the states dictated by the first premise, with quanta of energy (light) absorbed or released at frequencies given by the Planck relation.

Mathematically, Bohr's first postulate states that the angular momentum mvr of an electron of mass m, moving with linear velocity v in a circular orbit or radius r about the nucleus, is restricted to the values given by the equation

$$mvr = n\frac{h}{2\pi} \tag{6.2}$$

where $n = 1, 2, 3, \ldots$ (an integer). The integer n is known as the **principal quantum number;** it determines the radius and momentum of the orbital motion of the electron as well as its total energy. This can be shown by combining the classical

*The term *allowed* here means "allowed by the fundamental design or laws of nature."

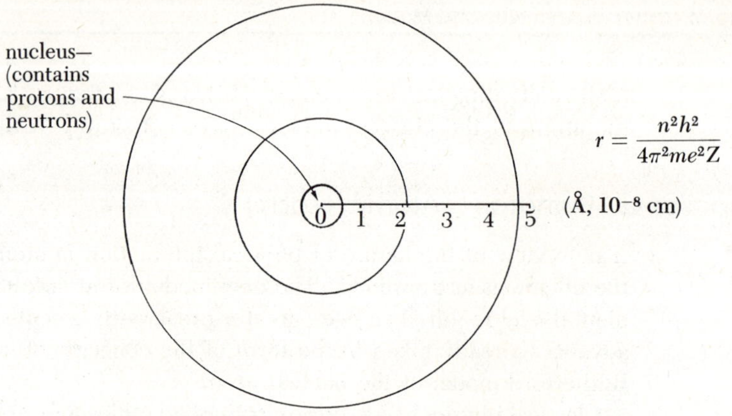

Figure 6.10
First three allowed orbits of the Bohr atom.

equations for the centrifugal and centripetal (electrostatic attractions) forces, with Bohr's angular momentum condition (Eq. 6.2).* The results of the mathematical development yield the following expression for the radius of orbital motion:

$$r = n^2\left(\frac{h^2}{4\pi^2me^2Z}\right) \tag{6.3}$$

where e is the electrical charge of the electron and Ze is the electrical charge (positive) of the nucleus. The symbol Z represents the number of protons in the nucleus —the **atomic number**. We see that r is proportional to the square of the principal quantum number and inversely proportional to the atomic number. Figure 6.10 is a representation of the orbits corresponding to the n values 1, 2, and 3.

As outlined above, several atomic models had been proposed, accepted for a time and then rejected, before Bohr proposed his model. Each of these earlier models, for example, those of the Greeks, Dalton, and Thomson, had been proposed originally because, it was believed, the model accounted for what were judged the important atomic features at the time. Likewise, each of these models was later discarded be-

*The "outward" force on the electron due to acceleration, which tends to force the electron into a wider orbit, is given by classical physics as mv^2/r. The centripetal (inward) force in this case is due to the electrostatic attraction of the electron (with charge $-e$) to the nucleus (with charge Ze); it is given by classical electrostatics as Ze^2/r^2. Hence, for a stable orbit, in which these two forces are balanced, we have

$$\frac{mv^2}{r} = \frac{Ze^2}{r^2}$$

or

$$v^2 = \frac{Ze^2}{mr}$$

From Eq. (6.2), we have

$$m^2v^2r^2 = \frac{n^2h^2}{4\pi^2}$$

or

$$v^2 = \frac{n^2h^2}{4\pi^2m^2r^2}$$

Substituting the expression for v^2 into the above equation, we get

$$\frac{n^2h^2}{4\pi^2m^2r^2} = \frac{Ze^2}{mr}$$

which gives

$$r = n^2\left(\frac{h^2}{4\pi^2me^2Z}\right)$$

cause it failed to explain some other observed property that was attributed to atoms on the basis of later experiments. The Greek model couldn't account for the great diversity of chemical behavior that was recorded over the years. Dalton's model did not answer the important question why the atoms of each different element showed the distinct chemical behavior that they did, and it did not account for subatomic particles (protons, neutrons, electrons) that were found in the discharge tube experiments. Thomson's model, while it included subatomic particles, could not be reconciled with the α scattering experiments, as pictured in Fig. 6.9. It is significant that as models of greater detail evolved, more complex observations and experiments were needed to disprove them. The Greek model could be rejected without resort to sophisticated, specialized experiments. Dalton's model virtually doomed itself by calling attention to the distinctness of the atoms of different elements, without explaining the origin of the distinctness, and yielded to the progress of research on subatomic particles. Rutherford's interpretation of the α scattering experiment, which overthrew Thomson's model, was a far more sophisticated and elegant development.

The experiments that challenged Rutherford's model, inspired Bohr's advance, and constituted the strongest proof of the value of the Bohr model were of the spectroscopic type. **Spectroscopy,** in its broadest sense, is the study of the interaction of electromagnetic radiation, light, for example, with matter. If a substance is subjected to light of the appropriate frequency, some of the light energy can be absorbed by the substance, leading to an excited state. Subsequently, the excited atoms or molecules of the substance may give up the extra energy by emitting light. Alternatively, there are other ways of providing enough energy to energize the substance into excited states; for example, it is often possible to excite atoms or molecules simply by raising the temperature of the substance several hundreds of degrees. Species that are excited in this way can then emit radiation as they fall from a high-energy state to a low-energy state; the lowest-energy state they can reach is the **ground state.**

When atoms or molecules absorb or emit energy in the form of electromagnetic radiation, only radiation of certain discrete frequencies is involved. Spectroscopy is largely concerned with determining which frequencies are absorbed or emitted, and in relating those frequencies to a fundamental understanding of the submicroscopic structures of the species involved. Figure 6.11 is a diagrammatic representation of spectroscopy, in terms of upward and downward transitions.

The experiments that Niels Bohr tried to account for, and which could not be reconciled with a classical interpretation of the Rutherford model, were studies of

Figure 6.11
Spectroscopic transitions. (Straight arrows represent change in energy of the pertinent species.)

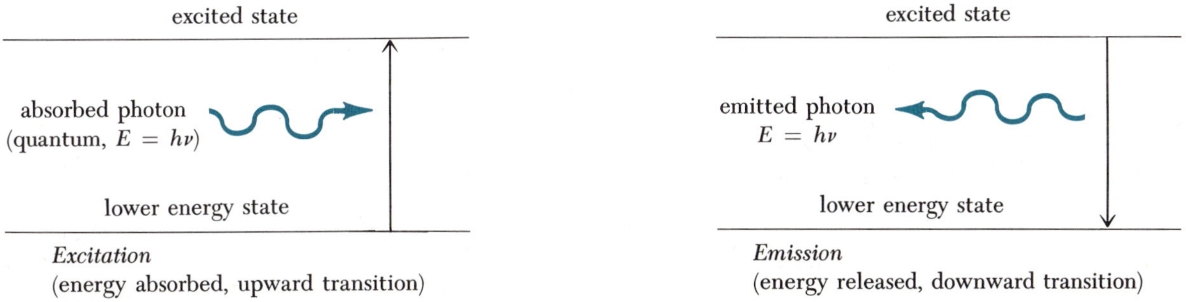

the emission spectra of atoms and ions. These spectra consisted of a series of discrete emissions (called lines); these emissions were at certain sharply defined frequencies, corresponding to the excited atoms or ions coming down in energy to states with lower energies. The classical concept of continuously variable energies, coupled with the basic ideas of the Rutherford model, could not account for emission lines.

During the nineteenth century the emission spectra of hydrogen atoms and certain ions, such as He^{+}, Li^{2+}, and Be^{3+}, had been reported. Until the Bohr development, however, there was no reasonable explanation for the specific patterns of discrete emission lines, representing frequencies, that were observed. Rydberg and Balmer had found that frequencies of the emission lines of a specific system, say hydrogen, could be calculated by the empirical relation given in Eq. (6.4):

$$\nu = R_{\mathrm{H}}\left(\frac{1}{n_2^{\,2}} - \frac{1}{n_1^{\,2}}\right) \tag{6.4}$$

In this equation ν represents the frequency of one of the discrete emissions, R_{H} is an empirical constant (called the Rydberg constant), and n_1 and n_2 are integers. Although this equation was reported in 1890, not until Bohr's publication of 1913 was a "reasonable" and successful model proposed to account for it.

Bohr recognized, in what must be one of the greatest examples of inspired, bold, and original thinking of twentieth-century science, that he could account for the Rydberg expression (Eq. 6.4) by making two dramatically novel assumptions—the two given above. Using these assumptions, Bohr derived an expression for the total energy of the electron in a hydrogen atom, kinetic energy plus potential energy, in terms of the principal quantum number introduced in Eqs. (6.2) and (6.3). Bohr's expression for the total energy is

$$E = -\frac{1}{n^2}\left(\frac{2\pi^2 mZ^2e^4}{h^2}\right) \tag{6.5}$$

where all of the algebraic symbols are as defined above.*

*The total energy is expressed as the sum of kinetic energy and potential energy (due to electrostatic attraction):

$$E = KE + PE = \frac{mv^2}{2} + \left(-\frac{Ze^2}{r}\right)$$

By recognizing that for a stable orbit, the centrifugal force must equal the centripetal force, we have

$$\frac{mv^2}{r} = \frac{Ze^2}{r^2}$$

$$mv^2 = \frac{Ze^2}{r}$$

Hence,

$$E = \frac{Ze^2}{2r} - \frac{Ze^2}{r} = -\frac{Ze^2}{2}\frac{1}{r}$$

Then, using the expression for r given in Eq. (6.3), we get

$$E = -\frac{Ze^2}{2}\cdot\frac{4\pi^2me^2Z}{n^2h^2} = -\frac{1}{n^2}\frac{2\pi^2mZ^2e^4}{h^2}$$

Bohr then reasoned that if the electrons were restricted to motion in orbits satisfying his first postulate, and if the only changes in energy corresponded to jumps from one of these allowed states to another, the observed emission spectrum of hydrogen could be predicted or explained. If we characterize the initial state of the system (before the energy jump) by the principal quantum number n_i and the final state (after the jump) by n_f, then we can represent the energies of the initial and final states by the formulas

$$E_{n_i} = -\frac{1}{n_i^2}\left(\frac{2\pi^2 mZ^2e^4}{h^2}\right) \quad \text{and} \quad E_{n_f} = -\frac{1}{n_f^2}\left(\frac{2\pi^2 mZ^2e^4}{h^2}\right)$$

Then, the energy difference corresponding to the jump is:

$$\Delta E = E_{n_f} - E_{n_i} = \left(\frac{1}{n_i^2} - \frac{1}{n_f^2}\right)\left(\frac{2\pi^2 mZ^2e^4}{h^2}\right) \tag{6.6}$$

Bohr then surmised that a quantum of energy having this magnitude would manifest itself as a photon with a frequency $\Delta E/h$, according to Eq. (6.1). Hence, it is predicted that the photons produced in the emission experiment on hydrogen have frequencies given by the equation

$$\nu = \frac{\Delta E}{h} = \left(\frac{1}{n_i^2} - \frac{1}{n_f^2}\right)\left(\frac{2\pi^2 me^4}{h^3}\right) \tag{6.7}$$

You recall that $Z = 1$ for hydrogen. When the values of π, m, e, and h are inserted into Eq. (6.7), it is found that the factor $2\pi^2me^4/h^3$ precisely equals the value of the constant R_H of Eq. (6.4). Hence, it is clear that the empirically based integers n_2 and n_1 in Eq. (6.4) are simply the principal quantum numbers of the lower and upper energy states, respectively, of the electron. Figure 6.12 gives a diagram of the first few allowed energy levels of the hydrogen atom. It is seen that highest energy available in the scheme is 0; all other energies are negative.* It is apparent from the figure that the gaps between successive levels are smaller as the principal quantum number is larger. The arrows of Fig. 6.12a represent possible emission transitions; in Fig. 6.12b some of the same transitions are represented schematically in terms of specific electron orbits.

EXAMPLE 6.1 Calculate the radii of the innermost three orbits of the electron in a hydrogen atom.

SOLUTION For the first orbit of hydrogen ($n = 1$, $Z = 1$),

$$r_1 = \frac{n^2h^2}{4\pi^2me^2Z} = \frac{(1)^2(6.6252 \times 10^{-27}\ \text{erg sec})^2}{4\pi^2(9.1083 \times 10^{-28}\ \text{g})(4.8029 \times 10^{-10}\ \text{abs esu})^2}$$

$$= 0.52937 \times 10^{-8}\ \text{cm} = 0.52937\ \text{Å}$$

*The energy value of zero for the highest orbit is an arbitrary but universally accepted assignment. Values of energy greater than zero then correspond to an electron in a state that is "not bound" in any way to the nucleus, one that is moving freely beyond the influence of the nucleus. The energy of the "free" electron is not governed by the quantized situation representative of the atom, and can have any value. Free electrons that are accelerated in the beams that produce images on television screens or that are used in electron microscopes are examples of such "positive energy" electrons.

Figure 6.12
Electronic transitions for the hydrogen atom. (a) An energy-level diagram for electronic transitions. (b) An "orbital transition" diagram to explain the interpretation of the experimental spectrum. (c) A band spectrum on a photographic plate.

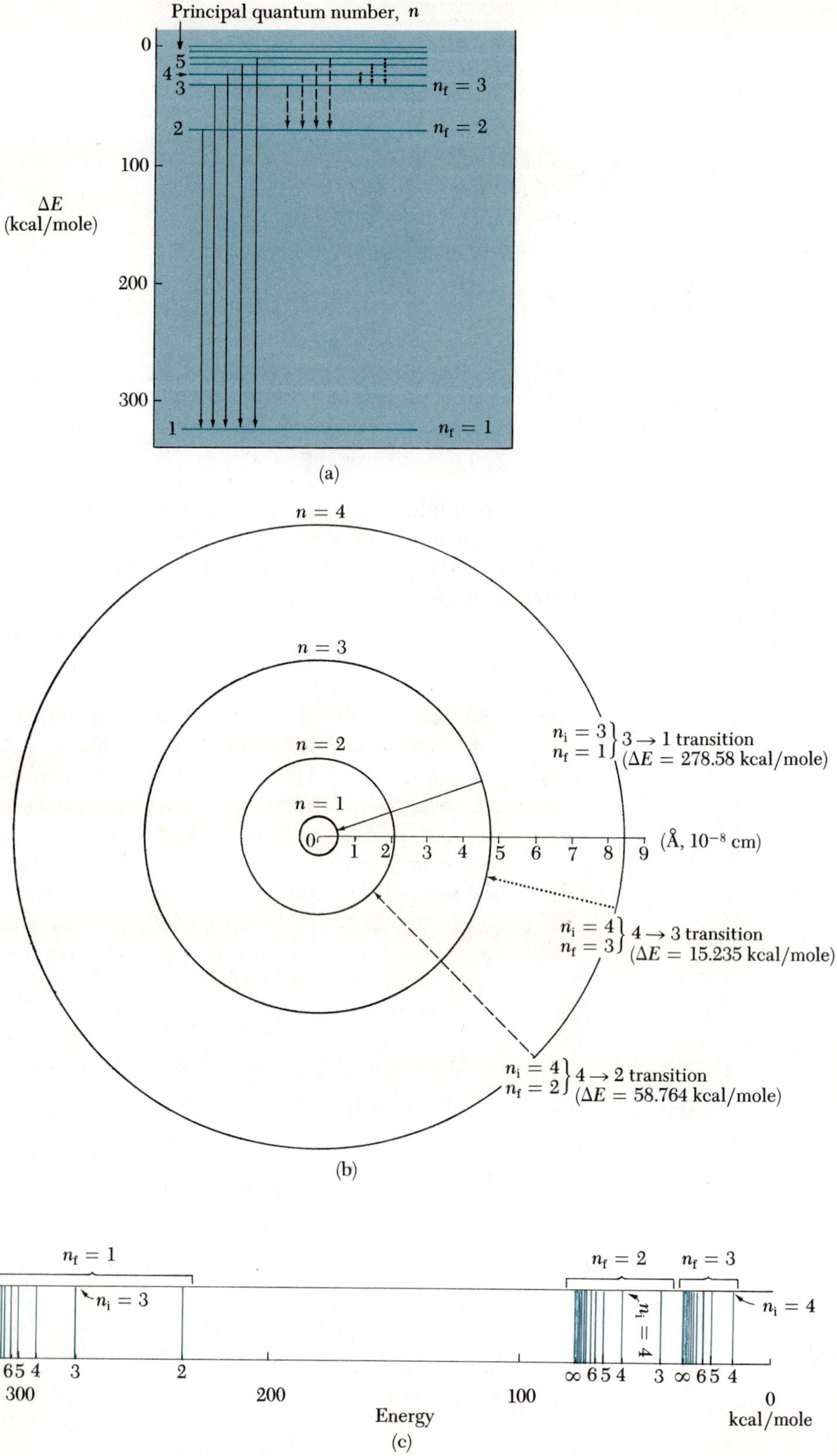

In calculating the second and third orbits, we note that the number of protons Z remains the same and that h, m, and e are all constants. Hence, the only factor that changes is n, the principal quantum number:

$$r_2 = n^2\left(\frac{h^2}{4\pi^2 me^2 Z}\right) = n^2(0.529\ \text{Å}) = 2^2(0.529\ \text{Å}) = 2.12\ \text{Å}$$

$$r_3 = n^2\left(\frac{h^2}{4\pi^2 me^2 Z}\right) = n^2(0.529\ \text{Å}) = 3^2(0.529\ \text{Å}) = 4.76\ \text{Å}$$

STUDY PROBLEM 6(a)

Calculate the innermost orbit radius for He^+.

EXAMPLE 6.2 Calculate the energy difference (in kcal/mol) between the two states of an electron (in a hydrogen atom) corresponding to $n = 1$ and $n = 2$, and compute the frequency of the light that is emitted if the hydrogen atom undergoes a transition from the $n = 2$ state to the $n = 1$ state.

SOLUTION

$$\Delta E = \frac{2\pi^2 mZ^2e^4}{h^2}\left(\frac{1}{n_f^{\,2}} - \frac{1}{n_i^{\,2}}\right)$$

The factor $2\pi^2 mZ^2e^4/h^2$ equals the Rydberg constant for the hydrogen atom R_H and has a value of 313.4 kcal/mol. Hence, for the $n = 2$ to $n = 1$ transition,

$$\begin{aligned}\Delta E &= 313.4\ \text{kcal/mol}(1/1^2 - 1/2^2)\\ &= 313.4\ \text{kcal/mol}(3/4) = 235.1\ \text{kcal/mol}\end{aligned}$$

The frequency of light is related to the energy by the equation

$$\Delta E = h\nu \qquad \text{or} \qquad \nu = \frac{\Delta E}{h}$$

The value h is given as 6.6256×10^{-34} J-sec. The abbreviation J-sec stands for joule-second, and a joule is the work done by a force of one newton acting for one meter. Hence, we shall convert the energy to the compatible units. We also want the frequency of each transition, (that is, per molecule), instead of the total for a mole.

$$\begin{aligned}\nu &= \frac{\left(235.1\dfrac{\text{kcal}}{\text{mol}}\right)\left(\dfrac{\text{mol}}{6.023 \times 10^{23}\ \text{molecule}}\right)\left(4.186\dfrac{\text{J}}{\text{cal}}\right)\left(\dfrac{10^3\ \text{cal}}{\text{kcal}}\right)}{6.6256 \times 10^{-34}\ \text{J-sec}}\\ &= 2.466 \times 10^{15}\ \text{sec}^{-1}\ \text{molecule}^{-1} \text{ or } 2.466 \times 10^{15} \text{ waves per second per hydrogen atom}\end{aligned}$$

The common units used for expressing frequency are sec^{-1}, or cycles per second (cps). Recently, the unit Hertz (Hz; after Franck Hertz) has been adopted in place of cps.

After the Bohr model was published, Sommerfeld suggested the refinement of including elliptical orbits as well as circular orbits. This model and the theoretical predictions based on it were quite successful in accounting for the available data from emission spectra of the hydrogen atom and various ions that contain only one electron, for instance, He^+, Li^{2+}, and Be^{3+}. This was considered a great victory of physical science, and stands as an important milestone of modern physics. It was soon recognized, however, that the Bohr model, with or without Sommerfeld's refinement, simply could not account for the spectra of atoms or ions with more than one electron. Although a truly giant step had been made in understanding physics and chemistry at the atomic and subatomic levels, an even better theory was needed.

The atomic theory that replaced the Bohr model retained the basic idea of a nuclear model with protons and neutrons, the heavy particles, contained in a small, central cluster, and electrons occupying a relatively large volume around this cluster. What is truly different about the new theory, which is still accepted, is the idea that one must abandon the approach that an electron can be viewed as a classical particle—a particle of clearly defined mass, charge, and size that occupies a specific point in space at any particular time. Intuitively this classical view of a particle seems entirely reasonable, but it turns out to be inconsistent with modern experiments on submicroscopic systems. A planetary orbit, as Bohr envisioned it, implies that one can exactly define the location and velocity of any particle of interest; modern theory requires us to abandon that assumption, as reasonable as it seems a priori.

6.2 Present View of Atomic Structure

Wave Nature of Electrons

Bohr's essentially classical view of an electron as a clearly defined charged particle moving in a precisely defined orbit soon came up against experimental data that challenged that philosophical foundation. It was found that electrons manifest a phenomenon that was previously attributable only to waves; the phenomenon is diffraction, a behavior not expected of particles.

When a beam of light impinges upon a sharply defined barrier, or set of sharp edges, the image that is cast upon a screen shows a series of patterns, for instance lines, called diffraction patterns, resulting from the interference phenomena characteristic of waves. Figure 6.13a depicts a typical experiment in which light waves are diffracted.

If a beam of electrons, like the scanning beam in a television tube, consisted simply of a stream of classical particles, the electrons would merely travel through space in a straight line, being scattered into other directions only if they were to collide with an obstacle. To the surprise of physicists in the first quarter of this century, beams of

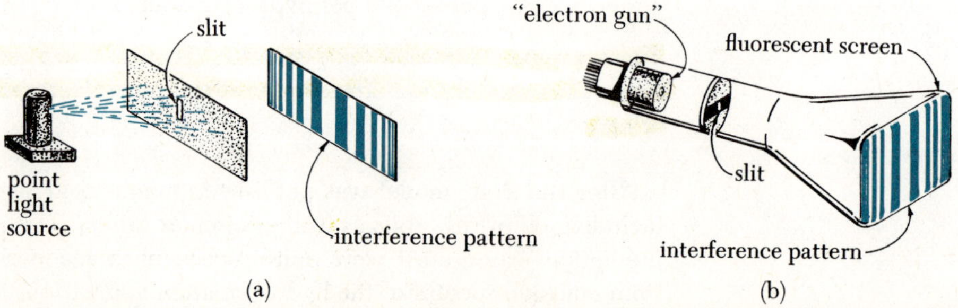

Figure 6.13 Diffraction behavior of (a) light waves and (b) an electron beam.

Spotlight

When waves encounter a slit, they flare out from the slit; this phenomenon is called **diffraction.** It was known before 1800 that light passing through two nearby slits would give a series of bright and dark areas on a screen—as shown in Fig. 6.13. Thomas Young, in 1803, interpreted these results by stating that waves could interfere with one another—either building each other up if they coincided properly or destroying each other if they were "out of phase." In his illustration shown herewith (from *Phil. Transactions*, 1803), the colored lines have been added to show lines of reinforcement. The *x*'s between lines show destructive interference.

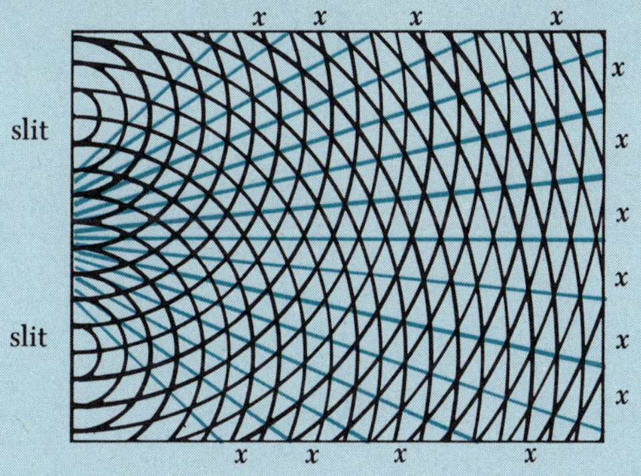

electrons were found to produce interference patterns of the type expected for diffraction phenomena of light waves; this observation is depicted in Fig. 6.13b. Other evidence for the wave character of electrons was secured in the observation that electrons were diffracted by metal foils in a fashion associated earlier with the wave nature of x rays.*

The logical conclusion for the interference experiments is that electrons are waves! But Crookes's pinwheel experiments and other evidence indicated that electrons are particles. The inevitable answer to the dilemma is that electrons manifest characteristics of both particles and waves! This was a radical idea in the 1920s, when it was first considered. Coupled with the fact that Bohr's model, based on a particle premise, could not account for the spectra of atoms with more than one electron, this apparently ambivalent view was at once unsettling and stimulating to atomic physics. The result was that some truly imaginative advances were made by exploring how the theory of wave motion could be incorporated in some way into an atomic model.

The behavior of confined wave motion—such as the motion of a plucked string of a stringed musical instrument—was well known to physicists before the beginning of the twentieth century. This kind of motion can provide useful guidelines or analogies for trying to understand how wave character can be incorporated into an atomic model. The form of vibration of a string with its two ends tied down is called **a standing wave.** Only waves with an integral number of half-wavelengths will fit between the points of attachment (see Fig. 6.14); hence, only such waves can be sustained and be standing waves. These integral numbers can be considered essentially quantum numbers, by which the motion of the string is "quantized." There are therefore only certain discrete, "allowed" wavelengths for the confined waves of a string; the system is quantized.

In 1926, Irwin Schrödinger put together the available ideas and information in a very novel way. Realizing that electrons have wavelike properties, he mathematically explored the possibility that the nucleus keeps the electron confined, exerting on it a confining influence that could lead to patterns of discrete, allowed wavelengths of the electron. He then proposed a theoretical model in which the

*A more detailed discussion of diffraction is included in Chap. 14.

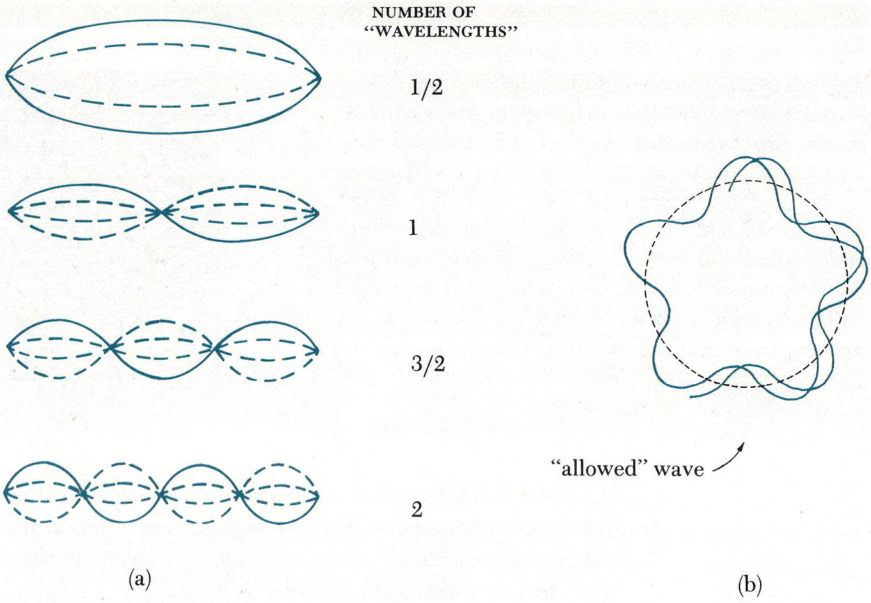

Figure 6.14 (a) "Time-lapse photos" of standing waves in a vibrating string. The number of possible wavelengths is limited to multiples of $\frac{1}{2}$. (A wave begins to repeat itself after one "wavelength.") (b) Wave representation for circular orbits of an electron about the nucleus of an atom. The "allowed" wave reinforces itself over several orbits to give a positive result. The wave shown by the dashed line does not reinforce itself and hence is destroyed by interference (since addition of the waves from several "trips" around the nucleus gives zero as a result).

allowed energies and the "whereabouts" of the electron were governed directly by the standing-wave character of the electron.

Schrödinger used a "wave equation" and a "wavefunction" to describe the behavior of the electrons in an atom, and developed a new type of mathematical physics to handle the problem. His approach is referred to as **wave mechanics**, and is one of the two essentially equivalent forms of modern quantum mechanics.* The mathematical details of Schrödinger's theory are not especially important for our present purposes.

Schrödinger's theory had profound philosophical consequences. These consequences are related to giving up, in accepting the postulates of wave mechanics, the intuitively comfortable assumption that one can define precisely the position and velocity of an electron in its motion about a nucleus, and that the motion can be described in terms of precisely definable orbits. This kind of classical exactness is, in wave mechanics, replaced by the concept of probability. In other words, instead of speaking with certainty about the exact location of an electron in an atom, we refer to the probability of its being at a particular location. This probability is given, according to the postulates of wave mechanics, by a function (mathematical expression) ψ^2, where ψ is the **wavefunction**; it is the key function in Schrödinger's theory.

*The other form was developed at about the same time by Heisenberg, and is referred to as matrix mechanics. Later it was shown by other scientists that wave mechanics and matrix mechanics are essentially two different forms of the same theory.

Spotlight

The basic mathematical form of Schrödinger's theory is represented by a famous equation that is known by his name:

$$H\psi = E\psi$$

In this equation, which can be considered a basic postulate of wave mechanics, E is the total energy of the system (one of the "allowed" energies if the equation is satisfied) and H stands for what is called a Hamiltonian operator. In the mathematical language of quantum mechanics, H stands for the sum of the kinetic energy (KE) and the potential energy (PE). According to the basic postulates, PE is simply the classical expression for the potential energy of the system ($-e^2/r$ for the special case of a hydrogen atom) and KE is given by a postulated "recipe." This recipe leads to

$$-\frac{h^2}{8\pi^2 m}\left(\frac{\partial^2}{\partial x^2}+\frac{\partial^2}{\partial y^2}+\frac{\partial^2}{\partial z^2}\right)$$

for the special case of an electron in a hydrogen atom, where m is the mass of the electron and $\partial^2/\partial x^2$ is the partial second derivative with respect to x. The symbol psi, ψ, stands for the wavefunction, which according to the basic postulates of wave mechanics provides a direct means of calculating the probability of finding an electron at a specific point in space.

The Schrödinger equation has solutions only for specific, discrete functions ψ and energies E, and only in certain combinations—like a lock and key arrangement. For each solution, that is, each combination of ψ and E that satisfies the equation, there exists an allowed state of the system—with its discrete allowed energy.

Spotlight

The untenable consequences of assuming that both position and velocity of a particle can be simultaneously determined exactly had been pointed out previously by Heisenberg, who recognized the fundamental limitations of making simultaneous measurements of position and velocity. He cast his discovery into the form of the equation:

$$\Delta x\,\Delta v \geq \frac{h}{m}$$

where h is Planck's constant, m is the mass of the particle in question, Δx represents the uncertainties in the particle's measured position, and Δv is the uncertainty in its velocity. This equation tells us that the product of these uncertainties must be at least as large as the value of h/m. This type of relation, in its generalized form, is known as the **Heisenberg uncertainty principle.**

An example of the magnitudes involved in such uncertainties can be found by considering a situation in which one knows the position of an electron at a particular instant to be at one Bohr radius (0.529×10^{-8} cm) from the nucleus with an uncertainty of 0.001×10^{-8} cm. The uncertainty on the velocity would then be

$$(1 \times 10^{-11}\text{ cm})\,\Delta v \geq \frac{6.6252 \times 10^{-27}\text{ erg sec}}{9.1083 \times 10^{-28}\text{ g}}$$

$$\Delta v \geq 7.2731 \times 10^{11}\text{ cm/sec}$$

Since the speed of light is 3.00×10^{10} cm/sec, we have essentially no idea of the speed if we can measure the position of the electron to 0.001 Å.

Would the Heisenberg uncertainty principle pose any problem in our macroscopic environment? Consider the case of an umpire who wished to judge the speed of a fastball weighing 250 g and traveling at 100 mi/hr to within 1 mi/hr (45 cm/sec) and call the strike at the same time. The uncertainty in determining the position would be:

$$\frac{45\text{ cm}}{\text{sec}}\,\Delta x \geq \frac{6.625 \times 10^{-27}\text{ erg sec}}{250\text{ g}}$$

$$\Delta x \geq 5.9 \times 10^{-31}\text{ cm, or } 6 \times 10^{-31}\text{ cm}$$

A simultaneous judgment of velocity and position of a fastball shouldn't affect the judgment of the strike zone!

Quantum Numbers from Wave Mechanics; Many-Electron Atoms

Applying Schrödinger's formalism to atoms disclosed that three quantum numbers were needed for specifying the state of an electron. For a one-electron atom (say H) or ion (say He^+ or Li^{2+}) only one of these quantum numbers n is required to specify the energy; however, for an atom or ion with more than one electron, a knowledge of just the quantum number n is not enough to determine the energy precisely. Three quantum numbers involved in the quantum mechanical description of a state of an atom are the principal quantum number n, which determines the gross, or approximate, energy of the state; a quantum number l, which is a measure of the angular momentum of an electron in its motion about the nucleus;* and a quantum number m_l, associated with the orientation of the angular momentum of the electron relative to specified directions in space (see Fig. 6.15). These quantum numbers have been shown to specify completely the energy of an electron in a specific state, and the electron probability distribution function ψ^2 of that state, denoted $\psi_{nlm_l}{}^2$. The subscripts of the probability function specify the quantum numbers of the electron,

*According to classical mechanics, the angular momentum of a particle of mass m, moving in a circular orbit of radius r and linear (tangential) velocity v, equals mvr.

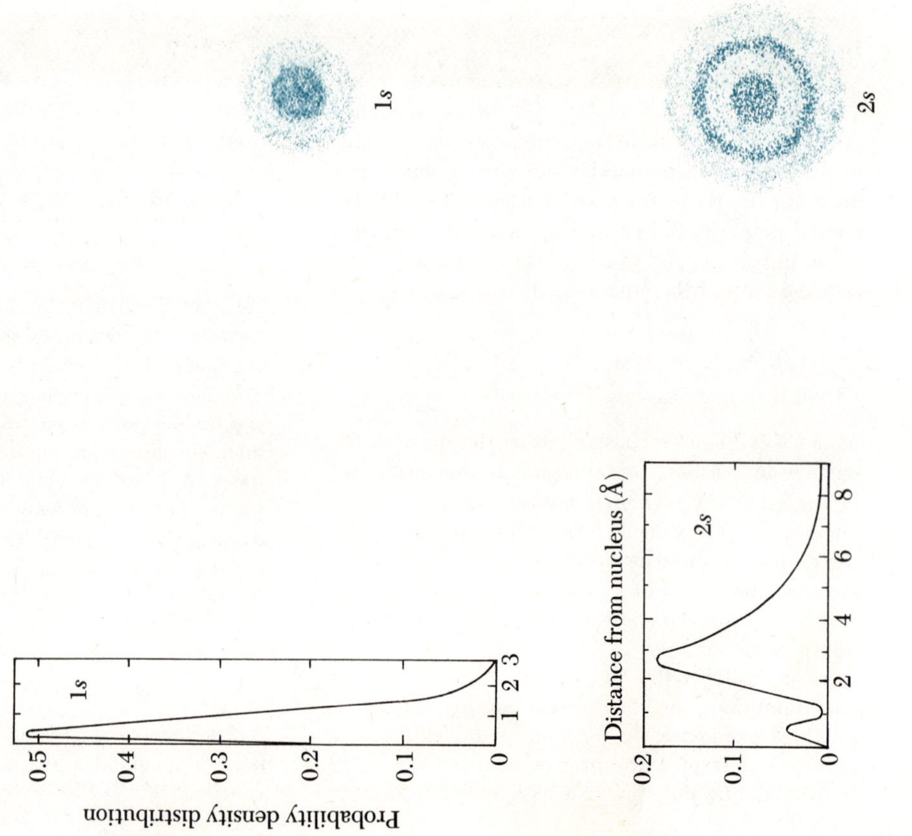

Figure 6.15 Representations of the probability density distribution of electrons about a nucleus. The representations shown here (turn book to read) are graphical and shaded drawings. The figures on the left show photos of probability density as a function of distance from the nucleus. The figures on the right represent the probability density by the darkness of the shading.

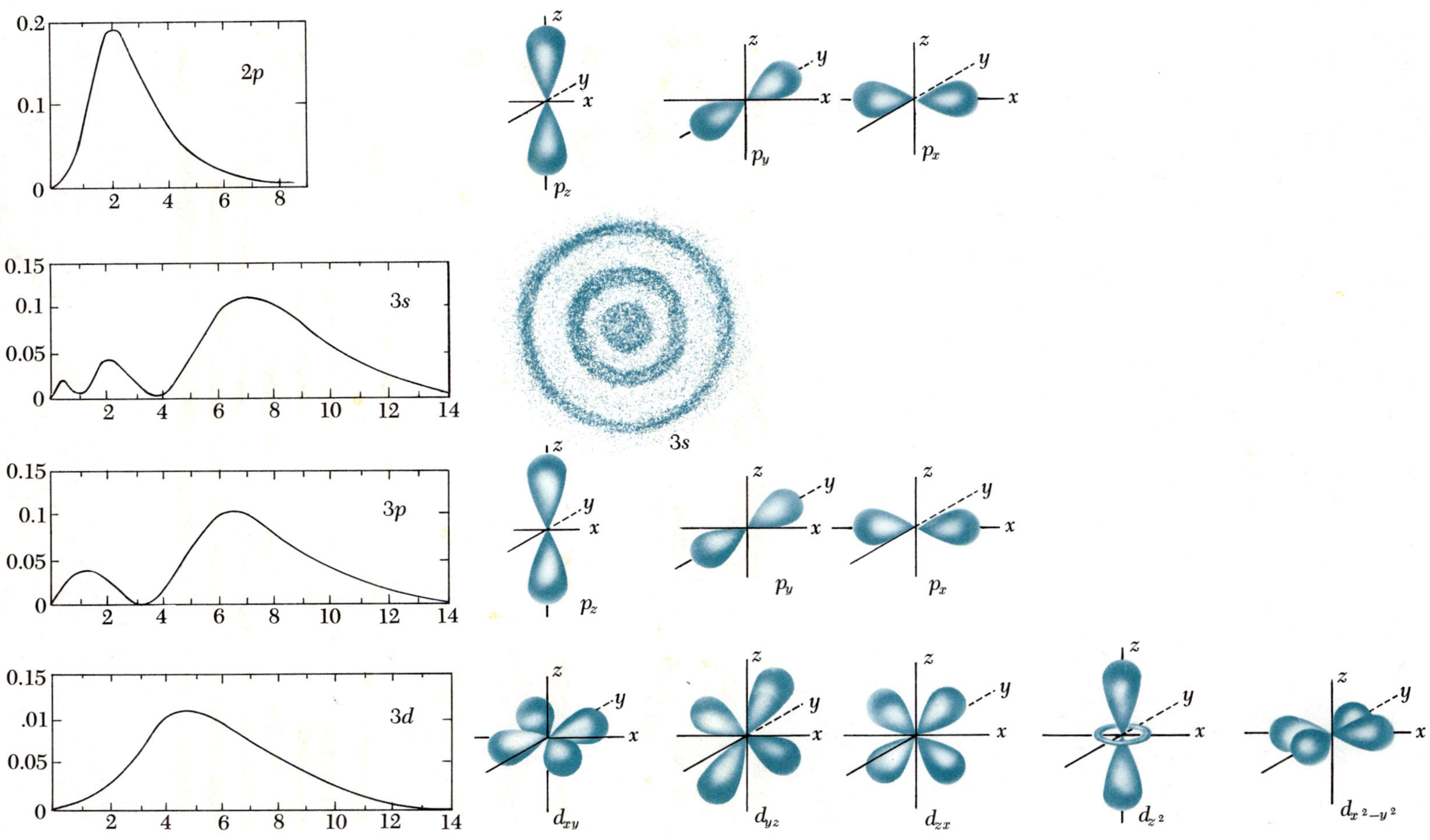

2p
3s
3p
3d
p_z
p_y
p_x
d_{xy}
d_{yz}
d_{zx}
d_{z^2}
$d_{x^2-y^2}$

and by specifying the three quantum numbers n, l, and m_l one specifies the state of the electron in an atom as completely as it can be specified, according to the philosophy and postulates of quantum mechanics. An electronic state, specified in this way, is often referred to as an **orbital**, by analogy to the orbit concept of Bohr.

The probability function $\psi_{nlm_l}{}^2$ is a precise mathematical expression; there are various ways of representing it—analytically, numerically, or graphically. Two of the common graphical methods are shown in Fig. 6.15. One of these methods is to show graphically how the probability density function varies with distance from the nucleus; this approach is shown on the left side of Fig. 6.15.* A second representation of the probability density function is shown on the right side of Fig. 6.15; the spatial dependence of $\psi_{nlm_l}{}^2$ is shown by shading in a three-dimensional picture. In this scheme, values of $\psi_{nlm_l}{}^2$ below some arbitrary value are not shown at all; this is why no probability density is shown in Fig. 6.15 beyond some finite distance from the origin in each case.

Much information about the electrons in an atom is available from knowing the quantum numbers. You can get this information if you know essentially what these numbers stand for, and what electronic features are associated with their specific values, even if you are not familiar with the precise mathematical techniques by which this information can be derived through quantum mechanics. The principal quantum number n is somewhat like the quantum number n of the Bohr theory. A knowledge of n gives one an idea of both the approximate energy of the electron in an atom and the range of distance from the nucleus in which the electron is most likely to be found, in other words, the range of radius of orbital motion having the highest probability. A knowledge of l is also necessary for pinpointing the energy exactly. The characteristics determined by a specific value of n define a **shell**. Thus, for each value of n (1, 2, 3 . . .), a shell is defined, which is roughly equivalent to the orbit defined by n in the Bohr theory. The highest value of n now known for an electron in a stable atom is 7.

The second quantum number that one needs to define the state, or energy, of an electron in an atom is l. This quantum number, while not primarily associated with specifying the most probable distances from the nucleus, tells one what the shape of the probability distribution is. A few of these shapes are given in Fig. 6.15. For a given value of n, it is known that l can be any one of the integers 0, 1, . . . , $n - 1$; l can never be as large as n. Thus, for an electron with $n = 4$, the value of l can be one of the following: 0, 1, 2, or 3. For reasons that are only of historical interest, another type of notation is in use, in which specific lowercase letters of the alphabet are associated with the l values. Thus, an electronic state with $l = 0$ is called an s state; if $l = 1$, it is called a p state; for $l = 2$, it is called a d state; for $l = 3$, an f state. For example, a state with $n = 3$ and $l = 2$ is called a $3d$ state, or a $3d$ orbital; a state with $n = 2$, $l = 0$ is referred to as a $2s$ orbital.

The third quantum number, m_l, designates the direction in which the maximum

*The function plotted on the vertical axes of the graphs in Fig. 6.15 is $4\pi\psi^2 r^2 dr$, where dr is a differential increment in radius (from differential calculus). The function ψ^2 represents the probability function at a point in space; the total probability of finding the electron in a small region of space is the produce of ψ^2 for that region and the volume of that region. In calculus, this product is given by $\psi^2 \cdot 4\pi r^2 dr$. It is this regional probability that is shown in Fig. 6.15. The maxima in the probability are found at Bohr radii.

concentration of electron probability density is oriented. It is well known that once the value of l is specified for an orbital, the only allowed values of m_l are the positive and negative integers ranging from $-l$ to l, that is, $-l, -l+1, \ldots, 0, \ldots, l-1, l$. Altogether there are $2l + 1$ possible values of m_l for a given value of l. Each of these corresponds to a distinct state, or orbital. All $2l + 1$ of them for a given n have the same energy value, for a free atom. From this, we see that for the case $l = 0$ (s case), m_l can only be zero; hence, once n is specified for the s case, a single orbital is defined; that is, there is only one orbital of the 1s type, only one of the 2s type, one of the 3s type, and so on. If $l = 1$ (p case), then m_l can have any one of the following values: -1, 0, 1. Once n is specified for the p case, *three* orbitals can be defined, all of them having the same energy value. That is, there are three orbitals of the 2p type, three of the 3p type, and so on. For $l = 2$ (d case), m_l can be -2, -1, 0, 1, or 2. Specifying n for the $l = 2$ case specifies five orbitals; hence, there are five 3d orbitals, five 4d orbitals, and so on. For $l = 3$ (f case), specifying n denotes seven orbitals, corresponding to the following possible values of m_l: $-3, -2, -1, 0, 1, 2, 3$; thus, there are seven 4f orbitals.

A given value of the principal quantum number n defines the shell, providing an approximate measure of the energy and average distance from the nucleus of an electron within that shell. Now we see that there may be more than one orbital or even more than one kind of orbital within a shell. Just as n is said to define a shell, a given value of l is said to define a **subshell.** Hence, we have s subshells (consisting of only one orbital), p subshells (consisting of three orbitals, p_x, p_y, and p_z; corresponding to $m_l = -1, 0,$ or 1), d subshells (consisting of five orbitals), and f subshells (consisting of seven orbitals). Figure 6.16 provides a summary of these relations, the "recipes" that we use to simplify electronic bookkeeping in atoms and molecules.

Figure 6.15 shows the directional features of a few specific atomic orbitals; it is seen that when there is more than one atomic orbital for a given subshell, each orbital corresponds to a different orientation of the electron probability density in space. For example, the three orbitals with $n = 2$ and $l = 1$ (p_x, p_y, and p_z) are seen to have probability distributions that are concentrated along three mutually perpendicular directions in space.* It is extremely useful to remember these directional characteristics of orbitals in considering the bonding characteristics of atoms; this is one reason why this second graphical means of representation, essentially pictorial, is so useful to chemists.

Electron Spin

So far, we have been concerned qualitatively with how the quantum numbers n, l, and m_l are related to the energies of electrons in atomic orbitals and to the distribution of electron probabilities in space. It was apparent even in the early days of the development of quantum mechanics, however, that another important feature of the

*All electrons in the three 2p orbitals in a free atom in space have the same energy. However, these three orbitals can be made to have different energies by rendering the three mutually perpendicular directions along which they are directed physically different—for instance, by applying a magnetic field along one of these directions. For this reason m_l is sometimes referred to as the magnetic quantum number.

Figure 6.16 Quantum numbers and orbitals

n (*determines shell*)	l (*determines subshell*)	m_l	*Common Notation*
1	0	0	1*s*
2	0	0	2*s*
	1	−1, 0, 1	2*p*
3	0	0	3*s*
	1	−1, 0, 1	3*p*
	2	−2, −1, 0, 1, 2	3*d*
4	0	0	4*s*
	1	−1, 0, 1	4*p*
	2	−2, −1, 0, 1, 2	4*d*
	3	−3, −2, −1, 0, 1, 2, 3	4*f*

electron should be included in any theory that would correctly describe electrons in atoms or molecules. This additional feature is referred to as electron spin, or more completely, intrinsic electron spin angular momentum.

Electron spin is a phenomenon for which strong experimental and theoretical evidence became available during the 1920s, and which is firmly established now in all modern theories of atomic and molecular structure.* Electron spin refers to the

*In 1925, Uhlenbeck and Goudsmit introduced the idea of electron spin to account for some of the subtle features of the emission spectra of some of the elements, and to account for the effect that an applied magnetic field has on the spectra. Soon after the development of quantum mechanics, P. A. M. Dirac formulated the theoretical basis supporting electron spin to the present.

universal behavior of electrons that one might try to explain within the framework of classical physics as spinning about an axis through the middle of the electron. Associated with this spinning is an intrinsic angular momentum and a magnetic moment; the moving charged particle gives rise to a magnetic field. Thus, in some ways, an electron acts like a tiny bar magnet. An interesting and highly pertinent feature of this spin and associated magnetic moment is that they behave as though there are only two situations available to the "spinning electron"; that is, this spin angular momentum is also quantized. It is as though the tiny bar magnet can point in only one of two possible directions (one the opposite of the other, say north and south). The way in which this behavior is described in quantum mechanics is by another quantum number, denoted m_s, which is restricted to only the value $\frac{1}{2}$ or the value $-\frac{1}{2}$. It is important to note that electron spin is a very interesting case of a "model." Chemists believe not that electrons are actually spinning like tops, but that the mathematical treatment of such a case is consistent with experimental results.

Taking electron spin into account, one sees that there are four quantum numbers, n, l, m_l, and m_s, which must be specified so that the state of an electron in an atom can be known.

Occupation of Orbitals

To understand how the various available electronic states of an atom are used, one needs to remember only a few simple rules. One rule, called the **Pauli exclusion principle,** states that *not more than one electron in an atom can occupy an allowed state,* a state corresponding to an acceptable set of the quantum numbers n, l, m_l, and m_s.* If we recall that the quantum numbers n and m_l specify an orbital, and that there are only two allowed values of m_s, we can readily see that the maximum number of electrons that can occupy a given orbital is two, and that if there are two, then they must have opposite spin; one of the two electrons has $m_s = \frac{1}{2}$, and the other has $m_s = -\frac{1}{2}$. Hence, according to the Pauli exclusion principle, the number of electrons occupying a specific atomic orbital is zero, one or two. A second important rule, the **Aufbau principle,** states that *the most stable state of an atom, the* **ground state,** *is the state of lowest energy,* in which the electrons occupy the lowest-energy orbitals available to them. Thus, if we wish to know how many electrons occupy each orbital in the ground state of an atom, we imagine "allocating" electrons to the various orbitals, filling them from the bottom up energetically (from the inside out spatially)—placing two electrons in each orbital until we have allotted all the electrons; for an odd number of electrons there will be only one electron in the highest-energy orbital that is used. For example, the ground state of the hydrogen atom, which has only one electron, consists of an electron in a 1s orbital, the lowest-energy orbital of all. For the ground state of beryllium, one expects two electrons of opposite spin in the lowest-energy orbital (1s) and two electrons of opposite spin in the orbital

*The original statement of this principle was made, in a slightly different form, by Wolfgang Pauli in 1925, before wave mechanics was truly developed.

of next lowest energy ($2s$); these two cases are denoted by the symbols $1s^1$ and $1s^2\,2s^2$, respectively. Similarly, the ground state of an oxygen atom would have orbitals occupied according to the scheme denoted $1s^22s^22p^4$. The knowledge of how many electrons occupy each kind of orbital constitutes knowledge of an **electronic configuration.** The notation just indicated for the ground states of hydrogen, beryllium, and oxygen atoms is a shorthand method of representing electronic configurations. A more complete way is to indicate each orbital by means of a clearly labeled horizontal line, and the electron occupation of the orbitals by means of arrows, which point up for a case with $m_s = \frac{1}{2}$ and down for a case with $m_s = -\frac{1}{2}$.

hydrogen atom $\quad \underset{1s}{\underline{\uparrow}}$

beryllium atom $\quad \underset{1s}{\underline{\uparrow\downarrow}}\ \underset{2s}{\underline{\uparrow\downarrow}}$

oxygen atom $\quad \underset{1s}{\underline{\uparrow\downarrow}}\ \underset{2s}{\underline{\uparrow\downarrow}}\ \underbrace{\underline{\uparrow\downarrow}\ \underline{\ \uparrow\ }\ \underline{\ \uparrow\ }}_{2p}$

The oxygen case above brings up a situation that was not properly anticipated in the Pauli and Aufbau principles. From those two principles, we know that the ground state of an oxygen atom has the configuration $1s^22s^22p^4$. The arrangement of four electrons in the three available $2p$ orbitals could, however, be (a) $\underline{\uparrow\downarrow}\ \underline{\ \uparrow\ }\ \underline{\ \uparrow\ }$ (as shown above), or (b) $\underline{\uparrow\downarrow}\ \underline{\uparrow\downarrow}\ \underline{\quad}$, or (c) $\underline{\uparrow\downarrow}\ \underline{\ \uparrow\ }\ \underline{\ \downarrow\ }$. There could be other arrangements equivalent to one of the latter two, like $\underline{\ \uparrow\ }\ \underline{\ \uparrow\ }\ \underline{\uparrow\downarrow}$, $\underline{\ \downarrow\ }\ \underline{\uparrow\downarrow}\ \underline{\ \downarrow\ }$, or $\underline{\quad}\ \underline{\uparrow\downarrow}\ \underline{\uparrow\downarrow}$. The point here is that all four of the $2p$ electrons can be paired as in (b)—with two in each of two orbitals—or two electrons can be unpaired and occupy separate orbitals as in (a). **Hund's rule** clarifies this by summarizing the pertinent results of both experimental and theoretical studies on this matter. Hund's rule tells us that when a given number of electrons can be arranged within a specific subshell in more than one way (affecting number of electrons occupying each orbital and relevant m_s values), the state of lowest energy corresponds to the case in which the electrons are spread out as much as possible among the different orbitals of the subshell, and for which all of the unpaired electrons have the same m_s value. Another way of stating Hund's rule is: *In the lowest-energy state, electrons fill orbitals of the same energy to give the maximum number of unpaired electrons.* Thus, the ground state of a carbon atom is

$\underset{1s}{\underline{\uparrow\downarrow}}\ \underset{2s}{\underline{\uparrow\downarrow}}\ \underbrace{\underline{\ \uparrow\ }\ \underline{\ \uparrow\ }\ \underline{\quad}}_{2p}$ instead of $\underset{1s}{\underline{\uparrow\downarrow}}\ \underset{2s}{\underline{\uparrow\downarrow}}\ \underbrace{\underline{\uparrow\downarrow}\ \underline{\quad}\ \underline{\quad}}_{2p}$; and the ground state of a nitrogen

atom is $\underset{1s}{\underline{\uparrow\downarrow}}\ \underset{2s}{\underline{\uparrow\downarrow}}\ \underbrace{\underline{\ \uparrow\ }\ \underline{\ \uparrow\ }\ \underline{\ \uparrow\ }}_{2p}$ instead of $\underset{1s}{\underline{\uparrow\downarrow}}\ \underset{2s}{\underline{\uparrow\downarrow}}\ \underbrace{\underline{\uparrow\downarrow}\ \underline{\ \uparrow\ }\ \underline{\quad}}_{2p}$. The spreading out of

electrons among the largest number of orbitals of the same energy is understandable on the grounds of classical electrostatics. Placing a second electron in a given orbital requires overcoming some forces of electrostatic repulsion, with a corresponding

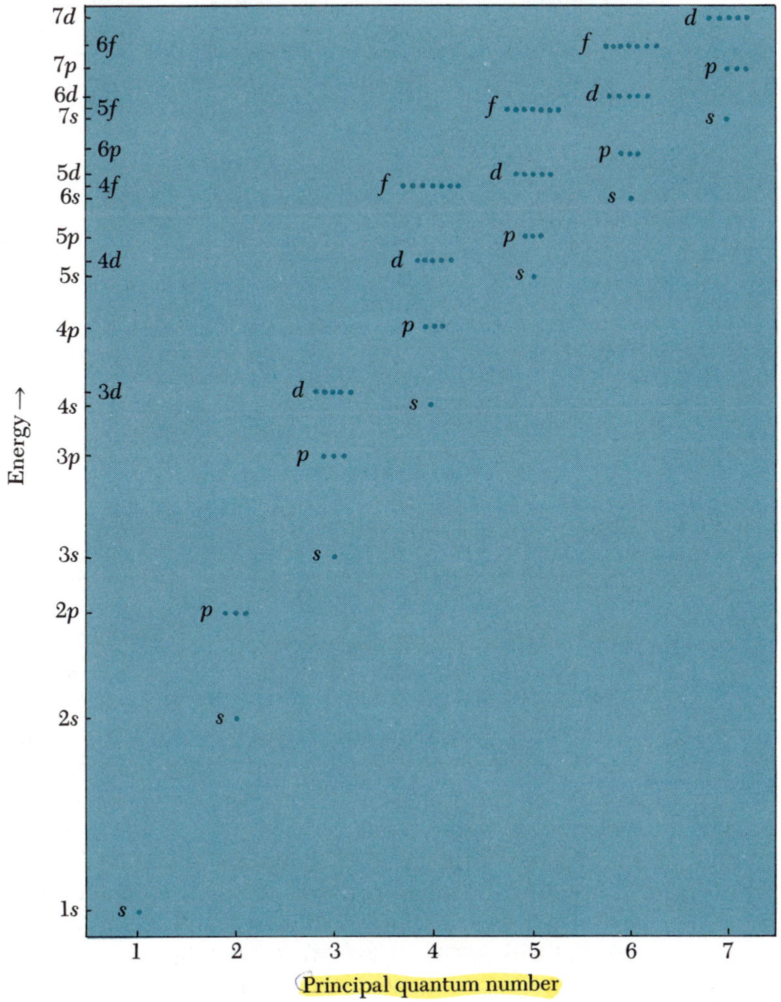

Figure 6.17 Relative energies of orbitals of neutral atoms in the ground state. Note: (1) the number of orbital-shape (l) quantum numbers = n; (2) the number of orbital-direction (m_l) quantum numbers = $2l + 1$; (3) although g, h, . . . subshells are possible, no known elements have electrons in these orbitals.

increase in total energy, and this additional energy can be reduced if two electrons occupy different orbitals.

In predicting electronic configurations of atomic ground states from the rules just cited, it is necessary to know how the energies of all the orbitals are related. The "energy order" of orbitals is determined experimentally and is illustrated in Fig. 6.17. Each circle in Fig. 6.17 represents an orbital, with its own set of n, l, and m_l quantum numbers. Each orbital can be occupied by a maximum of two electrons,

Atom	*Electronic Configuration*	*Orbital Diagram Representation*
H	$1s^1$	↑ (1s)
He	$1s^2$	↑↓ (1s)
Li	$1s^2 2s^1$	↑↓ (1s) ↑ (2s)
C	$1s^2 2s^2 2p^2$	↑↓ (1s) ↑↓ (2s) ↑ ↑ — (2p) (Hund's rule)
Cr	$1s^2 2s^2 2p^6 3s^2 3p^6 3d^5 4s^1$ or $[Ar]3d^5 4s^1$	↑↓ (1s) ↑↓ (2s) ↑↓ ↑↓ ↑↓ (2p) ↑↓ (3s) ↑↓ ↑↓ ↑↓ (3p) ↑ ↑ ↑ ↑ ↑ (3d) ↑ (4s)
Fe	$1s^2 2s^2 2p^6 3s^2 3p^6 3d^6 4s^2$ or $[Ar]\ 3d^6 4s^2$	↑↓ (1s) ↑↓ (2s) ↑↓ ↑↓ ↑↓ (2p) ↑↓ (3s) ↑↓ ↑↓ ↑↓ (3p) ↑↓ ↑ ↑ ↑ ↑ (3d) ↑↓ (4s)

one with spin "up" ($m_s = \frac{1}{2}$) and one with spin "down" ($m_s = -\frac{1}{2}$). From Fig. 6.17 the order of filling is seen to be 1*s*, 2*s*, 2*p*, 3*s*, 3*p*, 4*s*, 3*d*, 4*p*, 5*s*, 4*d*, 5*p*, etc. Some examples of electronic configurations of atoms are shown above. To simplify the symbolism for an electronic configuration, filled shells (filled orbitals with a given *n*) can be represented by the elemental symbol for the noble gas with the corresponding electronic configuration. Argon has the configuration $1s^2 2s^2 2p^6 3s^2 3p^6$, so the notation for the electronic configuration of the ground state of chromium can be shortened from $1s^2 2s^2 2p^6 3s^2 3p^6 3d^5 4s^1$ to $[Ar]3d^5 4s^1$.

In the example shown above, there is one exception to the above-stated rules for the electron populations of atomic ground states. We should expect Cr to have the ground state configuration $[Ar]3d^4 4s^2$, since the 4*s* orbital is of lower energy than the 3*d* orbitals. Instead, in the ground state that is observed experimentally, the 3*d* orbitals "borrow" one electron from the lower-energy 4*s* orbital. This same sort of thing occurs for the ground state of Cu, which we expect to have the configuration $[Ar]3d^9 4s^2$; but the observed ground state is $[Ar]3d^{10} 4s^1$. The characteristic that these two electronic configurations have in common is that the borrowed electron is used to complete a half-filled or filled inner subshell. An electronic configuration in which as many subshells as possible are either exactly half-filled or completely filled is especially stable; it has especially low energy. This tendency towards filled and half-filled subshells corresponds to a lowering of the total energy of the atom sufficiently to allow one electron to be borrowed from a lower-energy orbital. With this one exception, electronic configurations of most atoms can be predicted easily.*

Table 6.1 shows the electron configurations of the known elements.

*In very heavy elements, several orbitals have very nearly the same energy and slight deviations from our simple scheme are often found.

TABLE 6.1 Electronic Configurations of the Elements

Atomic Number	Element	1s	2s	2p	3s	3p	3d	4s	4p	4d	4f	5s	5p	5d	5f	6s	6p	6d	7s
1	H	1																	
2	He	2																	
3	Li	2	1																
4	Be	2	2																
5	B	2	2	1															
6	C	2	2	2															
7	N	2	2	3															
8	O	2	2	4															
9	F	2	2	5															
10	Ne	2	2	6															
11	Na	2	2	6	1														
12	Mg				2														
13	Al				2	1													
14	Si		Neon		2	2													
15	P		core		2	3													
16	S				2	4													
17	Cl				2	5													
18	Ar				2	6													
19	K	2	2	6	2	6		1											
20	Ca							2											
21	Sc						1	2											
22	Ti						2	2											
23	V						3	2											
24	Cr						5	①											
25	Mn						5	2											
26	Fe						6	2											
27	Co						7	2											
28	Ni			Argon core			8	2											
29	Cu						10	①											
30	Zn						10	2											
31	Ga						10	2	1										
32	Ge						10	2	2										
33	As						10	2	3										
34	Se						10	2	4										
35	Br						10	2	5										
36	Kr						10	2	6										
37	Rb	2	2	6	2	6	10	2	6			1							
38	Sr											2							
39	Y									1		2							
40	Zr									2		2							
41	Nb									4		①							
42	Mo									5		①							
43	Tc									5		2							
44	Ru									7		①							
45	Rh									8		①							
46	Pd				Krypton core					10		⓪							
47	Ag									10		①							
48	Cd									10		2							
49	In									10		2	1						
50	Sn									10		2	2						
51	Sb									10		2	3						
52	Te									10		2	4						
53	I									10		2	5						
54	Xe									10		2	6						

Continued on page 266.

TABLE 6.1 *(continued)*

Atomic Number	*Element*	*1s*	*2s*	*2p*	*3s*	*3p*	*3d*	*4s*	*4p*	*4d*	*4f*	*5s*	*5p*	*5d*	*5f*	*6s*	*6p*	*6d*	*7s*
55	Cs	2	2	6	2	6	10	2	6	10		2	6			1			
56	Ba															2			
57	La													①		2			
58	Ce										1			①		2			
59	Pr										2			①		2			
60	Nd										4					2			
61	Pm										5					2			
62	Sm										6					2			
63	Eu										7					2			
64	Gd										7			①		2			
65	Tb										8			①		2			
66	Dy										10					2			
67	Ho										11					2			
68	Er										12					2			
69	Tm										13					2			
70	Yb										14					2			
71	Lu										14			1		2			
72	Hf					Xenon core					14			2		2			
73	Ta										14			3		2			
74	W										14			4		2			
75	Re										14			5		2			
76	Os										14			6		2			
77	Ir										14			9		⓪			
78	Pt										14			9		①			
79	Au										14			10		①			
80	Hg										14			10		2			
81	Tl										14			10		2	1		
82	Pb										14			10		2	2		
83	Bi										14			10		2	3		
84	Po										14			10		2	4		
85	At										14			10		2	5		
86	Rn										14			10		2	6		
87	Fr	2	2	6	2	6	10	2	6	10	14	2	6	10		2	6		1
88	Ra																		2
89	Ac																	①	2
90	Th																	②	2
91	Pa														2			①	2
92	U														3			①	2
93	Np														5				2
94	Pu														6				2
95	Am														7				2
96	Cm							Radon core							7			①	2
97	Bk														8			①	2
98	Cf														9			①	2
99	Es														10			①	2
100	Fm														11			①	2
101	Md														12			①	2
102	No														13			①	2
103	Lw														14			1	2

NOTE: Exceptions to the "normal" pattern are circled. Note that there these exceptions involve orbitals that are close in energy (e.g., *3d* and *4s*, *4f* and *5d*).

EXAMPLE 6.3 Write the electronic configurations for aluminum, silicon, phosphorus, and arsenic. Which of the elements are in the same group?

SOLUTION

$$\text{Al} \quad 1s^22s^22p^63s^23p^1, \text{ or } [\text{Ne}]3s^23p^1$$

$$\text{Si} \quad 1s^22s^22p^63s^23p^2, \text{ or } [\text{Ne}]3s^23p^2$$

Silicon has the configuration of aluminum plus one more electron.

$$\text{P} \quad 1s^22s^22p^63s^23p^3, \text{ or } [\text{Ne}]3s^23p^3$$

Again, one electron has been added as the elements across a row are considered.

$$\text{As} \quad 1s^22s^22p^63s^23p^63d^{10}4s^24p^3, \text{ or } [\text{Ar}]4s^24p^3$$

Arsenic has a configuration like phosphorus plus an entire shell. Arsenic is in the same family as phosphorus, since it has the same number of electrons in the outermost shell.

STUDY PROBLEM 6(b)

Give the electronic configurations of Group IIA elements.

EXAMPLE 6.4 Write the electronic configurations of oxygen, sulfur, chromium, and copper. All these atoms could attain a filled or half-filled shell by promotion of one electron to the next higher energy orbital, but not all of them do so. Why?

SOLUTION

$$\begin{array}{rl} \text{O} & [\text{He}]2s^22p^4 \\ \text{S} & [\text{Ne}]3s^23p^4 \\ \text{Cr} & [\text{Ar}]3d^54s^1 \\ \text{Cu} & [\text{Ar}]3d^{10}4s^1 \end{array}$$

The first two species do not promote an electron because the next-highest orbital ($3s$ for O and $4s$ for S) is much higher in energy than the highest occupied sublevel ($2p$ for O and $3p$ for S). Such a large amount of energy cannot be compensated for by the extra stability of the half-filled shell. For chromium, and copper, the $4s$ orbital has a very similar energy to the $3d$ sublevel and the "promotion" energy is offset by extra stability of the half-filled shell.

6.3 Predictions of Properties of Atoms and Ions from Present View of Atomic Structure

Overview

In our journey from the Greeks to Schrödinger, we have referred to many philosophers and physicists, with only Dalton to represent the chemists. The experiments we have considered are largely the types that physicists have conducted; and for several years, physicists appreciated the results most acutely. After all, what do wavelike properties of electrons have to do with acids and bases or the structures and reactivities of molecules, or that some elements form ionic salts and others do not? In a word—everything! This claim will be substantiated dozens of times as we advance in our learning.

One of the most important features of atomic electronic configurations that has

Spotlight

Dimitrii Mendeleev (center, first row) was born in Tobolsk, Siberia in 1834 and received his chemical training in St. Petersburg. His ideas about the periodic table were apparently initiated after he attended the Chemical Congress at Karlsruhe in 1860, where Cannizarro clearly made the distinction between atoms and molecules. This distinction allowed the unambiguous determination of atomic weights, the basis of Mendeleev's table. It is interesting that the other discoverer of the periodic table, Lothar Meyer (left, first row), attended the same meeting, as did a close friend of Mendeleev, A. P. Borodin, the famous Russian composer who was chemistry professor at the Medico-Surgical Academy in St. Petersburg. Mendeleev was considerably bolder than Meyer in formulating the periodic table, predicting the properties of undiscovered elements which he named eka-aluminum, eka-born and eka-silicon. When these elements were discovered—gallium (1874), scandium (1879), and germanium (1885)—and were found to match the predicted properties, chemists became convinced of the value of Mendeleev's periodic table.

This picture was taken at a meeting of the British Association in Manchester in 1887. Standing are Wislicenus, Quincke, Schunck, Schorlemmer, and Joule. Meyer, Mendeleev, and Roscoe are seated.

The Manchester Literary and Philosophical Society

emerged from wave mechanics is the relation between the concept and the periodic table developed by Dmitri Mendeleev and Lothar Meyer in the mid 1800s. The periodic table was developed in a form quite similar to the one we use today, by considering similar chemical properties of elements and arranging elements by increasing atomic weight. Mendeleev arranged the periodic table so that Li, Na, and K were all in one column, because he knew that they undergo similar chemical reactions. From quantum mechanics, we see that Li, Na, and K have a similarity; for the ground state atoms, each has only one electron in its outermost shell, and in each case this electron is in an *s* orbital. As we shall see repeatedly, it is usually the outermost shell of electrons that determines chemical properties, for example, bonding characteristics; for this reason, the outermost shell is often referred to as **the valence shell.** Such chemical and electronic similarities are found in each column of elements in the periodic table, yet the table was constructed more than fifty years before a successful atomic model was developed.

If the quantum mechanical model of the atom is a good model, it should be consistent with the experimental properties that have been observed for atoms and ions, and with their classification according to the periodic table. We hope that the trends we can see for members of a group of elements in a particular period can be explained using the electronic configurations derived from our model.

Atomic Numbers

Since the scheme that we shall explore is the periodic table, let us investigate some of its information content. As we proceed across each row, there is a successive unit change in the positive charge of the nucleus. This charge is due to the protons in the nucleus. The number of protons in the nucleus is numerically equal to the positive charge in electronic units. This number is called the atomic number and is denoted by the symbol Z. The elements and their atomic numbers proceed as follows: H 1, He 2, Li 3, Be 4, B 5, C 6, N 7, O 8, F 9, Ne 10, and so on. As the atomic number increases by 1, one proton is added to the nucleus, and for a neutral atom in its ground state, one electron is added to the orbital available having the lowest energy. The number of protons in the nucleus, therefore, defines the elements, determining its atomic number. The element with eleven protons in the nucleus is sodium, Na. If we take one electron from this atom, it becomes the positively charged sodium ion (+1), denoted by the symbol Na^+, retaining its elemental designation.

EXAMPLE 6.5 What are the atoms or ions with (a) 10 protons; (b) 10 electrons; (c) an atomic number of 10?

SOLUTION
(a) Neon is the only element with 10 protons.
(b) Several elements could exist in a form with 10 electrons, such as sodium (Na^+), magnesium (Mg^{2+}), aluminum (Al^{3+}), fluorine (F^-), and oxygen (O^{2-}).
(c) Neon is the only element with atomic number 10, since it is the only element with 10 protons.

STUDY PROBLEM 6(c)

What are the atoms or ions with (a) 9 protons; (b) 9 electrons; and (c) atomic number 9?

Atomic Weight

Corresponding to each elemental symbol of the periodic table there is also a number shown which is larger than the atomic weight. The mass of an atom in atomic units nearly equals the number of protons plus the number of neutrons. These particles have about 2000 times the mass of an electron. Many other "elementary" particles have been found to reside in the nucleus, but they contribute very little to the atomic weight. (They are also unimportant in present chemical theory.) We can designate the approximate weight of an atom in the number of protons and neutrons with a notation like $^{19}_{9}F$, symbolizing a fluorine atom, with 9 protons and 10 neutrons in the nucleus. The upper left symbol is called the **mass number** and equals the sum of the protons and neutrons;

$$\text{mass number} = Z + \text{number of neutrons} \qquad (6.8)$$

EXAMPLE 6.6 How many neutrons are there in the nucleus of each of the following elements: hydrogen, lithium, silicon, phosphorus?

SOLUTION This question can't be answered readily because isotopes of the elements exist that differ by the number of neutrons. Thus, 99.84% of hydrogen atoms have one proton and no neutrons in the nucleus, whereas 0.0156% of hydrogen atoms, called deuterium, have one neutron; and a very small amount of hydrogen (about 1 atom in 10^{17}, called tritium) has two neutrons.

To answer such a question, one must know the identity of the isotope of the element.

EXAMPLE 6.7 What are the number of neutrons and protons in each atom of the following isotopes: $^{2}_{1}H$, $^{7}_{3}Li$, $^{6}_{3}Li$, $^{102}_{44}Ru$, sulfur of atomic weight 32, indium of atomic weight 109, indium of atomic weight 110?

SOLUTION The number of neutrons and protons can be obtained directly from the symbol for the isotope. The symbol $^{2}_{1}H$ means that the atomic weight is 2 and the atomic number is 1. The number of protons equals the atomic number, 1. The number of neutrons equals the atomic weight minus the atomic number $= 2 - 1 = 1$. For $^{7}_{3}Li$, there are 4 neutrons and 3 protons, for $^{6}_{3}Li$, 3 neutrons and 3 protons, and for $^{102}_{44}Ru$, 58 neutrons and 44 protons.

The last three examples require using the periodic table to obtain the atomic numbers. The isotope $^{32}_{16}S$ has 16 protons and 16 neutrons, $^{109}_{49}In$ has 60 neutrons and 49 protons, and $^{110}_{49}In$ has 61 neutrons and 49 protons.

STUDY PROBLEM 6(d)

What are the number of neutrons and protons in $^{31}_{15}P$, $^{125}_{53}I$, $^{127}_{53}I$?

Isotopes

Scientists have determined that an element can have atoms of different masses. That is, there may be two or more types of atoms with the same number of protons (fixed by the identity of the element) but a different number of neutrons. Atoms symbolized by $^{35}_{17}Cl$ have 17 protons and 18 neutrons, and atoms symbolized by $^{37}_{17}Cl$ have 17 protons and 20 neutrons. Since the two kinds of atoms have the same number of protons (same atomic number), they are defined to be the same element. Atoms of the same element but different atomic masses, (due to different numbers of neutrons) are called **isotopes.** Isotopes can be separated from one another because of their different weights, but this process is often very difficult, and also expensive. An isotope of hydrogen, called deuterium, $^{2}_{1}H$, occurs as 0.0156% of all hydrogen on this planet.* The substance $^{2}H_2O$, or D_2O, is called heavy water. The molecular weight of D_2O is $2 \times 2.0140 + 15.9994 = 20.0274$ amu, and of $^{1}H_2O$ is $2 \times 1.00797 + 15.9994 = 18.01534$ amu. Heavy water is used as a moderator in nuclear reactors and in isotopic tracer studies. Isotopes manifest the same chemistry, undergoing the same reactions and forming the same compounds, because chemistry is determined by electronic configuration and not by the number of neutrons.

If we now turn our attention back to the periodic table, we notice that the atomic weights are often quite different from integral numbers. For example, the atomic weight for chlorine given in the table is 35.453. How do we arrive at this result? For chlorine, two isotopes exist naturally. Natural chlorine consists of 75.53% of the isotope $^{35}_{17}Cl$ (75.53% "natural abundance") while $^{37}_{17}Cl$ has a natural abundance of 24.47%. By multiplying the fraction of each isotope that is present in the sample (normally the natural abundance) by the actual mass of the isotope (not the rounded-off isotope mass), we arrive at the atomic weight listed in the table. The weight of $^{35}_{17}Cl$ is actually 34.9688 amu, and $^{37}_{17}Cl$ weighs 36.9660 amu. For the average atomic weight, we must multiply the fraction of each isotope by its actual weight. For chlorine the equation is

$$\begin{aligned} \text{at. wt} &= (\text{fraction of } ^{35}\text{Cl}) \times (\text{wt of } ^{35}\text{Cl}) + (\text{fraction of } ^{37}\text{Cl}) \times (\text{wt of } ^{37}\text{Cl}) \\ &= (0.7553) \times (34.9688 \text{ amu}) + (0.2447) \times (36.9659 \text{ amu}) \\ &= 35.457, \text{ or } 35.46 \text{ amu} \end{aligned}$$

EXAMPLE 6.8 The *Handbook of Chemistry and Physics,* published by the Chemical Rubber Company, includes a table of isotopes that gives the natural abundance of isotopes. Using such a source of information, calculate the average atomic weight shown on periodic tables for silicon, sulfur, manganese, and iron.

SOLUTION (a) Silicon has three principal isotopes:

	% Abundance	*Atomic mass (in amu)*
$^{28}_{14}Si$	92.21	27.97693
$^{29}_{14}Si$	4.70	28.97649
$^{30}_{14}Si$	3.09	29.97376

$$\begin{aligned} \text{average mass} &= 0.9221\ (27.977) + 0.0470\ (28.976) + 0.0309\ (29.974) \\ &= 28.086, \text{ or } 28.09 \text{ amu} \end{aligned}$$

*Other planets may have different isotopic percentages.

(b) Sulfur has only one main isotope, $^{32}_{16}S$, with a mass of 32.064 amu.
(c) Manganese has only one main isotope, $^{55}_{25}Mn$, with a mass of 54.938 amu.
(d) Iron has four main isotopes:

	% Abundance	*Atomic mass (in amu)*
$^{56}_{26}Fe$	91.66	55.9349
$^{54}_{26}Fe$	5.82	53.9396
$^{57}_{26}Fe$	2.19	56.9354
$^{58}_{26}Fe$	0.33	57.9333

$$\text{average mass} = 0.9166\,(55.9349) + 0.0582\,(53.9396) + 0.0219\,(56.9354) + 0.0033\,(57.9333) = 55.847, \text{ or } 55.85 \text{ amu}$$

Spotlight

One of the areas in which a knowledge of isotopes is important is nuclear chemistry. In the "reactions" of nuclear chemistry, the composition and/or energy of the nucleus is altered. Some common examples of these reactions are:

(a) *Electron (β^- or β particle) emission.* For example:

$$^{14}_{6}C \rightarrow {}^{14}_{7}N + \beta^-$$

Note that when an electron (β^- or just β) is emitted from the $^{14}_{6}C$ nucleus, the overall process is equivalent to changing a neutron into a proton and an emitted electron. The addition of a proton in the nucleus of $^{14}_{6}C$ converts the neutral carbon atom to a nitrogen ion with a charge of +1. The emitted electron provides charge balance.

Carbon dating is a method used by archeologists to estimate the dates at which some carbon-containing material ceased living. Atmospheric CO_2 contains a small amount of $^{14}_{6}C$ and is continuously incorporated into living systems. Once a plant or animal dies, no more $^{14}_{6}C$ is incorporated into its tissue and the $^{14}_{6}C$ continues to decay by beta emission. In carbon dating, the number of beta particles emitted per minute per gram of total carbon dates the material. (For a quantitative discussion of carbon dating, see Chap. 13.)

(b) *Positron emission* (β^+, a particle with the mass of an electron, but the charge of a proton). For example:

$$^{11}_{6}C \rightarrow {}^{11}_{5}B + \beta^+$$

In this case, the overall effect in the carbon nucleus is the conversion of a proton into a neutron, with emission of a positron. This process is rarer than β^- emission.

(c) *Alpha particle ($^{4}_{2}He$) emission.* For example:

$$^{238}_{92}U \rightarrow {}^{234}_{90}Th + {}^{4}_{2}He$$

The emission of an α particle causes the atomic mass number to decrease by four and the atomic number to decrease by two. The charge on the emitted α particle is +2, but it generally picks up two electrons readily from some source to become a helium atom. The helium found in natural gas wells originated from radioactive decay.

(d) *Fission.* For example:

$$^{235}_{92}U + {}^{1}_{0}n \rightarrow {}^{139}_{56}Ba + {}^{94}_{36}Kr + 3\,{}^{1}_{0}n$$

In a fission (or fragmentation) reaction, a very large nucleus splits into two or more large nuclei. The case shown above occurs in a nuclear reactor such as those currently used in nuclear electrical generating plants. Since the neutrons that are released can cause fission of another $^{235}_{92}U$ nucleus, the reaction can be made to be a *chain* reaction, that is, the products of one step of the reaction cause the next step to occur.

(e) *Fusion.* For example:

$$^{2}_{1}H + {}^{3}_{1}H \rightarrow {}^{4}_{2}He + {}^{1}_{0}n$$

All the other processes described above result in the

fragmentation of the nucleus. In a fusion reaction, two nuclei are joined together. A very large amount of energy is required to bring the nuclei together, but energy may be released when they join. For the reaction of one tritium, ^3_1H, and one hydrogen, ^1_1H, nucleus (or of two deuterium nuclei, ^2_1H) to form a helium nucleus, the overall process is highly exothermic. Most fusion processes, however, are energetically unfavorable. Fusion occurs in stars, such as our sun, and may soon become a source of energy here on earth.

(f) *Neutron capture.* For example:

$$^{59}_{27}Co + ^1_0n \rightarrow ^{60}_{27}Co$$

This process is related to fusion in that the nuclear mass is increased. For neutron capture to occur, the reaction must "hit" the very small nucleus. Hence, a large number of neutrons per unit area (called the neutron flux) must be passed through the sample. Such reactions often produce radioactive products. The product in the process above, cobalt-60, is used in cancer therapy, because it gives off a γ as well as a β particle when it decays.

$$^{60}_{27}Co \rightarrow ^{60}_{28}Ni + \beta + \gamma\,(1.33\text{ MeV})$$

(g) *Gamma ray* (γ) emission. For example:

$$^{80m}_{35}Br \rightarrow ^{80}_{35}Br + \gamma\,(0.085\text{ MeV})$$

In this case, a *metastable* nucleus of bromine-80 (the "m" stands for metastable, meaning not the most stable, that is, in a nuclear excited state) gives off a burst of electromagnetic radiation called a gamma ray, with an energy of 0.085 million electron volts (or 2.0 million calories/mole of bromine atoms), and the nucleus is then in its ground state. The $^{80m}_{35}Br$ can be formed from $^{79}_{35}Br$ (which is 50.54% of natural bromine) by neutron capture. This phenomenon is used to analyze for very small amounts (trace levels) of bromine in a technique called neutron activation analysis. A flow of neutrons through the sample forms the $^{80m}_{35}Br$ by the reaction: $^{79}_{35}Br + ^1_0n \rightarrow ^{80m}_{35}Br$. The sample is removed from the neutron source and the number of gamma rays emitted with an energy of 0.085 MeV is counted to determine the bromine content. Neutron activation analysis can be used to determine the concentrations of a large number of elements.

The Atomic Mass Unit Scale

We have been using an atomic mass unit scale without having defined it precisely. Now that we have an idea of what isotopes are, the definition of this system should become clear. The most plentiful isotope of carbon is $^{12}_6C$ (98.89% natural abundance). The atomic weight of this isotope is defined to be exactly twelve atomic mass units, and other atomic weights are calculated from the weights of other atoms relative to the weight of $^{12}_6C$. Thus a hydrogen atom weighs 1.00797/12.0000 as much as $^{12}_6C$ and its weight is expressed as

$$\frac{1.00797}{12.0000} \times 12 = 1.00797\text{ amu}$$

Some of the related notations and definitions are summarized in Fig. 6.18.

Let us now investigate the atomic numbers and atomic weights of a row of the periodic table. As we proceed from left to right across a row of elements, the atomic numbers are larger by exactly one unit for each successive element. Since the atomic number gives for atoms the number of protons, and also the number of electrons since atoms are electrically neutral, moving from one element to the next corresponds to adding one proton and one electron to each atom to form the atom of the next element. The atomic weights, however, do not progress in such uniform steps. The atomic weight difference between N and C is 1.0052 amu, between Cd and Ag is 4.53 amu, and between Ni and Co is -0.22 amu! The number of neutrons added

Figure 6.18
A summary of atomic number and atomic weights.

SYMBOL $^{12}_{6}C$ *Atomic weight of isotope* (rounded to whole number) equals number of protons plus number of neutrons.
Atomic number equals number of protons; defines element.

ATOMIC MASS UNIT SYSTEM

$$\textit{Atomic weight of element X in atomic mass units} = \frac{\text{weight of atom of X}}{\text{weight of } ^{12}_{6}\text{C}} \times 12.0000 \text{ amu}$$

$$= \frac{\text{weight of mole of X}}{\text{weight of mole of } ^{12}_{6}\text{C}} \times 12.0000 \text{ amu}$$

Mass of electron = 0.0005487 amu
Mass of proton = 1.00797 amu
Mass of neutron = 1.00867 amu

Atomic weight of element X = sum of the products of fraction of each isotope times actual atomic weight of each isotope. The actual weight of an isotope cannot be determined by simply adding the sum of the weights of the protons and neutrons. The weight will be somewhat less than this sum. This "mass defect" is related to the forces responsible for holding the nucleus together.

For Li, which is 7.42% $^{6}_{3}Li$ and 92.58% $^{7}_{3}Li$, atomic weight = (0.0742) (6.01513) + (0.9258) (7.01601) = 6.942 amu.

Note: The value for the atomic weight can be determined most accurately by weighing a mole of Li; the results of this calculation are limited by our ability to determine natural abundance percentages.

from one element to another is not uniform. Each element has a specific number of protons, but the number of neutrons may vary; in other words, there can be different isotopes.

The basis for the periodic table is actually atomic number; it is not atomic weight, as Mendeleev and Meyer thought. The atomic number of an element defines its ground state electronic configuration, and it is this feature of atoms that is responsible for similarities of chemical properties.

The Sizes of Atoms

In 1870, Lothar Meyer determined the "atomic volume" of a large number of elements by dividing the atomic weight of an element by the density of a sample of the corresponding pure elemental substance (since density = mass/volume, then volume = mass/density). This procedure works reasonably well for those elements that can exist as pure substances, like metals, which form crystals composed of layers of separate atoms. However, some elements are never found as pure solid substances, and for these elements the size of individual atoms was computed as the relative amount of volume each type of atom occupies in compounds containing the element. Meyer's values were used to compute the atomic radius of each element, assuming that the atoms are spherical. His list has been expanded and refined by more recent work, especially that based on x-ray diffraction.* If a beam of x rays is diffracted

*The x-ray diffraction method uses the interference patterns associated with the diffraction of x rays from the arrays of atoms in crystals. The principle of the experiment is very similar to what is depicted in Fig. 6.13 and will be discussed in more detail in Chap. 14.

through a crystal, atomic positions can be determined and the radius of each type of atom can be calculated.

Two important types of radii that we shall consider by the atomic structure model are (1) atomic radii, the radii of atoms that are not chemically bound to other atoms, and (2) ionic radii, the radii of atoms from which one or more electrons have been abstracted (positively charged ions) or to which one or more electrons have been added (negatively charged ions).

Figure 6.19 shows atomic radii for the atoms of many elements. We can note two trends in atomic sizes from this figure. First, the atoms of elements in a group are progressively larger as we proceed down the group. This is readily understood, knowing that each succeeding element down the group differs from the one above it by one entire shell of electrons. Thus, the configuration of Li is $1s^22s^1$, of Na $1s^22s^22p^63s^1$, of K $[Ar]4s^1$, and so on. The principal quantum number, the shell number, largely determines the volume an outer electron occupies. For the simple Bohr model, the radius is proportional to n^2; and the volume of a sphere is proportional to r^3. Thus, the increase of n by unity greatly increases the volume of an atom. This is the case even though the number of protons and the positive charge of the nucleus is greater as one proceeds down a group. You might expect that the electron distribution would be contracted because of stronger electrostatic attractions. However, a shell of electrons between the nucleus and an "outer" electron (an electron with a higher n value) can effectively "shield" the outer electron from most of the attractive force of the positively charged nucleus. Each successive shell is successful enough in shielding the outer electrons from the increasing nuclear charge to allow those electrons to spend much time farther from the nucleus—their probability distributions are concentrated in regions farther from the nucleus.

The second trend which is evident from Figure 6.19 is the decrease in atomic size as we proceed across a row of the table. Moving across a row corresponds to adding electrons to the same shell. At the same time, the larger number of protons in the nucleus as one moves to the right along a row exert an increasing attraction for the electrons. Electrons in the same shell are not so effective as inner-shell electrons (those in orbitals with smaller n values) at shielding other electrons from the nuclear attraction. Hence, adding one electron and one proton increases the attractive force felt by all electrons in the outer shell, and they are all pulled slightly closer to the nucleus as one moves from left to right along a row.

EXAMPLE 6.9 In each of the following sets, rank the atomic or ionic species in order of increasing radius: (a) Li, N, C; (b) S, Mg, Si; (c) K, Ca, Mg; (d) P, Cl^-, Na^+.

SOLUTION

(a) N < C < Li. The size tends to decrease across the row.

(b) S < Si < Mg. Same trend as (a).

(c) Mg < Ca < K. Both period and group trends are involved here. Since an entire shell is added to Mg to form Ca, the radius of Ca is larger than the radius of Mg. Since there is a trend to smaller sizes across a period, the K radius is larger than the Ca radius. Therefore, Mg < Ca < K. One must be careful in size predictions when both period and group differences are involved since "period contraction" (across a row) can be greater than "group expansion" (down a column).

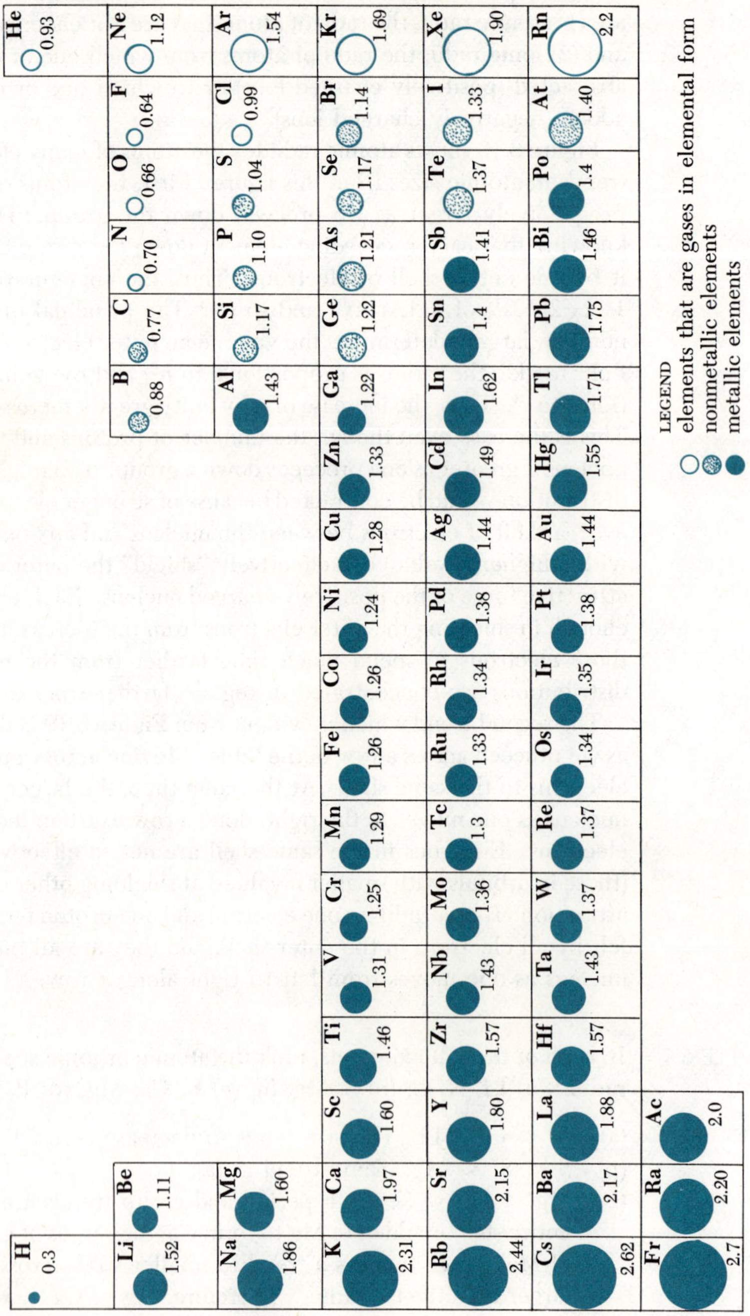

Figure 6.19 Relative atomic sizes (in angstrom units).

(d) $Na^+ < P < Cl^-$. These species are all in the same group. An electron has been removed from Na to form Na^+, greatly decreasing size (from 2.27 Å to 1.33 Å), and an electron has been added to Cl to form Cl^-, greatly increasing size (from 0.99 Å to 1.81 Å).

STUDY PROBLEM 6(e)

Rank these species in order of increasing radius: (a) Ca, Ge, Se; (b) S, Se, Te; (c) Br^-, Ge, Ca^{2+}.

Ion Formation and Ion Sizes

The sizes of ions and energies of ion formation can also be easily explained by our atomic structure model. Forming a positive ion from an atom requires abstracting an electron; the lowest possible expenditure of energy involves removing the electron from the highest-energy occupied orbital, for example, the 3*s* orbital in Na, as shown in Fig. 6.20. This process requires an amount of energy known as the ionization energy, or **ionization potential,** which can be measured accurately.* Since more than one electron can be removed stepwise from a many-electron atom, the ionization potentials are indicated: IP_1, for the abstraction of an electron from a neutral atom or molecule to form a $+1$ ion; IP_2, for the abstraction of an electron from a $+1$ ion to form a $+2$ ion; and so forth. Figure 6.21 shows the first ionization potential for many elements.

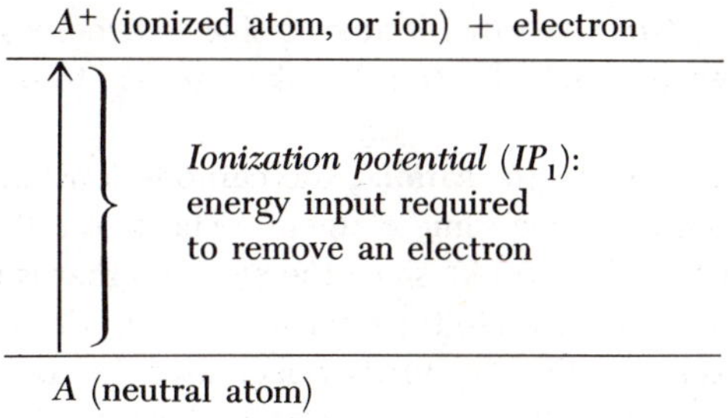

Figure 6.20 The ionization process and first ionization potential IP_1.

The main trend evident from Fig. 6.21 is that ionization potential is related to atomic size. In each row, the electron requiring least expenditure of energy to remove is one from a Group IA element, which has only one electron in the outer valence shell. As one electron and one proton are added stepwise in going across a row, the electrons are subject to greater nuclear attraction due to inefficient shielding by other electrons in the same shell. Hence, it becomes progressively harder to remove an electron from an atom. The highest IP_1 value of any atom in a given

*The ionization potential can be measured directly by observing how much energy one has to put into an atom to remove an electron. Two common ways of putting the requisite energy into the atom are electron impact and photoionization. The **electron impact method** involves firing a projectile electron of known energy into the atom from an electron gun, and observing how much energy the projectile electron must have in order to kick out an electron from the atom. **Photoionization** involves determining the minimum photon energy of radiation that should be applied to an atom to eject an electron from the atom.

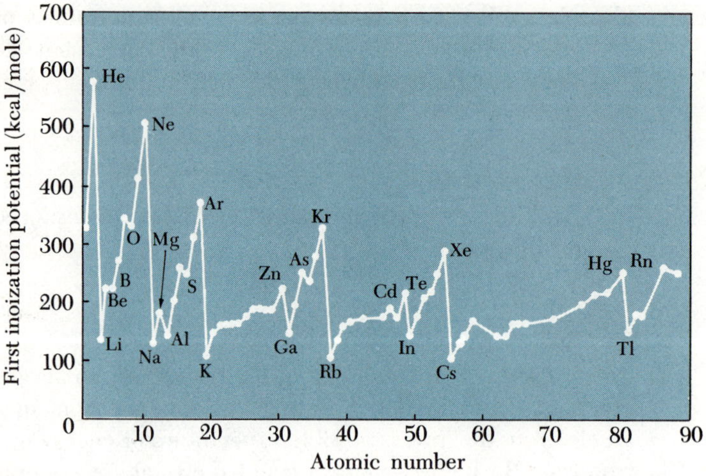

Figure 6.21 Relation between atomic number and ionization potential.

period is the one for the noble gas, with its filled-shell configuration. Also following a trend similar to what we saw for atomic size, it becomes easier to ionize atoms as we proceed down a group; this trend is likewise due to progressively larger atomic size.

Positively charged ions (or **cations**, the cathode ions) are much smaller than the corresponding atoms of the same element. The size difference is very noticeable for the alkali metals (Group IA), since the electron that is removed by ionization of the atom is the only electron in the outermost shell. Forming +2 and +3 ions in a second and third ionization step further decreases the size of an ionic species.

It is always more difficult to remove a second electron from an atom than the first electron, since negatively charged electrons must be pulled from a species that is already positively charged (+1 ion) instead of from a neutral atom. Just as IP_2 is greater than IP_1 for each element, succeeding *IP*'s always increase. For each element, the greatest incremental increase in an ionization potential (in going from atom to +1, to +2, to +3 ion, . . .) occurs when the electron must be removed from a filled shell. The IP_1 for Na is reasonably low; and the process involved is

$$\mathrm{Na([Ne]3}s^1) \xrightarrow{IP_1} \mathrm{Na^+ ([Ne])} + e^-$$

However, the IP_2 is very high, since an electron must be removed from an ion with a noble gas electron configuration:

$$\mathrm{Na^+ ([Ne])} \xrightarrow{IP_2} \mathrm{Na^{2+}} (1s^2 2s^2 2p^5) + e^-$$

The IP_1 and IP_2 for Ca are reasonably low; the processes are

$$\mathrm{Ca([Ar]4}s^2) \xrightarrow{IP_1} \mathrm{Ca^+ ([Ar]4}s^1) + e^-$$

and

$$\mathrm{Ca^+ ([Ar]4}s^1) \xrightarrow{IP_2} \mathrm{Ca^{2+} ([Ar])} + e^-$$

But the IP_3 is high. This kind of observation leads to a "noble gas configuration rule," which states that the ionizations that form configurations like those of the noble gases are more easily accomplished than those that require removing an electron from a species that already has a noble gas configuration.

EXAMPLE 6.10 For each of the following sets, rank the atomic or ionic species in order of increasing ionization potential: (a) Li, B, C; (b) Na, K, Li; (c) Li^+, Be, O^-.

SOLUTION

(a) Li < B < C. Within a period, the ionization energy tends to increase as the number of protons increases. Important exceptions to this trend occur when a half-filled or filled subshell configuration is disturbed. Thus $IP_{Be} > IP_B$, since the configuration of Be is $1s^22s^2$; and $IP_N > IP_O$, since the configuration of N is $1s^22p^22p^3$.

(b) K < Na < Li. Within a group, the ionization potential decreases as the size increases.

(c) $O^- < Be < Li^+$. In general, it is easier to remove an electron from a negatively charged species than from a neutral atom and easier to remove an electron from a neutral atom than from a positively charged ion.

STUDY PROBLEM **6(f)**

For each of the following sets, rank the atomic or ionic species in order of increasing ionization potential: (a) Mg, P, S; (b) P, As, Sb; (c) V^{2+}, Co^+, I^-.

Electron Affinity

The energy released when an electron is added to an atom, molecule, or negative ion is called the **electron affinity.** Figure 6.22 is a diagram that explains electron affinity. The symbolism—EA_1, EA_2, and so on—may be used like the *IP* symbolism.

The process that corresponds to the electron affinity concept produces negatively charged ions (**anions,** the anode ions).* Electron affinities are found to decrease as additional electrons are added to an atom, since the electron to be added will be

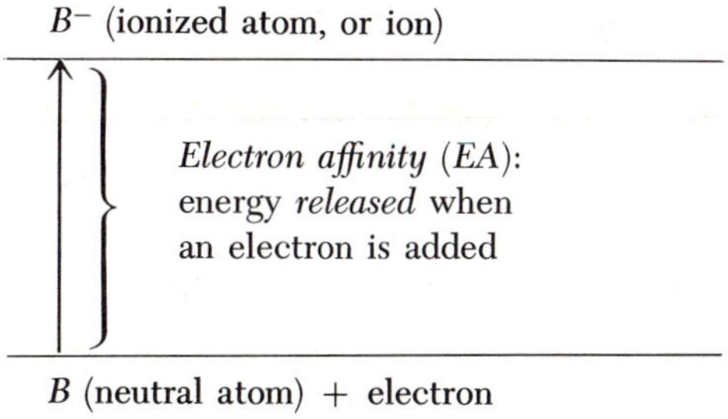

Figure 6.22
Accepting an electron and the first electron affinity, *EA*.

*The exception to this would be a process in which an electron is added to a positive ion, leading either to a neutral (from a +1 ion) or a positive ion of charge $n - 1$ (from a $+n$ ion). This literal exception is, however, not what one customarily means by the term *electron affinity,* which is usually applied to neutral species.

repelled by a negatively charged ion. Hence, it is generally more difficult to form -2 and -3 ions than -1 ions. If we inspect Group VIIA, the halogens, we note that these atoms need but one electron to attain a noble gas configuration. Atomic fluorine ([He]$2s^22p^5$) readily gains an electron to form F^- (electronic configuration [Ne]), and Cl ([Ne]$3s^23p^5$) readily gains an electron to form Cl^- ([Ar]). Group VIA atoms can gain two electrons to attain a noble gas configuration to form -2 ions, for example, O^{-2} ([Ne]) and S^{-2} ([Ar]) ions. Note that as a filled shell is reached, the next electron added must occupy a new shell—farther from the nucleus, and well shielded from the nuclear charge. The noble gas configuration is usually the limit to anion formation.

In the direction down a group, the electron is being added to shells of successively higher n values; for example, $n = 2$ for F, 3 for Cl, and 4 for Br. These valence shells are progressively farther from the nucleus. Electron affinities, then, are lower down a group (as atomic size is larger) and are higher across a row (as atomic size is less).

Negatively charged ions are much larger than corresponding atoms or cations. Even one of the smallest anions, Cl^- (radius = 1.81 Å) is larger than one of the largest cations, Cs^+ (radius = 1.67 Å). Negative ions of higher charge (O^{-2}, N^{-3}, . . .) are larger than the singly charged ions of the same element. Hence, nearly all anions are larger than any cations.*

EXAMPLE 6.11 For each of the following sets, arrange the atomic or ionic species in order of increasing electron affinity: (a) B, F, O; (b) S, O; (c) Li^+, C, F^-.

SOLUTION (a) $B < O < F$. The smaller size and larger number of protons causes a general trend of increasing electron affinity across a row. Note again that exceptions occur for half-filled subshells. Thus N is lower than C in electron affinity and the configuration of N is $1s^22s^22p^3$.

(b) $O < S$. In determining electron affinities, the increased number of protons of third-row group members appears to be more important than size, in contrast with determining ionization potential.

(c) $F^- < C < Li^+$. In general, positively charged ions attract electrons more effectively than neutral atoms do, and neutral atoms have higher affinities than negatively charged ions.

SUMMARY

From the fruits of many experiments, the model of the atom has progressed to a stage in which we can rationalize most properties of atoms and ions. The basic features of our model are:

1. A small central cluster consisting of nearly all the mass of the atom; this contains protons and neutrons, and is called the nucleus.
2. Identity of an element determined by the number of protons contained in the nucleus of the atom.
3. Very light, negatively charged particles, called electrons, occupying most of

*Adding an electron to a halogen atom increases the size by nearly as much as the opposite effect of "periodic contraction" across the row. For example, Cl^- (1.81 Å) is about the size of Na (1.86 Å), whereas the Cl radius is only 0.99 Å. Likewise, Br^- (1.95 Å) is about the size of K (2.27 Å) whereas the Br radius is only 1.14 Å.

the volume of the atom; these have wavelike properties and energies that can be described in terms of confined waves.

We have seen that the confined-wave description specifies orbitals, the regions of space in which the electron is most likely to be found. The orbitals are characterized by three quantum numbers, n, l, and m_l. A fourth quantum number m_s describes electron spin.

Electron configuration can be determined by placing electrons according to the Pauli exclusion principle and the Aufbau principle into orbitals of increasing energy, observing Hund's rule of maximum unpaired spins. These configurations can be used to explain atomic and ionic sizes, relative values of ionization potentials and electron affinities, and the maximum charges normally found for cations and anions.

We shall soon find that our atomic model can also be used to explain how atoms form ionic and molecular compounds and how these compounds interact.

STUDENT CHECKPOINTS

After studying this chapter, the student should be able to:

1. Outline the experimental observations and the reasoning that have led to our present concept of atomic structure.
2. Relate the dependence of the radius of one-electron atoms and ions to atomic charge and principal quantum number.
3. Use energy-level diagrams for electronic transitions.
4. Calculate transition energies by using the Rydberg equation.
5. Determine frequency when the energy of a photon is known, and vice versa.
6. Write electronic configurations for atoms and ions.
7. Write the quantum numbers for electrons in specified orbitals.
8. Calculate average atomic masses from isotope masses and abundance.
9. Determine trends in atomic and ionic sizes from the concepts that are relevant to these trends.
10. Define trends in ionization potentials and electron affinities and elaborate the reasoning that explains the trends.

EXERCISES

6.1. Briefly describe experiments that would lead to the following conclusions. State how the conclusion follows from the experimental results.
   (a) Neutral atoms can be decomposed into charged particles.
   (b) Electrons have particlelike properties.
   (c) Electrons have wavelike behavior.
   (d) The nucleus is much greater in mass than the electron.
   (e) The bulk of the mass of the atom resides in a small fraction of the total atomic volume.

6.2.* Calculate the radius of the fourth orbit of the hydrogen atom, using the fact that the radius of the first orbit is 0.529 Å.

6.3.* Calculate the radius of the first orbit of the ion Li^{2+}.

6.4. Using the energy diagram depicted in Fig. 6.12 indicate the $5 \rightarrow 2$ transition of the hydrogen atom.

6.5.* Calculate the energy of the transition referred to in Exercise 6.4, using the Rydberg equation (see Example 6.2).

6.6. Figure 6.12c depicts a strip of film from an emission spectrograph with atomic hydrogen in the source. Indicate the 5→2 transition. Find the series of lines that corresponds to transitions in which the ground state ($n = 1$) is the final state and the series in which the first excited state (n = 2) is the final state. The highest energy line occurs at an energy equal to the Rydberg constant. Use the Rydberg equation (Example 6.2) to show why this is so.

6.7.* What is the wavelength of the light emitted by the 5→2 transition of atomic hydrogen (Exercise 6.6)? What is the frequency?

6.8. The sodium "D" line, a line in the emission spectrum of sodium, is yellow (5890 Å). (a) What is the frequency of this emission? (b) How much energy (in joules) would be emitted at this wavelength by 1 g of excited sodium atoms?

6.9.* The highest energy line in the emission spectrum of an ion with one electron occurs at frequency 1.315×10^{16} Hz. Identify the ion.

6.10. The 2→1 transition in the emission spectrum of an ion with one electron occurs at frequency 1.315×10^{16} Hz. Identify the ion.

6.11.* Calculate the frequency and the energy (in kcal/mol) for the 3→2 transition of the ion Be^{3+}.

6.12. Can the electronic spectrum of the alpha particle, He^{2+}, be calculated using the Rydberg equation?

6.13.* How many electrons can occupy (a) the 3s orbital? (b) the 3d subshell? (c) the 4p subshell? (d) the $4p_x$ orbital? (e) the second shell? (f) the third shell?

6.14. What are the quantum numbers of the electron of highest energy in (a) the ground state nitrogen atom? (b) the ground state ion Mg^{2+}?

6.15.* What are the quantum numbers for the electron of lowest energy in (a) the ground state sulfur atom? (b) the ground state chloride ion?

6.16.* Give electronic configurations for the ground state atoms of the following elements: (a) potassium; (b) calcium; (c) scandium; (d) krypton.

6.17. Give electronic configurations for the ground states of the following ions: (a) K^+; (b) Ca^{2+}; (c) Sc^{3+}; (d) Br^-.

6.18.* Write ground state electronic configurations of the group known as the pnictides, of which nitrogen is the lightest member.

6.19. (a) What are the elements whose ground state atoms have the following electronic configurations: $1s^22s^22p^4$; [Ne]$3s^23p^3$; [Ar]$4s^23d^{10}4p^4$?

(b) Which of these elements are in the same group?

6.20.* Tell whether the following electronic configurations are ground state, excited state, or impossible configurations. For the possible configurations, identify the corresponding element:

(a) $1s^22s^22p^63s^13p^1$ (b) $1s^22s^22p^63s^23p^3$

(c) $1s^22s^22p^63s^23p^63d^54s^1$ (d) $1s^22s^22p^73s^23p^1$

6.21. Identify the following electronic configurations as ground state, excited state, or impossible. Identify the elements that correspond to possible configurations.

(a) $1s^22s^22p^33s^1$ (b) $1s^22s^22p^63s^23p^63d^{10}4s^1$

(c) $1s^22s^22p^63s^23p^63d^{11}$ (d) $1s^22s^12p^33s^2$

6.22.* Write ground state electronic configurations of fluorine, silver, and bromine atoms. Only one of these elements "promotes" an electron to attain a filled shell. Why?

6.23. Only one element has the atomic number 7, but several atomic species are found with seven electrons. Explain. Give some examples of species with seven electrons. (*Hint:* See Example 6.5.)

6.24.* (a) Atoms of two different masses (10.01618 amu and 11.01284 amu) are called boron. Explain.
(b) How many protons and neutrons are there in each of these types of atoms?
(c) If the natural abundance of the lighter isotope is 18.83% and only these two boron isotopes are known, what is the mass of a mole of boron?

6.25. Gallium exists as two isotopes, with mass numbers 69 and 71.
(a) How many protons are there in atoms of these two isotopes?
(b) Natural gallium is composed of 60.0% of the lighter isotope (68.956 amu). Atoms of the heavier isotope have an atomic mass of 70.954 amu. What is the average mass of an atom of gallium?

6.26.* Density is the mass of a substance per unit volume. If molecules of heavy water (deuterium oxide, ^2H_2O) occupy the same volume as molecules of ordinary water, and ordinary water has density 1.000 g/ml at 4 °C, what would the density of heavy water be at 4 °C? Is it reasonable to assume that molecules of heavy water occupy about the same volume as molecules of ordinary water? The actual density of ^2H_2O at 4 °C is 1.106 g/ml. Use this result to determine the volume occupied by a mole of ^2H_2O.

6.27. Which is a better guide to the chemistry of an atom, its atomic mass or its atomic number? Why?

6.28.* Which element of each pair is composed of larger atoms: (a) chromium or nickel? (b) chromium or molybdenum? (c) chromium or rhodium?

6.29. Which element of each pair is composed of larger atoms: (a) silicon or sodium? (b) silicon or germanium? (c) sodium or germanium?

6.30.* Which ion is larger in each of the following pairs: (a) Mg^{2+} or Ca^{2+}? (b) O^{2-} or F^-? (c) Na^+ or Mg^{2+}?

6.31. Using the definition of density (Exercise 6.26) and the atomic radii given in Fig. 6.20, decide for each of the following pairs which of the elemental solids is denser: (a) copper or chromium? (b) cobalt or rhodium? (c) osmium or copper?

6.32.* In each of the following pairs, which atom has the higher ionization potential: (a) aluminum or phosphorus? (b) silicon or germanium? (c) bromide ion or krypton?

6.33. In each of the following pairs, which atom has the higher ionization potential: (a) sodium or aluminum? (b) sodium or potassium? (c) sodium ion or magnesium?

6.34.* In each of the following pairs, which item has the greater electron affinity: (a) chlorine atom or aluminum atom? (b) sulfur atom or oxygen atom? (c) sodium ion or chloride ion?

6.35. Which item of each of the following pairs has the greater electron affinity: (a) chlorine atom or bromine atom? (b) sodium atom or chlorine atom? (c) Mg^{2+} or O^{2-}?

ANSWERS TO SELECTED EXERCISES

6.2 8.46 Å. **6.3** 0.265 Å. **6.5** 65.81 kcal/mol. **6.7** 4.34×10^{-5} cm; 6.90×10^{14} Hz. **6.9** Li^{2+}. **6.11** 4.109×10^{15} Hz; 391.8 kcal/mol. **6.13** (a) 2; (b) 10; (c) 6; (d) 2; (e) 8; (f) 18. **6.15** (a) and (b) $n = 1$; $l = 0$; or $1s$ orbital. **6.16** (a) $[Ar]4s^1$; (b) $[Ar]4s^2$; (c) $[Ar]4s^23d^1$; (d) $[Ar]3d^{10}4s^24p^6$. **6.18** N, $[He]2s^22p^3$; P, $[Ne]3s^23p^3$; As, $[Ar]3d^{10}4s^24p^3$; Sb, $[Kr]4d^{10}5s^25p^3$; Bi, $[Xe]4f^{14}5d^{10}6s^26p^3$. **6.20** (a) excited state of Mg; (b) ground state of P; (c) ground state of Cr; (d) impossible (seven p electrons). **6.22** F, $1s^22s^22p^5$; Ag, $[Kr]4d^{10}5s^1$; Br, $[Ar]3d^{10}4s^24p^5$. Only in the case of silver are the orbitals involved ($4d$ and $5s$) close enough in energy. **6.24** (a) The number of protons, not the mass, determines the identity of an element; (b) 5 protons, 5 neutrons and 5 protons, 6 neutrons; (c) 10.83 g. **6.26** (a) 1.112 g/ml; (b) yes, neutrons are very small and have no charge to attract the electron so atoms of isotopes are very nearly the same size; (c) 18.11 ml (20.0274 g/mol ÷ 1.106 g/ml = 18.11 ml/mol). **6.28** (a) chromium; (b) molybdenum; (c) rhodium. **6.30** (a) Ca^{2+}; (b) F^-; (c) Na^+. **6.32** (a) phosphorus; (b) silicon; (c) krypton. **6.34** (a) chlorine atom; (b) sulfur atom; (c) sodium ion.

ANSWERS TO STUDY PROBLEMS

6(a) 0.265 Å. **6(b)** Be, $[He]2s^2$; Mg, $[Ne]3s^2$; Ca, $[Ar]4s^2$; Sr, $[Kr]5s^2$; Ba, $[Xe]6s^2$; Ra, $[Rn]7s^2$. **6(c)** (a) F; (b) F, O^-, Ne^+, N^{2-}, Na^{2+}, etc.; (c) F. **6(d)** (a) 15 and 16; (b) 53 and 72; (c) 53 and 74. **6(e)** (a) Se < Ge < Ca; (b) S < Se < Te; (c) Ca^{2+} < Ge < Br^-. **6(f)** (a) Mg < S < P; (b) Sb < As < P; (c) I^- < Co^+ < V^{2+}.

7

Bonding and Molecular Structure

As we investigate chemical substances that occur naturally or that are made in chemists' laboratories, the variety of combinations of atoms that form compounds appears to be overwhelming. Nature offers us a nearly limitless array of different molecules, and the chemist's imagination can quickly devise schemes for making endless numbers of synthetic compounds. How can we possibly find a thread of consistency from all this diverse chemistry? How can we form some explanation for chemical reactivity?

Fortunately, there are some pronounced similarities and patterns in the types of compounds that any given element forms. Furthermore, molecules often contain certain groups of atoms, **functional groups** (for example, the —OH group or the $—NH_2$ group), which each react in a similar characteristic manner in many compounds. By understanding the patterns of the elements and of functional groups, we can organize chemistry in a very manageable form.

An important step in beginning an organization of chemistry is recognizing that compounds can be divided into essentially two classes: *ionic* compounds, composed of ionic parts and held together by electrostatic attractions between charges of opposite sign, and compounds composed of discrete neutral molecules, *covalent* compounds.

7.1 Ionic Compounds

We can imagine the simplest types of ionic compounds as compounds formed from two very different types of atoms—one atom that can easily lose one or more electrons to form a cation, and another atom that can readily accept one electron or more to form an anion. Thus, we can imagine the compound sodium chloride to be formed by the process

$$\mathrm{Na} + :\ddot{\underset{\cdot}{\mathrm{Cl}}}: \rightarrow \mathrm{Na}^+ :\ddot{\underset{\cdot}{\mathrm{Cl}}}:^-$$

in which the sodium atom gives up an electron to the chlorine atom, giving rise to a sodium ion, Na^+, and a chloride ion, Cl^-. The dots represent electrons in the outer (valence) shell of the atoms or ions of chlorine; the color dot represents the electron initially in the valence shell of the sodium atom.

Ionic compounds in the solid state are not found to consist of simple combinations of just one ion with another, for example, discrete pairs of ions. In solid ionic compounds, there are many positive and negative ions; they are arranged so that several negative ions are close to each positive ion and vice versa. In this way, nature tends to minimize the electrostatic energy of the solid system. Figure 7.1 shows the structure of solid NaCl and of another ionic compound, CsCl; such ionic compounds are often referred to as **salts**—ionic compounds formed from a cation other than H^+ and an anion other than OH^-. Since there are several equivalent chloride ions around each sodium ion, we cannot identify one particular chloride as "belonging" to a particular sodium ion. Therefore, there is no discrete ionic molecule in this ionic lattice. If we melt NaCl, forming a molten salt, the rigid structure of the solid is destroyed and the ions move about one another randomly; one particular chloride ion does not remain with any one particular sodium ion, but several chloride ions are moving in the vicinity of each sodium ion at any time, and they readily move from the region of one sodium ion to another.

Another important aspect of the structure of ionic compounds is the way in which the ions behave when ionic substances are dissolved. Ionic compounds are generally

Figure 7.1 (a) A portion of the NaCl crystal lattice (the ions Na^+ shown by smaller spheres). (b) One Cl^- is surrounded by 6 Na^+. (c) One Na^+ is surrounded by 6 Cl^-. (d) A portion of the CsCl crystal lattice (the ion Cs^+ shown by the color sphere).

(a) (b) (c) (d)

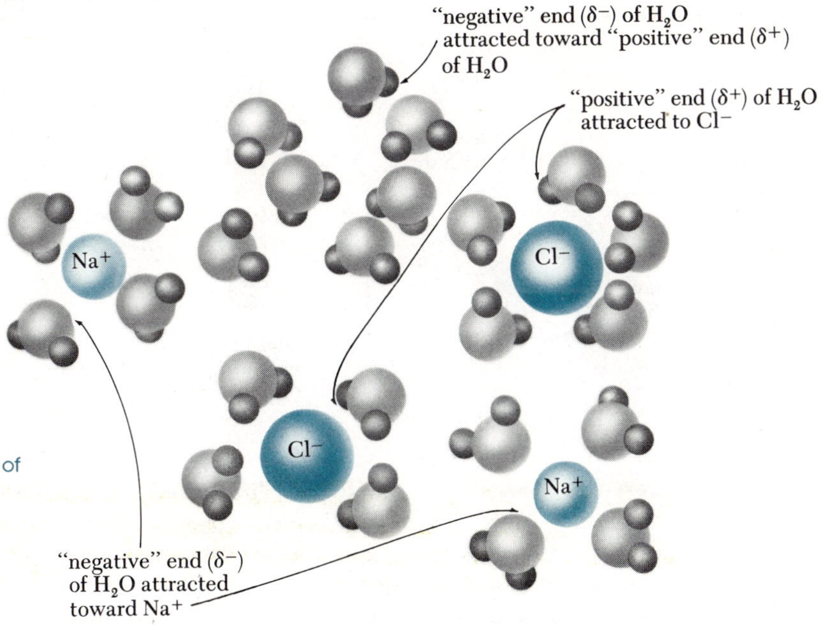

Figure 7.2
A portion of a solution of NaCl in H_2O.

soluble in solvents that are capable of forming an electrically insulating layer around the ions to keep them from coming together with ions of opposite charge (the "counter ions"), and re-forming a solid lattice. Such solvents are referred to as **polar compounds.*** The negative ends of several such solvent molecules will be attracted to a positive ion, and the positive ends of some other solvent molecules will be attracted to a negative ion. The manner in which the most common polar solvent, water, can shield the ions Na^+ from Cl^- is an important property of ocean water; it is shown in Fig. 7.2. Again, there is no discrete sodium chloride molecule in aqueous solution.

Determining That a Compound Is Ionic

For each state in which compounds are found, experiments can be performed to show that the compounds are indeed ionic and not molecular. For the solid state, we can determine the relative positions of the atoms or ions by x-ray diffraction. In this experiment, a very small beam of x radiation is passed through a crystal of the compound. The x-ray beam, manifesting the wave character of x rays, is diffracted by the atoms or ions in the crystal in a way that is analogous to what is described in Fig. 6.10. From the angles at which the ray is diffracted and the amount of the ray (the intensity) diffracted at each of the angles, the positions of the atoms or ions can be calculated. The experiment is shown schematically in Fig. 7.3 (also described in greater detail in Chap. 14). In an ionic compound, the observed positions of the atoms or ions show that there are several equivalent anions near each cation, and

*We shall soon deal with polar molecules in detail, but for the present we may think of a polar molecule as a species that is electrically neutral as a whole but in which there is a small amount of negative charge (δ−) and a small amount of positive charge (δ+) at different parts of the molecule.

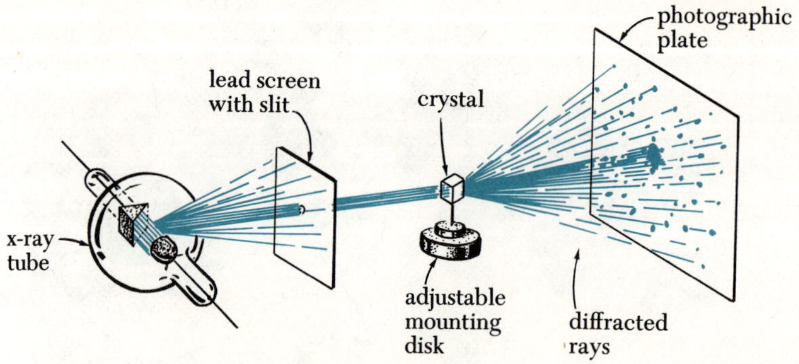

Figure 7.3 The x-ray diffraction experiment. Note that x rays are high-energy radiation (dangerous) and do not require an evacuated path, since "collisions" of the ray with air have little effect on the beam.

vice versa; there are no identifiable "molecules." Also, the cations are found by x-ray diffraction to be smaller than the corresponding atom and the anions larger than the corresponding atom, as expected due to abstracting one or more electrons from an atom to form a cation and adding one or more electrons to form an anion.

For the salts in a fluid state, either melted or dissolved, the existence of ions can be demonstrated by conduction of electric current. Thus, if a fluid medium is capable of passing current, a property referred to as **conductivity,** then one knows the fluid contains discrete, electrically charged particles or ions. If one places two electrodes in the salt solution or molten salt and applies a voltage across them, the cations will move toward the negative electrode, the cathode, while the anions will move toward the positive electrode, the anode, causing electrical current to flow. This demonstrates that the cations and anions can act separately, and are not bound together in a single, discrete molecule (Fig. 7.4).

Elements That Form Ionic Compounds

To decide which elements will form ionic compounds, we must consider the energies of the various processes that can be envisioned for their formation. The overall change in energy involved in converting elemental substances from their stable

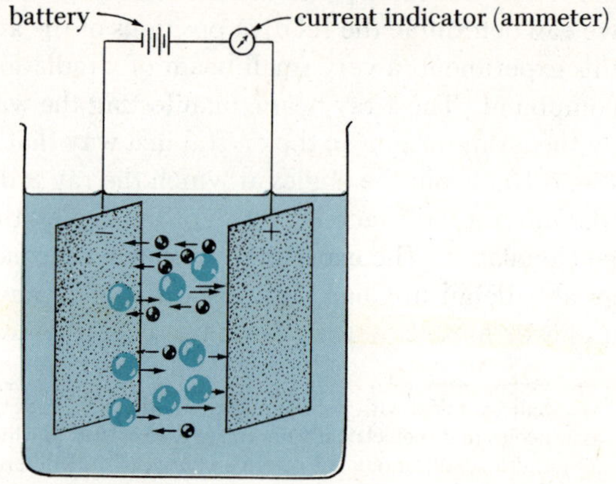

Figure 7.4 Use of conductance to demonstrate the presence of ions in solution. If a potential difference is applied to electrodes in an ionic solution (either a molten salt, or a solution of salt in a polar solvent), the cations (small spheres) will move toward the cathode and the anions (large spheres) will move toward the anode. This will cause a current flow, which is observed by a deflection of the ammeter needle.

Spotlight

The statement that one molecule is more stable than another means that the one is more likely to exist than the other. There are two general types of stability—thermodynamic and kinetic. If a molecule is thermodynamically more stable than another, it is more favored by the equilibrium constant for the expression that contains the concentrations of the two species. Molecules can also be kinetically stable, or "inert." In this case, they may not be so thermodynamically stable as other molecules, but reactions that can convert them into the more thermodynamically stable molecules are slow. The reaction $H^+ + NO_3^- + 4H_2(g) \rightleftharpoons NH_3 + 3H_2O$, for example, is exothermic due to the stability of the water produced; but it is slow. The molecule H_2 is then said to be kinetically stable in this situation. If no qualification is made, the statement that one molecule is more stable than another is taken to mean that it is more thermodynamically stable.

forms into compounds is called the **heat of formation.*** If energy is released in the formation of an ionic compound from the elements, the heat of formation is negative. This means that the end product of the processes has a lower energy, and tends to be more stable than the form of matter from which it was made, elements. The two most important classes of ionic substances for ordinary chemical purposes are ionic solids and ions in solution.

For the formation of an ionic solid from two elements, one can visualize the process in terms of a series of steps, the sequence known as the **Born-Haber cycle,** after Max Born (1882–1970) and Fritz Haber (1868–1954), both of whom contributed to its development. The Born-Haber cycle for the formation of NaCl is shown in Fig. 7.5. Going from the elements to the ionic compound NaCl by either path shown, one finds the total energy change must be the same:

$$\Delta H_f = \Delta H_s + \tfrac{1}{2}D + IP_1 - EA_1 + U \qquad (7.1)$$

For many ionic compounds the heat of formation ΔH_f is known. But to explain why ΔH_f for the formation of solid ionic compounds is negative for combinations of some elements and not for others, we must see how the several quantities in the stepwise path affect ΔH_f.

Table 7.1 gives the values of the energies of the various steps in the Born-Haber cycle for several compounds. Some of the entries are predicted values for compounds that have not been prepared, since combination of these elements as shown does not form a stable ionic compound. It is interesting to note which of the quantities are most important in determining which combinations of elements form ionic compounds. In general, the **lattice energy**, which is the energy decrease accompanying the formation of an ionic crystal from the gaseous ions, denoted as U in Eq. (7.1), must "overcome" the ionization potential; also the electron addition must not be a process that requires an excessively large energy input, or a large negative EA value.† Unless the positive and negative ions can be formed easily, with relatively favorable IP and EA values, an ionic compound is not formed. The lattice energy becomes

*The way in which heats of formation, and related quantities called free energies of formation, are related to compound stability is discussed in Chaps. 10 and 11.

†Remember that EA is defined as the energy *released* on electron addition, whereas all other quantities in Eq. (7.1) refer to the energy *required* for each process.

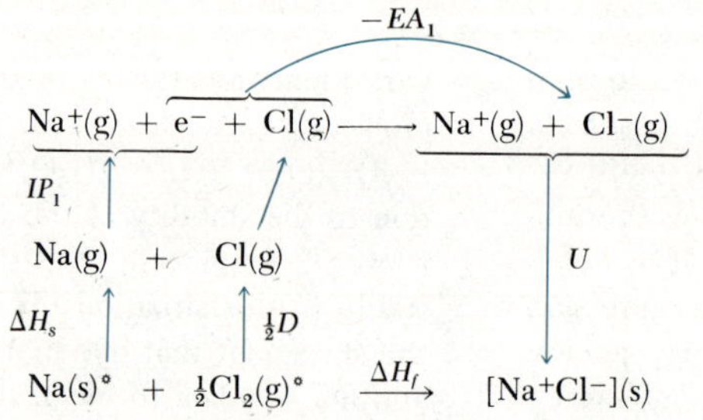

Figure 7.5
Born-Haber cycle for solid NaCl.

$$\begin{array}{c}
\text{Na}^+(g) + e^- + \text{Cl}(g) \xrightarrow{-EA_1} \text{Na}^+(g) + \text{Cl}^-(g) \\
IP_1 \uparrow \\
\text{Na}(g) + \text{Cl}(g) \\
\Delta H_s \uparrow \quad \uparrow \tfrac{1}{2}D \quad \quad \downarrow U \\
\text{Na}(s)° + \tfrac{1}{2}\text{Cl}_2(g)° \xrightarrow{\Delta H_f} [\text{Na}^+\text{Cl}^-](s)
\end{array}$$

Process	Energy Change	Definition
$\text{Na}(s)° \rightarrow \text{Na}(g)$	ΔH_s	*Heat of sublimation:* Energy necessary to vaporize a mole of Na(s)
$\text{Na}(g) \rightarrow \text{Na}^+(g) + e^-$	IP_1	*First ionization potential:* Energy necessary to remove a mole of electrons from a mole of Na(g)
$\tfrac{1}{2}\text{Cl}_2(g)° \rightarrow \text{Cl}(g)$	$\tfrac{1}{2}D$	*Dissociation energy:* Energy necessary to dissociate half a mole of $\text{Cl}_2(g)$
$\text{Cl}(g) + e^- \rightarrow \text{Cl}^-(g)$	$-EA_1$	*First electron affinity:* Energy released when a mole of electrons is added to a mole of Cl(g) is $+EA_1$.
$\text{Na}^+(g) + \text{Cl}^-(g) \rightarrow [\text{Na}^+\text{Cl}^-](s)$	U	*Lattice energy:* Energy of forming a lattice from a mole of $\text{Na}^+(g)$ and a mole of $\text{Cl}^-(g)$
$\text{Na}(s) + \tfrac{1}{2}\text{Cl}_2(g) \rightarrow [\text{Na}^+\text{Cl}^-](s)$	ΔH_f	*Heat of formation:* Energy necessary to form an ionic compound from the elements in their standard states

°Elements in the states in which they would be found at 25 °C, 1 atm pressure.

more favorable (more negative) if ions are more highly charged, for example, +2 or +3 instead of +1. Thus, the ionizations proceed to the point at which the next ionization potential would be unreasonably high, meaning until a noble gas configuration is reached, for example, to Al^{3+} for aluminum or to Mg^{2+} for magnesium.

A cycle analogous to the **Born-Haber cycle** can also be constructed for the formation of ionic compounds in a solution. For this cycle the last step would be transforming the relevant gaseous ions into ions in solution, or solvated ions; the energy associated with this step is called the **solvation energy**. For water and several other common polar solvent compounds, the values of the solvation energies of common ionic species are much like the corresponding lattice energies; hence, we expect to find the same ions in solution that we have in the solid state.° As a rough guideline,

°Whether or not a solid ionic compound dissolves in a given solvent depends largely on whether the solvation energy or the lattice energy has a larger magnitude (both are negative). If the solvation energy is much more negative, then the dissolved state tends to be more stable and the substance will dissolve well.

TABLE 7.1 Born-Haber Processes

Element	ΔH_s	IP_1	IP_2	D	$-EA_1$	$-EA_2$	U	ΔH_f	Compound Formed
SOME TYPICAL CASES OF ACTUAL COMPOUNDS									
Na	$Na(s) \rightarrow Na(g)$	$Na \rightarrow Na^+$							
Cl_2				$\frac{1}{2}Cl_2 \rightarrow Cl$	$Cl \rightarrow Cl^-$		$Na^+(g) + Cl^-(g) \rightarrow NaCl(s)$	$Na(s) + \frac{1}{2}Cl_2(g) \rightarrow NaCl(s)$	NaCl
	25	118.43		28.9	−83.1		−181.20	−92.0	
Ca	$Ca(s) \rightarrow Ca(g)$	$Ca \rightarrow Ca^+$	$Ca^+ \rightarrow Ca^{2+}$						
Br_2				$Br_2 \rightarrow 2Br$	$2Br \rightarrow 2Br^-$		$Ca^{2+}(g) + 2Br^-(g) \rightarrow CaBr_2(s)$	$Ca(s) + Br_2(g) \rightarrow CaBr_2(s)$	$CaBr_2$
	33	140.9	273.6	54.0	−156.2		−512.6	−167	
Ca	$Ca(s) \rightarrow Ca(g)$	$Ca \rightarrow Ca^+$	$Ca^+ \rightarrow Ca^{2+}$						
Cl_2				$Cl_2 \rightarrow 2Cl$	$2Cl \rightarrow 2Cl^-$		$Ca^{2+}(g) + 2Cl^-(g) \rightarrow CaCl_2(s)$	$Ca(s) + Cl_2(g) \rightarrow CaCl_2(s)$	$CaCl_2$[a]
	33	140.9	273.6	57.8	−166.2		−529.9	−191	
OTHER POSSIBILITIES									
Ca	$Ca(s) \rightarrow Ca(g)$	$Ca \rightarrow Ca^+$							
Cl_2				$\frac{1}{2}Cl_2 \rightarrow Cl$	$Cl \rightarrow Cl^-$		$Ca^+(g) + Cl^-(g) \rightarrow CaCl(s)$	$Ca(s) + \frac{1}{2}Cl_2(g) \rightarrow CaCl(s)$	CaCl[b] (Ca^+Cl^-)
	33	140.9		28.9	−83.1		−171.8	−52.1	
Na	$Na(s) \rightarrow Na(g)$	$Na \rightarrow Na^+$	$Na^+ \rightarrow Na^{2+}$						
Cl_2				$Cl_2 \rightarrow 2Cl$	$2Cl \rightarrow 2Cl^-$		$Na^{2+}(g) + 2Cl^-(g) \rightarrow NaCl_2(s)$	$Na(s) + Cl_2(g) \rightarrow NaCl_2(s)$	$NaCl_2$[c]
	25	118.43	1090	57.8	−166.2		−191.5	933	
Ca	$Ca(s) \rightarrow Ca(g)$	$Ca \rightarrow Ca^+$	$Ca^+ \rightarrow Ca^{2+}$						
Cl_2				$\frac{1}{2}Cl_2 \rightarrow Cl$	$Cl \rightarrow Cl^-$	$Cl^- \rightarrow Cl^{2-}$	$Ca^{2+}(g) + Cl^{2-}(g) \rightarrow CaCl(s)$	$Ca(s) + \frac{1}{2}Cl_2(g) \rightarrow CaCl(s)$	CaCl[d] ($Ca^{2+}Cl^{2-}$)
	33	140.9	273.6	28.9	−83.1	335	−707	21	

NOTE: All numerical values are in kcal/mole.

[a]The heat of formation is more favorable for $CaCl_2$ than for $CaBr_2$ due to both a more favorable electron affinity of Cl and a more favorable lattice energy.

[b]Formation of CaCl is favorable, but not as favorable as $CaCl_2$ formation. The big advantage of $CaCl_2$ formation is the more favorable lattice energy for a +2 ion over a +1 ion.

[c]Formation of $NaCl_2(s)$ is not favorable due to the very high IP_2 value for sodium.

[d]Formation of $Ca^{2+}Cl^{2-}(s)$ is not favorable mostly due to the unfavorable EA_2 value for chlorine.

Figure 7.6
Energy cycle for the formation of aqueous KI.

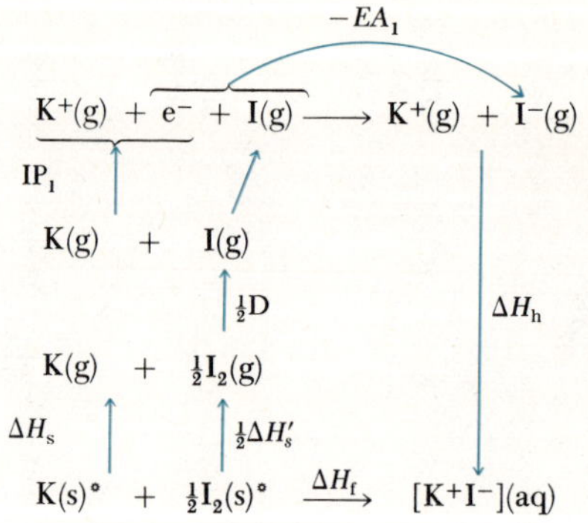

Process	*Energy Change*	*Definition*
$K(s) \rightarrow K(g)$	ΔH_s	*Heat of sublimation* (of K(s))
$K(g) \rightarrow K^+(g) + e^-$	IP_1	*First ionization potential* (of K(g))
$\frac{1}{2}I_2(s) \rightarrow \frac{1}{2}I_2(g)$	$\frac{1}{2}\Delta H_s'$	*Heat of sublimation* (of $\frac{1}{2}I_2(s)$)
$\frac{1}{2}I_2(g) \rightarrow I(g)$	$\frac{1}{2}D$	*Dissociation energy* (of $\frac{1}{2}I_2(g)$)
$I(g) + e^- \rightarrow I^-(g)$	$-EA_1$	*First electron affinity* (of I(g)) is $+EA_1$.
$K^+(g) + I^-(g) + nH_2O(l) \rightarrow [K^+I^-](aq)$	ΔH_h	*Hydration energy:* Energy necessary to dissolve one mole of gaseous ions K^+ and I^- in excess H_2O
$K(s) + \frac{1}{2}I_2(s) + nH_2O(l) \rightarrow [K^+I^-](aq)$	ΔH_f	*Heat of formation:* Energy necessary to form an aqueous KI solution from the elements in their standard states *and water.*

*Elements in the states in which they would be found at 25 °C, 1 atm pressure.

this generalization is useful. Figure 7.6 shows a cycle that includes the solvation step; in the case shown, the solvent is water, so the solvation step is called **hydration.**

Cycles of the type shown in Figs. 7.5 and 7.6 are merely mental constructions, useful for considering reasons for the properties of ionic substances. Such diagrams are not meant to represent the actual processes by which ionic solids or solutions are formed, which may be very different from these stepwise pathways. One can always compute the energy difference between any two specific states by considering any pathway between the states that is convenient for the calculation.

Typical Elemental Ions

Each of the alkali metals has one electron in its outermost shell. The first ionization potential is low for this group, and decreases with the increasing size of the atoms of the group. Lattice energies and hydration energies are large enough to allow formation of many ionic compounds of the alkali metals in the solid or solution state, so long as the electron addition to form the corresponding negative ion does not require too much energy. The second ionization potential for the alkali metals requires that an electron be abstracted from a filled shell, and it is therefore high. Consequently, the alkali metals form only +1 ions in ordinary chemical systems.

Atoms of the alkaline earth group (Group IIA) have two electrons in their outer shell in the ground state. The first and second ionization potentials for these elements (except for Be) can be adequately counterbalanced by lattice or hydration energies, and the alkaline earths usually exist as +2 ions in solids or solutions. The reason these elements are ordinarily not found as +1 ions is that the lattice energy, or hydration energy, of an ion is roughly proportional to the charge of the ion and hence is about twice as large for an Mg^{2+} case as for an Mg^{+} case. For the alkaline earths, this factor is great enough to overcome the energy requirement represented by the second ionization potential, and compounds containing +2 alkaline earth ions are far more stable than those containing the corresponding +1 ions would be. The third ionization potential corresponds to abstracting an electron from a closed shell, and is so high that it would overwhelm the corresponding increase in lattice energy for a +3 ion; hence, +3 ions of alkaline earth elements are very uncommon.

In considering the formation of anions, it is the electron affinity rather than the ionization potential that is important. Atoms that will form anions are those whose electron affinities either correspond to a release of energy (forming F^-, Cl^-, or Br^-, for example) or to a small enough input of energy (forming O^{2-} or S^{2-}, for example) that lattice energy can still render the heat of formation of an ionic compound negative.

The following three important points can be summarized from the discussion thus far: (1) Formation of an ionic compound requires that the heat of formation, which can be visualized in distinct steps by the Born-Haber cycle, is negative. This means that energy-requiring processes like ionization require compensation from energy-releasing processes—those corresponding to lattice energies, electron affinities, or solvation energies. (2) Only atoms or combinations of atoms that easily lose electrons to form cations (for example, the metals), together with atoms or combinations of atoms that readily gain electrons to form anions (for example, the halogens) will form ionic compounds. (3) The lattice energies are greatest for ions of highest charge, a situation that tends to favor formation of more highly charged ions; however, removal of an electron from a filled shell or addition of an electron to start a new shell requires addition of a large amount of energy. The net result of these two influences is that ions most readily formed are those that have a filled-shell configuration.* Thus, those commonly found have the following types of electronic configurations, indicated in parentheses:

$$Na^+ \text{ (Ne)}, Ca^{2+} \text{ (Ar)}, Cl^- \text{ (Ar)}, O^{2-} \text{ (Ne)}, \ldots$$

*The configurations of ions that are formed by removal of electrons from *d* or *f* orbitals—the ions of the transition metal series—also depend on ionization potentials and lattice or solvation energies; however, as we shall discuss in Chap. 18, the most stable ions in this series do not necessarily correspond to inert gas configuration. The atoms of the transition metal series (Sc through Zn) often have so many electrons that must be removed to form an inert gas configuration that the successive ionization potentials become too large to allow sufficient compensation by the lattice or solvation energies.

For very many elements, ionization energies are too high for the formation of ionic compounds to be common. Thus, for example, compounds of B^{3+} and carbon compounds of C^{4+}, or C^{4-}, are not common. These elements certainly combine with other elements, however, to form a wide variety of compounds. Indeed, the large portion of chemistry called **organic chemistry** is basically the study of carbon compounds. These compounds are not ionic, and are formed by what are called covalent bonds.

EXAMPLE 7.1 What are the ground state electronic configurations of the following ions: Mg^{2+}, Ga^{3+}, Te^{2-}, Te^{6+}, Br^{-}?

SOLUTION The electronic configurations of these species, all of which are in the "main groups" (the "A" groups) of the periodic table, are as follows: Mg^{2+}, [Ne]; Ga^{3+}, [Ar]$3d^{10}$; Te^{2-}, [Xe]; Te^{6+}, [Kr]$4d^{10}$; Br^{-}, [Kr].

STUDY PROBLEM 7(a)

Give the ground state electronic configurations of Ca^{2+}, Se^{2-}, Al^{3+}, Cs^{+}, and Zn^{2+}.

7.2 Covalent Compounds

General Features

It has long been known that many elements combine with each other in certain definite proportions to form compounds that do not have the properties of ionic substances. Such compounds have become known as **molecular compounds**, or **covalent compounds**. About 1910, G. N. Lewis introduced the concept that shared electrons are responsible for holding atoms together in these compounds. This development provided a way of exploring why only specific numbers of atoms could combine in molecular compounds. It was found that the atoms of each element tend to form a certain characteristic number of "connections," or **bonds**, to atoms of other elements. This number of bonds was often referred to as the **valence** of the element. The valence is determined by the number of electrons in the outer shell of an atom, often called the **valence shell**.

The carbon atom has four electrons in its valence shell, $1s^2 2s^2 2p^2$. If the atom could gain four electrons, it would have a filled-shell configuration (Ne). Similarly, if the atom could lose four electrons, it would assume the He filled-shell configuration of C^{4+}. However, the formation of either of these highly charged ions would require more energy than what could be compensated for by lattice or solvation energies. Lewis's development pointed out that if four atoms could be bonded to the carbon in a way in which each of the bonded atoms shared one of its electrons with the carbon atom, and if the carbon atom shared one of its valence electrons with each of the bonded atoms, then the carbon valence shell would be filled. These pairs of electrons were viewed as being shared by both atoms and were said to constitute a **covalent bond.**

For the case of methane, CH_4, the electrons that we initially associated with the valence shell of carbon are indicated by a small x and the electrons initially identified

Spotlight

Reproduced by permission of University Archivist, The Bancroft Library, University of California, Berkeley

G. N. Lewis was a powerful influence upon chemistry in America and throughout the world during the first half of the twentieth century. He was born in Weymouth, Massachusetts, in 1875, earned the A.B. degree from Harvard University in 1896 and the Ph.D. in 1899. After postdoctoral research in Leipzig and Göttingen and a year as head of a government laboratory in the Philippine Islands, he joined the faculty at the Massachusetts Institute of Technology in 1907. In 1912 he moved to a professorship at the University of California in Berkeley, where he remained until his death in 1946.

Lewis is especially well known for his formulation of early ideas of covalent bonding, his theory of acids and bases, and his contributions to thermodynamics. A textbook on thermodynamics which he coauthored is still in use (as a revision). During the first half of this century he was one of the dominant figures in physical chemistry.

with each hydrogen atom by a dot. Dashes are drawn around the pairs of shared electrons.

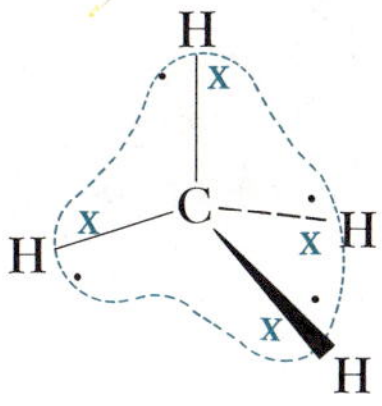

methane

In this example, each hydrogen atom shares its one valence shell electron with the carbon atom. The carbon atom, in turn, shares each of its valence shell electrons with a hydrogen atom to complete the 1*s* shell of hydrogen. Both of the atoms forming the covalent bond achieve a filled-shell configuration by sharing the bonding electrons.

The Covalent Bond

An extremely important and highly pertinent question is, How does the covalent bond make the combination of atoms more stable than the separate atoms? For ionic compounds, the attraction of the oppositely charged ions for one another, the lattice energy, is responsible for the formation of the compounds. In covalent compounds, an analogous yet distinctly different type of stabilization occurs. The atoms that participate in stable covalent bonds share the bonding electrons in such a way that there is a mutual attraction of both atomic nuclei for both shared electrons. Associated with this situation is the fact that the electrons in the bond "spend more time" (as represented in the electron probability distribution) between the nuclei of the atoms than they would for a case of atoms that do not bond. Fig. 7.7 shows the electron probability distribution for two helium atoms, which do not form a stable bond since each atom already has a filled valence shell, and for two hydrogen atoms, which form the H_2 molecule, with a stable covalent bond. The plot of electron density of two helium atoms that are brought close together is practically the same as a superposition of the electron-density plots of two independent helium atoms. Two hydrogen atoms, on the other hand, provide an excellent opportunity for covalent bonding, since each hydrogen atom is only "one electron away" from a filled shell, and this deficiency can be relieved by formation of a covalent bond. Figure 7.7 shows that the electron density for the resulting H_2 molecule is greater between the hydrogen nuclei than one would predict from simply superimposing the electron-density patterns of independent hydrogen atoms.

An analogous situation can be described for two fluorine atoms, in comparison with two neon atoms. Thus, a neon atom already has a filled shell and therefore

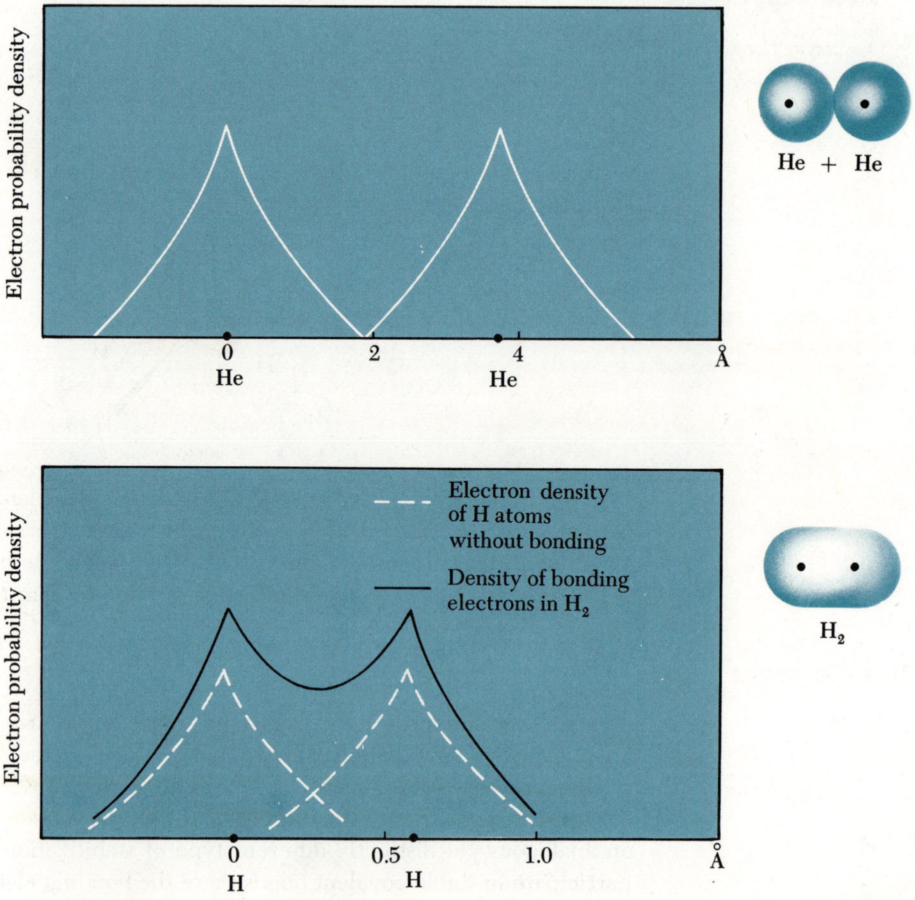

Figure 7.7 Electron probability densities for two unbound atoms and for a molecule with two bound atoms. Dots represent the positions of nuclei.

does not have a tendency to form Ne_2 by electron sharing. On the other hand, a fluorine atom, like a hydrogen atom, is only one electron away from a filled-shell configuration; and this deficiency can be eliminated by the sharing of an electron pair between the fluorine atoms, with each F atom contributing one electron to the pair in F_2.

In the bonding cases described for H_2 and F_2, the repulsive forces between the nuclei of the atoms are weaker than one might initially guess, since the increased electron density between the two nuclei somewhat shields the positive charges from each other. The bonding electrons, now shared, have strong attraction for both nuclei by being highly concentrated between them. Thus, the overall repulsive force, and corresponding energy contribution, are lessened and the attractive force, and energy contribution, increased relative to the uncombined atoms. The molecule is therefore more stable than the separated atoms.*

Bond Energy

In Sec. 7.1 we encountered the idea of "energy cycles" for visualizing the stabilities of ionic compounds. The Born–Haber cycle was devised for ionic solids, giving rise to the concept of lattice energy. A similar cycle, utilizing the concept of a hydration (solvation) energy was devised for ionic compounds in solution. An analogous energy cycle can be constructed for covalent compounds, and gives rise to the idea of a **bond energy.** The bond energy is defined as the energy required to break a specific type of covalent bond in a mole of gaseous molecules to form the corresponding gaseous atoms (not ions). For example, we define the bond energy, ΔH_b, of the hydrogen-iodine bond to be the energy *required* to dissociate HI(g) into gaseous, atomic hydrogen and iodine.

$$HI(g) \rightleftharpoons H(g) + I(g)$$

Alternately, we can say that the energy *required* when H(g) and I(g) combine to form HI(g) is then $-\Delta H_b$ (that is, $+\Delta H_b$ would be *released* in this process).

The formation of HI(g) from the stable forms of elemental hydrogen and iodine, $H_2(g)$ and $I_2(g)$, respectively, can be considered in the following steps: (1) the sublimation of $I_2(s)$ to $I_2(g)$, for which the required energy is ΔH_s, the sublimation energy of $I_2(s)$; (2) the dissociation of $I_2(g)$ into gaseous iodine atoms, for which the required energy is D', the *dissociation energy* of $I_2(g)$; (3) the dissociation of $H_2(g)$ into gaseous hydrogen atoms, for which the required energy is D, the dissociation energy of $H_2(g)$; and (4) the formation of HI(g) from H(g) and I(g), for which the energy *required* is $-\Delta H_b$. When *stable bonds* are formed, ΔH_b is a *positive* number and heat would actually be *released* by bond formation. The overall result of these four steps is the conversion of the stable elemental substances $I_2(s)$ and $H_2(g)$ into HI(g); the energy change for this entire stepwise conversion is denoted ΔH_f. Figure 7.8 summarizes this energy cycle in terms of individual steps in the process and the corresponding energy changes.

*An interaction that lessens the energy decrease the atoms experience in bond formation is the repulsion by each other of the bonding electrons. This interaction, however, is more than compensated for by the other favorable interactions.

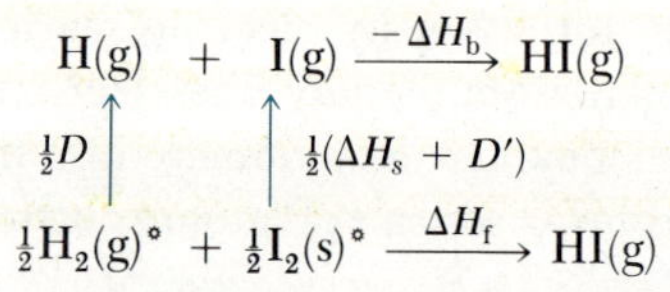

Figure 7.8
Energy cycle for the formation of HI(g).

Process	Energy Change	Definition
$\frac{1}{2}H_2(g)^* \rightarrow H(g)$	$\frac{1}{2}D$	*Dissociation energy* (of $H_2(g)$)
$\frac{1}{2}I_2(s) \rightarrow \frac{1}{2}I_2(g)$	$\frac{1}{2}\Delta H_s$	*Heat of sublimation* (of $\frac{1}{2}I_2(s)$)
$\frac{1}{2}I_2(g) \rightarrow I(g)$	$\frac{1}{2}D'$	*Dissociation energy* (of $\frac{1}{2}I_2(g)$)
$H(g) + I(g) \rightarrow HI(g)$	$-\Delta H_b$	*Bond energy:* Energy required to form HI(g) from H(g) and I(g). [ΔH_b is the energy required for the opposite process—making H(g) and I(g) from HI(g).]
$\frac{1}{2}H_2(g) + \frac{1}{2}I_2(s) \rightarrow HI(g)$	ΔH_f	*Heat of formation*

*Elements in the states in which they would be found at 25 °C and 1 atm pressure.

The bond energy is a direct measure of the strength of a covalent bond. A large number corresponds to a strong bond. Many correlations and patterns of chemical behavior can be accounted for in terms of bond energies, and we will make frequent use of this concept throughout the remainder of the text. The following chart summarizes the bond energies (in kcal/mol of bonds) of some single bonds between various elements.

	H	C	N	O	F	Cl	Br	I
H	104	99	93	111	135	103	88	71
C		83	70	84	116	79	66	57
N			38	53	65	48	58	–
O				33	44	49	48	48
F					37	60	60	67
Cl						58	52	50
Br							46	43
I								36

Covalent Bonding in NH_3 and CH_4

One finds that in many molecules more than one covalent bond is present. An example is the methane molecule, the main constituent of "marsh gas." In such cases, each bond contributes to the overall stability of the covalent molecule. A good guideline for deciding how many covalent bonds are to be expected in a given case is that atoms that form covalent compounds will tend to find other atoms with one electron or two to share until the valence shells are completed by the shared electrons. An example is carbon, which with four valence shell electrons forms bonds

Spotlight

Marsh gas, methane, is formed during the decomposition of plant and animal matter. It was one of the compounds that drew the interest of John Dalton, who developed the theory that each element was made up of atoms of a particular mass and that the atoms of different elements could bind together to form compounds (four hydrogen atoms to one carbon atom to form methane). The production of methane from decomposing waste is now being investigated as an energy source; the burning of methane is very exothermic:

$$CH_4 + 2O_2 \rightarrow CO_2 + 2H_2O$$

Mural by Ford Madox Brown in the Town Hall, Manchester, England. By permission of the Manchester City Council.

with four hydrogen atoms, each with one electron to share. The bonding of these elements thus yields completely filled valence shells. Another simple example is nitrogen [$1s^2 2s^2 2p^3$], which can share its three $2p$ electrons with three hydrogen atoms [$1s^1$] to form NH_3. Two common ways in which this example can be represented diagrammatically are an "orbital filling diagram"

$$N \underset{1s}{\uparrow\downarrow} \underset{2s}{\uparrow\downarrow} \underset{2p}{\overset{H\ H\ H}{\uparrow\downarrow\ \uparrow\downarrow\ \uparrow\downarrow}}$$

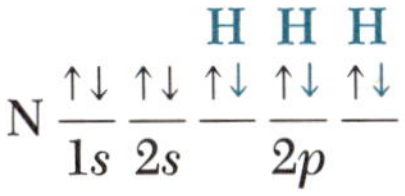

and a Lewis dot representation

$$:\!\overset{\cdot}{N}\!\cdot\ +\ 3H\cdot\ \rightarrow\ H\!:\!\overset{\cdot\cdot}{\underset{H}{N}}\!:\!H$$

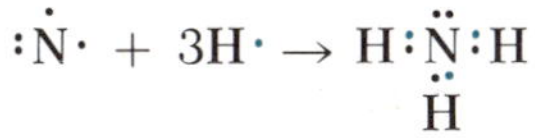

One should not read too many implications into these two kinds of representations. Such diagrams are useful only for gross electronic bookkeeping, not for providing great detail. In the orbital filling diagram, for example, we do not mean to imply that the hydrogen atoms provide electrons of a specific spin orientation (↓) and that they all are shared in nitrogen $2p$ orbitals. Actually, electrons in a molecule do not retain that kind of atomic identity; one can't speak realistically about which electron in a molecule came from which atom.° Nevertheless, so long as such unwarranted details are not read into these diagrams, this kind of symbolism can be useful in keeping track of the number of electrons in molecules in relation to the electron configurations of their constituent atoms. Similarly, the Lewis dot representation of NH_3 may seem to imply that the geometrical structure of the ammonia molecule is a

---

°As we shall see in some detail later in this chapter, each electron in a molecule belongs to the entire molecule—not to a specific atomic orbital. Furthermore, it will become clear that viewing electrons in a covalent molecule in terms of atomic orbitals is a gross oversimplification.

planar T. This kind of detail should not be read into Lewis dot formulas; this symbolism is useful only for showing how many covalent bonds to expect and not for showing the three-dimensional geometric structure of a species.

Why Carbon Forms Four Bonds

The ground state electronic configuration of carbon is $1s^2 2s^2 2p^2$ and there are therefore only two unpaired electrons, the $2p$ electrons. Yet carbon compounds generally form four bonds. Why not only two? The answer to this question lies in the additional stabilization (bonding energy) gained by forming four bonds to carbon rather than just two, and in electron promotion. This concept is associated with a hypothetical process that we can visualize rather than with a phenomenon having a physical reality. We find it attractive for keeping track of certain electron energy relations. Basically, we imagine promoting one of the paired $2s$ electrons of a ground state carbon atom up to a $2p$ orbital that was vacant in the ground state, yielding an excited carbon configuration in which there are four unpaired electrons. We can visualize the promotion as follows, where the hydrogens and related arrows are shown in color:

Promotion

$$\mathrm{C}\ \underset{1s^2}{\underline{\uparrow\downarrow}}\ \underset{2s^2}{\underline{\uparrow\downarrow}}\ \underbrace{\underline{\uparrow}\ \underline{\uparrow}\ \underline{\quad}}_{2p^2} \longrightarrow \mathrm{C}\ \underset{1s^2}{\underline{\uparrow\downarrow}}\ \underset{2s^1}{\underline{\uparrow}}\ \underbrace{\underline{\uparrow}\ \underline{\uparrow}\ \underline{\uparrow}}_{2p^3} \qquad \textit{requires energy input}$$

CH_4 *Formation*

$$\mathrm{C}\ \underset{1s}{\underline{\uparrow\downarrow}}\ \underset{2s}{\overset{\mathrm{H}}{\underline{\uparrow\downarrow}}}\ \underbrace{\overset{\mathrm{H}}{\underline{\uparrow\downarrow}}\ \overset{\mathrm{H}}{\underline{\uparrow\downarrow}}\ \overset{\mathrm{H}}{\underline{\uparrow\downarrow}}}_{2p}$$

The formation of CH_4 lessens the energy by four times the energy of a C–H bond. It is known from a variety of observations that the increased stabilization associated with each C–H bond is about 137 kcal/mol, and that the promotion energy corresponding to $C\ (1s^2 2s^2 2p^2) \rightarrow C\ (1s^2 2s^1 2p^3)$ is 151 kcal/mol. Thus, if we wish to compare the stabilization (energy) of CH_2 with CH_4, we should compute it as follows: To convert CH_2 to CH_4 requires adding promotion energy, and simultaneously increasing stabilization (lowering energy) in the amount of twice the C–H bond energy, due to the two additional covalent bonds. Thus, the total change expected for the process $CH_2 \rightarrow CH_4$ is:

$$151 + 2(-137) = 151 - 274 = -123\,\frac{\text{kcal}}{\text{mol}}$$

Hence, it is readily understood why CH_4 should be a common, highly stable species, whereas CH_2 can be detected only as an unstable species in sophisticated, specialized experiments. This example emphasizes the important guideline that the most stable chemical systems often tend to be the ones with the greatest number of covalent bonds possible for the system.

As a general rule, promotions within the valence shell are compensated for by the stabilizing influence of the additional bonds that can then be formed. Hence, *the* number of covalent bonds that an atom forms is generally the number needed to complete the filled-shell (noble gas) configuration. This statement is a form of a rule known as the **octet rule**, since the consequence for atoms in the second and third periods (Li to Ne, and Na to Ar) is eight electrons around each atom in the molecules.

EXAMPLE 7.2 How many covalent bonds are formed between the central atom or atoms and other atoms in the following molecules or ions: hydrazine, H_2NNH_2; nitrate, NO_3^-; silicon tetrafluoride, SiF_4?

SOLUTION

H_2NNH_2

$$\begin{array}{cc} H & H \\ H\!:\!\ddot{N}\!:\!\ddot{N}\!:\!H \end{array}$$

Each nitrogen atom forms three bonds with other atoms—two with hydrogen atoms and one with the other nitrogen atom.

NO_3^-

$$\begin{array}{c} :\ddot{O}: \\ \ddot{N}::O: \\ :\ddot{O}: \end{array}$$

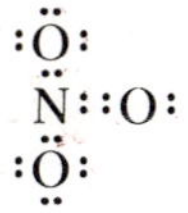

The nitrogen atom forms four bonds with the oxygen atoms.

SiF_4

$$\begin{array}{c} :\ddot{F}: \\ :\ddot{F}\!:\!\ddot{S}i\!:\!\ddot{F}: \\ :\ddot{F}: \end{array}$$

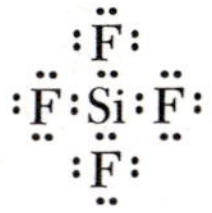

The silicon atom forms a bond with each fluorine atom for a total of four bonds.

STUDY PROBLEM 7(b)

How many covalent bonds are formed between the central atom or atoms and other atoms in the following species: (a) CH_3OH; (b) SO_4^{2-}; (c) $SnCl_5^-$? Draw the Lewis dot structure of each species.

Let us now use the octet rule to see how many hydrogen atoms are expected to combine with each of the following kinds of atoms to form stable species—carbon, nitrogen, oxygen, and fluorine:

$$\cdot \dot{\underset{\cdot}{C}} \cdot \quad + \; 4H \cdot \; \longrightarrow \; \begin{array}{c} H \\ H\!:\!\overset{\cdot\cdot}{\underset{\cdot\cdot}{C}}\!:\!H \\ H \end{array}$$

needs 4 electrons to complete octet

CH_4
methane

$$:\!\dot{N}\!\cdot \quad + \; 3H \cdot \; \longrightarrow \; \begin{array}{c} H \\ :\!N\!:\!H \\ H \end{array}$$

needs 3 electrons

NH_3
ammonia

$$:\!\ddot{O}\!\cdot \quad + \; 2H \cdot \; \longrightarrow \; \begin{array}{c} :\!\ddot{O}\!:\!H \\ H \end{array}$$

needs 2 electrons

H_2O
water

$$:\!\ddot{F}\!\cdot \quad + \; H \cdot \; \longrightarrow \; H\!:\!\ddot{F}\!:$$

needs 1 electron

HF
hydrogen fluoride

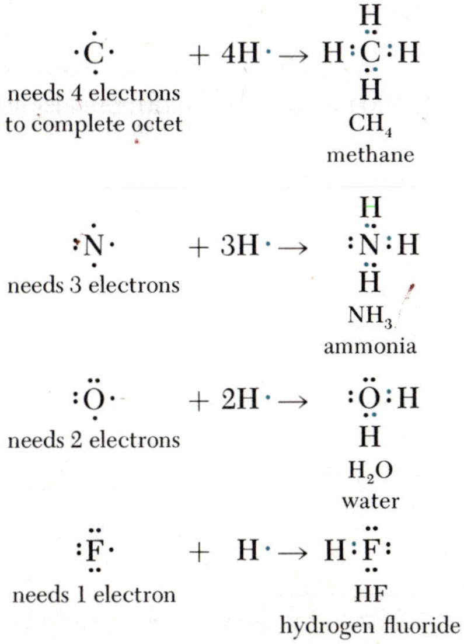

Some Atoms Combine with Different Numbers of Atoms of Another Element

We know that carbon and oxygen can combine to form carbon monoxide, CO, a poisonous gas that prevents our blood from using oxygen, and carbon dioxide, CO_2, which is a nontoxic product of expiration in animals and very useful in fire extinguishers or as the cooling agent "dry ice." Faced with this kind of apparent ambiguity in nature, we must be concerned whether our idea of valence is valid. Can we really apply the octet rule with confidence?

In answering these kinds of questions, let us consider the structure of CO. In terms of the ideas presented in the preceding section, we can visualize CO as

$$\cdot\dot{\underset{\cdot}{C}}\cdot + :\ddot{\underset{\cdot}{O}}\cdot \rightarrow :C::O: \qquad \text{or} \qquad :C{\equiv}O:$$

In the symbol on the right we have adopted the customary convention of representing a shared electron pair by a line, signifying one covalent bond. In this case, the carbon atom gains a share of four electrons from the oxygen atom by covalent sharing and the oxygen atom gains a share of the two electrons it needs to complete its octet. The net result is three shared electron pairs—three covalent bonds, represented by three lines in the formula, $:C{\equiv}O:$. (Often a formula of this type is drawn simply as $C{\equiv}O$, showing explicitly only those electrons counted in the covalent bonds.) Thus, the bond between the C and O atoms is made up of three pairs of electrons; and it is called a **triple bond.**

In CO_2, the electronic structure is written as

$$\cdot\dot{\underset{\cdot}{C}}\cdot + 2:\ddot{\underset{\cdot}{O}}\cdot \rightarrow :\ddot{O}::C::\ddot{O}: \qquad \text{or} \qquad O{=}C{=}O$$

Again, both the carbon and oxygen atoms can complete their valence shells by forming bonds consisting of more than just one pair of electrons. Both of the carbon-oxygen bonds contain two pairs of electrons and are therefore called double bonds.

EXAMPLE 7.3 What is the electronic structure of water?

SOLUTION Water, H_2O, can be visualized as arising from atomic oxygen, having six valence electrons, and two hydrogen atoms, each of which has only one electron. It is readily seen that the octet rule can be satisfied for oxygen if a covalent bond to each of two hydrogen atoms is formed:

$$:\ddot{\underset{\cdot\cdot}{O}}:H, \quad \underset{}{H} \qquad \text{or} \qquad \begin{array}{l} O{-}H \\ | \\ H \end{array}$$

EXAMPLE 7.4 What are the molecular and electronic structures of carbonic acid, H_2CO_3?

SOLUTION Although one can find exceptions, a generally useful rule of thumb in such cases is that a hydrogen atom will occupy a terminal position, since it is covalently bonded to only one atom,* and a carbon atom will be more central in the molecule, with four

*The main exceptions to this guideline come in boron compounds, as discussed in Chap. 9.

covalent bonds.° Then, recalling that atomic oxygen has six electrons in its valence shell and needs to participate in two covalent bonds to complete its octet, we are drawn to the following Lewis dot structure:

$$\text{H}:\ddot{\text{O}}:\overset{:\ddot{\text{O}}:}{\underset{}{\text{C}}}:\ddot{\text{O}}:\text{H}$$

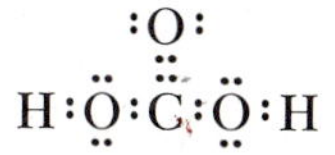

and the more concise formula,

$$\text{H}-\text{O}-\overset{\overset{\text{O}}{\|}}{\text{C}}-\text{OH}$$

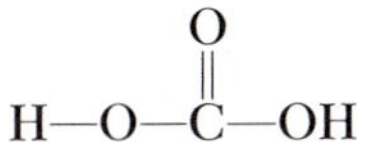

Some other structures that would also satisfy the octet rule for oxygen and carbon in Example 7.4 are

$$\text{H}-\text{O}-\text{O}-\overset{\overset{\text{H}}{|}}{\text{C}}=\text{O}, \quad \overset{\text{H}}{\underset{\text{H}}{\diagup}}\text{C}\overset{\text{O}}{\underset{\text{O}}{\diagup}} \quad \text{and} \quad \overset{\text{H}}{\underset{\text{H}-\text{O}}{\diagup}}\text{C}=\text{O}\diagdown\text{O}$$

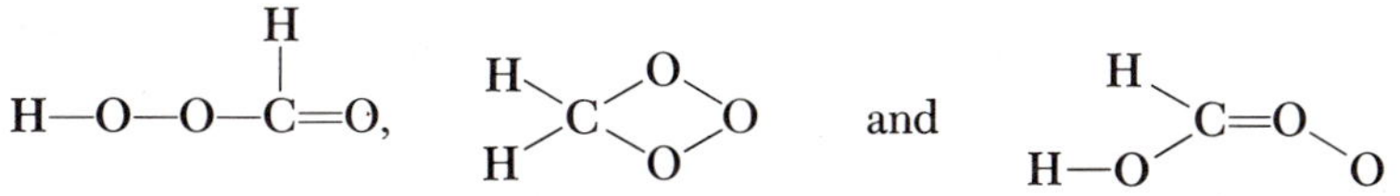

From experimental evidence, these possibilities can be ruled out. For example, oxygen-oxygen single bonds are especially weak.

Resonance

A third carbon-oxygen combination of great importance is the carbonate ion, CO_3^{2-}, one of the ions of importance in biological and geological systems. Noting that the carbonate ion has a -2 electric charge, corresponding to two more electrons than the total number available from a combination of neutral carbon and oxygen atoms, we can imagine the formation of CO_3^{2-} in the following fashion:†

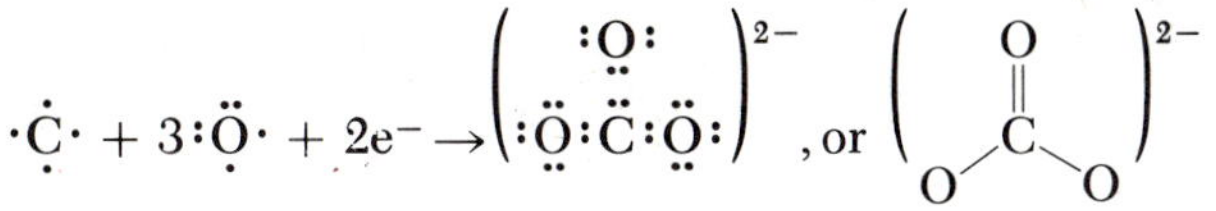

$$\cdot\dot{\text{C}}\cdot + 3\,:\!\ddot{\text{O}}\!\cdot + 2e^- \rightarrow \left(:\!\ddot{\text{O}}:\overset{:\ddot{\text{O}}:}{\text{C}}:\ddot{\text{O}}:\right)^{2-}, \text{ or } \left(\overset{\overset{\text{O}}{\|}}{\underset{\text{O}\diagup\diagdown\text{O}}{\text{C}}}\right)^{2-}$$

$$\text{I}$$

The carbon-oxygen bonds of the species consist of two single bonds and one double bond. <mark>A double or triple bond is stronger and shorter than a single bond</mark>, and it is

---

°The main exceptions to this rule of thumb are found in a class of compounds called isocyanides, or isonitriles. A typical example of these compounds is methyl isocyanide,

$$\text{H}:\overset{\overset{\text{H}}{|}}{\underset{\text{H}}{\text{C}}}:\ddot{\text{N}}:::\text{C}:$$

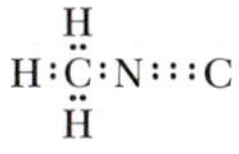

†In the CO_3^{2-} example, we have noted that for anions we must add an extra electron or electrons in the Lewis dot structure of the species. For cations (say NH_4^+), we must subtract one electron or more—corresponding to the electron deficiency (relative to total nuclear charges) implicit in the cation's charge. For the NH_4^+ case, we can write a hypothetical process for the formation of the species:

$$:\!\text{N}\cdot + 4\text{H}\cdot - 1e^- \rightarrow \text{H}:\overset{\overset{\text{H}}{|}}{\underset{\text{H}}{\ddot{\text{N}}}}:\text{H} \quad +$$

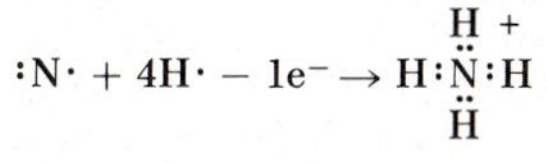

possible to distinguish these differences experimentally.* Experiments conducted on CO_3^{2-}, however, show that all three bonds are the same, contrary to what one expects from the above Lewis dot formula! Furthermore, these experiments show that these three bonds have properties intermediate between those expected for a C–O single bond and those for a C–O double bond. These observations are explained by resonance theory.

Quite possibly the reader may wonder at this point if there wasn't a considerable degree of arbitrariness in drawing structure I above for the carbonate ion. Why should the double bond point to the upper oxygen? Wouldn't it be just as reasonable to have the oxygen atom on the left or the one on the right be connected by a double bond? Indeed, there is no basis for making such a choice, and each of the following three structures is equally reasonable:

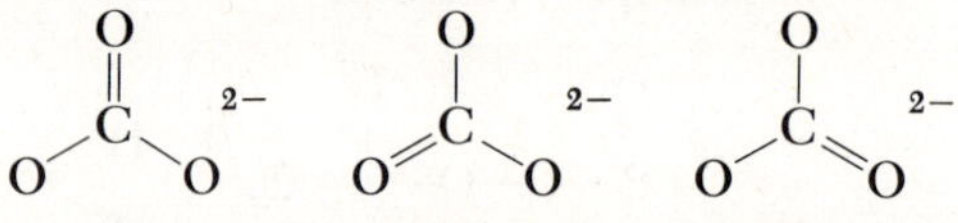

There is no reason for the carbon-oxygen double bond to point to any one of the three oxygen atoms instead of either of the others. Indeed, one finds, as indicated above, that the chemical and physical properties of the carbonate ion show that all three carbon-oxygen bonds are equivalent. Furthermore, these bonds appear intermediate in many respects between a C–O single bond and a C–O double bond. We can think of the situation as follows. The actual electronic structure of CO_3^{2-} doesn't correspond to any one of the three structures, but seems to be some kind of combination of all three simultaneously. This kind of situation is referred to as **resonance**, and is represented symbolically by

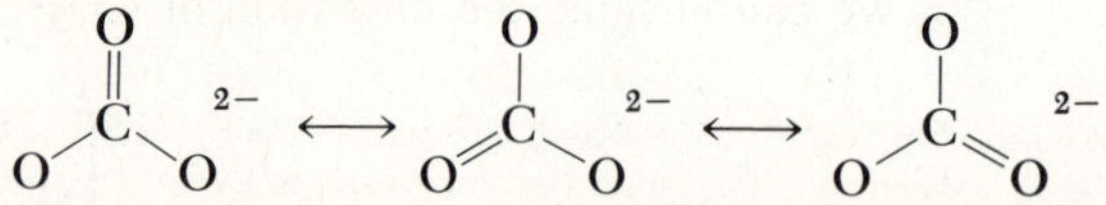

The structures I, II, and III are called **resonance forms** or **resonance contributors.** None exists as an entity, even for an instant, but for all three the character is contained in the actual species. That kind of species is referred to as a **resonance hybrid** of the contributing resonance forms.

An interesting and important feature of resonance is that whenever it occurs, it manifests itself as an increase in stability, a lowering of energy. For example, the carbonate ion, since it is a resonance hybrid, is more stable than one would have predicted by viewing it as only one of the contributing structures. This enhanced stability is especially large in cases like CO_3^{2-}, in which all the contributing resonance structures are equivalent. An example in which the resonance forms are not

*It is often possible to distinguish between single, double, and triple bonds linking two atoms by means of chemical reactivity differences, for example, the relative tendencies toward reaction with a specific reagent. Spectroscopic techniques that observe the vibrational characteristics of the bonds in molecules are also capable of making these distinctions (see Chap. 8).

equivalent is the azide ion, N_3^-, which is known to be linear. Recognizing that there are 16 valence electrons in this species (5 from each neutral N atom, plus 1 extra to give the -1 charge), we can write the following resonance structures:

$$:\!N{\equiv}N{-}\ddot{\underset{..}{N}}\!:^- \longleftrightarrow\ :\!\ddot{N}{=}N{=}\ddot{N}\!:^- \longleftrightarrow\ :\!\underset{..}{\ddot{N}}{-}N{\equiv}N\!:^-$$

IV V VI

EXAMPLE 7.5 Explain the great stability of benzene, a compound with formula C_6H_6 and constructed with the carbon atoms arranged in a regular hexagon, each with one hydrogen atom attached.

SOLUTION The basic outline of the molecule, according to the above information, is

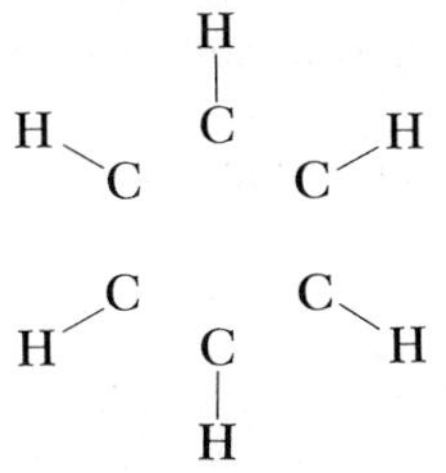

There are 30 valence electrons (4 from each carbon atom and 1 from each hydrogen) to be distributed among the bonds, and 12 have already been accounted for in the 6 C–H bonds. Of the remaining 18 electrons, we shall allot 12 of them to 6 C–C single bonds, yielding the following framework, with 6 electrons still not used:

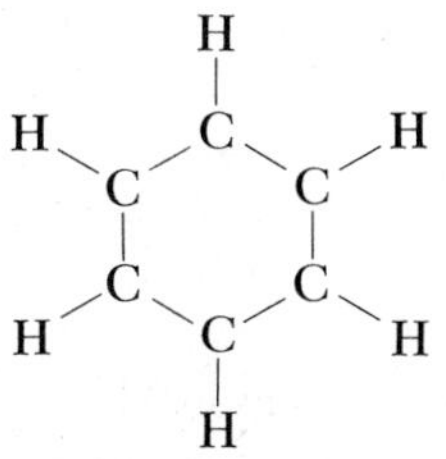

One notes that none of the carbon atoms in this framework meets the octet criterion (each has only 6 electrons in its valence shell, counting each shared electron pair with both atoms that share it); and we should like to allot the remaining 6 electrons so as to satisfy the octet rule for all carbon atoms. This can be achieved by including the 6 electrons as 3 additional covalent bonds, as follows:

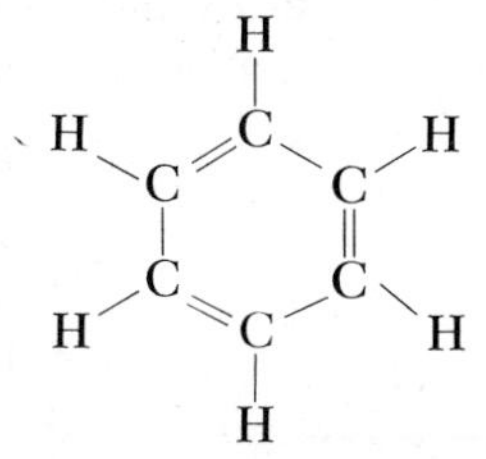

VII

It is readily seen that the octet rule is satisfied for all 6 carbon atoms in VII. However, an equally reasonable structure would have the following alternative arrangement of double bonds:

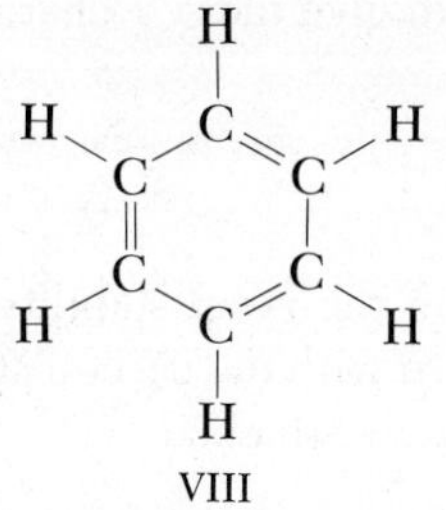

VIII

There is no reason to favor one of these structures over the other, and we conclude that this is a case of resonance: VII ↔ VIII.

EXAMPLE 7.6 Rationalize the acidity of acetic acid, which can be represented by the formula

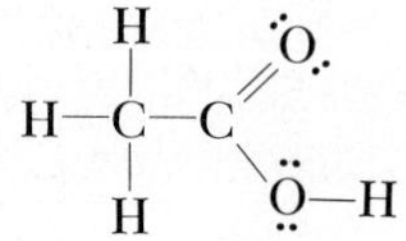

SOLUTION The acidic behavior of acetic acid can be represented by the chemical reaction $CH_3CO_2H \rightleftharpoons CH_3CO_2^- + H^+$ and the acidity of CH_3CO_2H is largely owing to the stability of the acetate ion $CH_3CO_2^-$. If one removes H^+ from the formula of acetic acid given above, one is left with

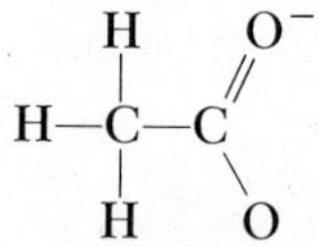

IX

An equally reasonable formula can be written if the choice of the C–O double and single bonds is reversed:

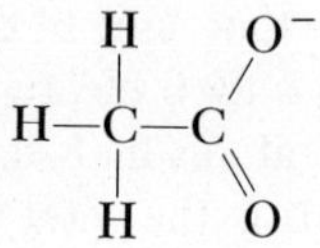

X

Since structures IX and X are equivalent, we recognize this as a prime example of resonance, described in terms of the resonance hybrid:*

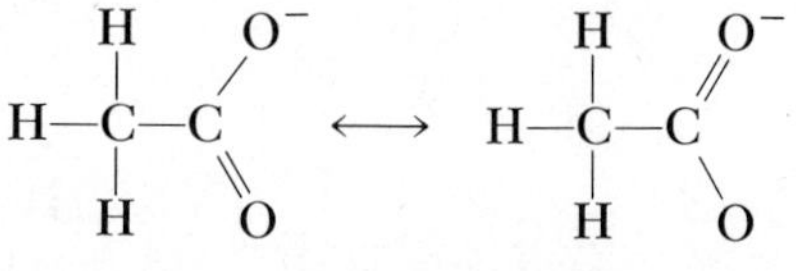

*The word *equivalent* in the sense of IX and X means essentially indistinguishable (without putting labels on each atom) and having equivalent energy. Thus, if IX and X existed separately (they don't!), they would have the same energy, chemical properties, and so on. Furthermore, one could not distinguish between them unless somehow the oxygen atoms could be labeled (note that if one rotates X it looks just like IX).

Associated with this resonance is a marked stability of the acetate ion, which is reflected in the acidity of the parent acid.

STUDY PROBLEM 7(c)

What are the resonance contributors of the nitrate ion, NO_3^-?

7.3 Importance of Molecular Geometry

General Shapes

Covalent molecules are characterized by their existence as discrete molecules; in methane gas, for example, we find four specific hydrogen atoms associated with one particular carbon atom. These discrete molecules are found to have definite shapes. The simplest type of covalent molecule is a diatomic molecule, for instance, HCl, N_2, O_2, in which the molecular shape is obvious. The next simplest arrangement occurs if there is one central atom to which several other atoms are attached. Other covalent molecules have other regular geometric figures. Presented herewith are some typical geometrical arrangements, in which an atom lies at the center of the

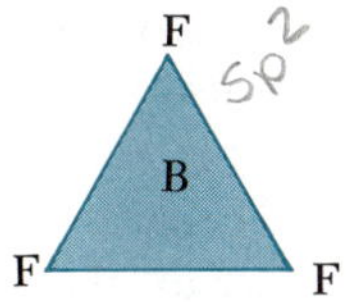

equilateral triangle

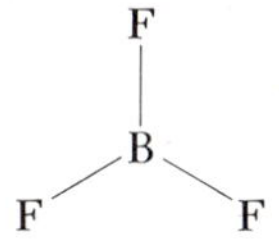

boron trifluoride

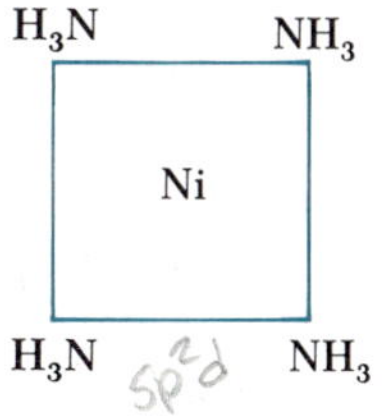

square

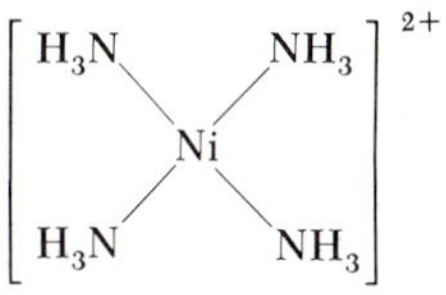

tetraamminenickel ion

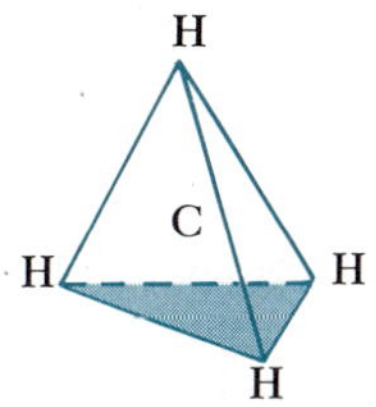

tetrahedron

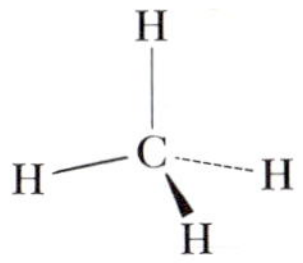

methane

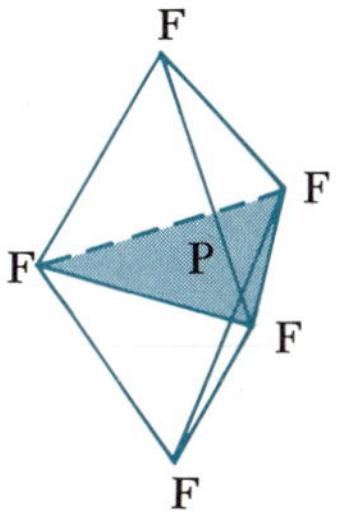

trigonal bipyramid

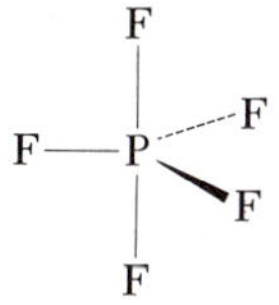

phosphorus pentachloride

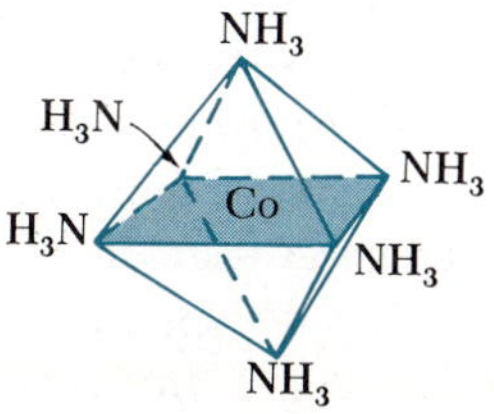

octahedron

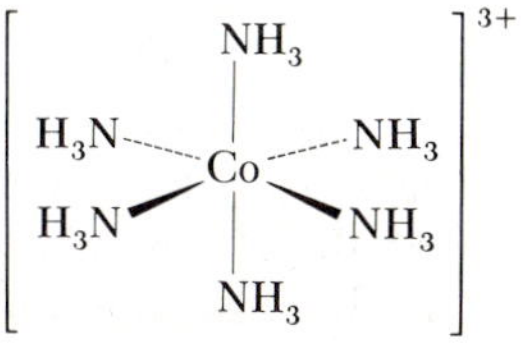

hexaamminecobalt ion

geometric figure and the other atoms lie at the vertices of the figure. Complicated cases, which may have some vertices unoccupied or which may be constructed by joining several simple geometric figures, will be investigated later.

Hybridization

The boron trifluoride molecule, BF_3, is an equilateral triangle, as illustrated above. The electronic structure of BF_3 can be viewed in terms of the hypothetical formation from the elements as follows:

$$\underset{2s}{\uparrow\uparrow}\ \underset{2p}{\uparrow\ \_\ \_}\ \xrightarrow{promotion}\ \underset{2s}{\uparrow}\ \underset{2p}{\uparrow\ \uparrow\ \_}$$

1 boron atom $[He]2s^22p^1$ → 1 boron atom $[He]2s^12p^2$

$$\underset{2s}{\uparrow}\ \underset{2p}{\uparrow\ \uparrow\ \_}\ +\ \underset{2s}{\uparrow\downarrow\ \uparrow\downarrow\ \uparrow\downarrow}\ \underset{2p}{\downarrow\ \_\ \_}\ \longrightarrow\ \underset{2s}{\overset{F\ F\ F}{B\uparrow\downarrow\ \uparrow\downarrow\ \uparrow\downarrow}}\ \underset{2p}{\_}$$

1 boron atom $[He]2s^12p^2$ + 3 fluorine atoms $[He]2s^22p^5$ → 1 boron trifluoride molecule BF_3

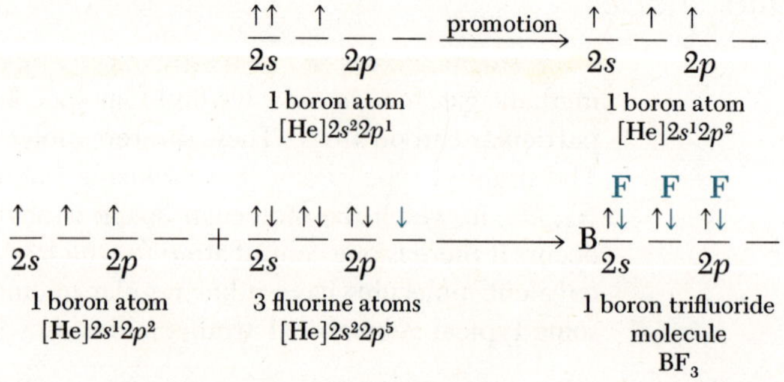

The boron promotion, an energy-consuming step, can be rationalized by the increased number of fluorine-boron covalent bonds that it makes possible (three bonds, in comparison to one that would be predicted in the absence of promotion). Each fluorine atom has one unpaired electron (a $2p$ electron) available for sharing with an unpaired electron of boron, three of which are available in the promoted state. It can be noted that the formula given for BF_3 shows that the octet rule is satisfied for each fluorine, but not for boron; the electron deficiency of boron in this compound is responsible for most of its chemical properties.

$$\cdot\ddot{B}\cdot\ +\ 3:\ddot{F}\cdot\ \longrightarrow\ \begin{matrix}:\ddot{F}:\\:\ddot{F}:B\\:\ddot{F}:\end{matrix}$$

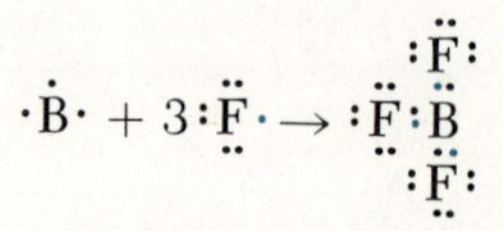

At this point, we have presented a method of explaining why boron compounds form more than one bond. Let us now investigate the shape of a boron compound such as BF_3. If we construct this molecule using the $2s$ orbital and two of the $2p$ orbitals on the boron atom, the shape of the molecule would be that shown in Figure 7.9b. The angle between two of the fluorine atoms would be 90° because they would be bound to the $2p$ orbitals of the boron atom, which are perpendicular to each other. The third fluorine atom would be bound to the $2s$ orbital. Since this orbital is spherical, the fluorine atom could bind anywhere about the boron and would be expected to get as far as possible from the other fluorine atoms (because of electron–electron repulsion), leading to the 135° angles shown in Figure 7.9b. This is *not* the shape of BF_3, however. The BF_3 molecule is a regular triangle; all F—B—F angles are 120°!

To explain this observation and clear up the apparent dilemma, a new concept is introduced, called **hybridization**. According to this concept, the three wave functions that describe the boron orbitals available for bonding in the promoted state

Figure 7.9
The structure of BF_3. (a) The experimentally determined arrangement. (b) A geometrical arrangement showing unhybridized orbitals and the predicted bonding to fluorine. (c) A geometrical arrangement showing boron atomic orbitals hybridized in the sp^2 fashion, which is also consistent with the experimentally determined bond orientations.

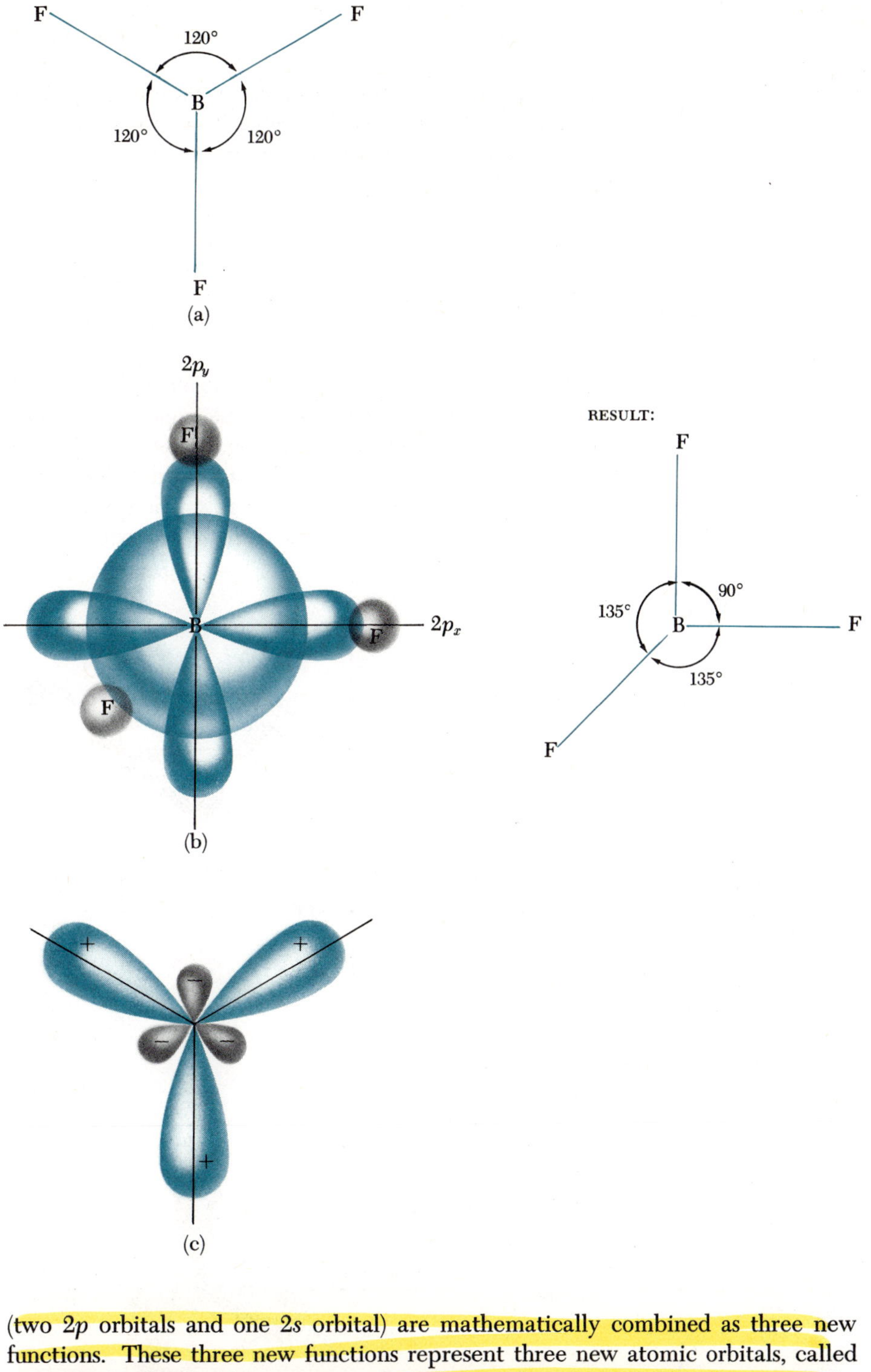

(two $2p$ orbitals and one $2s$ orbital) are mathematically combined as three new functions. These three new functions represent three new atomic orbitals, called sp^2 hybrid orbitals (the superscript 2 indicates that two $2p$ orbitals go into the hybridization); these three orbitals are equivalent (same size, shape, energy) and are directed so that their axes lie in a common plane, separated by 120° angles. These sp^2-hybridized orbitals are depicted in Fig. 7-9c. It is then readily seen how covalent bonding involving these three boron orbitals gives rise to the observed trigonal structure of Figure 7.9a. The situation can be summarized in the following hypothetical scheme for formation of BF_3 from atoms:

$$\mathrm{B}\underset{2s}{\underline{\uparrow\downarrow}}\ \underset{2p}{\underline{\uparrow}\ \underline{\ \ }\ \underline{\ \ }}\ \xrightarrow{\text{promotion}}\ \mathrm{B}\underset{2s}{\underline{\uparrow}}\ \underset{2p}{\underline{\uparrow}\ \underline{\uparrow}\ \underline{\ \ }}$$

$$\xrightarrow{\text{hybridization}}\ \mathrm{B}\underset{sp^2}{\underline{\uparrow}\ \underline{\uparrow}\ \underline{\uparrow}}\ +\ 3\mathrm{F}\underset{2s}{\underline{\uparrow\downarrow}}\ \underset{2p}{\underline{\uparrow\downarrow}\ \underline{\uparrow\downarrow}\ \underline{\downarrow}}\ \longrightarrow\ \mathrm{B}\overset{\mathrm{F}}{\underline{\uparrow\downarrow}}\ \overset{\mathrm{F}}{\underline{\uparrow\downarrow}}\ \overset{\mathrm{F}}{\underline{\uparrow\downarrow}}$$

The concept of hybridization is by no means limited to BF_3, or even to the trigonal sp^2 situation. It is a broadly applicable concept that finds its way into most qualitative discussions of covalent bonding. Indeed, we have already explored one case, CH_4, which is a classic example of hybridization. In that case, it was seen that promotion of one electron from the $2s$ orbital into a $2p$ orbital provided four unpaired electrons for covalent bonding to the hydrogen atoms. At this point we can

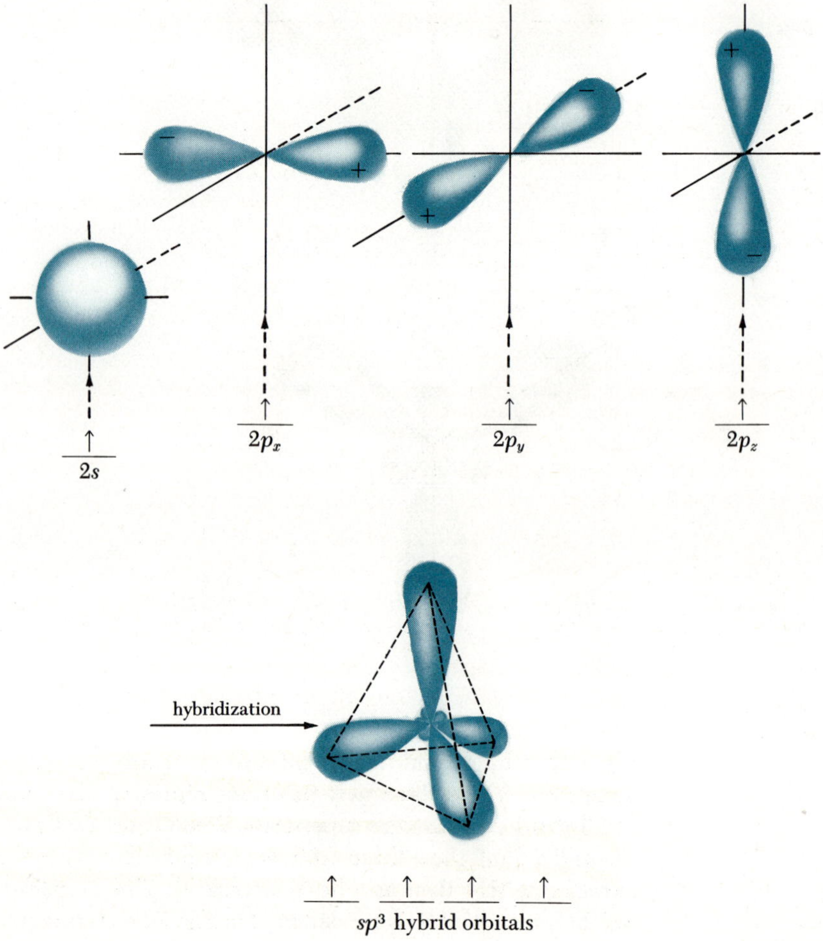

Figure 7.10
*sp*³ hybridization for the carbon atom in CH_4, showing four available hybrid orbitals (each containing one electron) for covalent bonding to four hydrogen atoms.

TABLE 7.2 **Common Hybridization Schemes to Produce Known Molecular Geometries**

Number of Electrons Usually Involved	*Orbitals Combined*	*Hybrid Orbital Type*	*Arrangement of Hybrid Orbitals*[a]	*Example*
2	s,p_z	sp	Linear	F—Be—F
3	s,p_x,p_y	sp^2	Equilateral-triangular (trigonal)	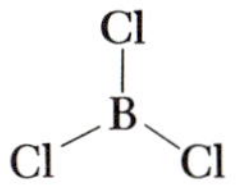
4	s,p_x,p_y,p_z	sp^3	Tetrahedral (Bent)	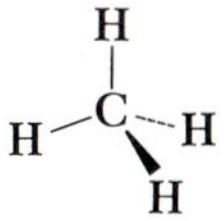
4	$s,p_x,p_y,d_{x^2-y^2}$	sp^2d	Square	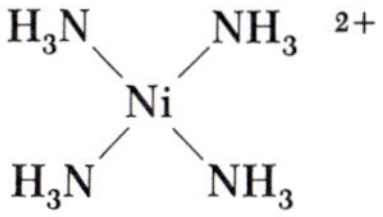
5	s,p_x,p_y,p_z,d_{z^2}	sp^3d	Trigonal bipyramidal	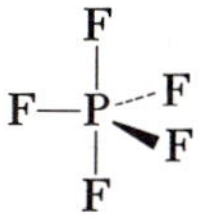
6	$s,p_x,p_y,p_z,d_{z^2},d_{x^2-y^2}$	sp^3d^2	Octahedral	

[a]This is not necessarily the molecular geometry, which is specified by the positions of the atoms and not by the positions of the electron pairs; for example, the oxygen atom in water has sp^3 hybridization, but the molecule is referred to as bent, not tetrahedral.

provide some additional details related to rationalizing the tetrahedral geometry observed for methane. If the wavefunctions describing the 2*s* and three 2*p* orbitals used in the promoted state are "mixed," four new functions describing what are called sp^3 hybrid orbitals can be formed; the superscript 3 indicates that three *p* orbitals go into the "mix". The four hybrid orbitals are equivalent in size, shape, and energy and are directed towards the four centers of a regular tetrahedron with carbon at its center. This is depicted in Fig. 7.10. In terms of these hybrid orbitals, it is easy to see that the structure of methane is tetrahedral, with one hydrogen atom involved in covalent bonding to each of the sp^3 hybrid orbitals.

Table 7.2 provides a summary of common hybridization schemes involving *s*, *p*, and *d* orbitals. This summary, which includes the BF_3 and CH_4 examples described above, shows the number and types of orbitals from which they are constructed and an example of an actual chemical species for each hybridization scheme. The most common use of hybridization theory is describing bonding in organic molecules.

EXAMPLE 7.7 What are the hybrid orbitals that would be assigned to bonds in the following molecules: SiH_4, NH_3, and phosphorus pentafluoride, PF_5?

SOLUTION The first thing to do is to construct Lewis dot formulas for each species. Then the orbitals of the valence shell can be "mixed" into hybrid orbitals.

SiH_4 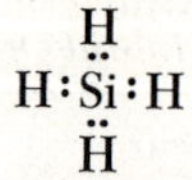 for Si, 4 valence orbitals: s, p_x, p_y, and p_z—sp^3 hybridization

NH_3 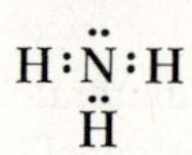 for N, 4 valence orbitals: s, p_x, p_y, and p_z—sp^3 hybridization

PF_5  for P, 5 valence orbitals: s, p_x, p_y, p_z and 1 d orbital—sp^3d hybridization

STUDY PROBLEM 7(d)

Assign hybrid orbitals to the central atom or atoms in hydrazine, H_2NNH_2; $SnCl_5^-$; arsine, AsH_3.

Predicting Molecular Geometries by the VSEPR Theory

The valence shell electron pair repulsion theory, or VSEPR, is a very successful model for predicting molecular shapes. Sidgwick and Powell suggested this model in 1940 and R. J. Gillespie has expanded and improved it in recent years. The VSEPR theory is based upon the hypothesis that the atoms in a covalent molecule are arranged in a manner that minimizes the electrostatic repulsion between electron pairs. Although there are some features of molecular structure that are not explained by this model, it is generally easier to use and provides more information than hybridization theory.

The VSEPR theory rests on the following several assumptions:

1. A molecule takes a shape that tends to minimize electrostatic repulsions (Table 7.3).
2. The electrostatic repulsions in the molecule (and hence its shape) will be determined by all electrons in the valence shell, including nonbonding pairs of electrons.
3. The repulsions of nonbonding pairs of electrons with other pairs of electrons will be stronger than the corresponding repulsions of bonding pairs. Thus, two nonbonding pairs will repel each other more than a nonbonding pair and a bonding pair and these, in turn, more than two bonding pairs of electrons.
4. The designation of the shape of the molecule will refer to the positions of the atoms of the molecule, not the nonbonding pairs (lone pairs) (see Table 7.3).

To see some examples of molecular shapes, as predicted by VSEPR theory, let us consider the covalently bonded compound of hydrogen with carbon, nitrogen, oxygen, and fluorine.

CH_4, Methane

All the carbon valence electrons are in bonding pairs. These electron pairs repel each other equally. The way in which these repulsions can be minimized is by placing the four electron pair bonds as far apart from each other as possible. A geometrical arrangement that places four bonding pairs as far apart as possible, as shown

TABLE 7.3 Shapes of Molecules as Determined by the VSEPR Method

Number of Electron Pairs	Arrangement of Least Repulsion	Shape of Electron Pairs	Examples	Shape of Molecule	
2		linear sp	BeH_2	linear	H—Be—H
3		triangular sp^2 (trigonal)	BF_3	triangular	F—B(—F)(—F)
4		tetrahedral sp^3	CH_4	tetrahedral	H—C(H)(H)(H)
			NH_3	pyramidal	H—N(H)(H)
			H_2O	bent	H—O—H
5		trigonal bipyramidal sp^3d	PF_5	trigonal bipyramidal	F—P(F)(F)(F)(F)
			SF_4	irregular	S(F)(F)(F)(F)
6		octahedral sp^3d^2	SF_6	octahedral	F—S(F)(F)(F)(F)(F)
			BrF_5	square pyramidal	Br(F)(F)(F)(F)
			XeF_4	square planar	Xe(F)(F)(F)(F)

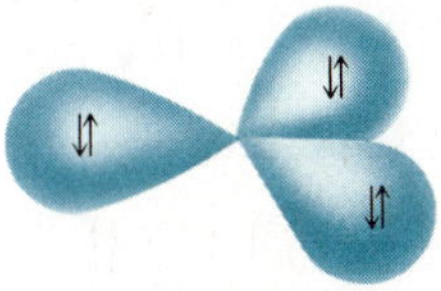

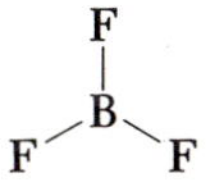

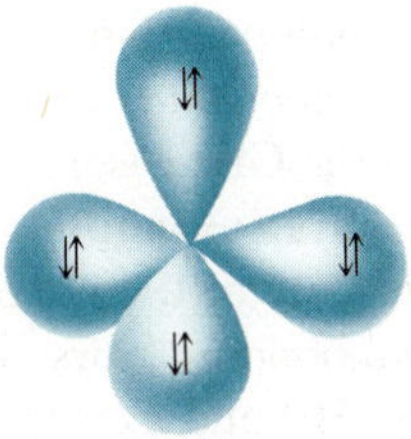

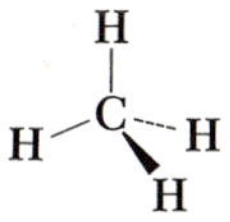

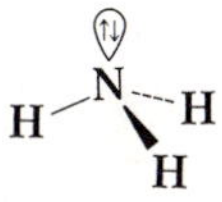

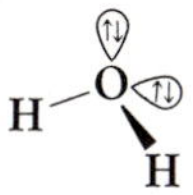

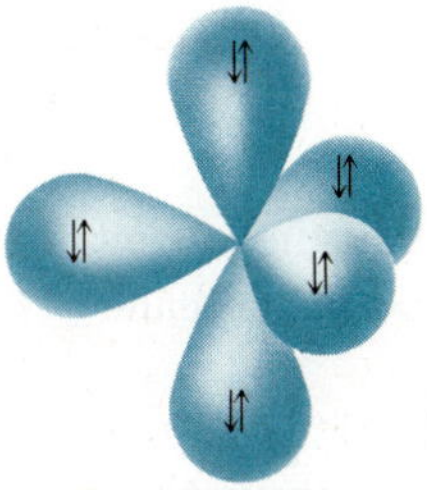

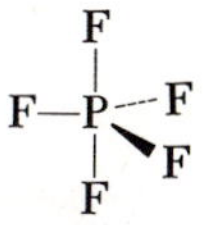

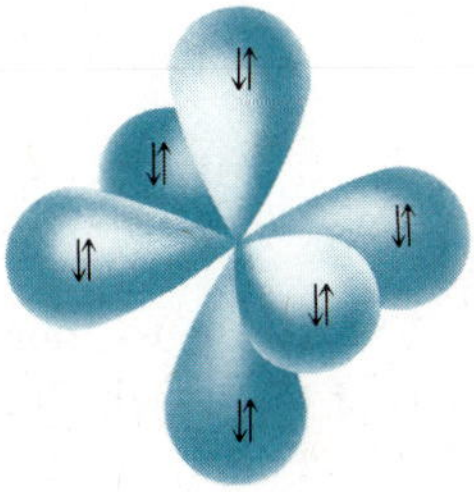

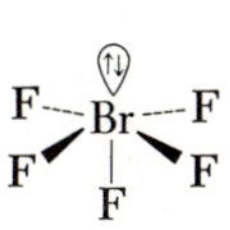

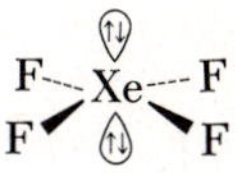

in Table 7.3, is the tetrahedron. Hence, this is the structural arrangement expected of methane, in agreement with the hybridization view. As you may recall from Fig. 3.1, the dotted line represents bond axis behind the plane of the paper, the wedged line represents a bond axis in front of the plane of the paper, and solid lines represent a bond axis in the plane of the paper.

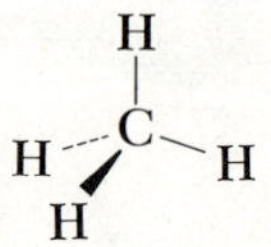

NH_3, Ammonia

Considering the Lewis dot structure,

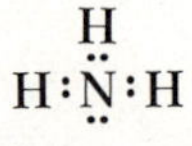

one sees that NH_3 has one lone pair and three bonding pairs on the nitrogen atom. Thus, there is a total of four electron pairs on nitrogen (the octet), and one can take a regular tetrahedron (as in the CH_4 case) as a first guess of the molecular geometry. We should recall, however, from assumption 3 above that the repulsions of the lone pair (nonbonding pair) with any of the bonding pairs are stronger than the mutual repulsions between any two bonding pairs. Hence, the angle between the lone pair axis and any one of the N–H bond axes (angle *a*) will be slightly larger than the H–N–H bond angle (angle *b*).

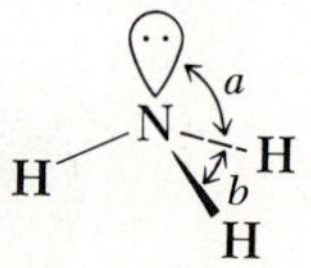

Experimentally, it is found that the angle *b* is 107.3°, slightly smaller than the tetrahedral angle of 109.5° found in methane.

H_2O, Water

From the Lewis dot formula H:Ö:H, one sees that there are two bonding electron pairs and two lone pairs on oxygen in this species. Hence, one chooses a tetrahedral structure as the first guess of a molecular arrangement.

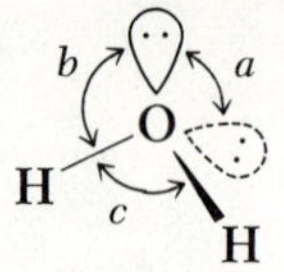

From assumption 3 of the VSEPR theory, one expects the angles to be in the order $a > b > c$. Hence, instead of the tetrahedral angle of 109.5°, the experimentally determined value of the H–O–H bond angle (c) is 105.2°. This angle is smaller than the H–N–H bond angle in ammonia, which is consistent with the additional lone pair repulsion present in H_2O, in comparison with NH_3.

HF, Hydrogen Fluoride

The Lewis dot structure, H:F̈: shows three lone pairs and one bonding pair, which according to the VSEPR theory would be situated in a tetrahedral arrangement.

BF₃, Boron Trifluoride

With only one bond in the molecule, however, the arrangement of atoms must be linear in any case.

The Lewis dot formula for boron trifluoride,

$$\text{F} : \ddot{\text{B}} : \text{F}$$
$$\text{F}$$

shows that the octet rule does not apply in this case. In this molecule there are three bonding electron pairs about the boron atom. For three pairs, as seen in Table 7.3, the optimum arrangement is an equilateral triangle, so the structure is the trigonal one that was also rationalized in terms of sp^2 hybridization.

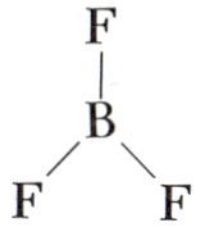

$$\begin{array}{c} \text{F} \\ | \\ \text{B} \\ / \quad \backslash \\ \text{F} \qquad \text{F} \end{array}$$

PF₅, Phosphorus Pentafluoride

From the point of view of the VSEPR theory, one recognizes that there are five bonding electron pairs in PF₅, and they (or the orbitals that contain them) should be arranged to minimize the electrostatic repulsions among the five electron pairs. The geometrical arrangement that accomplishes this is the trigonal bypyramid.

$$\begin{array}{c} \text{F}' \\ | \\ \text{F}-\text{P}---\text{F} \\ | \searrow \\ \text{F}' \quad \text{F} \end{array}$$

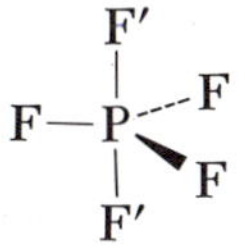

In this arrangement, there are two fluorine atoms (designated F′) which are distinctly different from the other three (designated F). Experimentally, it is found that the P–F′ bond length is greater than the P–F bond length.°

From the point of view of hybridization, one can think of the five atomic orbitals of phosphorus in a promoted state. Promotion results in one electron in each of five orbitals: one 3s orbital, three 3p orbitals and one 3d orbital. These orbitals are then hybridized to a set of five sp^3d orbitals and five fluorine atoms are then combined as shown below:

$$\text{P [Ne]} \underset{3s^2}{\uparrow\downarrow} \underset{3p^3}{\uparrow \quad \uparrow \quad \uparrow} \xrightarrow{\text{promotion}} \text{P [Ne]} \underset{3s^1}{\uparrow} \underset{3p^3}{\uparrow \quad \uparrow \quad \uparrow} \underset{3d^1}{\uparrow}$$

$$\text{P [Ne]} \underset{3s^1}{\uparrow} \underset{3p^3}{\uparrow \quad \uparrow \quad \uparrow} \underset{3d^1}{\uparrow} \xrightarrow{\text{hybridization}} \text{P [Ne]} \underset{sp^3d}{\underset{\text{F F F F F}}{\uparrow\downarrow \ \uparrow\downarrow \ \uparrow\downarrow \ \uparrow\downarrow \ \uparrow\downarrow}}$$

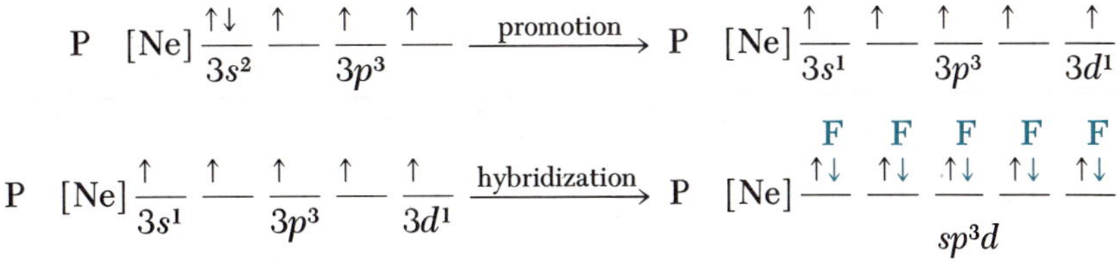

°This difference in bond lengths in the PF₅ case can be rationalized as follows. One expects that mutual electrostatic repulsions between pairs of electrons in two orbitals whose axes make a small angle with each other (like the 90° between the phosphorus orbitals involved in P–F′ and P–F bonds) should be stronger than the corresponding repulsions between pairs of electrons in orbitals with axes separated by a larger angle (like the 120° between phosphorus orbitals involved in two P–F bonds). Hence, the electron pair of each P–F′ bond, with three 90° F–P–F′ bond angles associated with it, is expected to have stronger electrostatic repulsions than each electron pair in a P–F bond, with only two 90° F–P–F′ bond angles. The net result of the increased repulsion of the electron pairs in the P–F′ bonds is to weaken those bonds, with attendant larger P–F′ bond length.

EXAMPLE 7.8 Using the VSEPR method, predict the geometries of $SnCl_5^-$, PF_6^-, BCl_3, OF_2, and HBr.

SOLUTION The starting point is a counting of valence electrons, which is followed by minimization of repulsion, according to Table 7.3.

$SnCl_5^-$

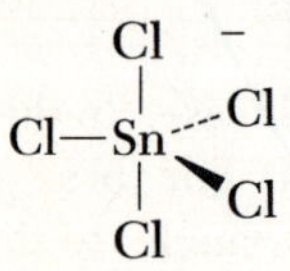

Four valence electrons from the neutral tin atom valence shell, plus 5 electrons provided by sharing with the 5 chlorine atoms and 1 extra electron from the net negative charge (assigned to the tin atom since the chlorine atoms are already assigned 8 electrons each) gives a total of 10 electron pairs. Using Table 7.3, one can predict the shape as a trigonal bipyramid.

PF_6^-

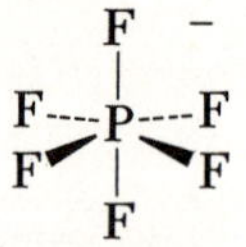

Five valence electrons from the neutral phosphorus atom, plus 6 electrons provided by sharing with the fluorine atoms, plus 1 extra electron from the net charge gives 12 electrons, or 6 pairs. According to Table 7.3, then, the shape is octahedral.

BCl_3

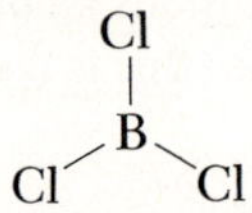

Three valence electrons from the neutral boron atom, plus 3 electrons provided by sharing with the chlorine atoms gives a total of 3 pairs. The shape is triangular.

OF_2

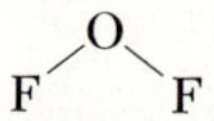

Six valence electrons from the neutral oxygen atom, plus 2 electrons provided by sharing with the fluorine atoms gives a total of 4 pairs. The electron pairs are, according to Table 7.3, arranged as a tetrahedron, but there are only 2 atoms attached to the oxygen central atom. The shape of the molecule, then, is bent.

HBr

Now here's an easy one. No matter how the electrons are arranged, a molecule with only two atoms must be linear.

STUDY PROBLEM 7(e)

Give the geometries of stilbine (SbH_3), methanol (CH_3OH), and PCl_4^+.

7.4 The Functional Groups Approach for Determining Molecular Structure

Overview

Complex molecules can be constructed from simpler molecules by attaching central atoms to one another. This is true in a literal sense; simple molecules can be combined in the laboratory to yield more complex molecules. But it is also true in a figurative sense; that is, complex molecules can be visualized as simpler molecules, or molecular fragments. What are the ideas, then, that make this conceptual method useful?

The conceptual method is analogous to visualizing an object constructed from Tinker Toys, an Erector Set, or Legos in terms of the basic parts of the particular

Spotlight

For many years it was believed that, except for a few simple carbon-containing salts like sodium bicarbonate, compounds containing carbon were all derived from living systems. For that reason, they were referred to as organic compounds, and all other substances were referred to as inorganic. It was believed that some vital force was necessary to produce an organic compound, and that its origin was in life itself. In 1828, the German chemist Wöhler shattered this doctrine by showing that urea, a compound that was universally accepted as organic, could be produced directly from a purely mineral substance, ammonium cyanate, which is inorganic:

$$\underset{\text{ammonium cyanate}}{NH_4OCN} \rightleftharpoons \underset{\text{urea}}{CO(NH_2)_2}$$

Although this evidence against the doctrine of a vital force was not accepted immediately, there is no longer any mystery associated with the conversion of inorganic compounds to organic; it is just a matter of knowledge and skill.

set. Some of the parts are functional as they apply to the structure, essentially holding the object together, whereas other parts are truly functional in a more active sense, like the wheels, pulleys, levers, and cranks. In any case, the entire object can be visualized by thinking of the functional parts from which it is constructed, and the geometrical arrangement of these parts. Similarly, molecules can be visualized by imagining the combination of certain recognizable groups of atoms, called **functional groups.** Some functional groups can be thought of as basically contributing only to the structural backbone of a molecule, whereas other functional groups can be associated with active functions—chemical reactions.*

A functional group is a recognizable group of atoms, recognizable in the sense that essentially the same group occurs in many different molecules; the functional group appears to give to a molecule containing it a particular set of relatively well-defined properties. For example, the carboxyl group imparts a weakly acidic character to any molecule containing this particular arrangement of atoms. Often it is convenient to define a class, or family, of compounds in terms of a common functional group they all contain. For example, all compounds containing the carboxyl group are called **carboxylic acids.** In a first approximation, which will be fairly good, one can say that the properties of a molecule, especially its chemical properties, result from a combination of the individual properties imparted by the functional groups it contains. These properties are often dominated by one functional group, or perhaps a few, of many contained in a complex molecule.

Organic Compounds

Basic Architecture

In the chemistry of carbon compounds, or organic chemistry, the concept of functional group has proved extremely useful. Most organic compounds can be thought of as constructed largely from a framework of carbon and hydrogen atoms, with various functional groups attached. Even the carbon-hydrogen (hydrocarbon) framework itself can be visualized as consisting of functional groups that are connected to each other. Some typical types of hydrocarbon architecture are here considered.

*Strictly speaking, a reaction is something that occurs to an entire molecule, not just to a specific atom or set of atoms within the molecule.

Alkanes

One of the simplest functional groups is the methyl group, $—CH_3$, in which the dash indicates the presence of a covalent bond to some other portion of the molecule that contains the methyl group, in other words, to some other functional group. If a hydrogen atom is attached to a methyl group, then one has methane, CH_4. The hybridization of the carbon atom in a methyl group is sp^3, so that it describes a tetrahedron with three covalent bonds from the central carbon to hydrogen atoms and the remaining orbital involved in a covalent bond to some atom in the rest of the molecule.

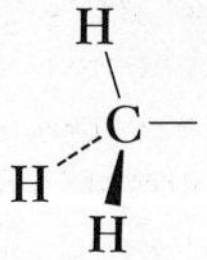

methyl

We can visualize combining this fragment with another methyl group to form a compound called ethane.

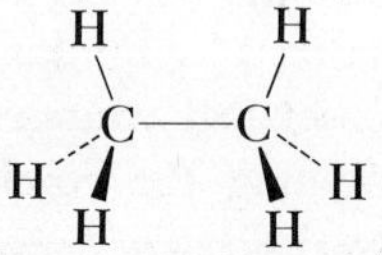

ethane

Just as we can visualize creating a methyl group by mentally removing a hydrogen atom from methane, we can visualize an ethyl group as an ethane molecule without one of its hydrogen atoms. Then, we can imagine combining an ethyl group with a methyl group to give a new molecule, propane.

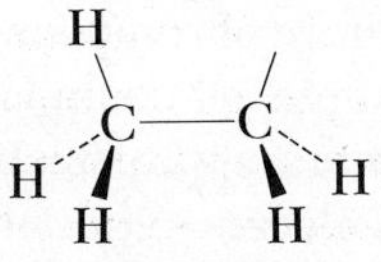

ethyl group

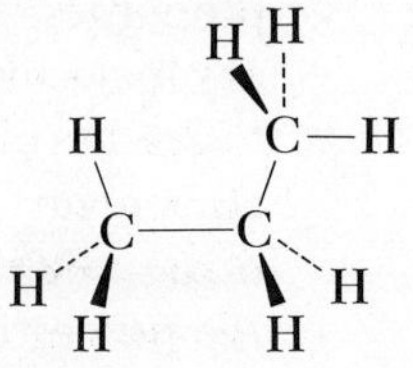

propane

In the same way, we can create a propyl group in our minds by taking away one hydrogen atom from propane; but in this case we perceive that we can visualize two very different types of propyl groups, depending on what carbon atom the hydrogen atom is removed from.

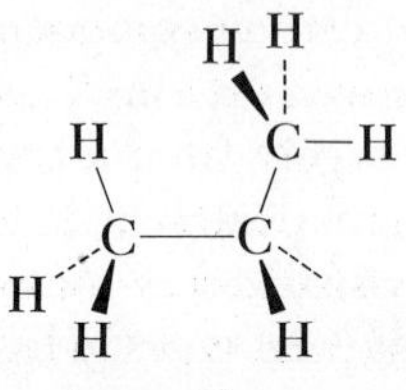

isopropyl group

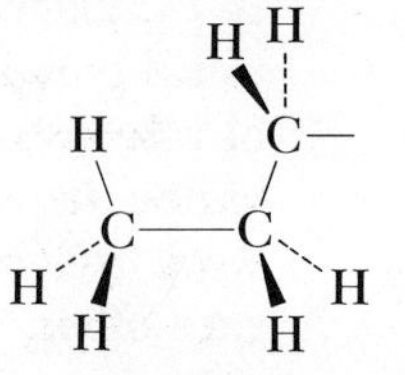

n-propyl group

Then, mentally combining these three-carbon groups, isopropyl and *n*-propyl, with methyl groups, we can imagine constructing two different four-carbon molecules, isobutane and *n*-butane.

[Structural diagrams of isobutane and *n*-butane]

By extrapolating this kind of molecule building to larger cases, by stepwise mental replacement of a hydrogen atom with functional groups (methyl, ethyl, and so on), we perceive how molecules of immense size and complexity are possible. The hydrocarbon molecules of any size that we design in this fashion are referred to as **alkanes**. The functional groups that one visualizes by eliminating a hydrogen atom from an alkane are called **alkyl groups** (for example, methyl, ethyl). In all alkanes, each carbon atom is involved in covalent bonds to four other atoms—hydrogens or other carbons. Many such compounds occur in nature, for example, components of petroleum; and nearly all others that one can imagine can be synthesized.

In considering alkanes, it is cumbersome to draw the detailed structural diagrams shown above. Some simpler kinds of notation have been developed, which serve many purposes. For example, in place of the structural diagram shown above for propane, one often writes

$$\text{H}-\overset{\overset{\displaystyle \text{H}}{|}}{\underset{\underset{\displaystyle \text{H}}{|}}{\text{C}}}-\overset{\overset{\displaystyle \text{H}}{|}}{\underset{\underset{\displaystyle \text{H}}{|}}{\text{C}}}-\overset{\overset{\displaystyle \text{H}}{|}}{\underset{\underset{\displaystyle \text{H}}{|}}{\text{C}}}-\text{H}$$

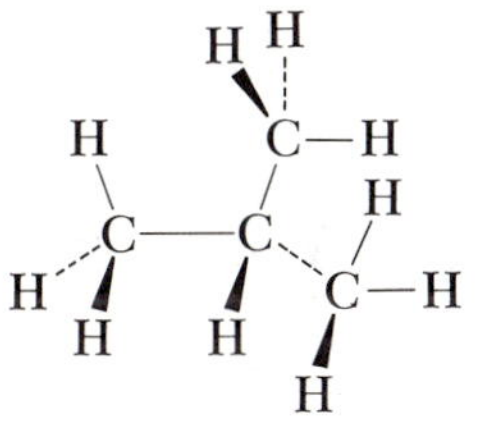

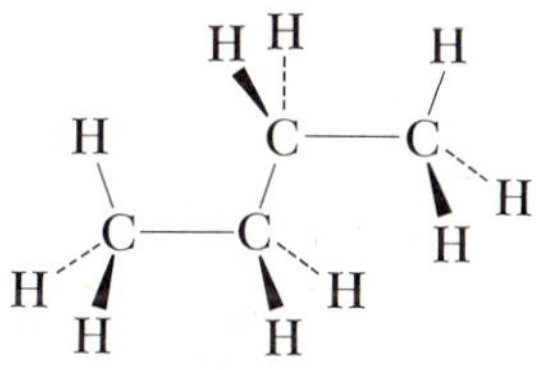

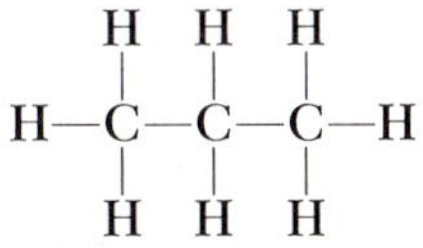

or the even simpler formula $CH_3CH_2CH_3$. These formulas still identify the functional groups that are the building blocks in these molecules, but the three-dimensional perspective is lost. Figure 7.11 summarizes the names, structures, and structural formulas of several alkanes.*

Alkenes

Another structural unit recognizable in many organic compounds is the carbon-carbon double bond. The simplest molecule containing this unit is **ethylene**, with

---

*There may be more than one species with the same molecular formula; such species are called **isomers**. Examples are *n*-butane and isobutane, both of which have the molecular formula C_4H_{10}. Another four-carbon alkane is cyclobutane with the structure

[Structural diagram of cyclobutane]

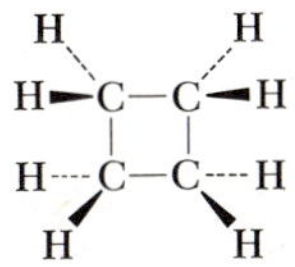

This compound is not an isomer of *n*-butane and isobutane, since it has a different molecular formula, C_4H_8.

methane
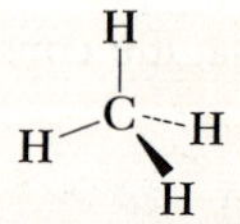

CH_4

ethane
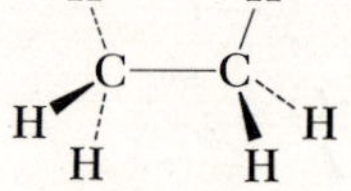

C_2H_6

propane
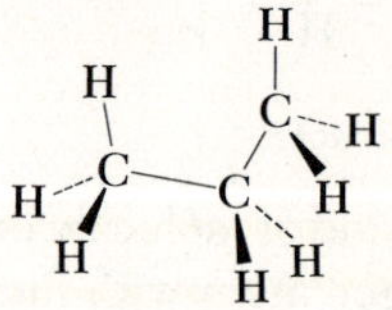

C_3H_8

cyclopropane
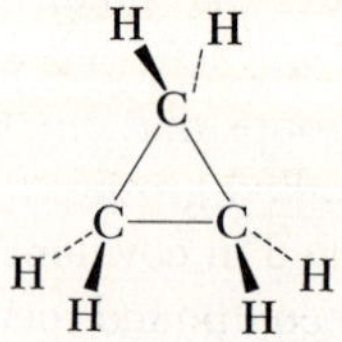

C_3H_6

butane
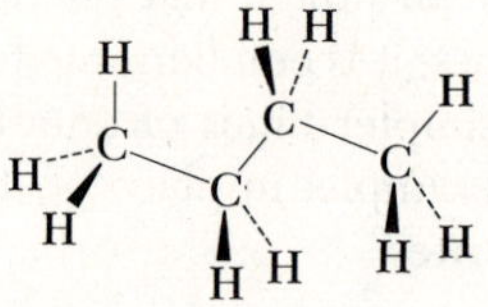

C_4H_{10}

isobutane
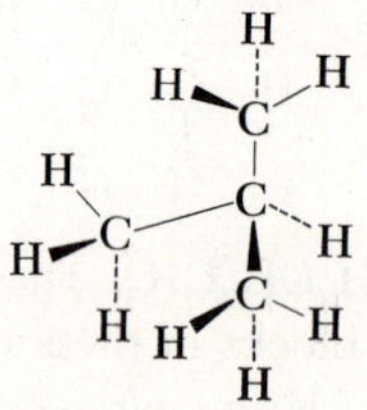

C_9H_{10} or $(CH_3)_3CH$

Figure 7.11
Names, structures, and structural formulas of alkanes.

the formula C_2H_4. Each of the carbon atoms in this molecule is visualized as sp^2-hybridized (Sec. 7.3). Two of the hybrid orbitals of each carbon are used in covalent bonding with hydrogen atoms, and the remaining pair of sp^2 orbitals (one from each carbon) overlap to form a C–C covalent bond. With an electron pair allotted to each covalent bond, one sees that there are two additional electrons still unaccounted

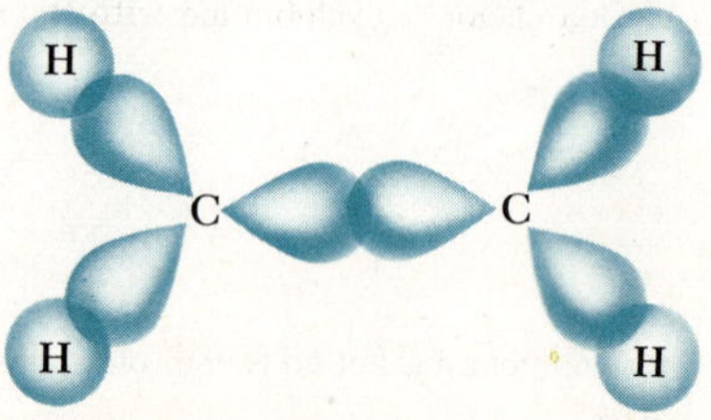

for. These two electrons are allotted to the remaining pair of carbon atomic orbitals, one $2p$ orbital of each carbon that is not involved in the sp^2 hybrids. The axes of these remaining $2p$ orbitals are each directed perpendicular to the plane that contains the three axes of the corresponding sp^2 hybrid orbitals.

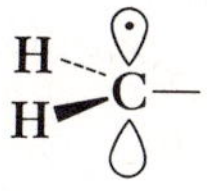

and

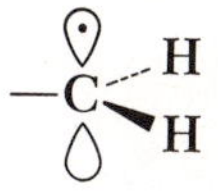

Furthermore, these $2p$ orbitals are arranged relative to each other in such a way that they can actually overlap.

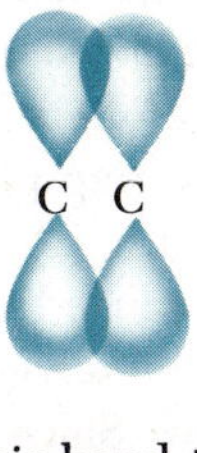

The net result is another electron pair bond that can be visualized as

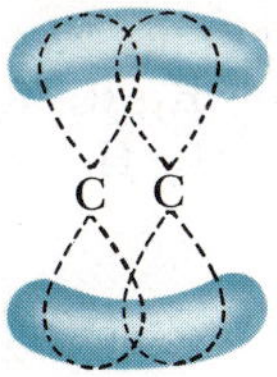

Having thus accounted for all electrons in C_2H_4, the structure of the compound can be drawn as

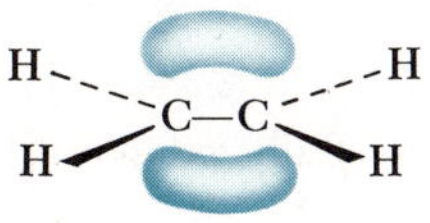

where the covalent bonds involving the sp^2 orbitals are shown as straight, dashed, or wedged lines, and the second C–C bond is shown in terms of a crude orbital overlap diagram. A simpler notation is

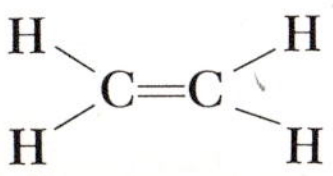

and an even simpler form is $CH_2{=}CH_2$. All these diagrams indicate the existence of two carbon-carbon covalent bonds—called a **double bond.** The simplest notation for a carbon-carbon double bond is C=C. A compound containing the C=C unit is called an **alkene.**

There are several interesting properties associated with C=C in a molecule. The C–C bond formed from the unhybridized $2p$ orbitals is called a **pi bond,** or **π bond.** It is considerably weaker than the C–C bond formed from the sp^2 orbitals, called a **sigma bond,** or **σ bond.** This is because the overlap in the former is not so large as in the latter. This difference in bond strength is readily manifested in the chemical properties of alkenes. For example, a reaction of the following type is quite common among organic compounds that contain the C=C units.

$$\begin{array}{c}H\\H\end{array}C=C\begin{array}{c}H\\H\end{array} + Br_2 \rightleftharpoons \begin{array}{cc}H&Br\\H-C-C-H\\Br&H\end{array}$$

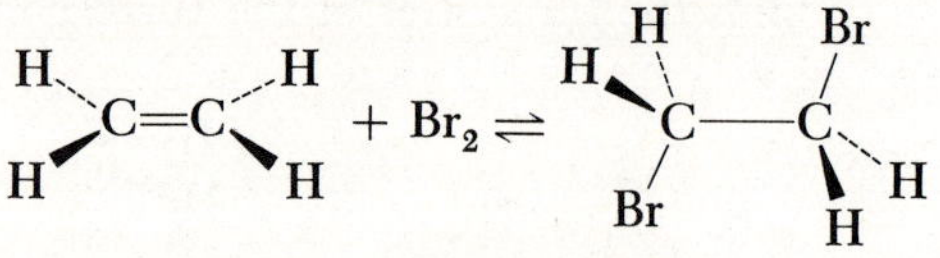

In this reaction, the C–C π bond is broken, not the C–C σ bond (the two new C–Br bonds that are formed make the reaction feasible).

Another interesting characteristic of a C–C double bond is that it is not rotationally flexible. That is, one side of the C=C cannot be rotated about the C–C axis without carrying the other end of the molecule around with it. This is easy to understand from the overlap pictures of the π bond drawn above; if one of the carbon atoms were rotated, carrying its $2p$ orbital with it, the overlap between the two $2p$ orbitals would be reduced or eliminated and the π bond would be broken. This could be avoided if the second carbon atom, with its $2p$ orbital, rotates in the same sense as the first one. Since much energy is required to break a C–C π bond, this kind of independent rotation is not observed (except under drastic conditions that would often lead to the decomposition of the compound). In the accompanying diagram, the symbols R_1, R_2, R_3, and R_4 represent functional groups that are structural, for example, CH_3, H, Br, and CH_2CH_3.

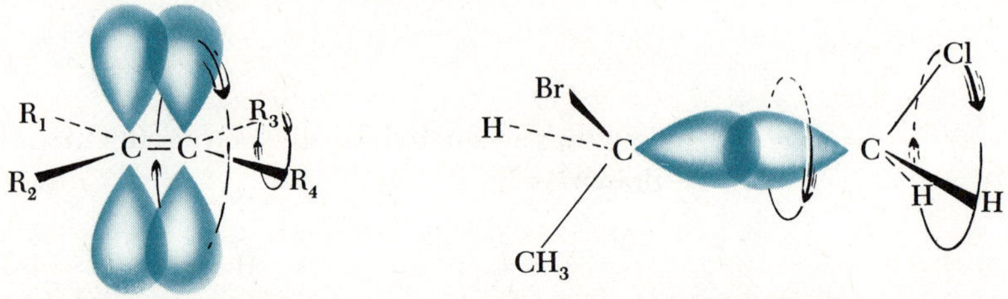

Essentially forbidden rotation Facile rotation

This rotational inflexibility of C–C double bonds is not found in C–C single bonds, since the overlap between the bonding orbitals is not interfered with by rotation of one carbon atom independent of the carbon to which it is attached. Rotation of the right side of this molecule, the

$$-C\begin{array}{c}Cl\\H\\H\end{array}$$

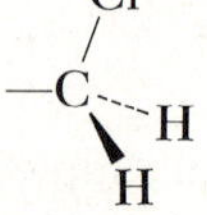

group, about the central C–C bond is facile, since the bonding network is not adversely affected.

The rotational inflexibility of C–C double bonds leads to the existence of geometrical isomers, compounds that differ from each other only in the placement of a given set of structural units about a C–C double bond. An example of this concept is provided by *cis-* and *trans-*diiodoethylene (discussed in Sec. 3.1).

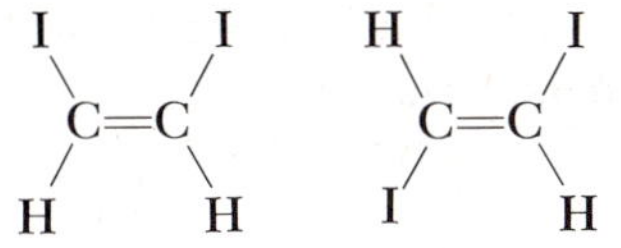

$$\underset{\text{cis}}{\overset{I}{\underset{H}{>}}C=C\overset{I}{\underset{H}{<}}} \qquad \underset{\text{trans}}{\overset{H}{\underset{I}{>}}C=C\overset{I}{\underset{H}{<}}}$$

The C=C unit can itself be considered a functional group, one to which four other structural units are attached in a real molecule. A specific example of this unit is the functional group that we can imagine deriving from ethylene by eliminating one hydrogen atom. It is called the vinyl group.

$$\overset{H}{\underset{H}{>}}C=C\overset{H}{\underset{}{<}}$$

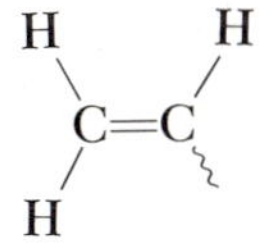

EXAMPLE 7.9 Write a structural formula for ethylvinylethylene.*

SOLUTION The name tells us that the molecule can be visualized as ethylene in which one hydrogen atom is replaced by a vinyl group and another is replaced by a methyl group. The following three possibilities immediately come to mind.

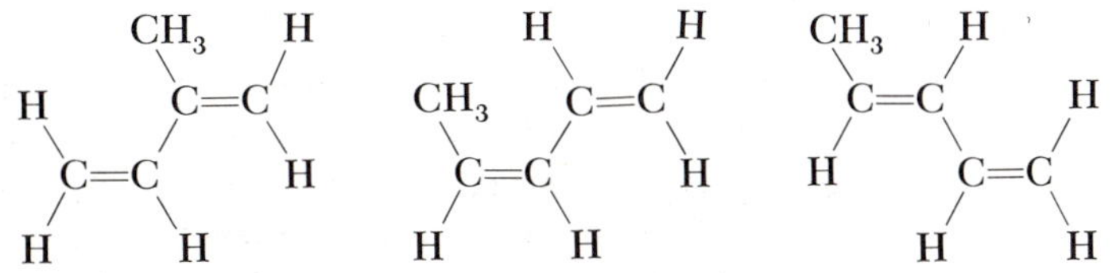

I   II   III

All other possibilities that we can imagine are related to these by a simple rotation about a C–C single bond.

The accompanying structure is equivalent to I in the sense that the two structures can be interconverted by rotation about a C–C single bond.

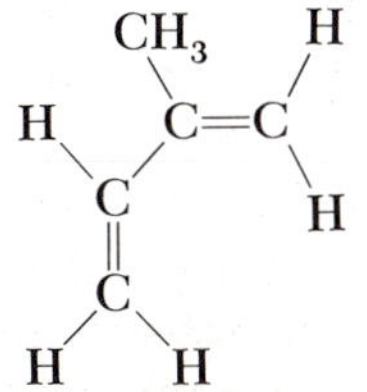

Alkynes

Another important structural unit that one finds in organic compounds is a carbon-carbon triple bond. The simplest molecule that contains this unit is called acetylene, C_2H_2, an important gaseous fuel used for welding torches. The simplest structural formula for acetylene is H—C≡C—H.

The structure of acetylene is readily understood in terms of sp hybridization of the carbon atoms. Each of the carbon hybrid orbitals is involved in a covalent σ bond,

---

*This is not the preferred nomenclature, but its meaning is unambiguous.

either with a hydrogen atom or with the other carbon atom. This gives the linear σ-bond structure:

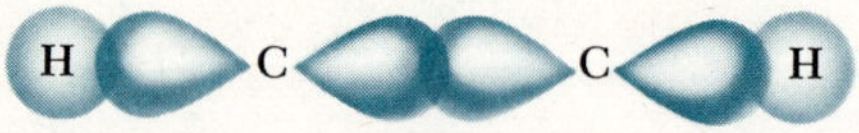

Counting the six electrons and the orbitals involved in the three electron-pair bonds of this σ framework, we see that two valence-shell electrons and two $2p$ orbitals remain for each carbon atom. These are arranged in two pairs that are properly set up for efficient overlap in the following way:

(a)
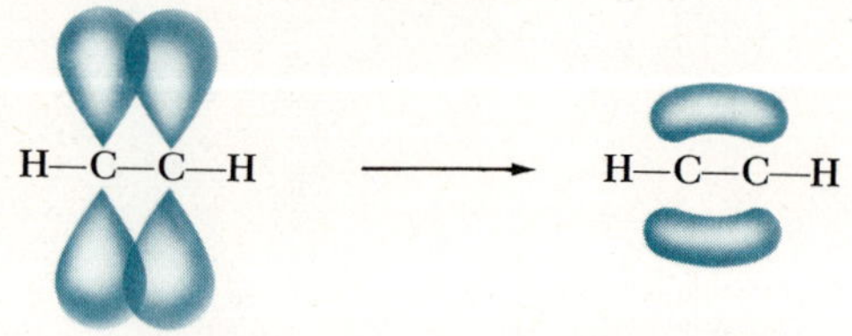

(b)
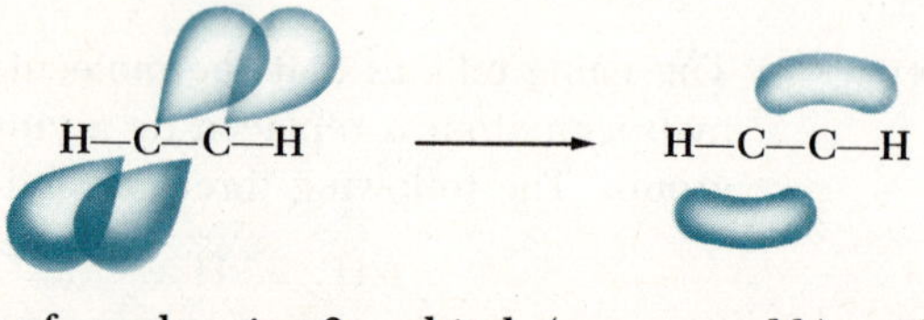

Each of these pairs of overlapping $2p$ orbitals (cases a and b), with a pair of electrons in each overlapping pair, constitutes a π-type electron-pair bond of the same general type described above for ethylene. In this case, however, there are two such bonds. Together with the C–C σ bond, this gives a total of three covalent bonds. The C≡C structural unit is an important functional group in organic chemistry. Compounds that contain it are called **alkynes**.

Other Functional Groups in Organic Chemistry

Only a few functional groups of organic chemistry have been described explicitly or referred to in passing. There are many others that are important in organic chemistry. To give you an early awareness and brief introduction to some of these groups, they are summarized in Table 7.4, which also gives the hybridizations of some of the central atoms involved, in cases where that would not be otherwise obvious at this stage.

Although it might seem from the emphasis in this section that the concept of the functional group is limited to organic chemistry, this is certainly not the case. Organic chemistry is a subject in which this concept finds perhaps its most extensive use, and it is a convenient vehicle for demonstrating the concept; but all branches of descriptive chemistry use it. The concept is a device that makes it possible for chemists to consider complicated problems in simpler parts, to focus attention on an especially pertinent small portion of a large, complex molecule when that portion is primarily associated with a property of interest. An automobile mechanic thinks of a car by its parts; if he is interested in how smoothly a particular car accelerates, he does not think about the chassis, the headlights, the bumpers, or the fenders; he focuses on the functional parts relevant to this particular property—the engine and perhaps the transmission. Chemists use the functional group in the same way. If one is interested in exploring the weakly acidic properties exhibited by G-penicillin,

TABLE 7.4 Some Typical Functional Groups in Organic Chemistry

Functional Group	*Name*	*Central Atom Hybridization*[a]
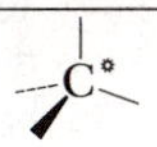	Alkyl	sp^3
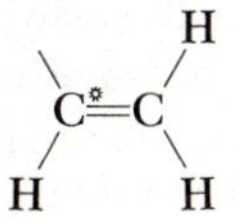	Vinyl	sp^2
$-\overset{*}{C}\equiv\overset{*}{C}-H$	Ethynyl	sp
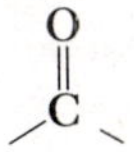	Carbonyl	sp^2
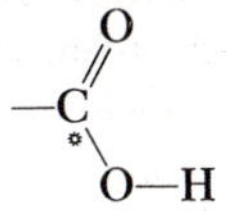	Carboxyl	sp^2
$-\overset{*}{C}\equiv N$	Cyano	sp
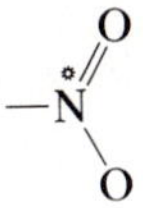	Nitro	sp^2
$-O-H$	Hydroxy	sp^3
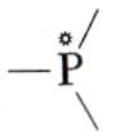	Phosphine	sp^3
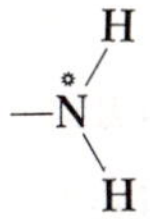	Amino	sp^2 or sp^3 (depends on the nature of carbon atom attached)

[a]Central atom indicated by asterisk.

it is reasonable, on the basis of the well-known acidic properties associated with the carboxyl group,

$$-\overset{\overset{\displaystyle O}{\|}}{C}OH$$

to focus attention on that portion of the molecule.

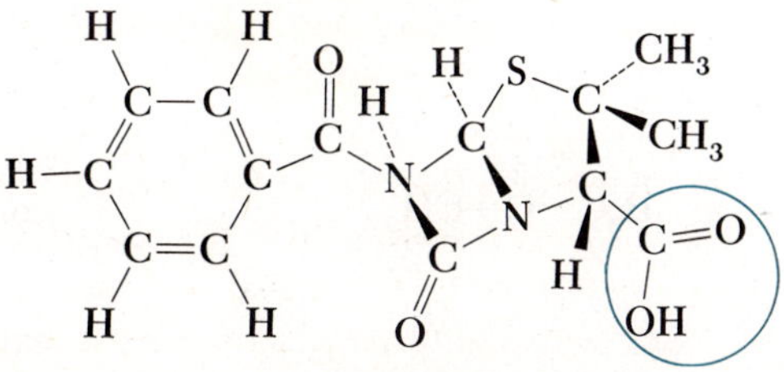

G-penicillin

Spotlight

Robert B. Woodward (1917–   ) of Harvard is one of the most accomplished molecular architects. Woodward has directed his students and postdoctoral associates in the synthesis of very complex molecules from small, common molecules by taking advantage of a knowledge of the chemistry of functional groups. Some of his accomplishments include the synthesis of cholesterol, cortisone, tetracycline, and chlorophyll (in the picture). In 1965 he was awarded the Nobel Prize. One of his latest achievements is the synthesis of Vitamin B_{12}, the most complex nonpolymeric natural molecule known. During this effort, which involved over 100 coworkers, Woodward collaborated with Roald Hoffman in formulating a general theory of organic reactions known as the Woodward–Hoffman rules; a theory which has been hailed as the most important advance in organic chemistry in the last twenty years.

Chemical and Engineering News

EXAMPLE 7.10 Identify the functional groups in the following molecules:

glycine, $H_2NCH_2\overset{\overset{O}{\|}}{C}-OH$,   acetone, $H_3C-\overset{\overset{O}{\|}}{C}-CH_3$;

hydroxylamine, H_2NOH; 1,3-hexadiene, $CH_2=CH-CH=CH-CH_2CH_3$.

SOLUTION $H_2N-CH_2-\overset{\overset{O}{\|}}{C}-OH$ consists of the functional groups amino ($-NH_2$), methylene ($-CH_2-$), and carboxylic acid $\left(-\overset{\overset{O}{\|}}{C}-OH\right)$.

At pH 7, in fact, the amine group acts as a base and the carboxyl acid group acts as an acid so that glycine is in the ionic form

$$^{(+)}H_3N-CH_2-C\overset{\displaystyle O}{\underset{\displaystyle O^{(-)}}{}}$$

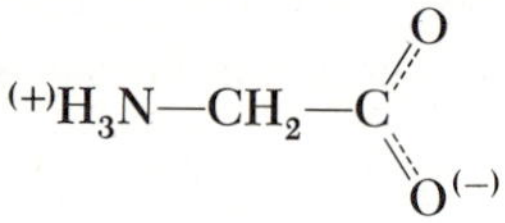

$H_3C-\overset{\overset{O}{\|}}{C}-CH_3$ consists of two methyl groups ($-CH_3$) and a carbonyl $\left(-\overset{\overset{O}{\|}}{C}-\right)$ group.

H$_2$NOH consists of an amino (—NH$_2$) and a hydroxyl group (—OH).

CH$_2$=CH—CH=CH—CH$_2$CH$_3$ consists of two ethylene groups $\left(\diagdown_{/}C=C_{\diagdown}^{/}\right)$, a methylene group (—CH$_2$—) and a methyl group (—CH$_3$).

STUDY PROBLEM 7(f)
Identify the functional groups in

(a) alanine, H$_2$N—CH(CH$_3$)—C(=O)OH;

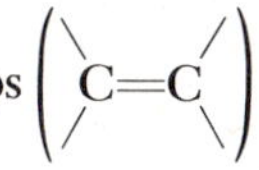

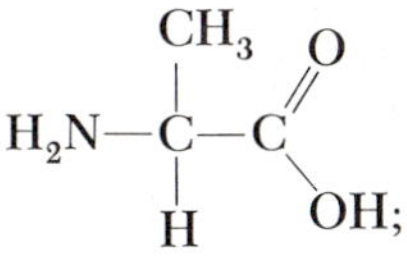

(b) methylethyl ketone, H$_3$C—C(=O)—CH$_2$CH$_3$;

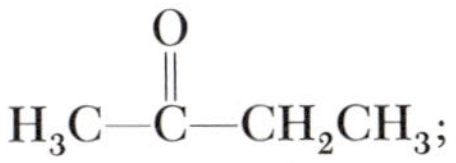

(c) hexaammine cobalt(III) trichloride, [Co(NH$_3$)$_6$]Cl$_3$.

7.5 Polar and Nonpolar Molecules—Electronegativity

Now that we have some idea of what covalent molecules are and what their shapes are, it is reasonable to begin exploring the general characteristics of covalent molecules. One highly pertinent characteristic is associated with the concept of polarity. As mentioned in Sec. 7.2, some liquids, such as water or liquid ammonia, are capable of dissolving ionic compounds because such liquids shield the oppositely charged ions from one another; that is, they reduce the tendency of the ions to recombine into a crystal lattice. Such solvents are said to be polar. Other solvents, which are said to be nonpolar, like carbon tetrachloride or benzene, do not manifest such a shielding effect. The shielding of electrical charge can be termed a **dielectric effect**, and a parameter called the dielectric constant is a measure of the effect. We can measure this parameter—and determine polarity—by using a set of oppositely

Spotlight

The charge stored between the plates at a capacitor per unit applied voltage across the plates is called the **capacitance** C of the apparatus. Then, the dielectric constant E of a substance is defined as the ratio of the capacitance of the system with the substance of interest between the plates to the corresponding capacitance of the system with nothing (a perfect vacuum) between the plates:

$$E = \frac{C}{C_{vacuum}}$$

If the liquid substance is a conductor (for example, a solution of a strong electrolyte or a molten salt), then applying a voltage difference across the plates results in a passage of charge (conduction) between the plates, rather than the storage of charge. The dielectric concept and the above defining equation apply to nonconducting fluids; however, once the concept of dielectric constant is defined, the associated idea of molecular polarity applies to solutions containing ions as well.

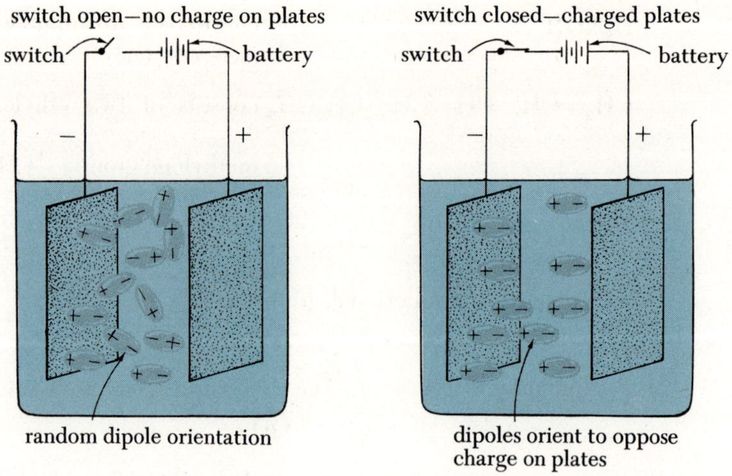

Figure 7.12
The insulating, or dielectric, effect of polar molecules.

charged plates (a capacitor) with the material of interest placed between them. The greater the electrical charge that can be stored on these plates for a given applied voltage across the plates, the higher the value of the dielectric constant of the substance between the plates, and the more polar the molecules of the substance.

As shown in Fig. 7.12, the effect of molecular polarity in a nonconducting fluid (insulator) can be explained by imagining that one end of the molecule is somewhat negative, while the other end is somewhat positive. The molecules are aligned so that an end of the molecule is attracted toward the capacitor plate of opposite charge. This effectively stabilizes the system by compensating for the negative charge of electrons stored on the negative plate with the positive end of the oriented molecules, thus increasing the amount of charge that can be stored on the plates for a given applied voltage.

So far we have said nothing to indicate why some substances have high dielectric constants and some substances have low values, that is, why some types of molecules are more polar than others. For example, why is H_2O polar and CCl_4 not? The polarity of molecules results from the fact that the electrons in covalent bonds need *not* be shared equally. The attraction of an additional electron by an atom varies from element to element. The ionization potential, or the energy needed to abstract an electron, also varies from one element to the next. We might imagine, then, that the electrons in a covalent bond between atoms of two different elements would not be shared equally. The tendency for an atom to pull the electrons in a covalent bond toward itself is called **electronegativity**.

Several scales of electronegativity have been developed, with some arbitrariness. Fundamentally, electronegativity is a powerful and extremely useful concept, but one that has been difficult to place on a quantitative basis for atoms in complex molecules. Electronegativity scales are relative scales; they compare the ability of the two atoms in a bond to attract the electrons. Fluorine is arbitrarily assigned an electronegativity value of about 4 and the electronegativities of other elements are usually scaled in comparison with fluorine.

One of the easiest electronegativity scales to define is one R. S. Mulliken proposed. His scale is based on the following equation:

$$EN = \text{electronegativity} = \frac{\text{ionization potential} + \text{electron affinity}}{2} \tag{7.2}$$

The ionization potential and electron affinity are weighed equally. The harder it is to take an electron away from an atom (the higher the ionization potential), the more the atom will keep its share of the bonding electrons. Also, the more an atom attracts an electron (the larger its electron affinity), the more it will tend to attract the electrons in the bonding pair. On this basis, we can use the periodic trends we have developed (last chapter) to rationalize some trends of electronegativity values.

EXAMPLE 7.11 Rank the electronegativities of atoms in the following molecules: H_2O, $AsCl_3$, TiO_2, $LiAlH_4$.

SOLUTION The ionization potentials and electron affinities (Chap. 6) of the atoms can be used to make the following predictions of relative electronegativities: O > H; Cl > As; O > Ti; H > Al > Li.

The electronegativities assigned to the elements from calculations of Allred and Rochow are summarized in Table 7.5. In moving across a row, we find that ionization to form cations is more difficult and electron affinity is larger. Moving down a column, we find that ionization potentials decrease and electron affinities also de-

TABLE 7.5 **Electronegativity of the Elements (Allred-Rochow Formula)**

H																
2.20																
Li	Be											B	C	N	O	F
0.97	1.47											2.01	2.50	3.07	3.50	4.10
Na	Mg											Al	Si	P	S	Cl
1.01	1.23											1.47	1.74	2.06	2.44	2.83
K	Ca	Sc	Ti	V	Cr	Mn	Fe	Co	Ni	Cu	Zn	Ga	Ge	As	Se	Br
0.91	1.01	1.20	1.32	1.45	1.56	1.60	1.64	1.70	1.75	1.75	1.66	1.82	2.02	2.20	2.48	2.74
Rb	Sr	Y	Zr	Nb	Mo	Tc	Ru	Rh	Pd	Ag	Cd	In	Sn	Sb	Te	I
0.89	0.99	1.11	1.22	1.23	1.30	1.36	1.42	1.45	1.35	1.42	1.46	1.49	1.72	1.82	2.01	2.21
Cs	Ba	*	Hf	Ta	W	Re	Os	Ir	Pt	Au	Hg	Tl	Pb	Bi	Po	At
0.86	0.97		1.23	1.33	1.40	1.46	1.52	1.55	1.44	1.42	1.44	1.44	1.55	1.67	1.76	1.96
Fr	Ra	**														
0.86	0.97															
		*La	Ce	Pr	Nd	Pm	Sm	Eu	Gd	Tb	Dy	Ho	Er	Tm	Yb	Lu
		1.08	1.06	1.07	1.07	1.07	1.07	1.01	1.11	1.10	1.10	1.10	1.11	1.11	1.06	1.14
		**Ac	Th	Pa	U	Np	Pu	Am	Cm	Bk	Cf	Es	Fm	Md		
		1.00	1.11	1.14	1.22	1.22	1.22			1.2 (estimated)						

SOURCE: *J. Inorg. Nucl. Chem.* 5:264 (1958).

crease. Thus, fluorine, in the upper right corner of the periodic table, should have the highest electronegativity, while francium (lower left corner) should have the lowest value. Elements in the left and lower portions of the table should have lower electronegativities than elements in the upper and right portions.

The Polar Bond

If we consider the molecule H—F, the fluorine atom has a much higher electronegativity than the H atom, so the probability density of the bonding electrons will be more highly concentrated near F than near H.

$$\delta^+ \longrightarrow \delta^-$$
$$\text{H}\text{——}\text{F}$$

bonding electrons in HF

Since each atom in the molecule would be electrically neutral if the sharing of the electron pair were exactly equal, the F is now somewhat negative, while the H end of the bond has lost some of its electron density and is somewhat positive. The amount of this effect, referred to as the **bond polarity**, depends on the relative electronegativities of the two atoms. This polarity is indicated above in terms of the "partial charges" δ^+ and δ^-.

If the electronegativities of the two atoms of a given bond are nearly the same (within about 0.3 unit on the Mulliken scale shown in Table 7.5), the sharing is nearly equal and the bond is essentially nonpolar. For the difference between 0.3 and 1.7 units approximately, the bond will be substantially polar. For a difference of more than about 1.7 units, one atom attracts electrons so much better than the other that the compound is generally ionic, and ions rather than discrete molecules exist.

EXAMPLE 7.12 Rank the bonds in the following species as nonpolar, polar, or ionic: CO_2, HBr, ClO_4^-, NH_3, O_2, RbCl, and $SbCl_3$.

SOLUTION The electronegativities are shown above each atom; ΔEN represents the electronegativity difference.

HBr   2.20 2.74 H—Br $\Delta EN = 0.54$; therefore, a polar bond

NH_3 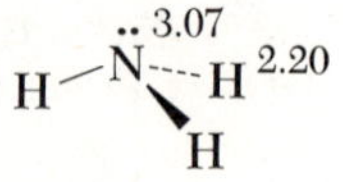

$\Delta EN = 0.87$; therefore polar bonds

O_2 3.50 3.50 O=O $\Delta EN = 0.00$; therefore a nonpolar bond

RbCl   0.89 2.83 Rb—Cl $\Delta EN = 1.94$; therefore, an ionic bond

$SbCl_3$ 1.82 2.83 Sb—Cl $\Delta EN = 1.01$; therefore polar bonds

STUDY PROBLEM 7(g)

Rank the bonds in the following list as nonpolar, polar, or ionic: phosphine, PH_3; phosgene,

$$Cl-\overset{\overset{O}{\|}}{C}-Cl$$

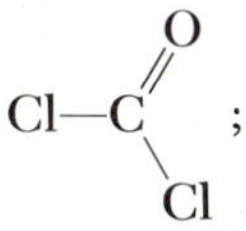

; and KBr.

Polar Molecules

From Table 7.5 one readily predicts that the bonds in both H_2O and in CCl_4 are all polar covalent bonds. Yet one finds that one of these molecules is polar, while the other is not. To determine the polarity of an entire molecule, not just a bond, one must know about both the polarity of the constituent bonds and the shape of the molecule, or how the bonds are arranged in space. For a molecule to be polar, one end of the molecule as a whole must be somewhat positive, while another part of the whole molecule must be somewhat negative. It is not sufficient that individual bonds are polar.

To consider this in detail, let us consider the molecules F_2, H_2O, CCl_4, and CO_2.

$$F-F \qquad H\overset{\nearrow\!\!\uparrow\!\!\nwarrow}{}H \qquad Cl\overset{\overset{Cl}{|\uparrow}}{\underset{Cl}{\searrow}}\!\!\overset{\leftarrow}{C}\!\!\overset{\rightarrow}{}Cl \qquad O\!\!=\!\!C\!\!=\!\!O$$

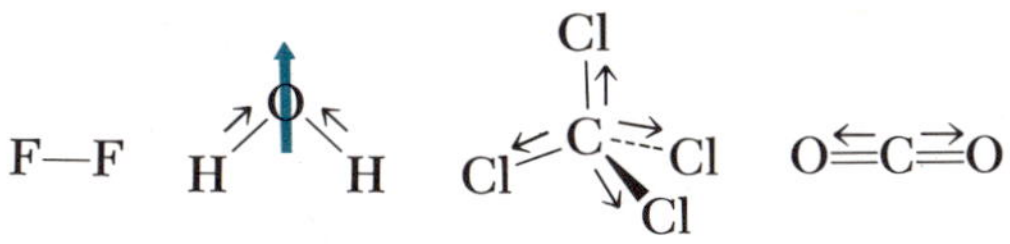

In the accompanying structural representations, an arrow is used to show a polar bond; the head of the arrow is in the direction of the negative end of the bond, towards the more electronegative atom. We can easily see that F_2 must be a nonpolar molecule, since it is made up of just one nonpolar bond. Also, CO_2 must be nonpolar, since the two ends of the molecule, each associated with a polar C=O, are equivalent and have a canceling effect; even though there are polar bonds, they add to each other in such a way that there is no "positive" end of the molecule. The molecule H_2O, on the other hand, is polar, because the two polar bonds are oriented so that the polarity of each bond is partly reinforced by the other. Each bond renders the oxygen end of the molecule somewhat negative and the H ends somewhat positive. The result of adding the two polar bond arrows together is shown by the colored arrow.

The arrows shown in the H_2O, CCl_4, and CO_2 diagrams are really vectors. A vector is a mathematical concept that conveys both magnitude and direction. It is often represented graphically by an arrow, the length of which is a measure of the magnitude of the vector quantity, the direction of the arrow representing the direction of the vector. A common example of a vector is velocity; a complete specification of velocity requires knowing both the rate of motion (speed) and the direction of motion. We can readily see that a bond polarity is a vector that can be represented by an arrow. The addition of these arrows according to the usual rules of vector algebra determines whether or not a molecule has a net polarity (nonzero resultant vector).

$$\nearrow\nwarrow = \uparrow \; ; \qquad \overset{\uparrow}{\underset{\downarrow}{}} = 0; \qquad \leftarrow \rightarrow = 0$$

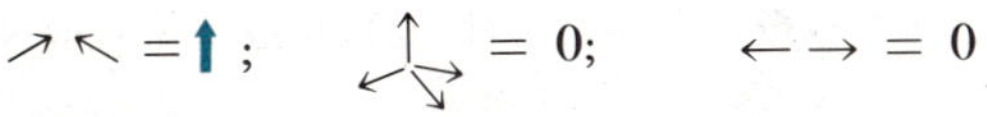

The situations for BF_3 and CO_2 are analogous; each BF bond is polar, but the

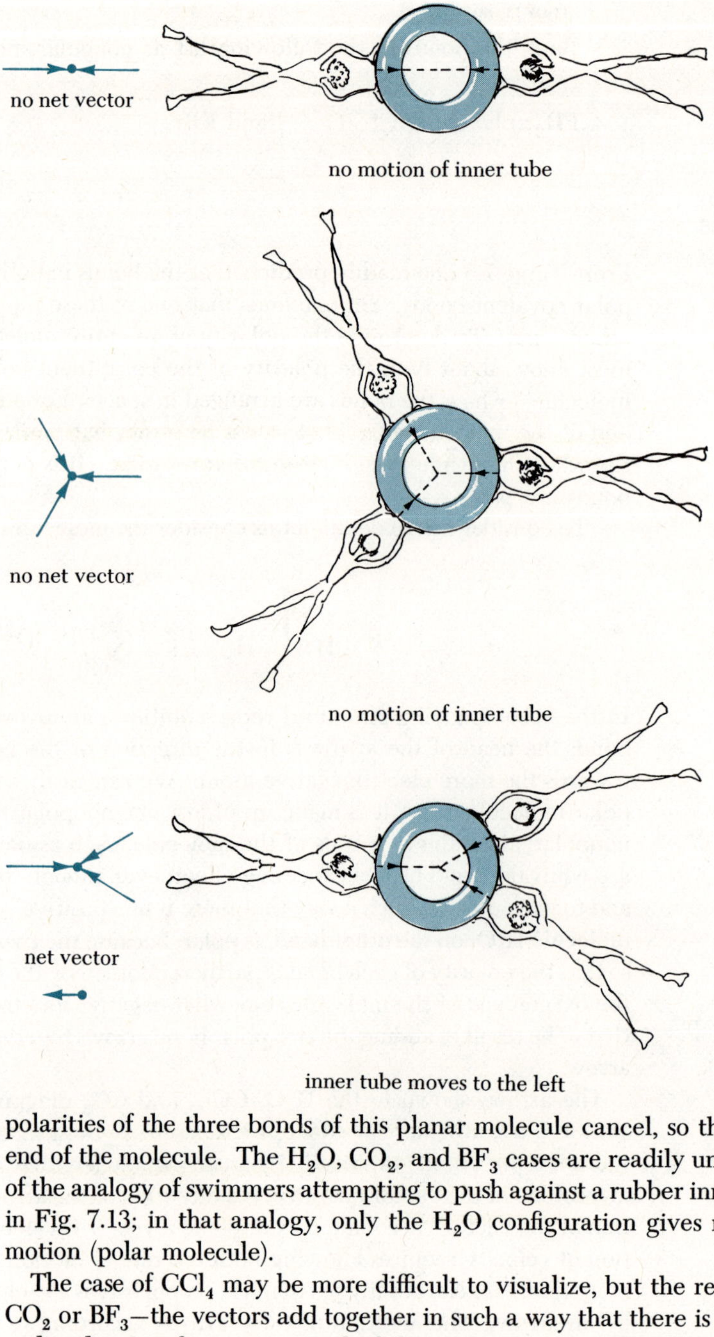

Figure 7.13 An illustration of vector addition. Swimmers attempt to move the inner tube by pushing against it. In the upper two cases, the swimmers are balanced and no movement of the inner tube occurs. There is no net force vector. In the bottom case, there is an imbalance and the two swimmers are able to move the inner tube (and the other swimmer!).

polarities of the three bonds of this planar molecule cancel, so there is no positive end of the molecule. The H_2O, CO_2, and BF_3 cases are readily understood in terms of the analogy of swimmers attempting to push against a rubber inner tube, as shown in Fig. 7.13; in that analogy, only the H_2O configuration gives rise to a resultant motion (polar molecule).

The case of CCl_4 may be more difficult to visualize, but the result is like that of CO_2 or BF_3—the vectors add together in such a way that there is no single positive end and no single negative end of the molecule, even though each C–Cl bond is somewhat polar (electronegativity difference = 0.33 unit). The molecule is nonpolar. If one of the Cl atoms is replaced by a fluorine atom, yielding fluorotrichloromethane, the bonds no longer all have the same degree of polarity (electronegativity difference for C—F = 1.6 units). This molecule can be represented as

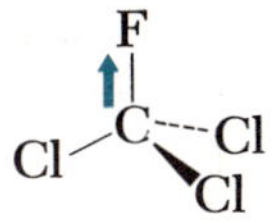

Adding the arrows gives a net dipole (a positive and negative charge separation) for the molecule as a whole, as shown by the colored arrow. The arrows are not balanced against one another as they were in the CCl_4 molecule, and $CFCl_3$ is polar.

Although it may not always be easy to visualize adding the polarity arrows, we can state some helpful guides for determining molecular polarity.

1. Molecules containing only nonpolar bonds (electronegativity difference of less than 0.3 unit) will be of low polarity. This is true no matter what the shape of the molecule might be.
2. For molecules in which all bonds are of the same type (for example, CO_2, BF_3, CCl_4, and PF_5), molecules that correspond to highly symmetrical figures, with one atom placed at its center, will be nonpolar. Some simple examples are shown herewith.

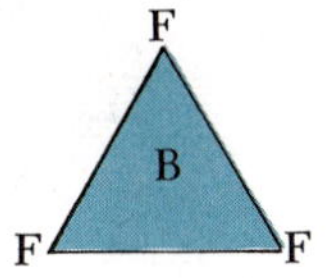

equilateral triangle

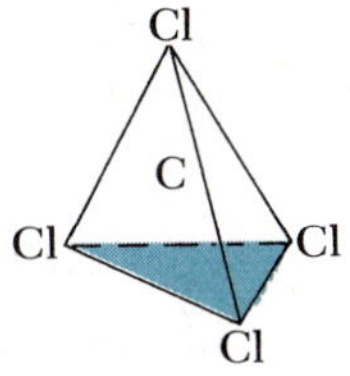

tetrahedron

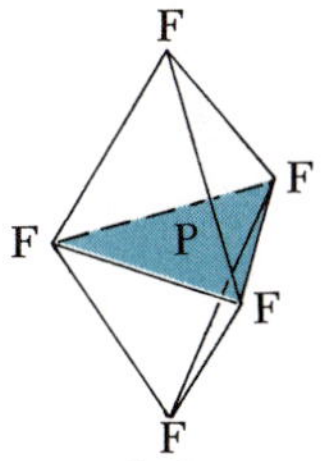

trigonal bipyramid

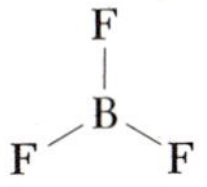

boron trifluoride

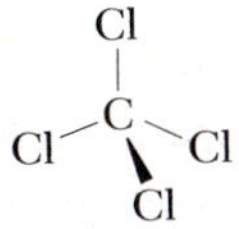

carbon tetrachloride

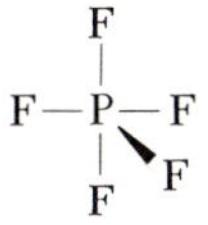

phosphorus pentafluoride

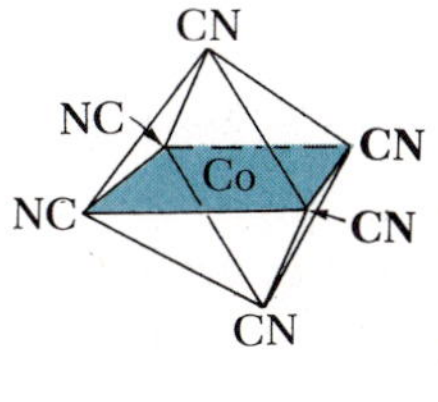

octahedron

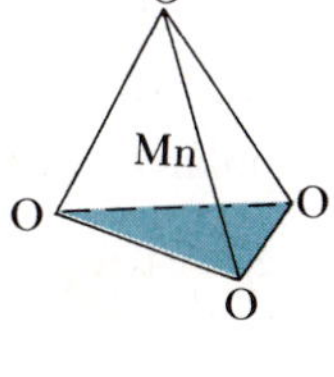

tetrahedron

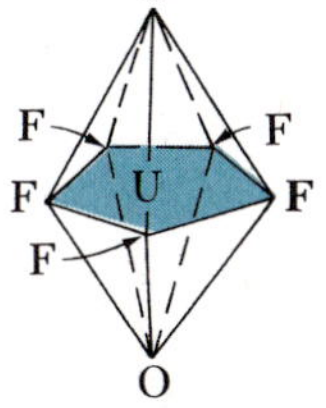

pentagonal bipyramid

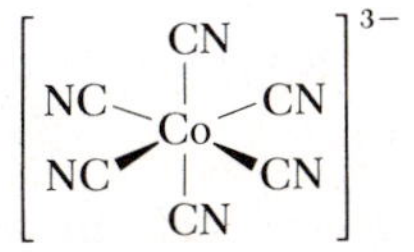

[hexacyano cobalt (III)]$^{3-}$ ion

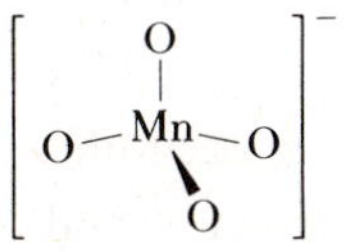

permanganate ion

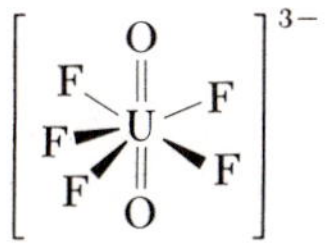

uranyl pentafluoride ion

3. Molecules with only polar bonds that are not highly symmetrical are polar molecules. Some simple examples are shown.

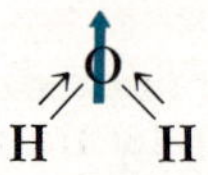
water

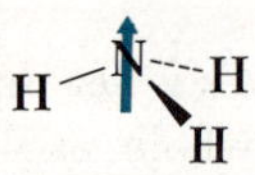
ammonia

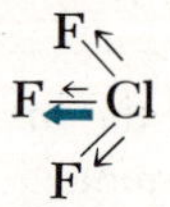
carbon trifluoride

sulfur tetrafluoride

4. Molecules with both polar and nonpolar bonds are not quite so easy to generalize. We must visualize the detailed geometrical structure of the molecule and determine whether the arrows for the polar bonds in the molecule cancel or reenforce each other. Nonpolar bonds can be ignored. Some examples are

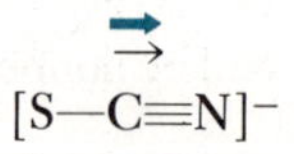
thiocyanate ion

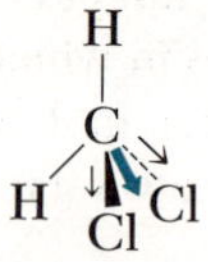
dichloromethane

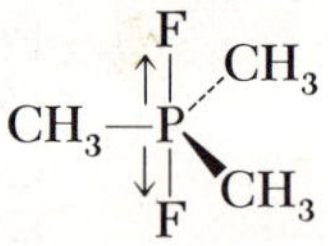
difluorotrimethylphosphorane (has no net dipole)

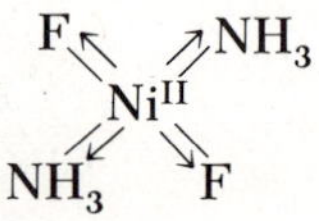
trans-difluorodiamminenickel (II) (has no net dipole)

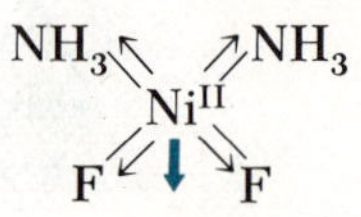
cis-difluorodiamminenickel (II)

EXAMPLE 7.13 Determine which of the following molecules are polar and which nonpolar: carbon tetrafluoride, CF_4; oxygen dichloride, OCl_2; nitric oxide, NO.

SOLUTION By first drawing the structure and noting the electronegativities, we can then find the electronegativity differences (Table 7.5) for each bond. The final step is to look at all the individual bonds and judge the polarity of the molecule as a whole.

CF_4 has the structure 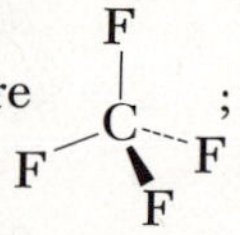 ; the bonds are each highly polar ($\Delta EN = 1.60$), but the structure is highly symmetric and the polar bonds cancel to give a nonpolar molecule.

OCl_2 has the structure 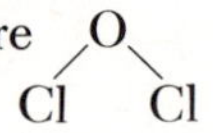 ; the bonds are somewhat polar ($\Delta EN = 0.67$), and they act in the same direction, reenforcing each other to give a polar molecule.

NO must be a linear molecule, so if the bond is polar, the molecule must be polar. The electronegativity difference is 0.43, so the molecule is polar.

STUDY PROBLEM 7(h)

Determine whether the following molecules are polar or nonpolar: SF_6, N_2, CO, BrF_5.

7.6 Atomic Charges

The atoms in covalent compounds do not have definite, integral charges, although the existence of bond polarity does mean that not all atoms in covalent molecules are strictly neutral (uncharged). This kind of situation can in principle be described as precisely as one wants by a quantum mechanical calculation, which can yield a complete description of the electron density distribution. However, such accurate calculations on complex molecules require immense amounts of electronic computer time, and a trained specialist for interpretation. Hence, some less sophisticated, less expensive, and more descriptive means of representing electron density in general, and bond polarity in particular, are desirable. There are various ways of approaching this need.

Percentage Ionic Character

One approach is a concept called **percentage ionic character.** This is an attempt to describe the polarity of a polar, covalent bond in terms of how much it resembles an ionic bond. A completely nonpolar covalent bond (exactly equal sharing) is said to have zero percent ionic character; a purely ionic bond (no sharing of electrons) is said to have 100 percent ionic character. A polar, covalent bond would have a percentage ionic character between 0 and 100; the larger the value, the more polar the bond. If we visualize a simple diatomic molecule AB, we can imagine any of these possibilities.

δ^+ 0 δ^- 0	δ^+ 0.1 δ^- 0.1	δ^+ 0.6 δ^- 0.6	$+$ $-$
A———B	A———B	A———B	A B
equal sharing; 0% ionic character	weakly polar bond; 10% ionic character	strongly polar bond; 60% ionic character	ionic; 100% ionic character
H—H	H—Br	H—F	$K^+\ F^-$

In these diagrams the symbol δ^- represents the "excess" electron density on the atom with higher electronegativity (B), and the symbol δ^+ represents the deficiency of electron density associated with the atom of lower electronegativity; δ^+ and δ^- would each be zero for a nonpolar bond.

Formal Charge

A less flexible and essentially less realistic approach, but one that has some use in keeping track of the number of electrons in species and reactions is based on the idea of formal charge. The **formal charge** is defined to be the charge that an atom seems to have in a simple structural formula, based upon the assumption that all electron pairs are shared evenly. Thus, in the structures

$$H-O-H, \qquad F-B(-F)-F, \qquad \text{and} \qquad O=C=O$$

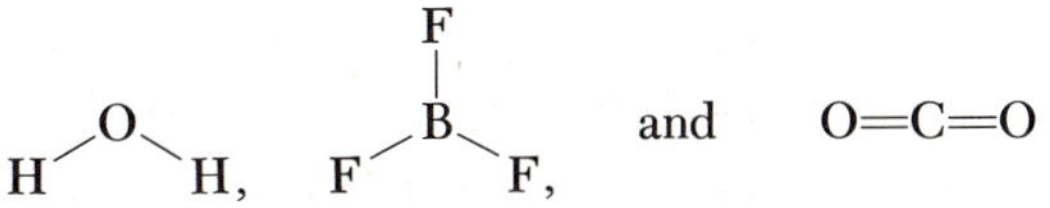

all the atoms have zero formal charge, even though we know that the bonds are polar.

Thus, the formal charge is really a property of a structural formula, not a property of a real species. In computing the formal charges in a structural formula, one simply counts up the total number of electrons associated with each atom in the formula and compares it with the charge of the **core** (inner electrons plus nucleus). Thus, writing Lewis dot formulas,

$$H:\ddot{O}:H \qquad F:\overset{F}{\underset{}{B}}:F \qquad :\ddot{O}::C::\ddot{O}:$$

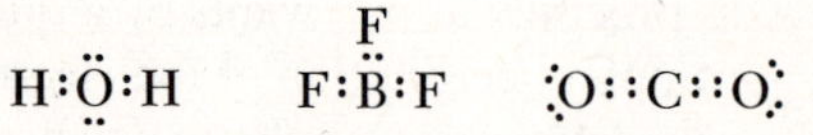

and allotting one electron of each covalent bond to each of the participating atoms, we see that the core charges match the number of associated electrons perfectly for each atom:

	H	O	B	F	C	O
Core charge	1	6	3	7	4	6
Associated valence electrons	1	6	3	7	4	6

For ammonia and the ammonium ion, we write

$$H:\overset{..}{\underset{H}{N}}:H \qquad \left[H:\overset{H}{\underset{H}{N}}:H\right]^{+}$$

$$H-\overset{H}{\underset{H}{N}}-H \qquad \left[H-\overset{H}{\underset{H}{N}}-H\right]^{+}$$

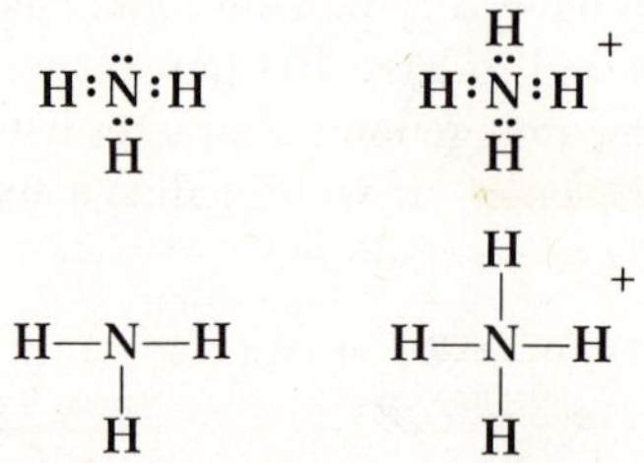

and assign a formal charge of zero to hydrogen in both species and a value of zero to nitrogen in ammonia. However, a value of $+1$ is assigned to nitrogen in NH_4^+, because the core charge of N is $+5$ and there are four valence electrons associated with it in this dot formula.

As indicated above, the formal charge is really more a property of a structural formula than of the species that the formula is meant to represent. For this reason, one uses the concept with the formulas that represent the contributing structure of resonance hybrids. For example, the azide ion, N_3^-, can be represented as

$$:\ddot{N}:N:::N:^- \longleftrightarrow :\dot{N}::N::\dot{N}:^- \longleftrightarrow :N:::N:\ddot{N}:^-$$

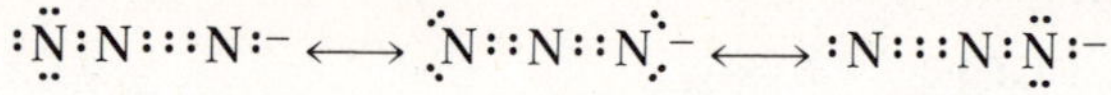

which, with the electrons allotted as above, yields the following pattern for formal charges:

$$\overset{2-}{N}-\overset{+}{N}\equiv N \longleftrightarrow \overset{-}{N}=\overset{+}{N}=\overset{-}{N} \longleftrightarrow N\equiv\overset{+}{N}-\overset{2-}{N}$$

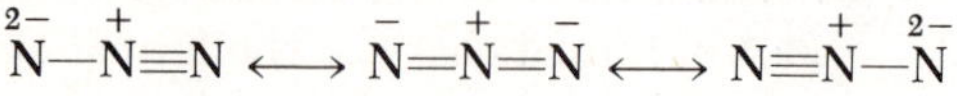

where the formal charges are given.

EXAMPLE 7.14 Write structural formulas, indicating any nonzero formal charges, for the following species: CN^-, CO, $CH_3CO_2^-$ (acetate), HCN, HNC, NO_3^-.

SOLUTION Starting with the Lewis electron dot formulas, and proceeding to the more abbreviated forms, we have

$$:C:::N:\quad :C:::O:\quad H:C:::N:\quad H:N:::C:$$
$$\overset{-}{C}\equiv N\quad \overset{-}{C}\overset{+}{=}O\quad H-C\equiv N\quad H-\overset{+}{N}\overset{-}{\equiv}C$$

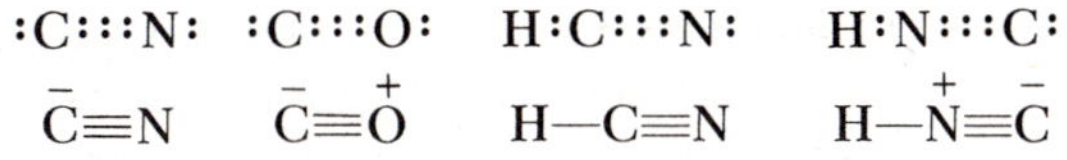

For acetate and nitrate ions, there are resonance hybrid structures to be represented, as shown in Sec. 7.3.

[Lewis dot and line structures of acetate resonance forms]

[Lewis dot and line structures of nitrate resonance forms]

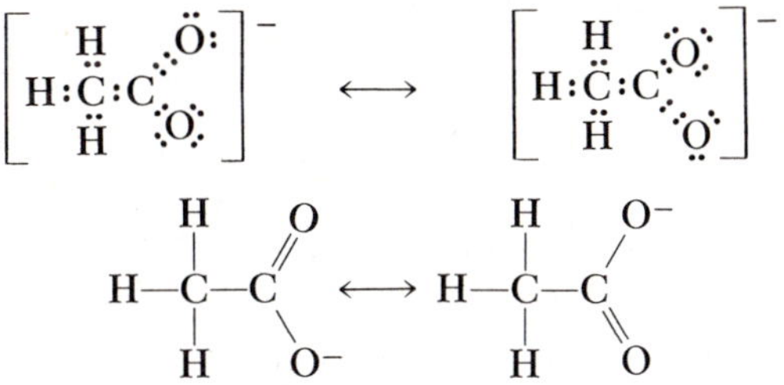

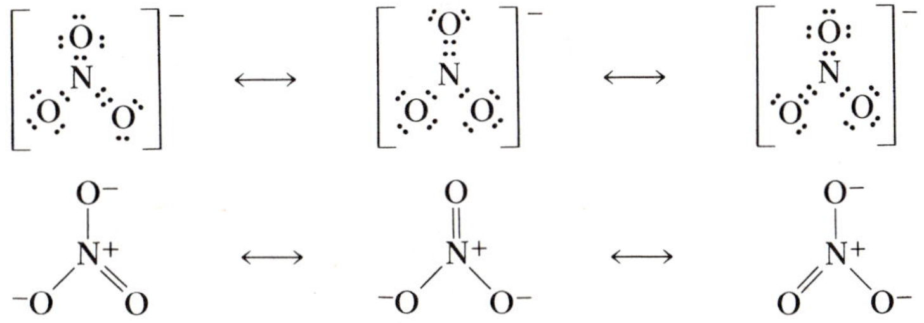

STUDY PROBLEM 7(i)

A correlation of formal charge with acid strength is sometimes made. The series of oxyacids Cl(OH), ClO(OH), ClO$_2$(OH), and ClO$_3$(OH) is ranked from weakest acid to strongest acid. What are the formal charges on the chlorine atom for this series? (In each case Cl is a central atom.)

As indicated symbolically in the electron dot formulas of the above example, the net charge on an ion "belongs to" the entire ion. The formal charge concept, however, places individual charges on individual atoms. The net charge on a species is the sum of the formal charges of all atoms in the species. For example, the total charge on the ion NO$_3^-$ is -1 (oxygen) plus -1 (oxygen) plus 0 (oxygen) plus $+1$ (nitrogen), which gives -1. The formal charge concept is useful both in matters of electronic bookkeeping and, to some extent, in visualizing gross charge distributions in molecular and ionic species.

7.7 Oxidation States

Another idea that is useful in keeping track of the total number of electrons in species and reactions is the concept of oxidation state. This concept provides another arbitrary means of assigning a specific integral number of electrons to each atom—hence, its usefulness in electronic bookkeeping. In this respect it is superficially similar to the concept of formal charge, but the two concepts are quite different. The oxidation state is useful for analyzing a class of reactions called **oxidation-reduction reactions**.

Oxidation is the removal of electrons from a species; **reduction** is the donation of electrons to a species. These two processes go on hand in hand in an electron transfer (oxidation-reduction) reaction. (Many important characteristics of oxidation-reduction reactions are discussed in Chap. 13.) A simple example of an oxidation-reduction reaction is the reduction of mercuric ion, Hg^{2+}, to metallic (liquid) mercury by metallic zinc:

$$Zn(s) + Hg^{2+} \rightleftharpoons Zn^{2+} + Hg(l) \tag{7.3}$$

Metallic Zn has the electronic configuration $[Ar]3d^{10}4s^2$, whereas the electronic configuration of Zn^{2+} is $[Ar]3d^{10}$. Thus, the zinc atom has lost two electrons in this process and has been oxidized, by Hg^{2+}. The electronic configuration of Hg^{2+} is $[Xe]5d^{10}$, and the electronic configuration of metallic mercury is $[Xe]5d^{10}6s^2$. Hence, mercury has gained two electrons and has been reduced in this process, by $Zn(s)$. Because the $Zn(s)$ has provided the electrons for donation, it is called the **reducing agent**; since the Hg^{2+} is responsible for removing electrons, it is called the **oxidizing agent** in the reaction.

Oxidation Number

Because of the simple nature of the species involved, the transfer of electrons from the reducing agent to the oxidizing agent in Eq. (7.3) is very easy to recognize. That is, however, not always the case. An example of a common oxidation-reduction reaction is the reaction of the permanganate ion, MnO_4^-, with the ferrous ion, Fe^{2+}, in acidic, aqueous solution:

$$8H^+ + MnO_4^- + 5Fe^{2+} \rightleftharpoons 5Fe^{3+} + Mn^{2+} + 4H_2O \tag{7.4}$$

This is a relatively complex chemical reaction, in the sense that it is not so easy to visualize in terms of an obvious electron transfer. It is in cases such as this that the concept of oxidation state becomes useful. To try to understand what is happening in the more complex reaction, it is advantageous to examine the dot structures of the species involved. The ion Fe^{2+} has the electronic configuration $[Ar]3d^6$ with the six $3d$ electrons as valence electrons, whereas Fe^{3+} has only five valence electrons. Focusing on the iron in the reaction shown in Eq. (7.4), we can write

$$\cdot\ddot{\underset{\cdot\cdot}{Fe}}\cdot^{2+} \rightarrow \cdot\ddot{\underset{\cdot\cdot}{Fe}}\cdot^{3+} + e^-$$

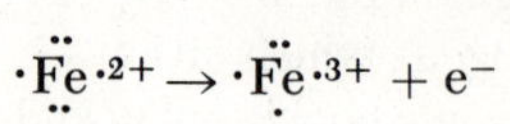

and recognize that the iron (in the form of Fe^{2+}) loses one electron in becoming Fe^{3+}. We say that Fe^{2+} and Fe^{3+} are in different **oxidation states**, and that the Fe^{2+} has been oxidized to Fe^{3+}. Focusing on the manganese in this reaction, we write

$$\begin{array}{c} :\ddot{O}: \\ :\ddot{O}:Mn:\ddot{O}:^- \\ :\ddot{O}: \end{array} \rightarrow \cdot\overset{\cdot}{Mn}\cdot^{2+}$$

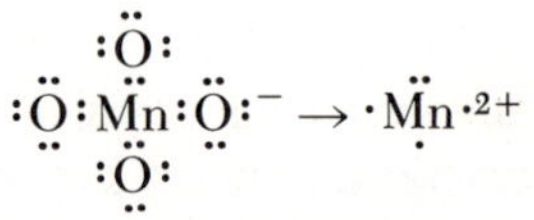

We can choose to identify each octet of electrons with oxygen in the MnO_4^- dot structure above, so that the manganese appears to gain five valence electrons in the transformation of MnO_4^- to Mn^{2+}; this situation can also be described by saying

that MnO_4^- and Mn^{2+} are in different oxidation states. The manganese appears to have been reduced from MnO_4^- to Mn^{2+} (in other words, that electrons have been donated to it). For making choices like identifying all the valence electrons in MnO_4^- with oxygen above, it is convenient to construct a self-consistent set of rules. Such rules allow one to develop a useful scheme for keeping track of electrons in electron-transfer reactions. A cornerstone of this scheme is characterizing the oxidation state by a variable called the **oxidation number.**

An exact value of an oxidation number can be defined for every atom in any species. Defining these numbers involves making arbitrary choices of which electrons "belong" to which atoms in a species; these choices are arbitrary in the sense that the rules on which they are based are arbitrary, but they are self-consistent in the sense that the same set of rules is always used to make the choices. Essentially the rules are based on assigning a pair of electrons shared in a covalent bond to the more electronegative atom of the bond. The following rules and guidelines are helpful in assigning oxidation numbers. The rules are given in order of precedence. That is, if in making oxidation number assignments on a particular ion or molecule, two of the rules conflict, use the one that appears first in the list.

1. The oxidation number of an elemental substance (for example, Cl_2, Hg(l), Na(s), O_2) is zero.
2. The oxidation number of a simple elemental ion equals the charge of the ion.
3. The oxidation number of any atom in any species can be computed by examining the electron dot formula, and assigning both electrons of a shared pair to the more electronegative atom in the bond. The oxidation number of an atom is then the charge that the atom appears to have on that basis (considering core charge and valence electrons allotted to it).
4. The sum of the oxidation numbers for all atoms in an ion or a molecule equals the net charge on the ion or molecule (zero for molecule).
5. The oxidation number of the ions of alkali metals (Group IA) is +1.
6. The oxidation number of the ions of alkaline earth metals (Group IIA) is +2.
7. The oxidation number of the hydrogen ion is +1, except when it is combined with metals, where it is −1.
8. The oxidation number of fluorine is always −1.
9. The oxidation number of oxygen is −2, unless rules 1 to 8 yield a contrary result (for instance, in H_2O_2, each O atom is assigned an oxidation number of −1, either from rule 3 or from rule 7).
10. The maximum positive oxidation number of an atom is its group number in the periodic table. The maximum negative oxidation number is (8 − group number).

The following are some examples of oxidation state assignments:

Simple Elemental Ions

Oxidation number equals charge.

ion	*oxidation number*
Ca^{+2}	+2
La^{+3}	+3
Cl^{-1}	−1

Polyatomic Ions Ions composed of several atoms that are covalently bound.

CARBONATE, CO_3^{2-}

1. Assign the oxidation number of each O as -2.
2. From the charge on the ion (and guideline 4), determine the "apparent" net charge x that carbon must have,

$$3(-2) + x = -2, x = -2 + 6 = 4$$

3. Therefore, the oxidation number of C is $+4$.

SULFATE, SO_4^{2-}

1. Assign each oxygen atom as -2.
2. Then, denote x as the oxidation number of sulfur; and apply rule 4,

$$4(-2) + x = -2, x -2 + 8 = 6$$

3. Therefore, the oxidation number of sulfur is $+6$.

THIOSULFATE, $S_2O_3^{2-}$*

1. Assign each oxygen atom as -2.
2. Denote the oxidation number of sulfur as x and apply rule 4,

$$3(-2) + 2x = -2, x = \tfrac{1}{2}(-2 + 6) = 2$$

3. Therefore, the oxidation number of sulfur is $+2$.

AMMONIUM ION, NH_4^+

1. Assign each H an oxidation number of $+1$.
2. Denote the oxidation number of N as x and apply rule 4,

$$4(1) + x = 1, x = -3$$

3. Therefore, the oxidation number of N is -3.

CYANIDE, CN^-

1. Write the electron dot formula.
2. Apply rule 3. :C:::N: All bonding electrons are assigned to N because it is the more electronegative atom. This results in eight electrons being assigned to the nitrogen atom and two electrons being assigned to the carbon atom.

   The oxidation number of nitrogen is

core charge	− valence electrons	= oxidation number
+5	−8	−3

*The prefix *thio* indicates that a sulfur atom has replaced an oxygen atom in the substance. Thus, thiosulfate is a sulfate ion with one oxygen atom replaced by a sulfur atom. Another example is a thioether; an ether contains a carbon-oxygen-carbon linkage (for example, CH_3—O—CH_3, dimethyl ether) while a thioether contains a C–S–C linkage (for example, CH_3—S—CH_3, dimethyl thioether).

Hence, the oxidation number of carbon is given by

$$\begin{array}{ccccc} \text{core charge} & - & \text{valence electrons} & = & \text{oxidation number} \\ +4 & & -2 & & +2 \end{array}$$

3. Note, as a check, that rule 4 is valid:

$$\begin{array}{ccccc} \text{ox. no. of C} & + & \text{ox. no. of N} & = & \text{total charge of the ion} \\ +2 & & -3 & & -1 \end{array}$$

IODINE MONOCHLORIDE, ICl

1. Write the electron dot formula: $:\ddot{\underset{..}{I}}:\ddot{\underset{..}{Cl}}:$

2. Applying rule 3, we conclude that eight electrons are associated with Cl and six with I. Hence, the oxidation number of I is $+7 - 6 = +1$. The oxidation number of Cl is $+7 - 8 = -1$.
3. Note that ox. no. I + ox. no. Cl =

$$+1 \quad + \quad -1 \quad = 0$$

which is consistent with the electrical neutrality of the species.

We can simplify the consideration of some complicated species by recognizing atomic ions of which they are composed. For example, in $NiCO_3$, the carbonate ion, CO_3^{2-}, has an ionic charge -2. For the compound to be neutral, the Ni atom must have a charge and an oxidation number $+2$.

In the ionic compound ammonium phosphate, $(NH_4)_3PO_4$, there are two polyatomic ions, the ammonium ion NH_4^+ and PO_4^{3-}. These two can be treated separately for calculations of the oxidation numbers of all the atoms. Thus, we know from above that the oxidation number of nitrogen in NH_4^+ is -3 and that of hydrogen is $+1$. Similarly, by applying either rule 3 or rule 4, we see that oxygen has oxidation number -2 in PO_4^{3-}; then, using rule 4 and denoting the oxidation number of phosphorus as x, we find that oxidation number to be 5: $x + 4(-2) = -3, x = 5$.

Complex Ions

Ions with transition metal atoms (Groups IIIB, IVB, VB, VIB, VIIB, VIII, IB, and IIB) as central atoms.

This is an important class of compounds that can be visualized readily as resulting from bonding between a so-called **donor** species, with an electron pair available for bonding, and a metallic ion, which is electron-deficient. The electron-deficient species is often referred to as an **acceptor,** or **Lewis acid.** The donor species is often called a **Lewis base,** or **ligand.** Viewing H_2O as a donor species and Cr^{3+} as an acceptor species, we can readily visualize the construction of a typical example, hexaaquochromium(III) cation, $[Cr(H_2O_6]^{3+}$:

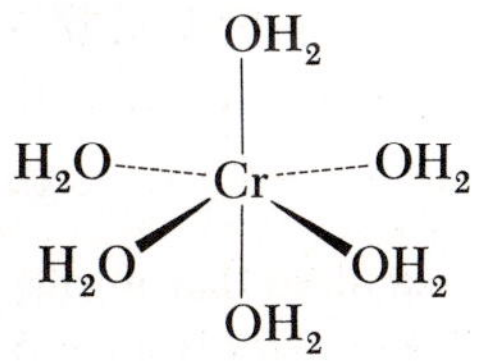

We view this species as simply a manifestation of the bonding interactions between an acceptor species and a donor species. If we view the ligand species (H_2O in this case) as retaining the same charge in the complex that it would have as a separate entity, H_2O, then the charge the chromium atom "appears" to have is +3, and that is the oxidation number that is assigned to it in this species. The same result would be obtained by directly applying rule 3 or 7 and 9.

CIS-DICHLORODIAMMINENICKEL(II), $Ni(NH_3)_2Cl_2$

1. This complex is viewed as being constructed from the ligands NH_3 and Cl^-.
2. If we imagine the ligands eliminated from the complex, with the electron pairs involved in donor-acceptor bonding to the metal, what is left is the ion Ni^{2+}. Thus, when the complex is viewed in terms of donors (ligands) and an acceptor (metal ion), the nickel atom appears to have a charge of +2. Hence, this is the assigned oxidation number; and the oxidation numbers of chlorine, nitrogen, and hydrogen are −1, −3, and +1.
3. The same conclusions would be reached by applying rule 3 directly.

Some Miscellaneous Cases

ClF_3

1. Rule 8 is the most straightforward starting point.
2. Then, if x denotes the oxidation number of chlorine, $x + 3(-1) = 0, x = +3$.
3. The oxidation number of chlorine is +3 in this species.

PCl_2F_3

1. Realizing that phosphorus is the central atom of this species, we use rule 3, and recognize that each of the three fluorine atoms and two chlorine atoms is allotted eight electrons.
2. Hence, the apparent local charge of each halogen atom is −1 and of the central atom is +5, which are the corresponding oxidation numbers.

PERMANGANATE ION, MnO_4^-

1. Applying rule 9, and denoting the oxidation number of the manganese atom in this species by the symbol x, then rule 4 gives $x + 4(-2) = -1, x = 7$.
2. The oxidation number of Mn is +7.

Balancing Oxidation-Reduction Reactions

Looking back at Eq. (7.4), one might legitimately wonder how to go about balancing an equation of this complexity. If one were to start with only a knowledge that MnO_4^- is transformed to Mn^{2+}, and Fe^{2+} into Fe^{3+}, what would be a reasonable way to proceed towards balancing the equation?

$$MnO_4^- + Fe^{2+} \rightarrow Mn^{2+} + Fe^{3+}$$

One recognizes that in this transformation, each Mn atom experiences a change in oxidation number from +5 to +2, and each Fe atom undergoes a change from +2 to +3. Within the bookkeeping scheme provided by the oxidation numbers, we say that the $MnO_4^- \rightarrow Mn^{2+}$ conversion requires five electrons from some source (the reducing agent). Each $Fe^{2+} \rightarrow Fe^{3+}$ conversion "releases" only one electron, which is available for reduction. Recognizing that the electrons required for the $MnO_4^- \rightarrow Mn^{2+}$ conversion can be provided only by the release of electrons in the $Fe^{2+} \rightarrow Fe^{3+}$ conversion, we see that five times as many iron atoms must be involved in this reaction as manganese atoms; each iron atom releases only one electron and each manganese atom requires five electrons.

		change in oxid. no.				*electrons transferred*
Fe^{2+}	$\rightarrow Fe^{3+}$	1	$\xrightarrow{\times 5}$	$5Fe^{2+}$	$\rightarrow 5Fe^{3+}$	5
MnO_4^-	$\rightarrow Mn^{2+}$	5	$\longrightarrow$	MnO_4^-	$\rightarrow Mn^{2+}$	5

Thus, as a first step in balancing the equation, we can write

$$MnO_4^- + 5Fe^{2+} \rightarrow Mn^{2+} + 5Fe^{3+} \tag{7.5}$$

This "equation" is balanced only in the sense that the number of electrons provided by the $Fe^{2+} \rightarrow Fe^{3+}$ transformation is the same as the number of electrons required by the $MnO_4^- \rightarrow Mn^{2+}$ conversion; it is not balanced in number of oxygen atoms or net electrical charge on each side (+9 left and +17 right, in the above expression). These additional balancing features can be accomplished from a knowledge of the solution in which the reaction occurs. In the present case the solution is acidic, so one knows that H^+ (H_3O^+) is present in significant concentrations. The H^+ can be used to balance the equation in electrical charge. We see that +8 is needed on the right side of Eq. (7.5) to equalize the charge on both sides.

$$8H^+ + MnO_4^- + 5Fe^{2+} \rightarrow Mn^{2+} + 5Fe^{3+} \tag{7.6}$$

Having balanced the charge, we now recognize that there is an excess of eight H atoms (in the form of $8H^+$) and four O atoms (in the form of MnO_4^-) on the left side of Eq. (7.6). This can be compensated for by adding 4 H_2O to the right side:

$$8H^+ + MnO_4^- + 5Fe^{2+} \rightleftharpoons Mn^{2+} + 5Fe^{3+} + 4H_2O \tag{7.7}$$

We recognize that this expression is balanced for both oxygen and hydrogen, and that it is simply the balanced form of Eq. (7.4).

Suppose that we were concerned with an analogous transformation occurring in basic solution instead of acidic. In this case, the appropriate manganese species for the +2 oxidation state is the hydroxide, $Mn(OH)_2$, and the pertinent +2 and +3 iron species would be $Fe(OH)_2$ and $Fe(OH)_3$. Then, the appropriate analog of Eq. (7.5) would be

$$MnO_4^- + 5Fe(OH)_2(s) \rightarrow Mn(OH)_2(s) + 5Fe(OH)_3(s) \tag{7.8}$$

To balance the charge for a basic solution, we recognize that OH^- instead of H^+ is the prevalent species, and an extra −1 charge is needed on the right side of Eq. (7.8). Therefore, we add OH^- on the right side; this yields

$$MnO_4^- + 5Fe(OH)_2(s) \rightarrow Mn(OH)_2(s) + 5Fe(OH)_3(s) + OH^- \tag{7.9}$$

We now have 14 oxygen atoms on the left side and 18 on the right, and we eliminate this imbalance by adding 4 H_2O to the left side:

$$MnO_4^- + 5Fe(OH)_2(s) + 4H_2O \rightleftharpoons Mn(OH)_2(s) + 5Fe(OH)_3(s) + OH^- \qquad (7.10)$$

Counting hydrogen atoms on each side, we recognize that both sides have 18 hydrogen atoms, and Eq. (7.10) is a balanced equation. As in other balanced equations involving strong electrolytes, we leave out the symbols of the spectator ions (Chap. 3).

Having worked through two related cases of balancing oxidation-reduction reactions by using oxidation numbers, it is now possible to formulate a recipe that can be applied generally to balancing such equations. This recipe can be given in terms of the following steps:

1. Identify which species in a chemical process contain atoms that undergo changes in oxidation number and determine the values of the oxidation numbers that change.
2. (a) Balance the electron transfer by introducing the appropriate stoichiometric coefficients to give an expression for which there is no net change in the sum of the oxidation numbers. (b) If appropriate, balance all other elements, except H or O, by introducing the appropriate species—not necessary in the above two examples.
3. Balance the charge by using a knowledge of the solution in which the reaction occurs; H^+ is used for acidic solutions and OH^- for basic solutions.
4. Correct the expression for any remaining imbalance in H or O atoms, or both, by introducing the appropriate number of H_2O's.

When each of these four steps is used in the examples below, it is indicated in the margin to the left. An example of a reaction in which step 2b is needed is the oxidation of metallic silver by concentrated nitric acid and hydrochloric acid to form AgCl(s) and NO(g). The species undergoing changes in oxidation states and the pertinent oxidation states are herewith summarized.

(1)
$$\overset{0}{Ag(s)} + \overset{+5,\,-2}{NO_3^-} \rightarrow \overset{+1,\,-1}{AgCl} + \overset{+2,\,-2}{NO(g)}$$

In step 2, one recognizes that there must be three $Ag \rightarrow AgCl$ transformations for each $NO_3^- \rightarrow NO$ transformation, and that one chloride ion must be available for each AgCl(s) unit formed. Hence, we write

(2a; 2b)
$$3Ag(s) + 3Cl^- + NO_3^- \rightarrow 3AgCl(s) + NO(g)$$

In step 3, we recognize a need for 4 H^+ on the left side to balance charge.

(3)
$$3Ag(s) + 3Cl^- + 4H^+ + NO_3^- \rightarrow 3AgCl(s) + NO(g) \qquad (7.11)$$

In the final step, we correct for the imbalance in H and O by adding two H_2O to the right side, yielding a balanced equation:

(4)
$$3Ag(s) + 3Cl^- + 4H^+ + NO_3^- \rightleftharpoons 3AgCl(s) + NO(g) + 2H_2O \qquad (7.12)$$

As a final check, one counts the number of atoms of each element on both sides and checks for charge balance.

At this point we may give another general definition of oxidation and reduction.

The **oxidation** of a species is a process in which the oxidation number of an atom in the species in question increases. **Reduction** is a process in which the oxidation number of an atom decreases. Although oxidation numbers give the appearance of localizing the oxidation-reduction process on specific atoms (the ones experiencing changes of oxidation number), these processes are inflicted on entire species, not individual atoms (except in the case of a simple monoatomic ion; for example, $Zn^{2+} \rightarrow Zn(s)$). Thus, in the oxidation-reduction reaction represented in Eq. (7.7), not just the manganese atom in MnO_4^- is reduced but the entire MnO_4^- species, even though the concept of oxidation numbers has the appearance of "blaming" the entire change on the Mn atom.

Oxidation-reduction reactions, also known as redox reactions, are an extremely important and interesting class of chemical process. These reactions account for many of the energy-providing schemes by which nature builds plants and animals, and keeps us all functioning. Molecular oxygen is commonly used to oxidize organic molecules to provide such energy. For example, the oxidation of glucose by molecular oxygen provides 686 kilocalories of energy for each mole of glucose oxidized to produce water and carbon dioxide. A **kilocalorie** is the amount of energy required to raise the temperature of one kg of water 1 °C. For the glucose oxidation, the "unbalanced equation," with pertinent oxidation numbers indicated, is

$$\text{(1)} \qquad \overset{0,\,+1,\,-2}{C_6H_{12}O_6} + \overset{0}{O_2} \rightarrow \overset{+1,\,-2}{H_2O} + \overset{+4,\,-2}{CO_2} \qquad (7.13)$$

From this we see that each carbon atom undergoes a change in oxidation number of +4, while each oxygen atom that is initially in O_2 undergoes a change of −2. Hence, for six carbon atoms involved in oxidation, there must be six O_2 molecules reduced. Thus, at step 2, we write

$$\text{(2)} \qquad C_6H_{12}O_6 + 6O_2 \rightarrow H_2O + 6CO_2 \qquad (7.14)$$

As both sides of the expression have zero net charge, the only remaining step is to correct the imbalance in hydrogen and oxygen, which can be accomplished by inserting the coefficient 6 before H_2O on the right side. This yields the balanced equation

$$\text{(4)} \qquad C_6H_{12}O_6 + 6O_2 \rightleftharpoons 6H_2O + 6CO_2 \qquad (7.15)$$

EXAMPLE 7.14 Write a balanced equation for the oxidation-reduction reaction

$$P_4 + OH^- \rightarrow PH_3 + H_2PO_2^-$$

SOLUTION (1) The oxidation states are

$$\overset{0}{P_4} + \overset{-2,+1}{OH^-} \rightarrow \overset{-3,+1}{PH_3} + \overset{+1,+1,-2}{H_2PO_2^-}$$

P has been both reduced (oxidation number change from 0 to −3) and oxidized (oxidation number change from 0 to +1). From the changes in oxidation number, three $H_2PO_2^-$ ions are formed for each PH_3 formed, giving

$$\text{(2)} \qquad P_4 + OH^- \rightarrow PH_3 + 3H_2PO_2^-$$

Next, the charge must be balanced.

(3) $$P_4 + 3OH^- \rightarrow PH_3 + 3H_2PO_2^-$$

Now, adding H_2O to balance H's, we obtain

(4) $$P_4 + 3OH^- + 3H_2O \rightleftharpoons PH_3 + 3H_2PO_2^-$$

Checking charge and mass balance, we find that the equation is totally balanced.

STUDY PROBLEM 7(j)

Write a balanced equation for

$$HgS(s) + NO_3^- + Cl^- \rightarrow [HgCl_4]^{2-} + NO_2(g) + S(s)$$

where the reaction is run in acidic solution.

7.8 Molecular Orbital Theory

Valence Bond Approaches

Our consideration of covalent bonding has thus far pivoted on one of two common general approaches. These are called the valence bond method and the molecular orbital method. In the valence bond method, one visualizes covalent bonds in terms of two atomic orbitals, which may be hybrid orbitals containing a total of two electrons (say, usually one electron in each orbital), and then "allowing" these occupied atomic orbitals to overlap in a way that results in the sharing of the electron pair over both atomic centers. This is clearly the approach we have used earlier. In some cases the approach was explicit, as when we visualized the formation of a π bond in ethylene by the overlap of two p orbitals, each containing an electron. In other cases, as in our discussion of resonance, that the overall model was based on the valence bond method may not have been so clear. However, whenever structural formulas are drawn with lines representing discrete covalent bonds, as in the resonance structures of Sec. 7.3, these lines conjure up the vision of a valence bond model of covalent bonding—two atomic orbitals, containing their two electrons, combining to form a covalent (valence) bond. This is an extremely useful and powerful approach for qualitative purposes, but has not been very popular in recent years as a basis for quantitative theoretical treatments.

The Molecular Orbital Method

The other important approach, the molecular orbital model, has a different tactic. Instead of picturing partially occupied atomic orbitals combined to form valence bonds, we imagine combining the available atomic orbitals to form orbitals that belong to the entire molecule—molecular orbitals. Once these molecular orbitals (abbreviated MO's) are derived, electrons are mentally allotted to them in a way that is entirely analogous to allotting electrons to atoms mentally in accounting for atomic structure. The mental or computational construction of orbitals for an entire molecule is in many ways analogous to the hybridization of orbitals (Sec. 7.3). In both

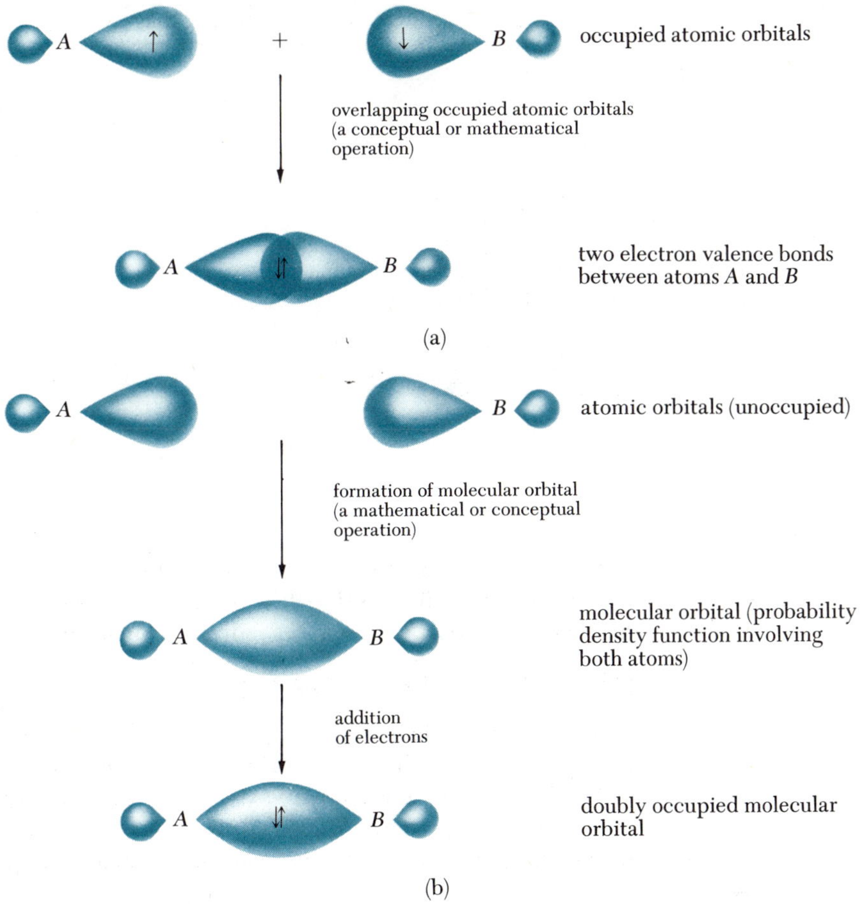

Figure 7.14 Comparison of (a) valence bond and (b) molecular orbital approaches for a diatomic system, *AB*.

cases, the mathematical justification of the approach is the combination of individual atomic orbitals into useful, new composite orbitals—hybrid atomic orbitals or molecular orbitals, depending on which case is at hand. This situation is described pictorially in Fig. 7.14, in which it is compared with the valence bond approach.

One experimental method that clearly shows how the molecular orbital approach is superior in some applications to the simple valence bond approach is measurement of magnetic properties. A substance is called **diamagnetic*** if it is slightly repelled in a magnetic field, and **paramagnetic** if it is attracted into a magnetic field. The extent to which a substance is diamagnetic or paramagnetic can be measured by the arrangement shown in Fig. 7.15. If a sample is diamagnetic, it will appear to be lighter when the electromagnet is turned on; it will appear heavier in the presence of the magnetic field if it is paramagnetic.

An intrinsic property of each electron is an associated magnetic moment, when each electron "acts" in a magnetic sense almost as if it were a little bar magnet

*Diamagnetism is associated with complex electronic circulations that are induced in a species by a magnetic field and with the magnetic field generated by those circulations.

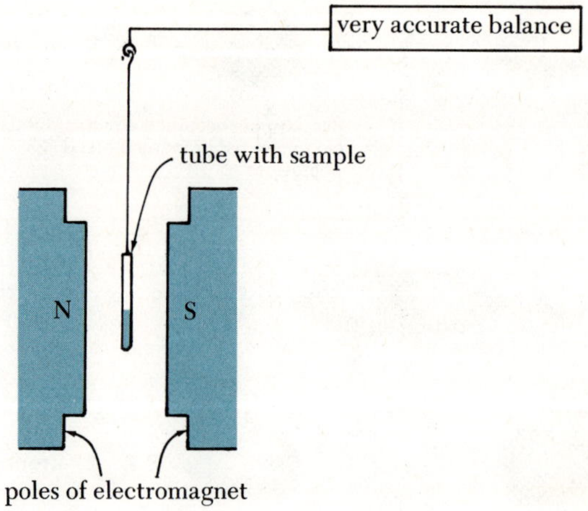

Figure 7.15 Measurement of magnetic properties (to determine diamagnetism and paramagnetism).

(Chap. 6). When electrons are paired as in covalent bonds or lone pairs—for instance, the nonbonded pair in NH_3—these magnetic moments cancel. If there is an odd number of electrons in a species, however, or if an even number is not perfectly paired $\left(\text{as in electron configuration } \underset{1s}{\underline{\uparrow\downarrow}}\ \underset{2s}{\underline{\uparrow\downarrow}}\ \underset{2p_x}{\underline{\uparrow\downarrow}}\ \underset{2p_y}{\underline{\uparrow}}\ \underset{2p_z}{\underline{\uparrow}}\right)$, then a net magnetic moment persists. This resultant nonzero magnetic moment gives rise to paramagnetism. The size of the effect can be related to the number of unpaired electrons.

The O_2 molecule provides a good example of how MO theory is consistent with observed magnetic properties in cases for which VB theory is inadequate. According to the VB approach, all the electrons in O_2 should be paired.

$$:\ddot{O}::\ddot{O}:$$

The molecule appears to be held together by a double bond, consisting of one π bond and one σ bond. Such a situation would clearly give rise to a diamagnetic substance. Experimentally one finds, however, that oxygen is paramagnetic, as one can clearly demonstrate in an apparatus analogous to that shown in Fig. 7.15 by using a sample of liquid or solid O_2. The magnitude of the paramagnetism indicates that there are two unpaired electrons per O_2 molecule.* The paramagnetism of O_2 can be accounted for in terms of molecular orbital theory. Let us, at this point, explore some of the simple details of this approach using the O_2 case as one of the vehicles.

In the molecular orbital model, orbitals that span the entire molecule or ion are constructed;† the most common approach is to construct them from atomic orbitals

*The magnitude of the paramagnetic effect due to an unpaired electron is usually at least 1000 times the diamagnetic effect of a pair of electrons. Hence, a single unpaired electron can be detected in a compound containing many pairs of electrons.

†The construction of molecular orbitals is not an arbitrary process. Precise mathematical procedures have been developed, based on the Schrödinger equation (Sec. 6.9). These procedures typically involve complex computational methods, with modern electronic computers. Nevertheless, the qualitative sense of the MO method is also of interest, without any computations.

as building blocks. Atomic orbitals can be combined in more than one way, forming molecular orbitals of more than one type. A mathematical analysis shows that the number of independent MO's that can be constructed from AO's equals the number of AO's from which they are constructed. If the energy of an electron in a given MO is lower than the energy of an electron in the AO's from which the MO is constructed, that MO is said to be a **bonding molecular orbital**; the electron is more stable in the orbital of the molecule than in the orbital of a separate atom, tending to favor the existence or formation of a bond. If on the other hand the energy of an electron in an MO is higher than the energy of an electron in the AO's from which it is constructed, then the MO is said to be **antibonding**; the electron is more stable in the orbitals of separate atoms than in the orbital of a molecule, tending to disfavor the existence or formation of a bond. In a case in which the MO and parent AO's are nearly the same in energy, the MO is often referred to as **nonbonding**. The hydrogen molecule is a simple example in which the construction of two MO's, one bonding and one antibonding, from two atomic orbitals can be illustrated. This case is shown diagrammatically in Fig. 7.16. The sign referred to is a consequence of the wave description. In that diagram it is seen that when two hydrogen atom 1*s* orbitals are combined with the same sign, the wavefunctions and consequently the probability density function reenforce in the overlap region, giving rise to a bonding MO. The energy of two electrons in this bonding MO is lower than if the electrons were in the two separate 1*s* orbitals of the two hydrogen atoms, H_a and

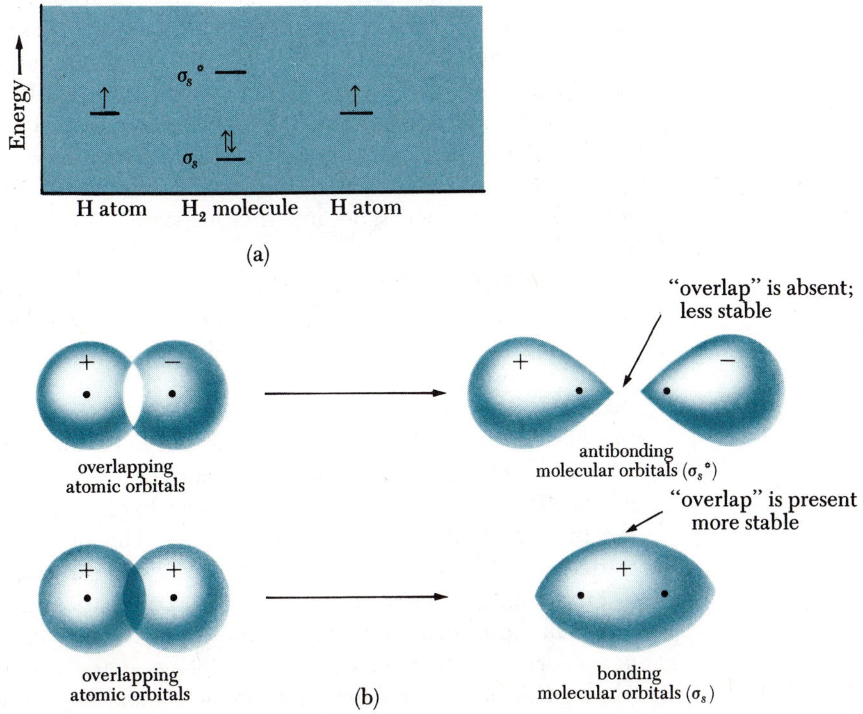

Figure 7.16
Molecular orbitals for H_2 derived from two 1*s* atomic orbitals. (a) Energy relations. (b) The role of the signs of atomic orbital functions. Note that like signs will add constructively (bonding) and unlike signs will add destructively (antibonding).

H_b. When two hydrogen atom 1*s* orbitals are combined with opposite sign, it is seen that they combine destructively, with a resulting absence of probability density at the point midway between the two nuclei; this region of zero value of the probability density function is called a **node.** For this case, the probability density is concentrated at the "outside" of the H_2 system, rather than between the positive nuclei, and the resulting energy of two electrons in such an MO, an antibonding MO, is higher than if the electrons were in two separate hydrogen 1*s* atomic orbitals. Both the bonding and antibonding MO's of H_2 are referred to as sigma (σ) MO's, meaning that they are cylindrically symmetrical about the axis of the associated bond, which is a line drawn between the centers of the two nuclei. The antibonding σ orbital for H_2 is indicated in Fig. 7.16 by the symbol "*," the asterisk indicating the antibonding character.

Allotting electrons mentally to the MO's of a molecule follows the same kind of rules described for the occupation of atomic orbitals in atoms (Chap. 6). To describe the ground state of a molecule, we imagine filling the molecular orbitals from the bottom up, a maximum of two electrons per MO and observing Hund's rule where appropriate. In the H_2 case, this brings us to the conclusion that the ground state is described as consisting of two electrons of opposite spin in the bonding MO.

The O_2 case is practically the same conceptually, but more complicated because it involves constructing MO's from not just 1*s* atomic orbitals but also from the 2*s* and 2*p* atomic orbitals of oxygen. This case is shown diagrammatically in Fig. 7.17.

As indicated in Fig. 7.17, the two 1*s* AO's combine to form two σ MO's—one bonding and one antibonding. Similarly, the two 2*s* AO's combine to form a σ_s and a $\sigma_s{}^*$ orbital (the subscript *s* indicating that the AO's from which these MO's are constructed are *s* orbitals). The sign combinations of the *s* AO's involved in constructing these σ_s orbitals and the general shapes of these MO's are similar to what is shown for the H_2 case in Figure 7.16. The two 2*p* atomic orbitals with axes directed along a line between the nuclei (let's call it the *z* axis) are combined to form one bonding MO and one antibonding MO, denoted σ_p and $\sigma_p{}^*$, respectively. The sign combinations of the $2p_z$ AO's involved in constructing σ_p and $\sigma_p{}^*$; and the resulting general shapes of these MO's are shown in Fig. 7.17b. These MO's like the σ_s and $\sigma_s{}^*$ MO's are all σ MO's, because they are symmetrical about an axis joining the two nuclei.

Another class of MO's is formed by combining the $2p_y$ orbitals of both oxygen atoms and by combining the $2p_x$ AO's of both oxygens. Depending on the combination of signs of the AO's, two general shapes of the resulting MO's result, as shown in Fig. 7.17b. One shape corresponds to the bonding MO, denoted π_p, and the other shape characterizes the antibonding MO, $\pi_p{}^*$. All these molecular orbitals are given the pi designation, denoting that they are antisymmetric relative to a plane of symmetry. By this we mean that if we could project these π MO's through a plane, we should see an image that is the same as what is on the opposite side of the plane, except that it would have the opposite sign. This situation is represented, and contrasted with the symmetric case, in Fig. 7.18. The construction of π orbitals depicted in Fig. 7.18b applies equally well to the combination of two $2p_x$ AO's or two $2p_y$ AO's. In the former case, the symmetry plane (that defines the designation) is the *y–z* plane; in the latter case it is the *x–z* plane. These cases are depicted in Fig. 7.19. The π_p molecular orbital formed from the combination of two $2p_x$ AO's has the same

Figure 7.17
Molecular orbitals of O_2.
(a) Energy relations.
(b) Representation of molecular orbitals from combinations of atomic orbitals.

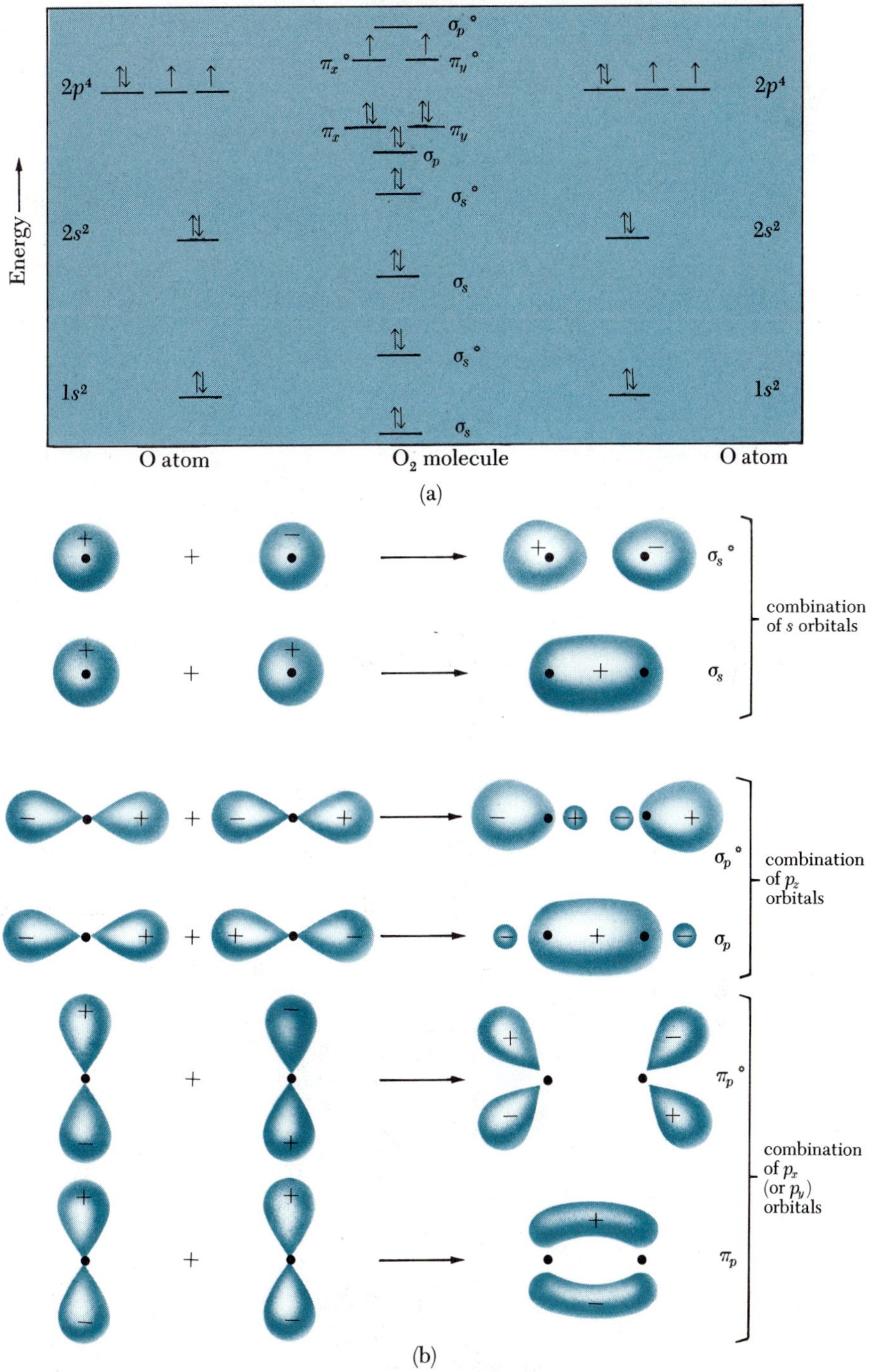

energy and shape as the π_p MO constructed from the two $2p_y$ AO's; only the orientation in space is different, by 90°. Similarly, the π_p^* molecular orbital constructed from the two $2p_x$ AO's has the same shape and energy as the π_p^* MO obtained from $2p_y$ AO's; only the orientation in space differs. This **degeneracy**, or identical energy, of two π_p molecular orbitals and of two π_p^* MO's is shown explicitly in Fig. 7.17a.

Figure 7.18 Symmetry property of a π orbital. (a) The antisymmetric case showing π symmetry. (b) The symmetric case (σ) for comparison.

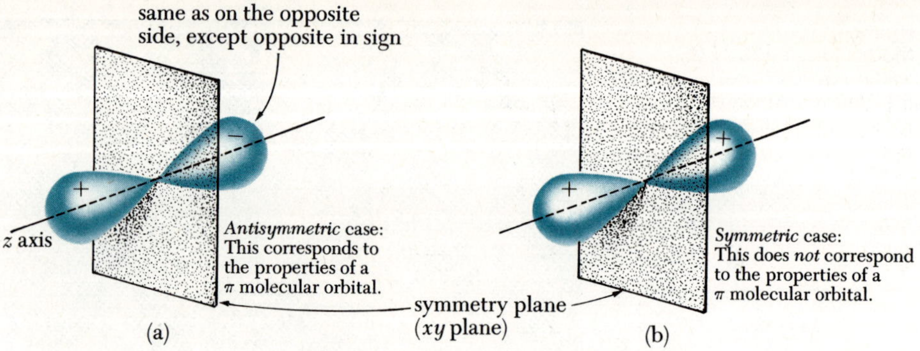

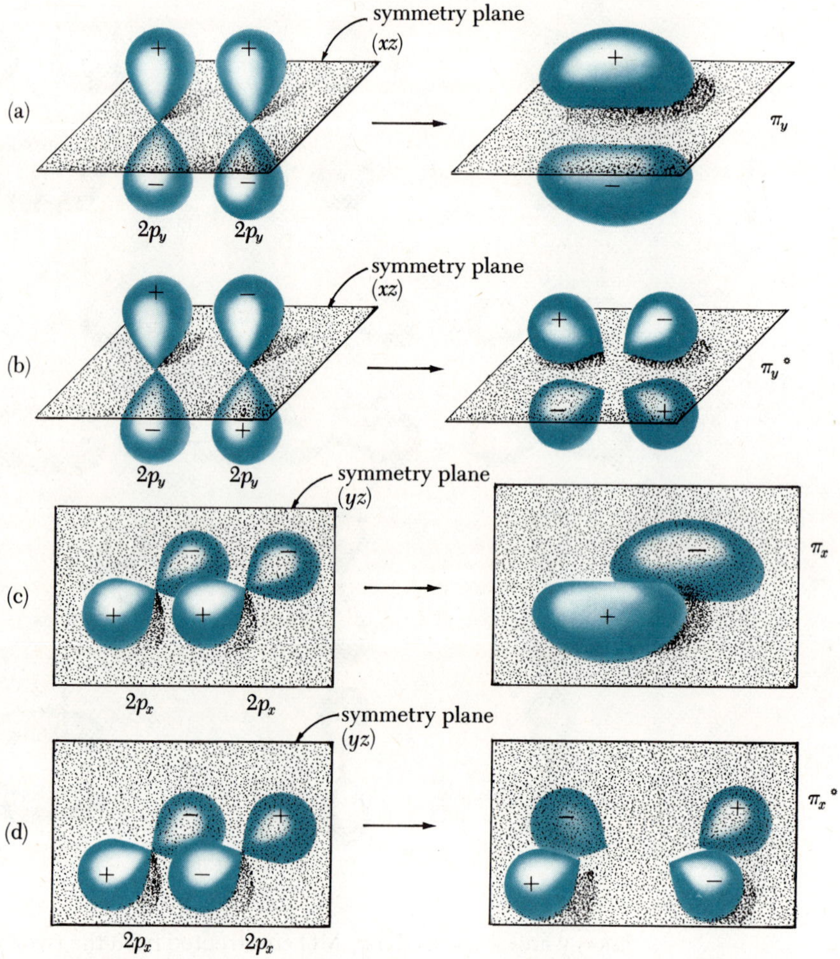

Figure 7.19 Symmetry relations of atomic and molecular orbitals relative to a reflection plane. (a) Construction of π_y MO from $2p_y$ AO's (xz symmetry plane). (b) Construction of π_y^* MO from two $2p_y$ AO's (xz symmetry plane). (c) Construction of π_x MO from two $2p_x$ AO's (yz symmetry plane). (d) Construction of π_x^* MO from $2p_x$ AO's (yz symmetry plane).

Using the diagram given in Fig. 7.17a, we can readily account for the ground state of O_2 by allotting electrons to each available orbital in the same way that was described for atoms (start from bottom, and use Pauli principle and Hund's rule). We see that the ground state of this 16-electron system has two electrons per orbital (paired, with opposite spin) up to the $\pi_p{}^*$ MO's. Then, we allot one electron to each of these degenerate MO's. According to Hund's rule both of these electrons have the same spin orientation. Recognizing these two unpaired electrons leads unequivocally to the prediction that O_2 should be a paramagnetic substance, in agreement with experiment.

Although the simple valence bond method could not account for the paramagnetism of molecular oxygen, it leads simply and directly to the conclusion that the oxygen atoms are held together by a double bond. The idea of single bonds, double bonds, and triple bonds is also relevant to the MO approach and enters MO theory through the concept of bond order. The **bond order** is a measure of the "extent of bonding"; it is defined to have its value unity for a single bond, 2 for a double bond, and 3 for a triple bond. Mathematically, it can be defined as half the number of electrons in bonding MO's minus half the number in antibonding MO's:*

$$\text{BO} = \tfrac{1}{2}(\text{no. electrons in bonding MO's} - \text{no. electrons in antibonding MO's}) \qquad (7.16)$$

This formula recognizes that electrons in bonding MO's contribute to the stability of a bond, whereas electrons in antibonding MO's detract from that stability. Applying Eq. (7.16) to the O_2 case shown in Fig. 7.14a, we obtain

$$\begin{aligned}\text{BO}_{O_2} &= \tfrac{1}{2}[2(\sigma_s) + 2(\sigma_s) + 2(\sigma_p) + 2(\pi_x) + 2(\pi_y) \\ &\qquad - 2(\sigma_s{}^*) - 2(\sigma_s{}^*) - 1(\pi_x{}^*) - 1(\pi_y{}^*)] \\ &= \tfrac{1}{2}(10 - 6) = 2\end{aligned}$$

where the occupied MO's are in parentheses.

The type of MO description given above for molecular oxygen can be extended easily to other simple diatomic species. Figure 7.20 provides a summary of the occupation of molecular orbitals for the ground states of B_2, C_2, N_2, O_2, and F_2. The number of unpaired electrons and bond orders corresponding to these MO occupations are also summarized in Fig. 7.20, together with relevant experimental data. A rough but recognizable correlation is seen to exist between the energy required to dissociate the molecule into atoms (bond strength) and the bond order. Also, that unpaired electrons give rise to paramagnetism is reflected in the table.

EXAMPLE 7.15 Using Figure 7.18a, predict the bond order and number of unpaired electrons for He_2, Li_2, N_2, and NO.

SOLUTION The molecular orbital diagrams and results are as follows.

He_2 $\quad \overset{\uparrow\downarrow}{—}\,\sigma_s{}^*$ $\quad \overset{\uparrow\downarrow}{—}\,\sigma_s$ $\quad$ BO $= \tfrac{1}{2}(2 - 2) = 0$; unstable molecule, no unpaired electrons.

*More detailed expressions, of which Eq. (7.16) is a special case, are defined for mathematical calculations based on MO theory.

Li_2

$\uparrow\downarrow$ σ_s
$\uparrow\downarrow$ σ_s^*
$\uparrow\downarrow$ σ_s

$BO = \frac{1}{2}(4 - 2) = 1$; single bond, no unpaired electrons.

N_2

$\uparrow\downarrow$ σ_p
$\uparrow\downarrow$ π_x $\uparrow\downarrow$ π_y
$\uparrow\downarrow$ σ_s^*
$\uparrow\downarrow$ σ_s
$\uparrow\downarrow$ σ_s^*
$\uparrow\downarrow$ σ_s

$BO = \frac{1}{2}(10–4) = 3$; triple bond, no unpaired electrons.

NO

$\uparrow$ π_x^* π_y^*
$\uparrow\downarrow$ σ_p
$\uparrow\downarrow$ π_x $\uparrow\downarrow$ π_y
$\uparrow\downarrow$ σ_s^*
$\uparrow\downarrow$ σ_s
$\uparrow\downarrow$ σ_s^*
$\uparrow\downarrow$ σ_s

$BO = \frac{1}{2}(10 - 5) = 2.5$; partial bond order (between a double and a triple bond), 1 unpaired electron.

SUMMARY

Compounds can be classified into two main types—ionic compounds and covalent, or molecular, compounds. By using a theoretical diagram for the formation of ionic compounds called the Born-Haber cycle, we can show why only certain combinations of elements give rise to stable ionic compounds. Two of the most important quantities that determine whether a particular ionic compound is stable are the ionization potential of the metallic element and the electron affinity of the nonmetallic element. Electron affinity and ionization potential are fundamental to electronegativity. The tendency of a combination of elements to form an ionic or covalent compound can be predicted from electronegativity values. Also, for covalent compounds, the polarity of bonds or of entire molecules can be predicted from electronegativity values.

Using either hybridization or VSEPR, we can correctly predict the shapes and polarity or nonpolarity of molecules and account for much of what is known about molecular architecture. In the VSEPR method, it is important to determine the number of electron pairs in the valence shell of the central atom of a molecule. In

Figure 7.20
A molecular orbital scheme for second row homonuclear molecules. Note that the π_p energy level is higher than the σ_p level for O_2 and F_2 (shown within color screen), whereas the reverse is true for Li_2 through N_2.

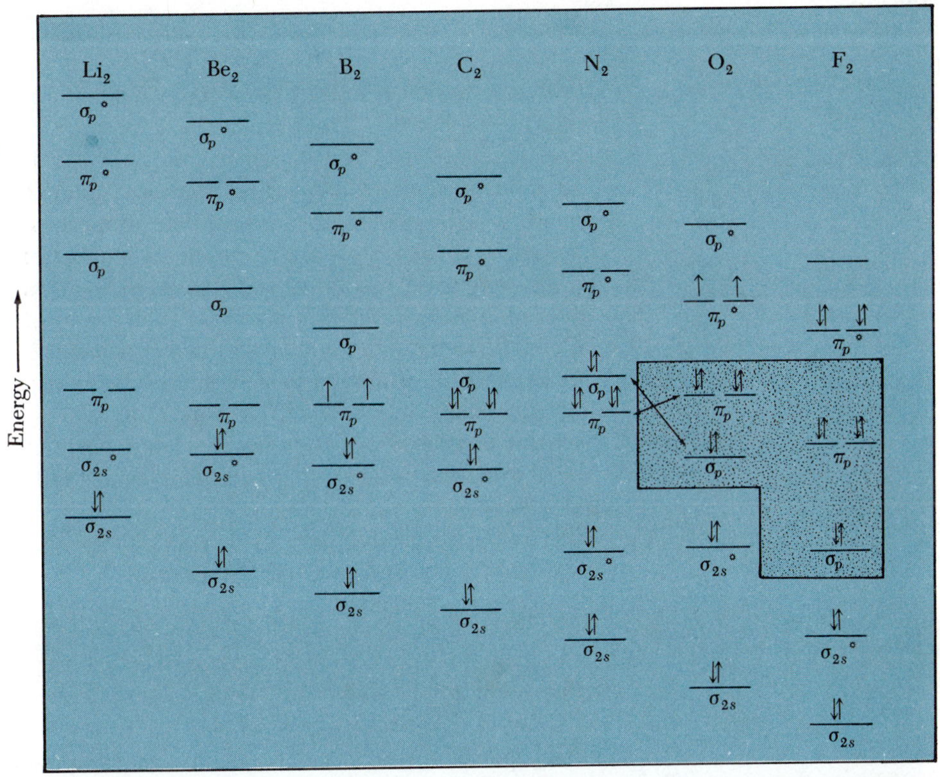

	Li_2	Be_2	B_2	C_2	N_2	O_2	F_2
Bond order	1	0	1	2	3	2	1
Bond strength (kcal/mole)	26	0	69	150	225	118	37
Number of unpaired electrons	0	0	2	0	0	2	0
Magnetism	diamagnetic	diamagnetic	paramagnetic	diamagnetic	diamagnetic	paramagnetic	diamagnetic

relation to functional groups, the structures of many parts of molecules are similar from one molecule to another, allowing us to picture building up a very large molecule by recognizing the functional groups of which it is composed.

Certain features of molecules are not obvious from their structure. A very important feature of atoms in molecules is their oxidation number, which can be used to understand a large class of reactions known as oxidation-reduction reactions. We can apply a systematic method for properly balancing such reactions.

STUDENT CHECKPOINTS

After studying this chapter, the student should be able to:

1. Explain how conductivity is used to indicate ionic behavior.
2. Explain why alkali metal ions are commonly of +1 charge while alkaline earth ions are commonly +2.
3. Recognize the similarity of electronic configurations within each group and explain how the electronic configuration affects the properties of the element.

4. Write electronic configurations of transition metal ions and also ions of representative elements.
5. Describe the distinction between ionic and covalent bonding, recognizing that in many chemical properties there is a gradual progression among compounds from nonpolar covalent to ionic properties.
6. Use Lewis dot structures to determine the number of bonds between atoms.
7. Define *resonance, resonance contributor,* and *resonance hybrid.*
8. Determine bond order in a molecule that is a resonance hybrid.
9. Discuss the hybridization assigned to a molecule.
10. Use the VSEPR approach to determine structures of compounds of the nontransition elements, those of the main group.
11. Determine polarity of bonds and molecules using electronegativity values.
12. Assign oxidation numbers to atoms in molecules and ions to balance oxidation-reduction reactions.
13. Use a molecular orbital diagram for homonuclear diatomic molecules to determine bond order and magnetic properties.

EXERCISES

7.1.* Explain how the conductivity experiment distinguishes between ionic and covalent compounds. Why does this difference exist?

7.2.* Using the Born-Haber cycle, choose the process that would cause the following species not to be the most favorable compounds formed from the constituent elements: (a) $CaCl$; (b) KCl_2; (c) Mg_2O; (d) $AlCl_2$; (e) Al_3Cl; (f) KBr_2.

7.3. Using the Born-Haber cycle and balancing equations, choose the most stable solid product of the following reactions; also choose one less stable possible product. Explain your choices. (a) $K(s) + Cl_2(g) \rightarrow ?$ (b) $Mg(s) + Br_2(g) \rightarrow ?$ (c) $Ca(s) + O_2(g) \rightarrow ?$ (d) $Al(s) + Cl_2(g) \rightarrow ?$ (e) $Al(s) + O_2(g) \rightarrow ?$

7.4.* Explain why lithium and fluorine have such different chemical properties.

7.5.* Explain the similar chemistry of strontium and calcium. (^{90}Sr is found in radioactive fallout and takes the place of calcium in bones and teeth.)

7.6. Why does lithium have a greater tendency for covalent bond formation than potassium does?

7.7.* Why does magnesium have a greater tendency for covalent bond formation than strontium does?

7.8. (a) Mixing elemental sodium and chlorine causes ionic sodium chloride to form, whereas mixing sodium and potassium leads to no chemical reaction. Why? (b) Chlorine and bromine can be made to combine, but the resulting species is not ionic. Why?

7.9.* What are the ground state electronic configurations of the following cations: (a) K^+; (b) Ca^{2+}; (c) Al^{3+}; (d) Tl^{3+}?

7.10. What are the ground state electronic configurations of the following anions: (a) Cl^-; (b) Se^{2-}; (c) N^{3-}?

7.11.* What are the ground state electronic configurations of the following transition metal cations: (a) Mn^{2+}; (b) Fe^{3+}; (c) Co^{3+}; (d) Pt^{2+}; (e) Ir^+?

7.12. What +1, +2, and −1 ions have the same ground state electronic configurations as the noble gases?

7.13.* What +2 ions have the following electronic configurations: (a) $[Ar]3d^7$; (b) $[Xe]4f^{14}5d^6$; (c) $[Kr]4d^6$; (d) $[Ar]3d^6$; (e) $[Kr]4d^7$?

7.14. What +3 ions have the following electronic configurations: (a) $[Ar]3d^5$; (b) $[Kr]4d^5$; (c) $[Ar]3d^6$; (d) $[Kr]4d^6$; (e) $[Xe]4f^{14}5d^6$?

7.15.* Write the ground state electronic configuration of boron. Why does boron form three bonds instead of just one?

7.16. How many covalent bonds are formed between the central atom and other atoms in the most stable products of the following reactions? Balance the equations appropriately. $B + F_2 \rightarrow$?; $H_2 + Cl_2 \rightarrow$?; $N_2 + H_2 \rightarrow$?

7.17.* How many bonds are formed between the central atom and other atoms in the following molecules or ions: (a) carbon dioxide; (b) the nitrogen molecule; (c) sulfate ion; (d) phosphorus pentafluoride?

7.18. Draw Lewis dot structures of (a) potassium iodide; (b) barium nitrate ($Ba(NO_3)_2$); (c) germanium chloride; (d) stannous fluoride.

7.19.* Draw Lewis dot structures of (a) sodium triiodide, NaI_3; (b) calcium nitrite, $Ca(NO_2)_2$; (c) dichlorotrifluorophosphorane, PCl_2F_3; (d) the oxygen molecule.

7.20. Draw the resonance contributors of sulfur dioxide, nitrogen dioxide, and benzene.

7.21.* In formaldehyde, 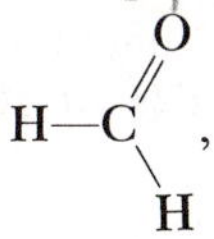, and carbonate ion, CO_3^{2-}, the average bond order is $1\frac{1}{3}$ in both bases. In the carbonate ion, all three bonds are the same, whereas in formaldehyde there is no evidence for any averaging. Explain.

7.22. The boron trifluoride molecule forms an equilateral triangle instead of a T-shaped molecule. How is this result rationalized?

7.23.* What is the shape of the hybridized orbitals that are termed sp, sp^2, and sp^3? Give examples of molecules for which these hybridizations are appropriate.

7.24. What is the shape of the hybridized orbital set formed from (a) s, three p, and a d orbital; and (b) an s, three p, and two d orbitals? Give examples of molecules for which these hybridizations are appropriate.

7.25.* What hybrid orbital combinations can be assigned to the central atom in the following species: (a) CO_2; (b) BCl_3; (c) NH_3; (d) PCl_5; (e) NO_3^-?

7.26. What hybrid orbital combination can be assigned to the central atom or atoms in the following species: (a) CoF_6^{3-}; (b) acetylene, HC≡CH; (c) HBr; (d) methanol, CH_3OH; (e) acetone, 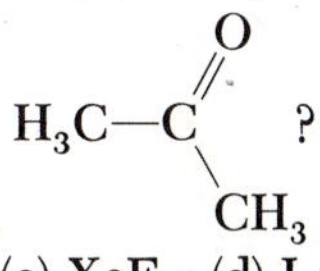 ?

7.27.* Predict the geometries of (a) Cl_2O; (b) PF_4^+; (c) XeF_4; (d) I_3^-; (e) CCl_4.

7.28. Predict the geometries of (a) ICl_3; (b) acetic acid, H_3CCOOH; (c) $CHCl_3$; (d) BrF_4^-.

7.29.* To which class of organic molecules does each of the following belong? Can you name them?

(a) $CH_3CH_2CH_2CH_3$; (b) CH_3NH_2; (c) $BrC\equiv CBr$;

(d) $H_3C-\underset{OH}{\overset{O}{\underset{\|}{C}}}$ (e) CH_3CH_2OH; (f) $\underset{CH_3}{\overset{Br}{\underset{|}{C}}}=CH_2$.

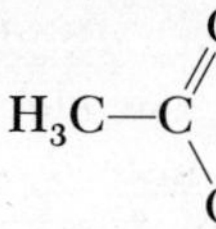

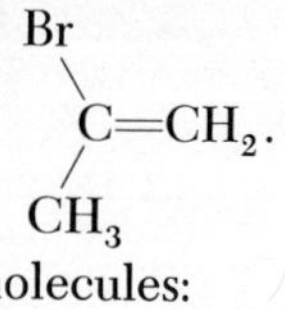

7.30. Indicate the functional groups in the following molecules:

(a) formic acid $H-\underset{OH}{\overset{O}{\underset{\|}{C}}}$; (b) diethylamine, $(CH_3CH_2)_2NH$;

(c) phenylalanine,

$$\begin{array}{c} HC-CH \\ HC \quad\quad C-CH_2 \quad O \\ HC=CH \quad C-C \\ H_2N \quad H \quad OH \end{array}$$

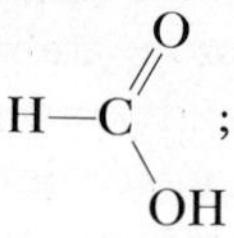

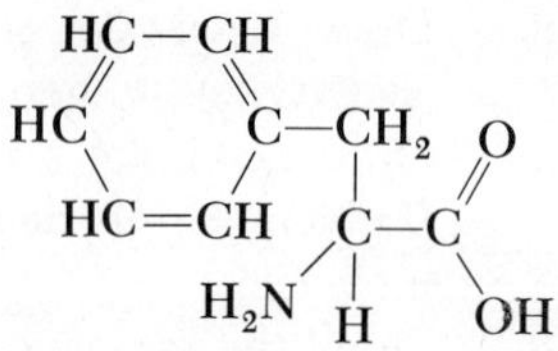

(d) *cis*-dichlorodiammineplatinum(II), $[Pt(NH_3)_2Cl_2]$.

7.31.* Indicate the functional groups in the following molecules: (a) butadiene, $CH_2=CH-CH=CH_2$; (b) dibromoacetylene, $BrC\equiv CBr$; (c) acetyl chloride, $CH_3-\underset{Cl}{\overset{O}{\underset{\|}{C}}}$.

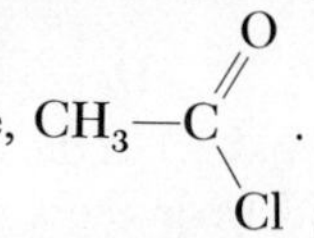

7.32. Using the Mulliken formulation of electronegativity, determine which atom is the more electronegative in each of the following species: CO_2, $SbCl_3$, ICl_3.

7.33. Another way of determining electronegativities, the Allred-Rochow method, uses the expression $Z_{eff}e^2/r^2$, where Z_{eff} is the effective nuclear charge that would affect an incoming electron. Do you expect the electronegativities calculated from this formula to be consistent with electronegativities determined by the Mulliken formula? Why or why not?

7.34. Classify the bonds in the following species as nonpolar, polar, or ionic: (a) HI; (b) N_2; (c) CCl_4; (d) $BiCl_3$; (e) TlCl; (f) CH_3OH.

7.35.* Which of the following molecules are polar: (a) carbon disulfide, CS_2; chloromethane, H_3CCl; the nitrogen molecule; acetone; $H_3C-\underset{CH_3}{\overset{O}{\underset{\|}{C}}}$; phosgene, $Cl-\underset{Cl}{\overset{O}{\underset{\|}{C}}}$? (b) Rank them in order of increasing polarity.

7.36. In an extraction (Fig. 2.12) in which carbon tetrachloride and water are used, indicate into which layer the following species will migrate predominantly: methanol, CH_3OH; carbon disulfide, CS_2; sodium chloride; diethyl ether, $CH_3CH_2OCH_2CH_3$.

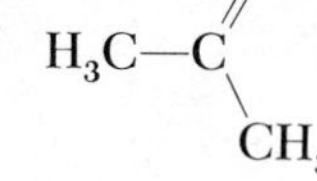

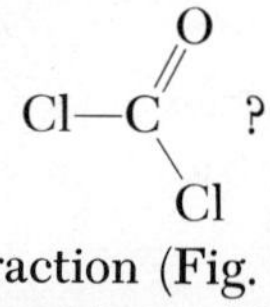

7.37.* Write Lewis dot formulas and indicate any nonzero formal charges for the following species: OCN^-, NO, NO_2^-, NH_4^+.

7.38. Give the oxidation number of each atom in MoO_3, Na_3VO_4, CH_3OH, and $Mg(ClO_4)_2$.

7.39.* Give the oxidation number of each atom in $Cr_2O_7^{2-}$, CrO_2^-, CoF_6^{3-}, H_2O_2.

7.40. Write balanced equations for the following: (a) $C_6H_5CHO + Cr_2O_7^{2-} \rightarrow C_6H_5COOH + Cr^{3+} + H_2O$ (acidic solution); (b) $Bi(OH)_3(s) + Sn(OH)_3^- \rightarrow Sn(OH)_6^{2-} + Bi(s)$ (basic solution); (c) $ClO^- + CrO_2^- \rightarrow CrO_4^{2-} + Cl^-$ (basic solution).

7.41.* Write balanced equations for the following: (a) $H_3PO_3 + Cr_2O_7^{2-} \rightarrow H_3PO_4 + 2Cr^{3+} + H_2O$ (acidic solution); (b) $Cu(NH_3)_4^{2+} + S_2O_4^{2-} \rightarrow SO_3^{2-} + Cu(s) + NH_3$ (basic solution); (c) $Fe_3O_4(s) + H_2O_2 \rightarrow Fe^{3+} + H_2O$ (acidic solution).

7.42. Predict the bond order and number of unpaired electrons for each of the following: (a) HeH; (b) He_2; (c) Na_2; (d) CO; (e) F_2.

ANSWERS TO SELECTED EXERCISES

7.1 Dissolved ionic compounds conduct a current and covalent compounds do not. Ionic compounds are held together by electrostatic forces. Polar solvent molecules can replace the counter ions by providing electrostatic attraction for the ions, shielding them from each other. **7.2** (a) Lattice energy would favor $CaCl_2$; (b) EA_2 of Cl; (c) more favorable lattice energy for MgO ($Mg^{2+}O^{2-}$); (d) more favorable lattice energy for $AlCl_3$ ($Al^{3+}Cl_3^-$); (e) unfavorable EA_3 for Cl; (f) unfavorable IP_2 for K. **7.4** Lithium is larger than fluorine with fewer protons to attract its electrons. Hence, it is much easier to remove an electron from lithium, which behaves as a metal, than from fluorine, which actually attracts an additional electron and acts as a nonmetal. **7.5** They have very similar electronic configurations. **7.7** Both commonly have the same charge (+2), and Mg^{2+} is much smaller than Sr^{2+}, polarizing electrons towards itself and sharing them to some extent. **7.9** (a) [Ar]; (b) [Ar]; (c) [Ne]; (d) $[Xe]4f^{14}5d^{10}$. **7.11** (a) $[Ar]3d^5$; (b) $[Ar]3d^5$; (c) $[Ar]3d^6$; (d) $[Xe]4f^{14}5d^8$; (e) $[Xe]4f^{14}5d^8$. **7.13** (a) Co^{2+}; (b) Os^{2+}; (c) Ru^{2+}; (d) Fe^{2+}; (e) Rh^{2+}. **7.15** (a) $1s^22s^22p^1$. The extra energy released by formation of two additional bonds more than compensates for the promotion energy required. **7.17** (a) 4; (b) 3; (c) 5; (d) 5. **7.19** (a) Na^+ + :Ï:Ï:Ï:⁻ ; (b) Ca^{2+} + 2 :Ö:N::Ö⁻ ; (c) 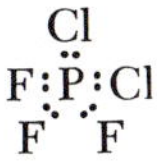; (d) :Ö::Ö: .

7.21 Carbonate has three equivalent resonance contributors. In formaldehyde, however, C–H double bonds cannot be drawn and there are no resonance contributors like those in carbonate ion. **7.23** *sp*, linear, BeH_2; sp^2, triangular, BF_3; sp^3, tetrahedral, CH_4. **7.25** (a) *sp*; (b) sp^2; (c) sp^3; (d) sp^3d; (e) sp^2. **7.27** (a) bent; (b) tetrahedral; (c) square planar; (d) T-shaped; (e) tetrahedral. **7.29** (a) alkanes, butane; (b) amine, methylamine; (c) alkyne, dibromoacetylene; (d) carboxylic acid, methylcarboxylic acid or acetic acid; (e) alcohol, ethyl alcohol, or ethanol; (f) alkene, 2-bromopropylene. **7.31** (a) two ethylene and one ethane; (b) two bromine atoms and one acetylene; (c) one chlorine atom, a carbonyl group, and a methyl group. **7.35** (a) all except CS_2 and N_2; (b) acetone > phosgene, but it is difficult to compare these with chloromethane because of the different angles involved. **7.37** :Ö::C::N̤̈⁻, N assigned −1; :Ṅ::Ö:;

$:\ddot{O}::\ddot{N}::\ddot{O}:^-$, N assigned -1; $H:\overset{\displaystyle H}{\underset{\displaystyle H}{\ddot{\underset{..}{N}}}}:H^+$, N assigned $+1$. **7.39** $Cr_2O_7^{2-}$: each Cr, $+6$; O, -2; CrO_2^-: Cr, $+3$; each O, -2; CoF_6^{3-}: Co, $+3$; each F, -1; H_2O_2: each H, $+1$; each O, -1. **7.41** (a) $8H^+ + 3H_3PO_3 + Cr_2O_7^{2-} \rightleftharpoons 3H_3PO_4 + 2Cr^{3+} + 4H_2O$; (b) $4OH^- + Cu(NH_3)_4^{2+} + S_2O_4^{2-} \rightleftharpoons 2SO_3^{2-} + Cu(s) + 4NH_3 + 2H_2O$; (c) $2Fe_3O_4(s) + H_2O_2 + 18H^+ \rightleftharpoons 6Fe^{3+} + 10H_2O$.

ANSWERS TO STUDY PROBLEMS

7(a) [Ar]; [Kr]; [Ne]; [Xe]; [Ar]$3d^{10}$. **7(b)** (a) 4 to C, 2 to O; (b) 4; (c) 5.

7(c) 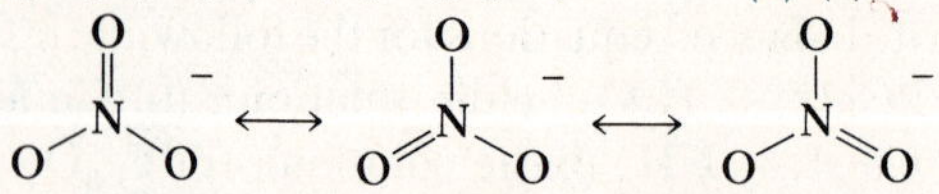

7(d) sp^3; sp^3d; sp^3. **7(e)** pyramidal; tetrahedral about C, bent about O; tetrahedral. **7(f)** (a) amine, alkyl, carboxylic acid; (b) two alkyl and one carbonyl; (c) amine, chloride, and cobalt(III). **7(g)** very slightly polar; C—O very polar; C—Cl slightly polar, ionic. **7(h)** nonpolar; nonpolar; polar; polar. **7(i)** 0; $+1$; $+2$; $+3$. **7(j)** $4H^+ + HgS(s) + 2NO_3^- + 4Cl \rightarrow [HgCl_4]^{2-} + 2NO_2(g) + S(s) + 2H_2O$.

8

Additional Aspects of Covalent Bonding and Spectroscopy

With our current background on covalent bonding, we can now investigate some important features of covalent compounds. These features influence chemical reactions that the compounds undergo and lead to interactions between molecules, changing their chemical or physical properties, or both. The importance of chemical reactions lies sometimes with specific functional groups and sometimes with specific types of interactions called Lewis acid-base interactions and hydrogen bonding.

Certain general methods used to investigate molecules involve obtaining information about the ways in which the molecules interact with electromagnetic radiation, absorbing or emitting radiant energy. The general title *spectroscopy* embraces the study of energy absorption and emission. Spectroscopy uses the principle of quantization of energy states (Chap. 6); and spectroscopic data show very clearly the need for the quantum theory to explain the realm of submicroscopic phenomena.

8.1 Reactivities of Functional Groups

The concept of the functional group is useful not only for visualizing known molecular structures but also for determining unknown molecular structures. This structure determination depends on the tendency of functional groups to retain not only the same basic (local) structure in many molecules, but also the same types of local chemical and spectroscopic characteristics. This concept allows us to envision a complicated molecule as a series of smaller groups, the structural characteristics of which are familiar and relatively constant from one molecule to another.

An important and powerful use of the concept of functional groups in understanding chemistry lies in identifying characteristic reactivities of individual portions of a molecule. Thus one can often predict what reactions the molecule can undergo. Technically, the entire molecule undergoes the reaction but it is often useful to localize the process to a specific functional group. An analogy is the availability of the whole body of the receiver to catch a pass in football, but we concentrate on the receiver's hands or arms for catching the football—forgetting about, say, his ear lobes. The first step in this approach is to become familiar with the reactivity characteristics of individual functional groups; we shall assume that their chemical properties can be considered independent of their environment. As one becomes better acquainted with the chemistry associated with various functional groups, it is possible to see how they interact in a molecule. In the football analogy, there is a working relation between fast legs and "good hands" in the receiver. Thus our chemical knowledge can be fine-tuned to a sophisticated state. This development constitutes much of the substance of organic chemistry, which is too extensive for us to cover completely here. We will concentrate on the individual characteristics of only a few functional groups.

The Hydroxyl Group and Acidity

The hydroxyl group, ~OH, occurs in several classes of molecules in which a common chemical characteristic is found; these molecules behave as acids, undergoing reactions in which the hydrogen atom can be removed from the ~OH group as a proton, H^+, and the remaining fragment is left negatively charged. According to the idea of formal charge, we associate the negative charge with the oxygen atom:

$$\text{base} + R\text{—}O\text{—}H \rightleftharpoons R\text{—}O^- + (\text{base})H^+ \qquad (8.1)$$

where R represents some atom or group of atoms. This type of dissociation is called a **heterolytic cleavage, because the bonding electrons are split unevenly**, both electrons remaining with the oxygen atom:

$$R\text{—}O\!:\!H \rightleftharpoons R\text{—}O\!:^- + H^+$$

Note the special case of water, in which the R group is a hydrogen atom, acting as a "hydroxy acid":

$$H\text{—}O\text{—}H \rightleftharpoons H\text{—}O^- + H^+$$

Alcohols

One class of molecules in which the hydroxyl group may act as an acid is the alcohol series.° An alcohol is composed of the ~OH functional group attached to an alkyl (hydrocarbon) or substituted alkyl group.

°The hydroxyl group can undergo other reactions. For example, alcohols can be dehydrated by strong acids to form alkenes:

$$\underset{\text{ethyl alcohol}}{\begin{array}{c}H\;\;\;H\\|\;\;\;\;\;|\\H\text{—}C\text{—}C\text{—}H\\|\;\;\;\;\;|\\H\;\;\;OH\end{array}} \xrightarrow[H_2SO_4]{\text{heat}} \underset{\text{ethylene}}{\begin{array}{c}H\;\;\;\;\;\;\;\;\;H\\\diagdown\;\;\;\;\;\diagup\\C\text{=}C\\\diagup\;\;\;\;\;\diagdown\\H\;\;\;\;\;\;\;\;\;H\end{array}} + H_2O$$

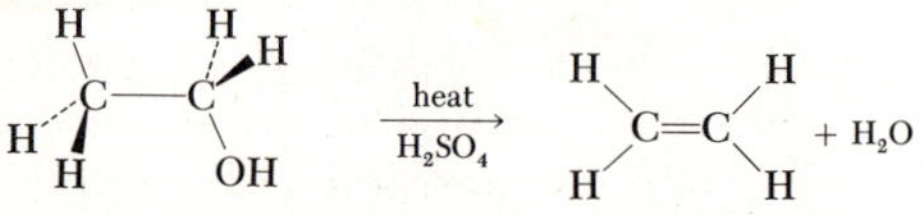

Methyl alcohol, or methanol, CH_3OH, is a highly poisonous substance sometimes referred to as wood alcohol, because it can often be obtained by heating wood. We can view this molecule as one composed of the methyl and hydroxyl functional groups. Ethyl alcohol, or ethanol, CH_3CH_2OH, is sometimes referred to as spirits, because it is the alcohol used in alcoholic beverages. We can view this molecule as composed of the methyl, $\sim CH_3$ methylene, $\sim CH_2$ and hydroxyl groups, CH_3-CH_2-OH, or as the ethyl and hydroxyl groups, C_2H_5-OH. A reaction in which an alcohol functions as an acid is a reaction with a hydride:

$$NaH(s) + CH_3OH \rightleftharpoons CH_3O^-Na^+ + H_2(g) \qquad (8.2)$$

where we can view the hydride ion H^- as the base. To "cleave" H^+ from an alcohol, very strong bases, such as the hydride ion, are needed. Hence, the alcohols are very weak acids.

It is now apparent how our knowledge of the way alcohols act like acids can be fine-tuned by determining how other structural groups on an alcohol affect the reactivity of the $\sim OH$ functional group. If we compare methyl alcohol with trifluoromethyl alcohol, CF_3OH, we find that the trifluoromethyl alcohol is a stronger acid—it requires less energy to remove a proton from trifluoromethyl alcohol than from methyl alcohol. We can explain this relative reactivity of the $\sim OH$ functional group in these two species by looking at the respective functional groups to which it is attached and determining how these functional groups might affect the reaction. The trifluoromethyl functional group, $\sim CF_3$, is effectively much more electronegative than the methyl functional group ($\sim CH_3$).° Its ability to pull electrons towards itself, to "withdraw electron density," is called an inductive effect. Since $\sim CF_3$ can withdraw electron density when the proton is removed,

$$F_3C-O-H \rightleftharpoons F_3C-O^- + H^+$$

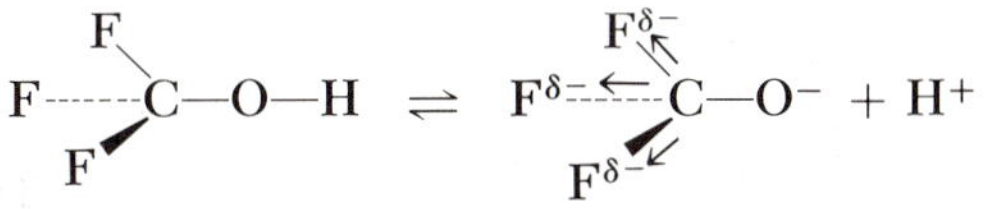

it tends, by spreading the excess negative charge over a larger region of the molecule, to stabilize the anion produced. This stabilization gives rise to a more favorable reaction than in the case of the CH_3 group:

$$H_3C-O-H \rightleftharpoons H_3C-O^- + H^+$$

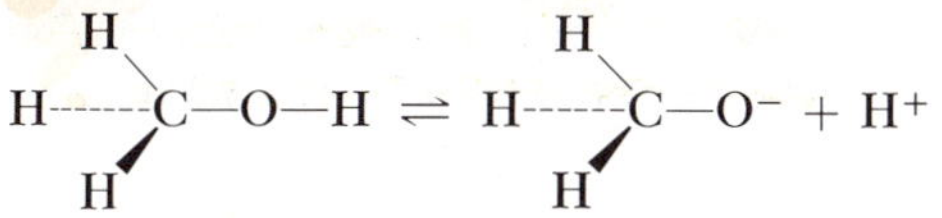

EXAMPLE 8.1 Rank the following in order of increasing acidity, and explain your reasoning: CHF_2OH, CH_2FOH, CF_3OH, CH_3OH.

SOLUTION The electronegativity of a substituted methyl group depends on the number and type of substituents. In this case, fluorine atoms are being substituted for hydrogen

---

°Electronegativity applied to atoms was introduced in Sec. 7.7. Here it is extended to functional groups, with the same general association of the word *electronegativity* with the ability to hold and attract electrons. This extension to functional groups provides a useful qualitative concept, but one that should be used with caution.

Carboxylic Acids

atoms. Each fluorine atom added will increase the electron-withdrawing ability of the substituted methyl groups and cause the alcohol to be a stronger acid. Hence, the order of increasing acidities is

$$CH_3OH < CH_2FOH < CHF_2OH < CF_3OH$$

Another class of molecules in which the hydroxyl group can function as an acid comprises the carboxylic acids. These molecules consist of an alkyl group or a substituted alkyl group and the **carboxyl group**,

$$\sim\!\!\overset{\displaystyle O}{\underset{}{\overset{\|}{C}}}\!\!-\!OH$$

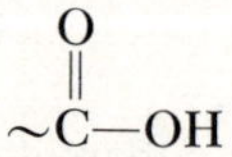

in which the hydroxyl group is included as an integral part. If a proton is removed from the hydroxyl group of a carboxylic acid by a base, an anion results, having the general name **carboxylate ion**. In this anion, according to resonance theory, the negative charge is spread over both oxygen atoms (see Example 7.4):

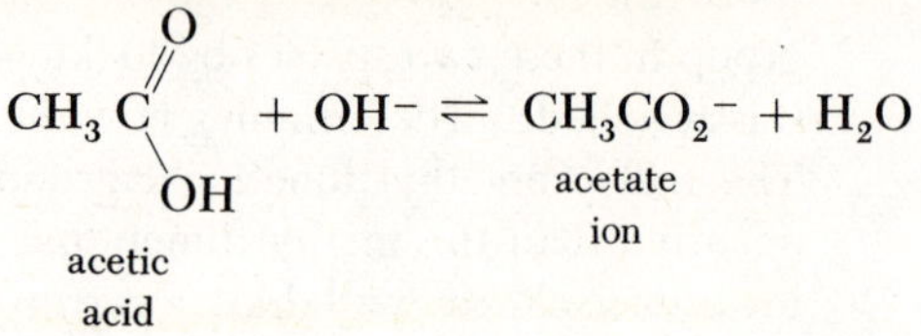

$$CH_3\underset{OH}{\overset{O}{\underset{}{\overset{\|}{C}}}} + OH^- \rightleftharpoons CH_3CO_2^- + H_2O \qquad (8.3)$$

acetic acid → acetate ion

In a shorthand notation, we can represent the resonance hybrid for $CH_3CO_2^-$ in the form

$$CH_3\!-\!C\!\!\begin{array}{c}{\scriptstyle O^-}\\[-2pt]{\scriptstyle O}\end{array}$$

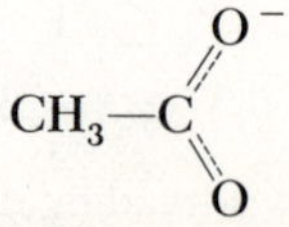

in which the dashed line represents partial double bond character (one-half here). Because the charge is spread over both oxygen atoms, the ion is stabilized, and has a lower energy than if the "delocalization" of charge were not possible. Thus, less energy is required for formation of the corresponding ion of a carboxylic acid. In this way, the $\sim\!\!\overset{\displaystyle O}{\underset{}{\overset{\|}{C}}}\!\!\sim$ functional group greatly increases the acidity of the $\sim\!OH$ functional group in a carboxyl group.

Substituted alkyl groups affect the relative acidities of carboxylic acids much as they affect alcohols. Thus, the trifluoromethyl carboxylic acid (trifluoroacetic acid), CF_3CO_2H, is a stronger acid than the methyl carboxylic acid (acetic acid):

$$H_3C\!-\!\underset{OH}{\overset{O}{\underset{}{\overset{\|}{C}}}} \rightleftharpoons H_3C\!-\!\overset{O}{\underset{O}{\overset{\|}{C}}}^- + H^+ \qquad F_3C\!-\!\underset{OH}{\overset{O}{\underset{}{\overset{\|}{C}}}} \rightleftharpoons F_3C\!-\!\overset{\delta-\ O}{\underset{\delta-\ O}{\overset{\|}{C}}}^- + H^+$$

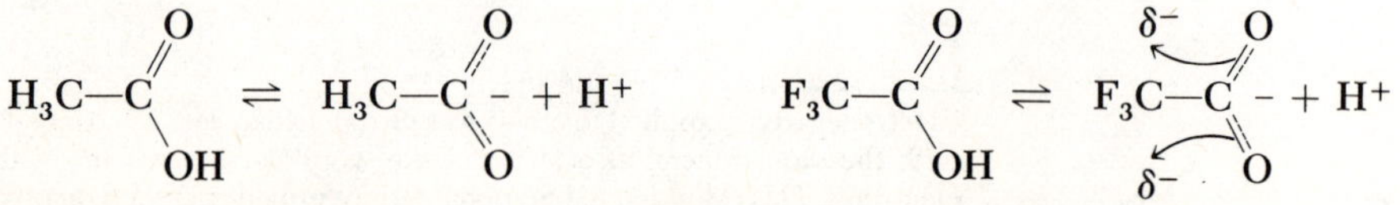

In the second case, because of the inductive effect of the CF_3 group, there is less negative charge concentrated on oxygen atoms, and a more stable ion results; hence, the acid is stronger.

The relative magnitudes of the substituent effects we have discussed are emphasized if we rank the four acids we have mentioned in order of decreasing acidity:

$$CF_3-\overset{\overset{\large O}{\|}}{C}-OH > CH_3-\overset{\overset{\large O}{\|}}{C}-OH > CF_3OH > CH_3OH$$

Compared with the fluorine atoms, the $\sim\overset{\overset{O}{\|}}{C}\sim$ group has the greater effect in determining anion stability and acid strength.

Water as an Acidic Functional Group; "Fine Tuning" by Metal Ions

When simple salts are dissolved in water, the metal ion is surrounded by water, forming an aquated ion, or "aquo complex." For example,

$$CoCl_2(s) + 6H_2O \rightleftharpoons Co(H_2O)_6{}^{2+} + 2Cl^-$$

The metal ion acts as a Lewis acid, forming a bond to the Lewis base, water. This can cause the water to give up a proton—breaking the water into a hydroxyl group, which is bound to the metal atom, and a proton, H^+:

$$Co(H_2O)_6{}^{2+} \rightleftharpoons (H_2O)_5Co(OH)^+ + H^+$$

Different metal ions provide a fine tuning that determines the tendency for the bound water to dissociate. The extent to which the proton is dissociated from the water molecule is given by the equilibrium constant K_a, for the reaction of the type illustrated above. The K_a value for $Co(H_2O)_6{}^{2+}$ is 1.25×10^{-10}; for $Ni(H_2O)_6{}^{2+}$, it is 3.98×10^{-10}; for $Zn(H_2O)_6{}^{2+}$, it is 1.26×10^{-9}; and for $Co(H_2O)_6{}^{3+}$, it is 2.00×10^{-3}. Two trends become evident from such data. First, the dissociation of water bound to a metal atom becomes slightly more favorable as one proceeds across the row of transition metal ions in the periodic table. This is because the ions become slightly smaller (Chap. 6). With ions of the same charge, then, the charge-to-radius ratio increases as the radius decreases. Relative to neutral H_2O, ions with a greater charge-to-radius ratio will have a stronger attraction for the hydroxide ion and will stabilize the product, the dissociated species. This effect is much greater if the charge is increased by a full unit. Thus, Co^{3+} is better, by a factor of 10^7, at causing bound water to dissociate than Co^{2+} is. The effect of increasing charge is illustrated by the series Li^+($K_a = 4.0 \times 10^{-15}$), Mg^{2+} ($K_a = 6.3 \times 10^{-13}$), and Sc^{3+} ($K_a = 1.3 \times 10^{-5}$).

EXAMPLE 8.2 For each of the following pairs, which substance would give the more acidic solution when dissolved in water: (a) $CoCl_2$ or $ZnCl_2$; (b) NaCl or $MgCl_2$; (c) $BaCl_2$ or $LaCl_3$?

SOLUTION The chloride salts of all of the metal ions were chosen, in this example, so that any effect due to the anion would be constant. (a) The ionic radius of Zn^{2+} is smaller than the ionic radius of Co^{2+} and the charge is the same, so Zn^{2+} would promote hydrolysis better and forms a more acidic solution. (b) Mg^{2+} has a higher charge than

Na⁺ does and forms a more acidic solution. (c) La³⁺ has a higher charge than Ba²⁺ and gives a more acidic solution. Note that highly charged ions often promote hydrolysis so well that they become predominantly hydroxides and often precipitate from solution.

STUDY PROBLEM 8(a)

For each of the following pairs, which member will give a more acidic solution when dissolved in water: (a) LiCl or $MgCl_2$; (b) $LaBr_3$ or $LuBr_3$; (c) $FeCl_3$ or $FeCl_2$?

The Amine Group, a Basic Functional Group

The amines are a class of molecules in which one or more alkyl groups R or substituted alkyl groups are attached to a nitrogen atom, taking the place of a hydrogen atom in NH_3. These compounds have the general formulas RHN_2, $RR'NH$, or $RR'R''N$, where R, R', and R'' represent alkyl groups or substituted alkyl groups and are called primary, secondary, or tertiary amines. An important characteristic reaction of the amines is the addition of a proton; that is, the amines act as bases. An example is the reaction of methylamine with a strong acid:

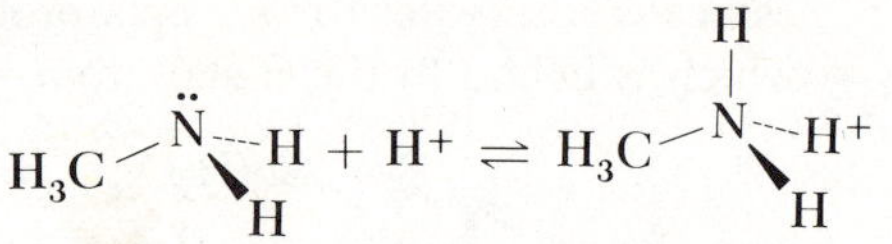

$$H_3C-\ddot{N}H_2 + H^+ \rightleftharpoons H_3C-NH_3^+ \tag{8.4}$$

Let's investigate the substituent effects of a functional group on amines. We consider the trifluoromethyl group. The trifluoromethyl group is a good electron-withdrawing group. This electron-withdrawing influence will tend to decrease the electron density of the attached nitrogen atom, which already has a formal charge of +1 in the protonated cation:

$$CF_3-\ddot{N}H_2 + H^+ \rightleftharpoons CF_3-NH_3^+$$

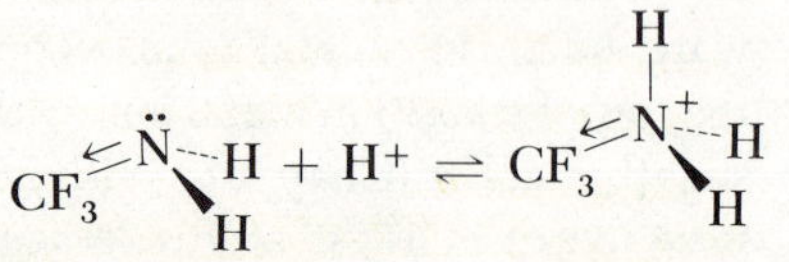

This electron-withdrawing influence exerted on a positive center in the substituted ammonium ion is a *destabilizing* influence on the ion; it represents an *increase* in energy associated with attempting to withdraw negative charge, electrons, from a positively charged center. The net result is a decreased tendency for formation of the cation, that is, a lower basicity of CF_3NH_2 than, say, $H-NH_2$ or CH_3-NH_2.

EXAMPLE 8.3   We have just seen that substituent effects causing changes in the basicities of amines are approximately additive, like the substituent effects causing changes in acidities of alcohols (Example 8.1). Rank the following in order of increasing basicity:

$F_3C-N(H)(CF_3)$, $F_2C(H)-N(H)(CH_3)$, $FC(H_2)-N(H)(CH_3)$, $F_3C-N(H)(CH_3)$

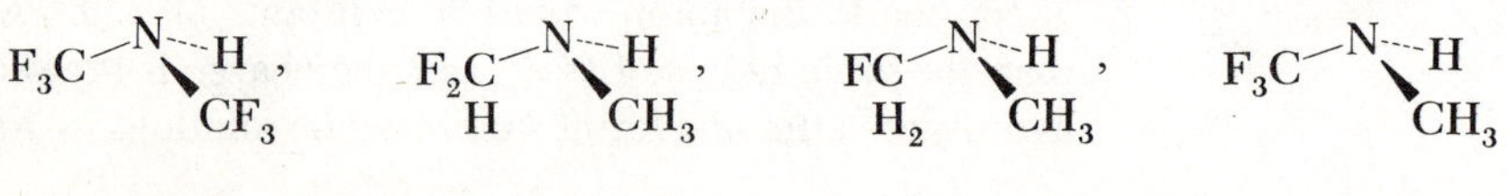

Rank the following in order of increasing acidity:

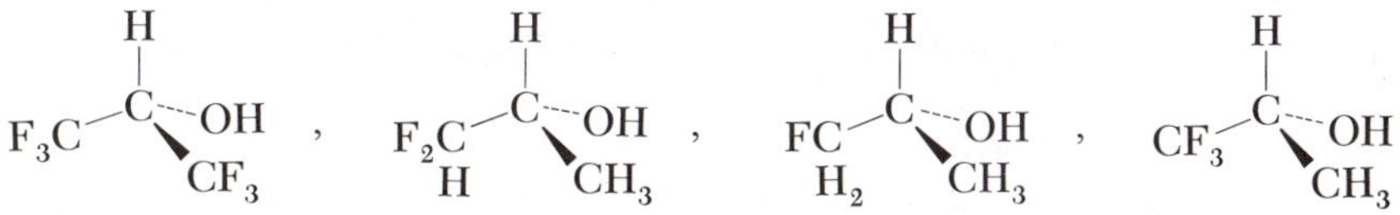

SOLUTION In basicity,

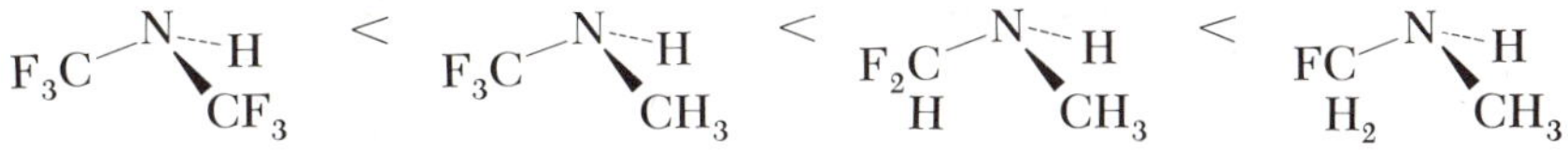

In acidity,

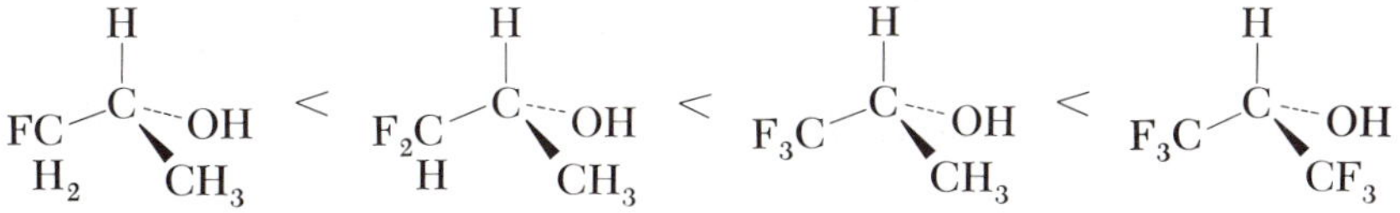

The trends are the opposite. The more complete the electron withdrawal by substituents, the weaker the base—since it will not have as great a "hold" on a proton. The acids become stronger (are able to shed a proton more easily), when a substituent of greater electron-withdrawing power is added. The trends therefore arise from the same phenomenon.

STUDY PROBLEM 8(b)

Between which acid-base pair among the following, do you suppose, would be the reaction with the largest equilibrium constant?

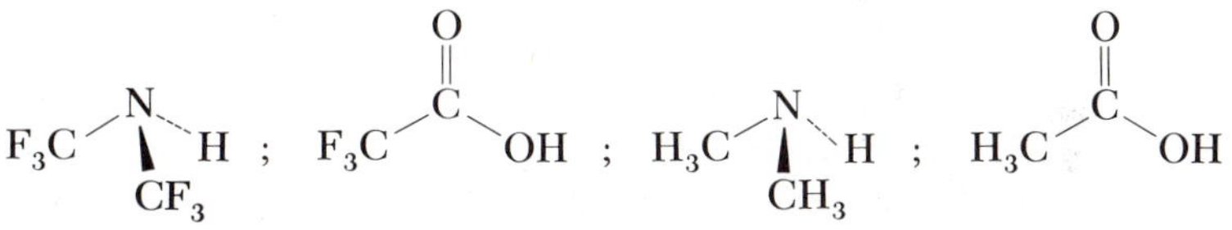

8.2 Lewis Acid–Base Bonds

We have found that the Arrhenius definitions of acids and bases are very specific—acids produce H^+ and bases produce OH^-. Brönsted and Lowry generalized the definition by proposing that acids are species that produce H^+ and bases are any species that accept H^+. G. N. Lewis, who was highly influential in developing the concept of chemical bond formation by electron sharing, generalized the acid-base concept still further. A Lewis acid is a species that accepts an *electron pair* in covalent bond formation (as H^+ does when it bonds to OH^-, for example) and a Lewis base is a species that donates an electron pair in covalent bond formation (as OH^- does when it bonds to H^+, for example). According to this view, acids are said to be **acceptor species** and bases **donor species.** We can say from this definition that electron-deficient compounds can act as Lewis acids and electron-rich compounds can act as Lewis bases.

We are now familiar with an important class of boron compounds in which there are three covalent bonds to boron, although the boron atom does not have an octet of valence electrons. Examples are BF_3 and BCl_3. We also know that many nitrogen compounds exist in which there is a lone pair of electrons in the valence shell and for which the octet rule is satisfied without all the nitrogen valence electrons shared in covalent bonds. Examples are NH_3, CH_3NH_2, H_2NNH_2. The boron atom in BF_3 or BCl_3 is electron-deficient and the nitrogen in any of the above examples is electron-rich. These types of compounds thus qualify as Lewis acids and Lewis bases, respectively; and a reaction between representative examples manifests these characteristics:

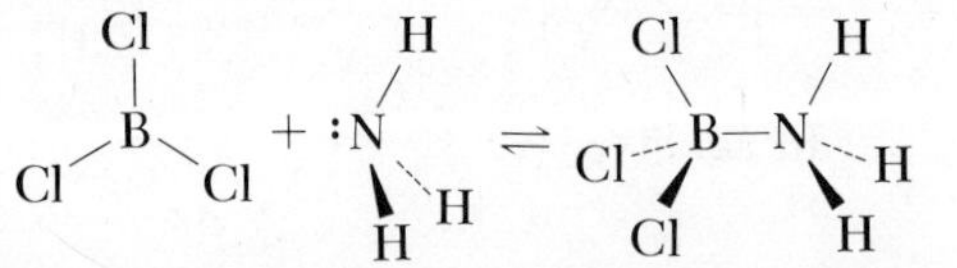

$$Cl_3B + :NH_3 \rightleftharpoons Cl_3B-NH_3 \tag{8.5}$$

Several features of the combined BCl_3–NH_3 species (called a Lewis acid–base adduct) should be noted. The arrangement of the atoms about boron is trigonal in BCl_3, but becomes tetrahedral when an electron pair is accepted from NH_3. We may justify this fact using VSEPR theory, considering the added lone pair to belong to the boron central atom as well as to the nitrogen central atom. Also, this is consistent with a change in hybridization of the boron atomic orbitals from sp^2 to sp^3 as the number of covalent bonds in which boron participates changes from three to four. There is no drastic change in the H—N—H bond angles associated with NH_3, as the B—N electron pair bond takes the place of the lone pair of ammonia. The B—N bond is what is sometimes called a **dative** or **coordinate** covalent bond. Both of the bonding electrons in the B—N bond can be viewed as having been donated by the nitrogen atom. The electron donor center in the adduct, the nitrogen atom, bears a formal charge of $+1$. The electron acceptor center in the adduct, the boron atom, bears a formal charge of -1.

Electron-deficient boron compounds can also undergo Lewis acid–base reactions with a wide range of other electron-rich species, such as F^- or Cl^-. In the case shown below, the F^- functions as a Lewis base:

$$BF_3 + F^- \rightleftharpoons [BF_4]^-$$

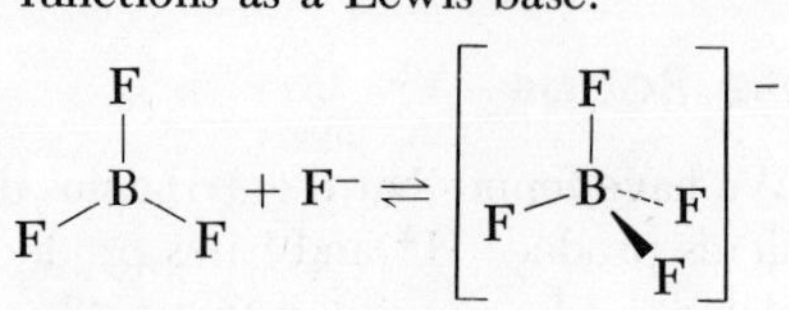

boron trifluoride   tetrafluoroborate

A variety of nitrogen compounds, including ammonia or amines, can act as Lewis bases toward H^+ (a Lewis acid) or other electron-deficient species:

$$CH_3NH_2 + H^+ \rightleftharpoons CH_3NH_3^+$$

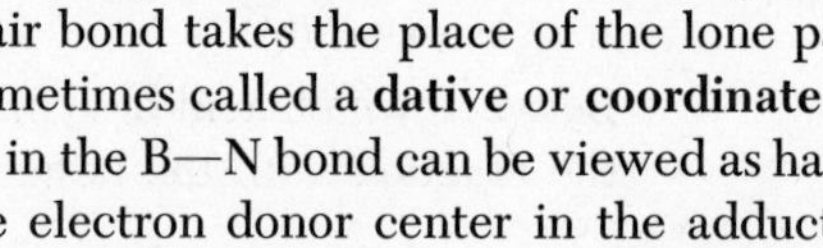

If the acidic reagent is gaseous HCl, instead of aqueous H^+, then the pertinent reaction is a displacement reaction, in which CH_3NH_3 displaces Cl^- as the Lewis base bonded to H^+:

$$CH_3-\underset{H}{\overset{\ddot{N}}{\vphantom{N}}}-H + HCl \rightleftharpoons CH_3-\underset{H}{\overset{H}{\underset{|}{\overset{|}{N^+}}}}-H + Cl^-$$

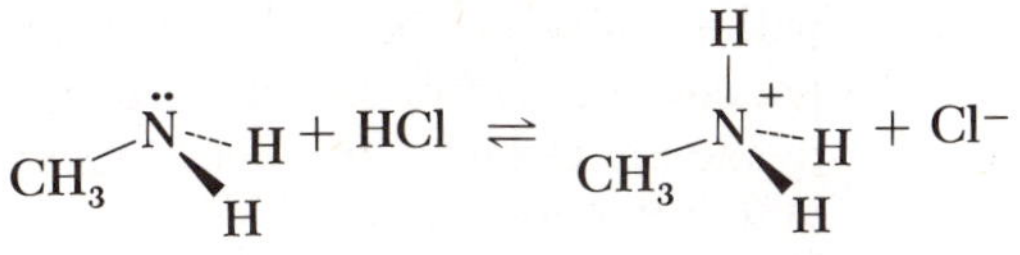

methylammonium chloride

Much of the chemistry of metal complexes can be viewed as manifestations of Lewis acid–base chemistry. For example, the existence of metal complexes can usually be understood descriptively in terms of Lewis acid–base bonds. The Lewis acid in these cases is a metal ion, and the Lewis base is the electron donor, the ligand, that is attached to it in the complex. Usually there are several ligands attached to a metal ion in a complex.

The ligands that bind to metal ions include both (1) neutral molecules, such as ammonia or water, which have lone pairs of electrons that can be donated to the metal, and (2) negatively charged ions, such as chloride ions or the hydroxide ion, which also have electrons to donate.

Some examples of Lewis acid–base bond complexes formed by metal ions° (arrows indicate formation of bonds)

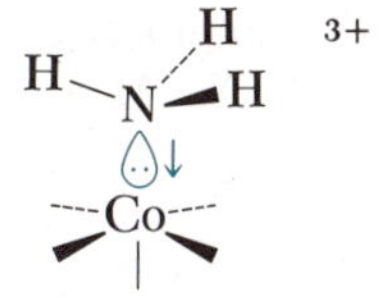

bond formation

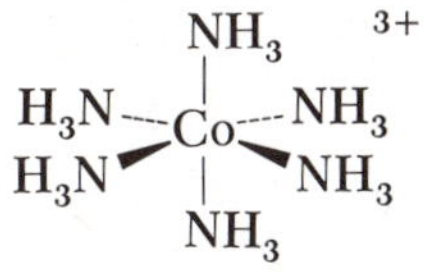

hexamminecobalt(III) cation

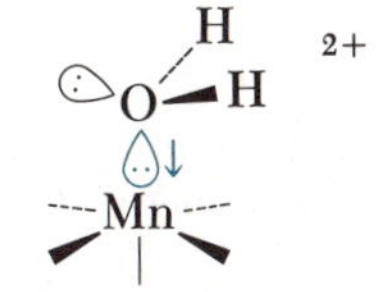

bond formation

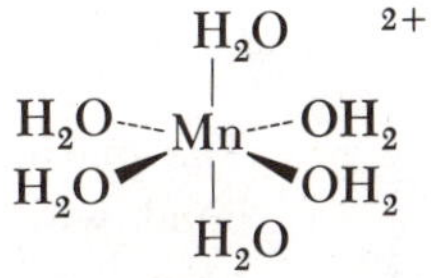

hexaaquomanganese(II) cation

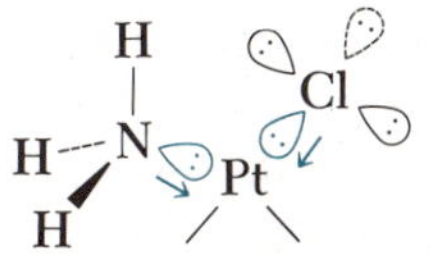

bond formation

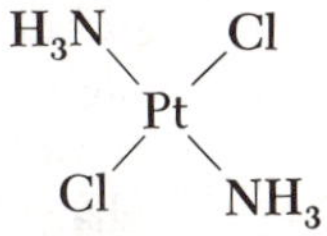

trans-dichlorodiammineplatinum(II)

By considering these examples, we can see that the total charge on the complex equals the charge on the metal ion plus the sum of the charges on the ligands (typically negative). Neutral molecules such as NH_3 or H_2O can bond to the metal ion

---

°The Lewis acid–base compounds formed by transition metal ions have geometries that generally cannot be determined by VSEPR theory. The theory currently used to predict these structures is too complex to be treated in this textbook, but some general patterns that are helpful guides are presented in Chap. 18.

without affecting the total charge of the complex. (The nomenclature for these compounds and further aspects of their chemistry will be treated in Chap. 18.) Each ligand is attached to the metal ion by a Lewis acid–base bond, with the ligands providing the electron pair.

8.3 The Hydrogen Bond

The hydrogen bond is a concept of wide-ranging importance in chemistry and molecular biology. The hydrogen bond is not a chemical bond in the usual sense. The interaction is primarily an electrostatic interaction, between species that are formally uncharged; some covalent character may also be present in the hydrogen bond. Numerous techniques have been applied to this interaction, and the consensus of the investigators is that the hydrogen bond is usually based mainly on a dipole-dipole interaction between a hydrogen atom that is bound to a highly electronegative atom X, (like F, N, or O) and another highly electronegative atom Y to which it is not formally bound. It is a $\mathrm{X{-}\overset{\delta+}{H}\cdots\overset{\delta-}{Y}\sim}$ interaction. In this situation, the positive end of the X—H dipole (for example, $\mathrm{\sim\overset{\delta-}{O}{-}\overset{\delta+}{H}}$) is on the hydrogen atom. This dipole can be attracted to the negative end of another polar bond, in which Y is formally involved; it is a $\mathrm{O{-}\overset{\delta+}{H}\ \ \overset{\delta-}{Y}\sim}$ interaction. For the specific case of water, the H_2O molecule can provide both the X—H unit ($\sim$O—H) and the electronegative Y, the oxygen atom:

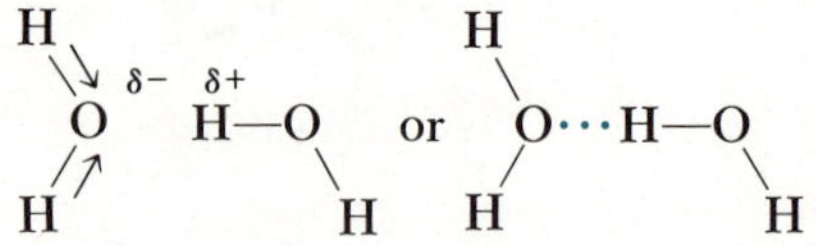

a hydrogen bond in water

The maximum magnitude observed for the energy associated with the hydrogen bond, which is the energy required to break it, is about 5%–10% of the energy of normal covalent bonds, for instance, carbon-carbon bonds. This means that the interacting molecules constitute a more stable situation by about 1–10 kcal/mol if the dipoles interact with one another than if the interactions are absent. This kind of interaction has its highest strength in the case of highly polar X—H bonds, and highly polar Y—R bonds (if X and Y are F, O, or N), as this maximizes the electrostatic attractions. Because of the small size of the hydrogen atom, the partial positive charge on the hydrogen atom can be situated very close to the partial negative charge on Y to which it is attracted; this causes an electrostatic attraction that is much larger than one would ordinarily expect of dipole-dipole interactions involving larger atoms.*

EXAMPLE 8.4 Choose the Lewis acids and Lewis bases from which one might expect to form the following compounds: $Rh(NH_3)_3Cl_3$, $ZnCl_4^{2-}$, HCl, $Cl_3B{:}NH_3$, NH_4^+, AgCl.

*Although the hydrogen bond has been described as based primarily on electrostatic dipole-dipole interactions, which is true in most cases, there are a few important examples in which one of the dipoles is replaced by an anion. An example is the interaction between fluoride ion and HF, $\mathrm{\overset{\delta-}{F}{-}\overset{\delta+}{H}{-}\overset{-}{F}}$, which is an extremely strong hydrogen bond (50 kcal/mol).

SOLUTION We can view electron acceptors as Lewis acids and electron donors as Lewis bases. In making these assignments, we imagine that we are putting the molecule together from "reasonable" fragments, or functional groups.

$Rh(NH_3)_3Cl_3$	can be viewed as Rh^{3+} bound to three NH_3 groups and three chloride ions. Rh^{3+} is the Lewis acid and the other species are Lewis bases.
$ZnCl_4{}^{2-}$	can be viewed as Zn^{2+} bound to four chloride ions. The Zn^{2+} is the Lewis acid and the four Cl^- are Lewis bases.
HCl	can be viewed as a combination of H^+ and Cl^-, where H^+ is the Lewis acid and Cl^- the Lewis base.
$Cl_3B{:}NH_3$	would consist of two fragments: Cl_3B and NH_3. The Cl_3B is electron-deficient (Lewis acid) and accepts a lone pair from the electron-rich NH_3 (Lewis base).
$NH_4{}^+$	can be viewed as the addition of H^+ (Lewis acid) to NH_3 (Lewis base).
AgCl	can be viewed as a combination of Ag^+ (Lewis acid) with Cl^- (Lewis base).

STUDY PROBLEM 8(c)

Choose the Lewis acids and bases from which one might expect to form the following compounds: HBr, $CuCl_2$, $BF_4{}^-$.

EXAMPLE 8.5 Show hydrogen-bonding structures that are possible among

$$\text{Acetic acid, } CH_3\overset{O}{\overset{\|}{C}}OH; \quad \text{ammonia;} \quad \text{and acetone, } CH_3\overset{O}{\overset{\|}{C}}CH_3.$$

SOLUTION Acetic acid can hydrogen-bond to other acetic acid molecules, to ammonia, and to acetone.

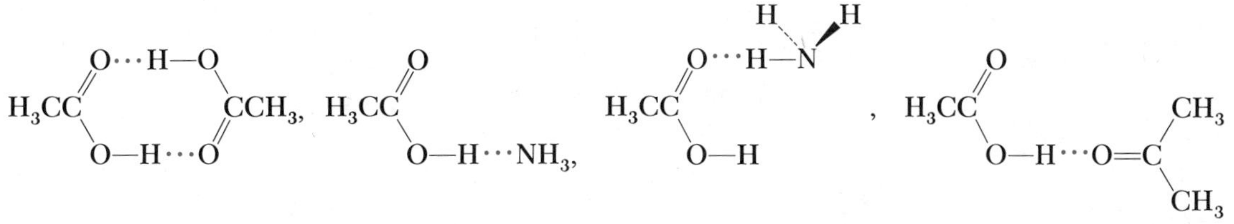

Ammonia can hydrogen-bond to other ammonia molecules, to acetic acid (above), and to acetone.

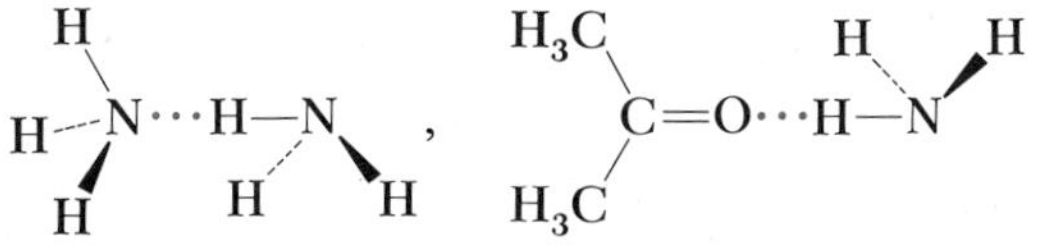

Acetone can hydrogen bond to acetic acid and ammonia, but not to other acetone molecules.

STUDY PROBLEM 8(d)

Show the hydrogen-bonding structures that are possible in a solution of the compound $NH_2CH_2{-}\overset{O}{\overset{\|}{C}}{-}H$ in water.

Figure 8.1
Boiling points of selected hydrides.

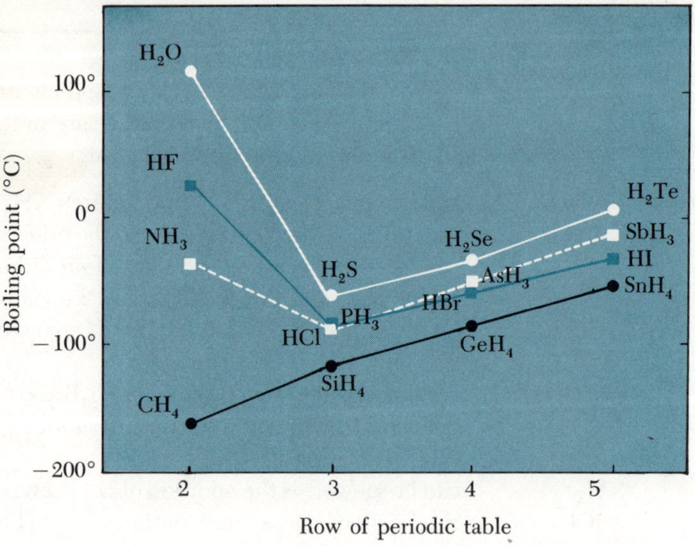

Hydrogen Bonding and Molecular Properties

Boiling Points of Hydrides

The hydrogen bond allows molecules to take a more stable configuration, one of lower energy, when they are properly associated with one another than when they are separated. This stabilization contributed by hydrogen bonding manifests itself in several important ways. For example, hydrogen-bonded molecules tend to require more energy to be melted or vaporized—they have higher melting and boiling points than we might otherwise expect. Figure 8.1 shows how hydrogen bonding affects boiling points of some simple compounds, the hydrides. Neglecting the points for NH_3, HF, and H_2O, one can see from the figure that normally the heavier hydrides within a given group of the periodic table boil at higher temperatures; this trend results from the size of the important intermolecular attractive forces, called dispersion interactions, which are generally greater for species with more electrons. (See Sec. 14.2.) The hydrides of the most electronegative elements, F, O, and N, however, boil at temperatures higher than one would expect from this trend, that is, higher than their heavier group neighbor. The reason for this deviation is hydrogen bonding. We note that methane, which contains carbon as the central atom, does not engage in hydrogen bonding. Carbon is relatively close to hydrogen in electronegativity, so methane cannot function as a hydrogen-bonding acid X—H. Also, CH_4 has no lone pairs of electrons and cannot function as a hydrogen-bonding base in a hydrogen bond with the hydrogen atom of another molecule. Similar arguments apply to the other hydride molecules in Fig. 8.1, which follow the simple trend just stated. By contrast, the H—F · · · H—F, H_2O · · · H—OH, and H_3N · · · H—NH_2 cases manifest very strong hydrogen bonds, and these are the cases that deviate markedly from the simple trends.

Molecular Structure

The hydrogen-bond stabilization also influences the particular ways in which two molecules can arrange themselves geometrically relative to each other. The interaction that occurs when the dipoles $\overset{\delta-}{X}—\overset{\delta+}{H}$ and $\overset{\delta-}{Y}—\overset{\delta+}{R}$ are closely placed and properly aligned with each other is the interaction that gives the most stable configuration, the one of lowest energy. This aspect of hydrogen bonding is extremely important in the structure of water and in the stabilization of particular arrangements of several types of molecules of vital biological importance.

Solid and Liquid H_2O

An interesting hydrogen-bonding phenomenon of great practical importance is the decrease in density of water as it freezes, causing ice to float on the surface of liquid water. This phenomenon is essential to maintaining aquatic life in cold regions of the earth, since ponds and lakes freeze from the top down, and thus life at the bottom of the water is happily allowed to survive. Most pure solids sink in the corresponding pure liquid, because most substances are less dense in the liquid state than in the solid state. Why is water so different? In the solid state there is less molecular motion than in the liquid state, and atoms, ions, or molecules can usually pack together more tightly than in the more mobile liquid state. Since there are more molecules per unit volume in the typical solid, it is denser. In ice, however, the evidence tells us that the molecules are less well packed than in liquid water. (See Fig. 8.2.)

As the temperature of water is decreased, the motion of the molecules becomes slower and they begin to align themselves in the lowest-energy configuration. For most pure substances, this would cause molecules to come very close together to

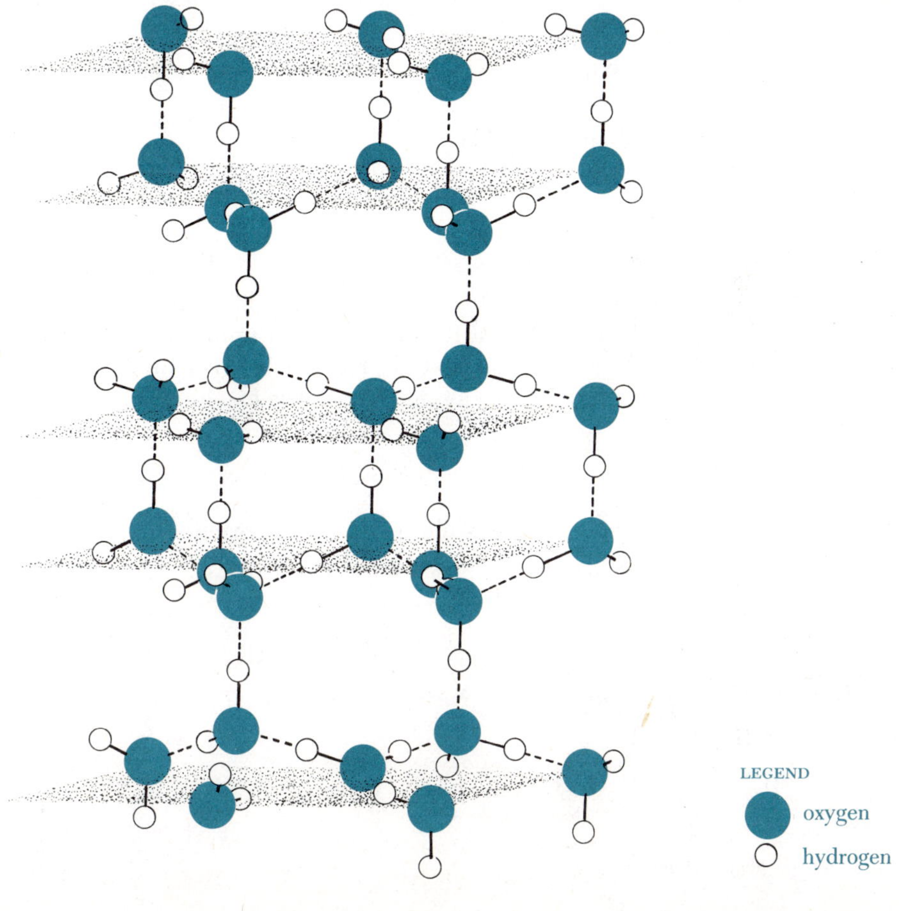

Figure 8.2 Hydrogen bonding in ice. Each oxygen atom forms two hydrogen bonds to hydrogen atoms on other water molecules, using both of its lone pairs of electrons. This is the structure of ice at 1 atm pressure. At higher pressures, some compression occurs and different structures are found.

maximize the various types of short-range attractive forces (van der Waals interactions; see Chap. 14) responsible for the existence of most liquids. The energy decrease associated with hydrogen-bonding in water, however, can overcome the tendency to closest-packing; to participate in as many strong hydrogen bonds as possible, the water molecule will align in an arrangement that allows each water molecule to form four nearly linear hydrogen bonds (Fig. 8.2). In this arrangement, there is much empty space in the lattice, resulting in a relatively low density.

OPTIONAL

Proteins and Nucleic Acids

Two of the most important hydrogen-bonding interactions in biological systems are those occurring between molecules of a type called peptides, and those between the class of molecules called nucleic acids. Let's investigate these two cases to see how such a subtle effect as hydrogen bonding can influence the structure of very large and very complex molecules.

Peptides; Proteins

The basic building block of proteins is the amino acid—a molecule containing both the amino, $\sim NH_2$, and carboxyl, $\sim C\underset{OH}{\overset{O}{\parallel}}$, functional groups. The molecular form of amino acids is

$$H_2N-\underset{H}{\overset{R}{\underset{|}{C}}}-\overset{O}{\overset{\parallel}{C}}OH$$

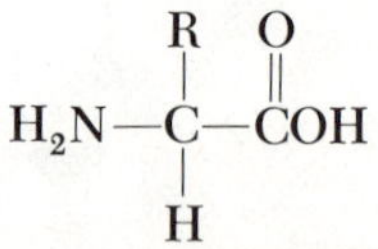

where R may be any of numerous different groups; hence each type of amino acid in a protein has its own particular properties. A peptide is a molecule that results from the "condensation" reaction between the amino group of one molecule and the carboxyl group of another molecule, eliminating a water molecule:

$$\sim\overset{O}{\overset{\parallel}{C}}-OH + H-N{\overset{\diagup}{\diagdown}} \rightleftharpoons \sim\overset{O}{\overset{\parallel}{C}}-N{\overset{\diagup}{\diagdown}} + H_2O \qquad (8.6)$$

This reaction is illustrated by the reaction of phenylalanine and glycine to form glycylphenylalanine dipeptide:

phenylalanine + glycine ⇌ glycylphenylalanine + H₂O

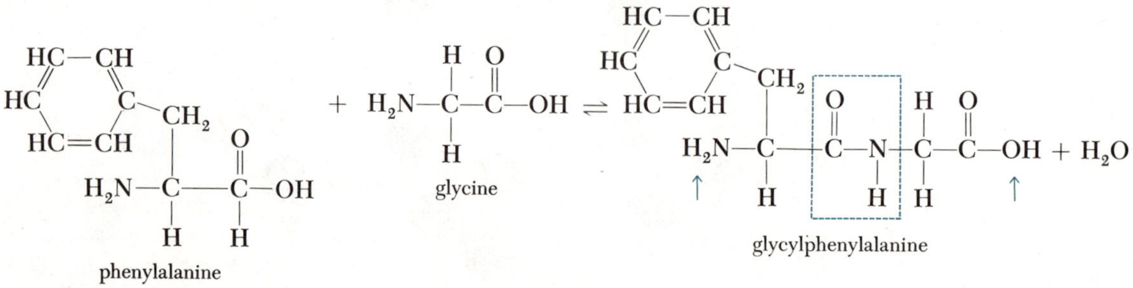

For this peptide, the structural components are recognizably derivable from the two amino acids. The peptide linkage is shown within the rectangular dashed line. From this structure, or that of any simple dipeptide (meaning built from two amino acids), it is seen that there is still one amino group and one carboxyl group, shown by arrows, that are available for attaching two more amino acids by peptide bonds. If these attachments are made, one then has four amino acids condensed into a polypeptide (*poly* meaning "many") but there will still remain an amino group and a carboxyl group available for additional attachments of amino acids. There is in principle no limit to the size of polypeptides that can be visualized in this way. The class of molecules called proteins, substances that are so important in the structure of biological organisms, consists of polypeptides with numbers of amino acid groups ranging from hundreds to many thousands. What characterizes a specific protein are the numbers and types of amino acid fragments and their relative arrangements—which are bonded to which, and in what order. An infinite variety of proteins can be imagined, based on the many known amino acids.

In proteins, the common hydrogen-bonding interaction of importance occurs between a carbonyl group in the peptide bond and the amine hydrogen of another peptide bond:

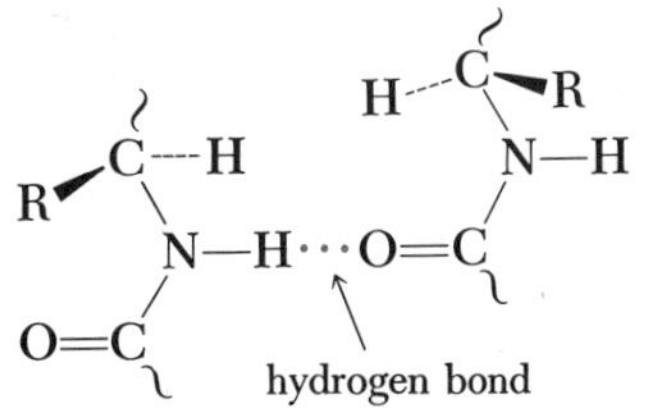

When this type of hydrogen bonding between the carbonyl groups ($\sim C(=O)\sim$) and amine groups of peptide bonds occurs within the same peptide chain (a given string of amino acid fragments), it can lead to the formation of what is called the alpha helix, as first postulated by Linus Pauling and R. B. Corey in 1946. Figure 8.3 shows a section of a protein in the α-helix configuration. Note that such a structure allows each carbonyl and amine group to participate in a hydrogen bond, and that all the hydrogen bonds are internal, existing within the same molecule. This leads to a very stable structure, which is fibrous. Such fibrous proteins are found in hair and wool. Polypeptides may also manifest hydrogen bonds between peptide chains—interchain hydrogen bonding. In this way, many chains can be joined to form sheets of protein instead of simple fibers. Again, a very stable structure results in which all possible carbonyl-amine hydrogen bonding is accomplished. In this case, however, the hydrogen bonding is not within the same chain, and the helical structure is not found. Figure 8.4 shows how the pleated sheet structures are formed by interchain hydrogen bonding. Pleated sheet structures are found for the proteins called fibroin, which occur in silk, and the α keratins, which occur in wool.

Nucleic Acids

Another well-known example of the structural effects of hydrogen bonding is found in the helical structure of the genetically important nucleic acids, especially a class of nucleic acids referred to as DNA, deoxyribonucleic acid. The **nucleic acids** are

OPTIONAL

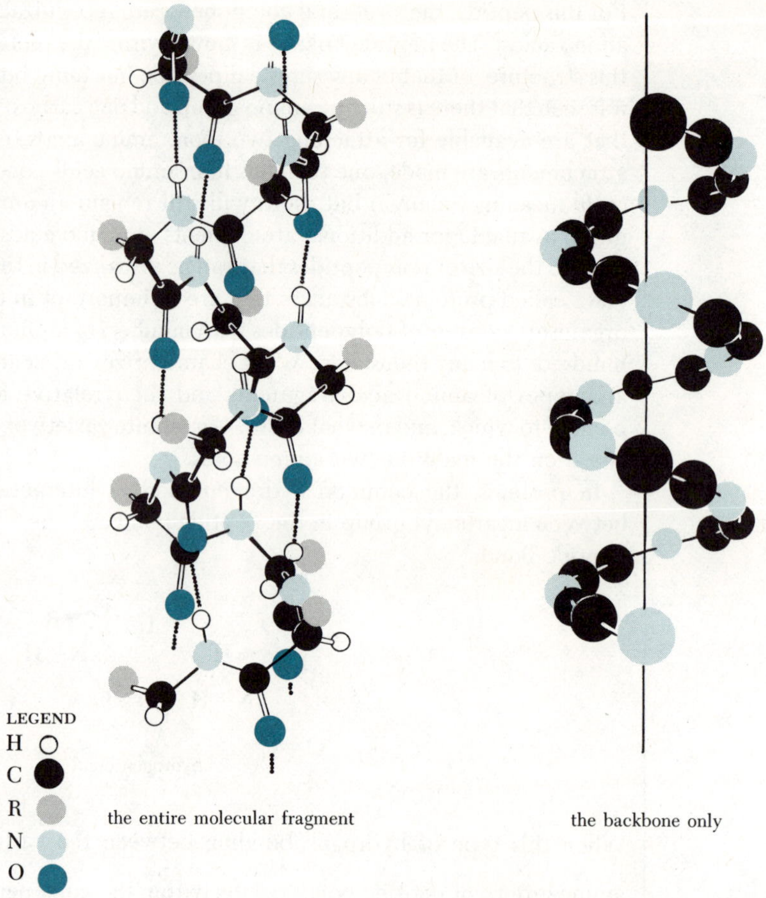

Figure 8.3
A section of a peptide chain in the α-helix configuration of a protein.

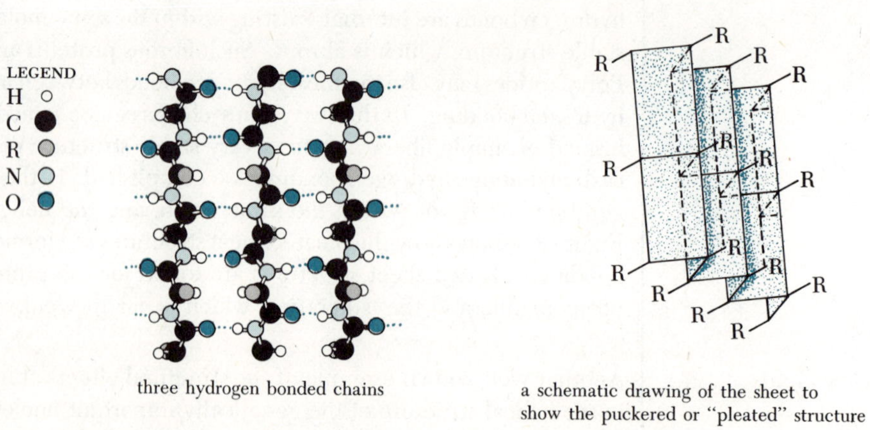

Figure 8.4
Formation of the pleated sheet structure of protein by interchain hydrogen bonding.

Spotlight

Linus Pauling (1900–), who has spent most of his scientific career at the California Institute of Technology, has contributed a great deal to modern chemistry by his experimental and theoretical developments concerning the nature of covalent bonding. Much of the experimental work involved the determination on molecular structures by x-ray crystallography. His fundamental contributions were recognized with the Nobel Prize in chemistry in 1954. His work has also included the areas of protein structure, immunochemistry, and, recently, the relationship between molecular abnormalities and disease. Pauling was active in bringing to public attention the dangers of radioactive fallout resulting from atmospheric tests of nuclear weapons. He was awarded the Nobel Peace Prize in 1963.

The California Institute of Technology, Pasadena, California

complex organic macromolecules, originally thought to be present principally in the cell nucleus. They transmit genetic information and control the synthesis of proteins. The nucleic acids are built from relatively small molecules called nucleotides, which are made up of three main parts—a nitrogen-containing cyclic portion, which is the base, a five-membered ring unit called ribose, and a phosphate group. Two representative nucleotides are shown herewith.

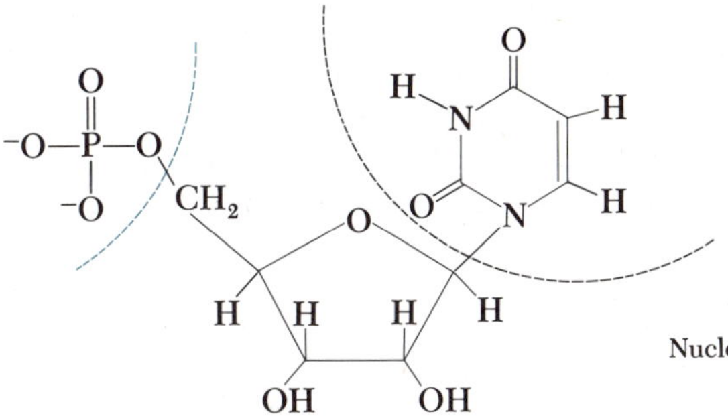

uridine monophosphate

Nucleotides

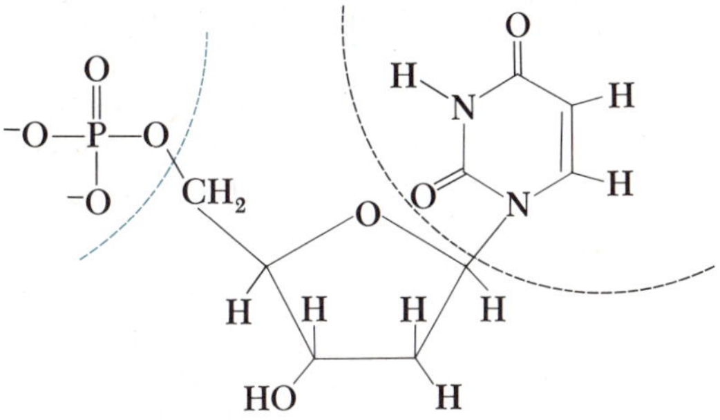

deoxyuridine monophosphate

In the structural formulas, the bases are shown within the black dashed lines and the phosphate groups within the colored dashed lines; the rest of the molecules correspond to the ribose unit. The deoxyribonucleic acids (DNA's) differ from ribonucleic acids (RNA's) in that the ribose rings have hydrogen atoms in DNA in place of the hydroxyl group that is present in RNA, as shown in the representations of the nucleo-

OPTIONAL

tides. The genetically important ribonucleic and deoxyribonucleic acids are macromolecules, formed when the nucleotide units are joined through "phosphate linkages," as shown in the accompanying structure.

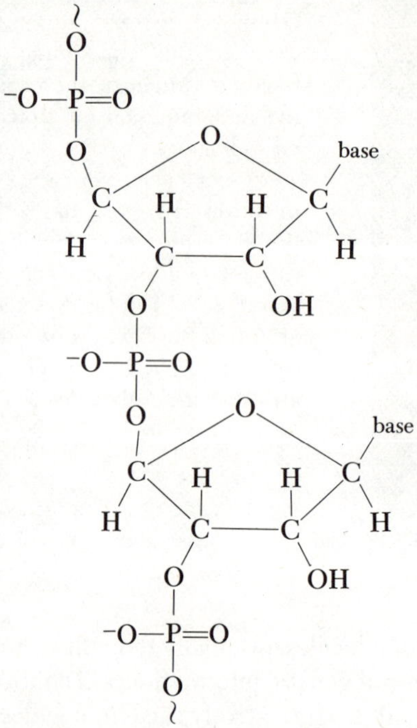

Macromolecules like these, which can be formed by combining many similar or identical subunits (here, the nucleotides), are often called **polymers.** The subunits are called **monomers.** For these nucleic acids and for proteins, the term *biopolymer* is often used.

An important feature of DNA structure is the existence of certain base pairs. The structures of the bases make it the most advantageous arrangement for one type of base, adenine, to be hydrogen-bonded with one called thymine and for cytidine to be hydrogen-bonded with guanine, as shown in the accompanying representations.

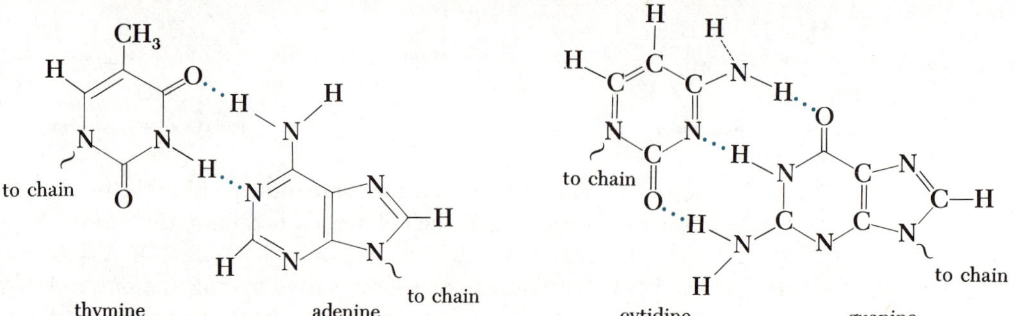

thymine adenine cytidine guanine

Such base-pairing by hydrogen bonding leads to the double helix structure of DNA, shown in Fig. 8.5. The deduction of this structure had fundamental importance in genetics. Watson and Crick received the Nobel Prize in 1962 for their contribution in deducing the double-helical nature of DNA. The double helix structure, it is proposed, is fundamentally important in the replication of DNA, with new strands of DNA formed by base matching of the old DNA strands. As a double helix unwraps, a new polymeric chain of nucleotides is formed on each of the original strands, so that when the process is completed, there are two new double helixes, each of which is like the original double helix. This is represented in Fig. 8.6.

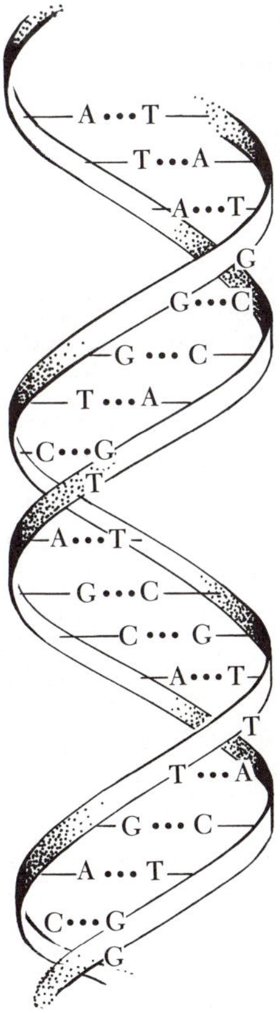

Figure 8.5
An illustration of the double helix structure of deoxyribonucleic acid (DNA). The ribbons represent the chain as shown on the facing page.

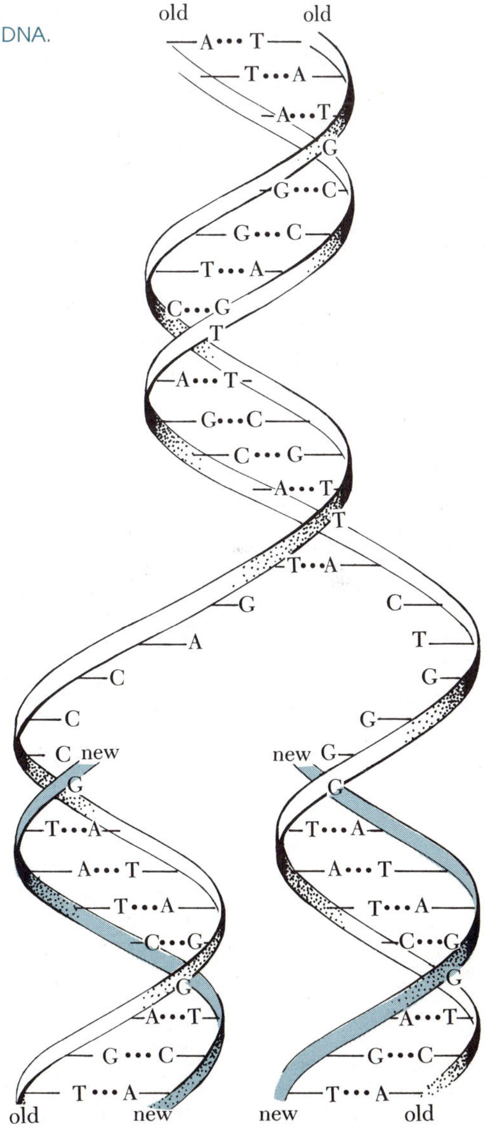

Figure 8.6
Replication of DNA.

8.4 Spectroscopy

Molecular structure, intermolecular interactions, chemical reactions, and most chemical and physical properties can be explained on the basis of the geometrical arrangements of atoms in molecules and the nature of the bonds that bind atoms. To investigate molecular structure and bonding, the chemist often uses the methods of spectroscopy. Just as atomic spectroscopy was seen to be useful in characterizing atomic structure, molecular spectroscopy is a powerful approach for studying molecular structure. Spectroscopy is the study of the interaction of electromagnetic radiation with matter. The absorption of radiation of the appropriate frequency leads to an excited state. When an excited state reverts to the ground state, electromagnetic radiation of a specific frequency is emitted.

The amount of energy that can be absorbed or emitted by a molecule in a spectroscopic study depends on the type of excited state involved; that is, it depends on the mode or modes of molecular energy affected by the energy absorbed or emitted. If the absorbed energy is taken up by the vibrational modes of the molecule, then the result of the absorption is called a vibrationally excited state, a state that differs from the ground state in that the vibrational energy has been increased by a specific amount; this kind of absorption of energy constitutes the basis of **vibrational spectroscopy.** If the absorbed energy causes an increase in the energy of rotation of the molecule, then the excited state differs from the ground state in having a higher rotational energy—in other words, a rotationally excited state. This is the basis for **rotational spectroscopy.** If the absorbed energy goes into increasing the energy associated with the electronic structure of the molecule, then the excited state is referred to as electronically excited; this is the basis for **electronic spectroscopy,** which is entirely analogous to the electronic spectroscopy of atoms.

The spectroscopic transitions, then, are the jumps from one energy level to another, induced by radiation. Studying them is useful to chemists in characterizing atomic and molecular species since good theoretical models are available for interpreting the results. These models relate the observed spectroscopic results to detailed information about molecular structure.

The Electromagnetic Spectrum

Energy is absorbed or emitted from radiation by matter in discrete bunches, called **quanta.** For light, these quanta are typically called **photons.** The energy content of each quantum of radiation energy is proportional to the frequency ν of the radiation; this is according to the Bohr-Planck relation, $E = h\nu$, where h is the proportionality constant **Planck's constant.** Thus, the energy is directly proportional to the frequency.

One finds that in nature there are various kinds of radiation sources that give off photons covering a wide range of frequencies, or energies. The entire range of these radiation types, covering the entire spread of frequencies, is called the **electromagnetic spectrum.** The frequency ranges of the electromagnetic spectrum can be categorized in many ways, for example by the sources giving rise to radiation of specific frequency ranges, (such as cosmic radiation), by the types of techniques used

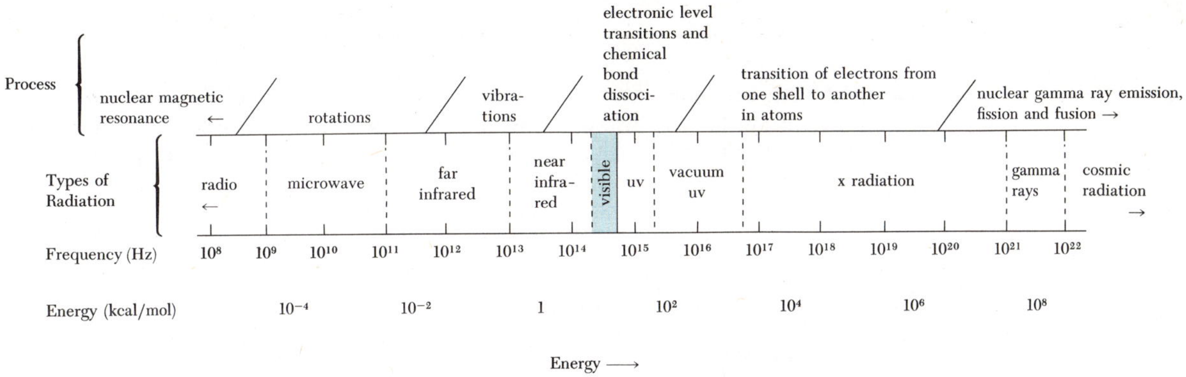

Figure 8.7 The electromagnetic spectrum, showing some of the approximate energy regions for various molecular, atomic, and nuclear processes.

to detect the radiation, (such as radio waves), or in terms of the modes of molecular energy that can be excited by radiation of a specific frequency range, such as vibrational spectroscopy. It has been found that the energy jumps involved in electronic spectroscopy, corresponding to transitions from a lower electronic energy level to a higher electronic energy level, require radiation at the higher frequency range of visible or ultraviolet light than vibrational excitations do (jumps from one vibrational energy level to another), which occur with radiation in the frequency range typically characterized as infrared. Transitions in rotational spectroscopy (jumps from one rotational energy level to another) occur with radiation in an even lower frequency range, the so-called microwave range. These frequency ranges of the electromagnetic spectrum and the types of spectroscopic transitions associated with them are summarized in Fig. 8.7.

EXAMPLE 8.6 Using Fig. 8.7, decide:
(a) which types of radiation are capable of severing chemical bonds, and
(b) which type of radiation can cause vibrational excitation.

SOLUTION (a) Radiation on the order of tens of kcal/mol and more, including visible, ultraviolet, x radiation, gamma rays, and cosmic radiation.
(b) Vibrational levels are quantized and are "excited" by infrared radiation.

It is relatively common to see specific regions of the electromagnetic spectrum represented by the wavelength λ of the radiation instead of frequency. The two are inversely related by the equation

$$\lambda = \frac{c}{\nu} \tag{8.7}$$

where c is the speed of light in a vacuum, 3.00×10^8 m/sec. Since the wavelength is inversely proportional to frequency, *short* wavelength radiation is of high energy, whereas *long* wavelength radiation is of low energy. Since the speed 3.00×10^8 m/sec applies to any type of electromagnetic radiation (not just what we call light), a characterization of spectroscopic information by wavelength is entirely equivalent to specifying the same information by frequency. That Eq. (8.7) is valid is readily seen by considering the distance traveled by a wave of frequency ν and wavelength λ in 1 sec. This distance is the number of cycles occurring in 1 sec ν times the length of each cycle λ:

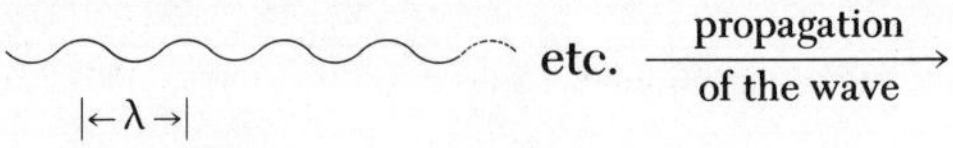

$$\text{distance per second} = \lambda\nu$$

But the distance traveled in 1 sec is also the speed, in the case of interest here, c. Thus

$$c = \lambda\nu$$

Electromagnetic radiation can be pictured as consisting of an alternating electromagnetic field, with the direction of the oscillating electric field perpendicular to the direction of the oscillating magnetic field. This is shown pictorially in Fig. 8.8. The oscillating electric field can interact with the charge distribution in a molecule or ion. For example, if a molecule executes a type of motion that causes its dipole moment to be altered by the motion, then the molecule can interact with the oscillating electric field of radiation of the correct frequency. The **dipole moment** in a diatomic molecule is the product of the charges q on each atom and the distance by which these charges are separated.

By virtue of this interaction, the energy of the molecule associated with this motion can be altered—the molecule can undergo a transition from one energy level to another. A necessary condition for this transition to occur is that the Bohr-Planck condition be satisfied. That is,

$$E = E_{\mathrm{f}} - E_{\mathrm{i}} = h\nu \tag{8.8}$$

where E is the change in energy, E_{f} is the final energy (after the transition), E_{i} is the initial energy (before the transition), and ν is the frequency of the radiation. Hence, the correct frequency is the one for which Eq. (8.8) is satisfied.

EXAMPLE 8.7 Calculate the energy in kcal/mol and wavelength in meters of x rays with frequency 3.0×10^{18} Hz, and of radio waves with frequency 3.0×10^{8} Hz.

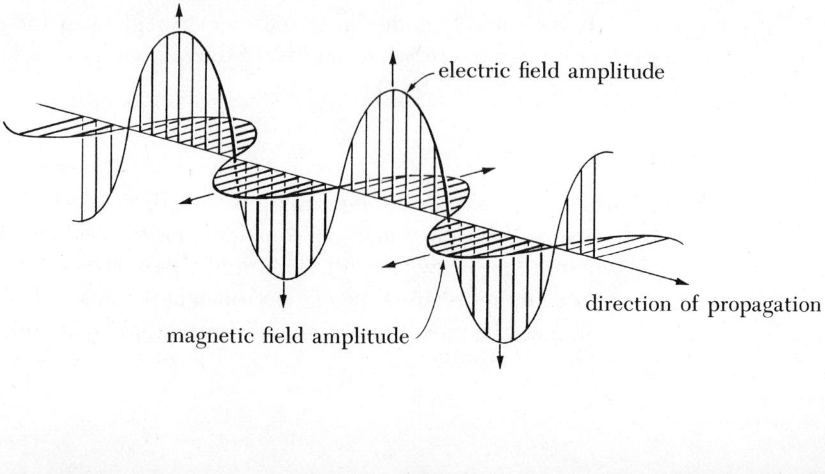

Figure 8.8
The oscillating electric and magnetic fields of electromagnetic radiation.

SOLUTION (a) X rays of 3.0×10^{18} Hz. Using the Bohr-Planck relation (Eq. 8.8) and the conversion factor from ergs to calories, we find

$$
\begin{aligned}
E &= h\nu \\
&= (6.6 \times 10^{-27} \text{ erg sec})(3.0 \times 10^{18} \text{ sec}^{-1}) \\
&= (19.8 \times 10^{-9} \text{ erg})\left(2.39 \times 10^{-8} \frac{\text{cal}}{\text{erg}}\right) \\
&= 4.73 \times 10^{-16} \text{ cal}
\end{aligned}
$$

But this is for one such photon. For a mole of these photons,

$$
\begin{aligned}
E &= 6.023 \times 10^{23} \times 4.73 \times 10^{-16} \\
&= 2.85 \times 10^{8} \text{ cal/mol, or } 2.85 \times 10^{5} \text{ kcal/mol}
\end{aligned}
$$

The wavelength of the radiation is

$$\lambda = \frac{c}{\nu} = \frac{3.0 \times 10^{8} \text{ m/Hz}}{3.0 \times 10^{18} \text{ Hz}^{-1}} = 1 \times 10^{-10} \text{ m, or 1 angstrom}$$

(b) Radio wave of 3.0×10^{8} Hz. Similarly,

$$
\begin{aligned}
E &= h\nu \\
&= (6.6 \times 10^{-27} \text{ erg sec})(3.0 \times 10^{8} \text{ sec}^{-1}) \\
&= (19.8 \times 10^{-19} \text{ erg})\left(2.39 \times 10^{-8} \frac{\text{cal}}{\text{erg}}\right) \\
&= 4.73 \times 10^{-26} \text{ cal}
\end{aligned}
$$

For a mole of these photons,

$$
\begin{aligned}
E &= 6.023 \times 10^{23} \times 4.73 \times 10^{-26} \\
&= 2.85 \times 10^{-2} \text{ cal/mol, or } 2.85 \times 10^{-5} \text{ kcal/mol}
\end{aligned}
$$

The wavelength of the radiation is

$$\lambda = \frac{c}{\nu} = \frac{3.0 \times 10^{8} \text{ m/sec}}{3.0 \times 10^{8} \text{ sec}^{-1}} = 1 \text{ m}$$

The Spectroscopic Experiment

The usual mode of operation in spectroscopy is to choose the frequency range appropriate to the mode of energy one wants to excite (say, molecular vibrations), and then irradiate the sample with a radiation source capable of providing the correct frequency or frequencies. This typically involves "scanning" the radiation frequency through the range of interest and then noting the particular frequency or frequencies causing transitions. Transitions are usually detected by a device that responds to the absorption of some of the radiation energy applied to the sample. For the special case of optical spectroscopy in the visible region of the electromagnetic spectrum, this is shown schematically in Fig. 8.9. In this scheme a lamp is used as a radiation source. The light emitted from a typical lamp covers a wide

Figure 8.9
Schematic diagram of an absorption spectrometer for the visible region of the electromagnetic spectrum. The movement of the prism permits one to scan the frequency range of interest. The response of the sample to this scan is displayed on the recorder.

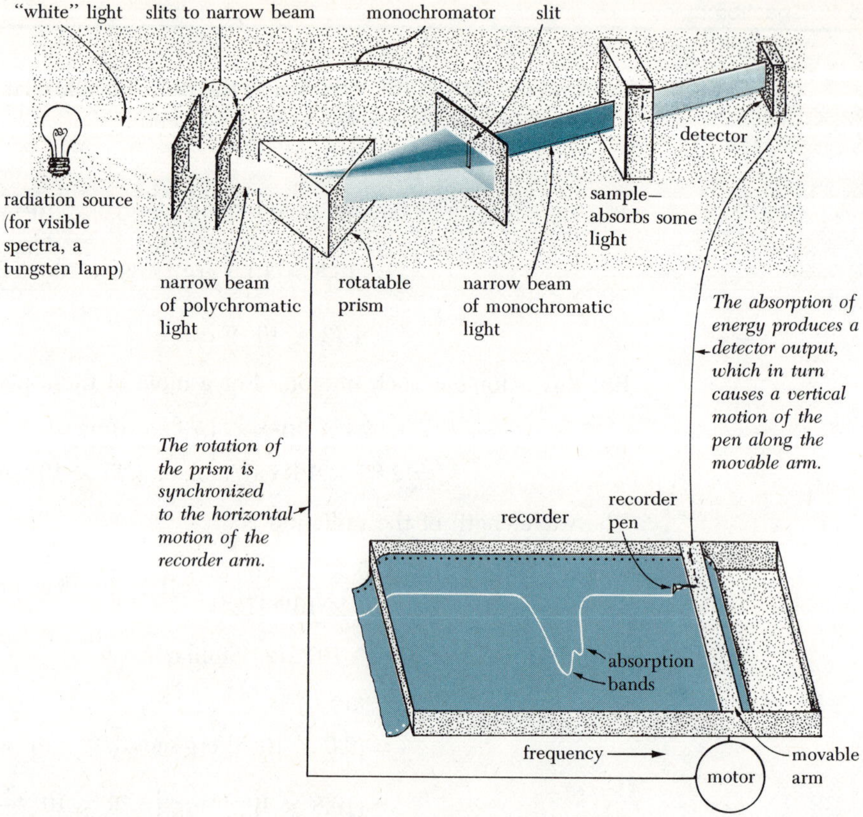

range of frequencies, but one is interested in knowing the response of the sample to radiation of each specific frequency over the same range (predetermined on the basis of what type of transitions one wishes to study). A prism is used in the scheme outlined in Fig. 8.9 to disperse the light from the lamp; light of different frequencies in the incident beam is dispersed into slightly different directions as it leaves the prism. By rotating the prism, one can select which frequency (or narrow range of frequencies) is directed to slit 2; only that frequency is then directed to the sample. By rotating the prism very slowly, one can "scan" the entire frequency range of interest (assuming that the source, a lamp in this case, emits radiation covering the entire frequency range of interest). Then one observes the response of the sample to each frequency, as the frequency is slowly scanned. This response is typically recorded on paper by the vertical response of a pen that moves in one direction (say, horizontally) in synchronization with the motion of the prism. If there is no response, the pen merely traces a straight horizontal line across the paper, a so-called "base line." If there is a response at a particular frequency, however, the pen shows an excursion from the base line, giving what is called a **peak,** an "absorption peak." This situation is also depicted in Fig. 8.9.

The response of the sample to the incident radiation is measured by an appropriate detection scheme. In the specific case depicted in Figure 8.9, absorption spectroscopy in the visible light region, the detection scheme consists of detecting the intensity of radiation that comes from the sample; if the sample absorbs radiation by undergoing transitions to higher energy levels, then the intensity of transmitted radiation is correspondingly reduced. Thus, in the case depicted in Fig. 8.9, the intensity of transmitted light dips at the frequency corresponding to a transition, giving the inverted peak as the prism rotates through the correct angle for providing the critical frequency through slit 2 to the sample. The frequency corresponding to the maximum excursion from the base line is identified as the **absorption frequency,** corresponding to a particular electronic transition in the sample (from a lower energy state to a higher energy state). The detector shown in Fig. 8.9 is a device capable of measuring the intensity of radiation incident upon it, a photocell with associated electronic circuitry.*

While the specific designs of spectroscopic apparatus used for the various types of molecular spectroscopy differ in many important details from what is described above for visible absorption spectroscopy, the overall strategy is basically the same in all cases. Each spectroscopic experiment requires a source of radiation, a way to select and vary the frequency, a device for detecting the occurrence of transitions, and a means of displaying the results in a useful fashion (for example, a recorder).

Electronic Spectra

One kind of spectrum that provides much information is the electronic spectrum. In this spectrum, the energy absorbed causes transitions from one electronic state to another. (This type of transition for atoms, viewed as an excitation of an electron from one atomic orbital to another, was described briefly in Chap. 6.) In a molecule, the transitions are viewed as arising from the excitation of an electron from one molecular orbital to another. For certain functional groups (for example, the $>C{=}O$ group) this type of transition occurs in a region of the electromagnetic spectrum that is conveniently accessible experimentally—in the visible range to the near-ultraviolet region of the spectrum. Since the energies of the molecular orbitals of a molecule depend not only on the functional group primarily identified with the transition (for example, $>C{=}O$) but also on the other groups bound to that functional group, accurate measurement of the transition frequencies can provide detailed information about molecular structure.

Electronic energy levels are generally separated by a much larger amount of energy than vibrational or rotational levels are. Most of the transitions between different electronic levels occur in the ultraviolet region of the spectrum. The electronic transitions of metal ions (Mn^{2+}, Co^{2+}, Ni^{2+}, Cu^{2+}, Fe^{3+}, and the like) or

*A photocell is a device containing a surface that is maintained at a negative potential; when this surface is acted on by radiation (photons), it emits electrons. These electrons are collected by a positive electrode, giving rise to current, which can be amplified appropriately, providing an electrical signal for driving the motor that positions the recorder pen vertically.

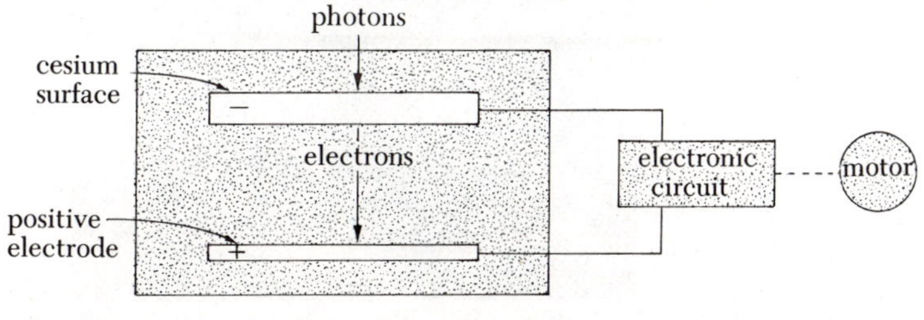

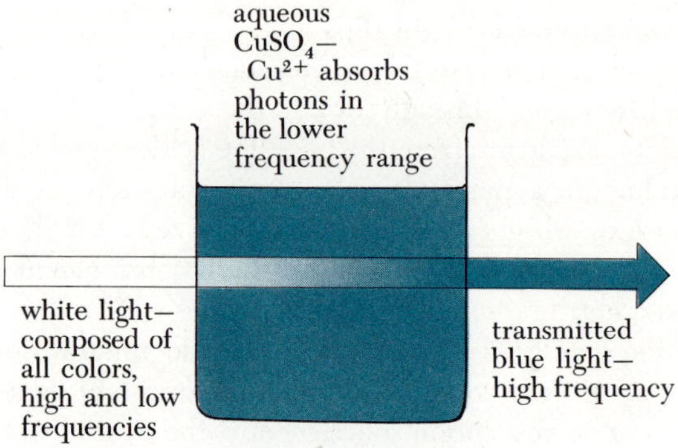

Figure 8.10
Origin of the blue color of aqueous Cu^{2+} solutions.

of compounds involving metal ions often have transition frequencies much lower than those of most organic molecules. These metal ion absorption frequencies generally lie in the near infrared and visible regions of the electromagnetic spectrum, giving rise to their characteristic colors. Most pure organic substances are colorless, because they have no available transitions that occur in the visible region of the electromagnetic spectrum. The ion Cu^{2+} in water appears blue, because it absorbs red light at the low frequency end of the visible range from a "white" light source like sunlight, leaving a proportionately higher intensity of radiation at the high-frequency end of the spectrum, the blue region. This is illustrated in Fig. 8.10. Another example is the visible spectrum of Ti^{3+} in aqueous solution, actually a complex containing six water molecules, which results from an electronic transition frequency in the center of the visible region, leaving a violet appearance from "white" light. This is depicted in Fig. 8.11.

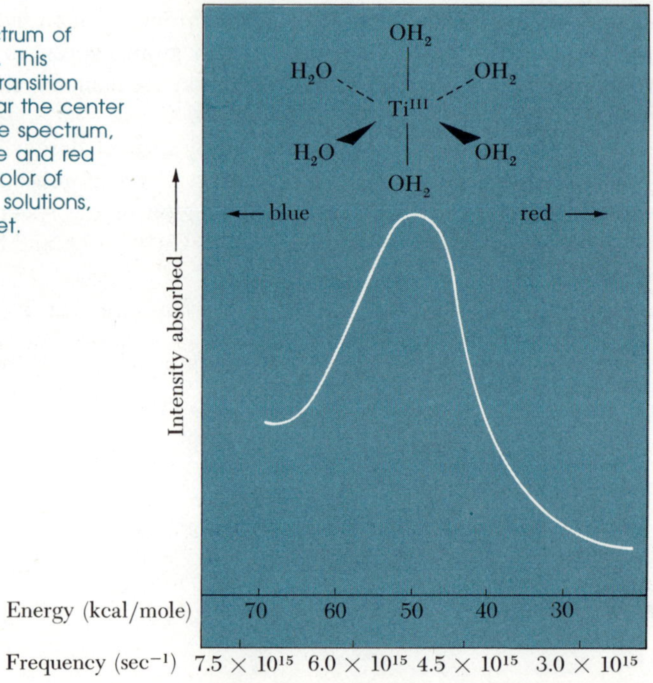

Figure 8.11
Visible spectrum of $[Ti(H_2O)_6]^{3+}$. This electronic transition absorbs near the center of the visible spectrum, leaving blue and red light. The color of $[Ti(H_2O)_6]^{3+}$ solutions, then, is violet.

Spotlight

An important and interesting phenomenon that has been investigated by visible spectroscopy is photosynthesis. Chlorophylls are molecules that contain a large planar cyclic system made up of four nitrogen atoms that bind the ion Mg^{2+}, as illustrated herewith. There are many resonance contributors for the ring system of chlorophylls. Largely because of the extensive resonance, these molecules absorb light in the visible region of the spectrum. Chlorophyll appears green to us because it can undergo an electronic transition that absorbs light in the red region of the spectrum (long-wavelength visible light, about 6500 Å, or 6.50×10^{-7} m) and also in the violet region (shorter wavelength, about 4000 Å). Green light, which occurs near the center of the visible spectrum, is not absorbed and gives the chloroplasts their color. When the chlorophyll molecule absorbs red and violet light, the electrons are excited to higher energy levels. As

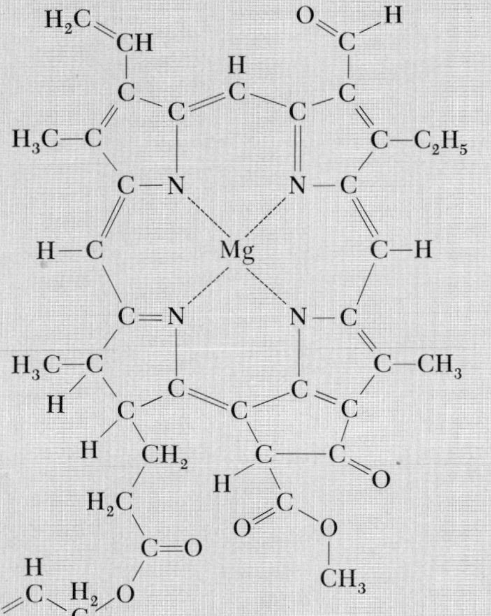

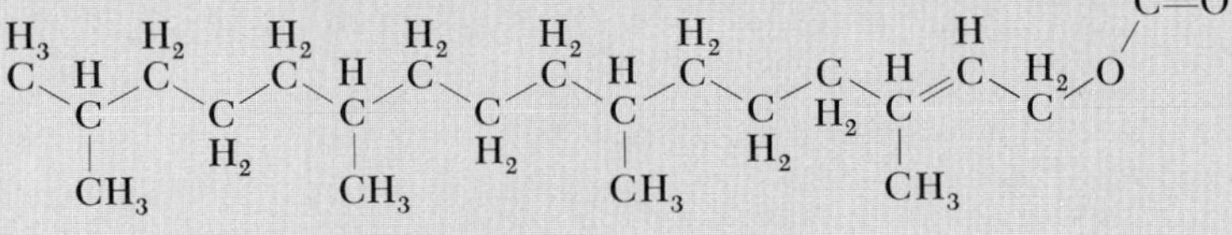

they return to their lower energy, or ground state, energy is released. This energy can be used by the plant to synthesize adenosine triphosphate, a molecule that "stores" energy. The energy stored by ATP is released in a large number of processes that take advantage of the favorable reaction:

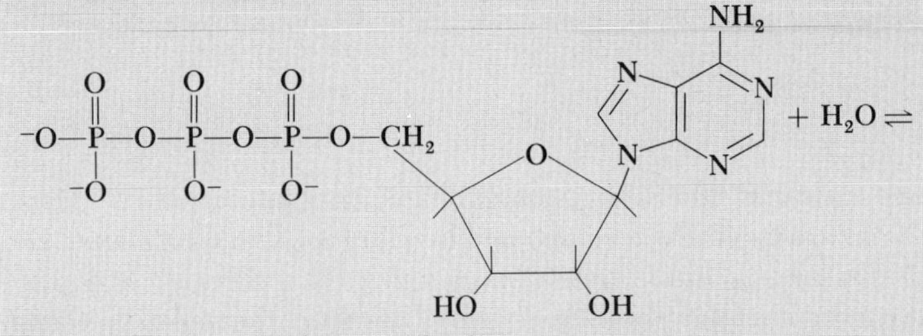

adenosine triphosphate

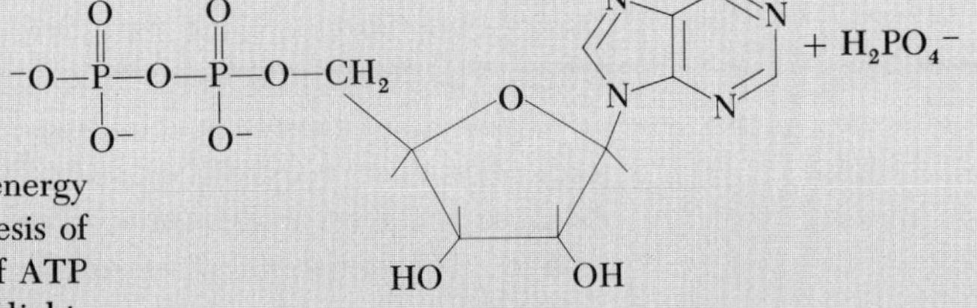

adenosine diphosphate

Once the plant has ATP, it can use the stored energy to maintain life processes, including the synthesis of chemicals needed for growth. The reservoir of ATP allows plants to function even in the absence of light; they needn't stop functioning at night.

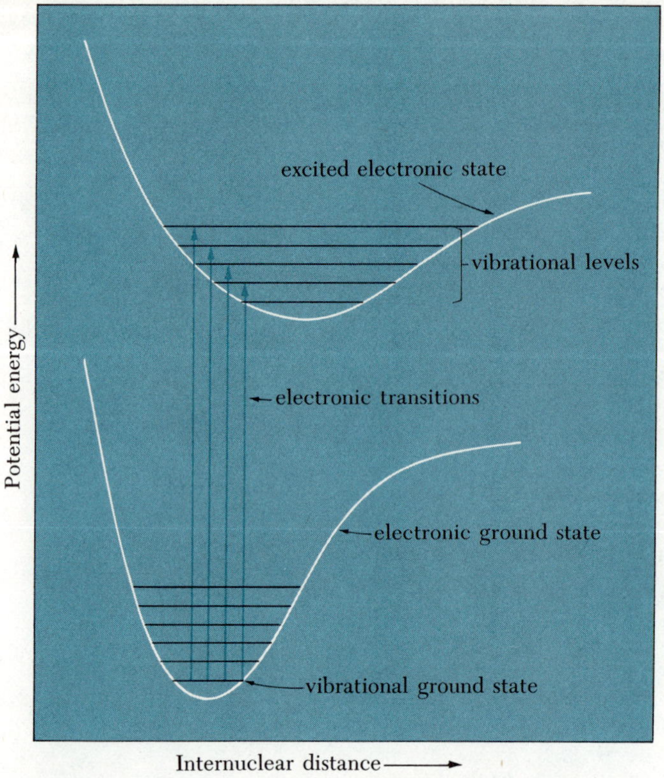

Figure 8.12 An energy level diagram showing electronic and vibrational energy levels. The curved lines show how the potential energy of the molecule changes as a function of the distance between the nuclei. Each curved line shows the energy for a particular electronic state, with the lower one being the more stable (ground) state. The horizontal lines show vibrational energy levels. The rotational levels are even more closely spaced and have been left out for clarity. The electronic excitation energy from the lower to the upper state can have a range of values, because of the several possible vibrational energy levels (and also rotational energy levels). The range of values leads to broad peaks in electronic absorption spectra.

Each electronic state has vibrational energy states associated with it. Hence, electronic transitions are typically accompanied by vibrational and rotational transitions, giving rise to many possible transition energies. This situation is depicted in Fig. 8.12. The curved lines show how the potential energy of the molecule changes as a function of the distance between the nuclei. Each curved line shows the energy for a particular electronic state, with the lower one being the more stable (ground) state. The horizontal lines show vibrational energy levels. The rotational levels are even more closely spaced and have been left out for clarity. The electronic excitation energy from the lower to the upper state can have a range of values, because of the several possible vibrational energy levels (and also rotational energy levels). The range of values leads to broad peaks in electronic absorption spectra. In addition, the peaks appear to be smooth rather than a series of spikes because of the close spacing of the energy levels.

Vibrational Spectra

Electromagnetic radiation can be absorbed when a dipole moment changes in the molecule. One type of motion for which the dipole moment changes in some molecules is vibrational motion, the oscillatory motion of atoms in molecules. Vibrational motion takes its simplest form in a diatomic molecule, in which case it is quite analogous to the oscillatory motion observed in the simple spring system depicted in Fig. 8.13. If such a mechanical arrangement is distorted from its resting configuration by a stretching or a compressing of the spring, it will oscillate with some characteristic frequency. Similarly, a diatomic molecule vibrates with a characteristic frequency. If the diatomic molecule is symmetrical, then it has no dipole moment, irrespective of whether it is oscillating. A symmetrical diatomic molecule is called homonuclear, meaning that both atoms are the same. For H_2 there is no change in dipole moment as the molecule vibrates, so electromagnetic radiation cannot induce vibrational transitions in the manner described. However, when H—Cl vibrates, the net dipole moment does change; hence vibrational transitions can be induced by radiation if its frequency is correct (according to Eq. 8.8). The vibration of HCl can be symbolized by the simple notation

$$\overset{\longleftarrow\text{\small 0000}\longrightarrow}{\text{H—Cl}}$$

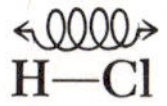

where the directions of the arrowheads indicate the relative motion of the atoms. Figure 8.14 is a pictorial representation of the absorption of radiation by the vibrational mode of HCl. The frequency of radiation giving rise to this transition is in the infrared region of the electromagnetic spectrum.

Now it is pertinent to ask what the study of vibrational transitions can tell us about molecules. To answer this question, we need a model of a vibrating molecule to relate absorption energies to pertinent characteristics of the vibrations. One of the

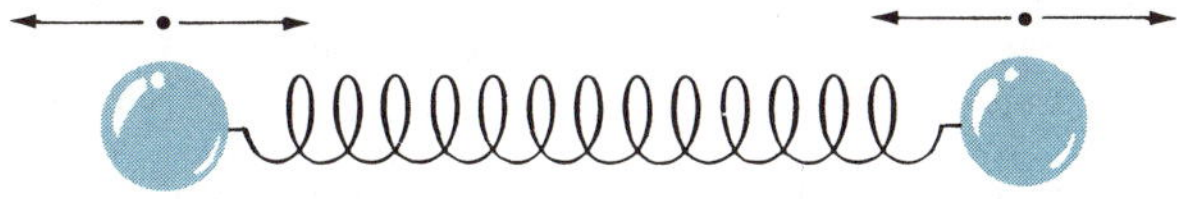

Figure 8.13
Vibrational motion in a simple spring system.

Figure 8.14
Vibrational transition in HCl.

photon input $\xrightarrow{h\nu}$

$h\nu = E_f - E_i$

excited state vibrational energy level, E_f

vibrational transition

ground state vibrational energy level, E_i

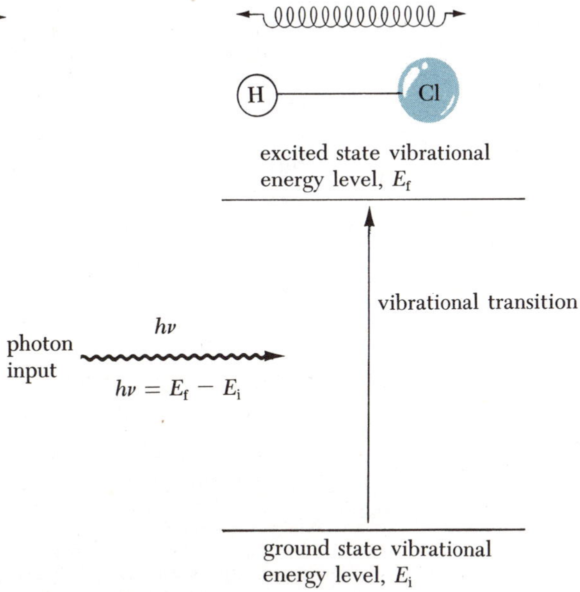

key features of the model is one which we have just considered—that the molecular dipole moments must change during a vibration for the molecule to be capable of undergoing vibrational transitions in the manner described. This feature, viewed as a criterion, allows us to assign the absorbancies, or peaks, to a set of vibrational motions that are "allowed"—that is, they can occur because of an associated dipole moment change. Another desirable feature of a model would be a way of predicting how many different vibrations are possible in a given molecule. It can easily be shown that a molecule containing N atoms has $3N - 5$ modes of vibration if it is linear, that is, having all atoms in a line, like H—Cl or H—C≡N.* These motions include translational motion of the molecule as a whole, rotational motion, and vibrational modes. All molecules have three translational modes of motion, since there are three independent directions of movement. Nonlinear molecules have three distinct rotational modes, corresponding to three mutually perpendicular axes about which rotation of a molecule occurs, as we visualize it. Linear molecules have two independent rotational motions; rotation about the bond axis does not change the molecule in any physically recognizable way, so it is not physically significant. The remaining modes of motion for a molecule are vibrations, $3N - 6$ for a nonlinear molecule and $3N - 5$ for a linear molecule.

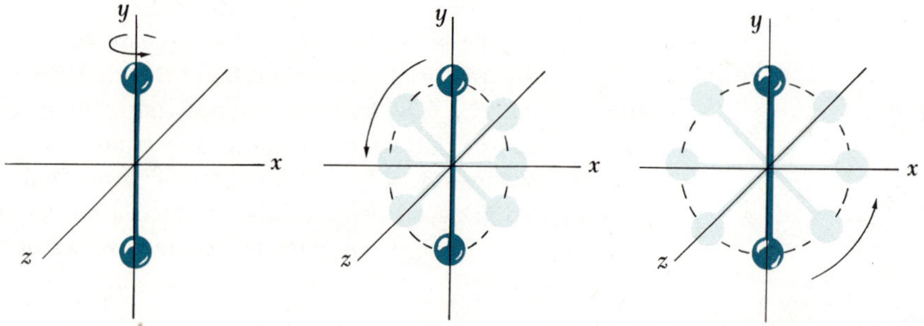

rotation about y (no change) rotation about x rotation about z

For a diatomic molecule such as HCl, the number of possible vibrational modes $= 3N -$ no. translations and rotations $= 3N - (3 + 2) = 3N - 5 = 6 - 5 = 1$. Thus, there is only one vibrational mode for diatomic molecules. This vibration can be excited by absorption of electromagnetic radiation if the molecule is polar.

EXAMPLE 8.8 Which of the following vibrational modes of CO_2 can be observed by infrared absorption?

$$\overset{\leftarrow\ \ \rightarrow}{\text{O=C=O}} \quad \text{or} \quad \overset{\leftarrow\ \leftarrow\ \rightarrow}{\text{O=C=O}}$$

(a) (b)

SOLUTION (a) This vibration is called a symmetric stretch. There is no net dipole moment at any stage during the vibration: hence, there is no change in the dipole moment and this vibration cannot lead to energy absorption.

*A molecule with N atoms has a total of $3N$ modes of motion (corresponding to the three independent coordinates needed to specify the position of any one atom).

(b) This vibration is an asymmetric stretch and gives rise to a change in dipole moment (from zero). This dipole moment manifests itself during the vibration; the two bond dipoles, $\overleftarrow{O{=}C}\overrightarrow{{=}O}$, will not be the same during the vibration. One of the dipoles is increased while the other is decreased. Electromagnetic radiation of the correct frequency can be absorbed as the energy of this vibration is increased.

EXAMPLE 8.9 How many vibrational modes are there for (a) CO_2; (b) H_2O; and (c) BF_3?

SOLUTION (a) CO_2; O=C=O, is linear; hence the number of vibrational modes is $3N$ − no. translations and rotations = (3)(3) − 5 = 4.

(b) H_2O; 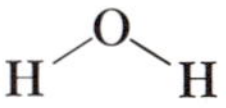, is nonlinear, and $3N$ − no. translations and rotations = (3)(3) − 6 = 3.

(c) BF_3; 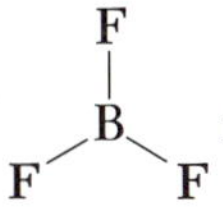 is nonlinear. $3N$ − no. translations and rotations = 3(4) − 6 = 6.

STUDY PROBLEM 8(e)

How many vibrational modes are there for formaldyhyde, 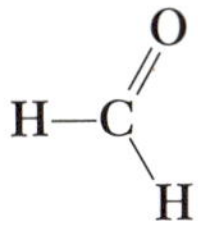 ; formic acid, 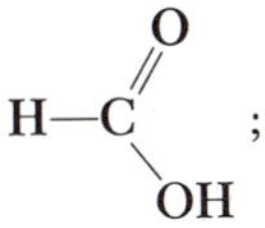 ; and $NiCl_4^{2-}$?

Structural Details from Vibrational Spectroscopy

OPTIONAL

There are six normal modes of vibration for a triangular molecule with four atoms ($3n - 6 = 3 \cdot 4 - 6 = 6$); these are shown here for the molecule boron trichloride.

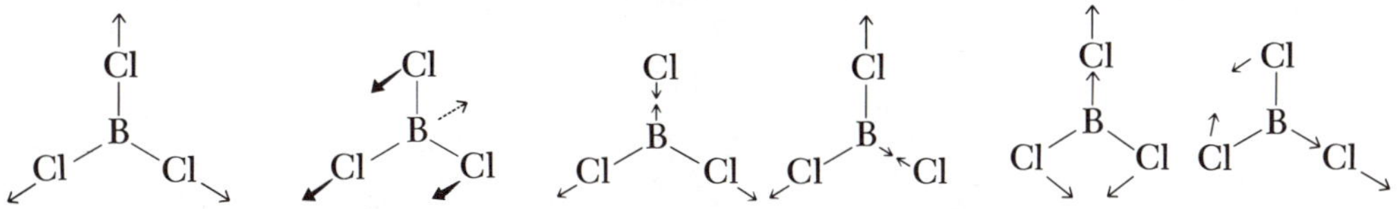

"symmetric" stretch bend stretching modes bending modes

These six modes show several features that are useful in interpreting infrared spectra. The "symmetric" motion on the left causes no change in the dipole moment and therefore is not activated by infrared radiation. The other modes, which are "infrared-active," can be stretching modes, bending modes, or a combination of stretching and bending. It is generally easier to cause the bending modes than a stretching mode, so lower-energy radiation is observed. An analogy is that the energy needed to bend a system composed of two balls at the end of a tight spring is

OPTIONAL

less than the energy needed to stretch the spring. Additionally, it can be seen that the pair of stretching modes that are not symmetric would require the same energy. Likewise, the pair of stretch and bend (combination) modes are alike and require the same energy. Thus, for BCl_3 there would be three infrared absorption frequencies instead of five.

When different atoms are bound to boron, however, this is not the case. For the vibrations of BCl_2Br, for example, one of the stretch and bend modes involves the stretching of a B–Br bond while the other involves the B–Cl bond. Hence, these modes require different energy. It also turns out that the "symmetric" motion is no longer symmetric since one corner of the triangle is now a B—Br stretch. So in BCl_2Br, all six modes are of different energy and all six absorb infrared energy. The number of infrared absorption peaks and their relative energies have been used to determine the precise geometrical structures of numerous inorganic compounds, especially ones that are very reactive and that are difficult to study by other methods. In practice, however, this application of vibrational spectroscopy is limited to relatively uncomplicated molecules. Its use requires a highly trained specialist.

A much more common use of infrared spectroscopy is in determining the gross structure of organic compounds by a functional group approach. Several organic functional groups have characteristic vibrational mode energies that tend to be similar from one compound to another. Figure 8.15 shows regions that are generally useful for assigning infrared absorption bands to particular pairs of atoms. One of the first things you probably notice about this figure is that the energy unit ordinarily used in infrared spectroscopy is the cm^{-1}, or "wavenumber". The value of the wavenumber is the inverse of the wavelength of the radiation, in centimeters. Radiation with a wavelength of 1×10^{-4} cm (one micrometer) has an energy of 1×10^{4} cm^{-1}.*

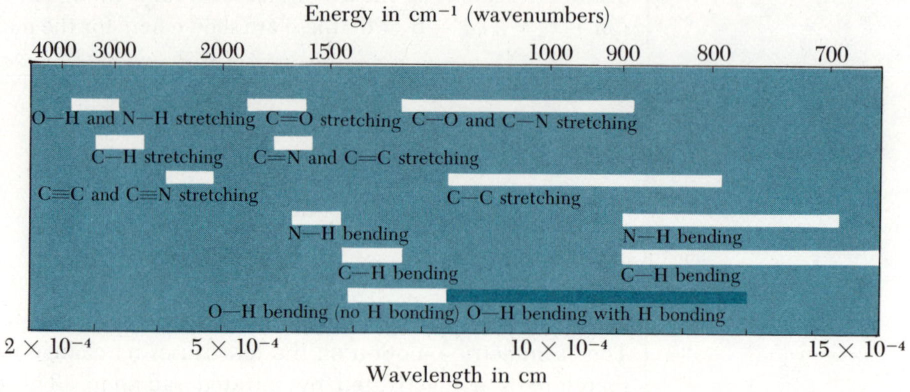

Figure 8.15
Some correlations of group vibrations to regions of infrared energy absorption.

*To convert the energy in units of cm^{-1} to more common units like J/mol or kcal/mol, obtain the wavelength λ (in cm) $=$ 1/wavenumbers (in cm)$^{-1}$. Then use this wavelength in the equation $\nu = c/\lambda$ to obtain frequency. Finally, obtain the energy from the proper choice of units for Planck's constant in $E = h\nu$.

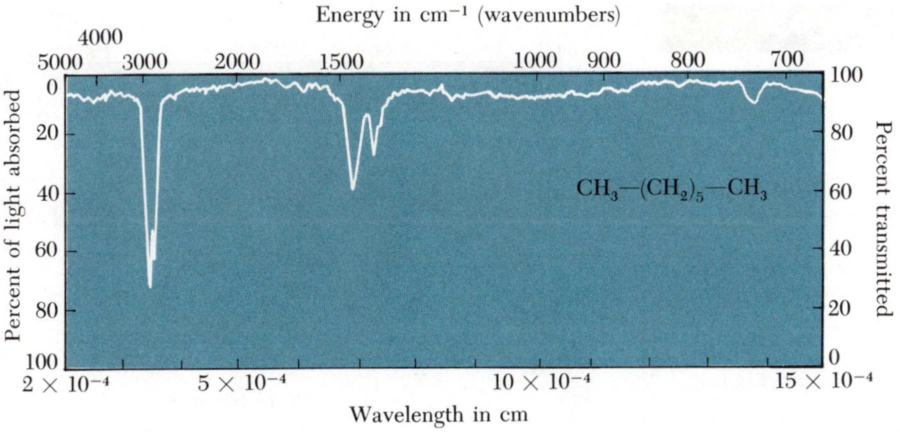

Figure 8.16 The infrared spectrum of an alkane: $CH_3(CH_2)_5—CH_3$ (*n*-heptane).

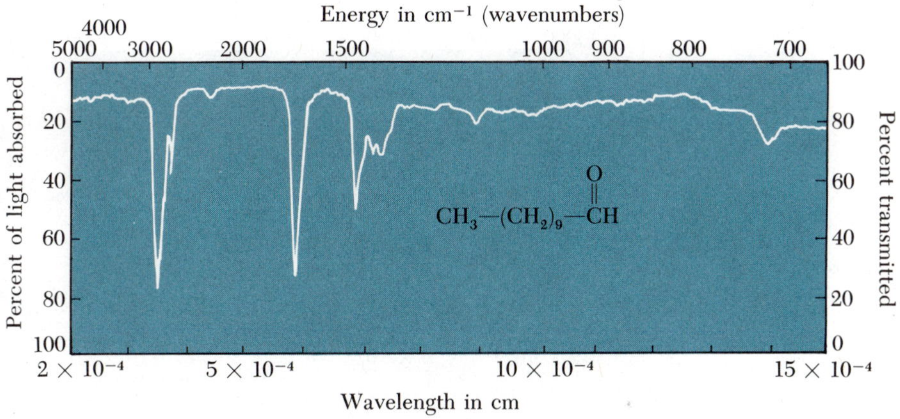

Figure 8.17 The infrared spectrum of an aldehyde: $CH_3—(CH_2)_9—C(=O)—H$ (*n*-decylaldehyde).

An important feature of the figure is that the higher the bond order, the higher the energy absorbed. Thus, the C≡C stretching mode is at a higher energy, and therefore more difficult to cause, than the C=C stretch, which is in turn higher in energy than the C—C stretch. Another important feature is that the stretching modes are more difficult than the bending modes.

Let us consider some infrared spectra to see how functional groups can be assigned. Figure 8.16 shows the spectrum of heptane, $CH_3—(CH_2)_5—CH_3$, which has only methyl and methylene functional groups. The C—H stretching energy is higher than the C—H bending energy. The peaks due to these modes are much more intense than the peak of the C—C stretch, which is hard to discern in this spectrum. The next spectrum (Fig. 8.17) shows the effect of the carbonyl group (>C=O) in the vibrational spectrum of decylaldehyde, $CH_3—(CH_2)_9—C(=O)H$. The sharp absorption peak near 1700 cm^{-1} is characteristic of the carbonyl group. The next spectrum (Fig. 8.18) shows the effect of combining a carbonyl group with an ether linkage (C—O—C)

OPTIONAL

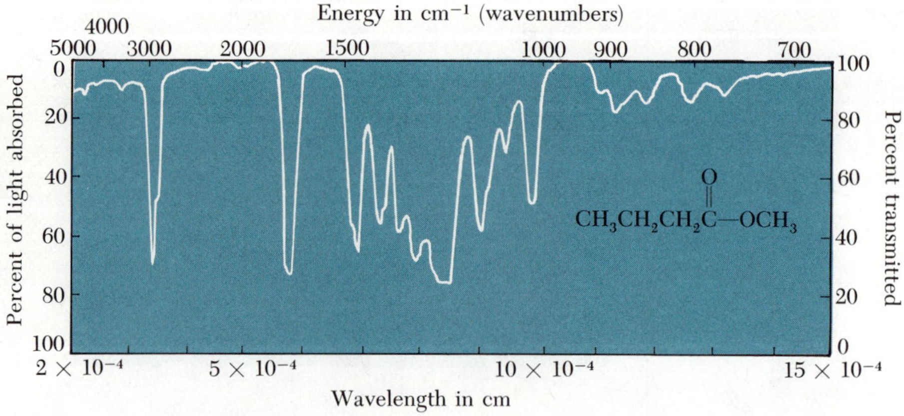

Figure 8.18 The infrared spectrum of an ester: $CH_3(CH_2)C(=O)—OCH_3$ (the methyl ester of butyric acid).

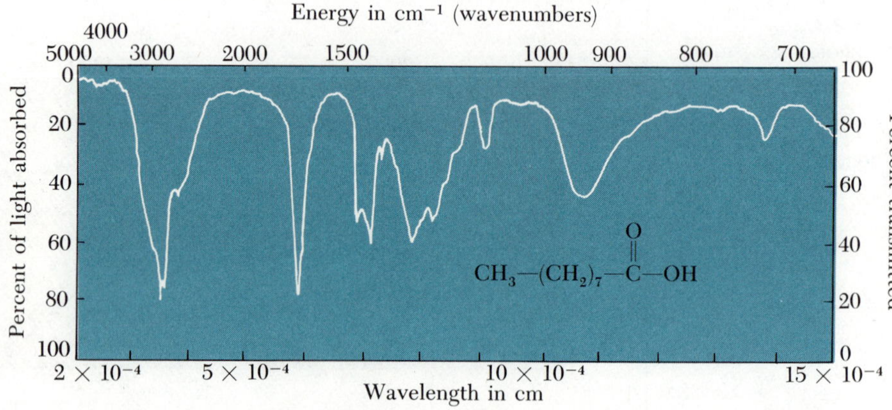

Figure 8.19 The infrared spectrum of a carboxylic acid: $CH_3(CH_2)_7C(=O)—OH$.

to form an ester ($\sim C(=O)—O—C\sim$), in this case $CH_3CH_2CH_2C(=O)OCH_3$. Note the sharp peak near 1700 cm^{-1} characteristic of the carbonyl group and the C—O stretching mode absorptions near 1300 and 1050 cm^{-1}. The presence of a hydroxyl group next to the carbonyl group gives the carboxyl group ($\sim C(=O)OH$). As Fig. 8.19 indicates, the O—H stretching mode overlaps with the C—H stretch near 3000 cm^{-1}. The absorption peak due to the O—H bending mode, which is generally broad, occurs near 950 cm^{-1}. Again, the carbonyl group absorption occurs near 1700 cm^{-1}.

EXAMPLE 8.10

Identify the possible functional groups X in the molecule $CH_3CH_2—CH_2—X$ from the infrared spectrum shown at the top of the next page. What information would tell you which possibility is correct?

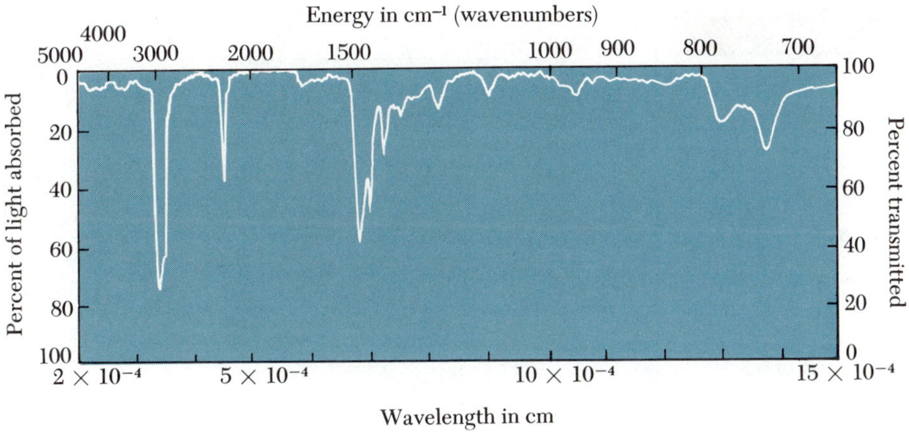

SOLUTION The infrared spectrum in this example is that of *n*-propylnitrile $CH_3—CH_2—CH_2—C≡N$. The band near 2200 cm^{-1} is due to the C≡N stretch, but could also be C≡O or C≡C. Elemental analysis would give the empirical formula and distinguish C≡N from C≡O and C≡C. Another experiment would be a mass spectrum analysis to give an accurate molecular weight, which would distinguish among the three possibilities.

STUDY PROBLEM 8(f)

A compound with molecular weight determined to be 58.08065 amu and with an elemental analysis of 62.2% C, 10.3% H, and 27.5% O has the infrared spectrum shown herewith. Identify the compound.

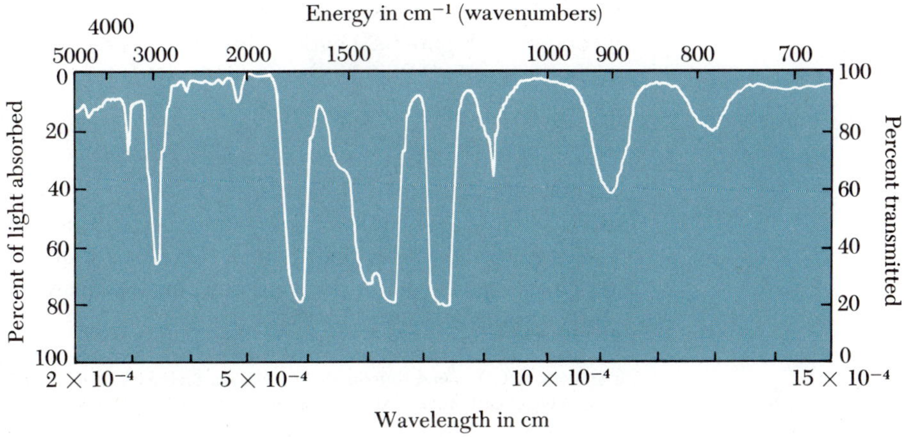

In another example of how vibrational spectroscopy can be used for identifying and studying specific functional groups, researchers have recently been able to study how oxygen is attached to hemoglobin in blood. By observing the stretching frequency corresponding to the O–O bond in O_2 bound to hemoglobin, they could

OPTIONAL

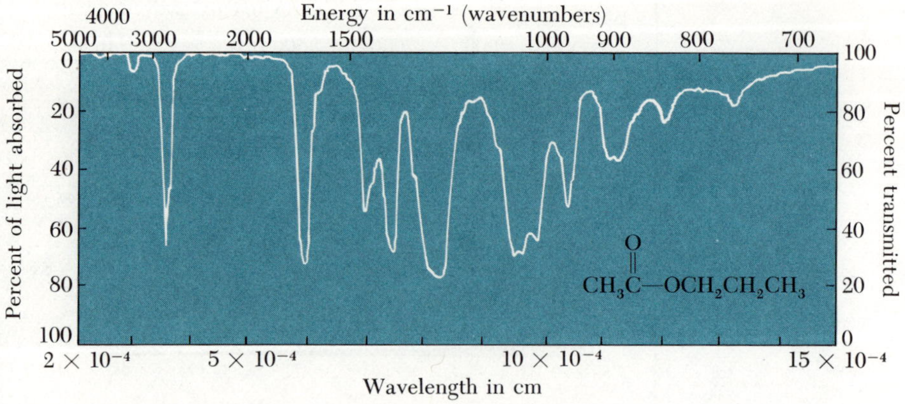

Figure 8.20 The infrared spectrum of an ester: $CH_3C(=O)OCH_2CH_2CH_3$ (the propyl ester of acetic acid).

study the characteristics of that binding.* Note that free O_2 would be inactive in infrared spectroscopy, since vibration does not change the dipole moment. But when O_2 becomes bound to the iron atom of hemoglobin, the bound oxygen atom is different from the other oxygen atom.

It is usually true of a complex organic compound, and Fig. 8.20 presents a good example, that not all the absorption frequencies can be accounted for simply by identifying them with individual bonds or functional groups. For such a case, the pattern of absorption frequencies can be counted on as characteristic of the particular compound—it is like a molecular fingerprint. This fact can be used to compare a compound of unknown structure with authentic samples of compounds of known structure that are considered to be the possible alternatives for the structure of the "unknown." If the vibrational spectrum of the unknown is compared with spectra of the authentic samples, a "fingerprint match" can be made. Such use can be seen by comparing Figure 8.20 with Figure 8.18. These spectra are those of the isomers $CH_3CH_2CH_2—C(=O)OCH_3$ and $CH_3C(=O)OCH_2CH_2CH_3$. As isomers, they have the same molecular formula and molecular weight but not the same structural formula. They even have the same functional groups, but their infrared spectra are different enough to distinguish them from one another. Such techniques are commonly used in the laboratory, and have also become useful in legal cases.†

*The binding of carbon monoxide to hemoglobin has been studied by observing the C≡O stretching frequency of CO bound to hemoglobin. This attachment of CO to hemoglobin is responsible for the toxic effect of carbon monoxide, a common and dangerous constituent of air that is polluted by internal combustion engines.

†Another very important aspect of the use of spectroscopy will be discussed later when we look at the rates of chemical reactions, or **reaction kinetics.** In many chemical reactions, reactants differ from products in one or more spectroscopic properties. In the reaction

$$2NO_2 \rightleftharpoons N_2O_4$$

which occurs in air polluted by automobile exhaust, the NO_2 is brown, absorbing energy in the visible region, while N_2O_4 is colorless. The infrared spectra of the two species are also quite different. These spectral differences can be used to find out how fast the conversion from reactant to product takes place.

Vibrational spectra, besides providing structural information, are useful in providing information about the strengths of bonds. In making this connection between bond strengths and vibrational spectra, it is customary to use the simple spring model, shown in Fig. 8.14, for the vibrations of the atoms joined by a chemical bond. We visualize vibration of the bond in correspondence to a stretching and a compressing of the spring. From classical physics one knows that the amount of energy necessary to stretch the spring depends on its strength; and the frequency with which the oscillations occur depends on both the strength (force constant) of the spring and the masses at the ends of the spring.*

From quantum mechanical developments it can be shown that the absorption frequency in the vibrational spectrum of a diatomic molecule is the same as the frequency of "mechanical" oscillation of the atoms. The same kind of relation exists for complex molecules, although it is often not straightforward to visualize the mechanical vibrations.

From several years of research in vibrational spectroscopy, including quantum mechanical treatments, a clear physical interpretation has emerged. Much of this can be summarized in Fig. 8.21, viewed most simply as describing vibrational motion

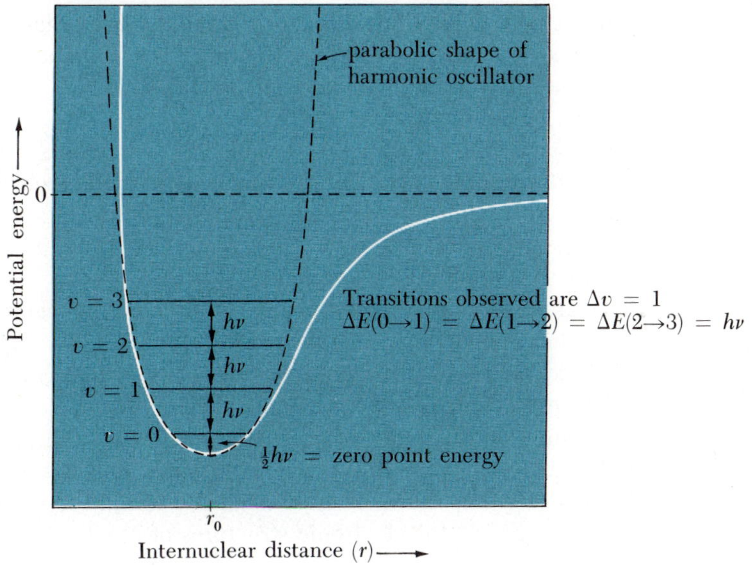

Figure 8.21 Use of harmonic oscillator energy levels, showing how only one energy is absorbed for vibration level transitions. The dashed line is the "idealized" harmonic oscillator potential curve. The white line is a typical potential curve for a molecule.

*If we use the model called a simple harmonic oscillator for a chemical bond vibration, the frequency of the vibration is related to the force constant of the bond and the masses m_1 and m_2 of the atoms constituting the bond according to the equation

$$\nu = \frac{1}{2\pi}\sqrt{\frac{k}{\mu}}$$

if we take the diatomic case for simplicity. Here ν is the frequency of the oscillation, $\mu = (1/m_1 + 1/m_2)^{-1}$ is the reduced mass, and k is called the "force constant." The value k is the proportionality constant that relates the force required to distort, by extending or compressing, the "spring" to the extent of the distortion x:

$$F = -kx$$

According to the classical development of the harmonic oscillator, the potential energy of the system is given as $\frac{1}{2}kx^2$. The force constant k is related to the strength of the bond ("spring"). Hence, one can get information about a bond by measuring its vibrational frequency.

OPTIONAL

in a diatomic molecule, although analogous diagrams can be drawn for the vibrations in more complex molecules. The dark line in Fig. 8.18 is a plot of how the potential energy of the diatomic molecule varies with the distance between the atoms.* From quantum mechanics, we learn that the vibrational energy is quantized, with the allowed energies given by the equation

$$E_v = h\nu_m(v + \tfrac{1}{2}) \tag{8.9}$$

In Eq. (8.9), ν_m is the classical mechanical frequency of vibration and v is a vibrational quantum number, an integer, 0, 1, 2, 3, . . . ; E_v is the vibrational energy value corresponding to the particular value of the vibrational quantum number. The numbered horizontal lines in Figure 8.18 represent the allowed energy levels.† The vibrational transitions responsible for the absorption of energy evident in a vibrational spectrum are jumps from one energy level to the next higher; expressed another way, these are changes in v of $+1$—from $v = 0$ to $v = 1$, or from $v = 1$ to $v = 2$, and so on. From Eq. (8.8), we see that the energy absorbed by the increase in vibrational energy is therefore

$$\Delta E = h\nu_m(v_f + \tfrac{1}{2}) - h\nu_m(v_i + \tfrac{1}{2}) = h\nu_m(v_f - v_i) = h\nu_m(1) = h\nu_m \tag{8.10}$$

This increase in energy must, according to the Bohr-Planck relation (Eq. 8.8), equal the energy of the absorbed photon $h\nu$, where ν is the frequency of the radiation absorbed. Hence

$$h\nu = h\nu_m, \qquad \text{or} \qquad \nu = \nu_m$$

Thus, the frequency of radiation absorbed equals the mechanical vibration frequency.

Other Energy Absorbing Processes; Magnetic Resonance

Several other energy-absorbing processes, or transitions, provide chemists with useful information about the structure and bonding of molecules. Some very widely used techniques involve magnetic resonance, which could more appropriately be called magnetic spectroscopy. These techniques include nuclear magnetic resonance (nmr) and electron spin resonance (esr, or sometimes epr—for electron paramagnetic resonance). The physical nature of the transitions has a bearing on the use of the magnetically dependent nuclear magnetic resonance spectroscopy.

In our models for electronic structures of atoms and molecules, we introduced

*The dashed line in Fig. 8.18 shows the potential energy function $\frac{1}{2}kx^2$, where $x = r - r_0$, for a simple harmonic oscillator. It is seen that the simple harmonic oscillator model is quite good at low vibrational energies but fails for very high values of the vibrational energy.

The potential energy curve in Figure 8.15 shows that as the bond is compressed greatly, that is, as the interatomic distance r becomes very small (approaching zero), the energy rises very rapidly. This corresponds to the repulsive forces that prevent atoms from coalescing. As the bond is stretched and extended greatly, the attractive forces of the bond are overcome, and the molecule dissociates into two atoms. At the point of dissociation, the energy of the system no longer depends on an increase in interatomic distance. The point of minimum potential energy $r = r_0$ is called the equilibrium bond distance. The energy difference D_e in Fig. 8.15 is called the dissociation energy: it is the energy difference between the energy of the system in its lowest level, $v = 0$, and the energy of the dissociated state.

†It is interesting to note that the lowest vibrational energy that is allowed is *not* zero, but is found to be the equation $(h/4\pi)(\sqrt{k/\mu})$, corresponding to $v = 0$.

the concept of intrinsic electron spin angular momentum. We find that the concept of intrinsic spin angular momentum also applies to the nucleus; some types of experiments indicate that certain types of nuclei behave as though they spin about an axis. We have learned that the spin angular momentum of an electron is characterized by the quantum number $\frac{1}{2}$ and that the spin angular momentum can be oriented to a magnetic field in only two possible ways, for which the quantum numbers $+\frac{1}{2}$ or $-\frac{1}{2}$ are used, or the symbols ↑ and ↓. The nucleus of each isotope has a characteristic nuclear spin angular momentum. It is characterized by a quantum number I, which is either an integer or half-integer—for example, $\frac{1}{2}$ or 1 or $\frac{3}{2}$ or 2. The nuclear spin for hydrogen, the isotope most commonly studied by nuclear magnetic resonance experiments, is $\frac{1}{2}$; that is, $I = \frac{1}{2}$.

Associated with intrinsic electron spin is a very small magnetic field (magnetic moment). Likewise, a nucleus with a nonzero spin also has a magnetic moment associated with it, like that of a tiny bar magnet. If a molecule containing a nucleus with a nonzero spin or a molecule with one or more unpaired electrons is placed in a magnetic field, any magnetic moment due to spin becomes aligned in quantized fashion relative to the applied magnetic field. For the case of a nucleus with $I = \frac{1}{2}$ (or as we have already seen, for an unpaired electron), the alignment of the intrinsic spin angular momentum and associated magnetic moment can be in only one of two possible orientations, identified by the quantum numbers $+\frac{1}{2}$ or $-\frac{1}{2}$. As with the roughly analogous case of a tiny bar magnet restricted to one of two possible orientations in a magnetic field, these two different alignments have somewhat different energies. This analogy is depicted in Fig. 8.22.

For either the $I = \frac{1}{2}$ case or the bar magnet model, the energy difference ΔE between the two possible states is directly proportional to the strength of the applied magnetic field H_0. For a nuclear spin case, the proportionality constant is given by the expression $\gamma h/2\pi$, where h is Planck's constant and γ is a factor called the **magnetogyric ratio.** This ratio is a constant characteristic of the specific nuclear isotope;

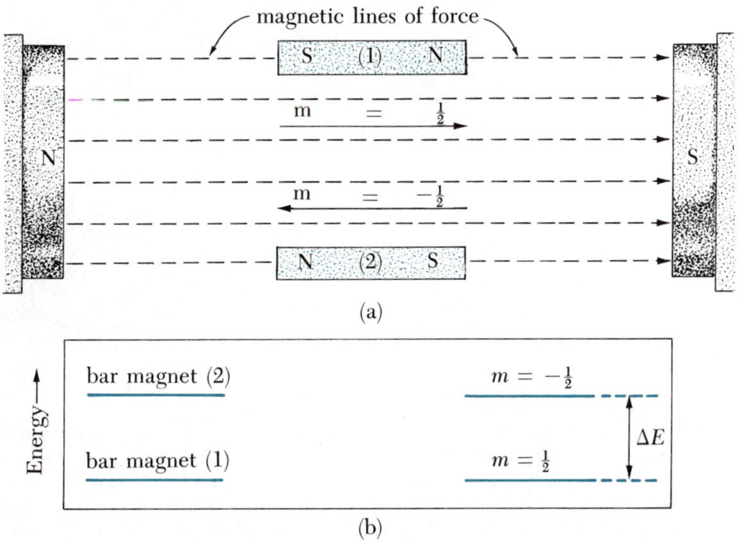

Figure 8.22
(a) Orientations of spin states (for $I = \frac{1}{2}$) in a magnetic field using bar magnets as an analogy. (b) Comparison of energy levels of spin states (and analogous bar magnets) in a magnetic field.

OPTIONAL

^{1}H has a specific γ value, ^{35}Cl has a specific γ value, and so on. The energy difference can now be expressed

$$\Delta E = \frac{\gamma h}{2\pi} H_0 \tag{8.11}$$

According to Eq. (8.11), the size of ΔE can be adjusted by varying H_0, essentially the strength of the magnetic field exerted by the magnet used in the experiment. This dependence of ΔE on H_0 is shown diagrammatically in Fig. 8.23.

In direct analogy to the other forms of spectroscopy we have discussed, it is possible to cause transitions between the energy levels shown in Figs. 8.22 and 8.23. For any particular value of H_0 represented in Fig. 8.23, there is a specific ΔE, defined by Eq. (8.11), that must be matched exactly by a quantum of electromagnetic radiation for a transition to occur. By the Bohr-Planck condition, this requires a frequency of $\Delta E/h$, which according to Eq. (8.11) is given by

$$\nu_o = \frac{\Delta E}{h} = \frac{\gamma H_o}{2\pi} \tag{8.12}$$

When this condition is satisfied, the oscillating magnetic field of the radiation of frequency ν_0 is able to interact with the magnetic moment of the nucleus, and energy of amount ΔE is absorbed as the nuclear (or electron) spins jump from the lower-energy state to the higher-energy state.* This condition, in which energy can be transferred between the electromagnetic radiation and the nuclear spins, is called **resonance**; this has no relation to the resonance referring to resonance contributors

Figure 8.23
Splitting of spin states (for $I = \frac{1}{2}$) by a magnetic field.

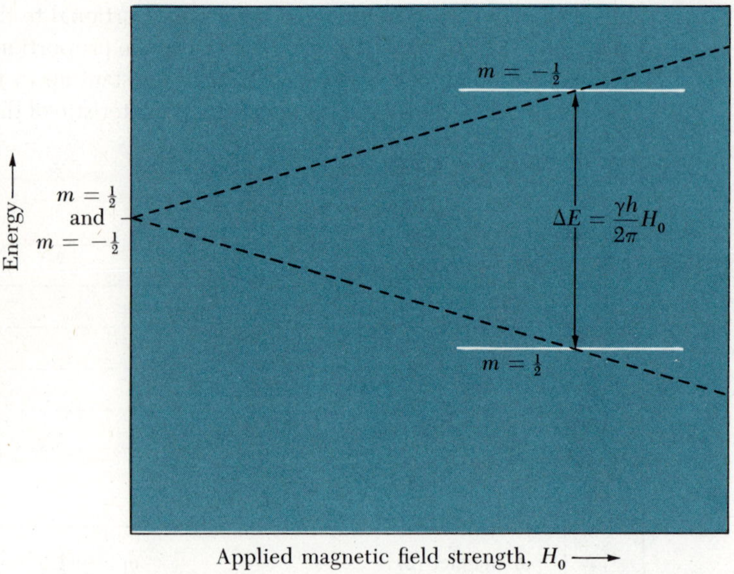

*The diagram given in Fig. 8.23 is for the simplest case, $I = \frac{1}{2}$. For larger values of I there are more than two energy levels. For example, for ^{23}Na, with $I = \frac{3}{2}$, there are four energy levels, corresponding to an orientation quantum number (analogous to the $+\frac{1}{2}$ and $-\frac{1}{2}$ of Fig. 8.23) that can have any of the values $\frac{3}{2}, \frac{1}{2}, -\frac{1}{2}$, or $-\frac{3}{2}$. However, even in such cases, the difference in energy between adjacent levels is given by Eq. (8.11), and the frequency of absorption of energy (resonance) is given by Eq. (8.12).

Figure 8.24
Schematic drawing of a simple nuclear magnetic resonance spectrometer.

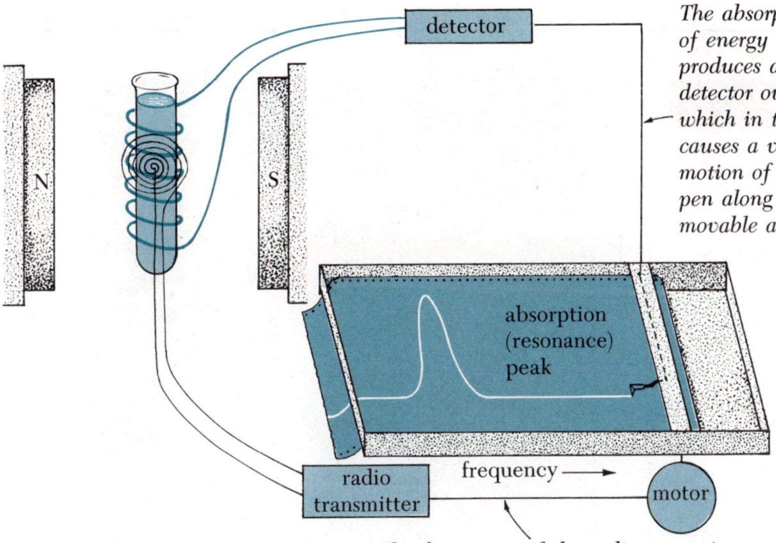

The absorption of energy produces a detector output, which in turn causes a vertical motion of the pen along the movable arm.

The frequency of the radio transmitter is synchronized to the horizontal motion of the arm.

in chemical bonding. The phenomenon is given the name magnetic resonance, but it is clearly just another example of spectroscopic transitions between energy levels. At the magnetic field strengths typically used (10^4 to 10^6 gauss), the nuclear resonance frequencies ν_0 occur in the range 10^6 to 10^9 Hz (the radiofrequency range), depending on the γ value for the nucleus of interest and the value of H_0 used.* A schematic diagram of a simple apparatus for performing a nuclear magnetic resonance experiment is shown in Fig. 8.24.

You may have been wondering what magnetic resonance spectroscopy has to do with chemistry. The determination of magnetogyric ratios for the stable nuclear isotopes and the corresponding factor for electrons is primarily within the realm of physics, and was completed many years ago. What makes magnetic resonance of interest in chemistry is that nuclear or unpaired electron spins in molecules do not respond in a magnetic resonance experiment in precisely the same way as if the spins were completely isolated. The difference arises from the sensitivity the spins have in molecules to the orbital motions of electrons and to the presence of other spins.

The electrons in a molecule, since they are moving charged particles, respond to the applied magnetic field by executing circulations that tend to cause a slight reduction of the net magnetic field from the full applied field H_0. Because of these circulations, nuclei in a molecule experience a net magnetic field that is slightly less intense than the applied field; we say that the nuclei are screened, or shielded, somewhat from the full strength of the applied field. The interesting and important point of this is that the extent of the shielding for a particular nucleus depends very sensitively on the electronic environment of that nucleus.

*There are some important practical reasons, of which one is signal strength, why the highest magnetic fields conveniently available are used. The 10^4 to 10^6 gauss range spans the highest fields usually attainable with permanent iron magnets, electromagnets, and superconducting magnets. One gauss is 10^{-4} newton/ampere-meter. The earth's magnetic field, which causes the deflection of a compass needle, is only a few tenths of one gauss.

OPTIONAL

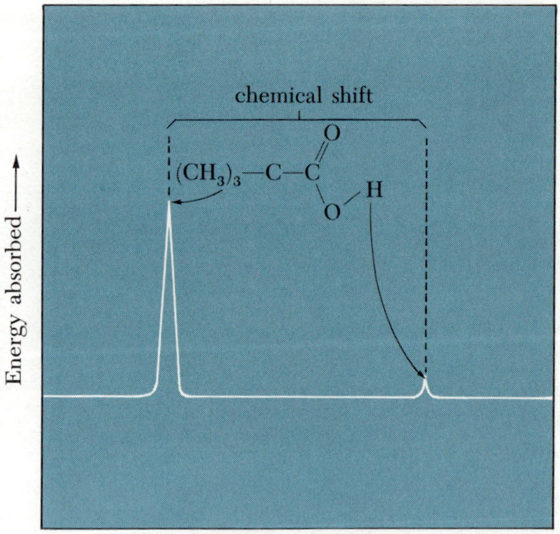

Figure 8.25 Symbolic representation of the resonance absorptions for the protons in pivalic acid.

If we place a molecule in a constant applied magnetic field, the stronger the shielding of a particular site in a molecule, the less intense the magnetic field experienced by a nucleus at that position, and according to Eqs. (8.11) and (8.12), the lower the resonance frequency. Hence, nuclei at chemically different positions in a molecule generally have resonance absorptions at different frequencies, assuming the applied field is fixed.

For ethyl chloride, CH_3—CH_2—Cl, the CH_3 protons are more strongly shielded than the CH_2 protons, and resonance absorption for the former is at a lower frequency than for the latter. This type of frequency difference is referred to as the **chemical shift.** It is represented symbolically in Fig. 8.25 for pivalic acid, $(CH_3)_3C$—C(=O)OH, for which the methyl protons are more shielded than the hydroxyl protons and hence absorb radiation at a lower frequency.

One feature of Fig. 8.25 is the great discrepancy between the sizes of the two absorption peaks. If one makes careful measurements, one finds that the area under the absorption peaks arising from the CH_3 protons is nine times greater than the area under the peak due to the hydroxyl proton. This is a manifestation of a very important feature of nmr, that the area of a peak is proportional to the number of nuclei giving rise to the peak; there are nine CH_3 protons and only one OH proton in each pivalic acid molecule.

Chemical shifts are generally determined and reported relative to some standard; for ^1H nmr the customary reference line is the absorption of $(CH_3)_4Si$, tetramethylsilane. Many thousands of different chemical shifts have been determined and published, especially for proton resonances. This information, like vibrational frequencies, is used to great advantage in determining the molecular structures of compounds. Some very important work of this sort has been carried out in recent years on the structures of proteins in solution.

EXAMPLE 8.11 Predict the general features of the ^1H nmr spectrum that one would obtain for a solution that is 0.5 *M* in CH_3OH, 2.0 *M* in $CHCl_3$, and 1.0 *M* in benzene (C_6H_6), with CCl_4 as the solvent.

SOLUTION There are four chemically distinct types of protons in this solution, two from CH_3OH and one from each of the other two solutes. Taking into consideration the molecular formulas and the concentrations of each species, we expect four different resonance peaks, with areas in the ratios (0.5×3) to (0.5×1) to (2.0×1) to (1.0×6), or 1.5:5:2:6, or 3:1:4:12.

The second "complicating" feature of magnetic resonance is that nuclear or unpaired electron spins in molecules also interact with, or "feel the presence of," other spins in the molecule. These interactions, often called **spin-spin interactions,** or **spin-spin coupling,** exert a definite influence on the total magnetic environment of each spin, altering it somewhat from what it would be if the spin were subjected only to the applied magnetic field. In this sense, these spin-spin interactions perturb the magnetic influence exerted on a nuclear (or unpaired electron) spin in a manner somewhat analogous to the perturbation exerted by electronic circulations—shielding. Because of spin-spin interactions, the net magnetic environment of a particular spin depends on the orientation of other spins (and their magnetic moments) in the same molecule.

As an example, let us consider the ^1H nmr spectrum of the compound dichloroacetaldehyde, $Cl_2CH—C(=O)H$. On the basis of the concept of shielding, we expect two different resonance peaks, one for the Cl_2CH proton and one for the $\sim C(=O)H$ proton. Now we wish to take account of the effect of spin-spin interactions. We note that the spin angular momentum and magnetic moment of the $\sim C(=O)H$ proton can be aligned in either the $+\frac{1}{2}$ (↑) or $-\frac{1}{2}$ (↓) sense relative to the applied magnetic field. Each of these orientations exerts its own specific perturbation on the magnetic environment of the Cl_2CH proton. Hence, the Cl_2CH proton will have two different resonance conditions (frequencies), one corresponding to the $+\frac{1}{2}$ orientation of the $\sim C(=O)H$ proton, and one corresponding to the $-\frac{1}{2}$ orientation. We say that the spin of the $\sim C(=O)H$ "splits" the resonance of the $CHCl_2$ proton into two peaks. Similarly,

OPTIONAL

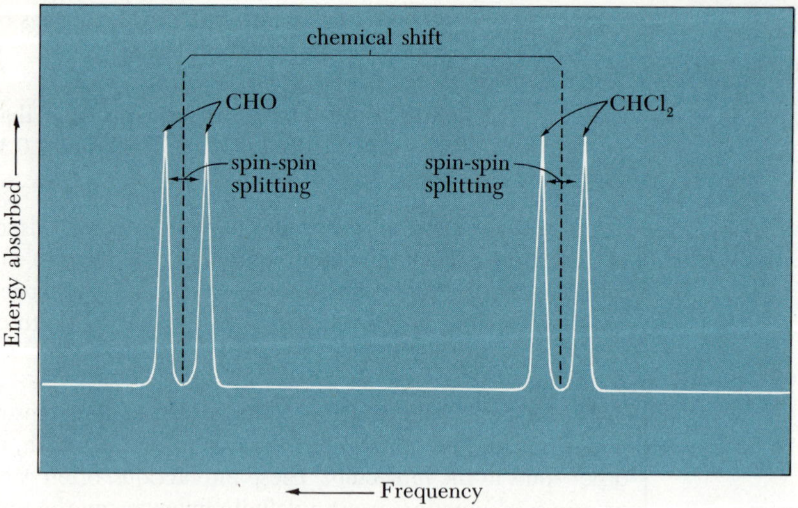

Figure 8.26
Proton nmr spectrum of $CHCl_2$—C(=O)—H, showing the chemical shift between the two chemically different types of protons and the spin-spin splitting (coupling) between them.

the $CHCl_2$ proton spin, because it has two possible spin orientations, can exert two different perturbations of the magnetic environment of the ~C(=O)H proton; hence, the proton spin of the $CHCl_2$ group splits the proton resonance of the ~C(=O)H group into two peaks. This case is described pictorially in Fig. 8.26.

The kind of spin-spin splitting represented by Fig. 8.26 is very common in the nmr spectra of molecules containing more than one nuclear spin. The splitting patterns are more complex if there is more than one chemically equivalent nucleus involved in the splitting, for example three equivalent protons in a methyl group. The resonance peaks of equivalent nuclei, for example, the protons of a CH_3 group, are not split by the spins of other equivalent nuclei. It is found that two equivalent protons, in a ~CH_2~ group for instance, cause a splitting pattern of three lines (a triplet), with peak intensity ratios 1:2:1. Three equivalent protons, in a ~CH_3~ group for instance, cause a quartet splitting (four lines) in the resonance of another nucleus in the molecule, with intensity ratios 1:3:3:1. These splittings are summarized in Fig. 8.27. A simple example of how these splittings are observed in actual spectra is illustrated for CH_3CH_2Cl in Fig. 8.28.

EXAMPLE 8.12 Predict the ^1H nmr spectrum of acetaldehyde, CH_3CHO.

SOLUTION There are two chemically different types of hydrogens in acetaldehyde, so one expects two resonance patterns, separated by a chemical shift. In addition, one expects the CHO proton resonance to be split into a quartet by interaction with the CH_3 proton spins, and the CH_3 proton resonance to be split into a doublet by

Spotlight

The number and intensities of the lines of the splitting patterns caused by spin-spin interaction with the protons of a CH_2 group or a CH_3 group are determined by the number of possible orientations of the proton spins of such groups. For example, the two protons of the CH_2 can have their spins lined up as ↑↑, ↑↓, ↓↑, or ↓↓, with the different resultant spin angular momentum projections M_z along the direction of the magnetic field. This is summarized in the diagram herewith.

The M_z value is an important factor in determining the perturbation applied to the magnetic environment of some other nuclear spin in the same molecule (for example, a CH_3 proton in CH_3CH_2Cl). Hence, we see that three different perturbations can be exerted, depending on which set of spin orientations within the CH_2 group is present, and there are two ways of arriving at the perturbation corresponding to $M_z = 0$. Thus, the resonance of another nucleus in the same molecule is split into three peaks, with the central peak indicating twice the intensity of either of the others.

Proton a	*Proton* b	M_z	
$\frac{1}{2}$	$\frac{1}{2}$	1	intensity 1
$\frac{1}{2}$	$-\frac{1}{2}$	0	intensity 2
$-\frac{1}{2}$	$\frac{1}{2}$	0	
$-\frac{1}{2}$	$-\frac{1}{2}$	-1	intensity 1

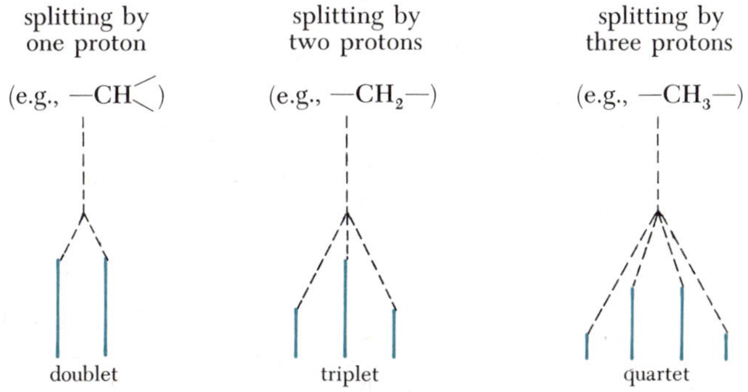

Figure 8.27 Spin-spin splitting patterns in magnetic resonance.

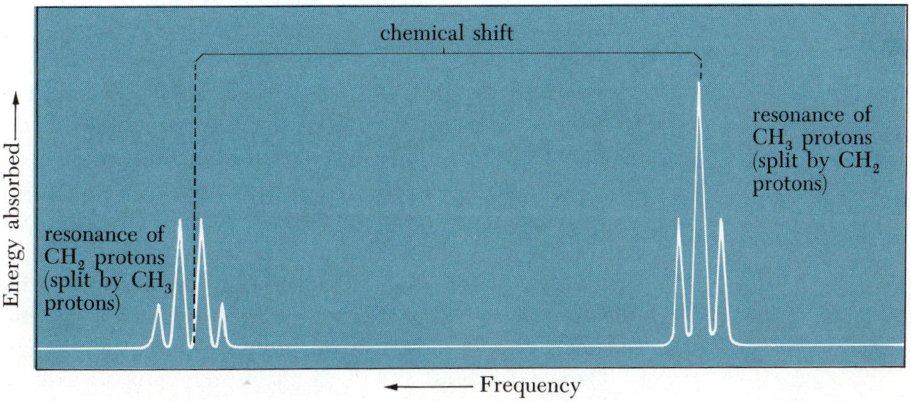

Figure 8.28 ^1H nmr spectrum of CH_3CH_2Cl showing the triplet pattern of the CH_3 resonance absorption and the quartet pattern of the CH_2 resonance absorption.

OPTIONAL

coupling with the CHO proton spin. The total areas of these peak patterns should be in the ratio 3:1, corresponding to the numbers of hydrogens. The predicted spectrum is

← Frequency

The kinds of splitting patterns explored above are very common in nmr spectra. As with some of the characteristics of other forms of spectroscopy, the splitting patterns are very useful in structure identification; a 1:3:3:1 quartet pattern tells one that a methyl group is almost certainly present. The magnitudes of the splittings (separation of the peaks in Hz) are also useful in providing details of molecular structure; a large body of information is now available in the literature, from both experimental and theoretical studies, that indicates how these splittings relate to molecular structural characteristics.

SUMMARY

An approach to understanding chemistry can be made by centering attention on functional groups. Certain groups of atoms tend to have similar behavior in a large number of molecules, often allowing us easily to predict the chemical reactivities of molecules by knowing their structure. It is often possible to improve the accuracy of our predictions by considering the effects of other functional groups on the group that is identified with the property, or reaction, in question.

We have also found that atoms can be bound by covalent bonds. The most straightforward visualization of the bond formation is the sharing of an electron pair by a species with one or more lone pairs of electrons, a Lewis base, with an electron-deficient species, a Lewis acid. This type of covalent bonding is called dative, or coordinate, covalent bonding. Common examples of dative or Lewis acid–base bonding are found in both the chemistry of the main group elements (for example, the ammonium ion, NH_4^+) and the transition elements (for example, coordination compounds such as the octahedral complex $[Co(NH_3)_3Cl_3]$).

Hydrogen bonding is another important type of interatomic interaction. It occurs when a hydrogen atom bound to a highly electronegative atom is attracted to an electronegative atom of a second species. Such interactions have profound effects on the physical and chemical properties of a wide range of substances, including such common simple molecules as water and ammonia. Hydrogen bonds also stabilize the structure of many complex molecules, such as proteins and polynucleic acids, that are basic to living systems.

Numerous experimental methods are used to study chemical bonding and hydrogen-bond interactions. Many of these methods use spectroscopic techniques. Spectroscopic methods involve observation of the absorption or emission of energy due to transitions between quantized energy states. The energy difference we observe may be associated with a change in electronic states (ultraviolet-visible spectroscopy), vibrational energy (infrared spectroscopy), or magnetic spin energy states (nuclear magnetic resonance spectroscopy). The study of spectroscopic transitions allows the chemist not only to learn about the transition processes themselves but also to use spectroscopy as a tool for identifying structural features of functional groups or entire molecules. For example, we can learn about bonding characteristics from the particular energies at which infrared light is absorbed. In addition, the absorption energy may characterize a particular functional group, allowing ready identification. Nuclear magnetic resonance spectroscopy is one of the most versatile and powerful techniques for identifying functional groups and deducing the structures of even very complex molecules.

Many other spectroscopic techniques besides those we have considered in this chapter are very useful to the modern chemist and to scientists in chemically related fields. By combining a fundamental knowledge of chemical bonding with properly chosen spectroscopic experiments, a chemist generally can readily deduce much information about molecular structure and even draw firm conclusions about many of the processes that occur at the molecular level.

STUDENT CHECKPOINTS

After studying this chapter, the student should be able to:

1. Identify some functional groups, defining their properties and the utility of the functional group approach.
2. Explain how by "fine tuning" the reactivity of one functional group can be modified by other functional groups in a molecule.
3. Apply the various definitions of acids and bases.
4. Identify Lewis acids and bases.
5. Identify hydrogen-bonding interactions.
6. Describe some of the effects of hydrogen bonding (for example, on boiling points, or on lowering the density of ice compared with water).
7. Explain the relation between energy-level changes and the absorption and emission of electromagnetic radiation.
8. Order the energies of electronic, vibrational, and rotational transitions.
9. Differentiate among frequency, wavelength, and energy.
10. Use a group frequency chart to analyze infrared spectra of small molecules.
11. Outline the basic features of molecular vibrations and infrared spectroscopy (that is, the idea of a harmonic oscillator, absorption of energy only if there is a change in dipole moment, and number of vibrational modes of a molecule).

EXERCISES

8.1. CF_3OH is a stronger acid than CH_3OH, but both are much weaker than $CH_3\overset{O}{\overset{\|}{C}}OH$. How are these facts explained?

8.2.* The K_a values for acetic acid, bromoacetic acid, $BrH_2\overset{O}{\overset{\|}{C}}COH$, and iodoacetic acid, $IH_2\overset{O}{\overset{\|}{C}}COH$, are 1.8×10^{-5}, 2.1×10^{-3}, and 7.5×10^{-4}, respectively. Explain this order of acidities.

8.3. The K_a values for acetic acid, chloroacetic acid, dichloroacetic acid, and trichloroacetic acid are 1.8×10^{-5}, 1.4×10^{-3}, 3.3×10^{-2}, and 2×10^{-1}. Explain this order of acidities.

8.4.* As noted in Exercise 8.3, $Cl_3CC(=O)OH$ is a stronger acid than $CH_3C(=O)OH$. From this information, which, do you predict, is the stronger base, $N(CCl_3)_3$ or $N(CH_3)_3$?

8.5. As noted in Exercise 8.3, $Cl_3CC(=O)OH$ is a stronger acid than $Cl_2CHC(=O)OH$, which is stronger than $ClCH_2C(=O)OH$. From this information, which, do you expect, is the stronger base: (a) $N(CHCl_2)_3$ or $N(CH_2Cl)_3$; (b) $N(CH_3)_2(CCl_3)$ or $N(CH_3)(CCl_3)_2$?

8.6.* Which salt of each of the following pairs would give the solution of lower pH when it is dissolved in water: (a) $FeCl_2$ or $FeCl_3$; (b) $ZnCl_2$ or TlCl; (c) $MnCl_2$ or $NiCl_2$?

8.7. Like water, ammonia can act as an acid as well as a base. If liquid ammonia is used as a solvent, what is the acid form (corresponding to H_3O^+ in aqueous solution) and what is the base form (corresponding to OH^- in water)?

8.8.* Like water, sulfuric acid can act as a base as well as an acid. If liquid sulfuric acid is used as a solvent, what is the base form (corresponding to OH^- in water) and what is the acid form (corresponding to H_3O^+ in aqueous solution)?

8.9. NH_3 acts as a Brönsted-Lowry or Lewis base, while NH_4^+ acts as an acid by any definition. Explain.

8.10.* Which forms a stronger Lewis acid-base adduct with BCl_3, NH_3 or $N(CF_3)_3$?

8.11. What is meant by the statement that the ion Cr^{3+} in water acts as a stronger Lewis acid towards Cl^- than H^+ does?

8.12.* What Lewis acids and bases can be used to produce each of these Lewis acid–base complexes: (a) $[CoCl_2(NH_3)_4]Br$; (b) BrF_4^-; (c) $CoCl_4^{2-}$; and (d) CaF_2?

8.13. What Lewis acids and bases can be used to produce each of the following Lewis acid–base complexes: (a) Sodium acetate; (b) $CoCl_2(H_2O)_2$; (c) $AlCl_4^-$; (d) NH_4Cl?

8.14.° Which member of the following pairs forms better hydrogen bonds with other molecules of the same compound: (a) NH_3 or PH_3; (b) HF or CH_3F;

(c) $CF_3\overset{\overset{O}{\|}}{C}-OH$ or $CH_3\overset{\overset{O}{\|}}{C}-OH$?

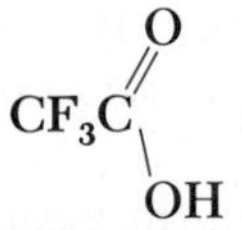

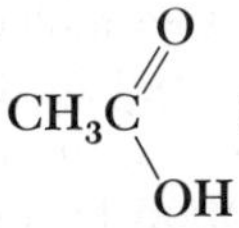

8.15. Which member of the following pairs forms better hydrogen bonds with other molecules of the same compound: (a) H_2O or H_2S; (b) CH_3OH or CF_3OH; (c) HF or HCl?

8.16.° What hydrogen bonds are possible in a solution of HCl dissolved in dimethylamine, $(CH_3)_2NH$?

8.17. What hydrogen bonds are possible in a solution of methanol, CH_3OH, and water?

8.18.° Acetic acid, $CH_3\overset{\overset{O}{\|}}{C}-OH$, is found as a dimer (two molecules linked together) in the gas phase. What is the structure of the dimer?

8.19. What is the role of hydrogen bonding in making ice float?

8.20.° The idea of base-pairing was essential to solving the structure of double helix DNA. What is meant by *base pairing*?

8.21. Both ammonia and arsine have higher boiling points than phosphine. Why? (*Hint:* Look for trends in Fig. 8.1.)

8.22.° Dark pigmentation of the skin can prevent burning of subcutaneous tissue by sunlight. Why?

8.23. What is the relative energy necessary for breaking chemical bonds, causing rotational excitation, and causing vibrational excitation? What are the relative frequencies of the radiation necessary for each process?

8.24.° Why is infrared spectroscopy generally very useful for deducing the structure of organic molecules? What other kinds of information aid in using infrared spectra to deduce structure?

8.25. How does the physical process of light emission differ from light absorption?

8.26.° Calculate the energy in kcal/mol and the frequency of microwaves with wavelength 0.100 cm.

8.27. Calculate the energy in kcal/mol, the frequency, and the wavelength of an infrared absorption band due to vibration of the carbonyl (CO) group in $Ni(CO)_4$, which is found at an energy of 2057 cm^{-1}.

8.28.° (a) How many vibrational modes are there for CS_2, HF, and NH_3? (b) How many rotational modes are there for each?

8.29. How many vibrational and rotational modes are there for (a) F_2; (b) OF_2; (c) PCl_3?

Which one of these molecules has no infrared absorption even though it has a vibrational mode?

8.30.° Why is the carbon-nitrogen stretching frequency higher in HCN than in $N(CH_3)_3$?

8.31. Why is the highest-energy stretching frequency of the CO group higher in

phosgene, $Cl-\overset{\overset{O}{\|}}{C}-Cl$, than it is in sodium acetate, $Na^+ CH_3-\overset{\overset{O}{\|}}{C}-O^-$?

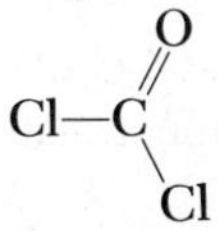

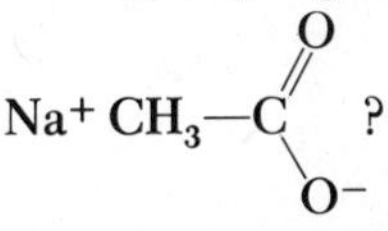

8.32.* Use the equation $\nu = \frac{1}{2\pi}\sqrt{\frac{k}{\mu}}$ to explain why the stretching frequencies for ^1HI and ^2HI are different. (Hint: What can you assume about the value of k for these two molecules?) This infrared "shift" caused by isotopic substitution is used by infrared spectroscopists to assign bands in complex infrared spectra to particular functional groups.

8.33. The absorption bands due to electronic transitions, which are generally found in the visible and ultraviolet regions of the spectrum, are typically much broader than vibrational bands in the infrared region. Why?

8.34.* The infrared spectrum shown below is that of a molecule with one functional group on an alkane chain: $CH_3(CH_2)_3X$. What is the functional group? How could you verify your choice?

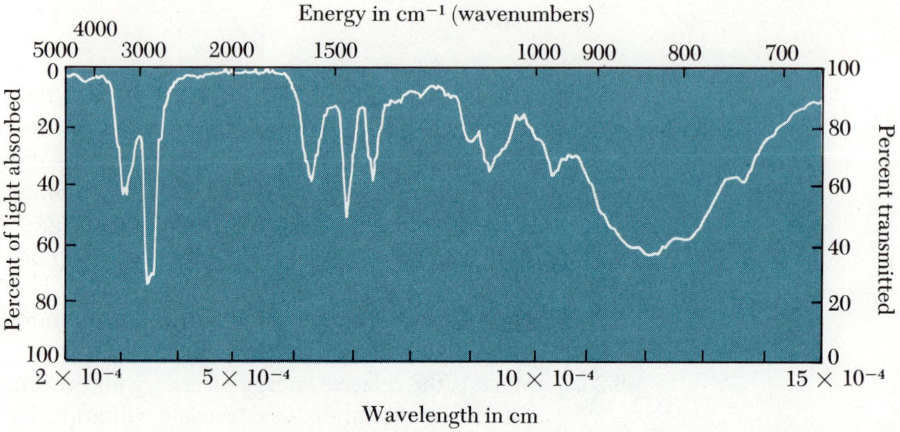

8.35. The infrared spectrum shown below is that of a molecule with one functional group within an alkane chain: $CH_3XCH_2CH_2CH_3$. The accurate mass of the molecule is 86.1349 amu. Identify the functional group. What should the elemental analysis results be?

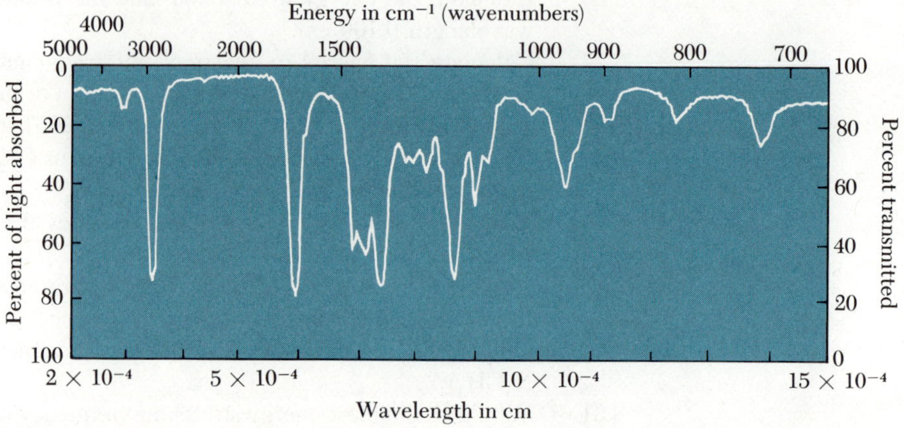

8.36.* Electronic spectra are generally not so useful as vibrational spectra in determining which functional groups are present in a particular molecule. Why?

8.37. What features would you predict for the ^{1}H nmr spectrum of (a) dimethyl ether, CH_3OCH_3; (b) diethyl ether, $CH_3CH_2OCH_2CH_3$; (c) acetone, $CH_3-\overset{\overset{O}{\|}}{C}-CH_3$?

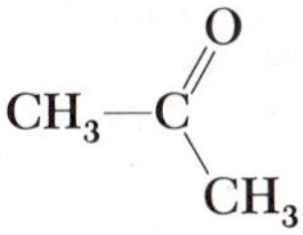

8.38.° What features would you expect to find in the ^{1}H nmr spectrum of (a) t-butylmethylketone, $(CH_3)_3C-\overset{\overset{O}{\|}}{C}-CH_3$; (b) ethyl alcohol, CH_3CH_2OH; (c) isopropyl alcohol, $(CH_3)_2\overset{\overset{H}{|}}{\underset{\underset{OH}{|}}{C}}$?

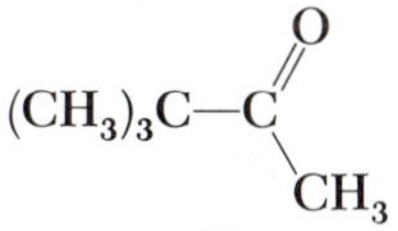

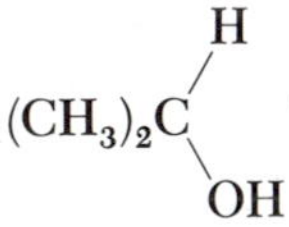

8.39. How would the ^{1}H nmr spectra of methanol, CH_3OH, and ethanol, CH_3CH_2OH, differ?

8.40.° What features would you find in the ^{1}H nmr spectrum of a solution that is 1.0 M in diethyl ether, $CH_3CH_2OCH_2CH_3$, 2.0 M in acetone, $CH_3\overset{\overset{O}{\|}}{C}-CH_3$, and 1.0 M in t-butyl alcohol, $(CH_3)_3COH$ in CCl_4 as solvent?

8.41. What do you predict as the ^{1}H nmr spectrum of a solution that is 1.0 M in trimethylamine, 1.0 M in trimethylphosphine, and 2.0 M in triethylamine, with deuterochloroform, $CDCl_3$, as the solvent?

ANSWERS TO SELECTED EXERCISES

8.2 The iodide group is better at stabilizing the anion (is more electron-withdrawing) than the CH_3 group. The bromide group is better yet. **8.4** $N(CH_3)_3$ (see Example 8.3). **8.6** (a) $FeCl_3$; (b) $ZnCl_2$; (c) $NiCl_2$. **8.8** HSO_4^- and $H_3SO_4^+$. **8.10** NH_3.
8.12 Lewis acid(s)   Lewis base(s)
   (a) Co^{3+} Cl^-, NH_3, Br^-
   (b) Br^{3+} F^-
   (c) Co^{2+} Cl^-
   (d) Ca^{2+} F^-

8.14 (a) NH_3; (b) HF; (c) $CH_3\overset{\overset{O}{\|}}{C}-OH$, **8.16** HCl with another HCl molecule (H—Cl $\cdots$ H—Cl) and with the amine (H—Cl $\cdots$ $NH(CH_3)_2$ and Cl—H $\cdots$ $NH(CH_3)_2$) and the amine with another amine molecule ($(CH_3)_2NH \cdots NH(CH_3)_2$).
8.18 Two molecules in hydrogen-bonded structures

$$CH_3-\overset{\overset{O \cdots H-O}{\|}}{\underset{\underset{O-H\cdots O}{\|}}{C}}\overset{}{\underset{}{C}}-CH_3$$

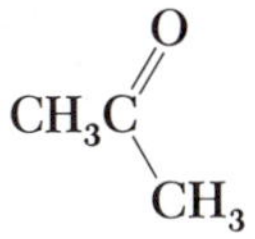

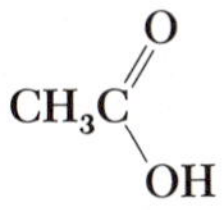

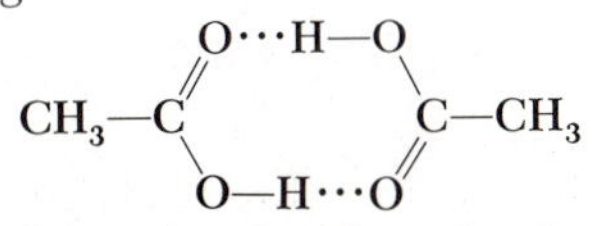

8.20 Pairs of the bases can hydrogen-bond with each other in a specific manner, guanine with cytidine and adenine with thymine. **8.22** The dark skin can absorb light that would otherwise break chemical bonds in exposed tissue, causing a burn. **8.24** Functional groups show characteristic absorption patterns in many molecules. In some molecules the number of bands and their energies can distinguish among possible structures. **8.26** 2.85×10^{-2} kcal/mol; 3.00×10^{11} Hz. **8.28** (a) 4,1,6; (b) 2,2,3. **8.30** The C–N bond in HCN is a triple bond—stronger than the C–N single bonds in $N(CH_3)_3$. **8.32** The bond strength, and hence the force constant k, can be assumed to be essentially the same for ^1HI and ^2HI. For these two molecules, then

$$\frac{\nu_{^1HI}}{\nu_{^2HI}} = \sqrt{\frac{\mu_{^2HI}}{\mu_{^1HI}}} = \sqrt{\frac{1.970}{1.000}} = 1.404$$

The ^1HI stretch is at a much higher frequency (energy) than the ^2HI stretch. **8.34** The functional group is $—NH_2$. Note the stretch, just above the C—H stretch (above 3000 cm^{-1}) and the broad N—H bond near 800 cm^{-1}. These features would be quite similar for an —OH group. Elemental analysis would distinguish the two possibilities. **8.36** Fewer groups give rise to electronic transitions and the bands are very broad, so different groups tend to have similar spectra. **8.38** (a) Two kinds of H atoms (two chemical shifts) with an area ratio of 2:1 (actually 6:3). These peaks could be split by spin-spin coupling because there are different types of protons in the molecule. (b) Three kinds of H atoms (three chemical shifts) with area ratios 3:2:1. These peaks would be split by spin-spin coupling. (c) Three kinds of H atoms (three chemical shifts) with areas 6:1:1. These peaks would be split by spin-spin coupling. **8.40** There would be two peaks due to CH_3 and CH_2 protons in diethyl ether, one peak from the CH_3 in acetone and two peaks due to CH_3 and H protons in *t*-butyl alcohol. The peak areas would be 6:4:12(2 × 6):9:1. Note that these ratios are achieved by multiplying the number of protons of each type by the molarity of the particular species. Each of these peaks would be split by spin-spin coupling because there are two or more types of protons in each compound.

ANSWERS TO STUDY PROBLEMS

8(a) (a) $MgCl_2$; (b) $LuBr_3$; (c) $FeCl_3$. **8(b)** the best base, $(CH_3)_2NH$, and the best acid, $CF_3C(=O)OH$. **8(c)** Lewis acid H^+, Lewis base Br^-; Lewis acid Cu^{2+}, Lewis base Cl^-; Lewis acid BF_3, Lewis base F^-. **8(d)** First considering the $H_2NCH_2C(=O)H$, the $\sim NH_2$ group on one molecule can use the hydrogen atom to bond to the nitrogen atom of an $\sim NH_2$ group on another molecule or to the $>C=O$ group of another molecule. Carbon is insufficiently electronegative for its attached hydrogen atoms to undergo significant hydrogen bonding. The water molecules can hydrogen-bond to other water molecules. They can also hydrogen-bond to the $H_2NCH_2C(=O)H$ molecule, with the water hydrogen atom bonding to the nitrogen atom of the $\sim NH_2$ group or to the oxygen atom of the $>C=O$ group and the water oxygen atom bonding to a hydrogen atom of the NH_2 group. **8(e)** 6; 9; 9. **8(f)** acetone, $CH_3—C(=O)—CH_3$.

9

Descriptive Chemistry: Trends and Patterns

9.1 Scope and Purpose

Certain trends in descriptive chemistry can be demonstrated on the basis of the periodic table. Particularly significant are variations in the chemical properties of the elements seen across a row or down a column of the periodic table. A truly comprehensive treatment of descriptive chemistry requires several volumes. To provide a glimpse into the great variety of processes and properties in chemistry and at the same time a view of periodic trends, we shall focus attention on the elements of the second period (Li → Ne), often referred to as the first short period. Some of the elements in this period are especially important in biology and industry and provide reasonable models for the diversity and patterns of chemical behavior of many other elements.

An overall view of "horizontal" and "vertical" trends of chemical properties in the periodic table is a very valuable tool for systematizing chemistry in one's own mind and for making predictions of reasonable value about chemical behavior. For example, who would have thought to look in bone for the strontium 90 isotope, a product of nuclear fission processes, if strontium had not been recognized as a close chemical relation of calcium? Another example occurs in the chemistry of lead, an important matter in current ecological questions regarding the fate of lead in the environment. Useful hints on the chemical behavior of tetravalent lead can be gained by viewing the better-characterized tetravalent members of the same group, carbon and silicon, and also germanium and tin. The elements within a group have many chemical and physical properties in common; they generally resemble each other more than they resemble the other elements in other groups. The similarities among the members of a group are expected on the basis of the similarities of their outer electronic structures. The groups are also known as "families" and some have been given the following common names: alkali metals (Group I), alkaline-earth metals (Group IIA), halogens (Group VIIA), and noble gases (sometimes called rare, or inert, gases) (Group VIIIA).

The elements lithium through neon are frequently considered prototypes for their congeners, the other group members. Although this is generally appropriate for lithium and fluorine, it is generally not appropriate for beryllium, boron, carbon, and especially nitrogen. For example, Li and F show properties that are strikingly similar to their heavier congeners, even as far down as the fifth period (Rb and I), but boron shows properties that are considerably different from those of even its next heavier congener, aluminum. In spite of this inconsistency between some of the elements of the second period and their heavier congeners, the elements of the third period are very similar to their heavier congeners. For example, at room temperature nitrogen is a diatomic gas, whereas P, As, and Sb are tetra-atomic solids. We are first concerned with considering how the properties of the elements of the second period change in the direction from left to right across the periodic table; subsequently we consider the heavier congeners.

9.2 Periodic Properties

Metals and Nonmetals

In any attempt to separate a large body of natural objects into two categories, one invariably finds that although most of the objects clearly belong to one or the other of the two categories, some belong to both and some to neither. An attempt to classify the chemical elements of the second period as either metals or nonmetals is not an exception (see definition in Sec. 5.7). Lithium is clearly a metal, fluorine is clearly a nonmetal, and neon is neither. Boron could be either, depending on the circumstances, a situation known as **amphoterism.** We distinguish here substances such as Ne that are not metals nor are they nonmetals, where the term *nonmetals* has come to mean a class of elements with a particular set of chemical properties that are usually the opposite of the properties of metals. Thus, in an attempt to classify the elements of the second period into metals or nonmetals, we must first clearly define the property around which the elements are to be classified, and then arrange the elements according to a sequence or scale that reflects the degree to which they demonstrate that property. There are several ways of defining metallic character. For example, if electrical conductivity is used as the criterion for metallic character, then boron, which is a semiconductor, is more metallic than nitrogen but less metallic than lithium.

Another basis for classification is acid or base behavior; the sequence can place the most acidic at one extreme and the most basic at the other. For example, since H^+ can be removed more easily from beryllium hydroxide, $Be(OH)_2$, it is an acid compared with lithium hydroxide, LiOH, but a base compared with the "hydroxide" of boron, $B(OH)_3$, which is boric acid. The oxides of the elements fall into a natural sequence descriptive of their acid-base behavior. The oxides of the elements at the left of the periodic table, the metals, are basic whereas the oxides of the elements at the right, the nonmetals, are acidic. The oxides of the elements in between show some of the properties of bases, but not necessarily all, and some of the properties of acids but not necessarily all. If the basic or acidic properties of the oxide of an element are used to determine the degree to which the element is a metal or a nonmetal, then neon is neither, since it forms no oxide.

Such correlations of metallic vs nonmetallic character and basic vs acidic properties of the oxides extend throughout the entire periodic classification. Thus, the second-period elements are arranged sequentially in order of decreasing metallic character.

The "Diagonal Relationship"

When the elements of the second period deviate markedly in properties from their congeners, they often show properties very similar to those of the heavier congeners in the next group. This is often referred to as the diagonal relationship. For example, in solubility, various lithium salts often resemble magnesium salts more than other alkali metal salts.

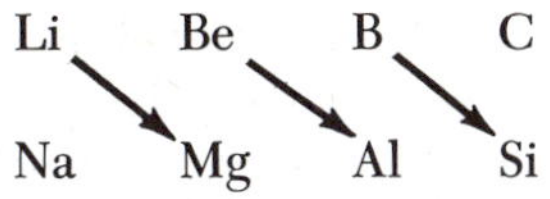

Diagonal Relationship

One of the best examples of the diagonal relationship is the striking similarity between the elements beryllium and aluminum and their most common oxides, BeO and Al_2O_3. Elemental Be and Al, and also BeO and Al_2O_3, dissolve in strong bases to form the so-called beryllate and aluminate ions; Mg and MgO do not behave similarly.

$$Be(s) + 2OH^- + 2H_2O \rightleftharpoons \underset{\text{beryllate ion}^*}{Be(OH)_4^{2-}} + H_2(g)$$

$$2Al(s) + 2OH^- + 6H_2O \rightleftharpoons \underset{\text{aluminate ion}^*}{2Al(OH)_4^-} + 3H_2(g)$$

$$BeO(s) + 2OH^- + H_2O \rightleftharpoons Be(OH)_4^{2-}$$

$$Al_2O_3(s) + 2OH^- + 3H_2O \rightleftharpoons 2Al(OH)_4^-$$

A possible explanation for this diagonal relationship is based on the ratio of charge to ionic radius. For Be^{2+} this ratio is 5.7; it is closer to that for Al^{3+} (5.9) than to that for Mg^{2+} (3.0).† On the whole, however, Li, Be, and B resemble their group con-

*As it exists in aqueous solution.

†Li and Mg and B and Si have charge-to-size ratios that are not nearly so close as those for Be and Al. This may explain why Be and Al provide the best example of the diagonal relationship.

geners more often than they resemble their diagonal congeners, and this diagonal relationship should not be overemphasized. It is generally only evident for the second row–third row diagonal elements. It is mainly helpful when these elements do not show properties similar to their respective group congeners; generally in such cases the element is more similar to its diagonal congener than to any other element.

Correlations among the Elements Li through Ne

Getting to know the most general aspects of the chemical properties of the second period elements will help us to understand and learn to predict how these properties change across the periodic table from Li to Ne. Especially pertinent are metallic character, bonding, and the most common simple ions for each element.

Lithium

The ground state electronic configuration of the Li atom is $1s^22s^1$. The low ionization potential, 5.39 eV, of the single 2*s* electron accounts for lithium's high degree of metallic behavior.* The chemistry of Li is relatively simple because it is mainly that of the ion Li^+ both in the solid state and in solution.

The Li^+'s small size results in a high charge-to-radius ratio, and consequently Li^+ has a powerful tendency to polarize the electron distribution of neighboring species, distorting it. The ion Li^+, it is held, can polarize negative anions to such an extent that an appreciable covalent electron density can exist between the nuclei. Thus, although most lithium compounds are largely ionic, lithium can also form bonds with considerable covalent character. Examples are the organolithium compounds, of which one is methyllithium, Li—CH_3.

STUDY PROBLEM 9(a)

Magnesium sometimes forms bonds with considerable covalent character, but sodium does not. Why is this so, do you think?

STUDY PROBLEM 9(b)

Lithium is the least dense (0.534 g/ml) of all the solid elements. Why is this so, do you think?

Beryllium

The ground state electronic configuration of the Be atom is $1s^22s^2$. In the direction left to right across the periodic table, Be is the first element that begins to show at least some nonmetallic character; it shows some amphoteric properties. Beryllium is still considered a metal, but it is more electronegative and less reactive than Li.

In contrast with Li^+, the discrete ion Be^{2+} does not exist in solution or lattices, even when combined with fluorine, the most electronegative element. The ionization potentials for the 2*s* electrons are high, 7.32 and 18.21 eV, and not easily offset by lattice or solvation energy. Thus, these two electrons occupy two *sp* hybrid orbitals in free molecules of the general type BeX_2, where X is some other atom. In these molecules Be has a coordination number of only 2, and to complete the

*The energy unit generally used for ionization potentials is the electron volt (eV). The electron volt is the kinetic energy an electron gains from being accelerated through a potential difference of 1 volt. If one electron volt of energy is gained by a mole of atoms, the total energy gain is 23.05 kcal/mol.

Spotlight

Few elements are obtained from such beautiful sources as beryllium. The element is found in a mineral that forms magnificent crystals—beryl, emerald, and aquamarine.

Boron is found as borax ($Na_2B_4O_7 \cdot 10H_2O$) and kernite ($Na_2B_4O_7 \cdot 7H_2O$) in deserts of the United States and as boronato-calcite ($CaB_4O_7 \cdot NaBO_2 \cdot 8H_2O$) in Chile. Borax hydrolyzes to form alkaline solutions:

$$B_4O_7^{2-} + 7H_2O \rightleftharpoons 4H_3BO_3 + 2OH^-$$

and is used to soften water, and in soaps. It acts like the polyphosphates, dissolving grease by saponification (adding —OH groups), which makes the grease soluble.

Boron nitrides, $(BN)_n$, are isoelectronic with carbon. Solid boron nitride has a structure like diamond and is extremely hard. It is used as an abrasive.

Boron hydrides (B_nH_m) have many interesting structures. One of the most fascinating features of these compounds is the ability of boron to form boron-hydrogen-boron linkages in which the hydrogen atom is two-coordinate rather than forming only one bond. The B–H–B system has only two electrons distributed in the two "bonds," rather than the four electrons expected for two covalent bonds. Some structures containing B—H—B linkages are diborane and pentaborane.

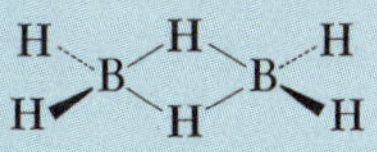
diborane

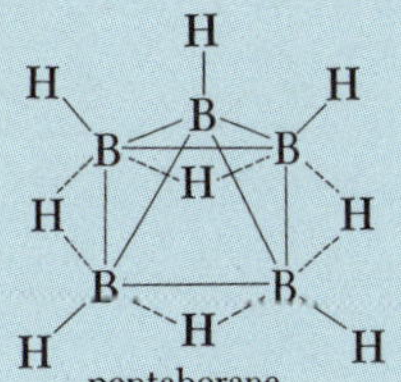
pentaborane

Professor William Lipscomb of Harvard University received the Nobel Prize in 1976 for his work in determining the structures of many boron hydrides and for formulating a theory for rationalizing and predicting such structures.

octet there is a strong tendency for Be to achieve fourfold coordination with Lewis bases. For example, alcohols and ethers form complexes such as $Be(HOR)_2Cl_2$ and $Be(OR_2)_2Cl_2$, where R is an atom or a hydrocarbon group such as CH_3.

Boron

The ground state outer electronic configuration of the boron atom is $2s^22p$. Nevertheless, it does not lose the single $2p$ electron in chemical processes in the liquid phase for the same reason that free ions Be^{2+} do not form; the first ionization potential is too high, 8.30 eV. Boron compounds contain predominantly covalent bonds, and boron is considered essentially nonmetallic even though its elemental properties are on the borderline between metals and nonmetals. It is a semiconductor, and can combine with most metals to form materials that are not easily classified by the terms so far employed here. (These materials will be discussed in more detail in Chap. 14.)

Like Be, boron has outer electrons that tend to occupy hybrid orbitals in trivalent compounds, such as BX_3, where X represents another atom or functional group, and the three sp^2 bonds leave B with an incomplete octet. To achieve maximum coordination, BX_3 compounds have a strong propensity for forming 1:1 complexes with Lewis bases. Some examples are $(CH_3)_3N{:}BCl_3$, $(CH_3)_3P{:}BH_3$, $(C_2H_5)_2O{:}BF_3$, and $(C_2H_5)_2S{:}B(CH_3)_3$. Complexes such as BH_4^- and BF_4^- are also readily formed. Consistent with the diagonal relation, boron frequently resembles Si more than it resembles its congeners. Boric acid, $B(OH)_3$, is remarkably similar to the silicic acids, H_2SiO_3 and H_4SiO_4, or $SiO(OH)_2$ and $Si(OH)_4$. Although boric acid is weak, it is definitely an acid and shows no amphoteric properties. On the other hand, $Al(OH)_3$ is mainly basic, with some amphoteric properties.

Although not to the same extent as carbon, boron has some tendency towards **catenation, an ability to form chains, or rings** with boron-boron bonds. A typical example is B_5H_{11}.

LEGEND
● B
○ H

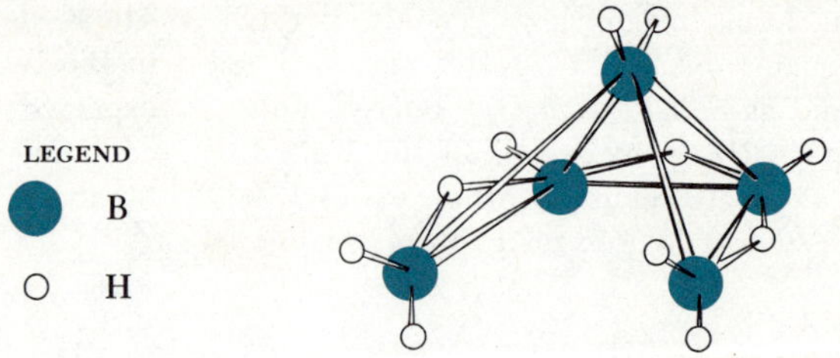

Structure of B_5H_{11}

Carbon

The ground state valence electronic configuration of carbon is $2s^2 2p^2$, and carbon even more readily hybridizes than boron, forming four equivalent sp^3 hybrid orbitals. Thus, carbon is completely **nonmetallic** and its chemistry is greatly dominated by formation of **covalent bonds**.

The ion C^{4+} does not form in chemical processes in solution; the first ionization potential is 11.22 eV. In fact, carbon is nonmetallic enough so that with some metals it forms crystalline carbides, in which C^{4-} may possibly exist. Thus, the carbides of the most reactive metals behave as essentially ionic compounds. Under the proper conditions, carbon can also form some cations, anions, and radicals, especially if there are functional groups bound to the carbon atom that can help absorb a charge through resonance delocalization. Some notable examples are $(C_6H_5)_3C^+$, $(C_6H_5)_3C^-$, and $(C_6H_5)_3C\cdot$, where C_6H_5 stands for a phenyl group, one resonance form of which is

$$\begin{array}{c} HC=CH \\ HC \qquad C- \\ HC-CH \end{array}$$

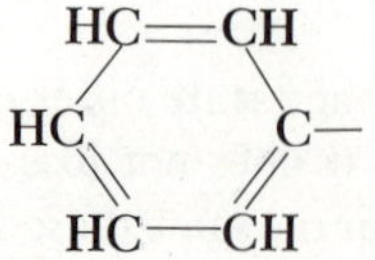

It should be borne in mind, however, that these charged or radical (odd-electron) species are all extremely reactive and react exceedingly rapidly in water. Covalent bond formation is the normal behavior for carbon.

Carbon also possesses a property that is not common with the preceding elements in the second period, namely, the ability to **form relatively stable multiple bon**ds. A typical example of a carbon compound that forms such bonds is ethylene (Chap. 7). In any attempt to construct an analogous molecule using boron instead of carbon, the boron octet would still be incomplete, because there are only three electrons per B atom available in the second shell. For beryllium, the situation for multiple bonds is even worse. The only possible structure that could be written with a double bond between two Be atoms is Be=Be, because there are not enough electrons in the second shell to accommodate another bond with any other atoms. In Chap. 7, molecular orbital theory was used to show why Be_2 is not stable (that is, the bond order of Be_2 is zero). Finally, Li does not even have enough electrons in the outer shell to form more than one bond between the two Li atoms.

Spotlight

Two of the elemental forms of carbon—graphite and diamond—have very different properties because of the arrangement of atoms within the solids. Diamond has a continuous network structure in which all atoms are tetrahedrally bound to four other carbon atoms. Since shearing a diamond would require breaking very many strong C–C bonds, the solid is very hard, and makes an excellent abrasive.

Graphite, on the other hand, is a very good lubricant. The graphite structure is layered. Since there are no covalent bonds between the layers, they slide easily over one another. Each carbon in the graphite structure is bound to three other carbon atoms (two single C–C bonds and one C–C double bond) rather than to four.

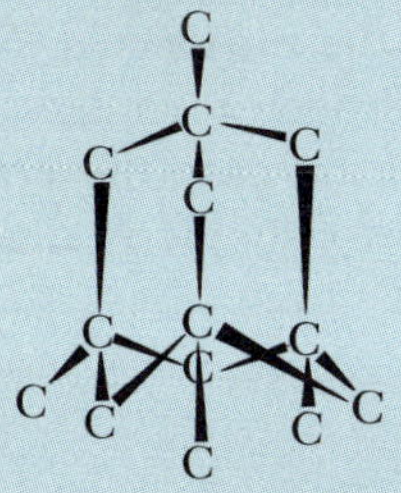

A portion of the diamond structure

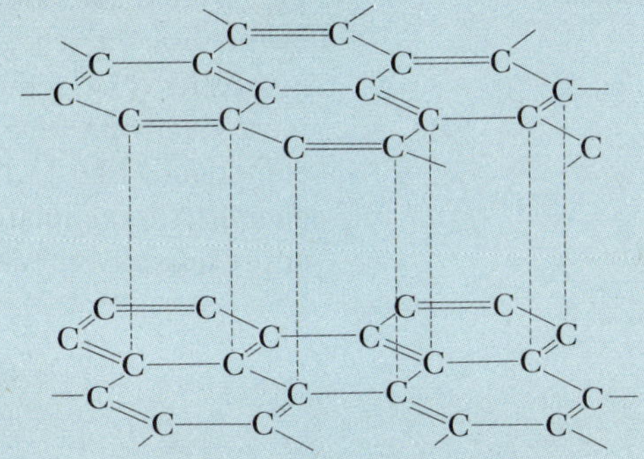

A portion of the graphite structure

Perhaps the most significant property of carbon is its ability to form long and stable chains; for some cases, such as the common plastics polyethylene (Tupperware, etc.), polyperfluoroethylene (Teflon), polymethylmethacrylate (Lucite, or Plexiglas) these chains have a "backbone" consisting entirely of carbon-carbon linkages, (see Chap. 14). In other cases, the proteins, for example, other atoms like nitrogen or oxygen can be involved. The reason for this ability to form chains may be, at least partly, the high bond energy of the carbon-carbon single bond, 83 kcal/mol. The only other single bond between two like atoms that has a higher bond energy is H_2 (104 kcal/mol), but hydrogen obviously cannot form chains because of its valence. The next highest bond energy is 58 kcal/mole, but this is for Cl_2, which again is prohibited from forming chains because its valence is saturated.

Associated with the ease with which carbon can form chains is its ability to form rings. If a carbon chain is progressively being lengthened, it can easily close on itself when it has attained a length of 5, 6, or 7 atoms.* This can occur for these ring sizes because the distance between the first and last carbons can be very close to the normal C–C bond distance (1.54 Å), if the molecule is bent in a ring without distorting the normal sp^3 C–C–C angle of about 109.5°. Smaller or larger rings are

*

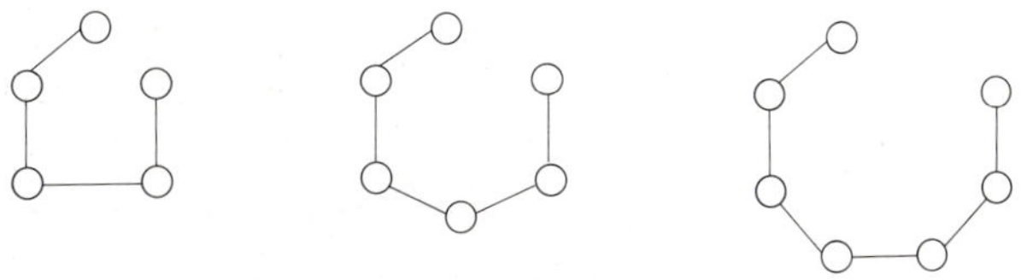

Representations of 5-, 6-, and 7-carbon chains about to form rings

possible, but the probability is less. Smaller rings introduce bond strain within the molecule, because the C–C–C bond angle will be significantly less than the optimum 109.5° for sp^3 orbitals. On the other hand, larger rings are less likely, because if the chain is eight or more atoms long, there is a severely reduced probability that the molecule will assume an orientation in which the first and last atoms can be close to 1.54 Å.

Nitrogen

The ground state outer electronic configuration of nitrogen is $2s^22p^3$. It is the first element of the second period that begins to show substantial electronegative characteristics (characteristics associated with high electronegativity). Thus, with the most electropositive elements, nitrogen forms the nitride ion, N^{3-}, in crystalline compounds like Li_3N. In general, however, nitrogen resembles carbon in that its chemistry is dominated by covalent bonding. In many cases the nitrogen atom adopts approximately sp^3 hybrid orbitals. The simplest example is ammonia, NH_3. The

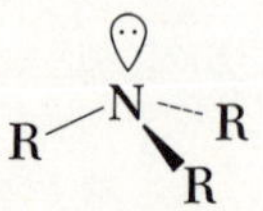

Structure of a typical NR_3 compound

ready availability of the lone pair of electrons in the projecting orbital of NR_3 compounds results in high basicity and a tendency to be attracted to a positively charged or electron-deficient site on another molecule. They are excellent Lewis bases and readily form covalent bonds with Lewis acids, including cations. For example, the Lewis acid BCl_3 forms the complex $(CH_3)_3N{:}BCl_3$ with trimethylamine, and the cations Ag^+ and H^+ form the diamminesilver(I) ion, $Ag(NH_3)_2{}^+$, and the ammonium ion, $NH_4{}^+$. Furthermore, under certain conditions, the lone pair can coordinate with hydrogen atoms contained in other molecules. This special type of bond is called a hydrogen bond. It has exceptionally important biological significance (Chap. 8).

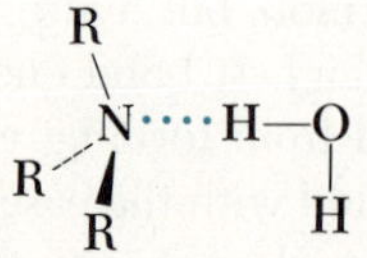

A molecule of H_2O associated with a molecule of NR_3 *via* a hydrogen bond

The formation of the diamminesilver(I) ion is the basis for the qualitative analysis of Ag^+. The silver ion is first precipitated with Cl^-, redissolved with NH_3 to separate it from other chloride ion precipitates, and reprecipitated with HNO_3:

$$Ag^+ + Cl^- \rightleftharpoons AgCl(s)$$

$$AgCl(s) + 2NH_3 \rightleftharpoons Ag(NH_3)_2{}^+ + Cl^-$$

$$Ag(NH_3)_2{}^+ + Cl^- + 2H^+ \rightleftharpoons AgCl(s) + 2NH_4{}^+$$

Salts of the ammonium ion, $NH_4{}^+$, are quite useful. Sal ammoniac, NH_4Cl, is used as a soldering flux because it decomposes at high temperatures to yield HCl:

$$NH_4Cl(s) \xrightleftharpoons{\text{heat}} NH_3(g) + HCl(s)$$

The HCl reacts with oxides on the metal surface, cleaning it:

$$CuO(s) + HCl(g) \rightleftharpoons Cu^{2+} + 2Cl^- + H_2O(g)$$

A further analogy with carbon is nitrogen's ability to form multiple bonds with other atoms of its species as well as with other elements. Some typical examples are elemental nitrogen (N≡N), nitriles (R—C≡N), cyanide ion (CN^-), imines

$$\left(\text{H—N=C}\begin{matrix}\text{R}\\ \text{R}\end{matrix}\right)$$

, azo compounds (R—N=N—R), and nitrous acid (HO—N=O).

Unlike carbon, nitrogen does not form stable, long chains of atoms, undoubtedly because the N—N bond energy (38 kcal/mol) is considerably less than the energy of the C—C bond (83 kcal/mol). Likewise, nitrogen has little tendency towards ring formation. Some compounds are known, however, that contain several nitrogen atoms consecutively linked in a chain or in a ring, but they are relatively unstable and generally explosive.

Oxygen

As expected from its ground state outer electronic configuration ($2s^22p^4$), oxygen is more electronegative than nitrogen. Adding an electron to the same shell and another proton to the nucleus causes a decrease in the atomic size and an increase in electron-attracting power. Many metallic oxides are known in which the oxide ion, O^{2-}, exists. Except at very high temperatures, such oxides are solids. High ionic charges on small ions (for example, the -2 charge on the one-atom ion of oxygen) lead to large lattice energies and therefore great stability of the solid state over the less ordered liquid state. The ionization potentials of nonmetals are high, however, so with elements like carbon and sulfur, oxygen forms covalent oxides. The transition metals, and many of the heavy metals, form oxides whose bonds may be conveniently regarded as intermediate between the two extremes.

When oxygen forms covalent single bonds, it is similar to nitrogen in the sense that it uses orbitals that seem to correspond nearly to sp^3 hybrid orbitals. As in the case of NR_3 compounds, where R is an atom or functional group, the lone pairs of electrons in OR_2 compounds are easily accessible and result in the formation of complexes with Lewis acids, including cations. For example, with the H_2O case (OR_2, with R = H), the cations H^+ and Cr^{3+} form the hydronium ion, H_3O^+, and the hexaaquochromium(III) ion, $Cr(OH_2)_6{}^{3+}$, respectively. The Lewis acid BF_3 forms the very stable boron trifluoride–etherate complex, $(C_2H_5)_2O{:}BF_3$, with diethyl ether, $(C_2H_5)_2O$. Hydrogen bonding is even more pronounced in OR_2 compounds than in NR_3 compounds. The compounds OR_2 and NR_3 have basic properties that are qualitatively similar.

Although OR_2 compounds have two pairs of unshared electrons, the oxygen atom can easily become three-coordinate, but very seldom four-coordinate. After one pair of electrons has become coordinated, as in H_3O^+, the oxygen atom will become partially charged in a positive sense. It is not very likely that an element as electronegative as oxygen would then donate another pair of electrons to another proton and become doubly charged, for example, the species H_4O^{2+} is very unlikely.

In rare cases oxygen can become four-coordinate, but in most of these examples the bonds have considerable ionic character and are not fully covalent. Four-coordination is achieved in solid metal oxides, and in certain hydrated salts such as $CuSO_4 \cdot 5H_2O$.

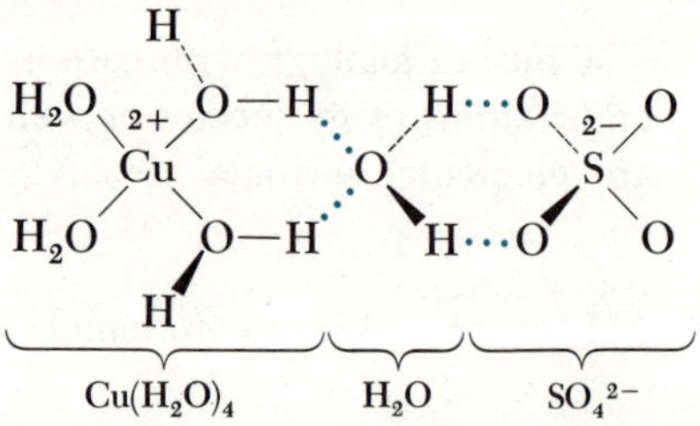

In the accompanying structure, four water molecules are associated with the cation and one is associated with the sulfate anion. The oxygen atom in the one water molecule that is not bound directly to the Cu^{2+} is four-coordinate.

Some of the most interesting types of compounds containing four-coordinate oxygen are beryllium complexes of the general formula $Be_4O(RCOO)_3$. They are volatile solids, soluble in nonpolar solvents (even hydrocarbons), and un-ionized in solution.

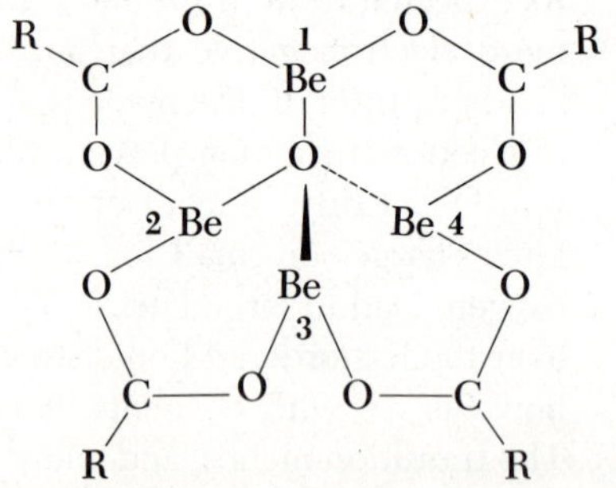

For clarity, only four carboxylate groups, RCO_2, are shown in the accompanying structure, but the fifth joins Be atoms 1 and 3, and the sixth joins Be atoms 2 and 4.

Like carbon and nitrogen, oxygen easily forms double bonds. Some typical examples are elemental oxygen (O=O), carbonyl compounds ($R_2C{=}O$), and oxides of carbon (for instance, O=C=O). The O—O bond energy (33 kcal/mol) is even less than the N—N bond energy. Consequently, oxygen has little tendency to form stable long chains or rings. Examples of compounds containing two linked oxygen atoms are peroxides, such as hydrogen peroxide, H—O—O—H, and peracetic acid,

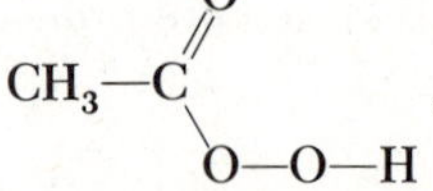

, and the superoxide ion, O_2^-. Ozone, O_3, contains three atoms in a cyclic chain. Because the bond energy is so low, these compounds are very unstable and powerful oxidizing agents.

Fluorine

Fluorine is the most electronegative, and in general, the most reactive of all the elements. Fluorine is the only element reactive enough to react with xenon, which before 1962 was considered completely inert. It is so electronegative that there are no compounds in which fluorine exhibits a positive oxidation state, even in the oxygen fluorides, and its chemistry is limited to the negative ion, F^-, and covalent bonding. Because of the high electronegativity of fluorine, we might not expect it to share its *p* electrons in hydrogen bonds. However, fluorine does show some tendency for hydrogen-bond formation. Indeed, it forms extremely strong hydrogen bonds in HF and in the ion HF_2^-, but in most situations the strength of hydrogen bonds to fluorine is considerably less than to nitrogen or oxygen. Fluorine does not form true multiple bonds. The valence of 1 for fluorine also prohibits it from forming chains or rings.

The reaction of HF with halogenated organic molecules produces aerosol propellants, often called Freons.

$$3HF + 2CCl_4 \rightleftharpoons CCl_2F_2 + CCl_3F + 3HCl$$

The product dichlorodifluoromethane (or Freon-12), CCl_2F_2, has a boiling point of -29.8 °C and is a very useful propellant. The other product, trichlorofluoromethane (or Freon-11), CCl_3F, boils at about room temperature and is also used as an aerosol propellant and in refrigeration. Unfortunately, some evidence suggests that fluorinated molecules may catalyze the decomposition of ozone in the upper atmosphere.

$$2O_3 \rightleftharpoons 3O_2$$

The ozone decomposition reaction is very slow in the absence of catalysts.

Neon

The valence shell of neon is a closed octet, and the high ionization potential of neon (21.6 eV) is exceeded only by the ionization potential of helium (24.6 eV). Accordingly the element is inert, and there are no known compounds of neon.

Correlations among the Oxides of Li through F

Oxides can be generally classified into four main groups:

1. Basic oxides (one extreme),
2. Acidic oxides (the other extreme),
3. Amphoteric oxides (oxides possessing some of the properties of both extremes but not all), and
4. Neutral oxides (oxides possessing none of the properties of either extreme).

These classifications all have one characteristic, which was mentioned earlier in this chapter in identifying metallic and nonmetallic character, namely, the acid-base behavior of oxides.

When an oxide is placed in water, it may dissolve, remain undissolved, or react chemically with water. The dissolution of many oxides in water is the result of a chemical reaction that produces a soluble product. Some examples are

$$Li_2O(s) + H_2O(l) \rightleftharpoons 2Li^+ + 2OH^-$$

$$SO_3(g) + H_2O(l) \rightleftharpoons 2H^+ + SO_4^{2-} \text{ (dilute sulfuric acid)}$$

In the first reaction above, Li_2O is an oxide of one of the least electronegative elements; hence it contains the oxide ion, O^{2-}. This ion is unstable in water, however, and rapidly hydrolyzes to the hydroxide ion:

$$O^{2-} + H_2O \rightleftharpoons 2OH^-$$

On the other hand, nonmetals have an ionization potential high enough that covalent bond formation with oxygen is the general rule. Many of these nonmetallic oxides contain an electron-deficient center, for example, and they can react with Lewis

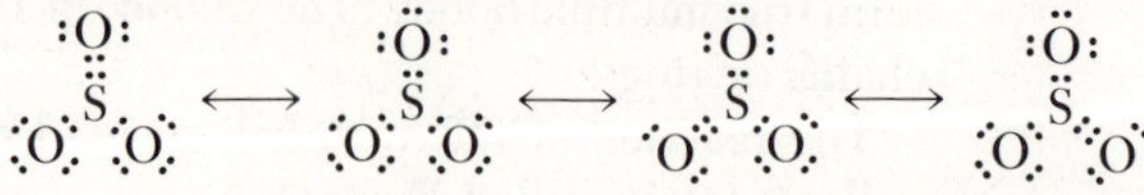

bases, particularly water. If a proton is removed from the resulting short-lived "water complexes," the remaining portion of the molecule can usually stabilize the resulting negative charge through resonance. Such complexes, then, are acidic.

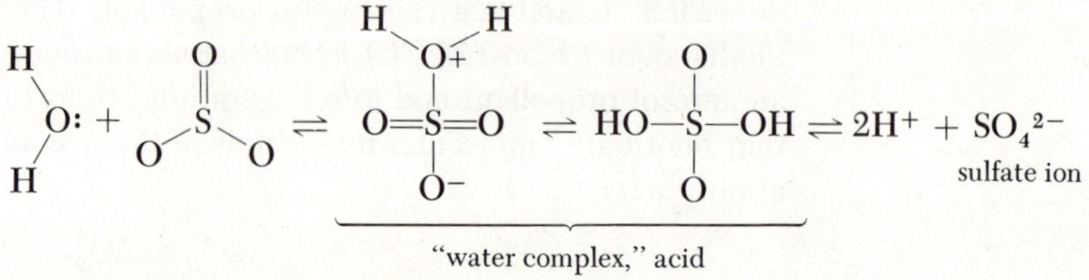

Thus, as shown in the reactions above, the oxide of a metal reacts with water to form a base, and the oxide of a nonmetal reacts to form an acid.

Metals are situated to the left and nonmetals to the right in the periodic table. If we consider the basic or acidic properties of the oxides of the second-period elements as we move from left to right across the periodic table, we see that although we occasionally find some neutral oxides (neither acidic nor basic character), the oxides are first basic, then amphoteric, and finally acidic. Let us now consider the properties of these oxides. Lithium oxide reacts directly with acids to form a salt and water. Thus, it is referred to as a **basic oxide.** Because it reacts with water to form a hydroxide, it is also called a **basic anhydride** (*anhydride,* meaning "without water").

Beryllium oxide is partially covalent. It is insoluble in water, but reacts with acids to form hydrated salts:

$$BeO(s) + 2HCl + 3H_2O \rightleftharpoons [Be(H_2O)_4]Cl_2$$

Beryllium oxide also reacts with bases to form the beryllate ion:

$$BeO + 2OH^- + H_2O \rightleftharpoons [Be(OH)_4]^{2-}$$

It is an example of an amphoteric oxide, because it reacts with both acids and bases.

Boron is sufficiently nonmetallic that its oxide is acidic. Boric oxide, B_2O_3, dissolves in water to form boric acid,

$$\underset{\text{boric oxide}}{B_2O_3(s)} + 3H_2O \rightleftharpoons \underset{\text{boric acid}}{2H_3BO_3} \text{ (or } 2B(OH)_3)$$

which, although weak, is definitely an acid and shows no amphoteric properties. For this reason, B_2O_3 and many other oxides of nonmetals are also called acidic anhydrides.

Carbon and nitrogen are similar to boron in that some of their oxides are acidic. For example:

$$CO_2(g) + NaOH(s) \rightleftharpoons \underset{\text{sodium bicarbonate}}{NaHCO_3(s)}$$

$$\underset{\text{dinitrogen pentoxide}}{N_2O_5(s)} + H_2O(l) \rightleftharpoons \underset{\text{dilute nitric acid}}{2H^+ + 2NO_3^-}$$

However, carbon and nitrogen also form oxides that are somewhat soluble in water—CO and N_2O, for example—but do not react with water to form either acids or bases. These oxides, then, are called **neutral oxides.**

Although fluorine forms compounds with oxygen, the products are more properly called oxygen fluorides, rather than fluorine oxides, because fluorine is the more electronegative element. Oxygen difluoride, OF_2, reacts with water to form oxygen and **hydrofluoric acid:**

$$\underset{\text{oxygen difluoride}}{OF_2(g)} + H_2O(l) \rightleftharpoons O_2(g) + \underset{\text{hydrofluoric acid}}{2HF(g)}$$

STUDY PROBLEM 9(c)

Write the equations for reactions of an acidic oxide, an amphoteric oxide, and a basic oxide that demonstrate their acid-base character. Choose examples from second row oxides.

Correlations among the Hydrides of Li through F

Only the most electropositive metals, like Li, will form saline hydrides, because of the very low electron affinity of hydrogen. The hydrides of such elements are ionic, containing the hydride ion, H^-. Very electronegative elements, like F, also form largely ionic species that can formally be called hydrides, although the polarities of the bonds are reversed from those of the saline hydrides; they contain hydrogen in the $+1$ oxidation state, instead of H^-. Whereas the hydrides of electronegative elements are acidic, the saline hydrides are basic, and vigorously react with water to form hydroxides and H_2 gas:

$$HF(g) + H_2O \rightleftharpoons H_3O^+ + F^- \qquad K_{eq} = 3.5 \times 10^{-4}$$

$$LiH(s) + H_2O \rightleftharpoons H_2(g) + Li^+ + OH^-$$

Elements that are neither very electropositive nor very electronegative form covalent hydrides that are relatively inert to hydrolysis, for example, methane CH_4. In summary, the hydrides are similar to the oxides in the sense that as one moves from left to right across the second period, the hydrides of the elements are first basic, then neutral, and finally acidic.

We shall now consider some of the detailed chemistry of the second period hydrides. The high reactivity of the saline hydride LiH can be explained by regarding H_2 as an extremely weak acid, or by regarding H^- as a very strong base:

$$LiH(s) + H_2O \rightleftharpoons H_2(g) + Li^+ + OH^-$$

or

$$H^- + H_2O \rightleftharpoons H_2(g) + OH^-$$

Lithium hydride is an excellent reducing agent, and is useful in synthetic organic chemistry. A simple example of this reducing ability is given in the following equation:

$$H_2C=O + LiH(s) \rightleftharpoons H_3C-O^-Li^+$$

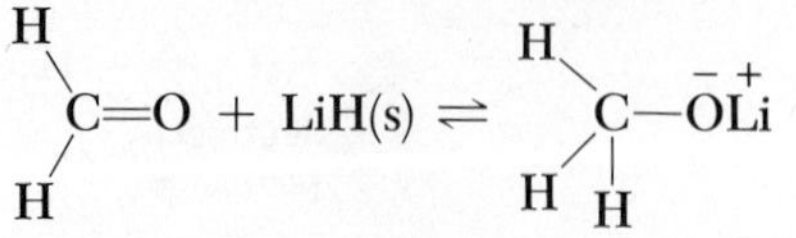

In this case the oxidation state of the carbon atom goes from 0 to -2.

Beryllium is not sufficiently electropositive to form a true saline hydride, and represents a transition between the ionic and covalent hydrides. The partially ionic character of the Be–H bond is consistent with its ability to react vigorously with water, and to act as a powerful reducing agent.

The low electron affinity of hydrogen, coupled with the more electronegative character of boron, causes the boron hydrides to be purely covalent; the bonds are essentially nonpolar. The simplest hydride, borane, BH_3, is too electron-deficient to be stable as such; however, a dimer (same composition) does exist. This dimer, called diborane, has a structure that includes hydrogen bridges (see structure on p. 417). Even though diborane contains no hydride ions, it is readily hydrolyzed to boric acid and hydrogen because water attacks the electron-deficient boron atom:

$$H_2O: + BH_3 \rightleftharpoons HO-BH_2 + H_2(g)$$

$$HO-BH_2 + 2H_2O \rightleftharpoons HO-B(OH)_2 + 2H_2(g)$$

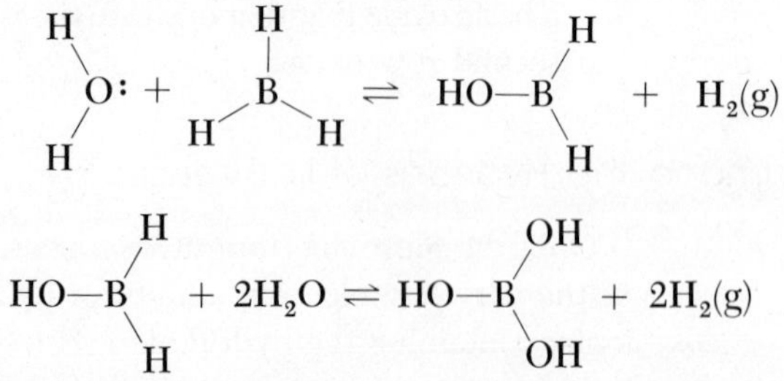

boric acid

Carbon hydrides are more commonly called hydrocarbons, of which methane is the simplest (see Fig. 7.11). The number of known and well-characterized hydrocarbons is enormous, and all classically belong in the realm of organic chemistry. As expected, the C–H bonds are purely covalent. The hydrocarbons have no tendency to hydrogen-bonding, and are completely inert towards water and many other reagents.

The most common hydride of nitrogen is ammonia, NH_3. In ammonia the nitrogen atom is sufficiently electronegative to make the N–H bond somewhat polar, resulting in a partial positive charge on the hydrogen atom, and some weakly acidic character. However, very strong bases are required to abstract a proton; for example;

$$H^- + NH_3(l) \rightleftharpoons NH_2^- + H_2(g)$$

Notice that although NH_3 is considered a base relative to H_2O (a weaker base), it is acidic relative to a stronger base (hydride ion).

Oxygen is more electronegative than nitrogen, and thus its hydride, H_2O, is more acidic than NH_3. Thus, the reaction

$$H_2O + NH_4^+ \rightleftharpoons H_3O^+ + NH_3$$

has an equilibrium constant 5.5×10^{-10}.* Actually, H_2O can be considered the middle of the acidity-basicity scale for the hydrides of the second period elements in the Brönsted-Lowry sense. (Further details of the acid-base properties of water and its hydrogen-bonding abilities are covered in Chap. 5 and 8.) Fluorine is the most electronegative of all the elements, and its hydride HF is the most acidic of the second-period hydrides.

STUDY PROBLEM 9(d)

Explain what is observed when the hydrides of the second period are added to water (using visual observation and a pH meter). Write the equations for the chemical reactions.

9.3 Trends Within the Main Groups

We have pointed out that whereas some of the elements of the second period frequently have properties that are considerably different from those of their heavier congeners, the elements of the third period are very similar to their heavier congeners. It is partly for this reason that we examined the trends across the second period first. We now turn our attention to trends within the main groups, in other words, down columns of the periodic table. In this section we will discuss only the main group (non-transition) elements. Hence, whenever a group designation is given, it should be understood that only A group elements are being considered.

Metallic Character

Just as the metallic character of the elements in any period tends to be greater as one moves from right to left across the periodic table, metallic characteristics tend to be more important as one moves down the table. Thus cesium, since it is in the lower left corner of the periodic table, is the most metallic element, and fluorine, in the upper right corner, is the most nonmetallic. This vertical trend, like the horizontal trend, is expected on the basis of the changes that occur in the electronic properties of the atoms (Chap. 6).

Correlations among the Elements

We are here also primarily concerned with how the chemical properties of the elements change as one moves down a specific column or group. Whenever appropriate, the following characteristic trends will be covered:

1. Metallic and nonmetallic characteristics
2. Ionic properties
3. Covalent properties, including catenation and organic derivatives

*The equilibrium constant given for the reaction between H_2O and NH_4^+ is for aqueous solution, for which solvation effects are very important. To be technically correct, one should evaluate these arguments by a value for this reaction in the gas phase if one is interested in the relative acidities of individual molecules.

4. Important differences between the second-row elements and their heavier congeners within a group when these differences are pronounced, as in the formation of multiple bonds.
5. Acceptor-donor tendencies, or Lewis acid–Lewis base properties.

Group I

The metals of Group I have only a single *s* electron in their outer shells, and removal of this electron leaves a noble gas core. Removal of a second electron is exceedingly difficult; the second ionization potentials for the elements Li through Cs are 75.62, 47.29, 31.81, 27.36, and 23.4 eV. Thus, these elements display very simple chemistries, since they are associated only with the +1 ions. All the isotopes of Fr are radioactively unstable, the longest half-life being about 20 min, and the chemistry of this element has not been extensively explored. The studies that have been done, however, show that its properties are consistent with its position in the periodic table.

It should be noted that several other ions chemically resemble the Group I ions; the most important is NH_4^+, with an ionic radius close to that of K^+ (1.43 Å for NH_4^+ and 1.33 Å for K^+). For example, the solubilities of the Group I and ammonium salts are quite similar.

It was previously pointed out that Li can form bonds with substantial covalent character. Other Group I metals also occasionally show this property, but as expected, the tendency for covalent bond formation decreases from Li to Cs.

Group II

In Group II, only Be shows predominantly covalent chemistry. Of the remaining elements, Mg (like Li, diagonal relationship) shows some covalent character, whereas the elements Ca, Sr, Ba, and Ra (which is radioactive) demonstrate predominantly ionic properties. For example, magnesium readily forms organomagnesium compounds, the so-called Grignard reagents, RMgX, of which $CH_3CH_2CH_2MgBr$ is an example.

The ionic chemistry of the elements in Group II is almost exclusively of the +2 oxidation state. These elements have oxidation number +2; and this is the first group in which one might naturally wonder about the stability of the +1 state. It suffices to say here that in this group only Mg has been shown to form a +1 state, and even Mg^+ is quite unstable and very short-lived. It forms only under unusual conditions; for example, Mg^+ has been electrically generated in solution in certain organic solvents.

Group III

The periodic trends in Group III bear a close relation to those in Group II. For example, boron's chemistry is predominantly covalent; Al shows some covalent character; and the remaining elements, Ga, In, and Tl, form predominantly ionic bonds. This is the first group in which anionic (negative-ion) species have some importance, but only boron forms borides; its heavier congeners are too metallic to form similar species.

This is also the first group in which stable multiple oxidation states occur. All the elements demonstrate a +3 state, but only boron does not form a +1 state. The stability of the +1 state increases from Al to Tl, but not until we reach Tl does this state become a dominant feature of the element's chemistry. In accordance with the general rule that was given earlier concerning the special stabilities of states containing completely filled subshells, one expects to find +1 and +3 states but not +2 states for elements in this group.

Like Mg, Al does not form a +1 state under ordinary conditions, but it can be electrically generated or formed at high temperatures; Al_2O and AlCl are known in the gaseous state. Similarly, GaCl exists in the gaseous state, and stable complexes of Ga^+ have been prepared by the reaction

$$2Ga(s) + 4GaCl_3(s) \rightleftharpoons 3Ga^+[GaCl_4]^-(s)$$

Although In^+ compounds appear to be more stable than Ga^+ compounds, the ion is not stable in solution. However, solid InCl, InBr, InI, and InO_2($In^+O_2^-$, indium superoxide) are known. The thallous ion, Tl^+, is quite stable, even more so than the thallic ion, Tl^{3+}.

Since thallium is at the bottom of its group, it is the most metallic element in Group III. The ionic radius of the ion Tl^+ is comparable to the radius of Rb^+, and in many ways Tl^+ closely resembles the Group I ions in chemistry. For example, the hydroxide is a water-soluble, strong base. Like the trivalent compounds of boron, BF_3 for example, the heavier congeners (with the possible exception of Tl) show marked properties of Lewis acids and this acceptor tendency decreases down the group. Aluminum chloride shows strong acceptor properties and is frequently used in certain types of syntheses of organic compounds, particularly in the famous Friedel-Crafts reaction:

$$CH_3CH_2Cl + AlCl_3(s) + C_6H_6 \rightleftharpoons C_6H_5CH_2CH_3 + AlCl_4^- + H^+$$

The other Group III elements show their acceptor properties in the dimerization of their halides, for example.

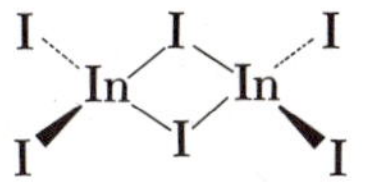

Structure of In_2I_4

STUDY PROBLEM 9(e)

Why is thallium much like rubidium in its chemistry? How do their chemistries differ?

Group IV

Group IV is in the middle of the periodic table, and on this basis it is not surprising that one observes the greatest group variation from nonmetallic to metallic characteristics as one moves down the group. For example, carbon is strictly nonmetallic, silicon is essentially nonmetallic, germanium is amphoteric, and tin is a metal, but not as metallic as lead. Only Sn and Pb have significant chemistries involving ionic states in aqueous solution; these ions are of the +2 and +4 types.* The chemistries of the remaining elements are dominated by covalent characteristics. All the elements in Group IV form tetravalent compounds with organic groups, compounds of the MR_4 type. For the carbon case, these simply correspond to organic compounds. For the other Group III elements, the most numerous and extensively studied substances are the organic derivatives of silicon.

Like Group III, this group also comprises anionic species, but since it is further to the right in the periodic table, the two lightest elements, carbon and silicon, form

*Ions of C, Si, and Ge can be important in the chemistry observed in nonaqueous solvents.

such species, in contrast to only the top element, boron, in the preceding group. For example, carbides and silicides are both known; $CaSi_2$, CaSi, and Ca_2Si are representative.

Carbon often participates in stable multiple bonds involving π bonds between p orbitals (Chap. 7); these are often referred to as $p\pi$–$p\pi$ bonds. The heavier congeners of carbon do not generally form such bonds, however, either to others of their species or to other Group IV elements. For Si, however, there is some evidence that it can form multiple bonds of the $d\pi$–$p\pi$ type, π bonds in which a $3d$ orbital of Si interacts with a $2p$ orbital of nitrogen, as in $(SiH_3)_3N$.*

While catenation is an exceedingly important feature of carbon chemistry, the tendency for catenation decreases for its heavier congeners. Silicon forms catenated hydrides, halides, and organic derivatives; Ge forms only catenated hydrides and organic derivatives, whereas Sn and Pb form only catenated organic derivatives, the organometallics.

In analogy to the trivalent compounds of Group III, the tetravalent compounds of Group IV elements, except for carbon, have some degree of Lewis acid character. The four heavier elements form complex halides, such as SiF_6^{2-}, $GeCl_6^{2-}$, $SnBr_6^{2-}$, and $PbCl_6^{2-}$.

Group V

Group V is similar to Group IV in showing a striking trend from nonmetallic to metallic character as one moves down the group. However, this is the first group in which all the elements show some ability to form negative species. The electronic structure of a noble gas could be achieved by the gain of three electrons, but this requires considerable energy; 350 kcal/mol are needed, for instance, to form P^{3-} from P. Consequently, the negative species are all relatively unstable. Nitrides, phosphides, arsenides, antimonides, and bismuthides are known, exemplified by Na_3N, Na_3P, Na_3As, and K_3Sb; but except for nitrides, the others are probably best classified as being between ionic compounds and alloys. The noble gas configuration could also be achieved by the loss of five electrons, but this also is energetically

*In $(SiH_3)_3N$, the $d\pi$–$p\pi$ bonding between N and Si causes the NSi_3 portion of the molecule to be in a planar configuration, as shown herewith. The overlap of a $3d$ orbital of Si and a $2p$ orbital of N leads to

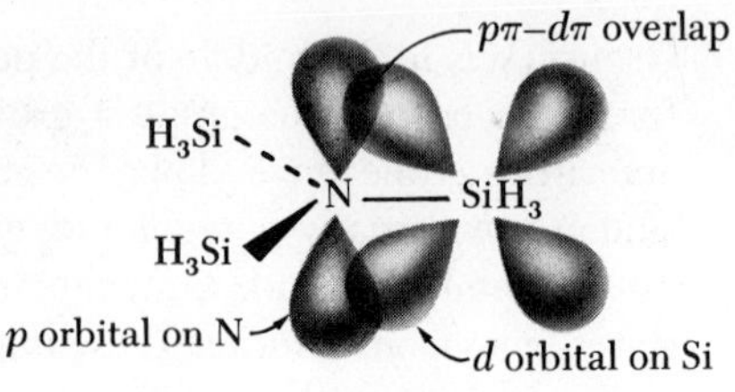

a π bond. If the 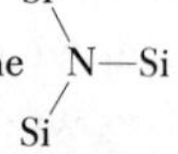

portion of the molecule were distorted so that it was no longer planar, this overlap would be lost and the molecule would be less stable. Consistent with this overlap view of $d\pi$–$p\pi$ bonding is the accompanying resonance structure.

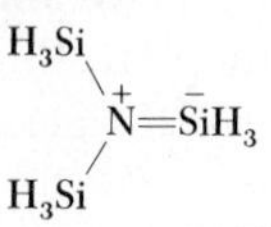

unfavorable. The +5 ions do not form, and only Bi is metallic enough to form compounds in which there is any real approach to a simple cation in solution, namely, Bi^{3+}, which has an electronic configuration in which all occupied subshells are filled. Thus, most of the chemistry associated with the elements of this group involves covalency. The most important oxidation numbers in these covalent compounds are −3 (NH_3) and +5 (P_2O_5). Nitrogen compounds show all the integer oxidation states from −3 to +5. All these elements form organic derivatives, with organonitrogen, organophosphorus, and organoarsenic compounds the most extensive of this type.

The main differences between the chemical properties of nitrogen and of phosphorus can be divided into two related categories:

1. Nitrogen can form $p\pi$–$p\pi$ bonds whereas phosphorus does not. In this difference, they are like carbon and silicon, of which carbon forms these bonds. On the other hand, phosphorus can provide $3d$ orbitals in $d\pi$–$p\pi$ multiple bonds, whereas nitrogen cannot. Thus, amine oxides have only one resonance structure, whereas phosphine oxides can have two resonance structures,

$$\underset{\text{amine oxide}}{R_3\overset{+}{N}\text{—}\overset{-}{O}} \qquad \underset{\text{phosphine oxide}}{R_3\overset{+}{P}\text{—}\overset{-}{O} \leftrightarrow R_3P\text{=}O}$$

which qualitatively accounts, at least partly, for the greater stability of the P–O bond vs the N–O bond in these types of molecules.

2. Nitrogen cannot expand its valence shell beyond eight electrons and cannot form compounds like NX_5. On the other hand, the existence of compounds such as PCl_5, PF_5, and PF_6^- suggests that one or two $3d$ orbitals, together with its $3s$ orbital and $3p$ orbitals are being hybridized to form bonding orbitals.

Group V, consistent with its position to the right of center in the periodic table, contains elements whose three-coordinate compounds (for example, NH_3) act as Lewis bases. It is found that the donor (Lewis base) tendencies in Group V compounds decrease in progressing down the group.* In fact, Bi does not appear to form any compounds in which the Bi atom functions as a Lewis base center. The triphenyl-phosphines, arsines, and stibines—$(C_6H_5)_3M$, where M is P, Ar, or Sb—react with organic halides to give the corresponding phosphonium, arsonium, and stibonium salts. For example, for the case M = P,

$$\underset{\text{triphenyl-phosphine}}{(C_6H_5)_3P} + \underset{\text{methyl bromide}}{CH_3Br} \rightleftharpoons \underset{\text{triphenylmethyl-phosphonium bromide}}{[(C_6H_5)_3PCH_3]^+Br^-(s)}$$

Group VI

Group VI, situated as it is, further to the right in the periodic table, contains a clear majority of elements showing almost purely nonmetallic chemistry. Only Po, and to a very slight extent Te, show metallic properties. Since polonium is radioactive, the only common Group VI element with metallic properties is tellurium. The oxides of these two heaviest elements appear to be ionic, and some salts of these

*With some transition metal cations, for example, Ag^+, as Lewis acids, the donor (Lewis base) order $N \gg P > As > Sb > Bi$ is changed to $P > As > Sb > N > Bi$. An approach to classifying these interactions is called hard and soft acid-base theory, and will be discussed in Chap. 18.

elements have been prepared but not well characterized. Clearly, however, nonmetallic behavior dominates this group. With the most metallic elements, O, S, Se, and Te form the oxide (O^{2-}), sulfide (S^{2-}), selenide (Se^{2-}), telluride (Te^{2-}), and polonide (Po^{2-}) ions.

Covalent bonding, as expected, is the dominant bond type in this group. Typical oxidation numbers in the largely covalent compounds are -2 (H_2S), $+4$ (SO_2), and $+6$ (SO_3). While oxygen has little tendency towards catenation, sulfur has a strong tendency, and this ability is lower the higher the atomic number in the group. For example, elemental sulfur and the polysulfide ion, S_n^{2-}, both contain relatively stable S–S bonds. Elemental Se contains Se_8 rings, and elemental Te contains Te–Te chains. Except for Po, all the elements form organic derivatives, for example, $(CH_3)_2O$, $(CH_3)_2S$, $(CH_3)_2Se$, and $(CH_3)_2Te$.

The main differences between oxygen and its congeners relate to the *d* orbitals. Oxygen does not have *d* orbitals available for bonding, but its congeners do. Thus, the heavier elements are not limited to a maximum coordination number 4: $SeBr_6^{2-}$, SF_5^-, $Te(OH)_6$, and PoI_6^{2-}. In addition, oxygen can form only $p\pi$–$p\pi$ type multiple bonds, such as O=O, but sulfur can use *d* orbitals for $d\pi$–$p\pi$ multiple bonds, as in the sulfate ion, SO_4^{2-}.

The donor properties of the divalent compounds of this group, and hence the tendency towards hydrogen bonding, decrease from oxygen to polonium. The tendency for catenation, as described above, is in the order $O < S > Se > Te > Po$.

Group VII

The chemistries of the elements of Group VII are almost completely nonmetallic, and it is difficult to speak of any sort of metallic character at all for these elements. However, one can always speak of lower tendency to nonmetallic character in the order down the group.

Little can be said about the chemistry of astatine, since all 20 isotopes of this element now known are unstable. The longest-lived isotope is At^{210}, with a half-life of only 8.3 hr. Thus, reasonable quantities have not been available for studying its chemistry.

Although the formation of positive ions is generally not the rule for these elements, there is definite evidence for the formation of such ions for I. For F, the most electronegative of all elements, there are no known compounds in which it has a positive ion state, even in the oxygen fluorides. Although there is no evidence for the existence of the ions Cl^+ or Br^+ as such, iodine is "metallic" enough to form ions such as I^+, I_3^+, and I_5^+. When I_2 is dissolved in 60% oleum (100% H_2SO_4 containing 60% dissolved SO_3), the ion I^+ seems to form:

$$I_2(s) + H_2S_2O_7(l) + 2SO_3 \rightleftharpoons 2I^+ + HS_2O_7^- + HSO_4^- + SO_2(g)$$

With 30% oleum, a 0.5 *M* solution appears to produce the ion I_3^+, and with excess I_2, the ion I_5^+.

More commonly, however, the elements Cl, Br, I, and At, like F, form the -1 ions. Even in these ions, however, one notes a decrease in nonmetallic character, in order down the column from F^- to At^-. Thus, Cl, more nonmetallic than I, oxidizes I^- in solution to I_2:

$$Cl_2(aq) + 2I^- \rightleftharpoons 2Cl^- + I_2(s)$$

In covalently bonded species, these elements have oxidation numbers 1, 3, 5, and 7; examples are Cl_2O, ClO_2^-, BrO_3^-, and IO_4^-. With the possible exception of astatine, only iodine forms stable catenated species—namely I_3^-, which is highly stable in aqueous solutions, and I_5^-, I_7^-, and I_9^-, which are loose aggregates. This tendency of iodine to catenation is presumably due to hybridization of d and p orbitals. This is not possible for fluorine, however. With chlorine it is unlikely, and with bromine it is not so favorable as with iodine. Thus, although the ion Br_3^- is known, it is not so stable as the ion I_3^-. For the elements F, Cl, Br, and I, numerous organic derivatives are known, again demonstrating the ability of these elements to form covalent bonds.

STUDY PROBLEM 9(f)

In addition to astatine, several other main-group (nontransition) elements are found only as radioactive isotopes. What are they?

The neutral halogen molecules Cl_2, Br_2, and I_2, act as acceptors. Some examples are the complexes formed from Cl_2 with benzene, C_6H_6, which functions as an electron donor by virtue of pi electrons, and Br_2 and I_2 with dioxane,

$$\begin{array}{c} H_2C\!\!-\!\!CH_2 \\ / \quad \quad \backslash \\ O \quad \quad O \\ \backslash \quad \quad / \\ H_2C\!\!-\!\!CH_2 \end{array}$$

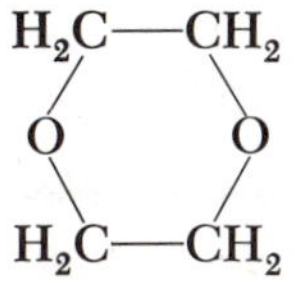

The stability of these complexes increases from Cl to I, indicating that the Lewis acid characters of the halogen molecules vary as $I_2 > Br_2 > Cl_2 > F_2$. The negative ions, F^-, Cl^-, Br^-, and I^-, however, act as donors (Lewis bases), and this tendency decreases from F^- to I^-. For example, the ion FHF^- is quite stable, but the salts of HBr_2^- and HI_2^- are unstable.

Group VIII

The chemical reactivity of the noble gases is limited; until 1962 these elements were believed to be totally inert. To date only Kr, Xe, and Rn have been made to react with other elements. Krypton represents the threshold of chemical reactivity for these gases, and not much is yet known about its compounds. Xenon is considerably more reactive and numerous compounds are known. Radon is radioactively unstable, and studies of its reactivity are therefore difficult. However, a radon fluoride has been prepared. (See Chap. 17 for the detailed chemistry of Kr and Xe compounds.)

Correlations among the Oxides

When we examine the trends in the chemical properties of the oxides in each group, we should remember the four general classifications of oxides—basic, acidic, amphoteric, and neutral (Sec. 9.2).

Group I

The oxides of all the elements of Group I are white solids, and all are such very strong bases that it is difficult to find trends in their properties. One property of basic anhydrides is their ability to react with water to form hydroxides; the main reaction is the hydrolysis of the oxide ion, O^{2-}, (as described in Sec. 9.2):

$$O^{2-} + H_2O \rightleftharpoons 2OH^-$$

The above reaction can be viewed as an acid-base reaction; the ion O^{2-} acts as the base and H_2O acts as the acid. One expects that the most basic metallic oxide will react the most vigorously with water, and that the heat liberated from the reaction of the oxide with water will be greater than for the least basic oxide. Table 9.1 shows the heat liberated for the general reaction

$$M_2O(s) + H_2O \rightleftharpoons 2M^+ + 2OH^-$$

where M represents a Group I metal.

TABLE 9.1 Heat Liberated for the Reaction $M_2O(s) + H_2O(l) \rightleftharpoons 2M^+ + 2OH^-$

Element	*Heat (kcal/mol)*
Li	32.3
Na	56.7
K	75.3
Rb	80.6
Cs	84.2

Group II

The tendency towards a larger basicity is clearly noticeable for the oxides in Group II. Beryllium oxide, BeO, is an amphoteric oxide, acidic enough to react with strong hydroxide bases (Sec. 9.2). Magnesium oxide, MgO, is more basic than BeO, and shows no amphoteric properties, but is still only weakly basic and only slightly soluble in water. The remaining oxides are more basic in increasing order, and evolve heat when mixed with water. A few drops of water placed on solid CaO produces clouds of steam, and BaO becomes visibly red hot. The oxides of Ca through Ra are all basic enough to absorb CO_2 (a Lewis acid) from the air to form the respective carbonates; that is,

$$\underset{\text{Lewis base}}{MO(s)} + \underset{\text{Lewis acid}}{CO_2(g)} \rightleftharpoons MCO_3(s)$$

where M represents Ca, Sr, Ba, or Ra.

Group III

Whereas only the oxide of the lightest element in Group II showed some acidic properties by being amphoteric, the oxide of the lightest element in Group III is definitely acidic, although only weakly, and the oxides and hydroxides of the next two elements are amphoteric. The oxides and hydroxides of the remaining two elements in this group are purely basic.

The hydroxide of boron, $B(OH)_3$, is boric acid, which can form the borate ion, $B(OH)_4^-$. Aluminum hydroxide and gallium hydroxide, when acting as acids, form the aluminate and gallate ions, which are sometimes formally written AlO_2^- and GaO_2^-, although they have more complicated structures, viz., $Al(OH)_4^-$, and $Ga(OH)_4^-$. These amphoteric hydroxides are less acidic than H_3BO_3:

$$H_3BO_3 + H_2O \rightleftharpoons H^+ + B(OH)_4^- \qquad K = 1 \times 10^{-9}$$
$$Al(OH)_3 + H_2O \rightleftharpoons H^+ + Al(OH)_4^- \qquad K = 4 \times 10^{-13}$$
$$Ga(OH)_3 + H_2O \rightleftharpoons H^+ + Ga(OH)_4^- \qquad K = 1 \times 10^{-15}$$

Indium and thallium form more than one oxide and hydroxide, whose basicities depend on the oxidation state of the metal. For example, In_2O, In_2O_3, Tl_2O, and Tl_2O_3 are known solids; and whereas thallous oxide, Tl_2O, is a strong base, thallic oxide, Tl_2O_3, is a weak base.

Group IV

Group IV, consistent with its central location in the periodic table, contains fewer clearly acidic or basic oxides than any other group. All the elements in this group form a monoxide and a dioxide of the type XO and XO_2, where X represents a Group IV element. As we shall see below, the acidities of the XO_2 compounds are lower with higher atomic numbers for elements in this group.

Carbon dioxide is a gas, under standard conditions of temperature and pressure. It and carbon monoxide are the only oxides of all the elements in this group or any preceding group that are not solids under standard conditions. Carbon dioxide is purely an acidic oxide, and readily reacts with an excess of hydroxides to form carbonates; for example,

$$2NaOH(s) + CO_2(g) \rightleftharpoons Na_2CO_3(s) + H_2O$$

Germanium dioxide also reacts with hydroxide bases to form the germanate ion, formally written GeO_4^{4-} but considerably more complex in structure. This oxide also begins to show some basic properties; it dissolves in hot concentrated hydrochloric and hydrofluoric acids to form $GeCl_4$ and GeF_4, but is not any more soluble in nitric or sulfuric acid than in water.

Tin dioxide readily dissolves in bases to form the stannate ion, formally written SnO_3^{2-}; the formula $Sn(OH)_6^{2-}$ is probably more accurate in reflecting its structure. The dioxide also readily dissolves in protic acids to form tin salts, and is thus definitely an amphoteric oxide:

$$SnO_2(s) + 2OH^- + 2H_2O \rightleftharpoons Sn(OH)_6^{2-}$$

$$SnO_2(s) + 4H^+ \rightleftharpoons Sn^{4+} + 2H_2O$$

Lead dioxide is relatively inert towards the ions H^+ and OH^- in dilute solution, but can react with acids and bases under more concentrated conditions. Lead dioxide and a concentrated hot alkali, say NaOH in water, form the plumbate ion, which like the stannate ion has a simplified formal representation, in this case PbO_3^{2-}; but $Pb(OH)_6^{2-}$ is probably more correct. With nitric acid, the oxide dissolves but does not form a stable ion Pb^{4+} in solution.

Group V

The elements of Group V, except for Bi, all form well-characterized trioxides and pentoxides of the type X_2O_3 and X_2O_5, where X represents a Group V element. All these oxides are solids under ordinary conditions, except N_2O_3, which is a gas. Bismuth trioxide has been well characterized, but the pentoxide is too unstable to have been obtained in a pure state. Some compounds having bismuth in the +5 oxidation state do exist, but like $NaBiO_3$, they are often very strong oxidants. This trend in the instability of the higher oxidation state with larger atomic number is also somewhat evident in other groups.

Both the trioxides and pentoxides of the three lightest elements are purely acidic anhydrides. The acidity of the pentoxides of these elements decreases from N to As, and is manifested by the ionization constants of the acids produced from the oxides when they react with water. Nitrogen pentoxide reacts to form nitric acid, HNO_3, a very strong acid. Phosphorus pentoxide forms the much weaker phosphoric acid, H_3PO_4, $K = 7.5 \times 10^{-3}$, and arsenic pentoxide forms the even weaker arsenic acid, H_3AsO_4, $K = 4.8 \times 10^{-3}$. These acids and oxides all react with hydroxide bases to form the corresponding anions, nitrate (NO_3^-), phosphate (PO_4^{3-}), and arsenate (AsO_4^{3-}).

Antimony pentoxide also forms an acid with water, namely, antimonic acid, formally written H_3SbO_3; but probably $HSb(OH)_6$ is more appropriate.

$$Sb_2O_5(s) + 7H_2O \rightleftharpoons 2H^+ + 2Sb(OH)_6^-$$

With hydroxide bases, the pentoxide reacts to form the antimonate ion, $Sb(OH)_6^-$. Antimony trioxide is definitely amphoteric, giving the antimonite ion, $Sb(OH)_4^-$, with hydroxide bases,

$$Sb_2O_3(s) + 2OH^- + 3H_2O \rightleftharpoons 2Sb(OH)_4^-$$

and antimony salts with acids. Solutions of these salts contain only a very small concentration of Sb^{3+}, and only the so-called basic antimony salts can be crystallized from these solutions. For example, Sb_2O_3 and H_2SO_4 give antimonious sulfate, which contains the antimonyl (SbO) group, and is appropriately written $(SbO)_2SO_4$.

$$Sb_2O_3(s) + 2H^+ + SO_4^{2-} \rightleftharpoons (SbO)_2SO_4(s) + H_2O$$

Bismuth trioxide is purely basic, forming bismuth hydroxide, $Bi(OH)_3$, with water, and normal bismuth salts with protic acids. In the absence of acids, these typical salts hydrolyze to basic salts in water; for instance, bismuth sulfate gives the basic sulfate $(BiO)_2SO_4$:

$$2Bi^{3+} + SO_4^{2-} + 2H_2O \rightleftharpoons (BiO)_2SO_4(s) + 4H^+$$

Note that the oxide described for Bi, the fifth element of this group, is more basic than the oxide described for the fifth element of the preceding group. This apparent discrepancy can be explained by noting that the comparison was not made between comparable degrees of oxidation states of the elements of these two oxides; PbO_2 (Pb in its highest state) was not compared with Bi_2O_5 (Bi in its highest state), but instead, with Bi_2O_3. The comparison with Bi_2O_5 could not be made because it has not been well characterized. A more valid comparison can be made between the highest oxidation states of the oxides of the fourth members of these two groups: SnO_2 is definitely amphoteric, while Sb_2O_5 is acidic.

It was previously mentioned that as one descends in a group, the stability of the higher oxidation state lessens. It is also true, however, that for any given oxidation state, the basicity of the oxide is higher as one descends in a group.

Group VI

The dioxides and trioxides of sulfur, selenium, and tellurium have been well characterized, but with polonium only the dioxide has been well characterized. The dioxides are much weaker acids than the trioxides, but the general trend toward weaker acidity in the direction down this group is apparent for either oxidation state.

The dioxides of S and Se are both quite soluble in water, producing solutions called sulfurous acid and selenious acid. Although sulfurous acid is formally written H_2SO_3, physical methods have shown that no such species exists to any measurable extent; instead, the solution contains hydrated SO_2, which like CO_2 can ionize water:

$$SO_2(aq) + H_2O \rightleftharpoons H^+ + HSO_3^-$$

Selenious acid, however, does contain H_2SeO_3. Sulfurous acid is somewhat stronger ($K \sim 10^{-2}$) than selenious ($K \sim 10^{-3}$), and both acids react with hydroxide bases to form stable anions. Sulfurous acid forms the sulfite (SO_3^{2-}) and bisulfite (HSO_3^-) ions and selenious acid forms the selenite (SeO_3^{2-}) and biselenite ($HSeO_3^-$) ions.

Tellurium dioxide is soluble in water only to the extent of ca 10^{-5} *M*, and while the resulting solution is called tellurous acid, essentially nothing is known about the actual species present. The dioxide does, however, dissolve in strong bases to form the tellurite (TeO_3^{2-}) and bitellurite ($HTeO_3^-$) ions. The emergence of weakly basic properties for TeO_2 is manifested by its dissolution in hydriodic acid, HI, to form TeI_4.

Polonium dioxide actually forms a basic hydroxide, $Po(OH)_4$; and some salts of this hydroxide have been reported, $Po(SO_4)_2$, for example.

The trioxides of S and Se dissolve in water to form the strong dibasic acids sulfuric acid, H_2SO_4, and selenic acid, H_2SeO_4, in which the first dissociation is essentially complete in dilute solutions. Sulfuric acid forms the sulfate (SO_4^{2-}) and bisulfate (HSO_4^-) ions, and selenic acid forms the selenate (SeO_4^{2-}) and biselenate ($HSeO_4^-$) ions. Tellurium trioxide does not react with water, but telluric acid, H_6TeO_6, can be prepared by the oxidation of tellurous acid. It is a much weaker dibasic acid (first *K* about 10^{-7}) than H_2SO_4 or H_2SeO_4, and forms the tellurate ($H_4TeO_6^{2-}$) and bitellurate ($H_5TeO_6^-$) ions.

Group VII

With the exception of iodine pentoxide, I_2O_5, the oxides of the elements of Group VII are all either unstable or of an uncertain nature. Thus, it is not appropriate to consider their acidities in terms of a direct reaction with hydroxide bases to form salts, or in terms of a direct reaction with water to form oxo acids. Many oxo acids and their salts are known, but they are not prepared by either of the two direct reactions mentioned above.

To find acidity trends in this group, then, we shall consider the ionization constants of some of the oxo acids with a fixed oxidation number, namely, those with formal oxidation state +1. Table 9.2 shows the known hypohalous acids, their ionization constants, and the ions they form with bases.

TABLE 9.2 **Hypohalous Acids and Some of Their Properties**

Name of Acid	*Formula*	*K*	*Anion*
Hypochlorous	HClO	3×10^{-8}	Hypochlorite, ClO^-
Hypobromous	HBrO	2×10^{-9}	Hypobromite, BrO^-
Hypoiodous	HIO	1×10^{-11}	Hypoiodite, IO^-
Hypoastatous	HAtO	?	Hypoastatite, AtO^-

Summarizing, then, we find that the basicity of the oxides is higher for elements in the direction right to left across the periodic table, and is higher towards the bottom for elements of a column. Thus, as expected, we find the amphoteric oxides in a diagonal area across the table, and starting with Be, this area proceeds to the right and diagonally down the table.

STUDY PROBLEM 9(g)

Both the formal charge (Chap. 7) and the electronegativity of the central atom of oxyacids are used in rationalizing trends in acid strength. Which method is most applicable to explaining the trend for the hypohalous acids?

Correlations among the Hydrides

Correlation of the basicity of the oxides is clear for both across and down the table, but it is unfortunate that a similar statement cannot be made concerning the hydrides. There are two important reasons for this. First, only some of the hydrides are stable enough to have been prepared so far, and even in these cases, enough information is not available for definite trends in their acidities to have been established. Second, many common definitions of acids or bases are ambiguous. For example, for substances that can donate protons when dissolved in water, one tends to use the Brönsted-Lowry concept, whereas for electron-deficient substances in nonaqueous systems one prefers the Lewis concept. Thus, in the Brönsted-Lowry sense, HF is a weaker acid in dilute aqueous solution than either HBr or HI, because HF is not nearly so dissociated in these systems. Recent measurements in the gas phase also indicate that HF is the weakest acid of the halogen halides. However, because the ion HF_2^- forms to a large extent in concentrated aqueous solutions (liberating H^+), whereas HBr and HI do not tend to form HBr_2^- or HI_2^- to such a large extent, HF in concentrated aqueous solution is more acidic than HBr or HI. The percentage ionic character (Chap. 7) for H—X (where X is the halogen atom) decreases from HF to HI, as expected from the electronegativity differences of the H—X bonds. The approximate percentage ionic character values for HF, HCl, HBr, and HI bonds are 60%, 20%, 10% and 5%, but the amount of ionic character of a bond in a molecule must not be confused with the tendency of the molecule to ionize in a solvent. The tendency to ionize in solution is determined by the relative stabilities of the un-ionized molecule and the separated ions in solution, not by the polarity of the bond. Thus, HF is a weak acid in dilute solutions, but becomes a strong acid in more concentrated solution (5 to 15 *M*) because stable ions more complex than F^- form; for example,

$$HF + H_2O \rightleftharpoons H_3O^+ + F^-$$
$$2HF \rightleftharpoons H_2F^+ + F^-$$
$$HF + F^- \rightleftharpoons HF_2^- \ (H_2F_3^-, H_3F_4^-, \ldots)$$

A general trend is apparent in the acidic and basic properties of the hydrides. As one moves down the groups III through VII, the stabilities of the hydrides are lower, and the acidities higher.

Groups I and II

With the exceptions of BeH_2 and MgH_2, which are solids that seem to represent a transition between the ionic and covalent hydrides, the hydrides of the Group I and II elements are all solids that react quite violently with water to form H_2 gas and hydroxides, as expected; but enough information is not available to observe clear trends in the relative basicities within either group.

Group III

In Group III, only B and Al form stable binary hydrides, gaseous diborane, B_2H_6, and solid polymeric alane, $(AlH_3)_n$. However, the complex ions BH_4^-, AlH_4^-, and GaH_4^- are stable and studies of these ions have shown that the Lewis acidities of the group MH_3, where M is B, Al, or Ga, are lower in the order B > Al > Ga.

Group IV

In Group IV, methane is purely covalent and the most stable of these hydrides. Silane, SiH_4, germane, GeH_4, and stannane, SnH_4, are gases under standard conditions, and their stabilities judged from decomposition reactions at high temperature are lower in the descending order $SiH_4 > GeH_4 > SnH_4$. Stannane decomposes slowly to its constituent elements at room temperature, and rapidly at 150 °C. Silane is a strong reducing agent and is rapidly hydrolyzed in the presence of compounds containing a hydroxide ion:

$$SiH_4(g) + 4OH^- \rightleftharpoons SiO_4^{4-} + 4H_2(g)$$

Germane and stannane are more stable to hydrolysis, even in the presence of a 15% solution of NaOH in water, and in this sense they are less basic, or more acidic, than silane. Plumbane, PbH_4, has been obtained only in small amounts and has not been well studied.

Group V

In Group V, ammonia, NH_3, phosphine, PH_3, arsine, AsH_3, and stibine, SbH_3, are gases under standard conditions, and their stabilities judged from decomposition reactions at high temperature are lower in the descending order $NH_3 > PH_3 > AsH_3 > SbH_3$. Stibine slowly decomposes to its constituent elements at room temperature and sometimes explodes when heated. Ammonia definitely forms small amounts of the hydroxide ion when dissolved in water, but pH measurements of aqueous solutions of phosphine show that these solutions are neither basic nor acidic. In this sense, then, phosphine is less basic than ammonia.

Group VI

Both the first and second ionization constants for the hydrides of S, Se, and Te (which are all gases under standard conditions) are greater from H_2S to H_2Te, indicating that H_2Te is a stronger Brönsted-Lowry acid than H_2S. These first ionization constants are consistent with the acidity trends obtained from recent measurements in the gas phase. Little is known about H_2Po, which has been obtained only in trace amounts.

SUMMARY

The elements in the second row of the periodic table (lithium through fluorine) show a number of trends because of the differing number of protons in the nuclei and differing electronic configurations. From trends such as the acid-base behavior of oxides and hydrides, the elements can be classified as metallic, nonmetallic, or amphoteric. The oxides of elements at the left of the periodic table are basic, classifying those elements as metallic, while the oxides of elements at the right side of the periodic table are acidic, classifying those elements as nonmetallic.

Comparison of the chemistry of the elements within each group reveals both trends and anomalies. Some vertical trends are the tendencies toward greater (a) metallic character, (2) stability of the lower oxidation states, (3) basicity of the oxides, and (4) acidity of the hydrides. Although the chemistry of elements within a group is normally quite similar, there are cases in which an element more closely resembles an element in another group than elements in its own group. This phenomenon is called the diagonal relationship and is exemplified by the chemical similarities of boron and silicon and those of lithium and magnesium. In addition, each element of the second row exhibits some chemical properties that are strikingly different from its heavier congeners. Among these properties are: (1) the ability of lithium to form covalent bonds in some cases, (2) the great tendency of beryllium to form covalent compounds, (3) formation of B—H—B bonds in compounds such as diborane, (4) the importance of catenation in the chemistry of carbon, (5) the maximum coordination number of four for nitrogen, (6) strong hydrogen bonding of the hydride of oxygen, water, and (7) the weak acidity of HF.

STUDENT CHECKPOINTS

After studying this chapter, the student should be able to:

1. Describe how metallic, nonmetallic, and amphoteric elements are distinguished from one another, and give examples.
2. Explain what is meant by *diagonal relationship*, and give examples of when the relation holds and when it does not.
3. Explain why the elements of the second period often differ from the heavier members of the same group, and cite examples.
4. Correlate the basicity of hydroxides and oxides.
5. Define *catenation* and how the process relates to the special property of carbon in forming long-chain compounds.
6. Discuss differences in chemistry of the second period elements relative to electronic configuration and size.
7. Describe the common oxidation states of the second period elements in the compounds they form and how the common oxidation states differ in the direction down a group.
8. Determine trends in metallic behavior within a period and a group.
9. Enumerate the properties of binary hydrides.

EXERCISES

9.1. Distinguish metallic, nonmetallic, and amphoteric elements. Note common properties of each and cite actual examples.

9.2. What is the diagonal relationship? Why does this relation apply better to beryllium and aluminum than to boron and silicon?

9.3. What are the beryllate and aluminate ions? How are they formed? Does the corresponding species form in the case of magnesium?

9.4. Lithium has a greater tendency to form compounds with organic species, for example, $LiCH_3$, which do not dissociate so readily as sodium does in polar organic solvents like acetone. Why is this true?

9.5.* Which element forms more ionic compounds—lithium or beryllium? Why?

9.6. Which is the least basic—LiOH, $B(OH)_3$, or $Al(OH)_3$?

9.7.* What is *catenation*? Which element has the greatest tendency to show this property? Why?

9.8. Why is carbon the lightest element that shows multiple bonding?

9.9. When carbon atoms are linked to form rings, the most likely sizes are those with 5, 6, or 7 atoms. Why?

9.10.* Draw the Lewis acid–base adduct formed from ammonia and boron trichloride.

9.11. Why are compounds containing long chains of linked nitrogen atoms generally unstable?

9.12. For carbon, nitrogen, and oxygen, give some examples of compounds of these elements that have multiple bonds.

9.13.* Peracetic acid is a useful reagent in bleaching wood pulp for paper. What kind of reaction is this?

9.14.* Rank the hydrogen bonds formed between the hydrogen atoms of ammonia and the electronegative atoms of CH_3F, $CH_3C(=O)H$, and CH_3NH_2 in order of increasing strength.

9.15. Rank the hydrogen bonds formed between water and CH_3F, $CH_3\overset{O}{\overset{\|}{C}}CH_3$, and $(CH_3)_3N$ in order of increasing strength.

9.16.* What is the only element of the first short period that forms no known compounds? Why is this so?

9.17. Describe the results of dissolving the oxide of a metal in water and the results of dissolving an oxide of a nonmetal in water.

9.18. Give examples of a basic oxide, an acidic oxide, and an amphoteric oxide.

9.19.* What is a "saline hydride"? Give an example.

9.20. Binary (two-element) compounds can form basic or acidic solutions. Illustrate this for reactions of oxides and hydrides in water.

9.21. Which is a better base—ammonia or water? Why?

9.22.* Assign oxidation states to the atoms in Ga_2Cl_4.

9.23. The chemistry of the thallium +1 ion is most like which other ion? Explain your reasoning.

9.24.* Which elements show the greatest range of oxidation states? Do these elements commonly have ionic or covalent bonding? Give some examples of compounds in which these elements exhibit their highest and lowest oxidation states.

9.25. One of the common halogens shows some metallic character. Illustrate its behavior as a metallic species.

9.26.° Describe the reaction between Cl_2 and KI in water. Is this reaction more favorable or less favorable if Br_2 is used?

9.27. When alkali metal oxides are dissolved in water, heat is released. What is the reaction? Which oxide causes the greatest heat release?

9.28.° The alkaline earth oxides are often used to remove water from other solvents. Which one is most effective? Is there any problem with using this oxide as a desiccant (drying agent)?

9.29. How can $SnCl_4$ be made from SnO_2?

9.30.° If one starts with elemental sulfur, how can sodium sulfite be obtained?

9.31. Within the following groups of acids, which is the strongest in aqueous solution: (a) sulfurous, selenious, or tellurous; (b) sulfuric, selenic, or telluric; (c) nitric, phosphoric, or arsenic?

9.32.° In each of the following sets, which pair of acids has a greater difference in acidity: (a) sulfurous and selenious, or selenious and tellurous; (b) sulfuric and selenic, or selenic and telluric; (c) nitric and phosphoric, or phosphoric and arsenic?

9.33. What reaction do calcium hydride and barium hydride undergo when added to water?

9.34. Alkali metal hydrides and alkaline earth hydrides generally show basic behavior, whereas Group III hydrides act as acids. Illustrate this statement with pertinent reactions.

9.35.° Why do the hydrides of nitrogen, oxygen, and lithium exist as a gas, a liquid and a solid, respectively, at room temperature?

9.36. Why do phosphine and silane exist as gases at room temperature, whereas beryllium hydride is a solid at the same temperature?

9.37.° Why is less known about the chemistry of polonium than of either tellurium or bismuth?

ANSWERS TO SELECTED EXERCISES

9.5 Lithium. The ionization potentials of beryllium are very high. **9.7** Ability to form long chains of the same element. Carbon; carbon-carbon bonds are especially strong.

9.10

$$\begin{array}{c} Cl \quad\quad H \\ \backslash \quad\quad / \\ B-N \\ /\,| \quad\; \backslash \\ Cl\; Cl \quad H \;\; H \end{array}$$

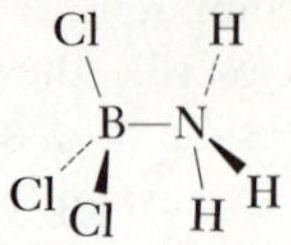

9.13 An oxidation-reduction reaction. Peroxide linkages (O–O single bonds) make molecules oxidizing agents. **9.14** The H···O hydrogen bond is stronger than H···N, and both are stronger than H···F. **9.16** Neon. It has an octet and therefore does not readily form covalent bonds. Its ionization potential and electron affinities are too poor for ionic compound formation. **9.19** A saline hydride is one that can be described as ionic, such as KH. **9.22** One Ga (+1), one Ga (+3), and 4 Cl's (−1). **9.24** The elements of Group VA, Group VIA, and chlorine. Covalent bonding mostly. For nitrogen, −3 in ammonia (NH_3) to +5 in nitrate ion (NO_3^-). For phosphorus, −3 in phosphine (PH_3) to +5 in phosphate (PO_4^{3-}). For sulfur, −2 in H_2S to +6 in SO_4^{2-}. For chlorine, −1 in NaCl to +7 in ClO_4^-. **9.26** $Cl_2(g) + 2I^- \rightleftharpoons 2Cl^- + I_2(s)$. Less favorable with Br_2. **9.28** RaO. Yes, it is radioactive; BaO would be the best choice for effectiveness (per mole) and safety. 9.30 $S(s) + O_2(g) \xrightarrow[\text{heat}]{} SO_2$; dissolve in water: $SO_2(g) + H_2O \rightleftharpoons H_2SO_3$ (actually $SO_2(aq)$); add sodium hydroxide: $(2Na^+) + 2OH^- + H_2SO_3^- \rightleftharpoons 2H_2O + SO_3^{2-} + (2Na^+)$; then evaporate the water to obtain $Na_2SO_3(s)$. 9.32 (a) selenious and tellurous; (b) selenic and telluric; (c) nitric and phosphoric. 9.35 NH_3 is a gas because it has a low molecular weight and the hydrogen bonding is not as stabilizing as for H_2O, which exists as a liquid because of the stronger hydrogen bonding (there are two hydrogen bonds per molecule). Lithium hydride is a solid because it is ionic. 9.37 Polonium is radioactive and there are no easily obtained isotopes with long half-lives (see Chap. 13).

ANSWERS TO STUDY PROBLEMS

9(a) Like lithium, magnesium as Mg^{2+} shows a high charge-to-radius ratio. This is an example of the diagonal relationship. 9(b) Only hydrogen and helium have fewer protons and neutrons in the nucleus than lithium does, and both of them are gaseous elements. Most of the mass of the atom is contained in the nucleus, whereas the size of the atom is determined by the electrons. The sizes of the atoms do not vary over as large a range as the masses (the radius of Li is 1.52 Å, and the radius of Cs is 2.66 Å; the molecular mass of Li is 6.939 amu and of Cs is 132.91 amu. Of all the solid elements, then, lithium has the lowest mass-to-volume ratio). 9(c) An acidic oxide (forms a solution with increased $[H^+]$): $B_2O_3(s) + 3H_2O \rightleftharpoons 2H_3BO_3 \rightleftharpoons 2H^+ + 2H_2BO_3^-$; an amphoteric oxide: $BeO(s) + 2OH^- + H_2O \rightleftharpoons Be(OH)_4^{2-}$, acidic (reacts with base); and $BeO(s) + 2H_3O^+ \rightleftharpoons Be^{2+} + 3H_2O$, basic (reacts with acid); a basic oxide (forms a solution with increased $[OH^-]$): $Li_2O(s) + H_2O \rightleftharpoons 2Li^+ + 2OH^-$. 9(d) Hydrogen gas is evolved when LiH and BeH_2 are added and the resulting solutions are basic. Hydrogen gas is also evolved when diborane is added to water, but the solution becomes acidic. When CH_4 is added to water a small amount dissolves, after which the rest just bubbles out again; there is no reaction. NH_3 forms a basic solution, while HF forms an acidic solution. The equations are all given in the chapter except that for BeH_2: $BeH_2(s) + 4H_2O \rightleftharpoons 2H_2(g) + Be(OH)_4^{2-} + 2H^+$. 9(e) The chemistry of thallium is mostly that of Tl^+, which is about the same size as Rb^+. The difference in their chemistries is due to the formation of Tl^{3+}, which is reasonably stable, whereas no other stable form of rubidium is found. 9(f) Elements 84–88, polonium, astatine, radon, francium, and radium. 9(g) Since the formal charges of all the halogens are the same in the hypohalous acids (+1), this method cannot explain any trend. Electronegativities decrease from chlorine to iodine, consistent with the greater acidity of hypochlorous acid.

10

Energy Changes in Chemistry

10.1 Chemical Reactions and Energy

Thus far we have considered chemical reactions and the chemical equations that represent them, and have focused our attention on the types and amounts of materials consumed and produced in chemical processes. From the law of conservation of mass and the other empirical laws describing weight relations in chemical reactions, the enormously useful rules of chemical stoichiometry have resulted. Having also explored the powerful implications of the law of chemical equilibrium, we might now view chemistry as the science concerned with the quantitative and qualitative aspects of the transformation that matter undergoes in chemical reactions, with emphasis on the amounts and properties of materials. At least one other aspect of chemical processes deserves our attention: the energy changes or transfers that accompany chemical processes.

Two basic reasons foster our interest in energy changes. First, success and development in a technologically oriented, industrialized civilization strongly depend on the ability of the culture to use the available energy sources efficiently. Our usual interest in the combustion of hydrocarbons in an automobile or jet airplane engine is not to produce the carbon dioxide, carbon monoxide, water, and so on that result from the combustion reactions, but to harness and use the energy liberated in the process. This is the aim for a large segment of the chemically oriented industry in a modern country. Clearly, the availability of energy and how it can or should be used to man's best advantage is one of the most important and most controversial issues in the Western world and in Japan today. This subject is likely to retain an important position in national and international policies for many years.

The second reason for our interest in energy is that energy relations in chemical reactions determine the extent to which reactions proceed and the rates at which reactions occur. The latter manifestations falls within the realm of theories of reaction rates, which is **kinetics**, whereas the former, the tendency of a reaction to occur, constitutes one of the applications of the science called **thermodynamics.**

Spotlight

The relation between energy consumption and energy resources in the world in general, and in the United States in particular, was until recently overlooked by most people. Recent events—increases in the price of crude oil from foreign producers and the harsh winter of 1976–1977, for instance—have brought this problem to every thinking citizen's attention. The problem essentially is that the United States and other industrialized countries have developed industrial systems and lifestyles that consume nonrenewable energy resources at a rapidly increasing rate, yet the resources we have been consuming are rapidly being depleted from the earth. The situation is clearly shown in the accompanying table.

This table indicates how the total energy consumed per year has increased so dramatically during the past century, and how the percentage provided by each primary energy source has changed. Until nearly 1850, most of the energy consumed in homes and industry in the United States came from burning wood; and this was at a rate for which the wood could be considered a renewable resource, since new trees could be grown rapidly enough to replace those being used for fuel. By the early 1900s, coal, a nonrenewable resource, had replaced wood as the main fuel. The rate of consumption of coal has remained relatively constant during the past few decades. Petroleum and natural gas, both also nonrenewable resources, have taken over as principal energy sources during the twentieth century. Water power has been of limited importance, initially as simple mechanical water wheels for, say, grinding grain, and now as hydroelectric power in water-powered turbines driving electrical generators. Nuclear energy has begun to exert an influence during the past dozen years.

Energy Consumption in the United States during the Nineteenth and Twentieth Centuries

	Wood		*Coal*		*Petroleum*		*Natural Gas*		*Water Power*		*Nuclear*	
1800	0.06	(99%)[a]	0.0006	(1%)	. . . .		. . . .		0.0001	(0.2%)	. . . .	
1850	0.3	(91%)	0.03	(9%)	. . . .		0.0006	(0.2%)	0.0003	(0.1%)	. . . .	
1900	0.3	(28%)	0.7	(67%)	0.02	(2%)	0.02	(2%)	0.001	(0.1%)	. . . .	
1920	0.2	(10%)	1.6	(74%)	0.3	(13%)	0.09	(4%)	0.009	(0.4%)	. . . .	
1930	0.2	(8%)	1.4	(61%)	0.6	(24%)	0.2	(9%)	0.01	(0.6%)	. . . .	
1940	0.2	(8%)	1.4	(52%)	0.8	(31%)	0.3	(11%)	0.02	(0.7%)	. . . .	
1950	0.2	(6%)	1.5	(39%)	1.4	(37%)	0.7	(19%)	0.04	(1%)	. . . .	
1960	0.1	(2%)	1.2	(24%)	2.1	(42%)	1.5	(31%)	0.05	(1%)	. . . .	
1970	. . . .		1.7	(23%)	3.1	(41%)	2.5	(33%)	0.2	(3%)	0.02	(0.2%)

NOTE: Figures except for percentages are in units of 10^{19} joules.

[a] The number in parentheses represents the percentage that each category contributes to the total energy consumed during the given year.

10.2 Thermodynamics

Scope and Definitions

In using thermodynamics, one focuses attention on some specific portion of the universe. This portion can be real or hypothetical; it is referred to as the **system.** The rest of the universe is considered the **environment,** or the **surroundings.**

In a sense the theoretical formalism of thermodynamics is more mathematics than science; the entire structure of the study is deduced logically, through standard methods of mathematical manipulations from a starting point of three basic postulates (mainly two), called the laws of thermodynamics. Thermodynamics depends in no way on any physical models or pictures we might use to visualize the nature of matter, nor on any submicroscopic theories of structure based on atomic or molecular concepts. Indeed, thermodynamics deals directly only with experimentally measurable, macroscopic properties of matter in bulk.

This macroscopic approach of thermodynamics can be considered both its greatest strength and its greatest weakness. It is a weakness in that classical thermodynamics does not provide any direct information on the exciting subject of the submicroscopic world. It is a strength in that it is not based on submicroscopic structural theories, which tend to change as time passes. Accordingly, the great American scientist, J. Willard Gibbs, who during the late 1800s contributed so much to the development of thermodynamics, once stated: "Certainly one is building on an insecure foundation who rests his work on hypotheses concerning the constitution of matter." The beauty of thermodynamics, then, is that from a very few postulates concerning macroscopic phenomena, the validity of which rest only on their success in agreeing with experimental facts, an enormous amount of information on the macroscopic behavior of matter can be correlated mathematically. Predictions of macroscopic behavior that are highly reliable can be made even on aspects that do not seem, on first glance, to be related directly to these concepts of energy that are at the root of thermodynamics. Furthermore, one need not despair that simply because thermodynamics does not rely on theories of submicroscopic structure it necessarily has no bearing on them. On the contrary, it provides a reliable framework against which to measure the validity of such theories. To be worthy of serious consideration, a theory of submicroscopic structure or behavior must be consistent with the predictions of thermodynamics. Thermodynamics often provides the most critical questions and tests for submicroscopic theories to treat.

The subject that provides the bridge between theories on the molecular level and the kind of properties thermodynamic formalism treats is called **statistical mechanics,** or **statistical thermodynamics.** It involves more sophisticated mathematics than we intend to develop; however, some of the essential ideas are used in Chap. 11.

An important characteristic of thermodynamics is that its focus is on **equilibrium systems,** that is, systems whose properties are stationary in time, and for which certain properties, such as temperature and pressure, must be uniform throughout the entire system.* Also for such systems, all properties must be constant

*The thermodynamics of equilibrium systems is called **equilibrium thermodynamics.** Recent theoretical developments have extended some of the features of thermodynamics to nonequilibrium systems; however, that area is far beyond our current study.

within each phase. Thus, in a three-phase system consisting of an ice cube floating in liquid water in a closed, vapor-filled jar, the temperature must be the same in all three phases for equilibrium to exist, and the density must be uniform throughout the ice; another value of density is uniform throughout the water, and a third density value is uniform throughout the vapor phase. However, the density will not be the same for all three phases. Systems that conform to such specifications are said to exist in **equilibrium states,** which we usually shall abbreviate by using the word **state.** The macroscopic properties used to specify the states of such a thermodynamic system are called the **state variables,** or **state functions.** These properties are usually temperature, pressure, volume, chemical composition, energy, and so forth; they provide all the information needed to reproduce exactly the state of the system in a new situation. That is, when the values of the state variables are known, the condition of the system is entirely specified, and for purposes of characterizing the system there is no need to know anything at all about how the state evolved or was constructed. Thus, the state variables depend only on the present condition of the system as it exists, and not in any way on its history. This property of state variables is extremely important, and is not by any means shown by all characteristics of a system. For example, the volume of a gas held in a balloon is a state variable of the gaseous system; but the shape of the gaseous sample, which is simply the shape of the balloon, is not a state variable since it does not characterize the system in a thermodynamic sense (even though it may in some other sense, like an esthetic context). The shape depends on how the balloon was constructed and how it was handled while it was being inflated, whereas the volume of the balloon depends only on the other thermodynamic, state variables of the systems, for example, temperature and pressure of the gas and amount of gas in the balloon.

Properties of State Functions

From our definition of a state function, it follows that for any process in which the state of the system changes, *the change in a state function must be independent of the method or pathway by which the state is changed.* That is, the change in the state function depends only on the initial and final states. If we use the symbol f to represent the state function and Δf to represent the change in the state function (where Δ stands for "change in"), then this can be summarized by the equation

$$\Delta f = f_{\mathrm{f}} - f_{\mathrm{i}} \tag{10.1}$$

and Δf is independent of path! In Eq. (10.1), Δf is the change of the state function that accompanies a process in which the system is transformed from the initial state to the final state, and f_{i} and f_{f} represent the values of the state function corresponding to the initial and final states. The symbol Δ always stands for "the difference" or the "change in," in the sense "final state minus initial state." So Δf stands for the change in f, the value of f in the final state minus the value of f in the initial state. The independence of path is no trivial property, and only properties of the system that qualify as state variables show it. (We shall see some examples of this distinction later in this chapter.)

Thermodynamics, then, is a science that deals primarily with systems in equilibrium states, and the changes that occur in properties of the system in transformations between these states. Knowing these changes, however, provides no positive information on the pathways by which the changes take place. Special methods are needed for obtaining such information (see Chap. 13).

Work

The act of performing work is usually thought of as making some change in the system against an opposing influence—for example, the movement of an object against some opposing force. Then, for a mechanical system, the amount of work performed in a given process is defined as the product of the force F against which the object must move and the distance s through which the force is applied:

$$w = Fs \tag{10.2}$$

For an electrical system, the work of displacing a change q through a change in electrical potential, or potential difference ΔV, is described by the equation

$$w = q\,\Delta V \tag{10.3}$$

10.3 What Is Energy?

Before we concentrate on the formulation and results of thermodynamics we should review what we mean by the term *energy*. The word itself represents a basic and important concept in science, and one which is unfortunately abstract. There is probably no good definition for the word in the sense of stating what energy is. However, as a working guide for using the concept, we can define it in the sense of how it manifests itself in our experience; in this way energy may be defined as the ability to do work. Applying this definition of energy, we find that it can take several forms. It is often convenient to consider the various forms of energy as belonging to either one of two categories, kinetic energy and potential energy.

Kinetic Energy

Kinetic energy is a property possessed by all bodies in a state of motion and is a direct result of the motion itself. The branch of physics called classical, or Newtonian, mechanics tells us that a body of mass m moving at a speed v has a kinetic energy equal to $\frac{1}{2}mv^2$. It is easy to see qualitatively how a body in motion can have the capacity to perform work simply because it is in motion. Figure 10.1 depicts a situation that demonstrates this property of moving bodies. The moving object (I), by virtue of its momentum mv, which can be at least partially transferred to a second object with which it collides, can displace the second object (II). The distance s of displacement will depend on the forces F restraining the displacement and the momentum of the moving body. The work performed by the moving object on the body it displaces over a small distance δs is then equal to $F\,\delta s$, if the magnitude of the force is assumed constant for this small displacement; this is only approximately true for a spring.

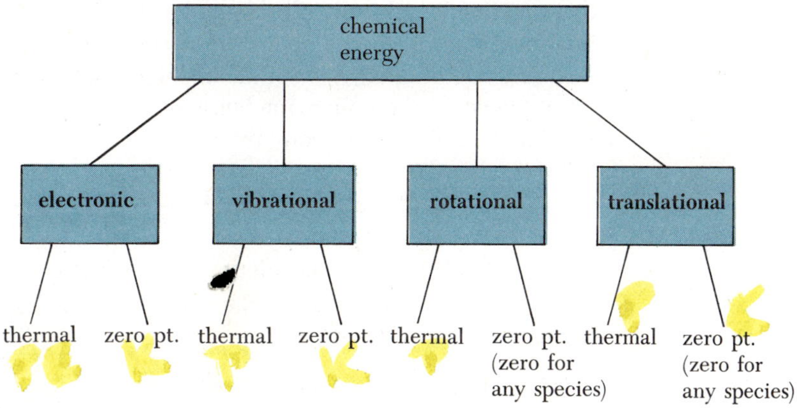

Figure 10.8 The decomposition of electronic, vibrational, rotational, and translational energies into thermal and zero point energy.

10.6 Energy in Transit: Heat and Work

Boltzmann Expression

We are now in a better position to understand the concept of heat relative to other classifications of energy. In Sec. 10.4 we referred to heat as a form of energy in transit, and now we are in a position to see what kind of energy can be involved. Equally important, we shall understand what ideas not to identify with heat.

Suppose that what we call heat (for example, from a flame or hot plate) is added to a system that is constituted in such a manner that it is unable to dissipate the added energy, say by performing work, emitting radiant energy, or the like. Then the only way in which the new energy can be accommodated is by an increase in the chemical energy of the system E; that is the only place for the new energy to go. According to the discussion of Sec. 10.5, the chemical energy is the sum of E_0 and E_{th}. Hence, an increase in E can occur in two ways or by a combination of them: by increasing the zero-point energy or by increasing the thermal energy, or both. The former depends only on the nature of the energy-level diagram, and not on the population of levels within the diagram. Hence E_0 is set by the chemical and physical (that is, solid, liquid, or gas) identity of the substance. The value E_0 is not associated explicitly with temperature changes, so long as such changes do not alter the identity of the system, in other words, alter the nature of the diagram. Increasing E_{th} depends on populating higher energy levels in the various modes of energy available to the molecules; this depends on temperature. The nature of the dependence of energy-level populations, and E_{th}, on temperature is given by an important relation called the Boltzmann equation.

In the late 1800s, this important formula was derived from largely statistical arguments to account for experimental measurements concerned with the "populations" of various energy levels, the number of molecules having each allowed energy. The **Boltzmann equation** gives the ratio of populations of two energy levels, say E_j and E_l, in terms of the absolute temperature and the energy difference $E_j - E_l$. The equation applies if the system is at equilibrium!

$$\frac{n_j}{n_l} = e^{-(E_j - E_l)/kT} \tag{10.5}$$

In this equation, graphically represented in Fig. 10.9, n_j is the population of the E_j energy level (that is, the number of molecular systems having energy E_j for some particular mode of energy), with an analogous meaning for n_l; k is a fundamental constant known as the Boltzmann constant, and it equals 8.315 J deg^{-1}mol^{-1}. For one molecule, Boltzmann's constant is

$$(8.315 \text{ J deg}^{-1}\text{mol}^{-1}) \frac{1}{6.022 \times 10^{23}} \frac{\text{mol}}{\text{molecule}}$$

or 1.381×10^{-23} J deg^{-1}molecule^{-1}. The symbol e stands for the base of natural logarithms, equal to 2.718, and T is the temperature in degrees K. Natural logarithms are defined by the relation

$$e^{\ln z} = z \tag{10.6a}$$

That is, e raised to the power "logarithm of a number z," gives the number z; $\ln z$ is the number to which e must be raised to give z as the result. Hence, if y is a number such that $e^y = z$, then y is the natural logarithm of z.

$$z = e^{\ln z} = e^y$$

$$\ln z = y = \ln(e^y) \tag{10.6b}$$

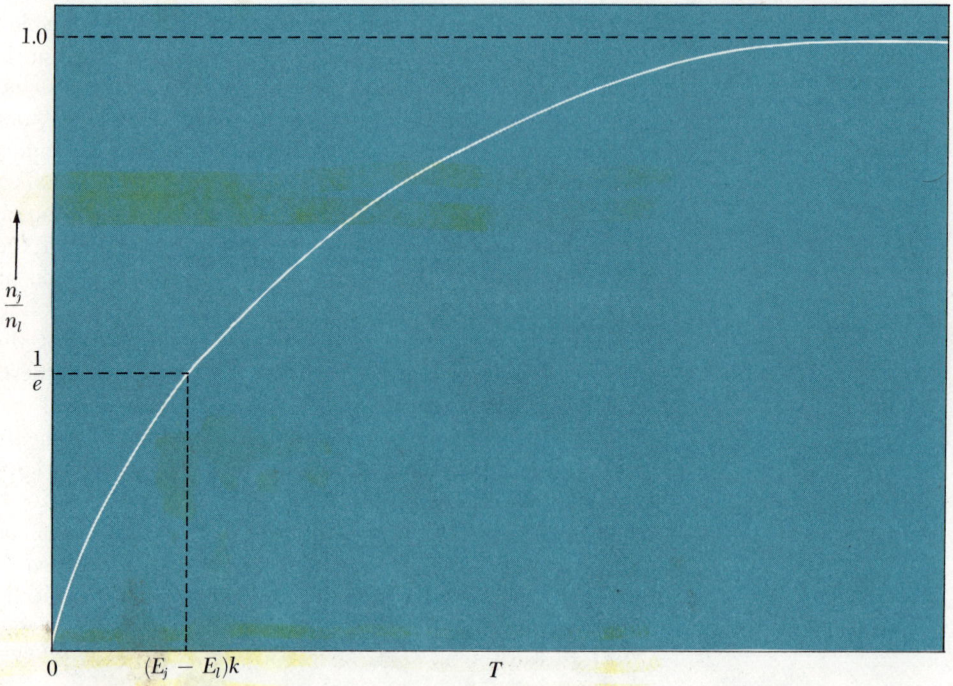

Figure 10.9 Plot of the relative populations of states with energies E_j and E_l (with $E_j > E_l$) as a function of temperature, showing the value for $T = (E_j - E_l)/k$ and the asymptotic limits as T approaches infinity.

The relation between the *natural* logarithm of a number z ($\ln z$) and the base 10 logarithm ($\log z$) is given by

$$e^{\ln z} = 10^{\log z} = z \qquad (10.6c)$$

$$\ln z = 2.303 \log z \qquad (10.7)$$

(For a more complete discussion of logarithms, see Appendix E.)

Equation (10.5) shows that the fraction of molecules occupying a higher energy level increases with the absolute temperature. To see this we observe that if $E_j > E_l$ in Eq. (10.5), the exponent on the right side is negative, rendering the exponential expression less than unity. If T is very small in comparison with $(E_j - E_l)/k$, then the exponent is a large negative number and n_j/n_l is very small (near zero). If T equals $(E_j - E_l)/k$, then the exponent is -1 and $n_j/n_l = e^{-1} = 0.37$; if T is very large (approaching ∞), then the exponent is a very small negative number and n_j/n_l is nearly as large as unity. These relations are demonstrated in Fig. 10.9, where n_j/n_l is plotted against T for a fixed, positive value of $E_j - E_l$. The overall behavior is *the higher the temperature, the higher the populations of the higher levels, and hence the higher the thermal energy.*

EXAMPLE 10.1 Calculate the ratio of the equilibrium populations at 25 °C of two electronic states with energies (energy levels) differing from each other by 1.0×10^{-18} J (per molecule).

SOLUTION Identifying the indices j and l of Eq. 10.5 with the higher and lower energy levels, respectively, we have

$$\frac{n_j}{n_l} = e^{-(1.0 \times 10^{-18} \text{J})/[(1.381 \times 10^{-23} \text{J deg}^{-1})(298 \text{deg})]}$$

$$= e^{-243}$$

Then, using Equation (10.6b), we get

$$\ln\left(\frac{n_j}{n_l}\right) = -243$$

and from Eq. (10.6d),

$$\ln\left(\frac{n_j}{n_l}\right) = 2.303 \log\left(\frac{n_j}{n_l}\right) = -243$$

$$\log\left(\frac{n_j}{n_l}\right) = \frac{-243}{2.303} = -105.5$$

$$\left(\frac{n_j}{n_l}\right) = \text{antilog}(-105.5) = \text{antilog}(0.5 - 106) = 3.2 \times 10^{-106}$$

This extremely small number indicates how very unlikely it is that higher electronic energy levels will be populated at room temperature if there are vacancies in electronic states at lower levels.

We can think of the result of Example 10.1 as a justification for the Aufbau principle introduced in connection with atomic orbitals and molecular orbitals (Chaps.

6 and 7). According to this principle, the ground states of atoms and molecules are accounted for by assigning the available electrons to the available orbitals, starting at the lowest energy and working up. We can now understand why that orbital filling order was adopted for ground states, that is, why we do not assign electrons to higher-energy orbitals so long as vacancies exist in lower-energy orbitals.

EXAMPLE 10.2 Calculate the ratio of the equilibrium populations at 25 °C of two vibrational states with energies differing from each other by 1.0×10^{-20} J.

SOLUTION Using the same approach as in the previous example, we get

$$\left(\frac{n_j}{n_l}\right) = e^{-(1.0\times10^{-20})/[(1.381\times10^{-23})(298)]} = 8.8 \times 10^{-2}$$

This number reflects a strong "preference" for the lower-energy vibrational state, but a much smaller preference than what was shown in Example 10.1 for the electronic case.

EXAMPLE 10.3 Calculate the temperature at which two rotational states with energies differing from each other by 1.0×10^{-22} J would have a ratio of populations of 0.976 at equilibrium.

SOLUTION Using the same approach as in the previous two examples gives

$$\frac{n_j}{n_l} = e^{-(1.0\times10^{-22})/[(1.381\times10^{-23})(T)]} = 0.976$$

$$\ln\left(\frac{n_j}{n_l}\right) = -\left[\frac{1.0 \times 10^{-22}}{(1.381 \times 10^{-23})(T)}\right]$$

$$= \ln(0.976) = 2.303 \log(0.976) = (2.303)(-0.0106)$$

$$T = -\left[\frac{1.0 \times 10^{-22}}{(1.381 \times 10^{-23})(2.303)(-0.0106)}\right] = 297\ ^\circ\text{K}$$

This tells us that at room temperature typical rotational states (with energy spacings about 1.0×10^{-22} J) have comparable populations (a ratio n_j/n_l near unity). This is in stark contrast to the electronic and vibrational cases treated in the previous two examples.

STUDY PROBLEM 10(a)

Calculate the ratio of the population of two vibrational energy states, with energies E_s and E_k, where $E_s - E_k = 2.8 \times 10^{-20}$ J. The system is in equilibrium at 100 °C.

Absorption of Heat

In some processes that are carried out at a fixed temperature, any added heat may go entirely into changing the zero-point energies of the substances making up the system, that is, changing the physical identities and energy-level diagrams of these substances. This is essentially what happens when heat is added to a pure substance

Spotlight

From another form of the Boltzmann relation, the fraction of molecules in a given level, say the j level, is given by the ratio of n_j to the sum of all the n's:

$$\frac{n_j}{\sum_i n_i} = \frac{n_j}{n_1 + n_2 + n_3 + \cdots + n_i} = \frac{e^{-E_j/kT}}{\sum_i e^{-E_i/kT}}$$

where the symbol $\sum_i n_i$ stands for the sum of all the n's, including n_j, and $\sum_i e^{-E_i/kT}$ stands for the sum of all of the expressions $e^{-E/kT}$ for each energy level.

Summation expressions in the general form $\sum_i f_i$ are common in the mathematical areas of chemistry. In such expressions i is called the index and generally takes on a range of integer values, for example, 1, 2, 3, ... The symbol f_i stands for the ith value of the quantity f, for example, f_1, f_2, f_3, ... Then, the symbol $\sum_i$ means "sum what follows for all values of i," for instance, $\sum_i f_i = f_1 + f_2 + f_3 + \cdots$. In the particular equation above, the subscript i refers to a particular level (E_i is the energy of that level and n_i is its population) and the symbol $\sum$ says "sum what follows for all values of the index i." In the left denominator above, this means "sum all n values" and on the right side it means "sum all values of $e^{-E_i/kT}$."

at its melting point or its boiling point. The heat goes into changing the physical state from solid to liquid, or from liquid to gas. Each physical state has its own characteristic energy-level diagram and zero-point energy. In other types of processes, especially those in which the temperature is allowed to increase, at least some of the added heat goes into an increase in the populations of higher energy levels that are associated with increased thermal energy and higher temperature. In some cases all the added heat goes into thermal energy; for example, if one raises the temperature of liquid water from 10 °C to 20 °C, all the added heat goes into an increase in thermal energy. If other avenues are available for using the added energy, then some of the heat may be used in activities like performing mechanical work in an engine or electrical work through an electrical motor. In general, the involvement of these other avenues will influence the amount of heat transferred. In any case, one fact should stand out clearly: *heat* is not synonymous with *thermal energy;* in some cases added heat may affect only the thermal energy, but in other cases it may manifest itself in other ways, such as changes in zero-point energy, or performing work, depending on the details of the process. Thus, heat cannot be a property of the system itself nor of the substances of which the system is composed, for it depends on other factors that hinge on details of the process. For clarity, we should speak of heat only in this sense of a *transfer* of energy, implying that the heat exists only in the process of being transferred. There is no meaning, therefore, in speaking of a system as "containing heat." The term *heat* is often used erroneously in this manner, when what is actually meant is *chemical energy* or *thermal energy.* Heat is analogous to work, which is another means of transferring energy from one object to another. Work, too, exists only at the time the transfer is being performed, and it is meaningless to speak of a system's "containing work," as though work were a property of the system itself. The system may be capable of providing work or heat, but it does not contain them. For many years it was believed erroneously that a system could contain heat, since heat was thought to be a mysterious fluid called **caloric,** which could flow from a hot body to a cold body. But, at the end of the eighteenth century, the caloric theory of heat was shown to be inconsistent with experimental evidence concerning the equivalence of heat and work.*

Heat and Work as Characteristics of an Operation

We now view *work* and *heat* in essentially the same sense, as methods of energy transfer. Work is a mechanical method that depends on some mechanical linkage by which one object or system can operate directly on another, thereby providing it with additional energy. Heat, on the other hand is a method of transfer that requires only a thermal link, that is, a means of contact by which the submicroscopic energy modes of one body can interact with the submicroscopic energy modes of another, allowing energy (heat) to flow from one body to the other. It is the result of overwhelming experimental evidence and the province of the second law of thermodynamics that the direction of spontaneous heat flow is always from a body of higher temperature to one of lower temperature. Thus, heat transfer is closely associated with temperature. With these points in mind, we can arbitrarily class all transfers of energy into or out of a system as either heat or work (in the absence of radiation), a step that will simplify the thermodynamic expressions we shall use. Accordingly, all transfers of energy into a system will be accounted for by a heat flow into the system or work performed on the system or both; and energy transfers from a system to its surroundings will be accounted for by a flow of heat out of the system or work performed by the system (on something outside the system), or both.

If the total chemical energy of the initial state is given the symbol E_i and the total chemical energy of the final state of a process is given the symbol E_f, then the change in E for a process can be written

$$\Delta E = E_f - E_i \qquad (10.8)$$

In these terms we see that transfers of energy into a system will correspond to an increase in E, that is, to a positive ΔE. Thus, a heat flow into a system from its surroundings or work performed on a system constitutes positive contributions to ΔE. Conversely, heat flow from a system to its surroundings and work performed by the system on the surroundings give negative contributions to ΔE; they tend to make E_f algebraically less than E_i.

Two assumptions have been more or less implicit in much of what has been considered so far about energy changes; first, the various forms of energy are, in a sense, equivalent and interconvertible, including energy in transit, or heat and work; and second, energy is conserved—neither created nor destroyed, but merely passed back and forth from one energy category to another. These assumptions, which may seem intuitively reasonable to us at this stage in history, have resulted from an evolutionary scientific development, and are well grounded in a wealth of experimental data.

*It is sometimes believed, erroneously, that since heat can be absorbed or given up by a system, the system must be able to contain it; however, we have seen that heat is only a manifestation of processes, or operations, and is not contained in any substance. In a sense heat can here be compared with light, or other forms of radiation. Under appropriate conditions light can be emitted or absorbed by a substance, but the substance does not *contain* light.

Mechanical Equivalent of Heat

Perhaps the most critical experiments in this development were carried out in 1798 by Count Rumford, minister of war in Bavaria, who was born in America as Benjamin Thompson. In boring brass for cannon barrels, he noted that heat was apparently generated in the process. He convinced himself that the heat was a manifestation of the work performed, and established a direct connection between heat and mechanical work. The establishment of work and heat simply as particular forms of energy was substantiated further by the work of the famous English scientist Sir Humphry Davy (1812), the Danish philosopher L. A. Colding (1843), the German physician J. R. Mayer (1814), and an English brewer with an avocation in science, J. P. Joule (1840–1850). Joule's work was particularly important; not only did he confirm that mechanical work could be converted to heat, but he also determined far more precisely than Rumford how much work had to be carried out to produce a given quantity of heat, namely, the **mechanical equivalent of heat.** In a typical experiment, the mechanical work was performed by a weight-and-pulley arrangement connected to a paddle that turned in a container of water; the resisting force was provided by the friction of the paddles moving through the water The heat produced by the motion of the paddles was measured in units of calories by the temperature rise of the water. A **calorie** is thus defined as the amount of heat required to raise the temperature of one gram of water one degree centigrade. In a variety of experiments with different weight-and-pulley arrangements and different amounts of water, Joule always found the same quantitative relation between the amount of work performed (computed in joules, as defined in Table 1.3) and the number of calories of heat produced. Stated in terms of modern units and higher precision, this relation is

$$1 \text{ cal} = 4.184 \text{ J} = 4.184 \times 10^7 \text{ erg} \tag{10.9}$$

Joule also was able to show that the same result was obtained if the transformation of work into heat was carried out through an intermediate stage of conversion to electrical energy, using a dynamo. He showed experimentally that when one form of energy disappears, an "equivalent" amount appears in some other form; this, if the units are appropriately converted, is seen to be identical to what disappeared. This principle, that *energy may be transformed or converted from one form into another, but neither created nor destroyed,* was also realized about the same time by the famous German physicist H. von Helmholtz. According to this principle, called the **law of conservation of energy,** whenever energy of any particular kind is produced, an equal amount of some other kind has been used up; in other words, energy is conserved. This is one form of stating the **first law of thermodynamics.**

10.7 First Law of Thermodynamics

Chemical Energy as a State Function

To formulate the first law of thermodynamics mathematically, let us suppose we have a system initially in a state A, and it undergoes a transformation to state B. If in this process the system absorbs heat from its surrounding by the amount q (thereby

Spotlight

In 1906, Albert Einstein (1879–1955) expanded the concept of the interconvertibility of different forms of energy and the law of conservation of energy to include mass. Einstein recognized that matter and energy are ultimately interconvertible and that every piece of matter is inherently equivalent to a specific quantity of energy. His famous equation relates the "energy equivalent" of a piece of matter to its mass m in terms of the speed of light, c (3.00×10^8 m/sec). Einstein's famous equation is

$$\text{energy} = mc^2$$

This is a mathematical statement of the interconvertibility of mass and energy, and gives rise to a combined law of conservation of mass-energy. This equation tells us that, if a piece of matter of mass m (say in kg) were converted entirely into energy, the amount of energy would be mc^2 (say in J). Alternatively, if an amount of energy E were converted into mass, the mass that would appear would be $m = E/c^2$. The conversion factor, $c^2 = 9.00 \times 10^{16}$ m^2/sec^2, is so large that very small mass conversions correspond to huge energies. Hence, within the framework of the energy amounts involved in ordinary chemical reactions, the changes in mass involved are negligible from the ordinary chemical point of view. However, in nuclear reactions, in which the elemental identities of atoms are changed, huge amounts of energy are involved (say 10^{12} J/mol). In this case, chemically significant mass changes occur and Einstein's equation relates the mass change quantitatively to the energy change involved.

gaining this amount of energy), and if it performs work on the surroundings by the amount w (thereby losing this amount of energy), then the net gain in energy of the system is given by

$$\Delta E = E_B - E_A = q - w \tag{10.10}$$

This is a mathematical statement of the law of conservation of energy and is considered a statement of the first law of thermodynamics. In this equation we have used the standard convention, which sets q positive when heat is actually absorbed by the system, and w positive when work is performed by the system. Naturally, if heat were given up by the system to the surroundings, q would be algebraically negative, and if work were performed on the system by the surroundings, w would be negative. In all cases, however, Eq. (10.10) is valid as it stands without changing the signs preceding the symbols q and w. It is only the signs of the numbers that q and w represent that depend on the direction of heat and work energy transfer; in applying Eq. (10.10), one must pay strict attention to the signs of these two numbers.

There will be an infinite number of ways of carrying out the transformation of the system from a specific state A to another specific state B, using different combinations of heat and work. For every combination, however, it is always found that $q - w$ equals the value of $E_B - E_A$, i.e., the value of ΔE for the A $\rightarrow$ B process. In other words, ΔE for the transformation from state A to state B is *independent of the path*. But this is the requirement set for the change of a state function. Thus, we conclude that the property E, for which a change is represented by the symbol ΔE, must be a state function. We have loosely referred to this interesting quantity as the chemical energy. It is more customary to designate this variable the **internal energy**; it is a characteristic of the system itself.

Conservation of Total Energy

Let us now note that the heat q absorbed by the system must be given up by the surroundings, so that the heat "absorbed" by the surroundings is $-q$. Also, the work w performed by the system must be performed on the surroundings, so that the work performed by the surroundings must be $-w$. If for the moment we consider the surroundings to be the system of interest, then applying the first law of thermodynamics to the surroundings, we conclude that

$$\Delta E_{\text{surroundings}} = \frac{\text{heat absorbed}}{\text{by surroundings}} - \frac{\text{work performed}}{\text{by surroundings}} = (-q) - (-w) \tag{10.11}$$

Combining this with Eq. (10.10), which refers to the system, we obtain the total ΔE for system plus surroundings (that is, for the universe):

$$\begin{aligned}\Delta E_{\text{total}} &= \Delta E_{\text{system}} + \Delta E_{\text{surroundings}} \\ &= q - w + (-q) - (-w) = 0\end{aligned} \tag{10.12}$$

Equation (10.11) is another way of expressing the first law. It states that ΔE_{total} is zero for any process when both the system and the surroundings are included in a calculation. Thus, no matter what processes may occur, when both systems and surroundings are considered, the total internal energy is unchanged. As Rudolf Clausius stated in 1850 in his presentation of the first law: "Die Energie der Welt ist konstant."

The system and surroundings together constitute the universe, an arrangement that is self-contained, by definition. There are no interactions—no heat flow or work —with anything outside this arrangement, because there is nothing outside this arrangement, since the surroundings include everything in the universe that we haven't initially labelled "the system." Such an arrangement, which interacts with nothing external to itself, is called an **isolated system.** By the same sort of reasoning that led to Eq. (10.12), we can state that ΔE is zero for any process carried out in an isolated system.

Distinction between State Functions and Other Quantities

At this point it is easy to see the distinction between state functions and other physical quantities. For this purpose let us imagine that the system we wish to consider is 1000 ml of water contained in a beaker. The beaker is arranged so that heat can be added to the system by means of an electric heating coil, and work can be performed on the system by an arrangement like an eggbeater. Such an arrangement is shown in Fig. 10.10. With such an apparatus it is possible to increase the internal energy of the water, as indicated by an increase in the temperature recorded by the thermometer, by adding heat through the electrical heater or by performing work with the beater or both. Suppose we start with the water at 10 °C (state A) and wish to raise its temperature to 20 °C (state B). The definition of a **calorie** is essentially the amount of heat required to raise the temperature of one gram of water one degree C. Hence, raising the temperature of 1000 g of water from 10 °C to 20 °C will require (1000 g)(1 cal deg^{-1}g^{-1})(10 deg) or 10,000 cal of energy input. So long as this combination of electrically generated heat q and mechanical work $-w$ is provided (a

Spotlight

One of the current approaches being explored to help solve the "energy" crisis of the United States and the world is nuclear fission. Fission is the fragmentation of a heavy nucleus into smaller nuclear particles. Such processes occur with a substantial increase in energy, as the total mass of the products of the reaction is less than that of the reactants. Hence, according to the Einstein equation ($E = mc^2$) and the principle of the conservation of mass-energy, this decrease in mass occurs with a corresponding increase in energy—energy that is available to perform work, or unfortunately, destruction. An important example of a nuclear fission reaction is the fission of the uranium 235 isotope, induced by bombardment of the $^{235}_{92}U$ nucleus with a neutron (symbol: ^1_0n). Two typical fission reactions of uranium 235 are

$$^1_0n + ^{235}_{92}U \rightarrow ^{137}_{52}Te + ^{97}_{40}Zr + 2^1_0n$$

$$^1_0n + ^{235}_{92}U \rightarrow ^{142}_{56}Ba + ^{91}_{36}Kr + 3^1_0n$$

The second process can be represented symbolically as

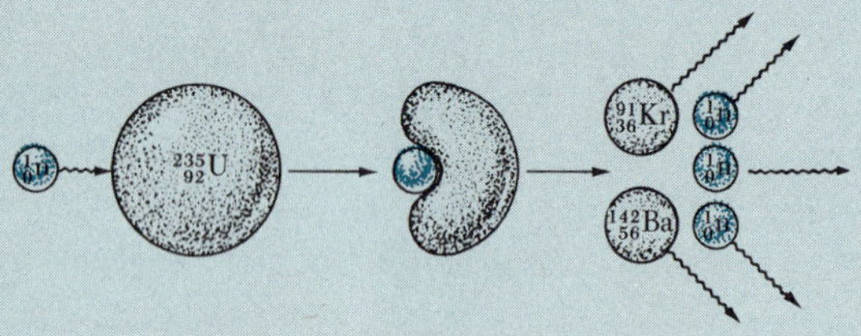

Most of the products of such fission reactions are themselves unstable, and more than 200 different isotopes are produced in the fission of $^{235}_{92}U$.

Each $^{235}_{92}U$ nucleus that undergoes fission produces, on the average, 2.4 neutrons. Thus, more neutrons are produced in the fission than are consumed initiating it. For example, taking the second example above, if one neutron induces a fission, three are produced; these three could induce three more fissions, which would produce nine neutrons, these could then induce nine new fissions, yielding 27 neutrons, which could induce 27 new fissions, and so on. Processes in which the species that initiates the process is also a product are called **chain reactions** (see Chap. 13). A neutron is such a species in the above cases. If the initiator is produced in greater numbers than the number consumed, then the chain reaction has the potential of being explosive. This is the basis of the atomic bomb.

In nuclear reactors, the chain fission process is moderated by inserting materials that absorb some of the neutrons, leaving enough to sustain the chain reaction but not enough to permit it to become explosive. If one computes the change in mass accompanying the fission of one gram-atom of uranium 235 (total mass of products minus total mass of reactants = about -0.25 g per g-atom of $^{235}_{92}U$), and calculates the corresponding energy released by Einstein's equation ($E = (0.00025)(3.00 \times 10^8)^2$), one finds that about 2×10^{13} J is the result. This is a huge amount of energy, more than 2 million times the amount of energy released when the same mass (235 g) of coal is burned. Furthermore, the production of energy by nuclear fission is "clean," in the sense of air and water pollution. However, one does have to consider serious problems, such as the disposal of the radioactive wastes, the possibility of dangerous accidents, and the possibility that the proliferation of a nuclear energy industry will lead to a proliferation of nuclear weapons. Hence, the use of nuclear fission as a standard means of producing energy remains a controversial issue throughout much of the world.

minus sign is before w in this case, as we are interested in work performed on the system, not by it), the state A will be transformed into the state B, with the characteristic increase in internal energy equal to $E_B - E_A$. Yet, depending on the amount of heat introduced or the amount of work performed on the system, q or $-w$ may individually range from 0 to 10,000 cal, so long as their sum equals 10,000 cal.* Thus, while ΔE depends only on the initial and final states, w and q depend on the

*To be more precise in describing the arrangement shown in Fig. 10.10, we should have to take into account that amount of energy that would be used in warming the heater, thermometer, and beater also. This type of correction will be considered in Sec. 10.8.

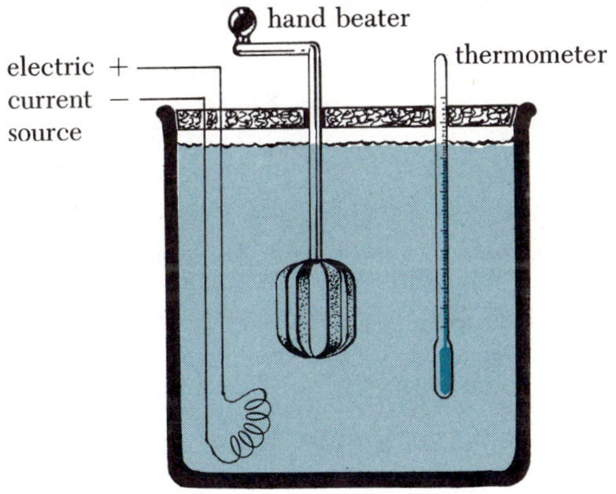

Figure 10.10
Apparatus for studying heat and work, consisting of 1000 g of water in a beaker, fitted with mechanical hand beater, electric heating coil, and thermometer.

detailed path by which the change in state is accomplished. Hence neither w nor q qualifies as a state variable.

EXAMPLE 10.4 Calculate the amount of work in joules that must be performed on the system shown in Fig. 10.10 in order to raise the temperature of the water from 10 °C to 15 °C, if the electrical heater provides 400 cal of heat.

SOLUTION The total amount of energy needed to accomplish this particular change of state is (1000 g)(1 cal $g^{-1}deg^{-1}$)(5 deg) = 5000 cal. Thus, the number of calories that must be provided by work is 5000 − 400 = 4600 cal. Hence, the work performed on the system is

$$-w = (4600 \text{ cal})\left(\frac{4.184 \text{ J}}{\text{cal}}\right) = 19.2 \times 10^3 \text{ J}$$

STUDY PROBLEM 10(b)

Calculate the heat in joules and in calories that must be added (via the electrical heating element) in the system shown in Fig. 10.10 to raise the temperature of the water 20 °C if 12.5×10^3 J work is performed by the beater.

10.8 Heat Capacity

The Concept

The principles described in Sec. 10.7 are based on scientific rigor. However, there is one oversight or oversimplification that was made in Example 10.4; this is associated with the fact that some of the energy transferred into the system from the heat and work would be used in raising the temperature of the various parts of the apparatus as well as the water. The systems to which thermodynamic treatments are directly applicable must be at equilibrium, and all parts of the system thus at the same temperature. If we were to include the thermometer, paddle, and so on as part of the system in Fig. 10.10, then we should have to recognize that these parts of the system would experience the same temperature change as the water and thus

Spotlight

Spectroscopic experiments on sunlight reveal that the composition of our sun is about 73% hydrogen, 26% helium, and 1% other elements. The energy produced in the sun comes from **fusion,** a nuclear reaction in which light nuclei combine to form a heavier nucleus; the combined mass of the "products" is slightly smaller than the combined mass of the "reactants." The overall fusion in the sun is shown as

$$4\,{}^{1}_{1}\text{H} \rightarrow {}^{4}_{2}\text{He} + 2 \text{ positrons}$$

A **positron** is a particle with the mass of an electron but with positive electrical charge. Such processes produce enormous amounts of energy, but require high-energy reactants—for example, high temperatures. The reactants must have enough energy to overcome repulsive forces between nuclei; the temperature of the sun is many millions of degrees C. In a hydrogen bomb, the requisite high temperature for nuclear fusion is generated by fission (basis of the atomic bomb). The fusion is shown as

$${}^{2}_{1}\text{H} + {}^{3}_{1}\text{H} \rightarrow {}^{4}_{2}\text{He} + {}^{1}_{0}\text{n}$$

For more than two decades, vigorous research has been under way. Scientists are trying to learn how to carry out the fusion in a controlled manner, so that the energy released can be harnessed for peaceful purposes. If one computes the change in mass accompanying the above process per gram-atom of consumed ${}^{2}_{1}\text{H}$ or ${}^{3}_{1}\text{H}$ (mass of ${}^{4}_{2}\text{He}$ + mass of ${}^{1}_{0}\text{n}$ − mass of ${}^{2}_{1}\text{H}$ − mass of ${}^{3}_{1}\text{H}$ = −0.0188 g), the energy released can be computed from Einstein's equation:

$$E = mc^2 = (1.88 \times 10^{-5}\text{ kg})(3.00 \times 10^{8}\text{ m/sec})^2$$

$$= (1.88 \times 10^{-5}\text{ kg})\left(9.00 \times 10^{16}\frac{\text{m}^2}{\text{sec}^2}\right)$$

$$= 1.70 \times 10^{12}\text{ J}$$

This is millions of times more energy than what is produced by burning an equivalent mass (5 g) of oil or coal. Fusion has an advantage over fission in the ready availability of large quantities of reactants (whereas known reserves of uranium are limited). Also, fusion is a clean process; the products are stable isotopes; so in contrast with nuclear fission, disposing of large amounts of the radioactive waste products is not a problem. The technical problems involved in carrying out *controlled* fusion are enormous, however. How does one, for example, physically "contain" a sample whose temperature is several million degrees C?

absorb part of the energy introduced into the system during heating and stirring. Alternatively, if we choose to define our system to be only the water, so that all parts of the apparatus are excluded from the system, then we must realize that some of the heat generated electrically would effectively not be passed into the system but used in warming part of the surroundings. Either approach to an analysis of this problem is acceptable; they are equivalent and lead to the same result. Both point to the need for understanding the relation between energy transfers and temperature changes in matter.

The first satisfactory approach to understanding this relation was taken by Joseph Black. In the late 1700s, he carried out a series of experiments and found, contrary to previously held opinion, that different quantities of heat were needed to raise the temperature of the same mass of different materials. He was unable to explain the reason for the differences.

The amount of heat required (and absorbed) when the temperature of one gram of any substance is increased by 1 °C (or °K) is called the **specific heat** of the substance. Of more interest to scientists, generally, is the **molar heat capacity.** This is frequently called simply the heat capacity, and is the amount of heat that must be absorbed by one mole of the substance to raise its temperature 1 °C. This quantity, given the symbol C, usually depends on not only the identity but also the temperature of the substance. Heat capacity is frequently diagnostic of details of submicroscopic structure and motions. For one mole of a given system, if the change in temperature ΔT results from the absorption of a given amount of heat q, then the molar heat capacity can be expressed formally as

$$C = \frac{q}{\Delta T} \tag{10.13}$$

Since q depends on the kind of process by which the temperature is increased by the amount ΔT (Sec. 10.7), it does not qualify as a state function. For the same reason, C also appears not to qualify as a state function; to make C unambiguous, it is necessary to specify the path by which a temperature increase is achieved. A simple and useful pathway to specify is one in which no work of any kind is performed, a specification that necessarily requires the volume of the system to remain constant during the energy transfer. (The need for this requirement will be discussed in the next section.) The heat capacity thus determined is called the **molar heat capacity at constant volume** and is designated by the symbol C_V.

Molar Heat Capacity C_V as a State Variable

With no work performed in a process, w equals zero, and Eq. (10.10) yields

$$\Delta E = q \text{ (constant volume)} \tag{10.14}$$

In this case we can write

$$C = C_V = \left(\frac{q}{\Delta T}\right) = \frac{\Delta E}{\Delta T} \tag{10.15}$$

Since both ΔE and ΔT in constant-volume equation (10.15) depend only on initial and final states, so must C_V; therefore this particular type of molar heat capacity qualifies as a state variable! The relation $C_V = \Delta E/\Delta T$ was derived by considering a constant-volume process. However, the result emerges as a relation among state variables, and is therefore independent of path (not restricted to constant-volume processes). Thus, for any process involving a temperature change ΔT, the corresponding change in the internal energy can be computed by Eq. (10.15) (assuming C_V can be considered constant):

$$\Delta E = C_V \Delta T \tag{10.16}$$

Because ΔE is a state variable, Eq. (10.16) is valid even if the volume of the system is not held constant during the process. If the volume is changed, then neither ΔE nor $C_V \Delta T$ will equal q; but Eq. (10.16) will still hold (assuming that C_V is constant throughout).

Much experimental and theoretical work has focused on the state function C_V, particularly because of its intimate relation to details of the submicroscopic world. In exploring this relation, one can think of the heat capacity as a measure of the capacity of a substance to absorb energy and store it by increasing the population of its higher energy levels—in other words, by increasing its thermal energy. If a substance has many different energy levels accessible (easily populated) at a given temperature, then it can, according to the Boltzmann expression, accommodate much energy while its temperature increases only slightly. In this case, the heat capacity will be large, because there is a place for the energy to go without a big increase in temperature. If, on the other hand, relatively few higher levels are low enough to be accessible at the initial temperature, a relatively large increase in temperature is needed, according to the Boltzmann expression, before there is a place for a substantial amount of energy to be accommodated, that is, a large temperature increase is needed before the higher levels can be populated. This gives a small ratio of q to ΔT, and results in a small heat capacity. Thus, large and complicated molecules with many modes of energy (rotational, vibrational, and so on) and many different accessible levels tend to be associated with large values for the heat capacity. We see therefore how measuring the dependence of C_V on temperature gives valuable insight into submicroscopic energy-level patterns.

EXAMPLE 10.5 If the molar heat capacity of pure liquid water is 18.0 cal $\text{mol}^{-1}\text{deg}^{-1}$ at about 20 °C, how much heat will be absorbed in raising the temperature of 30.0 g of water from 15.0 °C to 25.0 °C? What is ΔE for this process?

SOLUTION From Eq. (10.16), expressed in terms of 1 mol of substance, we can compute ΔE for the process directly. Recalling that the mol wt of H_2O is 18.02 amu, we write

$$\begin{aligned}\Delta E &= C_V \Delta T \\ &= \left[(30.0\ \text{g})\left(\frac{1}{18.02}\ \frac{\text{mol}}{\text{g}}\right)\right]\left(18.0\frac{\text{cal}}{\text{mol deg}}\right)(10.0\ \text{deg}) \\ &= 300\ \text{cal}\end{aligned}$$

If we assume, reasonably, that the change in volume of the system is negligible over this temperature range, then we can apply Eq. (10.14) to determine q:

$$q\ (\text{constant volume}) = \Delta E = 300\ \text{cal}$$

EXAMPLE 10.6 Knowing the value of the molar heat capacity (C_V) of He(g) to be 2.98 cal deg^{-1} mol^{-1}, calculate ΔE and q for a process in which 0.450 mol of helium is heated from 50 °K to 150 °K.

SOLUTION By Eq. (10.16),

$$\begin{aligned}\Delta E &= (0.450\ \text{mol})\left(2.98\frac{\text{cal}}{\text{deg mol}}\right)(100\ \text{deg}) \\ &= 134\ \text{cal}\end{aligned}$$

Since ΔE is independent of path, we were able to compute ΔE for this process by using Eq. (10.16), but we cannot determine q without knowing the path—q is not a state function. The path by which the temperature increase was achieved is not given, and therefore we cannot compute q.

STUDY PROBLEM 10(c)

Using the information given in Example 10.6, compute ΔE and q for a process in which 0.600 mol of He gas is cooled from 110 °C to −20 °C in a sealed glass tube.

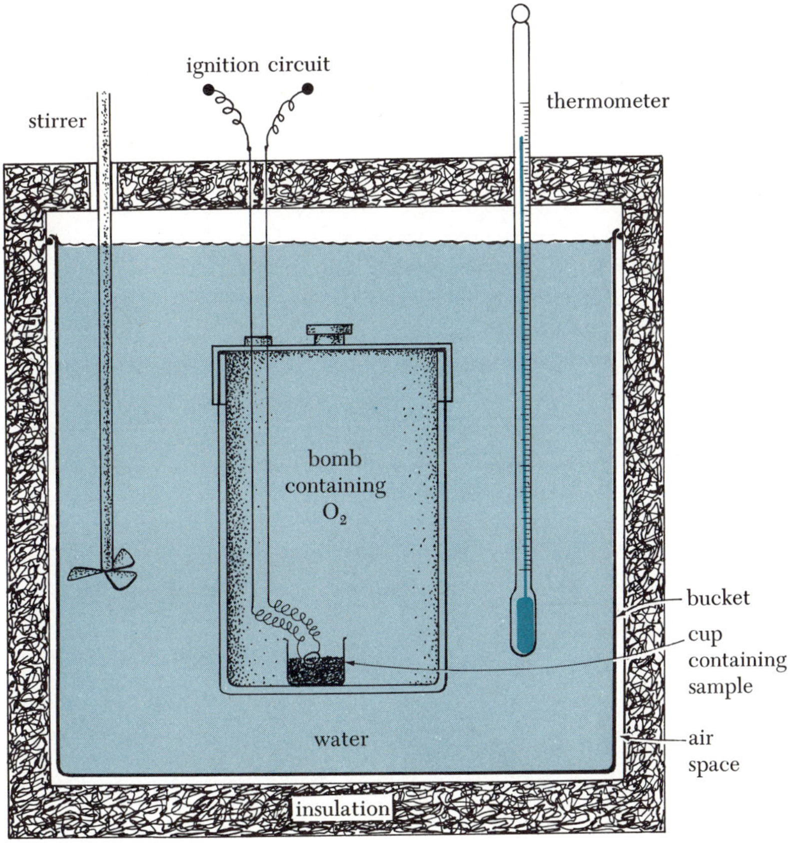

Figure 10.11 Constant-volume calorimeter, known as a bomb calorimeter. Ignition circuit initiates oxidation of samples; the heat from this combustion reaction raises the temperature of the stirred water a measurable amount.

Oxygen-Bomb Calorimeter

Before exploring more of the formalism of thermodynamics, we shall examine the means by which heat transfers are actually measured. This field of experimental science is referred to as **calorimetry** and the basic apparatus is a **calorimeter.** Various types of specialized calorimeters have been used, their designs depending on the detailed characteristics of the processes to be studied. Figure 10.11 is a schematic representation of one type of calorimeter, which is designed to measure q at constant volume for a combustion reaction, in which a substance of interest reacts with elemental oxygen. An example is the combustion of octane,

$$C_8H_{18}(l) + \tfrac{25}{2}O_2(g) \rightleftharpoons 8CO_2(g) + 9H_2O(l)$$

In this apparatus is a fixed-volume vessel, called a bomb, that is charged with a small, weighed sample of the combustible material and O_2 (gas) at a pressure of several atmospheres.* The charged bomb is placed in a water-filled bucket. That is, in

*The term *bomb* is used to describe the oxygen-bomb calorimeter because the oxidation that occurs within it usually occurs very rapidly, and "explosively." However, the muffled "explosion" is confined to the volume of the container, which is therefore not a bomb in the usual sense.

turn, placed in an insulated box, fitted with a stirrer to circulate the water in the bucket, a thermometer to measure its temperature, and an electrical system for igniting the mixture of oxygen and the other substances. All connections (stirrer, thermometer, ignition wires) between the contents in the interior of the insulating box and the surroundings are designed to eliminate or minimize heat flows into the box or out of it. After the contents of the bomb are ignited and the reaction is essentially instantaneously over, the contents of the box are allowed to reach thermal equilibrium, and therefore uniform temperature. The increase in temperature resulting from the heat evolved from the combustion reaction is measured by the thermometer; if the heat capacities of the contents of the box are known, the total q can be calculated. The value will be negative in this case, since heat is evolved by the system in the process. What is frequently done for convenience is to express the total capacity of the internal apparatus in terms of the so-called total water equivalent of the calorimeter. This total is the amount of water that would have the same capacity for absorbing heat as the actual combination of water plus bucket, plus bomb, and so forth.*

EXAMPLE 10.7 A 500-mg sample of liquid benzene, C_6H_6, was combusted in a bomb calorimeter of the type shown in Fig. 10.11 with a total water equivalent of 4147.0 g water. The temperature increase was found to be 1.208 °C. Calculate the amount of heat evolved in the combustion of one mole of benzene at constant volume.

SOLUTION Since the heat capacity of one gram of water is, to four significant figures, 0.9983 $\text{cal}^{-1}\text{deg}^{-1}$, the quantity of heat evolved in the reaction is

$$(0.9983\ \text{cal g}^{-1}\text{deg}^{-1})(4147.0\ \text{g})(1.208\ \text{deg}) = 5001\ \text{cal}$$

For benzene, of molecular weight 78.1 amu, combustion of one mole at constant volume would evolve the following amount of heat:

$$\left(\frac{5001\ \text{cal}}{0.500\ \text{g}}\right)\left(78.1\frac{\text{g}}{\text{mol}}\right) = 781 \times 10^3\frac{\text{cal}}{\text{mol}}$$

$$= \left(781 \times 10^3\frac{\text{cal}}{\text{mol}}\right)\left(4.184\frac{\text{J}}{\text{cal}}\right)$$

$$= 3.27 \times 10^6\frac{\text{J}}{\text{mol}}$$

STUDY PROBLEM **10(d)**

How much heat would be evolved (in calories and in joules) in the combustion of 62.4 g benzene?

STUDY PROBLEM **10(e)**

What would be the temperature increase accompanying the combustion of 341 mg of benzene in the calorimeter referred to in Example 10.7?

*Corrections must be made, in calorimeter experiments, for the heat introduced by the ignition circuit and by the stirring motion; these can be measured in separate experiments.

Spotlight

Coal, petroleum, and natural gas are **fossil fuels.** The basis of this name is that they are in a sense chemical fossils of life that existed on earth millions of years ago. The living forms have been converted to new chemical forms under conditions of high pressure and temperature. Natural gas consists of very simple hydrocarbon compounds, largely methane, CH_4, and ethane, CH_3CH_3. Petroleum consists primarily of large hydrocarbon compounds, comprising hundreds of them in a mixture; C_8H_{18} and C_9H_{20} are two examples of large hydrocarbons. Coal consists of a complicated mixture of very complex molecules, typical of which may be a "chicken-wire" type of structure based on benzene rings. The idealized structure shown below is really a guess, since the true nature of coal is still not well established and is a subject or intensive current research.

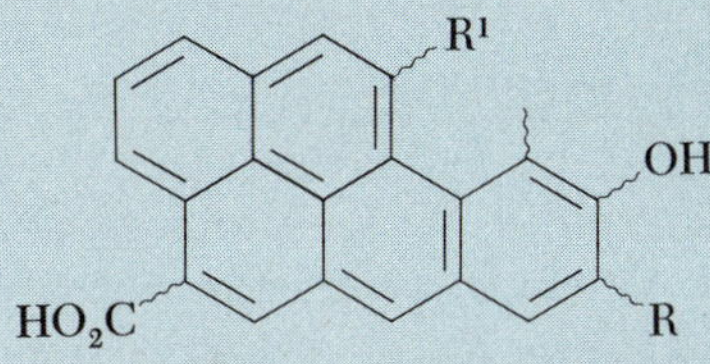

Another fossil fuel is **oil shale,** a type of "rock" that contains a substantial portion of a complex hydrocarbon material called *kerogen* (in amount, say, 10% to 25%). Kerogen is probably the most complex of these fossil fuels, and drastic measures, like heating to 500 °C, are necessary for extracting it from the rock. The accompanying table summarizes the fuel values of typical fossil fuels, which represent the amounts of energy available from combustion of one gram. Also given are the percentages of carbon, hydrogen, and oxygen in these fuels.

	C	*H*	*O*	*Fuel Value* kJ/g
Wood	50%	6%	44%	18
Anthracite coal	92%	3%	2%	31
Petroleum	85%	12%	0	32
Natural gas	75%	24%	0	29
Kerogen (from oil shale)	80%	10%	6%	29

The United States currently accounts for nearly a third of the energy consumption of the entire world. At the present rate of consumption, the U.S. reserves of oil and natural gas will be depleted before the typical reader of this textbook reaches middle age, and worldwide oil and gas supplies will not last much longer.

Coal is the most abundant fossil fuel on earth and constitutes about 80 percent of the fossil fuel reserves in the United States. Most of these reserves are contained in regions that are far from industrial centers; the Rocky Mountain states are one source. Serious environmental questions arise relative to using these coal reserves. Strip mining can leave intolerable scars on the landscape. Also, coal typically contains appreciable amounts of sulfur in various compounds, so that combustion of coal generates SO_2 as a gaseous pollutant. Much scientific and engineering research to minimize these problems is under way.

10.9 Enthalpy

Processes at Constant Pressure

We have just considered the consequences of keeping the volume constant in chemical processes. Such constant-volume conditions are referred to as **isochoric.** Another very common set of conditions under which chemical reactions are carried out calls for maintaining constant pressure. These are referred to as **isobaric** conditions. In such processes the volume generally changes, and one must take account of the work performed by the system or on it by reason of the change in volume. The need for this accounting can be visualized by considering a process in which the system is a gas contained in a piston-cylinder arrangement, as illustrated in Fig. 10.12. The

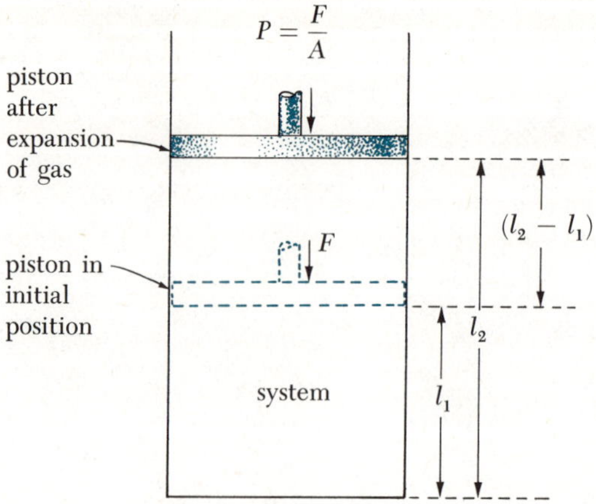

Figure 10.12 Cylinder-piston system for expansion of a gas against a pressure $P = F/A$, where F is the magnitude of the downward force exerted on the piston by some source and A is the area of the piston.

piston has a cross-sectional area A and exerts a constant pressure on the contained gas:

$$P = \frac{F}{A}$$

where F is the magnitude of the downward force that the piston exerts.

Let us imagine that a process is carried out that transforms the system contained in the cylinder from state 1 to state 2, and that in the process the system expands, whereupon its volume changes from $V_1 = (l_1)(A)$ to $V_2 = (l_2)(A)$. That is, as the system expands, it moves the piston a distance $l_2 - l_1$. During this expansion the system necessarily works against the force F; this work is given, according to Eq. (10.2), as $w = F(l_2 - l_1)$. But $l_1 = V_1/A$ and $l_2 = V_2/A$, so that

$$F(l_2 - l_1) = F\left(\frac{V_2}{A} - \frac{V_1}{A}\right) = \frac{F}{A}(V_2 - V_1) = P(V_2 - V_1)$$

Thus

$$w = P\,\Delta V \tag{10.17}$$

where ΔV is the change in volume for the process. Since this is the only form of work that the system can perform in the case under consideration, the heat absorbed in the process, the first law of thermodynamics tells us, is given by

$$q = \Delta E + w = \Delta E + P\,\Delta V \tag{10.18}$$

Thus, *for any process carried out at constant pressure, with only pressure-volume work performed, the heat absorbed by the system equals the change in internal energy plus $P\,\Delta V$ for the process.*

Definition of Enthalpy

It is instructive to rewrite Eq. (10.18) in the following form:

$$q = E_2 - E_1 + PV_2 - PV_1 \tag{10.19}$$

Now for an isobaric process, the pressure is the same at the beginning and end of the process; this can be written $P_1 = P_2 = P$. Hence Eq. (10.19) can be restated as

$$q = E_2 - E_1 + P_2V_2 - P_1V_1 = E_2 - E_1 + (PV)_2 - (PV)_1$$

Rearranging this equation, we obtain

$$q = E_2 + (PV)_2 - E_1 - (PV)_1 = (E + PV)_2 - (E + PV)_1 \qquad (10.20)$$

The symbols $(E + PV)_2$ and $(E + PV)_1$ stand for the values of $E + PV$ for states 2 and 1, respectively. Since expressions of this sort appear frequently in thermodynamic equations for isobaric considerations, it is convenient to give a special name and symbol to the expression $E + PV$; it is called the **enthalpy**, and given the symbol H.* Thus

$$H = E + PV \qquad (10.21)$$

Since E, P, and V are all state variables, this function H constructed from them will be a state function also, and the enthalpy thus qualifies. Equation (10.20) can be rewritten in terms of the new state variable:

$$q = H_2 - H_1 = \Delta H \qquad \text{(constant } P) \qquad (10.22)$$

which equates the heat absorbed in an isobaric process, where only pressure-volume work is performed, to the change in a simple state variable.

The enthalpy is a very "convenient" state variable, since ΔH can be measured directly. It can be measured as heat absorbed in a calorimetry experiment carried out at constant pressure, and not constant volume as described above.

According to the definition of the enthalpy in Eq. (10.21),

$$\Delta H = \Delta(E + PV) = \Delta E + \Delta(PV) \qquad (10.23)$$

Since all the variables involved in this equation are state variables, Eq. (10.23) will hold for any change in state and for all possible paths for these changes. For pathways in which the initial and final pressures are the same, the change in the product PV is due entirely to a change in V. In such cases, $\Delta(PV) = P\,\Delta V$, and Eq. (10.23) becomes

$$\Delta H = \Delta E + P\,\Delta V \qquad \text{(constant } P) \qquad (10.24)$$

Equation (10.24) can be combined with Eq. (10.22) to yield

$$q \text{ (constant pressure)} = \Delta E + P\,\Delta V \qquad (10.25)$$

Combining Eq. (10.25) with Eq. (10.14), we can write

$$q \text{ (constant pressure)} = q \text{ (constant volume)} + P\,\Delta V$$

This equation emphasizes that the heat absorbed in an isobaric process must differ from heat absorbed in an isochoronic process. The reason for this difference is the extra energy transferred into the system or out of it—extra energy resulting from pressure-volume work; the sign of $P\,\Delta V$ depends on the sign of ΔV. In chemical pro-

* *Heat content*, another name sometimes given to this expression, might be considered an unfortunate name. It implies that heat is contained in a substance; and whereas energy can be contained, heat cannot.

Spotlight

The adult human body, engaged in average work, expends about 2500 to 3000 kcal, or 10,000 to 13,000 kJ of energy a day. The energy requirements of our bodies are satisfied by chemical reactions involving the food we eat. The three main categories of food are *protein, fat,* and *carbohydrates.* **Proteins** consist of very large molecules built up of a variety of individual amino acids; food protein is used by the body largely as building material for muscle, skin, hair, and the like; protein is not used primarily as an energy source by a properly functioning body supplied with a balanced diet.

Fats are composed of carboxylic acids, with long hydrocarbon chains attached to the carboxyl group

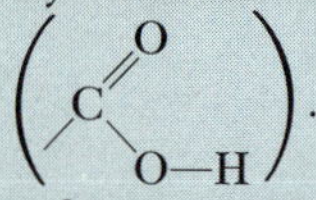

. In the body, as in an oxygen-bomb calorimeter (Fig. 10.11), fat molecules undergo combustion. A typical reaction is

$$2C_{57}H_{110}O_6(s) + 163O_2(g) \rightleftharpoons 114CO_2(g) + 110H_2O(l)$$

for which

$$\Delta H = -1.80 \times 10^4 \text{ kcal} = -7.55 \times 10^4 \text{ kJ}$$

The heat (energy) released by the combustion of one gram of fat, called its fuel value, is on the average about 38 kJ/g, or 9 kcal/g. Of course, the combustion in the body is a very complicated process, involving numerous steps. Fats are ideally suited for the storage of chemical energy in the body because of their high fuel value and because their insolubility in water is conducive to efficient storage.

Carbohydrates are a class of organic compounds with empirical formula CH_2O. They include starches and sugars. Carbohydrates undergo chemical transformations in the stomach, which convert them to glucose. Glucose, known as blood sugar, is soluble in blood and is transported by blood to cells, where it undergoes the following combustion reaction:

$$C_6H_{12}O + 6O_2(g) \rightleftharpoons 6CO_2(g) + 6H_2O$$

for which $\Delta H = -673.0$ kcal $= -2816$ kJ. Because of the rapid transformation of carbohydrates, they are not so suitable as fats for energy storage but more suitable for "quick energy." On the average, carbohydrates have a fuel value 17 kJ/g, or 4 kcal/g.

cesses involving only liquid and solid substances, volume changes are typically small, so that under conditions of constant pressure, ΔH and ΔE are nearly the same. For gaseous systems, however, $P\,\Delta V$ can be significant, and ΔH will differ from ΔE accordingly. To calculate the amount of this difference, one must know the quantitative relations between P, V, and T of the gaseous system, the equation of state of the gas. This is not known in detail for many gases, but as a guideline, we can take the equation of state of an ideal gas, $PV = nRT$, to see the magnitude of the difference (see Sec. 3.6). In this case,

$$H = E + PV = E + nRT$$

Thus, for one mole of an ideal gas at a fixed temperature T, we see that H and E differ by RT. Then, for a process involving a temperature change ΔT on an ideal gas system containing a fixed number of moles n,

$$\Delta H = \Delta E + \Delta(PV) = \Delta E + \Delta(nRT) = \Delta E + nR\,\Delta T \qquad (10.26)$$

Hence, for one mole of an ideal gas, ΔH and ΔE differ by $R\,\Delta T$ for any process in which the temperature change is ΔT, irrespective of path. In applying Eq. (10.26), it is usually convenient to express R as:

$$R = 1.987 \text{ cal mol}^{-1}\text{deg}^{-1} = 8.314 \text{ J mol}^{-1}\text{deg}^{-1}$$

Here R is expressed in different units from our earlier evaluation $R = 0.08205$ liter atm mol^{-1}deg^{-1} (Chap. 3).

EXAMPLE 10.8 Calculate the difference between ΔH and ΔE for a process in which 6.25 g of NH_3 gas is heated from 100 °C to 150 °C.

SOLUTION From Eq. (10.26), we write

$$\Delta H - \Delta E = nR\,\Delta T$$

Then, since the mol wt of NH_3 is 17.03 amu and R is 1.987 cal deg^{-1}mol^{-1},

$$\Delta H - \Delta E = \left[(6.25\text{ g})\left(\frac{1}{17.03}\frac{\text{mol}}{\text{g}}\right)\right]\left(1.987\frac{\text{cal}}{\text{deg mol}}\right)[(423 - 373)\text{deg}]$$

$$= 36.5\text{ cal} = (36.5\text{ cal})\left(4.184\frac{\text{J}}{\text{cal}}\right) = 152.6\text{ J}$$

STUDY PROBLEM 10(f)

Using Eq. (10.26) and the information given in Example 10.6, compute ΔH and q for a process in which 0.450 mol helium is heated from 50 °K to 150 °K.

Heat Capacity at Constant Pressure

Isobaric conditions provide another convenient path to specify for heat capacity measurements. Since q equals ΔH under these conditions (with only pressure-volume work involved), Eqs. (10.13) and (10.22) lead to an expression for the **molar heat capacity at constant pressure,** denoted by the symbol C_p:

$$C_p = \left(\frac{q}{\Delta T}\right)_{\text{const press}} = \frac{\Delta H}{\Delta T} \qquad (10.27a)$$

The heat capacity, C_p, is also a state variable; we note that it is constructed from ΔH and ΔT, both of which come under this classification. Accordingly, in any process for which C_p remains fixed, we can calculate ΔH for the process simply by knowing the temperature change, irrespective of the path, by rearranging Eq. (10.27a):*

$$\Delta H = C_p\,\Delta T \qquad (10.27b)$$

However, only in the special case of constant-pressure conditions will the equal quantities ΔH and $C_p\,\Delta T$ be the same as q for the process.

EXAMPLE 10.9 The heat capacity of solid zinc sulfide is fairly constant over the range 0 °C to 100 °C, with a value about 11.00 cal deg^{-1}mol^{-1}. Calculate q for the process in which 2.70 mol ZnS(s) is heated from 10 °C to 90 °C at constant pressure.

SOLUTION Using Eq. (10.27b), we have for this process, irrespective of path,

$$\Delta H = (2.70\text{ mol})(11.0\text{ cal deg}^{-1}\text{mol}^{-1})(80\text{ deg})$$

$$= 2.38 \times 10^3\text{ cal} = 2.38\text{ kcal}$$

*If C_p varies with temperature, then the change in enthalpy between two temperatures T_1 and T_2 is evaluated most conveniently by integral calculus:

$$\Delta H = H_2 - H_1 = \int_{T_1}^{T_2} C_p(T)\,dT$$

A similar relation exists between ΔE and C_V.

Since the process was carried out at constant pressure, this is also the amount of heat absorbed by the system (Eq. 10.22).

STUDY PROBLEM **10(g)**

Using the information given in Example 10.9, determine the temperature to which one would have to heat a 150-g sample of ZnS at constant pressure for ΔH of the process to equal 900 cal if the initial temperature is 15.0 °C. What is q for this process?

10.10 Heats of Reaction and Hess's Law

Enthalpy Changes of Typical Processes

The ultimate basis for acceptance of any laws in science is their ability to predict and agree with the results of experiments. One of the reasons we can be so confident of the predictive ability of thermodynamics in chemistry is that its predictions unfailingly agree with a great body of thermochemical measurements. Such measurements have been made on a wide variety of processes, using a great variety of calorimeters. Since the experimental chemist usually carries out reactions at constant pressure, the ambient atmospheric pressure, the enthalpy is a very important state variable. Consequently the thermodynamic data most frequently collected in reference tables are enthalpy changes for the processes of interest. For vaporization (vaporizing a liquid), fusion (melting), sublimation (vaporizing a solid), and dissolving a solute, these enthalpy changes are commonly called **heat of vaporization, heat of fusion, heat of sublimation,** and **heat of solution,** respectively. Typically they are reported as molar quantities, referring to one mole of substance. An example is the molar heat of vaporization. For chemical reactions, the ΔH values are frequently called the **heats of reaction:**

$$\text{heat of reaction} = \Delta H = H_{\text{product}} - H_{\text{reactant}} \tag{10.28}$$

The value of ΔH is positive for a reaction if the enthalpy of the products is higher than the enthalpy of the reactants:

$$\Delta H = H_{\text{prod}} - H_{\text{react}} > 0$$

Such a reaction is said to be **endothermic**, and heat is absorbed by the chemical system in a constant-pressure process with pressure-volume work and no other type of work. The value of ΔH for a reaction is negative if the enthalpy of the reactants exceeds the enthalpy of the products:

$$\Delta H = H_{\text{prod}} - H_{\text{react}} < 0$$

For such a reaction heat is released in an isobaric process, and the reaction is referred to as **exothermic.** A sample tabulation of enthalpy changes for a variety of processes is given in Table 10.1.

EXAMPLE 10.10 The C_p value of liquid water from 25 °C to 100 °C is nearly constant at 18.04 cal deg^{-1}mol^{-1} and the C_p value of water vapor from 25 °C to 125 °C is about 8.03 cal

$deg^{-1}mol^{-1}$. Calculate the change in enthalpy of 1 mol water heated at a pressure of 1 atm from the liquid state at 50 °C to the gaseous state at 110 °C.

SOLUTION As enthalpy is a state variable, the overall change in enthalpy is independent of path. Hence, we might as well compute the desired ΔH by viewing the overall change in terms of a pathway that is convenient for us to consider. Such a pathway is made up of three steps: (a) warming liquid water (at 1 atm) from 50 °C to 100 °C; (2) vaporizing water (at 1 atm) at 100 °C; and (3) warming water vapor (at 1 atm) from 100 °C to 110 °C. This is by no means the only pathway we could choose for our calculation; any path that leads from the given initial state to the given final state would give the same result. The pathway we have chosen just happens to be one for which we have all the information at hand for carrying out the calculation straightforwardly. For the first, second, and third steps, we have

$$\Delta H_1 = (18.04 \text{ cal deg}^{-1}\text{mol}^{-1})(50 \text{ deg}) = 902 \text{ cal/mol}$$

$$\Delta H_2 = 9.71 \times 10^3 \text{ cal/mol (from Table 10.1)}$$

$$\Delta H_3 = (8.03 \text{ cal deg}^{-1}\text{ mol}^{-1})(10 \text{ deg}) = 80.3 \text{ cal/mol}$$

$$\begin{aligned}\Delta H_{\text{tot}} &= \Delta H_1 + \Delta H_2 + \Delta H_3 \\ &= 10.69 \times 10^3 \text{ cal/mol} \\ &= (10.69 \times 10^3 \text{ cal})\left(4.184 \frac{\text{J}}{\text{cal}}\right) \Big/ \text{mol} \\ &= 44.7 \times 10^3 \text{ J/mol}\end{aligned}$$

TABLE 10.1 **ΔH° Values for Various Processes**

Process	$\Delta H°$, *kcal/mol*	T, °C
1. $Br_2(l) \rightleftharpoons Br_2(g)$	7.7	59
2. $C(dia) + O_2(g) \rightleftharpoons CO_2(g)$	−94.48	25
3. $C(gr) + O_2(g) \rightleftharpoons CO_2(g)$	−94.03	25
4. $C(gr) + 2H_2(g) \rightleftharpoons CH_4(g)$	−18.0	25
5. $CH_4(g) + 2O_2(g) \rightleftharpoons CO_2(g) + 2H_2O(l)$	−212.8	25
6. $C_2H_5OH(l) + 3O_2(g) \rightleftharpoons 2CO_2(g) + 3H_2O(l)$	−326.7	25
7. $CH_3CO_2H(l) + 2O_2(g) \rightleftharpoons 2CO_2(g) + 2H_2O(l)$	−207.9	25
8. $CO(g) + \frac{1}{2}O_2(g) \rightleftharpoons CO_2(g)$	−67.6	25
9. $CaCO_3(s) \rightleftharpoons CaO(s) + CO_2(g)$	43.4	25
10. $H_2(g) + \frac{1}{2}O_2(g) \rightleftharpoons H_2O(l)$	−68.3	25
11. $\frac{1}{2}H_2(g) + \frac{1}{2}Cl_2(g) \rightleftharpoons HCl(g)$	−22.1	25
12. $H_2O(l) \rightleftharpoons H_2O(g)$	9.71	100
13. $H_2O(s) \rightleftharpoons H_2O(l)$	1.44	0
14. $2NO_2(g) \rightleftharpoons N_2O_4(g)$	−13.9	25
15. $Pb(s) + PbO_2(s) + 2H_2SO_4(l) \rightleftharpoons 2PbSO_4(s) + 2H_2O(l)$	−121.7	25
16. $SO_2(g) + \frac{1}{2}O_2(g) \rightleftharpoons SO_3(g)$	−23.5	25
17. $SO_3(g) + H_2O(l) \rightleftharpoons H_2SO_4(l)$	−31.1	25

STUDY PROBLEM **10(h)**

Calculate the values of ΔH and q for a process in which 10.0 g of water is heated from 90 °C, where it is a liquid, to 120 °C, where it is a vapor, at 1 atm pressure.

Hess's Law

Some of the data presented in Table 10.1 can in practice be obtained by direct measurement. Others, for various reasons, cannot, and were calculated by a popular rule, called Hess's law. This law can be thought of as a manifestation that changes in state functions are independent of path. **Hess's law** states that *just as balanced chemical equations can be manipulated algebraically to provide new chemical equations, the corresponding heats of reaction can be manipulated in the same way*. This law, which Hess stated in 1840, actually predates the universal acceptance of the first law. Hess's law was based primarily on the recognition of a pattern in calorimetric data and was an important influence in the full acceptance of the first law. In illustration of the usefulness of Hess's law, we shall calculate the enthalpy change for the transformation of the graphite form of carbon to the diamond form of carbon:

$$C(gr) \rightleftharpoons C(dia) \qquad \Delta H = ?$$

From Table 10.1, we can obtain the heats of combustion of both C(gr) and C(dia), meaning the enthalpy changes for the combustion reactions with O_2. If we subtract both the equations and the ΔH values, according to Hess's law we obtain

$$\begin{array}{lll} & C(gr) + O_2(g) \rightleftharpoons CO_2(g) & \Delta H = -94.03\ \text{kcal/mol} \\ \textit{minus} & C(dia) + O_2(g) \rightleftharpoons CO_2(g) & \Delta H = -94.48 \\ \hline \multicolumn{2}{l}{C(gr) + \cancel{O_2(g)} - C(dia) - \cancel{O_2(g)} \rightleftharpoons \cancel{CO_2(g)} - \cancel{CO_2(g)}} & \Delta H = -94.03 - (-94.48) \\ & C(gr) \rightleftharpoons C(dia) & \Delta H = 0.45\ \text{kcal} \end{array}$$

There are at least two convenient ways of justifying this procedure. The first is to write the enthalpy change for the combustion of graphite in the form

$$H_{CO_2(g)} - H_{C(gr)} - H_{O_2(g)} = -94.03\ \text{kcal} \tag{10.29}$$

and the enthalpy change for the *reverse* of the combustion of diamond in the same form

$$H_{C(dia)} + H_{O_2(g)} - H_{CO_2(g)} = 94.48\ \text{kcal} \tag{10.30}$$

Since all the symbols in Eqs. (10.29) and (10.30) represent algebraic quantities, we can add these equations to obtain

$$H_{C(dia)} + \cancel{H_{O_2(g)}} - \cancel{H_{CO_2(g)}} + \cancel{H_{CO_2(g)}} - H_{C(gr)} - \cancel{H_{O_2(g)}} = 94.48 - 94.03$$

$$H_{C(dia)} - H_{C(gr)} = 0.45\ \text{kcal} = \Delta H$$

A second way to rationalize the use of Hess's law is in terms of the property of a state function that makes changes in it independent of path. Then, the process of interest, inaccessible as it is to direct measurement, can be thought of as the net result of two steps, each accessible to direct experiments:

$$\begin{array}{lll} \textit{Step 1} & C(gr) + O_2(g) \rightleftharpoons CO_2(g) & \Delta H = -94.03\ \text{kcal} \\ \textit{Step 2} & CO_2(g) \rightleftharpoons C(dia) + O_2(g) & \Delta H = 94.48\ \text{kcal} \end{array}$$

Clearly the total enthalpy change for a stepwise process must be the sum of the enthalpy changes for the steps; a similar statement can be made for any state function. Thus the desired ΔH is given as $-94.03 + 94.48 = 0.45$ kcal.

Hess's law provides a clear demonstration of one of the most powerful aspects of thermodynamics, namely its ability to provide exact information on processes that can be written although they may never have been carried out and possibly never will be. Many of the data that make these predictions possible are derived from combustion experiments as described in Sec. 10.8 and Fig. 10.11.

EXAMPLE 10.11 From the data in Table 10.1, calculate ΔH at 25 °C for the reaction

$$Pb(s) + PbO_2(s) + 2SO_3(g) \rightleftharpoons 2PbSO_4(s)$$

SOLUTION We note that this equation is simply the sum of the equations for process 15 and twice process 17 in Table 10.1. Thus, using Hess's law, we get

$$\Delta H = 2(-31.1) + (-121.7) = -183.9 \text{ kcal}$$

EXAMPLE 10.12 From the data of Table 10.1, calculate the enthalpy change at 25 °C for the process

$$C_2H_5OH(l) + O_2(g) \rightleftharpoons CH_3CO_2H(l) + H_2O(l)$$

SOLUTION From the table, we obtain the following information, considering the reverse of the combustion reaction for CH_3CO_2H.

$$C_2H_5OH(l) + 3O_2(g) \rightleftharpoons 2CO_2(g) + 3H_2O(l) \qquad \Delta H = -326.7 \text{ kcal}$$
$$2CO_2(g) + 2H_2O(l) \rightleftharpoons CH_3CO_2H(l) + 2O_2(g) \qquad \Delta H = 207.9 \text{ kcal}$$

Adding the second equation to the first gives the equation for the process of interest. Thus, according to Hess's law

$$\Delta H = -326.7 + 207.9 = -118.8 \text{ kcal}$$

STUDY PROBLEM **10(i)**

From the data of Table 10.1, calculate the enthalpy change at 25 °C for the reaction

$$CaO(s) + C(gr) + O_2(g) \rightleftharpoons CaCO_3(s)$$

The Need for References and Standards

Examples 10.11 and 10.12 demonstrate how thermodynamics, by applying data on processes that have been studied, makes predictions for processes that may not have been studied. Thus, tables of a few heats of reaction, such as heats of combustion, can be used to compute results for many chemical reactions. However, it would be unwieldy and inefficient to list thermochemical data for a large enough collection of chemical reactions to characterize thousands of compounds thermodynamically. It is far more efficient and convenient to list data for the individual compounds themselves—the enthalpies of compounds themselves, for instance. As discussed in Sec. 10.3, however, an absolute energy is not a straightforward concept, due to problems in assigning the zero of energy. Besides, thermodynamics deals primarily with energy *differences*. Thus, the enthalpy of a substance, or the internal energy (chemical), need only be compared with some reference state to be useful. Furthermore, to be useful, any reference state we might choose should be accessible experimentally, so that direct experimental comparisons can be made.

10.11 Standard Heat of Formation

Standard States

Basically there are two conventions to establish here: first, the definition of the reference state, the need for which was outlined above; and second, the precise specification of the states symbolized by the formulas in balanced chemical equations. The latter item must be settled so that we know precisely what reaction is to be assigned the specific ΔH value that has been reported, and what the conditions are. We adopt the convention that the *symbols in chemical equations to which thermodynamic parameters are assigned stand for species in states called standard states.* For a pure substance, the **standard state** is taken to be the physical state (solid, liquid, or gas) in which a substance exists at a pressure of 1 atm at some specified temperature. Thus, the standard state of pure water at 47 °C is liquid water at 47 °C under a pressure of 1 atm; and if the symbol $H_2O(l)$ appears in a chemical equation for a reaction at 47 °C, the symbol stands for 1 mol H_2O in this particular state. For N_2 in air at 30 °C, the standard state is pure gaseous N_2 at 30 °C and a pressure of 1 atm. For dissolved substances, the standard state is taken to mean the substance present in a concentration of 1 mole per liter under a pressure of 1 atm at a specified temperature.* Thus, by convention, the pressure assigned to standard states is fixed and need not be specified, whereas the temperature may be chosen to fit the requirements of each individual problem. If temperature is not specified, the value 25 °C is usually assumed. The value of a thermodynamic variable assigned to a substance in its standard state is indicated by a superscript zero placed after the symbol for the variable. Thus, the symbol $H^\circ_{Cl_2(g)}$ stands for the enthalpy of one mole of $Cl_2(g)$ in its standard state—at 1 atm and 298.15 °K (since temperature was implied rather than specified). The enthalpy change for a reaction is understood to correspond to the transformation of reactants in their standard state to products in their standard state, in other words, to the difference between the appropriate standard molar enthalpies. Such enthalpy changes are commonly referred to as **standard enthalpy changes,** or **standard heats of reaction,** and are symbolized ΔH°_T, where T stands for temperature. If one wishes to discuss the standard enthalpy change in the reaction

$$2H_2(g) + O_2(g) \rightleftharpoons 2H_2O(l)$$

at 25 °C, this is given as

$$\begin{aligned}\Delta H^\circ_{298} &= 2H^\circ_{298,H_2O(l)} - H^\circ_{298,O_2(g)} - 2H^\circ_{298,H_2(g)}\\ &= -136.6 \text{ kcal}\end{aligned}$$

*More precisely, the standard state of a solute is defined in terms of an activity of 1. See Chaps. 4 and 11 for discussions of concentrations and activities.

where the meaning of $\Delta H°$ is "the change in enthalpy accompanying the process by which 2 mol $H_2(g)$ in its standard state plus 1 mol $O_2(g)$ in its standard state are converted to 2 mol $H_2O(l)$ in its standard state (all standard states at 298 °K in this particular case)."

Reference States

With standard states now defined, it remains to specify reference states, the "chemical survey markers" to which substances in their standard states can be compared quantitatively. The reference state of an element is defined to be the element in its most stable form at 25 °C and 1 atm. Examples are $H_2(g)_{298°,1\,\mathrm{atm}}$, $Br_2(l)_{298°,1\,\mathrm{atm}}$, $Fe(s)_{298°,1\,\mathrm{atm}}$, and $C(gr)_{298°,1\,\mathrm{atm}}$. The enthalpies of substances in their standard states are referred to the enthalpies of these reference states. In a sense, this is like arbitrarily setting to zero the enthalpies of all elements in their reference states.

The main approach here is to consider the enthalpy of any substance in its standard state by how much of an enthalpy change accompanies the formation of the substance, in its standard state, from the corresponding elements in their reference states. The enthalpy change for the formation of a substance from its reference elements is called the **standard heat of formation,** symbolized $\Delta H°_{f,T}$; the subscript T denotes the temperature used in defining the standard state of the substance. If temperature is not specified, it is usually assumed to be 25 °C. Of course, the standard heat of formation of an element in its most stable form at 25 °C is zero by definition, since that is the reference state. An element in some other form at 25 °C or at some other temperature will, however, have a nonzero $\Delta H_f°$ value; this value would be the enthalpy change accompanying the formation of the nonreference state of the element from the reference state.

Let us explore these conventions for a specific example, the oxidation of sulfur dioxide to sulfur trioxide at 125 °C, which is 398 °K:

$$SO_2(g) + \tfrac{1}{2}O_2(g) \rightleftharpoons SO_3(g)$$

For this process,

$$\Delta H°_{398} = H°_{398,\,SO_3(g)} - (H°_{398,\,SO_2(g)} + \tfrac{1}{2}H°_{398,\,O_2(g)}) \tag{10.31}$$

For each species appearing in Eq. (10.31), we can write the enthalpy of the standard state in terms of the enthalpies of the elemental reference states by the concept of heat of formation. For $SO_3(g)$:

$$S(s)_{298} + \tfrac{3}{2}O_2(g)_{298} \rightleftharpoons SO_3(g)_{398}$$

$$\Delta H°_{f,398,SO_3(g)} = H°_{398,SO_3(g)} - (H°_{298,S(s)} + \tfrac{3}{2}H°_{298,O_2(g)}) \tag{10.32a}$$

Rearranging gives

$$H°_{398,\,SO_3(g)} = \Delta H°_{f,\,398,\,SO_3(g)} + H°_{298,\,S(s)} + \tfrac{3}{2}H°_{298,\,O_2(g)} \tag{10.32b}$$

For $SO_2(g)$,

$$S(s)_{298} + O_2(g)_{298} \rightleftharpoons SO_2(g)_{398}$$

$$\Delta H°_{f,\,398,\,SO_2(g)} = H°_{398,\,SO_2(g)} - (H°_{298,\,S(s)} + H°_{298,\,O_2(g)}) \tag{10.33a}$$

Rearranging gives

$$H^\circ_{398,\,SO_2(g)} = \Delta H^\circ_{f,398,\,SO_2(g)} + H^\circ_{298,\,S(s)} + H^\circ_{298,\,O_2(g)} \tag{10.33b}$$

For $O_2(g)$,

$$O_2(g)_{298} \rightleftharpoons O_2(g)_{398}$$

$$\Delta H^\circ_{f,398,\,O_2(g)} = H^\circ_{398,\,O_2(g)} - H^\circ_{298,\,O_2(g)} \tag{10.34a}$$

Rearranging gives

$$H^\circ_{398,\,O_2(g)} = \Delta H^\circ_{f,398,\,O_2(g)} + H^\circ_{298,\,O_2(g)} \tag{10.34b}$$

The symbols on the right and left sides of Eqs. (10.32 to 10.34) represent specific quantities with numerical values; they can be manipulated like any other algebraic variables. Hence, substituting into Eq. (10.31) the expressions for $H^\circ_{398,\,SO_3(g)}$, $H^\circ_{398,\,SO_2(g)}$, and $H^\circ_{398,\,O_2(g)}$, given by Eq. (10.32b), (10.33b), and (10.34b), we obtain

$$\begin{aligned}\Delta H^\circ_{398} &= \Delta H^\circ_{f,398,\,SO_3(g)} + H^\circ_{298,\,S(s)} + \tfrac{3}{2}H^\circ_{298,\,O_2(g)} \\ &\quad - [(\Delta H^\circ_{f,398,\,SO_2(g)} + H^\circ_{298,\,S(s)} + H^\circ_{298,\,O_2(g)}) \\ &\quad + \tfrac{1}{2}(\Delta H^\circ_{f,398,\,O_2(g)} + H^\circ_{298,\,O_2(g)})]\end{aligned} \tag{10.35}$$

After cancelling positive and negative values of the same magnitude, we obtain the following illuminating result:*

$$\Delta H^\circ_{398} = \Delta H^\circ_{f,398,\,SO_3(g)} - (\Delta H^\circ_{f,398,\,SO_2(g)} + \tfrac{1}{2}\Delta H^\circ_{f,398,\,O_2(g)}) \tag{10.36}$$

This equation tells us that the standard enthalpy change of the pertinent reaction at 125 °C is simply the difference between the standard heats of formation of reactants and products at 125 °C. This type of analysis can be carried out for any chemical process; hence, the essence of Eq. (10.36) applies to all chemical reactions at any specified temperature. This is the beauty and convenience of the concept of standard heats of formation: they can be manipulated as if they were absolute enthalpies.† Thus, to calculate the ΔH°_{398} value of Eq. (10.31), the only data needed are the $\Delta H^\circ_{f,398}$ values for the pertinent species. In many cases, these can be obtained directly from tabulated data available in the literature.

Standard heats of formation have been obtained by calorimetric methods, either directly or indirectly. The indirect approaches often involve using Hess's law. A brief sampling of these data are summarized in Table 10.2 for the special case of 25 °C.

Let us now consider a slightly simpler special case, namely, the same reaction as above, except at 25 °C instead of 125 °C. Considerations analogous to the above would lead to an equation similar to Eq. (10.36):

$$\Delta H^\circ_{298} = \Delta H^\circ_{f,298,\,SO_3(g)} - \Delta H^\circ_{f,298,\,SO_2(g)} - \tfrac{1}{2}\Delta H^\circ_{f,298,\,O_2(g)}$$

*From the literature, one finds that the $\Delta H^\circ_{f,398}$ values for $SO_3(g)$, $SO_2(g)$, and $O_2(g)$ have been determined to be −93.11, −69.95, and 0.72 kcal/mol. Hence, for Eq. (10.31), $\Delta H^\circ_{398} = -93.11 + 69.95 - 0.36 = -23.52$ kcal.

†Standard enthalpies of reference states appear with both signs in equations such as (10.35), and are cancelled in deriving equations of the type (10.36), which give ΔH° for a particular reaction of interest. Because of this cancellation, we could, for purposes of calculation, treat these standard enthalpies of reference states as if they were zero.

TABLE 10.2 **Standard Heats of Formation of Compounds at 25 °C**

Substance	H_f°, *kcal mol^{-1}*	*Substance*	H_f°, *kcal mol^{-1}*
$H_2O(l)$	−68.32	$Ag_2O(s)$	−7.31
$H_2O(g)$	−57.80	$CuO(s)$	−38.50
$HCl(g)$	−22.06	$FeS(s)$	−22.72
$HBr(g)$	−8.66	$FeO(s)$	−64.30
$HI(g)$	6.20	$Fe_2O_3(s)$	−196.50
$HNO_3(l)$	−41.40	$Fe_3O_4(s)$	−267.00
$H_2SO_4(l)$	−193.91	$NaCl(s)$	−98.23
$H_2S(g)$	−4.82	$KCl(s)$	−104.18
$CO(g)$	−26.42	$AgCl(s)$	−30.36
$CO_2(g)$	−94.05	$NaOH(s)$	−101.00
$NH_3(g)$	−11.04	$KOH(s)$	−101.78
$NO(g)$	21.60	$AgNO_3(s)$	−29.43
$NO_2(g)$	8.09	$Na_2SO_4(s)$	−330.50
$SO_2(g)$	−70.96	$PbSO_4(s)$	−219.50
$SO_3(g)$	−94.45	$Na_2CO_3(s)$	−270.30
$CuS(s)$	−11.6	$CaCO_3(s)$	−288.45
Methane(g), CH_4	−17.89	$Cu_2S(s)$	−19.0
Ethane(g), CH_3CH_3	−20.24	Acetylene(g), $HC\equiv CH$	54.19
Propane(g), C_3H_8	−24.82	Benzene(l), C_6H_6	11.72
n-Butane(g), C_4H_{10}	−29.81	Naphthalene(s), $C_{10}H_8$	14.40
n-Hexane(g), C_6H_{14}	−39.96	Methanol(l), CH_3OH	−57.04
n-Octane(g), C_8H_{18}	−49.82	Ethanol(g), CH_3CH_2OH	−56.30
Ethylene(g), $H_2C=CH_2$	12.50	Ethanol(l), CH_3CH_2OH	−66.36
$N_2O_4(g)$	2.31	Acetic Acid(l), CH_3CO_2H	−116.40

But the last term on the right side of this equation is zero, by definition, since $O_2(g)$ at 298 °K, 1 atm pressure, is the reference state. Hence, the standard heat of the reaction at 25 °C is given by

$$\Delta H^\circ_{298} = \Delta H^\circ_{f,298,SO_3(g)} - \Delta H^\circ_{f,298,SO_2(g)}$$

Then, using data from Table 10.2, we get

$$\Delta H^\circ_{298} = -94.45 + 70.96 = -23.49 \text{ kcal}$$

Applying the Concepts

As the following examples illustrate, the concepts of reference states and standard states and standard heats of formation provide chemists with a powerful means of systematizing and predicting energy relations in chemical processes.

EXAMPLE 10.13 Calculate ΔH° at 25 °C for the conversion of three moles of acetylene into benzene.

SOLUTION The reaction of interest is

$$3C_2H_2(g) \rightleftharpoons C_6H_6(l)$$

Spotlight

In addition to nuclear energy, other forms of energy are being studied as energy sources to replace fossil fuels. These replacements, it is hoped, will assume an important portion of the energy burden in time to save some of our fossil fuel resources. One important reason for conserving fossil fuels is that they are important for other purposes besides providing energy fuels. Petroleum, for example, is an important resource for the chemical industry, a starting material for such materials as plastics and synthetic fibers. It would be a shame if all the remaining petroleum, coal, and oil shale were burned up for their energy, leaving no convenient raw materials for other worthwhile purposes.

It seems likely that the long-range solution to the energy problem facing all of us will be a combination of primary conservation measures and developing alternative energy sources. Conservation measures will probably include improved home insulation, popularization of efficient transportation methods—trains, buses, small cars with good mileage characteristics—and perhaps other changes in lifestyle. Alternative energy sources will be chosen from nuclear fission and fusion, wind energy, tidal ocean energy, solar farms (raising plants or trees to be burned), geothermal energy (steam from geysers), and direct solar energy.

Solar energy is one of the most promising of the alternative energy sources, and many aspects of it are already practical and coming into widespread use. Each year the earth receives about 2×10^{24} J of energy from the sun. This is about 2.5×10^{4} times the total amount of energy consumed each year by the burning of wood and fossil fuels, nuclear energy, and hydroelectric sources. Hence, whenever the technology of harnessing even a small portion of total solar energy is perfected, much of the world's energy needs can be satisfied by that source.

There are several approaches under research and development for harnessing solar energy. One is the direct conversion of the energy from solar radiation into electricity; this involves solar cells, as used successfully in space satellites. Another is using solar energy to cause a transformation of H_2O into $H_2 + \frac{1}{2}O_2 \cdot H_2(g)$ can then be burned ($H_2(g) + \frac{1}{2}O_2(g) \rightarrow H_2O(g)$) to provide energy as heat. Probably the most straightforward and currently most popular application of solar energy is its collection by solar panels for heating buildings, and for a hot-water supply. The basic approach is shown in the accompanying figure.

In this system, the solar energy is collected in the solar panels from photons coming from a generally southern direction. These photons are absorbed and converted to heat on the black metal sheet, which is usually of copper or aluminum; a glass cover prevents this heat from being dispersed into the air outside, which is colder. The heat collected by the black metal sheets flows into the liquid circulating through the attached metal pipes, which are generally of copper. As this heated liquid—often a mixture of H_2O and antifreeze, $HOCH_2CH_2OH$—flows through the copper coils in the hot water storage tank, heat flows from the liquid into the water of the tank. This water then circulates through the plumbing and "radiators" of a normal hot-water heating system, the heat from the water being transferred to the air in the rooms by the radiators.

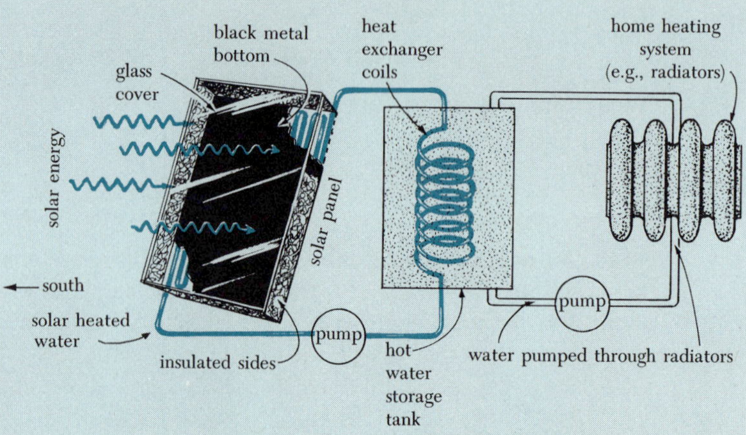

Using the information in Table 10.2, we find

$$\begin{aligned}\Delta H^\circ_{298} &= H^\circ_{f,298\ C_6H_6(l)} - 3H^\circ_{f,298\ C_2H_2(g)}\\ &= 11.72 - 3(54.19)\\ &= -150.85 \text{ kcal}\end{aligned}$$

EXAMPLE 10.14 Calculate ΔH° for the conversion of NO(g) to NO_2(g) at 25 °C.

SOLUTION The reaction of interest is

$$NO(g) + \tfrac{1}{2}O_2(g) \rightleftharpoons NO_2(g)$$

For this process,

$$\begin{aligned}\Delta H^\circ &= \Delta H^\circ_{f,298,NO_2(g)} - (\Delta H^\circ_{f,298,NO(g)} + \tfrac{1}{2}\Delta H^\circ_{f,298,O_2(g)})\\ &= 8.09 - 21.60 - 0 = -13.51 \text{ kcal}\end{aligned}$$

EXAMPLE 10.15 The heat of combustion of liquid toluene, C_7H_8, is -934.50 kcal/mol at 25 °C. Calculate $\Delta H^\circ_{f,298}$ for this compound.

SOLUTION The combustion reaction is

$$C_7H_8(l) + 9O_2(g) \rightleftharpoons 7CO_2(g) + 4H_2O(l) \qquad \Delta H^\circ = -934.50 \text{ kcal}$$

For this process,

$$\Delta H^\circ = 7\Delta H^\circ_{f,298,CO_2(g)} + 4\Delta H^\circ_{f,298,H_2O(l)} - \Delta H^\circ_{f,298,C_7H_8(l)} - 9\Delta H^\circ_{f,298,O_2(g)}$$

Then, using Table 10.2, and the convention for reference states, we get

$$-934.50 = 7(-94.05) + 4(-68.32) - \Delta H^\circ_{f,298,C_7H_8} - 0$$

Rearranging gives

$$H^\circ_{f,298,C_7H_8(l)} = 934.50 - 658.35 - 273.28 = 2.87 \text{ kcal}$$

STUDY PROBLEM **10(j)**

From the information in Table 10.2, compute ΔH° for the formation at 25 °C of *n*-octane from ethylene and H_2(g).

10.12 Bond Energies

Definitions

We have previously defined bond energy as the energy required to break a specific type of covalent bond in a mole of gaseous molecules to form the corresponding gaseous atoms. Having now developed a well-defined formalism for energy changes, we can be more specific about the energy referred to in that definition. Restating this definition in terms of a thermodynamic state variable, the **bond energy** is the **standard enthalpy change** of the reaction in which one mole of a specific type of bond in a gaseous compound is dissociated into gaseous atoms at 25 °C. Thus, for HI, the bond energy ΔH_b is the standard enthalpy change for the following reaction:

$$HI(g) \rightleftharpoons H(g) + I(g)$$

$$\Delta H_b = \Delta H^\circ_{f,298,H(g)} + \Delta H^\circ_{f,298,I(g)} - \Delta H^\circ_{f,298,HI(g)} \qquad (10.37)$$

In these terms, we can see that the ΔH_f value shown in Fig. 7.8 is actually ΔH_f°, the standard heat of formation of HI(g) at 25 °C. Similarly, the energy changes of the individual steps shown on the right side of that figure can now be identified as standard enthalpy changes for the individual steps.

In Fig. 7.8, the value $\frac{1}{2}D$ for the process $\frac{1}{2}H_2(g) \rightleftharpoons H(g)$ is simply ΔH° for that process, which incidentally is half the bond energy of the H–H bond. Similarly, $\frac{1}{2}D'$, half the dissociation energy of $I_2(g)$, is simply half the bond energy of the I–I bond. The quantity $\frac{1}{2}\Delta H_s$ is half the standard enthalpy of sublimation of $I_2(s)$, that is, half the standard enthalpy change for the process

$$I_2(s) \rightleftharpoons I_2(g)$$

Analogous identifications in terms of standard enthalpy changes can be made for Figs. 7.5 and 7.6.

For simple diatomic molecules, the bond energy is rigorously equal to the standard enthalpy change for the dissociation of the molecule into atoms; Eq. (10.37), for example, is rigorously exact. For molecules more complex than diatomic ones, however, the idea of a bond energy is not quite so straightforward to relate to a standard enthalpy change. Consider, for example, methane, CH_4. There are four C–H single bonds. There is no reason to believe that the standard enthalpy for dissociation of the "first" C–H bond

$$CH_4(g) \rightleftharpoons CH_3(g) + H(g) \qquad \Delta H_1^\circ$$

is exactly the same as that for dissociation of the "second" C–H bond

$$CH_3(g) \rightleftharpoons CH_2(g) + H(g) \qquad \Delta H_2^\circ$$

or the third, or the fourth. Furthermore, we cannot rigorously assume that the enthalpy change for dissociation of a C–H bond is the same for CH_4 as for some other compound, say CH_3OH. Hence, certain assumptions must be made in order to maintain the concept of bond energies at a relatively simple and workable level of complexity. With methane, for example, the bond energy of the C–H bond is *defined* to be the average value of the four C–H dissociation enthalpies. That is

$$\Delta H_b = \tfrac{1}{4}\Delta H^\circ$$

where ΔH° is the standard enthalpy change for the overall reaction

$$CH_4(g) \rightleftharpoons C(g) + 4H(g) \tag{10.38}$$

Experimentally this ΔH° value at 25 °C is found to be 397.2 kcal = 1662×10^3 J. Hence, the C–H bond energy is found to be 99.3 kcal/mol = 4.15 kJ/mol.

To derive a bond energy for ethane,

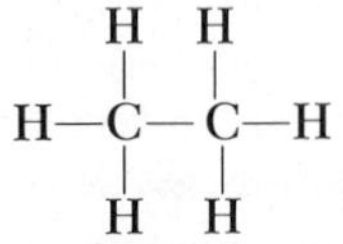

one can consider the complete dissociation of this gaseous substance into gas-phase atoms:

$$CH_3{-}CH_3 \rightleftharpoons 2C(g) + 6H(g) \qquad \Delta H^{\circ(\text{dis})}_{C_2H_6} \tag{10.39}$$

Within the concept of bond energies, this $\Delta H°$ value can be related to the bond energies of one C–C bond and six C–H bonds by the following approximate equation:

$$\Delta H°^{(dis)}_{C_2H_6} \cong \Delta H_{b,C-C} + 6\Delta H_{b,C-H} \tag{10.40}$$

Rearranging gives a value for the C–C bond energy

$$\Delta H_{b,C-C} \cong \Delta H°^{(dis)}_{C_2H_6} - 6\Delta H_{b,C-H} \tag{10.41}$$

Then, from the experimentally determined value of $\Delta H°^{(dis)}_{C_2H_6}$ for Eq. (10.39), and from the value of $\Delta H_{b,C-H}$ determined as indicated above, the value of $\Delta H_{b,C-C}$ is computed directly from Eq. (10.41).*

By stepwise reasoning of the kind just outlined, bond energies have been determined for a wide variety of covalent bond types. Some of these were summarized in Sec. 7.2. This collection is expanded and reexpressed in kilojoules in Tables 10.3 and 10.4.

EXAMPLE 10.16   From information in Sec. 7.2 and Table 10.2, calculate the bond energy for a carbon-carbon double bond.

SOLUTION   From the structure of ethylene, $\begin{array}{c} H \\ \diagdown \\ \end{array} C = C \begin{array}{c} H \\ \diagup \\ \end{array}$, we can consider the total energy of

dissociation of gaseous ethylene into gaseous atoms as follows:

$$C_2H_4(g) \rightleftharpoons 2C(g) + 4H(g) \qquad \Delta H°^{(dis)}_{C_2H_4}$$

$$\Delta H°^{(dis)}_{C_2H_4} \cong \Delta H_{b,C=C} + 4\Delta H_{b,C-H} \tag{A}$$

Since the value of $\Delta H°^{(dis)}_{C_2H_4}$ is not directly available to us from these tables, we take advantage of the fact the $\Delta H°_f$ values for both CH_3CH_3 and $H_2C=CH_2$ are given in Table 10.2. These two species are related by the following chemical equation:

$$H_2C=CH_2(g) + H_2(g) \rightleftharpoons CH_3CH_3(g) \tag{B}$$

for which the $\Delta H°$ value is given by   *product − reactants*

$$\Delta H° = \Delta H°_{f,CH_3CH_3} - \Delta H°_{f,CH_2=CH_2} - \Delta H°_{f,H_2} = -20.24 - 12.50 - 0$$
$$= -32.74 \text{ kcal} \tag{C}$$

In Eq. (C) the reference states for the heats of formation are the stable elemental substances at 25 °C. We could write an analogous equation based on some other arbitrary choice of reference states,

$$\Delta H° = -32.74 \text{ kcal} = \Delta H°'_{f,CH_3CH_3} - \Delta H°'_{f,H_2C=CH_2} - \Delta H°'_{f,H_2} \tag{D}$$

where the primes in Eq. (D) indicate that some alternative reference state has been used. Now, let this other choice of reference state be gaseous *atoms* at 25 °C. For example,

$$\Delta H°'_{f,H_2C=CH_2} = H°_{298,H_2C=CH_2} - 2H°_{298\,C(g)} - 4H°_{298\,H(g)} \tag{E}$$

---

*Equations (10.40) and (10.41) and any other equations relating bond energies to actual enthalpy changes for polyatomic molecules are only approximately valid. This is true because they are based on the assumption that a given type of bond has exactly the same bond energy in all molecules; and this assumption can be only approximately valid.

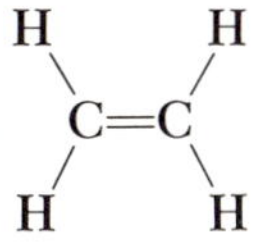

TABLE 10.3 **Bond Energy Values for Single Bonds**

H—H	436 kJ mol^{-1}	B—O	460 kJ mol^{-1}
Li—H	245	B—S	276
Na—H	202	B—F	582
K—H	182	B—Cl	388
Rb—H	167	B—Br	310
Cs—H	175	C—C	344
B—H	331	C—Si	290
C—H	415	C—N	292
Si—H	295	C—S	259
N—H	391	C—O	350
P—H	322	C—F	441
As—H	245	C—Cl	328
O—H	463	C—Br	276
S—H	368	C—I	240
Se—H	277	Si—Si	187
Te—H	241	Si—O	432
H—F	563	Si—S	227
H—Cl	432	Si—F	590
H—Br	366	Si—Cl	396
H—I	299	Si—Br	289
Li—Li	111	Si—I	213
N—N	159	Ge—Ge	157
O—O	138	Ge—Cl	208
N—O	222	Sn—Sn	143
Cs—Cs	45	N—F	270
B—B	225	F—F	156
B—C	312	Cl—Cl	243

TABLE 10.4 **Bond Energies for Some Multiple Bonds**

C=C	615 kJ mol^{-1}	C=S	477 kJ mol^{-1}
N=N	418	C≡C	812
O=O	402	N≡N	946
C=N	614	C≡N	890
C=O	725	P≡P	490

But an alternative $\Delta H_f^{\circ\prime}$ value defined in this way is simply the standard enthalpy change for formation of a gaseous species from gaseous atoms, which is simply the negative of the enthalpy of the dissociation reaction, that is, $-\Delta H^{\circ(\text{dis})}_{C_2H_4}$ of Eq. (A). Then the $\Delta H_f^{\circ\prime}$ values in Eq. (D) can be estimated from bond energies, as we have done above, say in Eq. (10.40):

$$\Delta H^{\circ\prime}_{f,CH_3CH_3} = -\Delta H_{b,C-C} - 6\Delta H_{b,C-H}$$

$$\Delta H^{\circ\prime}_{f,H_2C=CH_2} = -\Delta H_{b,C=C} - 4\Delta H_{b,C-H} = -\Delta H^{\circ(dis)}_{C_2H_4}$$

$$\Delta H^{\circ\prime}_{f,H_2} = -\Delta H_{b,H-H}$$

Substituting these equations into Eq. (D), using bond energies from Sec. 7.2,

$$-32.74 \text{ kcal} = -\Delta H_{b,C-C} - 6\Delta H_{b,C-H} - (-\Delta H_{b,C=C} - 4\Delta H_{b,C-H} - \Delta H_{b,H-H})$$

$$= -83 \text{ kcal} - (6)(99 \text{ kcal}) + \Delta H_{b,C=C} + (4)(99 \text{ kcal}) + 104 \text{ kcal}$$

$$\Delta H_{b,C=C} = (-33 + 83 + 594 - 396 - 104)\text{kcal} = 144 \text{ kcal}$$

$$= (144 \text{ kcal})\left(4.184\frac{\text{kJ}}{\text{kcal}}\right) = 602 \text{ kJ}$$

This example shows the approximate nature of the bond order concept for polyatomic compounds. The value obtained in Example 10.16 is a little different from what is given in Table 10.4 for C=C, because the latter is an average taken from several calculations of the general type outlined in Example 10.16.

STUDY PROBLEM **10(k)**

From information in Sec. 7.2 and Table 10.2, determine the bond energy for a carbon-carbon triple bond.

Applications

The concept of bond energy is useful in estimating the energetics of chemical reactions. Calculations based on bond energies are only estimates (often very good ones), because they are based on the fragile assumption that the bond energy of a given type of bond remains constant for all compounds containing that type of bond. If standard heats of formation are known for the species of interest, then exact calculations of the type represented by Examples 10.13–10.15 can be carried out, so one does not have to resort to using bond energies. If the pertinent ΔH_f° values are not known, then the approach of bond energies serves a useful predictive need. The following examples show the value and the limitations of the bond energy approach.

EXAMPLE 10.17 Estimate the standard molar heat of formation of NF_3(g).

SOLUTION We seek to calculate the ΔH° value for the following reaction:

$$\tfrac{1}{2}N_2(g) + \tfrac{3}{2}F_2(g) \rightleftharpoons NF_3(g)$$

NF_3(g) has an enthalpy of formation from gaseous nitrogen and fluorine atoms given by N—F bond energy:

$$\Delta H'^{(\text{from atoms})}_{f,NF_3} = -3\Delta H_{b,N-F}$$

The stable elements from which NF_3 would be formed, N_2(g) and F_2(g), have enthalpies of formation from the corresponding atoms that are simply the negative of the bond energies (the bond energy of a triple bond for N_2 and a single bond for F_2):

$$\Delta H'^{\text{(from atoms)}}_{f, N_2} = -\Delta H_{b, N \equiv N}$$

$$\Delta H'^{\text{(from atoms)}}_{f, F_2} = -\Delta H_{b, F-F}$$

Then,

$$\Delta H^{\circ}_{f, NF_3} = \Delta H'^{\text{(from atoms)}}_{f, NF_3} - \tfrac{1}{2}\Delta H'^{\text{(from atoms)}}_{f, N_2} - \tfrac{3}{2}\Delta H'^{\text{(from atoms)}}_{f, F_2}$$

$$= -3\Delta H_{b, N-F} - (-\tfrac{1}{2}\Delta H_{b, N \equiv N} - \tfrac{3}{2}\Delta H_{b, F-F})$$

$$= -3 \times 270 \text{ kJ} + (\tfrac{1}{2})(946 \text{ kJ}) + (\tfrac{3}{2})(156 \text{ kJ})$$

$$= -103 \text{ kJ}$$

EXAMPLE 10.18 Estimate the standard enthalpy change for the following reaction:

$$\begin{array}{c} H \\ \diagdown \\ C=O(g) \\ \diagup \\ H \end{array} + 2Cl_2(g) \rightleftharpoons 2HCl(g) + \begin{array}{c} Cl \\ \diagdown \\ C=O(g) \\ \diagup \\ Cl \end{array}$$

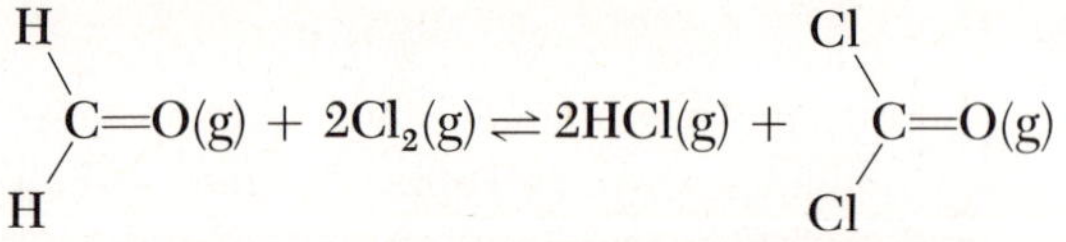

phosgene

SOLUTION As these species are not all represented in Table 10.2, we resort to using bond energies. Hence, we write ΔH° for this reaction in terms of heats of formation from the gaseous atoms:

$$\Delta H^{\circ} = \Delta H'^{\text{(from atoms)}}_{f, Cl_2CO} + 2\Delta H'^{\text{(from atoms)}}_{f, HCl} - \{\Delta H'^{\text{(from atoms)}}_{f, H_2CO} + 2\Delta H'^{\text{(from atoms)}}_{f, Cl_2}\}$$

Then, estimating these enthalpies of formation from gaseous atoms by means of bond energies, we have

$$\Delta H^{\circ} = (-\Delta H_{b, C=O} - 2\Delta H_{b, C-Cl}) - 2\Delta H_{b, HCl}$$
$$- [(-\Delta H_{b, C=O} - 2\Delta H_{b, C-H}) - 2\Delta H_{b, Cl-Cl}]$$

Using values from Tables 10.3 and 10.4 we get

$$\Delta H^{\circ} = (-725 \text{ kJ} - 2 \times 328 \text{ kJ}) - 2 \times 432 \text{ kJ}$$
$$+ (725 \text{ kJ} + 2 \times 415 \text{ kJ}) + 2 \times 243 \text{ kJ}$$

$$= -204 \text{ kJ} = (-204 \text{ kJ})\left(\frac{1}{4.184} \frac{\text{kcal}}{\text{kJ}}\right) = -49 \text{ kcal}$$

STUDY PROBLEM **10(l)**

Estimate the standard enthalpy change for the reaction

$$2H_2C{=}CH_2(g) \rightleftharpoons \begin{array}{c} H_2C-CH_2 \\ | \quad | \\ H_2C-CH_2 \end{array}$$

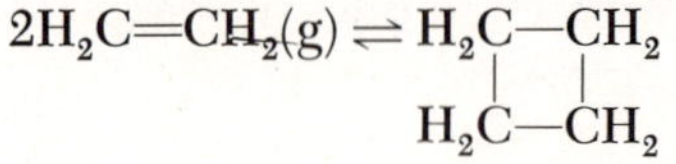

SUMMARY

Energy changes and energy transfers occur in chemical processes. The branch of chemistry that deals with this type of topic is thermodynamics, which focuses attention on a specific portion of the universe called the system. Of especial importance in thermodynamics are state variables; changes in state variables are independent of the pathway by which the system changes from one state to another.

Energy is expressed in various categories, for example, kinetic energy and potential energy. One important type of energy is chemical energy, the energy associated with the chemical form of an element or combination of elements. Chemical energy comprises electronic energy, vibrational energy, rotational energy, and translational energy, this order reflecting the successively smaller sizes of the spacings between the four types of energy levels; the translational levels are continuous. Another way in which chemical energy is considered is as the sum of zero-point energy and thermal energy. Zero-point energy changes when the chemical or physical state of the system changes; thermal energy, which is described by the Boltzmann equation, changes with a change in temperature.

Heat, like work, is seen to be a form of energy in transit, a way in which energy can be transferred to or from the system. The first law of thermodynamics, which is essentially a statement of the law of conservation of energy, relates the heat absorbed by the system in some process to the change in internal energy of the system and the work performed by the system for that process. For processes carried out at constant volume, where no work is performed, the heat absorbed by the system equals its change in internal energy (chemical). From processes carried out at constant pressure (in which only pressure-volume work is performed), the heat absorbed by the system equals the change in a state variable called enthalpy. A calorimeter is used to measure the amount of heat absorbed or released by the system in a chemical reaction. Hess's law, which is merely a manifestation of enthalpy as a state variable, permits one to combine enthalpy changes for the steps of multistep processes to obtain the ehthalpy change for the overall process.

To be precise and quantitative, thermodynamics requires the specification of standard states. Defining reference states provides a convenient means for computing differences of state variables. One such difference is the standard heat of formation of a substance; such quantities can be manipulated as if they were absolute enthalpy values in calculations of enthalpy differences for chemical processes.

The bond energy is defined as the energy required to rupture a covalent bond in a gaseous molecule, producing molecular fragments, for example, gaseous atoms from diatomic molecules. While bond energies are derived from and related to thermodynamic measurements, using bond energies to estimate "heats of reaction" involving complex molecules is based on an oversimplified assumption of bonding. Nevertheless, bond energies often provide very useful estimates of enthalpy changes in chemical reactions.

In examining the principle of the conservation of energy, we have emphasized how this principle relates to chemical processes and the energy transfers that accompany these processes. The formalism we have developed is essentially a method of bookkeeping—a recipe for keeping track of deposits and withdrawals of energy to and from a system.

This recipe has told us nothing about what types of processes will occur and what types will not occur; it has told us only what energy transfers must accompany any given process if that process does occur. On the basis of analogies in superficially similar mechanical situations, as with a ball rolling down a hill, we might be tempted at this point to predict that only processes in which ΔE (or perhaps ΔH) decreases will occur spontaneously in nature. For a period during the nineteenth century, some scientists held such views. However, overwhelming experimental evidence exists that shows this prediction clearly to be false; many spontaneous reactions are known for which $\Delta E°$ or $\Delta H°$ or both are positive. The first law is simply not equipped to handle spontaneity; that is, it is not able to tell us what determines the value of an equilibrium constant for a particular process.

STUDENT CHECKPOINTS

After studying this chapter, the student should be able to:

1. Define and specify a system.
2. Distinguish between state variables and other variables.
3. Define *energy, kinetic energy,* and *potential energy.*
4. Define *work* and how to compute work for simple mechanical processes.
5. Explain what chemical energy is, and the general magnitude of the energy gaps for its four contributions—electronic, vibrational, rotational, and translational.
6. Distinguish zero-point energy and thermal energy, and explain how the latter is related to the Boltzmann equation.
7. Discuss the transfer of energy as heat in terms of zero-point energies and thermal energies.
8. Relate the first law of thermodynamics and the law of conservation of energy.
9. Compute the change in internal energy of a system for a process in which the system that performs a known amount of work absorbs a known amount of heat.
10. Recognize E as a state variable, and calculate ΔE for a given process, knowing C_V and the change in temperature.
11. Describe the functioning of a calorimeter, and calculate the heat absorbed or released in a chemical process from data on a calorimetry experiment with a fully characterized calorimeter (that is, known "water equivalent").
12. Define *enthalpy* and discuss why enthalpy qualifies as a state variable.
13. Compute the enthalpy change from a knowledge of ΔE and a volume change in a constant-pressure process, and relate the change to the heat absorbed or released by the system.
14. Explain the need for reference states and standard states in thermodynamic formalism.
15. Apply Hess's law.
16. Discuss the meaning and practical utility of the standard heat of formation.
17. Compute standard enthalpy changes for reactions involving species for which standard heats of formation are known.
18. Define bond energy, and apply the concept of bond energy to estimating enthalpy changes in chemical reactions.

EXERCISES

10.1. Give five examples of state variables and five examples of properties that are not state variables.

10.2.* For the birth of a baby boy in a family initially consisting of a 30-year-old father, a 27-year-old mother, and a 3-year-old daughter, calculate Δf for each of the following "variables" (considering the birth as a "process" and the family as a "system"): (a) the number of family members; (b) the number of family income tax deductions; (c) the number of adults in the family; (d) the average age in the family.

10.3. Analyze the energy changes that take place as a skier skis down a slope.

10.4.* (a) How much work is performed when a 10-kg piece of steel is lifted 3 m? (b) What is the change in potential energy experienced by this system?

10.5. Explain how some of the categories into which chemical energy can be divided relate to each other, including cases in which categories in different sets overlap. (Three sets of categories have been mentioned in the text.)

10.6.* Suppose you have a sample of gaseous H_2 at a temperature of 5 °K, and you slowly increase the temperature. Indicate the order in which you expect the four modes of energy—electronic, vibrational, rotational, translational—to accept the added energy as the temperature increases. (*Hint:* Consider the sizes of the energy gaps for these four modes).

10.7. Is it zero-point energy or thermal energy that increases, or do both increase as the temperature of the H_2 is raised in Example 10.6?

10.8. Calculate the population ratio n_j/n_l for a system at equilibrium at 25 °C for the condition that the energy difference between the energy levels $E_j - E_l$ is 3.0×10^{-21} J (per molecule).

10.9.* Calculate the population ratio n_j/n_l for a system at equilibrium at 25 °C for an energy difference between the energy levels $E_j - E_l$ of (a) 100 cal/mol, (b) 100 kcal/mol.

10.10.* Calculate the population ratio n_j/n_l for a system at equilibrium at 100,000 °K for an energy difference between the energy levels $E_j - E_l$ of (a) 100 cal/mol, (b) 100 kcal/mol.

10.11. Calculate the change of internal energy of the system in a process in which the system performs 10.6 kJ of work and absorbs 6.2 kJ of heat.

10.12.* Calculate the heat absorbed by the system in a process with no work performed for the condition that the internal energy of the system decreases by 99.3 kcal.

10.13. How much work is performed by the system in a process in which the internal energy of the system remains constant and 33.7 kJ of heat is released by the system?

10.14. Calculate the temperature change of the water in the system shown in Fig. 10.10, if the beater carries out 11.2 kJ of work and the electrical heater provides 4.1 kJ of heat. (Neglect the heat capacity of everything but water in the apparatus.)

10.15.* Calculate the temperature change in an oxygen bomb calorimeter with a total water equivalent of 2662.5 g for the condition that ΔE for the combustion is -13.4 kJ.

10.16. A 0.0124-mol sample of a substance was combusted in an oxygen bomb calorimeter with a total water equivalent of 3922 g water. The temperature increase was found to be 1.92 °K. How much heat is evolved in the combustion of one mole of this substance?

10.17.* Calculate the change in the enthalpy of a system in a process carried out at a constant pressure for the conditions that 39.6 J of work was performed on the system by compressing it and ΔE for the system is 111.0 J.

10.18. Calculate the change in enthalpy that occurs in a 0.267-mol sample of a pure substance with a C_p value of 37.3 cal deg^{-1}mol^{-1} in the temperature range 50 °C to 80 °C if the sample is cooled from 65 °C to 55 °C.

10.19. If the combustion of glucose, $C_6H_{12}O_6$, proceeds with a decrease in enthalpy of 2.82×10^3 kJ per mole, what is q for the combustion of 16.2 g of glucose at constant pressure?

10.20.* (a) Using the information given in Example 10.6, calculate ΔE and ΔH for a process in which 3.26 g of He gas is cooled from 140 °C to 60 °C. (b) Under what conditions would either of these values equal q for the process?

10.21. Calculate the difference between ΔE and ΔH for the combustion of 2 mol $H_2(g)$, to produce $H_2O(g)$ at 150 °C.

10.22. Calculate the heat absorbed by 13.4 g of water, the system, when vapor at 150 °C and 1 atm pressure is cooled and condensed at 100 °C and the resulting liquid water is cooled to 72 °C, the entire process occurring at 1 atm.

10.23.* Calculate ΔE for the vaporization of one mole of water at 100 °C and 1 atm pressure. (*Hint:* Use the $\Delta H°$ value in Table 10.1 and neglect the volume of liquid water compared with the volume of gaseous water.)

10.24. Calculate ΔH, ΔE, and q for the dimerization of 0.10 mol $NO_2(g)$ at 25 °C and 1 atm, $2NO_2(g) \rightleftharpoons N_2O_4(g)$. (*Hint:* Use the $\Delta H°$ value in Table 10.1.)

10.25. How much heat is released as 5.00 g Pb(s) is converted to $PbSO_4(s)$ in the reaction that occurs in the lead "storage battery" of an automobile (reaction 15 in Table 10.1)?

10.26.* If one could find a catalyst that made possible the conversion of $CO_2(g)$ and $H_2O(l)$ to liquid enthanol (C_2H_5OH) at 25 °C, how much heat would be required to bring about the production of 2 mol ethanol under "standard conditions"?

10.27. Calculate ΔH and ΔE (assuming that the volume of liquid Br_2 is negligible compared with the volume of gaseous Br_2) for the condensation of 0.50 mol $Br_2(g)$ at 59 °C and 1 atm pressure.

10.28. Why is $\Delta H°$ larger, do you think, for the boiling of water than for the melting of ice?

10.29. (a) Judging strictly on energy availability, would you rather heat your home by burning graphite or by burning diamonds? (b) How would either of these compare (on a per gram basis) with methane as a fuel?

10.30.* Calculate $\Delta H°$ for the following process at 25 °C:

$$SO_2(g) + \tfrac{1}{2}O_2(g) + H_2O(l) \rightleftharpoons H_2SO_4(l)$$

10.31. Calculate $\Delta H°$ for the following process at 25 °C:

$$2HCl(g) + \tfrac{1}{2}O_2(g) \rightleftharpoons Cl_2(g) + H_2O(l)$$

10.32. Calculate $\Delta H°$ for the following process at 25 °C:

$$CH_4(g) + O_2(g) \rightleftharpoons CO_2(g) + 2H_2O(l)$$

10.33.* Calculate $\Delta H°$ for the following process at 25 °C:

$$2CaCO_3(s) \rightleftharpoons 2CaO(s) + 2CO(g) + O_2(g)$$

10.34. Suppose you calculate $\Delta H°$ for a certain reaction, using the method of standard heats of formation, as defined in the text. Then, you are told to repeat the calculation and to use a different set of reference states in defining your ΔH_f° values. How will the second result relate to the first result, and why?

10.35. Determine ΔH° for the following reaction at 25 °C:

$$2HBr(g) \rightleftharpoons H_2(g) + Br_2(l)$$

10.36.* Determine ΔH° for the following reaction at 25 °C:

$$2NO(g) + 5H_2(g) \rightleftharpoons 2NH_3(g) + 2H_2O(l)$$

10.37. Determine ΔH° for the following reaction at 25 °C:

$$Ag_2O + 2HCl(g) \rightleftharpoons 2AgCl(s) + H_2O(l)$$

10.38. Determine ΔH° for the following reaction at 25 °C:

$$HCl(g) + NaOH(s) \rightleftharpoons NaCl(s) + H_2O(g)$$

10.39.* Determine ΔH° for the following reaction at 25 °C:

$$Pb(s) + \tfrac{1}{2}O_2(g) + SO_3(g) \rightleftharpoons PbSO_4(s)$$

10.40. Determine ΔH° for the following hypothetical reaction of naphthalene at 25 °C:

$$C_{10}H_8(s) + H_2(g) \rightleftharpoons 5C_2H_2$$

10.41. What additional information is needed to compute $\Delta H_{b,\,N\text{–}H}$ from the value of ΔH_f° given for NH_3 in Table 10.2?

10.42.* Using bond energies, estimate ΔH° for the following reaction:

$$H\text{—}O\text{—}O\text{—}H(g) + NH_3(g) \rightleftharpoons H_2NOH(g) + H_2O(g)$$

10.43. (a) Using bond energies estimate ΔH° for the following reaction:

$$3CH_3\text{—}CH_3(g) \rightleftharpoons \underset{\text{benzene}}{C_6H_6(g)} + 6H_2(g)$$

(b) Why would you not expect this calculation to provide a very reliable estimate? (c) Would the estimate of ΔH° be too high or too low?

ANSWERS TO SELECTED EXERCISES

10.2 (a) 1; (b) 1; (c) 0; (d) −5 years. **10.4** (a) 294 J; (b) 294 J. **10.6** translational first, then rotational, then vibrational, and last, electronic (at very high temperatures—thousands of degrees K). **10.9** (a) 0.845; (b) 4.9×10^{-74}. **10.10** (a) 0.999; (b) 0.605. **10.12** −99.3 kcal. **10.15** $\Delta T = 1.20$ °K. **10.17** 71.4 J. **10.20** (a) $\Delta E = -194$ cal, $\Delta H = -324$ cal; (b) $q = \Delta E$ when V is constant, $q = \Delta H$ when P is constant. **10.23** 8.97 kcal. **10.26** 653.4 kcal. **10.30** −54.63 kcal. **10.33** 222 kcal. **10.36** −201.92 kcal. **10.39** −125.05 kcal. **10.42** −156 kJ.

ANSWERS TO STUDY PROBLEMS

10(a) $n_k/n_s = 230$. **10(b)** 71.1×10^3 J $= 17.0 \times 10^3$ cal **10(c)** $\Delta E = -232$ cal (which equals q, because volume was constant). **10(d)** 624×10^3 cal $= 2.61 \times 10^6$ J. **10(e)** 0.824 °K. **10(f)** $\Delta H = 224$ cal; q cannot be determined, as the path was not specified. **10(g)** 68.1 °C, $q = 900$ cal. **10(h)** $q = \Delta H = 5.58 \times 10^3$ cal. **10(i)** −137.4 kcal. **10(j)** −99.82 kcal. **10(k)** 197 kcal. **10(l)** −146 kJ.

11

The Driving Force of Chemical Equilibrium

In Chaps. 4 and 5 the law of chemical equilibrium and the existence of an equilibrium constant were presented essentially as empirical facts. Although the idea of equilibrium as a steady state has been explored above, the only rationale for the numerical value of the equilibrium constant was a qualitative discussion of the relative thermodynamic stabilities of reactants and products. From that general discussion and from analogy with simple mechanical systems (for example, the boulder shown in Fig. 4.15), we expect energy (specifically, enthalpy) to have an important part in determining the stability of a species. That is, we expect energy (enthalpy) to be important in determining the value of an equilibrium constant for a particular reaction. It is not difficult to see, however, that something else besides energy must be involved in determining the value of an equilibrium constant.

11.1 Entropy

Something More Than Energy

Let us imagine a system, as depicted in Fig. 11.1a, consisting of an evacuated chamber, to which two piston-cylinder chambers are attached by valves. Each of the chambers contains 0.4 mol Cl_2 gas. In the left chamber, the chlorine is the $^{35}_{17}Cl$ isotope, and in the right chamber, the chlorine is the $^{37}_{17}Cl$ isotope. Each 0.1 mol of $^{35}_{17}Cl—^{35}_{17}Cl$ is represented in the figure by the symbol 35–35 and each 0.1 mol of

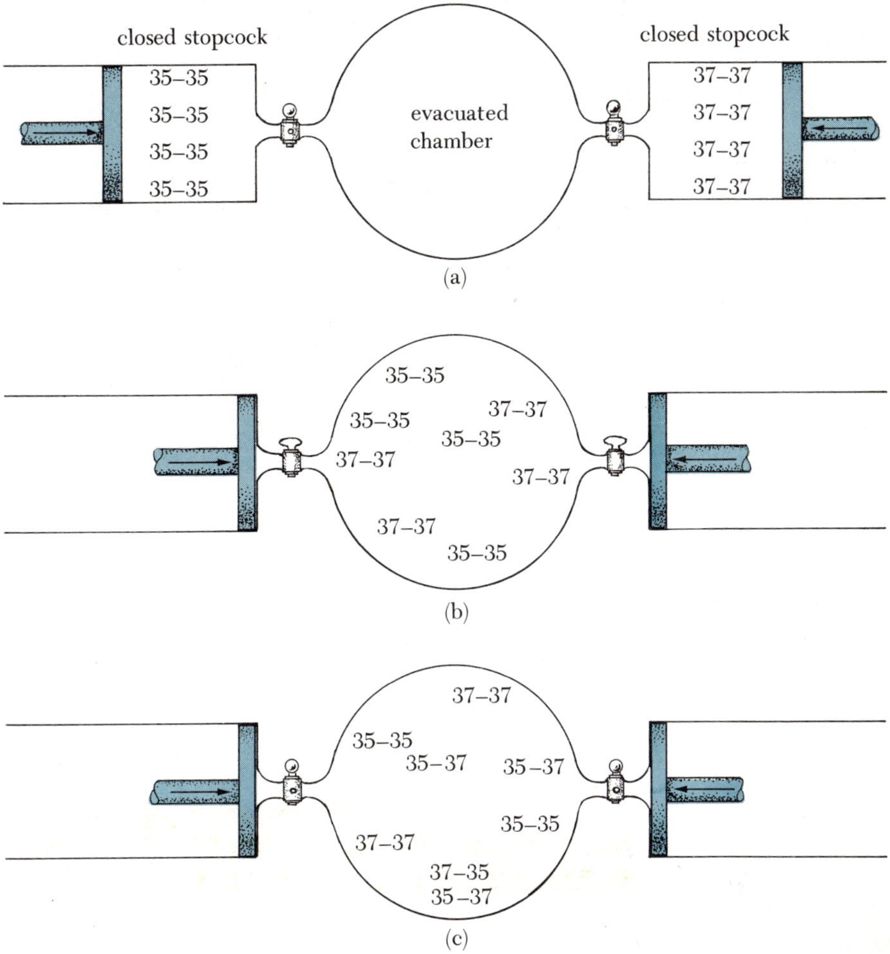

Figure 11.1 An experiment for studying the "exchange" reaction shown in equation (11.1). Each symbol of the type 35–35 represents 0.1 mol of the corresponding Cl_2 species (for example, $^{35}_{17}Cl—^{35}_{17}Cl$). (a) 0.4 mol $^{35}_{17}Cl—^{35}_{17}Cl$ initiallv in the left chamber, 0.4 mol $^{37}_{17}Cl—^{37}_{17}Cl$ initially in the right chamber, and the central chamber empty; the two stopcocks are closed. (b) The stopcocks are opened and the pistons moved towards the center chamber, forcing all Cl_2 molecules into the central chamber; shown at the instant of mixing. (c) The equilibrium mixture obtained after the reaction shown in Eq. (11.1) has been given sufficient time to occur.

$^{37}_{17}Cl—^{37}_{17}Cl$ is represented by 37–37. Now let us imagine opening the two valves and moving the pistons towards the central vessel, forcing the $^{35}_{17}Cl—^{35}_{17}Cl$ and $^{37}_{17}Cl—^{37}_{17}Cl$ into the central vessel. Figure 11.1b represents the system at the instant the two isotopes are mixed. Now if one waits long enough, the following equilibrium will be established *spontaneously* (although perhaps very slowly):

$$^{35}_{17}Cl—^{35}_{17}Cl + ^{37}_{17}Cl—^{37}_{17}Cl \rightleftharpoons 2\,^{37}_{17}Cl—^{35}_{17}Cl \tag{11.1}$$

For this process, we write the following equilibrium expression

$$K = \frac{[^{35}_{17}\text{Cl}{-}^{37}_{17}\text{Cl}]^2}{[^{37}_{17}\text{Cl}{-}^{37}_{17}\text{Cl}]\,[^{35}_{17}\text{Cl}{-}^{35}_{17}\text{Cl}]} \tag{11.2}$$

All three types of species represented in Eq. (11.1) are the Cl_2 type; all would be expected to have nearly the same internal energy or enthalpy.* Hence, so far as internal energy and enthalpy are concerned, all these species would be expected to have the same stabilities, and the equilibrium distribution of these species is what one would predict from simple statistics: half of the Cl_2 will be in the "mixed" form $^{35}_{17}\text{Cl}{-}^{37}_{17}\text{Cl}$, one-fourth will be $^{35}_{17}\text{Cl}{-}^{35}_{17}\text{Cl}$, and one-fourth will be $^{37}_{17}\text{Cl}{-}^{37}_{17}\text{Cl}$. This situation, depicted in Fig. 11.1c, corresponds to a K value for Eq. (11.2) of four, $(\frac{1}{2})^2/(\frac{1}{4})(\frac{1}{4})$. Now suppose that we consider this case in reverse, starting with the above-mentioned $\frac{1}{2}:\frac{1}{4}:\frac{1}{4}$ mixture, and we pull the pistons back away from the central vessel to their original positions of Fig. 11.1a. Do we expect the system to revert spontaneously to the mixture depicted in Fig. 11.1b, or to the two separated samples of $^{35}_{17}\text{Cl}{-}^{35}_{17}\text{Cl}$ and $^{37}_{17}\text{Cl}{-}^{37}_{17}\text{Cl}$ depicted in part a, which existed at the instant the valves were opened. The answer is *No*! Common sense tells us that the situation depicted in Figs. 11.1a or 11.1b will change spontaneously to what is shown in Fig. 11.1c, and that the reverse is not true. Hence, there is something inherently more stable about the $\frac{1}{2}:\frac{1}{4}:\frac{1}{4}$ mixture than the $0:\frac{1}{2}:\frac{1}{2}$ mixture, and internal energy and enthalpy have nothing to do with this difference in stabilities. Therefore another factor is at work here.

The other stability factor involved in the situation depicted in Fig. 11.1 is called entropy. **Entropy** is associated with the disorder, or randomness, of a system. Thus, the state in Fig. 11.1c has less order, or more randomness, than either of the states in Figs. 11.1a and 11.1b. We shall subsequently see that a larger entropy (disorder) is in favor of higher stability; and therefore the more disordered state (c) is more stable than the lower-entropy state (b).

Classical Definition of Entropy

During the later 1800s, great progress was made in the development of thermodynamics. One of the key concepts in developing this subject was the concept of a reversible process. A **reversible process** is one that proceeds against an opposing influence that is nearly strong enough to reverse the process. For example, a weight lifter midway in lifting precisely his maximum capacity is carrying out an essentially reversible process. If one placed one more kilogram of weight on the barbell, the athlete would not be able to lift, or even hold, the total weight, and the barbell would begin to fall. If one kilogram were removed from the maximum weight he can lift, he could lift the reduced load without question as the opposing force would not equal his strength. A chemical reaction that proceeds reversibly can also be reversed with only a slight increase in some opposing tendency. For example, if the reaction

$$3H_2(g) + N_2(g) \rightleftharpoons 2NH_3(g)$$

*A difference in a chemical property associated with different isotopes of the same element is called an **isotope effect**. Isotope effects on the thermodynamic variables of a set of species such as those in Eq. (11.1) are typically small.

is occurring reversibly, then a slight increase in the partial pressure (concentration) of NH_3 would cause the reaction to proceed from right to left. (We shall see in Chap. 12 that the maximum work that can be performed in a process can be achieved only if the process is carried out reversibly.)

From research aimed at maximizing the efficiency of engines running on heat, which were called heat engines, nineteenth-century scientists determined that an important thermodynamic quantity is *ratio of the heat absorbed by the system in a reversible process and the temperature at which the heat transfer occurs.*

$$\frac{q_{rev}}{T}$$

It was found that a state variable, called entropy, could be related to this quantity by the following equation:

$$\Delta S = \frac{q_{rev}}{T} = S_f - S_i \tag{11.3}$$

In this equation, ΔS represents the change in the entropy of the *system* associated with the particular process; S_f is the entropy of system in its final state; and S_i is the entropy of the system in its initial state.

Criteria for Spontaneity and Chemical Equilibrium

With the entropy change of any system defined by Eq. (11.3), it is worth while to consider the total entropy change of both the system and the surroundings:

$$\Delta S_{tot} = \Delta S + \Delta S_{surr} = \frac{q_{rev}}{T} + \frac{q_{rev,surr}}{T_{surr}}$$

The symbols ΔS, q_{rev}, and T refer to the system only. The symbols ΔS_{surr} and $q_{rev,surr}/T_{surr}$ come from applying Eq. (11.3) to the surroundings. Since the surroundings must provide the heat absorbed by the system, $q_{rev,surr}$, the heat absorbed by the surroundings in a reversible process, must equal $-q_{rev}$, the heat given up by the system. Also, we know that for any *reversible* process the temperature of system and surroundings must be the same—or irreversible heat transfer would occur. Thus, for a reversible process

$$\Delta S_{tot} = \frac{q_{rev}}{T} + \frac{-q_{rev}}{T} = 0 \qquad \text{(reversible process)} \tag{11.4}$$

The total entropy change of a reversible process is zero!

Now suppose that the temperature of the surroundings is higher than the temperature of the system, so that an irreversible spontaneous flow of heat occurs from the surroundings to the system. In that case,

$$\Delta S_{tot} = \frac{q_{rev}}{T} - \frac{q_{rev}}{T_{surr}} > 0 \qquad \text{(irreversible)}$$

because $T_{surr} > T$ and $1/T > 1/T_{surr}$, where q_{rev} represents the heat that would have been transferred from surroundings to system in a reversible process with the

Spotlight

The concept of entropy and its relation to criteria for spontaneity constitute the essence of what is called the **second law of thermodynamics.** This law, developed during the second half of the nineteenth century, has been stated in many ways, all of which can be shown to lead essentially to the idea that natural processes tend towards equilibrium. One often-quoted statement of the second law was made by Lord Kelvin (1824–1907): "It is impossible by a cyclic process to take heat from a reservoir and convert it into work without at the same time transferring heat from a hot to a cold reservoir." Another statement, made by Rudolf Clausius (1822–1888) is: "It is impossible to transfer heat from a cold to a hot reservoir without at the same time converting a certain amount of work into heat." Another pertinent and famous quotation from Clausius is: "Die Energie der Welt ist konstant; die Entropy der Welt strebt einem Maximum zu."

same initial and final states of the system as in the irreversible process under consideration. Hence,

$$\Delta S_{tot} > 0 \qquad \text{(irreversible)} \qquad (11.5)$$

The total entropy change of a spontaneous irreversible process is positive. Of course, various kinds of irreversible situations besides simply a temperature difference between system and surroundings can be imagined and demonstrated; nevertheless, the rigorous application of classical thermodynamics leads in all such cases to the result embodied in the expression (11.5)—the total entropy change of an irreversible process is positive! One can make a more generally useful statement by recognizing that all truly spontaneous processes are irreversible to some extent. Thus, the following general rule can be stated:

$$\Delta S_{tot} > 0 \qquad \text{(spontaneous)} \qquad (11.6)$$

for all spontaneous processes. This statement can be considered a fundamental rule of spontaneity in chemical and physical processes.

As powerful as the above spontaneity criterion is, it would be more convenient to have a criterion based entirely on the thermodynamic properties of the system itself, and not the system plus surroundings. Such a criterion is readily obtained for constant-pressure processes in which the only work performed is $P\,\Delta V$ work. In such cases, according to the first law of thermodynamics, $q_{rev} = \Delta H$. Then, by the statement (11.4), recalling that ΔS refers to the system only,

$$\Delta S_{tot} = \Delta S - \frac{\Delta H}{T} = 0 \qquad \text{(reversible processes)}$$

Multiplying both sides by T gives

$$T\,\Delta S_{tot} = T\,\Delta S - \Delta H = 0 \qquad \text{(reversible processes)}$$

Then, since $\Delta S_{tot} > 0$ for a spontaneous process, $T\,\Delta S_{tot} > 0$; hence,

$$T\,\Delta S_{tot} = T\,\Delta S - \Delta H > 0 \qquad \text{(spontaneous process)}$$

This can be rewritten as

$$\Delta H - T\,\Delta S < 0 \qquad \text{(spontaneous processes)} \qquad (11.7)$$

Now, we introduce a new thermodynamic state variable *for the system,* called the **Gibbs free energy,** symbolized G, and defined as

$$G = H - TS \tag{11.8}$$

where S is the entropy. For a given temperature, any change in TS is due to ΔS; hence, we write

$$\Delta G = \Delta H - T\,\Delta S \qquad \text{(constant } T\text{)} \tag{11.9}$$

Comparing Eqs. (11.7) and (11.9), we see that the following new statement for the condition of spontaneity can be written

$$\Delta G < 0 \tag{11.10}$$

(for any spontaneous process at some specific temperature with only PV work). This is the criterion for spontaneity that we were seeking based on the properties of the system.

For a process to occur spontaneously, it must proceed with a decrease in free energy of the system. If ΔG is positive, then the reverse process will occur spontaneously if the opportunity exists (for example, if the species of the right side of a chemical equation are present).

STUDY PROBLEM 11(a)

A certain process is under consideration for an industrial factory. For the system of interest, the ΔH and ΔS values for this process are -102.4 kJ and -0.084 kJ deg^{-1} at 100 °C. Will this process occur spontaneously at 100 °C?

Therefore, in chemical reactions the only situation that is not potentially subject to spontaneous change is one in which the free energy of the system is already at a minimum—and this must correspond to the condition of chemical equilibrium. This view is emphasized in Fig. 11.2. In this figure, the total free energy of the generalized reacting system

$$a\text{A} + b\text{B} + \cdots \rightleftharpoons m\text{M} + n\text{N} + \cdots \tag{11.11}$$

is plotted for conditions extending from the case with only left-side species (L) present to the condition in which all reactant species have been converted to right-side species (R). It is seen that in the case represented by this particular plot, the

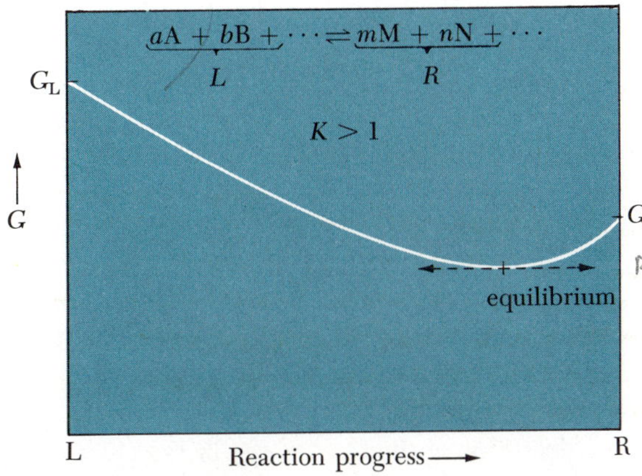

Figure 11.2 The dependence of the free energy of the reacting system (L ⇌ R) on progress from the left side (L) of the chemical equation to the right side (R).

value of G_R is lower than the value of G_L; hence, $\Delta G = G_R - G_L < 0$, and products are favored over reactants at equilibrium. In other words, K is greater than unity. Thus, it is seen that the position of minimum free energy occurs closer to the product (R) side than to the reactant (L) side.

It can be shown that the minimum free energy of the system always occurs somewhere between the two limits (L, G_L; R, G_R). If ΔG has a very large magnitude, then the minimum will occur very close to the side of lower G value, but the minimum is always displaced from it at least a little. The reason for this is that the entropy of the system is always larger in between the limits (L and R) than on either side, and according to Eq. (11.8), larger S values contribute to a lower G.

Another point of interest about Fig. 11.2 is that the curve is essentially flat over a very small (infinitesimal) shift to the right or left about the equilibrium point. This is represented by the dotted tangent line, and can be summarized by the notation

$$\delta G = 0 \qquad \text{(at equilibrium)} \tag{11.12}$$

which means that there is no change in G accompanying a small shift in the reactant-to-product ratio if the system is in equilibrium.

11.2 Mathematical Dependence of K on T, ΔS, ΔH, and ΔG

The Mathematical Function

Since the criterion for a potentially spontaneous process (for example, a chemical reaction with $K > 0$) can be stated in terms of ΔH, ΔS, and T (Eq. 11.7) or in terms of ΔG and T (Eq. 11.10), we have seen that at least in a qualitative sense changes in these thermodynamic state variables determine the value of the equilibrium constant at some specific temperature. Therefore, it should not be surprising for us to learn that a quantitative relation between K and ΔS, ΔH, and T or ΔG and T also exists. While a formal derivation of the quantitative relation will not be made, a simple statistical argument in the last section of this chapter makes the form of the expression plausible. At this point we simply state this important relation as follows:

$$K = e^{-\Delta G/RT} \tag{11.13}$$

In this equation, e is the base of natural logarithms (see Eq. 10.6) and R is the gas constant, in units compatible with ΔG (see Sec. 10.9). This equation is tremendously important, as it establishes the Gibbs free energy state variable G as the one that determines the values of the equilibrium constant at a particular temperature. We can note that Eq. (11.13) has a mathematical form that leads to the same conclusion as our spontaneity criterion (11.10). If ΔG is positive, then the exponent $(-\Delta G/RT)$ is negative, and $e^{-\Delta G/RT}$ is less than 1; therefore, K is less than 1 for $\Delta G > 0$, in agreement with criterion (11.10). Similarly, if ΔG in Eq. (11.13) is negative, then the exponent $(-\Delta G/RT)$ is positive, $e^{-\Delta G/RT}$ is greater than 1; therefore, K is greater than 1 for $\Delta G < 0$, in agreement with criterion (11.10). Therefore, the sign of ΔG determines whether a given reaction is potentially spontaneous in a specific direction at a particular temperature.

From the definition of the free energy in Eq. (11.8) and the expression for ΔG for

Spotlight

J. Willard Gibbs (1839–1903) has been considered by many to be the foremost American mathematical scientist of his day. Reared in an academic environment, he held the chair in mathematical physics at Yale University from 1871 until his death. During his scientific career he made monumental contributions to the field of thermodynamics, and was primarily responsible for developing chemical thermodynamics. His writings were couched in sophisticated and innovative mathematics and were not easily understood by his contemporaries; the breadth of their significance was largely unappreciated at the time they were published. Nevertheless, the great Ludwig Boltzmann, one of the pioneers in the field of statistical mechanics, recognized Gibbs as the greatest synthetic thinker since Isaac Newton. Gibbs, born near Yale University in 1839, was the first to receive a Ph.D. in science in America (Yale, 1863). A higher standard could not have been set for American science.

Culver Pictures

some specific temperature (Eq. 11.9), we can rewrite Eq. (11.13) in the following useful forms:

$$K = e^{-(\Delta H - T\Delta S)/RT} = e^{(-\Delta H/RT + \Delta S/R)}$$

Then making use of the properties of exponential expressions ($e^{a+b} = e^a \cdot e^b$), we write

$$K = e^{-\Delta H/RT} e^{\Delta S/R} \tag{11.14}$$

This equation shows explicitly that there are two different types of factors that determine the value of K at a specific temperature. One of these factors, $e^{-\Delta H/RT}$, clearly reflects the influence of energy, in the form of enthalpy, on the relative stabilities of reactants and products. The second factor, $e^{\Delta S/T}$, depends only on the entropy difference, and reflects the part of entropy in maintaining stability.

The Role of Standard States

Before we explore the consequences of Eq. (11.13) in any greater detail, we need to be specific about what we mean by the symbols representing the chemical transformations under consideration. When we write a balanced chemical equation of the type represented by Eq. (11.11), we necessarily attach certain meanings to symbols such as A, B, M, and N in addition to simply identifying the chemical formulae of the species represented. That is, we imply certain statements about the physical states of the species—whether they are solids, liquids, gases, or perhaps solvents or solutes in solution.

In many cases, especially if thermodynamic relations are involved, we should like a very specific set of conditions to be implied by the balanced chemical equation. A particularly important set of specific conditions are those that define what are called the standard states of the species involved. As defined in Chap. 10, these standard states are specified arbitrarily; for a gaseous species one chooses a partial pressure of 1 atm to define the standard state. For condensed phases, the pure solids or liquids are taken as the standard states, except with dilute solutions, for which the 1 *M* solution usually represents the standard states of solutes.* Since these standard states generally refer to whatever temperature happens to be of interest in the given problem, when we choose to interpret a balanced chemical equation such as (11.11) in terms of standard states, we imply, "*a* moles of species A in its standard state react with *b* moles of species B in its standard state, and so on, to form *m* moles of species M in its standard state, and so on, at some specific temperature." This is a purely formal statement; it may or may not correspond to a process that would actually be carried out. Even in instances for which the "written" process cannot be carried out directly, it still has meaning and is often useful in representing the process formally. In most actual cases, chemical reactions are not carried out under conditions in which the reactants are consumed in their standard states and the products produced in their standard states. Furthermore, at equilibrium, the reactants and products are generally not in their standard states.†

With this idea in mind, we emphasize that what is implied by the symbol ΔG for a reaction represented by the generalized equation (11.11) is the difference between free energies in standard states; we denote this difference by the symbol ΔG° to emphasize this convention. Thus,

$$\Delta G^\circ = mG^\circ_{\mathrm{M}} + nG^\circ_{\mathrm{N}} + \cdots - G^\circ_{\mathrm{A}} \cdots \tag{11.15}$$

where G°_{A} means the free energy of one mole of species A in its standard state, called its **standard molar free energy**; similar meanings apply to the analogous symbols for other species. Adopting the same symbolism for the entropy and enthalpy of standard states, we can write the following expressions for processes occurring at a particular temperature T and pressure:

$$\Delta G^\circ = \Delta H^\circ - T\,\Delta S^\circ \tag{11.16}$$

$$K = e^{-\Delta H^\circ/RT} e^{\Delta S^\circ/R} \tag{11.17}$$

$$K = e^{-\Delta G^\circ/RT} \tag{11.18}$$

*As discussed briefly in Chap. 4, the standard state for dilute solute species in solution is not simply a 1 *M* solution but rather a hypothetical solution, called a hypothetical one-molar solution. The hypothetical 1 *M* solution "acts" like a solution for which the value of an intensive solute property is n times as large as the value for a $1/n$ *M* solution, where n is a very large number (infinite dilution).

†It would be extremely coincidental for the equilibrium of a specific chemical reaction to exist when all reactants and products are present in their standard states. If such a highly coincidental case could be found, then for that case the equilibrium constant would have to equal unity, since each factor (activity) in the mass action expression defined in Eq. (4.21) would equal unity—by definition, the activity of any species is unity for the standard state.

From Eq. (10.6), we can rewrite Eq. (11.18) in the following useful form:

$$\ln K = 2.303 \log K = \frac{-\Delta G^\circ}{RT} \tag{11.19}$$

The standard states with respect to which Eqs. (11.15) through (11.19) are defined correspond to the same temperature T that occurs explicitly in these equations. Consequently, the temperature is often indicated as a subscript on the symbols ΔG°, ΔS°, and ΔH°. Therefore, one frequently encounters symbols such as ΔH°_{229}, which represents the value of ΔH° at 229 °K. If temperature is not specified, it is assumed to be 298 °K.

11.3 Standard Molar Free Energy Changes

Applications and the Algebra of Chemical Equations

Equation (11.17) provides an extremely valuable tool for understanding and predicting the conditions of equilibrium for a chemical reaction. For a given temperature, it gives quantitatively the relation between the equilibrium constant and the free energy that would accompany the transformation of reactants in their standard states to products in their standard states. By knowing ΔG° at a given temperature, one can calculate K for that temperature, even if for practical reasons K could not be measured directly; or if K is known for a given temperature, then ΔG° can be calculated for that temperature without recourse to any other measurement. The sign of ΔG° determines whether K is less than 1 or larger; however, the magnitude of ΔG° and the value of T must also be known to calculate the value of K.

In the following examples, we see how calculations involving ΔG° are carried out. In these examples we recognize that ΔG° values for individual processes can be manipulated in the same general ways that we used for ΔH° values (Chap. 10). Since we justified such manipulations for one state variable—enthalpy—we can now assume that they are valid for other state variables—G or S.

EXAMPLE 11.1 $\Delta G^\circ = 6.487$ kcal at 298.15 °K for the dissociation of acetic acid. Calculate K for this process.

SOLUTION From Eq. (11.9),

$$2.303 \log K = \frac{-\Delta G^\circ_{298}}{RT} = \frac{-6487 \text{ cal mol}^{-1}}{(1.987 \text{ cal deg}^{-1}\text{mol}^{-1})(298.15 \text{ deg})}$$

$$\log K = \frac{-6487}{(2.303)(1.987)(298.15)} = -4.755 = 0.245 - 5$$

$$K = [\text{antilog}\,(0.245)][\text{antilog}\,(-5)] = 1.76 \times 10^{-5}$$

EXAMPLE 11.2 $K = 3.24 \times 10^{-4}$ for the chemical reaction

$$CO_2(g) + H_2(g) \rightleftharpoons H_2O(l) + CO(g) \tag{A}$$

at 298.15 °K. Calculate ΔG° in kJ for this reaction at this temperature.

SOLUTION From Eq. (11.19), we see that for 298.15 °K

$$2.303 \log (3.24 \times 10^{-4}) = \frac{-\Delta G^\circ}{(RT)}$$

Then, multiplying both sides of the equation by RT, and substituting the appropriate values for these factors (298.15 deg K)(8.314 J deg^{-1}mol^{-1}), we get

$$\begin{aligned}\Delta G^\circ_{298} &= -(298.15 \text{ deg})(8.314 \text{ J deg}^{-1}\text{mol}^{-1})(2.303)[(\log 3.24) - 4] \\ &= -(298.15)(8.314)(2.303)(0.511 - 4)\frac{\text{J}}{\text{mol}} = 19.9 \times 10^3 \frac{\text{J}}{\text{mol}} \\ &= \left(19.9 \frac{\text{kJ}}{\text{mol}}\right)\left(\frac{1}{4.184}\frac{\text{kcal}}{\text{kJ}}\right) = 4.76 \text{ kcal (per mole of reaction)}\end{aligned}$$

By this we mean that at 298.15 °K

$$G^\circ_{\text{CO(g)}} + G^\circ_{\text{H}_2\text{O(l)}} - G^\circ_{\text{H}_2\text{(g)}} - G^\circ_{\text{CO}_2\text{(g)}} = 19.9 \text{ kJ} \tag{B}$$

EXAMPLE 11.3 $\Delta H^\circ = -9830$ cal at 25 °C for the reaction of Example 11.2. Calculate ΔS° for this process at 298.15 °K (25 °C).

SOLUTION From Eq. (11.16) and using the result of Example 11.2, we get

$$\Delta G^\circ = 4760 \text{ cal} = \Delta H^\circ_{298} - T\Delta S^\circ_{298} = -9830 \text{ cal} - (298.15 \text{ deg})\left(\Delta S^\circ_{298}\frac{\text{cal}}{\text{deg}}\right)$$

$$\begin{aligned}\Delta S^\circ_{298} &= \frac{4760 + 9830}{-298.15} = -48.9 \frac{\text{cal}}{\text{deg}} \\ &= S^\circ_{\text{CO(g)}} + S^\circ_{\text{H}_2\text{O(l)}} - S^\circ_{\text{CO}_2\text{(g)}} - S^\circ_{\text{H}_2\text{(g)}}\end{aligned}$$

EXAMPLE 11.4 $\Delta G^\circ = -31.26$ kcal at 298 °K for the reaction

$$CO_2(g) + 4H_2(g) \rightleftharpoons CH_4(g) + 2H_2O(l) \tag{C}$$

Using the results of Example 11.2, calculate ΔG° and K at 25 °C for the reaction

$$3H_2(g) + CO(g) \rightleftharpoons CH_4(g) + H_2O(g) \tag{D}$$

SOLUTION From Example 11.2, we know that ΔG° is 4.76 kcal for chemical equation (A), and hence ΔG° is -4.76 kcal for the reverse reaction, that is,

$$CO(g) + H_2O(l) \rightleftharpoons CO_2(g) + H_2(g) \tag{E}$$

This follows because the two different directions of a chemical reaction correspond to opposite signs of ΔG°. (Such a statement applies not only to ΔG°, but to any state variable.) But when chemical equation (E) is added to chemical equation (C), we obtain chemical equation (D), that is, the equation for the reaction with the unknown ΔG°. Thus, the unknown ΔG° must be the sum of these two ΔG°'s. We can summarize this as follows:

$$
\begin{array}{lr}
 & \Delta G^\circ\text{, kcal} \\
CO(g) + H_2O(l) \rightleftharpoons CO_2(g) + H_2(g) & -4.76 \\
CO_2(g) + 4H_2(g) \rightleftharpoons CH_4(g) + 2H_2O(l) & -31.26 \\
\hline
\begin{bmatrix} CO(g) + \cancel{H_2O}(l) \\ + \cancel{CO_2}(g) + 4H_2(g) \end{bmatrix} \rightleftharpoons \begin{bmatrix} \cancel{CO_2}(g) + H_2(g) \\ + CH_4(g) + \cancel{2}H_2O(l) \end{bmatrix} & -36.02
\end{array}
$$

$$CO(g) + 3H_2(g) \rightleftharpoons CH_4(g) + H_2O(l) \qquad \Delta G^\circ = -36.02\ \text{kcal}$$

Finally, from Eq. (11.19),

$$\log K = \frac{36.02 \times 10^3}{(2.303)(1.987)(298.15)} = 26.40$$

$$k = \text{antilog}\,(26.40) = 2.5 \times 10^{26}$$

STUDY PROBLEM 11(b)

The equilibrium constant for formation of $BaSO_4(s)$ from Ba^{2+} and $SO_4^{\,2-}$ in aqueous solution at 25 °C is 0.90×10^{10}. Calculate ΔG° in kJ for this process.

STUDY PROBLEM 11(c)

Calculate ΔH° in kJ for a chemical reaction in which ΔS° is 11.4 J deg^{-1} and ΔG° is 47.1 kJ at 36 °C.

STUDY PROBLEM 11(d)

Calculate ΔG° and the equilibrium constant for a chemical reaction at 100 °C for which ΔH° is -82.6 kcal and ΔS° is 29.1 cal deg^{-1}.

A characteristic of state functions is that changes in them depend only on the initial and final states and are independent of path; because of this, it is possible to calculate values of ΔG°, and hence K, for chemical reactions that have never been carried out (and may never even be attempted). For example, suppose no one had ever investigated the reaction represented by Eq. (D); because thermodynamic data, that is, ΔG° values, were available for the reactions given by Eqs. (A) and (C), we were able to calculate ΔG° and K for the reaction in question without performing any experiments on it directly. This is one of the great powers of thermodynamics, and has profound implications and applications in modern science and technology. We might be interested in making predictions about a chemical reaction that we suspect is occurring in a situation in which it is impractical or impossible for us to conduct certain experiments. Such a situation might be in outer space or inside a living cell, and it may not be experimentally possible to duplicate the system in the laboratory. However, if the reaction of interest can be related thermodynamically, in the sense of Example 11.4, to reactions that have been studied experimentally, then it will be possible to make reliable predictions on thermodynamic properties and equilibrium conditions for the reaction of interest.

It is reasonable to ask how one obtains the various G° values or ΔG° values required for calculations of the type demonstrated in Examples 11.2 to 11.4. In answering this, it is important to remember that our main interest here is in differences between free energies. Hence, as in the case of enthalpies, we are free to establish an arbitrary reference point, or reference substances. So long as we are consistent in this choice, then free energy *differences* calculated from this reference will be based on a self-consistent system of addition and subtraction, and will be correct. As we see how this system works, we notice the direct one-to-one correspondence with the scheme that has been outlined for $\Delta H°$.

The Role of Reference States: Standard Free Energies of Formation

In specifying this self-consistent system we must again call on the concept of the reference state of an element. We have defined the reference state as the element in its most stable form (for example, $H_2(g)$, $Br_2(l)$, Na(s)) at 25 °C and 1 atm. We then define the *standard molar free energy of formation* of a substance at a particular temperature T as the free energy change of the process by which elements in their reference states (at 25 °C) are converted to the substance of interest at the temperature T. For example, the standard molar free energy of formation of $CO_2(g)$ at 100 °C is defined by the following equation:

$$\underset{298}{C(gr)} + \underset{298}{O_2(g)} \rightleftharpoons \underset{373}{CO_2(g)} \qquad (11.20a)$$

and the corresponding equation

$$\Delta G^\circ_{f,373,CO_2(g)} = G^\circ_{373,CO_2(g)} - (G^\circ_{298,C(gr)} + G^\circ_{298,O_2(g)}) \qquad (11.20b)$$

Similarly, the standard molar free energy of formation of CO(g) at 100 °C is defined by the following two equations:

$$\underset{298}{C(gr)} + \underset{298}{\tfrac{1}{2}O_2(g)} \rightleftharpoons \underset{373}{CO(g)} \qquad (11.21a)$$

$$\Delta G^\circ_{f,373,CO(g)} = G^\circ_{373,CO(g)} - (G^\circ_{298,C(gr)} + \tfrac{1}{2}G^\circ_{298,O_2(g)}) \qquad (11.21b)$$

And the standard free energy of formation of $O_2(g)$ at 100 °C is defined as follows:

$$\underset{298}{O_2(g)} \rightleftharpoons \underset{373}{O_2(g)} \qquad (11.22a)$$

$$\Delta G^\circ_{f,373,O_2(g)} = G^\circ_{O_2(g),373} - G^\circ_{O_2(g),298} \qquad (11.22b)$$

By combining Eqs. (11.20b), (11.21b), and (11.22b), one can obtain an equation for the $\Delta G°$ value of the reaction

$$\underset{373}{CO(g)} + \underset{373}{\tfrac{1}{2}O_2(g)} \rightleftharpoons \underset{373}{CO_2(g)} \qquad \Delta G^\circ_{373} \qquad (11.23a)$$

After algebraic manipulation, Eqs. (11.20b), (11.21b), and (11.22b) yield the $\Delta G°$ value for Eq. (11.23a):

$$\Delta G^\circ_{373} = \Delta G^\circ_{f,373,CO_2(g)} - (\Delta G^\circ_{f,373,CO(g)} + \tfrac{1}{2}\Delta G^\circ_{f,373,O_2(g)}) \qquad (11.23b)$$

Equations (11.23a) and (11.23b) are an example of the following relationship: *The standard free energy change for any process at a particular temperature is the difference between the standard free energy of formation of product or products at that temperature and the standard free energy of formation of reactant or reactants at that temperature.* This is closely analogous to the relation encountered for $\Delta H°$ values (Chap. 10), and can be justified algebraically on the same kind of basis. Thus, to determine the $\Delta G°$ value of any reaction at any temperature, all that is needed are the values of $\Delta G^\circ_{f,T}$ for all relevant species at the temperature of interest.

Equation (11.23b) and the general rule that follows it can be justified as follows. Let us visualize the formation of $CO_2(g)$ at 100 °C (373 °K) from $O_2(g)$ and graphite, C(gr), both at 25 °C, in terms of $\Delta G°$ for the process

$$\underset{\substack{\text{reference states} \\ (298\ °\text{K})}}{C(gr) + O_2(g)} \rightleftharpoons \underset{\substack{\text{standard state} \\ (373\ °\text{K})}}{CO_2(g)} \qquad \Delta G^\circ_{f,373,CO_2(g)} \tag{F}$$

For this process

$$\Delta G° = \Delta G^\circ_{f,373,CO_2(g)} = G^\circ_{373,CO_2(g)} - G^\circ_{298,C(gr)} - G^\circ_{298,O_2(g)}$$

which rearranged gives

$$G^\circ_{373,CO_2(g)} = G^\circ_{298,C(gr)} + G^\circ_{298,O_2(g)} + \Delta G^\circ_{f,373,CO_2(g)} \tag{G}$$

$G^\circ_{298,C(gr)}$ and $G^\circ_{298,O_2(g)}$ are the standard molar free energies of carbon and oxygen in their reference states (graphite and one-atmosphere oxygen gas), and $\Delta G^\circ_{373,CO_2(g)}$ is the standard molar free energy of $CO_2(g)$ at 100 °C. Similarly, the standard free energy of formation of carbon monoxide at 100 °C would be defined in terms of the process

$$\underset{\substack{\text{reference states} \\ (298\ °\text{K})}}{C(gr) + \tfrac{1}{2}O_2(g)} \rightleftharpoons \underset{\substack{\text{standard state} \\ (373\ °\text{K})}}{CO(g)} \tag{H}$$

from which

$$\Delta G^\circ_{f,373,CO(g)} = G^\circ_{373,CO(g)} - [G^\circ_{298,C(gr)} + \tfrac{1}{2}G^\circ_{298,O_2(g)}]$$

$$G^\circ_{373,CO(g)} = \Delta G^\circ_{f,373,CO(g)} + G^\circ_{298,C(gr)} + \tfrac{1}{2}G^\circ_{298,O_2(g)} \tag{I}$$

Similarly, we represent the standard free energy of formation of $O_2(g)$ at 100 °C from $O_2(g)$ at 25 °C in terms of the process

$$\underset{\substack{\text{reference state} \\ (298\ °\text{K})}}{O_2(g)} \rightleftharpoons \underset{\substack{\text{standard state} \\ (373\ °\text{K})}}{O_2(g)} \tag{J}$$

Thus, we write

$$\Delta G^\circ_{f,373,O_2(g)} = G^\circ_{373,O_2(g)} - G^\circ_{298,O_2(g)}$$

which can be rearranged as

$$G^\circ_{373,O_2(g)} = \Delta G^\circ_{f,373,O_2(g)} + G^\circ_{298,O_2(g)} \tag{K}$$

We now wish to obtain an expression for the $\Delta G°$ value corresponding to Eq. (11.23a):

$$\Delta G° = \Delta G^\circ_{373} = G^\circ_{373,CO_2(g)} - (G^\circ_{373,CO(g)} + \tfrac{1}{2}G^\circ_{373,O_2(g)}) \tag{L}$$

TABLE 11.1 **Standard Free Energies of Formation at 25 °C**

Compound	*State*	$\Delta G^\circ_{f,298}$, kcal mol^{-1}	*Compound*	*State*	$\Delta G^\circ_{f,298}$, kcal mol^{-1}
Ag_2O	s	−2.59	FeS	s	−23.32
AgCl	s	−26.22	FeO	s	−58.4
AgBr	s	−22.93	H_2O	g	−55.64
AgI	s	−15.85	H_2O	l	−56.69
$AgNO_3$	s	−7.69	H_2O_2	g	−24.73
$CaCO_3$	s	−269.78	H_2O_2	l	−28.23
$CaSO_4$	s	−315.56	HCl	g	−22.77
CH_4	g	−12.14	HBr	g	−12.72
C_2H_2	g	50.0	HI	g	0.31
C_2H_4	g	16.28	H_2S	g	−7.892
C_2H_6	g	−7.86	H_2SO_4	aq	−197.67
C_6H_6(benzene)	g	30.99	NaCl	s	−91.78
C_6H_6(benzene)	l	29.7	Na_2CO_3	s	−250.4
C_8H_{18}(*n*-octane)	g	4.14	NaOH	s	−90.60
CO	g	−32.81	NH_3	g	−3.98
CH_3OH	l	−39.75	N_2O	g	24.76
CO_2	g	−94.26	NO	g	20.72
CuO	s	−30.40	NO_2	g	12.39
Cu_2O	s	−34.98	N_2O_4	g	23.49
$CuBr_2$	s	−30.3	$PbSO_4$	s	−193.89
CuS	s	−11.7	SO_2	g	−71.79
Cu_2S	s	−20.6	SO_3	g	−88.52
Fe_2O_3	s	−177.1			

Substituting the values of $G^\circ_{373,\,CO_2(g)}$, $G^\circ_{373,\,CO(g)}$, and $G^\circ_{373,\,O_2(g)}$ given by Eqs. (G), (I), and (K), we obtain the ΔG° value for equation (L):

$$\begin{aligned}\Delta G^\circ_{373} = {} & \Delta G^\circ_{f,373,\,CO_2(g)} + G^\circ_{298,\,C(gr)} + G^\circ_{298,\,O_2(g)} \\ & - (\Delta G^\circ_{f,373,\,CO(g)} + G^\circ_{298,\,C(gr)} + \tfrac{1}{2}G^\circ_{298,\,O_2(g)} \\ & + \tfrac{1}{2}\Delta G^\circ_{f,373,\,O_2(g)} + \tfrac{1}{2}G^\circ_{298,\,O_2(g)}) \qquad \text{(M)}\end{aligned}$$

After making the appropriate cancellations, we obtain Eq. (11.23b) directly. Clearly, the same process could be carried out for any process at any specific temperature (or for any state variable).

An immense body of data on $\Delta G^\circ_{f,T}$ values has been accumulated during this century. A brief abstract of this type of data on some simple compounds is given in Table 11.1. In the following examples, we see how these data can be practically applied.

EXAMPLE 11.5 From the data of Table 11.1, calculate ΔG° at 25 °C for the reaction $2NO(g) + O_2(g) \rightleftharpoons 2NO_2(g)$.

SOLUTION From the text discussion we know that

$$\begin{aligned}\Delta G^\circ_{298} &= 2\Delta G^\circ_{f,298,NO_2(g)} - (2\Delta G^\circ_{f,298,NO(g)} + \Delta G^\circ_{f,298,O_2(g)}) \\ &= 2 \times 12.39 - (2 \times 20.72 - 0) \\ &= -16.66 \text{ kcal}\end{aligned}$$

EXAMPLE 11.6 At 25 °C, $\Delta G^\circ = 25.82$ kcal for the reaction

$$2CuO(s) \rightleftharpoons Cu_2O(s) + \tfrac{1}{2}O_2(g)$$

Calculate ΔG°_f at 25 °C for $Cu_2O(s)$ in kJ from ΔG°_f for CuO(s) in Table 11.1.

SOLUTION

$$\begin{aligned}25.82 &= \Delta G^\circ_{298} \\ &= \Delta G^\circ_{f,298,Cu_2O(s)} + \tfrac{1}{2}\Delta G^\circ_{f,298,O_2(g)} - 2\Delta G^\circ_{f,298,CuO(s)} \\ &= \Delta G^\circ_{f,298,Cu_2O(s)} + \tfrac{1}{2}(0) - 2(-30.40)\end{aligned}$$

$$\begin{aligned}\Delta G^\circ_{f,298,Cu_2O(s)} &= 25.82 - 60.80 \\ &= -34.98\frac{\text{kcal}}{\text{mol}} = -34.98\left(\frac{\text{kcal}}{\text{mol}}\right)\left(4.184\frac{\text{kJ}}{\text{kcal}}\right) = -146.4\frac{\text{kJ}}{\text{mol}}\end{aligned}$$

STUDY PROBLEM 11(e)

Calculate ΔG° and K at 298 °K for the reaction

$$2NH_3(g) + \tfrac{5}{2}O_2(g) \rightleftharpoons 2NO(g) + 3H_2O(l)$$

Dependence of ΔG° and K on Temperature

From Eqs. (11.18) or (11.19), without reference to Eqs. (11.16) or (11.17), one might mistakenly assume that if K were known at any single temperature, then ΔG° could be calculated, and a new K could be computed for any other temperature. However, things are not quite so simple, since Eq. (11.16) tells us that ΔG° itself depends on the temperature. In fact, the dependence of ΔG°, or of K, on the temperature can be used to compute the value of ΔH° or ΔS° at any given temperature. The relations necessary to make these calculations can be obtained as follows.

From Eq. (11.16) and (11.19), we obtain, for any temperature, the equations

$$\frac{\Delta G^\circ}{T} = \Delta H^\circ\left(\frac{1}{T}\right) - \Delta S^\circ \qquad (11.24)$$

and

$$2.303 \log K = \ln K = \frac{-\Delta H^\circ}{R}\left(\frac{1}{T}\right) + \frac{\Delta S^\circ}{R} \qquad (11.25)$$

Equations (11.24) and (11.25) are both of the following general mathematical form:

$$y = mx + b \qquad (11.26)$$

In Eq. (11.24), y is $\Delta G^\circ/T$, m is ΔH°, x is $1/T$, and b is $-\Delta S^\circ$. In Eq. (11.25), y is $\ln K$, m is $-\Delta H^\circ/R$, x is $1/T$, and b is $\Delta S^\circ/R$. In equations of this general form, y is called the dependent variable; it depends on the independent variable, x, and also the constants m and b. Equation (11.26) is simply a mathematical representation of a straight line on a plot of y (vertical) vs x (horizontal). The y intercept is b, and the slope is m, as defined in Fig. 11.3. If in Eq. (11.24) we identify $\Delta G°/T$, $\Delta H°$, $-\Delta S°$, and $1/T$ with y, m, b, and x, respectively, in Eq. (11.26), then a plot of $\Delta G°/T$ vs $1/T$ will be a straight line with slope $\Delta H°$ and intercept $-\Delta S°$. Analogously, if we identify ln K (or 2.303 log K), $-\Delta H°/R$, $\Delta S°/R$, and $1/T$ with y, m, b, and x, then from Eq. (11.25), we can conclude, a plot of ln K (or 2.303 log K) vs $1/T$ will be a straight line with slope $-\Delta H°/R$ and intercept $\Delta S°/R$. The latter relation is shown in Fig. 11.4, where ln K is plotted against $1/T$ for the specific case of the following reaction:

$$H_2(g) + I_2(g) \rightleftharpoons 2HI(g) \tag{11.27}$$

Let us see how such a plot is used in typical calculations. We use the graph in Fig. 11.4 to calculate $\Delta H°$ and $\Delta S°$ for this specific reaction (Eq. 11.27). Since this plot is a straight line (which follows Eq. 11.25), $-\Delta H°/R$ and $\Delta S°/R$ must be constants, as expected. Hence, they can be obtained directly from the graph. The intercept, which is $\Delta S°/R$, is 1.20, and the slope, which is $-\Delta H°/R$, equals $2.89/(1.50 \times 10^{-3})$, or 1.93×10^3 deg. This gives

$$\begin{aligned}\Delta S° &= (1.987 \text{ cal deg}^{-1}\text{mol}^{-1})(1.20)\\ &= 2.38 \text{ cal deg}^{-1} \text{ (per "mole of reaction as written")}\\ \Delta H° &= -[1.987 \text{ cal deg}^{-1}\text{mol}^{-1}(1.93 \times 10^3 \text{ deg})]\\ &= -3.83 \times 10^3 \text{ cal (per "mole of reaction as written")}\end{aligned}$$

One can also derive mathematical expressions that relate the values of K at two temperatures, T_1 and T_2, to the values of $\Delta H°$ and $\Delta S°$ for the reaction; these expressions make it possible to determine $\Delta H°$ and $\Delta S°$ for a specific reaction from a determination of the equilibrium constant of that reaction at two temperatures.

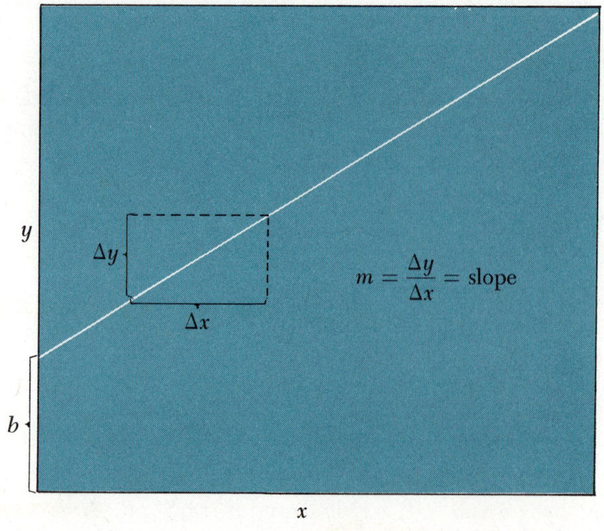

Figure 11.3 Plot of the function $y = mx + b$, with slope m and intercept b.

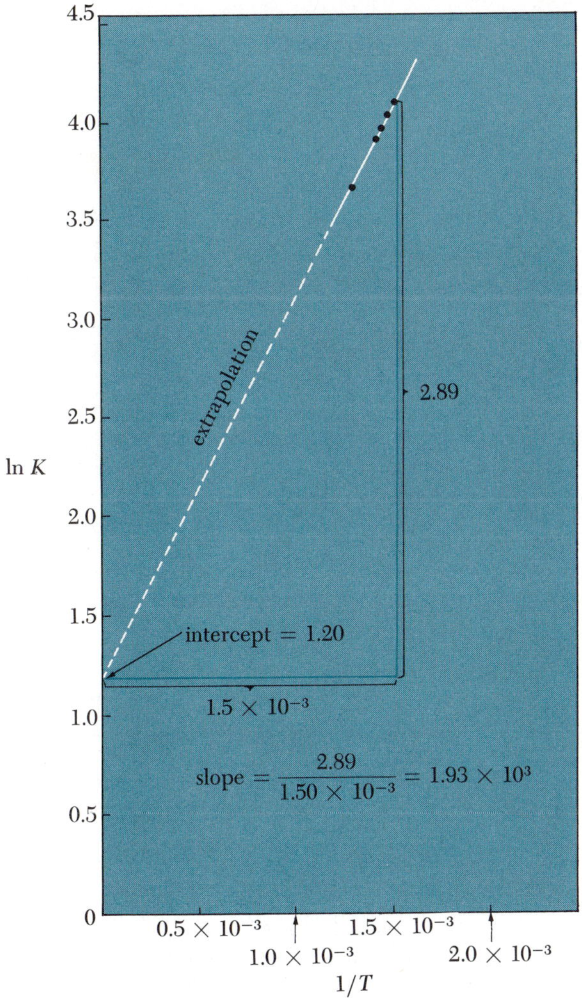

Figure 11.4 Plot of ln K vs $1/T$ for the reaction $H_2(g) + I_2(g) \rightleftharpoons 2HI(g)$. Data points correspond to the temperatures in °C: 393°, 410°, 427°, 443°, and 508°. Extrapolation of the solid line through the data points is shown as a dashed line. The distances used in computing the slope are also shown explicitly.

From Eq. (11.25), we have the following two relations to form a bridge between what is known and what is desired:

$$\ln K_1 = -\frac{\Delta H°}{R}\left(\frac{1}{T_1}\right) + \frac{\Delta S°}{R}$$

where K_1 is the value of K at T_1, and

$$\ln K_2 = -\frac{\Delta H°}{R}\left(\frac{1}{T_2}\right) + \frac{\Delta S°}{R}$$

where K_2 is the value of K at T_2. Then, subtracting the second equation from the first gives

$$\ln K_1 - \ln K_2 = -\frac{\Delta H°}{R}\left(\frac{1}{T_1} - \frac{1}{T_2}\right) = -\frac{\Delta H°}{R}\left(\frac{T_2 - T_1}{T_1 T_2}\right)$$

$$\ln\left(\frac{K_1}{K_2}\right) = \frac{\Delta H°}{R}\left(\frac{T_1 - T_2}{T_1 T_2}\right).$$

Using the relation between natural and base 10 logarithms (Eq. 10.7), we then get

$$2.303 \log\left(\frac{K_1}{K_2}\right) = \frac{\Delta H°}{R}\left(\frac{T_1 - T_2}{T_1 T_2}\right) \tag{11.28}$$

This is the desired relation. If one obtains $\Delta H°$ from Eq. (11.28), then $\Delta S°$ can be calculated by using this value in Eq. (11.25) with either T_1 and K_1, or T_2 and K_2.

EXAMPLE 11.7 The equilibrium constant of a certain reaction is found to be doubled when the temperature is increased from 50 °C to 60 °C. Calculate the value of $\Delta H°$ for this reaction.

SOLUTION Letting T_1 in Eq. (11.28) be 323 °K (50 °C) and T_2 be 333 °K (60 °C), we write

$$\ln\left(\frac{K_{323}}{K_{333}}\right) = 2.303 \log\left(\frac{K_{323}}{K_{333}}\right) = \frac{\Delta H°}{R}\left(\frac{323 - 333}{323 \times 333}\right)\frac{\text{deg}}{\text{deg}^2}$$

$$\ln\left(\frac{1}{2}\right) = 2.303 \log\left(\frac{1}{2}\right) = \frac{\Delta H°}{R}\left(\frac{-10}{10.76 \times 10^4 \text{ deg}}\right)$$

$$-0.693 = -\frac{\Delta H°}{R}\left(\frac{1}{10.76 \times 10^3 \text{ deg}}\right)$$

$$\begin{aligned}\Delta H° &= (0.693)(10.76 \times 10^3 \text{ deg})(1.987 \text{ cal deg}^{-1}\text{mol}^{-1})\\ &= 14.8 \times 10^3 \text{ cal (per mole of reaction)}\\ &= (14.8 \text{ kcal})\left(4.184 \frac{\text{kJ}}{\text{kcal}}\right)\\ &= 61.9 \text{ kJ (per mole of reaction)}\end{aligned}$$

EXAMPLE 11.8 A certain reaction is found to have an equilibrium constant of 2.64 at 10 °C and 18.4 at 100 °C. Calculate $\Delta S°$ for this reaction.

SOLUTION The strategy for solving this problem is to compute $\Delta H°$ from Eq. (11.28), and then compute $\Delta S°$ from $\Delta H°$ and K for either temperature from Eq. (11.25):

$$\ln\left(\frac{2.64}{18.4}\right) = 2.303 \log\left(\frac{2.64}{18.4}\right) = \frac{\Delta H°}{R}\left(\frac{283 - 373}{283 \times 373}\right)\frac{\text{deg}}{\text{deg}^2}$$

$$-1.94 = \frac{\Delta H°}{R}\left(\frac{-90}{10.56 \times 10^4 \text{ deg}}\right)$$

$$\frac{\Delta H°}{R} = (1.94)\left(\frac{10.56}{9.0} \times 10^3 \text{ deg}\right)$$

Then, focusing on 10 °C, we write Eq. (11.25) as follows:

$$2.303 \log K_{283} = \ln K_{283} = \frac{-\Delta H°}{R}\left(\frac{1}{283 \text{ deg}}\right) + \frac{\Delta S°}{R}$$

$$2.303 \log(2.64) = \ln(2.64) = -\left[(1.94)\left(\frac{10.56}{9.0} \times 10^3 \text{ deg}\right)\left(\frac{1}{283 \text{ deg}}\right)\right] + \frac{\Delta S°}{R}$$

$$0.971 = -8.04 + \frac{\Delta S°}{R}$$

$$\begin{aligned}\Delta S° &= (0.971 + 8.04)R = 9.01R \\ &= (9.01)(1.987 \text{ cal deg}^{-1}\text{mol}^{-1}) \\ &= 17.9 \text{ cal deg}^{-1} \text{ (per mole of reaction)}\end{aligned}$$

STUDY PROBLEM 11(f)

Calculate the $\Delta S°$ and $\Delta H°$ values of a reaction for which K is 0.28 at 100 °C and three times as large at 115 °C.

STUDY PROBLEM 11(g)

Calculate the value of the equilibrium constant of a specific reaction at 50 °C if $K = 1.3 \times 10^{-3}$ at 100 °C and $\Delta H° = 48.1$ kcal.

EXAMPLE 11.9 From the data of Tables 11.1 and 10.2 for the reaction $N_2(g) + O_2(g) \rightleftharpoons 2NO(g)$ calculate K for 25 °C and estimate it for a typical cylinder temperature of an internal combustion engine, say, 2500 °C. The results here are of interest relative to air pollution by nitrogen oxides and their derivatives in photochemical smog, which often gives the reddish brown color to polluted air.

SOLUTION The calculation of K for 25 °C is straightforward since the $\Delta G°_{f,298}$ values are available for all three species. Thus,

$$\begin{aligned}\Delta G°_{298} &= 2\Delta G°_{f,298,NO(g)} - (\Delta G°_{f,298,N_2(g)} + \Delta G°_{f,298,O_2(g)}) \\ &= 2(20.72) - 0 \\ &= 41.44 \text{ kcal}\end{aligned}$$

Then, from Eq. (11.19),

$$\begin{aligned}\log K &= \frac{-41.44 \times 10^3 \text{ cal mol}^{-1}}{(1.987 \text{ cal deg}^{-1}\text{mol}^{-1})(298.15 \text{ deg})(2.303)} \\ &= -30.37 \\ &= -31.0 + 0.63 \\ K &= 4.25 \times 10^{-31}\end{aligned}$$

For estimating K for 2500 °C, we can use Eq. (11.28), assuming $\Delta H°$ and $\Delta S°$ constant over the temperature range 25 °C to 2500 °C. (While this is not strictly a valid assumption to make in the present problem, it will suffice to provide an estimate of the desired quantity.) With this simplifying assumption, we can apply Eq. (11.28):

$$2.303 \log\left(\frac{K_{2773}}{K_{298}}\right) = \frac{\Delta H°}{1.987}\left(\frac{2773 - 298}{2773 \times 298}\right)$$

From the data of Table 10.2, we can calculate $\Delta H°$ at 25 °C for the reaction as follows:

$$\begin{aligned}\Delta H° &= 2\Delta H°_{f,298,NO(g)} - (\Delta H°_{f,298,N_2(g)} + \Delta H°_{f,298,O_2(g)}) \\ &= 2(21.60) - (0 + 0) \\ &= 43.20 \text{ kcal}\end{aligned}$$

Substituting this value of $\Delta H°$ into the above equation yields

$$2.303 \log\left(\frac{K_{2773}}{K_{298}}\right) = \frac{4.320 \times 10^4}{1.987}\left(\frac{2475}{2773 \times 298}\right)$$

$$\log\left(\frac{K_{2773}}{K_{298}}\right) = 28.27$$

$$\frac{K_{2773}}{K_{298}} = 1.9 \times 10^{28}$$

$$\begin{aligned} K_{2773} &= (1.9 \times 10^{28})K_{298} \\ &= (1.9 \times 10^{28})(4.25 \times 10^{-31}) \\ &= 8.1 \times 10^{-3} \end{aligned}$$

This is a value that could lead to significant concentrations of NO in a combustion cylinder of an internal combustion engine.

STUDY PROBLEM 11(h)

Estimate K and $\Delta G°$ for 400 °C for the reaction discussed in Example 11.9.

Examples 11.7, 11.8, and 11.9 are cases in which $\Delta H°$ and $\Delta S°$ values are assumed constant over the ranges of temperature of interest in each case for which data points are shown; that is why the solid line in Fig. 11.4, for example, is linear. However, in general, this is not always strictly the case, since both $\Delta H°$ and $\Delta S°$ may change somewhat as the temperature is varied.

Even if the thermodynamic quantities $\Delta H°$ and $\Delta S°$ do change somewhat as the temperature is varied, plots of ln K or $\Delta G°/T$ vs $1/T$ are still useful. These plots may take a curved form as shown in the accompanying diagram. In such a case, the slope (the change in $\Delta G°/T$ per unit change in $1/T$), can be obtained at any temperature, say T', by taking the slope of the straight line that runs tangent to the curve at the point corresponding to that temperature. This gives $\Delta H°$ for that temperature T', and $\Delta S°$ for the same temperature can be obtained by reading the value of $\Delta G°/T$ for the temperature T' from the graph and using Eq. (11.16).

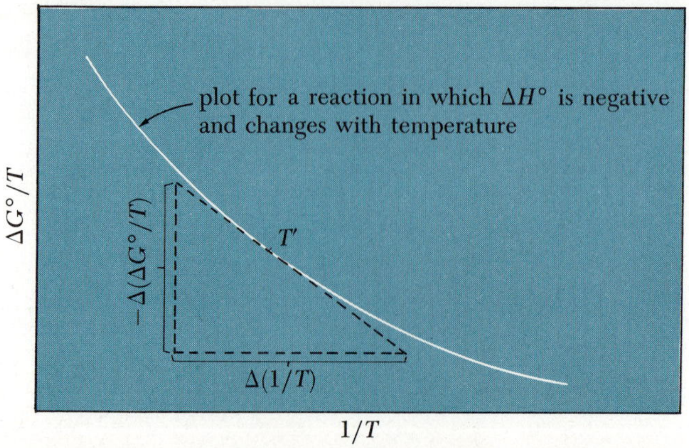

The Relations between ΔH_f°, ΔG_f°, and ΔS_f°

The definition and applications given above for $\Delta G^\circ_{f,T}$ are entirely consistent with the definition of $\Delta H^\circ_{f,T}$ in Chap. 10. In both cases attention is focused on the change in a thermodynamic variable that accompanies the transformation of matter in the form of certain elemental substances (in the reference states at 25 °C) to another substance in some particular standard state (at some specific temperature).

$$\begin{pmatrix}\text{matter in elemental}\\ \text{reference states at 25 °C}\end{pmatrix} \rightarrow \begin{pmatrix}\text{matter in standard state}\\ \text{substances at specified temperature}\end{pmatrix}$$

Using the NO(g) case of Example 11.9 as an example, 20.72 kcal is the free energy change, called the standard molar free energy of formation of NO(g) (at 25 °C), for the process

$$\tfrac{1}{2}O_2(g) + \tfrac{1}{2}N_2(g) \rightleftharpoons NO(g)$$

$$25\ °C,\ \text{reference states} \rightarrow 25\ °C,\ \text{standard state}$$

Correspondingly, 21.6 kcal is the enthalpy change accompanying this process, and is called the standard molar enthalpy of formation of NO(g) (at 25 °C) (see Table 10.2). With ΔG° and ΔH° defined for this "formation" reaction at the specific temperature 25 °C, ΔS° for this same process is precisely defined at this temperature by Eq. (11.16). We denote this corresponding entropy change by the analogous symbol $\Delta S^\circ_{f,298}$ for the standard entropy of formation of NO(g) (at 25 °C). Thus,

$$\Delta G^\circ_{f,298} = \Delta H^\circ_{f,298} - (298.15)(\Delta S^\circ_{f,298}) \qquad (11.29)$$

Of course, an expression like Eq. (11.29) is valid for any temperature. Hence,

$$\Delta G^\circ_{f,T} = \Delta H^\circ_{f,T} - T\,\Delta S^\circ_{f,T} \qquad (11.30)$$

EXAMPLE 11.10 From the data of Example 11.9, calculate $\Delta S^\circ_{f,298}$ for NO(g).

SOLUTION Using Eq. (11.29), we have

$$20.72 \times 10^3 = 21.60 \times 10^3 - (298.15\ \text{deg})(\Delta S^\circ_{f,298})$$

$$\Delta S^\circ_{f,298} = \frac{-8.8 \times 10^2\ \text{cal mol}^{-1}}{-298.15\ \text{deg}}$$

$$= 3.0\ \text{cal deg}^{-1}\text{mol}^{-1} = 12\ \text{J deg}^{-1}\text{mol}^{-1}$$

In Example 11.10, we have obtained $\Delta S^\circ_{f,298}$ without any need of measuring entropies directly. Obtaining this result is a typical manifestation of the power of thermodynamics; by applying thermodynamic logic rigorously, it is often possible, if the appropriately related thermodynamic properties have been measured, to get desired quantities without measuring them directly.

STUDY PROBLEM 11(i)

From the data given in Tables 10.2 and 11.1, calculate $\Delta S^\circ_{f,298}$ for $SO_3(g)$.

STUDY PROBLEM 11(j)

From the data given in Tables 10.2 and 11.1, calculate ΔS°_{298} for the reaction

$$SO_2(g) + \tfrac{1}{2}O_2(g) \rightleftharpoons SO_3(g)$$

OPTIONAL

11.4 A Statistical Rationale for Chemical Equilibrium

The equations defining ΔS and ΔG and relating them to the equilibrium constant of a reaction were stated without derivation, and with little justification at the beginning of this chapter. We here develop a simple statistical model that should make the general mathematical forms of Eqs. (11.14) and (11.17) plausible.

To introduce the statistical approach, we shall focus attention on a hypothetical mechanical model that bears little superficial similarity to a chemical system but has functional analogies that may prove illuminating. We shall visualize the mechanical model in a series of imaginary "experiments." (Experiments of this kind, which are designed to be carried out in the mind rather than in the laboratory, are sometimes called gedanken experiments from the German word *Gedanke,* for "thought.")

The Mechanical System

The mechanical system we wish to imagine is an incredibly huge box, containing an astronomical number of marbles, say 10^{23} of them; the box is constructed with a three-dimensional lattice of square "plateaus" and square "valleys," appearing as illustrated in Fig. 11.5. The areas of each plateau and each valley are all the same. The vertical distance between the valley floors and plateau tops is designated h, and is adjustable. The number of peaks and valleys is imagined to be even greater, much greater, than the number of marbles—say 10^{30} of each. The marbles have cross-sectional areas very much smaller than those of the individual peaks and valleys. Imagine furthermore that we can shake the entire box as vigorously as we want, thereby imparting to the marbles an amount of kinetic energy determined by how hard we shake the box. This kinetic energy would not be the same for all marbles. There would be an average value of kinetic energy (denoted here by $\overline{KE}$); some marbles would have more kinetic energy than the average, and some less. That is, there would be a distribution of kinetic energies. Most marbles would have kinetic energies near $\overline{KE}$, but a few would have much higher or much lower values.

Let us also imagine that the system consisting of the box and the marbles with average kinetic energy $\overline{KE}$ can be monitored by motion picture photography or some other means, so that the exact positions of each marble at any given time can be recorded. This would allow us to count the number of times that marbles make contact with plateau tops or valley floors during a short period of observation. We arbitrarily choose to label the situation of a marble in contact with the flat top of a plateau as a "U" state (U for "up") and the situation of a marble in contact with the floor of a valley as a "D" state (D for "down"). Thus, we specifically focus attention on two distinctly recognizable states of the marble, the U state and the D state. This method of observation then allows one to count the number of U states and D states that exist instantaneously during a short period of observation. We shall denote these numbers by the symbols $\langle U \rangle$ and $\langle D \rangle$, respectively.

The Analogy to a Chemical System

With this image and characterization of the hypothetical mechanical system in mind, we consider a series of straightforward gedanken experiments based on the

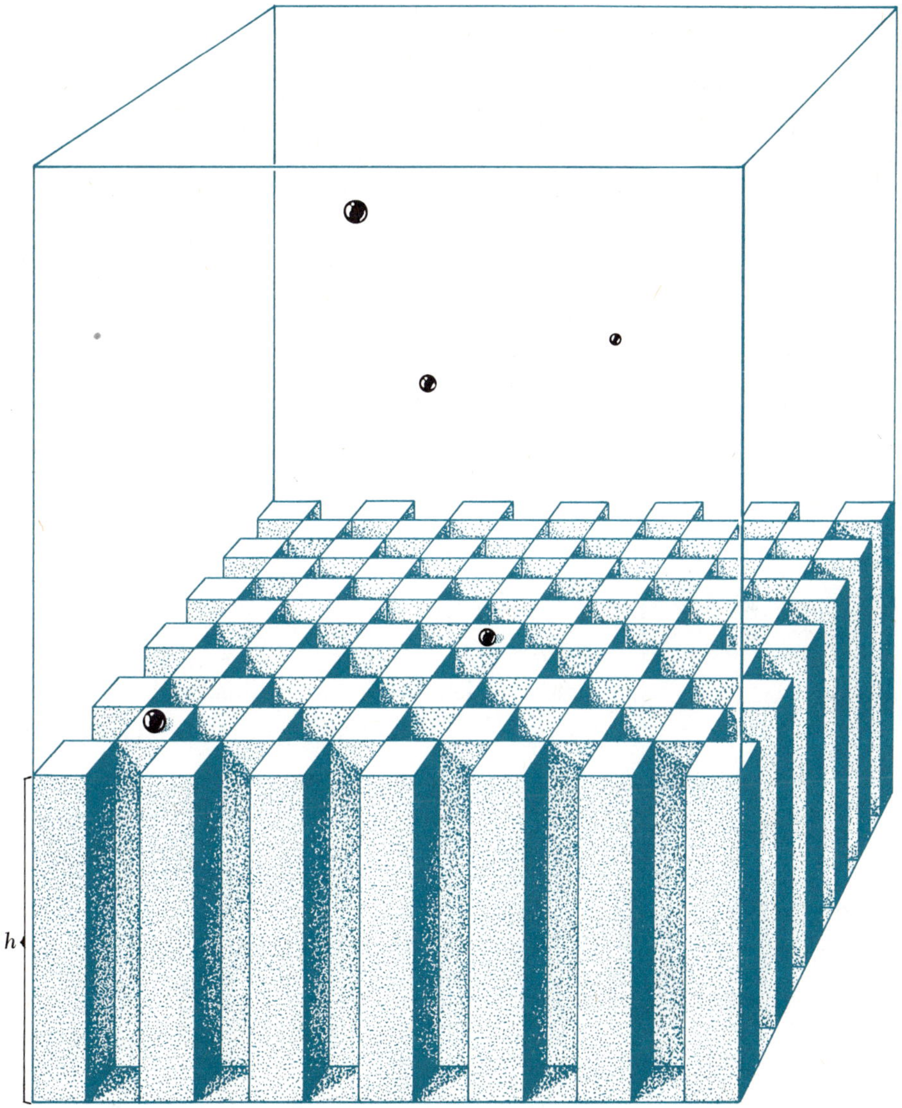

Figure 11.5
Representation of the type of construction of the "hypothetical" mechanical (box) system with alternating peaks and valleys.

system, hoping to develop insight into the nature of equilibrium in chemical systems. In bridging the gap between the chemical case and the mechanical model, we shall want to keep the following parallels in mind.

1. The average kinetic energy $\overline{KE}$ of the marbles in the mechanical system bears a direct analogy to the average thermal energy of the molecules in a chemical system. As we shall discuss in considerable detail in Chap. 15, this is proportional to the Kelvin temperature T of the chemical system.
2. The height h of the plateaus, which is directly proportional to the gravitational potential energy difference between the states U and D, can be thought of as analogous to the difference between the energies of two possible states in a dynamic chemical equilibrium.

OPTIONAL

We have seen in Sec. 10.9 that as far as the transfer of heat is concerned, the relevant energy difference for processes carried out at constant pressure is ΔH, the enthalpy of products minus the enthalpy of reactants for a chemical reaction. Since we are mainly interested in chemical equilibria under conditions of constant pressure, we shall conceptually identify h in the mechanical system with ΔH for the transformation of one chemical state into the other chemical state, or reactants into products.

These two states of a chemical system are represented by the formulas on the left and right sides of the balanced equation for the chemical reaction in question. We shall refer to these chemical states by the symbols "L" and "R." This means that the L and R chemical states are being viewed in direct analogy with U and D mechanical states. (The identification of either U or D with R, and the other with L, depends on which chemical state has the higher energy in the chemical reaction considered.) Thus, the mechanical process of transferring a marble from one position (state) to the other can be symbolized as either U $\rightarrow$ D or D $\rightarrow$ U, and is by analogy associated here with a chemical transformation that can be represented symbolically as

$$\mathrm{L} \rightarrow \mathrm{R} \tag{11.31}$$

A more complete symbolism for process (11.31) is given by

$$a\mathrm{A} + b\mathrm{B} + \cdots \rightarrow m\mathrm{M} + n\mathrm{N} + \cdots \tag{11.32}$$

where A and B represent reactant species and M and N represent product species. Thus, each of the two specified states (U and D) of the marble in the mechanical system bears an analogy either to aA molecules plus bB molecules, . . . , or to mM molecules plus nN molecules, . . . , in the chemical system searching for its equilibrium condition.

For chemical systems, we are familiar with the mass action quotient Q, which equals the equilibrium constant K at equilibrium:

$$Q = \frac{[\mathrm{M}]^m[\mathrm{N}]^n \ldots}{[\mathrm{A}]^a[\mathrm{B}]^b \ldots} = K \text{ at equilibrium} \tag{11.33}$$

We recall that K is a measure of the relative concentrations, or relative populations, of the states R and L at equilibrium (Chap. 4). We can emphasize this meaning by the condensed symbolism

$$K = \frac{<\mathrm{R}>}{<\mathrm{L}>} \tag{11.34}$$

where $<\mathrm{R}>/<\mathrm{L}>$ stands for the value of Q at equilibrium. In direct analogy to the chemical relation symbolized in Eq. (11.34), we shall be interested in a ratio of populations "measured" in the mechanical system and defined as

$$K' = \frac{<\mathrm{R}'>}{<\mathrm{L}'>} = \frac{<\mathrm{D}>}{<\mathrm{U}>} \text{ or } \frac{<\mathrm{U}>}{<\mathrm{D}>} \tag{11.35}$$

The choice depends on whether R is identified with D or with U. In this expression the populations $<\mathrm{R}'>$ and $<\mathrm{L}'>$ are each to be identified with either $<\mathrm{U}>$ or $<\mathrm{D}>$, the identities being determined by whether we wish to identify the higher-

TABLE 11.2 Parallel Relations between the Mechanical Model and Chemical Equilibrium. Corresponding Variables

Mechanical Case	*Chemical Case*
h	$\Delta H = H_R - H_L$
Marble	Equilibrating set of chemical species, e.g., aA + bB or mM + nN
KE of the marbles due to their motion	Thermal energy of molecules (Kelvin temperature)
U → D	L → R, exothermic reaction
D → U	L → R, endothermic reaction
Population ratio $\dfrac{<R'>}{<L'>} = K'$	$K = \dfrac{[M]^m[N]^n \ldots}{[A]^a[B]^b \ldots} = \dfrac{<R>}{<L>}$
$K'\,(\text{exo}) = \dfrac{<D>}{<U>}$	Exothermic reaction K
$K'\,(\text{endo}) = \dfrac{<U>}{<D>}$	Endothermic reaction K

energy mechanical state U with the right side R, or with the left side L of the particular chemical reaction in question. In other words, the identification of $<R'>$ and $<L'>$ in Eq. (11.35) in terms of U and D states of the mechanical system depends on whether we identify the D state with L and the U state with R, or the U state with L and the D state with R. This choice depends on whether the chemical process represented by the mechanical analogy occurs with a *decrease* in enthalpy (analogous to U → D) or an *increase* in enthalpy (analogous to D → U), that is, whether it is exothermic or endothermic. Some of these parallels are summarized in Table 11.2. With these parallels in mind, let us consider the following "experiments."

Hypothetical Experiments; The Energy Factor

Experiment 1

The experiment is begun by adjusting h to a particular value h_1, the subscript denoting the experiment number, and carefully placing each marble initially in the U state (on a plateau). Then the entire system is shaken for a while, and kinetic energy thereby imparted to the randomly bouncing marbles, so that a particular distribution of kinetic energies and a corresponding average kinetic energy $\overline{KE}$ is maintained. For this particular experiment, let us denote $\overline{KE}$ as $\overline{KE}_1$ and adjust it to equal about one-third the value of the potential energy difference between the U and D states determined by h_1.

The next step in the experiment is to determine the relative populations of the U and D state. Since we designed this experiment such that $\overline{KE} = \frac{1}{3}h_1$, the corresponding distribution of kinetic energies of the marbles would allow only a relatively small fraction of the marbles at any given instant to have enough kinetic energy to reach the top of a peak. The minimum kinetic energy required for a marble to reach the top of a peak would be h_1, and only a small fraction of the marbles in this experiment would have that much kinetic energy. Thus, at any instant, or over any period

OPTIONAL

of observation, many more marbles will exist in the state D than in the state U, and the ratio $<\mathrm{D}>/<\mathrm{U}>$ will be much greater than 1.

If we wish the mechanical model to represent an *exothermic* chemical in process, then the initial (L′) state is U and the final (R′) state is D; therefore L → R for the chemical system would be identified with U → D in the mechanical system. This is easily related to Experiment 1, in which all the marbles were *initially* in the U state. In such a case we identify $K = <\mathrm{R}>/<\mathrm{L}>$ with the mechanical population ratio

$$K'_{(\mathrm{exo})} = \frac{<\mathrm{D}>}{<\mathrm{U}>} \tag{11.36}$$

For this particular experiment we have concluded that this ratio is larger than unity; that is,

$$K'_{(\mathrm{exo})_1} = \frac{<\mathrm{D}>}{<\mathrm{U}>} \tag{11.36}$$

which is greater than one (>1). The subscript 1 again refers to Experiment 1.

Our common sense regarding the statistical nature of such an experiment tells us that if we repeat the experiment many times under exactly the same conditions, the same result would be obtained each time. Since the number of marbles involved is so large, the population ratios $<\mathrm{D}>/<\mathrm{U}>$ resulting from identical experiments would be "equal," at least to more significant figures than we should want to work with, even with computers. Furthermore, we should note that once the marbles have taken the energy distribution corresponding to $\overline{KE}$, then no matter how long we wait before making the population measurements, they will always lead to the same result, that is, the same $K'_{(\mathrm{exo})_1}$. Thus, we may logically conclude that $K'_{(\mathrm{exo})_1}$ is a constant for this experiment, any time it is carried out in the prescribed manner.

Experiment 2

Let us imagine conducting a second experiment, just the same as the first, with the same h and $\overline{KE}$, except that all the marbles will be initially placed on valley floors, or in D states. This would correspond in the analogous chemical system to starting out with the species in their lower-energy state; in other words, this mechanical experiment is closely identified with an *endothermic* chemical reaction. Thus, we shall conceptually identify the expression L → R for a chemical system with the expression D → U in the mechanical system.

Accordingly, the mechanical population ratio analogous to the chemical expression $K = <\mathrm{R}>/<\mathrm{L}>$ will now be

$$K_{(\mathrm{endo})} = \frac{<\mathrm{U}>}{<\mathrm{D}>} \tag{11.37}$$

Once the marbles have bounced about sufficiently to attain the kinetic energy distribution corresponding to some average kinetic energy $\overline{KE}$, then it is no longer of any consequence whether they started out in U or D states. Once $\overline{KE}$ reaches the steady value $\overline{KE}_1$, a particular distribution between U and D states is maintained that does not depend on the starting positions of the marbles. Thus, the value for the ratio $<\mathrm{U}>/<\mathrm{D}>$ that would be obtained in population measurements during this experiment would be the same as the value obtained in Experiment 1.

$$\left(\frac{<\mathrm{U}>}{<\mathrm{D}>}\right)_2 = \left(\frac{<\mathrm{U}>}{<\mathrm{D}>}\right)_1 = \left(\frac{<\mathrm{D}>}{<\mathrm{U}>}\right)_1^{-1}$$

which is less than 1. Thus,

$$K'_{(\mathrm{endo})_2} = \left(\frac{<\mathrm{U}>}{<\mathrm{D}>}\right)_2 = (K'_{(\mathrm{exo})_1})^{-1}$$

which is less than one (<1).

Experiment 3

In this experiment we should repeat the procedure of Experiment 1 (or 2, since we have seen that we get the same population distribution), except that only 10^{20} marbles would be used, instead of 10^{23}. This, in a chemical analog, would correspond to using fewer reacting systems, meaning fewer molecules or moles. The number of U states, $<\mathrm{U}>$, or of D states, $<\mathrm{D}>$, that would be counted in the measurements of this experiment would be smaller than the corresponding numbers from Experiments 1 and 2. However, common sense tells us that since the relative probabilities of occurrence of these two states will be independent of the number of marbles, the ratio $<\mathrm{U}>/<\mathrm{D}>$ will be the same for all three of these experiments:

$$\left(\frac{<\mathrm{U}>}{<\mathrm{D}>}\right)_1 = \left(\frac{<\mathrm{U}>}{<\mathrm{D}>}\right)_2 = \left(\frac{<\mathrm{U}>}{<\mathrm{D}>}\right)_3$$

Naturally, a similar result would have been obtained had a larger number of marbles, say 10^{24}, been used under the same conditions. Thus, the population distribution, defined in a specific manner, is the same for these three experiments.

Conclusions from Experiments 1, 2, and 3

If we accept the analogies drawn between the mechanical system and a chemical reaction, then the experiments described with the former lead us to conclusions pertinent to the latter. Thus for the mechanical system, we find that once h and $\overline{KE}$ are specified, $<\mathrm{U}>/<\mathrm{D}>$ is fixed, independent of the starting point (the direction of approach) or of the number of marbles. Similarly, it now seems plausible, even inescapable, that the value of Q for a given chemical reaction at a given temperature is, at equilibrium, independent of the direction of approach to the equilibrium, or of the amounts of materials employed. This means that whether we start with A and B or with M and N of Eq. (11.32) or with some mixture of all these species, and irrespective of the amounts of any of the species, the value of Q at equilibrium equals some constant K.[*] This K is characteristic of the particular reaction at the given temperature. In a sense, this is merely a rationalization of the law of chemical equilibrium (Sec. 4.2).

Experiment 4

This experiment would be carried out in almost exactly the same manner as any of the first three. The only difference is that the box would be shaken somewhat harder than in the previous cases, so that a distribution of kinetic energies would be imparted

[*] In these arguments we are assuming that the minimum number of atomic, molecular, or ionic building blocks required for the reaction are present. Thus A alone, or A and M alone, would not be capable of setting up the equilibrium indicated by Eq. (11.32).

OPTIONAL

to the marbles that would bring about a higher value of $\overline{KE}$. Thus, $\overline{KE}_4 > \overline{KE}_1$, and a larger percentage of marbles at any given instant in time would have sufficient kinetic energy to overcome the potential energy difference represented by h_1, and could thus attain the U state. (This is the same result we should get by reducing the value of h from the value h_1 to some smaller value, while maintaining $\overline{KE}$ equal to $\overline{KE}_1$.) In Experiment 4, the occurrence of the state U is more probable than the occurrence of this state in Experiment 1. Consequently, when the population measurements are made, the resulting ratio $<\mathrm{U}>/<\mathrm{D}>$ will be larger than it was in Experiments 1, 2, and 3. That is,

$$\left(\frac{<\mathrm{U}>}{<\mathrm{D}>}\right)_4 > \left(\frac{<\mathrm{U}>}{<\mathrm{D}>}\right)_1 = \left(\frac{<\mathrm{U}>}{<\mathrm{D}>}\right)_2 = \left(\frac{<\mathrm{U}>}{<\mathrm{D}>}\right)_3$$

On the basis of the parallel that we have drawn between chemical equilibrium and the mechanical system, the results of this experiment would imply that an increase in temperature of a chemical system will influence the equilibrium constant in a manner favorable to the higher energy state. This is consistent with Le Chatelier's principle (Chap. 4). We have learned that increasing the temperature decreases the the value of K for an exothermic chemical reaction and increases the value of K for an endothermic chemical reaction, the same predictions we should draw by comparing it with our mechanical analog.

Experiment 5

This experiment is carried out in the same fashion as any of the first three, except that we should try to impart an infinite amount of kinetic energy to the marbles. This is not a situation that could actually be achieved, even if we could construct the mechanical system. Nevertheless, we can imagine approaching it in a gedanken experiment, as we add more and more energy to the system. This would shift the distribution of kinetic energies so that the probability of occurrence of extremely high values would increase, with the average kinetic energy becoming arbitrarily higher and higher. As we imagine it, when $\overline{KE}$ becomes *much* higher than h_1, then most of the marbles have more than enough kinetic energy to overcome the potential energy difference between states D and U. If $\overline{KE}$ is increased towards infinity, essentially all the marbles achieve kinetic energies much greater than what is required to get from the D state to the U state. In this situation the potential energy difference represented by h is no longer of any consequence in determining the relative probabilities of occurrence of the states U and D. This is similar to the case of a multimillionaire in a candy store; a differential of a cent or two in the price of different varieties of candy will not be a factor in his choice. In other words, the effect of causing $\overline{KE}$ to approach an infinite value is equivalent to the effect of shrinking h to a value of zero, that is, equivalent to making the box have a flat bottom painted like a checkerboard, where the only distinction between the states U and D is analogous to the distinction between the red and black squares on an actual checkerboard.

The result of this experiment (or of one using a system with $h = 0$) would be that the populations of U states and D states occurring in the time of measurement would be equal. These results can be summarized symbolically by the equations

$$K'_{(\text{endo})_5} = \left(\frac{<\text{U}>}{<\text{D}>}\right)_{KE=\infty} = \left(\frac{<\text{U}>}{<\text{D}>}\right)_{h=0} = 1 = K'_{(\text{exo})_5} \tag{11.38}$$

Experiment 6

The procedure used here would be the same as in Experiments 1, 2, or 3 except that we should give a *lower* value of $\overline{KE}$ to the marbles. The result of such a change from the conditions of Experiment 1 would be simply the reverse of the results found in Experiment 5. That is, the occurrence of the U state would be even less probable than it was in Experiment 1, and the ratio $<\text{D}>/<\text{U}>$ measured in this case would be larger than $K'_{(\text{exo})_1}$.

As we imagined for Experiment 5, the effect of changing $\overline{KE}$ while maintaining h constant could be duplicated by changing h and maintaining $\overline{KE}$ constant. By either decreasing the average kinetic energy available to the marbles, or by increasing the value of h, we could cause the resulting population distribution to move towards the D state at the expense of the U state. The implications for chemical systems to be drawn from this experiment are analogous, but opposite in sense, to those from Experiment 5. This result implies that decreasing the temperature will tend to favor the lower energy state of a chemical system, that is, increase K for an exothermic chemical reaction and decrease K for an endothermic reaction. These results corroborate the conclusions drawn from Le Chatelier's principle.

Experiment 7

Let us follow up on the change in details from Experiment 1 to Experiment 6, and imagine the effect of continuing the reduction of KE all the way down to zero. We note that since no marbles can have less than zero kinetic energy, all marbles must have zero kinetic energy in this experiment for the average to be zero. However, we do not wish to consider an experimental procedure in which no kinetic energy at all is ever imparted to the marbles, for in such a case the marbles would be "trapped" in their initial configuration, whatever it is. That is, even marbles placed on top of plateaus wouldn't be able to roll off if they had no kinetic energy whatsoever. In such a case, marbles would be trapped in states with potential energies higher than what they "should" have according to their zero kinetic energies. A more appropriate experimental procedure would be to give the marbles initially enough kinetic energy to establish the appropriate population distribution (equilibrium), and then slowly reduce the average kinetic energy by shaking the box more and more slowly until a zero value of $\overline{KE}$ was approached. In this way, the marbles could maintain their equilibrium population distribution throughout the entire experiment.

For a mechanical system there is no particular problem in visualizing the system with $\overline{KE} = 0$; in that situation we expect all the marbles to be in the D state, so that the ratio $<\text{D}>/<\text{U}>$ is infinite, or the value of $<\text{U}>/<\text{D}>$ is zero. Of course, the same result would be obtained for any nonzero value of $\overline{KE}$ if h could somehow be made infinitely large.

The lesson to be learned for chemical equilibrium from Experiment 7 is that decreasing the temperature towards absolute zero shifts the equilibrium distribution to the lower energy state entirely (at absolute zero the thermal energy of molecules is zero). If one could reduce the temperature to zero, then K for an exothermic reaction would become infinite and K for an endothermic reaction would become zero. With chemical species, trapping in higher energy states would also be a

OPTIONAL

TABLE 11.3 Qualitative Relations among h, $\overline{KE}$, <U>, and <D>, Drawn from Experiments 4, 5, 6, and 7

Experimental Conditions		*Population Ratios*	
Variation	*Qualification*	<D>/<U>	<U>/<D>
Increases in $\overline{KE}$	h fixed	Decreases	Increases
Decreases in $\overline{KE}$	h fixed	Increases	Decreases
$\overline{KE} \rightarrow \infty$	h fixed	$\rightarrow 1$	$\rightarrow 1$
$\overline{KE} \rightarrow 0$	h fixed	$\rightarrow \infty$	$\rightarrow 0$
Smaller h	$\overline{KE}$ fixed	Smaller	Larger
Larger h	$\overline{KE}$ fixed	Larger	Smaller
$h = 0$	$\overline{KE}$ fixed	1	1
$h \rightarrow \infty$	$\overline{KE}$ fixed	$\rightarrow \infty$	$\rightarrow 0$

problem, so absolute zero would have to be approached gradually by a continuous removal of energy.*

Conclusions from Experiments 4, 5, 6, and 7

The pertinent results for the mechanical system are summarized in Table 11.3. There it is seen that the ratio <U>/<D> increases with increasing $\overline{KE}$, and with decreasing h, and that this ratio decreases with decreasing $\overline{KE}$ and with increasing h. This suggests that the feature that determines population distribution in the mechanical system may be something like the ratio of h and $\overline{KE}$. (This implication is explored in some detail later in this chapter.)

The Entropy Factor

All of our considerations in this exercise of relating mechanical to chemical models have been concerned with energy relations in mechanical population distributions or in chemical equilibrium. We have found that K' depends only on the values of h and $\overline{KE}$. This will be absolutely true in the mechanical case if the number of plateaus is precisely the same as the number of valleys, or alternatively expressed, if the number of ways in which a marble can attain the U state is exactly the same as the number of ways in which it can attain the D state. This is a special condition that we chose to impose upon the system for Experiments 1 through 7; the system need not have been designed that way, however. Indeed, in chemical systems the property analogous to the number of plateaus or peaks is in general not the same for the states represented by the two sides of a chemical equation. This avenue can be explored by considering some additional hypothetical experiments.

*For chemical systems there is an important reason why the absolute zero of temperature cannot quite be reached. How can one remove the "last traces" of thermal energy from chemical species that are already very near absolute zero? Although, as we have seen, heat flows spontaneously from a body at higher temperature to one at a lower temperature, we cannot find a body at a temperature lower than absolute zero. The problem is analogous to the futility of the direct, personal attempt of a trembling man to steady the hand of another trembling friend. Thus, we can imagine approaching absolute zero but not actually attaining it.

Experiment 8

The procedure to be followed in this experiment will be exactly the same as those used in Experiments 1 and 2; however, the system itself will be somewhat different. The box to be used will have 3×10^{30} valleys and 1×10^{30} plateaus. As $h = h_1$ and $\overline{KE} = \overline{KE}_1$ for this experiment, the modification introduced here would be expected to favor the D state over the U state by a factor 3 compared with what was found in Experiments 1 or 2. That is, when the measurement of populations is made, we should expect to find that

$$\left(\frac{<\mathrm{D}>}{<\mathrm{U}>}\right)_8 = 3\left(\frac{<\mathrm{D}>}{<\mathrm{U}>}\right)_1; \quad \text{that is,} \quad \left(\frac{<\mathrm{U}>}{<\mathrm{D}>}\right)_8 = \frac{1}{3}\left(\frac{<\mathrm{U}>}{<\mathrm{D}>}\right)_1 \tag{11.39}$$

Experiment 9

Again the procedure and $\overline{KE}$ and h will be the same as those used in the first two experiments, with the system itself further modified. The change in the apparatus here is a box constructed so that there are 100×10^{30} plateaus and 1×10^{30} valleys. In this case, we intuitively expect that the D state will be favored over the U state in the population distribution by a factor 100 less than the value for that factor in Experiments 1 or 2. Thus, we should expect to find that

$$\left(\frac{<\mathrm{D}>}{<\mathrm{U}>}\right)_9 = \frac{1}{100}\left(\frac{<\mathrm{D}>}{<\mathrm{U}>}\right)_1; \quad \text{that is,} \quad \left(\frac{<\mathrm{U}>}{<\mathrm{D}>}\right)_9 = 100\left(\frac{<\mathrm{U}>}{<\mathrm{D}>}\right)_1 \tag{11.40}$$

Conclusions from Experiments 8 and 9

We conclude that on the basis of these two experiments, some other factor besides the one depending on energy differences h and average available energy $\overline{KE}$ must influence the determination of K' for the mechanical system. Let us now try to explore these intuitive conclusions more analytically. We start by recognizing that there are basically two types of factors that determine the value of the population ratios in the mechanical system under discussion.

Factors That Determine the Mechanical Population Distribution

We shall denote one of the factors that determine the mechanical population distribution by the symbol K'_{E}. It is the ratio of the probability that a marble will come into contact with any *single* valley floor to the probability that a marble will come into contact with any *single* plateau top. This factor is concerned with only one plateau and one valley at a time—or one pair of alternatives; it is therefore independent of the number of peaks or the number of valleys available. It is determined solely by energy considerations, i.e., the values of h and $\overline{KE}$.

The second factor involved in determining the value of K' is independent of h and $\overline{KE}$, and depends only on the number of plateaus and the number of valleys. But this is simply the ratio of the number of possible D sites that may be attained (the number of ways of attaining the D state) to the number of possible U sites that may be attained (the number of ways of attaining the U state). We define the **degeneracy** of a state as the number of possible ways of attaining that state, and use the symbol Ω to stand for degeneracy. Then, the above statement is equivalent to the following mathematical relations:

$$\frac{\Omega_{\mathrm{D}}}{\Omega_{\mathrm{U}}} = \frac{\text{(degeneracy of D state)}}{\text{(degeneracy of U state)}} = \frac{\text{(no. of valleys)}}{\text{(no. of plateaus)}} \tag{11.41}$$

OPTIONAL

The explicit expressions for the degeneracy ratios for the mechanical analogs of exothermic and endothermic processes are then

$$K'_{\Omega(\text{exo})} = \frac{\Omega_D}{\Omega_U} \quad \text{and} \quad K'_{\Omega(\text{endo})} = \frac{\Omega_U}{\Omega_D} \tag{11.42}$$

The total ratio of probabilities of occurrence of the two states can be expressed as the product of the two factors (for the exothermic analog):

$$\begin{aligned} K' &= \frac{(\text{total prob. of occurrence of D states})}{(\text{total prob. of occurrence of U states})} \\ &= \frac{(\text{prob. of occurrence of any one D state})}{(\text{prob. of occurrence of any one U state})} \times \frac{(\text{no. of ways of attaining D})}{(\text{no. of ways of attaining U})} \\ &= K'_E \times K'_\Omega \end{aligned} \tag{11.43}$$

The first factor K'_E depends on energy relations, whereas the other, K'_Ω, depends on degeneracies. Apparently, if one wishes to understand the fundamental nature of chemical equilibrium, corresponding analogies to the factors K'_E and K'_Ω must both be considered.

Intuitive Rationalization of the Dependence of K on ΔH and T

Now let us extend our web of analogy connecting the chemical system and the hypothetical mechanical system by asserting that just as K' can be written as a product of two factors K_E and K_Ω, so the equilibrium constant K can be written:

$$K = K_E \cdot K_\Omega \tag{11.44}$$

where K_E and K_Ω have meanings that are analogous to the meanings of K'_E and K'_Ω in the mechanical system. Thus, we expect K_E to depend only on ΔH and T. The value K_E is the ratio of the probability of occurrence of a particular reactant state situation to the probability of occurrence of a specific product situation. This ratio of probabilities is independent of the number of ways of attaining each state, that is, the degeneracies. The degeneracies are involved in determining the ratio K_Ω, which is simply Ω_R/Ω_L, the ratio of the degeneracies of the right and left states of the chemical reaction.

It is now clear that the relations summarized in Table 11.3, and the experiments and reasoning that led up to them, were concerned only with the factor K_E, since in Experiments 1 through 7 the number of plateaus was equal to the number of valleys. That is, in those experiments, $K'_\Omega = \Omega_U/\Omega_D = \Omega_D/\Omega_U = 1$. Recognizing this and the chemical-vs-mechanical anlogies that were summarized in Table 11.2, one can qualitatively predict the behavior of K_E (the energy-dependent factor of a chemical equilibrium constant) with variations of ΔH and T. The similar statements given in Table 11.4 for the chemical case were obtained essentially by replacing the symbols h, $\overline{KE}$, and $<\text{U}>/<\text{D}>$ of Table 11.3 with the corresponding symbols ΔH, T, and K_E. Variations of temperature can be carried out readily for any chemical reaction. Variations of ΔH correspond to changing from one reaction to another (with a different enthalpy difference between reactants and products). The behavior

TABLE 11.4 Qualitative Relations for Chemical Systems Drawn from Table 11.3 by Analogy

Variation	*Qualification*	*"Energy factor" K_E of the Equilibrium Constant* — *Exothermic case*	*Endothermic case*	
Increase in T	ΔH fixed	Decreases	Increases	Changes in temperature
Decrease in T	ΔH fixed	Increases	Decreases	
$T \to \infty$	ΔH fixed	Would approach 1	Would approach 1	
$T \to 0$	ΔH fixed	Would approach ∞	Would approach 0	
Smaller ΔH	T fixed	Smaller	Larger	Changing ΔH, i.e., changing chemical reactions (comparing different chemical reactions)
Larger ΔH	T fixed	Larger	Smaller	
$\Delta H = 0$	T fixed	1	1	
Extremely large ΔH ($\Delta H \to \infty$)	T fixed	$\to \infty$	$\to 0$	

summarized in Table 11.4 depends, of course, on the sign and magnitude of ΔH. For example, if ΔH is positive ($H_R > H_L$, endothermic reaction), then as T is varied from near zero towards infinity, K_E varies from a value of zero towards unity. (Table 11.4 gives the behavior of K_E as the temperature is changed.) This is consistent with the view that only the lower energy state would be appreciably populated when almost no thermal energy is available to the system, whereas the population of the higher energy state increases as the molecules attain higher thermal energies. Finally, the higher energy state becomes as probable as the lower energy state when essentially an infinite amount of thermal energy is available. That $K_{E(\text{endo})}$ increases as T increases derives from the fact that the sign of ΔH is positive in this case. This mathematical relation is shown in Fig. 11.6.

If ΔH is negative ($H_R < H_L$, exothermic reaction), then the opposite sort of behavior is noted. As T is varied from near zero towards infinity, K_E changes from a value approaching infinity towards the number 1, for the same reasons on which the discussion of the previous example was based. Here, the fact that K_E decreases as T

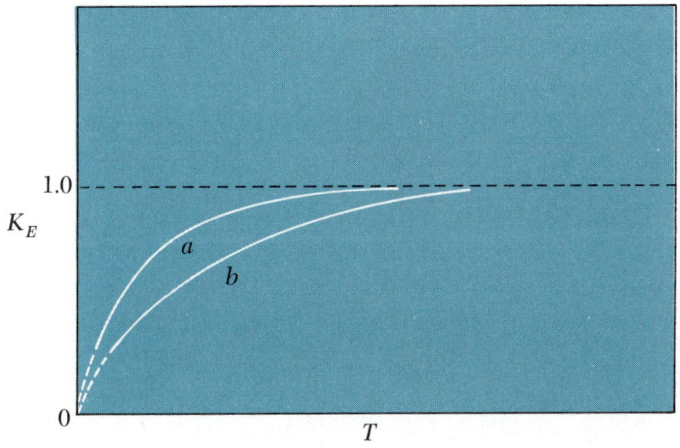

Figure 11.6 Plot of K_E vs T for two chemical reactions, a and b, with different positive ΔH values; $\Delta H_a < \Delta H_b$.

OPTIONAL

increases results from the fact that ΔH is negative. This behavior is shown in Fig. 11.7. In both this case and the endothermic case, the dependence of K_E on temperature parallels what would be predicted from Le Chatelier's principle.

We can also recall at this point that on the basis of the behavior discussed for the specific mechanical case with $h = 0$, K_E would equal 1 for all values of T if ΔH happened to equal zero. That is, if there is no enthalpy difference between the chemical states represented by the two sides of the chemical equation, the amount of thermal energy available to the molecules is of no consequence in determining K_E, or K itself. The plot for such a case would correspond to the dashed straight line in Fig. 11.7.

We want to be especially careful not to misrepresent the meaning of Figs. 11.6 and 11.7; they are plots of K_E, not K, against T. If we wish to consider K at a given temperature, we should have to multiply the value of K_E at that temperature times K_Ω. The value of K_Ω is, in our simple model, independent of T.

In Table 11.4 and Figs. 11.6 and 11.7, we described the qualitative dependence of K_E on ΔH and T. This dependence appears to be a function of the ratio $\Delta H/T$; i.e., a decrease in ΔH has the same effect as an increase in T. We know that if ΔH is positive (endothermic case), the larger the $\Delta H/T$ ratio, the smaller the value of K_E. Similarly, if ΔH is negative (exothermic case), the larger the magnitude of $\Delta H/T$, the larger the value of K_E.

We now seek a mathematical expression that describes these relations. While this can be derived by rigorous developments based on approaches called statistical thermodynamics or classical thermodynamics, we shall content ourselves here with discovering this expression on the basis of plausibility. Hence, without derivation or further justification, we note that a mathematical expression with the desired properties is of the exponential type,

$$K_E = e^{-c\Delta H/T} \tag{11.45}$$

where c is a constant, and e is the base of natural logarithms (Eq. 10.6).

We see that the expression given in Eq. (11.45) has the expected property of

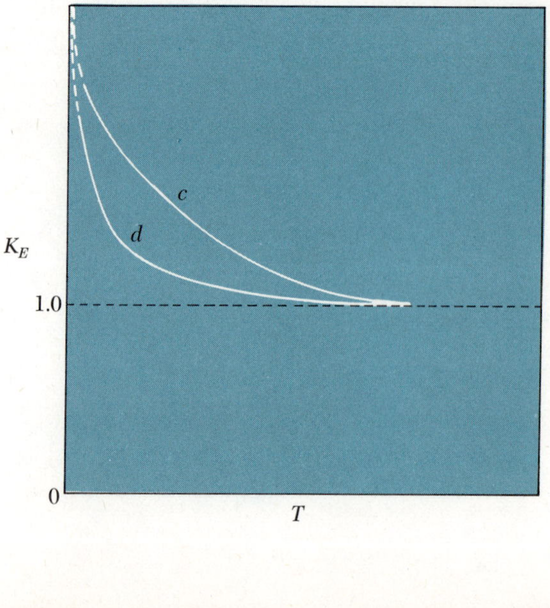

Figure 11.7
Plot of K_E vs T for two chemical reactions, c and d, with different negative ΔH values; $\Delta H_c < \Delta H_d$; that is, the magnitude of ΔH_c is greater than the magnitude of ΔH_d.

depending on ΔH and T through the ratio $\Delta H/T$. It is known from more formal developments that the constant that appears in Eq. (11.45) has the value $1/R$, where R is the gas constant met earlier. If ΔH is expressed in calories and T in degrees Kelvin, then R has the value 1.987 cal $\text{deg}^{-1}\text{mol}^{-1}$. If ΔH is expressed in joules, then R is 8.31 J $\text{deg}^{-1}\text{mol}^{-1}$:

$$K_E = e^{-\Delta H/RT} \tag{11.46}$$

We easily see that Eq. (11.46) provides an expression for K_E that is consistent with the qualitative requirements that we know it must meet. If $\Delta H = 0$, then $K_E = e^0 = 1$ for any temperature (neglecting the unattainable absolute zero of temperature). If ΔH is positive, then the exponent of e is negative, which indicates that K_E must be a number less than 1. Furthermore, Eq. (11.46) demands that for a given positive ΔH, the magnitude of the negative exponent $-\Delta H/RT$ decreases when T is increased, so that K_E increases. Ultimately, as T approaches the limit of infinity, the value of the exponent approaches zero, rendering K_E equal to e^0, or unity. This is the behavior that was predicted graphically in Fig. 11.6. For a chemical reaction with a negative ΔH, the behavior shown in Fig. 11.7 is contained in Eq. (11.46). In this case the exponent is positive, which means that according to Eq. (11.46), for any temperature (less than infinity) K_E must be greater than 1. This is as we know it should be from our conclusions from experiments 4, 5, 6, and 7 for the case where D is identified with R and U with L. Furthermore, the value of this positive exponent and the value of K_E itself decrease as T is increased, until as T approaches infinity, the exponent approaches zero and K_E approaches unity. Thus, Eq. (11.46) satisfies all our qualitative criteria; it is precisely what is obtained from formal thermodynamic developments.

The Dependence of K on ΔS

Our discussions of the mechanical system have indicated that a second factor, which does not depend on temperature and which we have denoted by K_Ω, is involved in K. For that system, Eqs. (11.41) and (11.42) tell us that K_Ω equals the ratio of the degeneracies of the two states involved, where the degeneracy is the number of ways the state can exist, the number of valleys and the number of plateaus in this mechanical case. We now assert that these same basic concepts apply to the case of chemical equilibrium. In this case we state the relation as

$$K_\Omega = \frac{\Omega_\text{R}}{\Omega_\text{L}} \tag{11.47}$$

where Ω_R now refers to the degeneracy of the chemical state represented by the right side of the chemical equation, and a corresponding meaning applies to Ω_L. Before we pursue this any further, it would be advantageous to explore the meaning of Ω from a different angle.

We have defined Ω to be the number of ways in which a particular state can be attained, or can exist. To clarify our meaning of this concept, let us consider a homely analogy that carries with it the crux of the idea. Suppose that the "system" in which we are interested is a student looking for a place to sit down in a building that houses

OPTIONAL

a 400-seat auditorium and a lobby with a telephone booth containing a stool. There are two conventional types of places to sit down in this building, and we shall identify each as providing a possible "state" for the "system." We use the label "*A* state" for a student seated in an auditorium seat, and the label "*B* state" for a student seated on the stool in the telephone booth. According to these specifications, Ω_A is 400, the number of ways in which the student can achieve the *A* state, that is, the number of auditorium seats from which to choose. Similarly Ω_B is unity, since there is only one way in which the *B* state can be achieved—if the student seats himself in the single telephone booth seat. With the system and states so specified, we can explore some other words or phrases often associated in chemical considerations with the concept of degeneracy; these are *randomness, disorder,* and *specification of detail,* or the amount of detail that can be specified for the system by simply knowing the state in which it exists.

It is straightforward to see that there is an inverse type of relation between specification of detail and degeneracy; if the former is large, the latter must be small. For example, since Ω_B is unity, we know in detail precisely where the student is seated if he is identified as being in a *B* state. By contrast, since Ω_A is 400, specifying that the student is in the *A* state tells us only that he is in one of 400 possible chairs—not very much detail. Associated with the idea of a limited specification of detail are the connotations of randomness and disorder. A precise specification of detail in a state brings to mind a highly ordered state—one with low disorder or low randomness. Thus, a low degeneracy is equivalent to considerable specification of detail, low randomness, and low disorder.

An example of these same concepts in a chemical system is provided by the hypothetical reaction of atomic hydrogen and carbon to form benzene in the gas phase, as shown in Eq. (11.48):

$6H + 6C \rightarrow$ 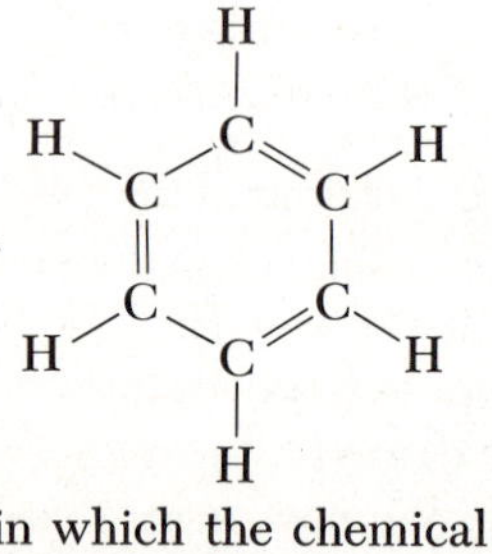 (11.48)

Clearly, there are more ways in which the chemical state represented by the left side of Eq. (11.48) can exist than the number of ways in which the benzene state can exist. In the former, the hydrogen and carbon atoms can be anywhere in the container; in the latter there are only a limited number of geometrical arrangements of atoms that are allowable, namely, those that correspond to the molecular structure of benzene. Thus, Ω_L must be greater than Ω_R, and we may logically consider the L state to be a state of higher disorder and randomness than the R state. Again, we see that the state of lower Ω (the R state) carries with it more specified detail about the system than the state of higher Ω (the L state). By knowing that the system is in the L state, we know only that all twelve atoms are in some specified container, which is not much detail. In contrast, by knowing that the system is in the R state, we know that all twelve atoms are located within a few angstroms of each other, that each carbon atom is bonded directly to one hydrogen atom and two other carbons, and that each hydrogen atom is directly bonded to a carbon atom, and so forth—considerable detail.

In chemistry, the word commonly associated with the concept of degeneracy and the words *randomness* and *disorder*, is *entropy*. This term is commonly thought of as a measure of the randomness or disorder of a physical or chemical system, and we want to retain this general meaning. However, we want our understanding of entropy to be more precise than a definition that depends only on a qualitative meaning. Accordingly, we define the property called entropy, symbolized by the letter S, in terms of the equation

$$S = R \ln \Omega \tag{11.49}$$

where R is the gas constant (Chaps. 3 and 10). This equation precisely relates what is called the entropy of a given state to the degeneracy of that state; Eq. (11.49) can be considered to be a fundamental postulate. In fact, Eq. (11.49) is frequently taken as the fundamental postulate of statistical thermodynamics, an elegant branch of science that deals with the relation between theoretical models of submicroscopic systems and the thermodynamic properties of macroscopic quantities of bulk materials. Like any postulate in science, Eq. (11.49) has justification only when its predictions and conclusions, properly applied, agree with all known experimental facts. Using Eq. (11.49) as the starting point, if a "correct" submicroscopic model of a certain physical or chemical system is used, the logical mathematical application of the formalism of statistical thermodynamics will lead to predictions of thermodynamic properties that agree with experimental results (to Eq. 11.3, for example). We shall now explore the consequences of Eq. (11.49) within the framework of our own statistical approach.

First let us note that the qualitative relation between degeneracy and entropy in the general discussion above is borne out by Eq. (11.49). A large value of Ω corresponds to a large value of S, and a small Ω corresponds to a small S. By the nature of the definition of S in Eq. (11.49), we can see readily that it is a property of a system that depends only on the state of the system and not on how the state was reached.

Now, by simply using the meaning of the natural logarithm (see Eq. 10.6a), we obtain from Eq. (11.49) the relation

$$\Omega = e^{\ln \Omega} = e^{S/R} \tag{11.50}$$

By substituting this result into Eq. (11.47), we obtain

$$K_\Omega = \frac{e^{S_R/R}}{e^{S_L/R}} = e^{(S_R - S_L)/R}$$

Then, noting that $\Delta S = S_R - S_L$, we obtain the result

$$K_\Omega = e^{\Delta S/R} \tag{11.51}$$

This now provides us with the relation that we sought, a mathematical expression for K_Ω in terms of ΔS.

At this point we can see clearly that Eqs. (11.49) and (11.51) are totally consistent with our picture of the relation between degeneracy and K_Ω. For example, when

OPTIONAL

$\Omega_R < \Omega_L$, then $S_R < S_L$ and ΔS is less than zero; hence $e^{\Delta S/R}$ has a negative exponent, which renders K_Ω less than 1. In the special case where $\Omega_R = \Omega_L$, then $\Delta S = 0$, and K_Ω becomes e^0, which is unity. If $\Omega_R > \Omega_L$, then $S_R > S_L$, and $\Delta S > 0$; hence $e^{\Delta S/R}$ has a positive exponent, so K_Ω is greater than 1.

If the results given for the two factors K_E and K_Ω in Eqs. (11.46) and (11.51) are substituted into Eq. (11.44), we obtain

$$K = e^{-\Delta H/RT}e^{\Delta S/R}$$

$$= e^{-(\Delta H - T\Delta S)/RT} = e^{-\Delta G/RT}$$

These relations are precisely the same as Eqs. (11.14) and (11.13). Hence, the statistical plausibility argument for the functional dependence of K on ΔH and ΔS is complete.

SUMMARY

The position of equilibrium and the equilibrium constant for a chemical reaction depend on both energy relations and purely statistical factors. Certain criteria for spontaneity and the relations between equilibrium constants and state variables exist. When the entropy change is defined in terms of the heat absorbed by the system in a reversible process, the relation $\Delta S > 0$ for the universe is found to be a condition for spontaneity. If we focus on just the system of interest, a useful spontaneity criterion can be stated in terms of a new state variable, the Gibbs free energy, $G = H - TS$; the spontaneity criterion is $\Delta G < 0$ for the system, which also gives new insight into the condition of chemical equilibrium ($\delta G = 0$ at equilibrium). The mathematical relations between K, ΔG, ΔS, ΔH, and T can be rationalized by a statistically oriented plausibility argument.

Standard free energies of formation and entropies of formation can be defined in direct correspondence to the standard heat of formation. From tables of ΔG_f° values, ΔG° values and equilibrium constants can be computed for reactions that may not have been studied directly. From the mathematical relations between K, ΔG°, ΔH°, ΔS°, and T one can compute ΔH° and ΔS° values from measurements of K at more than one temperature, or predict how an equilibrium constant will change when the temperature is changed.

STUDENT CHECKPOINTS

After studying this chapter, the student should be able to:

1. Explain why both statistical factors and energy relations are important in determining the value of K.
2. Describe a reversible process.
3. Give the classical definition of *entropy* in terms of a reversible process.
4. Explain the criterion for a spontaneous process in terms of the total entropy change.
5. Define *Gibbs free energy* in terms of enthalpy and entropy.
6. Compute ΔG°, ΔH°, or ΔS° from the other two values for a specific temperature.
7. Describe the criterion for a spontaneous process in terms of ΔG.

8. Compute K at a specific temperature from a knowledge of $\Delta G°$ or of $\Delta S°$ and $\Delta H°$.
9. Compute $\Delta G°$ values for any reaction if the pertinent $\Delta G_f°$ values are available.
10. Compute $\Delta H°$ and $\Delta S°$ for a reaction whose equilibrium constant is known at two temperatures.
11. Predict the value of K at some temperature for a reaction whose K value at another temperature and $\Delta H°$ are known.
12. Discuss the importance of thermodynamics in allowing calculations of quantities such as $\Delta G°$ for reactions that may never have been studied from other data on reactions that have been studied.

EXERCISES

11.1.* What would the final equilibrium distribution of Cl_2 species be if one started with 0.1 mol ${}^{35}_{17}Cl—{}^{37}_{17}Cl$ in the left chamber of Fig. 11.1 and 0.1 mol of ${}^{35}_{17}Cl—{}^{37}_{17}Cl$ in the right chamber?

11.2. Calculate the change in entropy that a system would undergo in a reversible process carried out at 150 °C if 16.7 J of heat is absorbed from the environment.

11.3.* Calculate the change in entropy that a system would undergo in a constant-pressure, reversible process at 100 °C if ΔE for the system is 47.9 kcal and the only work performed by the system, 14.8 kcal, is $P\,\Delta V$ work.

11.4. Determine whether each of the following processes would be spontaneous:
(a) ΔS for the system is 19.3 J deg^{-1} and ΔS for the surroundings is -17.8 J deg^{-1}.
(b) ΔS for the system is -19.3 J deg^{-1} and ΔS for the surroundings is 17.8 J deg^{-1}.
(c) ΔS for the system is -17.8 J deg^{-1} and ΔS for the surroundings is 19.3 J deg^{-1}.

11.5.* Calculate ΔH for a process at 130 °C for which ΔG is -19.2 kcal and ΔS is 16.5 cal deg^{-1}.

11.6. Calculate ΔS for a process at 13 °C for which ΔG is -39.4 J and ΔH is 6.2 J.

11.7. Calculate ΔG in kcal for a chemical reaction at 300 °K that has ΔH equal to -109.2 kcal and ΔS equal to 32.6 J deg^{-1}.

11.8.* Determine whether each of the following processes would occur spontaneously at 25 °C:
(a) ΔG for the system $= -3.5$ kJ.
(b) $\Delta H = -16.4$ kcal, and $\Delta S = 11.4$ cal deg^{-1}.
(c) $\Delta H = 42.2$ cal and $\Delta S = -8.1$ cal deg^{-1}.
(d) $\Delta H = 19.2$ kJ and $\Delta S = 41.6$ J deg^{-1}.
(e) $\Delta H = 86.4$ kcal and $\Delta S = 8.9$ cal deg^{-1}.

11.9. Calculate the value of $\Delta G°$ and the value of K for a chemical reaction at 40 °C for which $\Delta H°$ is 19.6 kcal and $\Delta S°$ is 22.3 cal deg^{-1}.

11.10.* Using Table 11.1, calculate $\Delta G°$ and K at 25 °C for the reaction

$$N_2O(g) + \tfrac{1}{2}O_2(g) \rightleftharpoons 2NO(g)$$

11.11. (a) Using Table 11.1, calculate $\Delta G°$ and K for the following reaction at 25 °C:

$$3C_2H_4(g) \rightleftharpoons C_6H_6(l) + 3H_2(g)$$

(b) Write the equilibrium expression for this reaction.

11.12.* (a) Using Table 11.1, calculate $\Delta G°$ and K at 25 °C for the reaction

$$H_2O_2(g) + CO(g) \rightleftharpoons H_2O(g) + CO_2(g)$$

(b) Write the equilibrium expression for this reaction.

11.13. (a) Calculate $\Delta G°$ and K for the following reaction at 25 °C:

$$Ca(s) + H_2SO_4(aq) \rightleftharpoons CaSO_4(s) + H_2(g)$$

(b) Write the equilibrium expression for this reaction.

11.14. (a) From the data in Table 11.1, calculate $\Delta G°$ and K for the following reaction at 25 °C:

$$SO_2(g) + \tfrac{1}{2}O_2(g) \rightleftharpoons SO_3(g)$$

(b) Write the equilibrium expression for this process.

11.15.* Indicate whether ΔG is 0, > 0, or < 0 for each of the following processes:
(a) $H_2O(l) \rightleftharpoons H_2O(g)$ at 100 °C, 1 atm.
(b) $H_2O(l) \rightleftharpoons H_2O(g)$ at 150 °C, 1 atm.
(c) $H_2O(l) \rightleftharpoons H_2O(g)$ at 50 °C, 1 atm.
(d) $H_2O(l) \rightleftharpoons H_2O(s)$ at 50 °C, 1 atm.
(e) $H_2O(l) \rightleftharpoons H_2O(s)$ at −50 °C, 1 atm.
(f) The process depicted in Fig. 11.1.

11.16.* Indicate whether ΔS is 0, > 0, or < 0 for the system in each of the following processes:
(a) $H_2O(l) \rightleftharpoons H_2O(g)$ at 100 °C, 1 atm.
(b) $H_2O(l) \rightleftharpoons H_2O(s)$ at 0 °C, 1 atm.
(c) $H_2O(g) \rightleftharpoons 2H(g) + O(g)$, 10,000 °C, 1 atm.
(d) The process depicted in Fig. 11.1.

11.17.* Indicate whether each of the following statements is true or false:
(a) If a process occurs spontaneously, ΔH must be less than 0 for the system in that process.
(b) If a process occurs spontaneously, ΔG must be less than 0 for the system in that process.
(c) If a process does not occur spontaneously, ΔG must be greater than 0 for the system in that process.
(d) If a process does not occur spontaneously, ΔS must be less than 0 for that system in the process.

11.18. When a protein is "denatured" (for example, by heating an egg), the complex array of interactions (for example, hydrogen bonding) that give the protein a specific and highly reproducible structure in its native state is overcome, and it assumes a "random coil" arrangement. Do you expect ΔS to be positive or negative for denaturation?

11.19.* Indicate each of the following statements as true or false:
(a) If a given reaction occurs spontaneously, then $\Delta H < T\Delta S$ for the system in this reaction.

(b) If a certain reaction occurs spontaneously, then ΔS for the system must be less than ΔS for the environment.
(c) If a certain reaction does not occur spontaneously, then $\Delta S°$ for the system may be greater than $\Delta H°/T$.

11.20. If a substance is to be thermodynamically stable relative to its elements (that is, ΔG for decomposition to the elements should be > 0), what can you say about its ΔG_f° value?

11.21. (a) From the data in Tables 10.2 and 11.1, calculate $\Delta H°$, $\Delta G°$, $\Delta S°$, and K for 25 °C for the following reaction:

$$3C_2H_2(g) \rightleftharpoons C_6H_6(l)$$

(b) Write the equilibrium expression for this process.

11.22. (a) From the data in Tables 10.2 and 11.1, calculate $\Delta H°$, $\Delta G°$, $\Delta S°$, and K for 25 °C for the following reaction:

$$2NO_2(g) \rightleftharpoons N_2O_4(g)$$

(b) Write the equilibrium expression for this reaction.

11.23. (a) Calculate $\Delta H°$, $\Delta G°$, $\Delta S°$, and K for 25 °C for the following reaction:

$$H_2(g) + Cl_2(g) \rightleftharpoons 2HCl(g)$$

(b) Write the equilibrium expression for this reaction.
(c) Calculate the equilibrium concentrations of all species present in a sample that is labeled pure HCl gas at 25 °C and 1.0 atm pressure.

11.24.* (a) Calculate $\Delta G°$ and K for the following reaction at 25 °C:

$$H_2S(g) + 2O_2(g) \rightleftharpoons 2H^+(aq) + SO_4^{2-}(aq)$$

(b) Write the equilibrium expression for this process.

11.25. (a) Calculate $\Delta H°$, $\Delta G°$, $\Delta S°$, and K for the following process at 25 °C:

$$Cu_2S(s) + O_2(g) \rightleftharpoons 2Cu(s) + SO_2(g)$$

(b) Write the equilibrium expression for this reaction.

11.26.* One of the main sources of energy for "running" the various functions of our bodies is the oxidation of glucose:

$$C_6H_{12}O_6 + 6O_2(g) \rightleftharpoons 6CO_2(g) + 6H_2O(l)$$

The $\Delta G°$ value for this process at 25 °C is -2.88×10^3 kJ. Calculate $\Delta G^\circ_{f,298}$ for $C_6H_{12}O_6(s)$.

11.27. In photosynthesis in plants, in which the energy of sunlight is stored, one main chemical reaction of interest is the reverse of what is shown in Example 11.26. (a) Could the photosynthesis reaction occur spontaneously if one bubbled CO_2 into water? (b) How is it possible for the photosynthesis reaction to occur in plants?

11.28. (a) From the data given in Chap. 10 and Exercise 11.26, calculate $\Delta S°$ for 25 °C for the reaction shown in Exercise 11.26. (b) How does this result relate to your intuitive understanding of entropy?

11.29.* According to the spontaneity criterion given in expression (11.6), the total entropy change for the universe in any spontaneous process is positive. What can you conclude about the total entropy of the universe with the passage of time?

11.30. Two important chemical reactions that are under study for the production of useful fuels are (a) one called the "water-gas" reaction, brought about by passing steam over hot charcoal or coal:

$$C(s) + H_2O(g) \rightleftharpoons CO(g) + H_2(g)$$

and (b) a process under study in attempts at "gasifying" coal:

$$CO(g) + 2H_2(g) \rightleftharpoons CH_3OH(l)$$

For 25 °C, calculate the $\Delta G°$ and K values for each of these processes (which determine how "easily" the products can be produced) and the $\Delta H°$ values for complete combustion of the products (these determine how much heat is available from burning these materials).

11.31. One of the main reasons for opposition to supersonic transports is based on the fear that NO emitted in the jet exhaust will deplete ozone, O_3, in the upper atmosphere. Ozone protects us from potentially dangerous ultraviolet radiation from the sun by absorbing this radiation. The controversy centers about the reaction

$$O_3(g) + NO(g) \rightleftharpoons O_2(g) + NO_2(g)$$

Calculate $\Delta G°$ and K for this process for 25 °C, knowing that $\Delta G^\circ_{f,298}$ is 39.1 kcal/mol for $O_3(g)$.

11.32.* Given the $\Delta G°$ values of the reactions (a) and (b) at 25°, calculate $\Delta G°$ for reaction (c) for 25 °C.
(a) $2HCl(g) + Ag_2O(s) \rightleftharpoons 2AgCl(s) + H_2O(l)$, $\Delta G° = -61.00$ kcal.
(b) $2HCl(g) + Ag_2CO_3(s) \rightleftharpoons H_2O(l) + 2AgCl(s) + CO_2(g)$, $\Delta G° = -53.37$ kcal.
(c) $Ag_2CO_3(s) \rightleftharpoons Ag_2O(s) + CO_2(g)$, $\Delta G° = ?$

11.33. From the data given in Table 11.1 and the fact that $\Delta G°$ at 25 °C is -9.72 kcal for the reaction

$$2NO(g) + Cl_2(g) \rightleftharpoons 2NOCl(g)$$

at 25 °C, calculate $\Delta G^\circ_{f,298}$ for NOCl.

11.34. Using the data in Tables 10.2 and 11.1, calculate $\Delta G°$, $\Delta S°$, and $\Delta H°$ for 25 °C for the complete combustion of (a) gaseous octane (C_8H_{18}), and (b) liquid benzene at 25 °C.

11.35. For the past few decades, the Western industrialized world has been relentlessly burning up all conveniently available fossil fuels to provide energy for its industrialized economy and society. When these energy resources are depleted, new energy sources like solar and nuclear energy may be available; but the kinds of compounds present in currently available fossil fuels may still be required as starting materials for the chemical industry, as in synthesizing plastics, pharmaceuticals, and fabrics. Taking octane and benzene as typical of the types of the desired structures, use the results of Exercise 11.34 to comment on the thermodynamics of using new energy sources to convert CO_2 and H_2O in the future to synthesize replacements for the raw materials we are now burning up.

11.36.* Calculate $\Delta H°$ and $\Delta S°$ (in J and J deg^{-1}) for a chemical reaction for which K is 1.3×10^{-6} at 110 °C and 9.3×10^{-3} at 150 °C.

11.37. Calculate the equilibrium constant for 50 °C for a reaction for which $\Delta H°$ is 46.6 kcal if K is 2.6×10^3 at 140 °C.

11.38. (a) From the data in Table 11.1, calculate K for the synthesis of NH_3 from H_2 and N_2 at 25 °C.
(b) Using the result from (a) and the data of Table 10.2, estimate K for the ammonia synthesis at 1000 °C.

11.39.* (a) Using the data in Tables 10.2 and 11.1, estimate the equilibrium constant of the following reaction at 25 °C, 1000 °C, and 2000 °C:

$$CO(g) + \tfrac{1}{2}O_2(g) \rightleftharpoons CO_2(g)$$

(b) Considering the results from part (a), if you were designing a furnace to minimize the production of CO (in other words, to provide complete combustion of some fuel), would you want a "cool" flame or a very hot flame?

11.40. From the data in Tables 10.2 and 11.1, calculate $\Delta S_f°$ for water at 25 °C.

11.41.* From the fact that for diamond at 25 °C, $\Delta H_f°$ is 0.45 kcal/mol and $\Delta S_f°$ is -0.78 cal $deg^{-1}mol^{-1}$, comment on the future of diamonds thermodynamically.

ANSWERS TO SELECTED EXERCISES

11.1 0.10 mol $^{35}_{17}Cl—^{37}_{17}Cl$:0.050 mol $^{35}_{17}Cl—^{35}_{17}Cl$:0.050 mol $^{37}_{17}Cl—^{37}_{17}Cl$. **11.3** 168.0 cal deg^{-1}. **11.5** -12.6 kcal. **11.8** (a) spontaneous; (b) spontaneous; (c) not spontaneous; (d) not spontaneous; (e) not spontaneous. **11.10** $\Delta G° = 16.68$ kcal; $K = 5.8 \times 10^{-13}$. **11.12** (a) $\Delta G° = -92.36$ kcal; $K = 5.5 \times 10^{67}$; (b) $K = [H_2O(g)][CO_2(g)]/[H_2O_2(g)][CO(g)]$. **11.15** (a) 0; (b) < 0; (c) > 0; (d) > 0; (e) < 0; (f) < 0. **11.16** (a) > 0; (b) < 0; (c) > 0; (d) > 0. **11.17** (a) false; (b) true; (c) true; (d) false. **11.19** (a) true; (b) false; (c) false. **11.24** (a) $\Delta G° = -189.78$ kcal; $K = 1.55 \times 10^{139}$; (b) $K = [H^+(aq)]^2[SO_4^{-2}]/[H_2S(g)][O_2]^2$. **11.26** -217.4 kcal. **11.29** increases. **11.32** 7.63 kcal. **11.36** $\Delta H° = 2.99 \times 10^5$ J; $\Delta S° = 668$ J deg^{-1}. **11.39** (a) $K = 1.166 \times 10^{45}$ at 25 °C; $K = 1.19 \times 10^7$ at 1000 °C; $K = 92.65$ at 2000 °C; (b) cool. **11.41** Because both $\Delta H_f°$ and $-T\Delta S_f°$ are positive, $\Delta G_f°$ must be positive for diamonds. Hence, they are thermodynamically unstable at 25 °C with respect to decomposing to graphite. However, this transformation is never observed, because diamonds are *kinetically* stable.

ANSWERS TO STUDY PROBLEMS

11(a) yes. **11(b)** -56.8 kJ. **11(c)** 50.6 kJ. **11(d)** $\Delta G° = -93.5$ kcal; $K = 5.8 \times 10^{54}$. **11(e)** $\Delta G° = -120.67$ kcal; $K = 3.19 \times 10^{88}$. 11(f) $\Delta H° = 21.1$ kcal; $\Delta S° = 54.0$ cal deg^{-1}. **11(g)** $K = 5.6 \times 10^{-8}$. **11(h)** $K = 1.92 \times 10^{-13}$; $\Delta G°_{400} = 39.16$ kcal. **11(i)** -19.89 cal $deg^{-1}mol^{-1}$. **11(j)** -22.67 cal deg^{-1}.

12

Electrochemistry

Chemistry is in many ways the science of how various collections of tiny charged particles behave. We can make that statement because matter is built out of nuclei and electrons. The essentially electrical character of chemistry is obvious in some situations, as in operating a flashlight from a "dry cell," or producing hydrogen and oxygen by applying a voltage to strips of metal in an aqueous solution. In most chemical processes, however, as in the burning of coal and the rusting of iron, the electrical foundations of the structure of matter are not so obvious. There are some aspects of chemical reactions that emphasize certain features of the electrical nature of matter, and these features can be summarized in concepts with wide applicability in chemistry. **Electrochemistry** is the study concerned with chemical changes that are brought about by electric current or that are capable of generating electric current.

12.1 Some Properties of Ionic Solutions

In Secs. 3.8 and 7.1, a general overview of electrolyte solutions was presented. In such solutions, the dissolved substance exists in an ionized form (to an extent that depends on whether the substance is a strong or a weak electrolyte). One clear-cut type of evidence for the existence of ions in solution is electrical conductivity.

The conductivity experiment is pictured in Figs. 3.7 and 7.4. When a voltage is applied across the electrodes in a conductivity apparatus, the positive ions, called **cations,** migrate towards the negative electrode, the **cathode,** and the negative ions, called **anions,** migrate toward the positive electrode, the **anode.** These migrations of ions constitute a net transfer of negative electrical charge through the solution from the cathode to the anode, which accompanies the transfer of negative charge, the electrons, from the anode to the cathode in the external circuit. This combination constitutes a completed circuit with its resulting electric current. Ordinarily only very small amounts of electric current are allowed to flow in a conductivity experiment, since one wishes to make such measurements on solutions that have not been drastically perturbed by electrical currents and the buildup of high concentrations of ions about the electrodes.

Some of the ions that do migrate to one of the electrodes in a conductivity apparatus undergo chemical transformation. For example, if current is allowed to pass in a conductivity experiment on an aqueous 3 *M* HCl solution, which contains the ions H^+ and Cl^-, one finds that bubbles of $H_2(g)$ appear at the cathode and bubbles of $Cl_2(g)$ appear at the anode, as shown in Fig. 12.1. Under the influence of an applied voltage, the ions H^+ migrate towards the cathode and are transformed into neutral H in the form of H_2, which makes the gas bubbles—that is, they become discharged; the ions Cl^- migrate to the anode and are discharged as Cl_2. (O_2 may also form, depending on conditions.) If the solution contained Cu^{2+} as the cations, instead of H^+, then the discharging at the cathode would produce copper atoms, which would plate out on the cathode. The discharging processes in these cases can be represented by the following symbolic chemical equations:

$$H^+ + e^- \rightleftharpoons \tfrac{1}{2}H_2(g) \qquad (12.1)$$

$$Cl^- \rightleftharpoons \tfrac{1}{2}Cl_2(g) + e^- \qquad (12.2)$$

$$Cu^{2+} + 2e^- \rightleftharpoons Cu(s) \qquad (12.3)$$

In each of these equations the symbol for an electron is included, underscoring the important fact that each of these discharging processes involves the transfer of an electron to or from an electrode. In the discharging of the cations H^+ and Cu^{2+}, electrons are transferred from the electrode (cathode) to the ions, making them

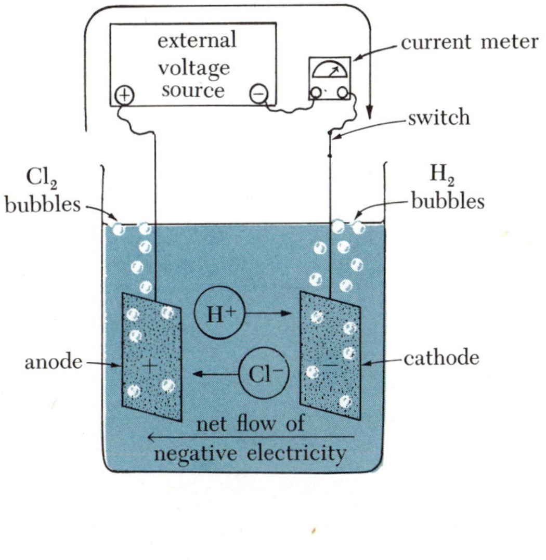

Figure 12.1 Discharging of the ions H^+ and Cl^- at the plates of a conductivity apparatus, which is being misused in that too much current is being allowed to flow.

neutral. For anions (for example, Cl⁻), discharging involves the transfer of electrons from ions to the electrode (anode). These electron transfers are simple examples of a wide class of chemical processes, oxidation-reduction reactions.

12.2 Oxidation-Reduction Reactions

The Concept of Electron Transfer

In Sec. 7.7, electron transfers were discussed in terms of the concepts of oxidation and reduction and oxidation states (or oxidation numbers). Definite rules were presented for determining the oxidation state of any atom in any molecular or ionic species. According to the defining rules, oxidation numbers are *apparent* atomic charges, when electrons are partitioned among atoms in a prescribed way. Oxidation numbers provide an arbitrary, but self-consistent, framework for electron bookkeeping.

In Chap. 7 we defined **oxidation** as a process by which electrons are removed from a species—an electron transfer, in which the oxidation number of the species is increased. Correspondingly, **reduction** is the addition of electrons to a species—an electron transfer in which the oxidation number of the species is reduced. According to these definitions, Eq. (12.1), when read normally from left to right, represents the reduction of H^+; it represents the transformation of H^+, a species with oxidation number 1, into H_2, a species with oxidation number 0. Equation (12.2) shows the oxidation of Cl^- (oxidation number -1) to Cl_2 (oxidation number 0). Equation (12.3) represents the reduction of Cu^{2+} to Cu.

Spotlight

There are three important compounds of lithium and hydrogen in which the oxidation number of hydrogen is -1. Because of this oxidation number of hydrogen, these compounds are called **hydrides**; they are very useful reducing agents, and the hydrogen is converted to the $+1$ oxidation state in their reducing actions. Lithium hydride, LiH, is prepared by the direct combination of the elements at high temperature.

$$2Li(s) + H_2(g) \rightleftharpoons 2LiH(s)$$

An example of the reducing capabilities of LiH is the reduction of a ketone. For example, acetone,

$$CH_3-C\!\!\underset{CH_3}{\overset{O}{\diagdown}}$$

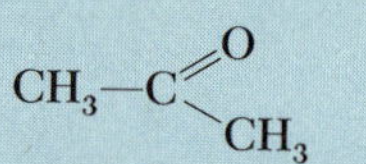

is smoothly and quantitatively reduced by lithium hydride to a lithium salt of 2-propanol, $CH_3-CH(OH)-CH_3$, which can be converted to the final product by reaction with water (hydrolysis):

$$LiH(s) + CH_3-\overset{O}{\underset{\|}{C}}-CH_3 \rightleftharpoons \underset{CH_3}{\overset{CH_3}{\diagup}}\!\!CH-O^-Li^+(s)$$

$$\underset{CH_3}{\overset{CH_3}{\diagup}}\!\!CH-O^-Li^+(s) + H_2O \rightleftharpoons CH_3-\underset{OH}{\overset{H}{\underset{|}{\overset{|}{C}}}}-CH_3 + Li^+ + OH^-$$

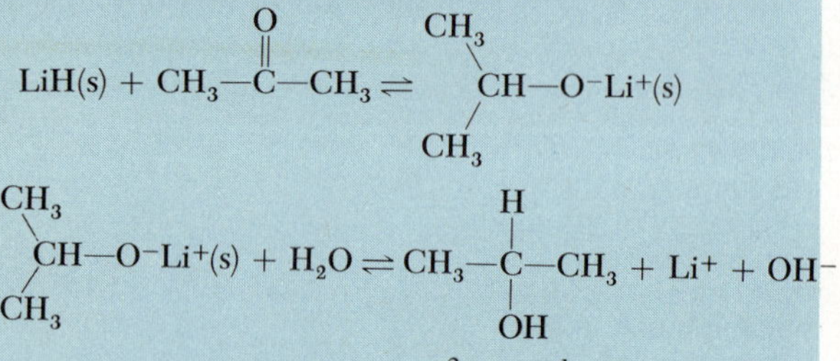

2-propanol

(Note the change in oxidation number of the carbonyl ($-\overset{O}{\underset{\|}{C}}-$) carbon atom.) In addition, lithium forms two important complex hydrides, lithium aluminum hydride, $LiAlH_4$, and lithium borohydride, $LiBH_4$. Both have excellent reducing properties and are extensively used for the reduction of carbonyl groups in organic compounds, analogous to the behavior of LiH.

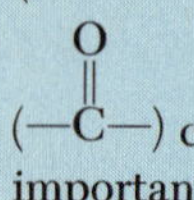

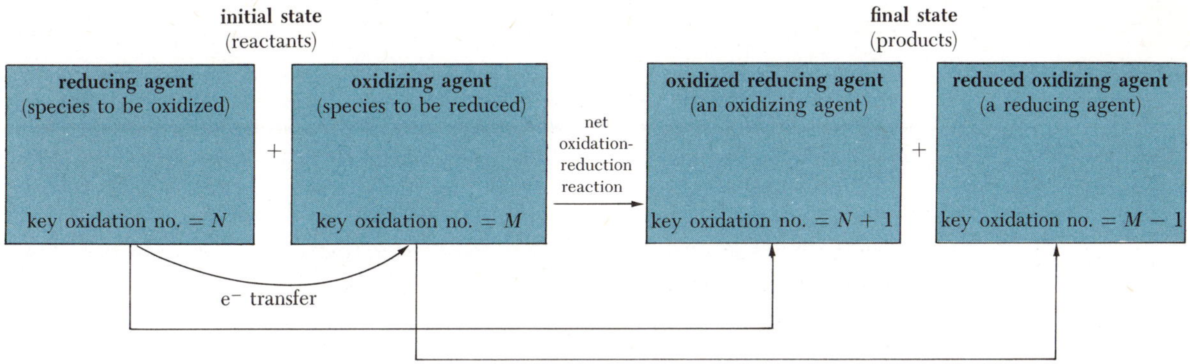

Figure 12.2 Diagrammatic representation of an oxidation-reduction reaction, showing a one-electron transfer and the resulting changes in oxidation numbers of the key elements of the reducing agent and the oxidizing agent; these are the ones undergoing changes in oxidation number.

Half-Reactions for Oxidation and Reduction

An oxidation-reduction reaction is a combination of oxidation and reduction. There can be no oxidation without reduction and no reduction without oxidation. In the half-reactions represented in Eqs. (12.1) to (12.3), only half of an oxidation-reduction is shown in each case. Clearly the reduction of Eq. (12.1) cannot occur unless there is some source of electrons from an oxidation, for example, the process shown in Eq. (12.2). It is in many ways convenient to think about an individual oxidation or an individual reduction, but one should be sure to remember that in practice these two processes always occur simultaneously. Nevertheless, they can be separated somewhat in space, as we shall see below.

The oxidation-reduction process could be compared to the process of buying and selling. One may find it convenient in a class on business management to focus on methods of selling. Or in a consumer science class, one may focus on methods of purchasing most effectively. Nevertheless, one recognizes that in an actual commercial transaction, there is neither buying nor selling without the companion process.

In the same sense, we can view oxidation-reduction chemistry as a "commerce in electrons." The basic idea is entirely analogous to viewing acid-base chemistry as a "commerce in protons," which we have already done (Sec. 5.2). The "purchasers" of electrons are the species being reduced, while the "sellers" of electrons are the species being oxidized. Clearly, the substance being reduced, called the **oxidizing agent,** takes electrons away from the species being oxidized. The substance being oxidized, called the **reducing agent,** gives up electrons to the substance being reduced (oxidizing agent). These relations are summarized in Fig. 12.2. From the figure, drawn for the case of a one-electron transfer, it is seen that after the electron transfer takes place, the species that has been reduced (originally the oxidizing agent) has an atom with a lower oxidation number due to an additional electron, which is potentially available for donation to some other species. In this sense, what was initially an oxidizing agent has been transformed into a reducing agent. Similarly, the species that was intially the reducing agent has an atom with a higher oxidation number since it has lost an electron; because of this electron deficiency this species can, in principle, function as an oxidizing agent. Thus, the oxidation-reduction process transforms a reducing agent into an oxidizing agent and an oxidizing agent into a reducing agent. This relation, presented by Eq. (12.4),

$$\text{ox. agent} + \text{red. agent} \rightleftharpoons \text{ox. agent}' + \text{red. agent}' \qquad (12.4)$$

is analogous to what we have seen in the conjugate acid-base relations (Sec. 5.2). The relation is also shown pictorially in Fig. 12.2. The practice of mentally separating an oxidation-reduction reaction into separate half-reactions, as in Eqs. (12.1) to (12.3) is shown pictorially in Figs. 12.3 and 12.4.

In the simple oxidation-reduction processes represented as half-reactions in Eqs. (12.1) to (12.3) or in related complete reactions such as those shown in Eqs. (12.5) and (12.6), the occurrence of electron transfer is relatively obvious from cursory considerations.

$$H^+ + Cl^- \rightleftharpoons \tfrac{1}{2}H_2(g) + \tfrac{1}{2}Cl_2(g) \qquad (12.5)$$

$$Cu^{2+} + 2Cl^- \rightleftharpoons Cu(s) + Cl_2(g) \qquad (12.6)$$

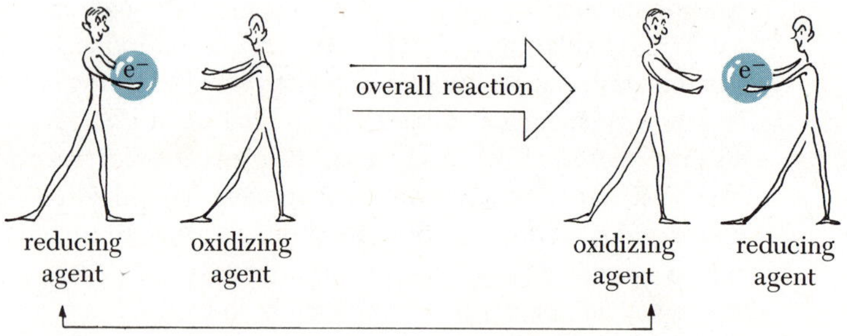

Figure 12.3 An oxidation-reduction reaction as a donation of an electron.

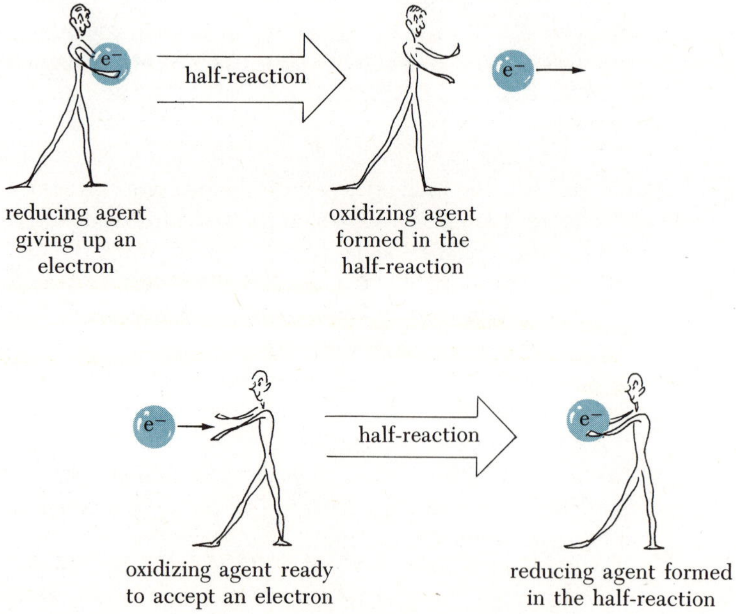

Figure 12.4 Artificially separating the steps of Fig. 12.3.

The electron-transfer feature is obvious in these cases because the structural entities are so simple.

We have also become familiar with some reactions in which the occurrence of electron transfer was not so obvious, that is, where the structural building blocks were sufficiently complex that the reaction could not easily be recognized as an oxidation-reduction (Sec. 7.7). The method that was described for balancing such equations is based on oxidation numbers. There is another method for balancing oxidation-reduction reactions—based on the idea of half-reactions.

Balancing Oxidation-Reduction Reactions by the Half-Reaction Method

Let us consider the same chemical transformation for which a balanced equation was derived in Sec. 7.7. An unbalanced representation of this transformation is

$$MnO_4^- + Fe^{2+} \longrightarrow Mn^{2+} + Fe^{3+} \qquad (12.7)$$

Let us try to separate the overall process in our minds into oxidation and reduction:

$$Fe^{2+} \longrightarrow Fe^{3+} \qquad (12.8)$$

$$MnO_4^- \longrightarrow Mn^{2+} \qquad (12.9)$$

Expression (12.8) is easily converted into a *balanced half-reaction equation,* that is, a balanced equation that corresponds to the oxidation or the reduction half of a total oxidation-reduction reaction. Only electrical charge is unbalanced in (12.8). Recognizing this, one can write

$$Fe^{2+} \rightleftharpoons Fe^{3+} + e^- \qquad (12.10)$$

In dealing with expression (12.9), we note that it is balanced for manganese, but not for oxygen. Oxygen balance can be accomplished by adding four water molecules to the right side of (12.9), giving

$$MnO_4^- \longrightarrow Mn^{2+} + 4H_2O$$

This expression, while balanced for manganese and oxygen, is unbalanced for hydrogen. This can be taken care of, in view of the acidic nature of the system, by adding $8H^+$ to the left side:

$$MnO_4^- + 8H^+ \longrightarrow Mn^{2+} + 4H_2O$$

This expression is now completely balanced so far as elemental balance is concerned. The only remaining imbalance is in electrical charge, and this can be eliminated by adding five electrons to the left side:

$$MnO_4^- + 8H^+ + 5e^- \rightleftharpoons Mn^{2+} + 4H_2O \qquad (12.11)$$

Equation (12.11) is completely balanced, a condition that is emphasized by the symbol sign. Equations (12.10) and (12.11) are balanced equations of the half-reactions involved in the process represented in unbalanced form by expression (12.7). If Eqs. (12.10) and (12.11) are now combined properly, one clearly must obtain the correct, balanced equation that represents the overall oxidation-reduction reaction. Since the number of electrons given up by the reducing agent, Fe^{2+}, must be the same as the number accepted by the oxidizing agent, MnO_4^-, it is seen that Eq. (12.10) must be multiplied by 5 before it is combined with Eq. (12.11). Then, one obtains

$$
\begin{array}{lr}
 & 5Fe^{2+} \rightleftharpoons 5Fe^{3+} + 5e^- \\
 & MnO_4^- + 8H^+ + 5e^- \rightleftharpoons Mn^{2+} + 4H_2O \\
\hline
\text{Sum} & MnO_4^- + 8H^+ + 5Fe^{2+} \rightleftharpoons 5Fe^{3+} + Mn^{2+} + 4H_2 \\
 & \text{ox.ag.} \qquad\qquad \text{red.ag.} \quad \text{ox.ag.} \qquad \text{red.ag.}
\end{array}
\qquad (12.12)
$$

Equation (12.12) is a properly balanced equation for the overall net reaction. Under this equation we have clearly indicated the "conjugate" oxidizing agent—reducing agent relation.

Now let us consider an analogous reaction, involving the same set of oxidation numbers, in a basic solution. In the presence of a basic solution, iron in its +2 oxidation state exists as $Fe(OH)_2(s)$ and in its +3 oxidation state as $Fe(OH)_3(s)$, and manganese in its +2 oxidation state exists as $Mn(OH)_2(s)$.* The transformations analogous to what are shown in expressions (12.8) and (12.9), are therefore

$$Fe(OH)_2(s) \rightarrow Fe(OH)_3(s) \qquad (12.13)$$

$$MnO_4^- \rightarrow Mn(OH)_2(s) \qquad (12.14)$$

Let us now generate balanced half-reaction equations from (12.13) and (12.14).

Clearly (12.13) can be balanced in its elemental stoichiometry by adding one hydrogen atom and one oxygen atom to the left side; for a reaction occurring in basic solution, this is done most readily by adding OH^- to the left side:

$$Fe(OH)_2(s) + OH^- \rightarrow Fe(OH)_3$$

This expression can be balanced in electrical charge by adding one electron to the right side:

$$Fe(OH)_2(s) + OH^- \rightleftharpoons Fe(OH)_3(s) + e^- \qquad (12.15)$$

Now working with (12.14), one can balance the expression for oxygen by adding two water molecules to the right side, yielding

$$MnO_4^- \rightarrow Mn(OH)_2(s) + 2H_2O$$

This expression, while balanced for oxygen, has an excess of six hydrogen atoms on the right side. The hydrogen imbalance can be eliminated by adding six H_2O molecules to the left side and six OH^- to the right side, giving an expression that is entirely balanced except for the electrical charge:

$$MnO_4^- + 6H_2O \rightarrow Mn(OH)_2(s) + 2H_2O + 6OH^-$$

After subtracting two water molecules from each side, removing thereby unnecessary

*Ordinarily, when permanganate, MnO_4^-, is reduced in basic solution, $MnO_2(s)$ is the manganese product, with a manganese oxidation number +4. Nevertheless, for those cases in which manganese is produced in the +2 oxidation state in basic solution, the species is $Mn(OH)_2(s)$.

redundancies, one can achieve electrical balance by adding five electrons to the left side:

$$MnO_4^- + 4H_2O + 5e^- \rightleftharpoons Mn(OH)_2(s) + 6OH^- \quad (12.16)$$

We must now adjust Eqs. (12.15) and (12.16), so they will represent the same number of electrons being transferred. After multiplying Eq. (12.15) by 5, the number of electrons transferred matches what is shown in Eq. (12.16); hence, the two equations can be added:

$$5Fe(OH)_2(s) + 5OH^- \rightleftharpoons 5Fe(OH)_3(s) + 5e^-$$
$$MnO_4^- + 4H_2O + 5e^- \rightleftharpoons Mn(OH)_2(s) + 6OH^-$$

$$\underset{\text{ox.ag.}}{MnO_4^-} + \underset{\text{red.ag.}}{5Fe(OH)_2(s)} + 4H_2O \rightleftharpoons \underset{\text{ox.ag.}}{5Fe(OH)_3(s)} + \underset{\text{red.ag.}}{Mn(OH)_2(s)} + OH^- \quad (12.17)$$

The balanced equations, (12.12) and (12.17), are precisely the same as those obtained in Sec. 7.7 by the method of oxidation numbers.

Having worked step by step through these examples, we can now write down a stepwise procedure for implementing the **half-reaction method of balancing oxidation-reduction reactions.** The steps are as listed herewith.

A. Identify the species being oxidized and reduced and their oxidation and reduction products; write down unbalanced expressions that represent these changes (for example, 12.8 and 12.9).

B. Obtain balanced half-reaction equations from the unbalanced expressions of step A by the following procedure:
   1. First balance each expression for all elements except hydrogen and oxygen by introducing the appropriate chemical symbols or coefficients or both; this can be done by inspection.
   2. Balance each expression for oxygen by introducing H_2O on either side, as needed.
   3. Balance each expression for hydrogen by:
      (a) Introducing H^+ to either side as needed (for acid solution).
      (b) Introducing H_2O and OH^- on opposite sides in equal numbers as needed (for basic solution).
   4. Balance each expression in electrical charge by adding electrons to the appropriate side, as needed.

C. Multiply each balanced half-reaction equation by a factor that will equalize the number of electrons produced in the oxidation half-reaction and the number of electrons consumed in the reduction half-reaction.

D. Add the two balanced half-reaction equations, and cancel identical entities on both sides.

The following examples include most types of manipulations represented in the rules just given.

EXAMPLE 12.1 Write a balanced equation for the oxidation of glucose, $C_6H_{12}O_6$, by O_2 to yield CO_2 and H_2O. This is a fundamental energy-producing process, carried out in animals in many complex steps.

SOLUTION In unbalanced notation, the basic transformation of interest is

$$C_6H_{12}O_6 + O_2 \rightarrow CO_2 + H_2O$$

The oxidizing agent is O_2, which has an oxidation number 0 and becomes H_2O (or CO_2) with oxygen oxidation number -2 in the products. The reducing agent is $C_6H_{12}O_6$, in which the carbon has an oxidation number 0, becoming 4 in CO_2. (We have taken the oxidation number of oxygen to be -2 and of hydrogen to be 1 in the glucose.) Then, the two processes of main interest can be represented in step A as

$$C_6H_{12}O_6 \rightarrow CO_2 \tag{12.18}$$

$$O_2 \rightarrow H_2O \tag{12.19}$$

Carrying out step B1 on (12.18), we have

$$C_6H_{12}O_6 \rightarrow 6CO_2 \tag{12.20}$$

Then, according to step B2, balancing for oxygen gives

$$6H_2O + C_6H_{12}O_6 \rightarrow 6CO_2 \tag{12.21}$$

Next, applying step B3 (arbitrarily choosing 3a) to balance for hydrogen gives

$$6H_2O + C_6H_{12}O_6 \rightarrow 6CO_2 + 24H^+ \tag{12.22}$$

Balancing with respect to charge according to step 4 yields a completely balanced half-reaction equation:

$$6H_2O + C_6H_{12}O_6 \rightleftharpoons 6CO_2 + 24H^+ + 24e^- \tag{12.23}$$

Next working with (12.19) in step B2, we obtain

$$O_2 \rightarrow 2H_2O \tag{12.24}$$

and applying step B3 for hydrogen balance,

$$4H^+ + O_2 \rightarrow 2H_2O \tag{12.25}$$

Implementing step B4 for electrical balance yields the balanced half-reaction equation

$$4H^+ + O_2 + 4e^- \rightleftharpoons 2H_2O \tag{12.26}$$

Finally, multiplying Eq. (12.26) by 6, according to step C, and adding, according to step D, we obtain

$$6H_2O + C_6H_{12}O_6 \rightleftharpoons 6CO_2 + 24H^+ + 24e^-$$

$$24H^+ + 6O_2 + 24e^- \rightleftharpoons 12H_2O$$

$$\text{Final equation} \quad C_6H_{12}O_6 + 6O_2 \rightleftharpoons 6CO_2 + 6H_2O \tag{12.27}$$

Equation (12.27) is the same net balanced equation that we obtained by the method of oxidation numbers in Sec. 7.7. In this particular example, the approach shown in Sec. 7.7 is probably more straightforward. Nevertheless, Example 12.1 does exhibit most of the steps listed above and demonstrates that the half-reaction method can be applied to reactions involving complex organic molecules, such as glucose, in which the half-reactions themselves are strictly mathematical conveniences.

EXAMPLE 12.2 Write a balanced equation for the reduction of triiodide ion to iodide ion by thiosulfate ion, $S_2O_3^{2-}$, which is oxidized to tetrathionate ion, $S_4O_6^{2-}$.

SOLUTION Separating the overall change into two processes according to step A, we have

$$I_3^- \rightarrow I^- \quad (12.28)$$

$$S_2O_3^{2-} \rightarrow S_4O_6^{2-} \quad (12.29)$$

A balanced equation can be written for (12.28) by implementing steps B1 and B4 (skipping steps B2 and B3 because they are not applicable):

$$I_3^- + 2e^- \rightleftharpoons 3I^- \quad (12.30)$$

Applying step B1 to (12.29), which simultaneously satisfies the goal of step B2, we obtain

$$2S_2O_3^{2-} \rightarrow S_4O_6^{2-}$$

Then, electrical balance gives the balanced half-reaction equation

$$2S_2O_3^{2-} \rightarrow S_4O_6^{2-} + 2e^- \quad (12.31)$$

Skipping step C, because it is unnecessary—two electrons are involved in both Eq. (12.30) and Eq. (12.31)—and adding according to step D gives the final result.

$$\begin{array}{l} I_3^- + 2e^- \rightleftharpoons 3I^- \\ 2S_2O_3^{2-} \rightleftharpoons 2e^- + S_4O_6^{2-} \\ \hline I_3^- + 2S_2O_3^{2-} \rightleftharpoons 3I^- + S_4O_6^{2-} \end{array} \quad (12.32)$$

EXAMPLE 12.3 Write a balanced equation for the oxidation of zinc metal to zinc hydroxide by MnO_4^{2-}, which is reduced to MnO_2, in basic solution.*

SOLUTION Separating this reaction into an oxidation process and a reduction process gives

$$Zn(s) \rightarrow Zn(OH)_2(s)$$

$$MnO_4^{2-} \rightarrow MnO_2(s)$$

Each of these expressions is already balanced for elements other than hydrogen and oxygen, so step B1 can be skipped. Balancing both expressions for oxygen according to step B2, we obtain

$$Zn(s) + 2H_2O \rightarrow Zn(OH)_2(s) \quad (12.33)$$

$$MnO_4^{2-} \rightarrow MnO_2(s) + 2H_2O \quad (12.34)$$

*$Zn(OH)_2(s)$ is **amphoteric,** that is, it functions as a base in the presence of excess acid:

$$Zn(OH)_2(s) + 2H^+ \rightleftharpoons Zn^{2+} + 2H_2O$$

and as an acid in the presence of excess base,

$$Zn(OH)_2(s) + 2OH^- \rightleftharpoons Zn(OH)_4^{2-}$$

Because of the latter process, reducing zinc metal in strongly basic media will give the zinc product $Zn(OH)_4^{2-}$ instead of $Zn(OH)_2(s)$.

Spotlight

The chemical reaction represented in Eq. (12.32) has important applications in analyzing some oxidizing agents. The unknown amount of oxidizing agent is introduced into a solution containing an excess of iodide ion, some of which is converted into I_3^- by oxidation. A titration is then carried out on this solution, containing I_3^-, with a standardized, or accurately analyzed, solution of sodium thiosulfate ($Na_2S_2O_3$, commonly called photographer's hypo), which converts the I_3^- back to I^-. As the solution containing I_3^- is orange violet and the solution without I_3^- present is colorless, it is possible to tell when just the correct amount of $S_2O_3^{2-}$ solution has been added to convert all the I^- to I_3^-; in other words, the titration endpoint is easily detectable.

For example, if the "unknown" oxidizing agent is hydrogen peroxide, then when it is added to a solution containing excess I^-, the following reaction will occur:

$$H_2O_2 + 2H^+ + 3I^- \rightleftharpoons I_3^- + 2H_2O$$

Then, the amount of I_3^- formed, and hence, the amount of H_2O_2 originally present, is determined quantitatively by titrating with $S_2O_3^{2-}$ solution, according to Eq. (12.32).

Ozone, O_3, a primary air-pollution concern, is a strong oxidizing agent. It is more reactive than O_2, typically reacting faster with reducing agents. It quantitatively oxidizes I^- to I_3^-, and the reaction is used for the quantitative analysis of ozone:

$$O_3(g) + 3I^- + H_2O \rightleftharpoons I_3^- + O_2(g) + 2OH^-$$

The method used for analyzing ozone in air consists of passing a known volume of air through a solution of potassium iodide or sodium iodide. The I_3^- thus formed is then titrated with a standardized solution of sodium thiosulfate. The net overall reaction for this titration is

$$I_3^- + 2S_2O_3^{2-} \rightleftharpoons 3I^- + \underset{\text{tetrathionate ion}}{S_4O_6^{2-}}$$

Suppose that 2.24×10^3 liters of air are passed through 100.0 ml of a 0.1 *M* solution of KI, and the I_3^- formed is found to require exactly 30.0 ml of a 0.0100 *M* $Na_2S_2O_3$ solution in a titration. What is the total amount of ozone in the air tested? The KI solution could have produced as much as

$$(0.100 \text{ liter KI soln})\frac{0.1 \text{ mol KI}}{\text{liter KI soln}}\,\frac{1 \text{ mol } I_3^-}{3 \text{ mol KI}} = 3.33 \times 10^{-3} \text{ mol } I_3^-$$

However, the amount of $S_2O_3^{2-}$ used was

$$(3.00 \times 10^{-2} \text{ liter})(1.00 \times 10^{-2} \text{ mol/liter}) = 3.00 \times 10^{-4} \text{ mol } S_2O_3^{2-}$$

The amount of I_3^- liberated then was only

$$(3.00 \times 10^{-4} \text{ mol } S_2O_3^{2-})\frac{1 \text{ mol } I_3^-}{2 \text{ mol } S_2O_3^{2-}} = 1.50 \times 10^{-4} \text{ mol } I_3^-$$

Thus, a sufficient amount of KI was available. As 1.50×10^{-4} mol I_3^- was produced, and the stoichiometry shows that an equal mole amount of O_3 was initially present, we conclude that 1.50×10^{-4} mol O_3 was present in the air sample.

Now, step B3b (basic solution) is implemented for (12.33) by adding $2OH^-$ to the left side and $2H_2O$ to the right side, then cancelling $2H_2O$ on both sides.

$$Zn(s) + 2H_2O + 2OH^- \rightarrow Zn(OH)_2(s) + 2H_2O$$
$$Zn(s) + 2OH^- \rightarrow Zn(OH)_2(s) \qquad (12.35)$$

With experience, one could probably write expression (12.35) directly, without this stepwise approach. Proceeding analogously with (12.34) to eliminate the deficiency of hydrogen on the left side, one obtains

$$MnO_4^{2-} + 4H_2O \rightarrow MnO_2(s) + 2H_2O + 4OH^-$$
$$MnO_4^{2-} + 2H_2O \rightarrow MnO_2(s) + 4OH^- \qquad (12.36)$$

Adding electrons appropriately to (12.35) and (12.36) by step B4, balanced half-reaction equations are obtained:

$$Zn(s) + 2OH^- \rightleftharpoons 2e^- + Zn(OH)_2(s) \tag{12.37}$$

$$MnO_4^{2-} + 2H_2O + 2e^- \rightleftharpoons MnO_2(s) + 4OH^- \tag{12.38}$$

Step C can be skipped because it is unnecessary in this case—two electrons are involved in both Eqs. (12.37) and (12.38). Then if these equations are added and two OH^- on each side of the result cancelled (step D), the final balanced equation is obtained:

$$Zn(s) + MnO_4^{2-} + 2H_2O \rightleftharpoons Zn(OH)_2(s) + MnO_2(s) + 2OH^-$$

STUDY PROBLEM **12(a)**

Using the half-reaction method, write a balanced chemical equation for the process

$$Cr_2O_7^{2-} + NO(g) \rightleftharpoons Cr^{3+} + NO_3^- \qquad \text{(acid solution)}$$

STUDY PROBLEM **12(b)**

Using the half-reaction method, write a balanced chemical equation for the reaction

$$Cl_2(g) + Ni(s) \rightleftharpoons Ni(OH)_2(s) + Cl^- \qquad \text{(basic solution)}$$

EXAMPLE 12.4 Calculate the number of grams of ferrous sulfate that can, in solution, be oxidized to the ferric state by 0.52 g $KMnO_4$ in acid solution.

SOLUTION Using the half-reaction method described in the preceding section, we can write the following half-reaction equations, and balanced net equation:

$$MnO_4^- + 8H^+ + 5e^- \rightleftharpoons Mn^{2+} + 4H_2O \tag{12.39}$$

$$5Fe^{2+} \rightleftharpoons 5Fe^{3+} + 5e^- \tag{12.40}$$

$$MnO_4^- + 8H^+ + 5Fe^{2+} \rightleftharpoons 5Fe^{3+} + Mn^{2+} + 4H_2O \tag{12.41}$$

It is also necessary to compute the number of moles of $KMnO_4$ in a 0.52-g sample. From the molecular weight 158.0 amu, we compute

$$(0.52 \text{ g KMnO}_4)\left(\frac{1}{158.0}\,\frac{\text{mol KMnO}_4}{\text{g KMnO}_4}\right) = 3.3 \times 10^{-3} \text{ mol KMnO}_4$$

Without even recognizing that Eq. (12.41) represents an oxidation-reduction reaction, we can use the methods described in Chap. 3 for stoichiometric calculations. From (12.41), we see that every mole of MnO_4^- reacts with 5 mol Fe^{2+}. Hence, the number of moles of Fe^{2+} that can be accommodated by 3.3×10^{-3} mol MnO_4^- is $3.3 \times 10^{-3} \times 5.0$, and the corresponding mass of $FeSO_4$ (mol wt = 151.9 amu) is given by

$$(3.3 \times 10^{-3} \times 5.0 \text{ mol}) \times \left(151.9\,\frac{\text{g}}{\text{mol}}\right) = 2.5 \text{ g}$$

Spotlight

One equivalent of an oxidizing agent in a reaction is defined as the amount of the substance that consumes Avogadro's number (6.022×10^{23}) of electrons, one mole of electrons, in a particular oxidation-reduction reaction. One equivalent of a reducing agent is defined as the amount of substance that provides one mole of electrons in a particular oxidation-reduction reaction. An equivalent can be defined only relative to a specific oxidation or reduction process; it is not uniquely defined by the oxidizing agent or the reducing agent of interest since the oxidation-reduction behavior of a particular species may depend significantly on the specific reaction system and conditions. For example, consider chlorine in aqueous solution; it often functions as an oxidizing agent, a behavior that can be represented by the following half-reaction:

$$Cl_2 + 2e^- \rightleftharpoons 2Cl^- \qquad \text{(a)}$$

However, under appropriate circumstances, aqueous Cl_2 can function as a reducing agent in a variety of ways, some of which are shown in the following half-reaction equations:

$$Cl_2 + 4OH^- \rightleftharpoons 2ClO^- + 2H_2O + 2e^- \qquad \text{(b)}$$

$$Cl_2 + 8OH^- \rightleftharpoons 2ClO_2^- + 4H_2O + 6e^- \qquad \text{(c)}$$

$$Cl_2 + 12OH^- \rightleftharpoons 2ClO_3^- + 6H_2O + 10e^- \qquad \text{(d)}$$

$$Cl_2 + 16OH^- \rightleftharpoons 2ClO_4^- + 8H_2O + 14e^- \qquad \text{(e)}$$

In a reaction system in which Cl_2 functions as an oxidizing agent in accordance with Eq. (a), one mole of Cl_2 consumes two moles of electrons; hence, in this instance one equivalent of Cl_2 as an oxidizing agent is one-half mole of Cl_2, or 35.5 g. If Cl_2 is involved in a reaction system in which it functions as a reducing agent according to Eq. (b), then one sees that one mole of Cl_2 produces two moles of electrons, so one-half mole of Cl_2 constitutes one equivalent of Cl_2 as a reducing agent in this instance. Jumping down to Eq. (e), one sees that if Cl_2 functions as a reducing agent by being converted to perchlorate ion, ClO_4^-, then one equivalent of Cl_2 is $\frac{1}{14}$ mol Cl_2. From this example, it is clear that one must know precisely what reaction is under consideration before the concept of an equivalent has any meaning. Once the system is defined, then one always knows that the number of equivalents of reducing agent consumed in a reaction is exactly the same as the number of equivalents of oxidizing agent consumed. This must always be true, because the number of moles of electrons provided by the reducing agent must be exactly the same as the number consumed by the oxidizing agent.

The concentrations of solutions of oxidizing agents and reducing agents are frequently specified in terms of normality. The **normality** of a solution of an oxidizing agent or of a reducing agent is defined to be the *number of equivalents* of the oxidizing agent or reducing agent *per liter* of solution.

Using the factor-label method in expanded form, we get the same result:

$$(3.3 \times 10^{-3} \text{ mol } MnO_4^-)\left(5.0 \frac{\text{mol } Fe^{2+}}{\text{mol } MnO_4^-}\right)\left(1 \frac{\text{mol } FeSO_4}{\text{mol } Fe^{2+}}\right)\left(151.9 \frac{\text{g } FeSO_4}{\text{mol } FeSO_4}\right) = 2.5 \text{ g } FeSO_4$$

STUDY PROBLEM 12(c)

Calculate the maximum amount of metallic iron that can be oxidized to $FeCl_3$ by 11.3 g Cl_2.

12.3 Electrochemical Cells

Physically Separating the Half-Reactions

The half-reaction is thus a device for mentally partitioning a complex oxidation-reduction reaction into separate, yet mutually dependent, oxidation and reduction processes. While the two half-reactions must always occur simultaneously, it is often useful to consider them separately—for example, in balancing chemical equations for oxidation-reduction reactions. In an ordinary oxidation-reduction in which one simply adds a reducing reagent to an oxidizing agent, the oxidation and reduction half-reactions occur at the same time and also at the same place, which means when and where the two reactant species come into contact with each other in the solution. This contact and resulting reaction gives the overall oxidation-reduction reaction. If aqueous $FeSO_4$ solution were our reducing agent and an aqueous chlorine solution our oxidizing agent, we should have the half-reactions

$$Fe^{2+} \rightleftharpoons Fe^{3+} + e^- \qquad \text{and} \qquad \tfrac{1}{2}Cl_2 + e^- \rightleftharpoons Cl^-$$

And the overall oxidation-reduction reaction would be

$$Fe^{2+} + \tfrac{1}{2}Cl_2 \rightleftharpoons Cl^- + Fe^{3+}$$

It is never possible to separate the oxidation and reduction half-reactions in time; they must always occur simultaneously, as emphasized above. It is possible, however, to separate them in space. This is precisely what is accomplished if an electrical current is allowed to flow in a conductivity experiment, as described in Sec. 12.1 and as depicted in Fig. 12.1. The oxidation half-reaction occurs at the anode and the reduction half-reaction occurs at the cathode. These types of anode and cathode reactions represent more of a nuisance than a focal point for conductivity experiments.* In other contexts, however, they constitute the basis of an important type of apparatus for carrying out oxidation-reduction reactions, electrochemical cells.

An **electrochemical cell** is a device in which an oxidation-reduction reaction is carried out, with the oxidation half-reaction occurring at the anode and the reduction half-reaction occurring at the cathode. Consider the apparatus pictured in Fig. 12.5. In this particular setup, zinc metal functions as a reducing agent, and also as the material out of which one of the electrodes (the anode) is constructed. On the other side of the cell, a platinum wire in contact with chlorine gas (being bubbled through the solution) serves as the other electrode, the cathode. The platinum plays no part except to conduct electrons and provide an inert surface on which an electrode reaction can occur. If the switch in the external circuit is left open, then there can be no electrical current flowing through the external circuit, and no significant reaction occurs in the apparatus. If the switch is closed, however, the meter registers a **voltage**, the electrical potential difference between the two electrodes, or a corresponding flow of electrical current. The voltage measured across the two electrodes, under conditions in which only a minute current is allowed to flow, is called the **electromotive force**, abbreviated emf.

An **electrical current** is a flow of electrically charged particles between points of different electrical potential, that is, across a voltage difference. In the case of a metallic conductor, like the copper wire in the external circuit of Fig. 12.5, the

*In a conductivity experiment, while oxidation and reduction processes occur at the plates when current is allowed to flow, these processes are of no direct interest in the experiment, which is concerned with measuring the mobilities of ions, not their reactivities. For this and other reasons, current flow is kept to a minimum in conductivity experiments. In fact, conductivity experiments are usually carried out by employing alternating current to minimize the effects of oxidation-reduction reactions and to avoid a buildup of ions near the electrodes; these problems would be more serious if the current were allowed to flow in one direction only.

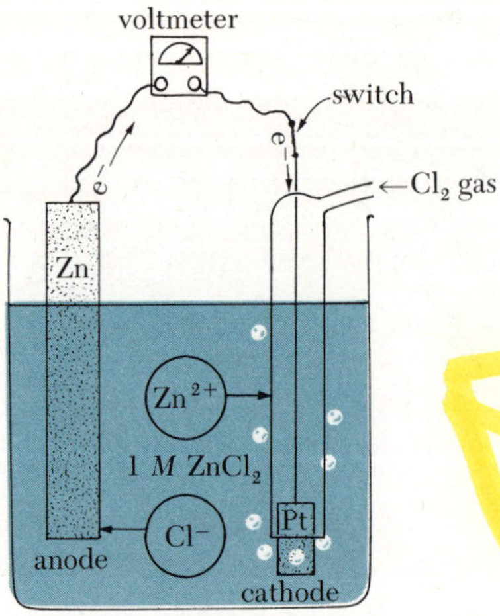

Figure 12.5 Electrochemical cell with $Zn|Zn^{2+}$ anode and $Cl^-|Cl_2(g)$ cathode, showing a glass bubbler through which $Cl_2(g)$ passes, providing contact for Cl^- and $Cl_2(g)$ with the platinum electrode.

particles that flow are electrons of the so-called conduction band of the metal (see Sec. 14.6). An electric current can flow through the external circuit only if electrons are being provided at the zinc electrode, flowing towards the platinum electrode and being consumed at the latter. Furthermore, to have a completed circuit, it is necessary that negative electricity be transferred from the Pt electrode to the Zn electrode through the solution; this can happen if negative ions migrate in solution from right to left and positive ions migrate from left to right in solution. These charge migrations are shown with arrows in Fig. 12.5.

The process that generates electrons at the Zn electrode can be represented by the following half-reaction equation:

$$Zn(s) \rightleftharpoons Zn^{2+} + 2e^- \tag{12.42}$$

As this is an oxidation process, where Zn is oxidized, the Zn electrode qualifies as the **anode,** the electrode at which oxidation occurs. At the platinum electrode, the electrons that are generated at the zinc electrode and carried to the Pt electrode by the external circuit are consumed in a reduction. Platinum, a very inert metal, participates in the reduction only in providing a conduit for the electrons being supplied and a surface on which the reduction can take place. The reduction is the conversion of Cl_2 to Cl^-, as represented by the following half-reaction equation:

$$Cl_2 + 2e^- \rightleftharpoons 2Cl^- \tag{12.43}$$

This is clearly a reduction, feeding on the electrons supplied by the oxidation of Zn at the anode. Hence, the Pt electrode is indeed the **cathode,** the electrode where reduction occurs.

From Eqs. (12.42) and (12.43), it is clear that the ions Zn^{2+} are generated at the anode and the ions Cl^- at the cathode. If it were not for the migration of ions in solution, these ion productions at the electrodes would lead to a substantial buildup of positive charge on the left side of the apparatus and of negative charge on the right side. These ion buildups cause the solution region near the cathode to appear negative and the solution region near the anode to appear positive. Because of this and because of the tendency in nature for the concentration of a given species to

Spotlight

From the point of view of the external circuit, that is, from measurements made on the electrical current and voltage manifested in the external wires, the negative electrode is the anode (Zn electrode of Fig. 12.5), as this is the one that is the source of negative electricity (electrons) for the external circuit. Similarly, the cathode (Pt electrode of Fig. 12.5) is the electrode towards which the electrons flow in the external circuit; so from the external point of view, the cathode is the positive electrode. The situation would appear different from a point of view within the cell itself. Within the cell of Fig. 12.5, the anode is a source of positive electricity (Zn^{2+}); hence, it appears to be the positive electrode. Similarly, inside the cell, the cathode is a source of negative electricity (Cl^-); hence, the cathode appears to be negative. Thus, designating an electrode as positive or negative depends on the point of view (external circuit or internal cell); unequivocal designations are *cathode* and *anode*, which are defined in terms of reduction and oxidation half reactions.

become equalized throughout a system, the ions Zn^{2+} produced at the anode tend to migrate from the anode towards the cathode.* Similarly, the ions Cl^- produced at the cathode tend to migrate towards the anode. The net result is a migration of positive ions from anode to cathode and negative ions from cathode to anode, that is, a net migration of negative electricity from cathode to anode. This complements the transfer of negative electricity, electrons, from the anode to the cathode in the external wiring to give a completed circuit. The amount of electrical current that the anions carry through the solution is generally different from the amount the cations carry; the fraction each carries depends on the ion mobilities. In the example under discussion, the only ions present in the system are the ones involved in the electrode processes, or Zn^{2+} and Cl^-. Often, ions that do not participate in the electrode reactions are present. Typically one adds to the cell an electrolyte that ionizes to give ions that are inert in the system under study; these ions are added to carry the electrical current through the solution. Such an electrolyte, because its ions "support" the electrical current, is called the **supporting electrolyte.**

Voltaic Cells

The chemical system pictured in Fig. 12.5 is an example of a voltaic cell, also called a galvanic cell. A **voltaic cell** is an apparatus in which an oxidation-reduction reaction occurs spontaneously, when the switch is closed, to produce an electric current. There are many such systems known in chemistry and an unlimited number can be imagined. Three such systems are pictured in Figs. 12.6 to 12.9. In each figure the anode reaction, cathode reaction, and net reaction are shown. Also given in each of the three cases is a standard shorthand notation for electrochemical cells. In this notation, the components of the anode compartment are listed on the left and the components of the cathode are listed on the right. Phase boundaries, the boundaries between regions of different phases (Chap. 2), are indicated by vertical lines. In this notation the cell shown in Fig. 12.5 is represented as $Zn(s)|ZnCl_2, 1M|(Pt)Cl_2$.

*The tendency for concentrations of a particular species to become equalized throughout a solution is implicit in the second law of thermodynamics, and the fact that $\delta G = 0$ for a system in equilibrium. A rough physical analogy is the tendency for a fluid to flow through a system in such a way that the pressure in the fluid is equalized throughout the system.

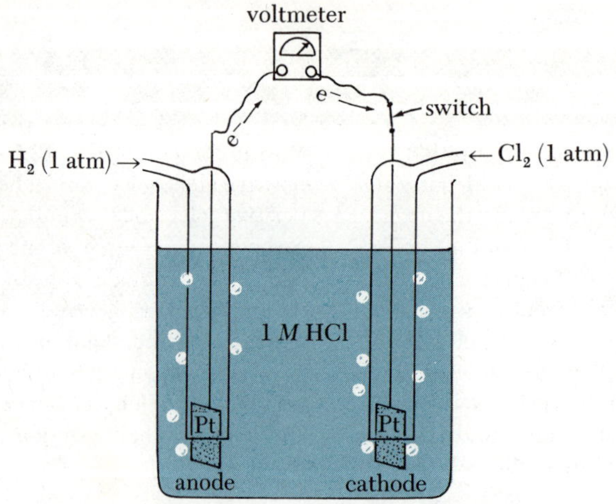

Figure 12.6
Voltaic cell: (Pt)H_2,1 atm|HCl,1 *M*|(Pt)Cl_2,1 atm.

anode reaction: $H_2(g) \rightleftharpoons 2H^+ + 2e^-$

cathode reaction: $Cl_2(g) + 2e^- \rightleftharpoons 2Cl^-$

net cell reaction: $H_2(g) + Cl_2(g) \rightleftharpoons 2H^+ + 2Cl^-$

In Fig. 12.6 a cell is shown in which H_2 is oxidized at the anode, while Cl_2 is reduced at the cathode, the corresponding half-reactions producing H^+ and Cl^-. In this cell, represented as (Pt)H_2,1 atm|HCl,1*M*|(Pt)Cl_2,1 atm, the platinum again functions only as a way of providing contact among the relevant species (for example, Cl_2 and electrons at the cathode). The electric current in solution is carried by the migration of the ions H^+ from the anode towards the cathode and the migration of the ions Cl^- from the cathode towards the anode.

In Fig. 12.7 a voltaic cell is shown that can be represented as Zn(s)|Zn^{2+},0.1 *M*|| Cu^{2+},0.1 *M*|Cu(s). In this notation the symbol "||" denotes the presence of a salt bridge, a gel (agar) containing a high concentration of KCl. The mobility of the ions K^+ and Cl^- in the gel provides for electrical continuity between the anode and cathode compartments of the cell.

Spotlight

An important type of voltaic cell currently under extensive study is the so-called fuel cell. A fuel cell is an electrochemical device designed to convert the energy of a combustible fuel *directly* into electrical energy as the combustion reaction, an oxidation-reduction, occurs. The usual procedure for extracting usable energy from a combustible fuel is to burn the fuel; then the heat generated is used to produce steam, which drives a turbine, which in turn powers an electrical generator. The fuel CH_4 in natural gas, for instance, would be oxidized in the reaction

$$CH_4 + 2O_2 \rightleftharpoons CO_2 + 2H_2O$$

This roundabout conversion of chemical energy to electrical energy seldom functions with an efficiency of more than about 40 percent conversion. If the combustion can be made to occur *directly* in a voltaic cell, the fuel cell, then a much higher conversion percentage can be expected. Fuel cells, when they become perfected, should be "clean" energy sources, producing few pollutants; they would be an important step towards reconciling the apparent conflict between providing safe, inexpensive energy, while protecting the environment at the same time. Of course, fuel cells would only be helpful as long as fuel resources last.

The voltaic cell shown in Fig. 12.8 is the basis of the "lead storage battery" common in automobiles. Electrical current within the cell is carried by the ions present in the sulfuric acid solution, H^+, HSO_4^-, and SO_4^{2-}. The cathode consists of a carbon plate designed in such a way that solid PbO_2 and $PbSO_4$ are retained on it and make good contact with both the carbon conductor and the solution. The anode is a metallic lead framework on which $PbSO_4(s)$ is imbedded.

The cell shown in Fig. 12.9 is a nickel-cadmium cell, represented as $Cd|Cd(OH)_2(s)|OH^-,1\ M|Ni(OH)_2(s)|NiO_2(s)$. This is the prototype of nickel cadmium batteries, which are rapidly replacing the common "dry cell" or the lead storage battery in many applications. The nickel-cadmium system is often preferred over the dry cell because the former is rechargeable (see the discussion below for the lead storage cell), whereas the latter is not. This has become increasingly attractive for applications such as small electronic calculators. Since no gases are involved

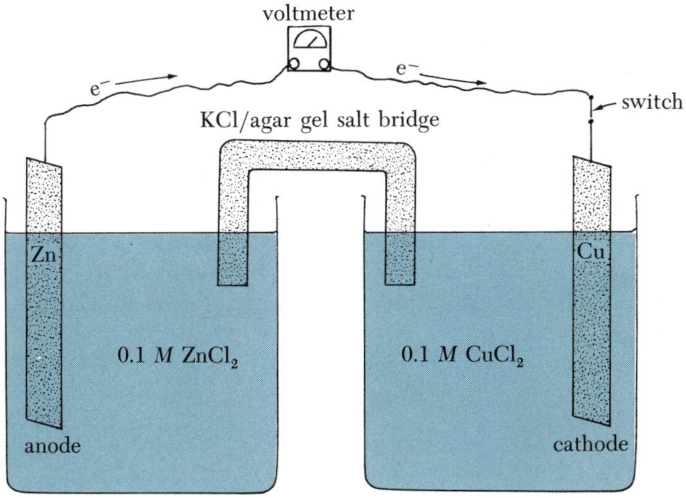

anode reaction: $Zn(s) \rightleftharpoons Zn^{2+} + 2e^-$

cathode reaction: $Cu^{2+} + 2e^- \rightleftharpoons Cu(s)$

net cell reaction: $Zn(s) + Cu^{2+} \rightleftharpoons Cu(s) + Zn^{2+}$

Figure 12.7 Voltaic cell: $Zn(s)|Zn^{2+},0.1\ M||Cu^{2+},0.1\ M|Cu(s)$.

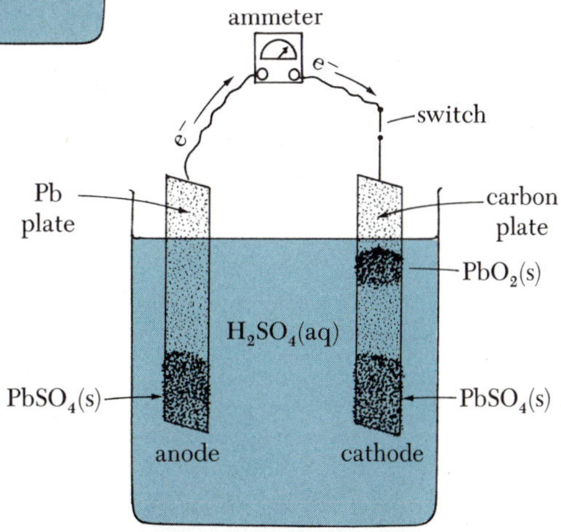

anode reaction: $Pb(s) + SO_4^{2-} \rightleftharpoons PbSO_4(s) + 2e^-$

cathode reaction: $PbO_2(s) + 4H^+ + SO_4^{2-} + 2e^- \rightleftharpoons PbSO_4(s) + 2H_2O$

net cell reaction: $Pb(s) + PbO_2(s) + 4H^+ + 2SO_4^{2-} \rightleftharpoons 2PbSO_4(s) + 2H_2O$

Figure 12.8 Voltaic cell, lead storage battery: $Pb(s)|PbSO_4(s)|H_2SO_4|PbO_2(s)|PbSO_4(s)$. The lead anode plate is coated with $PbSO_4(s)$. The carbon cathode plate is coated with $PbSO_4(s)$ and $PbO_2(s)$.

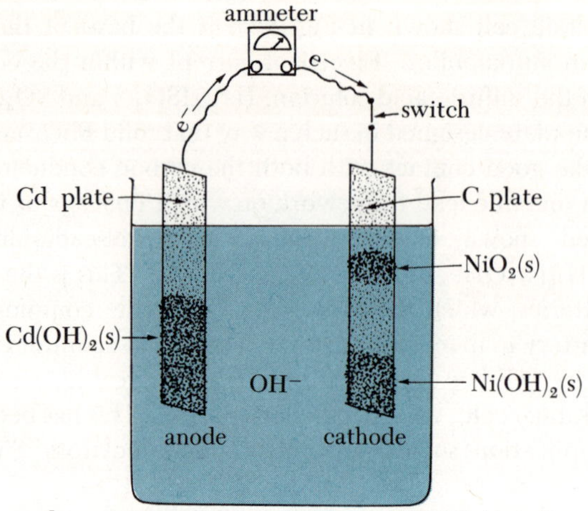

Figure 12.9 A nickel-cadmium cell; prototype for nickel-cadmium batteries: $Cd(s)|Cd(OH)_2(s)|OH^-, 1\ M|Ni(OH)_2(s)|NiO_2(s)|Ni$.

anode reaction: $Cd(s) + 2OH^- \rightleftharpoons Cd(OH)_2(s) + 2e^-$

cathode reaction: $NiO_2(s) + 2H_2O + 2e^- \rightleftharpoons Ni(OH)_2(s) + 2OH^-$

net cell reaction: $Cd(s) + NiO_2(s) + 2H_2O \rightleftharpoons Cd(OH)_2(s) + Ni(OH)_2(s)$

and the reactants and products readily adhere to the electrodes, the battery can be sealed into a compact device.

The so-called dry cell battery has been in use for many years for a multitude of purposes since its invention by LeClanche in 1866. Flashlight batteries are an example. While the detailed chemistry of this cell is complex, the electrode reactions can be idealized as follows:

anode reaction: $Zn(s) \rightleftharpoons Zn^{2+} + 2e^-$

cathode reaction: $2MnO_2(s) + 8NH_4^+ + 2e^- \rightleftharpoons 2Mn^{3+} + 4H_2O + 8NH_3$

The idealized net reaction is

$$Zn(s) + 2MnO_2(s) + 8NH_4^+ \rightleftharpoons Zn^{2+} + 2Mn^{3+} + 8NH_3 + 4H_2O$$

The anode reaction occurs at the zinc casing and the cathode reaction occurs at the carbon (graphite) rod. While inexpensive, this cell is not routinely reversible (rechargeable).

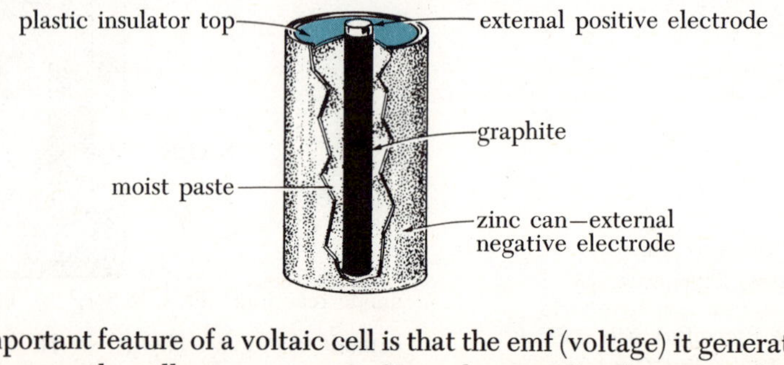

One important feature of a voltaic cell is that the emf (voltage) it generates gradually decreases as the cell passes current. Since the overall cell reaction for a voltaic cell occurs spontaneously (when the external circuit is completed), ΔG for the cell reaction must be negative. However, as the cell is operated, more and more reactants are converted to products and the ΔG value becomes less and less negative. Thus, the "driving force" of the cell reaction decreases, and so does the cell voltage. Eventually when chemical equilibrium is reached, the ΔG value becomes zero for further reaction (see Fig. 11.2). At this point there is no driving force for further reaction and the cell voltage drops to zero. For voltaic cells of any practical importance, one finds that the equilibrium constants are very large—say 10^n, where n is 5 or more. Hence, the point at which ΔG, and the cell voltage, reach zero occurs near the complete conversion of reactants to products (further to the right than shown in Fig. 11.2). (The qualitative relations between cell voltage ΔG and the concentrations of reactants and products are explored in greater detail in later sections of this chapter.)

Electrolytic Cells

Voltaic cells are based on chemical reactions that occur spontaneously, given the opportunity; closing the switch in an external circuit is an example. We have learned that reactions that have the tendency to occur spontaneously are reactions that have equilibrium constants greater than unity and $\Delta G°$ values that are negative. The tendency for the reactions to occur spontaneously is what provides the "driving force" causing electrons to flow through the external circuit from the anode to the cathode; this electrical current can be harnessed to operate an electrical apparatus or appliance. We might logically ask whether the reverse condition can be brought about. Can one "force" an "unfavorable reaction" to occur—a reaction with $K < 1$ and $\Delta G° > 0$? The answer is Yes, if one applies an external voltage to the electrodes of a cell constructed to provide a vehicle for the reaction. An external emf, from a direct current (dc) power supply or generator or from a battery can "pump" electrons into the cathode and out of the anode of an electrochemical cell. Such a "pumped" cell is called an electrolytic cell. An example is provided by the lead storage battery shown in Fig. 12.8.

Suppose one replaces the switch in the apparatus shown in Fig. 12.8 by a direct current power supply and connects the negative terminal, the source of electrons, to the $Pb(s)|PbSO_4(s)$ electrode of the lead storage cell. The electrons pumped into the $Pb(s)|PbSO_4(s)$ electrode are consumed in the reduction half-reaction

$$PbSO_4(s) + \underset{\substack{\text{from}\\\text{external}\\\text{circuit}}}{2e^-} \rightleftharpoons Pb(s) + SO_4^{2-} \qquad (12.44)$$

At the same time, with the positive terminal of the voltage supply connected to the $PbO_2(s)|PbSO_4(s)$ electrode, electrons are removed from that electrode, promoting the following oxidation half-reaction, which provides the electrons to the external circuit:

$$PbSO_4(s) + 2H_2O \rightleftharpoons \underset{\substack{\text{for}\\\text{external}\\\text{circuit}}}{2e^-} + PbO_2(s) + 4H^+ + SO_4^{2-} \qquad (12.45)$$

The net result of these two electrode reactions is

$$2PbSO_4(s) + 2H_2O \rightleftharpoons Pb(s) + PbO_2(s) + 4H^+ + 2SO_4^{2-} \qquad (12.46)$$

Spotlight

ALCOA

Aluminum was first isolated in 1828 by Wöhler. Until 1886, however, there was no commercially feasible method for producing the metal from the ore, bauxite, containing $Al_2O_3(s)$. It was known that it was possible to electrolyze molten Al_2O_3; however, the melting point of this substance is 2000 °C, which made the electrolysis impractical. In 1886, Charles Hall, a graduate student at Oberlin College, discovered that a mineral called cryolite, Na_3AlF_6. could be used in the molten state as a solvent for Al_2O_3 at about 1000 °C. This discovery made the process commercially feasible, and was responsible for the emergence of a huge industry. The overall process begins with a step in which some impurities are eliminated from the $Al_2O_3(s)$ ore by dissolving the $Al_2O_3(s)$ in a concentrated solution of sodium hydroxide:

$$Al_2O_3(s) + 6OH^- + 3H_2O \rightleftharpoons 2Al(OH)_6{}^{3-}$$

This is followed by filtration of the resulting solution, whereby solids that did not dissolve are eliminated. The $Al(OH)_6{}^{3-}$ solution is then acidified with CO_2 to reprecipitate the $Al_2O_3(s)$:

$$6CO_2(g) + 2Al(OH)_6{}^{3-} \rightleftharpoons Al_2O_3(s) + 6HCO_3{}^- + 3H_2O$$

The purified $Al_2O_3(s)$ is then mixed in a heated cell with cryolite, or even better materials that have been discovered recently. The resulting fused liquid, which contains the ions Al^{3+} and O^{2-}, is then electrolyzed:

$$\text{anode reaction:} \quad 2Al^{3+} + 6e^- \rightleftharpoons 2Al(s)$$

$$\text{cathode reaction:} \quad 3O^{2-} \rightleftharpoons \tfrac{3}{2}O_2(g) + 6e^-$$

$$\text{overall cell reaction:} \quad 2Al^{3+} + 3O^{2-} \rightleftharpoons 2Al(s) + \tfrac{3}{2}O_2(g)$$

A young Frenchman named Paul Heroult made the same discovery in France at nearly the same time as Hall made his discovery.

which is the reverse of the overall cell reaction given in Fig. 12.8. As the cell reaction of Fig. 12.8 is a spontaneous reaction capable of producing electrical energy in a voltaic cell, its reverse, represented by Eq. (12.46), would occur only if electrical energy were supplied to force the reaction to occur in the desired direction. This is precisely what is done when the lead storage battery is "charged up" by the generator or alternator of an automobile, or by a direct current voltage supply in a service station.*

An electrolytic cell is a chemical system in which a reaction that would not occur spontaneously is caused to occur by applying an external source of electrical energy.

*From Fig. 12.8 and Eq. (12.45), it is seen that when a lead storage battery "runs down," the concentration of sulfuric acid as the ions H^+ and $SO_4{}^{2-}$ decreases. By contrast, if the battery is charged, the sulfuric acid becomes more concentrated. Dilute sulfuric acid is less dense than concentrated sulfuric acid. Therefore, measuring density by a hydrometer is a convenient way for a service station attendant to check the condition of a lead storage battery. If a lead storage battery is operated for too long a period under conditions of high current flow, the characteristics of the two electrodes may be changed so drastically that no attempt at reversal of the net cell reaction will restore the cells to working order.

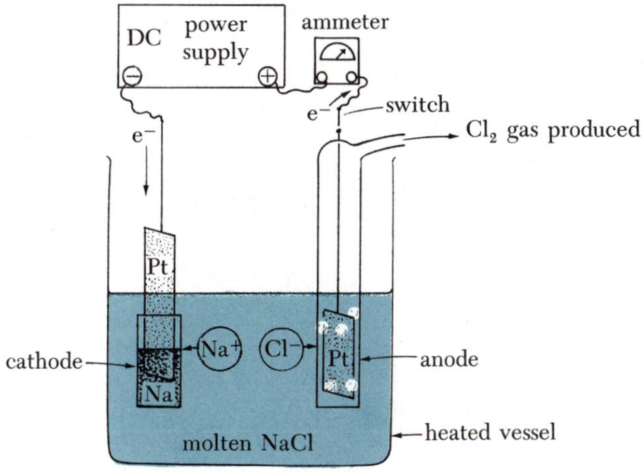

Figure 12.10
Cell for the electrolysis of molten NaCl (mNaCl), producing molten Na metal at the cathode and Cl_2 gas at the anode.

anode reaction: $Cl^-(mNaCl) \rightleftharpoons \frac{1}{2}Cl_2(g) + e^-$

cathode reaction: $Na^+(mNaCl) + e^- \rightleftharpoons Na(l)$

net cell reaction: $Na^+(mNaCl) + Cl^-(mNaCl) \rightleftharpoons Na(l) + \frac{1}{2}Cl_2(g)$

The process carried out in an electrolytic cell is referred to as **electrolysis**. The charging of a $Pb(s)|PbSO_4(s)|H_2SO_4|PbO_2(s)|PbSO_4(s)$ cell by forcing chemical reaction (12.46) to occur under the influence of a dc power source is an example of electrolysis. Figures 12.10 to 12.11 show some representative electrolysis reactions. The processes shown are of practical, and even commercial, importance.

Figure 12.10 depicts an electrolytic process in which metallic sodium and elemental chlorine, as the gas, are produced in a cell containing molten NaCl. At the cathode, Na^+ is reduced to liquid Na metal, which is solidified when allowed to cool; simultaneously at the anode the ions Cl^- are oxidized to Cl_2. Variations on this process are important in the industrial manufacture of sodium metal and chlorine gas.

Spotlight

Several thousand tons of metallic sodium are produced annually in the United States. This substance is used extensively in the production of industrially important organometallic compounds, i.e., compounds in which a metal atom replaces hydrogen in a hydrocarbon structure. The antiknock gasoline additive, tetraethyllead, has been an important case, although its use is declining because of environmental concerns. Metallic sodium is also used in the manufacture of sodium peroxide and sodium cyanide, and sodium vapor lights—the "yellow" lights often used to illuminate streets and highways.

Elemental chlorine, Cl_2, is used as a bleaching agent, in the paper industry, for instance, and as a germicide in water purification. Its use in the manufacture of a variety of pesticides, plastics, and industrial organic compounds has led to some serious environmental concerns. Examples of these are DDT (pesticide), polyvinyl chloride (PVC; Sec. 14.5), and polychlorinated biphenyls (PCB's), in which some of the hydrogen atoms on a

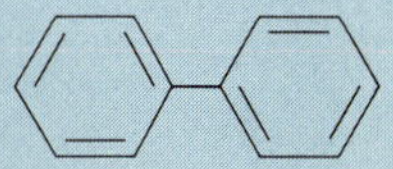

system are replaced by chlorine atoms.

Figure 12.11
Cell for the electrolysis of water in a dilute aqueous H_2SO_4 solution, producing $H_2(g)$ and $O_2(g)$.

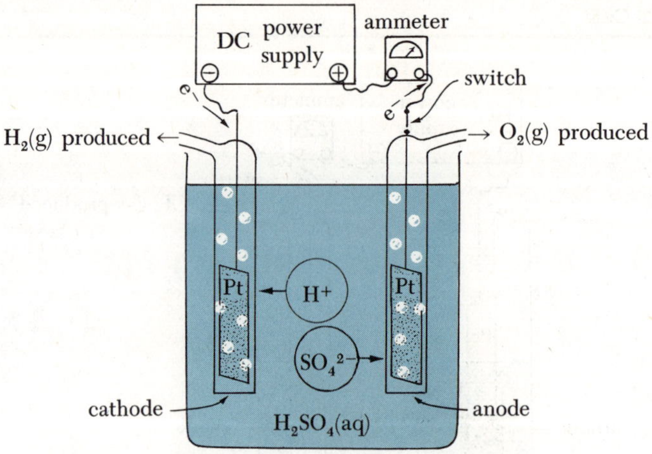

anode reaction: $H_2O \rightleftharpoons 2H^+ + \frac{1}{2}O_2(g) + 2e^-$

cathode reaction: $2H^+ + 2e^- \rightleftharpoons H_2(g)$

net cell reaction: $H_2O \rightleftharpoons H_2(g) + \frac{1}{2}O_2(g)$

Figure 12.11 shows the electrolysis of water to produce H_2 and O_2. This is a common experiment in an elementary chemistry laboratory, but it also has commercial application for the production of hydrogen gas. The ions H^+ and SO_4^{2-} of dissolved sulfuric acid carry the current within the cell. The ion SO_4^{2-} in this case is a good example of an ion that is not involved in either electrode reaction but participates in carrying current from one electrode to the other within the solution. It is easy to see the plausibility of these kinds of ion migration. The ions H^+ are being produced at the anode and consumed at the cathode. If there were no compensating processes, this would lead to an excess of positive ions (H^+) at the anode and a deficiency of H^+ (relative to SO_4^{2-}) at the cathode; this situation would yield an apparatus in which the left side of the solution is negatively charged (due to an excess of SO_4^{2-}) and the right side is positively charged. This untenable situation is avoided by two complementary processes, the migration of H^+ towards the cathode and the migration of SO_4^{2-} towards the anode.

In Fig. 12.12, we see an electrolytic process in which there is no net chemical change. There are well-defined electrode processes, however, the net result of which is the transfer of metallic silver from the silver-bar electrode to the copper spoon to be plated. In this way the metallic object, in this case the copper spoon, receives a coat of silver, the thickness of which depends on the strength of the current and the length of time it is allowed to flow. This general **electroplating** process is important commercially for a variety of metals.

If a copper spoon were placed in an $AgNO_3$ solution, there would be a tendency for the following reaction to occur:

$$2Ag^+ + Cu(s) \rightleftharpoons 2Ag(s) + Cu^{2+}$$

This would remove silver from the solution, producing Ag metal, but not in the form of a uniform layer on the spoon, the primary copper structure of which would be partly sacrificed in the process.

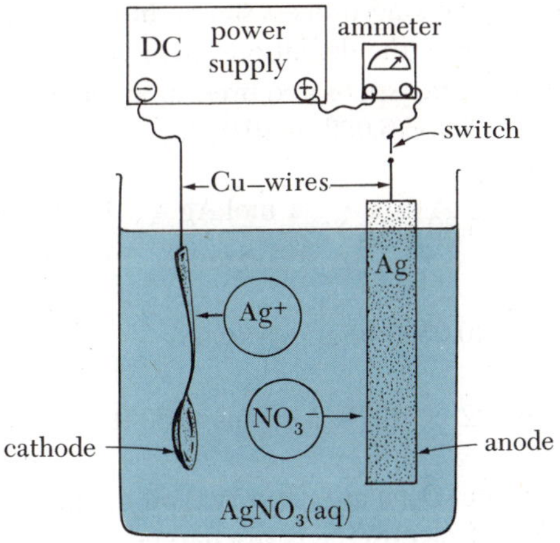

Figure 12.12 Electrolytic cell for plating silver on a copper spoon.*

anode reaction: $Ag(s) \rightleftharpoons Ag^+ + e^-$

cathode reaction: $Ag^+ + e^- \rightleftharpoons Ag(s)$

net cell reaction: $Ag\,(bar) \rightleftharpoons Ag\,(spoon)$

Faraday's Laws

Michael Faraday, an important British scientist of the nineteenth century (1791–1867), devoted the major portion of his research efforts to exploring fundamental relations in electrochemical systems. As with most early work, our hindsight may make his electrochemical discoveries seem obvious, but in his time they were important advances. One of the main findings of his work was that the amount of reaction taking place in any electrochemical process depends quantitatively on the amount of electricity passing through the system. This is shown by the mass of silver deposited on the copper spoon in Fig. 12.12. In modern terms, Faraday's laws can be condensed by using the following definition of a quantity of electricity called the faraday. One **faraday** of electricity is equivalent to one mole of electrons, 96,487 coulombs. Thus, if an external circuit passes one faraday of electricity, 6.022×10^{23} electrons flow through each cross-section of wire. In these terms, Faraday's laws can be condensed into the following statement: "During an electrochemical process (that is, electrolysis of an electrolytic cell, or the operation of a voltaic cell), the passage of one faraday through the circuit results in the transfer of one mole of electrons at the cathode and the transfer of one mole of electrons at the anode." This law is now entirely plausible, even predictable, on the basis of what we already know of electrochemical processes; but the development of Faraday's laws in the nineteenth century constituted an important advance in putting chemical weight relations on a quantitative basis, and in understanding electrochemical processes specifically. Examples 12.5 through 12.7 deal with manifestations of the law.

EXAMPLE 12.5 How many coulombs of electricity are required to deposit 5.39 g silver on a spoon in the arrangement depicted in Fig. 12.12, and how many electrons flow through the circuit in this process?

*The apparent inertness of NO_3^- in this figure and in Fig. 12.13 is due to the slowness of its reduction at the cathode.

SOLUTION In the electrochemical plating process shown in Fig. 12.12, the cathode reaction is $Ag^+ + e^- \rightleftharpoons Ag(s)$. From the definitions given in this section, it is clear that each mole of Ag(s) plated on the spoon requires an amount of electricity corresponding to one faraday of electricity (6.022×10^{23} electrons). Thus, since the atomic weight of silver is 107.87 amu,

$$\text{no. faradays} = 5.39 \text{ g Ag} \times \frac{1 \text{ mol Ag}}{107.8 \text{ g Ag}} \times \frac{1 \text{ faraday}}{1 \text{ mol Ag}} = 0.0500 \text{ faraday}$$

$$\text{no. coulombs} = (0.0500 \text{ faraday})\left(96{,}487 \frac{\text{coulomb}}{\text{faraday}}\right) = 4.82 \times 10^3 \text{ C}$$

This is $0.0500 \times 6.022 \times 10^{23}$ electrons, or 3.01×10^{22} electrons.

EXAMPLE 12.6 How many grams of $PbSO_4(s)$ are deposited on the anode of a lead storage battery when 0.015 faraday of electricity is discharged through the cell in operating an electrical appliance?

SOLUTION From the anode reaction given in Fig. 12.8, one sees that each mole of Pb(s) that is oxidized to $PbSO_4(s)$ at the anode requires the passage of two faradays of electricity. As the molecular weight of $PbSO_4(s)$ is $207.2 + 32.1 + 4 \times 16.0 = 303.3$ amu, 0.015 faraday of electricity generates $PbSO_4(s)$ in the amount of

$$(0.015 \text{ faraday})\left(0.5 \frac{\text{mol } PbSO_4}{\text{faraday}}\right)\left(303.3 \frac{\text{g } PbSO_4}{\text{mol } PbSO_4}\right) = 2.3 \text{ g } PbSO_4$$

EXAMPLE 12.7 Suppose an experiment is being carried out to isolate a metallic sample of silver of unknown isotopic distribution by the electrolysis of an aqueous $AgNO_3$ solution in the apparatus shown in Fig. 12.13. The silver nitrate solution contained in the cathode section (left side of Fig. 12.13) is the silver sample containing an unknown distribution of Ag isotopes; the silver sample was isolated from a silver ore in an

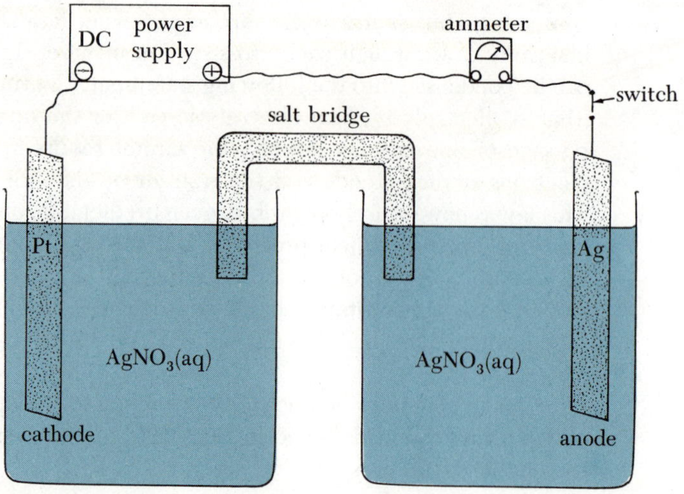

Figure 12.13 Electrochemical cell described in Example 12.7. An aqueous $AgNO_3$ solution in the cathode compartment contains silver with an unknown distribution of isotopes.

anode reaction: $Ag(s) \rightleftharpoons Ag^+ + e^-$
cathode reaction: $Ag^+ + e^- \rightleftharpoons Ag(s)$

Spotlight

Michael Faraday, born in 1791 in Surrey, England, was the son of a poor blacksmith. Young Michael was apprenticed to a bookbinder, and when he reached journeyman status, he educated himself in his spare time by reading books on science, principally in electricity and chemistry. He attended lectures by the famous English scientist, Sir Humphry Davy, in 1813 and secured Davy's help later to begin scientific work. After studying and working with Davy for some time, Faraday developed a reputation as a premier scientist on his own, and became a professor known for both the magnificent quality of his lectures and his important and timeless advances in research on chemistry and electricity.

Culver Pictures

unusual geological formation, and its isotopic distribution is of interest. If 15.25 mg of silver is deposited on the cathode, which is measured by weighing, when 1.401×10^{-4} faraday of electricity is passed through the cell, as monitored in the external circuit, what is the atomic weight of the silver of unknown isotopic distribution?

SOLUTION From the cathode reaction shown in Fig. 12.13, we see that each faraday of electricity passed through the circuit brings about the deposition of one mole, or gram-atom, of silver at the cathode. Thus,

$$1.401 \times 10^{-4} \text{ g-atom Ag} = 15.25 \times 10^{-3} \text{ g Ag}$$

$$1 \text{ g-atom Ag} = \frac{15.25 \times 10^{-3}}{1.401 \times 10^{-4}} \text{ g Ag} = 108.9 \text{ g Ag}$$

Hence, the atomic weight of this particular sample of silver is 108.9 amu, considerably different from the ordinary natural-abundance value 107.9 amu.

STUDY PROBLEM 12(d)

How many grams of silver can be plated on the spoon in the apparatus shown in Fig. 12.12 by the passage of 31.0×10^3 coulombs of electricity?

12.4 Standard Cell Potentials

To this point we have discussed emf only qualitatively. There are good reasons for concerning ourselves with cell emf's, or voltages, on a quantitative level. These parameters depend on the concentrations of reactant and product species in ways that can be useful for controlling the course of a reaction, or for measuring the concentration of a pertinent species, for example, H^+. Furthermore, the emf of a specific voltaic cell, or the voltage necessary to bring about a specific electrolysis, is a direct measure of the tendency of the corresponding net cell reaction to occur.

Cell Potentials, Free Energy Changes, and Equilibrium Constants

Voltaic cells, as described in Sec. 12.3 and depicted for a few representative cases in Figs. 12.5 through 12.9, clearly function because of their ability to convert chemical energy into electrical energy spontaneously. For a chemical system to be able to make this conversion and hence force electrons through an external circuit, it must be based on a chemical reaction with a substantial tendency to proceed spontaneously. Furthermore, since a voltage difference between two points is a direct measure of the tendency for electrical current to flow between those two points, the measured emf of a cell logically can be viewed as a measure of the tendency for the chemical reaction of an electrochemical cell to proceed. For example, if the concentrations of the pertinent species, that is, sulfuric acid and water, in the lead storage battery are such that there is a tendency for the net reaction shown in Fig. 12.8 to occur, then that tendency is translated into the voltage difference between the electrodes; and electrons will flow through the external circuit when the switch is closed. A cell based on a net reaction with a slight tendency to occur could be expected to generate a small emf; one with a high level of spontaneity, or "driving force," would be expected to develop a higher emf. An electrochemical system at equilibrium—with no tendency for any spontaneous reaction to occur—would generate no emf in an electrochemical cell. These qualitative arguments suggest a substantial correlation between cell emf and other measures of the driving force, or spontaneity, of a reaction—that is, ΔG or K.

A direct parallel exists between the emf of a cell and the free energy change ΔG of the cell reaction. As we shall justify quantitatively in Sec. 12.5, this parallel can be expressed in the following equation, where $\mathscr{E}$ stands for the emf generated by an electrochemical reaction:

$$\Delta G = -nF\mathscr{E} \tag{12.47}$$

In this equation, n stands for the number of electrons transferred from reducing agent to oxidizing agent (for example, two for Eq. 12.46), and F stands for the faraday, as defined above. An electrochemical reaction that generates a positive emf is seen, according to Eq. (12.47), to proceed with a negative ΔG, that is, with a decrease in free energy; such a reaction has an intrinsic tendency to occur spontaneously, as in a voltaic cell. According to the discussion in Sec. 11.1, a chemical reaction with a positive ΔG would occur spontaneously in the reverse direction if given the opportunity; for such a reaction, electrons must be "pumped" into the cathode in an electrolytic cell.

As useful and significant as Eq. (12.47) is, it leaves a considerable ambiguity that must be pinned down before it can be applied to specific cases. That ambiguity lies in the matter of concentrations and temperature. One cannot specify what the value of ΔG is for any given reaction without knowing the temperature of the system and the concentrations of all pertinent species, and the same is true of the $\mathscr{E}$ value of any electrochemical reaction. To be specific and consistent in defining values of ΔG or of other thermodynamic reaction variables such as ΔH and ΔS, we focus attention on reactions that transform reactants in their standard states to products in their standard states, the process generally being defined as taking place at 25 °C. For these conditions one uses the symbols $\Delta G°$, $\Delta H°$, and $\Delta S°$ for ΔG, ΔH, and ΔS. For these standard conditions, we also use the symbol $\mathscr{E}°$ for $\mathscr{E}$, and Eq. (12.47) takes the form

$$\Delta G° = -nF\mathscr{E}° \tag{12.48}$$

The $\mathscr{E}°$ values are called **standard cell potentials.**

The standard state for a liquid, say a solvent, is the pure liquid substance; for a solute it is a 1 molar solution; for a gaseous substance, it is the gas at a partial pressure of 1 atm; for a solid, it is the pure solid substance.*

Using these definitions, it is found that the emf of the standard cell ($\mathscr{E}°$ value) for the voltaic cells depicted in Figs. 12.5, 12.6, 12.7, and 12.8 are 2.12 V, 1.36 V, 1.10 V, and 2.05 V, respectively. For the electrolysis of water, which is represented in Fig. 12.11, the standard cell potential is -1.23 V. This pattern of positive $\mathscr{E}°$ values for voltaic cells and negative $\mathscr{E}°$ values for electrolysis systems is both general and expected. Voltaic cells are based on chemical reactions that proceed with negative $\Delta G°$ values, corresponding to a driving force to occur spontaneously in the direction of the left-to-right chemical equation, and a positive $\mathscr{E}°$. Electrolytic cells require the addition of electrical energy to force a chemical reaction to occur. They proceed with a positive ΔG; such reactions have a negative $\mathscr{E}°$ value, which is a measure of the voltage that must be applied externally to force the desired left-to-right reaction.

Another feature of Eq. (12.48) that is noteworthy is that it permits us to relate $\mathscr{E}°$ to the equilibrium constant of the electrochemical reaction. Rewriting Eq. (11.19), we obtain

$$\Delta G° = -RT \ln K$$

which, together with Eq. (12.48), yields the following powerful relation:

$$-nF\mathscr{E}° = -RT \ln K$$

Rearranging gives

$$\mathscr{E}° = \frac{RT}{nF} \ln K \tag{12.49}$$

This equation relates two experimentally accessible quantities that are measures of the driving force of an electrochemical reaction, $\mathscr{E}°$ and K. Since cell emf's can be measured accurately, if a chemical reaction can be carried out in a cell, then it is usually much easier to obtain K by measuring $\mathscr{E}°$ and using Eq. (12.49) than to determine K directly by measuring concentrations, partial pressures, and the like.

In applying Eq. (12.49), it is important to make sure that a self-consistent set of units is used. According to SI units, if we state F in coulombs (96.5×10^3 C), R should be used in units of joule deg^{-1} (8.31 J deg^{-1}). (Recall that one joule is one volt coulomb (VC), the energy required to pass one coulomb of electricity through a voltage

*Actually the standard state of a solute species is the "hypothetical 1 M solution", a solution with an activity that is 10^n times the activity of the solute in a 10^{-n} molar solution, where n is a very large number, or the value of a very dilute solution extrapolated to 1 M concentration.

Spotlight

Oxidation-reduction processes are involved in many important ways in living organisms. Electron-transfer mechanisms are important, for example, in a network of processes by which energy can be stored in "high-energy" chemical species (for example, ATP; see Sec. 17.5).

One very interesting class of electron-transport systems is the iron-sulfur proteins called **rubredoxins** and **ferredoxins.** Recent research showed that the latter proteins contain very interesting structural features that can be viewed as iron-sulfur "cages." In these cages the iron atoms are bound to sulfur atoms that are part of the polypeptide structure of the protein and also to what appears to be "inorganic sulfur," S^{2-}; an example of a sulfur-containing amino acid of protein is cysteine,

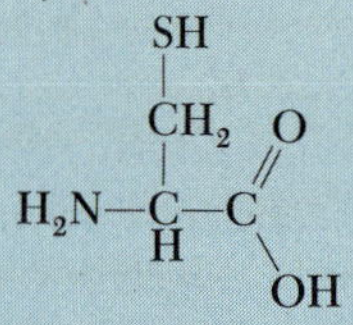

cysteine

The iron-sulfur cage in ferredoxin obtained from the bacterium *Clostridium* has been characterized as having the structure shown in the accompanying diagram. The iron atoms of this cage nearly form a tetrahedron, as they are placed at the alternate corners of an almost cubic box. The other vertices of this "cube" are occupied by the ions S^{2-}. Each iron atom is bound not only to three "inorganic sulfurs," but also to cysteine sulfurs of the protein structure. The molecular weight of the ferredoxin is 6200 amu. Since the iron atoms in this system can be viewed as being in +2 or +3 oxidation states, the entire Fe–S box may have a charge of −3, −2, or −1.

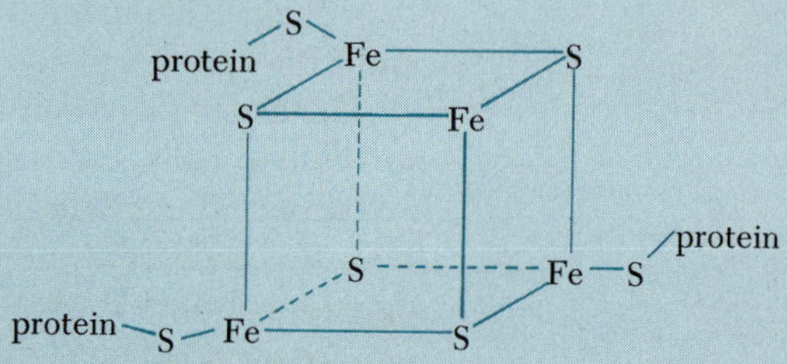

The reduction potentials of iron-sulfur proteins vary widely, from 0.35 V for the reduction of the −1 state to the −2 state of ferredoxin to −0.54 V for the reduction from the −2 to the −3 state of ferredoxin.

difference of one volt.) Then, as K and n are dimensionless, it is seen that the units are consistent.

$$\mathscr{E}^\circ\,(\mathrm{V}) = \frac{[(R\,(\mathrm{J\,deg^{-1}})][T\,(\mathrm{deg})]\ln K}{nF\,(\mathrm{C})}$$

$$= \frac{R(\mathrm{VC\,deg^{-1}})(T\,\mathrm{deg})}{nF(\mathrm{C})}\ln K$$

EXAMPLE 12.8 Calculate the equilibrium constant for the net reaction of the voltaic cell shown in Fig. 12.7.

SOLUTION The net reaction of interest is given by the equation

$$Zn(s) + Cu^{2+} \rightleftharpoons Cu(s) + Zn^{2+}$$

and the equilibrium constant is defined by the equation

$$K = \frac{[Zn^{2+}]}{[Cu^{2+}]}$$

signifying that the symbols representing solid zinc and copper do not appear in the expression. (See Secs. 4.4 and 4.6.) According to Eq. (12.49) and recalling from above that the $\mathscr{E}°$ value of this reaction is 1.10 V, we can write

$$\mathscr{E}° = \frac{RT}{nF} \ln K$$

$$1.10 = \frac{(8.31 \text{ J deg}^{-1})(298 \text{ deg})}{(2)(96.5 \times 10^3)} \ln K$$

In writing this equation, we recognize that the above chemical equation corresponds to the transfer of two electrons from reducing agent to oxidizing agent, and thus $n = 2$. Then,

$$\ln K = (1.10)\frac{(2)(96.5 \times 10^3)}{(8.31)(298)} = 2.303 \log K$$

$$\log K = \frac{(1.10)(2)(0.965 \times 10^5)}{(2.303)(8.31)(2.98 \times 10^2)} = 37.2$$

$$K = \text{antilog } 37.2 = [\text{antilog } (0.2)][10^{37}]$$
$$= 1.6 \times 10^{37}$$

EXAMPLE 12.9 Calculate the equilibrium constant for 25 °C for the electrolysis of water in an aqueous sulfuric acid solution, as depicted in Fig. 12.11.

SOLUTION The net reaction of interest and the corresponding pertinent equilibrium expression are

$$H_2O \rightleftharpoons H_2(g) + \tfrac{1}{2}O_2(g), \qquad K = [H_2(g)][O_2(g)]^{1/2}$$

where the symbol $[H_2O]$ has been omitted because the concentration of water in a dilute aqueous solution is nearly the same as the concentration of pure water. (See Sec. 4.4 and 4.6.) Using Eq. (12.49) and the above-stated $\mathscr{E}°$ value of -1.23 V, we write, recognizing $n = 2$ in this case,

$$2.303 \log K = \ln K = \frac{nF\mathscr{E}°}{RT} = \frac{(2)(96.5 \times 10^3)(-1.23)}{(8.31)(298)}$$

$$\log K = -\frac{(2)(0.965 \times 10^5)(1.23)}{(2.303)(8.31)(2.98 \times 10^2)} = -41.6$$

$$K = \text{antilog } (-41.6) = \text{antilog } (0.4 - 42)$$
$$= [\text{antilog } (0.4)]10^{-42} = 2.6 \times 10^{-42}$$

EXAMPLE 12.10 Calculate the equilibrium constant for 25 °C for the net cell reaction of the electrolytic cell shown in Fig. 12.14, noting that the emf of this cell is 0.577 V under standard conditions.

Spotlight

One of the most important types of chemical reactions, in a negative sense, is the corrosion of metals. The prime example is the formation of "rust", $Fe(OH)_3$, or Fe_2O_3, on iron-containing metals. The formation of rust is believed to proceed usually by a two-step process:

$$Fe(s) + \tfrac{1}{2}O_2(g) + H_2O \rightleftharpoons Fe(OH)_2(s)$$

$$2Fe(OH)_2(s) + \tfrac{1}{2}O_2(g) + H_2O \rightleftharpoons \underset{\text{rust}}{2Fe(OH)_3(s)}$$

So serious a problem is iron corrosion that the cost is estimated at billions of dollars annually in the U.S. economy, and about one-fifth of the annual steel production is required for replacing rusted iron items. The mechanisms of metal corrosion constitute a complex and sophisticated subject. Nevertheless, a substantial understanding of metal corrosion can be gained from a simple electrochemical model.

Both water and oxygen are usually necessary for the ordinary corrosion of iron. The oxygen is the oxidizing agent and the water is the solution of what is essentially a voltaic cell. Different parts of a metal object are found to have different tendencies to undergo oxidation. This can be due to nonuniformities in chemical composition of the metal—or to different elemental identities, as when one metal is joined to another—to different degrees of mechanical stress, and so on. These different regions can act as anode and cathode regions of a "minicell," with the metal object itself completing the circuit, acting as the external circuit. At the anode region, Fe(s) is oxidized to Fe^{2+}; at the cathode region, O_2 is reduced to OH^-. The Fe^{2+} generated at the anode migrates through the water (or moist medium) to the cathode region, where it reacts with OH^- to form $Fe(OH)_2(s)$; this substance then becomes oxidized to $Fe(OH)_3(s)$, rust. According to this mechanism, one might expect the rust to appear in the cathode region, while the "pitting" of the iron (due to consumption of Fe) occurs in the anode region. These regions can be a considerable distance apart, say several centimeters. According to this electrochemical model, it is not surprising that the presence of electrolytes in the water would accelerate rusting by enhancing the rate of current flow through the solution region of the "cell."

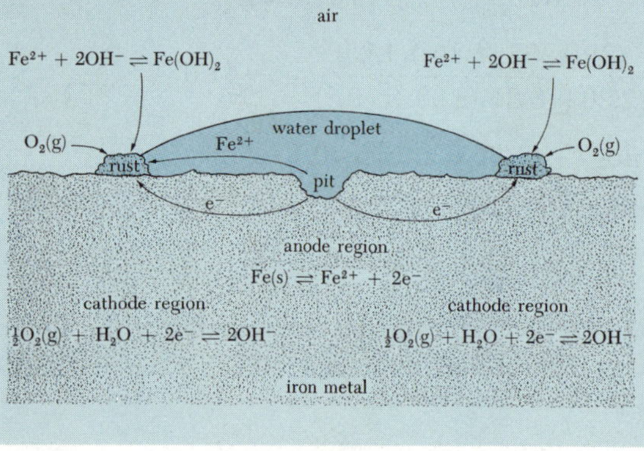

Different portions of a metal can have different tendencies to be oxidized. One instance of such selectivity occurs when the metal surface is covered with water droplets. The edge of a droplet has a higher concentration of dissolved O_2 than the center of the droplet does. Hence, the tendency for the reduction half-reaction of O_2 to occur is greater at the edges than near the center of the droplet-iron interface. The net result is again a small electrochemical cell.

Various methods are used to decrease the corrosion of iron, especially for objects in water or in moist earth. One method, called cathodic protection, involves connecting the iron object to a more active metal—one with a less positive reduction potential, like Zn or Mg. The more active metal then serves as the anode and the iron serves as the cathode, at which the O_2 reduction occurs rather than iron oxidation.

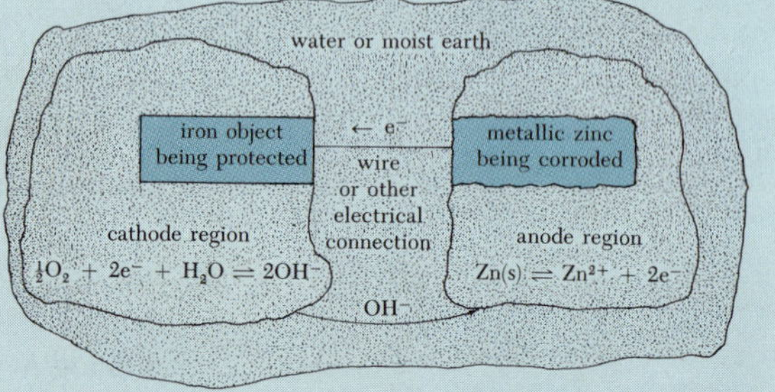

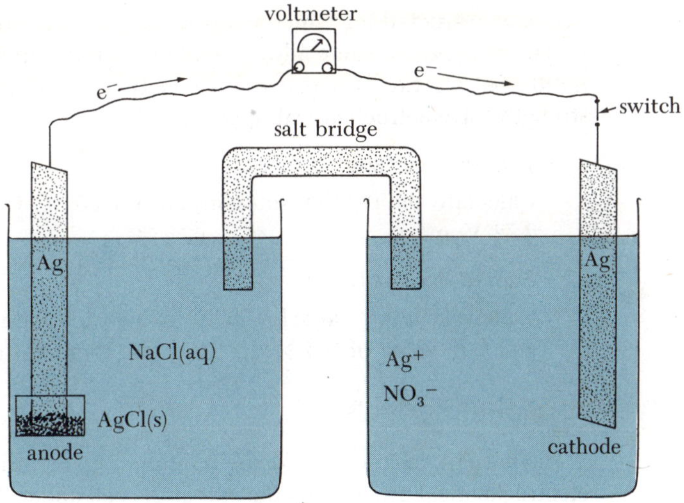

Figure 12.14 Electrochemical cell used for determining the solubility product of AgCl(s).

anode reaction: $Ag(s) + Cl^- \rightleftharpoons AgCl(s) + e^-$

cathode reaction: $Ag^+ + e^- \rightleftharpoons Ag(s)$

net cell reaction: $Ag^+ + Cl^- \rightleftharpoons AgCl(s)$

SOLUTION The net reaction and corresponding equilibrium constant for this system are

$$Ag^+ + Cl^- \rightleftharpoons AgCl(s), \qquad K = ([Ag^+][Cl^-])^{-1}$$

Recognizing that $n = 1$ for this case, from Eq. (12.49) we obtain

$$2.303 \log K = \frac{nF\mathscr{E}^\circ}{RT} = \frac{(1)(96.5 \times 10^3)(0.577)}{(8.31)(298)} = 22.5$$

$$\log K = \frac{22.5}{2.303} = 9.77$$

$$K = \text{antilog } 9.77 = [\text{antilog } (0.77)]10^9 = 5.9 \times 10^9$$

There are some important points to be derived from these three examples. One point is that for an electrochemical reaction to be capable of generating an emf of reasonable magnitude in a cell, say of the order of a volt, it must have a very large equilibrium constant; in Example 12.8 a reaction with an equilibrium constant of 2×10^{37} generates 1.1 V under standard conditions. Analogously, reactions with negative $\mathscr{E}^\circ$ values of reasonable magnitude have very small equilibrium constants, as evidenced by Example 12.9. The message in Example 12.10 is a little more subtle. The net reaction of the cell,

$$Ag^+ + Cl^- \rightleftharpoons AgCl(s)$$

is not an oxidation-reduction. Yet, a cell has been constructed for which an emf can be measured that corresponds to this net reaction, and this makes it possible for us to compute the equilibrium constant directly. In this particular case the equilibrium constant that has been determined is the inverse of K_{sp} for AgCl. Whenever a cell can be constructed for which the *net* reaction is the reaction of interest, the determination of the $\mathscr{E}°$ value allows one to obtain the equilibrium constant, whether or not the net reaction is an oxidation-reduction reaction.* This is one of the important strengths of electrochemical approaches.

STUDY PROBLEM 12(e)

Calculate the equilibrium constant for 25 °C for a reaction for which $\mathscr{E}°$ is 2.72 V and $n = 1$.

STUDY PROBLEM 12(f)

Calculate the $\mathscr{E}°$ value for 25 °C for an oxidation-reduction reaction with $n = 2$ and a K value of 3.8×10^{-18} at this temperature.

Half-Reaction $\mathscr{E}°$; Reduction Potentials

Having now characterized the relation between the equilibrium constant of an electrochemical process and the emf of the cell under standard conditions, we recognize that it is not necessary to have tables of both quantities in any collection of data on such systems. So far as we are now aware, if we need to know a value of $\mathscr{E}°$ (or K) for a specific electrochemical reaction, we should have to search for it in a table of data that would cover all possible combinations of reducing agents and oxidizing agents. Suppose one wishes to compile a table of $\mathscr{E}°$ values for all possible oxidation-reduction reactions among one hundred different oxidizing agents and one hundred different reducing agents; there are 10,000 combinations to be covered on the table! Fortunately, we find that this potentially overwhelming situation can be substantially simplified. This simplifying approach is directly analogous to what was developed in Sec. 5.7 and 5.8 for acid-base reactions.

Suppose we construct a series of cells for which the anode is a standard (Pt)H_2,1 atm$|H^+$,1M electrode (that is, a $H_2|H^+$ electrode under standard conditions) and the cathode consists of a platinum electrode at which any one of a variety of oxidizing agents can be reduced, only one for a given cell. The general arrangement is depicted in Fig. 12.15. At the cathode of this kind of arrangement, a process of the following type occurs:

$$\text{ox.ag.} + ne^- \rightleftharpoons \text{red. ag.}$$

The combination of an oxidizing agent and its "conjugate" reducing agent is referred to as a **couple.**† Thus, the cathode of Fig. 12.15 consists of some specific couple at a platinum electrode. In this system under consideration, all pertinent species of the cathode couple are in their standard states. If $\mathscr{E}°$ measurements are made on various

*That one can obtain an equilibrium constant from an $\mathscr{E}°$ value measured on a cell whose net reaction is not an oxidation-reduction is an example of the principle that changes in thermodynamic state variables (for example, $\Delta G°$, which is related to $\mathscr{E}°$ by Eq. 12.48) depends only on the initial and final states, not on the pathway. Hence, although the reaction of Ag^+ with Cl^- to yield AgCl(s) was conducted in this particular case through the combination of two oxidation-reduction half-reactions, the $\Delta G°$ and K values have not been altered for the overall net reaction.

†The Cu(II)–Cu(I) oxidation-reduction couple is involved in electron transport in biological systems. Some copper-containing proteins show the properties of a Cu(II)–Cu(I) couple; these are called **cuproproteins.** They include dehydrogenases (enzymes involved in removing hydrogen from a chemical species), the oxidoreductases, which electrons from certain organic species, the flavins and hydroquinones, to O_2.

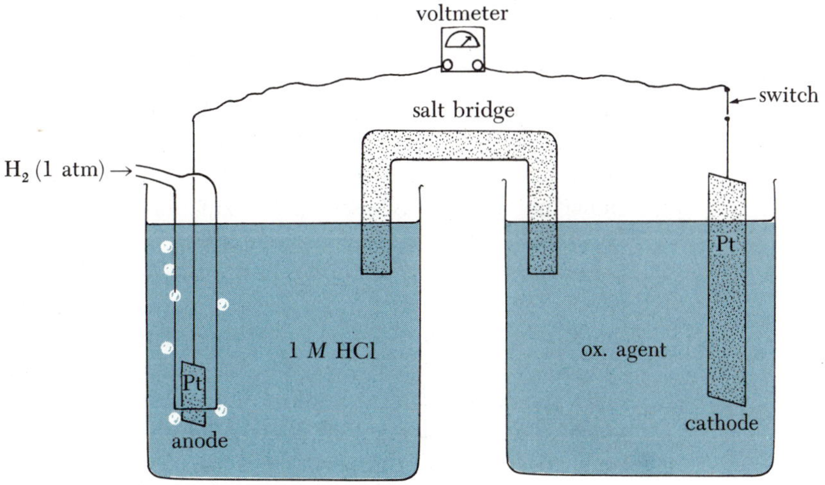

Figure 12.15 Generalized electrochemical cell in which the $(Pt)H_2$, 1 atm|HCl, 1 *M* electrode functions as the anode and some reduction, unspecified, takes place under standard conditions at the cathode.

cells of this type with a variety of different couples at the cathode, the differences in the $\mathscr{E}^\circ$'s measured can be attributed to differences in the cathode couple, since the anode system is the same in each case, $(Pt)H_2$,1 atm|H^+,1 *M*. Thus, the $\mathscr{E}^\circ$ values measured in such a series of experiments can be taken as measures of the strengths of the oxidizing agents used at the cathode. For example, if $F_2(g)$ is used at the cathode (at 1 atm pressure, together with a F^- concentration 1 *M*), an $\mathscr{E}^\circ$ value of 2.87 V is obtained. If $Cl_2(g)$ is used (at 1 atm, and with $[Cl^-] = 1$), $\mathscr{E}^\circ$ is found to be 1.36 V. For Br_2 (a pool of pure liquid in contact with the platinum and with the solution having $[Br^-] = 1$), $\mathscr{E}^\circ$ is determined to be 1.06 V. As the anode is the same in each case, these values mean that F_2 is a better oxidizing agent than Cl_2, which in turn is a better oxidizing agent than Br_2. That is, the half-reaction $\frac{1}{2}F_2 + e^- \rightleftharpoons F^-$ has a greater tendency to occur than the half-reaction $\frac{1}{2}Cl_2 + e^- \rightleftharpoons Cl^-$, and that half-reaction has a higher driving force than the half-reaction $\frac{1}{2}Br_2 + e^- \rightleftharpoons Br^-$. Also, if 1 *M* I_3^- is the oxidizing agent (with I^- in 1 *M* concentration), $\mathscr{E}^\circ$ is found to be 0.56 V, signifying that the half-reaction $I_3^- + 2e^- \rightleftharpoons 3I^-$ has even less driving force to proceed than the bromine half-reaction does. If the cathode has solid sulfur as the oxidizing agent (with $H_2S(g)$ present at 1 atm and H^+ at 1 *M*), the observed $\mathscr{E}^\circ$ value is 0.14 V; this shows that solid sulfur in this system is even a poorer oxidizing agent than I_3^- To summarize these points, the net cell reactions, pertinent cathode half-reactions, and measured $\mathscr{E}^\circ$ values are listed herewith.

$F_2(g) + 2e^- \rightleftharpoons 2F^-$,	$\mathscr{E}^\circ = 2.87$ V	$H_2(g) + F_2(g) \rightleftharpoons 2H^+ + 2F^-$
$Cl_2(g) + 2e^- \rightleftharpoons 2Cl^-$,	$\mathscr{E}^\circ = 1.36$ V	$H_2(g) + Cl_2(g) \rightleftharpoons 2H^+ + 2Cl^-$
$Br_2(l) + 2e^- \rightleftharpoons 2Br^-$,	$\mathscr{E}^\circ = 1.06$ V	$H_2(g) + Br_2(l) \rightleftharpoons 2H^+ + 2Br^-$
$I_3^- + 2e^- \rightleftharpoons 3I^-$,	$\mathscr{E}^\circ = 0.56$ V	$H_2(g) + I_2(s) \rightleftharpoons 2H^+ + 2I^-$
$S(s) + 2H^+ + 2e^- \rightleftharpoons H_2S(g)$,	$\mathscr{E}^\circ = 0.14$ V	$H_2(g) + S(s) \rightleftharpoons H_2S(g)$

Consider another set of cells, all having a fixed cathode reaction,

$$2H^+ + 2e^- \rightleftharpoons H_2(g)$$

These cells use as the cathode a standard hydrogen electrode identical to what is shown in Fig. 12.15 but operating in the reverse direction. This second series of cells would differ among themselves only in the anode reaction; the anode reactions are shown in Fig. 12.16 as based on the oxidation of a metal. Each anode reaction would be carried out under standard conditions; for example, each metal ion would be present in solution at 1 *M* concentration. Any number of couples could serve for the anode, but for simplicity, we limit consideration to the oxidation of the series of metals Mg, Al, Zn, Cd, and Pb. When these metal couples—$Mg|Mg^{2+}$, $Al|Al^{3+}$, $Zn|Zn^{2+}$, $Cd|Cd^{2+}$, and $Pb|Pb^{2+}$—are used individually at the anode, the following emf's are measured in the resulting five standard cells: 2.37, 1.66, 0.76, 0.40, and 0.13 volts. The total cell reactions for which these $\mathscr{E}°$ values are obtained are listed herewith, together with the pertinent anode half-reactions.

$Pb(s) + 2H^+ \rightleftharpoons H_2(g) + Pb^{2+}$	$Pb(s) \rightleftharpoons Pb^{2+} + 2e^-$,	$\mathscr{E}° = 0.13$ V
$Cd(s) + 2H^+ \rightleftharpoons H_2(g) + Cd^{2+}$	$Cd(s) \rightleftharpoons Cd^{2+} + 2e^-$,	$\mathscr{E}° = 0.40$ V
$Zn(s) + 2H^+ \rightleftharpoons H_2(g) + Zn^{2+}$	$Zn(s) \rightleftharpoons Zn^{2+} + 2e^-$,	$\mathscr{E}° = 0.76$ V
$Al(s) + 3H^+ \rightleftharpoons \frac{3}{2}H_2(g) + Al^{3+}$	$Al(s) \rightleftharpoons Al^{3+} + 3e^-$,	$\mathscr{E}° = 1.66$ V
$Mg(s) + 2H^+ \rightleftharpoons H_2(g) + Mg^{2+}$	$Mg(s) \rightleftharpoons Mg^{2+} + 2e^-$,	$\mathscr{E}° = 2.37$ V

As the cathode reaction in each of these five standard cells above is the reduction of H^+ to $H_2(g)$, it is clear that the measured $\mathscr{E}°$ values reflect the relative abilities of the five metals to function as reducing agents; Mg is the best of the five and Pb the worst.

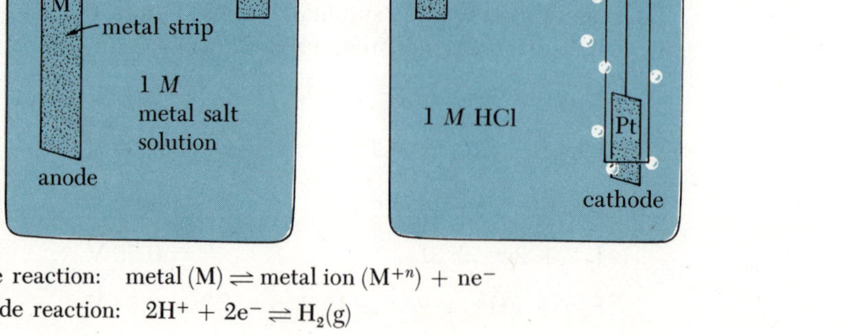

anode reaction: metal (M) $\rightleftharpoons$ metal ion (M^{+n}) + ne^-

cathode reaction: $2H^+ + 2e^- \rightleftharpoons H_2(g)$

Figure 12.16 Generalized electrochemical cell in which an $HCl, 1\ M|(Pt)H_2, 1$ atm electrode serves as the cathode and some oxidation, unspecified, occurs at the anode, under standard conditions.

Spotlight

Nitric acid, an oxidizing agent, is a very important chemical reagent. The resonance structures of the acid and its conjugate base, nitrate ion, are

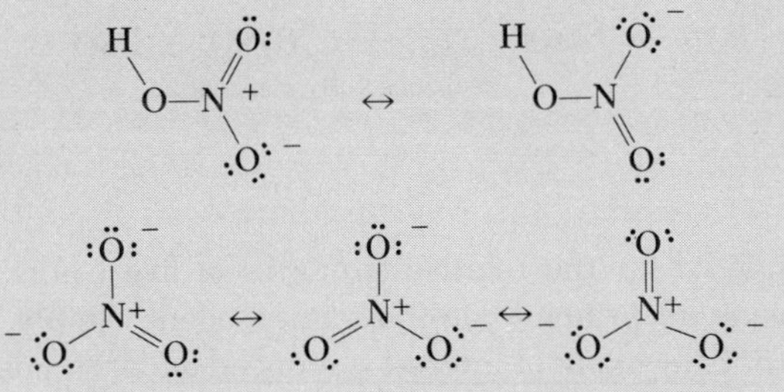

Nitric acid has three important chemical properties. It is (1) an acid, (2) an oxidizing agent, and (3) a nitrating agent.

1. As an acid, HNO_3 is one of the strongest known; it was called *aqua fortis* ("strong water") by the alchemists. It dissolves bases to form nitrates; lithium nitrate, $LiNO_3$, and beryllium nitrate, $Be(NO_3)_2$, can both be prepared in this way. For example, $LiOH(s) + H^+ + NO_3^- \rightleftharpoons LiNO_3(s) + H_2O$. When nitric acid is dilute, it attacks many metals to form their nitrate salts. The reaction of metals with HNO_3 is complicated by the oxidizing properties of the acid. When nitric acid is concentrated, it attacks all metals except Au, Pt, Rh, and Ir. (Some metals, such as Al, Cr, and Fe, in combination with HNO_3 form a "passive" state.) Gold and silver can be dissolved, however, by *aqua regia* (meaning "royal water," because it dissolves gold, the "royal metal"), which consists of three parts of concentrated HCl and one part of concentrated HNO_3. Aqua regia contains free Cl_2 and nitrosyl chloride, ClNO, the formation of which we describe by the following oxidation-reduction reaction:

$$NO_3^- + 3Cl^- + 4H^+ \rightleftharpoons Cl_2 + ClNO + 2H_2O$$

The dissolution of metallic silver or gold by aqua regia is attributable in part to the complexing action of the chloride ion:

$$Au(s) + 3NO_3^- + 4Cl^- + 6H^+ \rightleftharpoons AuCl_4^- + 3NO_2(g) + 3H_2O$$

2. As an oxidizing agent, HNO_3 is especially effective when it is concentrated. The dissolution of metals in HNO_3 is usually due to the oxidizing properties of NO_3^- in acid solution, and not to the oxidizing properties of H^+. Hence, when metals are treated with nitric acid, only Mg liberates only H_2. The reactions with other metals give varying products depending on the HNO_3 concentration and on conditions. For example, Cu, Ag, Bi, and Hg react with dilute HNO_3 forming NO:

$$3Cu(s) + 2NO_3^- + 8H^+ \rightleftharpoons 3Cu^{2+} + 2NO(g) + 4H_2O$$

With concentrated HNO_3, NO_2 is formed:

$$Cu(s) + 2NO_3^- + 4H^+ \rightleftharpoons Cu^{2+} + 2NO_2(g) + 2H_2O$$

With dilute HNO_3, Fe, Zn, and Cd give mainly N_2O:

$$4Zn(s) + 2NO_3^- + 10H^+ \rightleftharpoons 4Zn^{2+} + N_2O(g) + 5H_2O$$

but more concentrated HNO_3 gives NH_2OH or NH_4^+:

$$4Zn(s) + NO_3^- + 10H^+ \rightleftharpoons 4Zn^{2+} + NH_4^+ + 3H_2O$$

$$3Zn(s) + NO_3^- + 7H^+ \rightleftharpoons 3Zn^{2+} + NH_2OH + 2H_2O$$

The oxidation of Fe by nitric acid gives mainly the ferrous ion, Fe^{2+}, with 6% to 20% HNO_3, and mainly the ferric ion, Fe^{3+}, with higher than 20% HNO_3, because of further oxidation.

Metals that have been rendered "passive" by HNO_3 seem to lose many of their chemical properties. For example, iron dissolves in dilute HNO_3, but does not seem to react with HNO_3 in concentrations above about 77%; the iron, moreover, becomes passive since it will no longer dissolve in dilute HNO_3, nor will it reduce aqueous solutions of copper nitrate or silver nitrate to their respective metals. This passivity has been attributed to a thin film of oxide caused by the oxidizing properties of HNO_3, although its exact nature is unknown. Other strong oxidizing agents have the same effect; concentrated H_2SO_4, H_2O_2, and chromic acid, H_2CrO_4, also render iron passive, and studies have shown that a film of Fe_2O_3 exists. The passivity can be removed by striking the iron with a hard blow, scratching it, heating it in a reducing gas,

(Continued on page 578.)

or holding it in contact with metallic zinc while it is immersed in dilute HNO_3.

Nitric acid oxidizes many organic compounds, generally producing CO_2 and H_2O from hydrocarbons. For example, turpentine or even sawdust will burst into flames if added to warm fuming HNO_3, and glowing charcoal continues to burn even when submerged in the acid.

3. As a nitrating agent, nitric acid brings about the replacement of hydrogen atoms and —OH groups from organic molecules by $—NO_2$ groups, particularly when the nitric acid is mixed with concentrated H_2SO_4. The use of H_2SO_4 gives the ion NO_2^+, which is a very active nitrating agent:

$$3HSO_4^- + NO_3^- \rightleftharpoons NO_2^+ + 3SO_4^{2-} + H_3O^+$$

$$SO_4^{2-} + NO_2^+ + \underset{\text{benzene}}{C_6H_6} \rightleftharpoons \underset{\text{nitrobenzene}}{C_6H_5NO_2} + HSO_4^-$$

The kind of ranking just considered for the relative strengths of five oxidizing agents could be extended to include, in principle, any oxidizing agent, simply by constructing a cell in which the oxidizing agent of interest is reduced under standard conditions at the cathode and H_2 is oxidized under standard conditions at the anode. Hence, a giant table could be constructed in which the relative strengths of all oxidizing agents would be represented by the pertinent $\mathscr{E}°$ values. Similarly, the relative strengths of all reducing agents could be compared quantitatively in a huge table by listing in order the $\mathscr{E}°$ values obtained for all possible cells in which the reducing agent of interest is oxidized under standard conditions at the anode and H^+ is reduced under standard conditions at the cathode. While it may not be obvious at first sight, the information contained in these two tables can be combined into one unified table based on one set of concepts. This becomes clear when one considers the reverse of the net cell reactions for the five reducing agents that have been discussed; the $\mathscr{E}°$ values for the reversed reactions are simply reversed in sign.

For example, if $\mathscr{E}°$ is 2.37 V for the net cell reaction

$$Mg(s) + 2H^+ \rightleftharpoons H_2(g) + Mg^{2+} \qquad \mathscr{E}° = 2.37 \text{ V}$$

then $\mathscr{E}°$ must be -2.37 V for the reverse reaction

$$H_2(g) + Mg^{2+} \rightleftharpoons Mg(s) + 2H^+ \qquad \mathscr{E}° = -2.37 \text{ V}$$

This means that an applied external voltage would be required to force the reaction to occur. Accepting this view, let us now summarize the information given above for the five reducing agents, regarding the pertinent net cell reaction and half-reactions in the direction opposite to what was considered above.

$Pb^{2+} + 2e^- \rightleftharpoons Pb(s)$,	$\mathscr{E}° = -0.13$ V	$Pb^{2+} + H_2(g) \rightleftharpoons 2H^+ + Pb(s)$
$Cd^{2+} + 2e^- \rightleftharpoons Cd(s)$,	$\mathscr{E}° = -0.40$ V	$Cd^{2+} + H_2(g) \rightleftharpoons 2H^+ + Cd(s)$
$Zn^{2+} + 2e^- \rightleftharpoons Zn(s)$,	$\mathscr{E}° = -0.76$ V	$Zn^{2+} + H_2(g) \rightleftharpoons 2H^+ + Zn(s)$
$Al^{3+} + 3e^- \rightleftharpoons Al(s)$,	$\mathscr{E}° = -1.66$ V	$Al^{3+} + \frac{3}{2}H_2(g) \rightleftharpoons 3H^+ + Al(s)$
$Mg^{2+} + 2e^- \rightleftharpoons Mg(s)$,	$\mathscr{E}° = -2.37$ V	$Mg^{2+} + H_2(g) \rightleftharpoons 2H^+ + Mg(s)$

Viewing both five-member sets of reactions now as the reduction of an oxidizing agent, we can create a single table with ten entries as shown in Fig. 12.17. All ten entries on the left side of the figure have the same format, yielding a measure of the oxidizing strength of each oxidizing agent in the figure. The strongest oxidizing agent of the ten represented is seen to be $F_2(g)$, and the weakest is Mg^{2+}. By viewing the

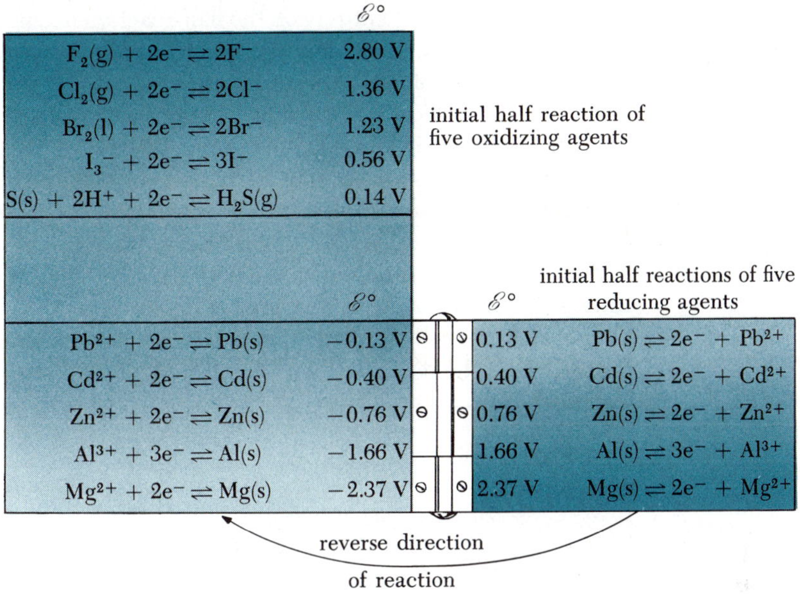

Figure 12.17
Evolution of a unified table of reduction potentials (standard potentials), representing reversing the sense of the lower reactions in terms of a gate swinging on hinges.

	$\mathscr{E}°$	
$F_2(g) + 2e^- \rightleftharpoons 2F^-$	2.80 V	
$Cl_2(g) + 2e^- \rightleftharpoons 2Cl^-$	1.36 V	
$Br_2(l) + 2e^- \rightleftharpoons 2Br^-$	1.23 V	initial half reaction of five oxidizing agents
$I_3^- + 2e^- \rightleftharpoons 3I^-$	0.56 V	
$S(s) + 2H^+ + 2e^- \rightleftharpoons H_2S(g)$	0.14 V	

	$\mathscr{E}°$	$\mathscr{E}°$	initial half reactions of five reducing agents
$Pb^{2+} + 2e^- \rightleftharpoons Pb(s)$	−0.13 V	0.13 V	$Pb(s) \rightleftharpoons 2e^- + Pb^{2+}$
$Cd^{2+} + 2e^- \rightleftharpoons Cd(s)$	−0.40 V	0.40 V	$Cd(s) \rightleftharpoons 2e^- + Cd^{2+}$
$Zn^{2+} + 2e^- \rightleftharpoons Zn(s)$	−0.76 V	0.76 V	$Zn(s) \rightleftharpoons 2e^- + Zn^{2+}$
$Al^{3+} + 3e^- \rightleftharpoons Al(s)$	−1.66 V	1.66 V	$Al(s) \rightleftharpoons 3e^- + Al^{3+}$
$Mg^{2+} + 2e^- \rightleftharpoons Mg(s)$	−2.37 V	2.37 V	$Mg(s) \rightleftharpoons 2e^- + Mg^{2+}$

reverse direction of reaction

five entries on the right side of Fig. 12.17, one sees that Mg(s) is the strongest reducing agent of those listed. As $F_2(g)$ is the strongest oxidizing agent represented in Fig. 12.17, F^- is the weakest reducing agent shown.

As the hydrogen electrode can be used as either the anode or the cathode (either $H_2(g)$ is oxidized or H^+ is reduced), we should expect that any oxidation-reduction couple can be placed on the kind of scale shown on the left side of Fig. 12.17. All that is necessary is to be able to determine the $\mathscr{E}°$ value of a cell in which one electrode is that at which the couple of interest functions and the other electrode is the standard hydrogen electrode. Extensive tables of this type have been constructed; a relatively small version is given as Table 12.1. In that table one sees that an $\mathscr{E}°$ value of zero is assigned to the half-reaction

$$2H^+ + 2e^- \rightleftharpoons H_2(g)$$

This is precisely what we should expect as the $\mathscr{E}°$ value for a cell of the general type shown in Fig. 12.15, in which the above half-reaction proceeds at the cathode, while the following "standard" reaction is occurring at the anode:

$$H_2(g) \rightleftharpoons 2H^+ + 2e^-$$

In this case one electrode simply balances the other, so there is no net cell emf, and $\mathscr{E}° = 0$.

The $\mathscr{E}°$ values in Table 12.1 correspond to cells for which the indicated half-reactions pertain to one electrode, while the other electrode is the standard hydrogen electrode. Recognizing this, it is convenient to assign these $\mathscr{E}°$ values for the complete cells to the indicated half-reactions. This yields a table that looks identical to what we should have constructed if we had assumed that the half-reaction occurring at the standard hydrogen electrode contributed nothing to the net cell emf ($\mathscr{E}°$), and had "blamed" $\mathscr{E}^\circ$ entirely on the half-reaction occurring at the other electrode. We know that this kind of unilateral blame cannot be rigorously valid. Nevertheless, it is quite permissible, and certainly convenient, to set up an emf scale in which we assign $\mathscr{E}^\circ$ values to individual half-reactions, relative to an arbitrary zero value of $\mathscr{E}^\circ$ for the standard hydrogen electrode. This (or any other) arbitrary zero for our emf scale is permissible because every cell has two half-reaction $\mathscr{E}^\circ$ values. These values are called **standard reduction potentials,** because the chemical equation describing each such half-reaction shows the reduction of an oxidizing agent. The reference reaction is the reduction of H^+ to H_2. Choosing this (or any other) arbitrary zero for our scale of standard reduction potentials is permissible for the following reason: Every cell reaction consists of two half-reactions and the $\mathscr{E}^\circ$ value for the cell is the difference between the two standard reduction potentials, as we see below.

TABLE 12.1 Standard Reduction Potentials

Half-Reaction (Couple)	*Half-reaction $\mathscr{E}^\circ$, volts*
A. Acid Solution	
$F_2(g) + 2e^- \rightleftharpoons 2F^-$	2.87
$O_3(g) + 2H^+ + 2e^- \rightleftharpoons O_2(g) + H_2O$	2.07
$S_2O_8^{2-} + 2e^- \rightleftharpoons 2SO_4^{2-}$	2.01
$H_2O_2 + 2H^+ + 2e^- \rightleftharpoons 2H_2O$	1.77
$MnO_4^- + 4H^+ + 3e^- \rightleftharpoons MnO_2(s) + 2H_2O$	1.69
$PbO_2(s) + 4H^+ + SO_4^{2-} + 2e^- \rightleftharpoons PbSO_4(s) + 2H_2O$	1.69
$NiO_2(s) + 4H^+ + 2e^- \rightleftharpoons Ni^{2+} + 2H_2O$	1.68
$HClO + H^+ + e^- \rightleftharpoons H_2O + \frac{1}{2}Cl_2(g)$	1.63
$HBrO + H^+ + e^- \rightleftharpoons \frac{1}{2}Br_2(l) + H_2O$	1.59
$BrO_3^- + 6H^+ + 5e^- \rightleftharpoons \frac{1}{2}Br_2(l) + 3H_2O$	1.52
$MnO_4^- + 8H^+ + 5e^- \rightleftharpoons Mn^{2+} + 4H_2O$	1.51
$PbO_2(s) + 4H^+ + 2e^- \rightleftharpoons Pb^{2+} + 2H_2O$	1.45
$Ce^{4+} + e^- \rightleftharpoons Ce^{3+}$	1.44
$Cl_2(g) + 2e^- \rightleftharpoons 2Cl^-$	1.36
$Cr_2O_7^{2-} + 14H^+ + 6e^- \rightleftharpoons 2Cr^{3+} + 7H_2O$	1.33
$ClO_2 + H^+ + e^- \rightleftharpoons HClO_2$	1.28
$MnO_2(s) + 4H^+ + 2e^- \rightleftharpoons Mn^{2+} + 2H_2O$	1.23
$O_2(g) + 4H^+ + 4e^- \rightleftharpoons 2H_2O$	1.23
$ClO_3^- + 3H^+ + 2e^- \rightleftharpoons HClO_2 + H_2O$	1.21
$ClO_4^- + 2H^+ + 2e^- \rightleftharpoons ClO_3^- + H_2O$	1.19
$Br_2(l) + 2e^- \rightleftharpoons 2Br^-$	1.06
$AuCl_4^- + 3e^- \rightleftharpoons Au(s) + 4Cl^-$	1.00
$NO_3^- + 4H^+ + 3e^- \rightleftharpoons NO(g) + 2H_2O$	0.96
$2Hg^{2+} + 2e^- \rightleftharpoons Hg_2^{2+}$	0.92
$Ag^+ + e^- \rightleftharpoons Ag(s)$	0.80
$Hg_2^{2+} + 2e^- \rightleftharpoons 2Hg(l)$	0.79

TABLE 12.1 **(continued)**

Half-Reaction (Couple)	*Half-Reaction $\mathscr{E}°$, volts*
$Fe^{3+} + e^- \rightleftharpoons Fe^{2+}$	0.77
$O_2(g) + 2H^+ + 2e^- \rightleftharpoons H_2O_2$	0.68
$I_3^- + 2e^- \rightleftharpoons 3I^-$	0.56
$I_2(s) + 2e^- \rightleftharpoons 2I^-$	0.54
$Cu^+ + e^- \rightleftharpoons Cu(s)$	0.52
$Cu^{2+} + 2e^- \rightleftharpoons Cu(s)$	0.34
$AgCl(s) + e^- \rightleftharpoons Ag(s) + Cl^-$	0.22
$SO_4^{2-} + 4H^+ + 2e^- \rightleftharpoons H_2SO_3 + H_2O$	0.17
$Cu^{2+} + e^- \rightleftharpoons Cu^+$	0.15
$S(s) + 2H^+ + 2e^- \rightleftharpoons H_2S(g)$	0.14
$AgBr(s) + e^- \rightleftharpoons Ag(s) + Br^-$	0.10
$2H^+ + 2e^- \rightleftharpoons H_2(g)$	0.00
$Pb^{2+} + 2e^- \rightleftharpoons Pb(s)$	−0.13
$Sn^{2+} + 2e^- \rightleftharpoons Sn(s)$	−0.14
$AgI(s) + e^- \rightleftharpoons Ag(s) + I^-$	−0.15
$Ni^{2+} + 2e^- \rightleftharpoons Ni(s)$	−0.25
$PbSO_4(s) + 2e^- \rightleftharpoons Pb(s) + SO_4^{2-}$	−0.36
$Cd^{2+} + 2e^- \rightleftharpoons Cd(s)$	−0.40
$Cr^{3+} + e^- \rightleftharpoons Cr^{2+}$	−0.41
$Fe^{2+} + 2e^- \rightleftharpoons Fe(s)$	−0.44
$Zn^{2+} + 2e^- \rightleftharpoons Zn(s)$	−0.76
$Mn^{2+} + 2e^- \rightleftharpoons Mn(s)$	−1.18
$Al^{3+} + 3e^- \rightleftharpoons Al(s)$	−1.66
$H_2(g) + 2e^- \rightleftharpoons 2H^-$	−2.25
$Mg^{2+} + 2e^- \rightleftharpoons Mg(s)$	−2.37
$La^{3+} + 3e^- \rightleftharpoons La(s)$	−2.52
$Na^+ + e^- \rightleftharpoons Na(s)$	−2.71
$Ca^{2+} + 2e^- \rightleftharpoons Ca(s)$	−2.87
$Ba^{2+} + 2e^- \rightleftharpoons Ba(s)$	−2.90
$K^+ + e^- \rightleftharpoons K(s)$	−2.93
$Li^+ + e^- \rightleftharpoons Li(s)$	−3.05
B. Basic Solution	
$O_3(g) + H_2O + 2e^- \rightleftharpoons O_2(g) + 2OH^-$	1.24
$ClO_2 + e^- \rightleftharpoons ClO_2^-$	1.16
$ClO^- + H_2O + 2e^- \rightleftharpoons Cl^- + 2OH^-$	0.89
$HO_2^- + H_2O + 2e^- \rightleftharpoons 3OH^-$	0.88
$ClO_2^- + H_2O + 2e^- \rightleftharpoons ClO^- + 2OH^-$	0.66
$MnO_4^{2-} + 2H_2O + 2e^- \rightleftharpoons MnO_2(s) + 4OH^-$	0.60
$NiO_2(s) + 2H_2O + 2e^- \rightleftharpoons Ni(OH)_2(s) + 2OH^-$	0.49

TABLE 12.1 **(continued)**

Half-Reaction (Couple)	*Half-Reaction* $\mathscr{E}°$, *volts*
$O_2(g) + 2H_2O + 4e^- \rightleftharpoons 4OH^-$	0.40
$ClO_4^- + H_2O + 2e^- \rightleftharpoons ClO_3^- + 2OH^-$	0.36
$ClO_3^- + H_2O + 2e^- \rightleftharpoons ClO_2^- + 2OH^-$	0.33
$IO_3^- + 3H_2O + 6e^- \rightleftharpoons I^- + 6OH^-$	0.26
$Co(NH_3)_6^{3+} + e^- \rightleftharpoons Co(NH_3)_6^{2+}$	0.10
$HgO(s) + H_2O + 2e^- \rightleftharpoons Hg(l) + 2OH^-$	0.10
$MnO_2(s) + 2H_2O + 2e^- \rightleftharpoons Mn(OH)_2 + 2OH^-$	−0.05
$O_2(g) + H_2O + 2e^- \rightleftharpoons HO_2^- + OH^-$	−0.08
$Cu(NH_3)_2^+ + e^- \rightleftharpoons Cu(s) + 2NH_3$	−0.12
$CrO_4^{2-} + 4H_2O + 3e^- \rightleftharpoons Cr(OH)_3(s) + 5OH^-$	−0.13
$Ag(CN)_2^- + e^- \rightleftharpoons Ag(s) + 2CN^-$	−0.31
$Hg(CN)_4^{2-} + 2e^- \rightleftharpoons Hg(s) + 4CN^-$	−0.37
$NiCO_3(s) + 2e^- \rightleftharpoons Ni(s) + CO_3^{2-}$	−0.45
$S(s) + 2e^- \rightleftharpoons S^{2-}$	−0.48
$Pb(OH)_3^- + 2e^- \rightleftharpoons Pb(s) + 3OH^-$	−0.54
$Fe(OH)_3(s) + e^- \rightleftharpoons Fe(OH)_2(s) + OH^-$	−0.56
$HgS(s) + 2e^- \rightleftharpoons Hg(s) + S^{2-}$	−0.72
$Cd(OH)_2(s) + 2e^- \rightleftharpoons Cd(s) + 2OH^-$	−0.81
$2H_2O + 2e^- \rightleftharpoons H_2(g) + 2OH^-$	−0.83
$SO_4^{2-} + H_2O + 2e^- \rightleftharpoons SO_3^{2-} + 2OH^-$	−0.93
$Zn(NH_3)_4^{2+} + 2e^- \rightleftharpoons Zn(s) + 4NH_3$	−1.03
$2SO_3^{2-} + 2H_2O + 2e^- \rightleftharpoons S_2O_4^{2-} + 4OH^-$	−1.12
$N_2(g) + 4H_2O + 4e^- \rightleftharpoons N_2H_4 + 4OH^-$	−1.16
$Zn(OH)_4^{2-} + 2e^- \rightleftharpoons Zn(s) + 4OH^-$	−1.22
$ZnS(s) + 2e^- \rightleftharpoons Zn(s) + S^{2-}$	−1.44
$Mn(OH)_2(s) + 2e^- \rightleftharpoons Mn(s) + 2OH^-$	−1.55
$H_2AlO_3^- + H_2O + 3e^- \rightleftharpoons Al(s) + 4OH^-$	−2.35
$Mg(OH)_2(s) + 2e^- \rightleftharpoons Mg(s) + 2OH^-$	−2.69
$Cu(OH)_2(s) + 2e^- \rightleftharpoons Cu(s) + 2OH^-$	−3.03

The information in Table 12.1 is separated into a set of cases relevant to systems that exist in acidic solutions and a set that pertains to basic solutions. This separation is made because the species of a given element in a particular oxidation state are often different in acidic and basic media; an example is $Cr_2O_7^{2-}$ in acidic solution vs CrO_4^{2-} in basic media. Another reason for the separation is that oxygen and hydrogen balance can involve H^+ in acidic solution and OH^- in basic solution. A key point is that there is in a sense a different hydrogen electrode in basic solution compared with the case discussed above ($2H^+ + 2e^- \rightleftharpoons H_2(g)$); in basic solution, the reaction is

$$2H_2O + 2e^- \rightleftharpoons H_2(g) + 2OH^-$$

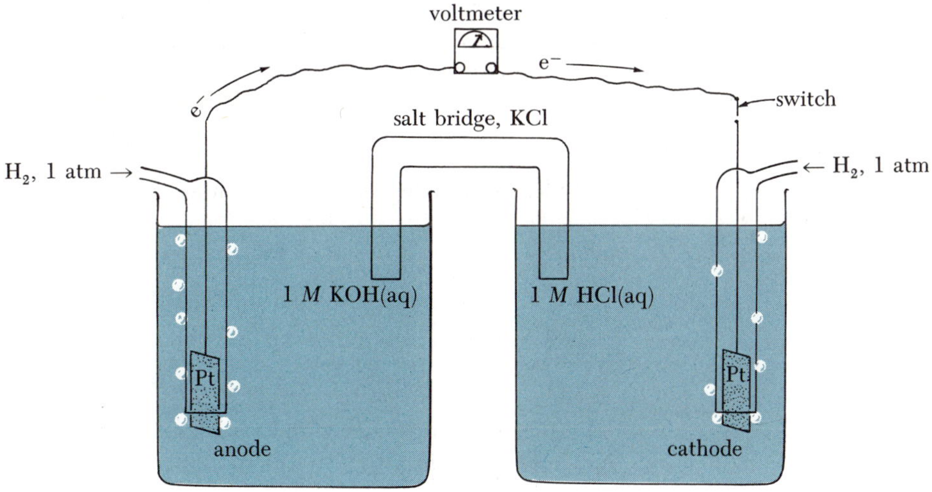

anode reaction: $H_2(g) + 2OH^- \rightleftharpoons 2H_2O + 2e^-$

cathode reaction: $2H^+ + 2e^- \rightleftharpoons H_2(g)$

net cell reaction: $2H^+ + 2OH^- \rightleftharpoons 2H_2O$

Figure 12.18 Electrochemical cell employing a standard basic hydrogen electrode at the anode and a standard acidic hydrogen electrode at the anode. The value for $\mathscr{E}°$ for the cell is 0.83 V.

The $\mathscr{E}°$ value of this half-reaction is -0.83 V relative to the standard (Pt)H_2(g), 1 atm|H^+,1 *M* electrode. This means that if a cell were constructed in which the standard acidic hydrogen electrode served as the cathode and the standard basic hydrogen electrode served as the anode, a cell emf of 0.83 V could be developed. This situation is depicted in Fig. 12.18. All the $\mathscr{E}°$ values summarized in both part A and part B of Table 12.1 are given relative to a zero value arbitrarily chosen for the standard acidic hydrogen electrode.

Applying Standard Reduction Potentials

A table of $\mathscr{E}°$ values for half-reactions, as represented by Table 12.1, is immensely useful in chemistry and chemical applications. Many of these applications are based on combining half-reaction $\mathscr{E}°$ values algebraically, like half-reaction equations, to give standard emf's corresponding to net cell reactions. Just as one combines two electrode couples to form a cell, adding two half-reactions to yield a net, overall reaction, one can add the corresponding half-reaction $\mathscr{E}°$ values to give the $\mathscr{E}°$ value for an entire cell. The procedure is straightforward, requiring that the sign of the half-reaction $\mathscr{E}°$ values used be consistent with the direction in which the half-reaction is written. Suppose, for example, that we wish to predict the standard emf of the voltaic cell shown in Fig. 12.7. The anode reaction is seen to be the reverse of the half-reaction $Zn^{2+} + 2e^- \rightleftharpoons Zn(s)$, which is shown in Table 12.1 to have a standard electrode potential, or reduction potential, of -0.76 V. Hence, we assign $+0.76$ V to the half-reaction with the direction shown at the bottom of Fig. 12.7. The cathode reaction,

$$Cu^{2+} + 2e^- \rightleftharpoons Cu(s)$$

is seen in Table 12.1 to have an assigned $\mathscr{E}^\circ$ value 0.34 V. Adding these, we obtain the standard emf of the net cell reaction:

$$\mathscr{E}^\circ_{\text{cell}} = \mathscr{E}^\circ_{\text{anode}} + \mathscr{E}^\circ_{\text{cathode}} \tag{12.50}$$

$$\mathscr{E}^\circ_{\text{cell}} = 0.76\text{ V} + 0.34\text{ V} = 1.10\text{ V}$$

Examples of applying Eq. (12.50) follow.

EXAMPLE 12.11 Calculate the standard emf of the lead storage battery described in Fig. 12.8.

SOLUTION From the information provided in Fig. 12.8 and Table 12.1, we see that $\mathscr{E}^\circ_{\text{anode}} = 0.36$ V and $\mathscr{E}^\circ_{\text{cathode}} = 1.68$ V. Hence

$$\mathscr{E}^\circ_{\text{cell}} = \mathscr{E}^\circ_{\text{anode}} + \mathscr{E}^\circ_{\text{cathode}} = 0.36\text{ V} + 1.68\text{ V} = 2.04\text{ V}$$

From this result, it is understandable that six such cells are used in series to provide the twelve volts in a typical automobile storage battery.

EXAMPLE 12.12 What is the minimum voltage that must be applied in the apparatus shown in Fig. 12.11 to bring about the electrolysis of water?

SOLUTION From Fig. 12.11 and Table 12.1, one finds that $\mathscr{E}^\circ_{\text{anode}} = -1.23$ V and $\mathscr{E}^\circ_{\text{cathode}} = 0$. Then, by Eq. (12.50), $\mathscr{E}^\circ_{\text{cell}} = -1.23$ V. Hence, the reverse reaction, $H_2(g) + \frac{1}{2}O_2(g) \rightleftharpoons H_2O$, would generate an emf of 1.23 V if the situation were favorable. Thus, about 1.23 V must be applied to reverse that tendency and bring about the desired process, the electrolysis.

The ability of the simple combining reaction of $H_2(g)$ with $O_2(g)$ to generate a substantial voltage has provided the basis of important fuel cells used in space capsules. Although the reaction of H_2 with O_2 to form water appears simple, developing a fuel cell based on this reaction has proved to be an extremely difficult technical problem. Part of the technical difficulty arises from the fact that the combination of H_2 with O_2 to form H_2O can be slow. Efficient catalysts had to be developed that would allow a convenient reaction rate—neither too fast, giving an explosion, nor too slow. This development was significant in the big successes of U.S. space efforts.

EXAMPLE 12.13 Calculate the equilibrium constant for the dissolving of silver chloride (i.e., the AgCl(s) solubility product), using data given in Table 12.1.

SOLUTION The way to approach this problem is suggested in Fig. 12.14, which depicts an electrochemical cell with a net reaction ($Ag^+ + Cl^- \rightleftharpoons AgCl(s)$) that is simply the reverse of the process of interest. From Table 12.1, the value of $\mathscr{E}^\circ_{\text{anode}}$ is -0.22 V and of $\mathscr{E}^\circ_{\text{cathode}}$, 0.80 V. Hence,

$$\mathscr{E}^\circ_{\text{cell}} = \mathscr{E}^\circ_{\text{anode}} + \mathscr{E}^\circ_{\text{cathode}} = -0.22\text{ V} + 0.80\text{ V} = 0.58\text{ V}$$

The reaction of interest, the reverse of the net cell reaction, is

$$AgCl(s) \rightleftharpoons Ag^+ + Cl^-$$

for which $\mathscr{E}^\circ$ is clearly the negative of the standard emf of the cell shown in Eq. (12.11). That is, the $\mathscr{E}^\circ$ value of interest is -0.58 V. Then, by Eq. (12.49),

Spotlight

The procedure embodied in Eq. (12.50) is justified on thermodynamic grounds by recalling Eq. (12.48), $\Delta G° = nF\mathscr{E}°$. We remember that the $\mathscr{E}°$ values quoted in Table 12.1 are actually for net cell reactions with the standard hydrogen electrode functioning as the anode. Thus, in the above example, $\mathscr{E}°_{\text{anode}}$ actually refers to the net reaction

$$Zn(s) + 2H^+ \rightleftharpoons H_2(g) + Zn^{2+}$$

to which we assign a free energy change, designated by the symbol $\Delta G°_{\text{anode}}$. Similarly, the $\mathscr{E}°_{\text{cathode}}$ in this example refers to the net reaction

$$Cu^{2+} + H_2(g) \rightleftharpoons 2H^+ + Cu(s),$$

for which the free energy change is designated $\Delta G°_{\text{cathode}}$. But the net reaction of the cell shown in Fig. 12.7,

$$Zn(s) + Cu^{2+} \rightleftharpoons Cu(s) + Zn^{2+} \qquad (\Delta G°_{\text{cell}} = ?)$$

is simply the sum of the above two reactions. Hence, on the basis of what we have learned about combining ΔG values, in Sec. 11.3, we see that for this net cell reaction, ΔG is given by

$$\Delta G°_{\text{cell}} = \Delta G°_{\text{anode}} + \Delta G°_{\text{cathode}}$$

From Eq. (12.48), we have

$$\Delta G°_{\text{cell}} = -nF\mathscr{E}°_{\text{cell}}$$

$$\Delta G°_{\text{anode}} = -nF\mathscr{E}°_{\text{anode}}$$

$$\Delta G°_{\text{cathode}} = -nF\mathscr{E}°_{\text{cathode}}$$

where a single n is used in these equations as a reflection of our having balanced the net cell reaction in number of electrons in the two electrode half-reactions.

Then, using these last three expressions for $\Delta G°_{\text{cell}}$, $\Delta G°_{\text{anode}}$, and $\Delta G°_{\text{cathode}}$ in the equation above for $\Delta G°_{\text{cell}}$, we obtain

$$-nF\mathscr{E}°_{\text{cell}} = -nF\mathscr{E}°_{\text{anode}} - nF\mathscr{E}°_{\text{cathode}}$$

On dividing both sides by $-nF$, we obtain

$$\mathscr{E}°_{\text{cell}} = \mathscr{E}°_{\text{anode}} + \mathscr{E}°_{\text{cathode}}$$

$$\ln K = \frac{nF}{RT}\mathscr{E}°$$

$$2.303 \log K = \frac{(1)(96.5 \times 10^3)}{(8.31)(298)}(-0.58)$$

$$\log K = -9.81 = 0.19 - 10$$

$$K = \text{antilog}\,(0.19 - 10) = [\text{antilog}\,(0.19)]10^{-10}$$

$$= 1.6 \times 10^{-10}$$

EXAMPLE 12.14 Predict what would happen if a voltage difference were applied across two platinum electrodes in an aqueous solution containing NaCl, $CuSO_4$, and HNO_3, with each solute species present at a concentration about 1 *M*, and if subsequently the voltage were increased until some electrolysis reaction began.

SOLUTION In considering this system, it is advantageous at the outset to enumerate the species to be dealt with; these are H_2O, Na^+, Cl^-, Cu^{2+}, $SO_4{}^{2-}$, H^+, and $NO_3{}^-$. A little thought convinces one that the species reduced most easily is the best oxidizing agent, having the highest $\mathscr{E}°$ value for its reduction; similarly, the species most easily oxidized is the best reducing agent, having the lowest $\mathscr{E}°$ for being formed from its corresponding oxidizing agent, so its oxidization has the highest $\mathscr{E}°$ value. The electrolysis reaction most easily brought about is the reaction occurring at the lowest applied voltage; this will be the one involving reduction of the best available oxidizing agent and oxidation of the best available reducing agent.* This is analogous to the principle used in Chap. 5—that the acid-base reaction most favored in a mixture of acids and bases is the one between the strongest acid available and the strongest base available. To proceed, it is convenient to list the available oxidizing agents and reducing agents with the $\mathscr{E}^\circ$ values of the corresponding couples from Table 12.1. Na^+, formally an oxidizing agent, is essentially inert and is not listed.

Available oxidizing agents	$\mathscr{E}^\circ$, V	*Available reducing agents*	$\mathscr{E}^\circ$, V
$NO_3^- + 4H^+ + 3e^- \rightleftharpoons NO(g) + 2H_2O$	0.96	$Cl_2(g) + 2e^- \rightleftharpoons 2Cl^-$	1.36
$Cu^{2+} + 2e^- \rightleftharpoons Cu(s)$	0.34		
$SO_4^{2-} + 4H^- + 2e^- \rightleftharpoons H_2SO_3 + H_2O$	0.17	$O_2(g) + 4H^+ + 4e^- \rightleftharpoons 2H_2O$	1.23
$2H^+ + 2e^- \rightleftharpoons H_2(g)$	0.0		

The species present in solution are underlined in these half-reactions. From this summary, it is clear that NO_3^- is the best oxidizing agent available in the solution, and H_2O is the best available reducing agent; hence, we expect NO_3^- to be reduced and H_2O to be oxidized in the electrolysis. The expected net reaction is then the sum of the two pertinent half-reactions, with appropriate coefficients to make the number of electrons transferred in both half reactions the same:

$$4NO_3^- + 16H^+ + 12e^- \rightleftharpoons 4NO(g) + 8H_2O \qquad \mathscr{E}^\circ = 0.96\text{ V}$$

$$6H_2O \rightleftharpoons 3O_2(g) + 12H^+ + 12e^- \qquad \mathscr{E}^\circ = -1.23\text{ V}$$

$$4NO_3^- + 4H^+ \rightleftharpoons 4NO(g) + 2H_2O + 3O_2(g) \qquad \mathscr{E}^\circ = -0.27\text{ V}$$

One sees that an applied voltage of about 0.27 V should be necessary to bring about this reaction. A higher applied voltage would be needed to bring about any other reactions in this system; such reactions would then occur along with the desired reaction.

In Example 12.14, $\mathscr{E}^\circ$ values were taken directly from Table 12.1, even though the half-reaction equations were multiplied by 4 and 3, because $\mathscr{E}^\circ$ values of a half-reaction, or a net reaction, are independent of the number of electrons appearing explicitly in the chemical equation. Thus, one can multiply a chemical equation for any half-reaction or net reaction, and the same $\mathscr{E}^\circ$ applies.† However, if one wishes to combine half-reactions to yield a new *half*-reaction, then the number of electrons

*The net reaction in which the best reducing agent is oxidized and the best oxidizing agent is reduced is the process that involves the smallest possible increase in free energy of the system.

†The fundamental basis for the rule that $\mathscr{E}^\circ$ is unchanged when the chemical equation is multiplied by some factor comes from Eq. (12.48), which can be rewritten $\mathscr{E}^\circ = -\Delta G^\circ/nF$. If it is desired to multiply both sides of a chemical equation by some number q, then the original ΔG° correspondingly becomes $q\,\Delta G^\circ$ and the original n becomes qn. Then, the "new" $\mathscr{E}^\circ$ is given by the equation above as

$$\mathscr{E}^\circ_{\text{new}} = -\frac{q\,\Delta G^\circ}{qnF} = -\frac{\Delta G^\circ}{nF}$$

which is the same as the original $\mathscr{E}^\circ$. Hence, $\mathscr{E}^\circ$ is unchanged. This result is in contrast to what happens to the values of ΔG° and the equilibrium constant under these circumstances; $K_{\text{new}} = (K_{\text{original}})^q$ (see Sec. 4.4), and $\Delta G^\circ_{\text{new}} = q\,\Delta G^\circ$.

Spotlight

As can be seen from Table 12.1, hydrogen peroxide, H_2O_2, can function as either an oxidizing agent or a reducing agent, under appropriate circumstances. In its use as an oxidizing agent, H_2O_2 has the unique advantage of forming only water as the by-product. Hydrogen peroxide is a reducing agent for species that are oxidizing agents stronger than O_2; for example, H_2O_2 reduces Ag_2O to Ag, Au_2O_3 to Au, PbO_2 to PbO, and O_3 to O_2; in all of these reactions, H_2O and O_2 are the other products.

$$H_2O_2 + PbO_2(s) \rightleftharpoons H_2O + PbO(s) + O_2(g)$$

Hydrogen peroxide is weakly acidic,

$$H_2O_2 \rightleftharpoons H^+ + HO_2^-, \qquad K = 1.5 \times 10^{-12}$$

and reacts with the hydroxides of the Group I and II metals to form the respective peroxides. In these reactions, H_2O_2 reacts in a manner consistent with its acidic properties. Barium peroxide hexahydrate, $BaO_2 \cdot 8H_2O$, for example, can be crystallized by mixing a solution of barium hydroxide with H_2O_2:

$$OH^- + Ba^{2+} + HO_2^- + 8H_2O \rightleftharpoons BaO_2 \cdot 8H_2O(s) + H_2O$$

The common 3% H_2O_2 solution is sold as an antiseptic and as a hair bleach. A 30% solution is used as a commercial bleach and as a laboratory reagent. In concentrations of 90% or more, H_2O_2 is used as a rocket fuel, either alone or mixed with an oxidizable substance. When used alone, it is catalytically decomposed to O_2 and H_2O, liberating a considerable amount of heat. In World War II, the German V1 and V2 rockets used H_2O_2 as an oxidizing agent for a mixture of methanol, CH_3OH, and hydrazine, $H_2N{-}NH_2$.

transferred in each and in the final half-reaction must be taken into account explicitly. Suppose, for example, that we wish to calculate $\mathscr{E}°$ for the half-reaction

$$O_3(g) + 6H^+ + 6e^- \rightleftharpoons 3H_2O \qquad (\mathscr{E}° = ?)$$

from the following half-reaction $\mathscr{E}°$ information taken from Table 12.1:

$$(1)\ O_3(g) + 2H^+ + 2e^- \rightleftharpoons O_2(g) + H_2O \qquad \mathscr{E}_1° = 2.07\ V$$

$$(2)\ O_2(g) + 4H^+ + 4e^- \rightleftharpoons 2H_2O \qquad \mathscr{E}_2° = 1.23\ V$$

We note that the half-reaction of interest is the sum of the half-reactions (1) and (2). In computing $\mathscr{E}°$ for this overall half-reaction, however, we must note that it is $\Delta G°$ values that we know how to combine; that is, $\Delta G° = \Delta G_1° + \Delta G_2°$ for this example. Using Eq. (12.48), we summarize our knowledge as follows:

$$\begin{array}{ll} O_3(g) + 2H^+ + 2e^- \rightleftharpoons O_2(g) + H_2O & \Delta G_1° = -n_1F\mathscr{E}_1° \\ O_2(g) + 4H^+ + 4e^- \rightleftharpoons 2H_2O & \Delta G_2° = -n_2F\mathscr{E}_2° \\ \hline O_3(g) + 6H^+ + 6e^- \rightleftharpoons 3H_2O & \Delta G° = \Delta G_1° + \Delta G_2° \end{array}$$

$$\Delta G° = -F(n_1\mathscr{E}_1° + n_2\mathscr{E}_2°)$$

Again, from Eq. (12.48),

$$\mathscr{E}° = -\frac{\Delta G°}{nF} = \frac{F(n_1\mathscr{E}_1° + n_2\mathscr{E}_2°)}{nF}$$

$$= \frac{n_1\mathscr{E}_1° + n_2\mathscr{E}_2°}{n} \qquad (12.51)$$

In the specific case under consideration,

$$\mathscr{E}° = \frac{2 \times 2.07\ V + 4 \times 1.23\ V}{6} = 1.51\ V$$

STUDY PROBLEM 12(g)

Calculate the $\mathscr{E}^\circ$ value of a cell for which the net cell reaction is

$$Cl_2(g) + 2Br^- \rightleftharpoons 2Cl^- + Br_2(l)$$

STUDY PROBLEM 12(h)

Calculate the equilibrium constant for the reaction shown in Study Problem 12(g).

STUDY PROBLEM 12(i)

From the data in Table 12.1, use Eq. (12.51) to compute $\mathscr{E}^\circ$ for the half-reaction

$$Hg^{2+} + 2e^- \rightleftharpoons Hg(l)$$

EXAMPLE 12.15 Predict the order in which metal ions would be reduced if a solution containing $Pb(NO_3)_2$, $Zn(NO_3)_2$, $Mn(NO_3)_2$, and $Fe(NO_3)_2$ in approximately 1 *M* concentrations were subjected to an increasing voltage in an electrolytic cell in which the standard hydrogen electrode functions at the anode and the cathode involves reduction of metal ions at some inert electrode, for example, Pt.

SOLUTION In this problem we were asked to determine the order in which the metal ions Pb^{2+}, Zn^{2+}, Mn^{2+}, and Fe^{2+} are reduced as the applied electrolytic voltage is increased. From the discussion above, we have seen that the first metal ion to be reduced, or the one most easily reduced, is the best oxidizing agent. Table 12.1 shows that of the four metal ions in this problem, Pb^{2+} is the best oxidizing agent, followed by Fe^{2+}, then Zn^{2+}, and finally Mn^{2+}. This is the order in which the four metal ion species would be reduced as the voltage applied by the power supply to the cell is increased. As the anode is a standard hydrogen electrode, $\mathscr{E}^\circ = 0$, the various reduction processes will, we expect, occur as the voltages indicated below are approached:

$$Pb^{2+} + 2e^- \rightleftharpoons Pb(s) \qquad -0.13\ V$$

$$Fe^{2+} + 2e^- \rightleftharpoons Fe(s) \qquad -0.44\ V$$

$$Zn^{2+} + 2e^- \rightleftharpoons Zn(s) \qquad -0.76\ V$$

$$Mn^{2+} + 2e^- \rightleftharpoons Mn(s) \qquad -1.18\ V$$

STUDY PROBLEM 12(j)

Determine the order in which the gases O_2, F_2, and Cl_2 would appear in an electrolysis experiment on an aqueous solution in which the concentrations of Cl^- and F^- are 1 *M*, the anode is a Pt wire, and the cathode is a hydrogen electrode. The voltage applied across the cell is slowly increased. The H^+ concentration is 1 *M* throughout the cell.

An interesting feature of the pattern of reducing voltages shown in Example 12.15 is that by carefully controlling the applied voltage, one can reduce these species selectively—one by one. Then, by measuring the current flowing through the circuit as each species is reduced in turn, the amount of each metal ion species present initially in the solution can be determined. This general approach constitutes the basis of an important class of analytical techniques.

Spotlight

Polarography is an important electrochemical method of quantitative and qualitative analysis developed by the Czech chemist, Jaroslav Heyrovsky (1890–1967). The polarographic method involves the determination of the electrical current that occurs when a solution containing oxidizable or reducible substances is electrolyzed in a special type of cell. One electrode of this cell consists of mercury falling dropwise from a fine-bore glass tube, regenerating a fresh mercury surface with each new drop that forms; the other electrode is typically a $Hg(l)|Hg_2Cl_2(s)$ electrode (calomel). As the potential applied across the cell is increased from zero, the current remains small until one of the species in solution begins to be reduced at the mercury cathode. This event causes a sharp rise in the current; the potential for which half of this sharp current increase occurs is called the **half-wave potential** and is characteristic of the species being reduced (hence, the use in qualitative analysis). The height of the steep current rise is proportional to the concentration of the species being reduced (quantitative analysis). The accompanying figure shows a **polarogram** of a solution containing approximately equal concentrations of Zn^{2+} and Tl^+; the oscillations in current that occur as each new drop is formed are represented. This technique works with very small concentrations of solute species, and so is very useful in studies of water pollution. Heyrovsky was awarded the Nobel Prize in Chemistry in 1959 for his discovery and development of polarography.

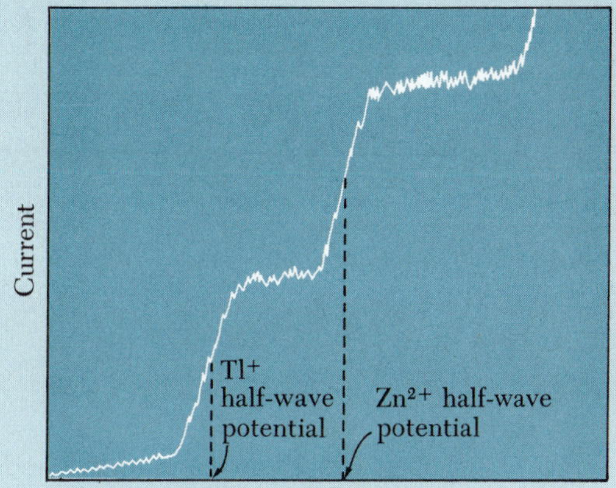

There is at least one point that has not yet been emphasized about Table 12.1, and yet is very important in taking maximum advantage of such compilations. This point is that applications of such data are not limited to electrochemical cells. The $\mathscr{E}^\circ$ values can be used for any oxidation-reduction reaction, or for any reaction that can conveniently be viewed as a combination of oxidation-reduction reactions, for example, Fig. 12.14. The ΔG° value, or equilibrium constant, of a particular oxidation-reduction reaction is the same whether the reaction is carried out in a cell or the components simply mixed directly in a flask. The ΔG° value can be related to the corresponding $\mathscr{E}^\circ$ value by Eq. (12.48) or Eq. (12.49). If the oxidation-reduction reaction is carried out by simply mixing the oxidizing and reducing agents instead of separating them by carrying out the overall reaction in a cell, then no emf is measured; nevertheless, the ΔG° and K values computed by Eqs. (12.48) and (12.49) from the $\mathscr{E}^\circ$ value that would be measured if the reaction were carried out in a cell are still valid.

EXAMPLE 12.16 Calculate the equilibrium constant for the reaction, if any, that occurs when granulated metallic iron is added to a solution of mercurous nitrate, $Hg_2(NO_3)_2$.

SOLUTION Using the information given in Table 12.1, we can see that Fe(s) is a stronger reducing agent than Hg(l); in other words, Fe(s) is capable of reducing Hg_2^{2+} to Hg(l) according to the equation

$$Fe(s) + Hg_2^{2+} \rightleftharpoons 2Hg(l) + Fe^{2+}$$

We can calculate the equilibrium constant for this reaction by computing the standard emf of a cell for which this would be the net cell reaction. The half-reactions and $\mathscr{E}^\circ$ values pertinent for such a cell would be

$$\begin{array}{lr} Hg_2{}^{2+} + 2e^- \rightleftharpoons 2Hg(l) & \mathscr{E}^\circ = 0.79 \\ Fe(s) \rightleftharpoons Fe^{2+} + 2e^- & \mathscr{E}^\circ = 0.44 \\ \hline Fe(s) + Hg_2{}^{2+} \rightleftharpoons 2Hg(l) + Fe^{2+} & \mathscr{E}^\circ = 1.23 \text{ V} \end{array}$$

Then, applying Eq. (12.49), we compute the equilibrium constant:

$$\ln K = \frac{nF\mathscr{E}^\circ}{RT}$$

$$= \frac{(2)(96.5 \times 10^3)(1.23)}{(8.31)(298)} = 2.303 \log K$$

$$\log K = \frac{(2)(96.5 \times 10^3)(1.23)}{(2.303)(8.31)(2.98 \times 10^2)} = 41.62$$

$$K = \text{antilog } 41.62$$

$$= \text{antilog } 0.62 \times 10^{41}$$

$$= 4.2 \times 10^{41}$$

From the examples and discussion above, it is possible to draw some useful generalizations concerning the $\mathscr{E}^\circ$ values of cells and of half-reactions. One generalization is that an oxidation-reduction reaction will have an equilibrium constant larger than unity, that is, a ΔG° value less than zero, if and only if the reactant oxidizing agent is a stronger oxidizing agent than the product oxidizing agent—the species produced by oxidation of the reactant reducing agent. Alternatively, we can say that the reactant reducing agent must be a stronger reducing agent than the one produced in the reaction by reduction of the reactant oxidizing agent. This same generalization can be stated in terms of Table 12.1 as follows:

An oxidation-reduction reaction will be spontaneous (for example, with $K > 1$ and $\Delta G^\circ < 0$) if and only if the half-reaction for the reactant oxidizing agent occurs higher on Table 12.1 than the half-reaction for the reactant reducing agent. This statement assumes reactants and products in their standard states. This is precisely the situation in which an oxidizing agent and a reducing agent are converted to products that are a weaker reducing agent and a weaker oxidizing agent. Reactions that meet these criteria are also the ones that if carried out in a cell produce positive $\mathscr{E}^\circ$ values. This pattern is precisely analogous to what we discussed in Sec. 5.3 for acid-base reactions, in which it was pointed out that the acid-base reactions with equilibrium constants larger than unity are those in which a stronger acid and stronger base react to produce a weaker base and a weaker acid.

The analogy with acid-base reactions can be taken further. If one draws a line on Table 12.1 between the reactant oxidizing agent and the reactant reducing agent, a *downhill* line, reading from left to right, corresponds to a spontaneous oxidation-reduction reaction, that is, one with $K > 1$ and $\Delta G^\circ < 0$ (and $\mathscr{E}^\circ > 0$ if the reaction is carried out in a cell). Using this rule, one can conclude, for example, that ozone, O_3, should not be stable in basic aqueous solution. Drawing a line in part B of Table

12.1 from $O_3(g)$ as an oxidizing agent to the OH^- as a reducing agent, one sees that it is downhill (left to right). This downhill line corresponds to the following favorable net reaction:

$$\begin{array}{rl} 2O_3(g) + 2H_2O + 4e^- \rightleftharpoons 2O_2(g) + 4OH^- & \mathscr{E}° = 1.24 \text{ V} \\ 4OH^- \rightleftharpoons 4e^- + O_2(g) + 2H_2O & \mathscr{E}° = -0.40 \text{ V} \\ \hline 2O_3(g) \rightleftharpoons 3O_2(g) & \mathscr{E}° = 0.84 \text{ V} \end{array}$$

This net reaction would provide a standard emf of 0.84 V if carried out in a cell, and has an equilibrium constant greater than unity—whether carried out in a cell or not!

OPTIONAL

Concentration Effect; The Nernst Equation

Throughout our discussions of cell potentials and half-reaction reduction potentials, care has been taken to limit the discussion to systems in which reactants and products occur in their standard states. It seldom happens, however, that chemical reactions or electrochemical cells are operated under standard conditions. Hence, it is important to consider the effects that deviations from standard states—say concentrations far from 1 *M* or partial pressures far from 1 atm—have on the concepts and relation so far considered.

That the emf of a cell should depend on the concentrations of reactant and product species (actually, on their activities; see Sec. 4.6) is easily seen. Suppose we set up a voltaic cell with all pertinent species in their standard states at 25 °C; the emf of the cell is, by definition, initially $\mathscr{E}°$. Now, suppose that the switch is closed in the external circuit, so electrons can flow through it and the cell reaction is allowed to proceed. If this reaction is allowed to proceed for a sufficiently long period, and the cell allowed to "run down," eventually enough of the reactant or reactants will be consumed and product or products produced so that chemical equilibrium is achieved. At this point there is no further tendency for any net reaction to occur, so there is no longer an emf generated by the cell; that is, $\mathscr{E} = 0$ at equilibrium. This change in emf from $\mathscr{E}°$ to 0 occurs continuously, the emf dropping as the concentrations of reactants decrease and of products increase as equilibrium is approached. Thus, we see that by merely allowing the concentrations of reactants and products to change, the emf of the cell changes dramatically; in this particular case, it changes from $\mathscr{E}°$ to zero.

The mathematical dependence of the emf $\mathscr{E}$ of an electrochemical cell on the concentrations of reactants and products is expressed in terms of the mass action quotient Q. The equation is

$$\mathscr{E} = \mathscr{E}° - \frac{RT}{nF} \ln Q \tag{12.52}$$

This important equation, or variations on it, is called the **Nernst equation.** It is widely useful in a variety of applications in oxidation-reduction chemistry. At equilibrium, $\mathscr{E} = 0$; and we write

$$\mathscr{E}° = \frac{RT}{nF} \ln Q_{\text{equil}} = \frac{RT}{nF} \ln K$$

Spotlight

The Bettmann Archive

Walter Nernst (1864–1941) was a leading German chemical physicist who made an enormous contribution to physical chemistry throughout his illustrious career. His many scientific contributions included important advances in thermodynamics and electrochemistry, and the relations between them. He was awarded the Nobel Prize in chemistry in 1920 for his work in thermodynamics, specifically on the third law of thermodynamics. This law states that the entropy of a perfect crystal approaches zero as the temperature is reduced towards absolute zero, and becomes zero at 0 °K.

One interesting and important equation that we can derive from the Nernst equation is the dependence of ΔG for a reaction on the activities (concentrations, and so on) of reactants and products. From Eqs. (12.52) and (12.47), we write

$$\Delta G = -nF\mathscr{E} = -nF\left(\mathscr{E}^\circ - \frac{RT}{nF}\ln Q\right)$$

$$= -nF\mathscr{E}^\circ + RT\ln Q$$

But from Eq. (12.48), $\Delta G^\circ = -nF\mathscr{E}^\circ$. Hence,

$$\Delta G = \Delta G^\circ + RT\ln Q$$

In formal thermodynamic developments, this equation would ordinarily be a starting point for the derivation of Eq. (12.52).

OPTIONAL

This is simply a restatement of Eq. (12.49). Under conditions in which all species are present in their standard states—say all solutes present at 1 *M* concentration—the value of Q is 1; and

$$\mathscr{E} = \mathscr{E}^\circ - \frac{RT}{nF}\ln(1) = \mathscr{E}^\circ - 0 = \mathscr{E}^\circ$$

That is, the emf of a standard cell is $\mathscr{E}^\circ$, which is consistent with our definition of $\mathscr{E}^\circ$.

More interestingly, Eq. (12.52) tells us precisely how the emf of a cell varies with the concentrations of pertinent species. Writing the equation as

$$\mathscr{E} = \mathscr{E}^\circ - \frac{RT}{nF}\ln\left(\frac{[M]^m[N]^m\ldots}{[A]^a[B]^b\ldots}\right)$$

we see that increasing the concentrations of products and decreasing the concentrations of reactants bring about a lowering in the cell emf, that is, subtracting the logarithm of a larger number from $\mathscr{E}^\circ$ in Eq. (12.52). This seems reasonable; if one increases the concentrations of products or decreases the concentrations of reactant species, one expects a decreased tendency for a net reaction in the forward direction, and that decreased tendency would be reflected in a lower value of $\mathscr{E}$.

Since many cases of practical interest are reactions carried out at 25 °C, and since base 10 logarithms are more popularly used than natural logarithms, it is convenient to convert Eq. (12.52) into a form specifically applicable to 25 °C, and restricted to that temperature. Substituting values of F, R, and 298 °K into Eq. (12.52), one obtains*

$$\mathscr{E} = \mathscr{E}^\circ - \frac{0.0591}{n} \log Q \tag{12.53}$$

EXAMPLE 12.17 Calculate the emf of a room-temperature cell of the type shown in Fig. 12.7 for
(a) $[Zn^{2+}] = [Cu^{2+}] = 1.0\ M$.
(b) $[Zn^{2+}] = 1.0\ M$ and $[Cu^{2+}] = 0.0010\ M$.

SOLUTION According to Table 12.1, $\mathscr{E}^\circ$ for this cell is 0.76 V + 0.34 = 1.10 V, the answer to part (a). Then, using the rules given in Chap. 4 for writing Q, we write Eq. (12.53) as follows for this case:

$$\mathscr{E} = 1.10 - \frac{0.0591}{2} \log \frac{[Zn^{2+}]}{[Cu^{2+}]}$$

For $[Zn^{2+}] = [Cu^{2+}] = 1.0$, we have for part (a)

$$\mathscr{E} = 1.10 - \frac{0.0591}{2} \log (1) = 1.10 - 0 = 1.10 = \mathscr{E}^\circ$$

For $[Zn^{2+}] = 1.0$ and $[Cu^{2+}] = 0.0010$, we have for part (a)

$$\begin{aligned}\mathscr{E} &= 1.10 - \frac{0.0591}{2} \log \left(\frac{1.0}{0.0010}\right)\\ &= 1.10 - \frac{0.0591}{2} \log (10^3)\\ &= 1.10 - \left(\frac{0.0591}{2}\right)(3)\\ &= 1.10 - 0.09 = 1.01\ \text{V}\end{aligned}$$

EXAMPLE 12.18 Suppose that we have an electrochemical cell of the general type shown in Fig. 12.8, except that the sulfuric acid is dilute; it is not concentrated as in a typical storage battery. Further, assume that the sulfate ion concentration in the cell is maintained exactly at 1 M but the H^+ concentration can be varied. Derive an expression that relates $\mathscr{E}$ for this cell to pH.

SOLUTION From the $\mathscr{E}^\circ$ values given in Table 12.1 for the half-reactions shown in Fig. 12.8, we compute $\mathscr{E}^\circ$ for this cell to be 0.36 + 1.69 = 2.05 V. Then, we write Eq. (12.53) for the net cell reaction of Fig. 12.8 as follows:

$$\mathscr{E} = 2.05 - \frac{0.0591}{2} \log \left(\frac{1}{[H^+]^4[SO_4{}^{2-}]^2}\right)$$

where the symbols corresponding to Pb(s), PbO_2(s), and $PbSO_4$(s), and H_2O have been replaced by unity in accordance with the rules given in Chap. 4 for writing Q.

*$0.0591 \log Q = (RT/F) \ln Q$, where $R = 8.31\ \text{J deg}^{-1}$, $F = 96.5 \times 10^3$ C, $T = 298$ °K, and $\ln Q = 2.303 \log Q$.

Spotlight

A precise description of the mechanism by which the glass electrode functions in a pH meter is far beyond our present interests. However, the basic design of this electrode and the cell in which it functions is simple, and is shown herewith. The glass electrode uses a couple such as $Ag(s)|AgCl(s)$ or $Hg(l)|Hg_2Cl_2(s)$ inside an enclosure that contains a solution of known pH, such as a 0.1 *M* HCl solution, and consists partly of a bulb made of special glass. This enclosure has a small pinhole in it to provide electrical contact—but not a flow of solution—between the 0.1 *M* HCl solution and the solution of unknown pH to be studied. The complete cell must contain two electrodes, and the second electrode is a **reference electrode;** this is typically the calomel electrode, for which the half-reaction is

$$Hg_2Cl_2(s) + 2e^- \rightleftharpoons 2Hg(l) + 2Cl^-$$

The reference electrode has a fixed potential; the glass electrode, however, is found to have an emf that depends on the pH of the solution on the outside of the special glass bulb. Furthermore the dependence of $\mathscr{E}$ on pH corresponds to the form of the Nernst equation, and can be used for a direct determination of pH. It is now common for the glass electrode and the reference electrode to both be encased in a single compartmentalized glass enclosure.

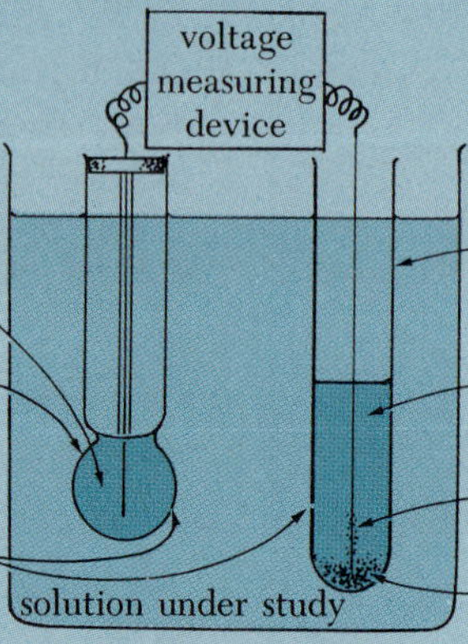

OPTIONAL

But in this case, we have specified that $[SO_4^{2-}] = 1$. Hence, the above equation can be simplified:

$$\mathscr{E} = 2.05 - \frac{0.0591}{2}\log\left(\frac{1}{[H^+]^4}\right)$$

Then, using the properties of logarithms gives

$$\begin{aligned}\mathscr{E} &= 2.05 + \frac{0.0591}{2}\log\,[H^+]^4 \\ &= 2.05 + \left(\frac{0.0591}{2}\right)(4)\log\,[H^+] \\ &= 2.05 + 0.118\log\,[H^+]\end{aligned}$$

Now, using the definition of pH ($-\log\,[H^+]$), we get

$$\begin{aligned}\mathscr{E} &= 2.05 - 0.118\,(-\log\,[H^+]) \\ &= 2.05 - 0.118\text{ pH}\end{aligned}$$

The final equation in Example 12.18 tells us that the pH increases as the emf in the system shown in Fig. 12.8 decreases. (This corresponds to the fact that the emf of a lead storage battery is high when the sulfuric acid concentration is high.) It also provides a specific example of how the emf that would be measured from a cell is quantitatively related to the concentration of a specific species (in this case H^+). Electrodes have been designed that make it possible to determine the concentration of a specific ion by measuring the emf of a cell in which such an electrode is incorporated. This kind of electrode is referred to as an **ion-specific electrode.** The most common example of an ion-specific electrode is the so-called *glass electrode,* which is incorporated into a cell for measurement of the pH. The combination of a suitable device for measuring voltages with the glass electrode in a cell for measuring pH is called a **pH meter.**

STUDY PROBLEM 12(k)

Determine the $\mathscr{E}$ value of the cell shown in Fig. 12.7 for the conditions that $[Zn^{2+}]$ is 0.30 *M*, $[Cu^{2+}]$ is 0.060 *M*, and the temperature is 25 °C.

STUDY PROBLEM 12(l)

Determine the $\mathscr{E}$ value of the cell shown in Fig. 12.8 for the conditions that the sulfuric acid concentration is 0.50 *M* and the temperature is 25 °C.

12.5 Electrochemistry and Thermodynamics

When we introduced the equation

$$\Delta G = -nF\mathscr{E}$$

which is numbered Eq. (12.47) in the text, we neither derived it nor substantially justified it. So far, we have based several important developments on this equation. Now, as we derive Eq. (12.47) on the basis of considering a readily visualized physical system, we shall obtain additional insight into the free energy and entropy concepts previously introduced (Chap. 11).

An Electrochemical-Mechanical System

Let us focus on a system consisting of an electrochemical cell maintained at some fixed temperature by a constant-temperature bath and connected with an electric motor capable of converting the electrical energy generated by the cell into mechanical energy. The system is pictured in Fig. 12.19. The cell depicted in this figure is designed so that the cell reaction is carried out at a constant pressure P and any change in volume associated with the reaction is accommodated at constant pressure by a movable piston arrangement. In this way an amount of work $P\,\Delta V$, which we shall refer to as "$P\,\Delta V$ work," can be performed by the system against the restraining pressure P—or alternatively, on the system, depending on the sign of ΔV. In addition to the $P\,\Delta V$ work, some "non-$P\,\Delta V$ work" can be performed by the cell system by lifting a lead weight that is coupled to a motor-pulley arrangement, the electric motor being driven by the current generated from the cell. The system, that is, the cell, is able to transfer energy to or from the surroundings by means of heat transfer due to its thermal contact with the constant temperature bath and by means of the $P\,\Delta V$ work and mechanical work of raising the lead weight.*

*We are assuming in the electromechanical model of Fig. 12.19 that the motor-pulley arrangement is perfectly efficient and that no energy is used up in frictional losses or electrical heating. This situation could not be realized in an actual apparatus; but it can be approached, and discussed as an ideal. Including these "energy losses" explicitly in our development would only make our equations more cumbersome, and they would not alter our important conclusions.

OPTIONAL

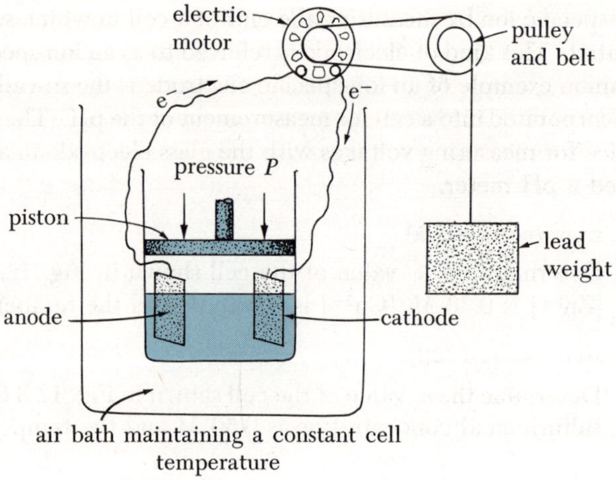

Figure 12.19
Electrochemical cell, maintained at constant temperature in an air bath; cell operating a piston against a constant pressure P and an electric motor driving a pulley that lifts a lead weight.

The first law of thermodynamics is stated as follows:

$$q = \Delta E + W$$

Since we want to focus on a specific reaction, the value of ΔE is fixed by our choice of reaction. If we neglect any frictional and electrical heating losses, the amount of energy q introduced into the cell through heat transfer from the bath as the cell reaction proceeds must exactly equal the energy "used" in increasing the internal energy of the reaction system ΔE and in performing work W by the system. That work is made up of both the $P\,\Delta V$ work, which takes place when the volume of the chemical system increases by an amount ΔV because, for example, of gaseous products formed, and the mechanical work of lifting the lead weight:*

$$W = P\,\Delta V + W_{\text{mech}}$$

Combining these last two equations, we obtain

$$q = \Delta E + P\,\Delta V + W_{\text{mech}}$$

or

$$W_{\text{mech}} = q - (\Delta E + P\,\Delta V)$$

Since the system under consideration functions at constant pressure, the expression $\Delta E + P\,\Delta V$ in the above equations can be replaced by ΔH (see Eq. 10.24). Hence, for the system shown in Fig. 12.19,

$$W_{\text{mech}} = q - \Delta H \qquad (12.54)$$

The mechanical work of this system is really useful work, work that could be used to perform a useful task. The value of W_{mech} in Eq. (12.54) depends on a variety of experimental variables because these variables determine the value of q (ΔE and ΔH are fixed by the initial and final states of the system). The maximum amount of

*Any of the quantities q, ΔE, $P\,\Delta V$, or W_{mech} could be either positive or negative, depending on the specific details of the cell reaction, temperature, pressure, mass of the lead weight, and so on. We are discussing this case as though all these quantities are positive. A negative sign for any of these quantities would not alter the reasoning or the basic conclusions.

useful work that can be performed by the system would be realized if the weight being lifted were of just the right size to render the process *reversible.* This would be the case if the emf generated by the cell were just barely large enough to drive the motor under the load of the lead weight to be lifted. If the weight were too heavy, the cell emf could not drive the motor against the restraining force and no useful work would be done by the system at all. Or if there were no weight at all, the motor (idealized, as it is) would move against no restraining force and again no useful work at all would be performed. Clearly, if the size of the lead weight were between these two extremes, some useful work would be done; and the amount could be maximized by using a weight that is infinitesimally smaller than one that would be just too large.* Hence, the maximum useful work would be obtained under *reversible* conditions:

$$(W_{\text{mech}})_{\max} = q_{\text{rev}} - \Delta H_{\text{rev}}$$

Enthalpy is a state function, hence as stated above, ΔH depends only on the initial and the final states of the system and is independent of pathway. Thus, ΔH is the same for a transformation between two given states whether the process is carried out reversibly or under irreversible conditions. Hence, for specific initial and final states of the electrochemical cell, Eq. (12.54) can be written

$$(W_{\text{mech}})_{\max} = q_{\text{rev}} - \Delta H \tag{12.55}$$

At this point it is useful to introduce the thermodynamic definition of the entropy charge for a process carried out at constant temperature. From Eq. (11.3), we see that

$$q_{\text{rev}} = T\,\Delta S$$

Substituting this expression for q_{rev} into Eq. (12.55), we obtain

$$(W_{\text{mech}})_{\max} = T\,\Delta S - \Delta H$$

or

$$(W_{\text{mech}})_{\max} = -\,(\Delta H - T\,\Delta S)$$

Recognizing from Eq. (11.9) that $\Delta H - T\,\Delta S$ is simply ΔG for a process at constant temperature, we can now write the following important equation:

$$\text{maximum useful work} = -\Delta G$$
$$(W_{\text{mech}})_{\max} = -\Delta G \tag{12.56}$$

This equation provides another important reason for having defined ΔG in Chap. 11 and for the name *free energy*. In Sec. 11.2, we found that the value of ΔG for reactants and products in their standard states, ΔG°, fixes the equilibrium constant for a given temperature. Now we see that ΔG represents the maximum amount of useful work that one can hope to obtain from a working cell. Fortunately this result is even more general than our example, which is based on an electrochemical cell. Thermodynamic developments show that the maximum useful work that can be extracted from any reaction equals $-\Delta G$ for the reaction. Any reaction with a negative ΔG can in principle be harnessed to perform useful work.

*In the idealized, or "frictionless," electromechanical system shown in Fig. 12.19, if the lead weight were just slightly heavier than the one that would yield a reversible system, the weight would fall, causing the motor to operate in the reverse direction; ideally, the motor could then function as a dynamo, generating an emf capable of reversing the direction of the cell current and cell reaction.

Spotlight

Hydrazine, $H_2N—NH_2$, as can be seen from Table 12.1, is an excellent reducing agent; $\mathscr{E}° = -1.16$ V. It is oxidized by O_2, burning with the evolution of a large quantity of heat, 149 kcal/mol; the gaseous products N_2 and H_2O result, and hydrazine has therefore been used in liquid-fuel rockets:

$$N_2H_4(l) + O_2 \rightleftharpoons N_2(g) + 2H_2O(g)$$

Hydrogen peroxide, H_2O_2, has also been used as the oxidizing agent in hydrazine liquid-fuel rockets. The reducing ability of hydrazine allows it to precipitate the respective metals from solutions of the salts of copper, silver, gold, and platinum; the overall reaction for Ag^+ is

$$4Ag^+ + N_2H_4 \rightleftharpoons 4Ag(s) + N_2(g) + 4H^+$$

Hydrazine also reduces ferric ion, Fe^{3+}, to ferrous ion, Fe^{2+}, and free halogens to halide ion. For bromine, the reaction is

$$2Br_2 + N_2H_4 \rightleftharpoons 4Br^- + N_2(g) + 4H^+$$

Hydrazine is conveniently prepared by the reaction of ammonia with sodium hypochlorite. The reaction proceeds in two steps via chloramine as an intermediate,

$$NH_3 + OCl^- \rightleftharpoons NH_2Cl + OH^-$$

$$NH_2Cl + OH^- + NH_3 \rightleftharpoons H_2N—NH_2 + H_2O + Cl^-$$

OPTIONAL

The Driving Force of a Reaction

What Eq. (12.56) tells us is that for a reaction

$$aA + bB + \cdots \rightleftharpoons mM + nN + \cdots$$

when a moles of A react with b moles of B to produce m moles of M and n moles of N, if reversible (ideal) conditions could be achieved, the maximum amount of useful work that could be obtained would be $-\Delta G$ for this process, where

$$\Delta G = mG_M + nG_N - aG_A - bG_B$$

and G_A is the value of the Gibbs free energy for one mole of species A in whatever the state specified in the conditions relating to the process in question.

There is another useful way of viewing the equal quantities W_{max} and $-\Delta G$. Focus attention again on a cell similar to the one represented in Fig. 12.19. We note that if this cell is capable of generating an emf equal to $\mathscr{E}$, then it can cause a current flow through any circuit with an opposing voltage that is less than $\mathscr{E}$. Let us denote this opposing voltage by the sumbol $\mathscr{E}'$. In the process of forcing n faradays (moles) of electrons through the external circuit, against the opposing voltage $\mathscr{E}'$, the cell will perform electrical work equal to the total charge transferred times the voltage difference across which it is transferred, that is, $F\mathscr{E}'$ joules per mole of electrons, or $nF\mathscr{E}'$ joules for n moles of electrons:

$$W = (nF)\mathscr{E}'$$

Since the maximum opposing voltage against which a current can be generated is $\mathscr{E}$, the maximum electrical work the cell can perform is equal to $nF\mathscr{E}$. Thus

$$(W_{mech})_{max} = nF\mathscr{E} \qquad (12.57)$$

Then, combining Eq. (12.56) and (12.57), we obtain the important relation

$$\Delta G = -nF\mathscr{E}$$

For a cell reaction with materials in their standard states, this becomes

$$\Delta G^\circ = -nF\mathscr{E}^\circ$$

These are Eqs. (12.47) and (12.48), stated in Sec. 12.4 without derivation. Joining these with Eq. (12.49), we can now summarize the relations as follows:

$$\mathscr{E}^\circ = \frac{-\Delta G^\circ}{nF} = \frac{RT}{nF}\ln K$$

This statement relates, for any oxidation-reduction reaction, three properties of the reaction that indicate quantitatively the driving force, $\mathscr{E}^\circ$, ΔG°, and K.

Rationale for the Nernst Equation

The relation between emf and concentrations can be pursued fruitfully if we recall the discussion of Secs. 4.2 and 4.6, in which the approach of a chemical system to equilibrium was presented as the approach of the mass action expression Q to its value at equilibrium K. For a generalized chemical reaction

$$a\text{A} + b\text{B} + \cdots \rightleftharpoons m\text{M} + n\text{N} + \cdots$$

Q is defined by

$$Q = \frac{A_M{}^m \cdot A_N{}^n \cdots}{A_A{}^a \cdot A_B{}^b \cdots} \tag{12.58}$$

where A_M represents the activity of species M, as defined in Sec. 4.6. The standard state of any species is, by definition, the state in which its activity is unity. Hence, the value of Q for a system in which all species are present in their standard states is unity, or

$$Q = \frac{(1)^m(1)^n \cdots}{(1)^a(1)^b \cdots} = 1 \qquad \text{(for all species in standard states)}$$

At equilibrium, the value of Q is the equilibrium constant, which is greater than unity for a reaction with a positive $\mathscr{E}^\circ$.

Let us consider a cell in which $K > 1$ and Q is initially 1 (all species in standard states). As the reaction is allowed to proceed, with passage of electrical current, and the cell correspondingly "runs down," some of the reactants initially present are converted to products. Therefore the concentrations or partial pressures, or both, of the reactants decrease, and the concentrations or partial pressures, or both, of the products increase; that is, the denominator of Eq. (12.58) decreases and the corresponding numerator increases. Eventually, equilibrium is reached, and $Q = K$, with $\mathscr{E} = 0$. Thus, as the emf is changing from the value $\mathscr{E}^\circ$ to zero, Q is increasing from its initial value of unity to its equilibrium value K.

It is convenient and, one might argue, intuitively reasonable to assume that the relation among $\mathscr{E}$, $\mathscr{E}^\circ$, and Q can be represented formally by an equation of the following general form:

$$\mathscr{E} = \mathscr{E}^\circ + f(Q,T) \tag{12.59}$$

where $f(Q,T)$ is some appropriate mathematical expression depending on both Q and T. The temperature dependence is included because we know from Eq. (12.47) that

OPTIONAL

$\mathscr{E}$ depends on temperature. Equation (12.59) is simply one way of stating that the emf of a cell can be expressed as the sum of two terms, one of which is what the emf would be under standard conditions ($\mathscr{E}°$) and the other the "correction term," $f(Q,T)$, which is a measure of how much the system deviates from standard conditions. From the form of Eq. (12.59), we see that $f(Q,T)$ must be zero if all species are present in their standard states ($Q = 1$), since $\mathscr{E} = \mathscr{E}°$ in that situation. It is also clear from Eq. (12.59) that at equilibrium, $f(Q,T)$ must equal $-\mathscr{E}°$, because $\mathscr{E} = 0$ at equilibrium. The changes that occur in the cell as it runs down from the assumed initial conditions to equilibrium can be summarized conveniently by the following three circumstances: initial (standard conditions), intermediate, and final (equilibrium).

Initial (standard states)	*Intermediate*	*Final (equilibrium)*
$Q = 1$	Q between 1 and K	$Q = K$
$\mathscr{E} = \mathscr{E}°$	$\mathscr{E}$ between $\mathscr{E}°$ and 0	$\mathscr{E} = 0$
$f(Q,T) = 0$	$f(Q,T)$ between 0 and $-\mathscr{E}°$	$f(Q,T) = -\mathscr{E}°$

It can be recalled from Eq. (12.49) that $\mathscr{E}°$ equals (RT/nF) ln Q at equilibrium, or when $Q = K$. Thus, at equilibrium $f(Q,T)$, which equals $-\mathscr{E}°$, also equals $-(RT/nF)$ ln Q. It is noteworthy that this same expression, $-(RT/nF)$ ln Q, is zero in a standard cell ($Q = 1$, ln $Q = 0$). Hence, the expression $-(RT/nF)$ ln Q meets the guidelines given above for the function $f(Q,T)$. Rigorous thermodynamic developments show that indeed $f(Q,T)$ of Eq. (12.59) is $-(RT/nF)$ ln Q, so that the proper statement of this equation is

$$\mathscr{E} = \mathscr{E}° - \frac{RT}{nF} \ln Q$$

This is the Nernst equation, stated without justification in Sec. 12.4.

SUMMARY

Electrochemistry is the study of electron-transfer reactions, that is, oxidation-reduction reactions. These reactions can be partitioned, either in our minds or in an electrochemical cell, into two half-reactions. One half-reaction is an electron donation by the reducing agent, which thereby becomes oxidized, and the other is an electron acceptance by the oxidation agent, which thereby becomes reduced. A method for writing balanced equations for oxidation reduction reactions is based on the half-reaction concept.

An electrochemical cell in which a reaction with negative ΔG is occurring generates an electromotive force, emf, and can be used as a voltage source. The emf of such a cell, which is a voltaic cell, is a direct measure of the driving force of the reaction. If reactants and products are all present in their standard states, then the cell emf is called the standard emf, $\mathscr{E}°$, and is mathematically related to $\Delta G°$ and K for the reaction. A reaction with a positive ΔG will not occur spontaneously in an electrochemical cell, but can be forced to occur if a sufficiently large external voltage is applied across the electrodes of the cell, an electrolytic cell.

By arbitrarily assigning a potential of 0 volts to the H_2(Pt)|H^+ electrode under standard conditions, one can establish a scale of standard reduction potentials for half-reactions. One can readily compute $\mathscr{E}°$ for any oxidation-reduction reaction for which the standard reduction potentials of the pertinent half-reactions are available in tables; hence, one can determine whether any proposed oxidation-reduction can or cannot occur spontaneously under standard conditions, and can compute the corresponding K values. The Nernst equation permits one to determine the emf of any cell for which the $\mathscr{E}°$ value, the concentrations of reactants and products, and the temperature are known.

STUDENT CHECKPOINTS

After studying this chapter, the student should be able to:

1. Explain an oxidation-reduction reaction in terms of half-reactions.
2. Balance equations for oxidation-reduction reactions by the half-reaction method.
3. Describe the functions of simple electrochemical cells in terms of half-reactions.
4. Apply Faraday's law.
5. Compute $\mathscr{E}°$ for any oxidation-reduction reaction from a knowledge of the pertinent standard reduction potentials.
6. Determine the minimum voltage that must be applied to bring about a specific electrolysis reaction, assuming all pertinent species are in their standard states.
7. Compute $\Delta G°$ and K for any oxidation-reduction reaction from a knowledge of the pertinent standard reduction potentials.
8. Compute the reduction potential of a half-reaction that is the sum of two other half-reactions from a knowledge of the standard reduction potentials of those two half-reactions.
9. Use the Nernst equation for computing $\mathscr{E}$ for any oxidation-reduction reaction at a specific temperature from a knowledge of $\mathscr{E}°$ for the reaction and the concentrations of all reactants and products.

EXERCISES

12.1.* Determine the oxidation number of each element in each species in the following equations:
(a) $Zn^{2+} + Fe(s) \rightleftharpoons Zn(s) + Fe^{2+}$
(b) $Cl_2(g) + 2Br^- \rightleftharpoons 2Cl^- + Br_2(l)$

12.2. Determine the oxidation number of each element in each species in the following equations:
(a) $Cr_2O_7^{2-} + 14H^+ + 6I^- \rightleftharpoons 2Cr^{3+} + 7H_2O + 3I_2(s)$
(b) $MnO_4^- + 8H^+ + 5Cl^- \rightleftharpoons Mn^{2+} + \frac{5}{2}Cl_2(g) + 4H_2O$

12.3.* Identify the oxidizing agent and reducing agent in each of the equations in Exercise 12.1.

12.4. Identify the oxidizing agent and reducing agent in each of the equations in Example 12.2.

12.5.* Identify (a) the element being oxidized, and (b) the element being reduced in each of the reactions shown in Exercise 12.1.

12.6. Identify (a) the element being oxidized, and (b) the element being reduced in each of the reactions shown in Exercise 12.2.

12.7.* Identify the half-reactions in each of the equations in Exercise 12.1.

12.8. Identify the half-reactions in each of the equations in Exercise 12.2.

12.9.* Indicate whether each of the following statements is true or false:
(a) The oxidation number of an elemental substance is always zero.
(b) The oxidation half-reaction and the reduction half-reaction can be separated in space.
(c) The oxidation half-reaction and reduction half-reaction can be separated in time.

12.10. Using the half-reaction method, obtain a balanced equation for each of the following reactions:
(a) $H_2(g) + Fe^{3+} \rightarrow H^+ + Fe^{2+}$ (acid solution)
(b) $Cl_2(g) + I^- \rightarrow Cl^- + I_3^-$ (acid solution)
(c) $Cr_2O_7^{2-} + I^- \rightarrow Cr^{3+} + I_2(s)$ (acid solution)

12.11.* Using the half-reaction method, obtain a balanced equation for each of the following reactions:
(a) $Cd(s) + NiO_2(s) \rightarrow Cd(OH)_2 + Ni(OH)_2$ (basic solution)
(b) $Mn(s) + OH^- \rightarrow Mn(OH)_2(s) + H_2(g)$ (basic solution)
(c) $Fe(OH)_2(s) + Pb(OH)_3^- \rightarrow Fe(OH)_3(s) + Pb(s)$ (basic solution)

12.12. Using the half-reaction method, obtain a balanced equation for each of the following reactions:
(a) $As(s) + NO_3^- \rightarrow AsO_4^{3-} + NO(g)$ (acid solution)
(b) $Br_2(l) \rightarrow BrO_4^- + Br^-$ (basic solution)
(c) $MnO_4^- + CN^- \rightarrow MnO_2(s) + CNO^-$ (basic solution)

12.13.* Using the half-reaction method, obtain a balanced equation for each of the following reactions:
(a) $C_3H_8(g) + O_2(g) \rightarrow H_2O(g) + CO_2(g)$ (basic solution)
(b) $MnO_4^- + HCO_2H \rightarrow Mn^{2+} + CO_2$ (acid solution)
(c) $P_4(s) \rightarrow HPO_3^{2-} + PH_3(g)$ (acid solution)

12.14. How many grams of $H_2(g)$ are consumed by 0.910 mole of Fe^{3+} in the following reaction?

$$H_2(g) + 2Fe^{3+} \rightleftharpoons 2H^+ + 2Fe^{2+}$$

12.15. How many moles of MnO_4^- are produced by 1.62 g of $NaBiO_3$ in the following unbalanced transformation in basic aqueous solution?

$$BiO_3^- + MnO_2(s) \rightarrow Bi^{3+} + MnO_4^{2-}$$

12.16.* How many grams of lead sulfate are produced in the lead storage battery as 754 mg of $PbO_2(s)$ are consumed?

12.17.* Indicate whether each of the following statements is true or false:
(a) In an electrochemical cell, anions migrate from the anode.
(b) In an electrochemical cell, the strongest oxidizing agent is reduced at the anode.
(c) In an electrochemical cell, the electrical current flowing between electrodes in the solution must be of exactly the same magnitude as the current in the external circuit.
(d) In an electrochemical cell, only the anode reaction can continue to occur if the external circuit is broken; the cathode reaction will stop.

12.18. Indicate whether each of the following statements is true or false:
(a) The emf generated by a cell will be doubled if the physical size of the cell is doubled, with all concentrations held fixed.

(b) The current flowing through the external circuit attached to an electrochemical cell can be doubled by doubling the physical size of the cell; all concentrations are held fixed.

12.19.* Indicate whether each of the following statements is true or false:
(a) In a voltaic cell, the strongest oxidizing agent in the system is reduced at the cathode.
(b) In a voltaic cell, the strongest reducing agent in the system is converted to a weaker reducing agent.
(c) In an electrolytic cell, a reducing agent and an oxidizing agent are converted into a weaker oxidizing agent and a weaker reducing agent.

12.20. (a) Draw a design for a cell for which the net reaction is

$$Zn(s) + 2H_3O^+ \rightleftharpoons Zn^{2+} + H_2(g) + 2H_2O$$

(b) Is this a voltaic cell or an electrolytic cell?

12.21. (a) Draw a diagram for a cell in which the net cell reaction is

$$Cl_2(g) + Cu(s) \rightleftharpoons 2Cl^- + Cu^{2+}$$

(b) Is this a voltaic cell or an electrolyic cell?

12.22. (a) Draw a diagram for a cell in which the net cell reaction is

$$4H_2O + 2MnO_2(s) + 3Cl_2(g) \rightleftharpoons 2MnO_4^- + 6Cl^- + 8H^+$$

(b) Is this a voltaic cell or an electrolytic cell?

12.23. How many grams of $PbSO_4(s)$ are produced in a lead storage battery when 0.0462 faraday of electricity is passed through the cell circuit in operating a lamp?

12.24.* How many coulombs of electricity are passed through the external circuit in the cell shown in Fig. 12.7 in order to deplete the anode of 0.592 g Zn(s)?

12.25. Recalling that one ampere of electrical current is one coulomb of electricity passed per second, calculate the amount of sodium metal produced in the cell shown in Fig. 12.10 in one hour if the current is 2.3 amperes (A).

12.26.* How long must a 0.12-A current be passed through the circuit shown in Fig. 12.12 in order to deposit 7.2 g silver on the spoon?

12.27. Recalling that one joule is the energy of moving one coulomb of electricity through a circuit with an emf of one volt (1 J = 1 C V), calculate the electrical energy consumed in an electroplating process in which the voltage applied to the cell is 2.6 V, and 2.60 faradays of electricity are passed in a 30 min procedure.

12.28.* (a) How many grams of cadmium are consumed in the operation of the nickel-cadmium cell shown in Fig. 12.9 if a current of 0.80 amperes is drawn for 25 min? (b) What is the total mass change of the system in this process?

12.29. Calculate ΔG for an oxidation-reduction reaction in an electrochemical cell with a measured emf of 1.33 V at 25 °C and $n = 1$.

12.30.* Calculate ΔG° and K for an oxidation-reduction reaction at 25 °C for which the standard emf is 0.832 V and $n = 2$.

12.31. Calculate ΔG° and $\mathscr{E}^\circ$ for an oxidation-reduction reaction with $n = 1$ and an equilibrium constant of 2.7×10^{-4} at 25 °C.

12.32.* Calculate $\Delta H°$ and $\Delta S°$ for an oxidation-reduction reaction with $n = 2$ and an equilibrium constant of 1.2×10^3 at 25 °C and 7.7×10^4 at 50 °C.

12.33.* Compute $\mathscr{E}°$ for 25 °C for each of the following reactions:
(a) The net cell reaction for the nickel-cadmium cell.
(b) The net cell reaction for the system shown in Fig. 12.5.
(c) $Cr_2O_7^{2-} + 14H^+ + 6Br^- \rightleftharpoons 2Cr^{3+} + 7H_2O + 3Br_2(l)$

12.34. Compute $\mathscr{E}°$, $\Delta G°$, and K for 25 °C for each of the following reactions:
(a) $HClO + Br^- + H^+ \rightleftharpoons H_2O + \frac{1}{2}Cl_2(g) + \frac{1}{2}Br_2(l)$
(b) The net cell reaction for the lead storage battery.
(c) $H_2(g) + ClO_4^- \rightleftharpoons ClO_3^- + H_2O$

12.35.* Which is the strongest reducing agent among the following species: $HClO_2$, NO(g), SO_4^{2-}, Cr^{2+}?

12.36. Which is the strongest oxidizing agent among the following species: NO_3^-, ClO_2, Fe^{2+}, La^{3+}, $NiO_2(s)$?

12.37.* Which is the strongest oxidizing agent and which the strongest reducing agent among the following species: Na^+, Cl^-, BrO_3^-, NO_3^-, Cr^{2+}, Cu(s)?

12.38. Which is the strongest oxidizing agent and which the strongest reducing agent among the following species: IO_3^-, SO_4^{2-}, Al(s), $Fe(OH)_2(s)$, $Co(NH_3)_6^{3+}$?

12.39. Which is the strongest oxidizing agent and which the strongest reducing agent among the following species: Ni^{2+}, Cr^{3+}, ClO_3^-, H_2O_2, SO_4^{2-}?

12.40. Indicate what reaction, if any, would have the greatest tendency to occur in an acidic aqueous solution containing each of the following species in 1 *M* concentrations: Na^+, SO_4^{2-}, MnO_4^-, Hg^{2+}.

12.41.* Indicate what reaction, if any, would have the greatest tendency to occur in a basic aqueous solution containing each of the following species in 1 *M* concentrations: Li^+, K^+, SO_3^{2-}, $H_2AlO_3^-$, HO_2^-, $S_2O_4^{2-}$, Cl^-.

12.42. Imagine that one has an acidic aqueous solution containing the following metal ions, each in about 1.0 *M* concentration: Na^+, Cu^{2+}, Pb^{2+}, Zn^{2+}, Ag^+. Suppose that one wishes to obtain from this solution a sample of copper metal. Explain how this could be done by electrolysis.

12.43.* Determine the order in which the following metal ions are reduced as an increasing voltage is applied to an electrolytic cell in which the standard hydrogen electrode functions at the anode: Ag^+, Mg^{2+}, Na^+, Cu^{2+}, Fe^{2+}.

12.44. A certain solution contains the ions Ag^+ and Cu^{2+}, and one wishes to determine how much of each is present. A cell is constructed for an electroplating process in which a platinum rod serves as the cathode and some other suitable electrode serves as the anode. A voltage is then applied to this cell that brings about the reduction of both Ag^+ and Cu^{2+}, plating out Ag and Cu on the surface of the platinum; it is found that 20.4×10^3 coulombs of electricity is required for this plating. Then the cell is "reversed," making the platinum electrode the anode, and a voltage is applied that brings about the oxidation of Cu to Cu^{2+}, but not Ag to Ag^+; 14.9×10^3 coulombs of electricity are required for this step. How many moles of Ag^+ and Cu^{2+} were present in the original solution? (This example is a simplified version of an important analytical technique called anodic stripping.)

12.45. Using standard reduction potentials, determine K_{sp} for AgBr(s) at 25 °C.

12.46.* Using standard reduction potentials, determine the equilibrium constant of the following reaction at 25 °C:

$$Zn^{2+} + 4NH_3 \rightleftharpoons Zn(NH_3)_4^{2+}$$

12.47 Knowing that $\mathscr{E}°$ at 25 °C is 0.15 V for the reaction

$$NiO_2(s) + H_2O + 2Ag(s) \rightleftharpoons Ag_2O(s) + Ni(OH)_2(s)$$

calculate the standard reduction potential corresponding to the following half-reaction:

$$Ag_2O(s) + H_2O + 2e^- \rightleftharpoons 2Ag(s) + 2OH^-$$

12.48.* From the data given in Table 12.1 and the known K_{sp} value of ZnS(s), calculate the standard reduction potential corresponding to the following half-reaction:

$$ZnS(s) + 2e^- \rightleftharpoons Zn(s) + S^{2-}$$

12.49. Calculate the emf of the cell shown in Fig. 12.7 if $[Zn^{2+}] = 0.010\ M$ and $[Cu^{2+}] = 0.030\ M$ at 25 °C.

12.50.* Calculate the emf of the cell shown in Fig. 12.9 if the OH^- concentration is 0.5 *M* at 25 °C.

12.51. (a) Write the net cell reaction for a cell constructed in two compartments, each of which is a hydrogen electrode; both electrodes have hydrogen being bubbled about a platinum wire at a pressure of 1 atm. In one hydrogen electrode, HCl is present at 1.0 *M*, and in the other it is present at 0.10 *M*. (b) What is the emf of this cell?

12.52. Calculate ΔG for the reaction

$$Zn(s) + Cu^{2+} \rightleftharpoons Zn^{2+} + Cu(s)$$

at 25 °C if $[Cu^{2+}] = 0.72\ M$ and $[Zn^{2+}] = 0.21\ M$.

12.53.* (a) Calculate $\mathscr{E}°$ for the following reaction:

$$2Al(s) + 3Pb^{2+} \rightleftharpoons 2Al^{3+} + 3Pb(s)$$

(b) Calculate $\mathscr{E}$ for this reaction at 25 °C if $[Al^{3+}] = 0.30\ M$ and $[Pb^{2+}] = 0.25\ M$.

ANSWERS TO SELECTED EXERCISES

12.1 (a) Zn^{+2}, +2; Fe(s), 0; Zn(s), 0; Fe^{+2}, +2; (b) $Cl_2(g)$, 0; Br^-, −1; Cl^-, −1; Br_2, 0. **12.3** ox. agent = Zn^{+2}, red. agent = Fe(s); ox. agent = $Cl_2(g)$, red. agent = Br^- **12.5** (a) Fe(s), Br^-; (b) Zn^{+2}, $Cl_2(g)$. **12.7** $Zn^{+2} + 2e^- \rightleftharpoons Zn(s)$, $Fe(s) \rightleftharpoons Fe^{+2} + 2e^-$, $Cl_2(g) + 2e^- \rightleftharpoons 2Cl^-$, $2Br^- \rightleftharpoons Br_2(l) + 2e^-$. **12.9** (a) true; (b) true; (c) false. **12.11** (a) $Cd(s) + NiO_2(s) + 2H_2O \rightleftharpoons Cd(OH)_2 + Ni(OH)_2$; (b) $Mn(s) + 2H_2O \rightleftharpoons Mn(OH)_2 + H_2(g)$; (c) $2Fe(OH)_2(s) + Pb(OH)_3^- \rightleftharpoons 2Fe(OH)_3(s) + Pb(s) + OH^-$. **12.13** (a) $C_3H_8(g) + 5O_2 \rightleftharpoons 3CO_2 + 4H_2O$; (b) $2MnO_4^- + 6H^+ + 5HCO_2H \rightleftharpoons 2Mn^{+2} + 8H_2O + 5CO_2$; (c) $P_4(s) + 6H_2O \rightleftharpoons 2HPO_3^{-2} + 2PH_3(g) + 4H^+$. **12.16** 1.91 g. **12.17** (a) false; (b) false; (c) true; (d) false. **12.19** (a) true; (b) false; (c) false. **12.24** 1.75×10^3 coulombs. **12.26** 14.9 hr. **12.28** (a) 0.699 g; (b) 0. **12.30** $\Delta G° = -161$ kJ; $K = 1.4 \times 10^{28}$.

12.32 $\Delta H° = 133.3$ kJ; $\Delta S° = 506$ J deg^{-1}. 12.33 (a) 1.3 V; (b) 2.12 V; (c) 0.27 V. 12.35 Cr^{+2}. 12.37 ox. agent = BrO_3^-; red. agent = Cr^{+2}. 12.41 $HO_2^- + S_2O_4^{-2} + OH^- \rightarrow 2SO_3^{-2} + H_2O$. 12.43 Ag^+, Cu^{+2}, Fe^{+2}, Mg^{+2}, Na^+. 12.46 1.34×10^9. 12.48 -1.43 V. 12.50 1.3 V. 12.53 (a) 1.53 V; (b) 1.52 V.

ANSWERS TO STUDY PROBLEMS

12(a) $Cr_2O_7^{-2} + 2NO(g) + 6H^+ \rightleftharpoons 2Cr^{+3} + 2NO_3^- + 3H_2O$. 12(b) $Cl_2(g) + Ni(s) + 2OH^- \rightleftharpoons 2Cl^- + Ni(OH)_2(s)$. 12(c) 5.93 g. 12(d) 34.7 g. 12(e) 9.4×10^{45}. 12(f) -0.515 V. 12(g) 0.30 V. 12(h) 1.4×10^{10}. 12(i) 0.86 V. 12(j) The Cl_2 and especially the F_2 would never appear so long as the solution remains as described, as O_2 is produced from the solvent (H_2O), of which there is essentially an infinite supply. 12(k) 1.08 V. 12(l) 2.03 V.

13

Kinetics and Mechanisms

13.1 Introduction

Chemical kinetics is the study of the **rates** of chemical reactions, how fast reactions occur. Some reactions are essentially complete in a few microseconds, such as explosions and many acid-base neutralizations. Others, such as the rusting of iron, may require days or even years to complete. The **mechanism** of a chemical reaction is a **pathway,** or sequence of specific steps, involved in the transformation of reactants, on the left side of a chemical equation, to products, on the right side of a chemical equation. Some reactions proceed on the simple collision of two molecules. Others require the formation of several intermediate species before the final product is obtained. Certain reactions proceed after simply breaking or making only one bond; others require extensive rearrangements of several atoms or groups of atoms within one molecule. In some cases each rearrangement requires at least one bond to be broken and at least one to be made, either simultaneously (concerted) or stepwise. The basic equation describing photosynthesis may be written

$$6CO_2(g) + 6H_2O \xrightarrow{\text{light}} C_6H_{12}O_6 + 6O_2(g)$$

Although this equation gives us the overall stoichiometry for the process, it tells us nothing of the mechanism, or pathway, by which it occurs. Actually, photosynthesis in plant cells is far more complicated than the overall equation suggests. Although the complete mechanism is not yet known, it is known that there are more than a hundred sequential chemical steps in the photosynthetic production of a molecule

Spotlight

Under certain conditions acetylene explosively decomposes to its constituent elements:

$$C_2H_2(g) \rightleftharpoons 2C(g) + H_2(g) \qquad \Delta H^\circ_{298} = -56 \text{ kcal}$$

It cannot be liquefied safely because of danger of explosion from shock, but procedures have been developed for safely handling the gas under pressures up to about 30 atm. At 12 atm, 300 volumes of the gas dissolve in one volume of acetone

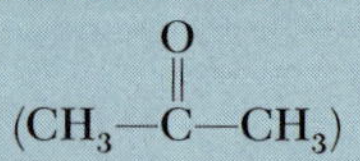

to form a stable solution. Consequently, for welding and other industrial uses, cylinders containing a porous material, for example, kapok, saturated with acetone are pressurized with acetylene.

of glucose from carbon dioxide and water, each step requiring a specific kind of agent called an enzyme.* Ideally, mechanism studies can give us a detailed motion picture of the entire sequence of events from the beginning to the end of a chemical reaction. We shall become aware that chemical kinetics can act to some extent like a camera in these studies.

In previous chapters we have been concerned primarily with only the initial and final states of a reaction or equilibrium; the specific steps involved or the time required for the reaction were largely ignored. Our previous studies of thermodynamics provide us with a powerful tool for treating and understanding reactions that are at equilibrium, but they give us no feeling for how long it takes before the equilibrium is actually achieved. The pitfalls one may encounter by relying solely on thermodynamics for practical predictions is dramatically demonstrated by the following equilibrium between acetylene, H—C≡C—H and its constituent atoms:

$$C_2H_2(g) \overset{K}{\rightleftharpoons} 2C(g) + H_2(g)$$

The value of the equilibrium constant at 25 °C is approximately 10^{37}.† If this were the only fact considered, one might be led to believe that acetylene would be too unstable even to exist at room temperature, since the equilibrium lies so far to the right. Indeed, in the thermodynamic sense, acetylene is unstable. The explanation for its existence lies in the fact that the value of the equilibrium constant does not tell us whether or not a pathway is available for the system to attain equilibrium—it tells us only the relative amounts of the species when they are at equilibrium; thus,

* 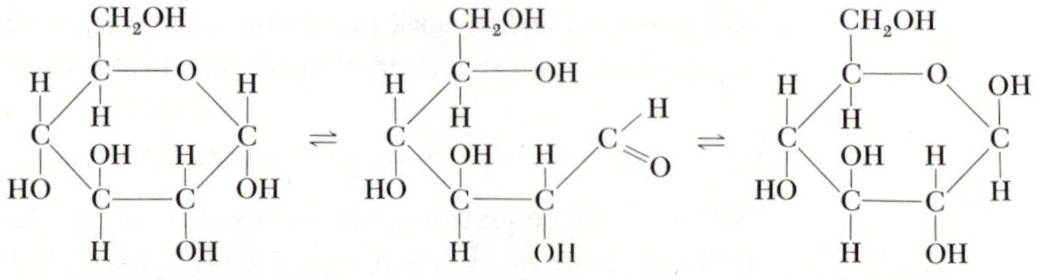

α—D—glucose   open form   β—D—glucose

In aqueous solution, glucose exists in three isomeric forms in dynamic equilibrium. It has been estimated that the open form is present to an extent of 0.024%.

†Calculated from the values given in Table 10.2.

if such a pathway is available and sufficient time elapses for an equilibrium to be established, then acetylene is decomposed into its constituent elements.

Because an equilibrium can be viewed as two opposing reactions proceeding at equal rates, there is sometimes a temptation to regard kinetics as more fundamental than thermodynamics.* A more realistic view is that kinetics and thermodynamics both relate to the same theoretical models as viewed at a submicroscopic level. Furthermore, kinetics by itself can provide only some of the information required to establish the detailed mechanistic course of a reaction. For a particular reaction, for example, it is quite common for a chemist to be able to propose at least two "reasonable" mechanisms that are kinetically indistinguishable; that is, they cannot be distinguished by rate studies. In these cases the chemist may also use several other kinds of techniques whenever possible. Two are the incorporation of specific functional groups or isotopes, which are commonly called "labels," and either the physical or chemical detection of possible intermediates.

To use rate studies for elucidating a mechanism, the chemist first undertakes an accurate product study under the same experimental conditions—of temperature, concentration, and the like—as the rate study. Without knowing precisely what the products are, he could never be sure of which reaction is under investigation. Often, two or more reactions occur simultaneously in the same chemical system. The next step is generally, but not always, using prior knowledge to list the most "reasonable" pathways by which the reaction may occur. These are guesses based on the individual's background, experience, and intuition. The chemist then conducts experimental studies to eliminate as many of the proposed mechanisms as possible, preferably all but one; more often, a choice of two or three remains. If the choices can finally be narrowed down to one mechanism, then it can rightfully be classified as a mechanistic theory. Experience has shown, however, that such theories are very susceptible to change as new knowledge is gained. The ultimate test is whether the proposed mechanism can successfully predict the outcome of experiments in which the reaction conditions are modified. If the proposed mechanistic theory is successful, it is accepted as the "correct" one and is often referred to as the "established mechanism." However, there can be no guarantee that the results of all future experiments will be predicted successfully; there is no scientific theory or law that can make such a guarantee. Should a contradictory experiment be found in the future, the mechanistic theory is modified to make it consistent with the experiment; in some cases an entirely different mechanism must be proposed.

This pattern of experimental studies is found in a wide variety of problems for which detailed insight into chemical processes must be obtained. Such problems are as diverse as air pollution, the manufacture of plastics, and chemotherapy. Most of the knowledge now available about smog has come from studies of this general type. In the case of chemotherapy, which is applying chemical principles and agents in treating disease, any progress besides that based on the results of trial-and-error approaches is likely to result only from experiments that give a fundamental understanding of the detailed modes of interaction of hormones, enzymes, toxins, carcinogens, and so on. Such understanding usually pivots on mechanism studies, and one might expect these studies to be vitally important in cancer research.

*See Chap. 4.

13.2 The Meaning of Rate

Reactions that proceed in one step are termed **elementary reactions,** and these can serve as stepping-stones in an overall pathway. The idea of an elementary reaction also serves as a mental stepping-stone in our visualization of overall mechanisms. For example, a reaction that produces an intermediate that subsequently gives the final product must proceed in at least two steps. Such a reaction can be represented as

reactant → intermediate	(elementary reaction)
intermediate → product	(elementary reaction)
reactant → product	(overall reaction)

As indicated above, an overall reaction is the sum of the elementary reactions of which it is constituted.

It is necessary to have a thorough grasp of the concepts of rates of elementary reactions before proceeding to overall rates. Consequently the rest of our discussions will be divided into two categories: elementary reactions and overall reactions.

13.3 Elementary Reactions

Factors influencing the Rate

Two general factors influence the rate of a particular elementary reaction—concentration and temperature.

Concentration Increased concentrations lead to increased rates. Five-molar HCl will react with active metals faster than 0.001 *M* HCl, and a wooden splint will glow in air, which is dilute oxygen, but will burst into flames in pure oxygen.

Temperature Like increased concentration, increased temperatures usually give higher reaction rates. The oxidation of compounds in food that leads to spoilage, for example, occurs faster at high temperatures—hence the advantage of refrigeration. A higher temperature leads to an increased energy for a collection of molecules. This is important because molecules must have a certain minimum energy before they can react; merely colliding may not be sufficient. This minimum energy is called the **activation energy** and represents a "barrier" that the molecules must overcome in order to react.

Rate Laws and Rate Constants

A **rate law,** sometimes referred to as a **rate expression** or a **rate equation,** is simply an equation that expresses the dependence of the rate of a reaction on the concentration of the reactants; it is always determined experimentally. This determination is accomplished by observing how the concentrations of the reactants change as time progresses.

For elementary reactions, the concentration of each of the reactants will appear as terms on the right side of the rate law. Each term will be raised to a power, expressed by an exponent, that equals the coefficient of that reactant in the written balanced chemical equation.

Let us consider two simple examples, namely, the abstraction of a hydrogen atom from an organic molecule by a methyl radical,*

$$CH_3 + HR \rightleftharpoons CH_4 + R$$

where HR stands for the organic molecule, and the direct combination of two methyl radicals to form ethane,

$$CH_3 + CH_3 \rightleftharpoons CH_3{-}CH_3$$

In the first example, doubling the concentration of CH_3, which means doubling the number of methyl radicals, leads to a twofold increase in the number of collisions between CH_3 and HR, and subsequently to a twofold increase in the rate of the reaction. The same result obtains if the concentration of only HR is doubled. Thus, the rate of the reaction is directly proportional to the concentration of each of the reactants and the rate will be given by

$$\text{rate} \propto [CH_3][HR]$$

$$CH_3 + HR \rightarrow CH_4 + R$$

Case A

$$\begin{matrix} CH_3 + HR & & CH_4 + R, & CH_3H + R, \\ + CH_3 + HR & \rightarrow & CH_3H + R, & CH_4 + R \end{matrix}$$

Case B (both concentrations doubled)

Here we see the effect of concentration on the collisional probability for forming CH_4 and R. The colored radicals are not meant to imply a difference between the radicals. The radicals are identical; the color is only used to show the origin of products when the concentrations are doubled. For Case A, only one collision is possible for CH_4 + R formation. For Case B, the total number of radicals has been doubled, but the number of possible ways to form CH_4 + R (any color) is increased by 2×2. Thus, the total probability for an effective collision is proportional to $[CH_3]$ times [HR].

In the ethane example, CH_3 plays the role of both CH_3 and HR in the example above. Doubling the CH_3 concentration, then, increases the rate fourfold. The rate is given by

$$\text{rate} \propto [CH_3]^2$$

The proportionality constant for any rate expression is normally referred to as the **rate constant** and is given the symbol k. Thus, for the reaction between two CH_3 radicals,

$$\text{rate} = k[CH_3]^2$$

This expression is the **rate law** for this process. The rate constant is a constant at any given temperature, but its value may change with temperature or under the influence of a catalyst.

*Free radicals are uncharged species produced by a "homolytic" cleavage of a bond, that is, a rupture in which one electron remains with each fragment; this is shown in the reaction $Br{:}Br \rightleftharpoons 2Br\cdot$. Free radicals are usually, but not always, very reactive.

Reaction Order and Molecularity

A **reaction order** for a reaction is defined as the sum of the exponents appearing in the rate law. For elementary reactions, the reaction order is equal to the sum of the coefficients of the left side of a written balanced chemical equation. If the sum is one, the reaction is referred to as a **first-order reaction.** Analogously, a sum of two is referred to as a **second-order reaction.**

The **molecularity** of an elementary reaction is defined to be the same as the reaction order. Terms such as *unimolecular, bimolecular,* and *termolecular* are used to describe molecularity. A distinction between molecularity and reaction order exists only with certain overall reactions.

First-Order Reactions

An elementary first-order reaction is defined as one whose rate is found by experiment to be directly proportional to the concentration of only one reactant; that is

$$\text{rate} = k[\text{A}] \tag{13.1}$$

where A represents the reactant. For reactions in solution, the unit of rate usually is mol/liter per unit of time, for example,

$$\frac{\text{mol}}{\text{liter}} \times \frac{1}{\text{sec}} \qquad \text{or} \qquad \frac{\text{mol}}{\text{liter} \cdot \text{sec}}$$

Since mol/liter are molar units, the units may be given as $M\ \text{sec}^{-1}$. This result is obtained because as the reaction proceeds, the concentration change (mol/liter) of the reactants is measured as a function of time. The units of a first-order rate constant are reciprocal time units, for example, sec^{-1}. The scheme is dimensionally self-consistent, as follows:

$$\text{rate}\left(\frac{\text{mol}}{\text{liter} \cdot \text{sec}}\right) = \text{rate}\ (\text{mol} \cdot \text{liter}^{-1}\text{sec}^{-1}) = [k(\text{sec}^{-1})][\text{A}(\text{mol} \cdot \text{liter}^{-1})]$$

The radioactive decay of the unstable nuclide ^{14}C to ^{14}N ($^{14}_{6}C \rightarrow {}^{14}_{7}N + \beta^-$) is a typical first-order reaction in which only one starting material is involved. It has been studied extensively and has been used widely in archaeology to determine the age of certain objects. We shall explore the details of this case as an illustrative example of first-order kinetics.

If we start at a time t_0 with a certain concentration of carbon 14 c_0, then at some later time t_1 we shall find that the concentration of carbon 14 c_1 is less than the original value.* Numerous determinations made at various times $t_1, t_2, t_3, \ldots$ will give various corresponding concentrations $c_1, c_2, c_3, \ldots$, from which the plot shown in Fig. 13.1 can be constructed. The plot shows that equal intervals of

*Measurements of the concentration of a chemical species at various time intervals can be accomplished in many ways, and some ingenious methods have been devised. The two most common methods are (1) withdrawal of a series of small measured samples or portions (frequently referred to as aliquots) of the reaction mixture, followed by a determination of the amount of the species present in the aliquots either by chemical methods (for example, titration) or by physical methods (for example, spectroscopy), and (2) continuous monitoring of the concentration of one of the species by physical methods.

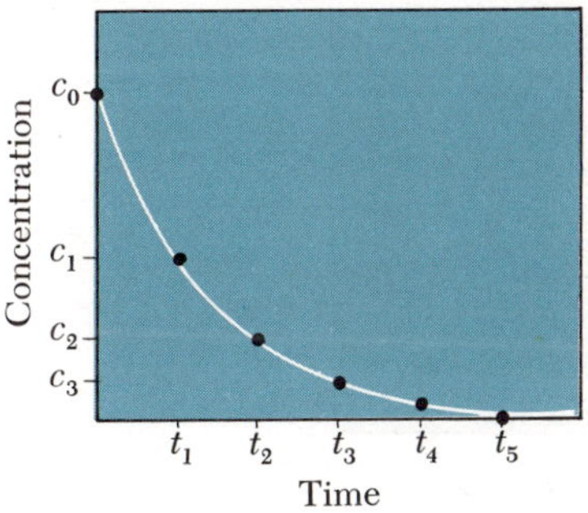

Figure 13.1 A plot of the concentration of a reactant vs time for a first-order reaction.

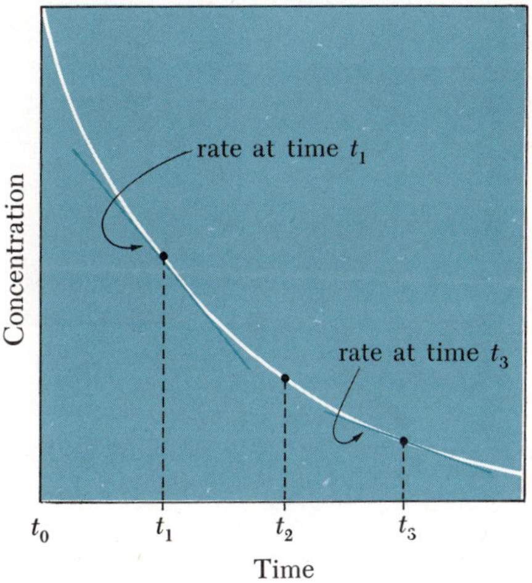

Figure 13.2 Instantaneous rates represented by slopes.

time do not give equal changes in the concentration. This is because the rate of decay decreases as time progresses. The rate is directly proportional to a continuously decreasing concentration of carbon 14. Thus, the plot shows graphically the same idea given in Eq. (13.1), and the form of the plot is typical of first-order processes.

The average rate during the first period is $(c_0 - c_1)/(t_1 - t_0)$, and for the second period $(c_1 - c_2)/(t_2 - t_1)$. The plot not only shows that these two average rates are not equal, but also shows that the rate is continuously decreasing during each infinitesimally small period. This continual decrease in rate can be seen by observing that the curved line connecting the experimental points is steeper during the earlier periods. The steepness of the line at any time equals the rate of the reaction at that time, and can be graphically represented by the slope of a line drawn tangent to the curve at that given point. The rates at times t_1 and t_3 are illustrated in Fig. 13.2 by slopes, and are called instantaneous rates.

The type of curve shown in Fig. 13.1 is treated easily by standard mathematical procedures.* Studies in analytic geometry show us that many lines, such as straight lines, circles, parabolas, and hyperbolas, can be represented by algebraic equations.

*For those who are familiar with calculus, the rate at which the concentration of carbon 14 is changing with time is $d[\mathrm{A}]/dt$. Then,

$$-\frac{d[\mathrm{A}]}{dt} = k[\mathrm{A}]$$

$$-\int_{c_0}^{c} \frac{d[\mathrm{A}]}{[\mathrm{A}]} = k\int_0^t dt$$

$$-\ln [\mathrm{A}]\Big|_{c_0}^{c} = kt\Big|_0^t$$

$$-\ln c + \ln c_0 = kt$$

This final equation can be rearranged to give any of those from (13.3) through (13.6) or it can be transformed into (13.2) by taking the antilogarithm.

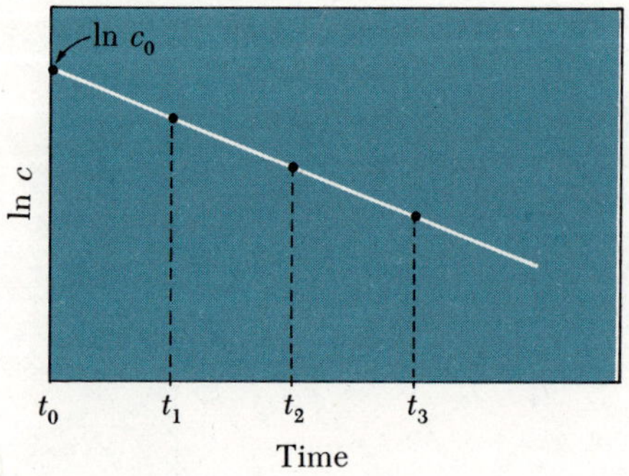

Figure 13.3
Graphical determination of a rate constant, using Eq. 13.4

Likewise, the curve in Fig. 13.1 represents an exponential function, and the equation for this line has the algebraic form

$$c = c_0 e^{-kt} \tag{13.2}$$

where e is the base of natural logarithms, k is a constant, c_0 is the y intercept in the plot, and c and t are the variables on the y and x axes, respectively. Taking the logarithm of both sides of Eq. (13.2) gives

$$\ln c = \ln c_0 - kt \tag{13.3}$$

and rearranging gives

$$\ln c - \ln c_0 = -kt \tag{13.4}$$

$$\ln \frac{c_0}{c} = kt \tag{13.5}$$

$$2.303 \log \frac{c_0}{c} = kt \tag{13.6}$$

Equation (13.3) has the form of a straight line; that is, $y = mx + b$, where x and y are the usual axes, m is the slope of the line, and b is the y intercept. If we now plot $\ln c$ (or $2.303 \log c$) on the y axis and t on the x axis, then we obtain the straight line shown in Fig. 13.3, where the slope of the line is equal to $-k$ and the y intercept is $\ln c_0$.

The value of k can thus be obtained from all the measurements, as all are used in choosing the best line. The k so determined can be used in any of Eqs. (13.3) through (13.6) to obtain the concentration at any chosen time. The concentration thus obtained can now be substituted for [A] in Eq. (13.1) to find the instantaneous rate at that particular time.

Note that to determine the value of k it is not necessary to know the original concentration. If the concentrations can be found at various times after the reaction has begun, and if the time that the reaction began is known, then a partial plot similar to Fig. 13.3, can be made. Extrapolation of the line back to t_0 will give $\ln c_0$. This is shown in Fig. 13.4. Conversely, if the original concentration is known but the time at which the reaction began is not, then t_0 can be found by extrapolation back to $\ln c_0$ and reading t_0 on the x axis. Alternatively, both of these extrapolations can be done mathematically rather than graphically.

In preparing the straight line plots, only two points are mathematically sufficient.

Spotlight

The procedure of plotting ln c vs time is the one archaeologists use to determine the age of carbonaceous materials. Carbon 14 is present in air in a very small but measurable amount. The isotope is continually being formed in the atmosphere by cosmic ray bombardment of ordinary nitrogen, ^{14}N, so that there is a constant percentage of carbon 14 in the carbon atoms of the CO_2 in air. Consequently, this same percentage of carbon 14 exists in the carbon atoms of all living organisms. When the organism dies, however, the incorporation of carbon 14 ceases, and the nuclei present at that time continue to decay in a first-order process without being replenished. Since the original concentration c_0 is known, the time of death t_0 can be found. The rate of decay for carbon 14 is very slow (see Table 13.1) and can accurately be used for estimating only long periods. Such measurements are the most accurate method of dating archaeological specimens, and the error may be as little as a few decades for specimens that are not too old. Materials older than about 40,000 years have too faint a radioactivity to be measured accurately; it is not possible to determine the age of coals in this way. Unfortunately, the wide dispersal of radioactive materials from man-made sources threatens to provide a background of radioactivity that will, in time, mask the effects of carbon 14.

In practice, however, one must be certain that there is not a large error in one of these two measurements, resulting in a large error in the value of k obtained. Hence, whenever possible, at least one more measurement should be taken as a check for accuracy. More important, more than two points are necessary to verify a straight-line relation regardless of the accuracy of the measurements, and an absolute minimum of three measured points is necessary; the point (c_0, t_0), if known, can be considered one of the three. It is customary, and considerably more accurate, to use many more than three. The best straight line can then be drawn; a typical example is shown in Fig. 13.5.

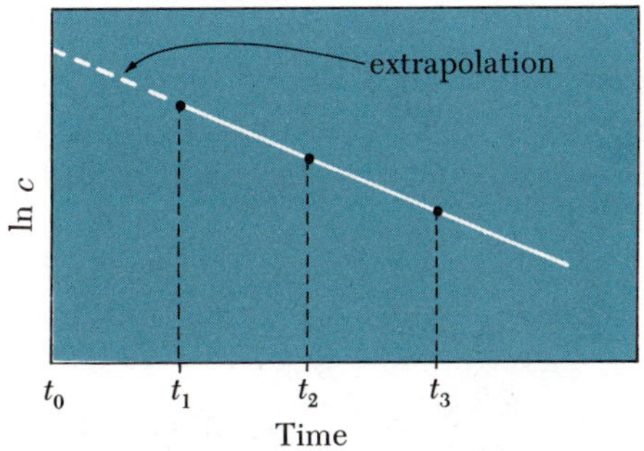

Figure 13.4 Extrapolation method to find c_0 or t_0.

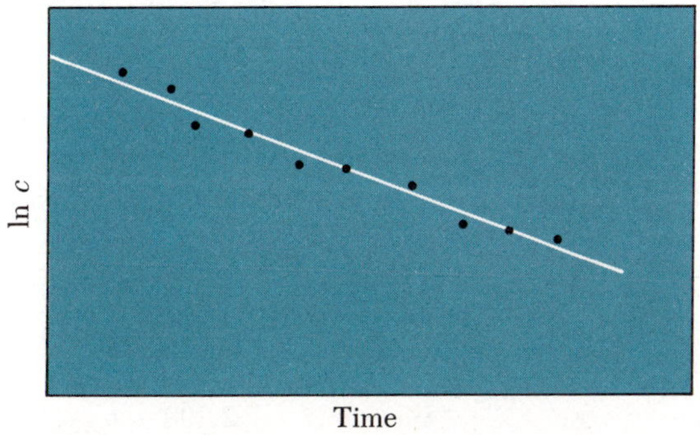

Figure 13.5 The "best" straight line through a set of scattered experimental points.

If the data obtained from a reaction do not give a straight line when plotted as described above, the reaction is definitely not simply first-order. Another common test is to use any of the equations (13.3) through (13.6) to calculate the value of k for each pair of concentration-time measurements, and then check for constancy in the values of k so obtained.

EXAMPLE 13.1

Sodium 24, in the form of a saline solution (NaCl), is sometimes used in medicine to pinpoint the location of blood clots. It radioactively decays to produce the stable isotope magnesium 24 and a beta particle (electron):

$$^{24}_{11}\text{Na} \rightarrow ^{24}_{12}\text{Mg} + \beta$$

A study of the decay of sodium 24 in a saline solution gave data as shown in the accompanying chart, where the concentrations are expressed in millimoles per liter.

Time, min	$[^{24}\text{Na}^+]$, *mmol/liter*
0	100.0
600	62.5
1200	39.3
1800	24.5
2400	15.4
3000	7.9

(a) Prove that the decay is a first-order process and calculate the rate constant, and (b) calculate the instantaneous rate of decay at $t = 1000$ min.

SOLUTION

(a) Use a rearranged form of Eq. (13.6) to calculate the value of k for each pair of concentration-time measurements, and then check for the constancy in the values of k. For the first time measurement, $t = 600$ min, $c = 62.5$ mmol/liter, and $c_0 = 100.0$ mmol/liter.

$$k = \frac{2.303}{t} \log \frac{c_0}{c}$$

$$= \frac{2.303}{600} \log \frac{100.0}{62.5} = 7.83 \times 10^{-4}\ \text{min}^{-1}$$

The other values of k obtained by this method are 7.78×10^{-4}, 7.82×10^{-4}, 7.80×10^{-4}, and 7.86×10^{-4} min^{-1}. Since the values obtained are constant within the probable experimental error, we conclude that the reaction is first-order.

(b) To find the instantaneous rate at $t = 1000$ min, we use Eq. (13.1); but first we need to know $[^{24}\text{Na}^+]$ at that time. This is obtained through using Eq. (13.6) again:

$$\log \frac{c_0}{c} = \frac{kt}{2.303}$$

$$\log \frac{100.0}{c} = \frac{7.82 \times 10^{-4} \times 10^3}{2.303} = 0.340$$

$$\frac{100.0}{c} = 2.19$$

$$c = 45.7 \text{ mmol/liter} = [^{24}\text{Na}^+]$$

at t = 1000 min. Then from Eq. (13.1),

$$\begin{aligned} \text{rate} &= k[^{24}\text{Na}^+] \\ &= 7.82 \times 10^{-4} \times 45.7 \\ &= 3.58 \times 10^{-2} \frac{\text{mmol}}{\text{liter} \times \text{min}} \end{aligned}$$

STUDY PROBLEM 13(a)

Determine the instantaneous rate of the reaction in Example 13.1 at 500 min.

Thus far, we have been concerned only with the rate of disappearance of a reactant. Since we are considering elementary reactions in this section, the rate of formation of product (or products) will equal the rate of disappearance of reactant. Because these two phrases are lengthy, a shorthand form is used to express the rates. For example, to denote the fact that a variable C, representing concentration, is changing with time, the symbol $\Delta C/\Delta t$ is frequently used.* Expressed in words, this symbol means "the change in C per change in time." Since C represents a quantity that must be positive, a minus or a plus sign precedes the symbol according to whether the variable C is decreasing or increasing as time passes. For the elementary reaction A $\rightarrow$ B, the rate of disappearance of A equals the rate of formation of B, and this is represented algebraically by

$$-\frac{\Delta[\text{A}]}{\Delta t} = \frac{\Delta[\text{B}]}{\Delta t} \qquad (13.7)$$

Hereafter, this notation will be used to express rates.

Introducing the concept expressed by Eq. (13.7) makes possible a modification of Fig. 13.2; a line for the product can be included, as shown in Fig. 13.6. Figure 13.6 shows that the slope at t_2 for the reactant is equal but opposite in sign to the slope for the product; the product is forming as fast as the reactant is disappearing. We notice one further important fact; the two curves intersect at one particular value of t (t_1 in Fig. 13.6). At this instant the concentrations are equal; therefore one-half the original reactant has been consumed. The time at which this occurs is conveniently called the **half-life** for the reaction, and it is designated $t_{1/2}$. It is very important that the definition of $t_{1/2}$ not be confused with "half the time required for the reaction to be complete," because strictly speaking the reaction is never entirely complete. This can be seen by observing in Eq. (13.5) that an infinite value of time is required for c to approach zero.

*In calculus, this concept is related to "the derivative of C with respect to time." It is conventional in calculus to denote an incremental change in time by Δt and the corresponding change in C by ΔC. The *average* rate of change of C relative to time over the interval Δt is, then, $\Delta C/\Delta t$. The *instantaneous* rate is the limit of the average rate as Δt becomes infinitesimal, and is denoted

$$\lim_{\Delta t \to 0} \left(\frac{\Delta C}{\Delta t}\right) = \frac{dC}{dt}$$

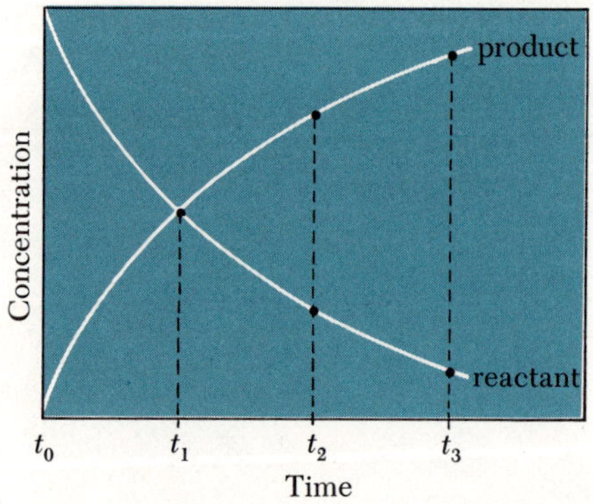

Figure 13.6
A plot of the concentrations of a reactant and a product vs time for a first-order reaction.

If the half-life for a first-order reaction is known, then the rate constant can be calculated. This is accomplished most readily through Eq. (13.6). At a value of t equal to the half-life, $c = \frac{1}{2}c_0$, and the equation becomes

$$2.303 \log 2 = kt_{1/2} \tag{13.8}$$

$$k = \frac{0.6932}{t_{1/2}} \tag{13.9}$$

The half-life for the radioactive decay of carbon 14 is 5760 yr; therefore the first-order rate constant is $12.0 \times 10^{-5}\ \text{yr}^{-1}$. Half-lives for the radioactive decay of some other nuclides are given in Table 13.1.

EXAMPLE 13.2 Since the number of beta particles emitted is proportional to the number of carbon 14 atoms, the number of counts per minute obtained from a Geiger-Müller counter will be proportional to the concentration of carbon 14. Immediately after being removed from the atmosphere, CO_2 normally gives a count of 12.5 beta particles per minute per gram of total carbon. A sample of wood taken from an Egyptian tomb was found, on combustion to CO_2 followed by precipitation of the product gas as $BaCO_3$, to give a count of 7.04 beta particles/(min × g C). How long ago was the wood part of a living tree?

SOLUTION Equation (13.6) relates the initial concentration to the concentration at any time; but to use this equation we need to know the first-order rate constant for the decay of carbon 14. This is conveniently obtained through the use of Eq. (13.9) and Table 13.1.

$$k = \frac{0.6932}{t_{1/2}} = \frac{0.6932}{5760\ \text{yr}} = 1.20 \times 10^{-4}\ \text{yr}^{-1}$$

$$t = \frac{2.303}{k} \log \frac{c_0}{c}$$

$$= \frac{2.303}{1.20 \times 10^{-4}\ \text{yr}^{-1}} \log \frac{12.5}{7.04} = 4780\ \text{yr}$$

TABLE 13.1 **Half-Lives of Several Unstable Nuclei**

Nucleus	*Half-Life*
^{3}H	12 yr
^{12}B	0.022 sec
^{14}C	5760 yr
^{24}Na	14.8 hr
^{214}Po	10^{-5} sec
^{228}Ra	1600 yr
^{234}Th	24 days
^{238}U	4.5×10^{9} yr

STUDY PROBLEM 13(b)

A wood sample treated in a manner similar to what is described in Example 13.2 gave a count of 11.8 beta particles/(min × g C). Approximately what date was the wood cut from a tree?

Second-Order Reactions

We have pointed out that for elementary reactions the molecularity is equal to the reaction order. An elementary second-order reaction, then, is **bimolecular,** involving two species, which may either be alike or different. The dimerization, or the union of two identical molecules, of borane, BH_3, and the oxidation of the chromium(II) ion by the cobalt(III) ion are examples of these two types of bimolecular reactions:

$$BH_3 + BH_3 \xrightarrow{k_1} B_2H_6$$

$$Cr^{2+} + Co^{3+} \xrightarrow{k_2} Cr^{3+} + Co^{2+}$$

The respective rate laws are

$$-\frac{\Delta[BH_3]}{\Delta t} = \frac{1}{2}\frac{\Delta[B_2H_6]}{\Delta t} = k_1[BH_3]^2 \qquad (13.10)$$

and

$$-\frac{\Delta[Cr^{2+}]}{\Delta t} = -\frac{\Delta[Co^{3+}]}{\Delta t} = \frac{\Delta[Cr^{3+}]}{\Delta t} = \frac{\Delta[Co^{2+}]}{\Delta t} = k_2[Cr^{2+}][Co^{3+}] \qquad (13.11)$$

If the units of $[BH_3]^2$ and $[Cr^{2+}][Co^{3+}]$ on the right sides of Eqs. (13.10) and (13.11) are $mol^2/liter^2$, the units of the second-order rate constants must be liter/(mol × unit time), for example, liter · $mol^{-1}sec^{-1}$, or $M^{-1}sec^{-1}$.

We have seen that for a first-order reaction a plot of ln c vs t gives the type of straight-line graph shown in Fig. 13.3. However, for the second-order rate law given by $-\Delta[A]/\Delta t = k[A]^2$, where A symbolically represents any reactant, it is found experimentally that a plot of $1/c$ vs t gives a straight line, as shown in Fig. 13.7.

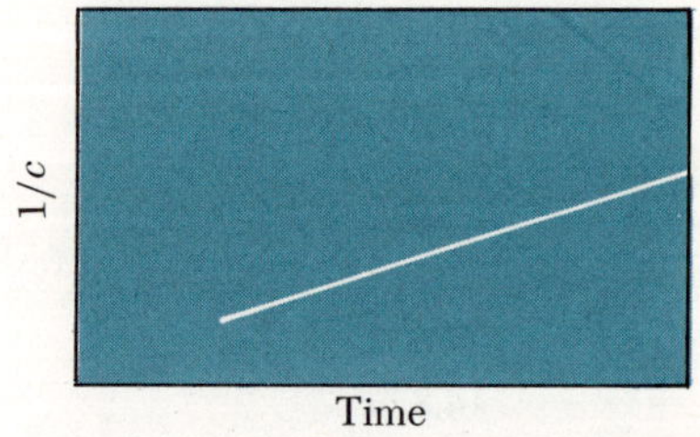

Figure 13.7
Plot of $1/c$ vs time for a bimolecular reaction involving a single reactant.

Any straight line can be represented by the general form $y = mx + b$, and it is found by the methods of calculus that the line in Fig. 13.7 can be represented by Eq. (13.12).* Hence, in Fig. 13.7, $1/c_0$ is the y intercept and k is the slope.

$$\frac{1}{c} = kt + \frac{1}{c_0} \tag{13.12}$$

The half-life expression can be determined from the fact that at $t_{1/2}$, $c = \frac{1}{2}c_0$; then

$$\frac{1}{\frac{1}{2}c_0} = kt_{1/2} + \frac{1}{c_0}$$

$$\frac{2}{c_0} = kt_{1/2} + \frac{1}{c_0}$$

or

$$\frac{1}{c_0} = kt_{1/2} \qquad \text{and} \qquad t_{1/2} = \frac{1}{kc_0}$$

Second-order reactions involving two different substances can be treated like first-order reactions, although the procedure is slightly more complicated.† The general expression is similar to Eq. (13.3):

$$\ln\frac{[\mathrm{A}]}{[\mathrm{B}]} = \ln\frac{[\mathrm{A}]_0}{[\mathrm{B}]_0} + k([\mathrm{A}]_0 - [\mathrm{B}]_0)t \tag{13.13}$$

*Equation (13.12) can be obtained from a second-order expression, such as Eq. (13.10), by the methods of calculus:

$$-\frac{d[\mathrm{A}]}{dt} = k[\mathrm{A}]^2 \qquad -\frac{d[\mathrm{A}]}{[\mathrm{A}]^2} = k\,dt$$

$$-\int_{c_0}^{c} [\mathrm{A}]^{-2}\,d\mathrm{A} = k\int_0^t dt \qquad \frac{1}{[\mathrm{A}]}\bigg]_{c_0}^{c} = kt\bigg]_0^t$$

$$\frac{1}{c} - \frac{1}{c_0} = kt$$

†In calculus, the integration of the rate law is simplified by letting x represent the amount of material reacted. The stoichiometry then leads to

$$\frac{dx}{dt} = k(a - x)(b - x)$$

where a and b are the initial concentrations of A and B. The integration is most easily carried out by first rearranging to the form $dx/dt = a + bx + cx^2$, whose closed-form solutions are available from a table of integrals.

where $[A]_0$ and $[B]_0$ are the initial concentrations of the reactants, $[A]$ and $[B]$ are the concentrations at any time t, and k is the second-order rate constant. Rearranging Eq. (13.13) gives the more common form

$$\frac{1}{[A]_0 - [B]_0} \ln \frac{[A][B]_0}{[A]_0[B]} = kt \qquad (13.14)$$

Equation (13.13), like Eq. (13.3), can be viewed as having the form $y = mx + b$. Hence a plot of ln ([A]/[B]) vs t produces a straight line. Such a plot is shown in Fig. 13.8. Notice the positive slope as predicted by the positive sign of k in Eq. (13.13), whereas the negative sign of $-k$ in Eq. (13.3) gave a negative slope. The y intercept in Fig. 13.8 is $\ln [A]_0/[B]_0$, and the slope is $k([A]_0 - [B]_0)$.

All the equations and figures given for the second-order reactions can be used for the same types of purposes described for the corresponding equations and figures of the first-order case.

EXAMPLE 13.3 Given: The formation of tetraethylammonium iodide from triethylamine in nitrobenzene at 100 °C is known to be bimolecular, with $k = 1383$ liter/(mol × sec), or 1383 $M^{-1}\text{sec}^{-1}$:

$$(C_2H_5)_3N\!: + CH_3{-}CH_2{-}I \rightleftharpoons (C_2H_5)_4N^+I^-$$

If the original concentration of the amine is exactly 2 M and of the iodide is exactly 1 M, how long will it take to form 0.5 M salt?

SOLUTION From the stoichiometry, when $[(C_2H_5)_4N^+I^-] = 0.5$ mol/liter then $[C_2H_5I] = 0.5$ mol/liter and $[(C_2H_5)_3N] = 1.5$ mol/liter. Using Eq. (13.14), we obtain

$$\frac{1}{2-1} \ln \frac{(1.5)(1)}{(2)(0.5)} = 1383\, t$$

$$2.303 \log 1.5 = 1383\, t$$

$$t = 2.932 \times 10^{-4} \text{ sec}$$

STUDY PROBLEM 13(c)

If the concentrations of the reactants were each 2×10^{-4} M, how long would it take to form 1×10^{-4} M salt? (Hint: When $A_0 = B_0$, treat the problem as though rate $= k[A]^2$.)

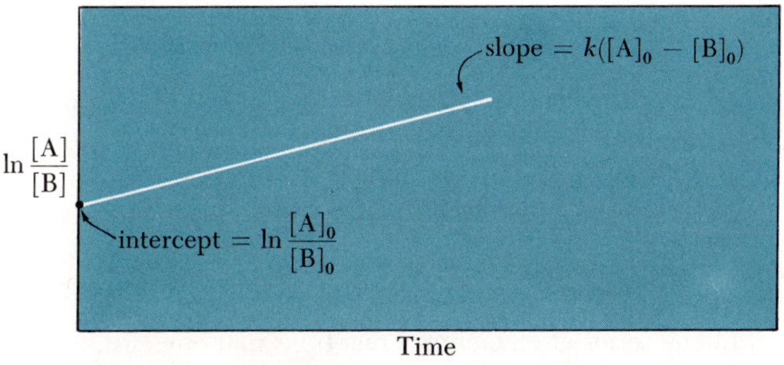

Figure 13.8 Plot of ln ([A]/[B]) vs time for a bimolecular reaction involving two different reactants.

Third-Order and Higher Reactions

The treatment of the rate laws for these kinds of reactions follows essentially along the same lines as the first two cases. We shall not consider this treatment in detail here, however, for three reasons: (1) No new principles would be introduced. (2) The increased complication is too complex for our current interests. (3) The probability that an elementary reaction will occur between more than two species drastically drops as the number of species increases; in fact, even termolecular reactions are very rare. Symbolically, third-order rate laws can be associated with elementary reactions as follows:

$A + B + C \rightarrow$ products	rate $= k[A][B][C]$
$A + 2B \rightarrow$ products	rate $= k[A][B]^2$
$3A \rightarrow$ products	rate $= k[A]^3$

13.4 Overall Reactions

Thus far we have concerned ourselves with chemical transformations that consist of only one elementary reaction—what we might call one-step reactions. Most chemical transformations consist, however, of a sequence of two or more elementary reactions; that is, most overall reactions proceed by mechanisms consisting of more than one step. A detailed experimental study designed to elucidate a mechanism generally pivots on studies of the overall rate (we shall explore this approach later). However, a complete mechanistic investigation of a reaction includes more than only a study of its overall rate. If the mechanism consists of a sequence of two or more elementary reactions, called a **consecutive mechanism**, one must identify each of them in detail.

Consider an overall reaction that can symbolically be represented by

$$A + B \rightleftharpoons \text{products}$$

The actual mechanism may simply be an elementary reaction of A with B to produce the products directly; or A and B may first produce an "intermediate," which then proceeds on to the final products. The former case corresponds to a second-order elementary reaction (covered in Sec. 13.3). If an intermediate is produced, then we could write at least two plausible mechanisms:

$$A + B \rightarrow \text{intermediate}$$
$$\text{intermediate} \rightarrow \text{products}$$

or

$$A \rightarrow \text{intermediate}$$
$$\text{intermediate} + B \rightarrow \text{product}$$

In both cases intermediates are involved, and one may be faced with the problem of deciding which mechanism—the direct one-step mechanism or one of the two-step mechanisms—is actually operating. Several theories and experimental procedures have been developed to help investigators make decisions like this.

Rate-Determining Step

In any series of elementary reactions that constitute an overall reaction, such as the two possibilities given above, the rates of the individual steps are generally not all equal, and there is usually one step that is slower than any of the others. This is an important point, because the overall rate can only be as fast as the slowest step, and no faster. It is analogous to a case in which water is flowing through an irregularly shaped pipe. If the pipe has varying diameters along its length, the maximum rate at which water is flowing is governed by the region of smallest pipe diameter—the "bottleneck." Similarly, the slowest step determines the rate of the overall reaction, and it is called the **rate-determining step.**

The concept of a rate-determining step is exceedingly important in connection with an overall rate, because the rate will depend on whichever elementary reaction is rate-determining. For example, the rate of the displacement of chloride ion from the *cis*-dichlorotetraamminecobalt(III) cation by bromide ion has been followed by several methods:

1. Rate of appearance of chloride ion by chemical analysis.
2. Rate of formation of products (that is, rate of incorporation of bromide ion in the new cobalt complexes) by chemical analysis.
3. Rate of disappearance of the initial cobalt complex by spectrophotometry (the spectra of the reactants and products in the visible region are different).

The rates of all the above are equal, and identical rates are obtained if nitrate or thiocyanate ion is used instead of bromide ion. This clearly indicates that the rate of the slow step is independent of the identity of the displacing anion. Furthermore, the rate is proportional only to the concentration of the initial cobalt complex,

$$-\frac{\Delta[\text{complex}]}{\Delta t} = k[\text{complex}]$$

Because the concentration of the displacing anion does not appear in the rate law, the anion must be involved in a fast step occurring after the slow step. A mechanism consistent with these results is:

$$cis\text{-CoCl}_2(\text{NH}_3)_4{}^+ \xrightarrow[\text{slow}]{k_1} \text{CoCl}(\text{NH}_3)_4{}^{2+} + \text{Cl}^-$$

$$\text{CoCl}(\text{NH}_3)_4{}^{2+} + \text{Br}^- \xrightarrow[\text{fast}]{k_2} cis\text{-CoClBr}(\text{NH}_3)_4{}^+ + trans\text{-CoClBr}(\text{NH}_3)_4{}^+$$

Spotlight

Isomers exist for transition metal compounds as well as for organic compounds (Chap. 7). The dichlorotetraamminecobalt(III) cation, $CoCl_2(NH_3)_4{}^+$, can exist in one form in which the chloride ligands are next to each other (*cis*) and in another form in which they are across from one another (*trans*).

These two forms can be separated from each other because of their different solubilities and they can be distinguished from each other because of their different colors.

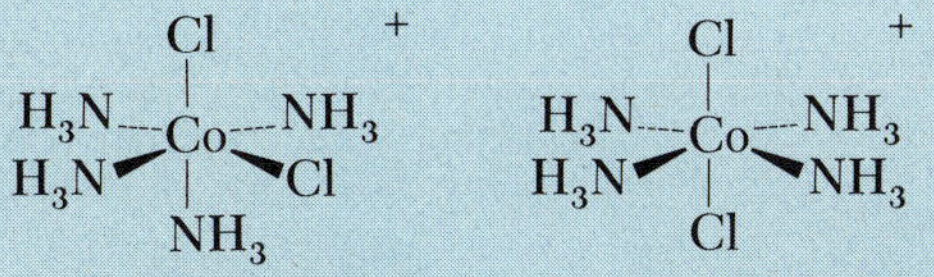

cis *trans*

In this reaction the ion Cl^- was replaced by the nucleophilic ion Br^- in a first-order reaction. The symbol S_N1 (substitution, nucleophilic, first-order) is used to represent processes occurring by such mechanisms.

A fundamental question concerning such reactions is, Why doesn't the *displaced* anion, as well as the *displacing* anion, attack the intermediate formed from the slow step? Indeed it does, and in this sense the first step is actually reversible. However, in the early stages at least, the concentration of the displaced anion is extremely small compared with the concentration of the displacing anion.* In the early stages then, the rate of the back reaction is very slow compared with the rate of the second fast step, and can in this case be neglected until an equilibrium is nearly approached.

An overall rate law depends on whichever elementary step is rate-determining. Hence, one very important point should now be clear. A rate law should never be deduced from only the stoichiometry of the reaction. The other important message that emerges from this discussion is that the experimental determination of the rate law can often permit the chemist to distinguish among the various proposed mechanisms. This kind of situation is the basis for applying kinetics to the study of reaction mechanisms.

Reaction Order

Like an elementary reaction, an overall reaction has an order that is still simply the sum of the exponents of the concentration terms, those that appear in the overall rate law. Occasionally, complicated rate laws arise in which fractional or negative exponents appear. These, however, do not pose a problem in determining the order from the rate expression; one simply adds the exponents. For example, if experiments on a particular reaction yield the rate law given in Eq. (13.15), then the overall order is 1:

$$\text{rate} = \frac{[\mathrm{A}][\mathrm{B}]^{3/2}}{[\mathrm{C}]^{1/2}[\mathrm{D}]} \tag{13.15}$$

Molecularity

The concept of a rate-determining step helps to clarify the distinction between reaction order and molecularity. Molecularity is defined as the number of molecules or species involved in the rate-determining step. Although the molecularity equals the reaction order for elementary reactions, such is not true for many overall reaction orders. Other terms such as *unimolecular* and *termolecular* refer to cases in which one and three molecules are involved in the slow step.

An example of the distinction results from the reaction of nitric oxide with oxygen

*For reactions in which the concentration of the displaced anion gradually builds up and increases the rate of the back reaction, the common procedure is to make measurements only during the beginning stages of the reaction, and then extrapolate back to zero time to give what is called the "initial" rate. The initial rate, then, theoretically represents the rate of disappearance of the reactant in the absence of any displaced anion.

to form nitrogen dioxide. In this reaction there is an equilibrium step before the rate-determining step:

$$2NO(g) \overset{K}{\rightleftharpoons} N_2O_2(g) \qquad (13.16)$$

$$N_2O_2(g) + O_2(g) \xrightarrow{k'} 2NO_2(g) \qquad (13.17)$$

The rate is given by $k'K[NO]^2[O_2]$, so the reaction is third-order. K is the equilibrium constant for Eq. (13.16). Only two molecules, N_2O_2 and O_2, react in the rate-determining step, however, so the reaction is bimolecular.

OPTIONAL

13.5 Additional Types of Reaction Mechanisms

Series First-Order

If a pure sample of a radioactive isotope is isolated, it will not remain pure, since radioactive decay causes a continuing decomposition. A decay product is thus formed, and if this species, called a "daughter" isotope, is also radioactively unstable, its concentration may be increasing or decreasing at any given moment, depending on the relative rates of its formation and decomposition. This situation is common in radioactive materials. Radioactive decay and growth curves typical of such systems are represented in Fig. 13.9, where line A represents the decay of a "parent" (the initial radioactive isotope), B the growth of a stable daughter, and C the growth and decay of a daughter that is itself unstable. This is a **consecutive mechanism.**

An example of the case represented by line C is the decay of bismuth 211, which produces an alpha particle and a daughter, thallium 207, whose half-life is of the same order of magnitude as that of the parent. The relevant nuclear reactions involved are

$$^{211}_{83}\text{Bi} \xrightarrow[t_{1/2} = 2.2\text{ min}]{k_1} {}^{207}_{81}\text{Tl} + \alpha \qquad (13.18)$$

$$^{207}_{81}\text{Tl} \xrightarrow[t_{1/2} = 4.8\text{ min}]{k_2} {}^{207}_{82}\text{Pb} + \beta \qquad (13.19)$$

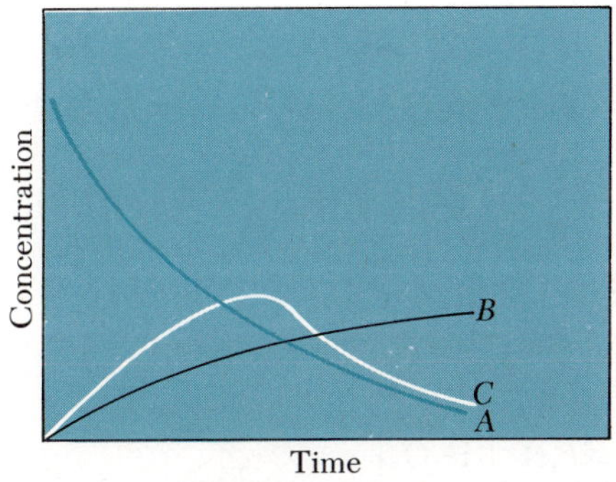

Figure 13.9 Growth and decay curves for some radioactive species.

OPTIONAL

where the symbol α represents an alpha particle, a helium nucleus, and β, a beta particle, or electron. The rate laws appropriate for these transformations are

$$-\frac{\Delta[\text{Bi}]}{\Delta t} = k_1[\text{Bi}] \tag{13.20}$$

$$\frac{\Delta[\text{Tl}]}{\Delta t} = k_1[\text{Bi}] - k_2[\text{Tl}] \tag{13.21}$$

$$\frac{\Delta[\text{Pb}]}{\Delta t} = k_2[\text{Tl}] \tag{13.22}$$

The relations between these rate laws and curves of the type shown in Fig. 13.9 can be handled by standard mathematical techniques.

Parallel Mechanisms

Under some conditions, a variety of reactions are possible for a given starting material, and a competitive situation can exist. In some cases parallel mechanisms lead to the same product or products. For example, iodide ion reacts with hydrogen peroxide under acidic conditions to produce iodine and water by two separate pathways. The overall net reaction is

$$2\text{I}^- + \text{H}_2\text{O}_2 + 2\text{H}^+ \rightleftharpoons 2\text{H}_2\text{O} + \text{I}_2(s) \tag{13.23}$$

Two plausible mechanisms for this reaction are

1.

$$\text{I}^- + \text{H}_2\text{O}_2 \xrightarrow[\text{slow}]{k_1} \text{H}_2\text{O} + \text{IO}^- \tag{13.24}$$

$$\text{IO}^- + \text{H}^+ \rightleftharpoons \text{HOI}$$

$$\text{I}^- + \text{H}^+ \rightleftharpoons \text{HI}$$

$$\text{HOI} + \text{HI} \rightleftharpoons \text{I}_2(s) + \text{H}_2\text{O}$$

2.

$$\text{I}^- + \text{H}^+ \overset{K}{\rightleftharpoons} \text{HI} \tag{13.25}$$

$$\text{HI} + \text{H}_2\text{O}_2 \xrightarrow[\text{slow}]{k_2} \text{H}_2\text{O} + \text{HOI} \tag{13.26}$$

$$\text{H}^+ + \text{I}^- + \text{HOI} \longrightarrow \text{I}_2(s) + \text{H}_2\text{O} \tag{13.27}$$

The total rate of disappearance of iodide ion due to both mechanisms is given by

$$-\frac{\Delta[\text{I}^-]}{\Delta t} = 2k_1[\text{I}^-][\text{H}_2\text{O}_2] + 2Kk_2[\text{H}^+][\text{I}^-][\text{H}_2\text{O}_2]$$

The appearance of two terms in the rate law is typical of reactions that proceed by two separate paths. The relative contributions that these two mechanisms make to the overall reaction is given by the ratio

$$\frac{2k_1[\text{I}^-][\text{H}_2\text{O}_2]}{2Kk_2[\text{H}^+][\text{I}^-][\text{H}_2\text{O}_2]}, \quad \text{which is simply} \quad \frac{k_1}{Kk_2[\text{H}^+]}$$

Thus, the relative importance of the mechanisms is determined by the H⁺ concentration.

EXAMPLE 13.4 The reaction of isobutyl bromide with ethoxide ion proceeds by two parallel paths as shown:

$$\begin{array}{c} H_3C \\ \diagdown \\ CH-CH_2-Br + EtO^- \\ \diagup \\ H_3C \end{array} \begin{array}{c} \xrightarrow{k_1} (CH_3)_2CHCH_2OCH_2CH_3 \\ \text{isobutyl ethyl ether} \\ \\ \xrightarrow{k_2} (CH_3)_2CH=CH_2 \end{array}$$

isobutyl bromide

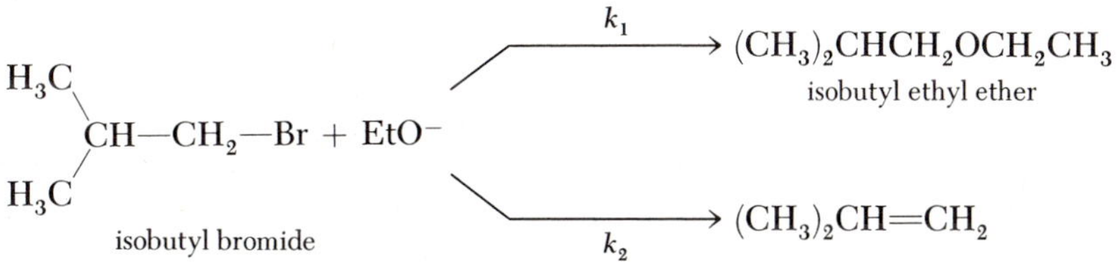

A mixture that is 1.00 M in isobutyl bromide and 2.00 M in sodium ethoxide is allowed to react at 55 °C, and then "quenched," or quickly stopped, by being rapidly cooled after 3.00 hr. Analysis of an aliquot shows that the mixture contains 80.00% unreacted starting material, 8.10% isobutyl ethyl ether, and 11.90% isobutylene. What are the values of k_1 and k_2?

SOLUTION Since the formations of the ether and the olefin require the same reactants and the same type of rate laws, the ratio of their yields at any time will equal the ratio of the respective rate constants. This fact provides us with one equation and two unknowns. The second equation relating these same unknowns obtains from the fact that the overall rate of disappearance of the isobutyl bromide due to both mechanisms is

$$-\frac{\Delta[\text{RBr}]}{\Delta t} = k_1[\text{RBr}][\text{EtO}^-] + k_2[\text{RBr}][\text{EtO}^-]$$
$$= (k_1 + k_2)[\text{RBr}][\text{EtO}^-]$$
$$= k[\text{RBr}][\text{EtO}^-]$$

where $k = k_1 + k_2$. We can now obtain k from Eq. (13.14):

$$k = \frac{1}{t([A]_0 - [B]_0)} \ln \frac{[A][B]_0}{[A]_0[B]}$$

$$= \frac{1}{3.00(2.00 - 1.00)} \ln \frac{(1.80)(1.00)}{(2.00)(0.80)}$$

$$= 0.0393 \frac{\text{liter}}{\text{mol} \times \text{hr}}$$

Now, $\dfrac{k_1}{k_2} = \dfrac{8.10}{11.90} = 0.681$

$$k_1 = 0.681 \, k_2$$

$$1.681 \, k_2 = 0.0393$$

$$k_2 = 0.0234 \frac{\text{liter}}{\text{mol} \times \text{hr}}$$

$$k_1 = 0.0393 - 0.0234 = 0.0159 \frac{\text{liter}}{\text{mol} \times \text{hr}}$$

OPTIONAL

STUDY PROBLEM 13(d)

What would the percentage of isobutyl ether, isobutylene, and unreacted isobutyl bromide be if the reaction described in Example 13.4 were quenched at 6.00 hr?

Chain Reactions

Many reactions of great practical importance occur by what are called chain mechanisms. These reactions, which are called **chain reactions**, occur in such diverse processes as the manufacture of plastics, the production of smog in the atmosphere, nuclear explosions, and nuclear reactor processes. To exploit, control, or manipulate such processes it is necessary to understand the mechanisms by which they occur. A **chain mechanism** consists of a series of identical or very similar reactions called **propagation steps.** They are triggered by a highly reactive species called an **initiator**, and will produce the same initiator or a species of similar reactivity as part of the sequence. The reaction will continue until the starting material is totally consumed or until a termination step occurs. In a termination step, a highly reactive species is diverted to an alternative route that proves to be a dead end. The initiator may be a cation, an anion, a nuclear particle, or a free radical (species with an unpaired electron), and may be added in catalytic amounts at the start of the reaction or generated in situ by irradiation of the reaction mixture.

Among the best-characterized reactions in chemistry is the hydrogen-bromine reaction,

$$H_2 + Br_2 \rightleftharpoons 2\ HBr$$

This reaction has been shown to follow the rate law

$$\frac{\Delta[HBr]}{\Delta t} = \frac{k[H_2][Br_2]^{1/2}}{\dfrac{k'[HBr]}{[Br_2]} + 1} \tag{13.28}$$

This complex rate expression is inconsistent with a "four-center" mechanism, which means one in which the transformation from reactants to products involves at some intermediate stage a nominal contact among four atoms in the species involved:

$$\begin{array}{c} H-H \\ + \\ Br-Br \end{array} \longrightarrow \begin{array}{c} H\text{-----}H \\ \vdots\ \ \ \ \vdots \\ Br\text{-----}Br \end{array} \longrightarrow \begin{array}{c} H\ \ \ H \\ | \ \ + \ | \\ Br\ \ \ Br \end{array}$$

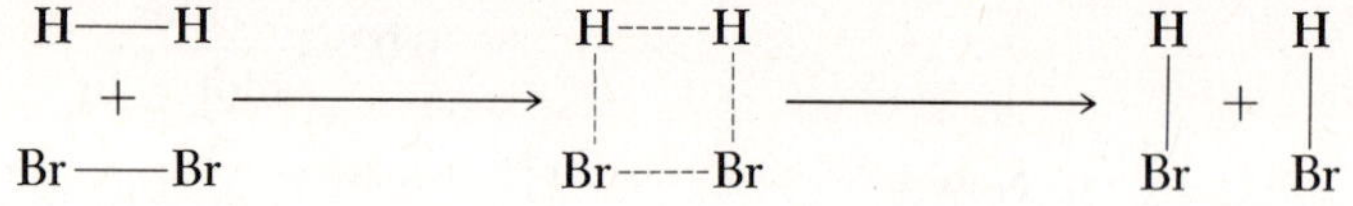

This mechanism would require a rate law that is first-order in both $[H_2]$ and $[Br_2]$. However, the experimentally observed rate law is consistent with the mechanism:

$$Br_2 \rightarrow 2\ Br \qquad \text{initiation}$$

$$\left.\begin{array}{l} Br + H_2 \rightarrow HBr + H \\ Br_2 + H \rightarrow HBr + Br \end{array}\right\} \text{propagation}$$

$$\left.\begin{array}{l} 2Br \rightarrow Br_2 \\ 2H \rightarrow H_2 \end{array}\right\} \text{termination}$$

In this scheme, H and Br are the **chain carriers.** The initiator is Br_2, but a different substance could be used instead. Note that the chain is maintained so long as one Br is produced for each Br consumed.

Some commercially important chain reactions involving organic molecules are those that produce the plastic polymers polyethylene, polystyrene, polypropylene (Tupperware), and polytetrafluoroethylene (Teflon®). The production of any of these, or of similar substances, proceeds by a mechanism of the following type: a free radical (R·) produced by the initiator (I) adds to one molecule of the olefin (monomer) to produce another reactive free radical in the propagation step. This larger free radical adds to another monomer unit and the process continues so that the polymer chain may become hundreds or thousands of monomer units long before the final termination steps.

$$I \rightarrow 2R\cdot \qquad \text{initiation}$$

$$\left.\begin{array}{ll} R\cdot + CH_2{=}CH_2 & \rightarrow RCH_2{-}CH_2\cdot \\ RCH_2{-}CH_2\cdot + CH_2{=}CH_2 & \rightarrow R(CH_2{-}CH_2)_2\cdot \\ R(CH_2{-}CH_2)_n\cdot + CH_2{=}CH_2 & \rightarrow R(CH_2{-}CH_2)_{n+1}\cdot \end{array}\right\} \quad \text{propagation}$$

$$\left.\begin{array}{ll} 2R\cdot & \rightarrow I \\ R(CH_2{-}CH_2)_n\cdot + R\cdot & \rightarrow R(CH_2{-}CH_2)_nR \\ 2R(CH_2{-}CH_2)_n\cdot & \rightarrow R(CH_2{-}CH_2)_{2n}R \end{array}\right\} \quad \text{termination}$$

The monomers for polyethylene, polystyrene, polypropylene, and polytetrafluoroethylene are $CH_2{=}CH_2$, $C_6H_5{-}CH{=}CH_2$, $CH_3{-}CH{=}CH_2$, and $CF_2{=}CF_2$.

In reactions of this general type there may also be present in the reaction mixture a foreign species that is very reactive towards free radicals. In these cases such a species may function as an inhibitor. The inhibitor (denoted here as HX) may react with a growing-chain radical to produce another radical (X·). If X· is relatively incapable of adding to a monomer, it will then represent a dead end so far as the polymerization is concerned, and this is how HX functions as an inhibitor. The action of the inhibitor explains the occurrence of an **induction period**—a period at the beginning of the reaction when no polymerization occurs. If the inhibitor is being consumed during this period, the "main" reaction is delayed and does not begin to accelerate until the inhibitor supply becomes exhausted.

Of all the chain reactions known, those involving nuclear reactions are the most spectacular, and according to their use, they present man with his greatest potential for progress or his greatest threat of destruction. In 1939, Otto Hahn and F. Strassmann, in Germany, reported that they had identified the element barium among the fragments produced by the bombardment of uranium 235 with neutrons. Excitement ran high among the physicists of the world and in 1942 Enrico Fermi successfully operated the first self-sustaining atomic pile under the west stand of the University of Chicago athletic field.

When uranium 235 is bombarded by neutrons, the first step is the formation of the unstable uranium 236, which immediately decays. More than 20 pairs of fragments have been identified. It has been established that approximately 200 MeV of energy is produced per fission, an enormous amount of energy, but more important than that, two or three neutrons are released per fission, the average being about 2.5.*

*One pound of uranium, or $\sim 10^{24}$ atoms, would theoretically produce 3×10^{13} J of work, an amount sufficient to keep a 100-W electric light bulb illuminated for 10,000 years.

OPTIONAL

If neutrons are available from any source, such as stray neutrons from cosmic rays or from spontaneous fission, they act as initiators. From this starting point, we can write a series of steps to represent one of the many possible nuclear chain reactions:

$$^{235}_{92}U + ^{1}_{0}n \rightarrow ^{236}_{92}U \tag{13.29}$$

$$^{236}_{92}U \rightarrow ^{141}_{56}Ba + ^{92}_{36}Kr + 3^{1}_{0}n + 200 \text{ MeV} \tag{13.30}$$

The neutrons produced from the uranium 236 not only will continue the propagation steps but will accelerate them because three neutrons are produced from each fission, which in turn was made possible by the capture of only one neutron. The termination step is the total depletion (or explosive removal) of the starting material.

13.6 Changing the Rate Constant

A rate constant is a constant at any given temperature but its value changes with temperature or the use of a catalyst. In actual practice, either of these methods may do more than just change the overall rate constant; in some cases rate changes by these methods are due to substantial alterations of the mechanism, especially when catalysts are employed.

Activation Barriers

Various theoretical models have been proposed to account for the effects of catalysts and temperature variation. A common feature of the modern views of this subject is the idea of a barrier to reaction. This view recognizes that reaction rates are in general substantially lower than what would be estimated from calculating the rate of collisions between molecules; not every collision leads to a transformation into the products. This discrepancy is usually visualized as a manifestation of a requirement that reacting systems must surmount an energy barrier. Such a barrier is understandable; as a chemical reaction proceeds, bonds must at least be partially broken before new ones can be entirely made, so that a "temporary" increase in potential energy must occur, even with exothermic reactions. The situation is represented pictorially in Figs. 13.10a for exothermic reactions and 13.10b for endothermic reactions.

The curves shown in Fig. 13.10 demonstrate the general relation between potential energy E and reaction progress. The situation is like rolling a ball over hilly terrain; if the ball is initially resting in a valley or depression of some sort, a certain minimum amount of energy must be applied to nudge it to the top of the hill so that it can roll into the next valley, even if the second valley is lower in altitude (or potential energy).

The highest point on a potential energy curve, such as that shown in Figs. 13.10a or 13.10b, corresponds to the arrangement of particles—atoms or ions—that has the largest amount of potential energy of any arrangement on the pathway from reactants to product. This particular geometrical configuration is called the **transition state,** and the energy of this configuration is represented as E^*. As shown in Fig. 13.10 the amount of energy required to overcome the barrier is the difference between E^* and the energy of the species on the left side of the chemical equation, E_L. This difference is referred to as the **activation energy** ΔE^*:

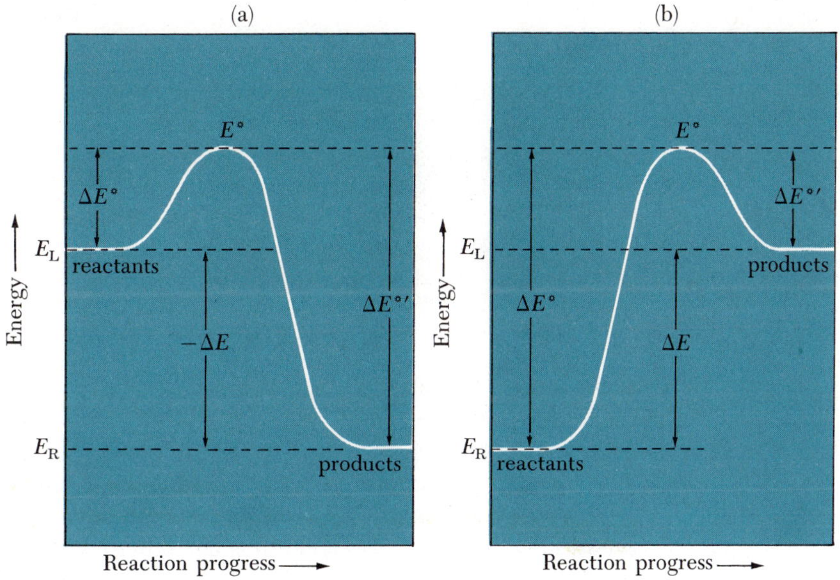

Figure 13.10 Energy barriers for (a) energy-releasing reaction and (b) energy-accepting reaction.

$$\Delta E^* = E^* - E_L \qquad (13.31)$$

Similarly, we see that the activation energy of the reverse reaction, $\Delta E^{*\prime}$ is given by

$$\Delta E^{*\prime} = E^* - E_R \qquad (13.32)$$

where E_R is the potential energy of the species on the right side of a chemical equation. The overall energy change, ΔE for the net reaction ($L \rightarrow R$) is seen to be the difference between the two activation energies:

$$\Delta E^* - \Delta E^{*\prime} = (E^* - E_L) - (E^* - E_R) = E_R - E_L = \Delta E \qquad (13.33)$$

As a specific example of these ideas let us consider the reaction of methyl bromide with hydroxide ion:

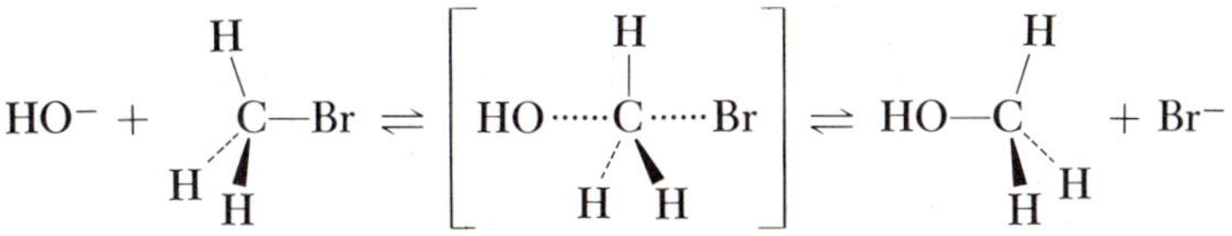
HO⁻ +
H
C—Br
H
H
HO······C······Br
H
H H
HO—C
H
H H
+ Br⁻

Figure 13.11 Formation of a transition state from methyl bromide and hydroxide ion.

As the hydroxide ion approaches the carbon atom from the back, it begins to form the carbon-oxygen bond while the carbon-bromine bond is breaking. At some point in this process a configuration is reached that has a higher energy than any other configuration encountered in the entire process. This point is called the **transition state.** Usually it is assumed that this configuration conforms approximately to the planar CH_3 arrangement implied in Fig. 13.11. From here, there are two routes available. The carbon-bromine bond may shorten again to its normal length, while the carbon-oxygen bond breaks, and the system will revert from the transition state to the configuration of the reactants. Alternatively, the carbon-oxygen bond may shorten and the system will proceed from the transition state to the configuration of the products. Thus, as the "old" carbon-bromine bond is being broken, and before the "new" carbon-oxygen bond substantially forms, the energy of the entire system increases and corresponds to moving up the energy barrier to the transition state

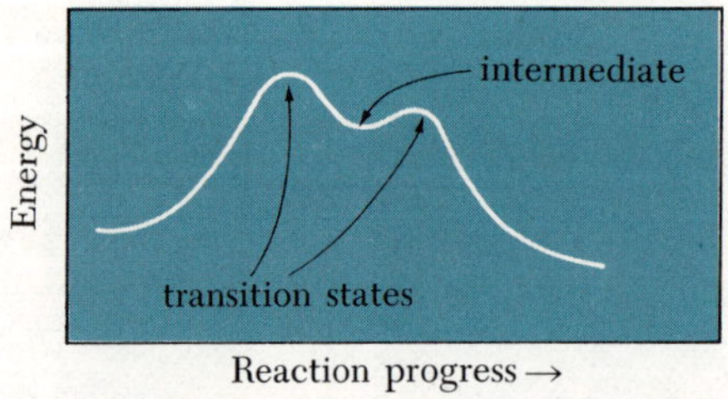

Figure 13.12 Relation between an intermediate state and a transition state.

at the "top of the hill." At this point the complex may "roll" either way; if it goes to the right, the new carbon-oxygen bond forms as the carbon-bromine bond breaks entirely, and the system rolls down the hill to the products.

We have frequently referred to an intermediate, defined as a species formed at some intermediate stage in a chemical transformation and eventually converted to product. How, then, does an intermediate differ from a transition state? An energy profile similar to Fig. 13.10 can be drawn showing the relation between an intermediate and the transition states of the two consecutive elementary reactions that lead to the intermediate and begin with it. This is shown in Fig. 13.12, where the lowest point of the dip represents the energy of the intermediate. The lower this minimum, the more stable the intermediate and the more likely the possibility that it can be "trapped," or isolated. If it is isolatable in principle, then the overall reaction can, in effect, be considered two separate consecutive reactions, each with its own transition state. In some cases, however, the dip is so slight that the distinction between the geometry of an intermediate and the transition state loses its significance.

The relation between the overall change in energy ΔE and the forward and reverse activation energies focuses attention on the fundamental relation between thermodynamics, or equilibrium, on the one hand and kinetics on the other. The thermodynamic parameter ΔE for a chemical reaction is intimately related to the equilibrium constant for that reaction, as the general approach of Chaps. 10 and 11 demonstrates. The rate of a reaction is determined, partly at least, by the value of its activation energy.

A related view of the relation between reaction rates and equilibrium was considered qualitatively in our introduction of equilibrium (Chap. 4). This view is simply that *at equilibrium the rates of forward and reverse reactions are equal.* Whether or not we have knowledge of the rate laws for the reaction viewed from right to left and from left to right, we know that the rates of these opposing processes are exactly equal and cancel each other when equilibrium is reached. This fact can be put in the form of a convenient quantitative relation by using the concept of a rate law.

To see how this comes about, let us consider as a specific example the following transformation, where the symbol "py" represents pyridine.

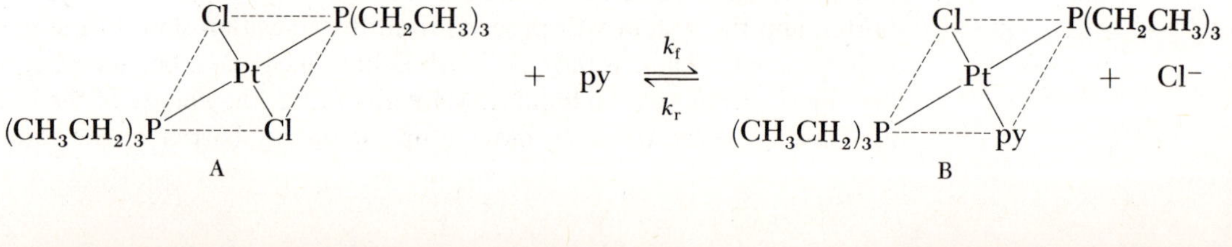

According to this scheme, the rates of forward and reverse reactions are given by

$$-\frac{\Delta[A]}{\Delta t} = k_f[A][py] = \frac{\Delta[B]}{\Delta t} \qquad \text{(forward)} \qquad (13.34)$$

$$-\frac{\Delta[B]}{\Delta t} = k_r[B][Cl^-] = \frac{\Delta[A]}{\Delta t} \qquad \text{(reverse)} \qquad (13.35)$$

Knowing that at equilibrium the rates of these opposing processes are equal, we can write

$$k_f[A]_{eq}[py]_{eq} = k_r[B]_{eq}[Cl^-]_{eq} \qquad (13.36)$$

where the subscript "eq" signifies that the indicated concentrations refer to values at equilibrium, since Eq. (13.36) is valid only for equilibrium. This equation can be rearranged to

$$\frac{[B]_{eq}[Cl^-]_{eq}}{[A]_{eq}[py]_{eq}} = \frac{k_f}{k_r} \qquad (13.37)$$

The left side of Eq. (13.37) is precisely the same as the value of the mass action expression Q at equilibrium. The value of Q at equilibrium is the equilibrium constant for the reaction; thus, we come to the interesting conclusion that *the equilibrium constant is the ratio of the rate constants for the forward and reverse reactions:*

$$K_{eq} = \frac{k_f}{k_r} \qquad (13.38)$$

The result embodied in Eq. (13.38) was derived here for the special case of a particular chemical reaction with a simple one-step mechanism; however, it can be shown that this relation between equilibrium constants and rate constants is valid for any chemical reaction, whether it is an elementary reaction or an overall reaction that proceeds by a complex mechanism. Example 13.5 shows the usefulness of the approach.

EXAMPLE 13.5 The rate of water formation from H^+ and OH^- at 25 °C is 1.4×10^{11} liter/(mol × sec). What is the rate constant for the dissociation of water at 25 °C?

SOLUTION

$$H^+ + HO^- \underset{k_r}{\overset{k_f}{\rightleftharpoons}} H_2O$$

$$K_{eq} = \frac{[H_2O]}{[H^+][HO^-]} = \frac{[H_2O]}{K_w}$$

$$K_{eq} = \frac{(55.5)}{10^{-14}} = 5.55 \times 10^{15}$$

$$\frac{k_f}{k_r} = \frac{1.4 \times 10^{11}}{k_r} = 5.55 \times 10^{15}$$

$$k_r = 2.5 \times 10^{-5}\ \text{sec}^{-1}$$

STUDY PROBLEM 13(e)

The rate constant for the oxidation of nitric oxide by oxygen,

$$2NO + O_2 \rightarrow 2NO_2$$

at 300 °C (conditions that might be found for exhaust gases from internal combustion engines) is $6.7 \times 10^5\ M^{-1}sec^{-1}$. If the rate constant for dissociation of nitrogen dioxide to form nitric oxide and oxygen at this temperature is 35 $M^{-1}sec^{-1}$, what is the equilibrium constant for the oxidation of nitric oxide at 300 °C?

Catalysts

A reaction is thermodynamically feasible if the ΔG is negative, but its rate may be so slow that it may appear not to occur at all. In such cases the use of a catalyst may be desirable. A **catalyst** is any substance that changes the rate of a reaction without being consumed by the reaction. Catalysts generally increase reaction rates, and are frequently referred to as promoters.* If a reaction produces a product that is itself a catalyst for the reaction, then the process is termed **autocatalysis.** Some catalysts, such as acids, are relatively effective for a large variety of reactions; others such as enzymes, vitamins, and hormones are spectacularly effective but mostly extremely specific. The last-mentioned substances act as catalysts for the biochemical reactions that are accelerated during the alteration of biological processes. Using a catalyst does not alter the thermodynamic state functions. Hence an equilibrium constant or the position of an equilibrium cannot be influenced by a catalyst.†

When chemical reactions that involve a catalyst are written, the name of the catalyst is customarily written above the arrow between the reactants and products. Some common examples are:

$$2KClO_3 \xrightarrow{MnO_2} 2KCl + 3O_2$$

$$2SO_2 + O_2 \xrightarrow{NO} 2SO_3$$

$$H_2C{=}CH_2 + H_2 \xrightarrow{Pt} H_3C{-}CH_3$$

$$CH_3CHO \xrightarrow{I_2} CH_4 + CO$$

$$H_2N{-}CO{-}NH_2 + H_2O \xrightarrow{\text{urease}} 2NH_3 + CO_2$$

$$(CH_3)_3\overset{+}{N}{-}(CH_2)_2{-}OOC{-}CH_3 + H_2O \xrightarrow[\text{cholinesterase}]{\text{acetyl-}} (CH_3)_3\overset{+}{N}{-}(CH_2)_2{-}OH + CH_3COOH$$

*There are cases in which a substance actually slows the rate of a reaction without being consumed. Such substances are usually termed *inhibitors* rather than *catalysts*. A common case of inhibition occurs when a substance can bind with an enzyme, preventing the molecule that normally reacts, or the substrate, from binding. The inhibition of enzymes has given much information about their binding sites and reaction mechanisms. It is often found that the inhibitor has a shape very much like the normal substrate (see the "lock-and-key" mechanism of enzymes in this section).

†Lowering the activation energy by using a catalyst permits the reactants to cross the barrier more easily, thus increasing the rate of the reaction. Note that it also affords the same advantage to the products in the case of an equilibrium; hence the equilibrium constant is unaffected in the presence of a catalyst. Both k_f and k_r are increased to the same degree. If the forward rate is increased by a factor of 1000, the reverse reaction is also increased by a factor of 1000.

A simple example of how a catalyst can provide an alternative mechanism with a lower activation energy is given by the oxidation of thallous ion by ceric ion,

$$2Ce^{4+} + Tl^{+} \rightleftharpoons 2Ce^{3+} + Tl^{3+} \tag{13.39}$$

Although the equilibrium for this reaction lies very much to the right, the reaction as written occurs very slowly (cf. the example given previously for acetylene). This is mainly because an easy pathway has not been provided. Note that for the reaction to occur as a one-step process would require the collision of three positive ions, and as we have mentioned, termolecular reactions are improbable. Manganous ions, Mn^{2+}, however, catalyze the reaction, since the following sequence of steps can occur:

$$Ce^{4+} + Mn^{2+} \rightarrow Ce^{3+} + Mn^{3+}$$
$$Mn^{3+} + Ce^{4+} \rightarrow Ce^{3+} + Mn^{4+}$$
$$Mn^{4+} + Tl^{+} \rightarrow Mn^{2+} + Tl^{3+}$$

The reaction can now proceed exclusively by a mechanism consisting of bimolecular collisions. This mechanism has a greater overall rate than the one-step route (without a catalyst). Note two important points: (1) The overall stoichiometry of the above sequence (that is, the sum of the three reactions after the manganese ions are cancelled from both sides) equals the overall stoichiometry of the reaction in the absence of the catalyst; and (2) the manganous ions are consumed in the first step but regenerated by the third step, and thereby qualify as catalysts.

In the areas of research dealing with catalysts, perhaps the most rapidly growing and most important concerns enzymatic action. Without the more than 500 different enzymes now known, life as we know it could not exist. Enzymes are much more effective than inorganic catalysts, frequently 100,000 to 1 million times more efficient. They are also very selective, and some are absolutely specific, operating for only one substance in only one reaction; an example is the hydrolysis of urea catalyzed by the enzyme urease. The rates of enzyme reactions are of paramount importance in biochemistry; reactions that are improperly controlled may cause serious metabolic disorders or diseases. Kinetic studies of enzyme reactions lead not only to an understanding of normal and abnormal biochemical processes but also to an elucidation of enzyme action.

Although the mechanisms of catalytic activities are complex and not fully understood in many cases, an appealing explanation has been offered for some enzymes; this is called the **lock and key theory.** The theory speculates that the geometrical shapes of the reacting substance, which is the substrate, and the enzyme are such that they fit together as a key fits into a specific lock. It is proposed that the enzyme and substrate molecules become attached to each other, with a specific mutual orientation, by various attractive interactions. The attractions include those due to the polarities of the chemical bonds within the molecules, hydrogen bonding, or the presence of an inorganic metal ion that binds the substrate. Within the enzyme there is a particular area, called the **active site,** which will then be situated adjacent to that portion of the substrate at which the principal reaction is to occur. It is the specific mutual orientation mentioned above that guarantees this proximity of active site and reaction site. After the activated substrate has reacted, the products will be different from the original substrate, and will be liberated subsequently from the enzyme. The regenerated enzyme is then free to act on other molecules of substrate.

In the scheme shown in the accompanying diagram for catalysis by an enzyme, the substrate bonds in a specific way to the enzyme, forming the enzyme-substrate complex. This complex then reacts to form products that cannot easily bind to the enzyme because of geometrical or bond polarity restrictions, or both.

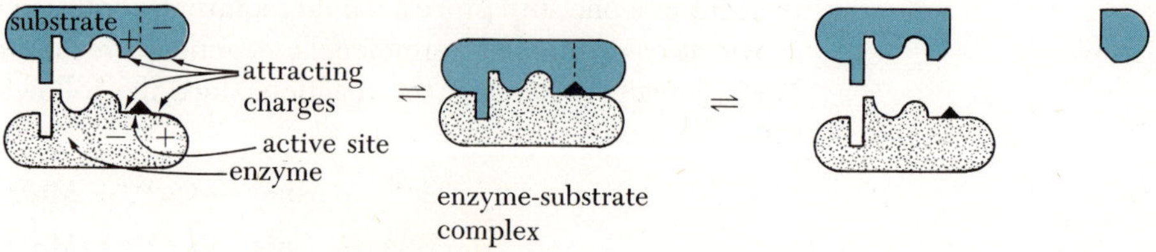

In 1913, Michaelis and Menten hypothesized that in enzyme reactions the reacting substance, or the substrate, and the enzyme first reversibly form an enzyme-substrate complex, and that this complex then proceeds to the product or products and the liberated enzyme in a rate-determining step. This second step is also reversible, but in many model enzyme reactions this equilibrium constant is so enormously large that only the forward direction need be considered; this step can be considered essentially irreversible. This scheme can be symbolically represented by

$$\mathrm{E} + \mathrm{S} \underset{k_{-1}}{\overset{k_1}{\rightleftharpoons}} \mathrm{ES} \xrightarrow[\text{slow}]{k_2} \mathrm{P} + \mathrm{E} \tag{13.40}$$

$$\text{rate} = \frac{\Delta[\mathrm{P}]}{\Delta t} = k_2[\mathrm{ES}] \tag{13.41}$$

where E, S, ES, and P stand for enzyme, substrate, enzyme-substrate complex, and product.

The steady state approximation is useful here in simplifying the kinetic analysis. The approximation, when applied to this case, states that the enzyme-substrate complex is sufficiently reactive to maintain its concentration at a very low level. Should there be a tendency for the concentration to increase, the rate of its decomposition would correspondingly increase, preventing any buildup. Similarly, if there should be a tendency for the concentration to decrease, the rate of the decomposition would decrease, offsetting the tendency. The influences tend to maintain the concentration of the enzyme-substrate complex at a low, constant level. This means that the rate at which the complex is forming equals its rate of consumption. Thus it is convenient to consider $\Delta[\mathrm{ES}]/\Delta t$ equal to zero, and $[\overline{\mathrm{ES}}]$ to mean the steady state concentration of ES. Furthermore, since substrate is consumed during the overall reaction, the steady state approximation is valid only in the early stages of reaction. Hence, in the considerations that follow, we are interested only in initial conditions, like initial rates and initial concentrations. Now, from the steady state approximation, the rate of formation of the complex equals its rate of consumption:

$$k_1[\mathrm{E}][\mathrm{S}] = k_{-1}[\overline{\mathrm{ES}}] + k_2[\overline{\mathrm{ES}}] \tag{13.42}$$

The concentration of free enzyme, [E], equals its initial concentration, $[E]_0$, minus the value bound in the complex, [ES], or

$$[E] = [E]_0 - [ES] \tag{13.43}$$

For the steady state,

$$[E] = [E]_0 - [\overline{ES}] \tag{13.44}$$

Substituting [E] given by Eq. (13.44) into Eq. (13.42) gives, after $[\overline{ES}]$ is factored on the right side,

$$k_1([E]_0 - [\overline{ES}])[S] = (k_{-1} + k_2)[\overline{ES}] \tag{13.45}$$

$$\frac{([E]_0 - [\overline{ES}])[S]}{[\overline{ES}]} = \frac{k_{-1} + k_2}{k_1} = K_m \tag{13.46}$$

The equality of the last two expressions in Eq. (13.46) constitutes the definition of K_m, called the **Michaelis-Menten constant.** The value of K_m is extremely useful in characterizing an enzyme. The variable k_2 is the first-order rate constant for the conversion of enzyme-substrate complex to products. We can write

$$\frac{\Delta P}{\Delta t} = k_2[ES] \tag{13.47}$$

$$\frac{\Delta[P]/\Delta t}{[ES]} = k_2 \tag{13.48}$$

Thus, k_2 is the ratio of the rate of change of concentration of product to the concentration of ES. But since [ES] is simply the concentration of active sites, k_2 can be interpreted as the number of substrate molecules converted to product molecules per second per enzyme active site. It represents a measure of how effective the enzyme is in catalyzing the reaction with the given substrate, and is referred to as the **turnover number.** (See Table 13.2.)

To use experimental data for determining K_m and k_2, one measures the reaction rate as a function of the initial substrate concentration $[S]_0$, and the enzyme concentration $[E]_0$, deriving a plot as shown in Fig. 13.13. The maximum rate, beyond

TABLE 13.2 Some Representative Turnover Numbers

Enzyme	*Turnover Number*, sec^{-1}
Papain	10
Peroxidase	10
Chymotrypsin	10^2
Ribonuclease	10^2
Dehydrogenases	10^3
Fumarase	10^3
Acetylcholinesterase	10^4
Urease	10^4
Carbonic anhydrase	10^5
Catalase	10^7

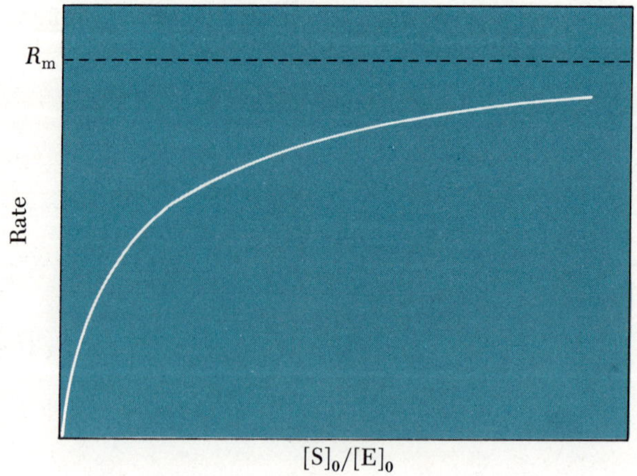

Figure 13.13
A plot of rate vs the $[S]_0/[E]_0$ ratio.

which increasing the substrate concentration no longer increases the rate, is called R_m. The relation between substrate concentration, rate, and the parameters R_m and K_m is given by

$$\frac{[S]}{R} = \frac{[S]}{R_m} + \frac{K_m}{R_m} \tag{13.49}$$

This can be derived from Eq. (13.46), using

$$R_m = k_2[E]_0$$

EXAMPLE 13.7 A study of the rate of hydrolysis of urea catalyzed by urease gave the following data.

Initial rate (mol/(liter × sec))	$[\text{Urea}]_0$, M
0.975	0.0167
1.03	0.0333

If the initial concentration of urease was 1.12×10^{-4} mol/liter in each experiment: (a) What is the Michaelis-Menten constant? (b) How many molecules of urea per second is each active site of urease capable of hydrolyzing?

SOLUTION From Eq. (13.49), we obtain

(a)
$$\frac{[S]}{R} = \frac{[S]}{R_m} + \frac{K_m}{R_m}$$

$$\frac{3.33 \times 10^{-2}}{1.03} = \frac{3.33 \times 10^{-2}}{R_m} + \frac{K_m}{R_m}$$

$$\frac{1.67 \times 10^{-2}}{0.975} = \frac{1.67 \times 10^{-2}}{R_m} + \frac{K_m}{R_m}$$

Taking the difference between the last two equations gives

$$\frac{3.33 \times 10^{-2}}{1.03} - \frac{1.67 \times 10^{-2}}{0.975} = \frac{(3.33 \times 10^{-2}) - (1.67 \times 10^{-2})}{R_m}$$

$$R_m = \frac{1.09 \text{ mol}}{\text{liter sec}}, \text{ or } 1.09\ M \text{ sec}^{-1}$$

Substituting this value of R_m into either of the two original equations gives $K_m = 1.91 \times 10^{-2}$ mol/liter, or $1.91 \times 10^{-2}\ M$.

(b)

$$R_m = k_2[\text{E}]_0$$

$$k_2 = \frac{1.09}{1.12 \times 10^{-4}}$$

$$= 9730 \text{ molecules/sec}$$

Hence, each active site of urease is capable of hydrolyzing 9730 molecules of urea per second, or on a macroscopic scale, each mole of active sites can hydrolyze 9730 mol urea per second.

STUDY PROBLEM 13(f)

For the same concentration of enzyme ($1.12 \times 10^{-4}\ M$) and urea ($0.0167\ M$ and $0.0333\ M$) as in Example 13.7, the initial rates at a slightly higher temperature were 1.95 and 2.00, respectively. At this temperature, what are the K_m and k_2 values?

Catalysts, in general, can be divided into two types, homogeneous and heterogeneous. **Homogeneous catalysis** occurs when both the catalyst and the reactants are in the same phase. Homogeneous catalysis may occur in either the liquid or the vapor phase. An important example of a liquid phase process occurs in the production of certain modern plastics.

Two additional examples of homogeneous catalysis are the acid-catalyzed hydrolysis of an ester in water,

$$\text{R—}\overset{\text{O}}{\overset{\|}{\text{C}}}\text{—O—R}' + H_2O \overset{H^+}{\rightleftharpoons} \text{R—}\overset{\text{O}}{\overset{\|}{\text{C}}}\text{—O—H} + \text{R}'\text{OH}$$

and the explosive combination of gaseous H_2 and Cl_2 catalyzed by water vapor. In **heterogeneous catalysis,** the catalyst is usually the surface of a solid. Two examples of heterogeneous catalysis are the Pt-catalyzed hydrogenation of unsaturated hydrocarbons with H_2, and the iron-catalyzed formation of ammonia from N_2 and H_2. The glass walls of ordinary laboratory equipment may frequently act as a catalyst, and in such cases the rate of the reaction can be significantly increased by introducing glass wool into the apparatus. An interesting aspect of surface catalysis, frequently referred to as **contact catalysis,** is that it may lead to "zero-order" reactions, that is, reactions whose rates are independent of the concentration of any reactant.

It is by no means uncommon for a small amount of a foreign substance to deactivate, or "inhibit," a catalyst grossly, if not completely. Such substances are called

Spotlight

The catalytic hydrogenation of unsaturated natural fats and oils is an important process in the food industry. Hydrogenation raises the melting points of unsaturated acids, and the process is called hardening. For example, oleic acid, $CH_3(CH_2)_7CH{=}CH(CH_2)_7COOH$, constitutes 60% of peanut oil and melts at 14 °C; after hydrogenation of the double bond, stearic acid, mp 69 °C, is produced. The unsaturated liquid fats and oils are useful for cooking and salad oils, whereas the hydrogenated solid products are useful in making soaps, lard substitutes, and certain margarines. Reduction of the reactive unsaturated centers eliminates odors and improves shelf life because the development of rancidity is associated with unsaturation. Unfortunately, removing these reactive centers also means that the human biochemical processes cannot easily degrade these fats; they are subsequently stored in the body.

catalytic poisons, and some of the most powerful biological poisons known act simply by inhibiting key enzymes. The inactivity may either be temporary or permanent. Temporary inhibition usually results from a preference of the catalyst for the poisonous substance over the normal substrate. This type of poisoning can usually be reversed if the concentration of the normal reactant is increased. Permanent poisoning, on the other hand, usually results from an essentially irreversible chemical reaction with the catalyst. Sulfur compounds are especially effective as permanent poisons for the common hydrogenation catalysts.

Temperature

A spectacular example of the effect of temperature on reaction rates is shown by the first-order decomposition of acetone dicarboxylic acid:

$$HOOC{-}CH_2{-}CO{-}CH_2{-}COOH \rightleftharpoons CH_3{-}CO{-}CH_3 + 2CO_2 \quad (13.50)$$

At 0 °C the rate constant is $2.5 \times 10^{-5}\text{sec}^{-1}$ and at 60 °C it is 5480 sec^{-1}. This means that the half-life at 0 °C is about 8 hr, but at 60 °C it is only 12 sec!

At present, there are several theories that try to describe the exact relation among the rate constant, the activation energy, and temperature for any reaction. These theories represent different approaches, but generally rate enhancements due to temperature increases are viewed in terms of increasing the fraction of molecules with sufficient energy to surmount the barrier, as shown in Fig. 13.10. Two very common theories are presented herewith.

Arrhenius Theory

Svante Arrhenius, in 1889, was the first to propose an analytical relation between rate constant and temperature,* namely,

*Arrhenius proposed that the rate at which ln k changes relative to temperature is given by

$$\frac{d \ln k}{dt} = \frac{E_a}{RT^2}$$

$$\int d \ln k = \frac{E_a}{R} \int T^{-2}\, dT$$

$$\ln k = \frac{E_a}{R}\left(-\frac{1}{T}\right) + C \text{ (constant of integration)}$$

$$k = Ze^{-E_a/RT}$$

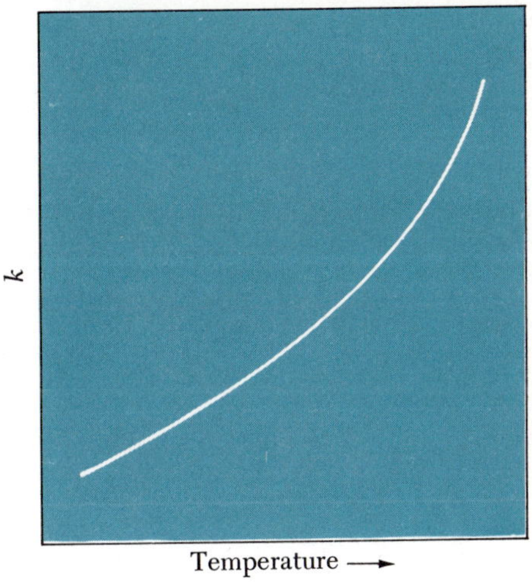

Figure 13.14
Arrhenius dependence of rate on temperature.

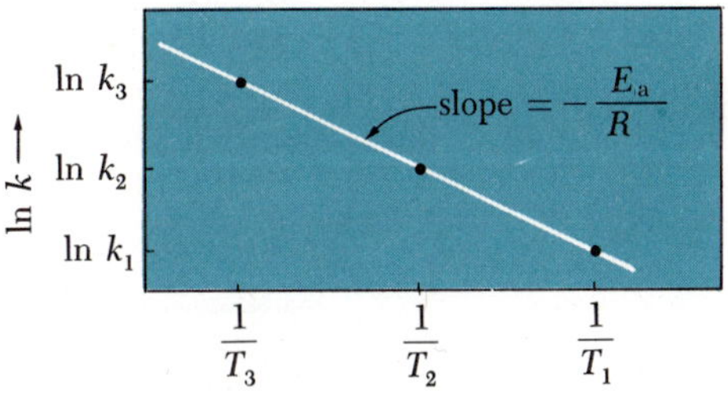

Figure 13.15 Graphical determination of Arrhenius activation energy. In this plot, T_3 is a higher temperature than T_1; thus, $1/T_3 < 1/T_1$.

$$k = Ze^{-E_a/RT} \tag{13.51}$$

where E_a is a constant called the **Arrhenius activation energy,** and Z is also a constant. A plot of how the rate constant for a "typical" reaction varies with temperature is shown in Fig. 13.14 and is called an **Arrhenius temperature dependence.** The rule of thumb that the rate of a reaction is doubled or trebled for each 10-deg rise in temperature originates from such a plot.

Taking the logarithm of both sides of Eq. (13.51) gives

$$\ln k = -\frac{E_a}{RT} + \ln Z \tag{13.52}$$

Equation (13.52) has the form of a straight line, and a plot of ln k vs $1/T$ (Fig. 13.15) yields $-E_a/R$ as the slope. It turns out that E_a is usually not a true constant and is somewhat temperature-dependent. Usually, however, E_a varies only very slightly with temperature, and the range of T normally covered in preparing a plot is small enough that a straight line is obtained.

Note that another frequently used equation can be derived from Eq. (13.52). If two different temperatures (T_1 and T_2) are used in Eq. (13.52), then one equation can be subtracted from the other to give Eq. (13.53), as follows:

$$\ln k_2 = -\frac{E_a}{RT_2} + \ln Z$$

$$-\left(\ln k_1 = -\frac{E_a}{RT_1} + \ln Z\right)$$

$$\ln k_2 - \ln k_1 = -\frac{E_a}{RT_2} + \frac{E_a}{RT_1}$$

$$\ln \frac{k_2}{k_1} = \frac{E_a}{R}\left(\frac{1}{T_1} - \frac{1}{T_2}\right) = \frac{E_a}{R}\left(\frac{T_2 - T_1}{T_1 T_2}\right) \tag{13.53}$$

From Eq. (13.53), the activation energy can be obtained from rate constants at two temperatures. Substitution back into Eq. (13.53) allows one to estimate k for any other temperature.

EXAMPLE 13.8 The formation of carbon disulfide and hydrogen sulfide from methane and sulfur has been found to obey second-order kinetics between 550 °C and 625 °C:

$$CH_4 + 2S_2 \rightleftharpoons CS_2 + 2H_2S$$

From the following data, estimate the rate constant for 600 °C.

T, °C	k, $M^{-1}\text{sec}^{-1}$
550	1.1
625	6.4

SOLUTION From the two rate constants at the two different temperatures, we can obtain E_a from Eq. (13.53). Note that the temperature must be expressed in degrees Kelvin, and that in this case if R is given as 1.99 cal/mol deg^{-1}, then E_a will be computed in calories.

$$\ln\frac{6.4}{1.1} = \frac{E_a(898 - 823)}{(1.99)(823)(898)}$$

$$\ln 5.8 = 1.76 = E_a(5.1 \times 10^{-5})$$

$$E_a = 34.5\frac{\text{kcal}}{\text{mol}}$$

Substituting this value of E_a back into Eq. (13.53) gives

$$\ln\frac{k}{1.1} = \frac{(3.45 \times 10^4)(873 - 823)}{(1.99)(823)(873)}$$

$$\ln\frac{k}{1.1} = 1.2$$

$$\frac{k}{1.1} = 3.3$$

$$k = 3.6\frac{\text{liter}}{\text{mol sec}}(\text{or } M^{-1}\text{sec}^{-1})$$

STUDY PROBLEM 13(g)

Estimate the rate constant for the reaction examined in Example 13.8 at 500 °C.

The observation that reaction rates generally increase with increasing temperature is explicable on the basis of the kinetic molecular theory. We shall see (Chap. 16) that for a given temperature the distribution of molecular velocities is given by the Maxwell distribution law. From this law, it is found that an increase in temperature

Spotlight

A typical example of an Arrhenius dependence of rate on temperature is shown in Figure 13.15. However, not all chemical reactions show this dependence. Some examples of non-Arrhenius dependence are shown in figures (a) and (b). Explosions generally have the dependence shown in (a), where the rate gradually increases with temperature until T_{ex} is reached, at which point the overall rate increases very rapidly and produces an explosion. The non-Arrhenius dependence of rate on temperature for the reaction

$$NO + NO \rightleftharpoons N_2O_2$$

$$N_2O_2 + O_2 \rightleftharpoons 2NO_2$$

is shown in (b). The overall rate actually decreases as the temperature increases, because the above equilibrium shifts to the left as the temperature increases.

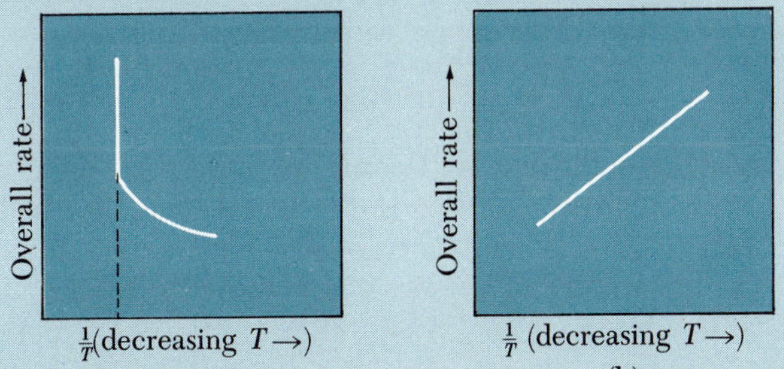

(a) (b)

increases the fraction of molecules that has a kinetic energy equal to or greater than the activation energy. Thus, there is an increase in the number of molecules that can react per unit time, increasing the reaction rate.

While it may be true that increasing the temperature must increase the k's of elementary reactions, there are some exceptions known for the k's of overall reactions. Consider, for example, an overall reaction that can symbolically be represented by

$$A \underset{k_{-1},\ \text{fast}}{\overset{k_1,\ \text{fast}}{\rightleftharpoons}} B$$

$$B \xrightarrow[\text{slow}]{k'} C$$

$$\frac{\Delta[C]}{\Delta t} = k[A]$$

where $k = k'K_{eq}$. Although increasing the temperature would increase all three elementary rate constants, it is possible for k_{-1} to increase considerably more than the other two rate constants. Hence, K_{eq} could decrease more than k' increases, and the overall first-order rate constant k may ultimately be lowered in spite of the higher temperature.

Absolute Rate Theory

The theoretical approach of the absolute rate theory probably has the most widespread use. This theory is sometimes called the transition state theory because it pivots on the concept of the transition state. It views the transition state almost as if it were an authentic state in the usual thermodynamic sense, at least to the extent that a law of chemical equilibrium can be applied to it. Ideally the absolute rate theory is treated by the methods of statistical mechanics, but a plausible case can be made on more intuitive grounds. The reasoning is as follows.

The rate of a reaction equals the number of activated complexes that pass over the potential barrier in a unit period; that is, the reaction rate is simply the rate at which the transition state passes on to products. Viewed in these terms, the reaction rate depends only on the "concentration" of the reacting system that is momentarily at the transition state, denoted here as $C^\ddagger$, and on the rate or frequency at which the transition state decomposes to products. Thus,

$$\text{rate} = (C^\ddagger)(\text{decomposition frequency of transition state}) \tag{13.54}$$

If, as suggested above, we treat $C^\ddagger$ in terms of an assumed equilibrium between the reactants and the transition state, then we can also define an equilibrium constant $K^\ddagger$. We can then define a free energy of activation from

$$\Delta G^\ddagger = -RT \ln K^\ddagger \tag{13.55}$$

analogous to the thermodynamic expression for a "normal" equilibrium. From the second law of thermodynamics, we can also define an entropy of activation, $\Delta S^\ddagger$, and an enthalpy of activation, $\Delta H^\ddagger$ since

$$\Delta G^\ddagger = \Delta H^\ddagger - T\Delta S^\ddagger \tag{13.56}$$

The rate constant can be expressed in transition state theory as

$$k = \frac{k_b T}{h} e^{-\Delta H^\ddagger/RT} e^{\Delta S^\ddagger/R} \tag{13.57}$$

Equation (13.57) gives the temperature dependence of the rate constant in terms of the fundamental constants k_b (the Boltzmann constant), h (Planck's constant), R (the gas constant), and the activation parameters $\Delta S^\ddagger$ and $\Delta H^\ddagger$. If this equation is rearranged somewhat, we can see how the parameters $\Delta S^\ddagger$ and $\Delta H^\ddagger$ can be extracted from experimental data. Dividing both sides of Eq. (13.57) by T and then taking the logarithm of each side, we obtain

$$\ln\left(\frac{k}{T}\right) = -\frac{\Delta H^\ddagger}{RT} + \ln\left(\frac{k_b}{h}\right) + \frac{\Delta S^\ddagger}{R} \tag{13.58}$$

This equation has the familiar form $y = mx + b$, where $(1/T)$ is viewed as the variable x, the expression $\ln (k_b/h) + \Delta S^\ddagger/R$ acts as the intercept b, the function $\ln (k/T)$ is the dependent variable y, and the ratio $-\Delta H^\ddagger/R$ is the slope m of the corresponding line. Accordingly, if the rate constant has been obtained experimentally at a variety of temperatures for a particular reaction, then a plot of 2.303 $\log (k/T) = \ln (k/T)$ vs the inverse of temperature gives a straight line with slope $(-\Delta H^\ddagger/R)$ and intercept $2.303 \log (k_b/h) + \Delta S^\ddagger/R$. Since k_b and h are fixed constants, the value of $\Delta S^\ddagger$ for the reaction of interest can be obtained directly from the intercept of the plot. This technique is outlined in the following example.

EXAMPLE 13.9 The rate of the reaction between N, N-dimethylaniline and methyl iodide to form phenyl trimethyl ammonium iodide,

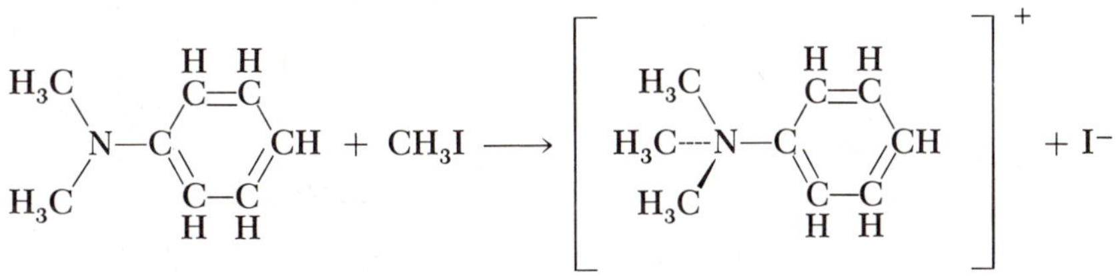

is 2.84×10^{-4} liter/(mol $\times$ sec) at 24.8 °C, 7.96×10^{-4} liter/(mol $\times$ sec) at 40.1 °C and 2.53×10^{-3} liter/(mol $\times$ sec) at 60.0 °C. Estimate (a) the enthalpy of activation and (b) the entropy of activation.

SOLUTION The values of $\Delta H^{\ddagger}$ and $\Delta S^{\ddagger}$ can be obtained by plotting those data in the form of $2.303 \log (k/T)$ plotted against $1/T$. Note that T must be in degrees Kelvin. (a) The plot is shown below. The slope $= -\Delta H^{\ddagger}/R$.

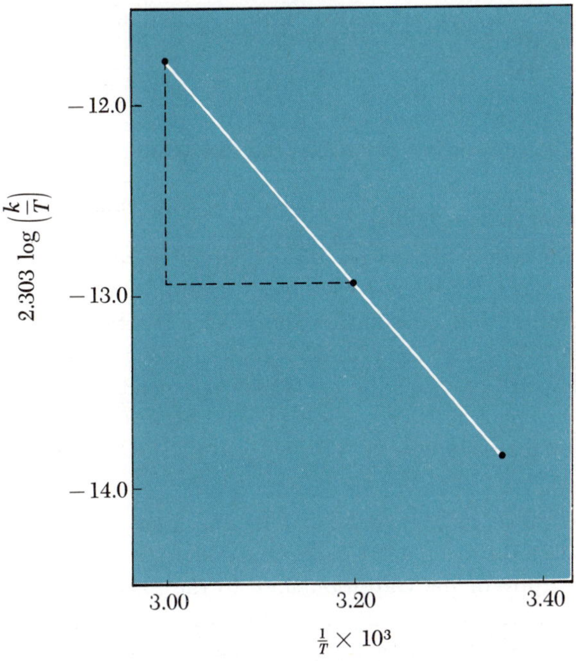

$$\Delta H^{\ddagger} = -R\left(\frac{\Delta Y}{\Delta X}\right)$$

$$= -R\left(\frac{-1.13}{0.200 \times 10^{-3}\ \text{deg}^{-1}}\right)$$

$$= -1.99\ \text{cal/(deg} \times \text{mol)} \times -5650\ \text{deg}^{-1}$$

$$= 11{,}243\ \text{cal/mol or } 11.2\ \text{kcal/mol}$$

(b) The intercept contains $\Delta S^\ddagger$. The equation for the line that has been plotted can be used to obtain the intercept:

$$y = mx + b$$

$$2.303 \log\left(\frac{2.84 \times 10^{-4}}{298}\right) = -5650 \text{ deg}^{-1}\left(\frac{1}{298 \text{ deg}}\right) + b$$

$$b = -13.86 + 18.96 = 5.10$$

but the intercept $= 2.303 \log (k_b/h) + \Delta S^\ddagger/R$:

$$5.10 = 2.303 \log\left(\frac{1.3806 \times 10^{-16}}{6.66262 \times 10^{-27}}\right) + \frac{\Delta S^\ddagger}{R}$$

$$\Delta S^\ddagger = (5.10 - 23.76)\,1.99 \frac{\text{cal}}{\text{deg mol}}$$

$$= -37.13 \frac{\text{cal}}{\text{deg mol}}$$

The units cal/deg are called entropy units (eu):

$$\Delta S^\ddagger = -37.1 \frac{\text{eu}}{\text{mol}}$$

Note that any value of y from the line could be used to obtain this result.

STUDY PROBLEM **13(h)**

The rate constant for the oxidation of nitric oxide by oxygen, $2NO + O_2 \rightarrow 2NO_2$ is $6.63 \times 10^5\ M^{-1}\text{sec}^{-1}$ at 327 °C and $6.52 \times 10^5\ M^{-1}\text{sec}^{-1}$ at 372 °C, and the rate constant for the reverse reaction is 83.9 $M^{-1}\text{sec}^{-1}$ at 327 °C and 407 $M^{-1}\text{sec}^{-1}$ at 372 °C. (a) What are the enthalpies and entropies for the forward reaction and for the reverse reaction at 372 °C? (b) What is the free energy of activation for each direction? (c) What is the equilibrium constant for the oxidation of NO by O_2 at 327 °C? at 372 °C?

The values of $\Delta H^\ddagger$ and $\Delta S^\ddagger$ obtained for a given reaction provide valuable insights into the mechanism of the reaction, to the extent that they relate the enthalpy and entropy of the transition state to the enthalpy and entropy of the reactants. For example, a highly negative value of $\Delta S^\ddagger$ would imply a large loss of randomness in progressing from reactants to transition state, giving a highly "restricted" transition state. The value of $\Delta H^\ddagger$, which is sometimes referred to as an activation energy, provides a measure of the enthalpy difference between the initial and transition states; however, it is important to note that $\Delta H^\ddagger$ is not the same as the activation energy E_a that is obtained according to the Arrhenius theory. To the extent that the two theories are directly comparable, it can be shown by the standard methods of calculus that $\Delta H^\ddagger$ and E_a are related by the expression

$$\Delta H^\ddagger = E_a - RT \tag{13.59}$$

Whereas the "activation energy" $\Delta H^{\ddagger}$ is obtained from a plot of ln (k/T), the activation energies, E_a and ΔE, referred to in Eqs. (13.52) and (13.58) are obtained from plots using ln k and ln $(k/\sqrt{T})$, respectively.

SUMMARY

A general approach to the study of the rates of chemical reactions shows that by changing the concentrations of the reactants, a rate law can be deduced. For a single-term rate law such as rate $= k[H_2][I_2]$, the order of the reaction is the sum of the exponents (in this case, two). The molecularity of a reaction is the number of molecules that react in the slowest, or rate-determining, step of a reaction. The way in which a reaction proceeds from reactants to products, the mechanism, must be consistent with the rate law. A knowledge of the chemical properties of the reactants is generally needed in addition to the rate law to distinguish among several possible mechanisms.

Graphical methods can be used for determining the reaction order. The rate constant is also of value; a known rate constant can be used for purposes such as ^{14}C dating. Parallel, or competitive, reactions are reactions that take place by two or more paths simultaneously, and consecutive reactions are those that take place in a series of steps. Chain reactions generally proceed by several rapid steps, which include both parallel and consecutive paths.

The concept of activation energy, an amount of energy required for a reaction to occur even if the overall reaction is exothermic, is essential to understanding reaction rates. The activation energy can be deduced from the change in a rate constant as a function of temperature change. Two methods of determining activation energy are the Arrhenius theory and the absolute rate theory. A catalyst can lower the activation energy of a process, giving a higher rate. Enzymes are catalysts that are often specific. A "lock and key" analogy is useful in discussing enzyme specificity. The equilibrium constant for a process is not changed by a catalyst; the forward and reverse reactions are accelerated to the same extent.

Understanding kinetics and mechanisms is essential to understanding biochemistry, as well as inorganic, organic, and nuclear chemistry.

STUDENT CHECKPOINTS

After studying this chapter, the student should be able to:

1. Determine whether a reaction is first-order or second-order by concentration vs time plots.
2. Determine the order of reaction from the change in rate as the initial concentration of a reactant is changed.
3. Determine the concentrations of product and of unreacted reactants at a given time for a first-order or second-order reaction.
4. Determine rate constants and express them in the proper units.
5. Relate the equilibrium constant to the forward and reverse rate constants.
6. Explain a catalyst as it affects the forward and reverse rates and the equilibrium constant in an equilibrium reaction.
7. Differentiate between an elementary reaction and an overall reaction.
8. Define *consecutive mechanism, parallel mechanism,* and *chain mechanism.*

9. Relate activation energy to rates of reaction.
10. Discuss the Arrhenius and absolute rate theory methods for obtaining activation parameters.
11. Discuss the effects of concentration, temperature, and catalysts on reaction rates, explaining how these effects are important in obtaining desired products in a reasonable amount of time in both synthetic and natural reactions.

EXERCISES

13.1.* Which of the following statements about rate laws are true?
(a) The rate law can be obtained directly from the balanced equation for any reaction.
(b) Concentrations of the species in the activated complex are found in the rate law.
(c) The power to which a particular concentration is raised is determined directly from the coefficient of that species in the balanced chemical equation.

13.2. Which of the statements are true?
(a) The mechanism of a reaction is established when the balanced equation for the reaction is established.
(b) A knowledge of the reaction products is essential to establishing a mechanism.
(c) Rate laws are essential in establishing a mechanism.
(d) An established mechanism must be consistent with all the data known about a reaction (reaction products, rate law, structure of the products if there are isomers, and the like).

13.3.* Which of the following statements are true?
(a) Catalysts speed up a reaction but do not take part in the reaction.
(b) Catalysts make a reaction more favorable by increasing the equilibrium constant.
(c) Enzymes are catalysts.
(d) Catalysts increase the reverse rate as well as the forward rate.

13.4. Define (a) an elementary reaction; (b) an overall reaction; (c) reaction order; (d) mechanism; (e) catalyst.

13.5.* What is the general equation for each of the following?
(a) The rate law of a first-order reaction (that is, $A \rightarrow B$).
(b) The rate law of a second-order reaction (that is, $A + B \rightarrow C$).
(c) The half-life of a first-order reaction.
(d) The half-life of a second-order reaction of the type $2A \rightarrow B + C$.

13.6 If the half-life of a first-order reaction is 6.9 min, and the initial concentration of the reactant is 0.10 M, what is the rate constant?

13.7.* If the half-life of a second-order reaction of the type $2A \rightarrow B + C$ is 6.9 min, and the initial concentration of A is 0.10 M, what is the rate constant?

13.8. Which is a straight line for a first-order reaction: the plot of concentration against time or the plot of the log of concentration against time?

13.9.* (a) What plot is linear for a second-order reaction of the type $A + B \rightarrow C + D$? (b) What is the intercept of such a plot?

13.10. If you have concentration vs time data for a first-order reaction but you do not know the initial concentration (c_0), how can you determine c_0?

13.11.* Technetium can be used to locate tumors because the gamma rays emitted by the isotope ${}^{99m}_{43}Tc$ are measurable even when the emitting nuclei are deep within the body. (In the symbol for the element, "m" stands for *metastable*, not the most stable form of ${}^{99}_{43}Tc$.) The process of emission is

$$ {}^{99m}_{43}\mathrm{Tc} \longrightarrow {}^{99}_{43}\mathrm{Tc} + 0.143\ \mathrm{MeV} $$

Given the data on the number of gamma rays emitted from a sample of ${}^{99m}_{43}Tc$, deduce (a) whether the process is first-order; (b) what the half-life of the decay of ${}^{99m}_{43}Tc$ is; and (c) what percentage of the radioactive isotope would be lost if the time required for preparing and administering a dose of ${}^{99m}_{43}Tc$ were 2 hr.

Time, hr	*Gamma rays emitted per sec*
0	10,000
4	6,300
8	4,000
12	2,500
16	1,575
20	990

13.12. A technique for determining very small amounts, or trace levels, of elements is called neutron activation analysis. The sample is bombarded with neutrons. Some atoms "capture" the neutrons, and radioactive species are formed. Ruthenium can be analyzed by this method. Ruthenium has several isotopes. One isotope undergoes the reaction shown herewith:

$$ {}^{100}_{44}\mathrm{Ru} + {}^{1}_{0}\mathrm{n} \longrightarrow {}^{101m}_{45}\mathrm{Rh} + \beta \xrightarrow{k} {}^{101}_{45}\mathrm{Rh} + 0.157\ \mathrm{MeV} $$

where the "m" means that this is not the most stable form of ${}^{101}_{45}Rh$. The counts measured per second at an energy of 0.157 MeV are given here. Deduce (a) whether the process is first-order; (b) what the half-life for the decay is; (c) how long the technician should wait to dispose of the sample if 10 counts per second is deemed safe.

Time, hr	*Counts per sec*
2	987
4	975
10	928
28	836
34	804
98	533

13.13* The isotope ${}^{100}_{44}Ru$ constitutes 12.62% of ruthenium. (a) If the number of counts per second of a standard sample containing 10.0 mg of ruthenium is 1987 at one hour, how much ruthenium is in the sample used to obtain the data in Exercise 13.12? (b) What is the concentration in units of parts per million (micrograms of ruthenium per gram of sample) if the sample weighs 1 g?

13.14. The isotope ${}^{90}_{38}Sr$ is a component of fallout that can replace calcium in bones and teeth. A collection of "baby teeth" has been maintained since the mid-1950s to monitor the effect of ${}^{90}_{38}Sr$. The rate constants for the reaction

$$ {}^{90}_{38}\mathrm{Sr} \xrightarrow{k_1} {}^{90m}_{39}\mathrm{Y} + \beta(0.546\ \mathrm{MeV}) \xrightarrow{k_2} {}^{90m}_{40}\mathrm{Zr} + \beta(2.95\ \mathrm{MeV}) + \gamma \xrightarrow{k_3} {}^{90}_{40}\mathrm{Zr} $$

(where β is a beta particle, an electron, and γ is a gamma ray, energy emission but no particle) are $k_1 = 2.47 \times 10^{-2}\ \mathrm{yr}^{-1}$, $k_2 = 2.23 \times 10^{-1}\ \mathrm{hr}^{-1}$, and $k_3 = 0.86\ \mathrm{sec}^{-1}$. If the beta particles at 0.546 MeV measured from a tooth in 1956 occurred at a rate of 100 min^{-1}, what would the measured count be in 1980?

13.15.* Plutonium forms extremely toxic compounds. One of the isotopes of plutonium produced in "breeder" reactors decays to less toxic, although still harmful, uranium:

$$ {}^{239}_{94}\mathrm{Pu} \xrightarrow{k_1} {}^{235}_{92}\mathrm{U} + \alpha \xrightarrow{k_2} {}^{231}_{90}\mathrm{Th} + \alpha \xrightarrow[\text{decay}]{\text{further}} $$

where the α particle is ${}_{2}^{4}He^{2+}$. If the "safe" level of a sample of buried ${}_{94}^{239}Pu$ is 0.01% of the original amount, and $k_1 = 2.84 \times 10^{-5}$ yr^{-1}, how long will it be before the waste material can be safely unearthed?

13.16. An archaeologist wants to determine by carbon dating whether a basket fashioned from reeds was made during the reign of Tutankhamen of Egypt ("King Tut") about 1350 B.C. or during the reign of Cleopatra, around 50 B.C. (a) Use the data in Table 13.1, the fact that the normal count for CO_2 in the atmosphere is 12.5 beta particles per minute per gram of carbon, and the fact that the measured count is now 8.37 counts min^{-1} per gram of C to determine the date at which the basket was made. (b) How accurate must this number be to distinguish between the two dates?

13.17.* The dissociation of $CrCl^{2+}$ to Cr^{3+} and Cl^- in water has a rate constant of 2.85×10^{-7} sec^{-1}. The process is first-order. If the chloride ion is removed from solution as it is formed so that it cannot recombine, how long will it take for 1.00 g of $CrCl^{2+}$ to become 90% dissociated?

13.18. The k for combination of Cr^{3+} and Cl^- to form $CrCl^{2+}$ is 3.0×10^{-8} liter/(mol $\times$ sec). If the $CrCl^{2+}$ is removed from solution as it is formed, how long will it take for one-half of the Cr^{3+} to be converted to $CrCl^{2+}$ with both Cr^{3+} and Cl^- at 1 M? Use the information given in Exercise 13.17 and this question to compute the equilibrium constant for the reaction:

$$Cr^{3+} + Cl^- \rightleftharpoons CrCl^{2+}$$

13.19.* A chemist decided to make Cr^{3+} from $CrCl^{2+}$ by removing the chloride ion with Hg^{2+}. The rate of dissociation of Cl^- from $CrCl^{2+}$ was increased by a factor of 200,000. The net reaction is

$$CrCl^{2+} + Hg^{2+} \underset{k_r}{\overset{k_f}{\rightleftharpoons}} Cr^{3+} + HgCl^+$$

The rate of the reverse reaction is 1.3×10^{-9} mol/(liter $\times$ sec). Use the data in Exercise 13.17 and this question to calculate the equilibrium constant.

13.20. What are the units for the indicated rate constant in the following examples?

(a) $RuCl^+ \underset{k_r}{\overset{k_f}{\rightleftharpoons}} Ru^{2+} + Cl^-$, forward reaction, rate $= k_f[RuCl^+]$.

(b) Reverse reaction, rate $= k_r[Ru_{(aq)}{}^{2+}][Cl^-]$.

(c) $2H_2 + 2NO \rightarrow 2H_2O + N_2$, rate $= k[H_2][NO]^2$.

13.21.* What are the units for the indicated rate constant in the following examples?

(a) ${}_{53}^{136}I \rightarrow {}_{54}^{136}Xe + \beta$, rate $= k[{}_{53}^{136}I]$.

(b) $2NO + Cl_2 \rightarrow 2ClNO$, rate $= k[Cl_2][NO]^2$.

(c) $NO_2 + 2HCl \rightarrow H_2O + NO + Cl_2$, rate $= k[HCl][NO_2]$.

13.22. In which of the following cases can the reaction as written be an elementary reaction?

(a) $H_2 + I_2 \rightarrow 2HI$, rate $= k[H_2][I_2]$.

(b) $H_2 + Br_2 \rightarrow 2HBr$, rate $= \dfrac{k[H_2][Br_2]^{3/2}}{k'[HBr] + k''[Br_2]}$

(c) $Cr^{2+} + Co^{3+} \rightarrow Cr^{3+} + Co^{2+}$, rate $= k[Cr^{2+}][Co^{3+}]$.

13.23.* In which of the following cases can the reaction as written possibly be an elementary reaction?
(a) $2I^- + H_2O_2 + 2H^+ \rightarrow 2H_2O + I_2$, rate $= k[I^-][H_2O_2] + k'[H^+][I^-][H_2O_2]$.
(b) $2BH_3 \rightarrow B_2H_6$, rate $= k[BH_3]^2$.
(c) $Cl_2 + CO \rightarrow Cl_2CO$, rate $= k[Cl_2]^{3/2}[CO]$.

13.24. For the reaction $PCl_3 + Cl_2 \rightarrow PCl_5$ in the gas phase, the rate doubles if the pressure of PCl_3 is doubled and also doubles if the pressure of Cl_2 is doubled. Write the rate law.

13.25.* For the reaction of chromium(II) with iron(III) compounds in acidic aqueous solution, the results in the accompanying chart are found for the reduction of a particular compound, with both reactants at 1.0 *M* concentration. (a) What is the rate law for this redox reaction? (b) What are the units of the rate constant?

[Cr(II)]	[Fe(III) compound]	*Relative rate*
1.0 *M*	1.0 *M*	1.0
0.5 *M*	1.0 *M*	0.5
0.5 *M*	0.5 *M*	0.25

13.26. The reaction of Cr(VI) with As(III) in mildly acidic aqueous solution is

$$8H^+ + 2HCrO_4^- + 3As(OH)_3 \rightleftharpoons 2Cr^{3+} + 3H_3AsO_4 + 5H_2O$$

The rate of $HCrO_4^-$ disappearance doubles when either the concentration of $As(OH)_3$ or $HCrO_4^-$ is doubled but is increased by a factor of 8 when the acid concentration is doubled. Write the rate law for the forward reaction.

13.27.* It is often desirable for an organic chemist to make aldehydes ($R{-}\overset{\overset{\displaystyle O}{\|}}{C}{-}H$) from alcohols ($RCH_2OH$) by using an oxidizing agent such as chromic acid. Formaldehyde, useful in laboratory preservation of tissue samples, can be obtained from methanol by this reaction:

$$6H^+ + 3CH_3OH + 2H_2CrO_4 \rightleftharpoons 3H{-}\overset{\overset{\displaystyle O}{\|}}{C}{-}H + 2Cr^{3+} + 8H_2O$$

The accompanying table gives the results of changing the methanol and chromic acid concentrations at constant acid concentration (3.5 *M* H_2SO_4).

$[CH_3OH]$	$[H_2CrO_4]$	*Relative rate of* $H\overset{\overset{\displaystyle O}{\|}}{C}H$ *formation*
0.010	0.010	1.0
0.020	0.010	2.0
0.010	0.005	0.5
0.020	0.005	1.0

What is the rate law at constant acid concentration? Determine the value of the rate constant at this acid concentration if the half-life with both reactants at $0.010\ M$ is 1.79×10^4 sec.

13.28. When the loss of chloride ion from $[Co(NH_3)_5Cl]^{2+}$ is monitored as a function of acid concentration, the rate increases with increasing pH to give the rate law

$$\frac{-\Delta[[Co(NH_3)_5Cl]^{2+}]}{\Delta t} = k[[Co(NH_3)_5Cl]^{2+}] + k'[[Co(NH_3)_5Cl]^{2+}][OH^-]$$

for this overall reaction:

$$H_2O + [Co(NH_3)_5Cl]^{2+} \rightarrow [Co(NH_3)_5(H_2O)]^{3+} + Cl^-$$

What type of mechanism is this?

13.29.* The rate law for the formation of $FeCl^{2+}$ from Fe^{3+} and Cl^- in aqueous solution is

$$\frac{\Delta[FeCl^{2+}]}{\Delta t} = k_1[Fe^{3+}][Cl^-] + k_2[Fe^{3+}][Cl^-][OH^-]$$

where $k_1 = 9.4\ M^{-1}sec^{-1}$ and $k_2 = 1.1 \times 10^4\ M^{-2}sec^{-1}$. At pH = 1.0, what fraction of the $FeCl^{2+}$ is formed by the direct combination of Fe^{3+} and Cl^-?

13.30 The reaction scheme here illustrated can be used to make polyethylene. The molecule R—O—O—R is a peroxide, where R is generally an alkyl group.

RO—OR → RO·

RO· + H₂C=CH₂ → RO—CH₂—CH₂·

2 RO—CH₂—CH₂· → RO—CH₂—CH₂—CH₂—CH₂—OR

RO—CH₂—CH₂· + H₂C=CH₂ → RO—CH₂—CH₂—CH₂—CH₂·

2 RO· → R—O—O—R

RO· + RO—CH₂—CH₂· → RO—CH₂—CH₂—OR

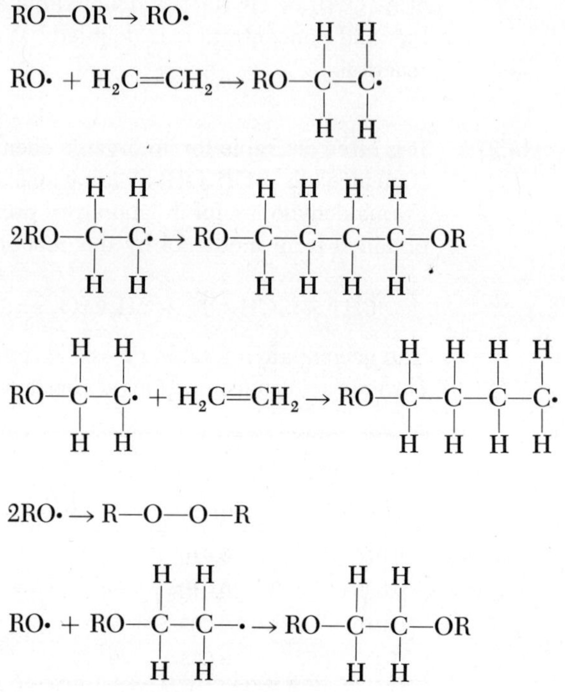

What type of reaction is this? What are each of the steps called? To be useful as a packaging film, the polyethylene molecules must be composed of very many ethylene units. What does this imply about the optimum concentrations of ROOR and $CH_2{=}CH_2$?

13.31.* Describe a reaction sequence for formation of Teflon® (polytetrafluoroethylene).

13.32. Draw energy barrier diagrams for a reaction in which ΔE is negative and for a case in which ΔE is positive. Give examples of a reaction of each type.

13.33.* How does a catalyst affect the rate of reaction? How does a catalyst affect an equilibrium constant? Use your answers to these questions and the definition of the equilibrium constant in terms of rates (13.18) to determine whether a catalyst makes forward and reverse rates faster by the same amount in an additive or multiplicative manner.

13.34. Platinum catalyzes the hydrogenation of many molecules, but urease catalyzes only the decomposition of urea. Why might this be?

13.35.* It is proposed that metal ions in enzymes have two different functions. It is held that in some cases they help the protein keep its shape and in other cases they are necessary components of the "active site." What is an "active site"?

13.36. The enzyme chymotrypsin causes the hydrolysis of groups of the type R—C(=O)—X to acids of the type R—C(=O)—OH. For a particular species of this type

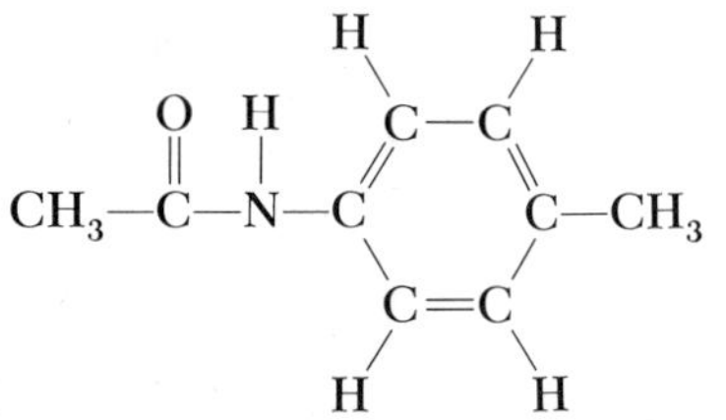

at pH 8.0, the turnover number is 8.7×10^{-2} sec^{-1} and the rate data in the accompanying chart were obtained with an initial chymotrypsin concentration of 1.0×10^{-4} *M*.

Initial rate (liter mol^{-1}sec^{-1})	[Substrate]$_0$ (*M*)
3.78×10^{-6}	1.00×10^{-2}
5.27×10^{-6}	2.00×10^{-2}
6.56×10^{-6}	4.00×10^{-2}

Calculate the Michaelis-Menten constant and the maximum rate under these conditions (R_m).

13.37.* Enzymes called ribonucleases decompose molecules of ribonucleic acids into the nucleotide units of which they are composed (Chap. 8). The final stage in this process is governed by Michaelis-Menten kinetics (Eq. 13.49). For the production of cytidine monophosphate from the intermediate stage (cyclic cytidine monophosphate) at pH = 7.0 and an enzyme concentration of 1.0×10^{-4} *M*, the data as shown in the accompanying chart were obtained. What is the value of the Michaelis-Menten constant and what is the turnover number?

Initial rate ($M^{-1}sec^{-1}$)	[Substrate]$_0$ (*M*)
4.10×10^{-4}	0.010
4.72×10^{-4}	0.020
5.08×10^{-4}	0.040

13.38. The production of cytidine monophosphate from RNA intermediates by ribonuclease from cattle pancreas (called bovine pancreatic ribonuclease) described in Exercise 13.37 has also been studied for pH 5.8. The data in the chart were obtained for the enzyme at 1.0×10^{-4} *M*.

Initial rate ($M^{-1}sec^{-1}$)	[Substrate]$_0$ (*M*)
1.92×10^{-4}	0.010
1.96×10^{-4}	0.020
1.98×10^{-4}	0.040

(a) Determine the values of K_m, R_m, and the turnover number from these data.

(b) Using the results of 13.37, which variable is affected most for this reaction, the Michaelis-Menten constant or the turnover number?

13.39.* The combination of CF_3 "free radicals"

$$2CF_3 \rightarrow C_2F_6$$

occurs with rate constants of 5.9×10^9 $M^{-1}sec^{-1}$ at 25 °C and 7.1×10^9 $M^{-1}sec^{-1}$ at 60 °C. (a) Use the Arrhenius form for the rate constant as a function of temperature to obtain the activation energy. (b) Predict the rate for −79 °C, where the product can be condensed to a liquid. (c) Would you term CF_3 a reactive species?

13.40. Dinitrogen pentoxide decomposes in a first-order manner to NO_2 and O_2 at rates shown in the accompanying table. Use these data to determine (a) the energy of activation using the Arrhenius equation; and (b) the enthalpy and entropy of activation from absolute rate theory.

k, sec^{-1}	*T*, °C
1.95×10^{-5}	20
7.25×10^{-5}	30
2.70×10^{-4}	40
9.05×10^{-4}	50
2.90×10^{-3}	60

13.41.* Using the Arrhenius equation, show what the activation energy of a reaction would be if its rate doubles when the temperature is increased by 10 °C. What is assumed in your calculation about the nonexponential term?

13.42. A reaction being run at 20 °C has an Arrhenius activation energy of 20 kcal/mol. How much will the rate at 20 °C be increased if a catalyst reduces the activation energy by 10 kcal/mol?

13.43.* Which of the following gas phase reactions would you expect to have the higher entropy of activation? Explain.

$$CO + NO_2 \rightarrow CO_2 + NO, \text{ rate} = k[CO][NO_2]$$

$$N_2O_5 \rightarrow 2NO_2 + \tfrac{1}{2}O_2, \text{ rate} = k[N_2O_5]$$

13.44. Biochemical reactions depend on the kinetic stability, or slowness in reacting, of many species that are not thermodynamically stable. Otherwise, proteins, nucleic acids, carbohydrates, and so on would become CO_2, H_2O, NO_3^-, and other very stable species. How does this explain the need for enzymes as specific catalysts?

13.45.* Catalysts are involved in maintaining the level of sugar in the blood, but the bicarbonate-carbonate buffer system in the blood is maintained without catalysts. What is the different feature in the chemical reactions that causes enzymes to be required in one case but not the other?

ANSWERS TO SELECTED EXERCISES

13.1 b. 13.3 c and d. 13.5 (a) rate = $k[A]$; (b) rate = $k[A][B]$; (c) $t_{1/2} = (2.303 \log 2)/k$; (d) $t_{1/2} = 1/C_0k$. 13.7 1.5 M^{-1}min^{-1}. 13.9 a plot of ln ([A]/[B]) against time; (b) ln $[A]_0/[B]_0$. 13.11 (a) yes; (b) 6 hr; (c) 21%. 13.13 (a) 5.0 mg; (b) 5000 ppm. 13.15 3.24×10^5 yr. 13.17 8.08×10^6 sec or 94 days. 13.19 4.4×10^7. 13.21 (a) sec^{-1}; (b) M^{-2}sec^{-1}; (c) M^{-1}sec^{-1}. 13.23 b. 13.25 (a) rate = k[Cr(II)][Fe(III)]; (b) M^{-1}sec^{-1}. 13.27 rate = $k[CH_3OH][H_2CrO_4]$; 5.60×10^{-3} M^{-1}sec^{-1}. 13.29 more than 99% by direct combination. 13.31 as in 13.30 but with $F_2C{=}CF_2$ in place of $H_2C{-}CH_2$. 13.33 (a) by decreasing the energy of activation; (b) it has no effect; (c) the rate constants are both multiplied by the same number. 13.35 An "active site" is the place in the enzyme at which reaction of a substrate occurs. 13.37 (a) 3.6×10^{-3} M; (b) 5.6 molecule/sec. 13.39 (a) 1.04 kcal/mol; (b) 2.3×10^9 M^{-1}sec^{-1}; (c) yes, it can undergo very fast reactions. 13.41 (a) 11 kcal/mol; (b) that is has no temperature dependence. 13.43 the first reaction, since two molecules must be brought together in the activated complex. 13.45 The activation energy required for ionic reactions in aqueous solution (the buffer system) is much lower than the activation energy for reactions involving covalent bond changes of molecules such as glucose.

ANSWERS TO STUDY PROBLEMS

13(a) 5.29×10^{-5} mol/(liter $\times$ min). 13(b) About A.D. 1500. An error of ± 0.1 counts/min per gram of carbon corresponds to about ± 70 yr. 13(c) 3.62 sec. 13(d) 65.2% unreacted isobutyl bromide; 14.1% isobutyl ether, and 20.6% isobutylene. 13(e) 1.9×10^4. 13(f) 8.32×10^{-4}; 18,303 molecules/sec. 13(g) 0.28 M^{-1}sec^{-1}. 13(h) (a) -1.48 kcal/mol and -35.8 eu/mol and 24.9 kcal/mol and -9.56 eu; (b) 20.0 kcal/mol and 30.7 kcal/mol; (c) 7.9×10^3, 1.60×10^3.

14

Solids

14.1 Condensed Phases

Condensed Phases vs Gases

The solid and liquid states commonly are referred to as condensed phases. They differ in many important ways from each other, but their common differences relative to the gaseous state are probably the most dramatic. Individual species of a gas can be viewed as moving about largely unhindered; there seems to be almost no correlation between the motions of different particles, and the interactions among the particles seem only transient. The particles of a gas, especially at temperatures substantially above the boiling point, spend most of the time moving freely through space, at large distances from other particles, and experiencing no strong forces of attraction or repulsion with other particles.

By contrast, molecular or ionic species in condensed phases are close together,[*] continuously experiencing forces of attraction—and to a lesser extent, repulsion. It is the forces of attraction that are responsible for the very existence of a condensed phase. These forces account for the relatively high density of solids and liquids, that

[*]The molecules or ions in a condensed phase are so close together that they are in direct physical contact. The concept of touching is one that is straightforward in the macroscopic world, but more elusive on the submicroscopic scale. A common approach is to *assign* molecular dimensions on the basis of the assumption that individual species in condensed phases are situated as close together as possible, essentially "touching." Packing them closer would require expending a great deal of energy to offset repulsive forces that operate if the individual particles are forced extremely close together.

is, the close packing of molecules and ionic species in condensed phases. Also resulting from the strong and persistent forces between the particles of a condensed phase and the associated close spacing is the restricted character of the motion of particles in a solid or liquid. The motion is especially constrained in the case of crystalline solids, in which movement is largely restricted to oscillations about fixed positions. A rough analogy to the different submicroscopic characteristics of gases and condensed phases is the comparison between the following two cases: (1) the eight or so players constituting the opposing offensive and defensive lines at the line of scrimmage in an American football game, and (2) the same players scattered and running randomly about the entire football field. Considering the amount of space that must be viewed in order to see all eight players, the "density" of players is much higher in case (1). Aside from an occasional collision, interactions between players in case (2) are absent and the movement is relatively free and random. By contrast, not only do the players at the line of scrimmage feel attractive interactions ("trapping" and maybe a little "holding") and repulsive forces (blocking), but also their motions are highly restricted and mutually correlated.

Gases are highly disordered collections of matter, and can accordingly be called high-entropy systems. The degree of order is far greater in condensed phases. Even liquids have a substantial level of short-range order (Chap. 16). Molecular and ionic species in a liquid tend, on the average, to line up relative to one another in specific ways that correspond to low-energy configurations—electrically positive regions next to electrically negative regions, for instance. A normal liquid, however, has essentially no long-range order. Knowing the geometrical orientation of a molecule at one point in a liquid tells one relatively little, if anything, about the orientation of another molecule a nanometer away. By contrast, true solids have both short-range and long-range order.

Interactions in Condensed Phases

Various kinds of attractive interactions are responsible for holding condensed phases together. For *covalent network crystals*, a very special type of solid, simple covalent bonds hold the substance together in its condensed form. For *metallic* solids, a very special kind of delocalized electron bonding is responsible for maintaining the integrity of the solid. The attractive forces in virtually all other types of condensed-phase systems can be described by a hierarchy of types of *electrostatic* attractions. These are shown schematically in Fig. 14.1. In this figure ions are represented by circles and molecules by an oblong symbol. Figure 14.1a represents the simple electrostatic attraction between two ions of opposite charge. Figure 14.1b represents the attraction of an ion for the oppositely charged portion of a molecule with a permanent electric dipole moment; δ^+ and δ^- represent the positive and negative ends of the dipole (see Sec. 7.5). In Fig. 14.1c, the α^- and α^+ symbols represent the oppositely charged ends of a dipole moment induced in a nonpolar molecule by the proximity of a nearby ion; the resulting induced dipole is attracted at one end by the charge of the ion responsible for inducing the dipole. Figure 14.1d represents the attractive interaction between oppositely charged ends of two molecules with permanent electric dipoles. In Fig. 14.1e is represented the induction of a dipole

Spotlight

The interactions between ions in solution and the polar regions of molecules in solution can in some cases be highly specific, favored for certain types of ions and largely nonexistent for others. For example, there are certain complex organic molecules that strongly associate (bind?) with specific metal ions and not others. One way in which this can occur is with a molecule that has polar groups and a cyclic, or doughnut-shaped, structure. The "hole" of this structure is just the right size for certain types of metal ions; larger ions don't fit in the hole, and smaller ions have their centers of positive charge too far away from the negative ends of the molecular dipoles for maximum attractive interactions. This mechanism of metal ion selectivity appears to operate in biological systems, allowing cell membranes, for example, to select sodium or potassium ions. For example, K^+ is ideally suited to fit into the hole of a valinomycin molecule, which is a species that has been identified with ion transport in living systems.

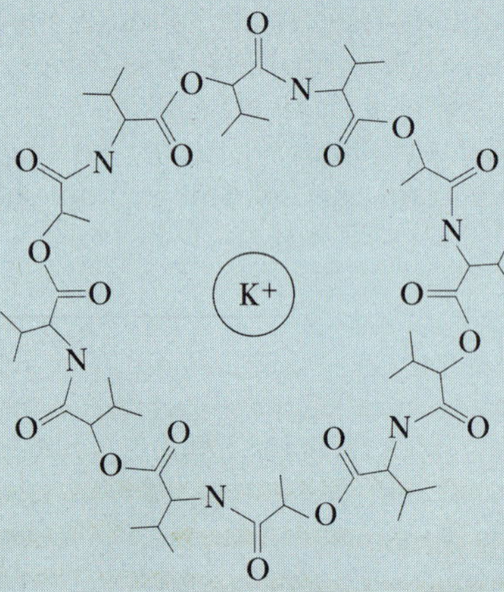

K^+-valinomycin complex

Symbolism

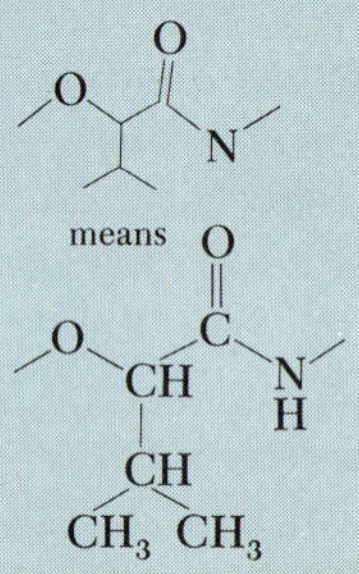

Such naturally occurring systems have recently been mimicked by chemists who have synthesized and studied a class of compounds known as crown ethers.

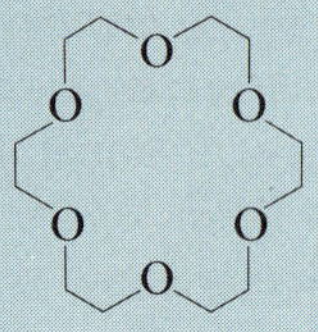

crown ether

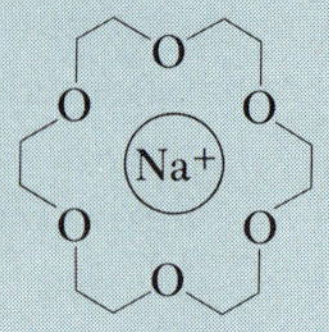

Na^+-crown ether complex

Symbolism

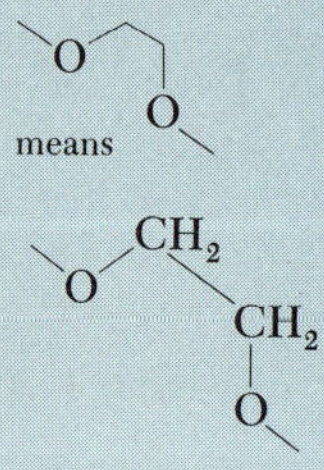

moment (with induced separated charges of α^+ and α^-) in a nonpolar molecule by the presence of a nearby molecule with a permanent dipole moment (with separated charges δ^+ and δ^-). The final part of Fig. 14.1 represents what are called *London*, or *dispersion*, forces between two nonpolar molecules.

We can visualize the basis of dispersion forces as follows. The various fluctuating influences to which a nonpolar molecule in a condensed phase is subjected cause a momentary displacement of the electron distribution of the molecule from what the normal distribution would be, or from the distribution that renders the molecule

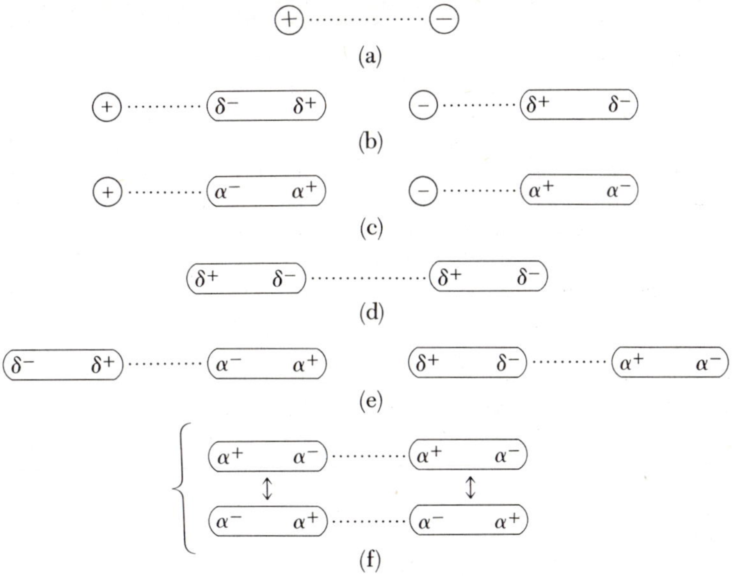

Figure 14.1 Various types of non-covalent attractive interactions. The dotted lines show the main attractive forces. (a) Simple ion—ion interaction. (b) Ion–permanent-dipole interaction. (c) Ion–induced-dipole interaction. (d) Permanent-dipole–permanent-dipole interaction. (e) Permanent-dipole–induced-dipole interaction. (f) Dispersion interaction (fluctuating mutually induced dipole–mutually induced dipole).

nonpolar on the average. This instantaneous displacement gives rise to an instantaneous dipole moment in the molecule (shown in terms of the symbols α^+ and α^- in Fig. 14.1f). This instantaneous dipole moment can induce an instantaneous dipole moment in some nearby molecule, and for an instant there is a mutual attraction between the opposite ends of these two dipoles. Very quickly the distortion that gave rise to the initial dipole moment in the first molecule fluctuates, like the electron distribution in the second molecule, and some new, often mutually attractive, situation arises. Because of the rapid and nearly random character of the fluctuations in electron distributions, the individual molecules remain nonpolar on the average. Nevertheless, these very short-lived distortions give rise to mutual attractions that can be quite substantial in magnitude. Such dispersion forces are very important in systems with large numbers of electrons. Other things being equal, the ease with which the fluctuations in the electronic distribution occur and the susceptibility to mutual polarizations (mutually induced dipole moments) increase as the total number of electrons in a molecule is larger. Dispersion forces are especially important in the condensed phases of substances for which other types of interactions, for example, ion–dipole, are absent. Even in the presence of ions or permanent dipoles, however, dispersion forces are significantly important in any situation in which the molecules are close enough to each other to experience any kind of mutual attraction.

The term *van der Waals interactions* generally refers to the combination of interactions represented in parts d, e, and f of Fig. 14.1, that is, 14.1 with ions removed. Van der Waals forces, including dispersion forces, are also present in systems containing ions; however, interactions involving ions, including those between ions, are typically stronger.

One should remember that various combinations of interactions of the various types can occur simultaneously in a specific situation. An example is hydrogen

Spotlight

The tendency for large molecules (usually with large dispersion forces) to be more likely to exist as solids at a given temperature than smaller molecules is used to advantage in the preparation of crystalline derivatives. It often happens in the laboratory that one wishes to identify a liquid sample, of which only a small quantity may be available. One useful property for identifying substances is the melting point. While the sample of interest may be a liquid at room temperature, it may be possible to convert it to a crystalline derivative, the melting point of which can be measured and compared with those of known compounds. For example, a simple ketone, like acetone, CH_3COCH_3, can be converted to a phenylhydrazone.

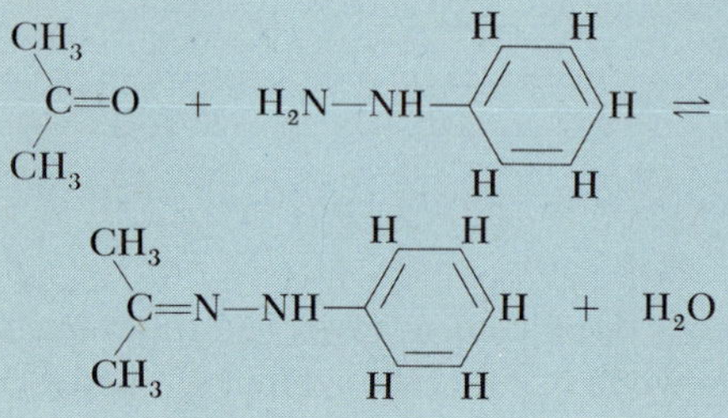

The acetone phenylhydrazone has a much higher molecular weight than acetone, and its correspondingly larger van der Waals interactions (including dispersion forces) lead to a much higher melting point (27 °C vs −95 °C), which is readily measured and compared with the melting points of suspected derivatives.

bonding (Sec. 8.3), which would generally be described as largely of the permanent dipole–permanent dipole type,

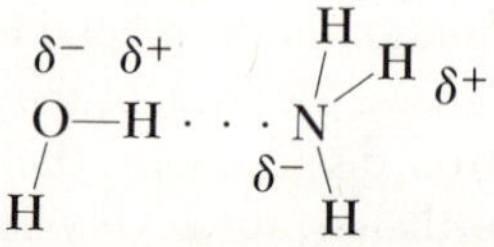

In the example above, the positive end of an O—H dipole in water is shown interacting with the negative end of a N—H dipole in ammonia. Theoretical calculations indicate that there is also a small amount of covalent character in a hydrogen bond.

Repulsive interactions are also present between molecules in close proximity. Such interactions are largely due to electron-electron repulsions that occur if molecules are too close together. These repulsions essentially determine how close the particles, the ions and the molecules, can pack together in a liquid or solid.

14.2 Solids

General Categories

In Sec. 2.1, solids were distinguished from liquids and gases by fluidity and the behavior of a substance when it is transferred to a new container. In these terms a **solid** was defined as a sample of matter that retains both its shape and its volume when transferred into a new container under the stress of gravity. When this definition is subjected to close scrutiny, it becomes clear that it is not very precise. For example, on a cold winter day, a piece of common roofing tar would appear to satisfy this definition of a solid. Yet, on a very hot summer day, the same piece of tar would slowly deform to conform to the shape of the container in which it is placed; it would flow very slowly, a characteristic generally associated with liquids. Furthermore, we should find no sharp dividing line between the two types of behavior; as the temperature is gradually increased, the apparent fluidity of the piece of tar gradually increases. Careful experimentation would disclose that even at lower temperatures, say 0 °C, the sample of tar would flow measurably under sufficient stress. In contrast, a solid substance like *t*-butyl alcohol, $(CH_3)_3COH$, at 0°C will fracture under stress rather than flow; and this behavior is retained as the temperature is increased to the melting point. At 25.5 °C, *t*-butyl alcohol melts sharply, and above that temperature the melted substance behaves as one expects a liquid to behave—when transferred from one container to another, it retains its volume but not its shape. Below its melting point, 25.5 °C, *t*-butyl alcohol does not exhibit gradual deformation like the tar sample; if one tries to produce such fluidity by applying a large stress, the sample will fracture.

The two types of materials just described, tar and *t*-butyl alcohol, are commonly characterized as amorphous solids and crystalline solids, respectively. In a scientific sense, the only solids are crystalline solids. A **crystalline solid** is one that is made up of macroscopic pieces that have regular shapes, generally appearing as well-defined polyhedra; they are called **crystals.** The regular shape of a crystal is due to its highly ordered submicroscopic structure. Such a solid, when heated slowly, does not gradually soften; rather, it sharply melts at a well-defined temperature, the **melting point.** The term *amorphous solid* is in a scientific sense a contradiction in terms. But the word *solid* is in such a widespread nonscientific usage for amorphous materials that we include a discussion of this type of matter in this chapter. Examination of an **amorphous solid** with a microscope shows that it is not composed of regularly shaped particles. Furthermore, when an amorphous solid is heated slowly, it is observed to soften gradually. There is no specific temperature at which melting occurs; strictly speaking it does not really melt but simply softens as the temperature is increased. Indeed, an amorphous solid is really not a solid at all in a true scientific sense; it does change its shape under stress and is found not to have the highly ordered submicroscopic structure characteristic of a true solid. What is commonly called an "amorphous solid" is really just a liquid of low mobility. Amorphous solids are sometimes referred to as *noncrystalline solids* (another scientific contradiction in terms). Some amorphous solids are *glasses,* or supercooled liquids (unstable liquids that are cooled below their freezing points); they have the same type of submicroscopic disorder as normal, highly mobile liquids, but not enough kinetic energy to manifest substantial fluidity at the given temperature. Some such substances can be **annealed** into crystalline solids by being carefully warmed and allowed to cool slowly; this provides sufficient mobility temporarily and enough time for the molecular or ionic species to assume the correct geometrical arrangements for a crystalline structure.

Polycrystalline solids

For many of the spectacular crystalline deposits found in underground caverns or for such common materials as rock salt, or granulated sugar, it is very easy to detect

TABLE 14.1 **General Types of "Solids"**

Crystalline Solids		
Single Crystals	*Polycrystalline*	*Amorphous "Solids"*[a]
Diamond	Copper, iron, etc.	Carbon black
Snowflake	Granite (mixture of substances)	Cut glass
Rock salt	Mothball	Tar
Sugar granule	Concrete (mixture of substances)	Typical plastic
Ruby	Graphite	Hair
Sapphire	Ice cube	Paper
Iodine crystal	Aspirin tablet	Obsidian
		Rubber
		Coal

[a]Not true solids in the scientific sense.

the crystalline nature of the solid. Superficial inspection of a crystalline solid does not always reveal, however, the regular crystal shapes that have been mentioned. In such cases, observation with a microscope shows that a macroscopic sample consists of a very large number of very small crystals, each of which does have a regular crystalline shape. Hence, such materials are not amorphous; they are referred to as *polycrystalline*. The polycrystalline designation applies to a wide variety of important materials. For example, pure metals, like solid copper or iron, are usually polycrystalline. The label can also apply to mixtures of pure substances, which can exist as a collection of very small crystals of different substances. Examples of each of these various categories are given in Table 14.1.

Crystalline Solids

General characteristics; Anisotropy

Crystalline solids thus are based on a submicroscopic structure characterized by a high level of ordering, which manifests itself macroscopically in a variety of ways, including sharp melting points and regular crystal shapes. Another manifestation is directional character, or **anisotropy**. By this, we mean that the response of a crystal to some external physical influence, say a mechanical force, a magnetic field, an electric field, or electromagnetic radiation, will generally depend on the specific orientation of the crystal in relation to the direction in which the external influence is applied. Thus, the susceptibility of a crystal to fracture under mechanical stress or the ability of a crystal to transmit light will in general depend on the specific orientation of the crystal in relation to the direction of the applied mechanical force or of the incident light. This is true, even if one has taken the trouble to grind the crystal to what appears superficially to be a spherical shape. Figure 14.2 shows that this kind of directional response is reasonable for a material whose macroscopic structure is based on a highly ordered submicroscopic structure. In this figure, an ordered array of oblong objects represents the ordered submicroscopic structure of a crystalline solid, and a small projectile represents an external physical influence. Clearly, the projectile will move through the array much more freely in the horizontal direction than in the vertical.

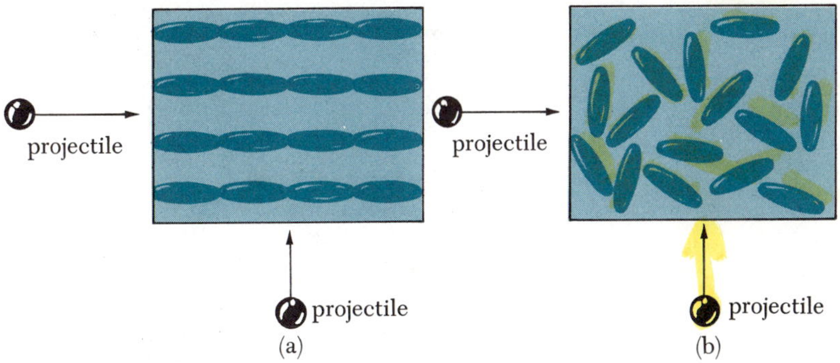

Figure 14.2 The response of an array of oblong objects to an external influence (in this case, represented by a small spherical projectile, moving in a specific direction). (a) An ordered array, representing a crystalline solid. (b) A disordered array, representing an amorphous solid.

The anisotropy exhibited by crystals is to be contrasted with the lack of anisotropy exhibited by amorphous solids. Amorphous materials respond to external physical influences in ways that are completely independent of the orientation of the sample relative to the direction characterizing the external influence. This independence of geometrical orientation is denoted by the word *isotropic* and is what one would expect from a material made up of molecular species that are arranged in a random instead of an ordered manner. The fate of the projectile in Fig. 14.2b as it impinges on the array of objects is, on the average, the same for the horizontal approach as for the vertical approach. The disordered array of objects is isotropic relative to the impinging projectile.

The regular shapes characteristic of crystalline solids are direct manifestations of the highly ordered nature of the submicroscopic structures and the anisotropies they exhibit.* When a crystal is formed, say from a saturated solution, these factors favor growth along certain directions, giving rise to specific crystal shapes.

Crystallography

Since early times, people have been fascinated by the regular and often beautiful shapes of natural crystals. The sparkling faces of minerals such as quartz, amethyst, and rubies were appreciated for their aesthetic features as well as their manifestation of order in nature. The study of crystals, called **crystallography,** has followed a history that is analogous to what has occurred in many other branches of science. First came the observation of many regular shapes in a wide variety of crystals; then wonderment and puzzlement; next, theoretical models were proposed to explain known patterns in the available data (observations); and to complete the cycle, better experiments were performed to test the proposed theories.

From numerous early observations and measurements on a wide variety of crystals, Abbé Haüy late in the eighteenth century proposed that the angles between corresponding faces of two crystals of the same substance are the same. In general, one finds that the appearance and dimensions of a crystal depend on the details of how it was prepared—conditions like rate of crystal growth and temperature and concentrations of species present in the solution from which the crystal was prepared. Nevertheless, according to Haüy, the angles between corresponding crystal faces

*The ordered anisotropic nature of crystals is responsible for the fact that a skilled diamond cutter can fracture a large diamond along a specific, predictable plane, whereas glass, an amorphous material, shatters or fractures in an unpredictable manner on impact.

Figure 14.3
Typical crystal shapes of quartz, SiO_2, and rock salt, NaCl, showing invariance of the angles between faces for a given substance.

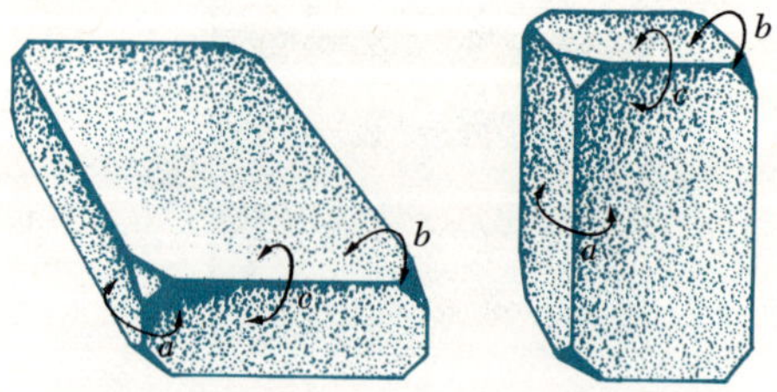

quartz crystals, with $a = c = 90°$, $b = 120°$

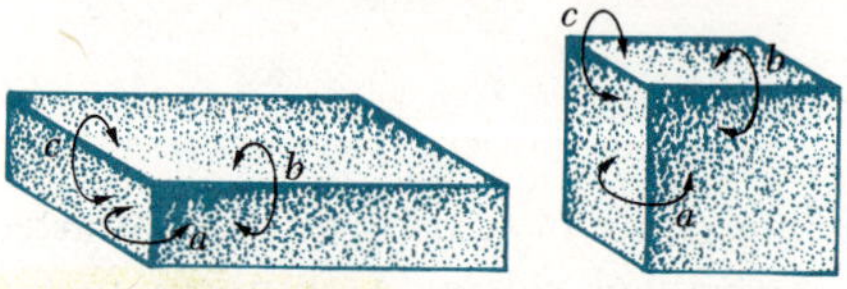

NaCl crystals, with $a = b = c = 90°$

should be the same for a given substance. Figure 14.3 shows typical shapes of quartz, SiO_2, and rock salt, NaCl, that can commonly be found. Although the two shapes shown for quartz are quite different, the interfacial angle $\angle a$ is 90° in both samples and likewise $\angle b$ is the same in both samples, 120°. Similarly, for each of the three NaCl crystals shown in Fig. 14.3, both interfacial angles shown are 90°.

The general features of a crystal's shape, including interfacial angles, is called the **crystal habit.** The study of crystal habits is referred to as **morphological crystallography,** and has been important in the development of geology. By cataloging crystal habits of specific minerals and correlating them with types of geological formations, geologists have obtained information that has helped them determine the geological histories of many areas.

While morphological crystallography has application in both geology and solid state chemistry, it does not provide chemists the sort of detailed, submicroscopic structural information typically of interest to them. The regularity of crystal shapes, especially the consistent patterns of interfacial angles for a specific substance, brought early scientists to the idea that crystals are built up in a regular manner from small, repetitive structural units that are characteristic of the substance. As early as 1611, Kepler had proposed an internal structural order in crystals as the basis of regular crystal shapes. In 1782, Haüy formalized the highly ordered submicroscopic structure of crystals as the basis of their regular and consistent geometrical features. That submicroscopic order should manifest itself in the form of regular crystalline shapes seems reasonable on the basis of simply considering building up a macroscopic structure from a particular type of repetitive submicroscopic structural unit. For example, it is easy to visualize how a six-sided object with 90° interfacial angles could be built from structural units that are rectangular on each side, as seen in Fig. 14.4a. Clearly, other shapes could be visualized for the macroscopic structure than what is shown in Fig. 14.4a. Nevertheless, the one shown is typical of what one might expect. Most important, this figure, together with the analogous case in Fig. 14.4b, emphasizes the point that only certain interfacial angles are expected for the macroscopic structure, the crystal, 90° in the case shown in Fig. 14.4a.

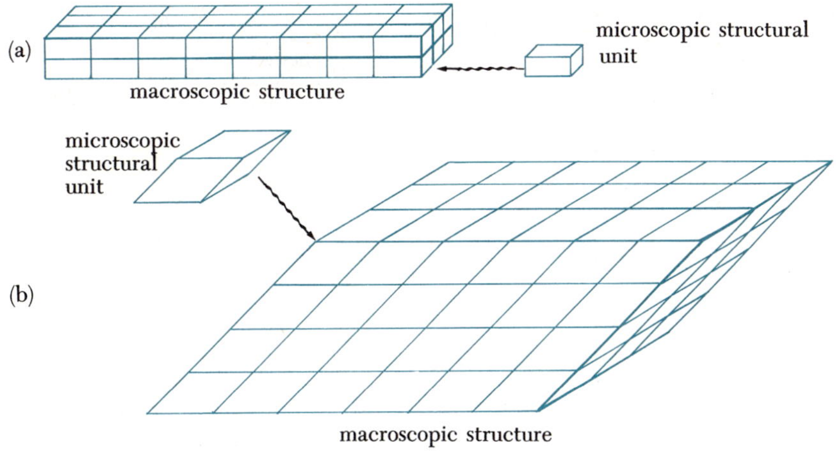

Figure 14.4 The building up of macroscopic structures of regular shapes—crystals—from repetitive microscopic structural units.

In crystallography, the repetitive submicroscopic structural units analogous to the small objects shown in Fig. 14.4 are called unit cells. A **unit cell** is a collection of molecules, ions, or atoms that if repeated in a regular manner (like the submicroscopic structural units of Fig. 14.4) constitutes the entire structure of a crystal; the unit cell is the smallest repeating unit. The crystal morphology can give some information about the shape of the unit cell, but not about its contents. The scientist needs to know the atomic-molecular makeup of the unit cell in order to understand crystal structure on a fundamental, submicroscopic level. This kind of understanding is an important component of fundamental chemistry and a necessary prerequisite to understanding a wide range of important chemical and physical properties of solids.

14.3 X-Ray Diffraction

Interference

The method chemists use to determine the detailed structure of the unit cell, and hence the detailed submicroscopic structure of a crystal, is **x-ray diffraction.** The diffraction phenomenon showed by traveling waves on encountering sharp boundaries (for example, narrow slits) was discussed briefly in Sec. 6.2 as one of the primary pieces of evidence for the wave character of electrons.

The physical basis of the diffraction phenomenon is the **interference** of electromagnetic waves—light, for example. Such waves, shown in Fig. 14.5, can be represented at a particular instant in time or at a particular point in space as **traveling sine waves** of varying electric field intensity, E.* As seen in Fig. 14.5, the period τ of a wave is defined as the time span between maxima (or minima) in E, as E oscillates about zero at a particular point in space; the wavelength λ is defined as the distance in space between the positions of the maxima (or minima) in E at a particular instant.

*In conjunction with the sinusoidally varying electric field intensity E is a correspondingly varying magnetic field B. Thus, a traveling wave of light, or any other electromagnetic radiation, is often represented as shown in Fig. 8.8.

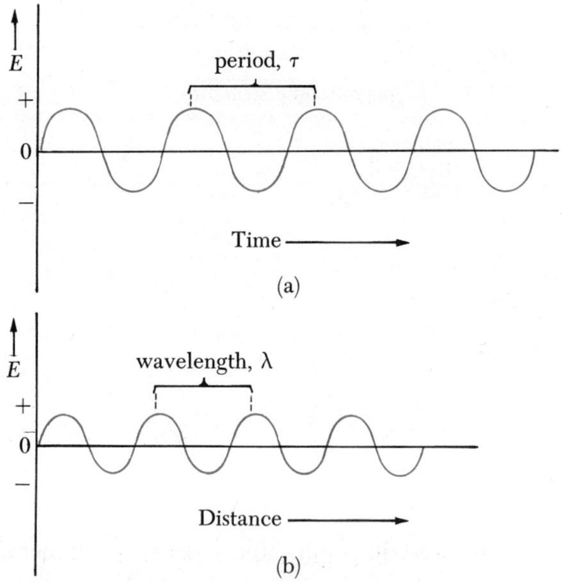

Figure 14.5 Oscillatory, or sinusoidal, behavior of the electric field intensity E of a traveling electromagnetic wave—light, for example. (a) Time dependence of E at a particular point in space. (b) Variation of E with increasing distance from the source at a given instant.

When two or more beams of light are superimposed at some point in space, the resultant light is characterized by an electric field intensity that is the simple sum of the intensities each of the beams would individually contribute. Thus, if two identical beams were superimposed at a given point, the light intensity at that point would be simply twice the intensity of a single beam, as shown schematically in Fig. 14.6a. This situation is referred to as **constructive interference.** If both light beams have identical periods, or wavelengths, and intensities but in one beam the propagation lags behind that of the other by one half-period, or one half-wavelength, then the resultant light intensity at the superposition point is zero; this is called **destructive interference,** and is shown schematically in Fig. 14.6b. Figure 14.6c shows the case in which the propagation lag, called the **phase difference,** is somewhere between one half-period and one period; the result is a diminished light intensity, but not zero light intensity. Inspection of Fig. 14.6, together with a little thought, leads one to conclude that if there were a phase difference of one period or any integral number of periods, the superposition of the two waves would give entirely constructive interference, just as if there were no phase difference at all.*

In terms of constructive and destructive interference, the diffraction experiment pictured in Fig. 6.13 and the approaches to be explored in this section are readily understandable. The basis for understanding such experiments is simply recognizing that the intensity of light at any point in a diffraction experiment is simply the sum of the intensities contributed by the various beams, or rays, at that point; and one must remember the interference associated with a given phase difference between different rays. In these diffraction experiments, the phase differences between two

*If two incident waves of the same intensity and period but with a phase difference that is an odd number of half-periods interfere at some point in space, the net result is completely destructive interference, just as if the phase difference were one half-period.

beams can be ascribed to different distances, or path lengths, traversed by the beams from the source to the common point P. For two beams, a difference in path lengths of exactly one wavelength, or any integral number of wavelengths, gives rise to a phase difference that is an integral number of periods τ; this, as we have seen, is essentially equivalent to having no phase difference at all, giving purely constructive interference at the point P. Similarly, any path difference that exactly equals one half-wavelength, or any integral multiple of $\lambda/2$, gives rise to a phase difference of one half-period, producing destructive interference between the two beams.

Figure 14.7 shows schematically the physical arrangement of a two-slit diffraction experiment. We need to know the path lengths between the source O and some point P on the detector surface. (Five specific points P are shown in Fig. 14.7—P_0, P_1, P_2, P_1', and P_2'.) For an experiment with visible light, the detector surface might, for example, be a photographic plate; for the diffraction of x rays, it could be a photographic plate or a movable radiation counter that could monitor the intensity of radiation at any point P. Since the distances $\overline{OS_1}$ and $\overline{OS_2}$ from the source to slit 1 and slit 2 are the same, differences in the total path lengths arise only from differences in the distances $\overline{S_1P}$ or $\overline{S_2P}$ (for example, where P is P_1, P_0, or P_2). The point P_0 is chosen so that the distance $\overline{S_1P_0}$ is exactly the same as $\overline{S_2P_0}$; hence, for P_0, both path lengths are the same, there is no phase difference, one has completely constructive

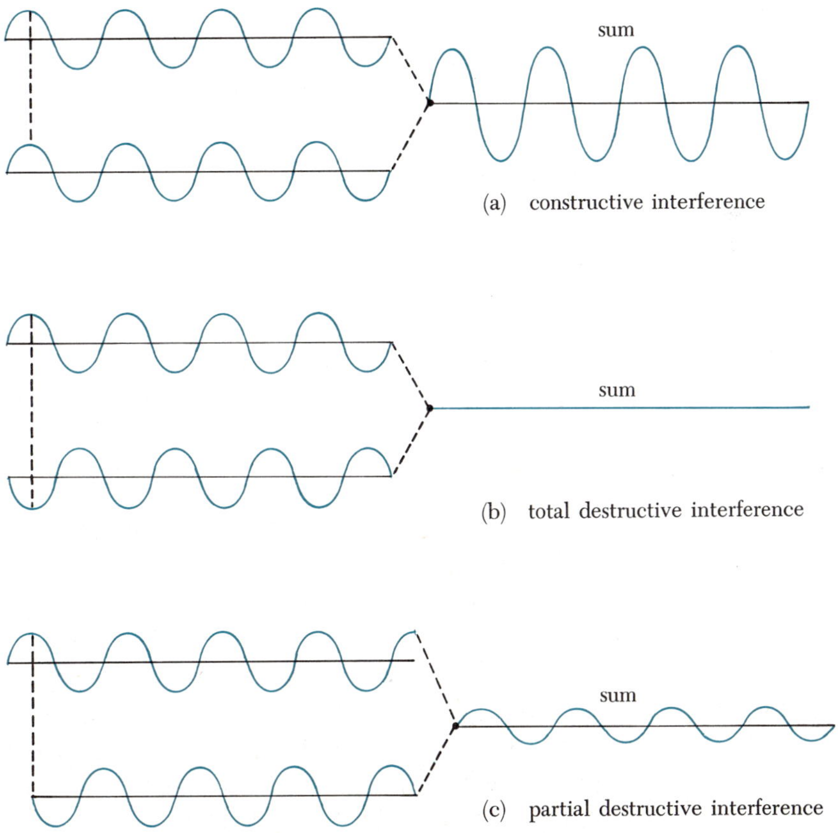

Figure 14.6 Interference patterns resulting from superposition of two beams of light of the same period, or wavelength, and intensity.

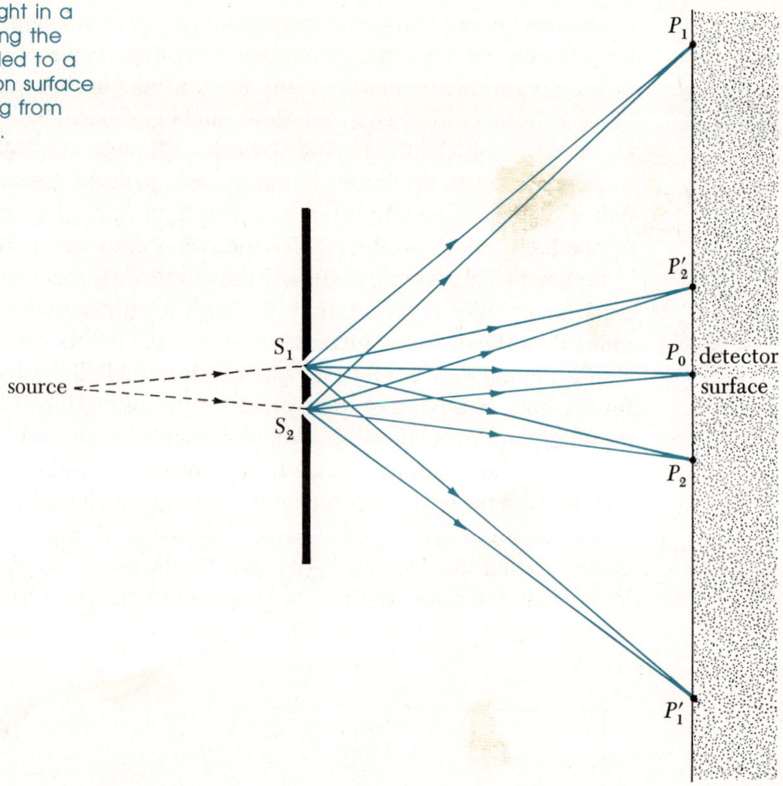

Figure 14.7 Diffraction of light in a two-slit arrangement, showing the different path lengths traveled to a given point on the detection surface by the two beams emerging from the slits at points S_1 and S_2.

interference, and the light intensity is high. If the distances $\overline{S_1P_1}$ and $\overline{S_2P_1}$ differ by an integral number of wavelengths, then the two beams interfere constructively at P_1, and at its counterpart, P_1', also resulting in a high intensity. Let us imagine that for point P_2, the distances $\overline{S_1P_2}$ and $\overline{S_2P_2}$ differ by an integral number of half-wavelengths; then the two beams interfere destructively at P_2, and at its counterpart, P_2', resulting in zero light intensity at that point. Moving along the detection surface away from its center point, one would find alternating regions of low intensity and high intensity, the distances between the regions depending on the geometry of the slit system and the distance of the detection surface from the slit system. Mathematical analysis of diffraction shows that the most dramatic diffraction effects, those on which measurements can be based most reliably, occur in slit systems for which the distance between the slits is of the same order of magnitude as the wavelength of the radiation being diffracted.

Diffraction of X-Rays by Crystals

The reason why diffraction is important for characterizing the structures of crystals is that the regular structure of a crystal, as an orderly repetition of the unit cell, provides an ordered array of sharp barriers, the atoms, which can diffract radiation of the proper wavelength. For crystals, with interatomic distances on the order of about 10^{-10} m (1 Å), radiation of about that wavelength is necessary for optimum diffraction. Such radiation is precisely what one obtains in x rays. X rays, a form of electromagnetic radiation, result from the emission of radiation by atoms in which an electron has been ejected from an inner shell, having been bombarded with a high-energy electron beam, and is replaced by an electron from an outer shell. The amount of energy emitted from the atom as the outer electron "drops" into the lower-energy, inner shell is just the energy difference between the two shells ΔE. Knowing these energies, and the Bohr-Planck frequency condition (Eq. 6.1, $\nu = \Delta E/h$) and the well-known relation between the frequency ν and wavelength λ of radiation (Eq. 14.1),* we can calculate the wavelengths of x rays.

$$\lambda\nu = c \tag{14.1}$$

where c = speed of light = 3.0×10^8 m/sec.

$$\lambda = \frac{c}{\nu} = \frac{c}{\Delta E/h} = \frac{hc}{\Delta E} \tag{14.2}$$

The wavelength is found to be typically of the order of 1 Å if the target materials, subjected to electron bombardment, are elements from scandium to zinc or from yttrium to cadmium, as arranged in the periodic table. Hence, x rays are ideal for diffraction by the ordered arrays of atoms in crystals. In 1912, Friedrich and Knipping discovered x-ray diffraction in crystals, a discovery that has affected chemistry profoundly.

EXAMPLE 14.1 Calculate the energy of the electron transition in copper that gives rise to x rays of wavelength 1.5×10^{-10} m. What are the frequency and energy quanta of this radiation?

SOLUTION From Eq. (14.2), we get

$$1.5 \times 10^{-10}\text{ m} = \frac{(6.626 \times 10^{-34}\text{ J sec})(2.998 \times 10^8\text{ m/sec})}{\Delta E}$$

$$\Delta E = \frac{(6.626 \times 10^{-34}\text{ J sec})(2.998 \times 10^8\text{ m/sec})}{1.5 \times 10^{-10}\text{ m}} = 1.3 \times 10^{-15}\text{ J}$$

The frequency is given by application of Eq. (14.2):

$$\nu = \frac{c}{\lambda} = \frac{2.998 \times 10^8\text{ m/sec}}{1.5 \times 10^{-10}\text{ m}} = 2.0 \times 10^{18}\text{ sec}^{-1}$$

STUDY PROBLEM 14(a)

Calculate the wavelength and frequency of x rays obtained by using manganese as the target material, for which an electronic energy change is 1.8×10^{-15} J.

The essence of x-ray diffraction in crystals can be seen in Fig. 14.8, which shows a cross section of an idealized crystal composed of equally spaced atoms. Let us focus attention initially on two beams of light, I and II, which hit neighboring atoms in

*Equation (14.1) is readily justified by noting the units of λ, ν, and c:

$$\lambda\text{ (m)} \times \nu\text{ (sec}^{-1}) = c\text{ (m/sec)}$$

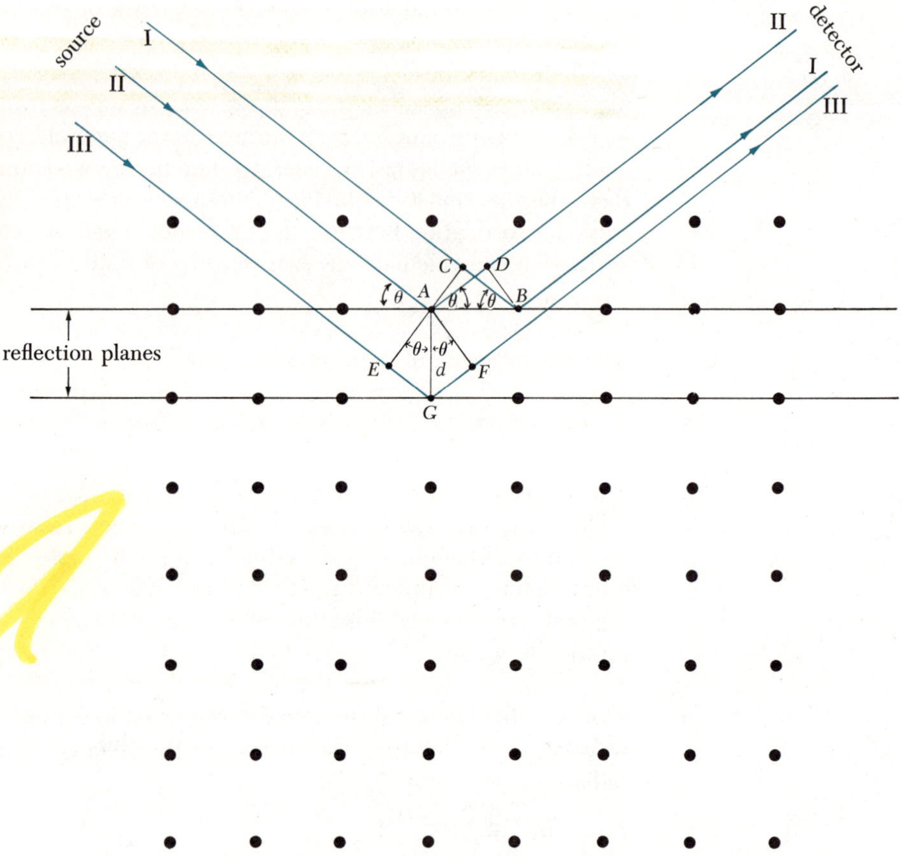

Figure 14.8
A cross section of an array of regularly spaced atoms, showing the "reflection" of three x rays. Rays I and II are "reflected" by two atoms within a given horizontal plane (perpendicular to the plane of the page). Ray III is reflected by an atom in another horizontal plane.

one horizontal row of the cross section shown; these beams are "reflected" at an angle θ, which equals the angle of incidence. From simple geometrical arguments, one sees that the triangles ABC and ABD are congruent and that the distances $\overline{AD}$ and $\overline{BC}$ are equal. Hence, the path lengths of beams I and II between the source and the detector are equal, and there is no phase difference at the detector for the radiation from those two beams. Therefore, constructive interference occurs. Similar analysis applies to all atoms in the row containing positions A and B. The net result is that there is constructive interference for radiation reflected from the atoms in a regular row (or plane, for a three-dimensional case) at a given angle that equals the angle the incident beam makes with the plane. In some respects, this reflection is similar to the reflection of light from an ordinary mirror.

The Bragg Equation

Of greater interest for determining submicroscopic crystal structure is the relation between two x rays impinging on atoms in different rows in the cross section of Fig. 14.8, for example, the rays II and III, which impinge on the atoms at points A and G, respectively, in Fig. 14.8. From the figure, one sees that the path length of ray III from source to detector is greater than the path length of ray II by the sum of distances $\overline{EG}$ and $\overline{GF}$. But from trigonometry, these distances are each equal to $d \sin \theta$, where d is the distance between the rows of atoms in the cross section and θ equals the angles of incidence and reflection described above. Thus, the difference in path length is $2d \sin \theta$. From the discussion above, rays II and III will interfere constructively at the detector only if this difference in path lengths equals an integral number of wavelengths of the radiation. Thus, the condition for constructive interference and a corresponding high intensity at the detector is

$$2d \sin \theta = n\lambda, \qquad n = 1,2,3, \ldots, \text{an integer} \tag{14.3}$$

This equation is called the **Bragg equation,** named after W. L. Bragg, who first derived it and applied it to determine structural features of crystalline solids. It tells us the following: If an x ray is incident upon a plane of regularly spaced atoms in a crystal with an angle of incidence θ, and if the detector is set to detect reflected x rays with a reflection angle equal to θ, then constructive interference occurs at the detector if the angle θ satisfies Eq. (14.3) for some integer value of n. Hence, the detector registers a strong response if the crystal is oriented so that the reflection angle is such that Eq. (14.3) is satisfied for n equal to some integer. In an actual x-ray diffraction experiment, the reflection angle θ is varied and the orientation of the crystal is varied to find those combinations that give rise to strong constructive interference. The kind of apparatus used is generally based on the sort of arrangement pictured in Fig. 14.9, called an **x-ray diffractometer.** The reflection angle and crystal orientation are determined by settings on the goniometer, which is a mechanical device for positioning the crystal, on its supporting device and by the position of the

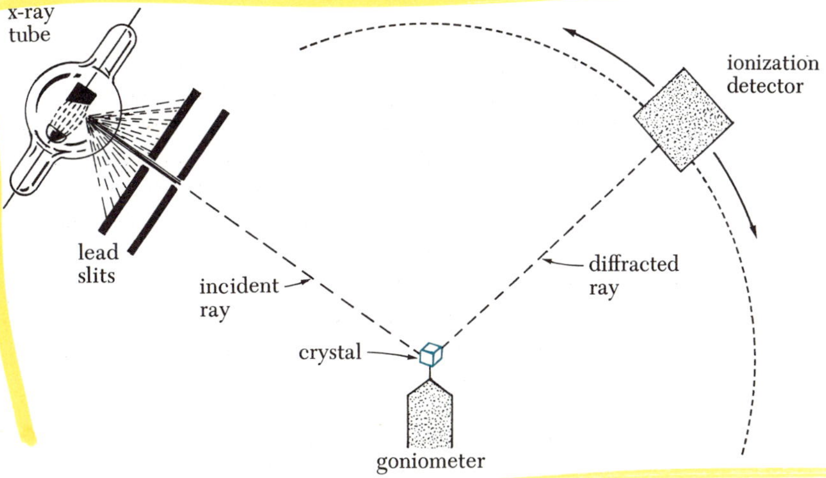

Figure 14.9 Schematic diagram of x-ray diffraction apparatus. The lead slits convert the x rays from the x-ray tube into a beam (the incident ray). A goniometer positions the crystal so that the incident ray is diffracted. An ionization detector, which is analogous to a Geiger counter, detects the diffracted ray. The detector can be moved along the indicated arc by a computer-controlled mechanism. The geometrical orientation of the crystal can be varied by the position and orientation of the goniometer.

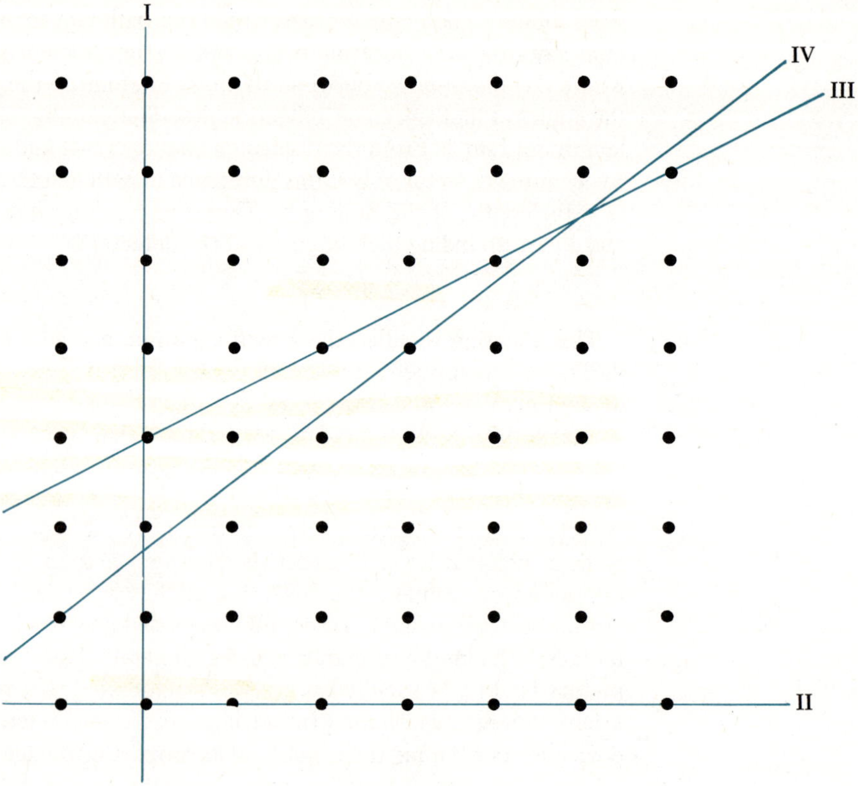

Figure 14.10
Cross section of an array of regularly spaced atoms, showing how the number of atoms in a plane depends on the orientation of the plane. (Lines I, II, III, and IV represent cross sections of planes.)

detector—generally an ionization chamber, a Geiger counter, for example.* As these variables are adjusted, certain conditions are found for which the proper angle θ exists between the incident x rays and some plane of atoms.

In the array of atoms of a crystal, any number of "planes of atoms" can be drawn, as we can see from Fig. 14.10, which shows a cross section of an array of regularly spaced atoms. Any horizontal or vertical line through the points representing atoms in this figure represents a plane containing a large number of atoms, for example, line I and line II. Line III represents a plane containing fewer atoms, and line IV, even fewer atoms. The planes that give rise to large x-ray intensities at the detector, for the appropriate angle of incidence, are the planes containing large numbers of atoms (for example, I and II in Fig. 14.10).

EXAMPLE 14.2 Calculate the first-order diffraction angle ($n = 1$) for which intense diffraction is obtained if x rays of 1.9×10^{-10} m wavelength are diffracted from a crystal in which successive layers of atoms are 2.3 Å (2.3×10^{-10} m) apart.

*In a modern x-ray diffractometer, the variations of the detector position and crystal orientation are controlled automatically by a small laboratory computer, which also monitors the radiation at the detector as these variations are carried out.

Spotlight

Another feature of the submicroscopic structure of a crystal that is important in determining the intensities of "reflected" x rays is the electron distribution within the crystal. Detailed theoretical analysis shows that the electron density is what determines the amount of radiation reflected at a particular place in the crystal array. We have discussed the reflecting atoms as though they were point charges; but in many cases, especially if covalent bonding is important, the electronic distributions about a particular atomic center may be far from spherical. Modern x-ray diffraction techniques have provided increasingly precise data. Hence, from the intensity patterns in x-ray diffraction experiments, it has become possible to derive "maps" of electron density—like contour maps of terrain, showing mountains, valleys, plateaus, and the like. This is one of the exciting areas of modern chemical research, and provides severe tests for the predictions of quantum mechanical calculations, for instance, the mathematical molecular orbital theory.

SOLUTION Using Eq. (14.3), we write

$$(2)(2.3 \times 10^{-10}\ \text{m})(\sin\theta) = (1)(1.9 \times 10^{-10}\ \text{m})$$

$$\sin\theta = \frac{1.9 \times 10^{-10}}{2 \times 2.3 \times 10^{-10}} = 0.413$$

Then finding in a table the angle θ that has $\sin\theta = 0.413$, we get $\theta = 24°$.

STUDY PROBLEM 14(b)

What is the d between planes of atoms in a simple crystalline array if the first-order ($n = 1$) diffraction maximum for x rays of 2.7×10^{-10}-m wavelength occurs at a diffraction angle of 46°?

In summary, there is constructive interference due to reflection of x rays by regularly spaced atoms within a plane, if the angle of incidence equals the angle of reflection, and constructive interference due to reflections of x rays by atoms in regularly spaced planes, at certain reflection angles satisfying the Bragg equation. And thus it is possible to relate the distance between reflecting planes d to the wavelength λ of the x rays. Early applications of the Bragg equation included the determination of x-ray wavelengths from diffraction experiments on simple crystals of known structure. The overwhelming emphasis in recent years is to use x rays of known wavelength in diffraction experiments designed to elucidate crystal structure. What x-ray diffraction experiments can provide is a complete, three-dimensional picture of where the atoms in the unit cell of a crystal are located; and if discrete molecules exist in the crystal, then these experiments can show the relative placement of atoms within individual molecules. Even if one is not primarily interested in the crystal structure of a system, and is more concerned with elucidating structural details of individual molecules, x-ray diffraction is often the best approach. This approach has been responsible for most of the accurate information now available on molecular structure; bond lengths and bond angles are examples of such information.

The high degree of order in a crystalline solid, expressed as the ordered repetition of the unit cell, is responsible for the patterns of constructive interference on which x-ray diffraction is based. Such order is not present in a liquid or an amorphous solid.

Hence, the usual x-ray diffraction approach simply does not apply to noncrystalline materials. If one has a sample of an amorphous solid or a liquid and wants to get the kind of detailed structural information available from x-ray diffraction experiments, the only recourse is to find an appropriate combination of conditions by which single crystals of the substance of interest can be prepared.* Once the single crystals are obtained (if they ever are!), the x-ray diffraction analysis can be carried out, and the detailed structural information is therefore available. In this way, a wide variety of chemical systems have been investigated, ranging from simple salts to complicated molecules of biological importance, of which enzymes, proteins, and DNA are notable. The famous Watson-Crick model of DNA, for which the Nobel Prize in Medicine and Physiology was awarded in 1962, was based on x-ray diffraction data.

14.4 Crystal Structure

The Unit Cell

The Seven Crystal Systems

The techniques of x-ray crystallography that have been outlined provide a detailed picture of the placement of the atoms in the unit cell. This unit cell is the fundamental building block of the crystal. From this unit, one can imagine building the entire crystal by just extending the unit cell in three dimensions with the contiguous attachment of additional unit cells. Figure 14.4 indicates this schematically. One might imagine that nature abounds in a wide variety of sizes and shapes of unit cells. Relatively few such shapes are actually found when crystal structures are analyzed. Sophisticated mathematical arguments show that any crystal can be represented as a repetition of a unit cell having one of only seven possible geometrical shapes. These shapes represent what are called the seven crystal systems, and are shown in Fig. 14.11. From Fig. 14.11, it can also be seen that each of these seven shapes can be defined by the following six factors: the three interfacial angles, α, β, and γ, and the three edge lengths, a, b, and c.

The reason that there are only seven basic shapes for unit cells is more geometric than physical. The definition of the unit cell carries with it the implication that the building up of a crystal by unit cells must completely fill the volume of the crystal with the individual volume elements defined by boundaries of the unit cells. There must be no gaps; if there were, we could not say that the volume of the crystal is built up completely in terms of unit cells. Detailed geometrical considerations show that only submicroscopic polyhedra of certain shapes are capable of filling a macroscopic three-dimensional space completely. A collection of many little spheres, for example, would not fill a macroscopic volume completely. If we imagine a beaker full of small glass spheres, we realize that one can still pour a considerable quantity of water into the beaker because of the gaps between spheres. The seven crystal systems conform to the requirement of filling the volume of a crystal entirely.

The shape of the unit cell ultimately determines the shape of the crystal, as implied

*Finding the conditions necessary to "grow" single crystals of a substance suitable for x-ray diffraction experiments is often more difficult than the x-ray diffraction experiments themselves. Variations of solvent (including solvent mixtures), temperature, and solute concentrations are explored in order to find the desired conditions.

in Fig. 14.4. Because crystal growth is different in different directions, the shape of a crystal need not resemble the shape of its unit cell; the interfacial angles must correspond, however. Hence, careful morphological studies of crystals can lead to a determination of the type of unit cell involved.

What morphological studies cannot provide is a detailed structural picture of the unit cell regarding placement of atoms, ions, or molecules. Generally, chemists really want to know how these basic structural units of chemistry are arranged in a unit cell, and this is precisely the kind of information x-ray diffraction data provide.

Figure 14.11 (a) Definition of the geometrical parameters, interfacial angles (α, β, γ), and edge lengths (a, b, c). (b) through (h) The seven unit cell shapes. (In some mathematical formulations, (e) and (f) are considered one basic type, giving a total of six "crystal systems.")

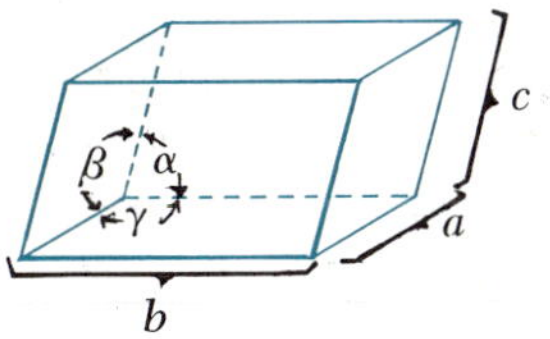

(a) definition of geometrical parameters

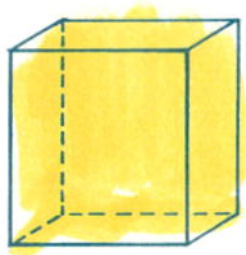

(b) cubic
$\alpha = \beta = \gamma = 90°$
$a = b = c$

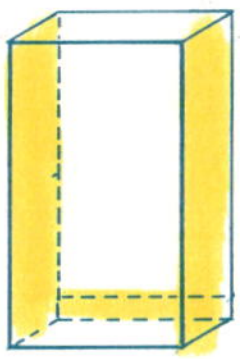

(c) tetragonal
$\alpha = \beta = \gamma = 90°$
$a = b \neq c$

(d) orthorhombic
$\alpha = \beta = \gamma = 90°$
$a \neq b \neq c$

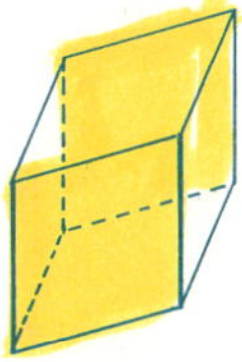

(e) rhombohedral
$\alpha = \beta = \gamma \neq 90°$
$a = b = c$

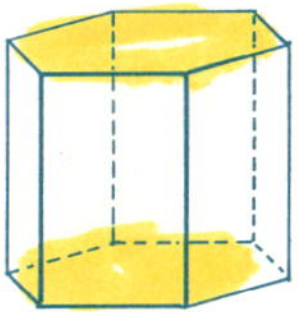

(f) hexagonal
$\alpha = \beta = 90°$, $\gamma = 120°$
$a = b \neq c$

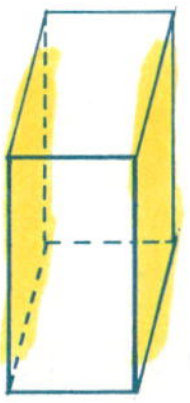

(g) monoclinic
$\alpha = \gamma = 90° \neq \beta$
$a \neq b \neq c$

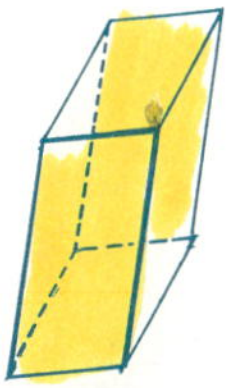

(h) triclinic
$\alpha \neq \beta \neq \gamma$
$a \neq b \neq c$

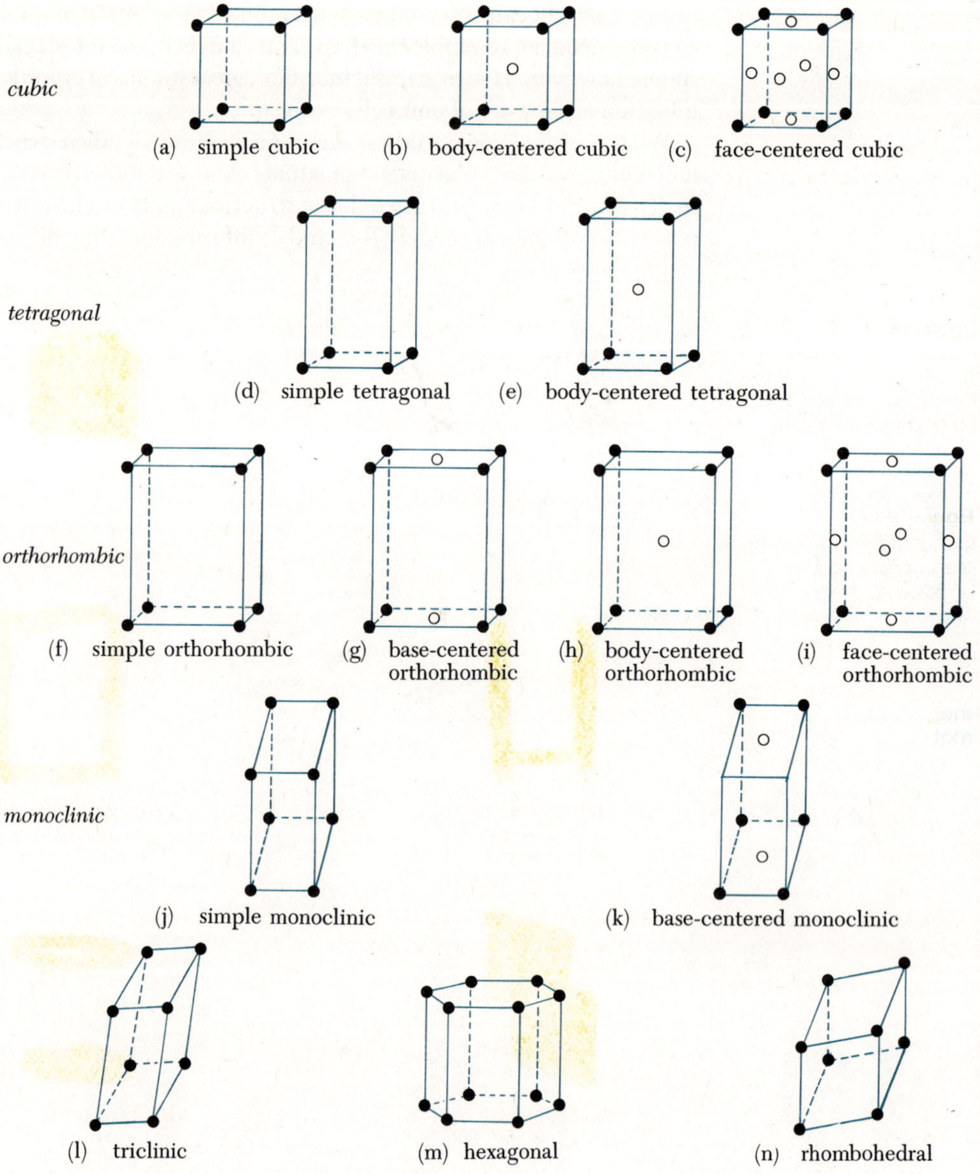

Figure 14.12 The fourteen Bravais crystal lattices. Corner lattice points shown as solid circles; body-centered and face-centered lattice points shown as open circles.

The Fourteen Bravais Lattices

With the sophisticated mathematical techniques used by scientists who have developed x-ray diffraction methods, one finds that crystal lattices can be characterized in greater detail than what is shown in Fig. 14.11. It is useful, besides specifying crystal shapes, to define *lattice points,* and by them, *crystal lattice systems.* By placing a lattice point at the corners or the middle of faces of each of the figures shown in Fig. 14.11, and in some cases at the center of a unit cell, one obtains the lattice types shown in Fig. 14.12. These fourteen lattice types are called the **Bravais crystal lattices;** they are named after A. Bravais, who deduced them in 1848. Each of these is derivable from the seven geometric forms of Fig. 14.11. A crystal lattice is said to be **body-centered** if within each unit cell there is a lattice point at the center. A lattice is called **face-centered** if there is a lattice point at the midpoint of each face. Those lattices with lattice points only at the corners are said to be **simple,** or **primitive, lattices.**

The unit cells and lattice systems of Fig. 14.11 and 14.12 do not represent the only unit cells or lattice systems one could call on to describe a particular crystal. One would find, however, that any others one might imagine using, like a body-centered hexagonal lattice, could be represented equally well by one of the fourteen Bravais lattices.

Lattice Points

In the simplest cases, lattice points of a crystal will actually be occupied by atoms, ions, or molecules. But this is definitely not the general case. In general, the lattice points are simply mathematical entities that one uses to characterize the shape of the unit cell and the symmetry and order of the crystal system. The actual structural units, say the molecules, are situated at specific places in each unit cell with definite orientations, and these positions and orientations are the same for all the unit cells in a given crystal; these positions are not necessarily lattice points. This situation is represented schematically in Fig. 14.13, in which an arrow represents a specific type of molecule, placed at a specific point in the unit cell and oriented in a specific direction. A lattice point is seen to be essentially a reference point—it is not necessarily where atoms, ions, or molecules are situated.

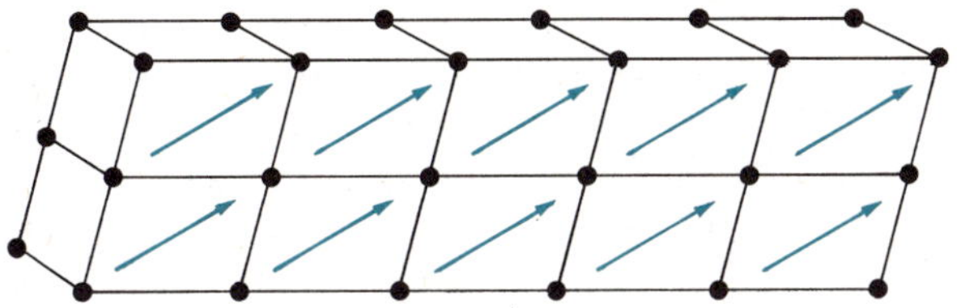

Figure 14.13 Schematic representation of the repetitive positioning and ordering of molecules in unit cells, with an arrow representing a molecule.

Types of Crystals; Bonding

The actual structure of a given crystal is characterized by whatever Bravais lattice it belongs to and by the values of the unit cell variables, α, β, γ, a, b, and c. These characteristics depend on a variety of fundamental considerations. The sizes of the structural units and the manner by which they are bonded are the most important factors. We shall subsequently examine some specific examples of how these factors interrelate.

From consideration of the different kinds of interactions, it is possible to classify crystals into the following four general types: (1) metallic, (2) ionic, (3) covalent-network, and (4) van der Waals (molecular). These categories are derived from the forces that hold the crystals together.

Spotlight

Beryllium is a silvery white metal, hard enough to cut glass. Beryllium-copper alloys (solid solutions of metals) possess unusual tensile strength and resiliency, thus finding use in springs and in delicate parts of instruments that must resist wear. The alloys of beryllium with aluminum, nickel, and cobalt sometimes find important use where resistance to corrosion is needed. The nonsparking properties of these alloys find use in contacts, slip-rings, and the like in motors. Each beryllium atom has only four electrons, and beryllium makes excellent x-ray windows, because the absorption of electromagnetic radiation, which must be avoided for windows, depends on the electron density in matter. Beryllium, when bombarded with α particles (He nuclei), is also an important source of neutrons for experimental purposes:

$$^{9}_{4}Be + ^{4}_{2}He \rightarrow ^{12}_{6}C + ^{1}_{0}n$$

The simple beryllium compounds are highly poisonous, but the mineral beryl, $3BeO \cdot Al_2O \cdot 6SiO_2$, is not considered hazardous. In this formula, the coefficients indicate the stoichiometry of the mineral.

Metallic Crystals

The "Electron Gas" Model

The bonding in metals is a special type of covalent bonding, in the broad sense of the *sharing* of electrons among atomic centers. Very roughly, the electronic structure of a metal in its solid state can be viewed as a "sea" of delocalized valence electrons, a so-called *electron gas,* in an array of metal cations. That is, one views one or more of the valence electrons of a metal atom as "ionized," leaving a metal cation that forms part of the lattice of such cations; and all the ionized electrons together constitute a mass of negative charge shared and attracted by all the ionic centers. The mutual attraction of the delocalized electrons for the metal cations is sufficiently strong to overcome the electrostatic repulsive forces between pairs of cations, and pairs of electrons, giving rise to the net structural stability of the crystal. In this view, a solid sample of sodium containing n atoms would be viewed as an array of n Na^+ ions, held together by the mutual attraction for n electrons, each one having originated from the 3*s* subshell of a sodium atom. This situation is represented in Fig. 14.14.

While this simple electron gas view of metallic structure is far from accurate, it does qualitatively conform to some important properties of metals, namely the conductivity of both electricity and heat. Since the electrons are represented as completely delocalized throughout the crystal network, it is reasonable that applying a

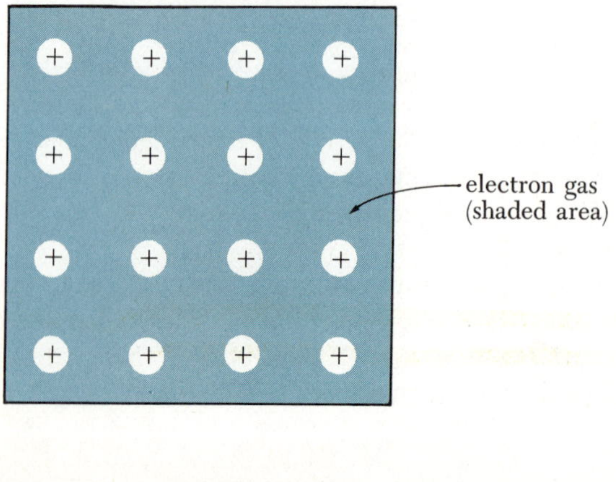

Figure 14.14
Simplified "electron gas" view of the metallic bonding in a metal such as sodium. The shaded area represents the electron gas, and the circles represent the metal ion obtained by removal of one or more valence electrons.

voltage difference across different ends of a piece of metal would give rise to a "drift" of electrons towards the electrically positive end. This drift of electrons constitutes **electrical conductivity**. Similarly, applying heat at one end of a piece of metal can cause agitation of the electrons in that region; this increase in average kinetic energy can then migrate to other regions of the piece, since the electrons are viewed as mobile in the electron gas. The net result would be an increase in temperature throughout the piece of metal, a manifestation of **thermal conductivity**. These two properties of metals will be subsequently examined in a more sophisticated and more accurate theoretical model.

Closest Packing

One important feature of many types of metallic crystals is what is called closest packing. This feature is common for crystals composed of identical particles, for example, the metal atoms of a metallic crystal, or small symmetrical molecules. Closest packed structures are the arrangements one would obtain by packing a collection of identical spheres to achieve maximum density. We can visualize these structures by reference to Fig. 14.15. First imagine placing six identical spheres in a contiguous hexagonal arrangement around a seventh sphere, all seven touching their neighbors, and with all seven centers in a single plane. (The color spheres in Fig. 14.15 represent this situation.) To add another dimension to this array, we imagine placing another layer of spheres, three for convenience, below the first layer. For maximum packing density, these three spheres would each be nestled into one of the six depressions (white in the figure) below the first layer. (These three spheres are represented in Fig. 14.15a by the black dashed circles.) To continue building this closest packed structure, add a third layer of spheres on top of the main, or middle, layer containing seven. These additional spheres, three again for convenience, will be placed into the depressions (white) on top of the middle layer. (The top layer of spheres is represented in Fig. 14.15 by gray spheres.) At this point one notes that there are two nonequivalent sets of three depressions in which to place the three spheres of the top layer. They can go into depressions that will place them directly above the three spheres in the bottom layer (Fig. 14.15b), or into the alternative set of depressions that are set off from the positions of the bottom set (Fig. 14.15c). The first of these alternatives is **hexagonal closest packing**; it corresponds to a rhombohedral lattice system, although this is by no means obvious from the figure. The second alternative, pictured in Fig. 14.15c, is called **cubic closest**

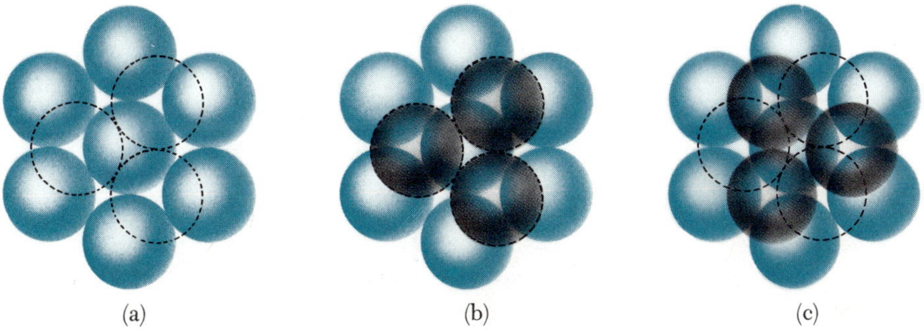

Figure 14.15 Visualization of closest packing structures. (a) Two layers of closest packed spheres: black dashed circles below; color spheres main level. (b) Hexagonal closest packed structure: gray spheres on top of (a). (c) Cubic closest packed structure: gray spheres on top of (a).

packing, and on close inspection of models is seen to correspond to the face-centered cubic lattice system; the unit cell is shown, along with the body-centered cubic system, in Fig. 14.16.

By focusing on the central sphere in the middle layer in Figs. 14.15b and c, which is representative of any sphere in the extended array, one can see that each sphere is in contact with twelve other spheres—six in the same layer, three in the layer below, and three in the layer above. Thus, atoms in these closest packed lattice structures are said to be 12-coordinate, or to have a coordination number 12. By noting that each corner of the face-centered cubic unit cell is shared by eight other unit cells (see Figs. 14.12c and 14.16a) and that each atom centered in a face "belongs to" two unit cells, we conclude that the total number of metal atoms in the unit cell of the face-centered cube, cubic closest packing, is four.

$$\underset{\text{corners}}{8 \times \tfrac{1}{8}} + \underset{\text{faces}}{6 \times \tfrac{1}{2}} = 4$$

Several metals, such as beryllium, magnesium, ruthenium, zinc, and cadmium, crystallize in the hexagonal closest packed structure. Several others crystallize in the cubic closest packed structure—strontium, aluminum, iron, cobalt, iridium, copper, silver, and gold. The third common crystal structure for metals is the body-centered cubic structure, shown in Fig. 14.12b, in which a metal atom occupies each lattice site. By focusing on the central lattice point in that structure, one notes that the coordination number of a metal atom in the body-centered cubic lattice is 8. Again recalling that each corner atom is shared by eight unit cells, one sees that only

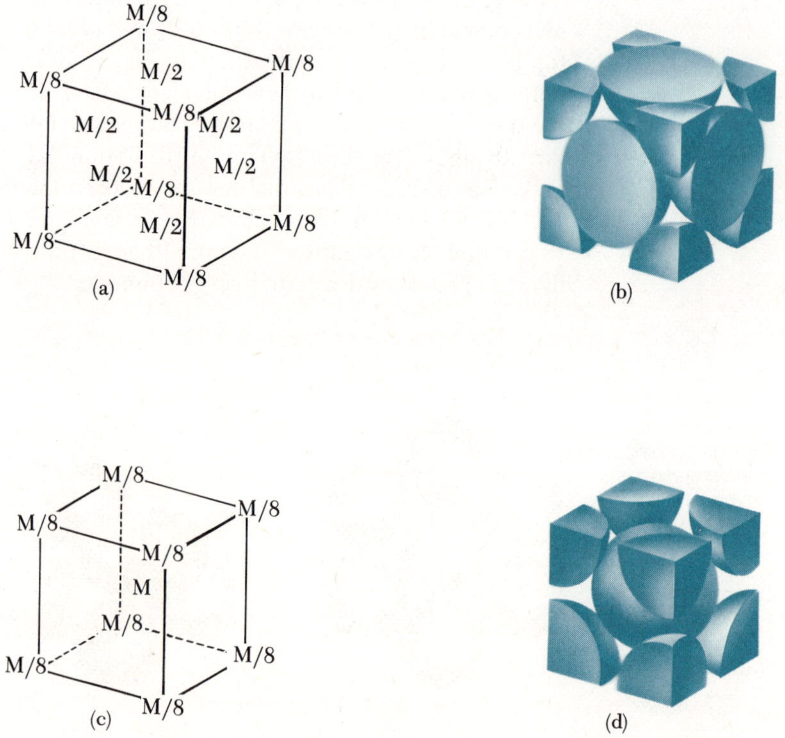

Figure 14.16 Cubic lattices. (a) The face-centered cubic lattice with an atom represented as M. (b) Another view of a face-centered cubic lattice. (c) The body-centered cubic lattice with an atom represented as M. (d) Another view of the body-centered cubic lattice.

TABLE 14.2 **Qualitative Guidelines on the Solubilities[a] of Common Ionic Solids**

Anions	*Soluble Compounds*	*Insoluble Compounds*	*Slightly Soluble Compounds*
NO_3^-, ClO_4^-, ClO_3^-, $CH_3CO_2^-$	Nearly all		$Ag(CH_3CO_2)$, $KClO_4$
Cl^-, Br^-, I^-	Nearly all	Ag^+, Pb^{2+}, Hg_2^{2+}	$PbCl_2$
O^{2-} (oxides)	Li^+, Na^+, K^+, Rb^+, Cs^+, Ba^{2+}	Nearly all	CaO, SrO
OH^-	Li^+, Na^+, K^+, Rb^+, Cs^+, Ba^{2+}, Sr^{2+}	Nearly all	$Ca(OH)_2$
SO_4^{2-}	Nearly all	Pb^{2+}, Sr^{2+}, Ba^{2+}	$CaSO_4$, Ag_2SO_4
CO_3^{2-}, S^{2-}, PO_4^{3-}, SO_3^{2-}	Na^+, K^+, Rb^+, Cs^+, NH_4^+	Nearly all	Li^+

[a]In water at 25 °C.

one-eighth of each corner atom "belongs to" the unit cell. Hence, we conclude that there are two metal atoms in the body-centered cubic cell:

$$\underset{\text{corners}}{8 \times \tfrac{1}{8}} + \underset{\text{body center}}{1} = 2$$

This structure is common for alkali metals and for several transition metals.

Ionic Crystals

An important class of crystals is made up of the many crystalline substances with ionic structures. The structural units of these substances are ions, simple or complex. The stable crystal lattices are those arrangements of ions in which the energy of electrostatic attraction between positive and negative ions outweighs the energy of repulsion between ions of the same electrical charge. This net stability requires lattice arrangements in which cations are surrounded by anions and anions are surrounded by cations.

There is a large variety of crystal structures that occur in this class of solids. Table 14.2 summarizes the common types of compounds one finds typically as ionic crystals and the solubilities one expects of these kinds of substances in water.*

The detailed lattice structure adopted by a particular ionic crystal depends on a variety of factors; the most important are the sizes and charges of the ions and the relative numbers of ions of each type present. The study of ionic crystals is an enor-

*The low solubility of the crystalline $BaCO_3$ constitutes the basis of a simple qualitative test for the existence of CO_2 in a sample of gas. A dilute aqueous solution of $Ba(OH)_2$ is suspended from the end of a stirring rod and held in the gas. A white cloudiness in the drop due to formation of the insoluble barium carbonate confirms the presence of CO_2:

$$Ba^{2+} + 2OH^- + CO_2(g) \rightleftharpoons BaCO_3(s) + H_2O$$

Figure 14.17
Unit cell of CsCl crystal. (a) The eight fold coordination of Cl^- (b) The coordination number of eight for Cs^+. (c) A composite of parts (a) and (b).

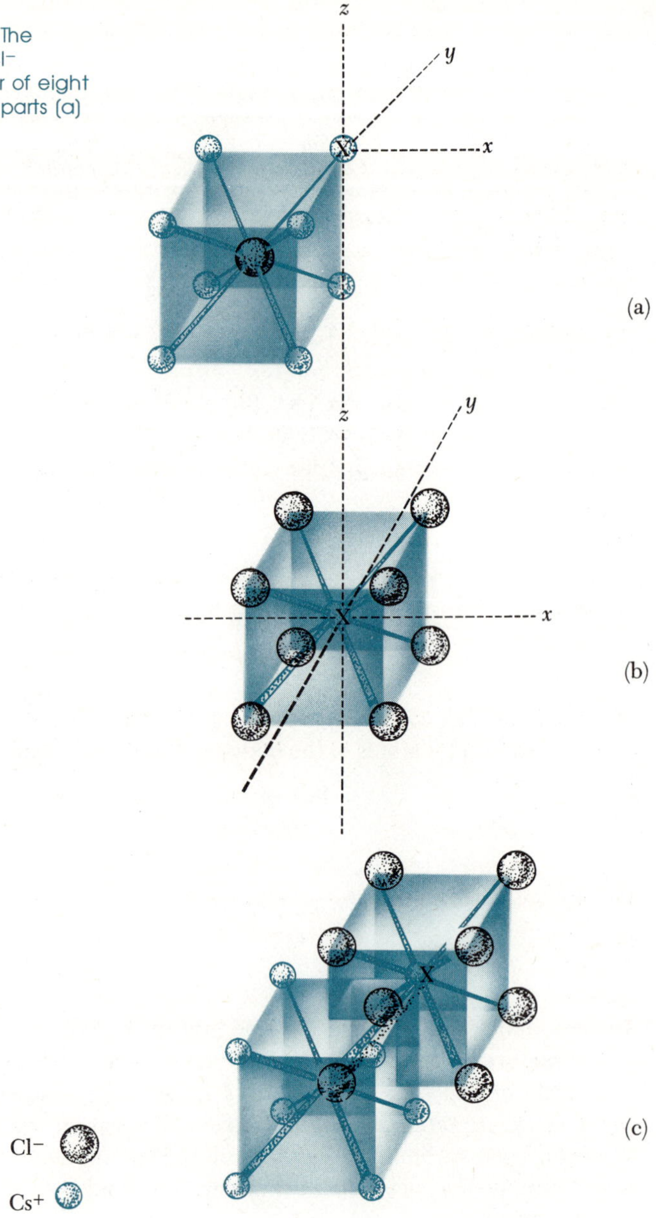

mously complex subject; nevertheless, there are simple examples of this class of substances.

The CsCl Structure

Crystalline cesium chloride is found in a crystal lattice with the type of unit cell shown in Fig. 14.17a. This unit cell has the appearance of a simple cube of cesium ions, with a chloride ion in its center. The color lines in Fig. 14.17a are meant to show interactions of the central chloride ion with its eight nearest neighbors. Accordingly, the coordination number of each chloride ion is 8 in this system. To recognize the coordination number of Cs^+, we can focus attention on the Cs^+ position marked in Fig. 14.17a with "X." One sees that there are eight cubes that share that point, each with an edge that shares one or more edges with the cube itself or with the three dashed axis lines shown. If one recognizes that each of these eight cubes, including the one shown explicitly in Fig. 14.17a, contains a Cl^- at its center, then it becomes clear that the Cs^+ position on which we are focusing is surrounded by eight chloride ions, or coordinated with them. This is emphasized in Fig. 14.17b by representing the specified Cs^+ with this arrangement of eight coordinated Cl^- around it. Clearly, any other Cs^+ can be considered in the same way. Thus, we see that either part a or part b of Fig. 14.17 is a valid way of representing the unit cell. The lattice can thus be described as interpenetrating simple cubic lattices of cesium ions and chloride ions; the corners of the Cl^- cubes penetrate to the centers of the Cs^+ cubes and vice versa. From either part a or b of Fig. 14.17, we can conclude that there is one Cs^+ and one Cl^- in each unit cell, recognizing that each corner ion is shared by eight cubes (so only one-eighth of the corner ion is counted with each cube).

The NaCl Structure

Another ionic crystal structure that is both relatively simple to visualize and representative as an important and common type is the "rock salt" structure, for which the unit cell is pictured in Fig. 14.18a. This unit cell can be represented as the result

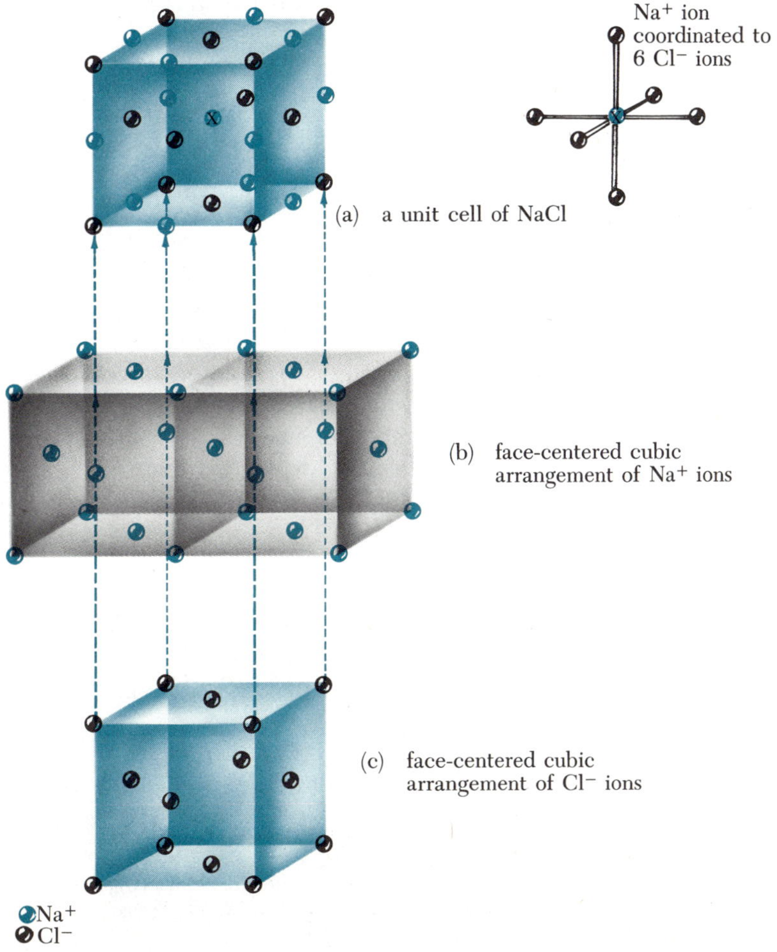

Figure 14.18 Crystal structure of NaCl. If the array of Cl^- ions in (c) is interpenetrated into the array of Na^+ ions in (b), then the overall arrangement produced is shown in (a).

of interpenetration of a face-centered cubic array of sodium ions (seen as two contiguous unit cells in Fig. 14.18b) with a face-centered cubic array of chloride ions (the unit cell shown in Fig. 14.18c). If one imagines bringing the array of part c into the middle of the array in part b, it is readily seen that the lattice depicted in Fig. 14.18a is the net result (the "east" and "west" faces of the array in Fig. 14.18b are neglected). Focusing attention upon the central Na^+ position in the unit cell shown in Fig. 14.18a (marked with "X"), we see, as demonstrated in the drawing on the right, that the ion Na^+ is coordinated to six Cl^-, and it is an octahedral coordination arrangement. This is true of all of the Na^+, which are structurally equivalent. Of course, one could equally well have drawn the unit cell with a chloride ion at the center, using the same general approach, from which it would also be obvious that each ion Cl^- is octahedrally coordinated to six Na^+.*

One additional point that can be extracted from Fig. 14.18a is the number of Na^+ and Cl^- ions in each unit cell. Recalling that one-eighth of each corner Cl^- and one-half of each face-centered Cl^- "belongs" to the unit cell, we calculate the total number of chloride ions in a unit all to be

$$8 \times \tfrac{1}{8} + 6 \times \tfrac{1}{2} = 4$$

Similarly, recognizing that one-fourth of each edge-centered Na^+ and all of the central Na^+ represented in Fig. 14.18a "belong" to the unit cell, we calculate the total number of sodium ions in the unit cell to be

$$12 \times \tfrac{1}{4} + 1 = 4$$

Thus, there are four sodium ions and four chloride ions in the unit cell.

The Ion Size Effect

One might legitimately ask why the crystal structure of CsCl involves a coordination number 8, while the crystal structure of NaCl manifests a coordination number of only 6. Clearly, one should expect that the stability of an ionic system would increase with the number of opposite charges packed around each ion. The answer lies in taking account of the relative sizes of the ions Na^+, Cs^+, and Cl^-. In the NaCl case, the Na^+ is simply too small to accommodate eight Cl^- (which are much larger than the Na^+) in a closely packed arrangement; six Cl^- can, however, pack closely. With the larger Cs^+ there is sufficient space available to pack eight Cl^- at Cs–Cl internuclear distances that are not appreciably larger than they would be if only six Cl^- were coordinated; hence, in this case, the eight-coordinate structure is favored.

Covalent-Network Crystals

In covalent-network crystals, there is a three-dimensional network of covalent bonds that extends throughout the entire crystal. This network forms a rigid framework that holds the crystal together, and generally gives it considerable strength.

The Diamond Structure

A brilliant example of a covalent-network crystal is provided by diamond (pure carbon), the unit cell of which is shown in Fig. 14.19a. Within this lattice is a com-

*One can also see the octahedral coordination of each Cl^- with six Na^+ directly in the unit cell representation of Fig. 14.18a. If we focus on the Cl^- position on the "eastern" face, and recognize that there would be an adjacent ion Na^+ directly east if a larger portion of the lattice were shown, the octahedral arrangement of sodium ions about the Cl^- becomes obvious.

Figure 14.19
Two covalent network crystals: (a) diamond and (b) cristobalite.

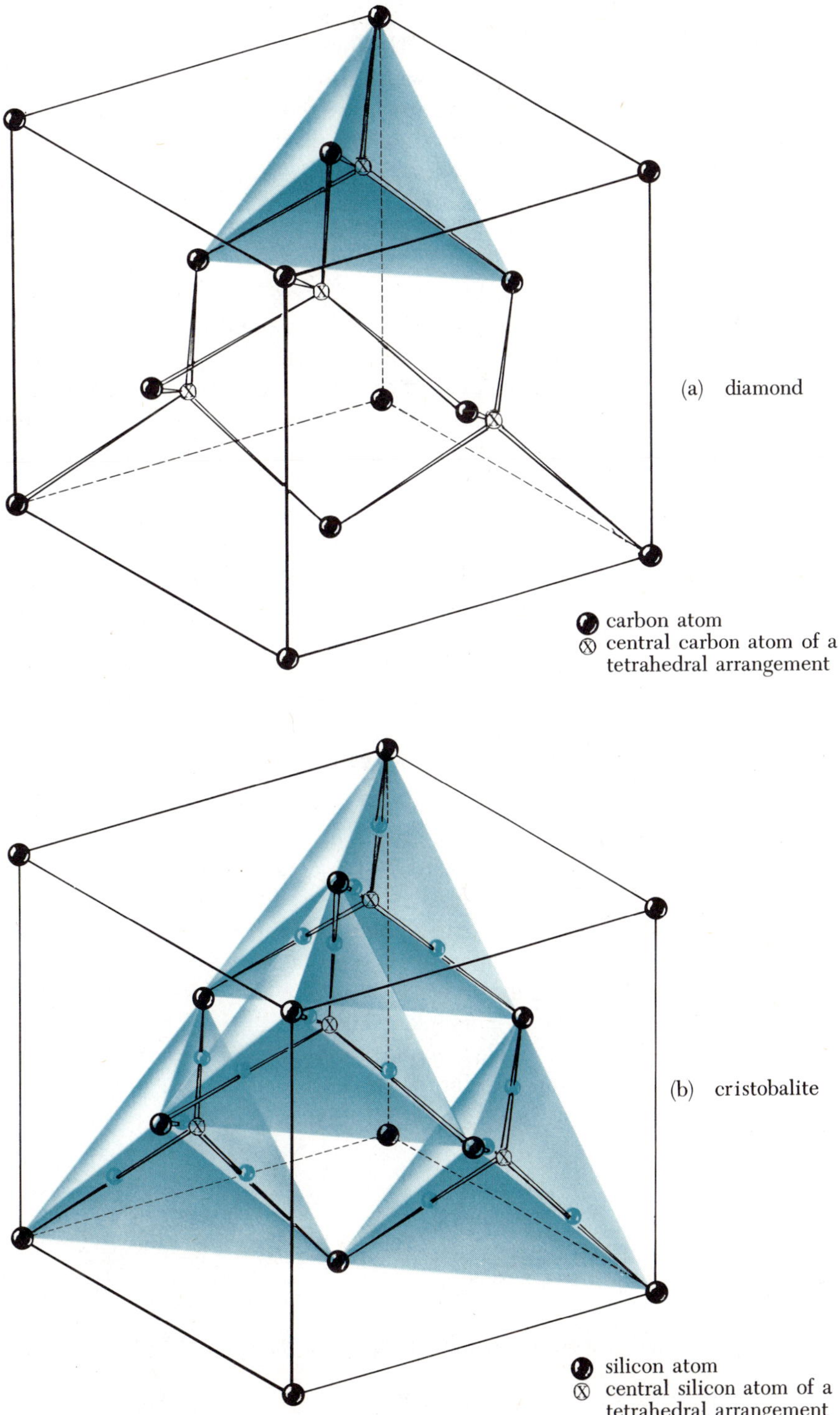

plex arrangement of bonds (only some of which are shown), wherein each carbon atom is connected by four bonds to other carbon atoms in a tetrahedral coordination scheme. Four such tetrahedral arrangements are identified in Fig. 14.19a. The central carbon atoms are shown with the mark "X" and with their four connected carbon atoms. Counting up the atoms in the unit cell, we find the number to be

$$\underset{\text{corner atoms}}{8 \times \tfrac{1}{8}} + \underset{\text{face-centered atoms}}{6 \times \tfrac{1}{2}} + \underset{\text{internal atoms}}{4} = 8$$

The Cristobalite Structure

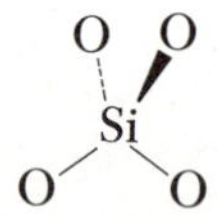

Another relatively simple covalent network crystal is cristobalite, a form of SiO_2; other forms are called quartz and tridymite.

The unit cell of cristobalite is shown in Fig. 14.19b. The relative positions of the silicon atoms are entirely analogous to the carbon atom positions of the diamond structure shown in Fig. 14.19a. Between each pair of silicon atoms that would be bonded if they were carbon atoms in a diamond structure is an oxygen atom in a Si—O—Si arrangement. Each silicon atom is then tetrahedrally bonded (see margin), with each of the four oxygen atoms bonded to another tetrahedral silicon atom. By analogy with the diamond structure, and by inspection of Fig. 14.19b, we see that the cristobalite unit cell contains eight silicon atoms. Recognizing that all the oxygen atoms in the picture of the unit cell are within the borders of the cell and not shared with any other unit cells, a simple count shows that there are sixteen oxygen atoms in the unit cell. This analysis yields the expected 1:2 silicon-to-oxygen ratio for SiO_2.

Van der Waals Crystals (Molecular Crystals)

Like covalent-network crystals, van der Waals crystals have the characteristic that the strong bonds present are covalent. However, in contrast with the covalent-network crystals, the covalent bonds of the van der Waals crystals are not directly involved in holding the crystal together. Covalent bonds in a van der Waals crystal are responsible for holding discrete molecules intact, and these molecules are maintained in a crystal lattice only by weak van der Waals interactions (Sec. 14.1). This is the kind of crystal structure that characterizes organic compounds and covalently bonded inorganic compounds in the solid state.

A simple example of a van der Waals crystal is solid pyrazine, $C_4N_2H_4$, an aromatic compound for which the molecular structure is

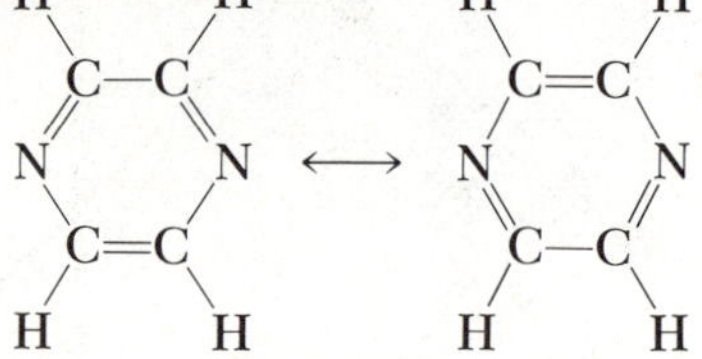

and for which we shall use the symbol . The unit cell for this crystal is shown in Fig. 14.20. The lattice type is rhombohedral and one sees that there is one molecule placed at each corner. As each of these is shared by eight unit cells, we see that there is one molecule ($8 \times \frac{1}{8}$) in a unit cell.

Figure 14.20 Unit cell of a pyrazine crystal showing a pyrazine molecule at each corner in a rhombohedral lattice; the plane of each molecule is inclined at about 30° to a vertical plane. Each C in this figure stands for a CH unit in the $C_4H_4N_2$ ring.

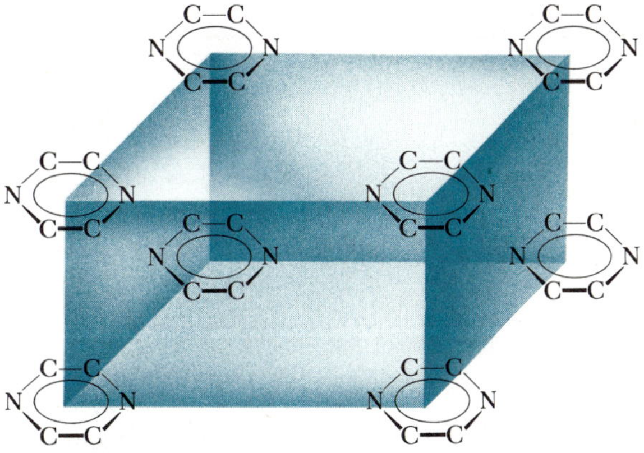

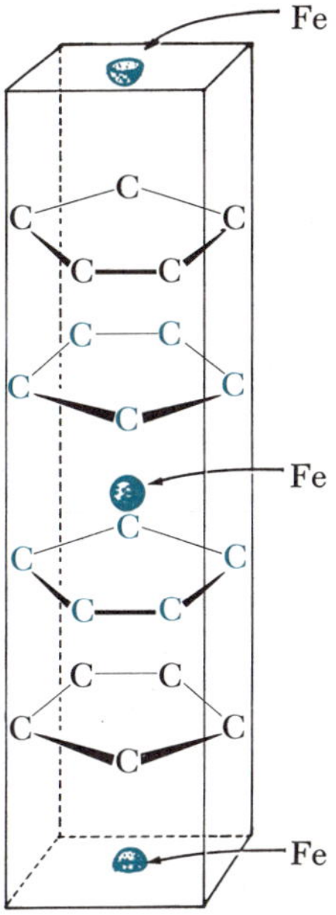

Figure 14.21 Monoclinic unit cell of the molecular crystal of ferrocene, $Fe(C_5H_5)_2$. C's shown in color represent CH portions of the C_5H_5 rings bonded to the central iron atom (shaded Fe). The unshaded Fe represent iron atoms at the faces of the unit cell, thus only "half-belonging" to the unit cell.

As mentioned above, the occurrence of a molecule, ion, or atom at a lattice point in a Bravais lattice is by no means general, and should perhaps be considered the exception rather than the rule. The examples we have shown, like pyrazine in Fig. 14.20, were chosen because of their relative simplicity in pictorial representation. Figure 14.21 shows the unit cell (monoclinic lattice) of a crystal of ferrocene, $Fe(C_5H_5)_2$. Clearly, in this case, the molecules do not lie at lattice points; this is the more common situation.

Spotlight

Ferrocene, $Fe(C_5H_5)_2$, is a "sandwich compound," which can be

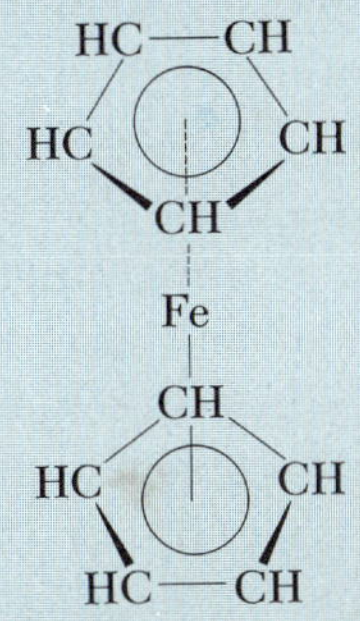

thought of as being constructed from Fe^{2+} and two $C_5H_5^-$:

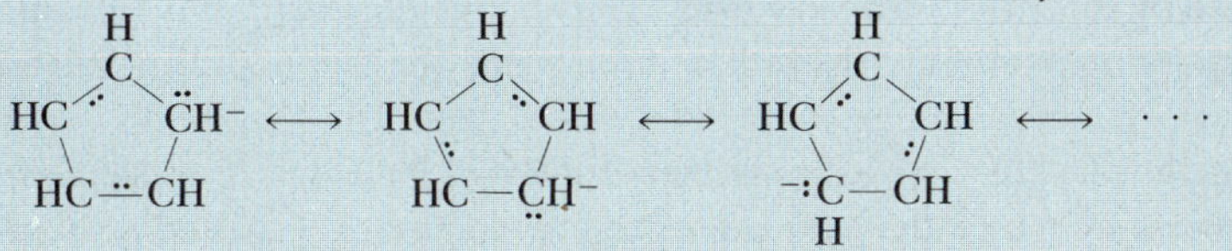

The electronic structure can be rationalized in terms of the six bonding electrons contributed by each of the ions $C_5H_5^-$ (those shown explicitly as dots in the resonance structure above) and the six valence-shell electrons contributed by Fe^{2+}, giving a rare-gas configuration,

$$[Ar]\ \underset{4s}{\underline{\uparrow\downarrow}}\quad \underset{3d}{\underline{\uparrow\downarrow}\ \underline{\uparrow\downarrow}\ \underline{\uparrow\downarrow}\ \underline{\uparrow\downarrow}\ \underline{}}\quad \underset{4p}{\underline{\uparrow\downarrow}\ \underline{\uparrow\downarrow}\ \underline{\uparrow\downarrow}\ \underline{\uparrow\downarrow}}$$

Each carbon atom bears the same bonding relation as any other carbon atom to the iron atom. This compound, discovered almost by accident in 1950, has proved to be the historical basis for a wide variety of interesting sandwich compounds that have been synthesized and studied since that time.

STUDY PROBLEM **14(c)**

On the basis of Fig. 14.21, determine the number of ferrocene molecules in the unit cell of a ferrocene crystal.

14.5 Polymers

Macromolecules

Fitting the popular concept of a solid, but usually not the scientific definition, is the important class of substances called **macromolecules,** meaning giant molecules. An important type of macromolecule is **polymers,** which are macromolecules that are made up of many repeating units. Strictly speaking, any molecule that can be represented as more than one repeating unit can be called a polymer (for example, dimer, two units; trimer, three units); but in common usage, the word *polymer* is typically used for substances with molecular weights in the range of thousands or millions of atomic mass units.

Clearly the most important polymers consist of biological macromolecules. The important classes of biological polymers include proteins, one of the main structural materials of animal bodies, nucleic acids, which determine genetic characteristics in organisms, and carbohydrate polymers, which are the structural materials of plants. (Two of these types of biological macromolecules were discussed in Sec. 8.3 in connection with hydrogen bonding interactions, which are very important in determining the geometrical configurations of these polymeric species.)

Another important class of polymers is the synthetic polymer. These polymers are substances that are manufactured commercially because of a need for their favorable structural properties. The materials that are popularly called plastics, synthetic rubber, and synthetic fabrics are synthetic polymers. Nylon, Orlon, and polyester are well-known synthetic fabrics.

Synthetic Polymers

Commercial Importance

There has been incredible progress in the chemical industry during the past few decades in the development and manufacture of synthetic polymers. These manufactured polymers have, over a period of years, replaced a variety of natural polymers—like natural rubber, cotton, wool, and wood—and also metals in thousands of manufactured products. In most cases the desired structural properties of a synthetic polymer surpass the properties of its natural counterpart in a particular product, and the product is generally less expensive to produce with the synthetic substance. The other side of that coin, however, is the fact that the nearly uncontrolled growth of the synthetic polymer industry during the past thirty years has brought with it some of the pollution problems, like the indestructible plastic containers, now faced by the industrialized world. The chemical industry is now turning some of its impressive capacities to the task of minimizing these undesirable features.

Polymerizations

The individual molecules from which a synthetic polymer is made are called **monomers** and the process by which they are joined to form polymers is called **polymeri-**

Spotlight

Wool is a natural polymer of the protein class (Sec. 8.3). Natural rubber is a polymer with the structure

$$-(CH_2-CH=C-CH_2)(CH_2-CH=C-CH_2)(CH_2-CH=C-CH_2)-$$
$$\qquad\qquad\quad CH_3 \qquad\qquad\quad CH_3 \qquad\qquad\quad CH_3$$

where three of the repeating units are shown; the parentheses indicate the identity of the repeating unit. Cotton and wood are carbohydrate polymers of the cellulose type, cotton being almost pure cellulose. The basic cellulose structure is shown below, with the repeating unit emphasized in brackets.

Natural polymers have been replaced in many applications by synthetic polymers. However, these natural polymers come from renewable resources (animals and plants), whereas synthetic polymers are made from nonrenewable resources (oil). The natural polymers are likely to experience a resurgence in popularity as nonrenewable resources become depleted.

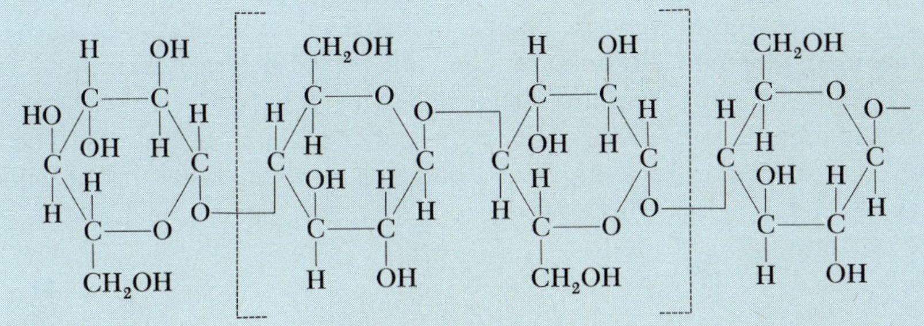

zation. An example of this process is the polymerization of the monomer ethylene to form the polymer polyethylene. We can visualize polymerization by the following steps.

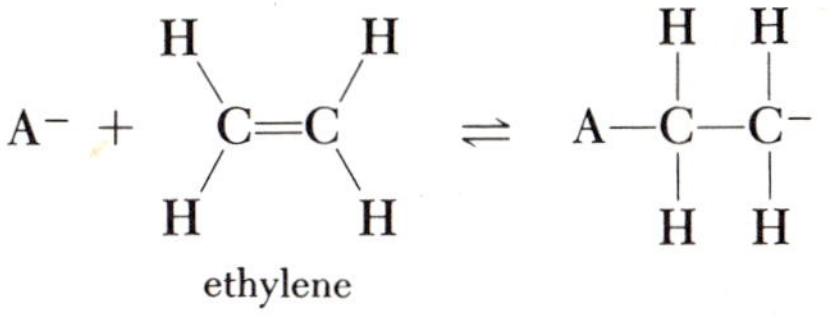

$$A^- + \underset{\text{ethylene}}{CH_2=CH_2} \rightleftharpoons A-CH_2-CH_2^- \qquad \text{(initiation)} \qquad (14.4a)$$

$$A-CH_2-CH_2^- + CH_2=CH_2 \rightleftharpoons$$
$$A-CH_2-CH_2-CH_2-CH_2^- \qquad \text{(propagation)} \qquad (14.4b)$$

$$A-CH_2-CH_2-CH_2-CH_2^- + CH_2=CH_2 \rightleftharpoons$$
$$A-CH_2-CH_2-CH_2-CH_2-CH_2-CH_2^- \qquad \text{(propagation)} \qquad (14.4c)$$

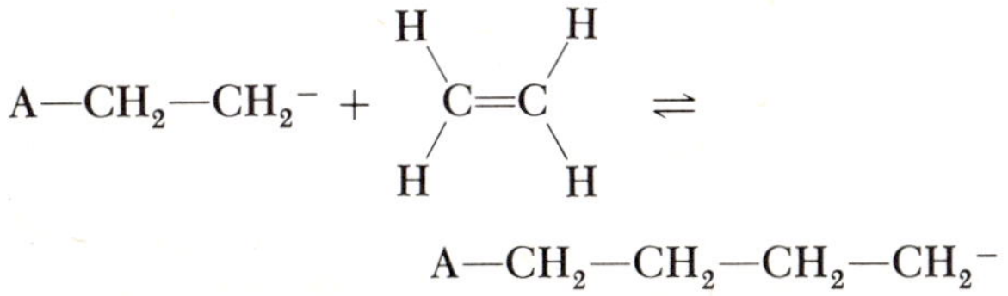

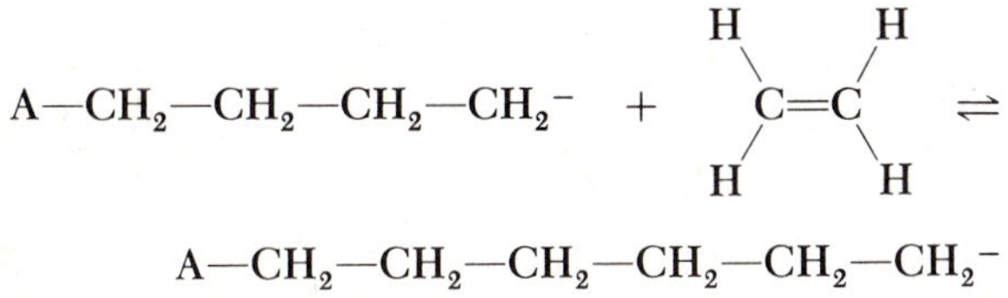

In this sequence the symbol "A⁻" stands for an **initiator,** the species that initiates the polymerization. In each step of the polymerization a $-CH_2-CH_2-$ unit is added; this unit corresponds in atomic composition to the monomer ethylene, $H_2C{=}CH_2$. After many monomer units have been added, say tens or hundreds of thousands, a process of the following type occurs:

$$A-(CH_2-CH_2-)_n^- + H_2C{=}CH_2 \rightleftharpoons (CH_2-CH_2)_n + A-CH_2-CH_2^- \qquad \text{(termination)} \quad (14.4d)^*$$

This step terminates the formation of one polyethylene molecule, $(CH_2-CH_2)_n$, and provides an initiator for making another one.

The polymerization represented by Eq. (14.4) is called **addition,** or **chain, polymerization,** because the initiation step (14.4a) starts a chain of reactions that is terminated for a particular growing polymer molecule in the step (14.4d). Much of the sophistication involved in the technology of producing these kinds of polymers is involved in choosing an appropriate initiator, A^-, and a set of conditions that will lead to molecular weights in a desired range, that is, n values in a desired range.†

There are many commercially important polymers of the general type represented by polyethylene. Some of them are shown in Table 14.3.

Not all synthetic polymers are based on the polymerization of just one type of monomer unit. An important example of a polymer involving two types of monomers, a **copolymer,** is nylon. This is produced in a process that can be depicted in a sequence of steps:

$$HO-\overset{\overset{\displaystyle O}{\|}}{C}-CH_2-CH_2-CH_2-CH_2-\overset{\overset{\displaystyle O}{\|}}{C}-OH + H_2N-CH_2-CH_2-CH_2-CH_2-CH_2-CH_2-NH_2 \rightleftharpoons$$
$$H-O-\overset{\overset{\displaystyle O}{\|}}{C}(CH_2)_4-\overset{\overset{\displaystyle O}{\|}}{C}-\overset{\overset{\displaystyle H}{|}}{N}(CH_2)_6-NH_2 + H_2O \qquad (14.5a)$$

$$HO-\overset{\overset{\displaystyle O}{\|}}{C}(CH_2)_4-\overset{\overset{\displaystyle O}{\|}}{C}-NH(CH_2)_6-NH_2 + HO-\overset{\overset{\displaystyle O}{\|}}{C}(CH_2)_4-\overset{\overset{\displaystyle O}{\|}}{C}-OH \rightleftharpoons$$
$$H_2O + HO-\overset{\overset{\displaystyle O}{\|}}{C}(CH_2)_4-\overset{\overset{\displaystyle O}{\|}}{C}-NH(CH_2)_6-NH-\overset{\overset{\displaystyle O}{\|}}{C}(CH_2)_4-\overset{\overset{\displaystyle O}{\|}}{C}-OH \qquad (14.5b)$$

---

*The formula $(CH_2-CH_2)_n$ for the final polymer in Eq. (14.4d) is not strictly correct as it would leave carbon atoms with only three covalent bonds on each end $\left(-C\begin{smallmatrix}H\\ \\H\end{smallmatrix}\right)$. Something like $CH_3-CH_2-(CH_2-CH_2)_{\overline{n-2}}CH{=}CH_2$ or $CH_3-CH_2-(CH_2-CH_2)_{\overline{n-2}}CH_2-CH_3$, where two hydrogen atoms have been abstracted from some source, would be more proper.

†The particular kind of chain polymerization represented in Eq. (14.4) is **anion polymerization,** because it is initiated and propagated by anions, for example, anions represented as A^- and $A-CH_2-CH^-$ in Eq. 14.4. Other common types of chain polymerization are **cation polymerization,** in which the initiation and propagation involve cations (say X^+ and $X-CH_2-CH_2^+$) and **radical polymerization,** involving initiation and propagation steps in which species with unpaired electrons (radicals) occur, for example, $R\cdot$ and $R-CH_2-CH_2\cdot$.

TABLE 14.3 **Some Common Polymers of the Polyethylene Type**

Monomer		*Polymer*		
Name	*Structure*	*Name*	*Structure*	*Common Use*
Ethylene	$H_2C{=}CH_2$	Polyethylene	$\text{+}CH_2{-}CH_2\text{+}_n$	Plastic containers, utensils
Propylene	$H_2C{=}C(CH_3)(H)$	Polypropylene	$(CH_2{-}\underset{H}{\overset{CH_3}{C}}\text{+}_n$	Plastic containers, and other items
Isobutylene	$H_2C{=}C(CH_3)(CH_3)$	Polyisobutylene	$\text{+}CH_2{-}\underset{CH_3}{\overset{CH_3}{C}}\text{+}_n$	Synthetic rubber
Vinylchloride	$H_2C{=}C(Cl)(H)$	Polyvinylchloride (PVC)	$\text{+}CH_2{-}\underset{H}{\overset{Cl}{C}}\text{+}_n$	Plastic pipe
Methylmethacrylate	$H_2C{=}C(CH_3)(C{=}O{-}O{-}CH_3)$	Polymethylmethacrylate	$\text{+}CH_2{-}\underset{C(=O){-}O{-}CH_3}{\overset{CH_3}{C}}\text{+}_n$	Plexiglas
Styrene	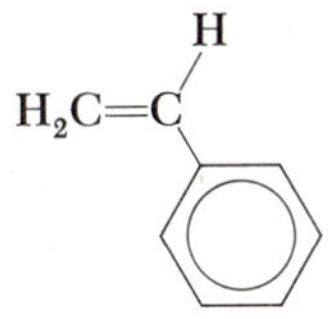	Polystyrene	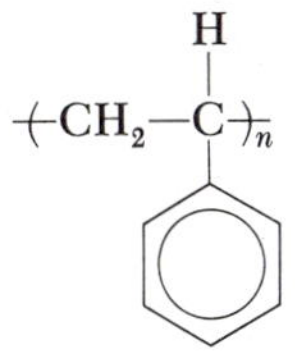	Foam plastic packing, adhesives
Tetrafluoroethylene	$F_2C{=}CF_2$	Polytetrafluoroethylene, Teflon®	$\text{+}CF_2{-}CF_2\text{+}_n$	Bearings, cooking surfaces

$$HO{-}\overset{O}{\overset{\|}{C}}(CH_2)_4{-}\overset{O}{\overset{\|}{C}}{-}NH(CH_2)_6{-}NH{-}\overset{O}{\overset{\|}{C}}(CH_2)_4{-}\overset{O}{\overset{\|}{C}}{-}OH + H_2N(CH_2)_6{-}NH_2 \rightleftharpoons$$

$$H_2O + HO{-}\overset{O}{\overset{\|}{C}}(CH_2)_4{-}\overset{O}{\overset{\|}{C}}{-}NH(CH_2)_6{-}NH{-}\overset{O}{\overset{\|}{C}}(CH_2)_4{-}\overset{O}{\overset{\|}{C}}{-}NH(CH_2)_6{-}NH_2 \qquad (14.5c)$$

$$\vdots$$

The two monomers used in the synthesis of this important copolymer are shown as the reactants in Eq. (14.5a).

The production of nylon is an example of a class of polymerization reactions called **condensation polymerization.** In these polymerizations two monomers are "condensed" to form a dimer and a small molecule (water, in the nylon case), which in turn is condensed with another monomer to form a trimer (and another small molecule, H_2O), and so on, until a large polymer is produced. The fibers used in synthetic fabrics are usually produced by this kind of polymerization.

Polymer Characteristics

How does one classify polymers in terms of the three states of matter? In many cases these materials qualify as amorphous solids (Sec. 14.2). In some cases, however, there is sufficient order in the polymer, or in specific regions of a polymer, that some degree of crystal-like behavior, **crystallinity**, is observed. Figure 14.22 shows pictorially how this can occur. To the extent that a polymer can be considered crystalline, it has the characteristics of both a covalent-network crystal, wherein covalent bonds hold each macromolecule intact, and of a van der Waals crystal, wherein the interactions responsible for holding molecules together in a van der Waals crystal also hold the macromolecules of a polymer together.

The physical and chemical properties of polymers are as varied as chemistry itself. Depending on the functional groups present in the monomers, the specific manner in which the monomers are put together in the macromolecules, and the overall configuration and size of the macromolecules, a huge variety of chemical and physical properties is possible. In biological systems, nature has had many millions of years in the evolutionary process to develop polymers with specific chemical and physical properties to meet specific needs. This has given rise, for example, to proteins of widely different properties, such as those that make up hair, cartilage, fingernails, and membranes. In the case of synthetic polymers, chemists have attempted to "design" polymers with a specific set of desired properties. This has largely been a successful effort. Most of the basic information that is available for correlating the chemical and physical properties of a polymer with its detailed structure has been developed during the past three decades.

Many fundamental structural characteristics contribute to the physical properties of polymers—to their degree of crystallinity, brittleness, strength, elasticity, and the like. Two of these are the degree of *branching*, and the degree of *cross-linking*. The cases discussed specifically above have been represented as strictly *linear* polymers,

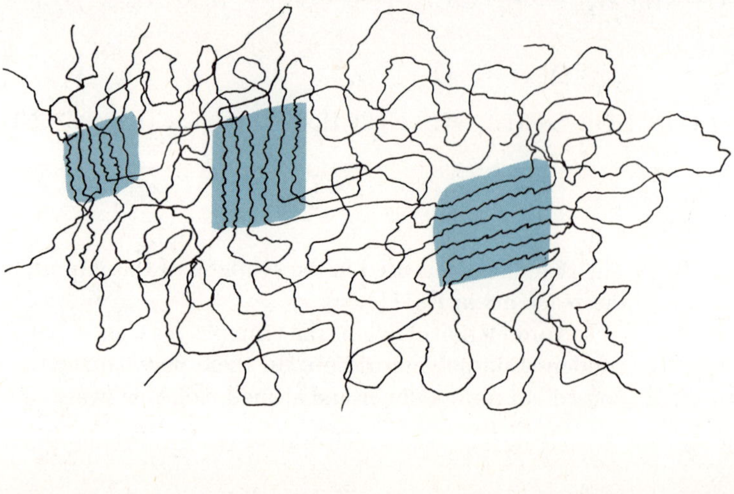

Figure 14.22
Representation of a solid polymer with three crystalline regions (shaded) in an otherwise amorphous structure.

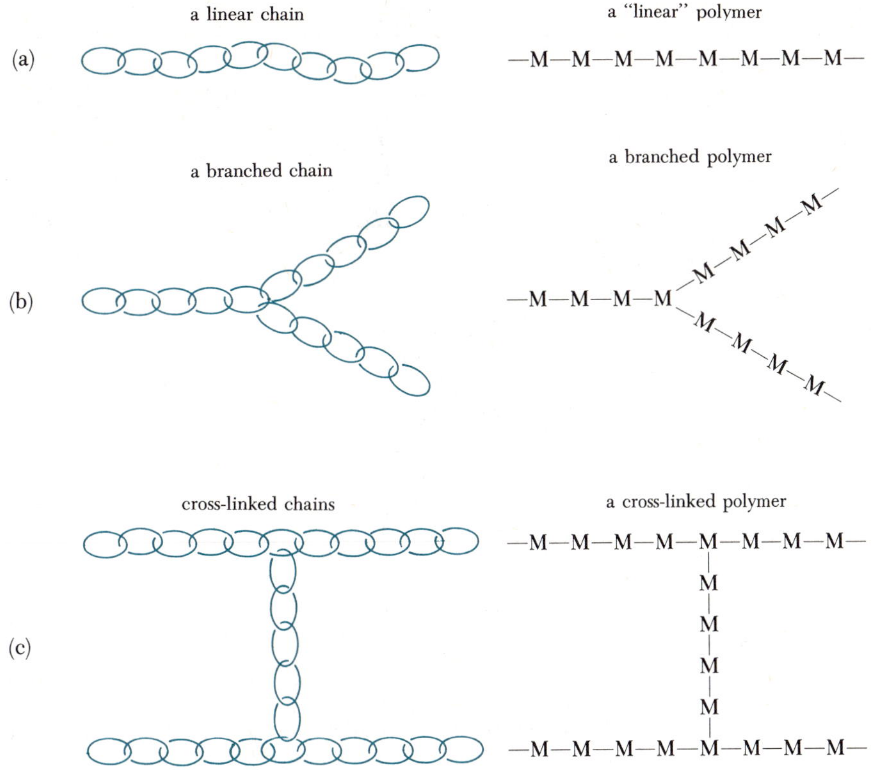

Figure 14.23
Branching and cross-linking in polymers; M represents the repeating unit in a polymer (related to the monomer).

(a) a linear chain   a "linear" polymer
—M—M—M—M—M—M—

(b) a branched chain   a branched polymer
—M—M—M—M(—M—M—M—M / —M—M—M—M)

(c) cross-linked chains   a cross-linked polymer
—M—M—M—M—M—M—M—M—
                    |
                    M
                    |
                    M
                    |
                    M
                    |
                    M
                    |
—M—M—M—M—M—M—M—M—

that is, polymers in which monomers are added one after another like the links in a simple steel chain (Fig. 14.23a). However, for many polymers it is possible to obtain structures in which branching occurs, analogous to attaching two steel chains to a single link of a third chain (Fig. 14.23b). In cross-linked polymers, two or more polymer chains are "linked" by additional chains between them, analogous to the case shown in Fig. 14.23c. The strength, brittleness, melting properties, and so on of a "plastic" are very dependent on how much branching and cross-linking are present in polymers.*

---

*It is very easy to see structurally how branching and cross-linking can occur in a polymer of the nylon type:

$\sim$C(CH$_2$)$_4$—C—NH(CH$_2$)$_6$—N$\langle$ C(CH$_2$)$_4$—C—NH(CH$_2$)$_6$—NH—C(CH$_2$)$_4$—C—NH(CH$_2$)$_6$—NH$\sim$ / C(CH$_2$)$_4$—C—NH(CH$_2$)$_6$—NH—C(CH$_2$)$_4$—C—NH(CH$_2$)$_6$—NH$\sim$

$\sim$C(CH$_2$)$_4$—C—NH(CH$_2$)$_6$—N—C(CH$_2$)$_4$—C—NH(CH$_2$)$_6$—NH$\sim$
                                           |
                                          C=O
                                           |
                                         (CH$_2$)$_4$
                                           |
                                          C=O
$\sim$C(CH$_2$)$_4$—C—NH(CH$_2$)$_6$—N—C(CH$_2$)$_4$—C—NH(CH$_2$)$_6$—NH$\sim$

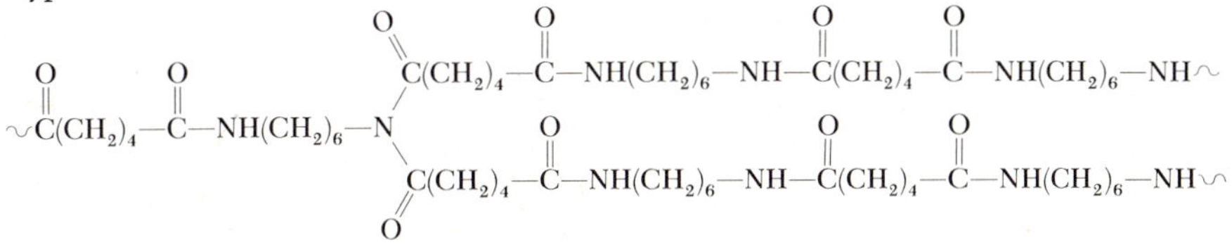

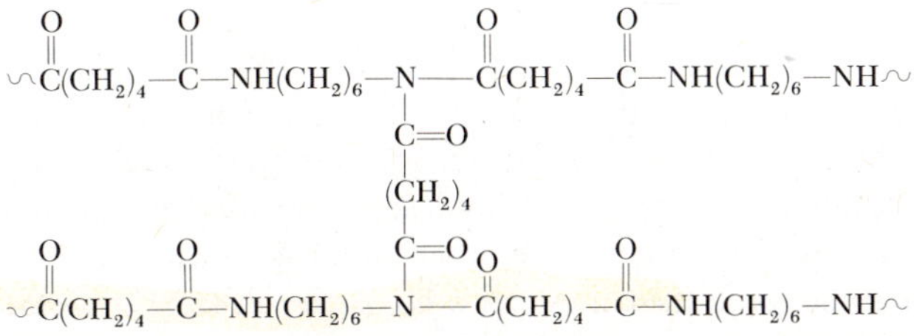

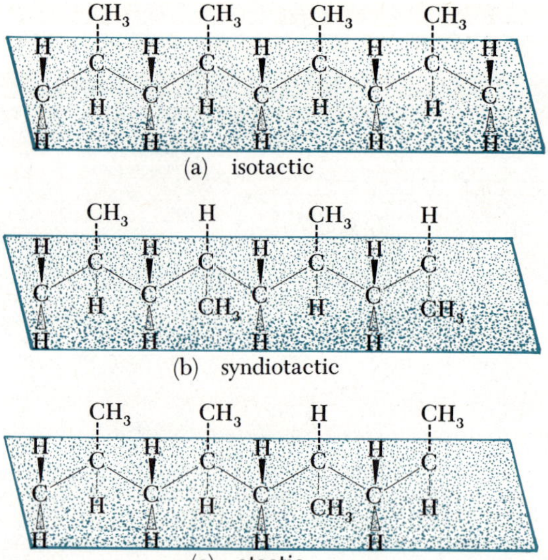

Figure 14.24 Representations of polypropylene macromolecule. (a) Isotactic. (b) Syndiotactic. (c) Atactic. Heavy dashed lines and black wedges are above the reference plane shown, light dashed line and shaded wedges below it. Wedges represent bonds oriented towards the reader; dashed lines represent bonds directed away from the reader.

Considerably more subtle structural details than branching and cross-linking can also have profound effects on the properties of a polymer. An important characteristic in many of the polymers of the type shown in Table 14.3 is the geometrical arrangement, or **stereochemistry,** of the polymer. As seen in Fig. 14.24 for polypropylene, there are three main types of stereochemical arrangements one can envision. In this figure, the main carbon chain network is represented in a single plane. In Figure 14.24a, the CH_3 groups are all shown on one side of the plane; this is called the **isotactic** arrangement. In part b, the CH_3 groups are seen to alternate regularly on both sides of the plane; this is the **syndiotactic** arrangement. In the arrangement depicted in part c, referred to as **atactic,** the methyl groups occur randomly on both sides. The physical properties of polypropylene strongly depend on which of these three arrangements predominates. Atactic polypropylene is a rubbery, soft elastic material. Both syndiotactic and isotactic polypropylenes are much less rubbery, as they are highly crystalline; the regularity in their structures permits the molecules to pack together efficiently in a regular crystalline lattice. The conditions of the polymerization, for example, the choice of initiator, largely govern the type of polypropylene that is produced.

14.6 Properties of Solids

The chemical and physical properties of solids and their relations to structural details largely constitute the subject matter of a complex and important field often referred to as materials science. The chemical properties of a solid usually reflect the properties of the corresponding liquid, and typically depend on the surface area of the solid available for contact with other molecules or ions with which reaction may occur. In some cases, the contact of other molecules on the surface of a solid, called **adsorp-**

tion, makes possible certain chemical reactions that would not occur without the influence of the surface of the solid.*

The physical properties of a solid depend in many interrelated ways on the submicroscopic structure of the solid. The detailed nature of these interrelations constitutes a very sophisticated subject, one in which materials scientists and engineers have made enormous strides during the past several years. Such progress has been an important factor not only in products manufactured for popular consumption but also in critical advances in space and medical technology.

Physical Properties

For many practical applications of solids, the most pertinent physical properties are essentially "mechanical" ones, for example, strength, brittleness, hardness, and melting point. From the nature of the bonding and forces that hold a solid together, it is possible to make some rough generalizations about these kinds of properties.

That a substance exists in the solid state under any conditions is evidence that appreciable attractive forces exist between the molecular or ionic species of the crystal under those conditions. As the temperature of a solid is raised, at some temperature the forces of attraction are overcome by the tendency for random motion of the particles, and the free energy of the liquid (or gaseous) phase becomes equal to the free energy of the solid phase. At this temperature, melting (or sublimation) occurs. This temperature, called the **melting point** (or sublimation point), is the temperature at which the two phases are in equilibrium with each other, and the free energy change for the melting (or sublimation) is zero:

$$\Delta G = \Delta H - T\,\Delta S = 0 \tag{14.6}$$

Thus, at the melting temperature, T is large enough that the term $T\,\Delta S$ equals the ΔH term. Hence, for melting

$$\Delta H = T\,\Delta S$$

$$T_{\text{melting}} = \left(\frac{\Delta H}{\Delta S}\right)_{\text{melting}} \tag{14.7}$$

From this we can see that a high melting point (or sublimation point) is favored by a large ΔH of fusion (or sublimation), that is, by strong attractive forces in the solid relative to the liquid (or relative to the gas). A high melting temperature is also favored by a small value of ΔS for the fusion (or sublimation) process. The ΔS values for the melting (fusion) or the sublimation of a solid and for the vaporization of a liquid are always positive, because the solid is a much more highly ordered state than the liquid, which in turn is less random than the corresponding gas. Solids that are held together by strong attractive forces requiring considerable energy to be disrupted for melting (or sublimation) have high values of ΔH_{fusion} (or ΔH_{subl});

*The surface of a solid can sometimes provide a reaction site, which makes possible a reaction pathway of lower energy than what would be available in the absence of the surface site. The availability of a lower energy pathway (Sec. 13.4) can increase the rate of an otherwise sluggish reaction to the point where it occurs rapidly.

TABLE 14.4 **Some Physical Properties of Representative Solids**

Substance	*Type of Solid*	*Strongest Interactions Available*	*Electrical Conductivity*	*Melting Point, °K*	ΔH_{fusion}, *kcal/mol*
CH_4	Molecular crystal	Dispersion	Very low	89	0.23
C_6H_6, benzene	Molecular crystal	Dispersion (highly polarizable)	Very low	278.5	2.37
HCN	Molecular crystal	Dipole-dipole (hydrogen bonding)	Very low	259	2.01
H_2O	Molecular crystal	Dipole-dipole (hydrogen bonding)	Very low	273.2	1.44
Hg	Metallic	Metallic bonding	Very high	234	0.57
Na	Metallic	Metallic bonding	Very high	371	0.63
Cu	Metallic	Metallic bonding	Very high	1356	3.11
NaCl	Ionic	Ionic bonding	Very low	1073	6.80
$MgSO_4$	Ionic	Ionic bonding	Very low	1468	3.50
SiO_2 (cristobalite)	Covalent network	Covalent bonding	Very low	2001	1.84
C (diamond)	Covalent network	Covalent bonding	Very low	4770	143
Polyethylene	Polymer	Covalent bonding and dispersion	Very low		

hence, such solids are expected to have high melting points; however, the fact that ΔS for the melting must also be taken into account precludes the validity of predicting trends in melting points from only a knowledge of the strengths of the attractions. The melting points and heats of fusion of a few representative solids are listed in Table 14.4.

EXAMPLE 14.3 Two crystalline forms of SiO_2, quartz and cristobalite, have melting points of 1610 °C and 1728 °C, and corresponding molar entropies of fusion ΔS_{fus} of 1.10 cal deg^{-1} and 0.92 cal deg^{-1}. Which form of SiO_2 has stronger crystal forces?

SOLUTION The strengths of the crystal forces are reflected in the molar heats of fusion ΔH_{fus}. These can be computed from Eq. (14.7), $\Delta H_{fus} = T\,\Delta S_{fus}$. For quartz

$$\Delta H_{fus} = (1610 + 273)\text{deg}(1.10 \text{ cal deg}^{-1}) = 2.07 \times 10^3 \text{ cal for one mole}$$

For cristobalite,

$$\Delta H_{fus} = (1728 + 273)\text{deg}(0.92 \text{ cal deg}^{-1}) = 1.8 \times 10^3 \text{ cal for one mole}$$

Hence, the crystal forces in quartz are stronger.

STUDY PROBLEM **14(d)**

Knowing that the heat of fusion of one gram of AgBr is 48.5 J and the melting point is 430 °C, calculate ΔS_{fus} (per mole, in J deg^{-1}).

Other properties that one expects to be correlated strongly with the strengths of the attractive interactions in solids are the macroscopic mechanical strength of the solid and its hardness. The solids with the weakest attractive forces are those that are held together only by van der Waals interactions. Such solids are found to be soft and mechanically weak—easily fractured under stress—and to have low melting points, as seen in Table 14.3. The types of solids with the strongest forces holding the structural entities together are the covalent-network solids, which are held together by a network of covalent bonds; this type of solid is characterized by very high melting points, great hardness, and high mechanical strength. Diamond is one example. High melting points and substantial mechanical strength and hardness are also characteristic of ionic crystals.

The class of solids manifesting the greatest variation in mechanical properties is the metallic solids. Their melting points range from below room temperature (mercury, −38.9 °C) to thousands of degrees C (tungsten, 3410 °C), with corresponding variation in mechanical strength and hardness. These variations can be attributed to variations in the sizes of the metal ions packed together in the crystal lattice and to the number of electrons involved in the bonding, as well as to the nature of the bonding orbitals. The arrangement of the metal ions in closest packed lattices is responsible for the very large densities of the heavy metals; in some cases they are about ten times as dense as many common van der Waals crystals.

Although polymers typically do not have sharp melting points, a high degree of crystallinity can give rise to rather narrow "melting ranges," which tend to fall in the same general temperature span as the melting points of van der Waals crystals. A typical synthetic polymer has, however, generally much greater mechanical strength than a van der Waals crystal does. The polymer owes its mechanical strength largely to the covalent bonds holding each macromolecule intact.

Electrical Conductivity

General Considerations

One of the most dramatic and important physical properties of metallic solids is their remarkable ability to conduct electricity. This phenomenon in metals depends on the flow of electrons from one end of a metal object, like a wire, to the other end when an electrical potential is applied across the object as part of a complete electrical circuit. The situation is represented in Fig. 14.25. The electrons depleted

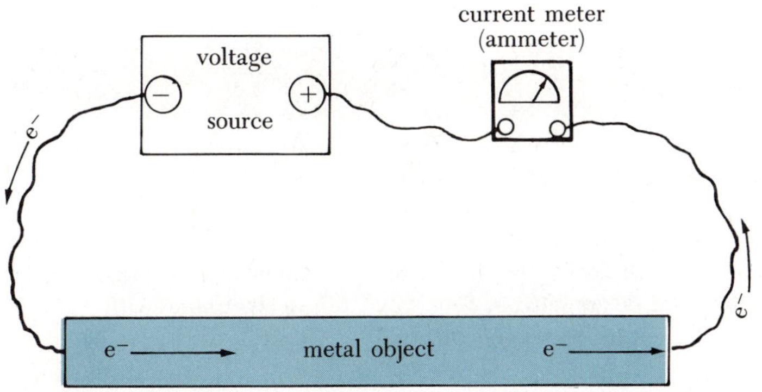

Figure 14.25 Conduction of electricity (electrons, e−) through a metal object under the influence of an applied electrical potential (from the voltage source).

Spotlight

Graphite is an important exception to the rule of thumb that molecular crystals are insulators; this material is one of two crystalline forms of elemental carbon; the other is diamond. The crystal structure of graphite is shown in Sec. 9.2. This substance is a soft, grayish blue solid with a metallic luster and a high electrical conductivity. The softness and high electrical conductivity are consistent with the layered structure of connected six-membered carbon rings, each one analogous in some respects to benzene. Each carbon atom is bound by single covalent bonds to three other carbon atoms. In addition, for every ring of six carbon atoms, there are three pairs of electrons used in resonance-type π bonding, analogous to the benzene case. These π electrons complete the octet for each carbon atom, and provide a framework of "loosely held" electrons that are similar to the conduction electrons of a metal, hence, the metallic lustre and electrical conductivity of graphite. Each layer of carbon atoms is tightly bound, but the forces between layers are weak, allowing them to "slip" over each other, which is the reason graphite is soft and a useful lubricant.

at the end from which electrons flow, the left side, are replaced by electrons supplied by the wire connected to the negative terminal of the voltage source in the external circuit, and the excess of electrons that would appear at the end to which they flow, the right side, are carried off into the wire connected to the positive terminal of the voltage source in the external circuit.

Nonmetallic solids can be characterized as insulators or semiconductors. Most nonmetallic solids, whether they are covalent-network, or ionic, or van der Waals crystals, fall into the former category; their measured conductivities are many orders of magnitude lower than the very high values characteristic of metals. A simple rationalization of this huge difference in conductivities is that the electrons in nonmetallic solids are localized in specific covalent bonds, or used as nonbonded electrons to complete an atomic subshell in a covalent or an ionic species in the crystal, for example, in the covalent bonds and on the oxygen atom of $CH_3—\ddot{\underset{..}{O}}—CH_3$, or on the sodium and chloride ions of Na^+Cl^-. In metallic solids, the bonding electrons are delocalized throughout the solid, and are thus capable of moving relatively freely throughout the solid, and hence carrying an electrical current.

The liquids obtained by melting van der Waals crystals have, like the solid form, very low conductivities. The liquids obtained by melting ionic crystals (at high temperatures) still consist of ions. These ions are quite mobile relative to the ions

Spotlight

Boron nitride, BN, is a solid prepared by heating boron with nitrogen or ammonia. It has four crystalline forms; two are layered and two have tetrahedral frameworks, analogous to the diamond structure (Fig. 14.19). The most common form is the "hexagonal layered" (see the structure of graphite in Sec. 9.2); it is a slippery white solid, which has been compared to graphite and is appropriately called "white graphite." Cubic BN is identical to diamond in hardness; it is stable in air but slowly hydrolyzes in water. It is interesting to note that the B—N unit has the same number of electrons as the C—C unit (that is, it is **isoelectronic** with C—C) and shows very similar chemical structures.

in the corresponding solid states, and their motion under the influence of an electrical potential difference permits the liquid to conduct electricity quite well. Metals, when melted, retain a high level of electrical conductivity.

Semiconductors, of which silicon and germanium are important examples, have conductivities that are intermediate between those of metals and insulators. Also, it is found that the conductivity of a semiconductor increases with increasing temperature, which is opposite to the behavior of a metallic conductor. To explain the means by which semiconductors manifest a limited conductivity, and to provide a unified view explaining both semiconductors and metallic conductors, it is necessary to introduce the band theory of conduction.

Band Theory

The band theory is essentially an extension of molecular orbital theory. In Sec. 7.8, it was seen that mathematical constructs called molecular orbitals can be obtained by combining atomic orbitals, and that the number of molecular orbitals obtained equals the number of atomic orbitals contributed to the development. In H_2, the simplest case, two 1*s* hydrogen atomic orbitals are combined to give two molecular orbitals, one called the bonding orbital (lower energy) and one the antibonding orbital (higher energy). Both molecular orbitals "belong" to the entire H_2 molecule, not to just one atom, and the two electrons that occupy the bonding molecular orbital in the ground state of H_2 are delocalized over the entire H_2 molecule.

A similar development applies to metallic solids. The big difference between the molecular orbitals of a metallic crystal solid and the molecular orbitals of individual molecules is that there are so many atomic orbitals to contribute in the solid. The orbitals belong to the entire crystal, and the number of contributing atomic orbitals is therefore on the order of, say, 10^{20}. Hence, there is also a very large number of molecular orbitals, say 10^{20}. With so many energy levels covering the energy range common to just the few molecular orbitals of a small nonmetallic molecule, they must be very close together, so close that they form bands of orbitals.

The pattern of bands and the distribution of electrons among the bands are the main features determining the conductivity characteristics of a solid. The kinds of patterns typical of a metallic conductor, an insulator, and a semiconductor are shown schematically in Fig. 14.26. In part a of that figure, there is one particular feature of the pattern of bands that is primarily responsible for the high conductivity in a metallic solid; this feature is that the highest energy band containing electrons, called the valence band, is only partially filled. Such a partially filled band is called a **conduction band,** and it gives rise to conductivity in the following way.

In any sample of a metal there is a distribution of electron velocities covering a wide range of speeds and all directions. In the absence of an applied voltage difference, the distribution of velocities within each occupied band is such that the net electron velocity is zero; in other words, there is no electron drift, or current, in any direction. When an electric field is applied, the electrons in a partially filled band can undergo a net drift in the direction towards the electrically positive end of the solid object. This drift is possible because there can be a net shift in the populations of the various states (orbitals of the band) to states in which electron velocities are in the "expected" current direction (towards the right side of Fig. 14.25) from states in which electron velocities are in the opposite direction (towards the left

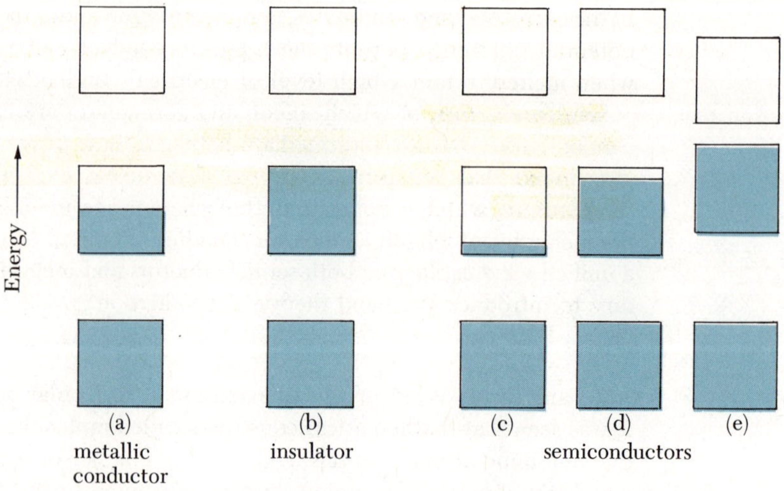

Figure 14.26
Patterns of bands of orbitals in (a) a metallic solid, (b) an insulator, and (c), (d), and (e) three types of semiconductors.

side of Fig. 14.25). This shift in the populations of the states gives rise to the net velocity in the expected direction, and the observed electrical current.

A solid insulator cannot respond to an applied electric field in the same way that a metallic conductor does. In an insulator, any band that is populated at all by electrons is completely filled, and the next higher band is too high in energy to be populated under ordinary circumstances. Hence, there is no opportunity for changes in the populations of states within the valence band, and the net electron velocity remains zero when the electric field is applied. The net result is very low conductivity.

There are three patterns of bands that one can visualize for semiconductor behavior. These are shown in Figs. 14.26c, d, and e. In the case shown in part c of the figure, the same kind of mechanism for conduction that was described for metallic solids applies, but there are few electrons available in the conduction band, so the conductivity is limited. In case d, the conduction band is almost filled, but not completely, so the redistribution of states (and electron velocities) within the band under the influence of an applied electric field is not very efficient. In Fig. 14.26e, we see a case in which the populated bands are entirely filled, but the lowest unoccupied band is not much higher in energy than the valence band. Hence, some electrons can be promoted from the latter band to the former in the presence of an applied electric field, and some conduction results.

The main effect that a temperature increase has on the conductivity of a metallic solid is to increase the resistance to the flow of electrons by increasing the amplitude of vibrations of the atomic centers (cores); these vibrations interfere with the movement of the electrons. Hence, an increase in temperature decreases the conductivity of a metal. In contrast to this, increasing the temperature of a semiconductor increases its conductivity, by promoting additional electrons into the conduction band, for example, across the gap between the filled and empty bands shown in Fig. 14.26e. The high thermal conductivity of metallic solids is also understandable in terms of band theory. The mechanism available for facile movement of electrons of the conduction band also makes possible the efficient transfer of energy by electrons from a warmer part of a metallic object to a cooler part of the object.

OPTIONAL

Doped Semiconductors

As the name *semiconductor* implies, in conductivity a semiconductor is between a metal and an insulator. More important for practical applications, conduction in semiconductors takes place through two different kinds of current, known as electron current and hole current. The electrons forming the covalent bonds in semiconductors are not strongly bound, and can be freed from these bonds by thermal or electrical means. When a fraction of the electrons are released from the covalent bonds, they are capable of supporting an electrical current of the same type occurring in metals, an electron current. However, the electron-depleted broken bonds constitute a net charge deficiency in the total bond network. These charge deficiencies are called **holes,** because they are formed from the lack of an electron in a covalent bond. Holes act as positively charged current carriers, and hole conduction can be thought of simply as the empty track of a moving electron. When a voltage is applied across a semiconductor, an electron from a nearby atom may leave and move to fill the hole. As a result, a new hole is formed. In turn, another electron may leave its atom and move, or "jump," to fill the new hole. This action continues so long as the voltage is applied, and can be thought of as electrons moving in one direction as holes move in the other. The mobile positive charge (hole) is capable of carrying an electric current, which is called hole current. In an extremely **pure,** or **intrinsic, semiconductor** the number of free electrons equals the number of holes, since an electron escaping from a covalent bond produces both a free electron and a hole. For germanium and silicon, the most common semiconductor materials used for electronic devices, the number of free electrons (and holes) per cm^3 at room temperature is 10^{13} and 10^{10}, respectively.

Because intrinsic semiconductors contain an equal number of both types of charge carriers, they are not useful materials for electronic devices. The practical usefulness of the semiconductor materials used in the electronics industry depends not only on their containing two types of charge carriers, but also on the potentiality for precisely varying the relative concentrations of electrons and holes during their manufacture. To control the relative concentrations precisely, certain impurities are added to very highly purified intrinsic semiconductor materials, producing materials called **doped,** or **extrinsic, semiconductors.**

Let us consider as an example of doped semiconductors the silicon case. Silicon exists as a covalent network crystal in which each atom is, in a tetrahedral arrangement, connected to four other atoms by covalent bonds. For our present purposes, which do not really depend on an accurate geometrical visualization, we represent the silicon crystalline network in the two-dimensional picture in Fig. 14.27a as though the silicon atoms could be squashed down into one plane. Each line in that figure represents one Si—Si covalent bond. All the valence-shell electrons in the system, meaning four electrons for each silicon, are involved in localized covalent bonds with sp^3 hybridization; thus, there is a sizable gap between the filled, valence band and the lowest unfilled, conduction band, as in Fig. 14.26e.

Figure 14.27 Impurity semiconductors. (a) A two-dimensional representation of a three-dimensional lattice of pure silicon. (b) Representation of an n-type semiconductor, with phosphorus as the donor atom; ⊙ represents the "extra" electron. (c) Representation of a p-type semiconductor, with indium as the acceptor atom; ○ represents the "electron hole."

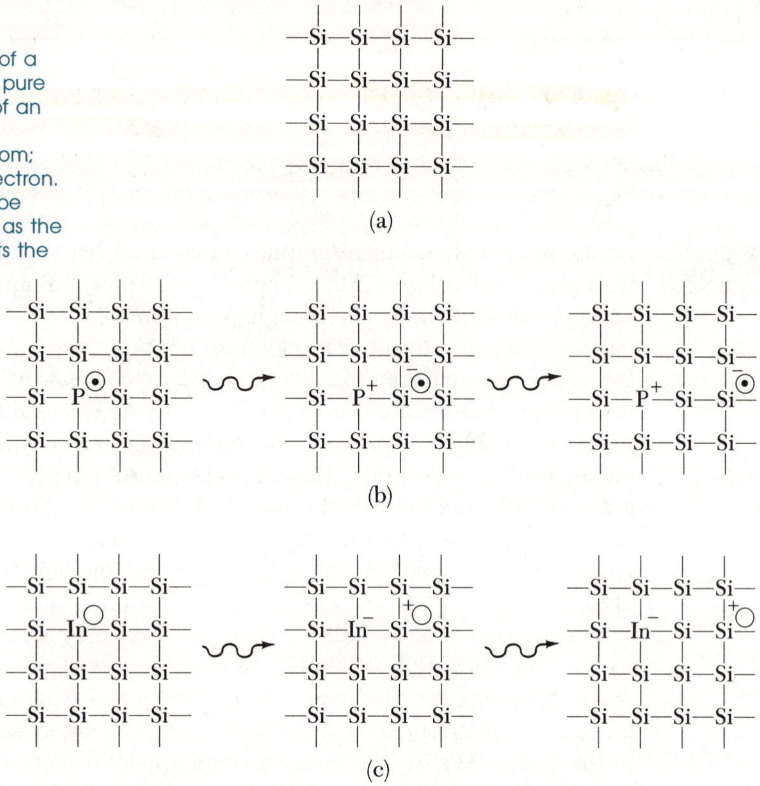

OPTIONAL

Let us imagine the effect of introducing a small amount of phosphorus as an impurity into a silicon crystal. The phosphorus atoms will simply replace silicon sites in the tetrahedral lattice, as represented in Fig. 14.27b.[*] The valence shell electrons of atomic phosphorus are two 3*s* and three 3*p* electrons; hence, there is one extra electron in the array shown in Fig. 14.27b, relative to that in part a of the figure. That is, there is one extra electron beyond what is needed to account for the simple picture of four localized covalent bonds to each atom. This extra electron clearly has a higher mobility than the electrons involved in the localized bonds; that is, it is more easily promoted into a conduction band and responds more effectively to the application of an electric field. The net result is effectively a reduction in the size of the energy gap between the valence band and the conduction band and a corresponding large increase in conductivity. In such cases as this, in which the impurity atom donates an "extra" electron to the system, the atom is referred to as a **donor;** it becomes a positive center in the lattice when the extra electron is conducted away, since the nuclear charge exceeds by 1 the number of electrons one would then assign to that atomic center. A semiconductor of this type is referred to as being of the **n type;** *n* stands for negative charge carrier. The extra electrons in this system behave much like the conduction electrons in metals. Moreover, there will still be some broken covalent bonds, each producing one electron and one hole.

[*]Impurities can be introduced into a crystal by including them in a melted sample that is allowed to crystallize under controlled conditions, or by allowing the impurity to diffuse slowly into an already formed crystal. The relative concentrations of the two types of carriers in a semiconductor are extremely sensitive to small quantities of impurities. To control these relative concentrations precisely, extreme care is taken to purify the germanium or silicon before the intentional "impurity" is added.

Thus, the material still contains both types of charge carriers, but more "free" electrons than holes. In this case, the free electrons are called the **majority carriers,** and the holes are called **minority carriers.** The donor impurities used to produce free electrons (phosphorus in the example shown) are found in Group V of the periodic table—phosphorus, arsenic, antimony, and bismuth.

Let us now imagine introducing indium as an impurity into a silicon lattice, to give the case depicted in Fig. 14.27c. For indium the number of valence shell electrons (three) is one short of what is required for four localized covalent bonds to each atom in the lattice. This constitutes a hole, which can be filled by an electron if one of the valence shell electrons originally associated with a silicon atom in the lattice migrates to that point; this forms a negative indium center in the lattice. Filling the hole on the indium atom in this manner creates a hole on the electron-depleted silicon atom, together with a local positive center on that silicon atom, as silicon's nuclear charge now exceeds the number of electrons associated with it. This hole can reside on any silicon atom, and migrates around the lattice much more readily than electrons would in a pure silicon crystal. Under the influence of an applied electric field, this migration of positive holes (by electron shifts from atom to atom) constitutes an electric current, which is considerably greater than what would be realized in the absence of the indium impurity. The impurity atom of this type, which accepts an electron from the remaining atoms of the lattice to form a negative center is called an **acceptor.** A semiconductor of this type is called a **p type;** *p* stands for positive charge migration, that is, migration of holes. The acceptor impurities, indium in the case described, are Group III elements.

Current Flow in n-Type and p-Type Semiconductors

The current flow through n-type and p-type materials is illustrated in Fig. 14.28. For an n-type semiconductor (Fig. 14.28a), an electrical potential applied across the material causes electrons released from the donor atoms to travel towards the positive terminal of the battery. Conduction of this type is similar to that in copper, but there are certain differences. For example, an increase in the temperature of copper increases the thermal agitation of the atoms and impedes current flow. In n-type material, increasing the temperature causes more carriers to become available, both by releasing more electrons from the donor atoms and by breaking more covalent bonds. Thus, an increase in temperature increases current flow.

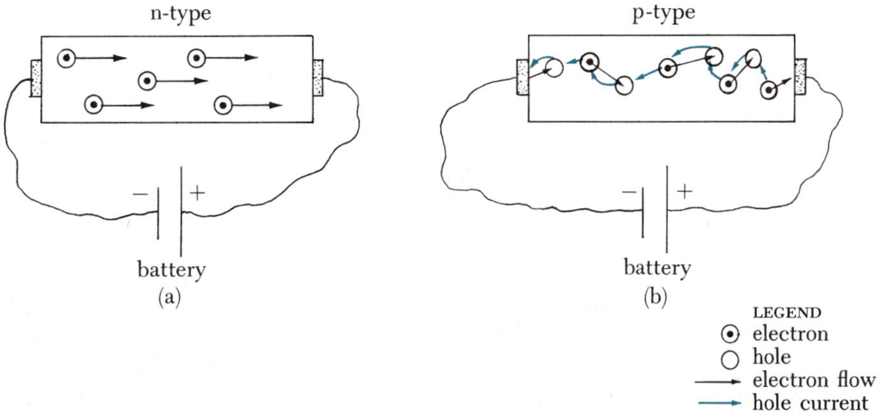

Figure 14.28
Current flow in semiconductors, with a battery as the voltage source. ⊙ represents an electron; ○ represents a hole. → represents electron flow; ⟶ represents hole current.

OPTIONAL

Current flow through a p-type material is illustrated in Fig. 14.28b. Electrons from the negative terminal of the battery cancel holes in the vicinity of the terminal. At the positive terminal, electrons are removed from covalent bonds, creating new holes that move towards the negative terminal. The process continues as a steady stream of holes, the hole current, moves towards the negative terminal.

In both n-type and p-type materials, electron flow in the external circuit is out of the negative terminal of the battery and into the positive terminal. Hole flow does not occur in the external wires, which are made of metal. Combinations of n-type and p-type semiconductors and other related devices form the basis of modern electronics equipment, such as high-fidelity sound equipment, television receivers, and computers.

SUMMARY

True solids are crystalline, and have both short-range order and long-range order. They are held together by a variety of essentially electrical forces, including in some cases covalent bonding. The morphology of a solid is determined by the shape and structural makeup of the unit cell, the smallest repeating unit of the crystal. The structure of the unit cell can be elucidated by x-ray diffraction experiments. There are fourteen Bravais crystal lattices, and all solids belong to one of these.

The gross properties of metals can be explained in terms of an "electron gas" model, in which metal ions are arranged in a closest packed array and are immersed in a sea of mobile valence electrons. In ionic crystals, simple electrostatic forces maintain the ions in their ordered lattice. In covalent network crystals, a giant network of covalent bonding holds the constituents of the crystal in its ordered arrangement. In van der Waals crystals, molecules are held together by weak electrical forces involving permanent or induced dipoles or both and dispersion forces.

Polymers constitute a class of substances made up of macromolecules. Most polymers are not strictly solids, but some have substantial crystalline character. Both natural and synthetic polymers have great practical importance; the latter are synthesized by the polymerization of monomers, of one or more kinds.

The properties of solids are intimately related to their submicroscopic structures. The conductivity of metals, insulators, and semiconductors can be accounted for by band theory, which is an expanded form of molecular orbital theory.

STUDENT CHECKPOINTS

After studying this chapter, the student should be able to:

1. Differentiate the types of attractive interactions that occur in condensed phases.
2. List the most important features of solids, whereby they can be distinguished from liquids.
3. Explain what anisotropy is in solids.
4. Define the "unit cell" concept, and explain how it relates in general terms to crystal shapes.
5. Elucidate the elementary principles of the x-ray diffraction experiment, and be able to apply the Bragg equation in simple cases.
6. Determine the number of molecules or ions of a given type in a simple unit cell.
7. Discuss the geometrical arrangements in closest packing.

8. Describe the four general types of crystalline solids.
9. Define the essential character of a polymerization of monomers to form a polymer.
10. Carry out calculations relating ΔH, ΔS, and T for the melting of a solid.
11. Explain the fundamentals of band theory and how it accounts for conductivity.

EXERCISES

14.1. What kind of attractive interactions holds each of the following crystals together: Mg, CCl_4, SiO_2, CH_3NO_2, Zn, CO_2? There may be more than one kind.

14.2.* What kind of attractive interactions, one or more, must be weakened to melt each of the following crystals: H_2, Kr, C (diamond), Co, KBr?

14.3. Give three examples of anisotropy in solids.

14.4. How would you explain why window glass is not really a solid in a scientific sense?

14.5.* Explain the difference between an amorphous "solid" and a polycrystalline solid.

14.6. What kind of fundamental information is available from morphological crystallography?

14.7.* (a) What are the main properties of x rays that make them suitable for determining the submicroscopic structure of crystals? (b) Why is visible light not used?

14.8. Describe constructive interference.

14.9.* Indicate whether each of the following statements is true or false:
(a) If two light beams converge on a given point with a phase difference corresponding to $\frac{8}{2}$ wavelengths, there will be destructive interference at that point.
(b) If two light beams converge on a given point with a phase difference of $\frac{5}{2}$ periods, there will be destructive interference at that point.
(c) The interference pattern in an apparatus such as that shown in Fig. 14.7 is independent of the wavelength of the radiation, but depends on the placement of the slits S_1 and S_2.

14.10. Indicate whether each of the following statements is true or false:
(a) From the point of view of diffraction, the distances $\overline{P_0P_1}$ and $\overline{P_0P_2}$ in Fig. 14.7 are more important than the distances $\overline{P_1S_1}$ and $\overline{P_1S_2}$.
(b) The distance $\overline{EG}$ in Fig. 14.8 is $2d \sin \theta$.
(c) The distance $\overline{AE}$ in Fig. 14.8 equals the distance $\overline{AF}$.

14.11.* Indicate whether each of the following statements is true or false:
(a) The purpose of a goniometer in an electron diffraction apparatus is to detect the x-ray intensity.
(b) The solid used in the x-ray diffraction determination of detailed unit cell structure must be a single crystal and not polycrystalline.
(c) The angle between the incident ray and the diffracted ray is always the same in an x-ray diffraction experiment.

14.12. Indicate whether each of the following statements is true or false:
(a) Only cubic lattices can have all three interfacial angles (α, β, γ) equal to 90°.

(b) A body-centered cubic crystal must have a single atom at each corner of the unit cell.

(c) The interfacial angles of a macroscopic triclinic crystal cannot be 90°.

14.13.* Suppose one had a van der Waals crystal in which small molecules occupied the lattice points of a base-centered monoclinic unit cell. How many molecules would there be per unit cell?

14.14. A simple cubic crystal can be visualized as eight spheres of radius r placed at the corners of the unit cell and just touching each other. (a) What is the volume of the cube in terms of r? (b) What fraction of the volume of the unit cell is "empty" in this model?

14.15.* Lithium bromide has the same type of unit cell as NaCl does. (a) How many Li^+ ions and how many Br^- ions are there per unit cell? The edge length of the LiBr unit cell is 5.501×10^{-10} m. (b) Calculate the density of the LiBr crystal.

14.16. Potassium chloride has the same type of crystal structure as NaCl. The density of KCl is 1.99 g/cm^3. Calculate (a) the length of the edge of the KCl unit cell; and (b) the mass of one KCl unit cell, in g.

14.17. The unit cell of solid nickel is face-centered cubic. The density is 8.90 g/cm^3. Assuming that the atoms "touch" in the crystal, calculate (a) the radius of a nickel atom, and (b) the fraction of the volume of the unit cell that is "empty."

14.18.* If one already knows the structural formula of a substance, why would one consider an x-ray diffraction study of the substance?

14.19. Metallic gold has a face-centered cubic structure. This element has an atomic radius 1.46×10^{-10} m. (a) Calculate the density of gold. (b) Suppose you had mistakenly assumed that the crystal structure was simple cubic; how much of an error would you have made in the computed density?

14.20.* If the edge of the unit cell of NaCl is 5.64×10^{-10} m, calculate (a) the volume of one mole of NaCl and (b) the density of NaCl.

14.21.* The body-centered cubic lattice of metallic lithium has a unit-cell edge length of 3.51×10^{-10} m. What is the density of this substance?

14.22. We know that the atomic weight of metallic copper is 63.55 amu, its density is 8.936 g/cm^3, and its crystal structure is face-centered cubic. From this knowledge, determine the edge length of the unit cell.

14.23. Discuss the strengths and limitations of x-ray diffraction methods from the point of view of types of samples that are suitable and type of information to be expected.

14.24.* What are the wavelength and the frequency of x rays generated when an inner-shell vacancy is filled by an electron from an outer shell with an energy that is 1.73×10^{-15} J higher than that of the inner-shell vacancy?

14.25. The x rays emitted from a metal target bombarded with electrons have a wavelength of 9.82×10^{-11} m. What is the energy of the x rays generated by the downward transitions of electrons in the metal, as the electrons fill inner-shell vacancies?

14.26.* What is the distance between neighboring planes of atoms in a simple crystal lattice like that shown in Fig. 14.8 if the first-order diffraction maximum occurs at 31° for x rays of wavelength 2.01×10^{-10} m?

14.27. What is the interatomic distance between reflecting planes of atoms in a simple crystal lattice like that shown in Fig. 14.8 if the second-order diffraction maximum ($n = 2$) occurs at 36° for x rays of wavelength 3.11×10^{-10} m?

14.28.* Calculate the wavelength of an x ray for which the first-order diffraction maximum occurs at an angle of 42° in a diffraction experiment for which $d = 2.22 \times 10^{-10}$ m.

14.29. Explain why SrS, BaS, and MgO have the NaCl type of crystal structure, while RbBr and CsBr have crystal structures of the CsCl type.

14.30.* Indicate whether each of the following statements is true or false:
(a) In a van der Waals crystal, it is not necessary to break covalent bonds in order to bring about melting.
(b) In a crystal of the diamond type, it is not necessary to break covalent bonds in order to bring about melting.
(c) One can melt a crystal of $(NH_4)_2SO_4$ without breaking any covalent bonds.

14.31. Explain how Avogadro's number could be determined from an x-ray diffraction experiment.

14.32.* Indicate whether each of the following statements is true or false:
(a) Ferrocene can be melted without breaking any covalent or ionic bonds.
(b) The crystal forces in pyrazine are stronger than those in graphite.
(c) Very few van der Waals crystals would remain unmelted at 25 °C.

14.33. Give two examples, other than polypropylene, in which isotactic, syndiotactic, and atactic isomers can be visualized.

14.34.* Polyester polymers are of the "condensation" type. Predict the structure of a polyester polymer made from the following monomers:

$HO—CH_2—CH_2—CH_2—OH$ and 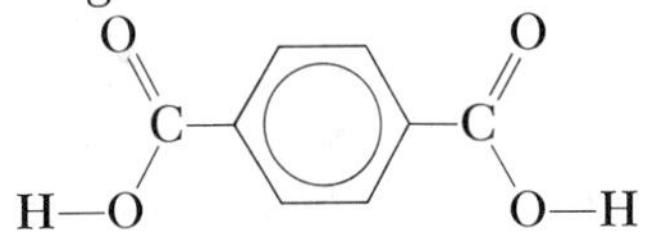

14.35. Epoxy resins result from the combination of a monomer, containing epoxide groups (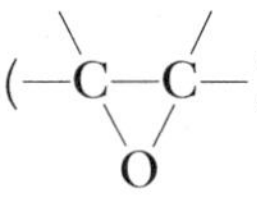), for example,

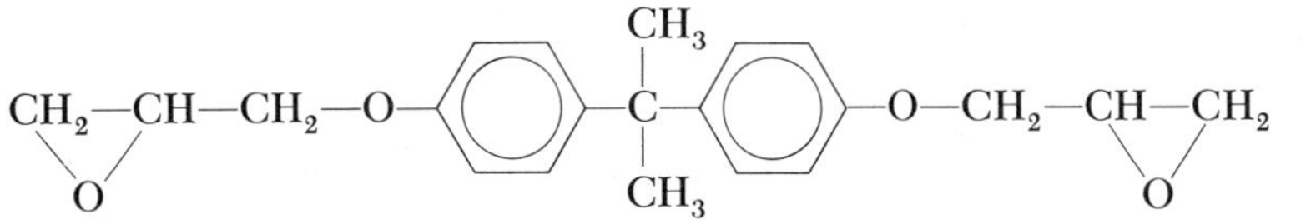

and a diamine compound, for example,

$$H_2N—CH_2—CH_2—NH_2$$

Knowing that one epoxide C—O bond of the epoxide is broken in the polymerization, predict the structure of an epoxy polymer resulting from the above two species.

14.36. Calculate the molar entropy of the melting of $CoCl_2$, for which the molar heat of fusion is 7.40 kcal and the melting point is 724 °C.

14.37.* Predict the melting point of CO_2, knowing that ΔH_{fus} is 8.32×10^3 J mol^{-1} and ΔS_{fus} is 38.3 J $deg^{-1}mol^{-1}$.

14.38. According to the band theory of metals, molecular orbitals encompassing the entire crystal are constructed from atomic orbitals, and these closely spaced "crystal orbitals" group together in bands. Assuming that a band of crystal orbitals comes from the 1*s* atomic orbitals, another from the 2*s* orbitals, another from the 2*p*, another from the 3*s*, and so on, draw a band energy diagram. (Assume that each of these bands is separated from the other by an energy gap, except for the 3*s* and 3*p* bands, which overlap.) On the basis of this diagram, discuss the conductivity of sodium and aluminum metals, keeping in mind whether a given band is completely filled, completely empty, or partially filled.

ANSWERS TO SELECTED EXERCISES

14.2 H_2: dispersion; Kr: dispersion; C: covalent bonds; Co: metallic bonding; KBr: ionic bonds. **14.5** A microscope shows that a polycrystalline solid is a collection of tiny crystals, which is not the case for an amorphous material. **14.7** (a) The x-ray wavelengths employed are comparable to interatomic distances in crystals; (b) The wavelengths of visible light are too large. **14.9** (a) false; (b) true; (c) false. **14.11** (a) false; (b) true; (c) true. **14.13** two. **14.15** (a) four of each per unit cell; (b) 3.46 g/cm³.
14.18 To determine precise geometrical details, for example, bond lengths and bond angles.
14.20 (a) 27.0 cm³; (b) 2.16 g/cm³. **14.21** 0.533 g/cm³. **14.24** $\nu = 2.61 \times 10^{18}$ sec^{-1}; $\lambda = 1.15 \times 10^{-10}$ m. **14.26** 1.95×10^{-10} m. **14.28** 2.97×10^{-10} m.
14.30 (a) true; (b) false; (c) true. **14.32** (a) true; (b) false; (c) false.
14.34

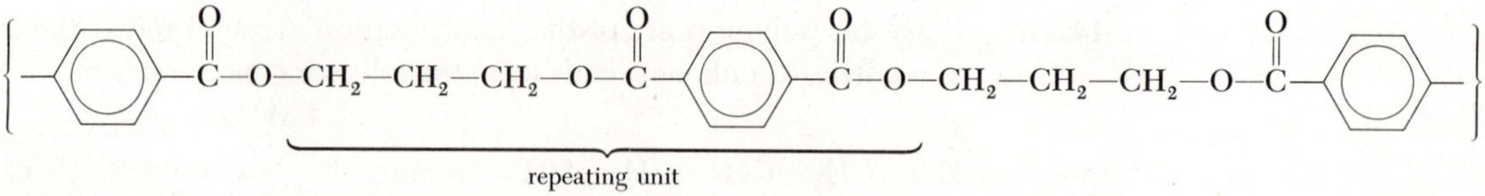

14.37 217 °K.

ANSWERS TO STUDY PROBLEMS

14(a) $\nu = 2.7 \times 10^{18}$ sec^{-1}; $\lambda = 1.1 \times 10^{-10}$ m. **14(b)** 1.9×10^{-10} m. **14(c)** two.
14(d) 13.0 J deg^{-1}.

15

Gases

15.1 Introduction

Our everyday experiences with the state of matter called a gas give us at least a qualitative understanding of the general physical characteristics of this state. Three common patterns of behavior that illustrate some of these characteristics are presented here, and we shall later investigate these properties in more detail in a quantitative sense.

1. When the volume of a given mass of a gas is decreased at a constant temperature, the pressure of the gas will increase. Conversely, if the volume of the gas is increased, the pressure will decrease. A balloon inflated with air is an example of a given mass of a gas, and if the volume is decreased by squeezing the balloon, say by submerging it in water, the pressure of the gas will increase.
2. When the temperature of a given mass of a gas is increased at a constant volume, the pressure will increase; conversely if the temperature is decreased, the pressure will decrease. If we took a balloon filled with gas from a refrigerator and held it firmly in our hands so that it could not expand, we should feel greater and greater pressure against our hands as the gas on the balloon warmed to our body temperature. If we could cool the gas again while we held it, we should notice a pressure decrease.
3. A gas will spread, or diffuse, into an empty space or into space containing another gas. Almost everyone has observed that if a gas with a distinct odor is released in a corner of a room, the gas will eventually diffuse through the air and spread itself entirely throughout the room. This is true even if there are no currents in the room, although such currents certainly accelerate the process.

Spotlight

Oxygen occurs in two allotropic forms; one is O_2 and the other is ozone, O_3, both gases under normal conditions. **Ozone** is a light blue, unpleasant-smelling gas at room temperature, but is a deep blue liquid and a black violet solid at low temperatures. Ozone is usually prepared by passing a silent electrical discharge through gaseous oxygen; concentrations of 3% to 10% O_3 can be obtained in this way. The peculiar smell evident in the vicinity of generators and ultraviolet lights, after lightning strikes, and near yellow phosphorus that has been exposed to air, is due to the presence of ozone. Lightning and ultraviolet rays from the sun produce ozone in the upper atmosphere, where it is important in partially screening ultraviolet radiation from the surface of the earth. The presence of ozone in the lower atmosphere is an important factor in air pollution.

Physical studies have shown ozone to be triangular, the O—O—O angle being about 117°. The stability of the ozone molecule can be visualized through the concept of resonance:

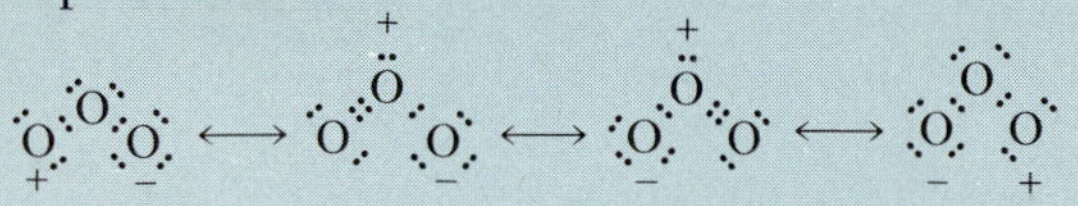

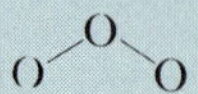

Ozone is well known to react with organic compounds at carbon-carbon double bonds, producing species called **ozonides.**

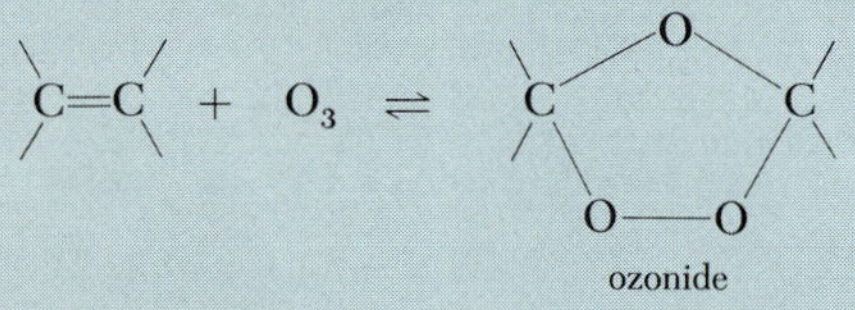

It is largely this reaction of O_3 with C=C bonds that partly accounts for the gradual deterioration of organic materials that contain double bonds, like cork and rubber, when exposed to air. Under certain conditions, for example, in the presence of ultraviolet light from the sun, saturated hydrocarbons can also react with O_3 in air to form organic peroxides, R—C(=O)—O—O—C(=O)—R, and hydroperoxides, R—C(=O)—O—O—H, and the presence of these products contributes significantly to air pollution. In urban areas, the main source of saturated hydrocarbons and also of compounds containing C=C bonds in the atmosphere is incomplete combustion of automotive and industrial fuels.

The first two types of gas behavior just described qualitatively were described quantitatively by the empirical gas laws introduced in Chaps. 2 and 3. In reviewing these gas laws along with additional gas laws, we shall develop a theoretical model able to account for all these empirical laws.

One important point to remember in considering the gas laws is that they work perfectly only for an ideal gas. We can define an **ideal gas** as a gas that obeys these laws exactly. No real gas is perfectly ideal, but many are very close to the ideal, especially at high temperatures and low pressures. Although gases at higher pressures and lower temperatures tend to deviate appreciably from the predictions of these laws, these predictions are still useful for providing semiquantitative descriptions of gaseous behavior. The behavioral difference between real gases and the ideal gas will be elaborated on in Sec. 15.6. The reason that gas behavior can be described quantitatively with such success has to do with the lack of order, the extreme randomness of gases. The solid state is amenable to detailed description because it is so highly ordered; gases can be described conveniently because the total lack of order is itself a well-defined condition. The intermediate degree of order in liquids renders the liquid state the most difficult to describe.

15.2 Gas Laws for an Ideal Gas

Boyle's Law

In 1662, John Boyle, an English chemist, physicist, and theologian, reported that when he confined a given quantity of a gas at constant temperature, its volume varied inversely with the pressure of the gas. As shown in Eq. (2.2), Boyle's law is stated mathematically as follows:

$$V = (\text{constant})\left(\frac{1}{P}\right) \qquad \text{(constant } T\text{, fixed mass)} \qquad (2.2)$$

or

$$V \propto \frac{1}{P} \qquad \text{(constant } T\text{, fixed mass)} \qquad (15.1)$$

where the symbol $\propto$ means "is proportional to."[*] The proportionality constant in Eq. (2.2) depends on the temperature and the weight of the gas. Another way of stating this inverse proportionality is

$$\frac{V_2}{V_1} = \frac{P_1}{P_2} \qquad \text{(fixed mass and temperature)} \qquad (15.2)$$

where the subscripts 1 and 2 stand for two different states of the system having the same temperature (and of course the same mass and chemical identity). These equations tell us, for example, that the volume of a gas sample having fixed temperature and mass will be doubled if the pressure is cut in half. In terms of Eq. (15.2), if $P_2 = \frac{1}{2}P_1$, then $V_2 = 2V_1$; that is,

$$\frac{V_2}{V_1} = \frac{P_1}{P_2} = \frac{P_1}{\frac{1}{2}P_1} = 2$$

Charles's Law

In 1787, Jacques Alexandre César Charles, a French physicist, observed that separate samples of air, H_2, CO_2, and O_2 expanded, each by the same volume ratio, when heated from 0 °C to 80 °C at a constant pressure. In 1802, Joseph Louis Gay-Lussac, a French chemist and physicist, found that when the temperature of a sample of gas was raised 1 °C at a constant pressure, the volume expanded by $\frac{1}{273}$ of the volume of the gas at 0 °C. It is now known that this fraction is more precisely $\frac{1}{273.15}$. Stated mathematically, these relations constitute *Charles's law* (often called the Charles–Gay-Lussac law):

$$V = V_0 + \frac{T}{273.15}V_0 \qquad \text{(constant P, fixed mass)} \qquad (15.3)$$

[*] A scuba diver must be cognizant of Boyle's law in order to avoid painful and dangerous air embolisms. If the diver inhales high-pressure air from the breathing apparatus at a 90-ft depth, which presents about 4 atm pressure, then in order to avoid an air embolism, he must exhale during his ascent. If he does not, the decreasing water pressure of the shallower depth causes expansion of the air in his lungs as the pressure within them tries to match the decreasing outside pressure of the water. The resulting increase in air volume within the lung brings about a rupturing of air sacs, an air embolism.

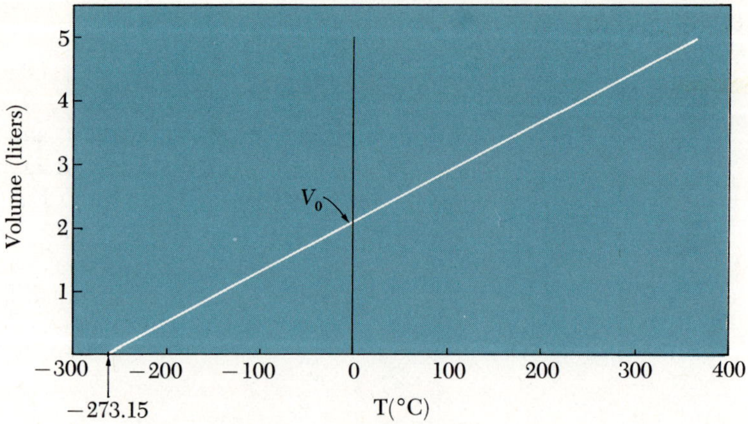

Figure 15.1
The dependence of the volume of a gas on temperature, at constant pressure. The volume of the gas at 0 °C is V_0 and at −273.15 °C the volume is zero.

where V_0 is the volume at 0 °C, and V is the volume at temperature T °C. A plot of this relation is seen in Fig. 15.1. From the plot and the parent equation (15.3), we see that if a gas is cooled to −273.15 °C, its volume would theoretically be reduced to zero. Below this temperature, the volume would apparently become negative, a physically unacceptable eventuality. Hence, −273.15 °C is considered the lowest attainable temperature—the **absolute zero** of temperature. We now make use of the Kelvin temperature scale, defined in Eq. (2.3):

$$\text{degrees K} = \text{degrees C} + 273.15 \tag{2.3}$$

On the Kelvin scale, 0 °K is equivalent to −273.15 °C, and each Kelvin degree is equal "in size" to one Celsius degree.

In terms of the Kelvin temperature scale, as stated in Chap. 2, Charles's law is expressible as a simple proportionality:

$$V \propto T \qquad \text{(constant } P\text{, fixed wt of gas)} \tag{15.4}$$

or

$$V = (\text{constant})(T) \qquad \text{(fixed } P \text{ and wt)} \tag{2.4}$$

The relation between V and T given by Eq. (2.4) is shown in Fig. 15.2. The constant in Eq. (2.4) depends on the pressure, the amount of gas, and the units of V.[*] For a given quantity of a specific gas at a constant pressure but two different temperatures, T_1 and T_2, we can express the simple proportionality as

$$\frac{V_1}{V_2} = \frac{T_1}{T_2} \qquad \text{(given wt of a specific gas at constant } P\text{)} \tag{15.5}$$

[*]A scuba diver who wants to obtain full value for money spent on compressed air should realize the consequences of the Charles–Gay-Lussac law when having air tanks filled. The diver who allows the temperature to increase markedly during the filling will be unhappy to note later, after the tanks and contained air have cooled, that the air pressure is appreciably lower than what was registered at the time of filling.

Spotlight

The gas that we now associate with the name carbon dioxide, CO_2, has been known for a long time, but Lavoisier was the first to show that it was an oxide of carbon. It was found to form whenever compounds containing carbon are burned in an adequate supply of air, and occurs naturally in air to the extent of about .03%; its concentration is remarkably stable as a result of the "carbon cycle," which involves the consumption of CO_2 by plants and the production of CO_2 by animals. The carbon cycle links the animal and plant kingdoms through photosynthesis.

Industrially, CO_2 is produced by heating limestone (calcium carbonate, $CaCO_3$), and as a by-product from alcoholic fermentation:

$$CaCO_3(s) \rightleftharpoons CaO(s) + CO_2(g)$$

$$\underset{\text{sugar}}{C_6H_{12}O_6} \rightleftharpoons 2CO_2 + \underset{\text{ethyl alcohol}}{2C_2H_5OH} \text{ (in wine)}$$

The laboratory preparation usually consists of acidifying a carbonate:

$$CO_3^{2-} + 2H^+ \rightleftharpoons H_2O + CO_2(g)$$

Carbon dioxide is a colorless gas, but it has a faint taste and smell. It is soluble in water, forming the mildly acidic solution known as carbonic acid, but with acidic properties primarily due to the reaction

$$CO_2 + 2H_2O \rightleftharpoons H_3O^+ + HCO_3^-$$

All natural sources of water contain dissolved CO_2 from the air. Commercially, a solution of CO_2 in water is sold as "soda water."

When CO_2 is pumped into steel cylinders, it liquefies. If the cylinder is turned upside down and the valve opened, part of the escaping CO_2 solidifies in the form of a white "snow." At normal atmospheric pressure solid CO_2 has a temperature of -78.5 °C or lower, and is commercially sold in the form of blocks. At ordinary room temperature and pressures, the material evaporates, or **sublimes**, directly to the gas without passing through a liquid phase. However, at a pressure of 5.2 atm or greater, a liquid state can form. Solid CO_2 is frequently used as a refrigerant and is called Dry Ice because it leaves no residue. When Dry Ice is placed under water, large bubbles of gaseous CO_2 are formed. The cold gas bubbling from the water condenses the water vapor in the surrounding air to a fog, which is dense enough to pour slowly down the outside of the vessel. The effect is quite eerie, especially if the water is colored, and makes an excellent theatrical prop.

The carbon dioxide molecule is linear and symmetrical, with a C–O bond distance of 1.163 Å; it has three significant resonance structures:

$$:\overset{+}{\ddot{O}}{\equiv}C{-}\overset{-}{\underset{..}{\ddot{O}}}: \leftrightarrow :\underset{..}{\ddot{O}}{=}C{=}\underset{..}{\ddot{O}}: \leftrightarrow :\overset{-}{\underset{..}{\ddot{O}}}{-}C{\equiv}\overset{+}{\ddot{O}}:$$

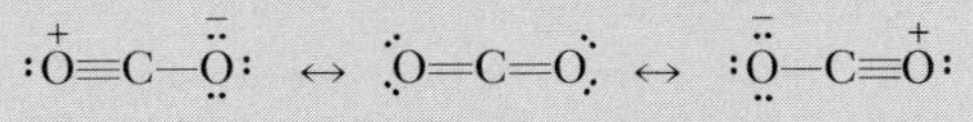

It does not burn and does not normally support combustion.

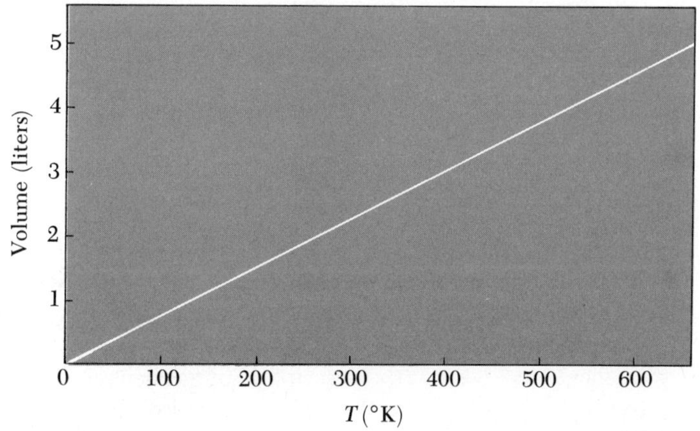

Figure 15.2
The dependence of the volume of a gas on temperature at constant pressure.

EXAMPLE 15.1 Suppose that a given mass of a certain gas occupies 4.0 liters at 27 °C and 1.0 atm, and that the pressure and temperature of the gas are changed to 2.0 atm and 327 °C. What is the new volume?

SOLUTION Let us designate the initial conditions by the subscript 1, and the final conditions by the subscript 2. We shall imagine the overall change to have occurred in the following two hypothetical steps:

$$\begin{pmatrix} 27\ ^\circ\text{C}, & 1.0\ \text{atm}, & 4.0\ \text{liters} \\ T_1 & P_1 & V_1 \end{pmatrix} \xrightarrow[\text{step}]{\text{first}} \begin{pmatrix} 27\ ^\circ\text{C}, & 2.0\ \text{atm}, & V' \\ T_1 & P_2 & V' \end{pmatrix}$$

$$\xrightarrow[\text{step}]{\text{second}} \begin{pmatrix} 327\ ^\circ\text{C}, & 2.0\ \text{atm}, & V_2 \\ T_2 & P_2 & V_2 \end{pmatrix}$$

In the first step, the gas is compressed from 1.0 to 2.0 atm with the temperature held constant. Under these conditions, the resulting volume V' would be given by Boyle's law (Eq. 15.2) as follows:

$$\frac{V'}{V_1} = \frac{P_1}{P_2}$$

$$\frac{V'}{4.0\ \text{liters}} = \frac{1.0\ \text{atm}}{2.0\ \text{atm}}$$

$$V' = \frac{(4.0\ \text{liters})(1.0\ \text{atm})}{2.0\ \text{atm}}$$

$$= 2.0\ \text{liters}$$

In the second step, the gas is heated at this pressure of 2.0 atm, from 27 °C (300 °K) to 327 °C (600 °K). The final volume will be given by Charles's law Eq. (15.5):

$$\frac{V_2}{V'} = \frac{T_2}{T_1}$$

$$\frac{V_2}{2.0\ \text{liters}} = \frac{600\ \text{deg}}{300\ \text{deg}}$$

$$V_2 = \frac{(2.0\ \text{liters})(600\ \text{deg})}{300\ \text{deg}}$$

$$= 4.0\ \text{liters}$$

Avogadro's Principle

In 1811, Amedeo Avogadro, an Italian physicist, stated the principle that equal volumes of all gases at the same pressure and temperature contain equal numbers of molecules (or moles). This principle was important historically in bringing a mathematical basis to the description of gas behavior, and provided a firm basis for testing concepts such as the mole. Avogadro's principle is shown schematically in Fig. 15.3.

Figure 15.3
A demonstration of Avogadro's principle.

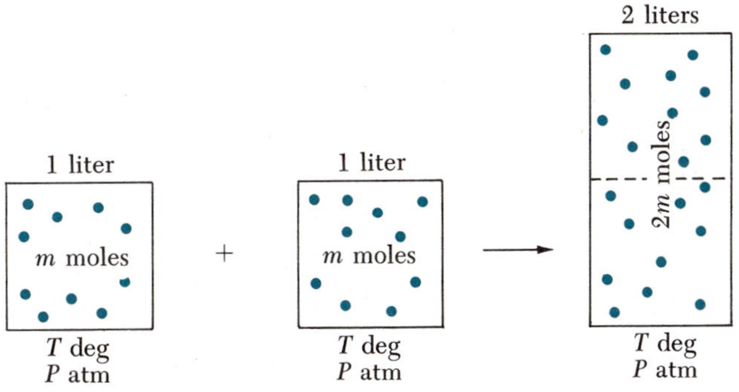

Equation of State of an Ideal Gas

As stated in Chaps. 2 and 3, the empirical gas laws for an ideal gas are expressible in a compact form called the ideal gas equation (or the equation of state of an ideal gas). This equation (Eq. 3.3) is

$$P = \frac{nRT}{V} \qquad \text{or} \qquad PV = nRT \tag{15.6}$$

In this equation, which is one of the most important equations in first-year chemistry, n represents the total number of moles of ideal gas in the system and R is a universal constant called the **gas constant.** The value of R depends on the units chosen for P and V. In the most common calculations using Eq. (3.3), V is expressed in liters and the pressure in atmospheres (760 torr); in this case

$$R = 0.08205 \text{ liter atm deg}^{-1}\text{mol}^{-1}$$

For all applications of gas laws, T is expressed in Kelvin degrees. In other units, R has the values 8.314 J $\text{deg}^{-1}\text{mol}^{-1}$, 8.314×10^7 erg $\text{deg}^{-1}\text{mol}^{-1}$, 62.36×10^3 ml torr $\text{deg}^{-1}\text{mol}^{-1}$, and 1.987 cal $\text{deg}^{-1}\text{mol}^{-1}$.

EXAMPLE 15.2 At 0 °C and 1 atm, 1 mol N_2 occupies 22.414 liters. Confirm the value of R in units of liter atm $\text{deg}^{-1}\text{mol}^{-1}$.

SOLUTION

$$PV = nRT \qquad \text{or} \qquad R = \frac{PV}{nT}$$

$$R = \frac{1\ (\text{atm}) \times 22.414\ (\text{liter})}{1\ (\text{mol}) \times 273.15\ (\text{deg})}$$

$$= 0.08205 \text{ liter atm deg}^{-1}\text{mol}^{-1}$$

EXAMPLE 15.3 The molecular weight of a gas was determined by filling an evacuated vessel with the dry gas at atmospheric pressure. From the following data, calculate the molecular weight.

Barometric pressure	= 756.9 torr
Volume of vessel	= 503.4 ml
Weight of gas	= 620.6 mg
Temperature	= 22.85 °C

SOLUTION

$$PV = nRT \qquad \text{or} \qquad n = \frac{PV}{RT}$$

$$n = \frac{(756.9 \text{ torr})(503.4 \text{ ml})}{(62.36 \times 10^3 \text{ ml torr mol}^{-1}\text{deg}^{-1})(296.0 \text{ deg})}$$

$$= 0.02064 \text{ mol}$$

$$0.02064 \text{ mol} = 0.6206 \text{ g}$$

$$1 \text{ mol} = \frac{0.6206 \text{ g}}{0.02064} = 30.07 \text{ g}$$

$$\text{MW} = 30.07 \text{ amu}$$

EXAMPLE 15.4 A sample of chlorine gas is contained in a 100-ml glass bulb at a pressure of 0.70 atm and a temperature of 0 °C. What will the pressure of this gas be if it is transferred to a previously evacuated 200-cm^3 vessel that is maintained at a temperature of 25.0 °C?

SOLUTION We assume that Eq. (15.6) is valid for both the initial and the final conditions. Hence, $n = PV/RT$. But the number of moles is the same in both states of the system, so we can write

$$\frac{P_1V_1}{RT_1} = \frac{P_2V_2}{RT_2} \qquad \text{or} \qquad \frac{P_1V_1}{T_1} = \frac{P_2V_2}{T_2}$$

$$\frac{(0.70 \text{ atm})(100 \text{ cm}^3)}{(273.2 \text{ deg})} = \frac{(P_2)(200 \text{ cm}^3)}{(298.2 \text{ deg})}$$

$$P_2 = \frac{100}{200}\,\frac{298.2}{273.2}(0.70 \text{ atm}) = 0.38 \text{ atm}$$

EXAMPLE 15.5 A 0.340-g sample of liquid benzene, C_6H_6, is introduced into an evacuated chamber of volume 250 cm^3 and maintained at a temperature of 100.0 °C. The benzene immediately vaporizes. What is the final pressure of the benzene gas in the chamber?

SOLUTION Since the molecular weight of benzene is 78.1 amu, a 0.340-g sample contains 0.340/78.1 mol. Then, by Eq. (15.6),

$$P = \frac{nRT}{V}$$

$$= \frac{(0.340/78.1 \text{ mol})(82.05 \text{ ml atm deg}^{-1}\text{mol}^{-1})(373.15 \text{ deg})}{250 \text{ ml}}$$

$$= 0.533 \text{ atm}$$

STUDY PROBLEM 15(a)

A certain sample of $N_2(g)$ occupies a volume of 200 cm^3 at 0.50 atm and 10 °C. What would the pressure of this sample be if its temperature were raised to 90 °C in the same 200-ml vessel?

Spotlight

Carbon monoxide, CO, is a poisonous gas formed to a very small extent by natural processes. For example, it has been detected in volcanic gases. It is usually formed whenever materials containing carbon are being burned in a limited supply of air or oxygen; the prime, and worrisome, example is the internal combustion in automobiles.

Industrially, CO is prepared in large quantities as "producer gas" and "water gas." Producer gas is a mixture of CO and N_2 formed by blowing air through a column of red hot coke (carbon) in a furnace called a producer. When the air first contacts the hot coke, CO_2 is produced; but as the CO_2 travels through the hot coke it becomes reduced to CO:

$$C(s) + O_2(g) \rightleftharpoons CO_2(g) \qquad \text{(initial contact with coke)}$$

$$C(s) + CO_2(g) \rightleftharpoons 2CO(g) \qquad \text{(continued contact with coke)}$$

Water gas is produced in much the same way as producer gas, but steam is used instead of air. Hydrogen, instead of nitrogen, is the other component:

$$C(s) + H_2O(g) \rightleftharpoons CO(g) + H_2(g)$$

In the laboratory, CO is generally prepared by heating concentrated sulfuric acid, a dehydrating agent, with either formic acid, HCO_2H, or oxalic acid, $HO_2C{-}CO_2H$:

$$H_2SO_4(l) + H{-}\overset{\overset{O}{\|}}{C}{-}O{-}H(l) \rightleftharpoons CO(g) + H_3O^+ + HSO_4^{2-}$$

$$H_2SO_4(l) + HO{-}\overset{\overset{O}{\|}}{C}{-}\overset{\overset{O}{\|}}{C}{-}OH(s) \rightleftharpoons CO(g) + CO_2(g) + H_3O^+ + HSO_4^{2-}$$

Carbon monoxide is a colorless, odorless, tasteless, and poisonous gas. It does not react with water, nor is it very soluble. It burns with a blue flame, forming CO_2, but does not support combustion. Hopcalite, a mixture of 50% manganese dioxide, MnO_2, 30% copper oxide, CuO, 15% cobaltic oxide, Co_2O_3, and 5% silver oxide, Ag_2O, will oxidize CO to CO_2 in air at ordinary temperatures. Hopcalite is therefore used in carbon monoxide respirators, devices designed to remove the poisonous species from air being breathed.

Carbon monoxide reacts with chlorine gas to form phosgene, carbonyl chloride, a poisonous gas used during World War I; phosgene is also a useful reagent for organic synthesis:

$$CO(g) + Cl_2(g) \rightleftharpoons \underset{\text{phosgene}}{Cl{-}\overset{\overset{O}{\|}}{C}{-}Cl(g)}$$

Carbon monoxide reduces many metallic oxides to the free metal when passed over the heated oxide. The reaction between CO and iron oxide is important in the production of iron from its ore in a blast furnace:

$$Fe_2O_3(s) + 3CO(g) \rightleftharpoons 2Fe(s) + 3CO_2(g)$$

When CO is mixed with hydrogen, as in water gas, and the mixture passed over a nickel catalyst at 300 °C, methane is formed:

$$CO(g) + 3H_2(g) \rightleftharpoons CH_4(g) + H_2O$$

However, if a mixture of zinc and chromium oxides is used as a catalyst at 350 °C–400 °C, then methanol is the product:

$$CO(g) + 2H_2(g) \rightleftharpoons CH_3OH(g)$$

All these reactions find extensive industrial applications.

STUDY PROBLEM 15(b)

A certain sample of $O_2(g)$ occupies a volume of 150 cm³ at 25 °C in a piston-cylinder arrangement in which the piston exerts a pressure of 1.20 atm. If the temperature of the system is raised to 175 °C and the piston is moved outward to decrease the pressure to 0.755 atm, what will the final volume be?

STUDY PROBLEM 15(c)

In an experiment designed to determine the molecular weight of an unknown gas, it is found that 427.3 mg of the gas occupies a volume of 181.3 cm³ at 78.8 °C and 1 atm. What is the molecular weight of the gas?

Dalton's Law of Partial Pressures

The law of partial pressures was discovered in 1781 by Henry Cavendish, a British chemist and physicist, but not announced until 1810, when John Dalton, who also developed the famous Dalton atomic theory, rediscovered it. This law states that *the total pressure exerted by a mixture of gases in a definite volume at constant temperature equals the sum of the individual pressures that each gas would exert if it occupied the same total volume alone at that temperature.* Mathematically, the law can be written

$$P = P_A + P_B + P_C + \cdots = \text{sum of partial pressures} \qquad (15.7)$$

where P is the total pressure, and P_A, P_B, P_C, . . . are the **partial pressures** of the respective gases in the mixture—the pressures that each would exert in the same container at the same pressure if none of the other species were present. Thus, if H_2 gas in a 1-liter vessel at 1 atm were forced at the same temperature into another 1-liter vessel already containing 2 atm of O_2, the total pressure would rise to 3 atm in the second vessel.

EXAMPLE 15.6 A particular breathing mixture for a deep-sea diver is to be made by introducing, at a constant temperature, 10.0 liters of O_2 under 2.0-atm pressure and 5.0 liters of He under 12.0-atm pressure into an evacuated 4.0-liter vessel. What would be the partial pressures of the O_2 and He, and what would be the total pressure?

SOLUTION By definition, the partial pressure of O_2, designated here as P_1, would be the pressure that the O_2 would exert if it alone occupied the total volume of 4.0 liters. From Boyle's law (Eq. 15.2),

$$\frac{2.0 \text{ atm}}{P_{O_2}} = \frac{4.0 \text{ liter}}{10.0 \text{ liter}}$$

$$P_{O_2} = 5.0 \text{ atm}$$

Spotlight

The composition of air at sea level, excluding water, is shown in the accompanying chart. With the exception of CO_2, the relative partial pressures of the other gases in air are essentially constant. The amount of CO_2 is quite variable and depends largely on where the sample of air is taken. Combustion processes in large cities tend to produce CO_2 and increase its partial pressure in the air over those cities. On the other hand, the process of photosynthesis consumes CO_2 and lowers its partial pressure in the air over forests.

The values in the table are average partial pressures, in atm, assuming that the total air pressure is exactly 1 atm.

Partial Pressures in Dry Air (in atm)

N_2	0.780	Ne	0.0000123
O_2	0.210	He	0.000004
Ar	0.0094	Kr	0.0000005
CO_2	0.0003	Xe	0.00000006
H_2	0.0001	Rn	trace

Similarly, for helium,

$$\frac{12.0 \text{ atm}}{P_{\text{He}}} = \frac{4.0 \text{ liter}}{5.0 \text{ liter}}$$

$$P_{\text{He}} = 15.0 \text{ atm}$$

Then, from Dalton's law of partial pressures,

$$P = P_A + P_B$$
$$= 5.0 + 15.0 = 20.0 \text{ atm}$$

Assuming that the ideal gas law is valid for both components of a two-component gas mixture, we can write

$$P_A = n_A \frac{RT}{V} \tag{15.8a}$$

$$P_B = n_B \frac{RT}{V} \tag{15.8b}$$

where P_A and P_B are the partial pressures, and where n_A and n_B are the numbers of moles of the gases, respectively. From Dalton's law,

$$P = P_A + P_B = (n_A + n_B)\frac{RT}{V} \tag{15.9}$$

Dividing P_A from Eq. (15.8a) by P from Eq. (15.9) gives the fraction of the total pressure due to component A:

$$\frac{P_A}{P} = \frac{n_A}{n_A + n_B} \tag{15.10a}$$

Similarly,

$$\frac{P_B}{P} = \frac{n_B}{n_A + n_B} \tag{15.10b}$$

It is now convenient to define the term *mole fraction;* **a mole fraction** is the ratio of the number of moles of a particular substance in a mixture to the total number of moles of all substances in that mixture. Letting X_A and X_B represent the mole fractions of the gases, we can now write

$$X_A = \frac{n_A}{n_A + n_B} \quad \text{and} \quad X_B = \frac{n_B}{n_A + n_B}$$

Then, from Eq. (15.10), we get

$$X_A = \frac{P_A}{P} \quad \text{and} \quad X_B = \frac{P_B}{P}$$

or

$$P_A = X_A P \quad \text{and} \quad P_B = X_B P \tag{15.11}$$

Thus, the *partial pressure of any individual component is the mole fraction of that component times the total pressure* of the gas mixture. We can note that the sum of the mole fractions equals 1.

$$X_A + X_B = \frac{n_A}{n_A + n_B} + \frac{n_B}{n_A + n_B} = \frac{n_A + n_B}{n_A + n_B} = 1$$

The sum of the parts equals the whole.

EXAMPLE 15.7 Determine the mole fractions, partial pressures, and total pressure in a 3.00-liter gas sample at 25 °C containing 0.0234 mol N_2 and 0.0162 mol O_2.

SOLUTION From the definition of mole fraction, we get

$$X_{N_2} = \frac{0.0234}{0.0234 + 0.0162} = \frac{0.0234}{0.0396} = 0.591$$

$$X_{O_2} = \frac{0.0162}{0.0234 + 0.0162} = \frac{0.0162}{0.0396} = 0.409$$

Now, assuming that this 0.0396-mol sample of gas is ideal, we write

$$\begin{aligned} P &= \frac{nRT}{V} \\ &= \frac{(0.0396 \text{ mol})(0.08205 \text{ liter atm deg}^{-1}\text{mol}^{-1})(298.15 \text{ deg})}{3.00 \text{ liter}} \\ &= 0.323 \text{ atm} \end{aligned}$$

Then, applying Eq. (15.11) gives

$$P_{N_2} = (0.591)(0.323 \text{ atm}) = 0.191 \text{ atm}$$

$$P_{O_2} = (0.409)(0.323 \text{ atm}) = 0.132 \text{ atm}$$

STUDY PROBLEM **15(d)**

Determine the mole fractions, partial pressures, and total pressure of a gas containing 1.84 g Cl_2 and 1.26 g N_2 at 50.0 °C in a 2.00-liter vessel.

Graham's Law of Diffusion

The process by which matter moves from one point to another, in the absence of macroscopic currents, is called **diffusion.** In 1829, Thomas Graham, a Scottish chemist, stated that at *some given temperature and pressure the rates of diffusion of various gases vary inversely as the square roots of their densities, d.* Mathematically, this law can be written

$$\frac{v_A}{v_B} = \frac{\sqrt{d_B}}{\sqrt{d_A}} \qquad \text{(constant } T, P\text{)} \qquad (15.12)$$

Spotlight

The dependence of the diffusion velocity of a molecule on its molecular weight has been applied to an important problem in the separation of the $^{235}_{92}U$ isotope from the heavier uranium isotopes, $^{237}_{92}U$ and $^{238}_{92}U$. Only the $^{235}_{92}U$ is useful for nuclear fission, that is, for nuclear reactors or nuclear bombs. Uranium hexafluoride, UF_6, is the only readily accessible, volatile compound of uranium; it has a vapor pressure of 115 torr at 25 °C. The hexafluoride of the lighter isotope, $^{235}_{92}U$, has a higher average molecular velocity, and is separated from the heavier isotopes by diffusion.

From the ideal gas equation, we see readily that the number of molecules per unit volume in a gas n/V is the same for all ideal gases at a given pressure and temperature:

$$\frac{n}{V} = \frac{P}{RT}$$

Hence, the ratio of the densities of two gaseous substances at the same T and P is simply the ratio of the corresponding molecular weights:

$$\frac{d_B}{d_A} = \frac{MW_B}{MW_A}$$

Hence, Graham's law can be expressed in terms of molecular weights as

$$\frac{v_A}{v_B} = \sqrt{\frac{MW_B}{MW_A}} \qquad \text{(for the same temperature and pressure)} \qquad (15.13)$$

EXAMPLE 15.8 The molecular weight of a pure gaseous substance was determined by allowing it to diffuse through an orifice. For calibration purposes, it was found that a given volume of N_2 required 50.0 sec to diffuse through the orifice. If the same volume of the unknown gas at the same temperature requires 79.6 sec, what is the molecular weight of the gas?

SOLUTION

$$\frac{v_A}{v_B} = \sqrt{\frac{MW_B}{MW_A}}$$

Recognizing that the velocities are inversely proportional to the time required to diffuse through the orifice, we write

$$\frac{v_{N_2}}{v} = \frac{(1/50.0)}{(1/79.6)} = \sqrt{\frac{MW}{MW_{N_2}}} = \sqrt{\frac{MW}{28.0\text{ amu}}}$$

$$\frac{MW}{28.0\text{ amu}} = \left(\frac{79.6}{50.0}\right)^2$$

$$MW = \left(\frac{79.6}{50.0}\right)^2 (28.0\text{ amu}) = 71.0\text{ amu}$$

STUDY PROBLEM 15(e)

In an apparatus designed to study gaseous diffusion for a particular gas temperature and pressure, it is found that an unknown gas diffuses 4.65 times more slowly than helium. What is the molecular weight of the unknown gas?

15.3 Kinetic Theory of Gases

The gas laws that we have just considered were discovered experimentally. These laws are obviously very useful for predictive purposes whether or not one can develop a theoretical model to account for them. Nevertheless, whenever we can develop an acceptable model that is consistent with our observations, we tend to feel that our understanding of nature has progressed appreciably. (This is what Dalton's model of the atom did for the orientation of scientists' thinking about atoms.) Furthermore, once one has a promising theory, then it can be tested by seeing whether it can predict results of experiments not yet performed. If it is again successful, we gain additional confidence that it contributes a higher level of knowledge of nature. The ideal gas laws are amenable to explanation by a straightforward theoretical model; it is called the kinetic theory of gases.

In 1738, Daniel Bernoulli, a Swiss mathematician, first proposed the **kinetic theory of gases.** It was later elaborated on by Clausius, Maxwell, Boltzmann, van der Waals, and Jeans. The theory rests on the following fundamental assumptions:

1. Gases consist of discrete particles of matter, called molecules, which are all the same—in mass and size, for instance—in any pure gaseous substance; the molecules differ, however, among different gaseous substances.
2. The molecules are in constant motion, colliding with each other and with the walls of vessels that contain them.
3. Molecular collisions with the vessel walls give rise to the phenomenon pressure.
4. Molecular collisions are perfectly elastic; that is, the molecules collide without loss of kinetic energy—there are no losses due to friction. Thus, molecular collisions are like collisions between idealized billiard balls.
5. At low pressures the diameter of any gaseous molecule is very much smaller than the spaces between molecules. Thus, molecules may be considered point masses, and it is assumed that the attractive forces between them are essentially negligible.
6. The absolute temperature of a gas is directly proportional to the average kinetic energy of the molecules of the gas.

From these assumptions one can construct a picture that is consistent with the ideal gas laws. Let us investigate how the theory can be used to derive them.

Consider a gaseous molecule with mass m and with velocity component (or speed) u_x in the x direction. Consider also that the molecule is inside a cubical container, as shown in Fig. 15.4. Velocity is a **vector** quantity; that is, it has both magnitude and direction. Any velocity vector, pointing in any direction, can be expressed in

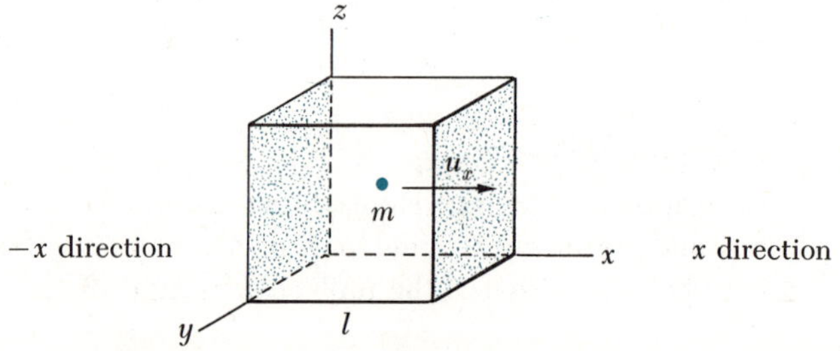

Figure 15.4 A gaseous molecule striking a wall of a cubical container.

terms of **components** along the x, y, and z directions. Choosing the x component for our illustration is essentially equivalent to merely considering motion that is directed along the x axis.

Then the momentum of the molecule in the x direction is mu_x. After the molecule collides with the wall of the cube, it rebounds in the $-x$ direction with momentum $-mu_x$; since the collision is perfectly elastic, there is no loss of kinetic energy, and no loss of momentum. The change in momentum per collision is therefore given by

$$\text{(momentum before collision)} - \text{(momentum after collision)} = mu_x - (-mu_x) = 2mu_x \qquad (15.14)$$

If the speed u_x of motion in the x direction is such that the length l of the cube is transversed in t seconds, then

$$u_x\left(\frac{\text{cm}}{\text{sec}}\right) = \frac{l\ (\text{cm})}{t\ (\text{sec})} \qquad (15.15)$$

The amount of time required for the particle to traverse the cube once is

$$(l\ \text{cm})\left(\frac{1}{u_x}\frac{\text{sec}}{\text{cm}}\right) = \frac{l}{u_x}\text{sec}$$

Thus, the frequency of collisions, expressing the number of collisions per second, is given by

$$\text{frequency of collisions} = \frac{1}{(l/u_x)\,\text{sec}} = \frac{u_x}{l}\,\text{sec}^{-1} \qquad (15.16)$$

Since the change in momentum along the x axis is $2mu_x$ (g cm/sec) for each wall collision, we write the following expression for the momentum change per second:

$$\text{(mom. change per collision)(no. collisions per sec)} = \{2mu_x\}\cdot\left\{\frac{u_x}{l}\right\} = \frac{2mu_x^2}{l}$$

So far we have been concerned with only the collisions of one molecule with the two shaded walls in Fig. 15.4. If there are n' molecules within the cube, all traveling with an x velocity component u_x, then the total momentum change per second due to collisions of all molecules with the shaded walls is

$$n'\frac{2mu_x^2}{l}$$

If we focus attention on just one of the shaded walls, then the total momentum change per second is half this number, that is, $n'mu_x^2/l$. Newton's second law of motion tells us that the rate of change of momentum, or momentum change per second, at a point is equal to the force exerted at that point by the phenomenon giving rise to the momentum change; in this case, the force is exerted against a wall by collisions of molecules against the wall. Hence, the force F_x exerted against either shaded wall by the motion of the molecules in the x direction is given by

$$F_x = \frac{n'mu_x^2}{l} \qquad (15.17)$$

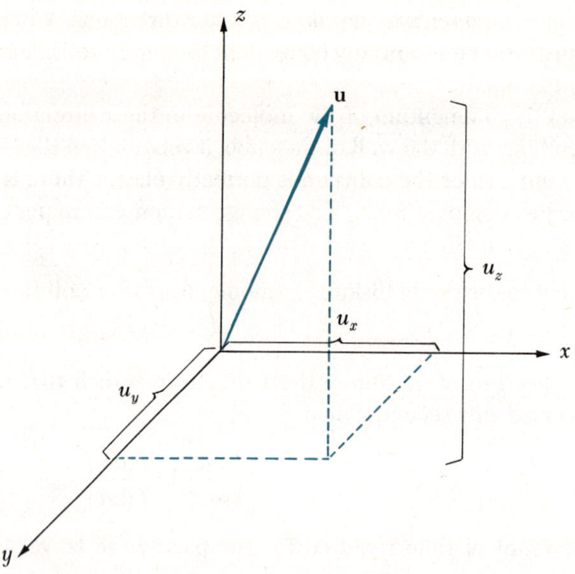

Figure 15.5
The factoring, or component analysis, of the velocity vector **u**. By trigonometry $u^2 = u_x^2 + u_y^2 + u_z^2$, where u is the length of the vector **u**.

In a real gaseous system, molecules move randomly in all directions, not just in the x direction. It can be shown that it is rigorously correct to "decompose," or factor, molecular motion into component motions along three mutually perpendicular axes, for example, the x, y, and z axes of Figs. 15.4 and 15.5. One can write the following equation that relates the x, y, and z components of velocity (u_x, u_y, u_z) to the total speed u:

$$u^2 = u_x^2 + u_y^2 + u_z^2 \tag{15.18}$$

We must also note that the molecules in a gas do not all move with just one speed; instead the speeds and velocity components cover a distribution of values. Hence, we define the *mean*, or average, values of the quantities given in Eq. (15.18). We indicate that we are dealing with averages by placing a bar over each of the symbols in Eq. (15.18):

$$\overline{u^2} = \overline{u_x^2} + \overline{u_y^2} + \overline{u_z^2}$$

Also, as no single direction of motion is favored over any other in a truly random system, we can write

$$\overline{u_x^2} = \overline{u_y^2} = \overline{u_z^2}$$

Thus,

$$3\overline{u_x^2} = \overline{u^2} \qquad \text{or} \qquad \overline{u_x^2} = \frac{\overline{u^2}}{3} \tag{15.19}$$

Using this relation between $\overline{u_x^2}$ and $\overline{u^2}$ and the expression for force given in Eq. (15.17), we can write

$$F_x = \frac{n'm\overline{u^2}}{3l}$$

Since the motion of molecules in a gas is random, we must conclude that the expression for F_x, which was derived for each of the shaded walls, must give the force exerted at any of the six walls of the cube:

$$F = \frac{n' m \overline{u^2}}{3l} \tag{15.20}$$

The pressure P exerted against any wall is by definition the force per unit area. As the force is given by Eq. (15.20) and the area of any single wall is l^2, we have

$$P = \frac{F}{l^2} = \frac{n' m \overline{u^2}}{3l \cdot l^2} = \frac{n' m \overline{u^2}}{3l^3}$$

But l^3 is simply the volume of the cube V. Hence

$$P = \frac{n' m \overline{u^2}}{3V} \tag{15.21}$$

This is a fundamental equation of the kinetic theory of gases. Although it was derived for a cubical container, it can be shown that the same result obtains for any other shape. The equation applies strictly to ideal gases; it is valid for real gases only to the extent that the conditions of assumptions 4 and 5 are valid.

Spotlight

There are four well-characterized gaseous oxides of nitrogen, NO, N_2O, NO_2, and N_2O_4.

Nitrous oxide, N_2O, is a gas first obtained by Priestley in 1772 from the reduction of nitric oxide, NO, with iron:

$$2NO(g) + Fe(s) + H_2O \rightleftharpoons N_2O(g) + Fe(OH)_2(s)$$

Pure N_2O is prepared by heating sodium nitrite, $NaNO_2$, and hydroxylamine hydrochloride in solution:

$$NO_2^- + NH_2OH + H^+ \rightleftharpoons N_2O(g) + 2H_2O$$

Nitrous oxide is known from spectroscopic measurements to be a linear molecule; it can be described in terms of two resonance structures:

$$:\overset{-}{\underset{\cdot\cdot}{N}}=\overset{+}{N}=\underset{\cdot\cdot}{\ddot{O}} \quad \leftrightarrow \quad :N\equiv\overset{+}{N}-\underset{\cdot\cdot}{\ddot{O}}:^-$$

It is a colorless gas that is soluble in water. The gas does not burn but will explode if detonated. It supports combustion because it decomposes to nitrogen and oxygen when heated; a glowing splint bursts into flames in N_2O.

Nitrous oxide produces a mild form of hysteria if inhaled in small quantities, hence the name "laughing gas." In larger doses it is an anesthetic and has found frequent use in dental surgery. It is used in whipped cream "bombs" as a propellant gas because of its solubility in cream.

Nitric oxide, NO, was known as early as 1600; however, Priestley is generally credited with the formal discovery in 1772. It occurs in nature during electrical storms. It is one of the main air pollution components and is formed from nitrogen and oxygen at the high temperatures found in the internal combustion engine. The commercial production involves the oxidation of ammonia,

$$4NH_3(g) + 5O_2(g) \rightleftharpoons 4NO(g) + 6H_2O$$

whereas the usual laboratory method is from copper and *dilute* nitric acid,

$$3Cu(s) + 8H^+ + 2NO_3^- \rightleftharpoons 2NO(g) + 3Cu^{2+} + 4H_2O$$

Nitric oxide contains an unpaired electron and is termed an odd molecule. The resonance structures are

$$:\dot{N}=\ddot{O}: \quad \leftrightarrow \quad :\ddot{N}=\dot{O}:$$

Continued on page 726

It is an unusual odd molecule in the sense that it does not readily form a dimer, even though it has an unpaired electron. In the liquid and solid states, however, some dimer exists, with the structure shown:

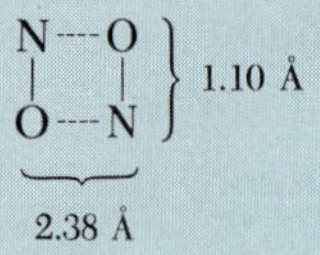

Nitric oxide is a colorless gas that is only slightly soluble in water. Its smell, taste, and other physiological properties have not been determined, because it immediately oxidizes in air to nitrogen dioxide,

$$2NO(g) + O_2(g) \rightleftharpoons 2NO_2(g)$$

At high pressures, NO decomposes at 30 °C to 50 °C:

$$3NO(g) \rightleftharpoons N_2O(g) + NO_2(g)$$

Aqueous oxidizing agents, such as I_2 or permanganate ion, MnO_4^-, in water, convert NO to the nitrate ion, NO_3^-:

$$2NO(g) + 3I_2(s) + 4H_2O \rightleftharpoons 2NO_3^- + 6I^- + 8H^+$$

$$10NO(g) + 6MnO_4^- + 8H^+ \rightleftharpoons 10NO_3^- + 6Mn^{2+} + 4H_2O$$

Nitric oxide can also act as an oxidizing agent; for example,

$$2NO(g) + 2H_2(g) \rightleftharpoons N_2(g) + 2H_2O$$

$$2NO(g) + 5H_2(g) \rightleftharpoons 2NH_3(g) + 2H_2O$$

$$4NO(g) + 2OH^- \rightleftharpoons N_2O(g) + \underset{\text{nitrite ion}}{2NO_2^-} + H_2O$$

$$2NO(g) + \underset{\text{sulfite ion}}{SO_3^{2-}} \rightleftharpoons N_2O(g) + \underset{\text{sulfate ion}}{SO_4^{2-}}$$

Nitrogen dioxide, NO_2, and dinitrogen tetroxide, N_2O_4, exist in equilibrium with each other at ordinary temperatures:

$$\underset{\text{brown, paramagnetic}}{2NO_2(g)} \rightleftharpoons \underset{\text{colorless, diamagnetic}}{N_2O_4(g)}$$

The mixture is sometimes referred to as nitrogen peroxide and can be prepared by mixing oxygen and nitric oxide,

$$2NO(g) + O_2(g) \rightleftharpoons 2NO_2(g)$$

or by a reaction of copper with *concentrated* nitric acid (dilute acid gives NO),

$$Cu(s) + 2NO_3^- + 4H^+ \rightleftharpoons 2NO_2(g) + Cu^{2+} + 2H_2O$$

Nitrogen dioxide has a bent structure

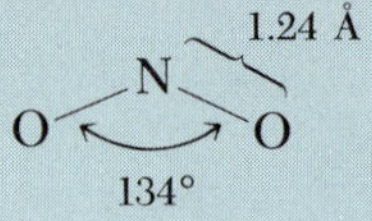

It is an odd-electron molecule, and has four main resonance structures.

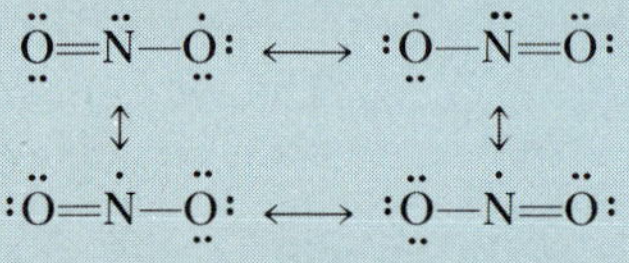

It can lose its odd electron to an oxidizing agent fairly easily to form the nitronium ion, NO_2^+,

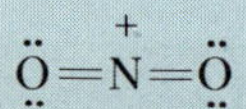

The ion NO_2^+ is linear and symmetrical with an N–O bond distance of 1.154 Å. It is reminiscent of the isoelectronic CO_2 molecule, which is also linear and symmetrical, with a C–O bond distance of 1.163 Å.

Dinitrogen tetroxide, the dimer, has three isomeric forms. The most stable form is the planar molecule whose structure is

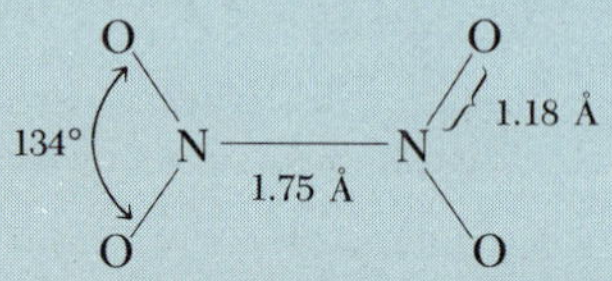

In anhydrous HNO_3 or H_2SO_4, N_2O_4 dissociates into the nitrosonium ion and the nitrate ion:

$$N_2O_4(g) \rightleftharpoons NO^+ + NO_3^-$$

The mixed oxides, NO_2 and N_2O_4, dissolve in water and react to form a mixture of nitrous and nitric acids:

$$2NO_2(g) + H_2O \rightleftharpoons HNO_2 + H^+ + NO_3^-$$

The mixture of acids, unless dilute or cold, is unstable and the HNO_2 decomposes:

$$3HNO_2 \rightleftharpoons 2NO(g) + H_2O + H^+ + NO_3^-$$

Use is made of this reaction in the commercial production of HNO_3.

From Eq. (15.21), we can derive the ideal gas law as follows. If n is the number of moles of gas in the cubical container, then

$$n' = nN$$

where N is Avogadro's number. Hence, from Eq. (15.21) we write

$$PV = \frac{nNm\overline{u^2}}{3} \tag{15.22a}$$

For the developments below, it is convenient to rewrite this equation as follows:

$$PV = \frac{2nN}{3} \times \frac{m\overline{u^2}}{2} \tag{15.22b}$$

Since we know from elementary physics that the kinetic energy of a particle (KE) is $\frac{1}{2}mu^2$, we can write from assumption 6 of the kinetic theory,

$$\tfrac{1}{2}m\overline{u^2} \propto \mathrm{T}$$

or

$$\tfrac{1}{2}m\overline{u^2} = (\text{const})\,T = \tfrac{3}{2}kT \tag{15.23}$$

where T is the absolute temperature and $3k/2$ is taken to be the proportionality constant. The constant k is called the Boltzmann constant; many years of experimentation have determined its value to be 1.381×10^{-16} erg deg^{-1} = 1.381×10^{-23} J deg^{-1}.

Using Eqs. (15.22) and (15.23), one can write

$$PV = \frac{2nN}{3} \times \frac{m\overline{u^2}}{2} = \left(n\frac{2N}{3}\right)\left(\frac{3kT}{2}\right) = n(Nk)T \tag{15.24}$$

Then, if we set Nk equal to the gas constant R, we obtain

$$PV = nRT \tag{15.10}$$

This is the **ideal gas law**, which we now see as derivable from a simple theoretical model. Furthermore, from the relationship $R = Nk$, we see that k can be thought of as the gas constant per molecule; that is,

$$k = \frac{R\ \text{cal deg}^{-1}\text{mol}^{-1}}{N\ \text{molecule mol}^{-1}} = \frac{R}{N}\text{cal deg}^{-1}\text{molecule}^{-1}$$

From Eq. (15.23), we see that for a mole of an ideal gas, the total translational kinetic energy is simply N times $\frac{3}{2}kT$.

$$E_{\text{tran}}(\text{per mole}) = N \times \tfrac{3}{2}kT = \tfrac{3}{2}NkT = \tfrac{3}{2}RT \tag{15.25}$$

For most actual gases, including many that obey the ideal gas laws, there will be other modes of energy (due, for instance, to rotational and vibrational motions) that are not included in this strictly translational picture; hence, in most cases, Eq. (15.25) does not account for all the important energy terms.

One can readily see that the above simple theoretical model successfully accounts for Graham's law of diffusion. Consider two different gases that are at the same pressure, having mean square speeds $\overline{u_1^2}$ and $\overline{u_2^2}$, and occupying volumes V_1 and V_2. As the pressures are the same, we can write from Eq. (15.21):

$$P_1 = \frac{n_1' m_1 \overline{u_1^2}}{3V_1} = \frac{n_2' m_2 \overline{u_2^2}}{3V_2} = P_2$$

or

$$\frac{\overline{u_1^2}}{\overline{u_2^2}} = \frac{n_2' m_2 / V_2}{n_1' m_1 / V_1}$$

But the numerator and denominator of the right side of this equation are the densities of gas 2 and gas 1, respectively. Hence,

$$\frac{\overline{u_1^2}}{\overline{u_2^2}} = \frac{d_2}{d_1}$$

or

$$\frac{(\overline{u_1^2})^{1/2}}{(\overline{u_2^2})^{1/2}} = \sqrt{\frac{d_2}{d_1}} \qquad (15.26)$$

which is a statement of Graham's law of diffusion. In Eq. (15.26), the quantity $(\overline{u^2})^{1/2}$ is the **root mean square** speed of molecules in gas 1, that is, the square root of the average of the squared speed of those molecules. The symbol "rms" is often employed to denote "root mean square."

15.4 Molecular Velocities

All molecules in a gas do not move with the same velocity. The random and frequent collisions between molecules give rise to a distribution of both energies and velocities. From consideration of probability relations, James Clark Maxwell, a Scottish physicist, showed in 1860 that the distribution of molecular speeds in a gas is as depicted in Fig. 15.6. This figure gives the distribution of speeds for three different temperatures, T_1, T_2, and T_3, where $T_1 < T_2 < T_3$. There are two important features to be noted in this figure. First, it demonstrates that it is highly improbable for a molecule to have either a very high speed or a very low one. The most probable speeds, corresponding to the largest fraction of molecules, occur at some intermediate value. The most probable speed, denoted by the symbol α, is the speed for which the curve reaches its peak, or maximum value. A second important feature of Fig. 15.6 is that the most probable speed increases as the temperature increases; that is,

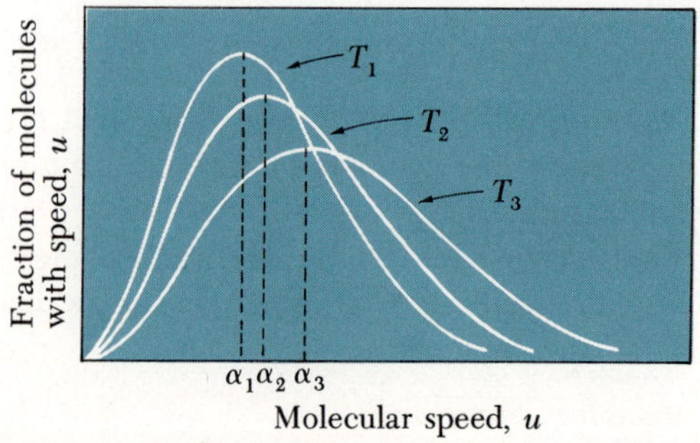

Figure 15.6 Distribution of molecular speeds in a gas at three different temperatures: $T_3 > T_2 > T_1$. The area under each curve represents all the molecules, and is the same for each curve; hence, the T_3 curve is shorter and fatter than the T_1 curve.

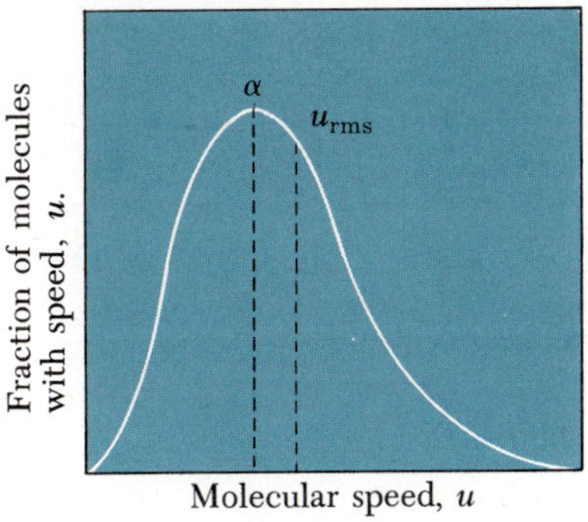

Figure 15.7
Distribution of molecular speeds in a gas.

$\alpha_1 < \alpha_2 < \alpha_3$, the subscripts referring to the three temperatures, T_1, T_2, and T_3. Indeed, a more general observation is that *higher speeds become more probable* (correspond to larger fractions of the molecules) *at higher temperatures.* Rearranging Eq. (15.23), we obtain the root mean square speed:

$$\overline{u^2} = \frac{3kT}{m}$$

$$\sqrt{\overline{u^2}} = u_{rms} = \sqrt{\frac{3kT}{m}} \tag{15.27}$$

where m is the molecular mass. From this result, we should expect the most probable speed to be proportional to the square root of T/m. Detailed treatments of the kinetic theory of gases show that the most probable speed is given by

$$\alpha = \sqrt{\frac{2kT}{m}} \tag{15.28}$$

where k is the Boltzmann constant, T is the absolute temperature, and m is the molecular mass.

Comparing Eqs. (15.27) and (15.28), we obtain

$$\frac{u_{rms}}{\alpha} = \frac{(3kT/m)^{1/2}}{(2kT/m)^{1/2}} = (\tfrac{3}{2})^{1/2} = 1.224$$

$$u_{rms} = 1.224\alpha$$

Thus, the rms speed is about 22% higher than the most probable speed. This relation is shown pictorially for a single temperature in Fig. 15.7.

EXAMPLE 15.9 For H_2 at 100 °C, calculate the root mean square velocity of a molecule in miles per hour.

SOLUTION To apply Eq. (15.27), we need the mass of H_2 in grams. From its molecular weight (2.016 amu), we write

$$2.016 \text{ g } H_2 = 6.022 \times 10^{23} \text{ } H_2 \text{ molecules}$$

$$\frac{2.016 \text{ g } H_2}{6.022 \times 10^{23}} = 1 \text{ } H_2 \text{ molecule} = 0.3347 \times 10^{-23} \text{ g} = 0.3347 \times 10^{-26} \text{ kg}$$

Then,

$$u_{rms} = \sqrt{\frac{3kT}{m}} = \sqrt{\frac{3(1.381 \times 10^{-23}\ \text{J deg}^{-1})(373.15\ \text{deg})}{0.3347 \times 10^{-26}\ \text{kg}}}$$

$$= 2.149 \times 10^3 \frac{\text{m}}{\text{sec}} = \left(2.149 \times 10^3 \frac{\text{m}}{\text{sec}}\right)\left(\frac{1}{1.094}\frac{\text{yd}}{\text{m}}\right)\left(\frac{1}{1760}\frac{\text{mi}}{\text{yd}}\right)$$

$$= 1.12\frac{\text{mi}}{\text{sec}} = \left(1.116\frac{\text{mi}}{\text{sec}}\right)\left(60 \times 60\frac{\text{sec}}{\text{hr}}\right) = 4018\frac{\text{mi}}{\text{hr}}$$

STUDY PROBLEM 15(f)

Calculate the most probable speed of a Cl_2 molecule at 200 °C.

15.5 Chemical Equilibrium in Gases

The law of chemical equilibrium has been stated in terms of the mass action expression (Eq. 4.5):

$$K = \frac{[\text{M}]^m[\text{N}]^n \ldots}{[\text{A}]^a[\text{B}]^b \ldots} \tag{4.5}$$

The quantities represented in brackets are taken to be molarities for solute species in liquid solutions. For the few cases in which gas systems have been discussed, molarities have been used for gaseous species also. Another common approach for gaseous species is to use the partial pressure species (in atm) in place of molarities in equilibrium expressions. Of course, we know that for ideal gases, molarity and partial pressure are proportional at a specific temperature:

$$P = \frac{nRT}{V} = \left(\frac{n}{V}\right)(RT) = (\text{molarity})(RT) \tag{15.29}$$

In this equation it is assumed that V is in liters.

If we denote the equilibrium constant in terms of molarities by the symbol K_c, then we write the following equilibrium expression for the synthesis of ammonia:

$$N_2(g) + 3H_2(g) \rightleftharpoons 2NH_3(g)$$

$$K_c = \frac{[NH_3]^2}{[N_2][H_2]^3} = \frac{(n_{NH_3}/V)^2}{(n_{N_2}/V)(n_{H_2}/V)^3} \tag{15.30}$$

Using Eqs. (15.29) and (15.30), one obtains

$$K_c = \frac{(P_{NH_3}/RT)^2}{(P_{N_2}/RT)(P_{H_2}/RT)^3} = \left[\frac{P_{NH_3}^2}{P_{N_2}P_{H_2}^3}\right](RT)^2 \tag{15.31}$$

where P_{NH_3} represents the partial pressure of ammonia in the system. The first expression in brackets on the right side of Eq. (15.31) has the same form (not the same value) as the Q expression in Eq. (15.30). Because K_c is a constant at a specific temperature, the expression in large brackets on the right side of Eq. (15.31) must be a constant. This expression is designated K_p.

Spotlight

Hydrogen fluoride, or hydrofluoric acid, HF, is a gas at room temperature and pressure (bp, 19.4 °C), and can be prepared by treating CaF_2 with concentrated H_2SO_4:

$$CaF_2(s) + H_2SO_4(l) \rightleftharpoons 2HF(g) + CaSO_4(s)$$

In the gas phase, HF is probably the most imperfect gas studied. It exists as monomer at high temperatures; at low temperatures, however, the molecules associate, and an equilibrium exists between HF, $(HF)_2$, $(HF)_4$, and $(HF)_6$. In the liquid state HF is electrically nonconducting and polymeric due to hydrogen bonding. The hydrogen bonds present in liquid HF account for its high boiling point relative to the other halogen hydrides (see Table 16.1). Solid HF (mp, −83 °C) consists of essentially infinite zigzag chains, as shown herewith.

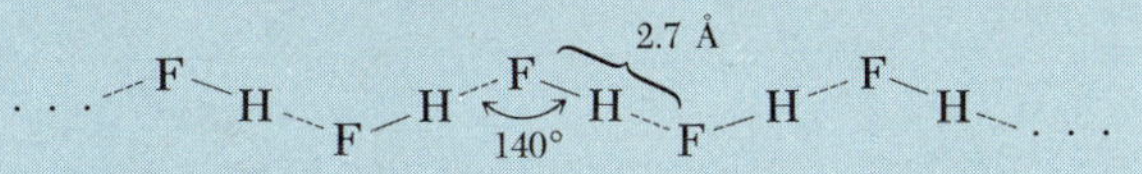

$$K_p = \frac{P_{NH_3}^2}{P_{N_2}P_{H_3}^3} \tag{15.32}$$

Equations of this form are what are often used in equilibrium calculations on gaseous systems. From Eqs. (15.31) and (15.32), we see that K_c and K_p are related for this case by the equation

$$K_c = K_p(RT)^2 \tag{15.33}$$

Equation (15.33) is just a special case of a general formula that we can derive for any gas phase reaction at equilibrium, using the ideal gas law (which is at least approximately valid). For the general case of a gas phase reaction one can write for the equilibrium condition

$$K_c = \frac{[M]^m[N]^n}{[A]^a[B]^b\cdots} = \frac{(P_M/RT)^m(P_N/RT)^n\cdots}{(P_A/RT)^a(P_B/RT)^b\cdots}$$

$$= \left[\frac{P_M{}^mP_N{}^n}{P_A{}^aP_B{}^b}\right]\frac{(RT)^{-m-n}\cdots}{(RT)^{-a-b}\cdots}$$

Then, if the expression in large brackets is identified as K_p, it follows that

$$K_c = K_p\frac{1}{(RT)^{\Delta n}} \tag{15.34}$$

where $\Delta n = m + n + \cdots - (a + b + \cdots)$, that is, the change in the total number of moles of gas associated with the chemical reaction as written (Eq. 4.3). For example, consider the chemical equation

$$N_2(g) + 2O_2(g) \rightleftharpoons NO_2(g)$$

for which $n = 2 - 3 = -1$. In this case, we write

$$K_c = \frac{[NO_2]^2}{[N_2][O_2]^2} = K_p\frac{1}{(RT)^{-1}} = K_p(RT)$$

where

$$K_p = \frac{P_{NO_2}^2}{P_{N_2}P_{O_2}^2}$$

Spotlight

The nitride of carbon is called cyanogen, $N{\equiv}C{-}C{\equiv}N$. It is a flammable gas that can be prepared directly from the elements, or by heating mercuric cyanide, $Hg(CN)_2$, either alone or with mercuric chloride, $HgCl_2$:

$$2C(s) + N_2(g) \rightleftharpoons C_2N_2(g)$$

$$Hg(CN)_2(s) \rightleftharpoons Hg(l) + C_2N_2(g)$$

$$Hg(CN)_2(s) + HgCl_2(s) \rightleftharpoons Hg_2Cl_2(s) + C_2N_2(g)$$

Cyanogen is a colorless, poisonous gas with a smell similar to that of peaches. It dissolves in water to give initially a substance called oxamide:

$$N{\equiv}C{-}C{\equiv}N + 2H_2O \rightleftharpoons \underset{\text{oxamide}}{H_2N{-}\overset{\overset{\displaystyle O}{\|}}{C}{-}\overset{\overset{\displaystyle O}{\|}}{C}{-}NH_2}$$

On standing for long periods, oxamide hydrolyzes to ammonium oxalate:

$$H_2N{-}\overset{\overset{\displaystyle O}{\|}}{C}{-}\overset{\overset{\displaystyle O}{\|}}{C}{-}NH_2 + 2H_2O \rightleftharpoons$$
$$2NH_4^+ + {}^-O_2C{-}CO_2^-$$

When cyanogen is mixed with pure oxygen, the mixture burns, producing one of the hottest flames (ca 4800 °C) known from a chemical reaction.

Cyanogen is referred to as a *pseudohalogen*, because some of its reactions are similar to reactions of the halogens; for example,

1. $2OH^- + (CN)_2(g) \rightleftharpoons CN^- + OCN^- + H_2O$
   $2OH^- + Cl_2(g) \rightleftharpoons Cl^- + OCl^- + H_2O$

2. $2M(s) + (CN)_2(g) \rightleftharpoons 2MCN(s)$
   $2M(s) + Br_2(l) \rightleftharpoons 2MBr(s)$

where M is a Group I metal, like Na or K.

It is important to remember that the same information is contained, in a somewhat different form, in equilibrium expressions based on molarities (and K_c) or equilibrium expressions based on partial pressures (and K_p).*

EXAMPLE 15.10 What is the ratio of K_c to K_p at 300 °C for the system described by Eq. (15.31)?

SOLUTION In that particular case,

$$K_c = K_p(RT)^2$$

As molarity is expressed in mol/liter, the appropriate value of R is 0.08205 liter atm deg^{-1}mol^{-1}. Then

$$\frac{K_c}{K_p} = (RT)^2 = [(0.08205 \text{ liter atm deg}^{-1}\text{mol}^{-1})(573.15 \text{ deg})]^2$$

$$= 2.21 \times 10^3 \frac{\text{atm}^2}{(\text{mol/liter})^2}$$

*The use of partial pressures instead of concentrations in equilibrium expressions for gases is in line with the concepts of activities and standard states. Thus, the activity in equilibrium expression (4.48) is, for gaseous species, expressed essentially as the ratio of the partial pressure of the species to the partial pressure of the species in its standard state, which is 1 atm. Hence, the activity of a gaseous species essentially equals the numerical value of the partial pressure in atm.

STUDY PROBLEM 15(g)

For the reaction

$$N_2O_4(g) \rightleftharpoons 2NO_2(g)$$

$K_c = 0.36$ at 100 °C. Calculate K_p for this temperature.

EXAMPLE 15.11 K_p at 25 °C is 2.5×10^{24} for the reaction

$$2SO_2(s) + O_2(g) \rightleftharpoons 2SO_3(g)$$

Suppose we are concerned with the efficiency of converting SO_2 to SO_3 under conditions in which we maintain the partial pressures of O_2 at 0.20 atm, approximately normal air. Calculate the ratio of the partial pressures of SO_3 and SO_2 in equilibrium with O_2 at a partial pressure of 0.20 atm.

SOLUTION For this reaction, the pertinent equilibrium expression is

$$2.5 \times 10^{24} = \frac{P_{SO_3}^2}{P_{SO_2}^2 P_{O_2}}$$

Then, substituting $P_{O_2} = 0.20$ gives

$$\frac{P_{SO_3}^2}{P_{SO_2}^2(0.20)} = 2.5 \times 10^{24}$$

$$\frac{P_{SO_3}}{P_{SO_2}} = \sqrt{(0.20)(2.5 \times 10^{24})} = 0.71 \times 10^{12}$$

Hence, as far as thermodynamics is concerned, the conversion should be fantastic.

STUDY PROBLEM 15(h)

For the case shown in Example 15.11, calculate what the partial pressure of O_2 would be at equilibrium if the ratio P_{SO_3}/P_{SO_2} is 1000.

15.6 Deviations from Ideal Behavior

So far, we have been concerned with a hypothetical state of matter called the *ideal gas*. No *real* gas is an ideal gas, but many gases under a wide range of conditions follow the ideal gas laws reasonably closely. Ideal gas behavior is approached most closely at low pressures and at temperatures well above the **critical temperature**, the highest temperature at which a gas can be liquefied by applying pressure.

A convenient manner of showing how much a real gas deviates from ideal gas behavior is to show how much the quantity PV/n varies at constant T as pressure is varied. According to the equation of state of an ideal gas (Eq. 15.6), this quantity equals RT for an ideal gas, and should be independent of pressure if temperature is held constant. Deviations from a constant value for a particular gas correspond to failure of the ideal gas formulation for that gas. Figure 15.8 illustrates the departure from ideal gas behavior for several gases. The ordinate is the quantity PV/n at a

Spotlight

Boron trifluoride, BF_3, is a colorless gas prepared by heating either B_2O_3 or borax, $Na_2B_4O_7 \cdot 10H_2O$ with CaF_2 and H_2SO_4:

$$CaF_2(s) + 2H_2SO_4(l) \rightleftharpoons 2HF(g) + CaSO_4(s)$$

$$Na_2B_4O_7 \cdot 10H_2O + 12HF(g) \rightleftharpoons 4BF_3(g) + 15H_2O + 2Na^+ + 2OH^-$$

Boron trifluoride, like the other boron halides, is known from spectroscopic studies to be planar:

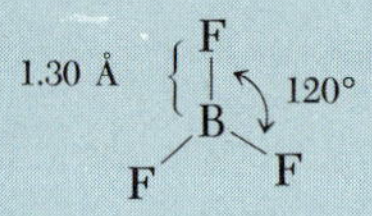

The shortness of the B–F bond has often been attributed to three resonance structures of the type

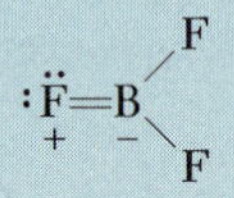

which are associated with a level of "double bond character" in each B–F bond.

Boron trifluoride is probably the most powerful neutral Lewis acid known. The formation of complexes with organic Lewis bases, such as amines, R_3N, phosphines, R_3P, ethers, R—O—R, and sulfides, R—S—R, is the main feature of its chemistry; for example,

$$(CH_3)_3N + BF_3 \rightleftharpoons (CH_3)_3\overset{+}{N}—\overset{-}{B}F_3$$

With BF_3, even the noble element argon forms coordination compounds that are stable above their melting points; six compounds are known, $ArBF_3$, $Ar(BF_3)_2$, $Ar(BF_3)_6$, $Ar(BF_3)_8$, and $Ar(BF_3)_{16}$. Hydrogen fluoride coordinates with BF_3 to form tetrafluoroboric acid, HBF_4, which ionizes to H^+ and BF_4^-.

Alkali and alkaline earth metals reduce BF_3 to elemental boron and the metal fluoride; for example,

$$3Na(s) + BF_3(g) \rightleftharpoons 3NaF(s) + B(s)$$

constant temperature 0 °C and the abscissa is *P*. For an ideal gas, the plot would simply give a horizontal straight line corresponding to the dashed lines in the figure.

Many equations of state have been developed to predict the behavior of real gases with more accuracy than what is provided by the ideal equation of state. Some of these equations are based on empirical observations, whereas others are based on theoretical considerations. One of the most common and most descriptive of such equations is the **van der Waals equation of state,** which was developed in 1873 by Johannes Diderik van der Waals, a Dutch physicist. His approach was to take into account the volume occupied by the molecules themselves and the attractive forces existing between them; both these entities are neglected in the model on which the ideal gas equation of state has been based.

According to the van der Waals approach, we let b represent the effective volume of the molecules in one mole of a gas. Then the volume that n moles will occupy in any total volume V will be nb. The "free space" available for compression of the gas is then $(V - nb)$, and the ideal equation of state can be modified as shown:

$$P(V - nb) = nRT \qquad (15.35)$$

To account for the attractive forces between the molecules, van der Waals made a correction in the pressure term. These forces effectively reduce the total force exerted on the walls of the container by the moving molecules, and the pressure on the walls is reduced. Van der Waals estimated that the pressure loss is

$$P' = \frac{n^2a}{V^2}$$

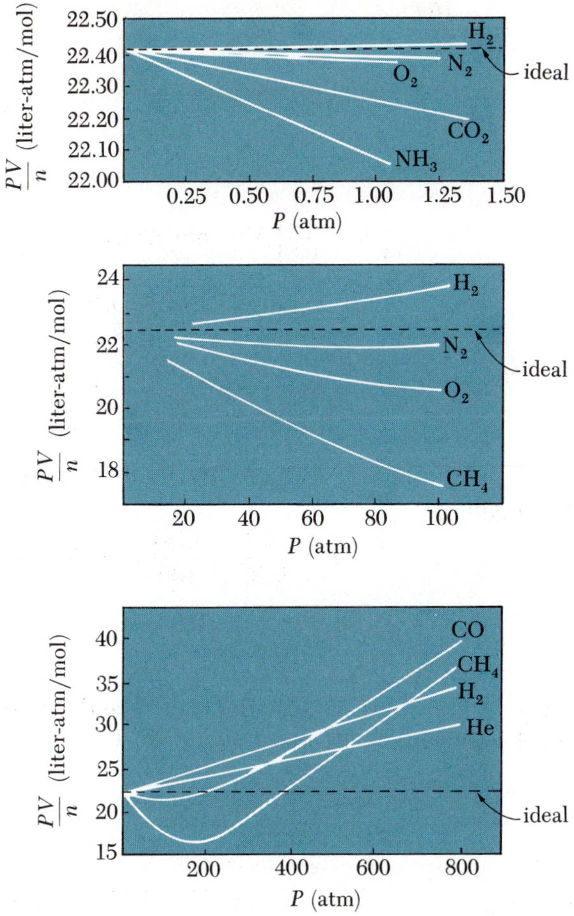

Figure 15.8 Deviations from ideal behavior for several real gases at 0 °C.

where P' is the reduction in pressure from ideality, and a is a constant characteristic of each gas and independent of pressure and temperature. The observed pressure will be the ideal pressure minus the pressure loss, or $P_{\text{obs}} = P_{\text{ideal}} - P'$; hence,

$$P_{\text{ideal}} = P_{\text{obs}} + \frac{n^2a}{V^2}$$

Applying this correction to Eq. (15.35) gives the van der Waals equation of state:

$$\left(P + \frac{n^2a}{V^2}\right)(V - nb) = nRT \qquad (15.36)$$

Table 15.1 gives values of the van der Waals constants a and b for some gases.

EXAMPLE 15.12 Calculate the pressure exerted in a 10.0-liter vessel containing 18.617 mol CO_2 at 100 °C, using (a) the ideal equation of state, and (b) the van der Waals equation of state.

SOLUTION (a)

$$P = \frac{nRT}{V}$$

$$= \frac{(18.617 \text{ mol})(0.08205 \text{ liter atm deg}^{-1}\text{mol}^{-1})(373.15 \text{ deg})}{10.0 \text{ liters}}$$

$$= 57.0 \text{ atm}$$

(b)

$$\left(P + \frac{n^2a}{V^2}\right)(V - nb) = nRT$$

$$P + \frac{n^2a}{V^2} = \frac{nRT}{V - nb}$$

$$P = \frac{nRT}{V - nb} - \frac{n^2a}{V^2}$$

Making the substitutions from Table 15.1 (a = 3.59 atm liter² mol⁻²; b = 0.0427 mol⁻¹), we get

$$P = \frac{570}{9.20} - \frac{1244}{100} = 49.5 \text{ atm}$$

The actual pressure of this system is 50.0 atm.

STUDY PROBLEM 15(i)

Using the van der Waals equation, calculate the pressure exerted in a 5.20-liter vessel containing 3.961 mol CO at 150 °C.

TABLE 15.1 Constants for the Van der Waals Equation

Gas	*a* atm liter^2 mol^{-2}	*b* mol^{-1}
Acetylene, C_2H_2	4.39	0.0514
Ammonia	4.17	0.0371
Carbon dioxide	3.59	0.0427
Carbon monoxide	1.49	0.0399
Chlorine	6.49	0.0562
Ethane, C_2H_6	5.49	0.0638
Ethylene, C_2H_4	4.47	0.0571
Helium	0.034	0.0237
Hydrogen	0.244	0.0266
Methane, CH_4	2.25	0.0428
Neon	0.211	0.0171
Nitric oxide, NO	1.34	0.0279
Nitrogen	0.39	0.0391
Nitrogen dioxide	5.28	0.0442
Nitrous oxide, N_2O	3.78	0.0442
Oxygen	1.36	0.0318
Sulfur dioxide	6.71	0.0564

OPTIONAL

15.7 Thermodynamic Relations for the Ideal Gas

Thermodynamic Properties and the Equation of State

An equation that mathematically relates the pressure, volume, temperature, and number of moles of a gas constitutes what is known as an **equation of state.** From such an equation, many thermodynamic properties can be derived. Equation (15.6) is the equation of state of an ideal gas. In pursuing some of the thermodynamic results that one can derive from this equation, we shall for simplicity adopt the very special case that the gas in question is not only ideal in translational motion, but also simple in that it has no modes of energy except the translational mode.* This would apply to a monatomic gas like one of the noble gases, say He or Ne. In this case, according to Eq. (15.25), the total internal energy of n mol of an ideal gas is $\frac{3}{2}nRT$. Thus, for a fixed number of moles, the total energy (internal) depends only on the temperature. To calculate ΔE for any process carried out on a given sample of this very special gas, all we need to know is the temperature change ΔT. We conclude that $\Delta E = 0$ for any process carried out on such an ideal gas at constant temperature, $\Delta T = 0$.

It is of substantial interest to know how to compute the change in enthalpy ΔH for a process carried out on the monatomic ideal gas. This general type of calculation would, for example, be useful in computing the heat losses involved in processing and transferring natural gas, the kind of problem that assumes special significance in light of current and projected energy shortages. Recalling from Eqs. (10.21) and (10.23) that

$$H = E + PV \qquad \text{or} \qquad \Delta H = \Delta E + \Delta(PV)$$

we see that one must know how to compute changes in the quantity PV in order to obtain ΔH. But this is an especially simple task for an ideal gas system, for we know that $PV = nRT$. Hence, since n is fixed for a particular sample, we see that changes in PV are given by the equation

$$\Delta(PV) = nR\,\Delta T \tag{15.37}$$

Then, using Eqs. (15.25) and (10.23), we see that *for each mole* of an ideal gas sample

$$\begin{aligned} \Delta H &= \Delta E + \Delta(PV) = \tfrac{3}{2}R\,\Delta T + R\,\Delta T \\ &= \tfrac{5}{2}R\,\Delta T \end{aligned} \tag{15.38}$$

We note from Chap. 10 that ΔH equals the heat absorbed by the system in any process carried out at constant pressure if no work besides $P\,\Delta V$ work is performed. For processes of this type, Eq. (15.38) provides a way of computing the amount of heat absorbed or released.

Recalling the concept of heat capacity from Chap. 10, we note that the heat capacity at constant volume C_V is given by the expression $\Delta E/\Delta T$. Then, noting that the relation between E and T for an ideal gas is given by Eq. (15.25), we see that

*Including other modes of energy, for example, rotational and vibrational motion, would introduce no new complications from the thermodynamic point of view if the equation of state is known. However, as these types of energy contributions are conspicuously quantized (see Chap. 8), simple, general equations of state for such systems are not available.

OPTIONAL

the change in E for one mole of an ideal monatomic gas associated with a change in temperature ΔT is $\frac{3}{2}R\ \Delta T$. Thus, for an ideal gas,

$$C_v = \frac{\Delta E}{\Delta T} = \frac{3}{2}R \qquad (15.39)$$

To compute ΔE for any process carried out on an ideal monatomic gas with a temperature change ΔT, one need use only the relation:

$$\Delta E = C_v\,\Delta T = \tfrac{3}{2}R\,\Delta T \qquad (15.40)$$

This equation also gives, according to the first law, the heat absorbed by one mole of an ideal gas during the process if there is no volume change and no work involved in the process.

Similarly, the heat capacity at constant pressure is given in Chap. 10 as $\Delta H/\Delta T$. Using Eq. (15.38), we obtain for an ideal monatomic gas

$$C_p = \tfrac{5}{2}R \qquad (15.41)$$

Then, to calculate ΔH for any process carried out on the ideal gas, we can simply use the equation

$$\Delta H = C_p\,\Delta T = \tfrac{5}{2}R\,\Delta T \qquad (15.42)$$

Isothermal Expansion of an Ideal Monatomic Gas

To see how the above equations pertain to real physical processes, and at the same time develop some facility with thermodynamic relations in simple physical systems, we explore in the following paragraphs the reversible, isothermal (constant-temperature) expansion of one mole of an ideal gas, for example in a piston-cylinder arrangement. For such a process, ΔT is by definition zero. Hence, from Eq. (15.40) and (15.42) we see that ΔE and ΔH are each zero for this process. Since the process is neither a constant-volume nor a constant-pressure process, however, neither ΔE nor ΔH is equal to the heat absorbed by the system. From the first law of thermodynamics, we know that the heat absorbed is given as the sum of ΔE and the work W performed by the system. As ΔE is zero in this case, we have

$$q = w \qquad (15.43)$$

It is convenient to visualize the entire expansion as occurring in infinitesimal stages, or steps. The work performed by the system (gas) against the restraining force holding the piston will be, *for each tiny step in the process*, given by the relation

$$w_{\text{step}} = (P\,\Delta V)_{\text{step}} \text{ where } P \text{ is the restraining pressure.} \qquad (15.44)$$

Since the temperature and number of moles of the gas are held constant while the volume is changed during the process, we can see from Eq. (15.6), that the pressure must change from step to step. Thus, we cannot use a single value of P in Eq. (15.44) for calculating w for all steps.

Indeed, if the process is truly reversible, then the restraining pressure of the piston must be just the same as the pressure of the gas—actually infinitesimally less than the gas pressure. But that is given by RT/V for one mole. Hence, the problem is really to compute

$$w_{\text{step}} = \left(\frac{RT}{V}\Delta V\right)_{\text{step}}$$

as V changes from one tiny step to the next; then we sum up the results for all the steps.

The techniques of calculus are required for solving this problem, and the result is

$$q = w = RT\ln\frac{V_2}{V_1} \tag{15.45}$$

for one mole of gas where V_2 is the final volume (after the expansion) and V_1 is the initial volume (before the expansion).* (The zero values of ΔE and ΔH computed for this process depend only on the initial and final states, not on the pathway, as E and H are state functions. Hence, ΔE and ΔH would be zero, so long as $\Delta T = 0$, for any path, whether the process were reversible or not. However, the calculation of q and w required a knowledge of the pathway, that is, knowledge that the process was reversible and isothermal.)

Changes in the thermodynamic quantity called entropy can be calculated simply for a reversible, isothermal process by the equation

$$\Delta S = \frac{q_{\text{rev}}}{T}$$

(Chap. 11). For the reversible, isothermal expansion we can apply this equation directly, since the process meets both the criteria; it is reversible and isothermal. Hence,

$$\Delta S = \frac{RT\ln(V_2/V_1)}{T} = R\ln\left(\frac{V_2}{V_1}\right) \tag{15.46}$$

Then, from Eq. (11.9), we can calculate the change in the Gibbs free energy for this process:

$$\Delta G = \Delta H - T\Delta S = 0 - RT\ln\frac{V_2}{V_1} \tag{15.47}$$

EXAMPLE 15.13 Calculate ΔE, ΔH, and q for a process in which 2.30 mol of an ideal monatomic gas is heated from 50 °C to 150 °C.

SOLUTION Since E and H are state functions, ΔE and ΔH can be determined directly from Eqs. (15.40) and (15.38), which are written for one mole, even though we have no knowledge of the path. For 2.30 mol,

*In calculus, the work in Eq. (15.45) is computed by evaluating the integral of $P\,dV$ from V_1 to V_2. For one mole of an ideal gas, $P = RT/V$; hence

$$w = \int_{V_1}^{V_2} P\,dV = \int_{V_1}^{V_2} RT\frac{dV}{V} = RT\int_{V_1}^{V_2} d\ln V = RT\ln\frac{V_2}{V_1}$$

OPTIONAL

$$\Delta E = (2.30)(\tfrac{3}{2}R\,\Delta T) = (2.30\text{ mol})(\tfrac{3}{2})(1.987\text{ cal deg}^{-1}\text{mol}^{-1})(150\text{ deg} - 50\text{ deg})$$
$$= 686\text{ cal}$$

$$\Delta H = (2.30)(\tfrac{5}{2}R\,\Delta T) = (2.30\text{ mol})(\tfrac{5}{2})(1.987\text{ cal deg}^{-1}\text{mol}^{-1})(150\text{ deg} - 50\text{ deg})$$
$$= 1.14\text{ kcal}$$

As the path is not specified, q cannot be computed.

EXAMPLE 15.14 Calculate ΔE, ΔH, ΔS, ΔG, and q for the reversible expansion of 0.160 mol of an ideal monatomic gas at a fixed temperature of 130 °C from a volume of 1.92 liters to 4.61 liters. No work besides pressure-volume expansion is involved in this particular process.

SOLUTION As the temperature is fixed, we see from Eqs. (15.38) and (15.40) that ΔH and ΔE are zero.

From Eq. (15.45) we compute q for this specified path as

$$q = (0.160)\left(RT \ln \frac{V_2}{V_1}\right)$$
$$= (0.160\text{ mol})(1.987\text{ cal deg}^{-1}\text{mol}^{-1})(403.15\text{ deg})\left(2.303 \log \frac{4.61}{1.92}\right)$$
$$= 112\text{ cal}$$

Using Eq. (15.46)

$$\Delta S = (0.160)\left(R \ln \frac{V_2}{V_1}\right)$$
$$= (0.160\text{ mol})(1.987\text{ cal deg}^{-1}\text{mol}^{-1})\left(2.303 \log \frac{4.61}{1.92}\right)$$
$$= 0.278\text{ cal deg}^{-1}$$

Using Eq. (15.47),

$$\Delta G = -(0.160)\left(RT \ln \frac{V_2}{V_1}\right) = -q = -112\text{ cal}$$

STUDY PROBLEM 15(j)

Calculate ΔE, ΔH, ΔS, and q for a process in which 0.842 mol of He gas is cooled from 80 °C to 20 °C.

STUDY PROBLEM 15(k)

Calculate ΔE, ΔH, ΔS, ΔG, and q in J for the reversible expansion of 0.242 mol of gaseous He at a fixed temperature of 45 °C from a volume of 3.32 liters to 9.14 liters.

SUMMARY

Of the three states of matter, gases occupy the extreme position of lowest order, highest randomness, and lowest density. A series of empirical laws, the gas laws, correlate the physical behavior of gases in terms of volume, pressure, temperature, and number of moles of the sample. The equation of state of an ideal gas relates all these variables and is obeyed reasonably well at high temperature and low pressure by most gases.

The kinetic theory of gases is based on a simple submicroscopic model of physical behavior in gases, treating gas molecules essentially like a collection of tiny billiard balls. This simple theory accounts successfully for the ideal gas law. A key feature of this theory is the assumed proportionality between the average molecular translational kinetic energy and the absolute temperature. Molecular speeds are described in terms of a distribution of speeds and the probability of occurrence of each speed.

Chemical equilibrium in gases is frequently described in terms of partial pressures instead of molarities. Partial pressures are proportional to mole fractions.

Deviations from ideal behavior occur because gases do not strictly obey the assumptions on which the kinetic theory of an ideal gas is based. The van der Waals equation provides a reasonable formulation of the deviation from ideal gas behavior and includes "correction" factors for the volume occupied by the gas molecules and for attractive interactions between gas molecules.

The ideal monatomic gas provides a convenient vehicle for exploring thermodynamic relations.

STUDENT CHECKPOINTS

After studying this chapter, the student should be able to:

1. Describe the general features of gases and how they differ from those of solids and liquids.
2. Relate P, V, T, and n in computations based on the gas laws, including the ideal gas equation of state.
3. Employ Graham's law of diffusion to relate the speeds of two types of molecules in terms of their molecular weights.
4. List key assumptions in the derivation of the ideal gas law by kinetic theory.
5. Explain each step in the derivation of the ideal gas equation of state by kinetic theory.
6. Interconvert K_c and K_p.
7. Work gas phase equilibrium problems using partial pressures.
8. Define and explain the quantities in the van der Waals equation, and apply the equation.

EXERCISES

Unless there is an indication to the contrary, assume that the gases in numerical problems are ideal.

15.1. Indicate whether each of the following statements is true or false:

(a) The entropy of a gas is always larger than the entropy of the corresponding liquid.

(b) The enthalpy of a gas is always larger than the enthalpy of the corresponding liquid.

(c) The free energy of a gas is always larger than the free energy of a liquid.

15.2.* Indicate whether each of the following statements is true or false:
(a) The average intermolecular attractions between molecules in a gas are weak.
(b) The intermolecular repulsions between molecules of a gas are always weak.
(c) The intermolecular distances between molecules are smaller for a gas than for the corresponding solid.

15.3. Indicate for each of the following statements whether it is true or false:
(a) Charles's law and Boyle's law cannot be simultaneously valid if the number of moles in the gas sample is held constant.
(b) If the temperature is held constant for a gaseous sample with a specific number of moles, the pressure and volume cannot be increased simultaneously.
(c) The equation $P_1 V_1 = P_2 V_2$ is valid for all gas samples at a given temperature.

15.4.* Describe the consequences of a temperature lower than -273.15 °C.

15.5. Describe what would happen to a balloon filled in San Francisco until it nearly bursts if it is brought to Denver without any gas leaking from it. Explain the basis of your prediction.

15.6.* Calculate the final volume of a 200-cm³ sample of N_2 gas at 25 °C if this sample is warmed to 300 °C at the same pressure.

15.7. What is the initial volume of a sample of Cl_2 gas at 25 °C if its final volume is 550 cm³ at 100 °C after the sample is warmed at constant pressure?

15.8.* A 0.350-liter sample of O_2 gas at 100 °C and 754 torr is heated to 200 °C in the same vessel. What is the final pressure?

15.9. A 742-ml sample of NH_3 gas at 50 °C and 1.00 atm is expanded to 942 ml at 1.00 atm. What is the final temperature?

15.10.* A 292-cm³ sample of CO_2 gas at 70 °C and 765 torr is expanded to 632 cm³ at the same temperature. What is the final pressure?

15.11. A 625-cm³ sample of CO gas at 0.984 atm and 25 °C is warmed to 75 °C while the sample volume is reduced to 395 cm³ (for example, in a piston-cylinder arrangement). What is the final pressure in the gas?

15.12. A 3.25-liter sample of NO(g) at 452 torr and 15 °C is warmed to 45 °C in a piston-cylinder arrangement in which the final pressure is 292 torr. What is the final volume?

15.13.* Calculate the number of moles of gas in a 340-cm³ vessel at 140 °C if the pressure of the gas is 730 torr.

15.14. Calculate the pressure of a 0.142-g sample of N_2 gas at 25 °C in a 0.311-liter vessel.

15.15.* Calculate the temperature of a 1.04-g sample of Cl_2 gas in a 300-ml vessel for a pressure of 0.75 atm.

15.16. Calculate the volume of a 0.179-mol sample of SO_2 gas that is at 35 °C and has a pressure 554 torr.

15.17.* Calculate the mole fractions, partial pressures, and total pressure for a 3.102-liter sample of gas containing 0.221 g He and 0.333 g Cl_2 at 30 °C.

15.18. For a sample of volume 2.15 liters and known to contain only N_2 and 128 mg of O_2, calculate the mole fractions and partial pressures of N_2 and O_2 at 37 °C and 0.915 atm.

15.19.* The air in a certain tunnel at 10 °C is found to contain a partial pressure of 0.016 atm CO_2. What is the molarity of CO_2 in the air in this tunnel?

15.20. The concentration of CO in a sample of exhaust collected from a poorly tuned engine is found to be 2.1×10^{-5} *M* at 25 °C. What is the partial pressure of CO in this sample?

15.21.* A 0.111-g sample of liquid benzene, C_6H_6, is introduced into an evacuated 300-cm³ chamber maintained at a temperature of 121 °C. What is the final pressure of the benzene vapor that results?

15.22. In an experiment designed to determine the molecular weight of a volatile substance, a 581.1-mg sample was found to generate a pressure of 741.3 torr in a previously evacuated vessel with a volume of 488 cm³. What is the molecular weight of the substance? (T = 25.0 °C)

15.23.* How many grams of methane, CH_4, must be added to a 400-ml flask at 30 °C to give a methane partial pressure of 322 torr?

15.24. A 0.640-g sample of a volatile liquid contains 102 mg benzene, C_6H_6, and 538 mg of an unknown substance. When this mixture is volatilized in a previously evacuated 500-ml vessel, the resulting pressure is found to be 722 torr at 120 °C. What is the molecular weight of the unknown substance?

15.25.* How many cm³ of HCl gas at a pressure of 370 torr and 50 °C will react exactly with a 520-cm³ sample of NH_3(g) at a pressure of 490 torr at 35 °C to produce NH_4Cl(s)?

15.26. How many cm³ of HCl(g) at a pressure of 740 torr and 25 °C will react completely with 2.61 g of Na_2CO_3 to produce NaCl?

15.27.* If 4.260 g of $BaCO_3$(s) is decomposed at a high temperature to form BaO(s) and CO_2(g), and the CO_2(g) is cooled to 25 °C and collected in a 500-cm³ container, what is the pressure of CO_2(g) in that container?

15.28. Indicate whether each of the following statements is true or false:
(a) The kinetic theory of gases is based on the assumption that a gas undergoes elastic collisions.
(b) The kinetic theory of gases requires an assumption regarding the relation between temperature and molecular energies.
(c) A vector quantity always has the same magnitude in all directions.

15.29.* Indicate whether each of the following statements is true or false:
(a) Graham's law of diffusion assumes that all molecules behave like vectors.
(b) The pressure exerted by a gas against the walls of a container is the same at every wall.
(c) The force exerted against the walls of a container is always the same at every wall.

15.30. Calculate the ratio of the diffusion rates of Cl_2 and F_2 for the conditions 100 °C and 711 torr.

15.31.* An unknown gas is found to diffuse 1.7 times more slowly than F_2 at the same temperature and pressure. What is the molecular weight of the unknown?

15.32. Indicate whether each of the following statements is true or false:
(a) The amount of time it takes for the molecule shown in Fig. 15.4 to travel from the right face of the cube to the left face is u_x/l.
(b) The root mean square of u_x is $\frac{1}{3}$ the root mean square of u.
(c) The pressure in an ideal gas is proportional to the average translational kinetic energy of the gas molecules.

15.33. Explain why Eq. (15.21) is intuitively reasonable.

15.34.* What is the key assumption that occurs in the kinetic theory derivation between Eqs. (15.14) and (15.24)?

15.35. Indicate whether each of the following statements is true or false:
(a) For figures such as Fig. 15.6, the higher the temperature, the greater the area under the curve.
(b) For figures such as Fig. 15.6 the higher the temperature, the greater the probability of higher speeds.
(c) The number of stationary molecules in a gas is negligible at any readily attainable temperature.

15.36.* Estimate the fraction of molecules that have a speed greater than α_3 for the T_3 curve in Fig. 15.6.

15.37. Calculate the ratio K_p/K_c for the following reaction at 120 °C:

$$NO(g) + \tfrac{1}{2}O_2(g) \rightleftharpoons NO_2(g)$$

15.38.* The value K_p at 25 °C is 1.2×10^{12} for the reaction

$$SO_2(g) + \tfrac{1}{2}O_2(g) \rightleftharpoons SO_3(g)$$

Calculate the partial pressure of O_2 that must exist for the partial pressure of SO_2 to be 2000 times less than the partial pressure of $SO_3(g)$ at equilibrium.

15.39. At 520 °C, K_p equals 1.6×10^{-2} for the reaction

$$H_2(g) + I_2(g) \rightleftharpoons 2HI(g)$$

Calculate the partial pressures of all species present at equilibrium in a sample prepared initially as 0.10 mol HI(g) in a 2.0-liter flask, and maintained at 520 °C.

15.40.* At 25 °C, K_p is 7.9×10^5 for the reaction

$$N_2(g) + 3H_2(g) \rightleftharpoons 2NH_3(g)$$

Calculate the partial pressure of H_2 in an equilibrium mixture in which P_{N_2} is 0.42 atm and P_{NH_3} is 1.2 atm.

15.41. Using the van der Waals equation of state, calculate the pressure of a sample of 1.04 mol ethane that is at 40 °C in a volume of 19.3 liters.

15.42.* Using the van der Waals equation, calculate the pressure of a one-mole sample of H_2 gas for a volume of 1.64 liters and 0 °C.

15.43. Calculate ΔE and ΔH for a process in which 0.0125 mol He gas is heated from 200 °C to 400 °C at a pressure of 0.055 atm.

15.44.* Calculate ΔE, ΔH, ΔS, and ΔG for the isothermal compression of 0.134 mol of a gaseous monatomic gas at a fixed temperature of 140 °C from a pressure of 0.30 atm to 0.90 atm.

ANSWERS TO SELECTED EXERCISES

15.2 (a) true; (b) false; (c) false. **15.4** negative volume; physically unreasonable. **15.6** 384 cm^3. **15.8** 956 torr. **15.10** 353 torr. **15.13** 0.00963 mole. **15.15** 187 °K. **15.17** $X_{He} = 0.922$; $X_{Cl_2} = 0.078$; $P_{He} = 0.443$ atm; $P_{Cl_2} = 0.0377$ atm; $P = 0.480$ atm. **15.19** 6.9×10^{-4} *M*. **15.21** 0.153 atm. **15.23** 0.109 g. **15.25** 722 cm^3. **15.27** 1.05 atm. **15.29** (a) true; (b) true; (c) false. **15.31** 1.1×10^2 amu. **15.34** assumption 6, embodied in Eq. (15.23). **15.36** about 0.6. **15.38** 2.8×10^{-18} atm. **15.40** 1.6×10^{-2} atm. **15.42** 13.8 atm. **15.44** $\Delta E = \Delta H = 0$; $\Delta S = -0.293$ cal deg^{-1}; $\Delta G = 121$ cal.

ANSWERS TO STUDY PROBLEMS

15(a) 0.64 atm. **15(b)** 358 cm^3. **15(c)** 68.06 amu. **15(d)** $X_{Cl_2} = 0.366$; $X_{N_2} = 0.634$; $P_{Cl_2} = 0.344$ atm; $P_{N_2} = 0.596$ atm; $P = 0.940$ atm. **15(e)** 86.5 amu. **15(f)** 3.33×10^2 m sec^{-1}. **15(g)** 11. **15(h)** 4.0×10^{-19} atm. **15(i)** 26.4 atm. **15(j)** $\Delta E = 151$ cal; $\Delta H = 251$ cal; insufficient information to compute ΔS or q (path not specified). **15(k)** $\Delta E = \Delta H = 0$; $\Delta S = 2.04$ J deg^{-1}, $\Delta G = -q = -648$ J.

16

The Liquid State

16.1 General Features

Observations

Our everyday experiences with liquids illustrate the following general characteristics of the liquid state.

1. Unlike a gas, a liquid sample has a volume that is independent of the volume of its container.
2. Like a gas, a liquid does not have a definite shape, and within the limits of the liquid's volume, it takes on the shape of its container.
3. Unlike a gas, a liquid is almost incompressible.*
4. Liquids generally have much greater densities and viscosities (roughly, the resistance to flow) than gases. For liquids, the densities are typically of the same order of magnitude as for solids.
5. While any gas will completely diffuse into and mix with any other gas in all proportions, only some combinations of liquids show this property. For example, water will mix with acetic acid in any proportion; they are **completely**

*That liquids are only slightly compressible accounts for their use in a variety of types of hydraulic systems, for example, hydraulic jacks and hydraulic valve lifters and brake systems in automobiles. In hydraulic systems the pressure exerted on a liquid is used to transmit mechanical forces in an advantageous manner.

miscible. Water will not mix with mercury to any appreciable extent; they are considered **immiscible.** Water will mix with only a certain amount of bromine; they are **partially miscible.**

6. At a specific pressure, when a liquid is heated to a certain temperature (a temperature characteristic of the specific liquid) it "boils" and is rapidly converted to a gas. The gas can be cooled and condensed and the substance returned to the liquid state; for example, the vapors from a pan of boiling water condense to liquid water when in contact with a cold window pane.
7. When a pure liquid is cooled to a certain temperature it may crystallize; the popular term is "freezing." The temperature at which this occurs is called the **freezing point** and is the same as the temperature at which the corresponding crystals would melt if warmed, called the **melting point** (abbreviated "mp").

Submicroscopic Structure of Liquids

A liquid is typically much closer in density to the solid state than to the corresponding gas. This is a manifestation of the fact that packing of molecules together in the liquid is roughly as close as the packing in the corresponding solid, in contrast to the nearly independent behavior of individual gas phase molecules. Attractive forces between molecules, essentially absent in a gaseous system, are in a pure liquid, of the same general type, and nearly as strong, as those in the corresponding crystalline solid. These attractive forces are not quite so strong in liquids, since the higher degree of random motion in the liquid (corresponding to a higher entropy) precludes the ordered arrangements of solids. These ordered arrangements make possible the enhanced mutual attractions and lower total energy of the solid. As indicated in Sec. 14.1, the solid's existing at temperatures below the melting point and the liquid's existing above the melting point is due, referring to Eq. (14.6), to the dominance of of the ΔH term of the equation at the lower temperature and of the $T\,\Delta S$ term at higher temperatures. The ΔH term reflects the dominant importance at lower temperatures of the lower energy of the solid, while the $T\,\Delta S$ term reflects the dominant importance at higher temperatures of the greater entropy of the liquid state.

In terms of order, liquids are "intermediate" between solids and gases. Molecules in a gas are essentially devoid of order, long-range or short-range. In a crystal, the submicroscopic structure shows a high level of both short-range and long-range order among the molecules, atoms, or ions. The molecules of a liquid maintain a certain level of short-range order, enough that they are properly situated for the intermolecular interactions responsible for the existence of the liquid state.* The motion of the molecules in a liquid is sufficiently vigorous and random that the order that does exist between pairs or small groups of molecules exists for only a short time; interactions between specific pairs of molecules at one instant are replaced by interactions between other combinations a short time later. The situation is analogous to a dance floor crowded with couples dancing to a lively beat, with the bandleader calling "Change partners" at frequent intervals.

*The order in a liquid diminishes rapidly with distance; that is, knowing the orientation of a specific molecule at a specific instant provides considerable information about the orientation of its neighbor, but very little information about the geometrical orientation of a molecule that is a few angstroms removed, meaning a few molecular diameters away.

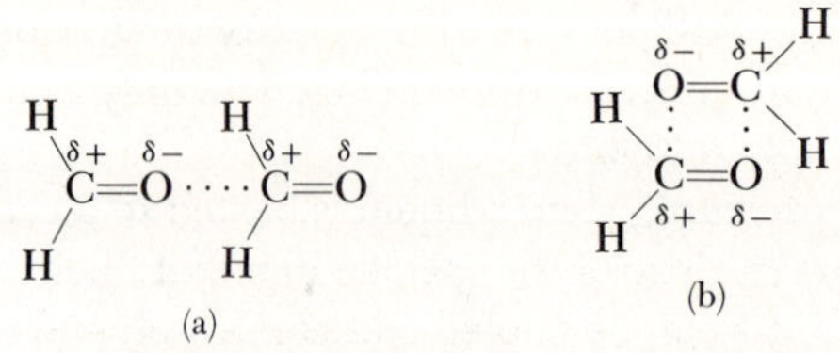

Figure 16.1 Two types of dipole-dipole interactions expected for formaldehyde, making use of the polarity of the C–O bond, showing "partial" negative and positive charges of the dipoles as δ− and δ+

The interactions between molecules in a liquid are qualitatively of the same type as those discussed in Sec. 14.1 and represented in Fig. 14.1 for condensed phases in general. Dispersion forces are nearly always important in liquids, especially in the absence of ions. If permanent dipoles are present, then dipole-dipole attractions (Fig. 16.1) or attractions between dipoles and induced dipoles also occur. For appropriately structured systems, as discussed in Sec. 8.3, the dipole-dipole interactions may be of the hydrogen-bonding type. Hydrogen bonding is especially important if there are —O—H groups and N—H groups present. For molecules without permanent dipole moments, dispersion forces are still present; these forces result from the fact that the electronic distribution in a molecule is not symmetrical at every instant. Instead, the negative electronic charge periodically fluctuates and gives rise to *transient* dipoles that are a result of the instantaneous positions of the electrons. These transient dipoles are averaged out on the whole, but their transient existence can induce a synchronized transient dipole in a nearby molecule, and a force of average mutual attraction between molecules. This dispersion effect, also called **London forces,** decreases sharply with increasing distance between the molecules, and increases with increasing polarizability of the molecules involved.

For liquids containing ions, for example, aqueous solutions of electrolytes, interactions among ions and between the ions and molecules with permanent dipole moments are important. An example is shown symbolically in Fig. 16.2 for an aqueous sodium chloride solution. The negative end of the water dipole is shown interacting with the Na^+ ion and the positive end of the O—H bond dipole is seen interacting with the Cl^- ion. When ions are present, interactions of the ion-induced dipole type (see Fig. 14.1) can also be important.

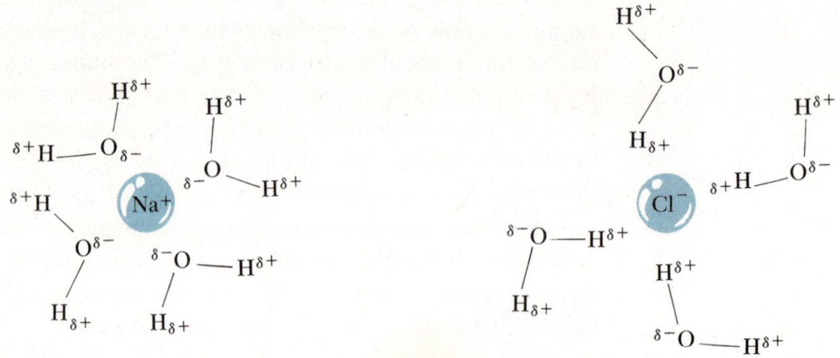

Figure 16.2 Symbolic representation of ion-dipole interactions responsible for the hydration of sodium and chloride ions in an aqueous NaCl solution, showing the "partial" positive and negative charges of the O–H dipoles as δ+ and δ− (where the magnitude of δ− is twice the magnitude of δ+). Actual geometrical relations are not implied in these drawings.

Spotlight

An important concept in describing solutions of organic molecules with polar functional groups is hydrophobic character. As shown in Figs. 14.1 and 16.2 for the interaction of water molecules with ionic or polar species, ionic or polar regions of molecules are able to interact favorably with water molecules by ion-dipole or dipole-dipole interactions of various types and by hydrogen bonding. These interactions are able to "compete" to some reasonable extent with the intermolecular interactions between water molecules. The loss of hydrogen bonding between water molecules whose space has been taken up by an organic molecule is at least partially compensated by the interactions between water and ionic or polar species. These ionic or polar regions of organic molecules, because of their favorable interactions with water molecules, are referred to as **hydrophilic** ("water-loving") regions, and are largely responsible for whatever solubility these molecules have in water. The hydrocarbonlike regions of organic molecules can interact with water molecules or with the hydrocarbon regions of other organic molecules only by London forces. These weaker forces are not able to compete with attractive forces between water molecules, and hence the hydrocarbon regions cannot easily take up the space otherwise occupied by water molecules in an aqueous solution. Such regions are referred to as **hydrophobic** ("water-hating"). Hydrophobic regions of separate molecules tend to cluster together, so that a minimum amount of the hydrogen bonding structure of the water is disrupted.

Another feature of liquids that is intermediate between gases and solids is the degree of difficulty in developing theoretical models to treat the states of matter. For solids, because of the high degree of order, successful models have been developed. (See Chap. 14.) Because of the general disorder and random motion that characterize a gas, the gaseous state is reasonably well explained by the kinetic theory; even nonideal gases can be treated theoretically by introducing corrections to the predictions of simple kinetic theory (for example, the van der Waals corrections of Eq. 15.43). But it is far more difficult to describe liquids in a theoretical model, because they cannot be approximated by either extreme order or disorder. One feature of liquids that has emerged from theoretical developments and that is useful to remember in discussions that follow is that some features of the kinetic theory of gases also apply to liquids. Clearly, the assumption that the molecules of a gas are independent and between collisions do not feel each other's presence cannot apply to liquids. But the idea of rapid, random motions with frequent intermolecular collisions carries over from gases to liquids. Perhaps most important for the discussions that follow is that average translational kinetic energy of gas-phase molecules is proportional to the absolute temperature; as given by Eq. (15.23), this principle also carries over to liquids. Thus, the distribution of translational speeds and kinetic energies in a liquid at a given temperature is described by the same type of plot that was shown for gases in Figs. 15.6 and 15.7.

16.2 Physical Properties of Pure Liquids

Phase Changes

Heating and Cooling Curves

Consider a sample of a pure substance at the temperature and pressure at which it exists as a solid. How does the sample behave when heat is added at constant pressure—and constant rate? Suppose that the temperature of the sample is monitored as time passes. Figure 16.3a shows how the temperature of the sample changes with time while heat is being added at a constant rate.

Figure 16.3
The variation of temperature with time as heat is added to the sample at a constant rate or removed from it. (a) Heating curve. (b) Cooling curve.

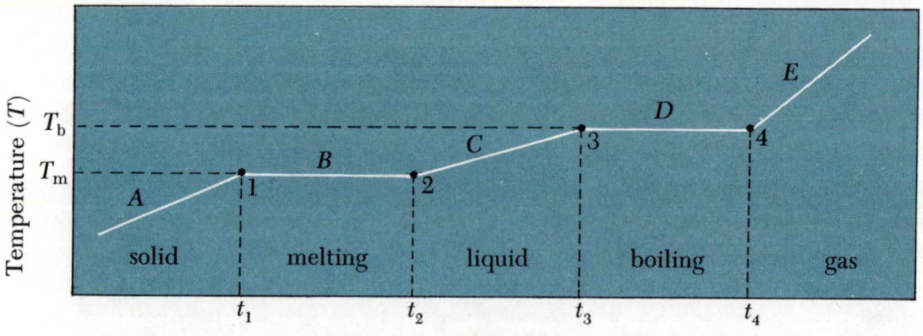

(a) Heating curve

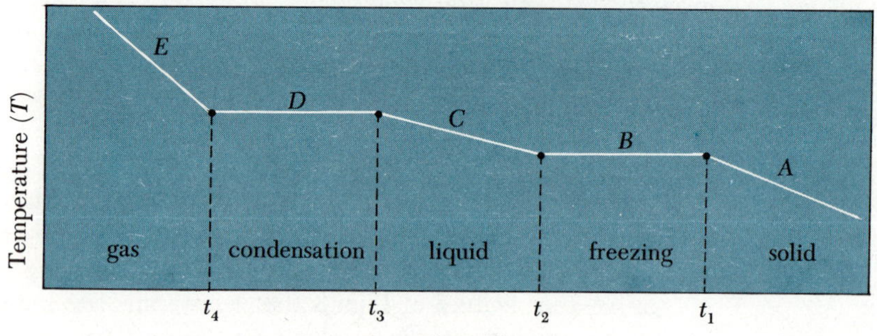

(b) Cooling curve

Segment A of the plot shown in Fig. 16.3a shows the increase in temperature that occurs as heat is added to the solid sample. The increase in temperature reflects the increase in average kinetic and potential energy that occurs in the vibrational motions of the solid. Thermodynamically this increase is accounted for by Eq. (10.27). It is found that when a certain point in time is reached (t_1 on the time axis, corresponding to point 1 on the plot), the substance begins to melt; the temperature remains constant at the value T_m for a time, corresponding to segment B of the plot, even though heat is continuously being added. This temperature T_m is called the **melting point.** During the melting, all the heat added to the sample is used up in transforming solid to liquid rather than increasing the average vibrational energy in the solid—and hence the temperature. This transformation involves a substantial increase in the average potential energy of the molecules, since the forces holding the crystalline lattice together must be overcome in the changeover to liquid state. When point 2 on the plot is reached (at time t_2), all the sample is melted, and the resulting liquid begins to warm up as the continuation of heating increases the average energy of molecules in the liquid state (segment C of the plot). This temperature increase continues, also governed by Eq. (10.27), until point 3 is reached, at which time the liquid begins to boil. The temperature at this point is T_b, called the **boiling point**, and it remains constant so long as any liquid remains to boil. Boiling is represented by the horizontal segment D. During vaporization, the added heat is used up in transforming liquid to vapor, rather than increasing the average energy in the liquid, which would correspond to a temperature increase. Much energy is required in overcoming the forces of attraction that hold molecules together in the liquid and in performing the $P\,\Delta V$ work involved in generating a gas.

It can be shown that, as heat is added to a substance at its melting point or at its boiling point, *all* of the added energy goes into increasing the potential energy of the molecules or ions as the phase change occurs (solid to liquid or liquid to gas). During these phase changes the average kinetic energy of the molecules or ions remains constant; hence, the temperature remains constant, as seen in the horizontal portions of the plot in Fig. 16.3a.

Point 4 of Fig. 16.3a represents the time t_4 at which enough heat has been added to convert all the liquid into gas. From that point on, the continued addition of heat is consumed in enhancing the average kinetic energy of the gas phase molecules, with a proportionate increase in temperature (segment E) governed by the value of C_p for the gas. The plot shown in Fig. 16.3a is called a **heating curve.**

The same kind of reasoning can be used to explain a cooling curve, that is, a plot of temperature vs time as heat is removed at a constant rate from a sample initially present as a gas. A cooling curve is shown in Fig. 16.3b, where segment E represents the cooling (rather than heating) of a gas, D represents condensation to a liquid (rather than vaporization of a liquid), C represents the cooling (rather than heating) of a liquid, B represents crystallization (rather than melting), and A represents the cooling (rather than heating) of a solid.

The amount of heat absorbed during melting (segment B) per mole is the enthalpy change for melting, usually referred to as the molar **heat of fusion,** ΔH_{fus}; this is also the quantity of heat released when one mole of liquid is crystallized at temperature T_m.*

$$\Delta H_{fus} = H_{liq} - H_{solid} \tag{16.1}$$

The amount of heat absorbed during boiling (segment D) is the enthalpy change for vaporization, usually referred to as the **molar heat of vaporization,** ΔH_{vap}; this is also the amount of heat *released* when one mole of vapor is condensed at temperature T_b.

$$\Delta H_{vap} = H_{gas} - H_{liq} \tag{16.2}$$

Heat of Vaporization

The heat of vaporization is a useful measure of the strength of the intermolecular forces present in a liquid, as it represents the amount of energy required to overcome these forces to create a gaseous system. From Fig. 8.3, it is evident that hydrogen bonding overrides the trend of increasing boiling point with increasing molecular weight of the hydrides in a given group of the periodic table. As discussed in Chap. 8, the boiling points of HF, H_2O, and NH_3 are anomalously high because of hydrogen bonding. Table 16.1 lists the heats of vaporization of the species represented in Fig. 8.3, together with the corresponding boiling points and molecular weights; also included are data on some representative hydrocarbons. From part A of the

*The temperature at which melting occurs T_m is such that the solid and liquid are in equilibrium, that is, $\Delta G = 0$ for the melting process, or $G_{solid} = G_{liquid}$. Then,

$$\Delta G_{fus} = 0 = \Delta H_{fus} - T_m\,\Delta S_{fus}$$

Similarly, at the boiling point T_b,

$$\Delta G_{vap} = 0 = \Delta H_{vap} - T_b\,\Delta S_{vap}$$

TABLE 16.1 Heats of Vaporization, Boiling Points (in parentheses), and Molecular Weights [in brackets] of Some Representative Liquids

A. Compounds from Fig. 8.3

CH_4 2.20 kcal/mol (−162° C) [16.0 amu]	NH_3 5.63 (−33°) [17.0]	H_2O 9.71 (100°) [18.0]	HF 7.21 (19.4°) [20.0]
SiH_4 2.95 (−112°) [32.1]	PH_3 3.49 (−88°) [34.0]	H_2S 4.49 (−61°) [33.1]	HCl 3.60 (−84°) [36.5]
GeH_4 3.58 (−90°) [76.6]	AsH_3 — (−55°) [78.0]	H_2Se — (−42°) [81.0]	HBr 3.90 (−70°) [80.9)
SnH_4 4.42 (−52°) [122.7]	SbH_3 — (−17°) [124.8]	H_2Te — (−4°) [129.6]	HI 4.34 (−36°) [127.9]

B. Hydrocarbons (straight chain, $CH_3(CH_2)_nCH_3$)

CH_4 2.20 (−162°) [16.0]	C_3H_8 4.32 (−30°) [44.1]	C_6H_{14} 8.10 (125°) [86.2]
C_2H_6 3.3 (−89°) [30.1]	C_4H_{10} 5.32 (0°) [58.1]	$C_{10}H_{22}$ 8.56 (160°) [142.3]

table one sees that, as noted earlier for the boiling points, the heats of vaporization of NH_3, H_2O, and HF are anomalously large in relation to the trend of increasing ΔH_{vap} with increasing molecular weight within a vertical group. For CH_4, in which hydrogen bonding is absent (carbon is not sufficiently electronegative), the usual trend of ΔH_{vap} with the molecular weight is maintained.

Spotlight

There is an interesting feature to note in the data given in Table 16.1. One might expect that even in the absence of appreciable hydrogen bonding, the polar substances in part A of the table would have larger boiling points and heats of vaporization than the nonpolar compounds of part B with comparable molecular weights. However, paired comparisons show that this kind of reasoning is not always so simple. For example, comparing the data for hexane, C_6H_{14}, of molecular weight 86.2 amu with the data for HBr (MW, 80.9 amu) or even HI (MW, 127.9 amu) shows what appears to be unreasonably large values of ΔH_{vap} and T_b for the hydrocarbon. Other such comparisons lead to the same kind of pattern, implying stronger-than-expected intermolecular forces for hydrocarbons. The explanation for this pattern lies in a closer examination of dispersion forces. For a molecule like HI, most of the electrons are not in the valence shell; most are inner-shell electrons and not very susceptible to the fluctuating polarizations that give rise to dispersion forces. By contrast, in a hydrocarbon only a small fraction of the total number of electrons is inner-shell electrons (the carbon 1*s* electrons); hence, most of the electrons contribute to dispersion interactions, which are therefore quite strong.

From part B of Table 16.1, the increase in ΔH_{vap} and boiling point with increasing molecular weights of hydrocarbons is clearly evident, reflecting the increased strength of dispersion interactions for molecules with larger numbers of electrons.

Under some circumstances it is also possible to vaporize a solid directly, without passing through the liquid state. This process is called **sublimation.** The amount of heat required for sublimation per mole of substance is called the **molar heat of sublimation,** ΔH_{subl}, where

$$\Delta H_{subl} = H_{gas} - H_{solid} \tag{16.3}$$

Vapor Pressure

The Dynamic Equilibrium

If a liquid at a constant temperature is placed in a previously evacuated container fitted with a pressure gauge, some of the molecules will escape from the liquid into the space above the liquid. The pressure gauge will initially show a continual increase in pressure as more and more molecules escape from the liquid, but after a time the pressure no longer increases, even if liquid is still present. As the pressure increases, the increased number of molecules in the vapor phase increases the probability that they will ricochet from other molecules or from the container walls and be recaptured on striking the surface of the liquid (see Fig. 2.1). When the rate of escape equals the rate of recapture, no further increase in pressure occurs. The pressure at this point is called the **saturated,** or **equilibrium, vapor pressure,** or simply the **vapor pressure.** This situation, in which the concentration of molecules in the gas phase increases until the rate of condensation (recapture) into the liquid equals the rate of evaporation into the gas phase, was discussed in Secs. 2.2 and 4.1 as an example of the steady state, or dynamic equilibrium.

Suppose, now, that we have a liquid in equilibrium with its vapor, and we try to increase the vapor pressure by reducing the volume of the container at a constant temperature. Such a process will initially increase the pressure and cause the rate of condensation to be temporarily higher than the rate of evaporation. As a result, some of the vapor will return to the liquid state, or condense, until the vapor pressure is lowered to its former value; at this point the two rates are again equal. Equilibrium is again established. Conversely, at a constant temperature, an increase in the volume of the container will initially decrease the pressure of the vapor and the rate of evaporation of some liquid will reestablish the same pressure (assuming that sufficient liquid is present). This behavior would be qualitatively predictable from Le Chatelier's principle (Sec. 4.1).

We have seen for gases that an increase in temperature leads to higher average molecular speeds (see Fig. 15.6). In a liquid, an increase in the temperature also results in an increase in the average molecular speed of molecules. This would give a corresponding increase in the rate of escape of molecules per unit area from the surface of the liquid as the temperature is increased, because an increasing fraction of the molecules will have enough kinetic energy to escape from attractive interactions in the liquid. This situation is depicted in Fig. 16.4.

To restore equilibrium after a sudden increase in temperature, a higher vapor pressure is necessary to increase the rate of recapture at this higher temperature.

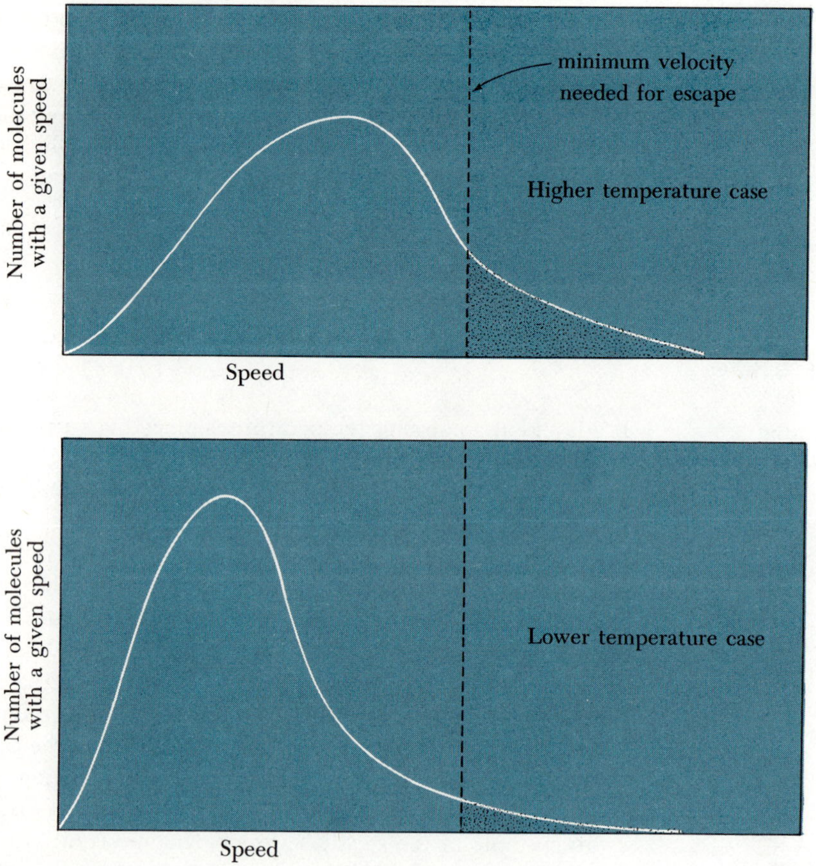

Figure 16.4 Distribution of molecular speeds at two different temperatures. Shaded area represents the fraction of molecules with speeds high enough to permit escape from the surface.

Thus, the saturated vapor pressure of a liquid increases with increasing temperature, and the magnitude of this pressure at any given temperature is a characteristic property of the liquid. Figure 16.5 shows the variation of the vapor pressures of several liquids as a function of temperature.

Solids, too, have characteristic vapor pressures. The disappearance of snow by sublimation, especially in high altitude climates of low humidity, is a practical consequence of the vapor pressure of crystalline water.

Evaporation

If an open container of a liquid is placed in a room, the liquid will begin to evaporate, and after a time the container will be dry. The time required for complete evaporation depends, among other factors, on the rate of escape of the molecules from the liquid phase, and the rate of recapture. Conditions that favor the rate of escape and reduce the rate of recapture will increase the net evaporation rate. For example, higher temperature and greater liquid surface area will increase the rate of escape.

Hence, a liquid in a large shallow pan on a summer day will evaporate faster than the same liquid in a narrow-necked bottle on a winter day. A liquid in a large shallow pan will evaporate faster on a windy day than on a calm day. On a calm day, some of the escaping molecules will remain near the liquid surface and rebound from molecules in the air to the liquid surface. A breeze blowing across the surface of the liquid reduces this tendency.

As shown in Fig. 16.4, not all the molecules in a liquid have the same velocity. Some have very low velocities, and some very high; and most have velocities that are between the two extremes. Molecules escaping from a liquid are those that have a high enough kinetic energy to overcome the forces that held them in the liquid state. Evaporation of a liquid causes the molecules with the highest energies to be lost. This lowers the average kinetic energy of the molecules remaining in the liquid phase, so the temperature of the liquid falls. For isothermal conditions to be maintained, heat must be supplied, and this heat comes from the surroundings. Hence, when alcohol is poured on the skin, a cooling sensation is experienced as the alcohol evaporates. For one mole of a given liquid at a given temperature, the amount of heat required to maintain isothermal conditions during vaporization is the molar heat of vaporization ΔH_{vap}.

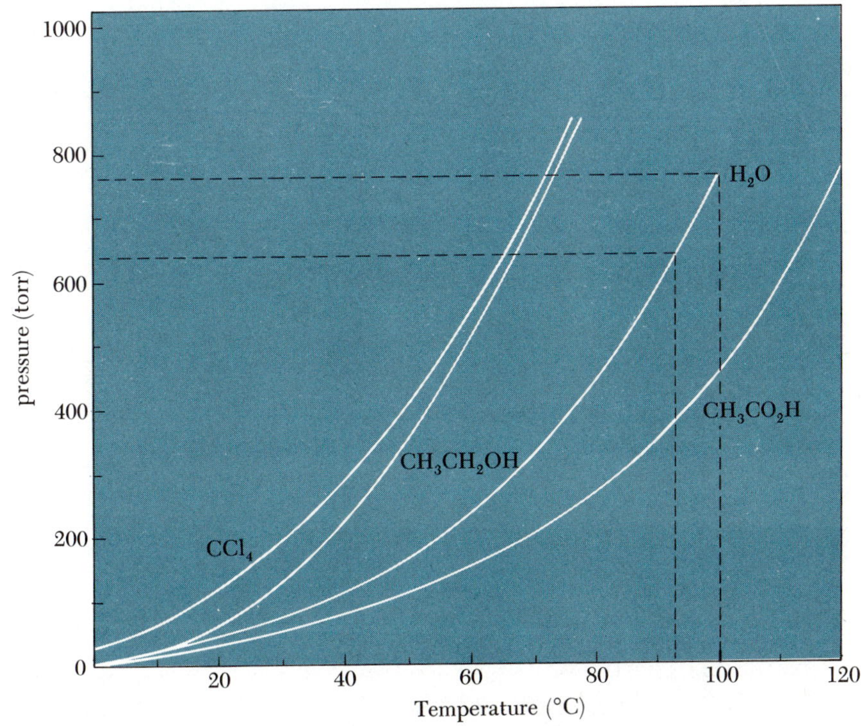

Figure 16.5 Variation of equilibrium vapor pressure with temperature for acetic acid, CH_3CO_2H, ΔH_v = 5.2 kcal/mol at 20 °C; ethanol, CH_3CH_2OH, ΔH_v = 10.0 kcal/mol at 20 °C; CCl_4, ΔH_v = 7.9 kcal/mol at 20 °C; and water, ΔH_v = 10.5 kcal/mol at 20 °C.

Spotlight

In 1834, Benoit-Pierre-Emile Clapeyron, a French civil engineer, derived an equation that applies to the equilibrium between any two phases having the same fixed composition. In 1850, Rudolf Julius Emmanuel Clausius, a German physicist and mathematician, modified the derivation. One form of this celebrated equation, called the **Clausius-Clapeyron equation,** is

$$\log P = -\frac{\Delta H}{2.303\ RT} + C$$

where C is a constant, and the other symbols have their usual significance. When this equation is applied to the variation of vapor pressures with temperature, P is the vapor pressure at the absolute temperature T, ΔH is ΔH_v, and R is the gas constant. The Clausius-Clapeyron equation has the form of a straight line, namely, $y = mx + b$, where m is the slope and b is the y intercept. Thus, a plot of the logarithm of the vapor pressure of a liquid vs the reciprocal of its absolute temperature will yield $-\Delta H/2.303R$ as the slope. For the liquids shown in Fig. 16.3, such a plot yields the straight lines shown in the accompanying illustration.

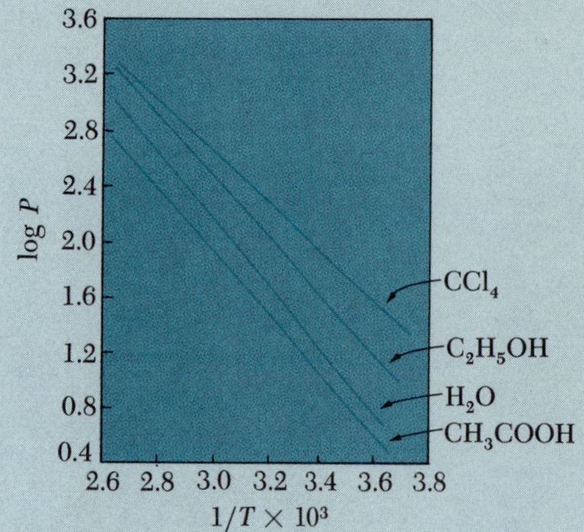

If the Clausius-Clapeyron equation is written for two different temperatures T_1 and T_2, we obtain

$$\log P_1 = -\frac{\Delta H_{\text{vap}}}{2.303RT_1} + C$$

$$\log P_2 = -\frac{\Delta H_{\text{vap}}}{2.303RT_2} + C$$

Subtraction of the second equation from the first gives

$$\log P_1 - \log P_2 = -\frac{\Delta H_{\text{vap}}}{2.303R}\left(\frac{1}{T_1} - \frac{1}{T_2}\right)$$

$$\log \frac{P_2}{P_1} = \frac{\Delta H_{\text{vap}}}{2.303R}\left(\frac{T_2 - T_1}{T_1T_2}\right)$$

For example, if the vapor pressure of carbon tetrachloride is 40.0 mm at 4.3 °C, the vapor pressure at 57.8 °C is found as follows.

ΔH_{vap} is 7900 cal/mol, and applying the above equation, we obtain

$$\log\left(\frac{P_2}{40.0}\right) = \frac{(7900)(53.5)}{(2.303)(1.987)(277)(331)} = 1.006$$

$$\frac{P_2}{40.0} = \text{antilog}\,(1.006) = 10.13$$

$$P_2 = (10.13)(40.0\ \text{mm}) = 405\ \text{mm}$$

EXAMPLE 16.1 The specific heat of water, that is, the calories required to raise 1 g of the substance 1 °C, is about 1.00 cal deg^{-1}g^{-1} and the specific heat of steam is 0.50 cal deg^{-1}g^{-1}; the molar heat of vaporization ΔH_v of water at 100 °C is 9.721 kcal/mol. How many calories are required to convert 10.0 g water at 20 °C to steam at 110 °C? (All quantities given and requested are for 1 atm pressure.)

SOLUTION 1. Calories required to raise 10.0 g water from 20 °C to 100 °C:

$$(100 - 20 \text{ deg})(1.00 \text{ cal deg}^{-1}\text{g}^{-1})(10.0 \text{ g}) = 800 \text{ cal}$$

2. Calories required to convert 10.0 g (0.555 mol) of water at 100 °C to steam at 100 °C:

$$(0.555 \text{ mol})(9.721 \text{ kcal/mol}) = 5395 \text{ cal}$$

3. Calories required to raise 10.0 g of steam from 100 °C to 110 °C:

$$(110 - 100 \text{ deg})(0.50 \text{ cal g}^{-1}\text{deg}^{-1})(10.0 \text{ g}) = 50 \text{ cal}$$

4. Total = 800 cal + 5395 cal + 50 cal = 6245 cal

STUDY PROBLEM 16(a)

How many kJ of heat are released by the system as 12.0 g of steam at 107 °C are converted to liquid water at 80 °C at 1 atm?

Boiling

As the temperature of a liquid increased, its vapor pressure rises. Eventually a temperature may be reached at which the vapor pressure exceeds the pressure in the liquid, and boiling occurs.

The internal pressure at any point in the liquid can be considered to be essentially that at the surface of the liquid, say about 1 atm if the liquid is in an open container on a typical day at sea level. When the vapor pressure slightly exceeds this internal pressure, the emerging vapor has enough pressure to "push away" the liquid and form bubbles. In this case, vaporization is not confined to the surface, for example, the interface between the liquid and the air, and bubbles can form anywhere in the interior of the liquid sample. This process is called **boiling,** and the temperature at which it occurs is the **boiling point** T_b. The temperature at which boiling occurs depends not only on the identity of the liquid, but also on the external pressure on the

Spotlight

Letting T_b represent the normal boiling point of a liquid—at 1 atm—we can rewrite the Clausius-Clapeyron equation for this case as

$$\log 1.0 = -\frac{\Delta H_{vap}}{2.303RT_b} + C$$

Since log 1.0 = 0, then

$$\frac{\Delta H_{vap}}{2.303RT_b} = C$$

$$\frac{\Delta H_{vap}}{T_b} = \text{a constant}$$

This is a statement of **Trouton's rule,** which F. T. Trouton proposed in 1884. When ΔH_v and T_b have the units calories/mole and °K, then the constant is taken as 21. For nonpolar liquids, such as hexane and benzene, the rule holds fairly well. The rule does not hold so well, however, for polar liquids, particularly for substances that show hydrogen bonding, such as water, alcohols, and carboxylic acids. As $\Delta H/T$ is equal to the entropy change of the system for a process carried out at constant temperature and pressure (Eq. 14.7), we see that Trouton's rule can be stated in terms of the entropy of vaporization ΔS_{vap}, the entropy change of the system for vaporization,

$$\Delta S_{vap} = 21 \text{ cal deg}^{-1}\text{mol}^{-1}$$

liquid. For example, at normal atmospheric pressure (760 mmHg), the boiling point of water is 100 °C. However, at atmospheric pressures typical of Colorado (600–690 mmHg), the boiling point is a few degrees below this (93 °C to 97 °C).*

Recognizing that boiling occurs when the vapor pressure of the liquid slightly exceeds the external pressure allows us to give another interpretation to plots of vapor pressure vs temperature of the type shown in Fig. 16.5. We can now think of the pressure represented in the figure as the external pressure at which boiling will occur at the corresponding temperature specified by the curve, because at that temperature the vapor pressure will just equal the particular external pressure under consideration. For example, consider the horizontal dashed line in Fig. 16.5 corresponding to a pressure of 760 torr. This dashed line intersects the water curve at a point corresponding to a vertical dashed line that represents a temperature of 100 °C. This means that water will boil at 100 °C if the external pressure is 760 torr, because that is precisely the vapor pressure water exerts at that temperature. Looking at the other set of dashed lines in Fig. 16.5, we see that at the temperature 93 °C water boils if the external pressure is 630 torr because that is the vapor pressure of water at 93 °C. Hence, such curves can be considered "boiling-point curves," giving the pressure-vs-temperature relation. Each curve in Fig. 16.5 represents the set of conditions (points on the line) for which the liquid and the vapor can exist in equilibrium. We can refer to such a curve as the liquid-vapor equilibrium line.

The same kind of analysis given above for the temperature dependence of the vapor pressure of a liquid can also be applied to a solid. Thus a *P*-vs-*T* curve can be drawn that represents either (1) the temperature dependence of vapor pressure of the solid, or (2) the relation between the temperature at which a solid can sublime and the external pressure on the system. These two plots, one for the solid and one for the liquid, are shown in part a of Fig. 16.6 for the specific case of water; analogous plots for other substances would look similar. The two curves in Fig. 16.5a are the solid-gas equilibrium line and the liquid-gas equilibrium line.

Phase Diagrams

Following the solid-gas equilibrium line of Fig. 16.6a to higher pressures and temperatures, we see that this line crosses the *P–V* line for the liquid-gas equilibrium at point *tp*. From this point on (the dashed color line) the vapor pressure of the solid is higher than that of the liquid at a given temperature. This is an indication that the "escaping tendency" of water in the solid state is greater than the escaping tendency of the liquid. Hence, the solid is less stable than the liquid; so at pressures and temperatures corresponding to the dashed portion of the solid-gas line, the solid would not continue to exist since it would be converted to the more stable liquid form.

Similarly, following the liquid-gas equilibrium line to temperatures below that of point *tp* (along the dashed white line), one finds that for the liquid, the vapor pressure and hence the escaping tendency is higher than for the solid. Thus, at the temperatures and pressures corresponding to the dashed white line, the liquid would not exist, as it would have been converted to the more stable solid.

At the point *tp*, the solid and the liquid have the same vapor pressure and escaping

*That the boiling point of water is lower at high elevations is responsible for the sometimes frustrating reality that cooking by boiling food takes longer in the mountains than at lower elevations. Since the temperature of the boiling water is lower, a longer time is required to bring about the desired chemical and physical changes involved in cooking.

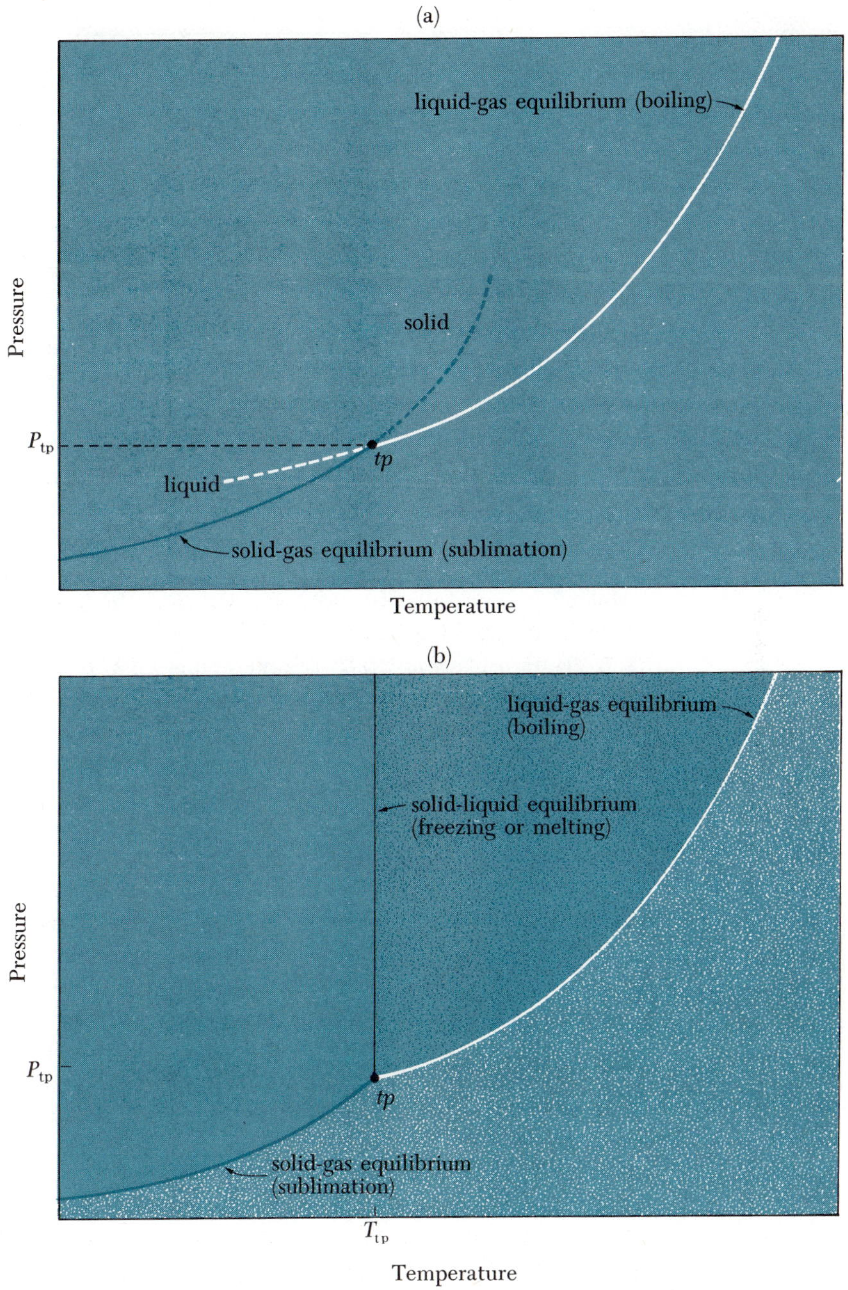

Figure 16.6 Pressure-temperature curves for water. (a) Solid-gas and liquid-gas equilibrium lines only. (b) Inclusion of solid-liquid equilibrium line.

tendencies; hence they can coexist in equilibrium. Thus, the temperature T_{tp} must be the temperature at which the solid would melt or the liquid would freeze at the external pressure P_{tp}. That is, T_{tp} is the melting point of water at a total pressure of P_{tp}. Furthermore, both the solid and the liquid can exist in equilibrium with the vapor if the external pressure is P_{tp}. As all three phases can exist in equilibrium with each other at this point, it is called the **triple point.** For water $T_{tp} = 0.01$ °C, and $P_{tp} = 4.58$ mmHg.

The triple point represents conditions under which the solid, liquid, and vapor phases can be in equilibrium with each other. At other external pressures, the solid-liquid equilibrium occurs at other temperatures, that is, other melting points. These can readily be determined experimentally, and are found generally to be not very far from the temperatures of the triple point. The solid-liquid equilibrium line is shown as the nearly vertical line passing through the triple point in part b of Fig. 16.6.

The reason that the solid-liquid equilibrium line in Fig. 16.6b slants in the direction it does is readily explained on the basis of Le Chatelier's principle, and the fact that ice has a lower density than liquid water. Suppose we have liquid and solid water in equilibrium with each other at the triple point, and the pressure on the system is suddenly increased. The way the system can attempt to minimize this sudden stress is to decrease its volume, which can be accomplished if the less dense solid is converted to the more dense liquid. The melting point is lowered to permit this. Water is somewhat unusual in having a solid phase that is less dense than the liquid phase, and most liquids would have a liquid-solid equilibrium line that would slant in the opposite direction. For them, an increase in external pressure would increase the melting point.

A diagram like Fig. 16.6b is a **phase diagram.** It provides much information on the behavior of a substance under variations of pressure and temperature. The solid lines shown in the diagram show the conditions under which two phases can coexist in equilibrium—or in the case of the triple point, three phases. The regions bounded by these solid lines represent combinations of pressure and temperature under which only one phase can exist. The evenly colored region of the figure shows combinations of pressure and temperature for which only the solid exists; every point in that area corresponds to a particular pressure and a particular temperature for which only the solid is stable. The region containing colored dots represents combinations of P and T at which only the liquid is stable. The region containing white dots represents combinations of P and T at which only the gas phase is stable.

To explore the meaning of a phase diagram in somewhat greater detail, let us focus on the water-phase diagram, reproduced again in Fig. 16.7. First, let us imagine a sample of ice at a temperature well below 0 °C and at a pressure of 760 torr, represented by point a in Fig. 16.7.* Let us now imagine heating this sample at constant pressure. As the pressure is held constant at 760 torr, we are concerned with moving horizontally across the diagram at the 760 level. As the temperature of the solid increases, the point representing the system moves to the right towards the intersection with the solid-liquid equilibrium line. That line is reached at point b, which is 0 °C, the melting point (as seen by following the dashed vertical line down to the temperature axis). As heat is added to the system and the solid melts, the temperature remains constant at 0° so long as any solid remains. When all the solid has melted, the temperature of the resulting liquid increases along the line b–c. When the point c is reached, the liquid begins to boil, as its vapor pressure equals the external pressure of 760 torr. The temperature at point c, the boiling point at 760 torr, is 100 °C. The continued application of heat at this pressure leads to the liquid-gas transformation at the constant temperature 100 °C so long as any liquid

*The phase diagram shown in Fig. 16.7 is purposely distorted in order to emphasize certain features that are present in the actual phase diagram but do not appear as dramatically as in this pictorial presentation.

remains. When all the sample has been converted to the gas phase (steam), the temperature of the steam rises, and the point describing the system moves to the right of point *c* along the same horizontal line.

If we had considered increasing the temperature of an ice sample at any fixed pressure above 4.58 torr, the same kind of behavior would have been observed qualitatively, except for different melting points and boiling points. For example, if the sample were heated at a pressure corresponding to the initial point *d*, then melting would occur at the temperature corresponding to point *e* (slightly above 0 °C) and the boiling point would be reached at point *f* (considerably below 100 °C). If the constant pressure of the experiment were chosen below the triple-point pressure, 4.58 torr, then the liquid phase would never be encountered. Thus, if the initial conditions of the solid are represented by point g, and the sample is heated at constant pressure, the temperature of the solid increases until point *h* on the solid-gas equilibrium line is reached. At this point the solid begins subliming, and the temperature remains constant until all the solid has been converted to gas. Continued heating of the gas at the same pressure simply raises its temperature according to the horizontal line to the right of point *h*. Other processes besides those carried out at constant pressure can also be represented by a phase diagram (cf. Exercise 16.19).

STUDY PROBLEM **16(b)**

What state of water is represented by the point *d* in the phase diagram of Fig. 16.7?

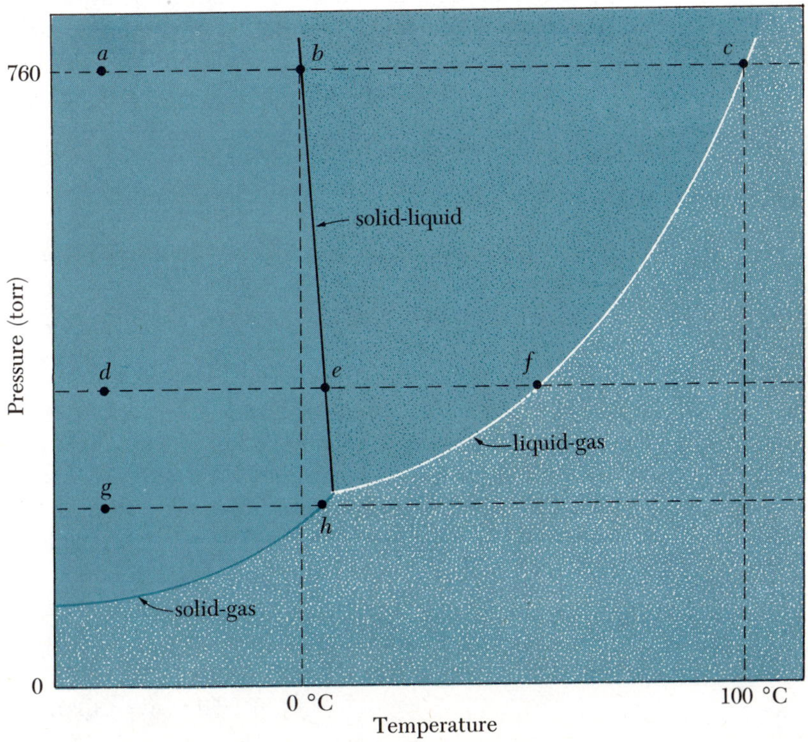

Figure 16.7
Phase diagram for water. Tilt of solid-liquid equilibrium line is exaggerated.

16.3 Special Role of Water

Distribution

Water is the most abundant compound on the earth, and it is widely distributed. Approximately three-fourths of the earth's surface is covered by water; it is estimated that the oceans and seas occupy a volume of 323,722 cubic miles. The human body contains approximately 70% water, aquatic plants 95%–99%, and even clay contains 10%–20% combined water. Water is present in the air up to altitudes of 40,000–45,000 ft, in amounts ranging from essentially 0% over some mountains and and deserts to 4% over oceans and seas. If all the atmospheric water condensed at once, it would cover the entire face of the earth with approximately one inch of rainfall. Small amounts are found even in outer space and on the moon.

Special Physical and Chemical Properties

Water freezes at 0 °C and boils at 100 °C at 1 atm; its high boiling point is attributable to a considerable extent of hydrogen bonding. When water is pure, it is a poor conductor of electricity, but it is the most generally used ionizing solvent. It can dissolve many common salts and form solutions that are highly conducting.

Water has a high heat capacity, which makes it highly efficient as a heat transfer agent, especially in automobile radiators and in home heating systems. It is this property that accounts for sea breezes and land breezes. Water has a higher heat capacity than land, and this causes the temperature of the air to be different over land compared with air over water. During the day, the air over the warm land rises, and cooler heavier air blows inland from the sea. At night, the sea, which has retained much of its daytime warmth because of its high heat capacity, warms the air over it. This air then rises and is replaced by the cooler heavier air blowing off the land, which has lost much of its heat.

At 3.98 °C, water has its highest density (1.000000 g/ml) and at 0 °C it has a higher density than ice at 0 °C. Thus, water expands when frozen, with a force of expansion that is enough to crack an automobile engine block. When the water on the surface of a lake is cooled by cold air, it sinks, and warmer water rises to the surface. When the whole body of water reaches 4 °C, no further circulation occurs, and the surface water finally cools to 0 °C; it freezes, and the resultant ice remains on the surface because of its lower density.

Water can act as a catalyst for some reactions. For example, gases that have been rigorously dried over solid P_4O_{10}, a common drying agent, for months or years show changes in their normal chemical reactivities: O_2 reacts with carbon only very slowly when heated, NH_3 and HCl gases do not form NH_4Cl, and neither a mixture of H_2 and Cl_2 nor a mixture of H_2 and O_2 will explode when sparked.*

*The dehydrating activity of $P_4O_{10}(s)$ is based on the following highly favored process (large negative ΔG value), producing phosphoric acid:

$$P_4O_{10}(s) + 6H_2O(g) \rightleftharpoons 4H_3PO_4(l)$$

Geological Aspects

The records of geology have been written mainly by the effects of water. For 2 billion years, wind, O_2, CO_2, and water in the form of rain, snow, and sleet, flowing streams and rivers, and expanding ice have slowly affected and changed the earth's upper crust. Falling rain eventually collects to form flowing rivers that carry sand and rolling rocks, and cut through and wear away the river beds. Rain seeps into cracked rocks, freezes, expands, and widens the cracks. Moving glaciers carry boulders and gouge out basins and widen valleys. These forces break down mountains and large rocks into smaller rocks, and the sand-blasting effect of driving rain and sleet breaks down these smaller rocks into soil.

The continuing geological effects of water result mainly from the water cycle. Water evaporating from oceans, seas, and lakes rises and eventually falls in the form of rain, snow, or sleet. Water evaporated from the warmer regions is transported by winds to cooler regions, where the heat liberated by the condensation of the water tends to extend the temperate zones nearer to the polar zones. Falling rain dissolves O_2, CO_2, N_2, NH_3, oxides of sulfur and nitrogen, and particulate matter. Rain contains approximately 0.013% dissolved N_2, 0.0064% O_2, and 0.0013% CO_2. Because rain contains dissolved CO_2, it is particularly effective in dissolving limestone as it percolates, that is, drains or seeps, into the ground:

$$CaCO_3(s) + H_2O(l) + CO_2 \rightleftharpoons Ca^{2+} + 2HCO_3^-$$

Water can dissolve large underground limestone beds to form sizeable caves. Within the caves, dripping solutions can redeposit the dissolved material to form stalactites and stalagmites; some form at the rate of 12 in. a month. Water percolating through the soil dissolves other minerals and plant nutrients, collects on impervious underground beds of rocks, and is eventually forced to the surface, where plant roots can reach the dissolved materials. Some water is forced to the surface as springs that feed rivers and streams, and eventually returns to the lakes, seas, and oceans to complete the never-ending water cycle. It has been estimated that 6524 cubic miles of water are discharged into the oceans per year. (Just try to imagine one cubic mile of water!)

Minerals leached from soil ultimately concentrate in the oceans and seas. Average seawater contains 96.5% water and 3.5% dissolved salts, principally consisting of 2.7% Na salts, 0.59% Mg salts, 0.14% Ca salts, and 0.07% K salts. The principal dissolved anions are Cl^-, Br^-, NO_3^-, SO_4^{2-}, and PO_4^{3-}. In addition, seawater contains 0.017% dissolved CO_2, 0.012% N_2, and 0.006% O_2. In land-locked seas and lakes, the percentage of dissolved salts is considerably higher; for example, 23% in the Dead Sea and the Great Salt Lake, and 27% in the Elton Lake in the USSR. From measurements of temperature, salinity, and dissolved O_2, movements of water can be traced. From such measurements, it is estimated that water from the Antarctic moves into the southern and tropical Atlantic at 200 miles per year, and that water at the bottom of the Atlantic Ocean was at the surface 150 years ago.

Biological Aspects

All forms of life, as we know it, contain and require water. It is generally believed that the warm waters of continental shelves, which are rich in nutrients, were the sources of all life. Body fluids have an overall electrolyte composition that is approximately the same as that of seawater. (See Table 2.2.) However, intracellular fluid (50% of body weight) has a markedly different composition from that of extracellular fluid (20% of body weight). The sea has continued to increase in salinity, while the composition of extracellular fluid has remained constant, and it has been suggested that in composition, extracellular fluid resembles the sea during the pre-Cambrian era, when animals with closed circulations came into existence.

The development of aquatic plants and animals is directly related to the abundance or scarcity of the substances dissolved in water. For example, brine shrimp thrive in very salty water; plankton, or microscopic plant and animal organisms that float or swim feebly in salt or fresh water, consume NO_3^- and PO_4^{3-} to produce organic substances containing nitrogen and phosphorus. The amount of dissolved O_2 is of special importance to marine life, and the amount dissolved depends on the past history of the water—how long since it was at the surface, how long it was in contact with air, and how plant and animal life have affected it. In general, the O_2 content of the oceans decreases in the direction from the surface to a depth of 1000 ft, increases again from 1000 to 4500 ft, and remains constant below 4500 ft.

A property of water that has important biological consequences is the decrease in density that accompanies freezing, that is, the lower density of ice compared with liquid water. In cold weather a layer of ice forms on the surface of a lake and remains there because of its lower density. This forms an insulating layer between the liquid and the cold air, helping to permit plant and animal life in the liquid water to survive. If ice were denser than liquid water, it would sink to the bottom, cutting off the O_2 supply to the lake bottom. Eventually the lake might freeze solid, destroying most forms of aquatic life.

The normal daily loss of water for humans is 1500 ml: 600 ml as perspiration, 400 ml in exhaled air, and 500 ml in the urine. Since the body produces approximately 300 ml of water through various oxidation reactions, the normal requirement for water intake is 1200 ml.

Water Pollution

The principal source of water pollution is of agricultural origin. Billions of tons of silt eroding from badly managed farm lands have covered and suffocated nestling beds of fish. The second most important source of water pollution is sewage from cities and wastes from mines and industrial plants; some of the important pollutants are heavy metals (such as Hg, Pb, As, and Cd), insecticides (such as DDT), oil, phosphates, and nitrates. Phosphates from detergents have received much attention because of **eutrophication,** which is the increase in the growth of algae and weeds from the addition of nutrients to the water. If aquatic plant life grows faster than the fish can consume it, then the bacteria that help decompose organic matter will thrive and use O_2 to such an extent that animal life cannot survive.* Thermal pollu-

*Because some substitutes for phosphates in detergents were found to be more harmful than the phosphates, and because an experiment on marine coastal waters showed that nitrogen contributed more to algal growth than phosphates, the surgeon general advised the American public, in 1972, to return to phosphate detergents.

tion, or abnormal increases in water temperature, also increase the growth of algae and bacteria, and in addition may make the temperature unsuitable for some fish.

Water Hardness and Purification

The processes required for purifying water are determined by the ultimate use of the water. For example, organic matter need not be removed from water used for irrigating soil; dissolved solids must be removed from water used in boilers; and contaminants, of which bacteria make up only one group, must be removed from drinking water.

For domestic purposes, water is purified by first removing the suspended matter. Aluminum sulfate and lime, $CaO(s)$, which can be considered a source of OH^-, are added to form a gelatinous precipitate of aluminum hydroxide, which entraps suspended matter and most bacteria:

$$Al^{3+} + 3OH^- \rightleftharpoons Al(OH)_3(s)$$

The $Al(OH)_3$, along with the suspended impurities, is removed by filtration through beds of gravel. Remaining bacteria are then destroyed with a germicide; Cl_2 is commonly used, although the effects of sunlight and O_2 from aeration are sometimes used. The remaining dissolved solids are removed at this point if desired.

"Hard" water contains salts of calcium, magnesium, and sometimes iron, usually as bicarbonates or sulfates. In small amounts, these salts are beneficial in drinking water, but they appreciably reduce the effectiveness of soaps, for example, sodium stearate, $CH_3(CH_2)_{16}\overset{\overset{\displaystyle O}{\|}}{C}ONa$. Sodium stearate is water-soluble, but calcium and magnesium stearate are not. Thus, an insoluble soap scum forms in hard water:

$$2CH_3(CH_2)_{16}CO_2^- + Ca^{2+} \rightleftharpoons [CH_3(CH_2)_{16}CO_2]_2Ca(s)$$

Detergents are water-soluble sodium salts of alkyl or aryl sulfuric or sulfonic acid, for example, sodium lauryl sulfate, $CH_3(CH_2)_{10}OSO_3Na$. Detergents do not form insoluble calcium salts or magnesium salts, and are suitable for use in hard water. Some detergents also contain phosphates, such as hexasodium hexaphosphate, $Na_6P_6O_{18}$, which keep the Ca^{2+} in solution as a complex ion, for example, $Ca_2P_6O_{18}^{2-}$.

Spotlight

Soaps, like sodium stearate, and synthetic detergents function as detergents because the two "ends" of the molecule have such different properties. One end is polar, or ionic, like the $—CO_2^-$ group of stearate, and the other is similar to a hydrocarbon (the $CH_3(CH_2)_{16}—$ portion of stearate). The ionic groups, which interact with water molecules and are responsible for the solubility of detergent molecules in water, cluster together. The resulting formations, called **micelles,** have a polar region and a hydrocarbon region. The latter is responsible for the ability of a detergent to dissolve organic "dirt" like oils and dyes, because of the tendency of "like to dissolve like" (Sec. 16.4). (See the Spotlight in Sec. 16.1 describing the related concepts of hydrophobic and hydrophilic character.)

Water hardness is either temporary or permanent. Temporary hardness results from the presence of bicarbonates of calcium or magnesium, whereas permanent hardness results from the sulfates or sometimes the chlorides of these elements. Temporary hardness can be removed simply by boiling, which precipitates the cations as carbonates:

$$M^{2+} + 2HCO_3^- \rightleftharpoons MCO_3(s) + H_2O + CO_2(g)$$

where M^{2+} is Ca^{2+} or Mg^{2+}. These carbonates can form residues in boilers, teakettles, and steam irons. Temporary hardness can also be removed by adding ammonia or NaOH,

$$Ca^{2+} + 2HCO_3^- + 2NH_3 \rightleftharpoons CaCO_3(s) + 2NH_4^+ + CO_3^{2-}$$

$$Ca^{2+} + 2HCO_3^- + 2OH^- \rightleftharpoons CaCO_3(s) + CO_3^{2-} + 2H_2O$$

or by adding $Ca(OH)_2$ in an amount determined by analyzing the Ca^{2+} and Mg^{2+} contents,

$$Ca^{2+} + HCO_3^- + OH^- \rightleftharpoons CaCO_3(s) + H_2O$$

$$Mg^{2+} + 2HCO_3^- + 2Ca^{2+} + 4OH^- \rightleftharpoons 2CaCO_3(s) + Mg(OH)_2(s) + 2H_2O$$

Permanent hardness can be removed only by chemical means, one of which is by using zeolites (see Sec. 17.4). It can also be removed by adding Na_2CO_3 and $Ca(OH)_2$ in amounts determined by analyzing the Ca^{2+} and Mg^{2+} contents:

$$Ca^{2+} + CO_3^{2-} \rightleftharpoons CaCO_3(s)$$

$$Mg^{2+} + CO_3^{2-} + Ca^{2+} + 2OH^- \rightleftharpoons CaCO_3(s) + Mg(OH)_2(s)$$

16.4 Physical Properties of Solutions

Solubility

Definitions

When a solute is dissolved in a solvent, the resulting mixture is referred to as being any one of the following: unsaturated, saturated, or supersaturated. An **unsaturated solution** is a solution that is capable of dissolving additional solute at the same temperature. A **saturated solution** is a solution that is incapable of dissolving additional solute at the same temperature when undissolved solute is present. A **supersaturated solution** is a solution that contains more dissolved solute than a saturated solution at the same temperature when no undissolved solute is present. Such a solution is unstable, and if solute is added, it will not dissolve; instead, dissolved solute will precipitate, forming a saturated solution in the presence of undissolved solute.

Supersaturated solutions can be formed by preparing a saturated solution at a given temperature, removing undissolved solute, and then slowly lowering the temperature without any mechanical disturbances. Because a supersaturated solution is unstable in the presence of undissolved solute, failure to remove undissolved solute before cooling a saturated solution will result in precipitation of dissolved solute, and a saturated solution will form at the lower temperature instead of a supersaturated one. Rapid cooling or mechanical disturbances like jarring or stirring may also cause precipitation of the excess dissolved solute.

Spotlight

The important concept of the ionization of electrolytes in water solutions was developed in an 1887 paper by the Swedish chemist Svante Arrhenius (1859–1927). During the years that followed, Arrhenius, the famous German chemist Wilhelm Ostwald, and also the brilliant Dutch chemist J. H. Van't Hoff developed many of the important concepts now identified with electrolyte solutions. These three scientists are often considered the "fathers" of physical chemistry, since their efforts set the stage for recognizing the importance of systematic and fundamental studies of the physical properties of chemical systems.

Culver Pictures

Arrhenius received a barely passing evaluation of his thesis in 1884; but his work on the ionization, or dissociation, of electrolytes earned him the Nobel Prize in Chemistry in 1903. Recognition did not come immediately to Arrhenius and his theory. However, Ostwald, already with an established reputation as a first-class scientist, espoused the Arrhenius theory and accelerated its acceptance. Arrhenius worked with Ostwald in the latter's laboratory in Riga, and then again in Leipzig. For a time, Ostwald and Arrhenius were joined by Van't Hoff, who made important contributions to the understanding of geometrical arrangements in molecules and the effect of temperature on equilibrium, in addition to his studies on solutions. In 1901, Van't Hoff was awarded the first Nobel Prize in Chemistry for his work on osmotic pressure.

Culver Pictures

Ostwald's famous laboratory in Leipzig became a mecca for students and young scientists interested in the emerging field of physical chemistry. In 1909, Ostwald was awarded the Nobel Prize in Chemistry for his important contributions in catalysis, chemical equilibrium, and reaction rates.

United Press International Photo

To determine whether a solution is unsaturated, saturated, or supersaturated, one can simply add additional solute. If the additional solute dissolves, the solution was unsaturated; if no further dissolution occurs, the solution was saturated; if precipitation of solute occurs, the solution was supersaturated.

The **solubility** of a particular solute in a given solvent is defined as the amount of solute that will dissolve for a given amount of solvent. There are many ways of stating solubility. For example, in the tables of solubility data of some handbooks, solubility is stated simply as the number of grams of solute that will dissolve in 100 g of the solvent. Frequently, solubility is stated in terms of the concentration of the solute in a saturated solution.

Concentration, in turn, can be represented in a variety of ways. To say that a given solution of benzene in chloroform is 20% benzene by weight means that 100 g of the solution contains 20 g of benzene and 80 g of chloroform. To say that a given water sample contains 55 parts per million (ppm) of lead means to the environmental scientist that 1.0 g of the sample contains 55 μg of lead in some form, say an ionized Pb(II) salt or finely divided, solid $PbCO_3$ particles.

The most common way of representing concentrations in chemistry is in molarity, denoted by M, the number of moles of solute per 1000 ml of solution (Sec. 3.9). While molarity is by far the most common convention, other fairly common terms are also used. One of these is **molality**, the number of moles of solute per 1000 g of solvent. Molality is identified by the symbol m, for example, a 2.1 m NaCl solution in water. As the density of water is essentially 1 g/ml, 1000 g water is almost exactly 1000 ml. Hence, for very dilute aqueous solutions, molarity and molality are nearly the same numerically. However, for concentrated aqueous solutions, or for solutions in some solvent with a density substantially different from unity, these two measures of concentration may be numerically much different. In such cases, 1000 ml of solution may contain an amount of solvent that is far different from 1000 g. In any case, if one knows the densities of the solutions, it is always possible to relate one unit of concentration to the other precisely.

Another way of expressing concentration is in terms of the mole fraction, as already described for gases (Sec. 15.2). The **mole fraction** of any compound in solution is defined as the ratio of the number of moles of the particular component to the total number of moles of all components present in solution. Thus, the mole fraction of a compound is the fraction that compound constitutes of the total number of moles of all compounds present in the solution. If X_a represents the mole fraction of a species indexed by the letter "a" in a solution in which the various species are present in mole fractions N_a, N_b, N_c, . . . , then X_a is defined:

$$X_a = \frac{N_a}{N_a + N_b + N_c + \cdots} = \frac{N_a}{\text{total no. moles}} \tag{16.4}$$

Similarly,

$$X_b = \frac{N_b}{\text{total no. moles}}$$

$$X_c = \frac{N_c}{\text{total no. moles}}, \ldots$$

Spotlight

Professor Joel H. Hildebrand (University of California, Berkeley), is shown here addressing a national meeting of the American Chemical Society in 1976, at the age of 94. Professor Hildebrand has made important contributions to the understanding of solutions and has remained active in research throughout his "retirement." His scientific contributions have advanced the state of knowledge of intermolecular forces and how they influence solubility. His work has helped elucidate general structural features of liquids. He has also been a major influence in chemical education for much of this century.

Chemical and Engineering News

It is easy to confirm that the sum of the mole fractions of all the compounds present in a solution should be unity:

$$X_a + X_b + X_c + \cdots = \frac{N_a + N_b + N_c + \cdots}{N_a + N_b + N_c + \cdots} = 1$$

EXAMPLE 16.2 The density of a 20.0% by weight solution of H_2SO_4 in water is 1.139 g/cm³. What are the molarity and molality of H_2SO_4 in this solution? (The relation between density and concentration for the sulfuric acid–water system is used to advantage in the method common in a gasoline service station for determining the status of a lead storage battery (see Chap. 12). The density is measured with a hydrometer; the density is a measure of the H_2SO_4 concentration—hence, the condition of the cell.)

SOLUTION A sample of this solution containing 1000 g water would contain $(\frac{20}{80})$(1000 g) of H_2SO_4. Then, as the MW of H_2SO_4 is 98.08 amu, a sample of the solution containing 1000 g of H_2O would contain

$$\left(\frac{20}{80}\right)(1000 \text{ g } H_2SO_4)\left(\frac{1}{98.08}\ \frac{\text{mol } H_2SO_4}{\text{g } H_2SO_4}\right) = 2.55 \text{ mol } H_2SO_4$$

Hence, this solution is 2.55 m in H_2SO_4. One liter of this sulfuric acid solution would have a mass (1.139 g/cm³)(1000 cm³) = 1139 g. Of this, (0.20)(1139 g) would be sulfuric acid. Thus, the number of moles of H_2SO_4 in a 1000-cm³ sample would be

$$[(0.20)(1139) \text{ g } H_2SO_4]\left(\frac{1}{98.08}\ \frac{\text{mol } H_2SO_4}{\text{g } H_2SO_4}\right) = 2.32 \text{ mol } H_2SO_4$$

Hence, this solution is 2.32 M in H_2SO_4.

EXAMPLE 16.3 What is the molality of benzene, C_6H_6, in a chloroform solution, $CHCl_3$, in which its mole fraction is 0.30?

SOLUTION From what is given, we know that the ratio of the number of moles of benzene to the number of moles of chloroform is

$$\frac{N_{C_6H_6}}{N_{CHCl_3}} = \frac{0.30}{0.70}$$

Since molality is defined in terms of 1000 g of solvent, it is of interest to determine the number of moles of chloroform (MW 119.4 amu) in a 1000-g sample:

$$(1000 \text{ g } CHCl_3)\left(\frac{1}{119.4}\ \frac{\text{mol } CHCl_3}{\text{g } CHCl_3}\right) = 8.38 \text{ mol } CHCl_3$$

Then, the number of moles of benzene in a solution of the specified type containing 1000 g of $CHCl_3$ would be

$$(8.38 \text{ mol } CHCl_3)\left(\frac{0.30 \text{ mol } C_6H_6}{0.70 \text{ mol } CHCl_3}\right) = 3.59 \text{ mol } C_6H_6$$

Hence, the solution is 3.59 m C_6H_6 in chloroform.

EXAMPLE 16.4 In practical applications, the degree of the hardness of water is often stated in equivalents of $CaCO_3$, that is, the amount of $CaCO_3$ that would contain the same number of moles of metal ion, expressed in terms of ppm. The total amount of Ca(II) and Mg(II) salts are then expressed as though all the Ca(II) and Mg(II) present were only Ca(II). A sample of water contains 712 ppm of $MgSO_4$ and no Ca(II) salts. What is its degree of hardness?

SOLUTION One mole of $MgSO_4$ (120 g) is, within the context of the above stated convention, "equivalent" to 1 mol $CaCO_3$ (100 g). Thus, 712 ppm of $MgSO_4$, that is, 712 g $MgSO_4$ in 1.0×10^6 g of sample, is equivalent to

$$(712 \text{ ppm } MgSO_4)\left(\frac{100 \text{ g } CaCO_3/\text{mol}}{120 \text{ g } MgSO_4/\text{mol}}\right) = 593 \text{ ppm } CaCO_3$$

STUDY PROBLEM **16(c)**

The density of a 10.0% by weight solution of sulfuric acid in water is 1.066 g/cm^3. What is the molality and what is the molarity of H_2SO_4 in the solution?

STUDY PROBLEM **16(d)**

What is the mole fraction of HCl in a 2.3 m hydrochloric acid solution?

Concepts

The limit of solubility of any particular solute in a given solvent is governed by the law of chemical equilibrium. In Chap. 4 we encountered the methodology by which this law can be used to formulate solubility relations for solids, with molarity as the measure of concentration for the dissolved species.

The amount of solute needed for preparing a saturated solution depends on the nature of the solute and the solvent, the temperature, and particularly for gases, the

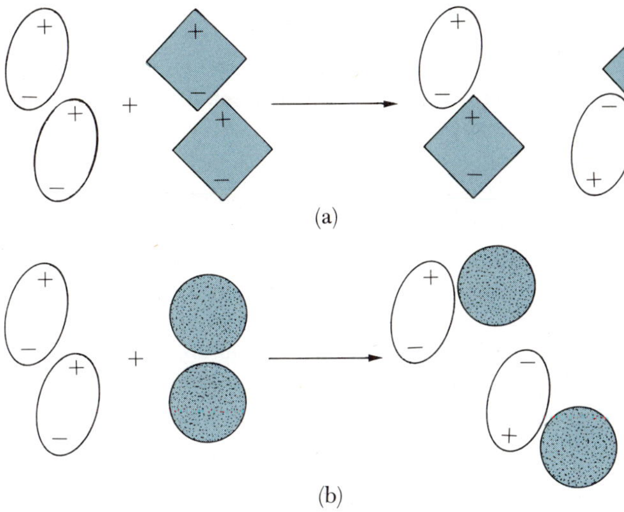

Figure 16.8 A symbolic rationalization for the rule of thumb "Like dissolves like." Separate components on the left; solution on the right. (a) Two polar components, showing how dipole-dipole attractions between different polar molecules in the solution replace dipole-dipole attractions between polar molecules in the individual components. (b) One polar and one nonpolar component, showing how the dipole-dipole attraction between the molecules of the pure polar component is not suitably compensated for in the solution.

pressure. At present, our knowledge of the forces between the solute molecules and the solvent molecules is insufficient for us to predict quantitatively from a fundamental theoretical basis the solubility of one substance in another at any temperature. Nevertheless, chemists have developed several empirical rules and plausible theoretical models that function reasonably well at a qualitative level.

Intermolecular attractions are, of course, much greater in solids and liquids than in gases; and for solids and liquids to dissolve in a solvent, the solvent molecules must have a sufficient attraction for the solute molecules or ions to separate, come between, and surround them. At the same time, the interactions between the solute molecules being dissolved and the solvent molecules must compensate for the solvent-solvent attractions that are to some degree sacrificed when solute species become inserted between solvent molecules. Thus, in dissolution, the factors affecting the intermolecular interactions between the solute molecules or ions with their own kind and with the solvent molecules are important. It is not surprising, then, that we find that substances composed of polar molecules or ions generally require polar solvents for dissolution to occur. Similarly, nonpolar solute substances generally require nonpolar solvents, because the nonpolar nature of the solute molecules prohibits them from coming between molecules of a polar solvent, which are strongly attracted to each other. To dissolve in a polar solvent, nonpolar solute molecules would have to become inserted between the solvent molecules, disrupting the attractive dipole-dipole interactions between solvent molecules. As these solvent-solvent interactions would not be replaced with comparable solute-solvent attractions, the dissolution is energetically unfavorable. This situation is depicted in Fig. 16.8. Thus, the phrase "Like dissolves like" is a useful and common guideline, and is readily rationalized.

Temperature Dependence

Like other equilibrium phenomena, the temperature dependence of the value of an equilibrium constant governing a solubility relation can be computed from Eq. (11.28). The appropriate $\Delta H°$ value to use in such a computation is the **molar heat solution,** ΔH_{sol}, defined as the enthalpy change accompanying the dissolution of one mole of solute in the solvent of interest.* Depending on the interplay of intermolecular interactions of the solvent-solvent, solute-solvent, and solute-solute type discussed above, the ΔH_{sol} value may be positive or negative. That is, dissolution may be either endothermic or exothermic, depending on the identities of the solute and solvent compound. Hence, the solubility may either increase (if ΔH_{sol} is positive) or decrease (if ΔH_{sol} is negative) as the temperature is increased. Figure 16.9 shows how the solubilities of some substances in water vary with temperature.

One factor that works in favor of the dissolution of a solute, even if ΔH_{sol} is unfavorable, (that is, positive), is the entropy of the dissolution process, ΔS_{sol}.

$$\Delta S_{sol} = S_{sol} - (S_{solvent} + S_{solute})$$

The components of a solution are "mixed up" and hence less ordered in the solution than as separate components. Hence, the entropy (randomness) of a system consisting of two separate, pure liquids is lower than the entropy of a single solution of

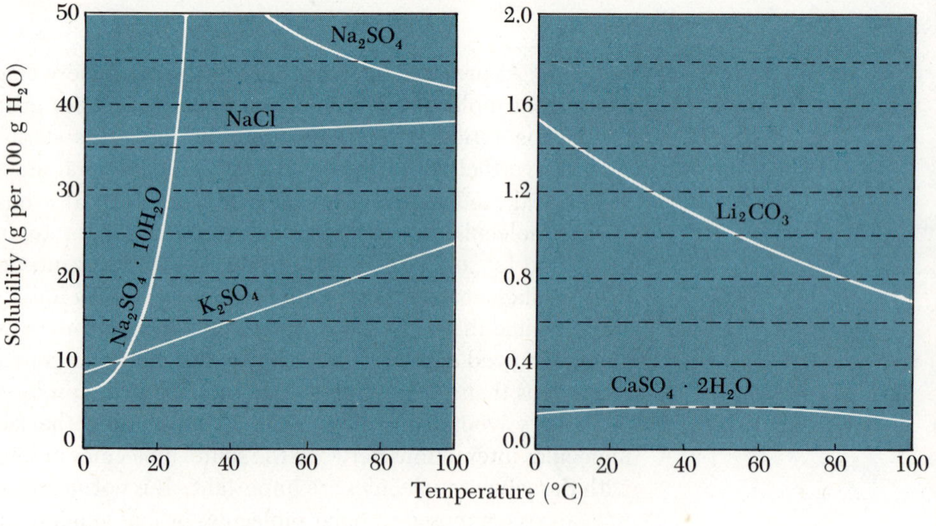

Figure 16.9
Effect of temperature on solubilities of some substances.

*Since there is essentially an absence of intermolecular interactions in a gas, dissolution of a gas in a liquid solvent typically is accompanied by a net increase of intermolecular attractions and hence negative ΔH_{sol}. Therefore, the equilibrium constant governing the dissolution decreases with an increase in temperature, and the solubility of the gas decreases as the temperature is raised. The formation of air bubbles at the bottom of water in a pan being heated is evidence of this effect. As the temperature of the water is increased, the solubilities of the dissolved gases (components of air) decrease, and the gases begin coming out of solution, forming bubbles. That the solubility values of gases in liquids are generally small, even though the ΔH_{sol} values are typically negative, is due to the decrease in entropy that accompanies the dissolution of a gas. A large fraction of the randomness of the gas molecules is lost when they enter a solution.

Spotlight

In 1808, William Henry, an English chemist, discovered that at equilibrium the solubility of a gas in a liquid is directly proportional to the pressure of the gas above the liquid; that is,

$$C = kP$$

The constant k is called Henry's law constant; it depends on the nature of the gas and liquid, the temperature, and the units of C and P, which represent the concentration of the gas in the liquid and the pressure of the gas above the solution. When two or more gases are dissolved, the law applies for each gas if C is the concentration and P is the partial pressure. This law applies only when the molecular species of the gas above the solution is the same as the gas dissolved in the solution. For example, the law does not apply to the gas HCl dissolved in water, for in water HCl ionizes, whereas in the gas phase HCl is molecular. Similarly, the law does not apply to gaseous SO_3 dissolved in water, since it reacts with water to form H_2SO_4.

For an example of the application of Henry's law, consider a 1.000-liter vessel containing dry O_2 at a pressure of 142.9 torr at 25.0 °C, into which is placed 500.0 ml of "degassed" water (water from which dissolved gases have been expelled) also at 25.0 °C. At equilibrium the pressure above the solution is 299.8 torr. Knowing that the vapor pressure of water at 25.0 °C is 23.8 torr, we can find the Henry's law constant and the percentage of O_2 dissolved in the water. From the ideal gas law, the number of moles of O_2 present in the vessel is

$$n = \frac{PV}{RT} = \frac{142.9 \times 1.00}{62.630 \times 298.2} = 7.651 \times 10^{-3} \text{ mol}$$

The pressure of O_2 above the solution after equilibration is 299.8 − 23.8 = 276.0 torr, and the number of moles undissolved is

$$n = \frac{PV}{RT} = \frac{276.0 \times 0.5000}{62.360 \times 298.2} = 7.421 \times 10^{-3} \text{ mol}$$

The number of moles dissolved is

$$(7.651 \times 10^{-3}) - (7.421 \times 10^{-3}) = 2.30 \times 10^{-4} \text{ mol}$$

Henry's law constant is therefore

$$\frac{2.30 \times 10^{-4} \text{ mol}/0.500 \text{ liter}}{276 \text{ torr}} = 1.67 \times 10^{-6} \text{ mol/liter}^{-1}\text{torr}^{-1}$$

the same two substances. The difference in entropy between these two cases—exclusive of any specific ordering that may occur in the separate components or the solution—is called the "entropy of mixing", and favors formation of a solution. The positive ΔS value for the dissolution, ΔS_{sol}, tends to make the value of $\Delta G_{sol} = \Delta H_{sol} - T\Delta S_{sol}$ more favorable, that is, more negative, than it would be otherwise if entropy effects did not exist.

Colligative Properties

Definitions

When a nonvolatile solute is dissolved in a solute, the resulting solution shows changes in its physical properties. Some of these changes are found to depend not on the nature or identity of the solute particles in solution but only on their number; such properties are referred to as **colligative properties.** Familiar examples are the boiling point, the vapor pressure, and the freezing point (melting point) of the solution; a less familiar example is osmotic pressure, a property of great importance in understanding living systems.

Altered Phase Diagram; Vapor Pressure Lowering

The way in which the colligative properties of a solvent are affected by the addition of a nonvolatile solute can be understood at a qualitative level by considering phase diagrams of the type shown in Fig. 16.7. In Sec. 16.2, we saw that the equilibrium

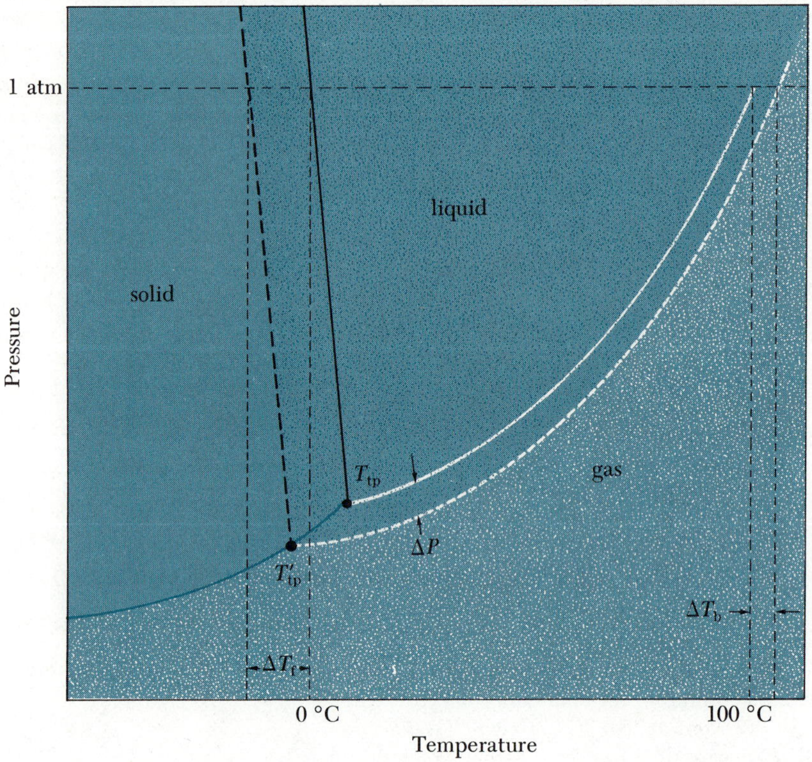

Figure 16.10 Phase diagram for water as altered by the presence of a nonvolatile solute. The solid black line and the solid white line represent the solid-liquid equilibrium and the liquid-gas equilibrium for pure water. The corresponding black and white dashed lines are for the aqueous solution.

vapor pressure of a pure liquid is attained only when the rate of escape of molecules from the liquid surface to the vapor phase equals the rate of condensation of the molecules from the vapor phase to the liquid surface. Adding a nonvolatile solute will reduce the rate of escape per unit area at the surface of the liquid, because a portion of the surface of the liquid will be occupied by particles of the solute instead of molecules of the solvent. Only a small fraction of the surface molecules has large enough kinetic energy to escape the liquid, and that fraction is an even smaller number when a portion of the surface is occupied by solute molecules. For equilibrium to become reestablished after a solute is added, the rate of condensation must also decrease, and this reduction can be accomplished only by reducing the pressure of the vapor above the solution. Thus, the equilibrium vapor pressure of a liquid will be lowered on the addition of a nonvolatile solute. This can be represented for an aqueous solution by altering the phase diagram of Fig. 16.7, as indicated by the dashed liquid-gas line in Fig. 16.10. The length of the line ΔP represents the amount by which the vapor pressure is lowered. The solid-gas equilibrium line in this altered phase diagram is the same as that for pure water, since for the case of interest here, the solid that exists in equilibrium with gas or liquid in this type of system is *pure* ice. Hence, the vapor pressure curve for the solid, which is the same as the solid-gas equilibrium line, is unchanged when the nonvolatile solute is added.

Figure 16.10 shows that the temperature (T'_{tp}) for which the solid and the liquid solution have the same vapor pressure—and escaping tendency—is lower than the temperature (T_{tp}) of the corresponding situation for pure water. Thus, the triple point has a lower temperature, T'_{tp}. Now, the water vapor pressure of the liquid is reduced by the presence of the solute, but the water vapor pressure of the solid is unaffected. Hence, there is a displacement of the solid-liquid equilibrium, not just at the triple point but at any external pressure. A new solid-liquid equilibrium line (dashed) is shown for the solution case in Fig. 16.10.

The changes in the water-phase diagram (Fig. 16.10) brought about by adding the nonvolatile solute have two readily measured consequences besides the reduction in water vapor pressure ΔP. We see from the diagram that relative to the pure water case, a higher temperature is now required for the solution to bring the water vapor pressure up to 1 atm, that is, to bring about boiling at 1 atm. The magnitude of the **boiling-point elevation** for an external pressure of 1 atm is shown as ΔT_b in Fig. 16.10. The other readily measured consequence of the alterations of the phase diagram in Fig. 16.10 is the lowering of the temperature at which liquid (solution) and solid (ice) can coexist in equilibrium at a specific pressure; that is, the melting point (or freezing point) is lowered. This **freezing-point lowering** is shown for an external pressure of 1 atm as ΔT_f in the figure.*

Freezing-Point Lowering

All three colligative properties represented in Fig. 16.10, namely P, T_b, and T_f, can be expressed quantitatively. Thermodynamic developments show that the freezing-point lowering ΔT_f is proportional to the molality of solute species in solution. This is expressed

$$\Delta T_f = T - T' = K_f m \tag{16.5}$$

where T is the freezing point of pure solvent, T' is the freezing point of solvent from the solution of interest, m is the molality of all solute species in solution, and K_f is a constant characteristic of the type of solution, meaning the solvent mainly. The factor K_f is called the **molal freezing point depression constant,** or sometimes, the cryoscopic constant. Table 16.2 summarizes K_f values for some important solvents.

TABLE 16.2 **Some Representative Cryoscopic Constants**

Solvent	K_{fus}[a]	*Freezing Point, °C*
Acetic acid	3.9	16.7
Benzene	5.12	5.5
Camphor	37.7	178.4
Cyclohexane	20.0	6.5
Tribromophenol	20.4	96.0
Water	1.86	0.0

[a]The molal freezing point depression, in units of deg $\left(\frac{\text{mol}}{1000\text{ g solv}}\right)^{-1}$.

*Since adding a nonvolatile solute raises the boiling point of a liquid and lowers the freezing point, ethylene glycol, $HOCH_2CH_2OH$, which is much less volatile than water, can function as an antifreeze in engines in winter and also render summer boiling over less likely.

EXAMPLE 16.5 When a 6.25-g sample of an unknown molecular compound is dissolved in 100 g of camphor, the resultant solution is found to have a melting point of 173.6 °C. What is the molecular weight of the compound?

SOLUTION The freezing-point lowering is 176.4 °C − 173.6 °C = 2.8 deg. Using the K_f value for camphor in Table 16.2 and Eq. (16.5), we obtain

$$\Delta T_f = 2.8 \text{ deg} = 37.7 \text{ deg}\left(\frac{\text{mol}}{1000 \text{ g solv}}\right)^{-1} m\left(\frac{\text{mol}}{1000 \text{ g solv}}\right)$$

$$m = \left(\frac{2.8}{37.7}\right)\frac{\text{mol}}{1000 \text{ g solv}} = 0.074\frac{\text{mol}}{1000 \text{ g solv}}$$

Thus, the molality of the unknown compound in that solution is 0.074 *m*. But the actual solution studied contained 100 g, not 1000 g, of solvent:

$$0.074\frac{\text{mol}}{1000 \text{ g solv}} = 0.0074\frac{\text{mol}}{100 \text{ g solv}}$$

Hence, the 6.25-g sample is 0.0074 mol:

$$6.25 \text{ g} = 0.0074 \text{ mol}$$

$$1 \text{ mole} = \frac{6.25}{0.0074} \text{ g} = 845 \text{ g}$$

$$\text{MW} = 845 \text{ amu}$$

Boiling Point Elevation

Resembling the freezing-point depression, the change in boiling point brought about by introducing a nonvolatile solvent should be proportional to the concentration of solute particules in solution. Thermodynamic developments show this to be the case, and one obtains the following equation for the boiling-point elevation, ΔT_b:

$$\Delta T_b = T' - T = K_b m \qquad (16.6)$$

where T is the boiling point of pure solvent, T' is the boiling point of the solution, m is the molality of all solute species in solution, and K_b is called the **molal boiling-point elevation constant,** or the **ebullioscopic constant.** Table 16.3 summarizes K_b values of some representative solvents.

EXAMPLE 16.6 From the data in Table 16.3 calculate the boiling point of an 0.20-*m* solution of sucrose in water.

SOLUTION

$$\Delta T_b = K_b m$$

$$= \left[0.512 \text{ deg}\left(\frac{\text{mol}}{1000 \text{ g solv}}\right)^{-1}\right]\left(0.20\frac{\text{mol}}{1000 \text{ g solv}}\right)$$

$$= 0.10 \text{ deg}$$

$$\text{Boiling point} = 100.0 + 0.10 = 100.10 \text{ °C}$$

TABLE 16.3 Some Representative Ebullioscopic Constants

Solvent	K_b[a]	*Boiling Point, °C*
Acetic acid	3.07	118.1
Acetone	1.71	56.5
Benzene	2.53	80.1
Carbon tetrachloride	5.03	76.8
Cyclohexane	2.79	81.4
Ethanol	1.22	78.5
Methanol	0.83	64.6
Water	0.512	100.0

[a]The molal boiling point elevation, in units of deg $\left(\frac{\text{mol}}{1000\text{ g solv}}\right)^{-1}$.

EXAMPLE 16.7 Calculate the boiling point of a sample of an aqueous NaCl solution that is 9.35% NaCl.

SOLUTION A sample of this solution containing 1000 g water would contain the following amount of NaCl.

$$\left(\frac{9.35\text{ g NaCl}}{90.65\text{ g H}_2\text{O}}\right)(1000\text{ g H}_2\text{O}) = \frac{9.35 \times 1000}{90.65} = \text{g NaCl}$$

In terms of moles of NaCl, this is

$$\left(\frac{9.35 \times 1000}{90.65}\text{ g NaCl}\right)\left(\frac{1}{58.5}\frac{\text{mol NaCl}}{\text{g NaCl}}\right) = 1.76\text{ mol NaCl}$$

As NaCl is a strong electrolyte, the total number of moles of solute particles in a sample containing 1000 g H_2O would be $2 \times 1.76 = 3.52$ mol. Hence, the value of m in Eq. (16.5) is 3.52, and

$$\Delta T_b = (0.512)(3.52) = 1.80\text{ deg}$$

Thus, the boiling point is 101.80 °C.

EXAMPLE 16.8 A 3.650-g sample of a certain monoprotic, nonvolatile weak acid with molecular weight 101.2 amu is dissolved in 180 g water, and the resulting solution has a boiling point of 100.11 °C. Determine the percentage ionization and the ionization constant of this acid.

SOLUTION Using Eq. (16.6), we find

$$0.11\text{ deg} = \left[0.512\text{ deg}\left(\frac{\text{mol}}{1000\text{ g solv}}\right)^{-1}\right](m) = \frac{0.11}{0.512}\frac{\text{mol}}{1000\text{ g solv}} = 0.22\ m$$

Thus, the total concentration of solute species is 0.22 mol/1000 g H_2O, or

$$\left(\frac{180}{1000}\right)\left(0.22\frac{\text{mol}}{180\text{ g H}_2\text{O}}\right) = 0.039\frac{\text{mol}}{180\text{ g H}_2\text{O}}$$

If the acid were completely non-ionized, the number of moles of solute species in 180 g H$_2$O would be

$$(3.650 \text{ g})\left(\frac{1}{101.2} \frac{\text{mol}}{\text{g}}\right) = 0.0361 \text{ mol}$$

Clearly, some of the acid has ionized. We assume that some fraction α of the acid ionizes according to the following equation:

$$\text{HA} + \text{H}_2\text{O} \rightleftharpoons \text{H}_3\text{O}^+ + \text{A}^-$$

Then $(0.0361)(1 - \alpha)$ mol of acid remains non-ionized and $2(0.0361)(\alpha)$ is the total of moles of ions produced. Hence the total number of moles of solute particles in the solution containing 180 g of water would be

$$(0.0361)(1 - \alpha + 2\alpha) = (0.0361)(1 + \alpha) \text{ mol}$$

But from the boiling-point elevation, we found the total number of moles of solvent to be 0.0394. Thus $(0.0361)(1 + \alpha) = 0.039$.

$$1 + \alpha = \frac{0.039}{0.0361} = 1.1$$

$$\alpha = 0.1$$

Thus, the fraction of the acid ionized is 0.1 and the fraction left non-ionized is $1.0 - 0.1 = 0.9$. Hence, the acid is 10% ionized. The molality of non-ionized acid is $(0.22)(0.9)$ and the molality of H$^+$ and of A$^-$ are each $(0.22)(0.1)$. Then, setting molality equal to molarity for such dilute aqueous solutions, we write

$$[\text{HA}] = (0.22)(0.9) = 0.2$$

$$[\text{A}^-] = [\text{H}_3\text{O}^+] = (0.22)(0.1) = 0.02$$

$$K_a = \frac{[\text{H}_3\text{O}^+][\text{A}^-]}{[\text{HA}]} = \frac{(0.02)(0.02)}{(0.2)} = 2 \times 10^{-3}$$

EXAMPLE 16.9 A certain nonvolatile carboxylic acid, R—CO$_2$H, where R is an organic group, has a molecular weight of 130.22 amu. It is suspected that like many carboxylic acids, this acid in solution in a hydrocarbon solvent undergoes some degree of association to form a dimer held together by hydrogen bonds:

$$2\text{R}-\text{C}\begin{matrix}\diagup\text{O}\\ \diagdown\text{O}-\text{H}\end{matrix} \rightleftharpoons \text{R}-\text{C}\begin{matrix}\diagup\text{O}\cdots\text{H}-\text{O}\diagdown\\ \diagdown\text{O}-\text{H}\cdots\text{O}\diagup\end{matrix}\text{C}-\text{R}$$
dimer

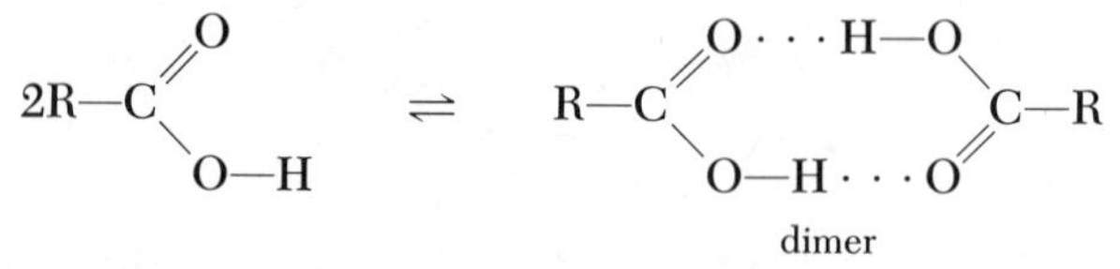

Experimentally it is found that a solution containing 11.63 g of this acid and 100 g of benzene has a boiling point of 82.0 °C. What fraction of the carboxylic acid exists as a hydrogen-bonded dimer at 82.0 °C?

SOLUTION Using Eq. (16.6) and the data from Table 16.3 gives

$$\Delta T_b = 82.0 - 80.1 = (2.53)\,m$$

$$m = \frac{1.9}{2.53} = 0.75\frac{\text{mol}}{1000\text{ g}}$$

$$0.75\frac{\text{mol}}{1000\text{ g benzene}} = 0.75\left(\frac{100}{1000}\right)\frac{\text{mol}}{100\text{ g benzene}} = 0.075\frac{\text{mol}}{100\text{ g benzene}}$$

If the carboxylic acid remained completely undimerized, the number of moles of solute species in 100 g benzene would be

$$(11.63\text{ g})\left(\frac{1}{130.22}\frac{\text{mol}}{\text{g}}\right) = 0.0893\text{ mol}$$

If a fraction β of the acid dimerizes, then $(0.0893)(1 - \beta)$ mole remains undimerized and $(0.0893)(\beta/2)$ mol of dimer forms; the total number of moles of solute species in the 100 ml of benzene would be $(0.0893)(1 - \beta + \beta/2)$. Then, $(0.0893) \cdot (1 - \beta/2) = 0.075$.

$$1 - \frac{\beta}{2} = \frac{0.075}{0.0893} = 0.84$$

$$\frac{\beta}{2} = 1.00 - 0.84 = 0.16$$

$$\beta = 0.32$$

which is the fraction existing as the hydrogen-bonded dimer.

STUDY PROBLEM 16(e)

What is the boiling point of a 1.32 m solution of KOH in methanol, CH_3OH?

STUDY PROBLEM 16(f)

When 2.625 g of a molecular compound is dissolved in 100 g of camphor, the melting point of the resulting solution is found to be 174.9 °C. What is the molecular weight of the solute?

OPTIONAL

Osmotic Pressure

In 1748, Abbé Mollet reported the phenomenon called **osmosis,** a process in which a solvent passes through special types of membranes from a region of low solute concentration to a region of high solute concentration. For an example of this phenomenon, consider a solution of sucrose in water that is separated from pure water by a **semipermeable membrane;** *semipermeable* in this example means permitting the passage of molecules of water but not molecules of sucrose. Water will pass from the pure water side, through the membrane, and into the sucrose solution as pictured in Fig. 16.11a. The tendency for water to pass through the membrane, that is, to **osmose,** can be measured by applying an amount of external pressure on

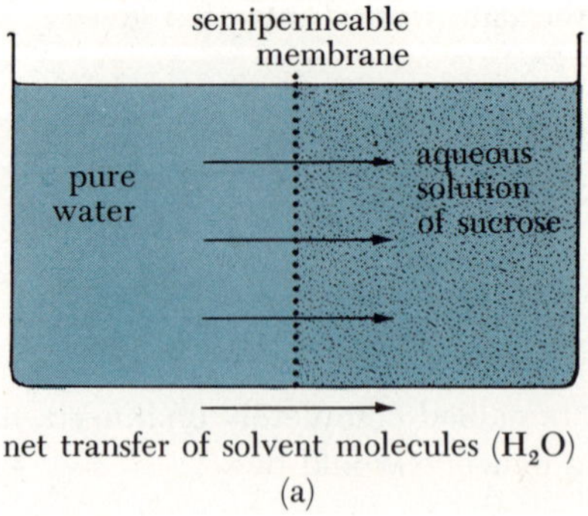

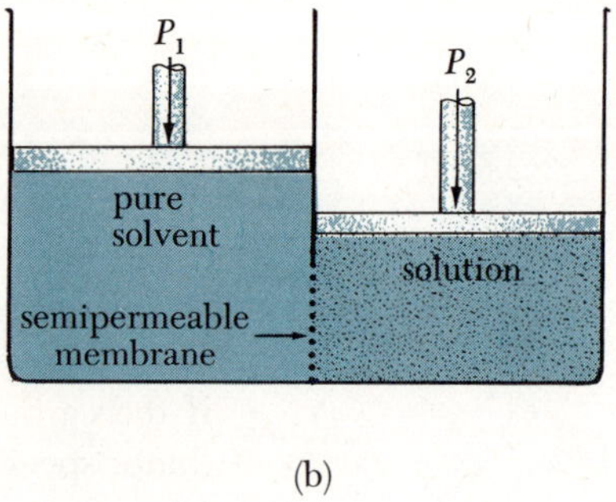

Figure 16.11 Osmosis. (a) Flow of solvent (water) through a semipermeable membrane. (b) Osmotic pressure, $P_2 - P_1$, required to stop the flow of solvent across the membrane, showing pistons pressing down on the compartments with pressures P_1 and P_2.

OPTIONAL

the solution side sufficient to prevent the osmosis of water (see Fig. 16.11b). This just counterbalances the tendency for osmosis. The pressure difference ($P_2 - P_1$ of Fig. 16.11b), which just accomplishes this balance, is called the osmotic pressure of the solution; while it is independent of the nature of a truly semipermeable membrane or of the identity of the solute species, it is dependent on the number of particles in solution, the nature of the solvent, and the temperature. Hence, osmotic pressure is also a colligative property.

To explain osmosis, one can recall that the solvent vapor pressure of the solution is lower than the vapor pressure of the pure solvent. Hence, the escaping tendency of solvent is greater from the pure solvent than from the solution. The pressure exerted on the membrane by solvent molecules in the solution, therefore, must be less than the pressure exerted by the pure solvent, and pure solvent will tend to pass through the membrane and into the solution until the two pressures become equal. However, if as soon as the osmosis began, an external pressure differential $P_2 - P_1$ that was equal to the initial vapor pressure differential were applied to the system with the higher pressure on the solution side, then the passage of water and the concomitant phenomenon of osmosis would cease. Thus, the external pressure differential required is a measure of the osmotic pressure.

A phenomenon analogous to osmosis is observed when a container of pure solvent and a container of a solution of a nonvolatile solute dissolved in the same solvent are placed in a sealed enclosure, as shown herewith. Under these conditions solvent from the container of pure solvent, whose vapor pressure is higher than that of the solution, will evaporate, or "distill," into the container of the solution.

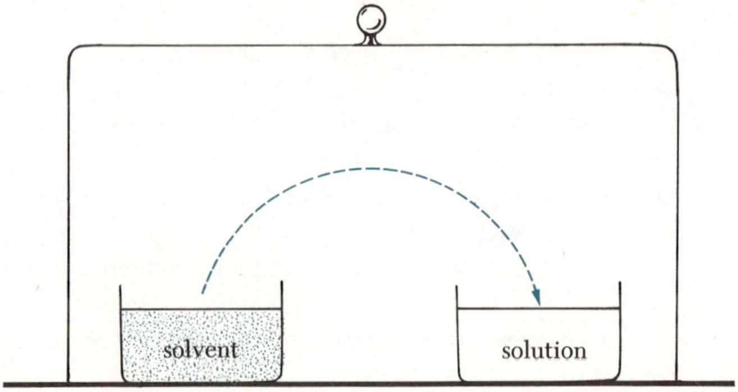

It is found experimentally, and can be derived from thermodynamics, that the osmotic pressure is directly proportional to the concentration. This is expressed as*

$$\pi = k_{osm} m \tag{16.7}$$

where $\pi = P_2 - P_1$ is the osmotic pressure, k_{osm} is the molal osmotic constant, and m is the molality of solute particles in the solution. For dilute aqueous solutions at 25 °C, if π is expressed in atm, then k_{osm} is about 24 atm (mol/1000 g solv)$^{-1}$. Hence, appreciable osmotic pressures can be generated in aqueous solutions even by very small solute concentrations. For a solute molality of 0.010 mol/1000 g H_2O

$$\pi = \left[24 \text{ atm}\left(\frac{\text{mol}}{1000 \text{ g } H_2O}\right)^{-1}\right]\left(0.010\frac{\text{mol}}{1000 \text{ g } H_2O}\right)$$
$$= 0.24 \text{ atm}$$

This pressure is large enough to support a column of water about 8 ft high.†

The tendency for water to pass through membranes is of great physiological importance. For example, if red blood cells are placed in pure water, the cells will swell and rupture, or **hemolyze.** If they are placed instead in a strong saline solution, they will shrink, in the process **crenation.** These effects are explicable in terms of osmosis.‡ The fluid inside the cell is called the **intracellular fluid,** and contains dissolved inorganic electrolytes, organic electrolytes, and nonelectrolytes. When pure water surrounds the cells, water will osmose into the cell and cause hemolysis. When the cell is surrounded by a strong saline solution, intracellular water will osmose out of the cell and cause crenation. However, a 0.9% solution of NaCl in water does not show either of these effects, and is called a **physiological saline solution.** Thus, solutions that are given intravenously must contain a specific concentration of added salt to prevent both hemolysis and crenation.

*Thermodynamic developments show that another form of Eq. (16.7) is

$$\pi = (kT)\left(\frac{n}{V}\right)$$

where kT is the analog of k_{osm} of Eq. (16.7), and concentration has been expressed as the number of moles of solute particles per liter, n/V. Rearranging gives

$$\pi = \frac{nkT}{V}$$

which is very similar to the ideal gas equation of state

$$P = \frac{nRT}{V}$$

In fact, k in the above equation is found to have nearly the same value as R.

†One can imagine osmotic pressure exerted over the membranes in tree roots, between the water in the soil and the aqueous solution in the living tree. This pressure is partly responsible for the ability of trees to transport water from their roots to their uppermost extremities.

‡The normal osmotic pressure of intracellular fluid cannot be greatly changed without irreversible and lethal changes. Thus, if the intake of water or Na^+ or both is either abnormally high or abnormally low, a complex series of feedback devices controls both the volume and the Na^+ concentration of the urine. In addition, the normal osmotic pressure is maintained by the regulation of water intake through the thirst mechanism, which operates with even the slightest increase in the tonicity of extracellular fluid.

OPTIONAL

Solution of Liquids in Liquids

Raoult's Law

In 1884, François M. Raoult, a French physical chemist, stated that the partial pressure of any volatile constituent of a solution equals the vapor pressure of the pure constituent multiplied by the mole fraction of that constituent in the solution. Stated mathematically for a two-component system, Raoult's law is

$$P_A = P_A{}^0 X_A \tag{16.8a}$$

$$P_B = P_B{}^0 X_B \tag{16.8b}$$

where P_A and P_B are the partial pressures for the volatile components A and B, $P_A{}^0$ and $P_B{}^0$ are the vapor pressures of *pure* A and B, and X_A and X_B are the mole fractions of A and B. A solution that obeys Eq. (16.8) is called **ideal.**

Dalton's law of partial pressures (see Sec. 15.2) tells us that the total pressure of the vapor equals the sum of the contributions (partial pressures) from both components; that is,

$$P = P_A + P_B$$

where P is the total pressure above the two-component mixture. Substitution of Eq. (16.8) into Dalton's law gives

$$P = P_A{}^0 X_A + P_B{}^0 X_B$$

Then, since $X_A + X_B = 1$,

$$\begin{aligned} P &= P_A{}^0 X_A + P_B{}^0(1 - X_A) \\ &= (P_A{}^0 - P_B{}^0)X_A + P_B{}^0 \end{aligned} \tag{16.9}$$

This last equation has the form of a straight line, $y = mx + b$. Thus, for a two-component system that obeys Raoult's and Dalton's laws, a plot of P vs X_A will give a straight line with a slope equal to $P_A{}^0 - P_B{}^0$ and a y intercept equal to $P_B{}^0$. Figure 16.12 shows a plot, at a constant temperature, for a system in which $P_B{}^0 > P_A{}^0$. In this figure, the dashed lines represent the partial pressures of A and B as a function of X_A. The sum of these partial pressures at any value of X_A equals P, according to Dalton's law, and is given by the solid line in Fig. 16.12.

Equation (16.9) gives the relation between the total and partial pressures as a function of the mole fractions of A and B in solution; these mole fractions may not be the same as the mole fractions of A and B in the vapor above the solution. To obtain the latter relation, we use Dalton's law. Let x_A and x_B represent the mole fractions of A and B in the vapor (small x represents vapor; large X, liquid). Then, from Dalton's law, we can write

$$x_A = \frac{P_A}{P} \tag{16.10}$$

and substituting Eqs. (16.8a) and (16.9) into Eq. (16.10) gives

$$x_A = \frac{P_A{}^0 X_A}{(P_A{}^0 - P_B{}^0)X_A + P_B{}^0} \tag{16.11}$$

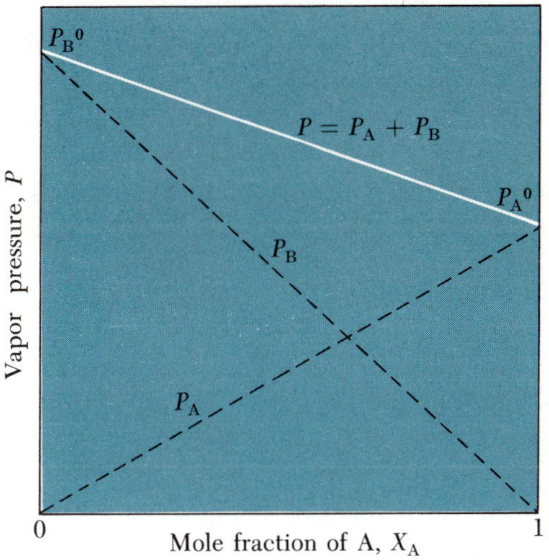

Figure 16.12
Pressure vs mole fraction X_A for an ideal mixture of two miscible liquids, A and B.

This is the relation between the mole fraction of A in the vapor, the mole fraction of A in the solution, and the vapor pressures of pure A and B. From Eq. (16.11), we can see that x_A equals X_A only when $P_A{}^0$ equals $P_B{}^0$. In general, however, $P_A{}^0$ does not equal $P_B{}^0$; and in these cases the relation between x_A and X_A is most easily seen by reference to a graphical representation of Eq. (16.11), a plot called a liquid-vapor composition plot. Figure 16.13 is such a plot, where the abscissa (horizontal axis) represents X_A or x_A. That is, there are really two plots on the same figure. For the plot of pressure vs the mole fraction of A in the liquid, the abscissa is X_A and the relation is given by the line labeled "liquid"; but for pressure vs the mole fraction of A in the vapor, the abscissa is x_A and the relation is given by the line labeled "vapor." The information shown in Fig. 16.13 was calculated using Eq. (16.11) and is for the specific case of an ideal solution of A and B, whose vapor pressures at 25 °C in the pure liquid states are 300.0 and 100.0 torr, respectively. A horizontal line in

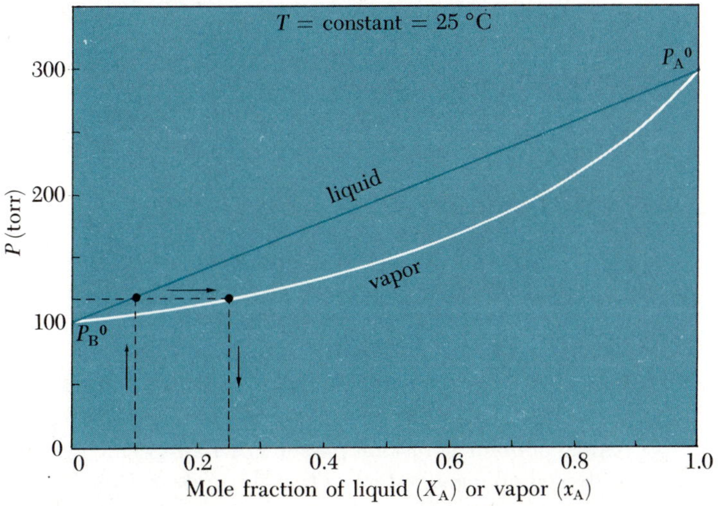

Figure 16.13
Plot of pressure vs liquid and vapor compositions (X and x) for an ideal mixture of liquids for which $P_A{}^0$ = 300 torr and $P_B{}^0$ = 100 torr.

OPTIONAL

the figure corresponds to a particular total vapor pressure ($P_A + P_B$); and its intersections with the liquid and vapor lines give the compositions of the liquid and vapor for that total vapor pressure (when the vertical lines, the dashed lines, are dropped from those points of intersection). The dashed lines shown in Fig. 16.13 show that for the liquid represented, with X_A equal to 0.1, the total vapor pressure above the liquid is 120 torr (given by the horizontal line intersecting the line labeled "liquid" at this composition); at this total pressure, the mole fraction of the equilibrium vapor x_A is 0.25. Thus, the vapor is richer in the more volatile component A that the liquid is. It is precisely this relation that provides the basis for the purification of liquids by distillation.

EXAMPLE 16.10 (a) Assuming that Raoult's law is valid, as in Fig. 16.13, calculate P_A, P_B, and P (the total vapor pressure) for a solution of A and B with a mole ratio of 1:1 at the temperature for which that figure is valid. (b) What is the composition of the vapor in equilibrium with this 1:1 mixture?

SOLUTION (a) Using Eq. (16.8) and the $P_A{}^0$ and $P_B{}^0$ values given with Fig. 16.13, we get

$$P_A = (0.500)(300 \text{ torr}) = 150 \text{ torr}$$

$$P_B = (0.500)(100 \text{ torr}) = 50 \text{ torr}$$

Then, by Dalton's law of partial pressures,

$$P = 150 \text{ torr} + 50 \text{ torr} = 200 \text{ torr}$$

(b) The vapor composition is given by the partial pressures, as we know from the ideal gas equation that the pressure exerted by a gas is proportional to the number of moles of the gas (for a given T and V).

Hence,

$$\text{percentage A} = \frac{P_A}{P_A + P_B} \times 100 = \frac{150 \text{ torr}}{200 \text{ torr}} \times 100 = 75\%$$

$$\text{percentage B} = \frac{P_B}{P_A + P_B} \times 100 = \frac{50 \text{ torr}}{200 \text{ torr}} \times 100 = 25\%$$

STUDY PROBLEM 16(g)

For the system shown in Fig. 16.13, determine P_A, P_B, P, and the vapor composition in equilibrium with a mixture of 80% A and 20% B.

Distillation

Figure 16.13 is a plot of pressure vs mole fraction at a constant temperature. One can construct an analogous plot of temperature (boiling point) vs mole fraction at a constant pressure. To construct this plot, we must recall that the boiling point of a liquid is the temperature for which the vapor pressure of the liquid equals the external pressure, and that the vapor pressure above a mixture of two liquids is a function of the composition of the mixture and the vapor pressures of the pure constituents (Eq. 16.9). From these considerations, one can construct a plot of the boiling point at a constant pressure of 1 atm vs the mole fraction. For the same kind of two-component system as that illustrated in Fig. 16.13, this plot of boiling point vs mole fraction is shown by the line labeled "liquid" in Fig. 16.14 (further assuming

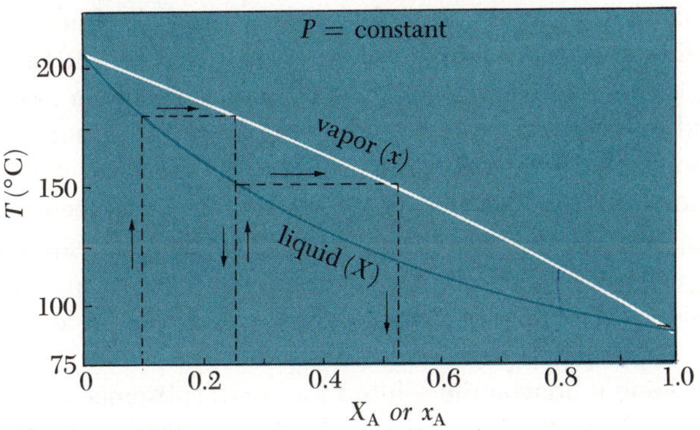

Figure 16.14
Plot of temperature vs liquid and vapor compositions for a two-component system. (Boiling point of B is 206 °C; boiling point of A is 90 °C.)

the special case that the normal boiling points of pure A and pure B are 87 °C and 206 °C, respectively).*

Note that the boiling point decreases as X_A approaches 1, because $P_A^\circ > P_B^\circ$ at any temperature and a liquid with a higher vapor pressure boils at a lower temperature. From Fig. 16.13, we see that at any temperature the vapor above the mixture is richer in the more volatile, or lower-boiling component. Thus, the line labeled "vapor" in Fig. 16.14 lies above the "liquid" line, and points on the vapor line lie to the right (larger fraction of A) than the corresponding points on the liquid line for the same temperature (horizontal line). The values shown for the vapor line in this figure were calculated from Eq. (16.12). The dashed lines in this figure show that a mixture with a mole fraction of A (X_A) equal to 0.1 boils at approximately 180 °C (follow the first arrow), and that the mole fraction of A in the vapor (x_A) is 0.26 (second arrow). If this vapor is condensed to a liquid, which is called the **distillate**, the condensed liquid (follow third arrow) will be richer in A than the original liquid was ($x_A > X_A$); and the residual liquid being boiled will therefore be richer in B than the original liquid. If this distillate is brought to boiling at approximately 150 °C (fourth arrow), the mole fraction of A in the vapor phase will be about 0.53 (fifth arrow), that is, even richer in A. These steps can be repeated until the distillate is essentially pure A, with a residue approaching pure B.†

A distillation apparatus that condenses the vapor above the original mixture directly into a receiving vessel, is said to have one "theoretical plate," that is, it performs only the first step of the previously described stepwise procedure (first three arrows of Fig. 16.14). Such an apparatus is shown in Fig. 2.10. Some types of apparatus can perform many condensations and vaporizations while in operation; and if the distillate shows an enrichment in the lower-boiling component that corresponds to n of the type of steps shown in Fig. 16.14, then the apparatus is said to have n theoretical plates. Distillations carried out with an apparatus providing many theoretical plates can provide an excellent means of purification for a liquid.

*The "liquid" curve shown in Fig. 16.14 can be obtained by using the Clausius-Clapeyron equation to determine $P_A{}^0$ and $P_B{}^0$ as functions of temperature, so one can find the combinations of temperature and composition that render P in Eq. (16.9) equal to 1 atm. The plot given in Figure 16.14 assumes that the molar heats of vaporization of the two components are equal and constant over the temperature range considered.

†It must be emphasized that Figs. 16.13 and 16.14 were constructed from equations applicable to ideal gases and solutions. While some mixtures demonstrate properties that are consistent with ideal laws, many do not.

SUMMARY

In many respects—say in order, and the strength of intermolecular interactions—the liquid state is intermediate between the solid state and the gaseous state. The general kinds of attractive interactions present are the same in liquids as in solids. When a solid is melted, the forces holding the crystal together are overcome as the liquid is formed; this requires an input of energy, the heat of fusion. The attractive interactions of the liquid must be overcome for gas molecules to be formed in vaporization; this requires an energy input, the heat of vaporization. A phase diagram summarizes the conditions (temperature and pressure) under which the solid, the liquid, or the vapor is stable, and the conditions under which any two, or all three (the triple point) can coexist at equilibrium.

Adding a nonvolatile solute to a liquid substance reduces its vapor pressure; the way in which the addition alters the phase diagram for the substance is the basis for understanding boiling-point elevations, freezing-point depressions, and in general, colligative properties. From Raoult's law, one can compute the vapor pressure contributed by each component in an ideal solution. This permits one to understand the manner in which a distillation can be used to separate a liquid solution into pure components (assuming they have substantially different boiling points and vapor pressure).

Water is a unique and important substance in our world. Water plays central roles in biology and geology. Achieving and maintaining water purity requires keeping in mind some simple but important chemistry.

STUDENT CHECKPOINTS

After studying this chapter, the student should be able to:

1. Describe the nature of liquids, in general, in terms of intermolecular interactions, order, and randomness.
2. Explain the general features of heating or cooling curves and why they have the general forms they have.
3. Describe boiling in terms of pressures.
4. Outline the general features of the phase diagram of a pure substance and describe the significance of each region and line in the diagram.
5. Describe what would occur in a sample if it were subjected to a process corresponding to moving across a horizontal line or up a vertical line in a phase diagram.
6. Define *molality* and *mole fraction,* and relate these values to molarity provided that enough information is given.
7. Predict what qualitative effect adding a nonvolatile solute has on a phase diagram.
8. Compute the freezing-point depression or boiling-point elevation of a solution containing nonvolatile solute particles at a known concentration.
9. Apply the computational methods of colligative properties to determine chemically useful results—for example, solute molecular weight and percentage ionization of a solute electrolyte.

EXERCISES

16.1. Indicate whether each of the following statements is true or false:

(a) A liquid is approximately as dense as the corresponding solid and approximately as compressible as a gas.

(b) All liquids, like gases, are completely miscible.
(c) The entropy of one mole of a liquid is less than the entropy of one mole of corresponding gas.

16.2.* Indicate whether each of the following statements is true or false:
(a) The entropy of one mole of a liquid is greater than the entropy of one mole of the corresponding solid.
(b) At the melting point, $\Delta H_{fus} = T\,\Delta S_{fus}$.
(c) A liquid has long-range order, but no short-range order.

16.3. Describe the types of attractive forces present in liquids.

16.4. Explain what is meant by short-range order in liquids.

16.5. Discuss the relative densities of solids, liquids, and gases.

16.6. Describe London forces.

16.7.* Why is it more difficult to describe liquids theoretically than to describe solids or gases?

16.8. Explain why NH_3 and H_2O have much higher boiling points than N_2 and O_2, which have higher molecular weights.

16.9.* Which of the following two isomers would you expect to have a higher boiling point? Why?

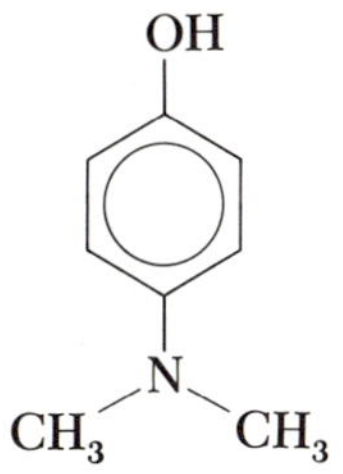

A

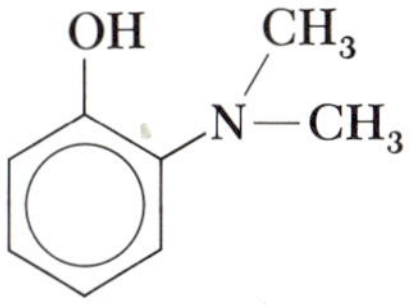

B

16.10. Consider the following two substances: molten KCl, liquid CH_4.
(a) Which has the higher electrical conductivity?
(b) Which has the stronger intermolecular attractions?
(c) Which has the higher boiling point?

16.11.* Consider the following three substances: molten NaCl, liquid Cl_2, liquid Na.
(a) Rank these liquids in decreasing order of electrical conductivity.
(b) Which of these liquids would have the weakest intermolecular attractions?
(c) Which would have the lowest boiling point?

16.12. Consider the following three liquids:

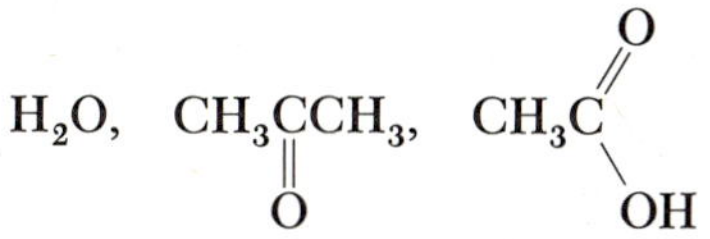

(a) In which of these is hydrogen bonding least important?
(b) Which would be the best solvent for LiCl?
(c) Which would be the best solvent for benzene, C_6H_6?

16.13.* Indicate whether each of the following statements is true or false:
(a) The heat of vaporization of a liquid is always greater for compounds containing hydrogen than for compounds that do not.
(b) The heat of fusion of a solid is always large if ΔS_{fus} is large.
(c) A large ΔH_{fus} value and a small ΔS_{fus} value lead to a high melting point.

16.14. Indicate whether each of the following statements is true or false:
(a) A large value of ΔH_{vap} and a small value of ΔS_{vap} lead to a high boiling point.
(b) The heat of vaporization of a liquid reflects only the difference between the potential energies of molecules in the liquid and the vapor states.
(c) The liquid state of a substance has a lower free energy than the solid state of the same amount of the same substance at the same temperature and pressure.

16.15. Describe the interactions represented pictorially in Fig. 16.2.

16.16.* Indicate whether each of the following statements is true or false:
(a) The steepness of the line A in Fig. 16.3a would be greater if the heat capacity of the solid were higher.
(b) The temperature corresponding to D in Fig. 16.3a is greater than that corresponding to D in Fig. 16.3b.
(c) Between the beginning and the end (left and right) of the line shown in Fig. 16.3a, an amount of heat exactly equal to the sum $\Delta H_{fus} + \Delta H_{vap}$ would be added to one mole of the substance in question.

16.17.* Using Fig. 16.5, estimate the vapor pressure of CCl_4, CH_3CH_2OH, H_2O, and CH_3CO_2H for 60 °C to the nearest 10 torr.

16.18. Using the Clausius-Clapeyron equation, compute the vapor pressure of CCl_4 for 30 °C, knowing that it is 40.0 mm at 4.3 °C.

16.19. Describe in detail what would happen to a sample of water at a temperature of 55 °C and a pressure of 50 torr if the sample were compressed to a pressure of 800 torr, with the temperature maintained constant.

16.20.* Explain how Dry Ice, which is solid CO_2, can sublime directly to CO_2 gas without passing through the liquid state.

16.21. Describe the meaning of the liquid-gas line in Fig. 16.7.

16.22. How does the direction of slant of the solid-liquid line in a phase diagram relate to the melting point of the corresponding solid?

16.23. Give five examples of the following rule of thumb in solubility: Like dissolves like.

16.24.* Calculate the molality of acetone (CH_3COCH_3) in a solution that is 25.0% acetone and 75.0% benzene (C_6H_6) by weight.

16.25. Calculate the weight percentage of CCl_4 in a solution that is 1.7 *m* CCl_4 in methanol, CH_3OH.

16.26.* Calculate the mole fraction of $KMnO_4$ in an 0.84 *m* solution of $KMnO_4$ in water.

16.27. Calculate the molality of triphenylphosphine, $(C_6H_5)_3P$, in a solution in benzene, where the benzene mole fraction is 0.962.

16.28.* Calculate the percentage iodine by weight in a 0.15 *m* I_2 solution in CCl_4.

16.29. The density of a 23.0% by weight solution of sulfuric acid in water is 1.163 g/cm^3. Calculate the molarity and molality of H_2SO_4 in this solution.

16.30.* A 25.0% by weight solution of H_2SO_4 in water is found to have a sulfuric acid concentration of 3.00 *M*. What is the density of this solution?

16.31. A certain solution of sucrose, $C_6H_{12}O_6$, in water contains 216.2 g of sucrose per liter of solution. This solution is 20.0% sucrose by weight. What are (a) the molarity, (b) the molality, (c) the density of this solution?

16.32.* Calculate the boiling point of benzene in an 0.83 *m* solution of a nonvolatile organic compound in benzene at 1 atm pressure.

16.33. Calculate the freezing point of water in a 1.14 *m* $Mg(NO_3)_2$ solution.

16.34.* Calculate the boiling point of a 1.43 *m* solution of $CaCl_2$ in water.

16.35. Calculate the boiling point of a solution made up by dissolving 1.62 mol of a nonvolatile weak monobasic acid in 1000 g of water for the condition that the acid is 5.0% ionized in that solution at about 100 °C.

16.36.* Determine the percentage of ionization and the ionization constant of a nonvolatile, weak monobasic acid, given the following information. When 1.18 mol of the acid is dissolved in 1000 g of water, the resulting solution is found to have a boiling point of 100.72 °C.

16.37.* Determine the molecular weight of a compound if a solution containing 7.24 g of the compound in 125 g of camphor was found to have a melting point of 159.9 °C.

16.38. A mixture of two compounds, 2.92 g A and 4.61 g B, is dissolved in 500 g tribromophenol. The resulting solution is found to have a melting point (freezing point) of 92.1 °C. The molecular weight of A is 145.0 amu. What is the molecular weight of compound B?

16.39. At 150.6 °C, the vapor pressure of pure decane, $CH_3(CH_2)_8CH_3$, is 400 torr. What would be the partial pressure of decane over a solution containing 80% decane and 20% octane, $CH_3(CH_2)_6CH_3$, by weight at 150.6 °C?

16.40.* The boiling point of pure benzene at 1 atm is 80.1 °C. What would be the vapor pressure of benzene over a solution containing 1 mol of a nonvolatile hydrocarbon and 12 mol benzene at 80.1 °C?

16.41. Using Fig. 16.5, estimate the partial pressures and total pressure of the equilibrium vapor over a solution containing 12.4 g CCl_4 and 3.24 g CH_3CH_2OH at 50 °C.

16.42. Explain how a liquid can be separated into pure components by distillation.

16.43.* One heats a liquid mixture of 80% A and 20% B for the constant-pressure system shown in Fig. 16.14. Estimate (a) at what temperature the solution will boil; (b) what the composition of the vapor first obtained will be; (c) what the boiling point of the liquid that could be condensed from that vapor will be.

16.44. For the particular case shown in Fig. 16.14, how many distillation "steps" (plates) would be required to obtain 95% pure B, starting from a liquid mixture that is 40% A and 60% B?

16.45. What are the main metal ions responsible for water "hardness"?

ANSWERS TO SELECTED EXERCISES

16.2 (a) true; (b) true; (c) false. **16.7** Solids and gases represent the extremes of a high degree of order and of no order, respectively. **16.9** Compound A would have a higher boiling point. It would engage in *inter*molecular hydrogen bonding, effectively giving a higher molecular weight, whereas the hydrogen-bonding tendency of B can be satisfied at least partially by *intra*molecular hydrogen bonds. **16.11** (a) liquid Na > molten NaCl > liquid Cl_2; (b) liquid Cl_2; (c) liquid Cl_2. **16.13** (a) false; (b) false; (c) true. **16.16** (a) false; (b) false; (c) false. **16.17** CCl_4, 550 torr; CH_3CH_2OH, 500 torr; H_2O, 230 torr; CH_3CO_2H, 140 torr. **16.20** By proceeding from left to right across a phase diagram at a pressure lower than that of the triple point (like the dashed line passing through *g* and *h* in Fig. 16.7). **16.24** 5.74 *m*. **16.26** 0.015. **16.28** 3.7%. **16.30** 1.18 g/cm³. **16.32** 82.2 °C. **16.34** 102.2 °C. **16.36** 19% in the specified solution; $K = 5.4 \times 10^{-2}$. **16.37** 118 amu. **16.40** 0.92 atm. **16.43** (a) 100 °C; (b) $x_A = 0.94$, $x_B = 0.06$; (c) 88 °C.

ANSWERS TO STUDY PROBLEMS

16(a) 28.3 kJ. **16(b)** solid. **16(c)** 1.13 *m*; 1.09 *M*. **16(d)** 0.040. **16(e)** 66.8 °C. **16(f)** 2.8×10^2 amu. **16(g)** $P_A = 240$ torr; $P_B = 20$ torr; $P = 260$ torr; about 92% A and 8% B.

17

Descriptive Chemistry. Vertical Trends for the Heavier Main Group Elements

In this chapter we shall consider the heavier main group elements found in Groups IA to VIIIA of the periodic table, beginning with sodium. Main group elements have completely empty or completely filled *d* and *f* sublevels in all their common oxidation states, whereas transition elements have partially filled *d* or *f* sublevels or both in at least one of their common oxidation states. The transition elements, groups IB to VIIIB, will be considered in Chap. 18.

Although much familiar chemistry involves the lighter elements, hydrogen to fluorine, the chemistry of heavy elements is also very important. Sodium, potassium, and calcium are found throughout cellular plants and animals and are necessary for very many biological processes. Phosphorus, sulfur, chlorine, and iodine are also essential components of living organisms. Silicon makes up nearly one-quarter of the earth's crust and is found in many important minerals. Aluminum is of great commercial value. Hundreds of such statements dealing with heavier main group elements could be made to emphasize their importance.

Although there are a considerable number of heavy main group elements (34 of them, sodium through radium), much of their chemistry can be understood by knowing the chemistry of the first-row elements and some important vertical trends. A survey of the chemistry of the heavy elements quickly reveals that the properties of one member of a group correspond well to the properties of the other members, with one general exception—the lightest element of the group, that in row 2. One of the principal aims of this chapter is to explain this discontinuity and to show how vertical trends can be used to understand much of the chemistry of the heavy elements. In contrast to Chap. 9, which is based on type of compounds formed by each element, this chapter is arranged by groups.

17.1 Group 1A: Li, Na, K, Rb, Cs, and Fr. The Alkali Metals

General Considerations

The most significant chemistry of the alkali metals (Li, Na, K, Rb, Cs, and Fr) is that of the +1 ions. In the neutral atoms, all these elements have one *s* electron outside a noble gas core. The first ionization potential (energy required to remove one electron from the neutral atom in the gas phase) of each of these elements is low enough to allow relatively easy formation of the +1 ion. Removing the next electron requires disturbing the noble gas type of electron configuration and requires so much energy (the second ionization potential) that the chemistry of ions of greater than +1 charge is not known for the alkali metals under usual circumstances, say in stable solutions or crystals.

The +1 ions of the alkali metals are, in general, small enough to have little tendency to be polarized and are too large to cause significant polarization of other species—negative ions or solvent, for example; such mutual perturbations of the electron distribution of one atomic center with another are generally associated with covalent bonding.* Since these +1 ions are neither polarized easily nor the cause of substantial polarization, they generally form bonds without any significant covalency. The exception to this general pattern is Li^+, which is small enough to cause significant polarization of anions. For this ion, because of its large charge-to-radius ratio—meaning a high concentration of charge per unit volume due to the small size of Li^+—a significant distortion of the electron distribution of an anion can be caused; this gives the bond of lithium to more electronegative elements some covalent character. That is, the higher charge-to-radius ratio of the ion Li^+ polarizes the electron cloud of a nearby anion effectively, and even shares a portion of the electron density. The other alkali metal ions are large enough—the charge-to-radius ratios are low enough—to preclude covalency in the bonding.

If we consider the overall chemistry of the alkali metals, then we find that the general consequence of the small size of lithium is to render it an exception. The chemical properties of the other alkali metals are more easily related to one another. In general, trends for a given property among them are caused by the increasing size of the atoms and + 1 ions in the series Na < K < Rb < Cs.† As the atomic size is larger, the outer electron is more easily removed since it is farther from the nucleus and increasingly shielded (Chap. 6); hence the ionization potential of the alkali metals

*One can think of the formation of a purely covalent bond as such a complete perturbation of a pair of electrons by two atomic centers that neither electron can be assigned more to one atomic center than to the other.

†All known isotopes of francium are short-lived radioactive species and the chemistry of francium is relatively unimportant.

Spotlight

Since the alkali metals have low first ionization potentials, they are chemically very reactive. They never occur in nature in an elemental form, but are found as salts that are quite water-soluble. The leaching action of rain tends to concentrate these salts in the oceans and in landlocked lakes and seas. Some notable deposits of salts are borax, $Na_2B_4O_7 \cdot 10H_2O$, in Death Valley, California; NaCl in Salt Lake, Utah; Chilean saltpeter, $NaNO_3$, in Chile; and KCl in Searles Lake, California. The pure metals are generally obtained by electrolysis of the fused chlorides; for example,

$$2NaCl(l) \rightleftharpoons 2Na(s) + Cl_2(g)$$

The low ionization potentials also give rise to weak binding energies in the metal lattices (Chap. 14); as a result, the alkali metals are soft enough to cut with a knife and are good conductors of heat and electricity. They have a silvery appearance when freshly cut. Sodium is so efficient as a heat transfer medium that it is used as a coolant in the valves of some aircraft engines; the initially hollow valves are partially filled with metallic sodium. Sodium is also used as the heat transfer medium in "breeder" reactors. Rubidium and cesium lose their electrons so easily that electrons are ejected when light energy strikes a polished surface of the metal; hence, these metals find important applications in photoelectric cells.

The weakly held outer electrons can also be solvated. One of the most important examples is a solution of Na in liquid NH_3, which has been shown in dilute solutions to consist of Na^+ and "solvated electrons." The solution is deep blue and is widely used in organic synthesis as a reducing agent.

TABLE 17.1 Some Fundamental Properties of the Alkali Metals (Group IA)

Element	*Electronic Configuration*	*Metal Radius, Å*	*+1 Ion Radius, Å*	*Ionization Potentials, eV*	
				1st	*2nd*
Li	[He]2*s*	1.52	0.60	5.390	75.62
Na	[Ne]3*s*	1.86	0.95	5.138	47.29
K	[Ar]4*s*	2.27	1.33	4.339	31.81
Rb	[Kr]5*s*	2.48	1.48	4.176	27.36
Cs	[Xe]6*s*	2.65	1.69	3.893	23.4
Fr	[Rn]7*s*				

is lower from lithium to cesium; this can be seen in Table 17.1, which summarizes some fundamental properties of alkali metals. As a consequence of the easier ionization, many reactions in which the alkali metals are ionized will proceed much more easily with the heavier alkali metals. If the reaction is exothermic with sodium, we should expect it to release even more heat with potassium, rubidium, and cesium. A familiar example is the reaction of alkali metals with water to form the metal hydroxide and hydrogen gas.*

$$M + H_2O \rightleftharpoons MOH(aq) + \tfrac{1}{2}H_2(g) \qquad (17.1)$$

where M is an alkali metal and MOH stands for the corresponding metal hydroxide. Sodium reacts vigorously with water; small chunks are propelled rapidly about the

*Note that this is an oxidation-reduction reaction since the metal has been oxidized from an oxidation state of zero (element) to +1, while the oxidation state of hydrogen in H_2O has been reduced from +1 to zero.

surface of the H_2O and quickly consumed in the formation of the NaOH solution and H_2:

$$Na + H_2O \rightleftharpoons NaOH + \frac{1}{2}H_2$$

Adding even small pieces of potassium to water causes flames, and in a similar process rubidium and cesium react explosively. Sodium barely reacts with Br_2 ($Na + \frac{1}{2}Br \rightarrow NaBr$, again an oxidation-reduction reaction), whereas rubidium and cesium react very rapidly. The alkali metals tarnish rapidly in air, reacting with the O_2; rubidium and cesium again are the most reactive, so reactive that they must be handled in an inert atmosphere, one without O_2, to prevent combustion.

An interesting feature of the combustion of alkali metals with O_2 is that the types of products of the reaction, all solids under normal circumstances, are different for the various metals. Lithium forms the oxide Li_2O, whereas sodium forms the peroxide, Na_2O_2; and potassium, rubidium, and cesium form superoxides, KO_2, RbO_2, and CsO_2. Again this feature of alkali metal chemistry can be attributed to the increasing size of the heavier alkali metal ions. The lattice energy of an ionic compound is important in determining whether or not a solid ionic compound is easily formed (Chap. 7). One of the ways that a large, or favorable, lattice energy is achieved is for the sizes of the cation and anion to be very similar. In this case the ions can pack very efficiently in the crystal (Sec. 14.4). The oxide ion, O^{2-}, is much smaller than the peroxide ion, O_2^{2-}. Two sodium ions pack with each peroxide ion in Na_2O_2, and there are two lithium ions for each oxide ion in Li_2O. The larger alkali metal ions—potassium, rubidium, and cesium—form a very different crystal lattice, in which there is one superoxide ion, O_2^-, for each metal ion. Thus, the size variation of oxygen anions O^{2-}, O_2^{2-}, O_2^- parallels the size variation of alkali metal cations, Li^+, Na^+, K^+, Rb^+, and Cs^+, and accounts for the crystalline products formed by the combustion of alkali metals with O_2.

Analysis

Nearly all salts of the alkali metals are water-soluble; hence, the number of precipitation reactions that can be used as a qualitative test for these cations is few. This is one of the reasons that the alkali metals are distinguished by flame tests rather than precipitation reaction in many schemes for qualitative analysis. The characteristic emission colors of the most common alkali metals, sodium (yellow flame) and potassium (red violet flame), are very intense, allowing visual determination of their presence by observing the sample in a flame. Using blue glass allows one to see the potassium coloration in the presence of sodium since the blue glass absorbs the color produced by the sodium.

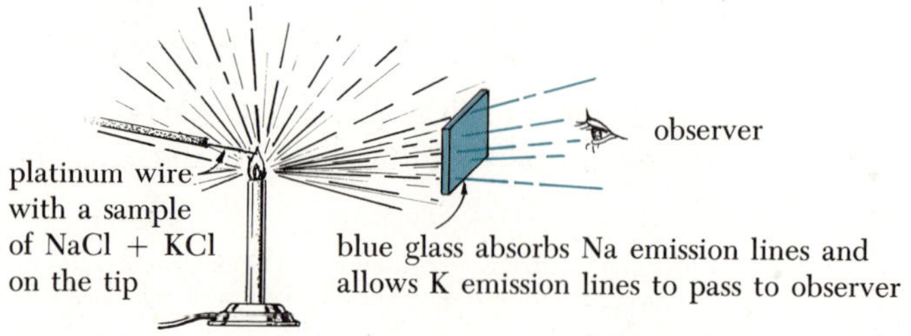

Rubidium and cesium are very similar to potassium in the way they color flame, and their precipitation reactions are also very similar. For example, the perchlorate salts, $MClO_4$, of each of these cations are all nearly insoluble in water; also, these alkali cations are all precipitated as tetraphenylborate salts, $MB(C_6H_5)_4$, by the tetraphenylborate anion, $B(C_6H_5)_4^-$. To distinguish these ions from one another, the chemist often uses an instrumental method such as emission spectroscopy, in which the spectral lines resulting from the downward transitions of atoms in excited states are recorded. This is the same phenomenon used to produce spectra of hydrogenlike atoms and ions (Sec. 6.8). These emission lines are recorded much more accurately than we can distinguish emission colors with our eyes.

Even though Na^+ and K^+ are very similar, our cellular systems recognize the difference between them in biologically important ways. In biological systems, ionic size once again appears to be the dominant feature in the difference between Na^+ and K^+ chemistry.

17.2 Group IIA: Be, Mg, Ca, Sr, Ba, and Ra. The Alkaline Earths

General Properties

Like the alkali metals, the alkaline earths (Be, Mg, Ca, Sr, Ba, and Ra) exhibit chemistry predominantly due to one type of species. The chemistry of the alkaline earths, except for beryllium, is that of the +2 ions. Neutral atoms of these elements all have two *s* electrons outside a noble gas core. By now it should be clear why these elements do not form +3 ions as part of their normal chemistry—but why don't they have +1 ions as well as +2 ions? This question may be answered by considering lattice energy (of solids) or solvation energy (of aqueous solutions). As we discussed in Sec. 7.4, a highly charged ion, all other things equal, has a greater lattice energy or solvation energy than an ion of lower charge. If the lattice energy difference between the two ions is large enough to overcome the additional ionization potential to form the more highly charged ion, that ion will be more stable. This situation is seen diagrammatically in Fig. 17.1. By investigating Table 17.2, which summarizes some pertinent properties of alkaline earths, one can see that the second ionization potentials for the alkaline earths are not extremely high. They are much lower than

TABLE 17.2 Some Fundamental Properties of the Alkaline Earths (Group IIA)

Element	*Electronic Configuration*	*Metal Radius, Å*	*+2 Ion Radius, Å*	*Ionization Potentials, eV*	
				1st	*2nd*
Be	[He]$2s^2$	1.11	0.31	9.32	18.21
Mg	[Ne]$3s^2$	1.60	0.65	7.64	15.03
Ca	[Ar]$4s^2$	1.97	0.99	6.11	11.87
Sr	[Kr]$5s^2$	2.15	1.13	5.69	10.98
Ba	[Xe]$6s^2$	2.17	1.35	5.21	9.95
Ra	[Rn]$7s^2$	2.20	1.57	5.28	10.10

Figure 17.1
Energy diagram showing how the energy required to ionize a +1 metal ion to a +2 metal ion can be more than compensated for by a larger lattice energy (or hydration energy) of the M^{2+} species, a crystalline solid or an aquated metal ion.

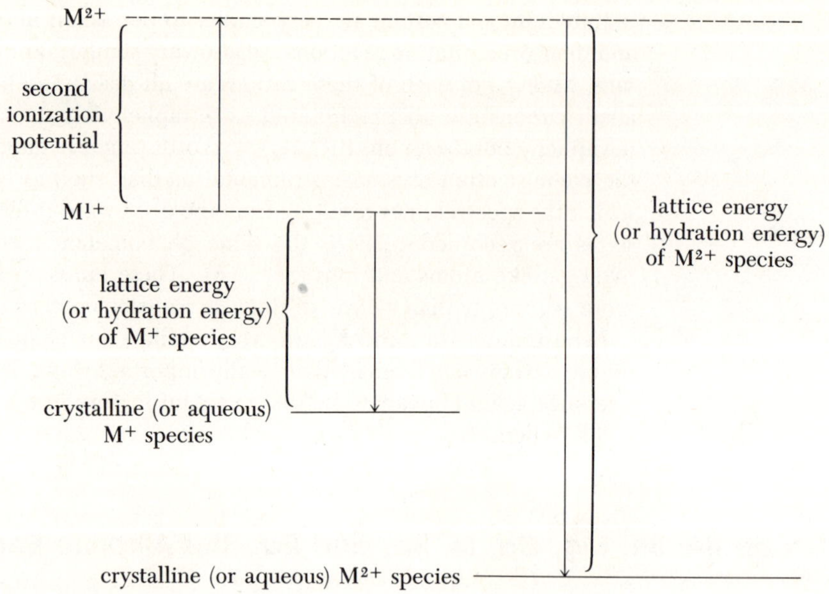

the second ionization potentials of the alkali metals shown in Table 17.1. In fact, for any system involving a +1 alkaline earth ion that we predict to be stable, the +2 ion is even more stable. (See the Born-Haber cycle for Ca^+Cl^- and $Ca^+Cl_2^-$, Table 7.1.)

In the discussion of the alkali metals, we mentioned that Li^+ has some tendency to form bonds with significant covalent character, due to its relatively large charge-to-radius ratio. This feature that is so important in Li^+ chemistry is even more important for Be^{2+}. The ion Be^{2+} is smaller than Li^+, since one more electron has been removed, and there is an additional proton in the nucleus; and the charge of

Spotlight

Like the alkali metals, the alkaline earth metals have low ionization potentials. Therefore, they are sufficiently chemically reactive that they do not occur in nature in an elemental form. Except for Be, which is hard enough to cut glass, the Group IIA metals are also soft enough to be cut with a knife to give a silvery surface, are good conductors of heat and electricity, dissolve in liquid NH_3 to give blue solutions with reducing properties, and are generally obtained by electrolysis of the fused halides.

In an elemental form, only magnesium finds extensive industrial use, mainly for light, strong alloys such as those used in engines and aircraft construction. Metallic magnesium is important in the laboratory for the preparation of **Grignard reagents,** written RMgX, where R is an organic group and X is a halide. The reagent and some of its reactions were reported in 1900 by the French chemist Victor Grignard, and it proved to be so extraordinarily useful that a Nobel Prize was awarded to Grignard in 1912. Grignard reagents are of more practical importance for the laboratory synthesis of organic compounds than any other single reagent. Typically, a Grignard reagent is prepared by the direct action of an alkyl halide or an aryl halide on the metal in the presence of an organic solvent. For convenience, the formula written for the reagent is RMgX, even though the structures of Grignard reagents in solution are more complex.

Spotlight

The basic properties of the oxides and hydroxides of magnesium and calcium find several important uses. The substances MgO and $Mg(OH)_2$, called milk of magnesia, are used medicinally as antacids. Calcium hydroxide, $Ca(OH)_2$, called slaked lime, and CaO, called quicklime, are extensively used in mortar and Portland cement, respectively. Mortar gradually hardens due to the reaction of $Ca(OH)_2$ with CO_2 in the air to form $CaCO_3$. This reaction also occurs extensively in nature during the formation of limestone caverns, and $CaCO_3$ occurs naturally in the form of marble and in the shells of marine animals. The essential constituents of Portland cement are believed to be tricalcium silicate, $3CaO \cdot SiO_2$, and tricalcium aluminate, $3CaO \cdot Al_2O_3$. Hydration occurs when these anhydrous substances are mixed with water, forming hydrated calcium silicate, $2CaSiO_3 \cdot 5H_2O$, tetracalcium aluminate, $4CaO \cdot Al_2O_3 \cdot 12H_2O$, and $Ca(OH)_2$ in a mass of interlacing needles that give strength and hardness to set cement.

Be^{2+} is twice the charge of Li^+. Hence, Be^{2+} has a charge-to-radius ratio several times as great as that of Li^+. (Charge-to-radius ratio for $Be^{2+} = 6.5$, for $Li^+ = 1.7$; given in units of electronic charge per Å.)

In the following discussion, we shall consider the elements magnesium, calcium, strontium, barium, and radium. The charge-to-radius ratio of beryllium is so high that its chemistry is very much different from the chemistry of the remaining elements of Group IIA. Therefore, the chemistry of beryllium will not be discussed in this treatment of Group IIA.

The Hydroxides

The alkaline earths all form basic hydroxides. A tendency within the hydroxides readily explained by periodic trends is that $Mg(OH)_2$ is a weaker base than the heavier alkaline earth hydroxides. In the Arrhenius view (see Sec. 5.2), the tendency for a hydroxide compound to act as a base in aqueous solution can be symbolized by the equation

$$M(OH)_x(aq) \rightleftharpoons M(OH)^+_{x-1}(aq) + OH^-(aq) \qquad (17.2)$$

According to the Brönsted-Lowry approach, we may think of the stronger base as the compound that can best compete with the free hydroxide ion for a proton, H^+:

$$\underset{\text{base}}{M(OH)_x} + \underset{\text{acid}}{H_2O} \rightleftharpoons \underset{\text{base}}{OH^-} + \underset{\text{acid}}{M(OH)H^+} \qquad (17.3)$$

Smaller, more highly charged metal ions can better polarize the negatively charged hydroxide group in $M(OH)_x$, leaving a less negative center, meaning a less basic center, to attract a proton from another acid, which would be H_2O in the case shown. In a given group of metal hydroxides therefore, the weakest attraction for the proton and therefore the weakest base will be the hydroxide of the lightest, smallest element. For this reason, $Mg(OH)_2$ is a weaker base than the heavier alkaline earth hydroxides. By contrast, all the alkali metal hydroxides—NaOH through CsOH—are strong bases since the corresponding metal ions are only singly charged and the interaction with OH^- is in all cases so weak that the metal hydroxides are quite soluble in water and highly ionized in aqueous solution.

The Halides

Salts of the alkaline earth metal ions with negative ions of the halogen series (F^-, Cl^-, Br^-, and I^-) are common. One of the notable properties of these salts is the lower solubility of salts of the heavier alkaline earth metals. How can the solubility differences of alkaline earth halides be explained? To understand this phenomenon, we can envision the process that occurs when a compound dissolves. An ionic compound dissolves by becoming solvated; its constituent ions become surrounded by solvent. The solvent needs to be polar to stabilize the ions in solution—that is, to shield the ions from each other so that they do not recombine and crystallize. Water is an excellent polar solvent. In considering the solubility of a substance, we are necessarily concerned with the equilibrium constant, or $\Delta G°$ value, for the dissolution:

$$\text{solid} \rightleftharpoons \text{dissolved species}, \; \Delta G^\circ_{dis} = \Delta H^\circ_{dis} - T\,\Delta S^\circ_{dis} \tag{17.4}$$

A large part of the ΔG°_{dis} value is determined by the change in enthalpy for the dissolution; that is, $\Delta H^\circ_{dis} = H^\circ_{sol} - \Delta H^\circ_{cryst}$, where ΔH°_{sol} represents the molar enthalpy change for dissolution of gaseous species (ions) and ΔH°_{cryst} represents the molar enthalpy change for crystallization of the gaseous ions. The difference in enthalpies, ΔH°_{dis}, is seen diagramatically in Fig. 17.2. From that figure it is clear that

$$\Delta H_{dis} = \text{hydration energy} - \text{lattice energy} \tag{17.5}$$

The ΔS°_{dis} factor in Eq. (17.4), which is $S^\circ_{sol} - S^\circ_{cryst}$, is always favorable (positive) for dissolution, because the crystalline state is much more highly ordered than the dissolved state. To see how ΔH°_{dis} varies for various alkali metal halides, one can examine Fig. 17.3 and Fig. 17.4, which show how the lattice energies for a representative class of alkali halides (chlorides) and the metal ion hydration energies, respectively, vary with the metal identity. From Eq. (17.5), it is clear that to favor solubility, one wants a larger hydration energy and a smaller lattice energy—to give

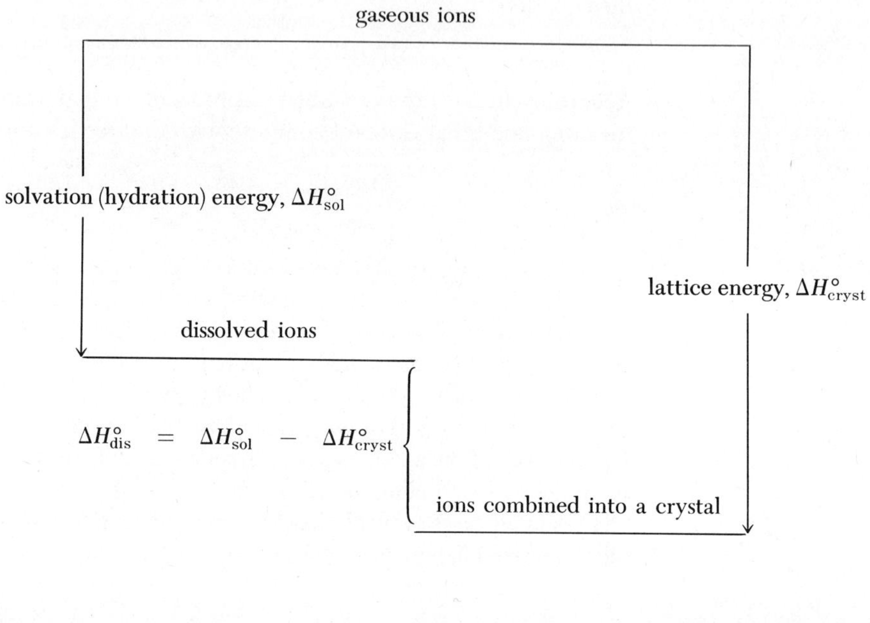

Figure 17.2 Energy (enthalpy) diagram showing the relation between the $\Delta H^\circ_{dissolution}$ (enthalpy change for dissolution) and the lattice energy and hydration energy for an electrolyte. The case shown is a sparingly soluble salt. (The magnitude of the lattice energy is larger than the magnitude of the hydration energy.)

Figure 17.3
Lattice energies of the alkali halides.

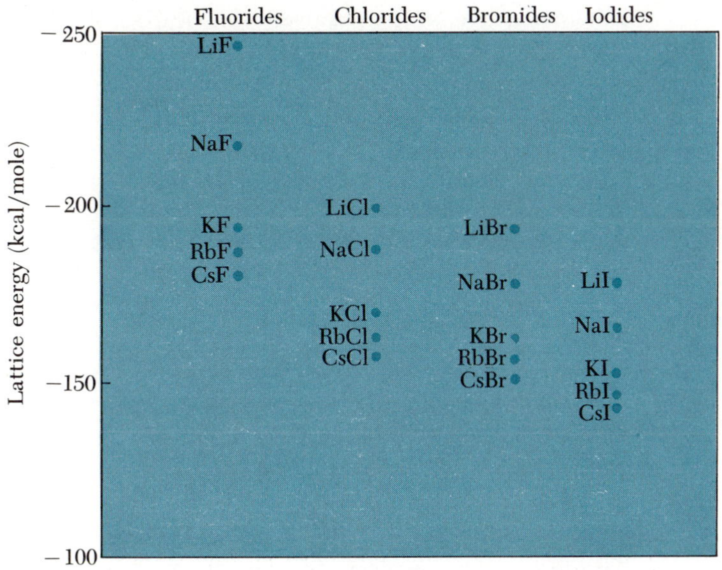

Figure 17.4
Hydration energies of the alkali metal ions and halide ions.

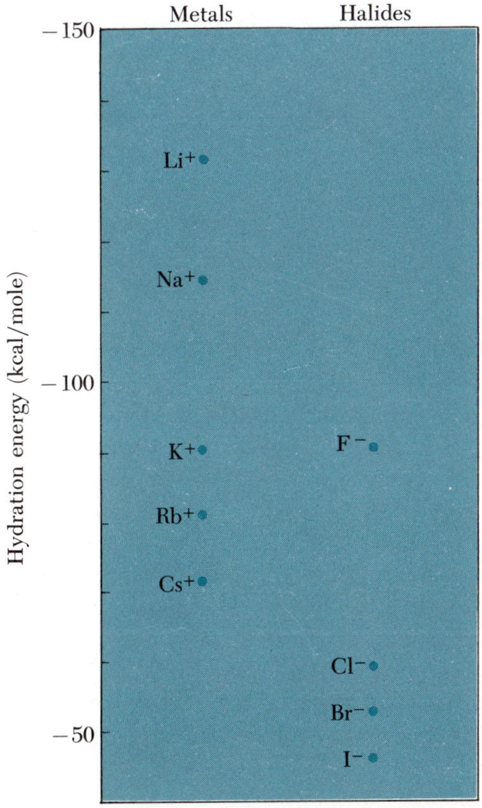

a more negative, or less positive, value of ΔH°_{dis} and hence a more negative, or less positive, value of ΔG°_{dis}. Figures 17.3 and 17.4 show that the hydration energies drop off more rapidly than the lattice energies do in the direction towards heavier alkali metal ions. Thus, ΔH°_{dis} becomes less favorable in the direction from $MgCl_2$ towards $BaCl_2$; hence, the solubility decreases in that order. The phenomenon is again explained by the difference in the size of the ions M^{2+} in the direction down the group. The hydration energy has a somewhat greater dependence on size than the lattice energy and therefore is sharply lower as the ionic size is greater.

Carbonates

The carbonates of alkaline earth metals, especially $MgCO_3$ and $CaCO_3$, are familiar to us as components of hardwater scale that deposits in pipes, boilers, and teakettles.* These carbonates are not very soluble in water, and like the halides, they have less solubility as the metal ions are heavier.

The substance $CaCO_3(s)$ is formed when water containing Ca^{2+} and carbonate ions is boiled or heated. In cold samples, much of the carbonate is present in the form of bicarbonate, the hydrogen carbonate ion, HCO_3^-. Calcium bicarbonate, $Ca(HCO_3)_2$, is reasonably soluble. The ion HCO_3^- is in equilibrium with dissolved CO_2 (or carbonic acid) and the carbonate ion (see Sec. 5.2):

$$2HCO_3^- \rightleftharpoons CO_3^{2-} + CO_2 + H_2O \qquad (17.6)$$

Since CO_2 can be driven off by heating, this equilibrium can readily be shifted towards formation of more carbonate ion as bicarbonate ion is consumed. Calcium carbonate is much less soluble than calcium bicarbonate, and $CaCO_3(s)$ precipitates as the concentration of CO_2^{2-} builds up.

17.3 Group IIIA: B, Al, Ga, In, and Tl

General Considerations

Group IIIA tends to form compounds in which the bonding is not accurately described as either purely ionic or purely covalent. The most stable oxidation state for most of the metals in this group is +3 (the stability of the +1 oxidation state increases down the group, and Tl^+ is especially stable). For a stable +3 ion to form in a crystal or in solution, energy sufficient to overcome the first three ionization potentials would have to be supplied from sources such as lattice energy or solvation energy. Even if this energy were available, the ions formed would be highly charged (+3) and reasonably small. Thus, they would be good polarizing ions and would cause distortion and some sharing of the electron clouds of anions or solvent molecules, leading to some covalency. The importance of this covalency decreases as the ions are larger, and less effective at interacting with the electrons of the anion or solvent. Thus, boron is essentially "purely covalent" in its bonding with other atoms; aluminum and gallium form polar covalent bonds with very large covalent character; indium is more ionic; and thallium tends to be the most ionic.

*Salts of iron also contribute markedly to the hardness of water.

TABLE 17.3 **Some Fundamental Properties of Group IIIA Elements**

Element	Electronic Configuration	Metal Radius, Å	+3 Ion Radius, Å	Ionization Potentials, eV 1st	2nd	3rd
B	$[He]2s^22p$	0.80	.....	8.30	25.15	37.92
Al	$[Ne]3s^23p$	1.43	0.50	5.98	18.82	28.44
Ga	$[Ar]3d^{10}4s^24p$	1.22	0.62	6.00	20.43	30.6
In	$[Kr]4d^{10}5s^25p$	1.63	0.81	5.79	18.79	27.9
Tl	$[Xe]4f^{14}5d^{10}6s^26p$	1.70	0.95	6.11	20.32	29.7

Why does thallium form a stable +1 ion while its congeners do not? An inspection of the ionization potentials of Group IIIA, given in Table 17.3, along with some other pertinent data on Group IIIA elements, shows that the ionization potentials of thallium are greater than the corresponding ionization potentials of indium, whereas our general rule about ionization potentials (Sec. 6.3) states that larger atoms should have lower ionization potentials. How can we explain this? In Chap. 6 we argued that adding an extra shell of electrons while adding an equal number of protons to the nucleus should make electron removal easier. This statement relies on the concept that inner shells of electrons shield the outer electrons from the effect of an increased number of protons. In going from indium to thallium, we add 32 protons; also electrons in number fourteen 4*f*, ten 5*d*, two 6*s*, and one 6*p*. The *f* electrons do not shield outer shells so well as *s*, *p*, or even *d* electrons. Hence, the greater number of protons in the thallium nucleus can attract the 6*s* and 6*p* electrons more effectively than the nucleus in indium can attract its 5*s* and 5*p* electrons. Thus, thallium has higher ionization potentials, which tends to stabilize the +1 ions. But what about the stabilizing influences of lattice energy for ionic bond formation, hydration energy for formation of aqueous solutions, or bond energies for covalent bond formation? How do these factors influence the relative stability of Tl^+? Each of these factors is less important for Tl^{3+} than for In^{3+}, because of the larger size of Tl^{3+}. Hence, the tendency to favor the +3 oxidation state is less for thallium than for any of the other Group IIIA elements.

Occurrence and Properties of Elements

Aluminum is the most common metallic element in the earth's crust. One of the most common commercial sources is the ore bauxite, containing Al_2O_3. It is purified by being dissolved in aqueous NaOH, with subsequent reprecipitation caused by bubbling CO_2 through the filtered $Al(OH)_6^{3-}$ solution:

$$Al_2O_3(s) + 6OH^- + 3H_2O \rightleftharpoons 2Al(OH)_6^{3-} \qquad (17.7)$$

after which solid impurities are filtered off:

$$2Al(OH)_6^{3-} + 6CO_2 \rightleftharpoons Al_2O_3(s) + 6HCO_3^- + 3H_2O \qquad (17.8)$$

The reprecipitated $Al_2O_3(s)$ is then dissolved in molten cryolite, Na_3AlF_6, at 800 °C to 1000 °C, giving the ions Al^{3+} and O^{2-}; the solution is then electrolyzed.* The relatively great stability of the Al(III) oxidation state in these ores requires a large electrical energy input for conversion to the metal (zero oxidation state).

$$2Al^{3+} + 3O^{2-} \rightleftharpoons 2Al(l) + \tfrac{3}{2}O_2(g) \quad (17.9)$$

This energy input is several times the energy required to melt aluminum metal and work it into the desired form. Hence, aluminum recycling is a great energy-saving process.

Aluminum is highly electropositive, yet it does not generally corrode badly. The reason for aluminum's resistance to corrosion is the tough oxide film that rapidly develops on a clean aluminum metal surface, preventing further reaction of the metal.

Gallium, indium, and thallium are found only in trace quantities in nature. They are generally obtained in the elemental state by electrolysis of aqueous solutions of their salts. An amazing and as yet inadequately explained property of gallium is its abnormally low melting point, 30 °C. Since gallium has a boiling point in the same range as the boiling points of the other Group IIIA metals, 2070 °C, it has the longest liquid range of any known substance, and like mercury, finds use as a thermometer liquid.

Compounds of Group IIIA Metals. Oxides

Oxides of aluminum are some of the most useful and beautiful of nature's compounds.† Aluminum oxide, Al_2O_3, exists in two crystalline forms: α-Al_2O_3, a hexagonally close packed array (Sec. 14.3) with Al^{3+} distributed symmetrically among the octahedral sites in the lattice; and β-Al_2O_3, which has Al^{3+} in both octahedral and tetrahedral sites in the lattice. The α-Al_2O_3 is a mineral called **corundum,** which is very useful as an abrasive. Powdered aluminum oxide is called **alumina.** Because of the highly polar structure of alumina, other polar molecules that come in contact with it tend to attach themselves to it. The strength and lifetime of such an attachment depends on the polarity of the polar molecule and other structural details. Different types of molecules have different tendencies to attach to alumina, and these differences can be used to separate different types of compounds from each other in a solution. This type of separation approach is the basis for an important class of techniques called chromatography.‡

---

*The electrode reactions for the electrolysis of Al_2O_3 in molten cryolite are

$$\text{anode:} \quad 2O^{2-} \rightleftharpoons O_2(g) + 4e^-$$
$$\text{cathode:} \quad Al^{3+} + 3e^- \rightleftharpoons Al(l)$$

†Pure, large-crystal corundum is called white sapphire. If trace amounts of Te^{2+}, Fe^{3+}, or Ti^{4+} were present as impurities in the corundum when it crystallized, the crystals formed are what are called blue sapphires. Similarly, a ruby results from Cr^{+3} as an impurity. All these gems can now be crystallized in laboratories to make the synthetic gems that have the same chemical compositions as the corresponding natural gems.

‡In chromatography a solution (gaseous or liquid) consisting of a solvent (gaseous or liquid) and a mixture of compounds to be separated is passed through a column containing a material such as alumina (called the stationary phase) to which molecules of the substances to be separated attach to different degrees. As the solution passes through the column, the substance with the stronger attachment proceeds more slowly, while the substance with weaker attachments progresses through the column faster and emerges first at the outlet of the column. In this way separation of the substances can be achieved. Depending on the nature of the carrier or the stationary phase, a wide variety of useful chromatographic techniques can be designed.

Group IIIA Compounds as Lewis Acids

In Chap. 8, the concept of Lewis acids and bases was presented. Lewis acids are substances that are electron-deficient, like H^+ and BF_3, while Lewis bases are electron-rich, like OH^- and NH_3. Compounds of trivalent Group IIIA elements, such as $AlCl_3$ and BCl_3, are electron-deficient and often act as Lewis acids. Thus, if solid NaCl is added to molten $AlCl_3$, the species $AlCl_4^-$ is readily formed. In this ion, the chloride ion from NaCl has donated an electron pair to be shared with the aluminum atom, completing the valence octet of the latter.

$$Cl^- + AlCl_3 \rightleftharpoons AlCl_4^-$$

The tendency to act as a Lewis acid depends on the ability of the electron-deficient atom to attract an additional electron pair. The stability of the Lewis acid-base adduct is greater the smaller the metal ion, that is, the greater the charge-to-radius ratio of the corresponding metal ion. Lewis acid-base adducts are particularly stable for compounds of aluminum and gallium (and of course beryllium) and are also known for indium and thallium. Formation of Lewis acid-base adducts is especially significant in an important reaction in synthetic organic chemistry, called the **Friedel-Crafts reaction**. This reaction is catalyzed by $AlCl_3$, or $GaCl_3$, and is believed to occur in the following steps:

$$R-\underset{\underset{O}{\|}}{C}-Cl + AlCl_3 \rightleftharpoons R-\underset{\underset{O}{\|}}{C}^+ + AlCl_4^-$$

$$R-\underset{\underset{O}{\|}}{C}^+ + C_6H_6 \rightleftharpoons C_6H_5-\underset{\underset{O}{\|}}{C}-R + H^+$$

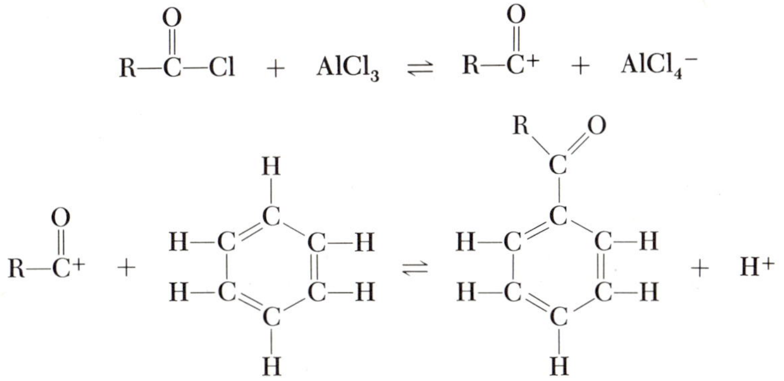

$$AlCl_4^- + H^+ \rightleftharpoons AlCl_3 + H^+ + Cl^-$$

The catalyst therefore is regenerated.°

Another way besides the Lewis acid-base complex formation in which compounds of Group IIIA metals become four-coordinate, satisfying the octet rule, is the linking of two MX_3 units, where X is a halogen; thus halogen atoms, and their electrons, are shared between two metal atoms. An example of this kind of compound is Al_2Br_6:

$$\begin{array}{c} Br \quad\quad Br \quad\quad Br \\ \diagdown \;\; \diagup \;\; \diagdown \;\; \diagup \\ Al \quad\quad Al \\ \diagup \;\; \diagdown \;\; \diagup \;\; \diagdown \\ Br \quad\quad Br \quad\quad Br \end{array}$$

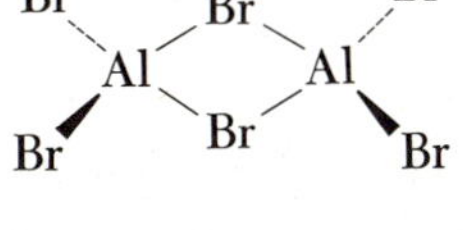

°A catalyst creates a reaction path that allows the reaction to proceed more rapidly. Although the equilibrium constant for the reaction is not changed, the equilibrium is achieved more rapidly (see Chap. 13).

In this structure each aluminum is bound to four bromine atoms and the geometry of each aluminum center is tetrahedral. Such dimers will generally dissociate in the presence of excess halide anion to form monomeric tetrahedral ions, such as the $AlCl_4^-$ mentioned above:

$$Al_2Br_6 + Br^- \rightleftharpoons 2AlBr_4^-$$

An important class of Lewis acid-base adducts that are synthetically useful are the complex hydrides of Group III metals. These hydrides have the formula MH_4^- and are often used as the lithium salts, for example $LiAlH_4$, lithium aluminum hydride. These compounds can be visualized as resulting from the addition of the hydride ion, H^-, which acts as a Lewis base, to the neutral hydride of the Group IIIA metal, for example, AlH_3, which acts as a Lewis acid. Since the ability of these elements in trivalent compounds to act as electron-pair acceptors decreases (because of atomic size) in the order B > Al > Ga, we expect that formation of BH_4^- would be the most favorable, and hence, that BH_4^- would be the most stable towards dissociation (into BH_3 and H^-), AlH_4^- next, and GaH_4^- least. In fact, BH_4^- is quite stable, existing as a discrete ion even in water, whereas AlH_4^- reacts rapidly in water, but is stable as solid $LiAlH_4$ up to a temperature of 120 °C; and $GaAlH_4$ reacts explosively with H_2O and decomposes slowly at 25 °C even as a solid. The most generally useful of those hydrides is AlH_4^-, generally used as $LiAlH_4$, since this compound is reasonably reactive but stable enough to keep in the laboratory for long periods. The MH_4^- species such as AlH_4^- are capable of providing a hydride ion in a reaction with some electron-deficient species, represented in the following symbolic equation by the symbol "ϕ":

$$AlH_4^- + \phi \rightleftharpoons AlH_3 + \phi H^-$$

Since the hydride ion carries with it two electrons that are used in bonding to ϕ in the species ϕH^-, the net result is a reduction in some atom in ϕ. Hence, the MH_4^- species are considered reducing agents.

The reduction involves addition of H^- (oxidation state -1) to a more electronegative element, changing the assigned oxidation state of the hydrogen atom to $+1$ and causing a corresponding decrease in the oxidation state of the more electronegative atom, for example, C, N, or O. A useful application of reduction by these metal hydrides species is the reduction of carboxylic acids to alcohols. The net reaction, a manifold process, is

$$2R\overset{\overset{\displaystyle O}{\|}}{C}{-}OH + LiAlH_4 + 2H_2O \rightleftharpoons Al(OH)_3 + LiOH + 2R{-}CH_2OH \qquad (17.10)$$

Note that the oxidation state of carbon in a carboxylic acid, such as formic acid, $H\overset{\overset{\displaystyle O}{\|}}{C}{-}OH$, is 2, whereas for the corresponding alcohol, CH_3OH, the oxidation state of carbon is -2. The change in oxidation state for each carbon is -4, corresponding to a reduction.

Spotlight

Carbon, silicon, and tin exist in allotropic forms, carbon as diamond and graphite, and silicon as amorphous or crystalline silicon. The ordinary form of tin is called white tin, or β-tin. When β-tin is cooled below 13 °C, it changes to gray tin, or α-tin, the rate of change having its maximum at −48 °C. Gray tin is interesting because it exists as a gray powder, and numerous cases have been reported in which tin objects have crumbled to a gray powder in winter. The change is called tin pest, and the "disease" manifests itself first as a tarnishing of the tin, followed by wartlike formations, and finally a crumbling to powder. Tin disease seems to be "contagious" and the change is more rapidly induced when α-tin touches β-tin.

17.4 Group IVA: C, Si, Ge, Sn, and Pb

General Considerations

Since about half of all chemists are organic chemists, many chemists are most familiar with the chemistry of carbon compounds. Group IVA is thus the most commonly cited example of the discontinuity in chemical properties of the first long period relative to subsequent periods. The chemistries of silicon, germanium, tin, and lead vary smoothly and more or less predictably, but the chemistries of these elements vary radically from the chemistry of carbon.*

Metallic Character

Group IVA elements, like other groups, have ionization potentials that are lower as the size of ionic radius is larger. (See Table 17.4.) Hence, a greater metallic character is manifested by the lower members of the group. Silicon is almost totally

TABLE 17.4 Some Fundamental Properties of Group IVA Elements

Element	*Electronic Configuration*	*Covalent Radius, Å*	*Ionic Radius,*[a] *Å*	*Ionization Potentials, eV*			
				1st	*2nd*	*3rd*	*4th*
C	$[He]2s^2 2p^2$	0.77	.....	11.264	24.37	47.86	64.48
Si	$[Ne]2s^2 3p^2$	1.18	.....	8.149	16.34	33.46	45.13
Ge	$[Ar]3d^{10}4s^2 4p^2$	1.23	Ge^{4+} 0.54	7.809	15.86	34.07	45.5
Sn	$[Kr]4d^{10}4s^2 5p^2$	1.41	Sn^{2+} ? Sn^{4+} 0.71	7.332	14.63	30.6	39.6
Pb	$[Xe]4f^{14}5d^{10}6s^2 6p^2$	1.75	Pb^{2+} 1.12 Pb^{4+} 0.78	7.415	15.03	32.0	42.3

[a]In most cases, these "ions" have significant covalent character. The most ionic of these is Pb^{2+}.

*Mendeleev found several blank spaces in the periodic table he developed, and in 1869 he predicted the existence and properties of germanium, which he called eka-silicon. The element was discovered 19 years later, and because its properties were found to be those Mendeleev had predicted, confidence in the validity of the periodic table was strengthened.

nonmetallic, germanium is a metalloid, acting as either a metal or a nonmetal; and tin and lead are metallic.

An important characteristic of Group IVA elements, which is also a persistent periodic trend of several groups, is the increasingly greater stability of lower oxidation states with higher atomic number. At first this may seem contradictory, since it is known that ionization potentials are lower with higher atomic number within a group. The important factor to consider, however, is the difference between *oxidation state* and *ionic charge.* In compounds of silicon and germanium with oxidation state +4, the bonds are very much covalent. Four valence electrons of silicon and germanium are shared with other atoms, and the ionic character is not high. Hence, the +4 oxidation state of the Si or Ge does not mean that a +4 charge resides on these atoms; and we need not be concerned with the first four ionization potentials of these elements in rationalizing the stability of these compounds. Although the bonds in compounds in which tin and lead have a +4 oxidation state may be higher in ionic character than the corresponding bonds in analogous silicon and germanium compounds, they have significant covalent character nonetheless. There are compounds in which tin or lead is in the +2 oxidation state, and in which the bonds to these elements are mostly ionic. In such cases the +2 oxidation number also represents an ionic charge, not just an oxidation state.

As indicated above, it is easy enough to ionize Sn and Pb to the +2 state that these elements can act essentially as +2 ions. It is significant that in Group IVA, as well as other groups, the difference in ionization potentials between the *p* subshell and the *s* subshell within the valence shell is great enough that only the *p* electrons are removed readily to produce stable ions. This is why tin and lead can exist as stable +2 ions (two 5*p* electrons removed) but not as stable +4 ions (two 5*p* and two 5*s* electrons removed). Thus, unlike the elements of Groups IA and IIA, these ions do not achieve the noble gas configuration.

Stable ions of the transition metals also generally fail to show a noble gas configuration. The concept of the noble gas configuration is still useful, however, since it defines the limiting ionic charge or oxidation state as the highest ionic charge or oxidation number that we might even wish to consider. That is, for stable compounds we do not find an ionic charge or oxidation state so large that it would involve removing electrons from inside the valence shell. The maximum oxidation number found for Group IVA is +4.

Spotlight

Metallic tin finds extensive use in the manufacture of "tin cans" (tin-plated steel) and in important alloys such as solder and pewter. An allow of tin and niobium promises to be important in manufacturing "superconducting" magnets, that is, magnets that can generate and maintain enormous magnetic fields but use practically no power. A superconducting magnet weighing only a few pounds that is started with a small battery and then disconnected from the battery, is comparable to a 100-ton electromagnet operated continuously from a large power supply.

Metallic lead is an important constituent of "lead-acid" batteries, and is used in the manufacture of alloys such as solder. Compounds of lead were used widely in paints, but their use has been discontinued because of their poisonous nature.

A Special Property of Silicon and Germanium

Silicon and germanium have some very special properties that are due to their position in Group IV, where they are neither so covalent as carbon nor so metallic as tin and lead. The most important of these properties is that they are semiconductors (Chap. 14). That is, the electrical conductivity of silicon and germanium increases exponentially with temperature. The conductivity is very sensitive to trace impurities in the solid forms of the elements. The revolution in electronics that allows almost incredible miniaturization and very low current requirements has been made possible by semiconductor solid state devices.

Structures of Group IVA Compounds

Most of the structures of Group IVA compounds are predicted correctly using VSEPR Theory (Sec. 7.3). Thus, covalent compounds of silicon in the oxidation state, for example, $Si(CH_3)_4$ and $SiCl_4$, have four pairs of electrons in the silicon valence shell and a tetrahedral structure—like that found for corresponding carbon compounds. Unlike carbon, however, the heavier elements of Group IVA have low-lying *d* orbitals that can accommodate additional electrons. Thus, a Cl^- can donate a pair of electrons into a covalent bond with tin in $SnCl_4$, giving rise to $SnCl_5^-$. In this compound, there are five pairs of electrons in the valence shell, which gives rise to a trigonal bipyramidal structure

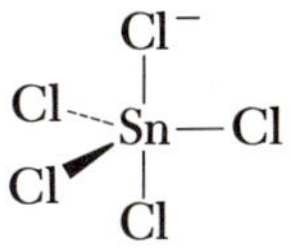

The ion $SnCl_3^-$ is also known. Its structure is pyramidal with one tetrahedral site occupied by a lone pair of electrons

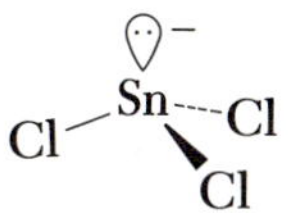

Octahedral geometry is found for ions such as SiF_6^{2-} and $SnCl_6^{2-}$,

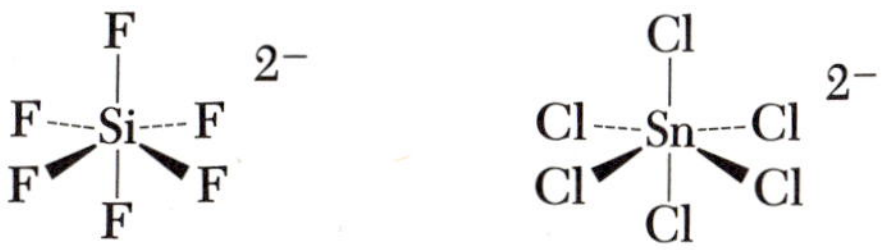

as there are six pairs of electrons in the valence shell.

Multiple Bonding

A very important feature in carbon chemistry is carbon's ability to form stable $p\pi$–$p\pi$ double bonds with other carbon atoms, and with oxygen or nitrogen atoms. The strength of the *p* bond depends, of course, on the amount of overlap of the *p* orbitals of the two atoms forming the bond. Silicon, germanium, tin, and lead show

Figure 17.5
An illustration of the reason that π bonding doesn't occur for the heavier Group IV elements—namely, the very small p orbital overlap.

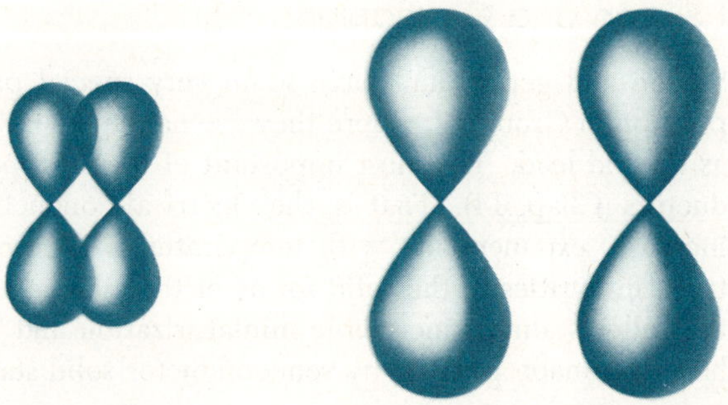

very little or no tendency to form $p\pi$ bonds, either with other atoms of the same element or with atoms of other elements. This may be understood in terms of the greater size of these elements, making p orbital overlap with other elements too small to allow stable bond formation. This situation is represented in Fig. 17.5. The more diffuse empty d orbitals of these elements are, however, large enough to provide effective overlap for π bonding. In compounds of these elements in which $d\pi$–$p\pi$ bonding is exhibited, another element, such as N, donates a pair of electrons from a filled p orbital into the empty d orbital of the Group IVA element. The resulting bond is reasonably stable because the d orbitals of these elements are of low energy; in contrast, there are no d orbitals available in the valence shell of carbon, nitrogen, or oxygen.

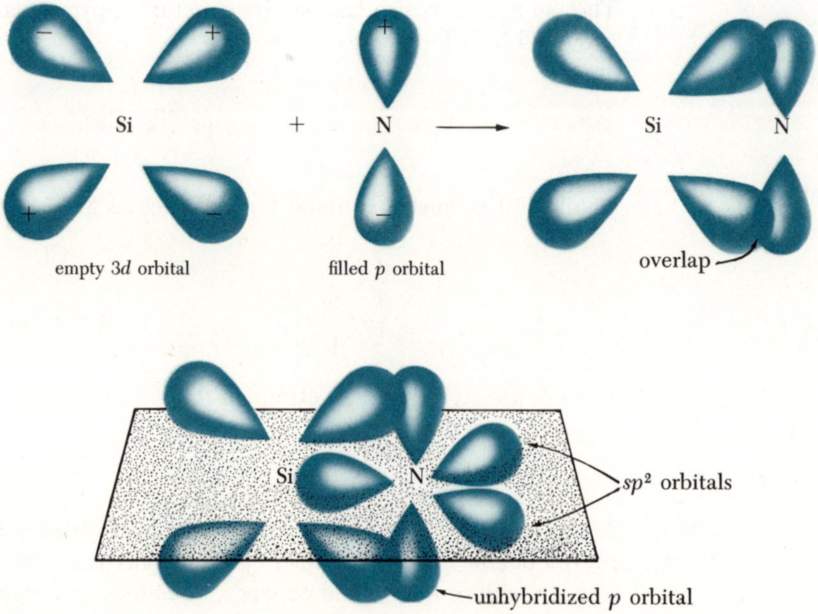

A compound in which such bonding is found is trisilylamine,

$$
\begin{array}{cc}
(Me)_3Si & (Me)_3 \\
\diagdown & \diagup \\
N-Si & \\
| & | \\
Si-N & \\
\diagup & \diagdown \\
(Me)_3 & Si(Me)_3
\end{array}
$$

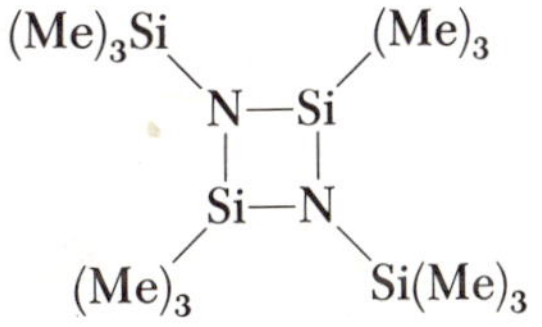

which has a planar configuration about the nitrogen atoms (sp^2 hybridization, with the remaining filled p orbital used in $d\pi$–$p\pi$ bond formation with silicon) rather than pyramidal geometry (sp^3 hybridization) as found in $N(CH_3)_3$. The Si—N bond lengths are shorter in compounds that are proposed to have $d\pi$–$p\pi$ bonding than in compounds that exhibit only Si—N single bonds.

Compounds of the Group IVA Elements

Silicon is one of the predominant elements of the earth's crust, about 28 percent; and it is found mainly in silica, SiO_2, and silicate compounds, that is, species containing silicate ions (SiO_4^{4-}, $Si_2O_7^{6-}$, and so on). The common forms of silica are quartz and cristobalite. In both of these compounds, the silicon atom is tetrahedrally coordinated to four oxygen atoms and the oxygen atoms are shared by two silicon atoms. Cristobalite has a diamondlike structure, with silicon atoms forming a tetrahedron roughly analogous to the carbon arrangement in diamond (Fig. 17.6), but with an oxygen atom between each pair of silicon atoms. Quartz has helices of Si—O linkages. The helices may be "right-handed" or "left-handed" and crystals of each type can be easily separated from one another.

The separation of crystals of "right-handed" and "left-handed" substances from one another was first accomplished in 1848 by Louis Pasteur. In the course of a crystallographic study of tartaric acid, he observed two different kinds of crystals. Using a magnifying lens and a pair of tweezers, he separated the crystals into two

Spotlight

The other Group IVA elements also tend to occur in nature in the form of their oxides, except for lead, which occurs generally as the sulfide. All these elements are usually prepared by the reduction of their oxides with carbon, for example,

$$SiO_2(s) + 2C(s) \rightleftharpoons Si(s) + 2CO(g)$$

The oxides and hydroxides of Ge, Sn, and Pb are amphoteric (see Sec. 9.3) but SiO_2 is almost purely acidic. Hydrated silicon dioxide, $SiO_2 \cdot 2H_2O$, acts as the acid orthosilicic acid, H_4SiO_4; but SiO_2 reacts with concentrated HF to form SiF_4, which is a gas. HF is the only common acid that attacks glass, which contains silicates.

In addition to glass, SiO_2 is extensively used in the manufacture of cement, ceramics (which include pottery, bricks, tile, and porcelain), and "silicones."

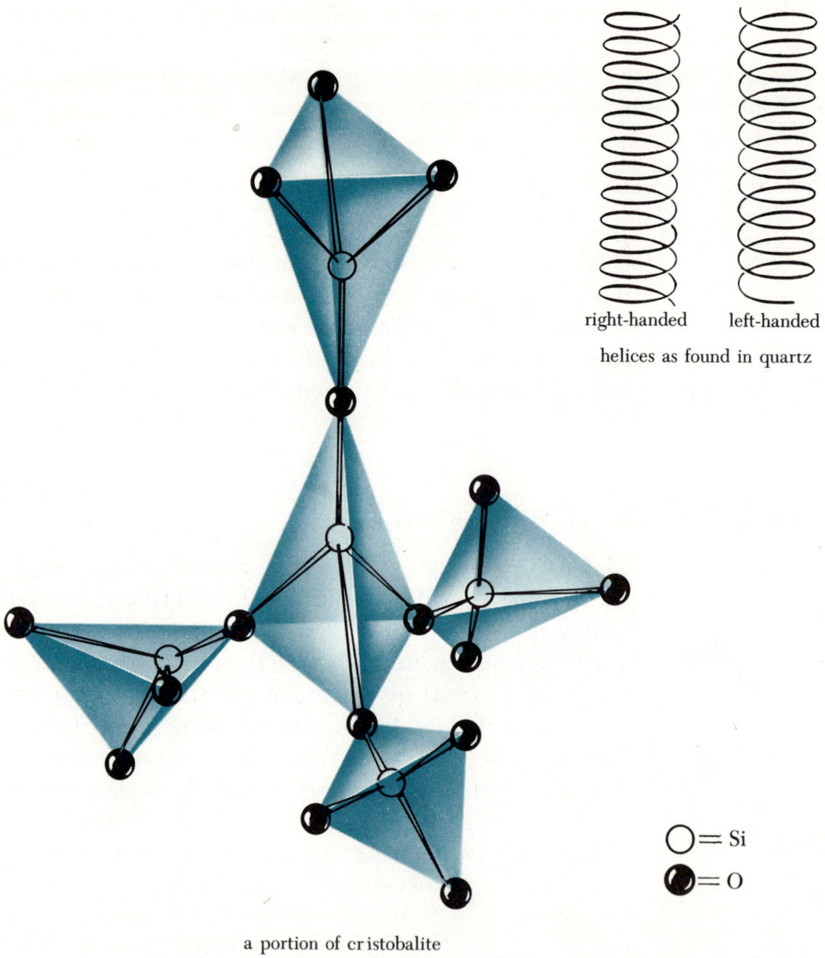

Figure 17.6
Cristobalite and quartz structures.

groups. When he dissolved one of the lots of crystals in water, the solution caused rotation of plane-polarized light in a right-handed direction. A solution of the second lot of crystals caused a rotation of equal magnitude but in the left-handed direction.

Silicates vary greatly in structure. The basic structural unit is the SiO_4^{4-} tetrahedron. This simple building block may form compounds by occurring singly as a polyatomic ion, by sharing oxygen atoms to form small groups, such as $Si_2O_7^{6-}$, or by forming many-unit polymers that are chainlike, cyclic, or in sheets. These three situations are shown in Fig. 17.7.

The various structures of silicate minerals have striking effects on the physical properties of the minerals. For example, a linear chain silicate should have strength preferentially along one direction, the direction along the chain, like a fiber. This type of structure is found in minerals known as amphiboles, such as various asbestos minerals that are indeed fibrous, an unusual feature for minerals in general. A sheet type of silicate structure with cations between the sheets should be easily cleaved into thin sheets. Mica is a well-known example of a mineral with a sheet silicate structure.

Figure 17.7
Silicate anion structures.

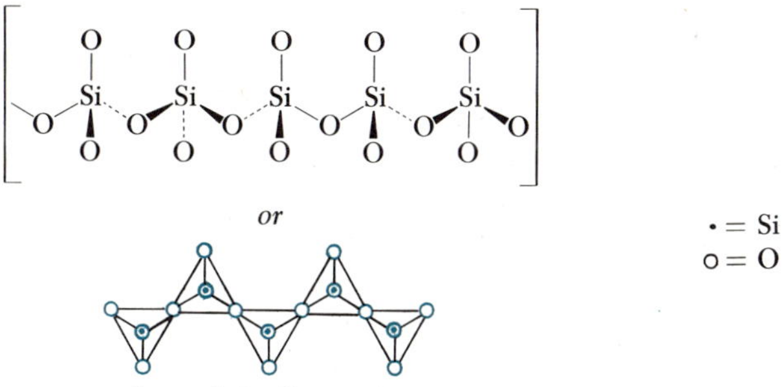

a linear chain silicate anion
found in the class of minerals called pyroxenes

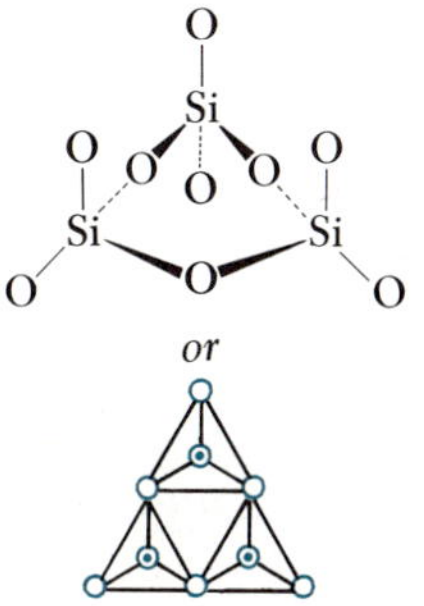

$Si_3O_9{}^{6-}$

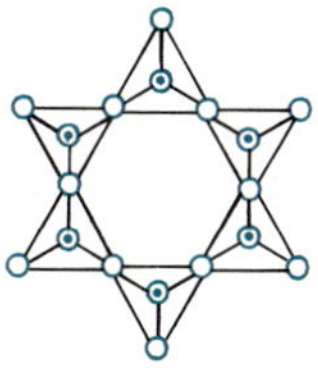

$Si_6O_{18}{}^{12-}$

cyclic silicate anions
found in benitoite ($BaTiSi_3O_9$) *and beryl* ($Be_3Al_2Si_6O_{18}$)

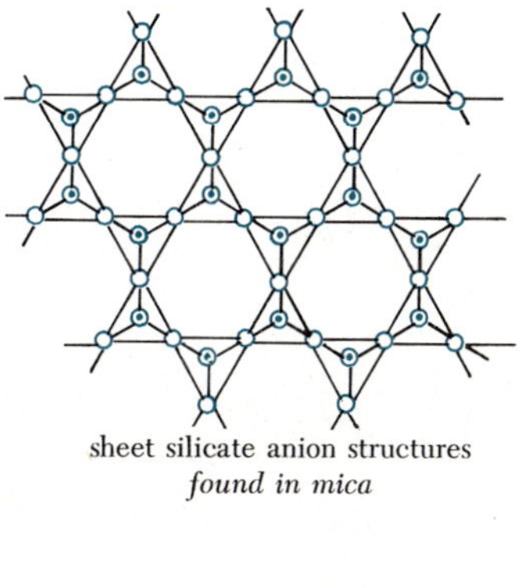

sheet silicate anion structures
found in mica

The next step in complexity of linking of silicon-oxygen bonds is a three-dimensional, or "framework," structure. The mineral silica has a fully repetitious three-dimensional structure of this kind. In many minerals, there are interruptions in the silica-type structure, leaving negatively charged oxygen atoms, so that cations must be present to balance the charge. Some very common minerals of this type are the feldspars and zeolites. The zeolites contain holes, channels, or cavities in which cations reside. The cations in the zeolite channels can often be rapidly replaced by other cations, leading to the usefulness of the zeolites as ion-exchangers. The cavities can also hold small molecules. A zeolite with cavities of the proper size can be used as a "molecular sieve" to selectively absorb particular gases or liquids. Sieves with a cavity diameter of about 4 Å will selectively absorb water from other solvents. This property makes these sieves very useful in chemistry, since many reactions must be undertaken without any water present and other methods of water removal are often inconvenient or inefficient. Many of the zeolites can now be made synthetically, and experimental conditions can be carefully controlled to make a compound with the particular desired properties—like proper cavity size.

Of the many other types of compounds formed by Group IVA elements, perhaps the most familiar are the halides of tin, including stannous fluoride, SnF_2, an important active ingredient in toothpastes,* and tetraethyllead, $Pb(CH_2CH_3)_4$, the organometallic compound of lead, which is the controversial antiknock additive in gasoline.

Both tetraethyl lead and tetramethyl lead, $Pb(CH_3)_4$, are used as antiknock agents in gasoline. A principal commercial synthesis is given by the reaction

$$\underset{\text{lead-sodium alloy}}{4NaPb} + \underset{\text{ethyl chloride}}{4CH_3CH_2Cl} \rightleftharpoons Pb(CH_2CH_3)_4 + 3Pb + 4NaCl \qquad (17.11)$$

The tetralkyllead compounds $Pb(CH_2CH_3)_4$ and $Pb(CH_3)_4$ are covalent, nonpolar, highly toxic liquids. They constitute the lead component of gasoline that is responsible for the lead pollution caused by automobile emission.

*Stannous fluoride is one of the most effective cavity-preventing compounds known. Its beneficial effect arises from its ability to ionize in aqueous solution to give F^-. The fluoride ion combines with tooth enamel to reduce the deleterious effect of acids (lactic, acetic, and pyruvic) found in the mouth and to reduce effects of bacteria on tooth enamel.

Spotlight

It is often desirable to exchange one ion for another in a solution. For example, hard water is caused by the presence of such ions as Mg^{2+}, Ca^{2+}, and Fe^{3+}. The Na^+, in contrast, does not lead to formation of salt deposits on pipes or reduce the effectiveness of laundry detergents. Zeolites allow one to "trade" Na^+ for Mg^{2+}, Ca^{2+}, and Fe^{3+} in a water supply. We need merely fill the zeolite cavities with Na^+, then allow the hard water to pass through the solid mineral. The ions Mg^{2+}, Ca^{2+}, and Fe^{3+}, since they are more highly charged, are generally attracted much more strongly to the negatively charged cavities in the zeolite than Na^+ is. Hence, these ions take the place of the Na^+ in the zeolite and release Na^+ into the water. The net effect is the "softening" of water.

Spotlight

The main reason for the use of tetralkyllead compounds is that they improve the combustion properties of gasolines with lower "octane" ratings. Gasoline is composed of several hydrocarbons that result from the distillation of crude oil. Smooth combustion is promoted by several branched hydrocarbons, such as iso-octane:

$$CH_3CH_2CH_2CH_2CH_2CH_2CH_2CH_3CH$$

n-octane
no branching

$$\begin{array}{c} H_3C \\ \diagdown \\ CHCH_2CH_2CH_2CH_2CH_3 \\ \diagup \\ H_3C \end{array}$$

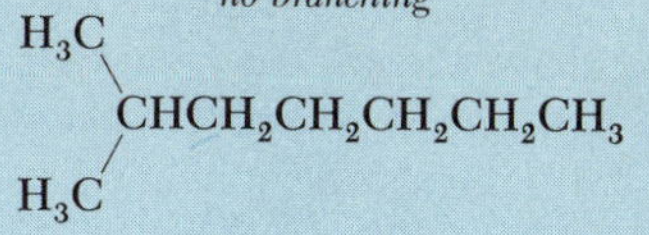

iso-octane
single branching

$$\begin{array}{c} H_3C CH_2 \\ \diagdown | \\ CH-CH-CH_2CH_2CH_3 \\ \diagup \\ H_3C \end{array}$$

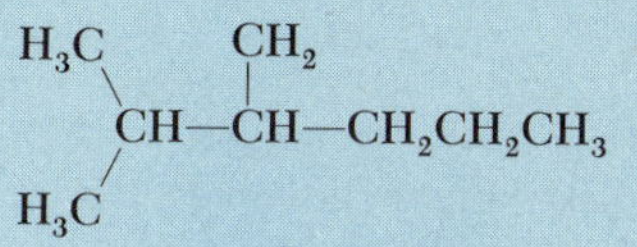

2,3-dimethylhexane (or 3-methylisoheptane)
multiple branching

The synthesis of branched hydrocarbons to improve the combustion properties of a gasoline solution is significantly more expensive than the addition of tetralkyllead (which is added in amounts of about 3–5 g per gallon of gasoline). Smooth combustion is more critical in engines in which there is a large compression of the gasoline-air mixture before ignition, in other words, high-compression engines. Thus, the largest expenses involved in the move toward lead-free automobile fuels is in the premium grades of gasolines, which are required by high-compression engines.

17.5 Group VA: N, P, As, Sb, and Bi. The Pnictides

General Considerations

The differences in the chemistries of nitrogen and the other Group VA elements are analogous to the differences between carbon and the other Group IVA elements. Similar trends in some properties are evident in comparing Table 17.5 with Tables 17.3 and 17.4. As we have seen, carbon atoms forms $p\pi$–$p\pi$ bonds with other carbon atoms and with the atoms of other second row elements, for example, N and O. Nitrogen also forms π bonds by p-orbital overlap. The larger elements of Group VA,

TABLE 17.5 Some Fundamental Properties of the Pnictides (Group VA)

Element	Electronic Configuration	Covalent Radius, Å	Ionic Radius, Å, −3 Ion	Ionization Potentials, eV 1st	2nd	3rd	4th
N	$[He]2s^22p^2$	0.73	.....	14.53	29.62	47.01	77.45
P	$[Ne]3s^23p^3$	1.10	.....	11.01	19.65	30.16	51.34
As	$[Ar]3d^{10}4s^24p^3$	1.21	.....	9.80	20.21	28.32	50.30
Sb	$[Kr]4d^{10}5s^25p^3$	1.41	0.92	8.64	16.48	25.28	44.23
Bi	$[Xe]4f^{14}5d^{10}6s^26p^3$	1.52	1.08	7.29	16.68	25.56	45.10

Spotlight

Arsenic, antimony, and bismuth occur free in nature, but only to a small degree. Like the Group IVA metals, they are prepared by the reduction of their oxides with carbon. Phosphorus, however, does not occur free in nature, and is prepared by the reduction of phosphate rock with coke and silica:

$$2Ca_3(PO_4)_2(s) + 6SiO_2(s) + 10C(s) \rightleftharpoons P_4(g) + 10CO(g) + 6CaSiO_3(s)$$

Phosphorus, arsenic, and antimony form allotropes. The two common allotropes of phosphorus are the "yellow" and "red" forms. Yellow phosphorus is highly poisonous and chemically very reactive. Because it is spontaneously flammable in air above 30 °C, it is usually kept under water. In air in the dark, it emits an eerie pale green light; the glow has been the subject of much investigation, but is not fully understood. Red phosphorus is much less reactive and not poisonous; it is used in the striking surfaces for "safety" matches.

Arsenic is used in the manufacture of insecticides, for example, lead arsenate, $Pb_3(AsO_4)_2$, and medicinals, like Salvarsan. Antimony is an important constituent of alloys and is used in certain medicinals, for example, "tartar emetic," or potassium antimonyl tartrate, $K(SbO)C_4H_4O_6$. Bismuth forms important low-melting alloys; the principal one melts at 100 °C and is used in the construction of water sprinklers in automatic fire extinguishing systems.

however, have such poor *p*-orbital overlap that we do not find $p\pi$–$p\pi$ double-bonding in compounds of these elements. Like silicon, however, the larger elements, phosphorus, arsenic, antimony, and bismuth, may have their low-energy *d* orbitals used in forming $d\pi$–$p\pi$ bonds.

The presence of empty *d* orbitals in the valence shells of the large Group VA elements, like what is also found in the larger elements of Groups III-VIIA, in general leads to another important consequence—the larger Group VA elements can bond to more than four other atoms. The maximum number of bonds to nitrogen in stable species is found to be four—for example, in NH_4^+ and $N(CH_3)_4^+$. Phosphorus is found to form five (PF_5) or even six (PF_6^-) stable bonds. How can we explain such a "valence expansion" for the heavier elements? To answer this question, let us first look at the electron configurations of neutral nitrogen and phosphorus atoms in their ground states:

$$\text{nitrogen: } [\text{He}]2s^22p^3 \quad \text{or} \quad [\text{He}]\ \underset{2s}{\underline{\uparrow\downarrow}}\ \underbrace{\underline{\uparrow}\ \underline{\uparrow}\ \underline{\uparrow}}_{2p}$$

$$\text{phosphorus: } [\text{Ne}]3s^23p^3 \quad \text{or} \quad [\text{Ne}]\ \underset{3s}{\underline{\uparrow\downarrow}}\ \underbrace{\underline{\uparrow}\ \underline{\uparrow}\ \underline{\uparrow}}_{3p}$$

It appears that each of these elements could form only four bonds by using the *s* orbital and the three *p* orbitals. They could form three bonds by sharing one electron in each of three orbitals and one additional bond by donating a pair of electrons to another atom, say to form a two-electron bond with H^+. Phosphorus, however, has another set of orbitals in its valence shell. Thus, the valence shell of a neutral phosphorus atom in its ground state should be represented as

$$\text{phosphorus: } [\text{Ne}]\ \underset{3s}{\underline{\uparrow\downarrow}}\ \underbrace{\underline{\uparrow}\ \underline{\uparrow}\ \underline{\uparrow}}_{3p}\ \underbrace{\underline{\ }\ \underline{\ }\ \underline{\ }\ \underline{\ }\ \underline{\ }}_{3d}$$

If more stabilization (a reduction in energy) can be gained by forming additional bonds, in PF_5, for example, the $3d$ orbitals can become involved. Thus, considering the situation of a neutral phosphorus atom, we can imagine that an electron can be "promoted" from the $3s$ orbital to a $3d$ orbital:

$$\text{phosphorus:} \quad [\text{Ne}] \quad \underset{3s}{\uparrow} \quad \underset{3p}{\uparrow \uparrow \uparrow} \quad \underset{3d}{\uparrow \_\_\_\_}$$

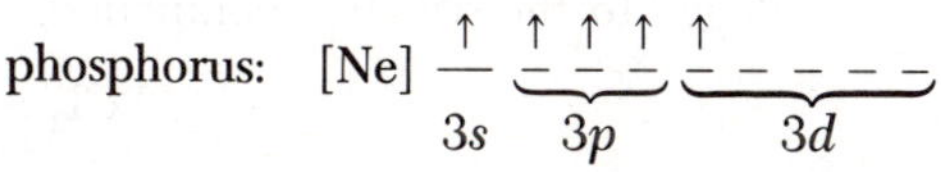

giving five orbitals that are only half-filled and available for forming five covalent bonds. This option is not open to nitrogen since it has no empty orbitals in its valence shell, and the energy required for promotion to the next higher shell is too great to be compensated for by the formation of two additional bonds.

For this reason, compounds and polyatomic ions such as PF_5, PF_4Cl, $P(OCH_3)_5$, $SbBr_6^-$, $P(CH_3)_6$, SbF_5, and $Sb(C_3H_8)_5$ are well known, but corresponding stable species with five or six bonds to a nitrogen atom are not found. This tendency for valence expansion of heavier elements is, of course, not limited to Group VA. The availability of low-energy d orbitals can be used to explain the existence of a variety of compounds in Group IIIA through Group VIIIA, such as AlF_6^{3-}, $SnCl_5^-$, SF_6, BrF_5, and XeF_6.

Compounds of the Group VA Elements

Hydrides

All the Group VA elements form trihydride compounds that are gases at 25 °C. All except BiH_3 are known by trivial names: NH_3, ammonia; PH_3, phosphine; AsH_3, arsine; SbH_3, stibine. The compounds PH_3, AsH_3, and SbH_3 are relatively stable and very poisonous. They are pyramidal with H—P—H, H—As—H, and H—Sb—H angles of 93.5°, 91.8°, and 91.3°, respectively.

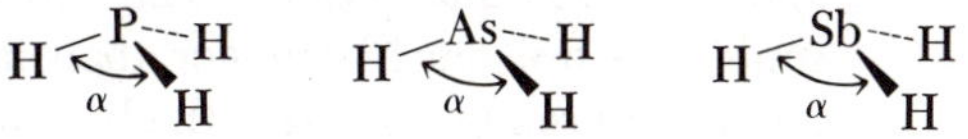

$$\alpha = 93.5° \qquad \alpha = 91.8° \qquad \alpha = 91.3°$$

Phosphorus, arsenic, and antimony are significantly less electronegative than nitrogen so that the central atoms on P, As, or Sb in PH_3, AsH_3, or SbH_3 have a smaller excess electron density than nitrogen does in ammonia. Hence, PH_3, AsH_3, and SbH_3 are much weaker bases than NH_3.

Halides

The heavier Group VA elements form both trihalides, of which PF_3, $AsCl_3$, and SbI_3 are examples, and pentahalides, of which PF_5, PF_3Cl_2, and AsF_2Cl_3 are examples. The trihalides can be made by adding elemental halogen to an excess of the Group VA element. These compounds are volatile and are pyramidal in the gas phase, with a lone pair of electrons occupying one tetrahedral site.

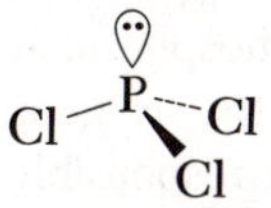

They are hydrolyzed rapidly by water, producing HCl, H_3PO_3, and $H_4P_2O_5$; the relative amounts of the acids generated, or the exact stoichiometry of the hydrolysis, depend on reaction conditions. The pentahalides are formed by adding an excess of the elemental halogen to the Group VA element. They can also be made by adding halogen to the trihalide compound:

$$PCl_3 + F_2 \rightleftharpoons PCl_3F_2 \tag{17.12}$$

The pentahalides have a trigonal bipyramidal geometry. If the compound is a mixed halide, the more electronegative (and smaller) halide will occupy an axial position, as shown herewith for the compounds PCl_4F, PCl_3F_2, and PCl_2F_3:

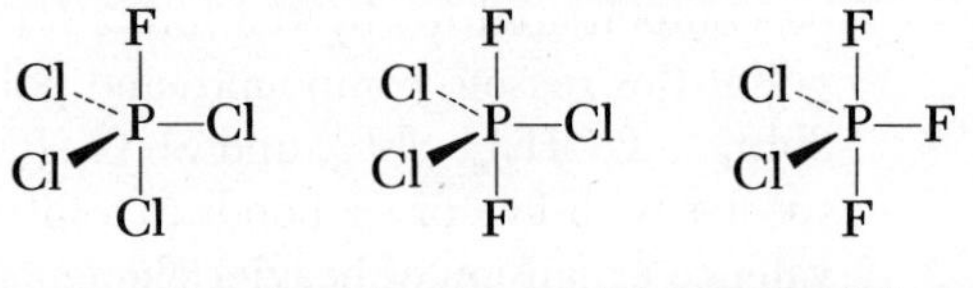

You may recall from Sec. 7.3 that the trigonal bipyramidal structure is predicted by VSEPR, since in the pentahalide compounds there are five pairs of electrons in the valence shell of the Group VA element; we can think of this (for bookkeeping purposes) as arising from five valence-shell "promoted" phosphorus atoms and one from each of the five halogen atoms.

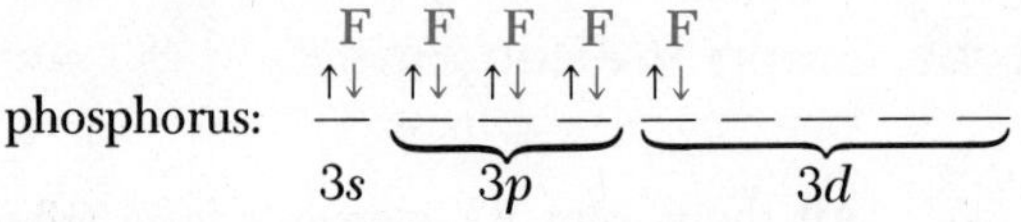

If there is more than one type of halogen, the more electronegative element occupies the axial position, because electronic charge in this position experiences more repulsion than at the equatorial positions. If we consider F—P—F angles, an axial group forms three 90° angles, whereas an equatorial group is involved in two 90° and two 120° interactions. The VSEPR scheme predicts that smaller angle (90°) repulsions are much stronger than larger (120°) repulsions. An electronegative ligand will tend to polarize the bonding electrons away from the central atom and will tend to decrease these larger repulsions; clearly a more electronegative ligand will be more effective in decreasing these repulsions than a less electronegative group.

Oxo Anions

The oxo acids and anions of phosphorus, meaning species containing only oxygen and phosphorus and in some cases hydrogen, are important compounds. Perhaps the most familiar species is phosphoric acid, a main component present in Coca-Cola and other soft drinks in the form of both H_3PO_4 and $H_2PO_4^-$. The phosphate ion, PO_4^{3-}, derived from phosphoric acid, is well known as an important and required biochemical species, which in excess causes pollution problems by stimulating overgrowth of algae in bodies of water.

Orthophosphoric acid, H_3PO_4, is tetrahedral, with three hydrogen atoms that are bound to the oxygen. These three hydrogens can be readily removed as ions, and hence are responsible for the ionization and acidic character of H_3PO_4.

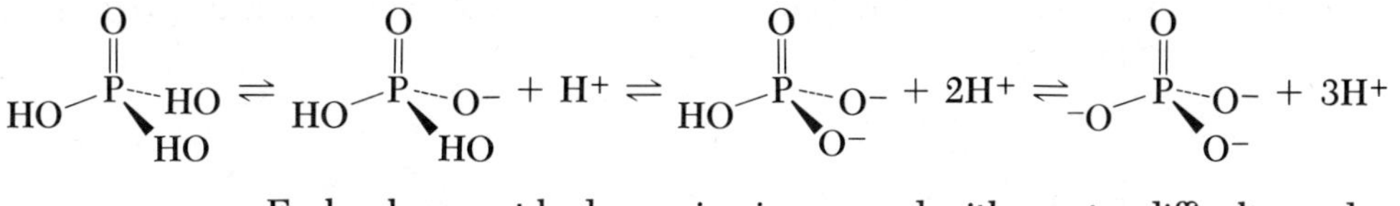

SEC. 17.5 Group VA: N, P, As, Sb, and Bi

$$\text{HO-P(=O)(OH)(HO)} \rightleftharpoons \text{HO-P(=O)(OH)(O}^-) + \text{H}^+ \rightleftharpoons \text{HO-P(=O)(O}^-)(\text{O}^-) + 2\text{H}^+ \rightleftharpoons {}^-\text{O-P(=O)(O}^-)(\text{O}^-) + 3\text{H}^+ \quad (17.13)$$

Each subsequent hydrogen ion is removed with greater difficulty, and concomitant greater addition of energy, since it must leave a more negatively charged ion to which it is electrostatically attracted. Each anion that is produced in the ionization of orthophosphoric acid is stabilized by resonance (Sec. 7.2). The ion PO_4^{3-} is an example:

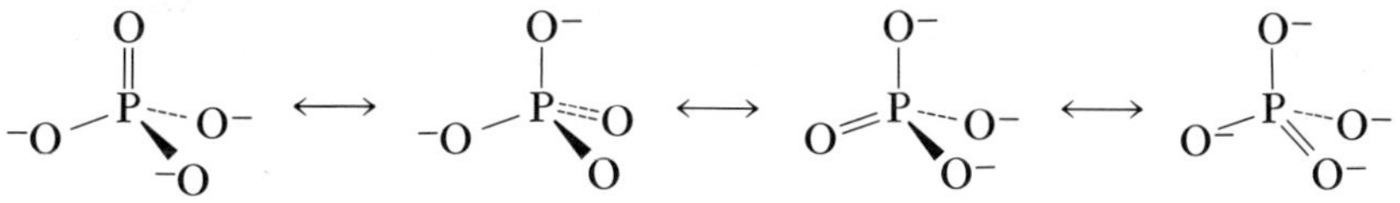

resonance contributors

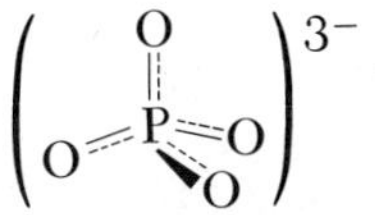

resonance hybrid

Some other phosphorus-containing acids have hydrogen atoms that are bound to the phosphorus atoms rather than to oxygen atoms, for example, phosphorous acid, H_3PO_3:

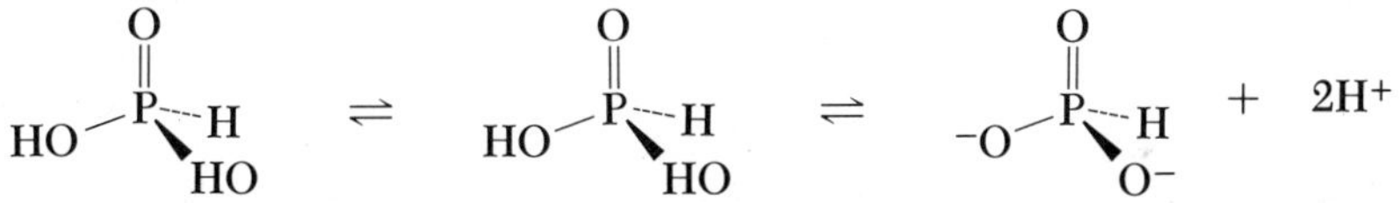

$$\text{HO-P(=O)(H)(HO)} \rightleftharpoons \text{HO-P(=O)(H)(O}^-) \rightleftharpoons {}^-\text{O-P(=O)(H)(O}^-) + 2\text{H}^+ \quad (17.14)$$

The hydrogen atom bond to the phosphorus atom in a phosphorous acid molecule is not easily removed as H^+, and therefore it is not an "acidic hydrogen." The anion produced by ionization of this acid is also resonance-stabilized.

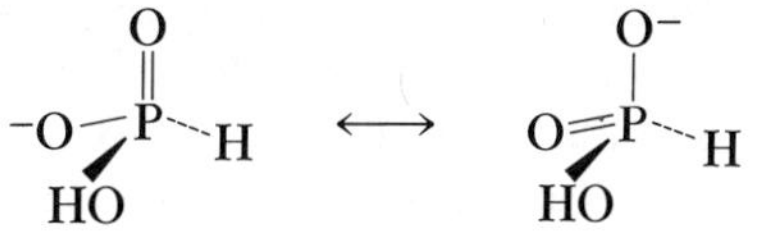

Phosphate groups can be linked to form diphosphates or triphosphates:

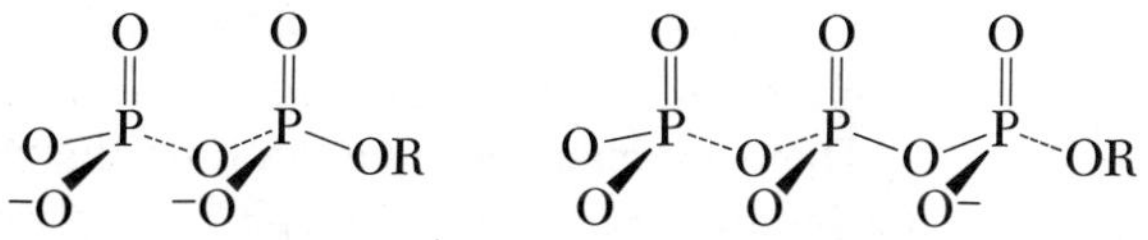

where R is a chemical group, for example, a nucleoside; see Eq. (17.15). Such species can undergo reactions in which the phosphate groups are separated from each other (cleaved), with each piece becoming solvated, a process that releases energy. Thus, the phosphate chain can be thought of as an energy storage group or energy reservoir. Many biochemical processes derive the energy they require from phosphate cleavage of adenosine triphosphate (ATP) to form adenosine diphosphate (ADP) and the phosphate ion:

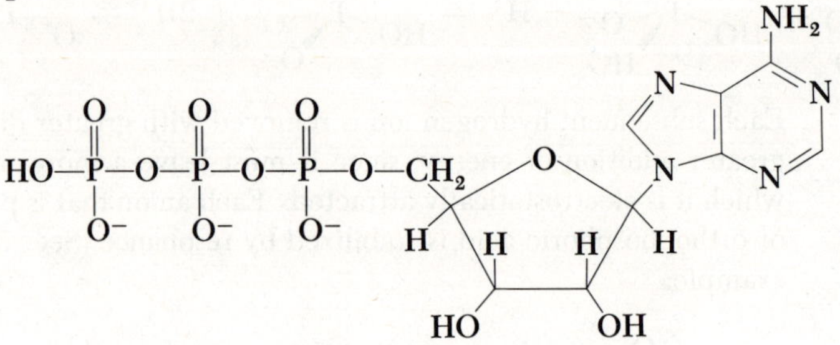

$+ \quad H_2O \rightleftharpoons$

ATP

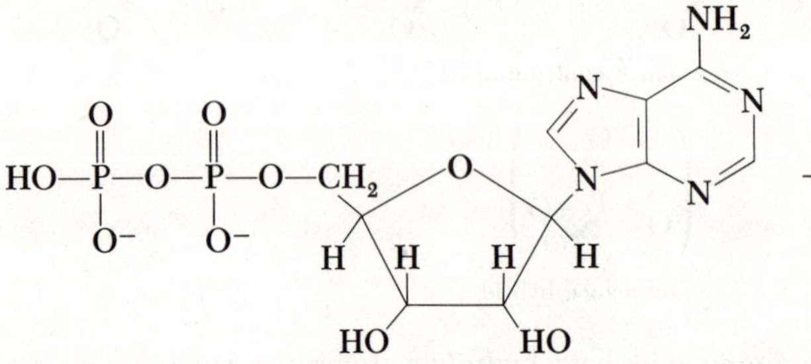

$$+ \quad H_2PO_4^- \qquad \Delta G° = -7 \text{ kcal/mol} \tag{17.15}$$

ADP

Phosphates as Pollutants; Eutrophication

The term *eutrophication* refers to the processes that occur in lakes and rivers when the supply of nutrients or foods increases from outside and usually unnatural sources. This increase in the supply of growth-supporting materials can lead to a large increase in the amount of algae and plankton that the aquatic ecosystem supports. In turn, this may lead to an increase in the populations of organisms that feed on the algae and plankton. Organisms consume oxygen to convert simple compounds into the complex substances involved in constructing structures. The decomposition of dead organisms, wherein organic compounds are oxidized, also consumes oxygen. Eventually, the combination of these processes may cause serious depletion of the oxygen content of the body of water, especially at depths where circulation and replenishing of O_2 from the surface is inefficient. The O_2 concentration may become so low that many organisms will die, and the ecosystems become greatly disturbed.

There are many nutrients that may be limiting factors in a particular ecosystem—for instance, phosphorus and nitrogen compounds, trace metal ions, and sulfur compounds. The Group VA elements nitrogen and phosphorus are often the limiting "growth factors" and may be supplied by fertilizer runoff or by detergents in sewer effluent. The phosphorus species that is easily assimilated into growth processes is the phosphate ion, which commonly is a large component in detergents, since it acts as a water-softener by forming soluble complexes with the ions Ca^{2+} and Mg^{2+}. The common phosphorus oxyanion salt used in detergents is sodium tripolyphos-

Spotlight

One of the oldest chemical processes used by man for synthesizing a useful product is soap making. Soap factories for large-scale production were common in the Middle Ages. The same process was used in these factories that had been used in homes for hundreds of years. Fats, or glycerides, are hydrolyzed by a base to produce a glycerol, which is water-soluble, and soap. The reaction is

The R in the formula below is typically a large organic group, containing several carbon atoms. Soap molecules can "surround" oil droplets with the organic end of the molecule next to the oil and the polar end extended into the water. Strong interactions between the polar ends of soap molecules and water allow the soap molecule itself to be water-soluble and soap solutions to support large quantities of fats and oils in solution or in a dispersed state.

$$Na^+ + OH^- + \begin{array}{c} CH_2-O-C-R \\ \| \\ O \\ CH-O-C-R' \\ \| \\ O \\ CH_2-O-C-R'' \\ \| \\ O \end{array} \rightleftharpoons \begin{array}{c} CH_2-OH \\ CH-OH \\ CH_2OH \end{array} + \begin{array}{c} RCOO^-Na^+ \\ R'COO^-Na^+ \\ R''COO^-Na^+ \end{array}$$

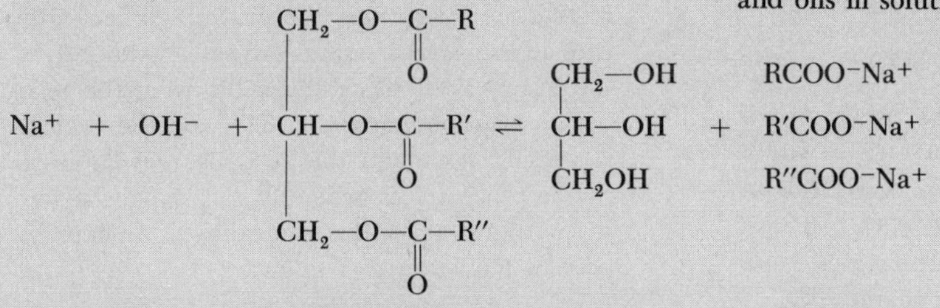

a glyceride (fat)   glycerol   soap

Ashes from various sources provided the base; for example, potash is KOH, marine alkali is mostly NaOH, and burnt wine lees give K_2CO_3. Animal tallow provided the fat for making soaps.

---

phate, $Na_5P_3O_{10}$, which also manifests the useful feature of hydrolyzing to form a basic solution:

$$P_3O_{10}^{-5} + H_2O \rightleftharpoons HP_3O_{10}^{4-} + OH^- \tag{17.16}$$

The basic character of the resulting solution neutralizes acids that may be found on the surfaces of grease and of solid matter one wants to remove with the detergent. Also, the excess OH^- converts fats ("fatty acids") to soaps by **saponification**. Unfortunately, the tripolyphosphate provides phosphate for use in an organism's metabolism (in algae, for instance) by a slow but persistent hydrolysis:

$$P_3O_{10}^{5-} + 4OH^- \rightleftharpoons 3PO_4^{3-} + 2H_2O \tag{17.17}$$

Spotlight

Effluents can be treated to remove phosphorus in several ways:

1. Bacterial treatment (secondary treatment) of sewage removes about 30% of the phosphorus. In this treatment, bacteria metabolize some phosphorus-containing species.
2. Insoluble phosphate species can be formed by using several readily available compounds; the insoluble phosphate compounds are then allowed to settle to the bottom of a holding pond, contributing to the sludge.
   a. Aluminum sulfate, $Al_2(SO_4)_3 \cdot 18H_2O$ is common for this purpose. It functions by first producing Al^{3+} by ionization, followed by the reaction

$$Al^{3+} + PO_4^{3-} \rightleftharpoons AlPO_4(s)$$

The product is the insoluble $AlPO_4(s)$, which settles out.

Continued on p. 820

b. Sodium aluminate, $NaAlO_2$, is also used:

$$H_2O + H^+ + AlO_2^- \rightleftharpoons Al^{3+} + 3OH^-$$

$$Al^{3+} + PO_4^{3-} \rightleftharpoons AlPO_4(s)$$

c. Calcium oxide (lime), which reacts as follows, is also used.

$$CaO(s) + H_2O \rightleftharpoons Ca(OH)_2(s)$$

$$Ca(OH)_2(s) \rightleftharpoons Ca^{2+} + 2OH^-$$

$$3Ca^{2+} + 2PO_4^{3-} \rightleftharpoons Ca_3(PO_4)_2(s)$$

3. An anion exchange resin can be used. Anion exchange resins consist of large polymers to which positively charged groups such as $R—\overset{+}{N}H_3$ (R is an organic group) are bound covalently. These positive centers of the resin can form strong ionic bonds to anions of high negative charge (PO_4^{3-}, for instance) that impinge on it from a solution with which it is in contact. By virtue of these ionic bonds, the resin can remove a high percentage of PO_4^{3-} from solution. As with zeolites (Sec. 17.4), the ion bound to the resin can be replaced by another ion. Thus, the anion exchange is saturated with an innocuous anion such as Cl^-, that PO_4^{3-} will replace when the resin is placed in a phosphate-containing solution. The ion Cl^- is then released into the water. When after prolonged use, the resin is saturated with PO_4^{3-}, the PO_4^{3-} must be removed by using the resin with a solution containing Cl^-; and the cycle is repeated. Although the resin is recyclable in such a process, it is generally very expensive and PO_4^{3-} removal by this procedure is tedious by present techniques.

17.6 Group VIA: O, S, Se, Te, and Po. The Chalcogens

General Considerations

Neutral atoms of the elements of Group VIA need obtain only two electrons to attain a noble gas electron configuration. Some properties of these elements are given in Table 17.6. The electronegativities of the elements of this group, called the chalcogens, are 2.44 for sulfur, 2.48 for selenium, 2.01 for tellurium, and 1.76 for polonium. These are large enough to allow formation of a stable -2 ion for sulfur, selenium, and tellurium, but generally not for polonium. Polonium is comparable in electronegativity to tin or copper and its chemistry is mainly metallic. While sulfur, selenium, and tellurium can form stable negative ionic species and do so principally in compounds with the alkali metals and alkaline earths, most of their compounds are covalent. In many of these compounds, the oxidation number of

TABLE 17.6 **Some Fundamental Properties of the Chalcogens (Group VIA)**

Element	*Electronic Configuration*	*Covalent Radius, Å*	*Ionic Radius, Å, −2 Ion*	*Ionization Potentials, eV*			
				1st	*2nd*	*3rd*	*4th*
O	$[He]2s^22p^4$	0.73	1.40	13.62	35.13	54.94	77.45
S	$[Ne]3s^23p^4$	1.03	1.84	10.36	23.42	35.00	47.31
Se	$[Ar]3d^{10}4s^24p^4$	1.16	1.98	9.76	21.51	32.00	42.89
Te	$[Kr]4d^{10}5s^25p^4$	1.43	2.21	9.02	18.60	31.22	38.16
Po	$[Xe]4f^{14}5d^{10}6s^26p^4$	1.67	2.30	8.41			

TABLE 17.7 Structures of Typical Chalcogen Species Observed Experimentally and As Predicted by VSEPR Theory

Compound Formula	*No. of Pairs of Valence Electrons about the Central Atom (Electrons Contributed)*[a]	*Geometry*
H_2S	4 (6 from S, 1 from each H)	Angular
$(CH_3)_2Se$	4 (6 from Se, 1 from each CH_3)	Angular
$(CH_3)_3S^+$	4 (5 from S, 1 from each CH_3)	Pyramidal
SO_2	4 (6 from S, 2 from one O)	Angular
TeF_4	5 (6 from Te, 1 from each F) note: promotion is required $[Te]5s^25p^4 \rightarrow [Te]5s^25p^35d^1$	Irregular
SeF_5^-	6 (7 from Se, 1 from each F) promotion required $[Se^-]4s^24p^5 \rightarrow [Se^-]4s^24p^34d^2$	Square pyramidal
$Te(OH)_6$	6 (6 from Te, 1 from each OH) promotion required $[Te]5s^25p^4 \rightarrow [Te]5s^15p^35d^2$	Octahedral

[a] For purposes of electron bookkeeping, we shall regard the net charge on an ionic species to affect the central atom. Electrons used in π-type bonding are not included in the valence electron count (see Chap. 6).

the Group IVA atom is -2. Examples are H_2S, which has the odor of rotten eggs, and dimethyl sulfide, $(CH_3)_2S$, an odorous material found in onions.

Compounds with the oxidation states $+4$ (for example, SeO_2, TeO_2, $SeBr_4$, and SF_4) and $+6$ (for example, SF_6 and SeF_6) are also quite common.

Chalcogen compounds have a variety of geometries. The number of groups bound to the chalcogen atom can range from two to six. Using the VSEPR method (Chap. 7), one can rationalize the many geometries associated with the varied number of bound groups, as shown in Table 17.7.

The group properties of the chalcogen compounds follow patterns similar to those found for compounds of Groups IIA through VA. Ionization of the heavier elements becomes gradually easier with larger size (down the group in the periodic table), allowing the heaviest of the chalcogens, polonium, to show metallic character. As indicated above, coordination numbers of the heavier elements are not limited to two, but may be greater; this increase can be explained by invoking "promotion" of some valence electrons. Such promotion is energetically feasible because there are empty *d* orbitals that have nearly the same energy as the *p* orbitals in the valence shell.

Properties of the Elements and Their Compounds

Sulfur is found in nature in its elemental form. Such natural occurrence of an elemental species generally indicates that atoms of the element can form stable bonds to other atoms of the same element; oxygen as O_2 is an example, and carbon as graphite or diamond is another. There are also the metals, which form strong metallically bound crystal lattices (Sec. 14.4). Solid elemental sulfur is very commonly

Spotlight

Sulfur occurs in its elemental form in large deposits in the southern United States, mainly Texas and Louisiana, and also in Italy, Spain, Japan, Ireland, and Mexico. Sulfide ores such as galena, PbS, and sphalerite, ZnS, are also sources of sulfur, although the metals in these cases are the primary objective of the mining operations. Familiar minerals that contain sulfur but are not important sources of sulfur are gypsum, $CaSO_4 \cdot 2H_2O$, and epsom salts, $MgSO_4 \cdot 7H_2O$.

The occurrence of large deposits of free sulfur is in a sense puzzling, since elemental sulfur is relatively easy to oxidize, with O_2 from the atmosphere. One possible explanation is that the elemental sulfur was deposited when there were significant quantities of reducing compounds in the atmosphere, H_2S for instance, and these deposits have not been exposed for long periods to an oxidizing atmosphere.

Selenium and tellurium occur in their elemental forms in nature, but generally they occur as selenides and tellurides. Polonium occurs in uranium ores as a product of radioactive decay, but it is so rare, with one ton of uranium ore possibly containing 100 μg Po, that it is now prepared by the neutron bombardment of natural bismuth, ^{209}Bi. The reaction produces ^{210}Bi, the radioactively unstable parent of Po:

$$^{209}_{83}Bi + {}^{1}_{0}n \rightarrow {}^{210}_{83}Bi \rightarrow {}^{210}_{84}Po + \beta$$

An additional source is from the decay of radon used in hospitals for the treatment of cancer:

$$^{222}_{86}Rn \rightarrow {}^{218}_{84}Po + {}^{4}_{2}He$$

Selenium has two important uses because of its special electrical properties. Because its electrical conductivity increases in direct proportion to the intensity of light striking its surface, it is used in light-measuring devices and other optical instruments. Secondly, when selenium is in contact with the surfaces of other metals, it forms devices called semiconductor diodes, which have the property of passing electrons from selenium to the metal but not from the metal to selenium (see Chap. 14).

Tellurium is used in semiconductor devices, and in alloys to give them a more durable surface.

Polonium radioactively decays to lead, and emits as many alpha particles per unit time as 5000 times its weight of radium. This has an interesting application, because the energy released is so large (140 watts/g) that 0.5 Po reaches a temperature of 500 °C and exhibits a blue glow. The alpha irradiation is stopped within the metal and its container, and thus Po has attracted attention for uses as a lightweight heat source for thermoelectric power in space satellites.

found in a unique structure—an eight-membered ring made up of only one element, S_8. The structure shown herewith is present in the most stable crystalline form of S_8, known as S_α, which occurs as large yellow crystals in volcanic deposits. Another

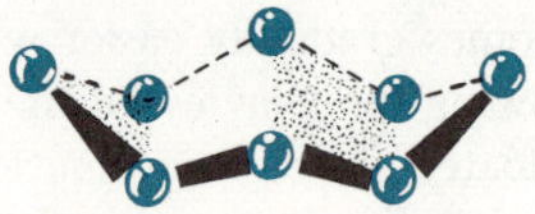

crystalline form, S_β, is the stable form above 95.5 °C. The S_β form melts at 119 °C. The difference between the S_α and S_β forms is simply in the way the S_8 molecules pack together in the solid state. Since S_α is converted only slowly to S_β above 95.5 °C, rapid heating of the S_α form will "bypass" the S_α-to-S_β structural change and lead to melting of S_α at 112.8 °C. In such a rapid heating process, S_α can exist between 95.5 °C and 112.8 °C, a region in which the S_β form is actually thermodynamically more stable. A species existing in such a state is said to be **metastable.** These processes are illustrated in Fig. 17.8.

Elemental sulfur is known to exist in several other structural forms besides the S_8 cyclic ring, including S_6, S_7, S_9, S_{10}, and S_{12}. Molten sulfur, formed by melting S_8 or the other less common structures, consists of several species, including linear poly-

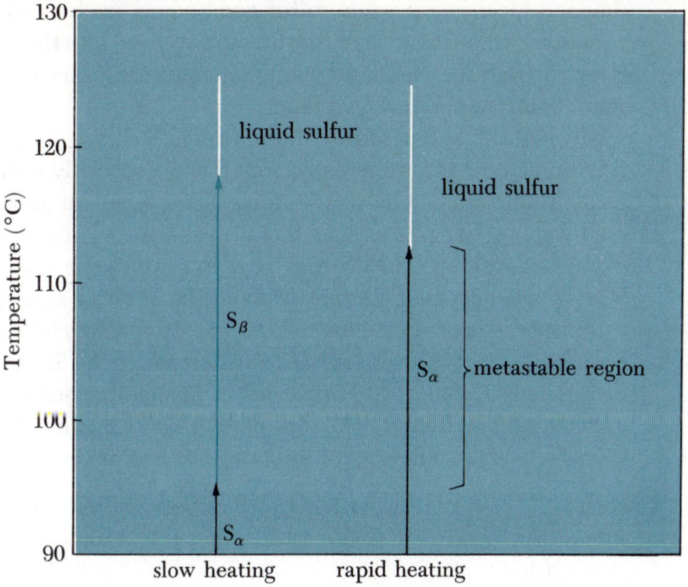

Figure 17.8 Structural changes of sulfur on being heated.

mers. These polymers are responsible for pronounced changes in the viscosity of liquid sulfur as the temperature of the melt is increased.

Although a comparable selenium species, Se_8, is known, it is not as stable as a solid structure that contains infinite spiral chains of selenium atoms. Tellurium does not tend to form chains as easily as sulfur or selenium. The solid state structure of polonium is metallic; that is, the structure does not contain individual molecules and the solid conducts electricity (see Sec. 14.4).

Hydrides

The hydrides H_2S, H_2Se, and H_2Te are obtained by the reactions of acids with metal salts of S, Se, and Te. They are all weak acids. Since the bond strength for a series of covalent compounds of hydrogen generally is lower as the size of the bound atom is greater, due to diminished overlap, the H–Te bond is weaker than the H–Se bond, which is in turn weaker than the bonds in H_2S. Thus, H_2Te is the strongest acid of the three but still classed as a weak acid, since only a small portion of the H_2Te molecules are dissociated in aqueous solution. The K_a for the process $H_2X \rightleftharpoons H^+ + HX^-$ is 1.9×10^{-8} for H_2S, 1.3×10^{-4} for H_2Se, and 2.3×10^{-3} for H_2Te.

Oxides of the Heavy Chalcogens

All the chalcogens except for polonium form dioxides and trioxides; for polonium, only the dioxide is known. The dioxides can be obtained simply by burning the element in air. Although SeO_2 and TeO_2 are both solids at ordinary temperatures and pressures, SO_2 is a gas. The existence of SO_2 as a gas at standard temperature and pressure gives rise to some serious pollution problems. Instead of being easily retained for disposal when sulfur compounds are burned, for example, as a sulfur-containing impurity in coal, the SO_2 is dispersed into the atmosphere. Sulfur dioxide is very irritating, readily attacking mucous membranes. It also promotes the corrosion of steel and damages plants.

Much of the SO_2 pollution in the United States occurs from the burning of sulfur-containing coal and some petroleum that is especially sulfur-rich. A large electrical generation plant can burn 8000 tons of coal a day, or about 3 million tons a year. If high-sulfur coal, such as that found in the Midwest with an average of 4% sulfur, were used, about 120,000 tons of sulfur would be converted to 240,000 tons of SO_2—a tremendous pollution problem for a city's atmosphere.

Since pressure is being exerted to find alternatives to the use of our limited petroleum supply, coal burning is likely to increase, so the SO_2 pollution problem becomes increasingly important. Fortunately, chemical means exist to remove SO_2. These include the reaction of SO_2 with lime, CaO. Lime is formed by heating limestone, $CaCO_3$, to drive off CO_2. Lime and sulfur dioxide react as follows:

$$CaO(s) + SO_2(g) \rightleftharpoons \underset{\text{calcium sulfite}}{CaSO_3(s)} \tag{17.18}$$

Also, SO_2 can be oxidized to SO_3 and dissolved in water to form sulfuric acid:

$$2SO_2(g) + O_2(g) \rightleftharpoons 2SO_3(g) \tag{17.19}$$

$$SO_3(g) + H_2O(l) \rightleftharpoons H_2SO_4(l) \tag{17.20}$$

In fact, SO_2 is converted into SO_3 in the atmosphere quite readily by the action of sunlight or airborne catalysts in concert with O_2. The SO_3 is absorbed and converted to sulfuric acid by water vapor, according to the last step of the reaction sequence above (Eq. 17.20). If the SO_3 or sulfuric acid remains airborne, the result is a very harmful form of smog, which attacks the respiratory system. It can also be carried down in rain to cause the increased corrosion and damage typical of strong acids.

Oxo Acids

As we have seen, SO_3 forms sulfuric acid when dissolved in water. Solutions of SO_2 in H_2O are also acidic. The acidity is often attributed to a species called sulfurous acid, H_2SO_3, which is ostensibly formed from SO_2 and H_2O:

$$H_2O + SO_2 \rightleftharpoons H_2SO_3 \tag{17.21}$$

$$H_2SO_3 + H_2O \rightleftharpoons HSO_3^- + H_3O^+ \tag{17.22}$$

The existence of the sulfurous acid molecule has not been demonstrated. The acidic behavior of SO_2, dissolved in water, is better represented as

$$SO_2 + H_2O \rightleftharpoons HSO_3^- + H_3O^+ \tag{17.23}$$

When SeO_2 dissolves in water, H_2SeO_3 is indeed formed. It is a very weak acid. Tellurium dioxide is essentially insoluble in water and there is no evidence for the formation of an H_2TeO_3 species.

The dioxides of sulfur, selenium, and tellurium can be oxidized in the presence of water to form sulfuric, selenic, and telluric acids. Sulfuric acid, H_2SO_4, and selenic

Spotlight

The preparation of sulfuric acid from SO_2 forms the basis of the "contact process," one of two industrially important methods for producing sulfuric acid:

$$2SO_2(g) + O_2(g) \xrightarrow[\text{catalyst } (Fe_2O_3,\ CuO,\text{ or } V_2O_5)]{450\ °C} 2SO_3(g)$$

$$H_2SO_4(l) + SO_3(g) \rightleftharpoons \underset{\text{"oleum"}}{H_2S_2O_7(l)}$$

$$H_2S_2O_7(l) + H_2O(l) \rightleftharpoons 2H_2SO_4(l)$$

The oleum is prepared first because SO_3 dissolves much faster in sulfuric acid than it does in water.

Besides being a strong acid, H_2SO_4 is also a strong oxidizing agent,

$$Cu(s) + 2H_2SO_4(l) \rightleftharpoons CuSO_4(s) + 2H_2O(l) + SO_2(g)$$

and a strong dehydrating agent. It can even remove the elements of water from substances that do not contain water per se. Dry sugar when mixed with H_2SO_4 first turns brown, darkens, and finally turns to a charred, swollen mass, with the evolution of steam:

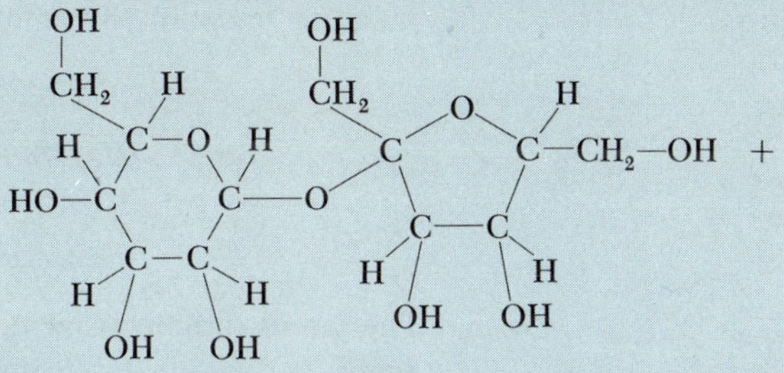

sucrose

$$11H_2SO_4 \rightleftharpoons 12C + 11H_2SO_4 \cdot H_2O$$

Sulfuric acid is the world's most widely used industrial chemical. It is used by industries involved in fertilizers, petroleum products, dyes, drugs, metals, textiles, paper, and other chemicals. It is so important that even in 1843 the great German chemist Justus von Liebig wrote, "We may fairly judge the commercial prosperity of a country from the amount of sulfuric acid it consumes."

acid, H_2SeO_4, are strong acids, readily forming salts of the bisulfate anion, HSO_4^-, and the biselenate anion, $HSeO_4^-$, as well as the sulfate anion, SO_4^{2-}, and the selenate anion, SeO_4^{2-}. Telluric acid has a much different structure from the sulfuric and selenic acids. It is octahedral, rather than tetrahedral, with the formula $Te(OH)_6$. The oxidation

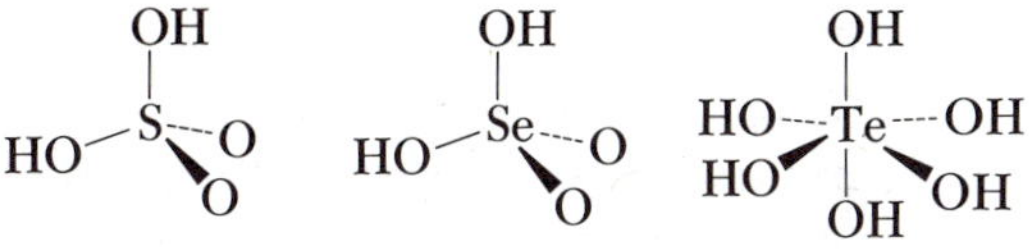

sulfuric acid selenic acid telluric acid

state of Te in telluric acid is 6, and S and Se in sulfuric and selenic acids also are of oxidation state 6. Telluric acid is not a strong acid, as can be seen from the equilibrium constants given in the accompanying chart.

Equilibrium	K_a
$H_2SO_4 + H_2O \rightleftharpoons HSO_4^- + H_3O^+$	1
$HSO_4^- + H_2O \rightleftharpoons SO_4^{2-} + H_3O^+$	1.2×10^{-2}
$H_2SeO + H_2O \rightleftharpoons HSeO_4^- + H_3O^+$	1
$HSeO_4^- + H_2O \rightleftharpoons SeO_4^{2-} + H_3O^+$	1.2×10^{-2}
$Te(OH)_6 + H_2O \rightleftharpoons Te(O)(OH)_5^- + H_3O^+$	2.1×10^{-8}
$Te(O)(OH)_5^- + H_2O \rightleftharpoons Te(O)_2(OH)_4 + H_3O^+$	6.5×10^{-12}

Oxo acids of sulfur can also have two structural features for which analogs have not been found for the oxo acids of selenium and tellurium. These features are S–S bonds and peroxo functional groups. The greater tendency for catenation (sec. 9.3) in sulfur undoubtedly contributes to the stability of such acids as

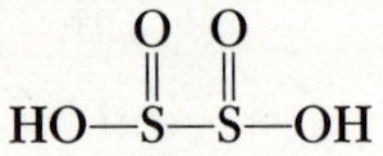

dithionous acid

and

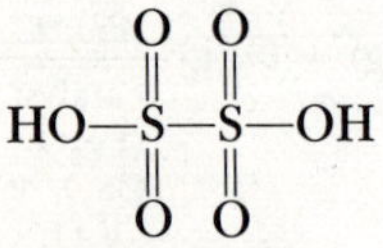

dithionic acid

The ion of dithionous acid, $S_2O_4^{2-}$, is called dithionite and is useful as a reducing agent; it can donate electrons to some other species:

$$S_2O_4^{2-} + 4OH^- \rightleftharpoons 2SO_3^{2-} + 2H_2O + 2e^- \qquad \mathscr{E}° = 1.12\text{ V} \quad (17.24)$$

Examples of peroxo acids (acids containing the $—O—O—^{2-}$ functional group) are

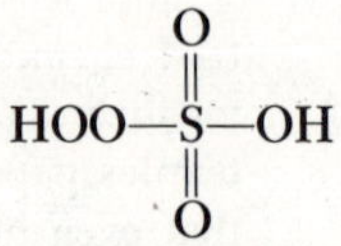

peroxomonosulfuric acid

and

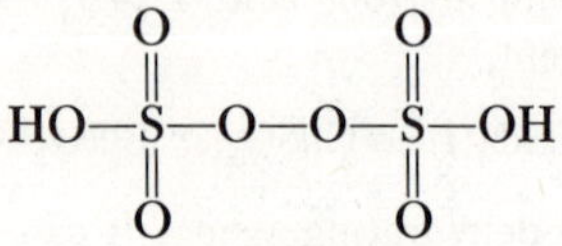

peroxodisulfuric acid

Salts of peroxomonosulfuric acid are useful as oxidizing agents, due to the very large reduction potential for the oxidation state change from +7 to +6.

$$S_2O_8^{2-} + 2e^- \rightleftharpoons 2SO_4^{2-} \qquad \mathscr{E}° = 2.01\text{ V} \qquad (17.25)$$

17.7 Group VIIA: F, Cl, Br, I, and At. The Halogens

General Considerations

Group VIIA is known as the halogen group. With a valence shell configuration of ns^2np^5, where n is the principal quantum number of the valence shell, neutral atoms of these elements need gain only one electron to attain a noble gas configuration. Some properties of these elements are given in Table 17.8. The monovalent ions, Cl^-, Br^-, and I^-, all are significant in halogen chemistry. Astatine (from the Greek word for "unstable") is found only as radioactive isotopes, the most common of which

TABLE 17.8 **Some Fundamental Properties of the Halogens (Group VIIA)**

Element	*Electronic Configuration*	*Covalent Radius, Å*	*Ionic Radius, Å, −1 Ion*	*Ionization Potential, eV*	*Electron Affinities, eV*
F	$[He]2s^22p^5$	0.71	1.36	17.43	3.62
Cl	$[Ne]3s^23p^5$	0.99	1.81	13.01	3.69
Br	$[Ar]4s^24p^5$	1.14	1.95	11.84	3.45
I	$[Kr]5s^25p^5$	1.33	2.16	10.45	3.15
At	$[Xe]6s^26p^5$	.....	.....	.....	.....

has a half-life of only 8.3 hr; half of it disintegrates in 8.3 hr, in other words. (See Chap. 13.) The chemistry of astatine has not been extensively investigated, but those studies that have been performed indicate that it resembles the other halogens, readily forming At^- but also At^+. Like iodine, astatine shows more nonmetallic than metallic properties.

The halogens, besides their ionic chemistry, form stable compounds in which covalent bonding gives a complete valence octet. In general, the covalent molecules have structures in which the halogens are monovalent, or form only one covalent bond. The familiar molecules Cl_2, Br_2, HCl, HBr, HI, and so on show halogen monovalence. Two principal exceptions to monovalency occur. One of these is the series of interhalogen compounds, such as ClF_3, ClF_5, BrF_5, and IF_7. Another exception is the capacity of halogens for forming "bridges" between two or even three atoms in such compounds as the polymeric beryllium chloride,

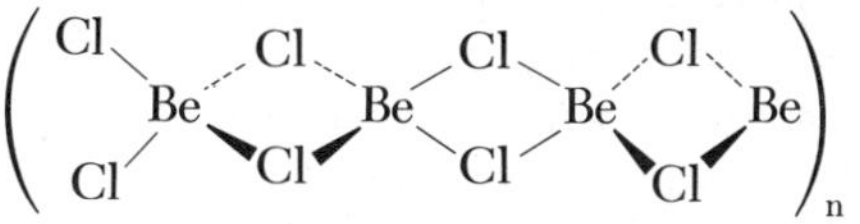

and the dimer of $AlCl_3$,

The heavier elements, in accord with our findings for previously cited groups, are less electronegative, and the second row element, in this case fluorine, differs in many respects from the subsequent congeners. However, in contrast with the preceding groups, the elements towards the bottom of the group do not tend towards metallic behavior, and exceptions associated with the lightest element of the group are not so pronounced. Even astatine appears to have predominantly nonmetallic chemistry. Although there are a significant number of exceptions in the chemistry of fluorine compared with the other halogens, the effect of the small size of its atom and ion and its high electronegativity does not generally cause gross discontinuities from trends within the group.

All the halogens are sufficiently reactive in their elemental state to be found in nature only as compounds. Fluorine is especially unstable as its elemental species, F_2. The F–F bond strength, 37 kcal/mol, is very low compared with the strengths of fluorine bonds to many other elements or with the lattice energies that result from ionic compound formation with many metals. The exceptionally low bond energy of F_2 and large amounts of energy released on formation of compounds with other elements make F_2 one of the most universally reactive species, able to oxidize even some of the noble gases!

The elemental halogens are generally obtained by electrolysis of naturally occurring salts of the halogens, called halides. Fluorine can be obtained from molten mixtures of KF and HF. Hydrogen fluoride is used to lower the temperature at which KF melts. A solution of one part KF to two to three parts HF melts at 70 °C to 100 °C; a 1:1 solution melts at 250 °C, whereas KF itself melts at 880 °C. Fluorine cannot be formed from F^- by electrolysis in aqueous solution because a competitive

Spotlight

Elemental chlorine is a greenish yellow gas, and is an excellent oxidizing agent; the oxidizing properties account for its largest use in bleaching wood pulp for paper. It also finds important uses as a germicide for water, and in the manufacture of HCl.

Bromine is a dense reddish brown liquid that is efficiently produced from seawater because of its enormous demand in the production of antiknock gasolines. Seawater is first acidified and then treated with chlorine gas:

$$Cl_2(g) + 2Br^- \rightleftharpoons Br_2(l) + 2Cl^-$$

This oxidation-reduction is typical of other halogens; that is, a halide ion can be oxidized by any other halogen above it in the periodic table.

Iodine is a dark bluish black crystalline solid that is manufactured from the $NaIO_3$ contained in Chilean saltpeter beds:

$$2IO_3^- + 5HSO_3^- \rightleftharpoons I_2 + 5SO_4^{2-} + 3H^+ + H_2O$$

It has a vapor pressure of 3 mm at 55 °C, and can therefore be purified easily by sublimation.

The most important use of iodine is in medicine. The radioactive isotope ^{131}I is used in treating tumors in the thyroid gland, where iodine accumulates. For external wounds, a solution of I_2 and KI in alcohol is sometimes used; the solution is called tincture of iodine. Iodoform, CHI_3, is another important antiseptic.

Astatine is so rare that less than one ounce exists in the earth's crust. It is another element that was predicted to exist by Mendeleev in 1869, who called it eka-iodine. It was finally prepared in 1940 by the bombardment of bismuth with alpha particles. Only 0.05 μg has been prepared to date.

reaction, the generation of $O_2(g)$ from water, takes place instead. This $\mathscr{E}°$ value is less negative than the $\mathscr{E}°$ for generation of F_2:

$$2F^- \rightleftharpoons F_2(g) + 2e^- \qquad \mathscr{E}° = -0.82 \text{ V} \qquad (17.26)$$

The F_2 produced from the KF–HF melt is extremely reactive and consequently corrosive. Fortunately, some metals are rapidly coated with a layer of the metal fluoride, which is resistant to further attack—much as the oxide coating formed on aluminum metal protects it from further corrosion.

Chlorine is abundant in nature as sodium chloride; sea water is one natural vehicle of this substance. Elemental chlorine can be obtained from salt water by electrolysis:

$$Na^+ + Cl^- + H_2O \rightleftharpoons Na^+ + OH^- + \tfrac{1}{2}H_2(g) + \tfrac{1}{2}Cl_2(g) \qquad (17.27)$$

This is possible even though the chloride-chlorine potential is more negative than the H_2O–O_2 potential. In the case of chloride, the rates of reaction favor the formation of the Cl_2 and do not permit the thermodynamically favored process to occur.

$$2Cl^- \rightleftharpoons Cl_2 + 2e^- \qquad \mathscr{E}° = -1.36 \qquad (17.28)$$

The liberation of O_2 from H_2O is slow unless a high potential (1.4 V) is supplied. At the potential sufficient to produce Cl_2 rapidly (1.36 V), little O_2 is produced.

The bromine case is similar to that of Cl_2 production:

$$2Br^- \rightleftharpoons Br_2(aq) + 2e^- \qquad \mathscr{E}° = -1.09 \text{ V} \qquad (17.29)$$

With iodides, the potential is lower than the H_2O–O_2 potential:

$$2I^- \rightleftharpoons I_2(aq) + 2e^- \qquad \mathscr{E}° = -0.54 \text{ V} \qquad (17.30)$$

In both cases, straightforward electrolysis in aqueous solutions can be carried out.

Halides

The only elements in the periodic table that do not form halides—compounds in which the halogen is in the -1 oxidation state—with chlorine, bromine, or iodine are helium, neon, and argon. The noble gases xenon, krypton, and radon form fluorides but no other halides. The bonds in halides range from essentially purely ionic, like CsF, which has an electronegativity difference, ΔEN, of 3.1, and intermediate, like $MgBr_2$, which has a value ΔEN of 1.3, to purely covalent, like F_2, Cl_2, Br_2 of $\Delta EN = 0$. Most pure metal halides occur as essentially ionic crystals. When these are dissolved in water, the resulting aqueous solutions conduct electricity and are hence deemed predominantly ionic. Very many organic molecules contain chlorine atoms that are definitely covalently bound. Carbon-chlorine bonds in such compounds are only very slightly polar ($\Delta EN = 0.3$) and in most cases show no tendency towards ionic dissociation. One simple chemical test for determining qualitatively the degree of ionic character of a halide, or more correctly, the tendency of a halide to ionize in solution, is to dissolve the halide in an aqueous solution, sometimes with alcohol added for enhancing the solubility of an organic halide. The next step is to see whether a precipitate forms when a silver nitrate solution is added. If the halide compound ionizes in solution, the resulting halide ion will react with Ag^+ to form a precipitate of silver halide. For example, if NaCl is examined in this way, AgCl will precipitate when aqueous $AgNO_3$ is added, but no precipitate forms if CH_3—Cl is the halide, since the C–Cl bond is largely covalent and does not ionize in solution. Molecular halides with substantial ionic character often react with water to liberate the hydrohalogen acid and an acid of the other element. An example is phosphorus tribromide:

$$\underset{\text{phosphorus tribromide}}{PBr_3} + 3H_2O \rightleftharpoons \underset{\text{phosphorous acid}}{HOP(OH)_2} + 3H^+ + 3Br^- \tag{17.31}$$

Halides can often be formed by direct reaction of the elemental halogen with an element; for example,

$$P + \tfrac{3}{2}Cl_2 \rightleftharpoons PCl_3 \tag{17.32}$$

The stoichiometry is often controlled by the relative amounts of the elements and experimental conditions such as temperature and pressure. Thus, if excess Cl_2 is present, the above reaction is replaced by

$$P + \tfrac{5}{2}Cl_2(g) \rightleftharpoons PCl_5(g) \tag{17.33}$$

Variable stoichiometry is quite common for the heavier members of the nonmetallic groups, in which valence expansion, for example, "promotion," is well known.

Metallic halides can often be obtained by removing water from salts of hydrated metal ions and halide ions, either by heating in a vacuum or by treating with a powerful dehydrating agent; this is an agent such as $SOCl_2$ that reacts with water, or otherwise removes it. The technique is common for obtaining the halide salts of transition metals. For example, manganese chloride is prepared, at high temperature and low pressure, by the reaction

$$[Mn(H_2O)_6]Cl_2 + 6SOCl_2 \rightleftharpoons MnCl_2 + 12HCl(g) + 6SO_2(g) \tag{17.34}$$

The SO_2 and HCl produced are removed during the reaction by a vacuum or by sweeping the system with an inert gas like N_2 or Ar.

Organic halides can be obtained by the reaction of the halogen with an alkene functional group (Chap. 7), for example,

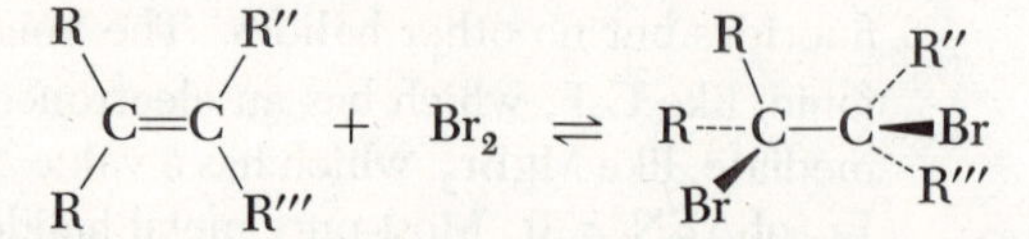

$$\underset{R}{\overset{R}{>}}C=C\underset{R'''}{\overset{R''}{<}} + Br_2 \rightleftharpoons R-\underset{Br}{\overset{R}{C}}-\underset{R'''}{\overset{R''}{C}}-Br \quad (17.35)$$

Equation (17.35) represents the reaction that constitutes the principal use of bromine, which is in the production of ethylene dibromide, an additive to gasoline to provide antiknock properties:

$$C_2H_4(g) + Br_2(l) \rightleftharpoons BrCH_2CH_2Br(l)$$

Numerous other means of introducing one or more halogens have also been used in organic chemistry.

Hydrides

The binary acids of the halogens, hydrofluoric (HF), hydrochloric (HCl), hydrobromic (HBr), and hydriodic (HI), all act as acids. The order of acid strength is HI > HBr > HCl > HF. This order parallels trends for acids of other series of structurally related acids involving the various elements of a group, since larger atoms have weaker covalent bonding with H, allowing more facile ionization. It may seem somewhat surprising, however, that HF is such a weak acid.

$$HF + H_2O \rightleftharpoons H_3O^+ + F^- \quad K_A = 6.9 \times 10^{-4} \quad (17.36)$$

This is due to the very strong H–F bond, having a bond energy of 136 kcal/mol, which can be compared with 103 kcal/mol for HCl, 87 for HBr, and 71 for HI. In the chalcogens, a similar large variation in acidity between H_2O and the other chalcogen hydrides is found; the dissociation constants for H_2O and H_2S are 10^{-14} and 10^{-8}, respectively.

$$2H_2O \rightleftharpoons OH^- + H_3O^+ \quad K = 1.0 \times 10^{-14} \quad (17.37)$$

$$H_2S + H_2O \rightleftharpoons HS^- + H_3O^+ \quad K = 1.9 \times 10^{-8} \quad (17.38)$$

$$\frac{K_{H_2S}}{K_{H_2O}} = 10^6$$

Oxo Acids

Although only one relatively stable oxo acid is known for fluorine, FOH, several are known for the other halogens. Typically, the nomenclature of the multiple oxo acids of the halogens has been a bane to beginning chemistry students. But the subject is really not very complicated.

The **hypohalous acids** have the general formula XOH, where X is the halogen. All four hypohalous acids—hypofluorous, hypochlorous, hypobromous, and hypoiodous acid—are known, although only FOH can be obtained in a pure state. In these compounds, with perhaps the exception of FOH, the oxidation number of the halogen

Spotlight

In general, the naming of the oxo acids follows the rule that the suffix *ic* is used for the most common acid. The acid with the next lowest oxidation state has the suffix *ous,* and the next lowest state is given the prefix *hypo* with the suffix *ous.* Finally, the acid with a higher oxidation state is given the prefix *per* with the suffix *ic.* For the salts of oxo acids, *ous* changes to *ite,* and *ic* changes to *ate.* It should be noted that binary acids have the prefix *hydro* with the suffix *ic;* for their salts, *ic* changes to *ide* and *hydro* is dropped. For example:

Acid	*Name*	*Salt*
HCl	Hydrochloric	Chloride
$HClO$	Hypochlorous	Hypochlorite
$HClO_2$	Chlorous	Chlorite
$HClO_3$	Chloric (most common)	Chlorate
$HClO_4$	Perchloric	Perchlorate
$H_2N_2O_2$	Hyponitrous	Hyponitrite
HNO_2	Nitrous	Nitrite
HNO_3	Nitric (most common)	Nitrate
H_2S	Hydrosulfuric	Sulfide
H_2SO_3	Sulfurous	Sulfite
H_2SO_4	Sulfuric (most common)	Sulfate
H_3PO_3	Phosphorous	Phosphite
H_3PO_4	Phosphoric (most common)	Phosphate

is $+1$, and they are readily reduced to the 0 or -1 oxidation states; examples are HOCl to Cl_2 or Cl^-. Hence, they are good oxidizing agents. In basic solution the hypohalous acids are converted to the corresponding hypohalite ions, OCl^-, OBr^-, and OI^-. These species undergo a kind of reaction in which they function both as an oxidizing agent and as a reducing agent, yielding halogen species as products that have halogen oxidation numbers both larger ($+5$) and lower (-1) than that of the halogen atom in the reactant ($+1$). Such oxidation-reduction reactions are called **disproportionation reactions.** An example is the disproportionation of IO^-, which reacts with itself so rapidly that it has not been definitely identified in solution:

$$3IO^- \rightleftharpoons 2I^- + IO_3^- \qquad (17.39)$$

iodine oxidation number: $(+1)$ (-1) $(+5)$

A halogen species with the formula HXO_2 is called a **halous acid,** with the halogen oxidation number $+3$. Only one halous acid has been definitely demonstrated, namely chlorous acid, $HClO_2$. Salts of chlorous acid, such as $NaClO_2$, are known as chlorites and are used in bleaching agents because of the oxidation power of the chlorite ion, ClO_2^-. Their action in whitening clothes is by causing oxidation of electron-rich functional groups that are responsible for color; these groups absorb light in the visible region. Since hypochlorites are also good oxidizing agents, they are also used as bleaches.* The reactions of chlorites and hypochlorites as oxidants can be quite favorable due to the potentials shown here:

$$ClO_2^- + 4e^- + 2H_2O \rightleftharpoons Cl^- + 4OH^- \qquad \mathscr{E}^\circ = 0.77\ V \qquad (17.40)$$

$$ClO^- + 2e^- + H_2O \rightleftharpoons Cl^- + 2OH^- \qquad \mathscr{E}^\circ = 0.88\ V \qquad (17.41)$$

These potentials are for the reactions in alkaline solution, since bleaches are generally used in conjunction with soaps or detergents, which produce alkaline solutions.

*A common bleaching powder is calcium hypochlorite, which is prepared as follows:

$$3Ca(OH)_2(s) + 2Cl_2(g) \rightleftharpoons Ca(OCl)_2(s) + CaCl_2 \cdot Ca(OH)_2 \cdot H_2O(s) + H_2O(l)$$

Halic acids are acidic species in which the oxidation number of the halogen is +5. Only iodic acid is actually isolated as the pure acid itself; its formula is $H_2I_2O_6$. Both chloric and bromic acids, $HClO_3$ and $HBrO_3$, can be obtained in solution—in ionized form. They are very powerful oxidizing agents.* The ions derived from the halic acids on ionization are halate ions, XO_3^-. Iodate, IO_3^-, is quite common in nature—as a minor constituent in seawater, and as $CaIO_3(s)$ and $NaIO_3(s)$ in mineral deposits of a sedimentary character.

The **perhalic acids,** with general formula HXO_4, contain halogen atoms in the +7 oxidation state. These acids are potentially very strong oxidizing agents, but often react slowly. The most common of these, perchloric acid, $HClO_4$, is a very strong acid, and can be explosive in its pure state. The salts of the halic acids are called perhalates, for example, sodium perchlorate, $NaClO_4$. Such salts, while also potentially strong oxidizing agents, usually are stable because of slow reactions. However, reactions between perchlorates and organic species may occur very rapidly—sometimes causing severe explosions—and many chemists have learned to respect perchlorates through unfortunate experiences.

Perbromic acid was recently synthesized by the oxidation of the bromate ion.

$$BrO_3^- + H_2O \rightleftharpoons BrO_4^- + 2H^+ + 2e^- \qquad \mathscr{E}° = -1.76\text{ V} \tag{17.42}$$

This oxidation is so difficult that early attempts with even very strong oxidants were unsuccessful. In fact, several chemists offered theories to show why the perbromate ion could not exist! Use of the very strong oxidizing agent F_2 has recently been shown to produce perbromate:

$$BrO_3^- + F_2 + 2OH^- \rightleftharpoons BrO_4^- + 2F^- + H_2O \tag{17.43}$$

Periodic acid is also known. Like perchlorate and perbromate, periodate is a tetrahedral ion.

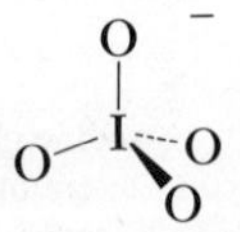

Periodiate ion is not so strong an oxidizing agent as perchlorate or perbromate and undergoes reduction at reasonable rates. It is useful for oxidizing organic compounds. The formation of crystalline salts of HIO_4 often leads to conversion from tetrahedral coordination about the iodine to an octahedral form, for instance as in $Na^+[H_4IO_6]^-$:

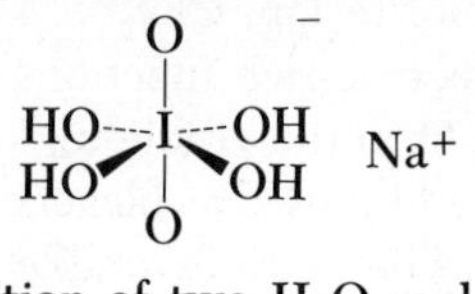

This can be imagined as the addition of two H_2O molecules to the iodate ion.

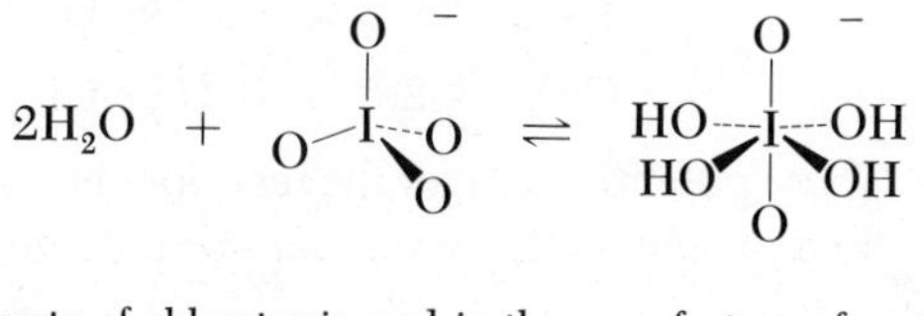

(17.44)

*The oxidizing property of chlorates is used in the manufacture of matches and explosives. Sodium chlorate also finds important uses as a weed killer.

Interhalogens

Different halogens may react with each other to form numerous compounds called interhalogens, with varying stoichiometry. All halides except fluorine are capable of valency expansion—the use of orbitals beyond those normally considered the valence shell. This is responsible for the fact that some of the interhalogen compounds involve coordination numbers larger than 1. The interhalogen compounds are generally formed from one atom of a heavy halogen and one or several fluorine atoms. The exceptions are BrCl, ICl, ICl_3, and IBr. The stoichiometries commonly found for interhalogen compounds are 1:1 (BrCl, ClF, BrF, and so on), 1:3 (ClF_3, BrF_3, IF_3), 1:5 (ClF_5, BrF_5, IF_5), and 1:7 (IF_7). The odd-number stoichiometries can be understood by the promotion-hybridization model developed in Chap. 7. We can regard the 1:1 interhalogens as the simplest case; sharing an electron from each halogen atoms leads to completion of the valence octet of each. No promotion and hybridization scheme is necessary. For the 1:3 compounds, we can imagine promotion of one electron in the heavier halogen; for example, for BrF_3,

$$\text{bromine:}\quad \underset{4s}{\underline{\uparrow\downarrow}}\ \underbrace{\underline{\uparrow\downarrow}\ \underline{\uparrow\downarrow}\ \underline{\uparrow}}_{4p}\ \underbrace{\underline{\ \ }\ \underline{\ \ }\ \underline{\ \ }\ \underline{\ \ }\ \underline{\ \ }}_{4d}\ \xrightarrow{\text{promotion}}\ \underset{4s}{\underline{\uparrow\downarrow}}\ \underbrace{\underline{\uparrow\downarrow}\ \underline{\uparrow}\ \underline{\uparrow}}_{4p}\ \underbrace{\underline{\uparrow}\ \underline{\ \ }\ \underline{\ \ }\ \underline{\ \ }\ \underline{\ \ }}_{4d}$$

Now three electrons in three orbitals are available for bonding. Each of three fluorine atoms can share one electron to form complete pairing. Bromine difluoride would not be stable, since the same amount of promotional energy would be expended (one electron promoted); but only one extra bond is formed. Formation of two extra Br–F bonds provides more net stabilization (lowering of energy) and is hence much more favorable. Bromine pentafluoride can result from the promotion of two electrons

$$\text{bromine:}\quad \underset{4s}{\underline{\uparrow\downarrow}}\ \underbrace{\underline{\uparrow\downarrow}\ \underline{\uparrow\downarrow}\ \underline{\uparrow}}_{4p}\ \underbrace{\underline{\ \ }\ \underline{\ \ }\ \underline{\ \ }\ \underline{\ \ }\ \underline{\ \ }}_{4d}\ \xrightarrow{\text{promotion}}\ \underset{4s}{\underline{\uparrow\downarrow}}\ \underbrace{\underline{\uparrow}\ \underline{\uparrow}\ \underline{\uparrow}}_{4p}\ \underbrace{\underline{\uparrow}\ \underline{\uparrow}\ \underline{\ \ }\ \underline{\ \ }\ \underline{\ \ }}_{4d}$$

allowing formation of five covalent bonds. Formation of only four Br–F bonds would not be so favorable, as it would also require the promotion of two electrons; hence, BrF_4 is not found. It should be noted that these arguments have all been directed towards neutral compounds. The case for interhalogen ions is somewhat different. The ion BrF_4^+, in which the "extra promoted electron" is absent, is found to be a relatively stable species. The ion BrF_6^- is also known.

The structures of interhalogen compounds and ions are consistent with the VSEPR theory. By counting the number of valence electrons about the central atom, including those donated by bound atoms, and placing electron pairs as far from one another as possible, we arrive at the structures shown in Table 17.9.

An important class of interhalogen ions is typified by I_3^-, and comprises the trihalide ions. As indicated in Table 17.9, these ions are found to be linear. The ion I_3^- is formed easily by dissolving $I_2(s)$ in an aqueous solution of an iodide salt:

$$I^- + I_2(s) \rightleftharpoons I_3^- \qquad (17.45)$$

In a similar manner, ICl_2^- can be formed,

$$I^- + Cl_2 \rightleftharpoons ICl_2^- \qquad (17.46)$$

TABLE 17.9 Structures Observed Experimentally and Predicted from VSEPR Theory for Some Interhalogen Compounds and Ions

		Number of Electron Pairs	*Structure (Predicted and Experimental)*
1:1	ClF	4 (7 electrons from Cl, one from F)	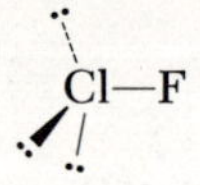 linear
1:2	ClF_2^-	5 (7 from Cl, one from each F, one from net charge)	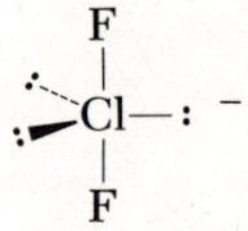 linear
1:2	ClF_2^+	4 (7 from Cl, one from each F minus one from net charge)	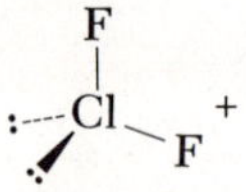 bent
1:3	BrF_3	5 (7 from Br, one each from three F's)	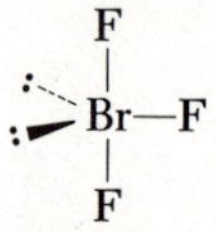 T-shaped
1:4	BrF_4^-	6 (7 from Br, one from each F, one from net charge)	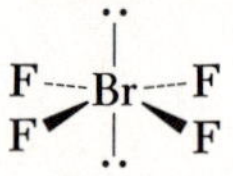 square planar
1:4	ClF_4^+	5 (7 from Cl, one each from F, minus one from net charge)	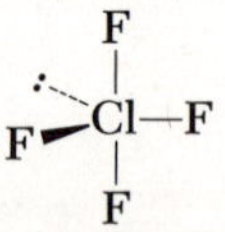 irregular
1:5	IF_5	6 (7 from I, one from each FO)	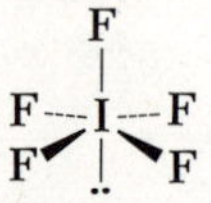 square pyramidal
1:6	IF_6^-	7 (7 from I, one from each F, one from net charge)	 Predicted: distorted octahedral. Found: *distorted octahedral*—exact structure has not been determined
1:6	IF_6^+	6 (7 from I, one from each F minus one from net charge)	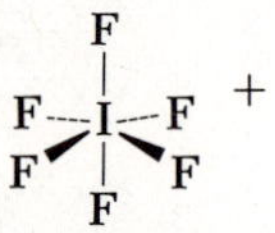 octahedral

TABLE 17.9 **(continued)**

		Number of Electron Pairs	*Structure (Predicted and Experimental)*
1:7	IF_7	7 (7 from I, one from each F)	 monocapped octahedral

No stable monatomic halogen cations, such as Cl^+, Br^+, or I^+ are known; At^+ may have been made but the evidence is not conclusive. Several polyatomic halogen cations do exist. For example, ICl_2^+ can be produced in oleum, H_2SO_4 with about 30% dissolved SO_3:

$$ICl_3 + H^+ \rightleftharpoons ICl_2^+ + H^+ + Cl^- \qquad (17.47)$$

The structure predicted by VSEPR theory for ICl_2^+ is bent rather than linear.

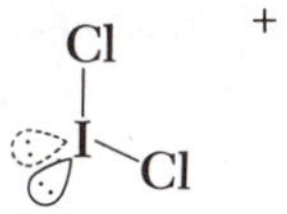

This is quite different from the case of the triatomic anions such as ICl_2^-, which are linear.

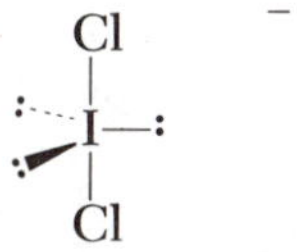

17.8 Group VIIIA: He, Ne, Ar, Kr, Xe, and Rn. The Noble Gases

General Considerations

Until about 1962, the Group VIIIA elements were known as the inert gases. However, from the time that Neil Bartlett was successful in achieving a reaction of Xe with the extremely strong oxidant, PtF_6, that title became inappropriate and the title *noble gases* superseded it. Like the noble metals, exemplified by Au, Pd, Pt, Ph, and Ir, which are reluctant to react, due to a great stability of the metallic state (zero oxidation number), so also the noble gases resist reaction. Only one known source of oxidizing power is enough to bring about chemical reactions with noble gas elements; that source is fluorine in some appropriate chemical form, like PtF_6 or F_2. Only three of the noble gases have shown a tendency to react, all by reactions that increase the oxidation number of the noble gas atom. Those three are xenon, krypton, and radon. Xenon is the lightest of the "reactive" noble gases; the ionization potentials of argon, neon, and helium are too high (see Table 17.10) for these elements to react with known chemical species, even F_2.

Spotlight

The noble gases are all colorless, and together constitute approximately 0.9% of the atmosphere, as shown in the accompanying table.

Noble Gas	*% in air*
He	0.0005
Ne	0.0015
Ar	0.9
Kr	0.0001
Rn	0.00001

It should be noted, however, that He is the second most abundant element in the *universe;* hydrogen is the most abundant.

In 1785, Henry Cavendish, an English chemist and physicist, found that when the oxygen, nitrogen, carbon dioxide, and water were removed from a sample of air, a small bubble remained that represented $\frac{1}{120}$ of the whole. This measurement was remarkably accurate. Over 100 years later, in 1892, Lord Rayleigh measured the density of the residual gas obtained in this manner, and postulated that it was an allotrope of nitrogen. Not until 1894 did William Ramsay show it to be a hitherto unrecognized gas, and wrote to one of his colleagues, "Has it occurred to you that there is room for gaseous elements at the end of the first column of the periodic table?" The gas was called argon, from the Greek word for "inert," and these investigations soon led to the discovery of the whole family of noble gases.

Helium occurs in natural gas wells in the United States, which contain practically the world's supply. Its occurrence in these wells undoubtedly originated from the decay of radioactive elements. It is still used in inflating navigable airships because it has a lifting power equal to 98% of that of hydrogen but is not flammable, like hydrogen. Another Hindenburg type of disaster cannot result with helium-filled airships. One of the recent largest uses for He has been for pressuring liquid fuel rockets; a Saturn booster requires 13 million cubic feet for a firing. Helium is not as soluble in blood as nitrogen is. Thus, deep-sea divers breathe a mixture of helium and oxygen before and during their descents to deplete N_2 from their blood. This prevents N_2 from being released as small bubbles in the blood when the divers surface, thereby preventing the "bends," which can be fatal.

Helium is also used in cryoscopy. It boils at 4 °K and is an unusual liquid below this temperature. For example, it has such a low viscosity that it forms films only a few hundred atoms thick, and flows apparently without friction, even up over the edges of the vessel that contains it. There is as yet no satisfactory explanation for this phenomenon.

Neon was also discovered by Ramsay. It is extensively used in discharge tubes for neon signs.

Argon is sometimes used instead of N_2 in electric light bulbs. It is also used in welding when an inert atmosphere is required, say in welding Mg to prevent the formation of a coating of MgO.

Krypton, as well as xenon, was also discovered by Ramsay. Krypton is sometimes used in place of argon in light bulbs, and is used in photographic flash lamps where brilliant illumination is required.

Xenon is used in making vacuum tubes, and in stroboscopic lamps, bactericidal lamps, and lamps used to excite ruby lasers for generating coherent light. It is potentially useful as a gas for ion engines.

Radon's most important application is for the radioactive treatment of cancer in hospitals. It is sealed in tiny tubes, called needles or seeds, for insertion into the cancerous area. At ordinary temperatures the gas is colorless, but when cooled below its freezing point of −71 °C, it exhibits a brilliant yellow phosphorescence; at the temperature of liquid air, −196 °C, it is orange red.

Compounds of the Noble Gases

The compounds of Xe are all covalent, and the range of this element in oxidation number in known compounds is from +2 (in XeF_2) to +8 (in XeO_4 and $XeO_6{}^{4-}$). The fluorides XeF_2, XeF_4, and XeF_6, all gases at standard temperature and pressure, are prepared by direct reaction of Xe gas with F_2 gas, generally at tempcratures of 250 °C or above in nickel reaction vessels. The ratios of the three gaseous products

TABLE 17.10 Some Fundamental Properties of Noble Gas Elements (Group VIIIA)

Element	*Electronic Configuration*	*1st IP, kcal/mol*	*BP, °K, 1 atm*	*Radius, Å*	*Known to Form Compounds*
He	$1s^2$	566.8	4.2	0.93	No
Ne	$[He]2s^22p^6$	497.2	27.1	1.12	No
Ar	$[Ne]3s^23p^6$	363.4	87.5	1.54	No
Kr	$[Ar]3d^{10}4s^24p^6$	322.8	120.9	1.69	Yes
Xe	$[Kr]4d^{10}5s^25p^6$	279.7	166.2	1.90	Yes
Rn	$[Xe]4f^{14}5d^{10}6s^26p^6$	247.9	211.4	2.20	Yes

produced depend on temperature, pressure, and the ratio of F_2 to Xe in the mixture of reactants. The structures of these three gases are those predicted by VSEPR. Xenon difluoride is linear, because of five electron pairs,

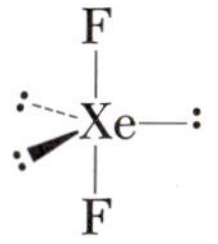

XeF_4 is square (the only square molecule formed by representative elements),

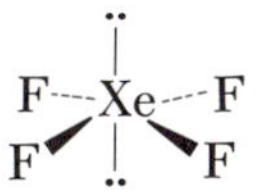

and XeF_6 is a distorted octahedron, because of valence pairs about the Xe center. The xenon fluorides are powerful oxidizing agents, as indicated by the following half-reaction potential:

$$XeF_2 + 2H^+ + e^- \rightleftharpoons Xe + 2HF \qquad \mathscr{E}° = 2.64\ V \qquad (17.48)$$

It was through using xenon fluorides that the first chemical synthesis of the powerful

Spotlight

In the 1930s Linus Pauling suggested that xenon halides, if they could be made, should be stable, but initial attempts to produce xenon fluoride using an arc discharge were unsuccessful. The discovery of xenon fluorides did not occur until three decades later and took a much more circuitous route than the initial arc discharge attempts. Neil Bartlett was attempting to prepare PtF_2 from SF_4 and PtF_4, when he noticed that a red solid formed on the inside of the reaction vessel. It was fortunate that the fluorine-nitrogen gas mixture that was being used in the reaction vessel contained a small amount of oxygen as an impurity. The red compound was found to be $O_2{}^+PtF_6{}^-$. Rather than dismiss this finding as an annoying interference, Bartlett made the observation that the first ionization potentials of O_2 and Xe are very similar (278.5 and 279.7 kcal/mol). He reasoned that if $O_2{}^+PtF_6{}^-$ could be formed, perhaps $XePtF_6$ could also! An attempt to bring about a reaction of hexafluoroplatinum(IV) with xenon was indeed successful, and the "inert gases" became the "noble gases."

oxidant perbromate, BrO_4^-, was achieved. The xenon fluorides form crystalline solids with rather low vapor pressures. One reason for these low vapor pressures is that in the xenon fluorides, unlike many other fluorine-containing compounds, the lone pairs of electrons about the center (xenon) atom appear to be quite chemically active, available as Lewis base centers that allow formation of complexes (Lewis acid-base adducts) of one xenon fluoride molecule with another in the solid state. X-ray diffraction studies of the crystalline state support this view.

Xenon fluorides are hydrolyzed by water to give the explosive trioxide, XeO_3, in which Xe has an oxidation state +6:

$$XeF_6 + 3H_2O \rightleftharpoons XeO_3 + 6HF \tag{17.49}$$

Formation of XeO_3 in basic solution leads to $HXeO_4^-$, which undergoes disproportionation to give Xe species in oxidation states of +8 and zero.

$$XeO_3 + OH^- \rightleftharpoons HXeO_4^- \tag{17.50}$$

$$2HXeO_4^- + 2OH^- \rightleftharpoons XeO_6^{4-} + Xe + O_2(g) + 2H_2O \tag{17.51}$$

The ion XeO_6^{4-} is called the perxenate ion, having a xenon oxidation number +8. Stable perxenate salts can be obtained by precipitating the XeO_6^{4-} from basic solution. A salt such as Na_4XeO_6 is reasonably stable in the solid state. It is an extremely strong oxidant; for example, it readily produces the permanganate ion from Mn^{2+}:

$$4OH^- + 8Mn^+ + 5XeO_6^{4-} \rightleftharpoons 8MnO_4^- + 5Xe + 2H_2O \tag{17.52}$$

Four compounds of krypton have been made. These include KrF_2 and KrF_4, which have structures analogous to those of XeF_2 and XeF_4.

The longest-lived isotope of radon is ^{222}Rn, which has a half-life of only 3.823 days and is difficult to handle because of its radioactivity. The experiments in which oxidation by F_2 have been attempted indicate that F_2 oxidation produces Rn^{2+} (the solution conducts electricity), and there is no reported evidence for covalent species. The tendency towards metallic behavior and lower oxidation states for the heavier member or members of a group thus holds in the noble gas series, as we have seen also for the other groups of the representative elements.

SUMMARY

In the descriptive chemistry of the elements in the third and succeeding periods of the periodic table, it is evident that there is a general discontinuity in the chemistry of these elements and the elements of the second period. The three exceptions to this rule are the alkali metals, the halogens, and the noble gases. Even with these elements, however, there are some very special features of the heavier elements that differ from lithium, fluorine, and neon. For example, lithium shows some tendency towards covalent bond formation, F_2 has a much weaker bond than the other halogen molecules, and the heavier noble gases do exhibit some chemistry, unlike neon.

Vertical trends have been emphasized in terms of structure, with valency expansion common for the heavier elements; metallic character, which is greater in the direction down a group; and acid-base character, in which trends were discussed for both oxides and hydrides.

Some of the special properties of various elements, such as the catenation of sulfur, the semiconductor properties of germanium and silicon, and the great oxidizing power of fluorine, have been examined. Many of these special properties, as well as the properties that are more easily predicted from vertical trends, make the heavier elements important in industrial application and essential in biochemistry.

STUDENT CHECKPOINTS

After studying this chapter, the student should be able to:

1. Discuss the discontinuities in the chemistry of the elements of the second period in relation to the heavier elements of the same group.
2. Cite examples that illustrate the discontinuity between the chemistry of the element of the second period and succeeding periods.
3. Detail the trends in base strength of hydroxides and oxides and acid strength of hydrides.
4. Detail the trends in common oxidation state within groups.
5. Cite examples of donor-acceptor compounds (Lewis acid-base compounds) and trends in their stability.
6. List trends in coordination number within a group and account for these trends.
7. Give names and formulas for the oxo acids of the halogens.
8. Predict structures of compounds of the main group elements.
9. Discuss the importance of the heavier main group elements to industry and to the chemistry of living systems.

EXERCISES

17.1.* What is the difference between Na^+ and K^+ that distinguishes them chemically?

17.2. Which are most alike: Li^+ and Na^+, Na^+ and K^+, or Li^+ and K^+? Explain.

17.3.* Why is the +2 ion the most common form of calcium?

17.4. Beryllium shows an even greater tendency toward covalency than lithium. Why?

17.5.* Which are most alike: Li^+ and Be^{2+}, Li^+ and Mg^{2+}, or Li^+ and Na^+? Justify your choice.

17.6. Why is metallic beryllium harder than metallic calcium?

17.7.* Why is $Mg(OH)_2$ a weaker base than $Sr(OH)_2$?

17.8.* Why is radioactive ^{90}Sr an especially dangerous component of fallout for young children?

17.9. Why is NaOH a stronger base than $Ca(OH)_2$?

17.10. Which is more soluble, $BaCl_2$ or $CaCl_2$? Why?

17.11. Why are Ca^{2+} compounds often more soluble in cold water than in hot water?

17.12.* What disagreeable features are characteristic of "hard" water? What are the chemical reasons for these phenomena?

17.13. Explain the differences in chemical patterns in Group IIIA—especially the difference between boron and the rest of the group and thallium and the rest of the group.

17.14.* Why is the recycling of aluminum energy-efficient?

17.15.* In molten aluminum chloride solutions, the "melt" is made less acidic by adding another chloride salt, such as NaCl. Why is the "melt" less acidic?

17.16. How can electron-deficient compounds with empirical formulas $BeCl_2$ and $AlBr_3$ satisfy the "octet rule"?

17.17. Why is $LiAlH_4$ called a reducing agent, even though the oxidation state of the aluminum atom does change in most of its reactions? Illustrate using a chemical reaction.

17.18.* Pb(II) compounds are generally ionic, while Pb(IV) compounds are covalent. Why?

17.19. Ionization potentials of atoms generally decrease down a group; but while tin is commonly found in a +2 oxidation state, silicon is commonly found in a +4 oxidation state. Explain.

17.20.* The amount of lead in a sample can often be determined quantitatively by reaction with ethylenediaminetetraacetic acid (EDTA), but this method does not work with the lead compound used in gasoline. Explain.

17.21. How are the submicroscopic (molecular scale) structures of various silicates reflected in their macroscopic properties?

17.22.* Why does PCl_3 react readily with F_2, while NCl_3 does not? Why could the reaction be termed oxidative addition?

17.23. Three protons can be titrated from H_3PO_4 with a NaOH solution, but only two from H_3PO_3. Why?

17.24.* Why is the chemistry of polonium more like that of tin than silicon? (For an interesting story, find out how this element was named.)

17.25. Arrange the hydrides of Group IVA in order of (a) increasing acidity, and (b) increasing boiling point. Which order is more easily predicted?

17.26.* Molten sulfur can exist in an elastic form called amorphous sulfur. What might the structure of the sulfur molecules be when sulfur is in this state?

17.27. Scandinavia and the Eastern United States are experiencing problems due to the increasing acidity of rain with each succeeding decade. To what do you ascribe this phenomenon?

17.28.* Give examples of the violation of "monovalency" of halogen bonding. Does the halogen atom act as a Lewis acid or a Lewis base in such species?

17.29. Write the chemical formulas for hypochlorous acid, sodium hypochlorite, perbromic acid, chloric acid, hydrofluoric acid, and calcium perchlorate.

17.30.* Name KBrO, $HClO_4$, $HBrO_2$, $Mg(ClO)_2$, FOH, HI, $NaIO_3$.

17.31. Write the chemical formulas for hyponitrous acid, sodium nitrite, sulfurous acid, magnesium sulfite, potassium phosphite.

17.32. Name Na_2SO_3, $(Ca)_3(PO_4)_2$, $NaNO_2$, H_2SO_3, MgS.

17.33. Which of the following pairs is a better oxidizing agent: (a) NaCl or $NaClO_2$? (b) $HClO_3$ or $HClO_2$? (c) HClO or $HClO_2$?

17.34.* What are the structures of $H_4IO_6^-$, $HBrO_4$, ClO_2^-, and $HClO_3$?

17.35. What are the structures of telluric acid, hypobromous acid, sodium triiodide, iodine trichloride?

17.36. Why are neutral interhalogen compounds only found with odd-number stoichiometries? Illustrate with some examples.

17.37.* Why is ICl_2^+ bent and ICl_2^- linear?

17.38. Although the noble gases are inert under normal conditions, numerous applications have been found for them. Discuss at least one use for each of the noble gases.

17.39.* Why are there stable compounds known for the noble gases xenon, radon, and krypton, but not for argon or neon? Do any of the "normal" periodic trends associated with other groups in the periodic table apply to the noble gases?

17.40. Xenon tetrafluoride is the only square planar neutral molecule in which all the atoms are non–transition elements. What are some examples of square planar ions in which the atoms are all non–transition elements?

ANSWERS TO SELECTED EXERCISES

17.1 The size and hence the charge-to-radius ratio. **17.3** Because this oxidation state provides the highest lattice energy or solvation energy without requiring extremely large ionization energies. **17.5** Li^+ and Mg^+. Both ions have covalent and ionic chemistry. **17.7** The higher charge-to-radius ratio of Mg^{2+} causes Mg^{2+} to attract OH^- groups more strongly. **17.8** The chemistry of Sr is much like Ca and Sr can replace Ca in bones and teeth. It is therefore retained by the body, where it can cause damage due to radioactive decay. **17.12** It forms precipitates in boilers and hot-water heaters and interferes with the action of soap. The $CaCO_3$ and $MgCO_3$ in hard water precipitate when the water is heated. The ions Ca^{2+} and Mg^{2+} bind well to the ionic groups of soap molecules. **17.14** It takes much less energy to melt and reform metallic aluminum than it does to reduce Al(III) in ores. **17.15** $AlCl_3$ is a Lewis acid and Cl^- is a Lewis base. Addition of Cl^- causes $AlCl_4^-$ to be formed (better represented in these melts as $Al_2Cl_7^-$), reducing the concentration of Lewis acid. **17.18** The ionization potentials of lead can be compensated for by lattice energy or solvation energy to give Pb^{2+} but they are too high for ionic compounds of Pb^{4+} to exist. **17.20** To form a compound with EDTA the lead must be ionic (for example, $Pb(ClO_4)_2$ dissolved in water forms $Pb^{2+} + 2ClO_4^-$ and the Pb^{2+} can react with EDTA). The lead in gasoline is in the compound tetraethyllead, which is covalent and does not readily dissociate in solution. **17.22** PCl_3 can undergo "valency expansion." Since the oxidation state of phosphorus is +3 in PCl_3 but +5 in PCl_5, it is oxidized at the same time that two chlorine atoms are added, hence "oxidative addition." **17.24** Polonium can be ionized readily to the +2 state, like tin. Silicon shows almost exclusively covalent chemistry of the +4 oxidation state and is not readily ionized. **17.26** The sulfur atoms form chains of varying length rather than identical S_8 cyclic units. **17.28** Halogens can form two bonds when they bridge two atoms, for example, aluminum atoms in Al_2Br_6. They are also multivalent in many interhalogen compounds, such as BrF_5 and ICl_3. They can act as either Lewis acids or Lewis bases. In the interhalogen compounds, the central atom (Br in BrF_5) is less electronegative than the other atoms and is therefore regarded as a Lewis acid. **17.30** Potassium hypobromite, perchloric acid, bromous acid, magnesium hypochlorite, hypofluorous acid, hydroiodic acid, sodium iodate. **17.34** Octahedral with hydrogen atoms on four of the oxygen atoms; tetrahedral with a hydrogen atom on one oxygen atom; bent; pyramidal with a hydrogen atom on one oxygen atom. **17.37** There are four pairs of electrons in the valence shell of the iodine atom in ICl_2^+ but five pairs in ICl_2^-. **17.39** The ionization potentials and promotion energies of neon and argon are too unfavorable to be overcome by bond formation. Yes, the trend towards stability of lower oxidation states (radon shows only the +2 state so far) and increasing metallic character (radon acts as a +2 ion).

18

Coordination Chemistry of the Transition Elements

18.1 Introduction

In this final chapter you will have the opportunity to apply the chemical principles you have become familiar with to the chemistry of coordination compounds of the transition elements. A proper discussion of coordination chemistry involves electrostatic and covalent bonding concepts, Lewis acid-base chemistry, thermodynamics, electrochemistry, kinetics, spectroscopy, and descriptive chemistry. The field of coordination chemistry has expanded greatly in recent years due both to improved experimental methods for investigating the coordination complexes and to more useful conceptual models for their physical and chemical behavior. Crystal field theory (Sec. 18.5) and "hard and soft acid-base theory" (Sec. 18.3) are two examples of such models. Coordination compounds have some interesting properties and models have been developed for explaining them. In considering them, we shall use many of the concepts we have so far developed.

Most of the chemical compounds that we have so far considered involve elements of Groups IA through VIIIA. In stable structural situations and in their neutral states, atoms of these elements have either totally empty or totally filled *d* and *f* sublevels. The 44 elements of the A groups are referred to as the representative elements or the main group elements. An element whose atoms typically have partially filled *d* or *f* sublevels in any of its common oxidation states is termed a **transition element.** Zinc, cadmium, and mercury are generally included with the transition elements due to many chemical similarities, even though these three elements ordinarily exist in situations in which their *d* and *f* subshells are completely filled. The total number of transition elements now known is 61.* All the transition elements are metals. They are good conductors of electricity (or heat) and are lustrous in the elemental state. In compounds, they have positive oxidation numbers rather than negative oxidation numbers with very few exceptions.

The bonding in which transition elements engage is generally Lewis acid-base bonding (Sec. 8.2) with a substantial amount of ionic character. One can view the Lewis acid-base bonding as follows. A positively charged transition metal ion acts as a Lewis acid, accepting and sharing pairs of electrons from Lewis bases, such as ammonia (NH_3), H_2O, and Cl^-. The Lewis bases, called **ligands,** donate both of the electrons to be shared in the covalent bond. This type of bond, in which both shared electrons are viewed as being donated by just one of the partners in the bond, is referred to as a **dative bond** (from the Latin, "to give"), or **coordinate covalent bond.**† The compounds thus formed are termed **coordination compounds,** or **transition metal complexes.** It is very important to note that the transition metal atom shares the electrons from the ligands to some extent in the complex, and is not a simple ion in a coordination compound. Nevertheless, it is often convenient to view the metal atom almost as though it really exists as an ion; the amount of ionic character that exists in the complex is very significant and is reflected in the common tendency of chemists to speak of the metal atoms in transition metal complexes as metal ions. Strictly speaking, one should refer to these metal atoms simply as atoms of a certain assigned oxidation number.

Much chemistry of the transition metals exists besides coordination chemistry.

*It is conceivable that some new transition elements may be produced synthetically in the near future by the methods of high-energy nuclear physics.

†We should recognize that the assignment of both electrons in what we have called a dative bond to the ligand does not describe the fate of electrons in the molecule. The electrons belong to the bond (actually, to the entire molecule), irrespective of where they came from.

Spotlight

Nitric oxide, NO, is an interesting ligand, with an odd number of electrons. An important qualitative test for nitrates, known as the brown ring test, is based on the formation of a complex of NO with Fe(II). In this test, the material suspected of containing NO_3^- is mixed with concentrated sulfuric acid, H_2SO_4. A solution of ferrous sulfate, $FeSO_4$, is then slowly poured down the side of the container so that it forms a layer on top of the denser H_2SO_4 layer. If NO_3^- is present, Fe^{2+} will reduce it to NO. Additional Fe^{2+} then complexes with the NO to form $[Fe(H_2O)_5NO]^{2+}$, which appears as a brown ring or layer at the interface of the two layers:

$$[Fe(H_2O)_6]^{2+} + NO \rightleftharpoons [Fe(H_2O)_5NO]^{2+} + H_2O$$

The removal of the odd electron from NO is not very difficult. Stable solid compounds containing the nitrosonium ion, NO^+, have been isolated. For example, nitrosonium bisulfate, $NO^+HSO_4^-$, is an isolatable intermediate in the lead chamber process used for the commercial production of sulfuric acid, in which NO functions as a catalyst for the oxidation of SO_2 by O_2:

$$NO(g) + \tfrac{1}{2}O_2(g) \rightleftharpoons NO_2(g)$$

$$\underline{NO_2(g) + SO_2(g) + H_2O(l) \rightleftharpoons H_2SO_4(l) + NO(g)}$$

$$SO_2(g) + \tfrac{1}{2}O_2(g) + H_2O(l) \rightleftharpoons H_2SO_4(l)$$

The ion NO^+ is isoelectronic with CO and can similarly form bonds with metals; $[Ni(CO)_4]$ and the isoelectronic $[Co(CO)_3NO]$ are typical examples.

Of importance are oxides such as permanganate and chromate ions (MnO_4^- and CrO_4^{2-}), hydrides, and carbides. Comprehensive treatment of transition metal chemistry can be found in numerous current monographs. Here we are chiefly concerned strictly with coordination chemistry of the transition elements.

18.2 Nomenclature of Coordination Compounds

At first glance the formulas and nomenclature of coordination compounds may appear to be quite strange and unnecessarily complicated. Brackets and parentheses, as well as odd-sounding suffixes like *-ito*, *-ato*, and *-ido* abound in coordination chemistry. Fortunately, the rules are reasonably straightforward and quickly mastered.

Formulas

First let us consider some typical formulas. The coordination ion or molecule itself is enclosed within brackets. The symbol for the central atom is placed first, followed by the symbols for the ligands in the following order: anionic, neutral, and, in rare instances, cationic. Examples of this part of the formula are $[PtCl_4]^{2-}$, $[CoCl_2(NH_3)_4]^+$, in which the ligands are Cl^- and NH_3; and $[RuBr_3(H_2O)_3]$, the ligands being Br^- and H_2O. If the compound is a solid that contains cations in the crystal lattice that compensate for the negative charge of the complex ion and are not attached to the central atom itself, the symbols for the cations are placed before the brackets; an example is $K_2[PtCl_4]$, an ionic solid in which the ions are K^+ and $[PtCl_4]^{2-}$. Charge-compensating anions that are not bound to the central atom are shown *after* the bracket; $[CoCl_2(NH_3)_4]ClO_4$ is an example of a solid complex containing the ions $[CoCl_2(NH_3)_4]^+$ and ClO_4^-.

The order in which information is presented in the name of a coordination compound is

1. Charge-compensating cations.
2. Anionic ligands (word representation ending in "*o*"; for example, nitrate becomes *nitrato* and carbonate becomes *carbonato*). The following anions are very common ligands:

H^-	hydrido	F^-	fluoro	NO_3^-	nitrato
Br^-	bromo	OH^-	hydroxo	NO_2^-	nitrito
Cl^-	chloro	I^-	iodo	CO_3^{2-}	carbonato
CN^-	cyano	O^{2-}	oxo	OCN^-	cyanato

The number of times a particular "inorganic" anion occurs as a ligand (e.g., the examples listed above) in a coordination species is generally indicated by the following prefixes: *di* (two), *tri* (three), *tetra* (four), and so on. A multiple occurrence of an organic anionic ligand, such as

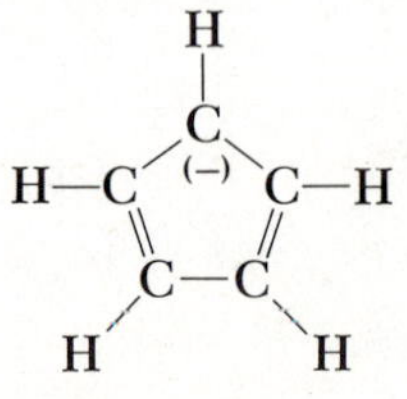

cyclopentadienyl ion

is denoted by using the prefixes *bis* (two), *tris* (three), *tetrakis* (four), and so on, before the name of the ligand. The ligand name is then enclosed in parentheses, as in bis(cyclopentadienyl).

3. Neutral ligands are presented in the order: *aquo* (H_2O), ammine (NH_3), other neutral inorganic ligands in alphabetical order, then neutral organic ligands in alphabetical order. Multiple occurrences are denoted by *di*, *tri*, *tetra*, and so on for the inorganic ligands, and *bis*, *tris*, *tetrakis*, and so on for the neutral organic ligands.
4. The central metal atom. The oxidation number of the central metal atom is given as a Roman numeral in parentheses after the name of the metal. If the coordination complex is an anion, the central atom name ends in *-ate;* $K_2[PtCl_4]$, for instance, is potassium tetrachloroplatinate(II), indicating the presence of a minus charge on $[PtCl_4]^{2-}$.
5. Charge-compensating anions.

By the guidelines just listed, we name some representative coordination compounds as follows.*

Formula	*Name*
$[Co(SCN)(NH_3)_5]Cl_2$ Note that the metal atom symbol is *first* in the formula, giving the order metal atom, anionic ligand, neutral ligands, compensating anion (Cl^-)	thiocyanotopentaamminecobalt(III) chloride Order: anionic ligand, neutral ligands, metal atom, compensating anion
$[Rh(OH)_2(en)_2]ClO_4$ (where "en" stands for $H_2NCH_2CH_2NH_2$, ethylenediamine)	dihydroxobis(ethylenediamine)rhodium(III) perchlorate Note the prefixes of hydroxo and ethylenediamine ligands
$[Rh(OH)_2(en)_2]ClO_4$ In this case there are charge-compensating cations and two types of anionic ligands (in alphabetical order)	rubidium pentacyanonitrosylferrate(III) Note the prefix *-ate* on ferrate, since this is a negatively charged coordination ion. Nitrosyl stands for the NO ligand
$[Fe(C_5H_5)_2]$	bis(cyclopentadienyl)iron(II) As in organic nomenclature, some coordination compounds also have trivial, or nonsystematic, names; this complex is popularly called ferrocene
$[CoBr_2((CH_3)_2NH)_2]$ In this case, the ligand $(CH_3)_2NH$ occurs twice.	dibromobis(dimethylamine)cobalt(II)

EXAMPLE 18.1 Name the following complex compounds:

(a) $[Pt(NH_3)_6]Cl_4$ (b) $K_2[CuCl_4]$
(c) $K_4[Fe(CN)_6]$ (d) $[PtBr_2(en_2](ClO_4)_2$
(e) $[RuCl_2(NH_3)_4]$

*Often an exception is made to the "normal" order, with H_2O or NH_3 preceding the anionic ligands and and other neutral ligands.

Spotlight

The nitrite ion, NO_2^-, acts as a ligand with several transition metal cations.

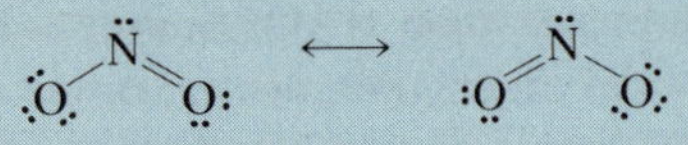

The hexanitritocobaltate(III) ion, $[Co(NO_2)_6]^{3-}$, is important in the qualitative analysis of cobalt. The ion NO_2^- complexes with the cobaltous ion, Co^{2+}, and then additional NO_2^- oxidizes the cobalt(II) to cobalt(III); in the presence of K^+, a yellow precipitate of potassium hexanitrocobaltate(III) forms.

$$Co^{2+} + 6NO_2^- \rightleftharpoons [Co(NO_3)_6]^{4-}$$

$$[Co(NO_2)_6]^{4-} + NO_2^- + 2H^+ \rightleftharpoons [Co(NO_2)_6]^{3-} + NO(g) + H_2O$$

$$[Co(NO_2)_6]^{3-} + 3K^+ \rightleftharpoons K_3[Co(NO_2)_6](s)$$

The ion NO_2^- can be distinguished from the NO_3^- in the reaction with Fe^{2+} in acidic solution to produce NO. The same brown color forms as in the brown-ring test for nitrates, except that a weakly acidic solution, dilute acetic acid, is required to form a brown color in the case of nitrites, whereas a strongly acidic solution, concentrated sulfuric acid, is required in the case of nitrates.

SOLUTION (a) Hexaammineplatinum(IV) chloride. The order of naming is neutral ligands (prefix *hexa-*, since they are inorganic); metal atom (with oxidation number in parentheses); and the compensating anion.
(b) Potassium tetrachlorocuprate(II). Note that the compensating cation is given first and the ending *-ate* is used in the name of the complex metal ion, since the coordination species is anionic.
(c) Potassium hexacyanoferrate(II) (or potassium ferrocyanide). Ferrocyanide, $[Fe(CN)_6]^{4-}$, and ferricyanide, $[Fe(CN)_6]^{3-}$, are commonly used coordination ions that are generally known by their trivial names.
(d) Dibromobis(ethylenediamine)platinum(IV) perchlorate. In this example the prefix *di* is used for the inorganic ligand while *bis* is used for the organic ligand.
(e) Dichlorotetraammineruthenium(II). This is an example of a neutral coordination compound, rather than an ion; therefore no compensating cation or anion appears in the formula. The prefixes in this case both refer to inorganic ligands—one anionic and one neutral.

STUDY PROBLEM 18(a)

Name (a) $[PdCl_2(NH_3)_2]$; (b) $Na_2[ZnCl_4]$; and (c) $[Ru(en_2)Cl_2]Br$.

EXAMPLE 18.2 Give the proper formula for each of the following compounds.
(a) hydroxopentaaquoaluminum(III) perchlorate.
(b) bromochlorodiaquodiamminerhodium(III) nitrate.
(c) potassium tetrachlorocuprate(II).
(d) carbonatoidobis(ethylenediamine)chromium(III).

SOLUTION (a) $[Al(OH)(H_2O)_5](ClO_4)_2$. Note the order: central metal atom, anionic ligand, neutral ligand, then compensating anion. Also note that as this case shows, the

nomenclature of transition metal complexes is sometimes extended to the compounds of representative elements, for example, aluminum.
(b) $[RhBrCl(H_2O)_2(NH_3)_2]NO_3$. The anions are shown first in alphabetical order. Water (*aquo*) takes precedence over ammonia (*ammine*) when the terms for neutral ligand are arranged in the formula.
(c) $K_2[CuCl_4]$. The symbol for the charge-compensating cation appears first in the formula.
(d) $[Cr(CO_3)I(en)_2]$. Abbreviations are quite common in the formulas of coordination compounds; in this case, "en" is the abbreviation used for ethylenediamine.

STUDY PROBLEM 18(b)

Give the formula for (a) chloropentaaquochromium(III) perchlorate; (b) sodium hexafluorocobaltate(III).

18.3 Bonding in Coordination Compounds

An interesting and important feature of coordination compounds is that the bonds have some properties characteristic of electrostatic attraction and some features that are typical of covalent bonding. Electrostatic bonding can be considered by viewing each interacting entity as possessing a particular charge, an electric dipole, and examining the net effect of electrostatic attractions and repulsions on the energy, or stability, of a particular arrangement of the interacting entities. In this way, we can rationalize the lattice energy of NaCl, MgF_2, and other ionic solids (Chap. 14). Similarly, a large fraction of the total stability of transition metal complexes can be accounted for by a simple electrostatic model, based on assuming an electrostatic attraction between a positively charged metal ion and the polar or ionic Lewis bases, the ligands. Predictions of thermodynamic stability based entirely on electrostatic consideration, however, are not consistent with measured equilibrium constants, so further consideration of bonding in the compounds is necessary.

Some properties of coordination compounds are much different from what is typically found in ionic compounds. For example, when coordination compounds are dissolved in water, it is often found that the complex does not undergo a dissociation of the ligands from the metal atom. In such cases, the dissolved coordination complex may act as an ion with a net charge that is very different from that assigned to the central metal ion in a purely ionic model. For example, the complex bromochlorotetraamminecobalt(III) chloride acts, in aqueous solution, as a combination of +1 ions (bromochlorotetraammine(III) cation) and −1 ions (chloride), that is, it dissociates in solution:

$$[CoBrCl(NH_3)_4]Cl(s) \rightleftharpoons [CoBrCl(NH_3)_4]^+ + Cl^-$$

A second important property that transition metal complexes have in which they resemble covalent bonding types is their ability to retain particular shapes. Various methods have been developed to explore or predict the nature of the bonding in transition metal complexes, for example, to determine the extent of covalency in the bonds between the metal atom and the ligands and to compare the properties of a complex with what one would expect for simple ionic bonding.

Spotlight

A biologically important cobalt-containing complex is Vitamin B_{12}, the structure of which is shown herewith.

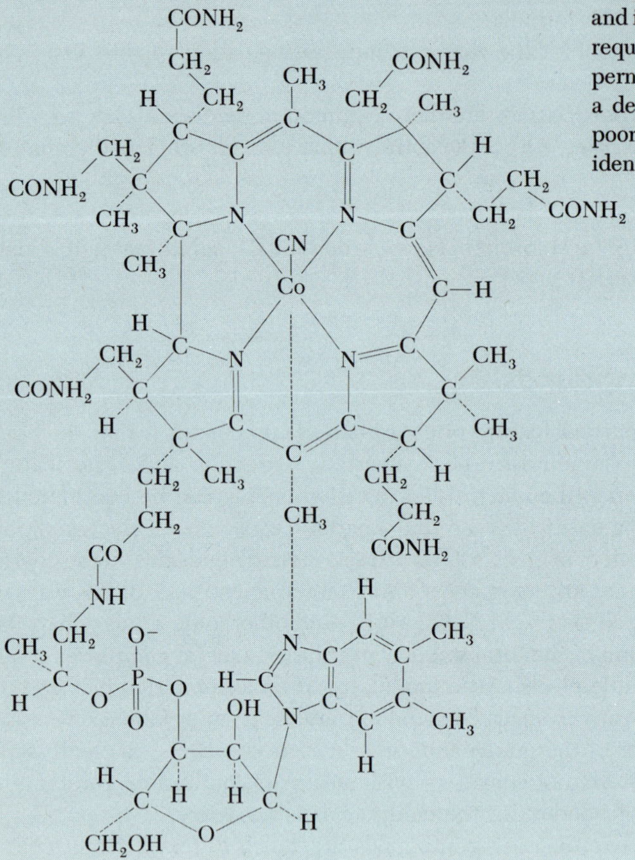

Vitamin B_{12} acts as an agent for transferring hydride ion, H^-, causing the reduction of other species. This vitamin is synthesized in our lower intestinal tract and is lost by excretion. Very small amounts of B_{12} are required for good health, but deficiencies can cause pernicious anemia. This affliction manifests itself by a deficiency of hemoglobin in the blood, leading to poor respiration. Before the B_{12} requirement was identified, pernicious anemia was fatal.

Stability Constants; Equilibria of Coordination Compounds

Equilibrium constants allow us to calculate the concentrations of various species in solution when the system has reached equilibrium.* The equilibrium constants for coordination compounds are often called stability constants, and several compila-

*It is important to realize throughout this discussion that some reactions approach equilibrium much faster than others. As a result, chemists often make and use chemical species that are predicted to be in low concentration because of equilibrium constant considerations (thermodynamic stability) but whose reactions are so slow (giving them kinetic stability) that they can exist in high concentrations for long periods.

tions are available in most libraries. These stability constants can be written as a series of equilibrium constants:

$$Cu^{2+}(aq) + NH_3 \rightleftharpoons [CuNH_3]^{2+} \qquad K_1 = \frac{[ML]}{[M][L]} \tag{18.1}$$

$$[CuNH_3]^{2+} + NH_3 \rightleftharpoons [Cu(NH_3)_2]^{2+} \qquad K_2 = \frac{[ML_2]}{[ML][L]} \tag{18.2}$$

$$[Cu(NH_3)_2]^{2+} + NH_3 \rightleftharpoons [Cu(NH_3)_3]^{2+} \qquad K_3 = \frac{[ML_3]}{[ML_2][L]} \tag{18.3}$$

$$[Cu(NH_3)_3]^{2+} + NH_3 \rightleftharpoons [Cu(NH_3)_4]^{2+} \qquad K_4 = \frac{[ML_4]}{[ML_3][L]} \tag{18.4}$$

where M is Cu^{2+} and L is NH_3. The stability constants are also written as products of the equilibrium constants for individual steps. The symbol "β" is sometimes used to denote the equilibrium constant for a particular species written in terms of the concentrations of metal ion (M) and ligand (L) in solution at equilibrium.

$$\beta_1 = \frac{[ML]}{[M][L]} = K_1 \tag{18.5}$$

$$\beta_2 = \frac{[ML_2]}{[M][L]^2} = \frac{[ML_2]}{[ML][L]} \cdot \frac{[ML]}{[M][L]} = K_1K_2 \tag{18.6}$$

$$\beta_3 = \frac{[ML_3]}{[M][L]^3} = \frac{[ML_3]}{[ML_2][L]} \cdot \frac{[M_2]}{[M][ML]} \cdot \frac{[ML]}{[M][L]} = K_1K_2K_3 \tag{18.7}$$

$$\beta_4 = \frac{[ML_4]}{[M]]L]^4} = K_1K_2K_3K_4 \tag{18.8}$$

The stability constants of most complexes decrease successively. For the complexes of copper and ammonia at 25 °C, the values are $K_1 = 10^{4.13}$, $K_2 = 10^{3.53}$, $K_3 = 10^{2.87}$, and $K_4 = 10^{2.15}$. There are several reasons for this decrease. One of the most common and important reasons is that as ammonia ligands are added they take the place of a water molecule that was attached to the metal atom. As more ammonia ligands are added, the odds that another ammonia ligand will replace a water ligand rather than change places with an ammonia ligand already on the metal atom become less favorable. A reaction in which identical ligands change places,

$$[Cu(NH_3)_3{}^{2+}] + NH_3 \rightleftharpoons [Cu(NH_3)_3]^{2+} + NH_3 \tag{18.9}$$

does not affect the equilibrium, since no observable change has taken place.

For many years, inorganic chemists have relied on their own knowledge of the ability of various ligands to bind to particular metal ions or they have looked up values in a compendium that lists values for stability constants. Predicting stability constants in any general way was not very successful. Several chemists have recently made progress in prediction, however. Professor Ralph Pearson has developed one of the most widely used predictive methods, the hard and soft acid base theory.

TABLE 18.1 **Hard and Soft Lewis Bases**

Hard	*Soft*[a]
H_2O, OH^-, F^-	R_2S, RSH, RS^-
$CH_3CO_2^-$, PO_4^{3-}, SO_4^{2-}	I^-, SCN^-, $S_2O_3^{2-}$
CO_3^{2-}, ClO_4^-, NO_3^-	CN^-, RNC, CO
ROH, RO^-, R_2O	C_2H_4, C_6H_6
NH_3, RNH_2, N_2H_4	H^-, R^-
Borderline	
pyridine, N_3^-, Cl^-, Br^-, NO_2^-, SO_3^{2-}, N_2	

[a]R stands for an alkyl or aryl group (see Chap. 8).

TABLE 18.2 **Hard and Soft Lewis Acids**

Hard	*Soft*
H^+, Li^+, Na^+, K^+	Cu^+, Ag^+, Au^+, Tl^+, Hg^+
Be^{2+}, Mg^{2+}, Ca^{2+}, Sr^{2+}, Mn^{2+}	Pd^{2+}, Cd^{2+}, Pt^{2+}, Hg^{2+}
Al^{3+}, Sc^{3+}, Ga^{3+}, In^{3+}, La^{3+}	Tl^{3+}, $Tl(CH_3)_3$, BH_3
Cr^{3+}, Co^{3+}, Fe^{3+}, As^{3+}	$GaCl_3$, GaI_3, $InCl_3$
Si^{4+}, Ti^{4+}, Zr^{4+}, Th^{4+}, U^{4+}	I_2, Br_2, ICN
$Al(CH_3)_3$, $AlCl_3$, AlH_3, BF_3	M^0 (metal atoms)
Borderline	
Fe^{2+}, Co^{2+}, Cu^{2+}, Zn^{2+}, Pb^{2+}, In^{2+}, Sb^{3+}, Bi^{3+}, Rh^{3+}, Ir^{3+}, Ru^{2+}, Os^{2+}, SO_2, GaH_3	

According to this approach, bases that are "hard," generally having high electronegativity and other tendencies related to their ability to hold their electrons tightly, will tend to bind most strongly to "hard" Lewis acids. "Softer" bases, in turn, bind more strongly to "softer" acids. One way to look at these trends is that the interaction of a hard base with a hard acid will emphasize strong electrostatic attraction, whereas the interaction of a soft base and a soft acid will favor strong covalent bonding. Some species are neither very hard nor very soft and therefore sometimes have higher stability constants for combination with hard partners and sometimes with soft partners. These species are termed "borderline" cases. Tables 18.1 and 18.2, taken from Professor Pearson's work, illustrate the hard and soft classification. In Table 18.3 it is evident that the highest stability constants are found for hard-hard and soft-soft combinations. The stability constants of species formed with species whose hardness is borderline cover a large range and are less easily predicted.

EXAMPLE 18.3 For each of the following, determine the more favorable free energy value at 27 °C for: (a) binding of F^- or I^- to Fe^{3+}; and (b) binding of F^- or I^- to Hg^{2+}.

TABLE 18.3 **Some Examples of the Correlation of Hardness and Softness with Equilibrium Constants**

Hard	K_{eq}	*Soft*	K_{eq}
		ACID	
$Fe^{3+} + F^-$ (Hard)	$10^{5.2}$	$Hg^{2+} + F^-$ (Hard)	$10^{1.0}$
$Fe^{3+} + Cl^-$ (Borderline)	$10^{1.5}$	$Hg^{2+} + Cl^-$ (Borderline)	$10^{6.7}$
$Fe^{3+} + I^-$ (Soft)	$10^{1.3}$	$Hg^{2+} + I^-$ (Soft)	10^{13}
		BASE	
$NH_3 + Co^{3+}$ (Hard)	$10^{7.3}$	$SCN^- + Mn^{2+}$ (Hard)	$10^{1.2}$
$NH_3 + Cu^{2+}$ (Borderline)	$10^{4.3}$	$SCN^- + Cu^{2+}$ (Borderline)	$10^{2.3}$
$NH_3 + Cd^{2+}$ (Soft)	$10^{2.7}$	$SCN^- + Hg^{2+}$ (Soft)	10^{16}

SOLUTION Using the K_{eq} values in Table 18.3 and the relation $\Delta G = -RT \ln K_{eq}$ (or $\Delta G = -2.303\ RT \log K_{eq}$), we get

(a) $\Delta G_{F^-} = -2.303(596) \log 10^{5.2} = -1373(5.2) = -7139$ cal/mol

$\Delta G_{I^-} = -2.303(596) \log 10^{1.3} = -1373(1.3) = -1785$ cal/mol

The ΔG for binding of F^- to Fe^{3+} is 5.4 kcal/mol more favorable.

(b) $\Delta G_{F^-} = -2.303(596) \log 10^{1.0} = -1373$ cal/mol

$\Delta G_{I^-} = -2.303(596) \log 10^{13} = -13(1373) = -17{,}849$ cal/mol

The ΔG for binding of I^- to Hg^{2+} is more favorable by 16.5 kcal/mol.

STUDY PROBLEM 18(c)

The K_{eq} value for the replacement of NH_3 by Cl^- in $[Co(NH_3)_6]^{3+}$ at 25 °C is 32. What is the free energy of this reaction?

Paramagnetism and Diamagnetism in Transition Metal Complexes

If the chromium atom in $[CrCl_2(H_2O)_4]Cl$ were simply a +3 ion, that is, if the bonding in the transition metal complex were simply ionic, then we should expect that the electronic configuration of the chromium in this substance would be $[Ar]3d^3$ (Chap. 6). Such a species would have three unpaired electrons. The existence of unpaired electrons, and even the number of unpaired electrons, can be determined for a substance by measuring the extent to which a sample of the substance is paramagnetic or diamagnetic. A compound is said to be **paramagnetic** if it weighs more in the presence of a magnetic field than out of a magnetic field, that is, if it is "drawn into" a magnetic field.* The results of an experiment using an apparatus such as that schematically illustrated in Fig. 18.1 would be, for a paramagnetic substance, a higher weight reading when the electromagnet is turned on than when

*The basis for paramagnetism is the magnetic moments associated with unpaired electrons (see Chaps. 6 and 8). These moments can align themselves either with or against a magnetic field to which they are subjected. There is a slight (temperature-dependent) excess of the alignment that leads to a mutual attraction of the electron magnetic moments with the applied magnetic field; this is responsible for the observed "increase in weight."

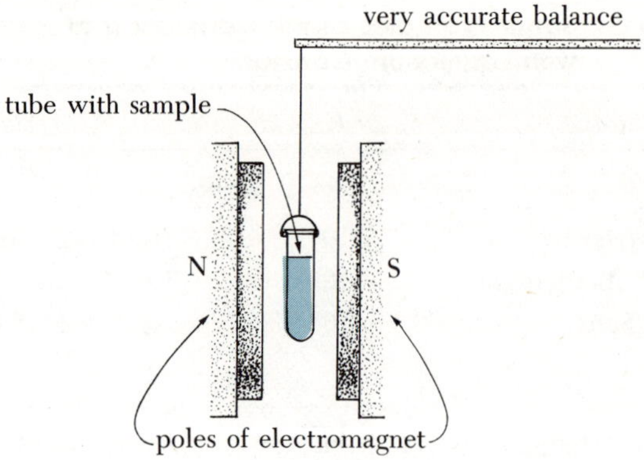

Figure 18.1
A simple apparatus for measuring paramagnetism or diamagnetism in a sample, showing the sample in an electromagnet, and attached to the arm of a balance.

it is off. A greater number of unpaired electrons per gram of sample results in a greater increase in weight in the presence of the magnetic field. The weight increases in weight per mole of the complex $[CrCl_2(H_2O)_4]Cl$, which is found to have three unpaired electrons, is greater than for $[VCl_2(H_2O)_4]Cl$, with two unpaired electrons; this is in turn greater than for $[TiCl_2(H_2O)_4]Cl$, with one unpaired electron. The electronic configurations that one would assign to the metal atoms in these complexes, taking the ionic view, are $[Ar]3d^3$ for Cr(III), $[Ar]3d^2$ for V(III), and $[Ar]3d^1$ for Ti(III). In this series of compounds, the experimental facts regarding the number of unpaired electrons are in accord with what one would predict directly from the electronic configurations of the metal ions of a purely ionic view; in this sense, these data can be viewed as support for an ionic interpretation of the bonding in these complexes.

Not all species experience a weight increase on being placed in a magnetic field. A species such as $K_2[TiCl_6]$, in which titanium is predicted to have the electronic configuration [Ar] in a strictly ionic view, has no unpaired electrons and actually weighs less on being introduced into a magnetic field than in the absence of a field. Compounds that weigh less in a magnetic field than outside the field are said to be **diamagnetic.*** Diamagnetism indicates a lack of unpaired electrons, and its presence is important to know in judging the nature of bonding in a transition metal complex. In the $[TiCl_6]^{2-}$ complex, the observation of diamagnetism agrees with the prediction of a simple ionic model that there are no unpaired electrons.

Conductivity and the Nature of Bonding

The paramagnetism-diamagnetism data cited above, and many other forms of experimental evidence, point to the importance of ionic character in the structures of transition metal complexes. Indeed the large portion of the stability of these complexes can be accounted for by the electrostatic attraction of the metal "ion"

*Diamagnetism occurs because electrons experience currents that are induced by the presence of the applied magnetic field. These induced electronic circulations themselves generate magnetic fields, and these induced magnetic fields give rise to a repulsion by the applied magnetic field. These induced electron currents and the associated diamagnetism are also present in paramagnetic species, but the paramagnetic effect is generally much larger.

for electron pairs of the ligand. Nevertheless, some experiments clearly suggest significant covalent character in bonding between a transition metal atom and ligands. One of these is the conductivity experiment, in which the ability of a species to carry electric current between two electrodes is measured. Ions are required to carry the current in a solution (Sec. 12.1). The number and identity of the ions determine how much current is carried when a given voltage is applied across the electrodes of a specific cell. When coordination compounds are dissolved in water, it is often found that the ligands do not dissociate from the metal to yield a metal ion. In such cases, the dissolved coordination complex will (unless perhaps all of the ligands are neutral) conduct less electric current than would be expected if the complex dissociated into ions. If the total negative charge of the ligands just balances the positive charge assigned to the central metal ion, the complex itself is neutral. A nondissociating transition metal complex will not conduct current in solution. The compound $[CoCl_3(NH_3)_3]$ is an example of such a substance; it does not contribute to the conductivity when dissolved in water. Its behavior is much different from what one would expect if the ions Co^{3+} and Cl^- were present due to a dissociation of $[CoCl_3(NH)_3]$. Thus, in behavior, this complex is much different from a compound like NaCl, which immediately dissociates in water, allowing the water molecules to surround the Na^+ and Cl^-, and providing ionic species capable of conducting electrical current.

Of course, not all transition metal complexes are neutral. Since the conductivity experiment provides quantitative information about the number and identity of ions, one can also investigate charged complex ions. The ion $[CoCl_2(NH_3)_4]^+$, together with a counteranion like ClO_4^-, has an effect on conductivity roughly like that of a singly charged nonatomic cation, say Cs^+, also with a counteranion, but with differences due to the sizes, shapes, and internal charge distributions of the two ions. This is much different from the conductivity one would expect for the ions that would be present if the complex dissociated—Co^{3+} and two Cl^-. The ion $[CoCl_4(NH_3)_2]^-$, on the other hand, has conductivity characteristics roughly like those of a -1 ion, much like ClO_4^-. The conductivity is definitely not that of a Co^{3+} and four Cl^-. Thus, the conductivity evidence in these two cases suggests that there is something about the bonding in these complexes, namely convalency, that causes them to behave differently in solution from purely ionic compounds, the latter typically ionizing in aqueous solution.

Studies of Structural Geometry

Another characteristic in which transition metal coordination complexes resemble covalent compounds is the ability to retain particular shapes, their structural integrity. X-ray diffraction studies of simple ionic compounds (Sec. 14.3) show that the ions tend to pack as efficiently as possible with positive and negative ions alternating throughout the crystal. In compounds such as Na_3PO_4, with polyatomic ions, the polyatomic ions, which have atoms covalent bound to each other, are easily identified in x-ray studies. In the case of Na_3PO_4, the ions PO_4^{3-} are found as tetrahedral units with four oxygen atoms in close proximity and a phosphorus atom at the center. Coordination complexes act like polyatomic ions in x-ray studies. In the

Spotlight

In 1893, Alfred Werner in Zurich developed the concept of directed valency in coordination compounds. The existence of geometrical isomers, for example, cis and trans isomers, and a large body of previously puzzling data led Werner to the idea of specific geometrical orientations of metal ligand bonds, and specifically to the concept of octahedral complexes. Werner's ideas of metal ligand coordination were further advanced by G. N. Lewis (1875–1946) in Berkeley in 1916. Lewis, who was born not far from Harvard University and earned his A.B. and Ph.D. degrees there, visualized the bond between a central metal ion and each of the attached ligands in terms of electron-pair bonds. This concept of an electron-pair bond, and the associated acid-base concept that bears Lewis's name, have remained important features of modern chemical ideas. Professor Lewis also made important contributions to chemical thermodynamics and was a significant force in physical chemistry during the first half of the twentieth century.

crystal structure of $[CoCl(NH_3)_5]SO_4$, for example, the ion $[CoCl(NH_3)_5]^{2+}$ is revealed as a definite structural entity with the structure

$$\begin{array}{c} Cl \\ | \\ H_3N\text{---}Co\text{---}NH_3 \\ H_3N \quad | \quad NH_3 \\ Cl \end{array}^+$$

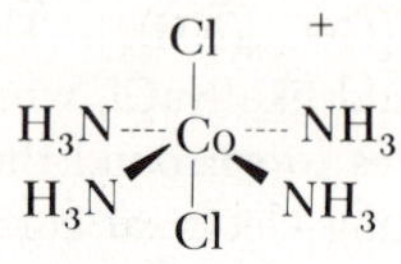

This polyatomic transition metal complex ion is found to alternate in the crystal structure with the polyatomic sulfate ion.

Another feature of the structures of transition metal complexes that indicates covalent character is the occurrence of complexes whose shapes are not what would be predicted as the most efficient based on electrostatic interactions. Such shapes can in some cases be shown experimentally, through various forms of spectroscopy, to persist in solution as well as in the solid state. Two means of packing ligands about a positively charged metal ion that would maximize favorable electrostatic interactions—minimizing the total electrostatic energy—are:

1. Forming an octahedral arrangement like that found in the NaCl crystal structure (Fig. 7.1) or in $[CoCl(NH_3)_5]^{2+}$.
2. Forming a tetrahedral arrangement, which is found in the solid state structure of ZnS and in the complex $[CoCl_4]^{2-}$.

Many transition metal complexes do indeed have such geometrical arrangements. Transition metal complexes are also found in other shapes, however—geometrical arrangements that are not known for typical ionic species. Two notable shapes that indicate covalent character in the bonding of coordination complexes are the linear shape and the square planar configuration.

Linear complexes are generally found for metal "ions" with a d^{10} configuration and a low oxidation number (+1, and in some cases, +2). Some representative complex ions known to be linear in the solid state are the diamminesilver(I) ion,

$$\begin{array}{c} H \quad H \qquad\qquad H \\ \diagdown | \qquad\qquad | \diagup \\ N\text{---}Ag\text{---}N \\ \diagup \qquad\qquad \diagdown \\ H \qquad\qquad H \end{array}^+ \quad II$$

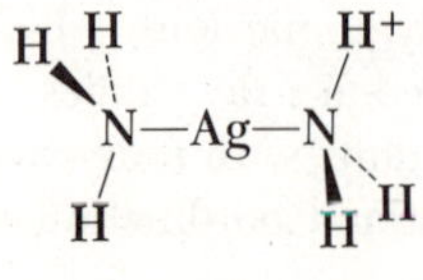

the dichlorocuprate(I) ion, Cl—Cu—Cl⁻, and dimethylmercury(II),

$$\begin{array}{c} H \\ H \diagdown | \\ C-Hg-C \\ / | \diagdown \\ H \; H \; H \end{array}$$

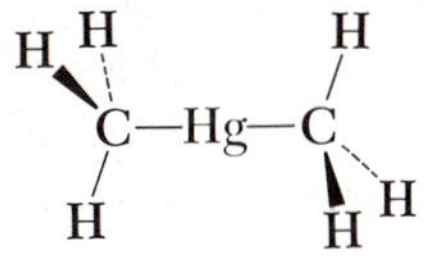

If only electrostatic interactions were important, about six electron donors would be packed about these relatively large metal ions in the crystal. The ion Ag^+, for example, is approximately the size of Rb^+, which has six Cl^- as nearest neighbors in the crystalline state of RbCl.

The square planar geometry is found for complexes in which the central metal atom (ion) has a d^8 electronic configuration (and for some d^9 metal atoms). Examples of d^8 configurations are Pt(II), Pd(II), Rh(I), Au(III), and Ni(II).° In square planar compounds of these metals, the "top" and "bottom" of the metal ion are exposed, seemingly poised for convenient attraction of electron-pair-donating species. Although one might expect chloride ions to be strongly attracted to such a species, for instance to $[Pt(NH_3)_4]^{2+}$, this is not the case. In the crystalline structure of $[Pt(NH_3)_4]Cl_2$, the chloride counterions are not bound to the platinum atom. The four ammonia molecules are bound to the platinum atom to form a square complex ion:

$$\begin{array}{ccc} H_3N & & NH_3 \\ & \diagdown \; / & \\ & Pt & \\ & / \; \diagdown & \\ H_3N & & NH_3 \end{array}^{2+}$$

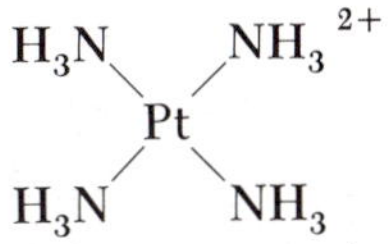

Apparently, one has to be somewhat more sophisticated than to adopt a strictly ionic approach if one is to understand bonding and geometries of the complexes of transition metals. A theoretical model is needed for this purpose. Some additional background and skills are needed before we develop a suitable theory.

The Metal Ion

The Ionic View

Many chemical and physical properties of transition metal complexes can be interpreted by an ionic system. Thus, much of Co(II) chemistry is discussed in terms of Co^{2+} and much of Co(III) chemistry in terms of Co^{3+}. For this reason it is very important to be able to characterize in detail just what the metal ion *would be* if the complex really were strictly ionic. Two of the most significant characteristics of this metal atom are its oxidation number and the number of d electrons associated with it.

The Oxidation Number of the Metal

The properties of transition metal complexes depend strongly on the oxidation state of the central metal ion. The chemistry of Co(II) complexes, for example, differs greatly from the chemistry of Co(III) complexes. Cobalt(II) chemistry generally has much more in common with the chemistry of Fe(II), Mn(II), or Ni(II) than with

---

°Ni(II) complexes may also have other shapes besides square planar. Many are octahedral.

Co(III) chemistry. Similarly, Co(III) behaves more like Cr(III) in its chemistry than like Co(II). Hence, it is important to be able to recognize readily the oxidation number of a metal ion in a complex simply by seeing the chemical formula of the complex. The oxidation number is indeed the charge that we assign to the metal ion in the sample ionic model.

How can one determine the oxidation number of cobalt, for example, in $[CoCl_3(NO_3)_3]$ or $[CoCl_4]^{2-}$ or $[Co(NO_3)_6]^{3+}$? We begin by assigning the identities and electronic structures to the ligands. With the help of a few familiar principles, it is not difficult to learn the identities and net charges of common ligands, say H_2O, NH_3, Cl^-, F^-, CN^-, or CO. For unfamiliar ligands, it is a straightforward task to figure whether the ligand is neutral or charged. A pair of electrons in an "electron-rich" species is necessary for that species to act as a Lewis base. If the octet of a potential ligand is complete for the neutral species and at least one lone pair of electrons is available, the species can act as a *neutral* ligand. Examples of these ligands are ammonia, NH_3, $\mathrm{H{:}\ddot{N}{:}H}$ (with H below N); H_2O, $\mathrm{H{:}\ddot{\underset{..}{O}}{:}H}$; carbon monoxide, $\mathrm{:C{:::}O:}$; and trimethylphosphine, $\mathrm{:P(CH_3)_3}$. Species that need, to complete the octet, one or two additional electrons beyond what would be available in a neutral molecule or atom will bind as negatively charged ligands, for example, $\mathrm{:\ddot{\underset{..}{Cl}}:^-}$; cyanide, $\mathrm{:C{:::}N:^-}$; carbonate, $\mathrm{:\ddot{\underset{..}{O}}{::}\overset{:\ddot{O}:}{\ddot{C}}{:}\ddot{\underset{..}{O}}:^{2-}}$; and nitrite, $\mathrm{:\ddot{O}{::}N{:}\ddot{\underset{..}{O}}:^-}$.

In $[CoCl_3(NH_3)_3]$, therefore, the three NH_3 groups are neutral, while each of the three chlorides has a -1 charge. Since the net charge of the species is zero, applying Rule 4 of Sec. 7.7 gives

$$\text{oxidation number of cobalt} + 3(0) + 3(-1) = 0$$
$$\text{oxidation number of cobalt} = +3$$

Similarly, for $[CoCl_4]^{2-}$,

$$\text{oxidation number of cobalt} + 4(-1) = -2$$
$$\text{oxidation number of cobalt} = +2$$

While for $[Co(NH_3)_6]^{3+}$,

$$\text{oxidation number of cobalt} + 6(0) = +3$$
$$\text{oxidation number of cobalt} = +3$$

EXAMPLE 18.3 Determine the oxidation number of the metal ion in each of the following complexes: (a) $[PtBr_2(NH_3)_2]$; (b) $[Ru(NH_3)_5N_2]^{2+}$; (c) $[Ni(CO)_4]$; (d) $[MnCl_2(\text{pyridine})_4]$.

SOLUTION (a) charge of NH_3 = 0; charge of Br^- = -1; net charge = 0.

$$\text{oxidation number of Pt} + 2(0) + 2(-1) = 0$$
$$\text{oxidation number of Pt} = +2$$

(b) charge of NH_3 = 0; charge of N_2 = 0; net charge = +2.

$$\text{oxidation number of Ru} + 5(0) + 1(0) = +2$$
$$\text{oxidation number of Ru} = +2$$

(c) charge of CO = 0; total charge = 0.

$$\text{oxidation number of Ni} + 4(0) = 0$$
$$\text{oxidation number of Ni} = 0$$

(d) charge of pyridine = 0,

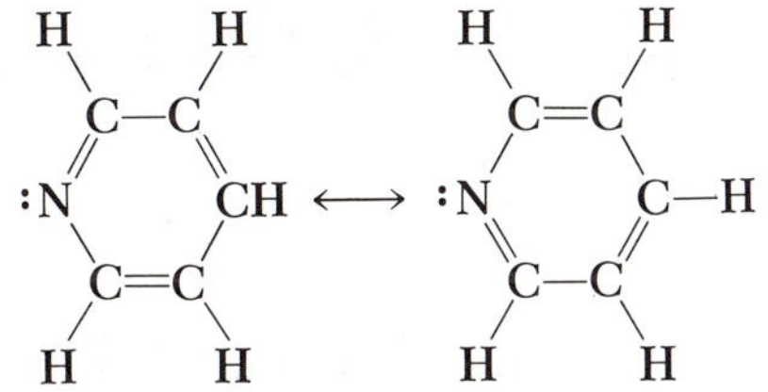

charge on Cl^- = −1; net charge = 0.

$$\text{oxidation number of Mn} + 4(0) + 2(-1) = 0$$
$$\text{oxidation number of Mn} = +2$$

STUDY PROBLEM 8(d)

Determine the oxidation number of the metal ion in (a) $[CrCl(H_2O)_5](ClO_4)_2$; (b) RuO_4; (c) $Fe(CO)_5$.

The Number of d *Electrons on the Metal Atom*

Since we later develop a simple theoretical model based essentially on the ionic view, the electronic configuration of the metal ion, especially its number of *d* electrons, is important. The electron configuration of a neutral atom of the element Ni is $[Ar]3d^84s^2$; but for Ni(II) it is $[Ar]3d^8$, and not $[Ar]3d^64s^2$. For neutral cobalt atoms, the electron configuration is $[Ar]3d^74s^2$, and for Co(III) it is $[Ar]3d^6$, and not $[Ar]3d^44s^2$. In transition metal ions—not neutral atoms—the *d* orbitals thus appear to be of lower energy than the *s* orbital of the next-higher shell. This situation occurs because the energy of an electron in a particular orbital depends on both its attraction by the nucleus and its repulsion by the other electrons. This balance of forces is often different in cations, in which there is a deficiency of electrons relative to the number of protons of the nucleus, from the balance that exists in the corresponding neutral atoms. It is an experimental fact, determined especially by spectroscopic and magnetic studies, that the relative positions of the 3*d* and 4*s* (or 4*d* and 5*s* or 5*d* and 6*s*) orbitals in the filling order appear to switch when a transition metal atom is ionized. The net result is, for example, that the configuration $[Ar]3d^6$ is more stable than the configuration $[Ar]3d^44s^2$.

EXAMPLE 18.4 Determine the number of *d* electrons identified with the central metal atom of each of the following complexes: (a) $[Cr(NH_3)_6]^{3+}$; (b) $[Ni(CN)_4]^{2-}$; (c) $[ZnCl_4]^{2-}$.

SOLUTION To determine the number of *d* electrons, we must first determine the oxidation number of the metal ion in the complex. It is then customary to determine the number of *d* electrons outside the noble gas core for the neutral metal atom and deduct the number of electrons needed to give the apparent charge on the assumed metal ion.

(a) $[Cr(NH_3)_6]^{3+}$

$$\text{oxidation number of Cr} + 6(0) = +3$$
$$\text{oxidation number of Cr} = +3$$

The electron configuration of Cr is [Ar]$3d^5 4s^1$, or 6 electrons outside the [Ar] core. Thus, Cr(III) has $6 - 3 = 3$ electrons outside [Ar]. This leads us to [Ar]$3d^3$ (three *d* electrons), since occupancy of the 3*d* orbitals is favored over the 4*s* orbitals for cationic species, as indicated above.

An alternative way of determining the number of *d* electrons uses atomic numbers directly. To use this method, subtract the atomic number of the preceding noble gas and the apparent charge. For Cr(III), this gives: $27 - 18 - 3 = 6$.

(b) $[Ni(CN)_4]^{2-}$

$$\text{oxidation number of Ni} + 4(-1) = -2$$
$$\text{oxidation number of Ni} = +2$$

For neutral atoms the Ni electronic configuration is [Ar]$3d^8 4s^2$, or 10 electrons outside the [Ar] core; Ni(II) has $10 - 2 = 8d$ electrons, corresponding to [Ar]$3d^8$. Or by the atomic number method, $28 - 18 - 2 = 8d$ electrons.

(c) $[ZnCl_4]^{2-}$

$$\text{oxidation number of Zn} + 4(-1) = -2$$
$$\text{oxidation number of Zn} = +2$$

For neutral atoms, the Zn electron configuration is [Ar]$3d^{10} 4s^2$, or 12 electrons outside the [Ar] core; Zn(II) has $12 - 2 = 10$ electrons outside [Ar], or [Ar]$3d^{10}$. Thus, there are 10 *d* electrons allotted to Zn in this formulation.

STUDY PROBLEM 18(e)

Determine the number of *d* electrons associated with the central metal atom in (a) $[Co(NH_3)_6]^{2+}$; (b) $[Fe(CN)_6]^{3-}$; (c) $[Fe(CN)_6]^{4-}$.

Geometry of the Complex

Typical Geometries

We shall not examine the detailed reasons for the relative stabilities of particular geometries of coordination complexes. Many of the structural features can be rationalized, however, by some of the simple ideas already introduced (Chap. 7) and by the simple theory we shall soon discuss to explain the spectral and magnetic properties of these complexes. The shapes of coordination complexes cannot confidently be predicted by the VSEPR bonding theory, which was found to be so useful for predicting structures of covalent compounds of the nontransition elements. It is often also the case that the structures of transition metal complexes are not explained in terms of simple electrostatic interactions.

To begin our discussion of geometry, we note that the following shapes are common:

1. Linear. With two ligands; generally found for d^{10} metal ions of +1 or +2 oxidation state. Examples are $[CuCl_2]^-$ and $[Ag(NH_3)_2]^+$.
2. Square planar. With four ligands; found for many d^8 and some d^9 ions.* Examples are $[Pt(H_2O)_2Cl_2]$, $[Ni(CN)_4]^{2-}$, $[AuCl_4]^-$, and $[RhCl_2(CO)(P(CH_3)_3)]$.

*For a d^9 metal ion with four Lewis base donors, VSEPR theory would focus on four nonbonding pairs and four bonding pairs, that is, eight pairs of electrons, leading to the prediction of a square antiprismal arrangement of electrons with ligands on alternate corners—giving a tetrahedral geometry.

3. Tetrahedral. Also four ligands; metal ions other than d^8 ions (including some d^9 ions). Examples are $[CoCl_4]^{2-}$, $[ZnCl_4]^{2-}$, and $[Ni(CO)_4]$.
4. Octahedral. Six ligands; very common. Examples are $[Cr(NH_3)_3Cl_3]$; $[Fe(H_2O)_6]^{2+}$, $[Ni(NH_3)_6]^{2+}$, $[CoF_6]^{3-}$.

These geometries are the most common. Many cases of other shapes are known and the particular guidelines given above are sometimes violated; for example, some d^8 complexes are tetrahedral. Some of these violations occur because the ligands attached to the metal atom are particularly bulky and interfere with each other in a structure that we should otherwise predict to be stable. Such interference is called **steric interference,** and decreases the stability of an otherwise favorable arrangement.

Hybridization

The concept of hybridization (Chap. 7), while less than perfect, can be used to some extent in rationalizing the geometries of transition metal complexes. As seen in the summary in Table 7.1, a set of octahedrally arranged hybrid orbitals can be "built" from two *d* orbitals, an *s* orbital, and three *p* orbitals; these six hybrid orbitals are referred to as d^2sp^3, or sp^3d^2. The structure of $[Co(H_2O)_6]^{3+}$ is readily rationalized in these terms. Thus, for electronic bookkeeping, we note that if we didn't know or care anything about Hund's rule, we might write the electron population in Co^{3+} as follows:*

$$[\mathrm{Ar}]\ \underset{4s}{\underline{\quad}}\ \underbrace{\underline{\uparrow\downarrow}\ \underline{\uparrow\downarrow}\ \underline{\uparrow\downarrow}\ \underline{\quad}\ \underline{\quad}}_{3d}\ \underbrace{\underline{\quad}\ \underline{\quad}\ \underline{\quad}}_{4p}$$

We can then imagine using the vacant 4*s*, two 3*d*, and three 4*p* orbitals to form six octahedrally arranged d^2sp^3 hybrid orbitals. These d^2sp^3 hybrids could then be used to form six bonds to the H_2O ligands, using the electron pairs available from the ligands. These additional six electron pairs (represented below by color arrows) would completely fill up the 4*s*, 3*d*, 4*p* orbital population diagram for cobalt:

$$[\mathrm{Ar}]\ \underset{4s}{\underline{\uparrow\downarrow}}\ \underbrace{\underline{\uparrow\downarrow}\ \underline{\uparrow\downarrow}\ \underline{\uparrow\downarrow}\ \underline{\uparrow\downarrow}\ \underline{\uparrow\downarrow}}_{3d}\ \underbrace{\underline{\uparrow\downarrow}\ \underline{\uparrow\downarrow}\ \underline{\uparrow\downarrow}}_{4p}$$

Hence, d^2sp^3 hybridization, with its six octahedrally arranged hybrid orbitals, can be used to rationalize the existence of the octahedral $[Co(H_2O)_6]^{3+}$ complex.

The tetrahedral arrangement of the ligands in $Ni(CO)_4$ is similarly rationalized in terms of sp^3 hybridization. If for bookkeeping purposes only, we choose to write the electronic configuration of atomic nickel

$$[\mathrm{Ar}]\ \underset{4s}{\underline{\quad}}\ \underbrace{\underline{\uparrow\downarrow}\ \underline{\uparrow\downarrow}\ \underline{\uparrow\downarrow}\ \underline{\uparrow\downarrow}\ \underline{\uparrow\downarrow}}_{3d}\ \underbrace{\underline{\quad}\ \underline{\quad}\ \underline{\quad}}_{4p}$$

*Overlooking Hund's rule in setting up a hybridization argument is of no consequence. Hund's rule applies to the electronic configurations of actual species, but we are concerned, in writing down the population diagrams of the metal atom orbitals, only with electronic bookkeeping. Strictly speaking, in considering hybridization schemes, we don't really need to consider how many electrons are available. We could imagine a hybridization scheme to be constructed from the available atomic orbitals, and *then* consider how many electrons are available to fill the hybridized and unhybridized orbitals.

we recognize the availability of the 4*s* and three 4*p* orbitals for hybridization and bonding. Invoking four tetrahedrally oriented sp^3 hybrid orbitals of nickel and four electron pairs (again represented by color arrows) from the four CO ligands to form these bonds, we see that all the atomic orbitals for nickel are then completely accounted for:

	4s	3d	4p
[Ar]	↑↓	↑↓ ↑↓ ↑↓ ↑↓ ↑↓	↑↓ ↑↓ ↑↓

Similarly, the square planar geometry of a $[Ni(CN)_4]^{2-}$ complex, in which the oxidation number of nickel is +2, can be rationalized in terms of dsp^2 hybridization. Again, neglecting Hund's rule, and writing the Ni^{2+} configuration,

	4s	3d	4p
[Ar]	—	↑↓ ↑↓ ↑↓ ↑↓ —	— — —

we recognize that the 4*s* orbital, a 3*d* orbital, and two 4*p* orbitals can be used to build four dsp^2 (or sp^2d) hybrids, which are oriented in a square planar way. Then, using these four hybrid orbitals and four electron pairs from the ligands (represented below by color arrows), we rationalize the square planar structure, and the following diagram:

	4s	3d	4p
[Ar]	↑↓	↑↓ ↑↓ ↑↓ ↑↓ ↑↓	↑↓ ↑↓ —

The linear arrangement of $[Ag(NH_3)_2]^+$ is also readily rationalized in terms of hybridization, in this case *sp* hybridization (see Table 7.1). Considering the electronic configuration of Ag^+ to be

	4s	3d	4p
[Ar]	—	↑↓ ↑↓ ↑↓ ↑↓ ↑↓	— — —

and invoking the use of the 4*s* orbital and one 4*p* orbital, leads to two, collinear *sp* orbitals. Then, with one electron pair contributed by each of the two NH_3 ligands, these two *sp* hybrid orbitals lead to a rationalization of the linear structure and the diagram

	4s	3d	4p
[Ar]	↑↓	↑↓ ↑↓ ↑↓ ↑↓ ↑↓	↑↓ — —

STUDY PROBLEM 18(f)

Describe the hybridization scheme for (a) $[Ru(NH_3)_6]^{3+}$; (b) $[PtCl_4]^{2-}$; (c) $[Hg(NH_3)_2]^{2+}$.

Chelates

The term *chelate* is derived from the Greek word for "claw." It denotes a ligand with more than one functional group, which may therefore form more than one bond to the metal ion. For example, using more than one functional group, ethylenediamine, $H_2NCH_2CH_2NH_2$, can bond to the metal ion at either or both of

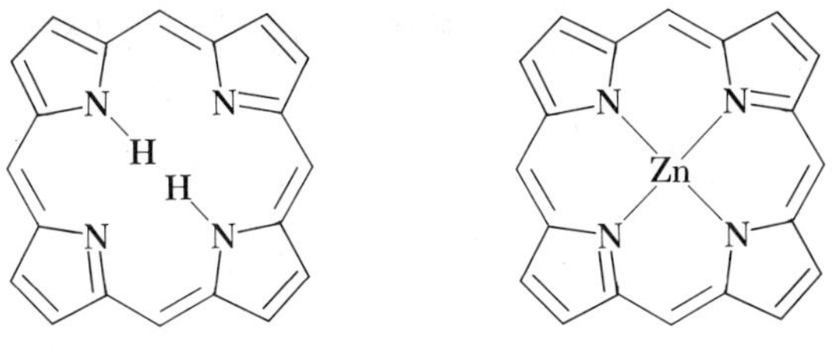

(a) (b)

Figure 18.2
(a) Basic porphyrin "parent molecule," from which the quadridentate ligand of −2 charge is formed by removal of two H^+. (b) Zn(II)-porphyrin complex.

its nitrogen atoms. The compound dichloro(ethylenediamine)palladium(II), $[PdCl_2(en)]$, is an example. It has the structure

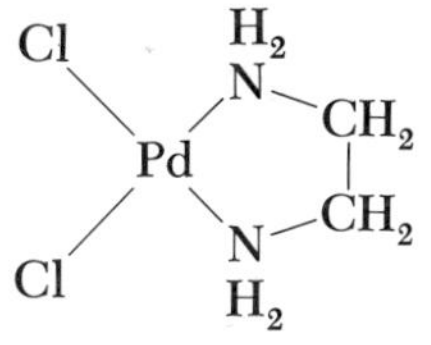

This structure is square planar, like most d^8 compounds. An obvious restriction here is that the chlorides must be next to each other (*cis*), whereas for the compound $[PdCl_2(NH_3)_2]$, there are two isomers, cis and trans.*

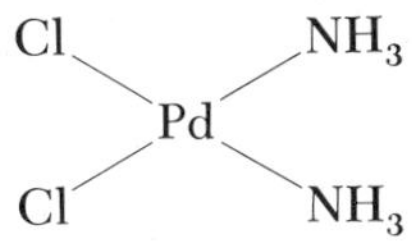

and

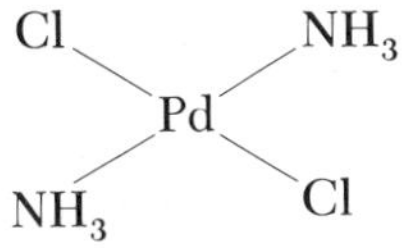

cis trans

Some other chelates are even more restrictive in their geometrical requirements than the cis constraint of ethylenediamine. A class of biologically important chelates called porphyrins constitutes an example of such chelates (Fig. 18.2a). Porphyrins and slightly modified porphyrins are essential in the function of hemoglobin, in which an Fe(II) porphyrin complex carries O_2 from the lungs to the cells; of myoglobin, an Fe(II) porphyrin complex that provides O_2 transport in muscles; of chlorophyll, an Mg(II) complex of a porphyrinlike molecule; and of many other species. The porphyrin ligand is constrained by the various covalent bonds in its framework to be square. Thus, a Zn(II) porphyrin complex (Fig. 18.2b) is found to be square, instead of having the tetrahedral shape that four-coordinate Zn(II) complexes commonly would have with "unrestricted" ligands, for example, $[ZnCl_4]^{2-}$. Chelating ligands bind much more strongly to transition metal ions than comparable unidentate, or one-toothed, ligands, those that form only one bond to the metal atom. This is evident since the equilibrium constant for the combination of ethylenediamine with Fe(II), given by the equation

$$[Fe(H_2O)_6]^{2+} + H_2NCH_2CH_2NH_2 \rightleftharpoons [Fe(H_2O)_4(en)]^{2+} + 2H_2O \qquad (18.10)$$

is 2.19×10^4, while for NH_3 complexation

$$[Fe(H_2O)_6]^{2+} + 2NH_3 \rightleftharpoons [Fe(H_2O)_4(NH_3)_2]^{2+} + 2H_2O \qquad (18.11)$$

*The cis isomer of dichlorodiammineplatinum(II) is effective in arresting the growth of certain types of malignant tumors and is undergoing clinical trials. The trans isomer is totally ineffective. Knowing about isomers is important!

Figure 18.3 (a) Ethylenediaminetetraacetate ligand. (b) Structure of the chelate (complex) formed between the $EDTA^{4-}$ and Pb^{2+}.

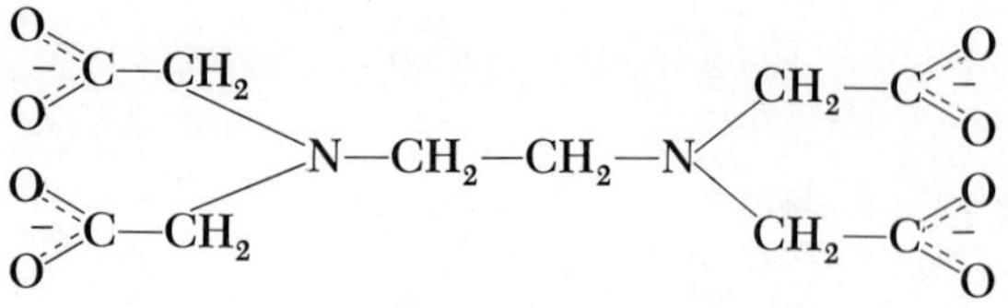

(a) ethylenediaminetetraacetate

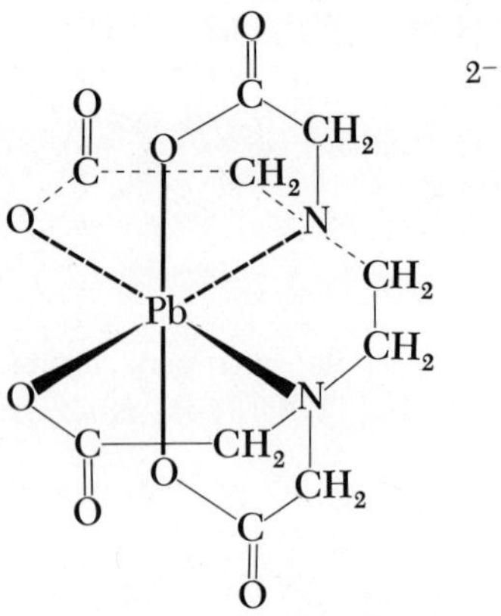

(b) $[Pb(EDTA)]^{2-}$ complex

the equilibrium constant is only 1.6×10^2. This preference for multidentate ligands is utilized by both chemists and nature. Many molecules in biological systems whose functions depend on maintaining the availability of a metal ion are able to accomplish this by providing a chelate site for binding the metal ion. This chelating site can bind better to the metal ion than can other Lewis bases that may be present—binding with the chelating agent has a larger equilibrium constant. One can also take advantage of the strong binding property of chelates to remove an excess of metal ions from systems in which they are not wanted. This approach is the basis for important methods of treatment for heavy metal poisoning, like lead poisoning. Large concentrations of Pb^{2+} can be removed from the bloodstream by using the hexadentate chelate, ethylenediaminetetraacetate acid (EDTA).* The resulting chelate complex has the formula $[Pb(EDTA)]^{2-}$ and the structure shown in Fig. 18.3. This species has a much higher stability constant than complexes of Pb^{2+} with other ligands that are available in the bloodstream.

EXAMPLE 18.5 The reaction of hexaamminenickel(II) with ethylenediamine illustrates the chelate effect. For this reaction at 25 °C:

$$Ni(NH_3)_6{}^{2+} + 3en \rightleftharpoons Ni(en)_3{}^{2+} + 6NH_3$$

$K_{eq} = 10^{9.7}$ and $\Delta H° = -2.9$ kcal/mol. How much does the entropy term contribute to the free energy difference between the chelate complex and the hexaammine complex?

*Entropy is the main reason that the formation of a chelate complex generally is governed by a larger equilibrium constant than the formation of a corresponding complex with monodentate ligands. In the formation of the chelate complex, fewer ligand species are removed from the free solvated state and "tied down" to the metal ion than when equivalent monodentate ligands are complexed. Hence, there is a smaller decrease in randomness, or entropy, in the formation of the chelate complex, resulting in a more favorable equilibrium constant.

Spotlight

"Optical" or "stereo" isomers have been important in the study of the reaction mechanisms of coordination compounds. Optical isomers occur in pairs in which one isomer is the mirror image of the other. Structures are only isomers if they are different—not "superimposable" on each other. An example of a pair of optical isomers is furnished by the two forms of *cis*-dichlorobis(ethylenediamine)cobalt(III) cation. As a pair of optical isomers, they are called **enantiomers.**

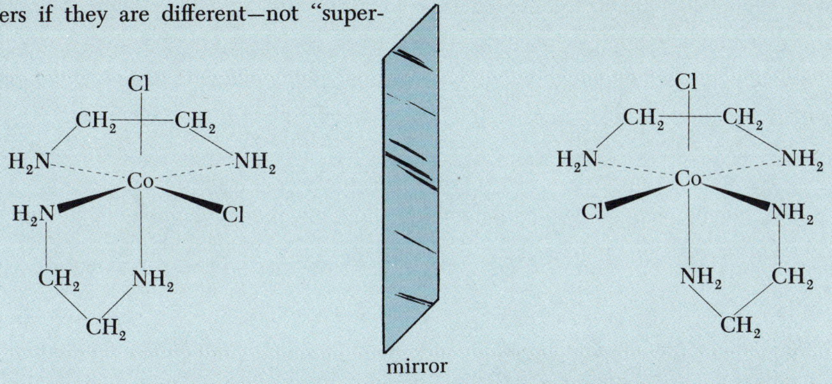

If the isomer on the right is rotated so that the chloride ligands coincide with the isomer on the left, the ethylenediamine ligands are not oriented the same way:

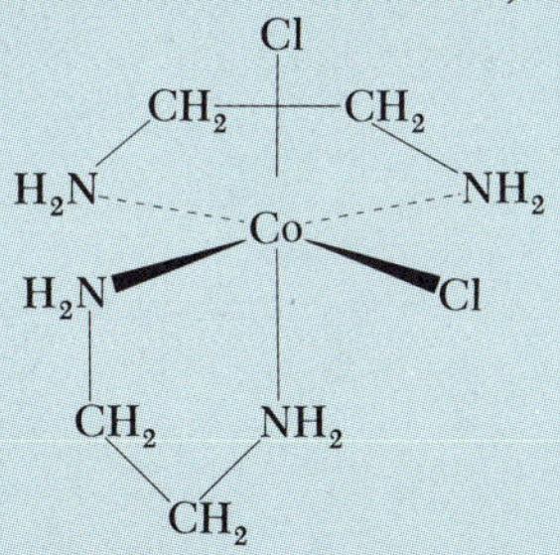

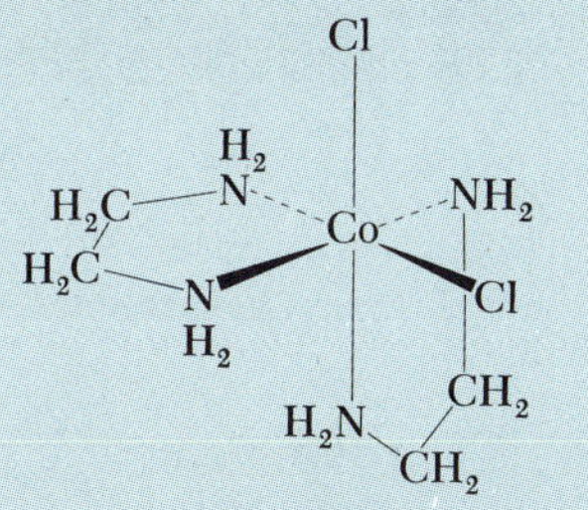

This effect can be visualized using your hands. Your right hand is the mirror image of your left hand but they are different from each other. These species are called **optical isomers** because they have the property of rotating a beam of light that has been properly polarized in one orientation in different directions; one rotates the beam clockwise and the other counterclockwise. The *trans*-dichlorobis(ethylenediamine) cobalt(II) cation does not exist as optical isomers because the molecule and its mirror image are the same. (This effect is discussed in most organic chemistry textbooks.)

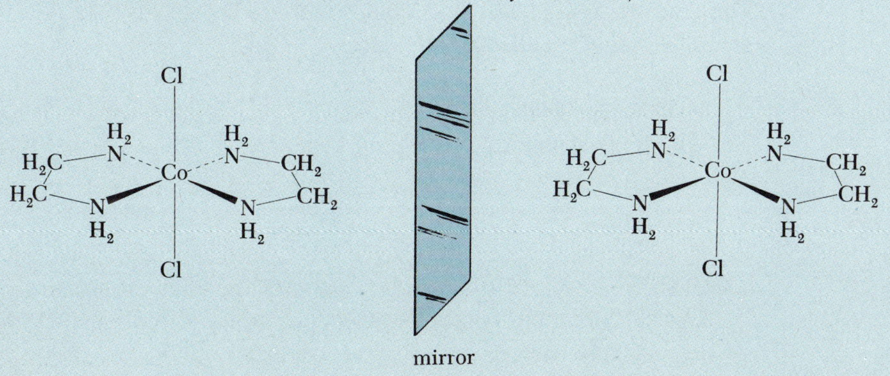

Spotlight

Professor John C. Bailar (1904–) has done extensive research involving isomers of coordination compounds. The work of Bailar and his many outstanding graduate students has been instrumental in the development of modern coordination chemistry. Professor Bailar has received many honors not only in recognition of his research but for his contributions to chemical education as well.

Chemical and Engineering News

SOLUTION First we need to determine $\Delta G°$. Then, by difference, we can obtain the entropy term $-T\,\Delta S°$.

$$\Delta G° = -RT \ln K_{eq} = -2.303\,(592) \log 10^{9.7}$$
$$= -13.2 \text{ kcal/mol}$$
$$= \Delta H° - T\,\Delta S°$$
$$-T\,\Delta S° = \Delta G° - \Delta H° = -13.2 + 2.9 = -10.3 \text{ kcal/mol}$$

The entropy term clearly contributes more to the relative stability of the chelate complex than the enthalpy term does.

STUDY PROBLEM 18(g)

The formation constants for the reactions

$$Cu^{2+} + en \rightleftharpoons Cu(en)^{2+} \qquad K_1$$
$$Cu(en)^{2+} + en \rightleftharpoons Cu(en)_2{}^{2+} \qquad K_2$$

at 25 °C are $K_1 = 10^{10.66}$ and $K_2 = 10^{9.33}$. By using the values given in Sec. 18.3 for the formation constants of copper amine complexes, determine the free energy stabilization of $Cu(en)_2{}^{2+}$ relative to $Cu(NH_3)_4{}^{2+}$. (*Hint:* Use the β value for the appropriate complex.)

Rates of Ligand Substitution

To this point no attention has been directed to the possibility that ligands may not be permanently attached to the metal ion in a complex. Generally one finds neither that the formation of a ligand-metal bond is instantaneous, nor that the lifetime of the bond is infinite. The dynamics or rates of the various bond-forming and bond-breaking processes are of considerable interest. The substitution reactions given by the equations above for complexation of Fe(II) with ethylenediamine and ammonia are rapid. There is a very large range of rates for ligand exchange of transition metal complexes. Many of the metal ions in our examples undergo slow substitution, and laboratory manipulations are conveniently possible with such species without any threat to their structural integrity. Common examples of this situation are most of the complexes of Co(III), Cr(III), Pt(II), Ru(II), Ru(III), Rh(III), and Ir(III). For some other complexes, ligand substitution is rapid; these include most complexes of Cr(II), Co(II), Ni(II), and Ag(I); such species are called **labile.** If small amounts of labile complexes are added to water, they are quickly converted to "aquated" species—complexes with water serving as the ligands:

$$[CoCl_4]^{2-} + 6H_2O \rightleftharpoons [Co(H_2O)_6]^{2+} + 4Cl^- \quad (18.12)$$

$$[Ni(NH_3)_6]^{2+} + 6H_2O \rightleftharpoons [Ni(H_2O)_6]^{2+} + 6NH_3 \quad (18.13)$$

To be sure what species are present when working with transition metal complexes, one must know whether the complex is inert and able to maintain its bonding structure throughout the experiment, or whether it is "labile" and will quickly be converted to some other species. The ultimate structural fate of a labile complex, with enough time allowed for all pertinent processes to occur, depends on the equilibrium constants for binding with whatever Lewis bases are present and the amount of each Lewis base. Two good rules of thumb for predicting the rate at which one ligand can be replaced by another are: (1) A particular ligand is exchanged much more slowly when the metal atom has a d^3, d^6, or d^8 configuration (except for Ni(II) and high-spin Fe(II)) than when other electronic configurations are involved. (2) Chelating ligands are replaced more slowly than unidentate ligands.

The detailed theory of explaining the rates of ligand substitution reactions, which we shall omit, has a great deal conceptually in common with the theory we shall now develop for explaining spectral and magnetic properties.

18.4 Spectral and Magnetic Properties of Transition Metal Complexes

Many transition metal complexes are colored. This observation means that energy in the visible region is absorbed or emitted, as it is in atomic absorption and emission spectra. By a spectrometer, one can scan the visible region of the spectrum and discover that certain energies (frequencies) of light are indeed absorbed. The color that we see is due to the combination of all the energies of light that are not absorbed. Table 18.4 gives the relation among energies, wavelengths, and colors of light in the visible region.

Transitions in the visible region at the electromagnetic spectrum (4000–7000 Å wavelengths) are too low in energy (about 40–80 kcal/mol) to be the result of

TABLE 18.4 **Relations among Energies, Wavelengths, and Colors in Light Absorption Spectra**

Energy of Photons, kcal/mol	*Approximate Wavelength, Å*	*Color*	*Color Observed*[a]
65	4400	Violet	Yellow
62	4600	Blue	Orange
60	4800	Blue green	Red
53	5400	Yellow green	Purple
50	5700	Yellow	Violet
46	6200	Orange	Blue
43	6600	Red orange	Green
41	7000	Red	Blue green

[a] If light of the given energy has been absorbed.

exciting electrons of the transition metal ion from one subshell to another, say from 2*s* to 2*p* or from 3*d* to 4*s*. A specific example of such a process would be exciting Cr(III) from the electronic configuration [Ar]$3d^3$ to [Ar]$3d^24s^1$. Since processes involving two different subshells are not involved in the absorption of visible light by transition metal complexes, a theory has been developed that involves electron excitation within a subshell, that is, from one *d* orbital to another.* *This requires the d orbitals of the metal to be of different energies.* But so far we have always assumed that all orbitals in a given sublevel—all three *p* orbitals or all five *d* orbitals or all seven *f* orbitals—had the same energy. This assumption, a theoretical result of quantum mechanics, was based on our concern with free ions, which had no covalent attachments to other atoms or ions and no electrostatic perturbations by other ions, as in a crystal. The properties of transition metal ions, on the other hand, definitely show some covalency in the attachment of the Lewis bases, ligands, to the metal ion. As we shall see, the presence of the ligands can be thought of as giving rise to energy differences among the orbitals of the *d* subshell.

Crystal Field Theory

The theoretical model we shall use to explain the optical and magnetic properties of transition metal complexes is called **crystal field theory.** Its basic premise is that although we know that there is a substantial amount of covalent character in the metal ligand bonds, we can overlook that aspect of the electronic structures of transition metal complexes if we are primarily interested in accounting for the main features of optical and magnetic properties. Having neglected covalency, we then focus attention on what appears to be a metal ion surrounded by an array of ligands with which it interacts electrostatically (as in an ionic *crystal*). The crystal field

*Strictly speaking, it is not precisely correct to discuss the absorption of light by transition metal complexes in terms of the individual atomic orbitals of the metal. The orbitals of the complex belong to the entire complex—they are *molecular* orbitals. Nevertheless, to the extent that we can view the structure of a complex as though there is a metal ion at the center, it is reasonable to focus attention on the orbitals in that metal ion.

approach then considers the influences that these interactions have on the energies of the five d orbitals in the valence shell of the metal.

How do the energies of the d orbitals become different because of the perturbations by the Lewis bases? This question can be answered by looking closely at the shapes of the d orbitals and considering the orientation of the ligands in a transition metal complex. For an example, we compare in Fig. 18.4 the orbital orientation with ligand positions in an octahedral complex. Let us consider the effect of six ligands that are oriented in an octahedral arrangement on the cartesian axes shown in Fig. 18.4a. The question of interest is whether the presence of these ligands alters any of the orbital energies, the energy of an electron in any one of the five d orbitals in part a of that figure. This problem can be met by asking the question: In terms of energy, does it make a difference whether an electron is placed in an orbital with a "lobe" directed along an axis on which a ligand lies (d_{z^2} or $d_{x^2-y^2}$) or in an orbital with lobes directed between such axes (d_{xy}, d_{xz}, d_{yz})? If an electron is placed in the $d_{x^2-y^2}$ or d_{z^2} orbital, it will have a high probability of occupying a region of space that is directly facing the pairs of electrons being donated by the Lewis bases on the x, y, and z axes; this is a relatively high-energy situation because of electron-electron repulsions. The d_{xy}, d_{xz}, and d_{yz} orbitals, however, lie *between*

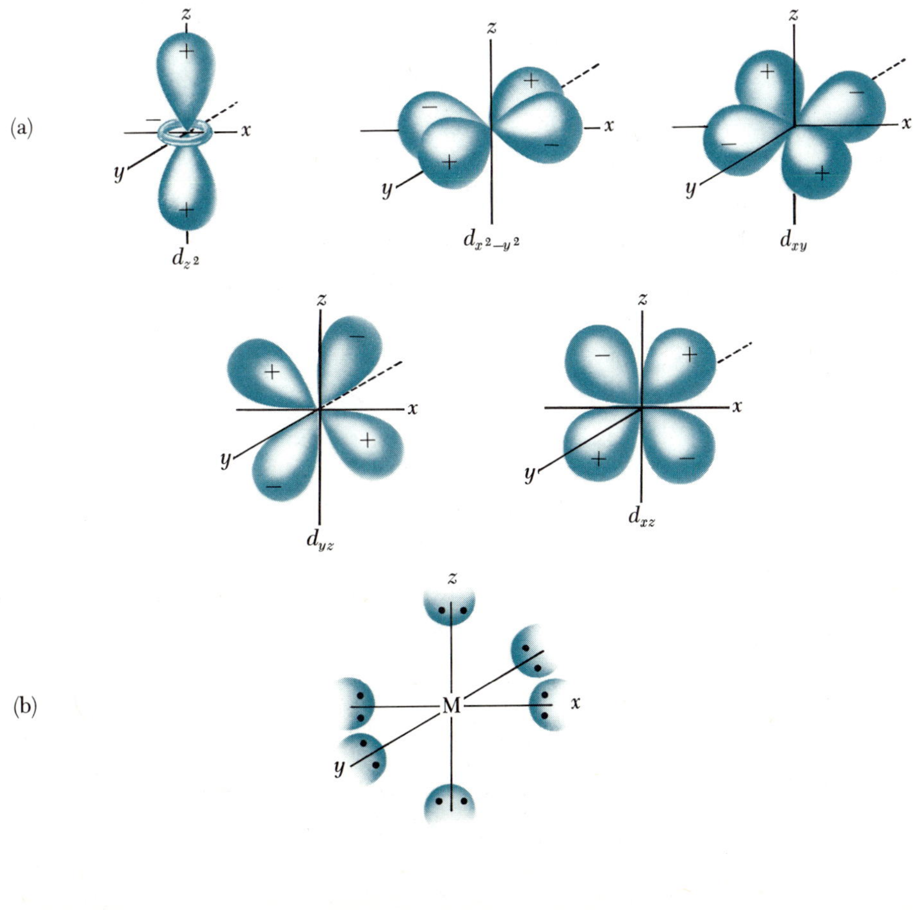

Figure 18.4
(a) Arrangement of d orbitals relative to x, y, z axes. (b) Six ligands about a metal ion (M), in an octahedral arrangement.

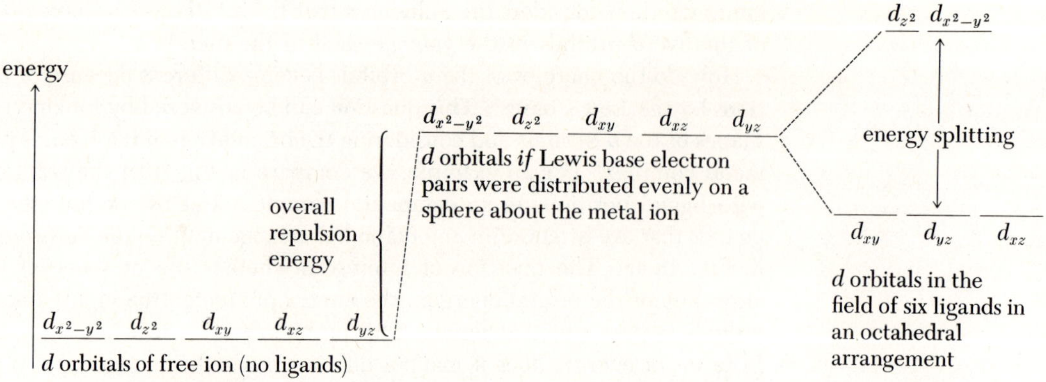

Figure 18.5 The influence of six ligands, arranged octahedrally about a metal ion, on the energies of electrons in the *d* orbitals of the metal ion (showing also what the effect of the overall repulsion would be if there were no directional character—if the ligand electron pairs were evenly distributed on a sphere, although they are not.)

the axes that bear electron pairs from Lewis bases; and an electron in one of these orbitals experiences a lower level of electron-electron repulsion than an electron in a d_{z^2} or $d_{x^2-y^2}$ orbital. Hence, it is much more favorable, less energy being needed, for an electron to occupy the d_{xy}, d_{xz}, or d_{yz} orbital than the d_{z^2} or $d_{x^2-y^2}$ orbitals.

More sophisticated developments along these same lines show that the d_{xy}, d_{xz}, and d_{yz} orbitals in an octahedral complex all have the same energy and a lower energy (more favorable) than the energy of the d_{z^2} and $d_{x^2-y^2}$ orbitals.* This leads to the energy diagram of the *d* orbitals in an octahedral complex shown in Fig. 18.5. In considering energy diagrams like Fig. 18.5, it must be remembered that the energy represented is only the energy of an electron in the *d* orbitals. Even though this energy is raised by adding ligands, the *overall* energy of the entire coordination compound is much lower (more favorable) by addition of ligands, because of the very large energy stabilization provided by electrostatic attraction and the partial covalency of the Lewis acid-base bonds.

Visible Spectra

Now we can see how these different energies of *d* orbitals relate to the magnetic and optical properties of transition metal compounds. For the simple case of a d^1 complex (a complex in which the central metal is viewed as an ion having only one *d* electron), we can picture the nature of a color-producing absorption as shown in Fig. 18.6. As the correct frequency for the absorption is given by $\Delta E/h$, where ΔE is the difference in energy between the two pertinent levels and h is Planck's constant (see Sec. 6.5), a photon of radiation with this frequency is absorbed by the system as the electron undergoes a jump in energy ΔE from the d_{xy}, d_{xz}, d_{yz} level to the d_{z^2}, $d_{x^2-y^2}$ level. The resulting higher energy state is called the **excited state.** The

*Mathematical treatments of crystal field theory for an octahedral complex show that the energy of an electron is the same in either the d_{z^2} or $d_{x^2-y^2}$ orbital, even though it is not obvious from the pictorial representation of these orbitals.

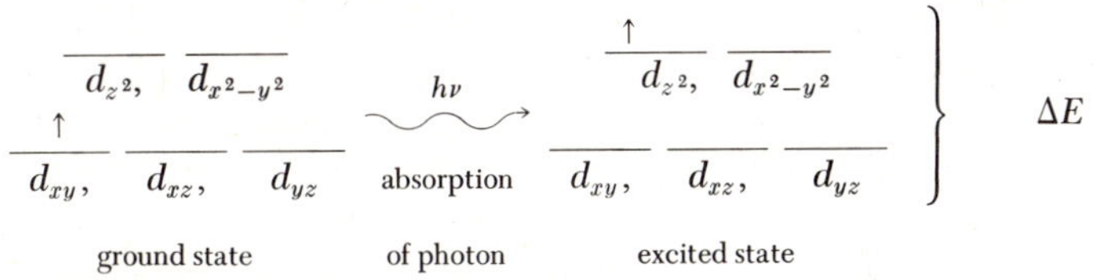

Figure 18.6 Absorption of a photon with frequency ν, where $\nu = \Delta E/h$, as a d^1 metal ion in an octahedral crystal field undergoes a transition from the d_{xy}, d_{xz}, d_{yz} energy level to the energy level of d_{z^2} and $d_{x^2-y^2}$.

excited state may lose its extra energy by various means, thereby returning to the more stable ground state. These means include energy loss to the vibrational and translational motions of the molecule, without light emission.

The energy of the light absorbed in a process of the type depicted in Fig. 18.6 is the energy difference, or energy "gap," between the two types of d orbitals—those between ligand electron pairs and those facing electron pairs. As an example, the absorption spectrum of $[Ti(H_2O)_6]^{3+}$ is shown in Fig. 18.7. The wavelength at which maximum absorption occurs in the spectrum shown in Fig. 18.7 is 4975 Å. From this we can calculate the energy difference ΔE between the pertinent levels. We use $\Delta E = h\nu$, expressed in terms of ΔE, the wavelength λ, the frequency ν, and appropriate conversion factors. From Eq. (8.8),

$$\Delta E = h\nu$$

$$= \frac{hc}{\lambda}$$

$$= \frac{1.58 \times 10^{-37} \text{ kcal-sec/molecule} \times 6.022 \times 10^{23} \text{ molecule/mol} \times 3.00 \times 10^{10} \text{ cm/sec}}{4975 \text{ Å} \times 10^{-8} \text{ Å/cm}}$$

$$= 57.4 \text{ kcal/mol}$$

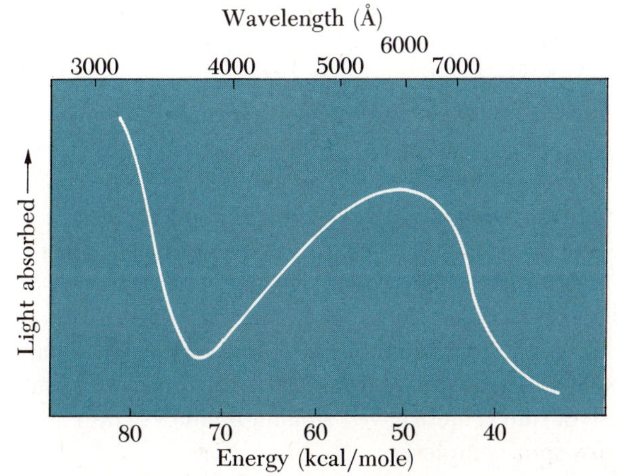

Figure 18.7 Absorption spectrum of the ion $[Ti(H_2O)_6]^{3+}$ in aqueous solution, showing the relation between the light absorbed and the wavelength of the light or the energy of the transition responsible for the absorption.

As one can see from the spectrum shown in Fig. 18.7, the energy absorption occurs over a wide range of frequencies. This does not destroy our notions of quantized energy states. The wide energy range merely results from the fact that at a given instant the energy gap for a particular ion $[Ti(H_2O)_6]^{3+}$ can be different from that of some other ion $[Ti(H_2O)_6]^{3+}$ due to such factors as different instantaneous bond lengths for some of the H_2O–Ti bonds from vibrations. The electron transition occurs very rapidly. If a bond happens to be stretched when the electron transition occurs, the energy of the *d* orbital facing the H_2O electron pair would be different from the energy in the situation in which the bond is at its average distance, or compressed.

In $[Ti(H_2O)_6]^{3+}$, the absorption bond in the visible region is so broad that only for the region from about 4200 Å to 3600 Å is energy not absorbed. The color of this complex ion is violet, since this is the only color range that survives as light passes through with the sample; frequencies in the blue through red color ranges are at least partly absorbed. Table 18.4 shows the relation between the energy and the wavelength of absorbed light and the color of a solution containing the absorbing species.

Magnetic Properties; High Spin and Low Spin

The magnetic properties of transition metal compounds and ions are also compatible with this picture of an energy gap between *d* orbitals due to the "crystal field." An experimental determination of the extent of paramagnetism in a substance provides a measure of the number of unpaired electrons in each molecule or ion. If we imagine placing electrons, one by one, in the *d* orbitals of octahedral complexes of transition metals, we can make predictions about the number of unpaired electrons that one would expect to find for each case. These predictions are based on the following two principles

1. *Other considerations being the same, electrons occupy the lowest energy levels available to them—a maximum of two per orbital, with opposite spin orientation.*
2. *Other considerations being the same, electrons of the same spin have lower energy than electrons of opposite spin, according to Hund's rule.*

Applying these rules to the d^1, d^2, and d^3 cases, we obtain the predictions shown in Fig. 18.8. For the d^1, d^2, or d^3 cases, one predicts the same number of unpaired electrons that would be predicted if there were no difference in energy of different *d* orbitals. The predictions of the crystal field model, however, are important for the d^4, d^5, d^6, and d^7 cases. These cases are depicted in Fig. 18.9. In these cases, one sees that the two rules stated above can lead to conflicting predictions, so that more than one orbital population diagram may seem reasonable. To choose between them, one needs to know which rule has priority. On the left side of Fig. 18.9, we see the electron configurations that one would predict on the basis of emphasizing rule 2; these configurations, compared with the corresponding configurations on the right, have a larger number of unpaired electrons and are called **high-spin.** On the right side of the figure are listed the electronic configurations one would predict on the basis of rule 1; these have a smaller number of unpaired electrons and are referred to as **low-spin** complexes.

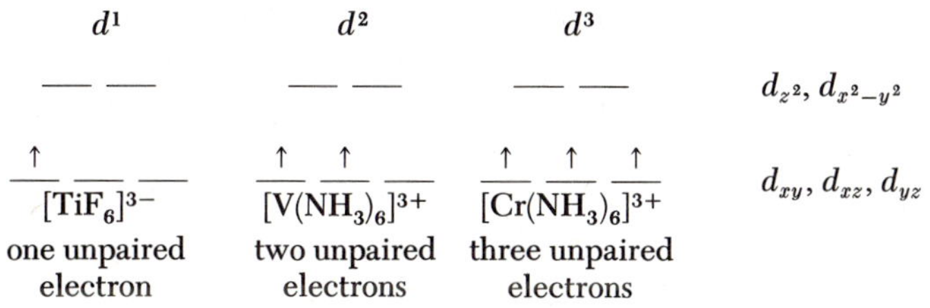

Figure 18.8 Electron populations of d orbitals for d^1, d^2, and d^3 octahedral complexes, with specific examples cited.

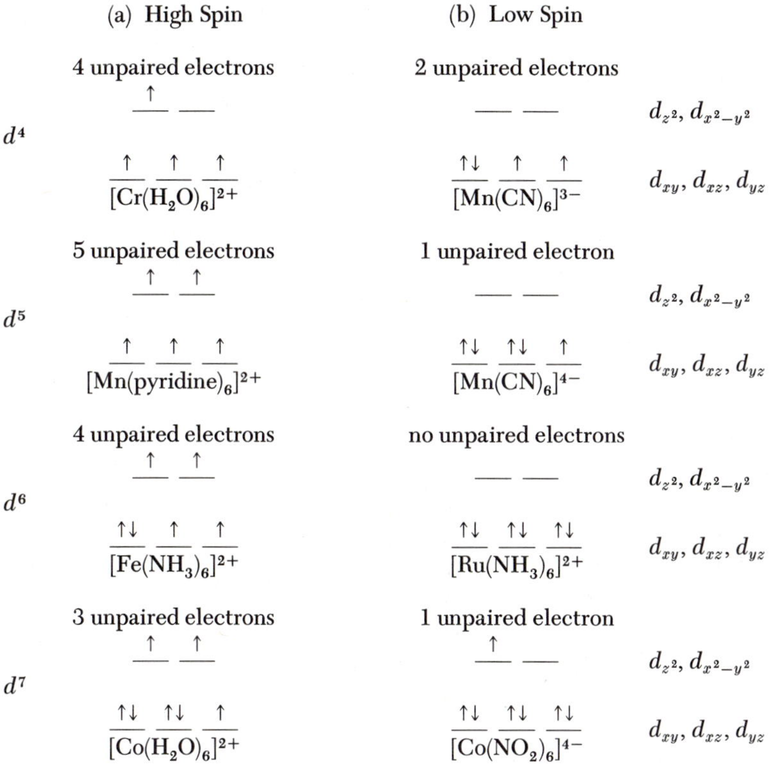

Figure 18.9 Populations of d orbitals for d^4, d^5, d^6, and d^7 metal ions in an octahedral crystal field, showing (a) high-spin and (b) low-spin cases, with specific examples cited.

The amount of weight increase caused by unpaired electrons when a complex is subjected to a magnetic field is called the **magnetic susceptibility** of the complex. Although the actual weight increase per mole will depend on the magnetic field employed, the relative weight increases of two species in the same magnetic field can be readily predicted. For the transition elements, except for lanthanide ions, which require a more sophisticated treatment, the relative values can be predicted quite well by the equation for the magnetic moment per mole—generally within ± 20%.

$$\mu = 2\sqrt{S(S+1)} \tag{18.14}$$

where S is the total number of unpaired electrons divided by 2. In actual experiments, diamagnetic effects from nontransition elements in the compound, for example, the ions K^+ and Cl^- in $K_2[CoCl_4]$, are subtracted out to compare μ values. For compounds with few other atoms besides the transition metal, however, this correction is very small.

EXAMPLE 18.6 If the weight increase for $K_3[CoF_6]$, a high-spin complex of Co(III), is 0.290 g per mol, what weight increase is expected for $[Fe(NH_3)_6]F_2$, a high-spin complex?

SOLUTION High-spin Co(III), a d^6 species, has 4 unpaired electrons. The ion $[Fe(NH_3)_6]^{2+}$ is also a high-spin complex and it too is d^6 with 4 unpaired electrons. The relative weight increases, therefore, are predicted by Eq. (18.14) to be the same:

$$\mu = 2\sqrt{\tfrac{4}{2}(\tfrac{4}{2} + 1)} = 2\sqrt{2(3)} = 4.90$$

The weight increase of $K_3[CoF_6]$ is 0.290 g per mol and the molecular weight is 289.9 so the weight increase is 1.00×10^{-3} times the molecular weight. For $[Fe(NH_3)_6]F_2$, the molecular weight is 177.9 and the predicted weight increase in the same magnetic field (and at the same temperature) is 0.178 g weight increase per mole.

EXAMPLE 18.7 In the same magnetic field used in Example 18.6, what weight increase is expected for $K_3[Mn(CN)_6]$, a low-spin complex?

SOLUTION Mn(III) has 4 d electrons and in the low-spin case 2 are paired, leaving only 2 unpaired electrons. Therefore,

$$\mu = 2\sqrt{S(S + 1)} = 2\sqrt{\tfrac{2}{2}(\tfrac{2}{2} + 1)} = 2\sqrt{2} = 2.83$$

The relative weight increases for this complex would be 2.83/4.90 of the weight increase for $K_3[CoF_6]$, which was 1.00×10^{-3} times the molecular weight. For $K_3[MnCN)_6]$, a weight increase of 5.78×10^{-4} times the molecular weight, or a 0.189-g weight increase per mole, is predicted.

STUDY PROBLEM 18(h)

For the same magnetic field used in Examples 18.6 and 18.7, predict the weight increases for (a) $K_3[Cr(H_2O)_6]$; (b) $K_2[Cr(H_2O)_6]$, high-spin; (c) $[Co(NH_3)_6]Cl_3$, low-spin.

To emphasize the factors involved in determining whether a high-spin or low-spin configuration will exist in a particular complex, let us imagine the fate of an electron about to occupy the most favorable orbital. In the d^4 case, the fourth electron can minimize electrostatic interactions with other electrons by jumping to a higher-energy orbital (following Hund's rule) to form a high-spin complex. This would happen only if jumping cost less energy than pairing with another electron (violating Hund's rule). Thus, high-spin complexes result when the energy gap is smaller than the pairing energy. The ligands that give rise to the small gaps characteristic of high-spin complexes exert weak crystal fields and are called **weak-field ligands.** Conversely, low-spin complexes result when it costs less energy to form a pair than to make the jump. Low-spin complexes have large energy gaps between

Spotlight

The most familiar of the oxygen-carrying molecules in animals is hemoglobin. **Hemoglobin** is a protein of molecular weight about 66,000 amu; it has four ferroprotoporphyrin IX groups in each molecule. The

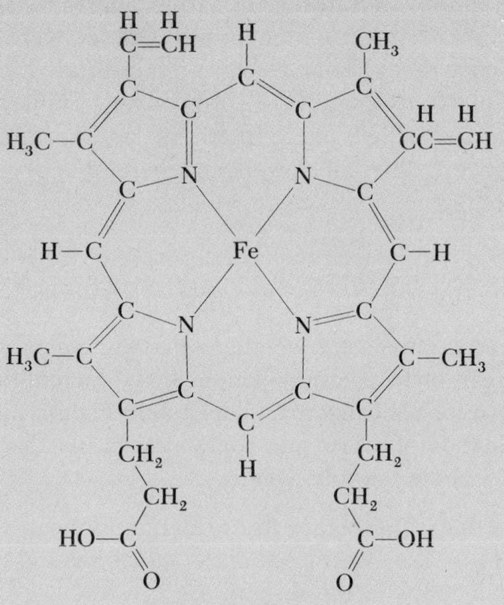

ferroprotoporphyrin IX

binding of oxygen in hemoglobin is said to be "cooperative"; that is, each successive O_2 molecule that attaches to the Fe^{2+} of a ferroprotoporphyrin IX group is more easily accommodated than the one before! The easiest by far is the fourth O_2 molecule. This situation can occur because the Fe^{2+} without bound O_2 is "high-spin" with four unpaired electrons. One of these electrons occupies the $d_{x^2-y^2}$ orbital, causing a large repulsion with the four nitrogen atoms of the porphyrin ring (Fig. 18.2a). This causes the ion Fe^{2+} to be pushed up from the plane of the four nitrogen atoms. When oxygen displaces the weaker-field H_2O molecule from its site of bonding to iron, it causes the Fe^{2+} to become low-spin, removing an electron from the $d_{x^2-y^2}$ orbital. The ion Fe^{2+} then inserts into the plane of the four nitrogens, as shown herewith

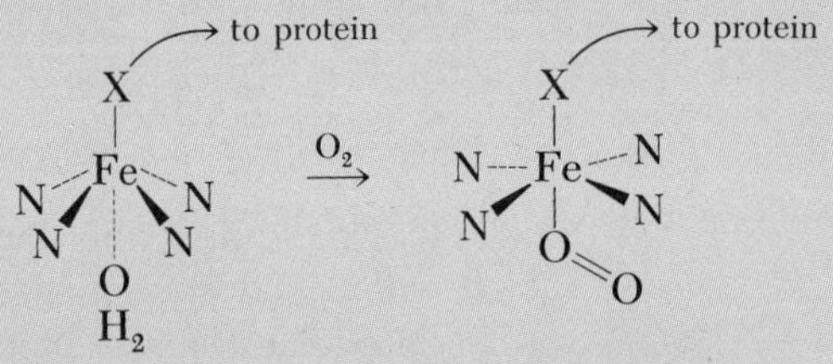

It has been proposed that this shift in position of the Fe^{2+}, which is bound to the protein, can cause structural changes in the rest of the protein. Thus, the other ferroprotoporphyrin IX groups "know" that O_2 has been added to another such group. The main function of the cooperative effort is probably not to encourage addition of O_2 in the lungs, but to make O_2 removal from the hemoglobin in red cells specific. The effect of cooperation when *removing* O_2 from the four ferroprotoporphyrin IX groups is that the first O_2 is held most strongly, and is thus hardest to remove. Once the first O_2 is removed, the other three will come off much more easily.

the two sets of *d* orbitals; the ligands that give rise to the large gaps characteristic of low-spin complexes are called **strong-field ligands.**

Up to this point, the d^8 and d^9 cases have not yet been considered. The reason that discussion of these cases has been delayed is that d^8 and d^9 ions represent cases in which octahedral complexes are neither "high-spin" nor "low-spin." The terms *high spin* and *low spin* make sense only if there is a choice, based on applying the two rules given above for the possible number of unpaired electrons. For an octahedral complex, the electronic configurations of the d^8 and d^9 ions are

d^8 d^9

To cause a change in the number of unpaired electrons in the d^8 case would require moving an electron from one of the upper orbitals to the other, and pairing the spins, in violation of Hund's rule; in the ground state, an electron will not occupy an orbital with another electron already in it if there is an empty orbital of the same energy available. In the d^9 case, no other electronic configuration can be written for these five *d* orbitals.

Of course, the concepts of high spin and low spin are not at all applicable to systems in which the central metal is described as d^0 or d^{10}. In the former case, there are no electrons in the *d* orbitals of the valence shell; hence, electronic transitions of the type discussed above, or unpaired *d* electrons, simply are not involved. Similarly, in a d^{10} case, the *d* subshell is completely filled; hence, there are no unpaired *d* electrons, no electronic transitions between different *d* levels, and therefore no color due to *d*–*d* transitions.

Factors Influencing the *d* Orbital Energy Gap

It is important to be able to predict and to understand whether a particular complex will be low-spin or high-spin. One can make such predictions if the factors that determine the size of the energy gap, or splitting, are understood. Factors that increase this gap will favor the formation of a low-spin compound rather than a high-spin compound. These factors include the following:

1. The oxidation state of the metal ion. The higher this oxidation number, the stronger the attraction of the Lewis base electron pairs the metal ion and the greater the repulsion affecting electrons in the d_{z^2} and $d_{x^2-y^2}$ orbitals. This increases the gap; that is, for a given set of ligands, the splitting for Co(III) is greater than for Co(II).
2. The row of the periodic table in which the transition metal ion is found. The second-row ions, for example, Pd(II), are larger than the first-row ions of the same charge, for example, Ni(II); hence the d_{z^2} and $d_{x^2-y^2}$ orbitals extend farther towards the electron pairs of the Lewis bases. Hence, the lower the row of the periodic table, the larger the splitting.
3. The nature of the ligands attached to the metal ion. The interaction of the Lewis base itself with the *d* orbital is, of course, important in determining the repulsion. The strength of this interaction is gauged by an empirical variable called the crystal field strength. The order of the crystal field strengths of ligands is called the **spectrochemical series,** which is based primarily on spectroscopic determinations of the influence of ligands in the energy gap between *d* orbitals. For some common ligands, this series is, with CO exerting the strongest and I^- the weakest field,

$$CO > CN^- > NO_2^- > \text{pyridine} \approx NH_3 > H_2O > F^- > Cl^- > Br^- > I^-$$

The reason this is called the spectrochemical series is that the energy gap not only influences whether a complex is high-spin or low-spin, but also determines the color, meaning the spectroscopic properties, of the complex. Table 18.5 provides some examples that show the consequences of these features in determining the high-spin or low-spin characteristic of a transition metal complex.

TABLE 18.5 Some Examples of High-Spin and Low-Spin Cases

	High-Spin	*Low-Spin*	*Reason*
d^4	$[Cr(H_2O)_6]^{2+}$	$[Mn(CN)_6]^{3-}$	Two effects: (1) CN^- gives a larger gap than H_2O, and (s) higher charge of Mn^{3+} causes a larger gap than Cr^{2+} does.
d^5	$[Mn(pyridine)_6]^{2+}$	$[Mn(CN)_6]^{4-}$	CN^- causes a larger gap than pyridine.
d^6	$[Fe(NH_3)_6]^{2+}$	$[Ru(NH_3)_6]^{2+}$	Ru(II) is in the second row of transition metals and is larger than Fe(II).
		$[Co(NH_3)_6]^{3+}$	Higher oxidation state of Co(III) causes larger gap.
d^7	$[Co(H_2O)_6]^{2+}$	$[Co(NO_2)_6]^{4-}$	NO^- causes a larger gap than H_2O.

Using the qualitative principles just outlined, one can predict which of two complexes is more likely to be low-spin, but one cannot predict absolutely whether or not a complex is low-spin unless the actual size of the energy gap and the pairing energy for a particular metal ion are known. Such numbers are usually well known. The energy gap is principally determined from visible absorption spectra; and pairing energies, which are quite characteristic for each *d* electron configuration, have been evaluated from a combination of experiments and calculations.

The kinds of concepts described in this section also apply to other coordination geometries besides the octahedral. Crystal field splitting diagrams for the square planar and tetrahedral cases are shown in Figs. 18.10 and 18.11.

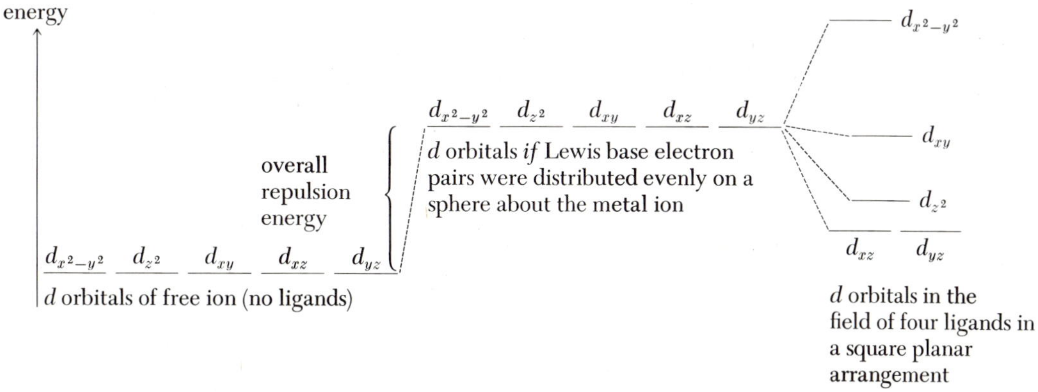

Figure 18.10 The influence of four ligands, arranged in a square planar configuration about a metal ion, on the energies of electrons in the *d* orbitals of the metal ion. We see what the effect of the overall repulsion would be if there were no directional character—if the ligand electron pairs were evenly distributed on a sphere, although they are not.

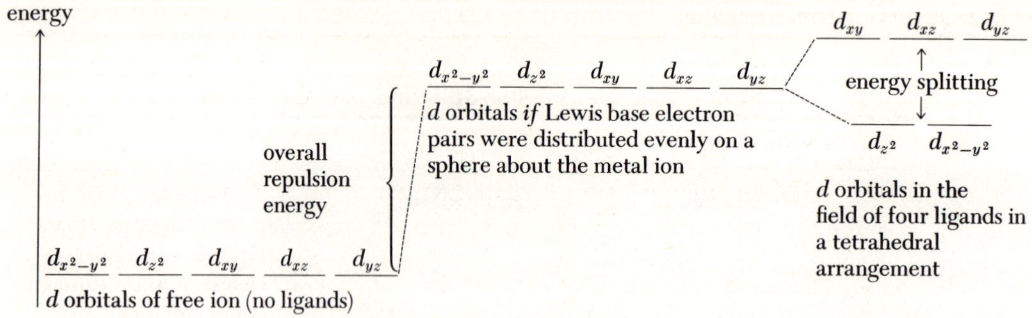

Figure 18.11 The influence of four ligands, arranged tetrahedrally about a metal ion on the energies of electrons in the *d* orbitals of the metal ion. This shows also what the effect of the overall repulsion would be if there were no directional character—if the ligand electron pairs were evenly distributed on a sphere (which they aren't). The energy splitting for a tetrahedral complex is only four-ninths the octahedral energy splitting for the same metal atom and ligands.

18.5 Reactions of Transition Metal Complexes

Selectivity of Transition Metal Ions

A knowledge of the bonding and of the rates of reactions of transition metal complexes allows chemists to optimize several important processes in which transition metals are essential. Nature also finds ways to use transition metal ions for very specific reactions that require their unique properties.

Spotlight

An interesting and important case of biologically important transition metal complexes is found in the behavior of the "nitrogen-fixing" enzymes, called **nitrogenases.** They contain both iron and molybdenum and make possible the production of NH_3 from atmospheric nitrogen, N_2. Enzymes called hydrogenases make the hydrogen available. Nitrogenases, with molecular weights of about 200,000 amu, are found in the bacteria that infect root nodules of legumes and account for the prodigious production of amino acids from these plants. This is why beans are such a good protein source. Models have been suggested for the way in which these enzymes work. The model proposed by R. W. F. Hardy shows how the bonding of N_2 to the metal ion site would lead to NH_3 production. The net reaction is $N_2 + 3H_2 \rightleftharpoons 2NH_3$.

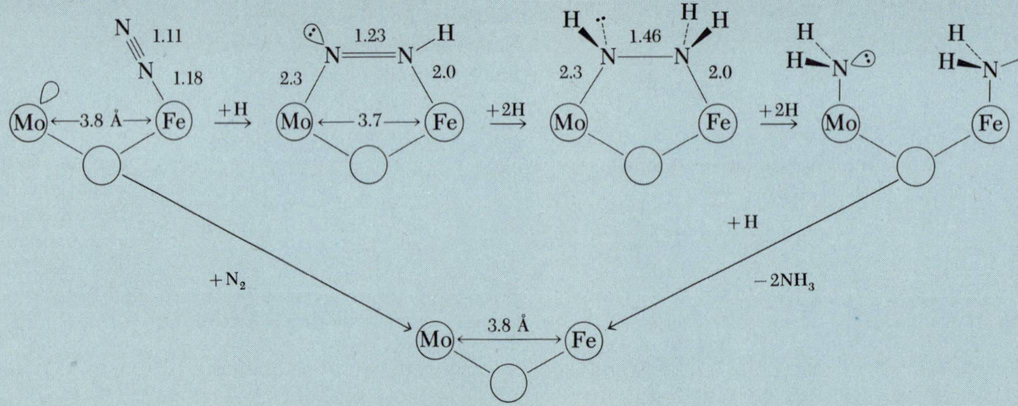

Most reactions involving transition metal ions take advantage of one or both of the following two properties: (1) their ability to function as good Lewis acids, often with reaction rates that are more or less roughly predictable for a given metal ion, and (2) they have two or more reasonably stable oxidation states.

The Lewis acid properties are often used in the purification, separation, and analysis of metal ions. The ability of the Lewis acids to use covalency in bonding to Lewis bases and the particular stability of certain geometries cause a large variation in equilibrium constants for the reactions of different transition metal ions with a particular Lewis base. In this manner reactions with transition metal ions can be specific and selective; that is, one can design experiments in which only one type of transition metal ion in a solution containing several will be caused to react with a specific Lewis base.

A characteristic example of this selectivity is the quantitative complexation of Ni(II) ion by dimethylglyoxime to form bis(dimethylglyoximato)nickel(II):

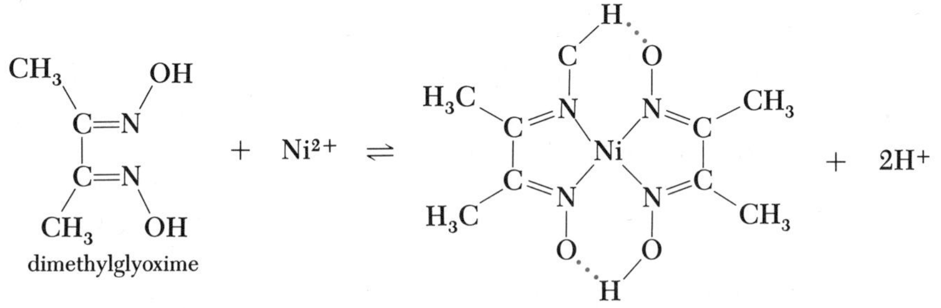

(18.15)

The dotted lines indicate hydrogen bonds (see Chap. 8), which help stabilize the structure. The formation of this solid, brick red compound is the basis of a precipitation method used to analyze for nickel.* The bright yellow solid Pd(II) complex of dimethylglyoxime is useful in analyzing for the Pd(II) ion.

The chemical form in which a transition metal exists can often be determined by using differences in rates of reaction; the rates may vary greatly for different oxidation states of a transition metal ion. For example, reactions in which one Lewis base is replaced by another can occur some 10^{10} times faster for Co(II) than for Co(III). Such rate differences allow one to selectively cause a reaction with the labile ions—those that react rapidly—and leave the slow, inert ions behind. In this way, one distinguishes between the ions and also separates them. If we knew that we had both Co(II) and Co(III) compounds in a sample, the total cobalt content could be determined by a spectroscopic method based on the element, such as atomic absorption (Chap. 6). The amount of Co(II) could then be determined by titration with ethylenediaminetetraacetic acid. A Co(II) species will react rapidly with the EDTA, producing a soluble species analogous to the Pb(II) complex shown in Fig. 18.3. The Co(III) species do not react rapidly in this way. Cobalt(III) would then be determined by the difference of the total cobalt and Co(II) results:

$$\text{moles Co(III)} = \underset{\text{(atomic absorption)}}{\text{total moles cobalt}} - \underset{\text{(EDTA analysis)}}{\text{moles Co(II)}}$$

*In analyzing for Ni^{2+} or Pd^{2+} in a solution, one adds an excess of the dimethylglyoxime reagent, and a precipitate forms, which is the solid complex. This precipitate is filtered, dried, and weighed, providing an absolute measure of the amount of Ni^{2+} or Pd^{2+} present in the solution of interest.

Enzyme Systems

Metal ions acting as Lewis acids are known in numerous enzymatic systems. **Enzymes** are complex molecules that function in biological systems as catalysts, enhancing the rates at which specific reactions occur. Zinc(II), because it is in many ways similar in its chemistry to the transition metal ions, is often included in discussions of this subject, although Zn (d^{10}) is actually not a transition metal. In many enzymes, Zn(II) and Co(II) will act very similarly, because of their similar properties (size, charge, and rate of ligand replacement), whereas Ni(II) and Co(III) are totally inactive, due to the different geometry applicable to Ni(II), and to the slow reaction rates of Co(III). The similarity in properties of Zn(II) and Co(II) in enzymatic systems has been used in the study of a type of enzyme called alkaline phosphatases. The **alkaline phosphatases** are a group of several enzymes that catalyze the following general reaction:

$$(O_3POR)^{2-} + OH^- \rightleftharpoons PO_4^{3-} + ROH \qquad (18.16)$$

This reaction occurs under alkaline conditions in organisms ranging from bacteria to mammals. The R group is a ribonucleotide or deoxyribonucleotide polymer, such as a fragment of RNA or DNA. One of these enzymes from the bacterium *Escherichia coli* has four Zn(II) "ions" in each molecule. Zinc(II) has very few spectroscopic properties that can be used as probes for the detailed study of these important enzymes. Thus, Zn(II) is d^{10}, with no useful optical spectrum (that would result from transitions among *d* orbital levels), and it is diamagnetic. A much more useful probe for laboratory studies is the colored, paramagnetic d^7 Co(II) ion. When Co(II) is substituted for Zn(II) in the enzyme, the catalytic activity is retained, and the system is then more amenable to studies concerned with the environment of the metal ion. Both Co(II) and Zn(II) are labile and often form tetrahedral complexes; and they have about the same Lewis acidity since they are of similar size and have

Spotlight

The ion Zn^{2+} is much more effective in functioning as a Lewis acid catalyst than many other metal ions, for instance Mg^{2+}. At least part of the reason for this is the specific bonding Zn^{2+} shows, because like transition metal ions, it has some covalency in its bonding. Zinc(II) is found in a large number of enzymes. Modern analytical methods are becoming more and more sensitive, steadily increasing the number of known metal-containing biological molecules. Zinc is readily abundant and easily able to be incorporated into enzymes, since it undergoes very rapid ligand replacement. The enzyme is able to retain such labile metal ions as zinc by "encapsulating" them within its protein structure and preventing "attack" on the metal ion by other good ligands that are present in the biological system, like Cl^- or H_2O. Often there is only one approach to the metal ion (often called a channel, cleft, or pocket) allowed by the encapsulating protein. The function of the zinc may be only to hold the enzyme in a particular shape to allow the enzyme to perform its catalytic function. The Zn^{2+} may bind to a Lewis base region of the substrate, such as a lone pair of electrons on nitrogen or oxygen, and thereby aid in the catalysis by its activity as a Lewis acid.

Nutritional zinc deficiency is rare since zinc is abundant in most soil and is readily obtained from both animal and plant food sources.

Spotlight

Due to the size, complexity, and frequent instability of enzymes and other metal complexes involved in electron and oxygen transport, much remains to be learned of the ways in which these molecules act on the molecular scale. In many cases the actual structure of the molecule around the metal ion is not known. Even when this information is available, there is often a lack of other, analogous molecules that might be used to test a hypothesis about the mode of reaction of the biologically active molecule. The frequent necessity of additional molecules for "activating" the catalyst complicates the problem further. This area of research, in which transition metal complexes are so important, is indeed challenging, and promises to remain so for quite a while. Continued development in spectroscopic methods, synthetic studies to make better model systems, and the general advancement of chemical theory are all combining to make such difficult problems more tractable.

the same charge. Nickel(II) often will not function in enzymes in place of Zn(II), probably because it requires different geometries (octahedral or square planar), even though it is relatively labile and has similar Lewis acidity. Alkaline phosphatase is inactive with Co(III) that may be substituted for Zn(II), probably because it is not labile, is inert to ligand substitution, and requires octahedral geometry. (Further consideration of catalysis in general and enzymes in particular is in Chap. 13.)

Multiple Oxidation States

Another important property of transition metals is their ability to exist in several oxidation states. One can predict the maximum oxidation number that is to be found for stable species; this is given by the number of electrons outside the noble gas core in a neutral atom in its ground state. For example, with ground state electronic configuration for Co, $[Ar]3d^7 4s^2$, the maximum oxidation number is 9. Species having the maximum predicted oxidation numbers are not found to be stable for all elements, even though they lead to the noble gas configuration. Such absences are due to the large amounts of energy that would be needed to form these ions—to remove all the valence electrons. The most commonly formed ions of transition metals are $+2$ and $+3$, which in terms of lattice energy and ionization potential are relatively easy to achieve. Accordingly, $+2$ and $+3$ are the most common oxidation states.

Reasonable stability of more than one oxidation state is an important feature of the chemistry of pairs of ions such as Fe(II)–Fe(III) and Cu(II)–Cu(I) and their complexes in biological systems. For such species, oxidation-reduction reactions allow these transition metal ions to function as catalysts.

EXAMPLE 18.8 The stability of a particular oxidation state of a metal ion depends markedly on the other species in solution available for reaction. Using Table 12.1, determine the species that are stable in acidic, oxygen-free aqueous solutions under the following conditions: (a) Fe^{3+} with excess $Cl_2(g)$ present; (b) Fe^{2+} with excess $Cl_2(g)$ present; (c) Fe^{3+} with excess $I_2(g)$ present; (d) Fe^{3+} with excess I^- present.

SOLUTION Consulting Table 12.1, we can decide whether an ion can react with another species present in solution to give a more stable ion, in other words, whether there is a reaction possible with an overall positive $\mathscr{E}°$.

(a) $Cl_2(g)$ is an oxidizing agent and Fe^{3+} is in the oxidized state, so it is stable.

(b) $Cl_2(g)$ oxidizes Fe^{2+} to Fe^{3+}:

	$\mathscr{E}°$ (volts)	$n\mathscr{E}°$ (volts)
$Cl_2(g) + 2e^- \rightleftharpoons 2Cl^-$	1.36	2.72
$2Fe^{2+} \rightleftharpoons 2e^- + 2Fe^{3+}$	−0.77	−1.54
$Cl_2(g) + 2Fe^{2+} \rightleftharpoons 2Cl^- + 2Fe^{3+}$		1.18 volts

Therefore, Fe^{3+} would be more stable in this solution because Fe^{2+} is readily oxidized by Cl_2.

(c) $I_2(g)$ is not capable of oxidizing Fe^{2+} under these conditions:

	$\mathscr{E}°$	$n\mathscr{E}°$
$I_2(s) + 2e^- \rightleftharpoons 2I^-$	0.54	1.08
$2Fe^{2+} \rightleftharpoons 2e^- + 2Fe^{3+}$	−0.77	−1.54
$I_2(s) + 2Fe^{2+} \rightleftharpoons 2I^- + 2Fe^{3+}$		0.46

Therefore, either Fe^{2+} or Fe^{3+} can exist in a solution with excess $I_2(s)$.

(d) From (b) above, I^- reduces Fe^{3+}:

$$2Fe^{3+} + 2I^- \rightleftharpoons I_2(s) + 2Fe^{2+} \qquad n\mathscr{E}° = +0.46$$

Therefore Fe^{2+} is stable but Fe^{3+} is reduced. (For this reason, Fe^{3+} complexes with I^- are not isolated from highly acidic solution. In less acidic solution, the $\mathscr{E}°$ value changes to favor Fe^{3+} (see Table 12.1) and Fe^{3+} and I^- can coexist in solution.)

STUDY PROBLEM 18(i)

Determine the stable species in acidic, oxygen-free aqueous solution under the following conditions: (a) Ce^{4+} in the presence of excess Fe^{2+}; (b) Cu^{2+} in the presence of excess Fe^{2+}; (c) Ag(s); (d) Ag^+; (e) Zn(s).

Many examples of useful oxidation-reduction catalysts are found in laboratory procedures. If one wants to dissolve chromium(III) chloride in water, for example, the process can be facilitated by adding a small amount of chromium(II) chloride. Chromium(II) chloride is very soluble in H_2O. However, $CrCl_3$ does not appear to dissolve at all when added to water, even though it is actually quite soluble once in solution. If one waits long enough, quite a bit will eventually dissolve. The sluggishness of $CrCl_3$ dissolution is due to the very slow ligand exchange rate for Cr(III). Added Cr^{2+} facilitates this exchange catalytically:

$$CrCl_3(s) + Cr^{2+}(aq) \xrightarrow{\text{rapid}} CrCl_2(s) + Cr^{3+}(aq) + Cl^- \qquad (18.17)$$

$$CrCl_2(s) \xrightarrow{\text{rapid}} Cr^{2+}(aq) + 2Cl^- \qquad (18.18)$$

$$\text{net reaction:} \quad CrCl_3(s) \xrightarrow{\text{rapid}} Cr^{3+} + 3Cl^- \qquad (18.19)$$

A second example involves the use of As_2O_3 as a reducing agent in an oxidation-reduction titration used in analytical procedures. As shown in the equations that follow, osmium tetroxide, OsO_4, is used as a catalyst for reactions in which As_2O_3 is the titrant. For example, if one wishes to find the concentration of Ce(IV), a strong oxidizing agent, a titration with As_2O_3 can be used.

$$2Ce(IV) + As_2O_3 \rightarrow As(V) + 2Ce(III) \qquad (18.20)$$

By itself, the reaction would take months. A little OsO_4, however, speeds up the reaction to such an extent that it is over in less than a second. The reaction scheme (unbalanced) is

$$As_2O_3 + OsO_4 \rightarrow As(V) + Os(IV) \qquad (18.21)$$

$$2Ce(IV) + Os(VI) \rightarrow 2Ce(III) + OsO_4 \qquad (18.22)$$

Since OsO_4 is consumed in the first step and regenerated in the second step, it is not part of the net reaction:

$$2Ce(IV) + As_2O_3 \xrightarrow{OsO_4} As(V) + 2Ce(III) \qquad (18.23)$$

Important industrial processes also use oxidation-reduction catalysis. An example of this is the Wacker process, which converts ethylene, a by-product of petroleum refining, into a useful industrial chemical, acetaldehyde, $\mathrm{CH_3C}\begin{smallmatrix}\diagup\!\!\mathrm{O}\\ \diagdown\mathrm{H}\end{smallmatrix}$. The reaction sequence, which uses $PdCl_2$ and $CuCl_2$ as catalysts in solution, can be written as

$$\begin{matrix}\mathrm{H} & & & & \mathrm{H}\\ & \diagdown & \mathrm{C{=}C} & \diagup & \\ \mathrm{H} & \diagup & & \diagdown & \mathrm{H}\end{matrix} + Pd^{2+} + H_2O \rightleftharpoons \mathrm{CH_3\overset{\overset{\displaystyle O}{\|}}{C}H} + Pd(s) + 2H^+ \qquad (18.24)$$

$$Pd(s) + 2Cu^{2+} \rightleftharpoons Pd^{2+} + 2Cu^+ \qquad (18.25)$$

$$2H^+ + 2Cu^+ + \tfrac{1}{2}O_2(g) \rightleftharpoons 2Cu^{2+} + H_2O \qquad (18.25)$$

The overall reaction is then

$$\begin{matrix}\mathrm{H} & & & & \mathrm{H}\\ & \diagdown & \mathrm{C{=}C} & \diagup & \\ \mathrm{H} & \diagup & & \diagdown & \mathrm{H}\end{matrix} + \tfrac{1}{2}O_2(g) \rightleftharpoons \mathrm{CH_3\overset{\overset{\displaystyle O}{\|}}{C}H} \qquad (18.27)$$

The fact that copper can exist in both the +1 and +2 oxidation states is critical in the reaction sequence.

SUMMARY

We have outlined some of the chemical and physical characteristics of coordination compounds of the transition elements. These compounds can be viewed as if formed from a transition metal atom that acts as a Lewis acid and a number of ligands (generally 2, 4, or 6) that act as Lewis bases. The bonding in transition-metal coordination compounds shows characteristics of both ionic and covalent bonding. The relative importance of each of these types of bonding is often predictable by a scale of hardness and softness. The hard Lewis acids and bases are generally smaller and more electronegative (showing greater electrostatic or "ionic" character) than the softer Lewis acids and bases, which are generally larger and more polarizable, thus showing greater covalent character.

Many coordination compounds undergo dissociation slowly and retain their bonding in solution. The identity of such species can sometimes be deduced by the extent to which they will conduct current when dissolved in water. Covalent character in these compounds is also evident from their tendency to retain a characteristic shape—linear Ag(I), Hg(II), and Cu(I) species; square Pd(II) and Pt(II) complexes; and octahedral Cr(III) and Co(III) complexes, for example.

An important property of coordination compounds is their tendency to be drawn into a magnetic field (paramagnetism) or to be rejected from a magnetic field (diamagnetism). From the extent of paramagnetism, the number of unpaired electrons and the oxidation state of the metal ion can generally be deduced. A transition metal ion with 4 to 8 *d*-level electrons can show different degrees of electron pairing. This effect can be understood by crystal field theory, and the degrees of electron pairing are called high spin (maximum number of unpaired electrons) and low spin (minimum number of unpaired electrons).

Coordination compounds are important in oxygen binding, the reactions of Vitamin B_{12}, and biological oxidation-reduction reactions. They are also useful as catalysts. In addition, they have a rich and diverse chemistry that continues to fascinate many chemists.

STUDENT CHECKPOINTS

After studying this chapter, the student should be able to:

1. Determine which elements are transition elements.
2. Name and write formulas of transition metal compounds.
3. Determine the oxidation state assigned to the transition metal atom in a given compound or ion.
4. Explain what is meant by a stability constant and by hard and soft Lewis acids and bases.
5. Explain how conductivity is used to study transition metal compounds.
6. Discuss how the rates of reactions of transition metal compounds span a wide range.
7. Define *diamagnetism* and *paramagnetism* and explain how to use magnetic properties of transition metal compounds in determining the number of unpaired electrons.
8. Cite examples of linear, tetrahedral, square, and octahedral transition metal species.
9. Review the importance of transition metal ions in catalysis, oxidation-reduction reactions, and chemical analysis.

EXERCISES

18.1. What features of transition elements are consistent with their classification as metallic elements?

18.2.* Name the following compounds, which contain transition elements: $[Cr(NH_3)_6]Cl_2$, $[CrCl_2(en)_2]Br$, $FeSO_4$, $K[PtCl_3NH_3]$, $Ca[ZnCl_4]$.

18.3. Name the following compounds: $[CoCl_3(NH_3)_3]$, $Mn(Cl_4)_2$, $[RuClNO(en)_2]Br$, $Na_4[Fe(CN)_6]$.

18.4.* Write formulas for the following compounds: *trans*-dichlorobis(ethylenediamine)cobalt(III) chloride, sodium tetrachlorocuprate, *cis*-dichlorotetraammineruthenium(II).

18.5. Write formulas for the following compounds: ferric perchlorate, potassium pentacyanoaquoferrate(III), bis(cyclopentadienyl)ruthenium(II) (also known as ruthenocene).

18.6.* The stability constants for the combination of silver(I) and ammonia are $K_1 = 2.3 \times 10^3$ and $K_2 = 6.9 \times 10^3$, whereas the stability constant for $AgNO_3$ is 0.50; that is, $AgNO_3$ formation is negligible in the presence of NH_3. For an aqueous solution 1.0×10^{-3} *M* in $AgNO_3$ and 1.0 *M* in NH_3, calculate:
(a) the ratio $[Ag(NH_3)_2^+]/[Ag(NH_3)^+]$;
(b) the ratio $[Ag(NH_3)^+]/[Ag(aq)^+]$;
(c) the concentration of each species.

18.7. The stability constant for the combination of copper(II) with perchlorate ion is less than 1. Using data given in the chapter, calculate the following for an aqueous solution 1.0×10^{-3} *M* in $Cu(ClO_4)_2$ and 1.0 *M* in NH_3:
(a) the value of β_4;
(b) the ratio of the concentrations of $Cu(NH_3)_4^{2+}$ and Cu^{2+} in the solution. (*Hint:* To two significant figures $[NH_3]$ remains 1.0 *M*);
(c) the ratio of concentrations of $Cu(NH_3)_3^{2+}$, $Cu(NH_3)_2^{2+}$, and $Cu(NH_3)^{2+}$ to Cu^{2+} based on the values of β_3, β_2, and β_1;
(d) the actual concentrations of each species.

18.8. By a method that distinguishes aqueous silver(I) from $Ag(NH_3)^+$ or $Ag(NH_3)_2^{2+}$ and uncomplexed ammonia from ammonia bound to silver, the following values were determined: $[Ag^+] = 2.75 \times 10^{-4}$ *M* and $[NH_3] = 1.42 \times 10^{-3}$ *M*. The solution had been made up by dissolving 1.70 g of silver nitrate in 100 ml of 0.020 *M* aqueous ammonia. What are the concentrations of $Ag(NH_3)^{2+}$, and $Ag(NH_3)^+$ in the solution? (Use the stability constants given in Exercise 18.6.)

18.9.* The stability constants for the progressive combination of mercury(II) with ammonia show a sharp decrease,

$$K_1 = 6.3 \times 10^8, \quad K_2 = 5.0 \times 10^8, \quad K_3 = 10, \quad K_4 = 6.0$$

whereas those for cobalt(II) and ammonia decrease gradually,

$$K_1 = 97, \quad K_2 = 32, \quad K_3 = 8.5, \quad K_4 = 4.4, \quad K_5 = 1.1, \quad K_6 = 0.18$$

Why does this happen, do you think?

18.10. Why is F^- said to be harder than Br^-?

18.11.* Which in each of the following pairs has a higher stability constant (for loss of one halide ion):
(a) $AlCl_4^-$ or $AlCl_3I^-$?
(b) ZnI^+ or HgI^+?
(c) $FeCl_3$ or FeI_3?

18.12. Which in each of the following pairs has a higher stability constant:
(a) $[Co(NH_3)_6Cl]^{2+}$ or $[Co(NH_3)_6I]^{2+}$ (loss of halide)?
(b) $[Pt(NH_3)_3Cl]^+$ or $[Pt(NH_3)_3I]^+$ (loss of halide)?
(c) $HgSCN^+$ or $FeSCN^{2+}$?

18.13.* When some of the following are dissolved in water, the solution will immediately conduct a current. Indicate the species for which this is observed and show what ions are formed. $[RuBr(NH_3)_5]Cl_2$; $K_2[Fe(CN)_5NO]$; $Ni(CO)_4$; *cis*-$[PtCl_2(NH_3)_2]$.

18.14. When $[RhCl_2(NH_3)_4]Cl$ is dissolved in water and $AgClO_4$ is added, a white precipitate forms immediately. If the solution is then filtered, the white precipitate continues to be formed much more slowly in the filtrate. What is happening? Write out and balance the appropriate equations.

18.15.° (a) When $[Co(CO_3)I(NH_3)_4]$ is added to acidic solutions, bubbles form. Show what is happening. (b) In neutral solutions, a yellow precipitate slowly forms if $Pb(NO_3)_2$ is added. Show the equation for this process. (c) Why does the yellow precipitate form faster when $[Co(CO_3)(NH_3)_5]I$ is dissolved than when $[Co(CO_3)I(NH_3)_4]$ is dissolved?

18.16. Which complex of each of the following pairs would show greater paramagnetism:
(a) $[CrCl_3(NH_3)]$ or $[TiCl_2(H_2O)_4]Cl$?
(b) $K_3[Fe(CN)_6]$ or $K_2[ZnCl_4]$?
(c) $[Mn(H_2O)_6]Cl_2$ or $KMnO_4$?

18.17.° Which complex of each of the following pairs would show greater paramagnetism (*Hint:* See Fig. 18.9.):
(a) $[Fe(NH_3)_6]Cl_2$ or $[Ru(NH_3)_6]Cl_2$?
(b) $[Mn(pyridine)_6]Br_2$ or $[Zn(NH_3)_6](ClO_4)_2$?

18.18. What are the isomers of $[PtCl_2(NH_3)_2]$?

18.19.° What are the isomers of (a) $[CoCl_2(NH_3)_4]$? (b) $[CoCl_3(NH_3)_3]$?

18.20. Give an example of transition metal species that are (a) linear, (b) square, and (c) octahedral.

18.21.° Determine the oxidation number of the metal atom in each of the following complexes: (a) $[PtI_2(H_2O)_2]$; (b) $[Fe(CO)_5]$; (c) $K_3[Fe(CN)_6]$.

18.22. Determine the oxidation number of the metal atom in each of the following complexes:
(a) $[RhCl_2(NH_3)_4]Cl$;
(b) $[RhCl(CO)_2]_2$, which has the structure

$$\begin{array}{ccc} OC & Cl & CO \\ \diagdown & \diagdown & \diagup \\ Rh & & Rh \\ \diagup & \diagup & \diagdown \\ OC & Cl & CO \end{array}$$

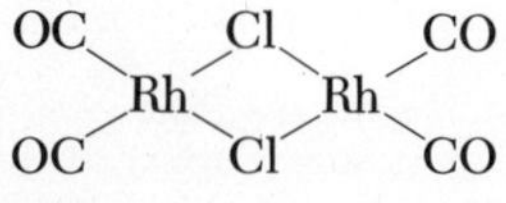

(c) $KMnO_4$.

18.23.° Why are there isomers of $[PdBr_2(NH_3)_2]$ but not of $[PdBr_2(ethylenediamine)]$?

18.24. If 1.0 millimole of $[Fe(H_2O)_6](ClO_4)_2$, 2.0 millimoles of ethylenediamine, and 6 millimoles of methylamine are dissolved in 1.0 liter of water, what is the predominant species? Why?

18.25.° Describe the process by which light in the visible region of the spectrum is absorbed by the ion $[Ti(H_2O)_6]^{3+}$.

18.26. What is meant by low spin and high spin? What experiment can be performed to distinguish whether a species such as $[Co(NH_3)_6]^{3+}$ is high-spin or low-spin? For some transition metal ions in a particular oxidation state, there is only one spin state possible. Give an example.

18.27.° Given that one of each set of the following pairs is high-spin and one low-spin, tell which is high-spin and determine the number of unpaired electrons in each:
(a) $[Co(NH_3)_6]F_3$ or $K_3[CoF_6]$;
(b) $[Fe(NH_3)_6]Cl_2$ or $[Ru(NH_3)_6]Cl_2$;
(c) $Co^{3+}(aq)$ or $Co^{2+}(aq)$.

18.28. What is the approximate energy of the light absorbed by a complex with only one absorption band that gives (a) an orange solution? (b) a blue solution? (c) a red solution? What are the approximate wavelengths and frequencies of the absorbed light for those species?

18.29. How does Cu^{2+} function in the catalytic production of acetaldehyde from ethylene (the Wacker process)?

18.30.* "Hydrated" chromic chloride, $Cr(H_2O)_6 \cdot Cl_3$, dissolves in water much faster than anhydrous chromic chloride, $CrCl_3$. Why is this so, do you think?

ANSWERS TO SELECTED EXERCISES

18.2 Hexaamminechromium(III) chloride; dichlorobis(ethylenediamine)chromium(III) bromide; ferrous sulfate; potassium trichloroammineplatinate(II); calcium tetrachlorozincate(II). 18.4 *trans*-$[CoCl_2(en)_2]Cl$; $Na_2[CuCl_4]$; *cis*-$[RuCl_2(NH_3)_4]$. 18.6 Note that so little NH_3 is used up that $[NH_3] \approx 1.0\,M$ in the presence of the Ag^+. (a) 6.9×10^3; (b) 2.3×10^3; (c) $[Ag^+] = 6.3 \times 10^{-11}\,M$; $[Ag(NH_3)^+] = 1.4 \times 10^{-7}\,M$; $[Ag(NH_3)_2{}^+] = 1.0 \times 10^{-3}\,M$. 18.9 Mercury forms stable linear complexes such as $Hg(NH_3)_2$ whereas cobalt(II) can form six-coordinate octahedral complexes such as $[Co(NH_3)_6]^{2+}$. 18.11 (a) $AlCl_4{}^-$; (b) HgI^+; (c) $FeCl_3$. 18.13 $[RuBr(NH_3)_5]^{2+}$ and $2Cl^-$; $2K^+$ and $[Fe(CN)_5NO]^{2-}$. 18.15 (a) $[Co(CO_3)I(NH_3)_4] + 2H^+ + H_2O \rightarrow [Co(I)(H_2O)(NH_3)_4] + H_2CO_3$, $H_2CO_3 \rightleftharpoons H_2O + CO_2(g)$; (b) $2[Co(CO_3)I(NH_3)_4] + Pb^{2+} + 2H_2O \rightleftharpoons 2[Co(CO_3)(H_2O)(NH_3)_4] + PbI_2(s)$; (c) In $[Co(CO_3)I(NH_3)_4]$, the iodide atom is bound to some degree covalently to the Co(III) atom and dissociates slowly. In $[Co(CO_3)(NH_3)_5]I$, the iodide ion is ionic and will be dissociated in aqueous solution. 18.17 (a) $[Fe(NH_3)_6]Cl$; (b) $[Mn(py)_6]Br_2$. 18.19 (a) cis and trans chloride ions; (b) chloride ions disposed as a T and on one face of the octahedron. 18.21 (a) +2; (b) 0; (c) +3. 18.23 $[PdBr_2(NH_3)_2]$ is found as cis and trans isomers but ethylenediamine is not long enough to form a trans isomer. 18.25 This is a *d–d* transition in which the excitation mainly involves the *d* electron on the Ti^{3+} atom. 18.27 (a) $K_3[CoF_6]$, four; (b) $[Fe(NH_3)_6]Cl_2$, four; (c) $Co^{2+}(aq)$, three. 18.30 In the hydrated form the chloride ions are already dissociated from the chromium atom, so only electrostatic interactions are present. In $CrCl_3$ the Cr–Cl bonds must be broken for dissolution to occur and the process is slow.

ANSWERS TO STUDY PROBLEMS

18(a) dichlorodiamminepalladium(II); (b) sodium tetrachlorozincate(II); (c) dichlorobis(ethylenediamine)ruthenium(III) bromide. 18(b) (a) $[CrCl(H_2O)_5](ClO_4)_2$; (b) $Na_3[CoF_6]$. 18(c) 2.05 kcal/mol. 18(d) (a) 3; (b) 8; (c) 0. 18(e) (a) 7; (b) 5; (c) 6. 18(f) (a) d^2sp^3, with the six *d* electrons of Ru(II) occupying three of the five 4*d* orbitals; (b) dsp^2, with the eight *d* electrons of Pt(II) occupying four of the five 5*d* orbitals; (c) *sp*, with the eight *d* electrons of Hg(II) occupying all five of the five *d* orbitals. 18(g) 9.97 kcal/mol. 18(h) (a) 0.219 g weight increase per mole; (b) 0.238 g weight increase per mole; (c) no unpaired electrons, no weight increase. Actually some slight weight decrease due to diamagnetism. 18(i) (a) Ce^{4+} reacts readily to give Fe^{3+}, not stable; (b) Cu^{2+}; (c) Ag(s) is stable in acid; (d) Ag^+ is stable in acid since H^+ is an oxidizing agent; Ag^+ can be reduced by $H_2(g)$; (e) Zn(s) reacts to form Zn^{2+} in acid.

Appendixes

Appendix A

Index to Important Reference Tables

Index to Important Reference Tables (continued)

Appendix B

Association Constants for Complex Ions (25 °C)

Equilibrium	*log K*
$Cd^{2+} + NH_3 \rightleftharpoons [Cd(NH_3)]^{2+}$	2.74
$[Cd(NH_3)]^{2+} + NH_3 \rightleftharpoons [Cd(NH_3)_2]^{2+}$	2.18
$[Cd(NH_3)_2]^{2+} + NH_3 \rightleftharpoons [Cd(NH_3)_3]^{2+}$	1.45
$[Cd(NH_3)_3]^{2+} + NH_3 \rightleftharpoons [Cd(NH_3)_4]^{2+}$	1.00
$Cd^{2+} + 4NH_3 \rightleftharpoons [Cd(NH_3)_4]^{2+}$	7.37
$Co^{3+} + NH_3 \rightleftharpoons [Co(NH_3)]^{3+}$	7.3
$Co^{2+} + NH_3 \rightleftharpoons [Co(NH_3)]^{2+}$	1.98
$[Co(NH_3)]^{2+} + NH_3 \rightleftharpoons [Co(NH_3)_2]^{2+}$	1.51
$[Co(NH_3)_2]^{2+} + NH_3 \rightleftharpoons [Co(NH_3)_3]^{2+}$	0.93
$[Co(NH_3)_3]^{2+} + NH_3 \rightleftharpoons [Co(NH_3)_4]^{2+}$	0.64
$[Co(NH_3)_4]^{2+} + NH_3 \rightleftharpoons [Co(NH_3)_5]^{2+}$	0.41
$[Co(NH_3)_5]^{2+} + NH_3 \rightleftharpoons [Co(NH_3)_6]^{2+}$	−0.74
$Cu^{2+} + NH_3 \rightleftharpoons [Cu(NH_3)]^{2+}$	4.13
$[Cu(NH_3)]^{2+} + NH_3 \rightleftharpoons [Cu(NH_3)_2]^{2+}$	3.53
$[Cu(NH_3)_2]^{2+} + NH_3 \rightleftharpoons [Cu(NH_3)_3]^{2+}$	2.87
$[Cu(NH_3)_3]^{2+} + NH_3 \rightleftharpoons [Cu(NH_3)_4]^{2+}$	2.15
$Cu^{2+} + SCN^- \rightleftharpoons [CuSCN]^+$	2.3
$Fe^{3+} + F^- \rightleftharpoons [FeF]^{2+}$	5.2
$Fe^{3+} + Cl^- \rightleftharpoons [FeCl]^{2+}$	1.5
$Fe^{3+} + I^- \rightleftharpoons [FeI]^{2+}$	1.3
$[Fe(CN)_5]^- + CN^- \rightleftharpoons [Fe(CN)_6]^{2-}$	9
$Fe^{2+} + 6CN^- \rightleftharpoons [Fe(CN)_6]^{4-}$	24
$Fe^{3+} + 6CN^- \rightleftharpoons [Fe(CN)_6]^{3-}$	31
$Fe^{3+} + SCN^- \rightleftharpoons [Fe(SCN)]^{2+}$	2.3
$[Fe(SCN)]^{2+} + SCN^- \rightleftharpoons [Fe(SCN)_2]^+$	1.6
$[Fe(SCN)_2]^+ + SCN^- \rightleftharpoons [Fe(SCN)_3]$	1.4
$[Fe(SCN)_3] + SCN^- \rightleftharpoons [Fe(SCN)_4]^-$	0.8
$Mn^{2+} + SCN^- \rightleftharpoons [MnSCN]^+$	1.2
$Hg^{2+} + SCN^- \rightleftharpoons [Hg(SCN)]^+$	16
$Hg^{2+} + NH_3 \rightleftharpoons [Hg(NH_3)]^{2+}$	8.80
$[Hg(NH_3)]^{2+} + NH_3 \rightleftharpoons [Hg(NH_3)_2]^{2+}$	8.70
$[Hg(NH_3)_2]^{2+} + NH_3 \rightleftharpoons [Hg(NH_3)_3]^{2+}$	1.00
$[Hg(NH_3)_3]^{2+} + NH_3 \rightleftharpoons [Hg(NH_3)_4]^{2+}$	0.78
$Hg^{2+} + F^- \rightleftharpoons [HgF^+]$	1.0
$Hg^{2+} + Cl^- \rightleftharpoons [HgCl]^+$	6.7
$Hg^{2+} + I^- \rightleftharpoons [HgI]^+$	13
$Ag^+ + NH_3 \rightleftharpoons [Ag(NH_3)]^+$	3.36
$[Ag(NH_3)]^+ + NH_3 \rightleftharpoons [Ag(NH_3)_2]^+$	3.84
$AgCl + Cl^- \rightleftharpoons [AgCl_2]^-$	1.67
$Tl^+ + Cl^- \rightleftharpoons TlCl$	0.50
$Tl^{3+} + Cl^- \rightleftharpoons [TlCl]^{2+}$	8.1
$[TlCl]^{2+} + Cl^- \rightleftharpoons [TlCl_2]^+$	5.5
$[TlCl_2]^+ + Cl^- \rightleftharpoons TlCl_3$	2.2
$Zn^{2+} + 4NH_3 \rightleftharpoons [Zn(NH_3)_4]^{2+}$	8.70
$Zn^{2+} + Cl^- \rightleftharpoons [ZnCl]^+$	0.72
$[ZnCl]^+ + Cl^- \rightleftharpoons ZnCl_2$	−0.23
$ZnCl_2 + Cl^- \rightleftharpoons [ZnCl_3]^-$	−0.68
$[ZnCl_3]^- + Cl^- \rightleftharpoons [ZnCl_4]^{2-}$	0.37
$Zn^{2+} + 4OH^- \rightleftharpoons [Zn(OH)_4]^{2-}$	20.2

Appendix C

Solubility Products

Substance	K_{sp} *at 25°*	*Substance*	K_{sp} *at 25°*
Aluminum		Iron	
$Al(OH)_3$	1.9×10^{-33}	$Fe(OH)_2$	7.9×10^{-15}
Barium		$FeCO_3$	2.11×10^{-11}
$BaCO_3$	8.1×10^{-9}	FeS	1×10^{-19}
$BaC_2O_4 \cdot 2H_2O$	1.1×10^{-7}	$Fe(OH)_3$	1.1×10^{-36}
$BaSO_4$	1.08×10^{-10}	Lead	
$BaCrO_4$	2×10^{-10}	$Pb(OH)_2$	2.8×10^{-16}
BaF_2	1.7×10^{-6}	PbF_2	3.7×10^{-8}
$Ba(OH)_2 \cdot 8H_2O$	5.0×10^{-3}	$PbCl_2$	1.7×10^{-5}
$Ba_3(PO_4)_2$	1.3×10^{-29}	$PbBr_2$	6.3×10^{-6}
$Ba_3(AsO_4)_2$	1.1×10^{-13}	PbI_2	8.7×10^{-9}
Bismuth		$PbCO_3$	1.5×10^{-13}
BiO(OH)	1×10^{-12}	PbS	8.4×10^{-28}
BiOCl	7×10^{-9}	$PbCrO_4$	1.8×10^{-14}
Bi_2S_3	1.6×10^{-72}	$PbSO_4$	1.8×10^{-8}
Cadmium		$Pb_3(PO_4)_2$	3×10^{-44}
$Cd(OH)_2$	1.2×10^{-14}	Magnesium	
CdS	3.6×10^{-29}	$Mg(OH)_2$	1.5×10^{-11}
$CdCO_3$	2.5×10^{-14}	$MgCO_3 \cdot 3H_2O$	*ca.* 1×10^{-5}
Calcium		$MgNH_4PO_4$	2.5×10^{-13}
$Ca(OH)_2$	7.9×10^{-6}	MgF_2	6.4×10^{-9}
$CaCO_3$	4.8×10^{-9}	MgC_2O_4	8.6×10^{-5}
$CaSO_4 \cdot 2H_2O$	2.4×10^{-5}	Manganese	
$CaC_2O_4 \cdot H_2O$	2.27×10^{-9}	$Mn(OH)_2$	4.5×10^{-14}
$Ca_3(PO_4)_2$	1×10^{-25}	$MnCO_3$	8.8×10^{-11}
$CaHPO_4$	5×10^{-6}	MnS	5.6×10^{-16}
CaF_2	3.9×10^{-11}	Mercury	
Chromium		$Hg_2O \cdot H_2O$	1.6×10^{-23}
$Cr(OH)_3$	6.7×10^{-31}	Hg_2Cl_2	1.1×10^{-18}
Cobalt		Hg_2Br_2	1.26×10^{-22}
$Co(OH)_2$	2×10^{-16}	Hg_2I_2	4.5×10^{-29}
$CoS(\alpha)$	5.9×10^{-21}	Hg_2CO_3	9×10^{-17}
$CoS(\beta)$	8.7×10^{-23}	Hg_2SO_4	6.2×10^{-7}
$CoCO_3$	1.0×10^{-12}	Hg_2S	1×10^{-45}
$Co(OH)_3$	2.5×10^{-43}	Hg_2CrO_4	2×10^{-9}
Copper		HgS	3×10^{-53}
CuCl	1.85×10^{-7}	Nickel	
CuBr	5.3×10^{-9}	$Ni(OH)_2$	1.6×10^{-14}
CuI	5.1×10^{-12}	$NiCO_3$	1.36×10^{-7}
CuCNS	4×10^{-14}	$NiS(\alpha)$	3×10^{-21}
Cu_2S	1.6×10^{-48}	$NiS(\beta)$	1×10^{-26}
$Cu(OH)_2$	5.6×10^{-20}	$NiS(\gamma)$	2×10^{-28}
CuS	8.7×10^{-36}	Potassium	
$CuCO_3$	1.37×10^{-10}	$KClO_4$	1.07×10^{-2}
		K_2PtCl_6	1.1×10^{-5}
		$KHC_4H_4O_6$	3×10^{-4}

Solubility Products (continued)

Substance	*K_{sp} at 25°*	*Substance*	*K_{sp} at 25°*
Silver		Thallium	
$\frac{1}{2}Ag_2O(Ag^+ + OH^-)$	2×10^{-8}	TlCl	1.9×10^{-4}
AgCl	1.8×10^{-10}	TlCNS	5.8×10^{-4}
AgBr	3.3×10^{-13}	Tl_2S	1.2×10^{-24}
AgI	1.5×10^{-16}	$Tl(OH)_3$	1.5×10^{-44}
AgCN	1.2×10^{-16}	Tin	
AgCNS	1.0×10^{-12}	$Sn(OH)_2$	5×10^{-26}
Ag_2S	1.0×10^{-51}	SnS	8×10^{-29}
Ag_2CO_3	8.2×10^{-12}	$Sn(OH)_4$	*ca.* 1×10^{-56}
Ag_2CrO_4	9×10^{-12}	Zinc	
$Ag_4Fe(CN)_6$	1.55×10^{-41}	$ZnCO_3$	6×10^{-11}
Ag_2SO_4	1.18×10^{-5}	$Zn(OH)_2$	4.5×10^{-17}
Ag_3PO_4	1.8×10^{-18}	ZnS	1.6×10^{-23}
Strontium			
$Sr(OH)_2 \cdot 8H_2O$	3.2×10^{-4}		
$SrCO_3$	9.42×10^{-10}		
$SrCrO_4$	3.6×10^{-5}		
$SrSO_4$	2.8×10^{-7}		
$SrC_2O_4 \cdot H_2O$	5.61×10^{-8}		

Appendix D

Table of Common Logarithms

No.	*0*	*1*	*2*	*3*	*4*	*5*	*6*	*7*	*8*	*9*
1.0	0000	0043	0086	0128	0170	0212	0253	0294	0334	0374
1.1	0414	0453	0492	0531	0569	0607	0645	0682	0719	0755
1.2	0792	0828	0864	0899	0934	0969	1004	1038	1072	1106
1.3	1139	1173	1206	1239	1271	1303	1335	1367	1399	1430
1.4	1461	1492	1523	1553	1584	1614	1644	1673	1703	1732
1.5	1761	1790	1818	1847	1875	1903	1931	1959	1987	2014
1.6	2041	2068	2095	2122	2148	2175	2201	2227	2253	2279
1.7	2304	2330	2355	2380	2405	2430	2455	2480	2504	2529
1.8	2553	2577	2601	2625	2648	2672	2695	2718	2742	2765
1.9	2788	2810	2833	2856	2878	2900	2923	2945	2967	2989
2.0	3010	3032	3054	3075	3096	3118	3139	3160	3181	3201
2.1	3222	3243	3263	3284	3304	3324	3345	3365	3385	3404
2.2	3424	3444	3464	3483	3502	3522	3541	3560	3579	3598
2.3	3617	3636	3655	3674	3692	3711	3729	3747	3766	3784
2.4	3802	3820	3838	3856	3874	3892	3909	3927	3945	3962
	0	*1*	*2*	*3*	*4*	*5*	*6*	*7*	*8*	*9*

Table of Common Logarithms (continued)

No.	*0*	*1*	*2*	*3*	*4*	*5*	*6*	*7*	*8*	*9*
2.5	3979	3997	4014	4031	4048	4065	4082	4099	4116	4133
2.6	4150	4166	4183	4200	4216	4232	4249	4265	4281	4298
2.7	4314	4330	4346	4362	4378	4393	4409	4425	4440	4456
2.8	4472	4487	4502	4518	4533	4548	4564	4579	4594	4609
2.9	4624	4639	4654	4669	4683	4698	4713	4728	4742	4757
3.0	4771	4786	4800	4814	4829	4843	4857	4871	4886	4900
3.1	4914	4928	4942	4955	4969	4983	4997	5011	5024	5038
3.2	5051	5065	5079	5092	5105	5119	5132	5145	5159	5172
3.3	5185	5198	5211	5224	5237	5250	5263	5276	5289	5302
3.4	5315	5328	5340	5353	5366	5378	5391	5403	5416	5428
3.5	5441	5453	5465	5478	5490	5502	5514	5527	5539	5551
3.6	5563	5575	5587	5599	5611	5623	5635	5647	5658	5670
3.7	5682	5694	5705	5717	5729	5740	5752	5763	5775	5786
3.8	5798	5809	5821	5832	5843	5855	5866	5877	5888	5899
3.9	5911	5922	5933	5944	5955	5966	5977	5988	5999	6010
4.0	6021	6031	6042	6053	6064	6075	6085	6096	6107	6117
4.1	6128	6138	6149	6160	6170	6180	6191	6201	6212	6222
4.2	6232	6243	6253	6263	6274	6284	6294	6304	6314	6325
4.3	6335	6345	6355	6365	6375	6386	6395	6405	6415	6425
4.4	6435	6444	6454	6464	6474	6484	6493	6503	6513	6522
4.5	6532	6542	6551	6561	6571	6580	6590	6599	6609	6618
4.6	6628	6637	6646	6656	6665	6675	6684	6693	6702	6712
4.7	6721	6730	6739	6749	6758	6767	6776	6785	6794	6803
4.8	6812	6821	6830	6839	6848	6857	6866	6875	6884	6893
4.9	6902	6911	6920	6928	6937	6946	6955	6964	6972	6981
5.0	6990	6998	7007	7016	7024	7033	7042	7050	7059	7067
5.1	7076	7084	7093	7101	7110	7118	7126	7135	7143	7152
5.2	7160	7168	7177	7185	7193	7202	7210	7218	7226	7235
5.3	7243	7251	7259	7267	7275	7284	7292	7300	7308	7316
5.4	7324	7332	7340	7348	7356	7364	7372	7380	7388	7396
5.5	7404	7412	7419	7427	7435	7443	7451	7459	7466	7474
5.6	7482	7490	7497	7505	7513	7520	7528	7536	7543	7551
5.7	7559	7566	7574	7582	7589	7597	7604	7612	7619	7627
5.8	7634	7642	7649	7657	7664	7672	7679	7686	7694	7701
5.9	7709	7716	7723	7731	7738	7745	7752	7760	7767	7774
6.0	7782	7789	7796	7803	7810	7818	7825	7832	7839	7846
6.1	7853	7860	7868	7875	7882	7889	7896	7903	7910	7917
6.2	7924	7931	7938	7945	7952	7959	7966	7973	7980	7987
6.3	7992	8000	8007	8014	8021	8028	8035	8041	8048	8055
6.4	8062	8069	8075	8082	8089	8096	8102	8109	8116	8122
	0	*1*	*2*	*3*	*4*	*5*	*6*	*7*	*8*	*9*

Table of Common Logarithms (continued)

No.	0	1	2	3	4	5	6	7	8	9
6.5	8129	8136	8142	8149	8156	8162	8169	8176	8182	8189
6.6	8195	8202	8209	8215	8222	8228	8235	8241	8248	8254
6.7	8261	8267	8274	8280	8287	8293	8299	8306	8312	8319
6.8	8325	8331	8338	8344	8351	8357	8363	8370	8376	8382
6.9	8388	8395	8401	8407	8414	8420	8426	8432	8439	8445
7.0	8451	8457	8463	8470	8476	8482	8488	8494	8500	8506
7.1	8513	8519	8525	8531	8537	8543	8549	8555	8561	8567
7.2	8573	8579	8585	8591	8597	8603	8609	8615	8621	8627
7.3	8633	8639	8645	8651	8657	8663	8669	8675	8681	8686
7.4	8692	8698	8704	8710	8716	8722	8727	8733	8739	8745
7.5	8751	8756	8762	8768	8774	8779	8785	8791	8797	8802
7.6	8808	8814	8820	8825	8831	8837	8842	8848	8854	8859
7.7	8865	8871	8876	8882	8887	8893	8899	8904	8910	8915
7.8	8921	8927	8932	8938	8943	8949	8954	8960	8965	8971
7.9	8976	8982	8987	8993	8998	9004	9009	9015	9020	9025
8.0	9031	9036	9042	9047	9053	9058	9063	9069	9074	9079
8.1	9085	9090	9096	9101	9106	9112	9117	9122	9128	9133
8.2	9138	9143	9149	9154	9159	9165	9170	9175	9180	9186
8.3	9191	9196	9201	9206	9212	9217	9222	9227	9232	9238
8.4	9243	9248	9253	9258	9263	9269	9274	9279	9284	9289
8.5	9294	9299	9304	9309	9315	9320	9325	9330	9335	9340
8.6	9345	9350	9355	9360	9365	9370	9375	9380	9385	9390
8.7	9395	9400	9405	9410	9415	9420	9425	9430	9435	9440
8.8	9445	9450	9455	9460	9465	9469	9474	9479	9484	9489
8.9	9494	9499	9504	9509	9513	9518	9523	9528	9533	9538
9.0	9542	9547	9552	9557	9562	9566	9571	9576	9581	9586
9.1	9590	9595	9600	9605	9609	9614	9619	9624	9628	9633
9.2	9638	9643	9647	9652	9657	9661	9666	9671	9675	9680
9.3	9685	9689	9694	9699	9703	9708	9713	9717	9722	9727
9.4	9731	9736	9741	9745	9750	9754	9759	9763	9768	9773
9.5	9777	9782	9786	9791	9795	9800	9805	9809	9814	9818
9.6	9823	9827	9832	9836	9841	9845	9850	9854	9859	9863
9.7	9868	9872	9877	9881	9886	9890	9894	9899	9903	9908
9.8	9912	9917	9921	9926	9930	9934	9939	9943	9948	9952
9.9	9956	9961	9965	9969	9974	9978	9983	9987	9991	9996
	0	1	2	3	4	5	6	7	8	9

Appendix E Essential Mathematical Background

1. Algebra

The level of algebra required for using this text is very basic. A very brief review of the essentials follows.

(a) The algebraic methods employed in this text are based upon the concept of the **algebraic equation,** a statement that the numerical value of what is on the left side of an equal sign is the same as the numerical value of what appears on the right. The most important rule we use to manipulate or solve algebraic equations is: *The validity of an algebraic equation remains intact in any manipulation in which the same operations are performed on both sides of the equation.*

For example, if we start with the equation

$$x^2 = 4.62$$

then we can also write

$$\sqrt{x^2} = \sqrt{4.62} \quad \text{(taking the square root of both sides)}$$

or $$\frac{1}{x^2} = \frac{1}{4.62} \quad \text{(inverting both sides)}$$

or $$(x^2)^{5/2} = (4.65)^{5/2} \quad \text{(raising both sides to the } \tfrac{5}{2} \text{ power)}$$

or $$\log(x^2) = \log(4.65) \quad \text{(taking the logarithm of both sides)}$$

or $$\frac{x^2 + 3.21}{11.1} = \frac{4.65 + 3.21}{11.1} \quad \text{(adding 3.21 to both sides and then dividing the result of both sides by 11.1)}$$

and so forth.

An example of how this can be applied follows:

$$(1.03)(y^2 + 14.2) = 62.6$$

$$y^2 + 1.42 = \frac{62.6}{1.03} \quad \text{(divide both sides by 1.03)}$$

$$y^2 = \frac{62.6}{1.03} - 14.2 \quad \text{(subtract 14.2 from both sides)}$$

$$= 61.4 - 14.2 = 47.2$$

$$y = \sqrt{47.2} = \pm 6.87 \quad \text{(taking the square root of both sides)}$$

(b) *Quadratic equations.* Very often in solving chemical problems, one encounters algebraic equations of the following form:

$$ax^2 + bx + c = 0$$

where a, b, and c are distinct constants and x is the algebraic variable. An equation of this general form is called a **quadratic equation.** The solutions to the quadratic equation are given by the formula

$$x = \frac{-b \pm \sqrt{b^2 - 4ac}}{2a}$$

An example of such an equation might arise in the solution of a particular equilibrium problem, where the equilibrium expression gives rise, say, to the following equation:

$$\frac{y^2}{0.031 - y} = 1.6 \times 10^{-2}$$

(where y might, for example, represent the value of $[H_3O^+]$ resulting from the ionization of a weak acid). Rearranging this equation (twice using the general rule stated above),

$$y^2 = (1.6 \times 10^{-2})(0.031 - y) = (1.6 \times 10^{-2})(0.031) - (1.6 \times 10^{-2})y$$

$$y^2 + (1.6 \times 10^{-2}y) - (0.031 \times 1.6 \times 10^{-2}) = 0$$

In this case, the algebraic variable is y, and the constants are:

$$a = 1, b = 1.6 \times 10^{-2}, c = -0.031 \times 1.6 \times 10^{-2} = -4.96 \times 10^{-4}$$

Then, according to the formula above,

$$y = \frac{-1.6 \times 10^{-2} \pm \sqrt{(1.6 \times 10^{-2})^2 - 4(1)(-4.96 \times 10^{-4})}}{2(1)}$$

$$= \frac{-1.6 \times 10^{-2} \pm \sqrt{(2.56 \times 10^{-4}) + (19.84 \times 10^{-4})}}{2}$$

$$= \frac{-1.6 \times 10^{-2} \pm \sqrt{22.4 \times 10^{-4}}}{2} = \frac{(-1.6 \times 10^{-2}) \pm (4.7 \times 10^{-2})}{2}$$

$$= \frac{3.1 \times 10^{-2}}{2} = 1.6 \times 10^{-2}$$ (The minus option was disregarded, as it would lead to the unreasonable result of a negative concentration.)

2. Exponents and Exponential Notation

In algebraic notation, the number of times a number is multiplied together is given by an exponent. For example,

$$a^5 = a \cdot a \cdot a \cdot a \cdot a$$

$$2^4 = 2 \cdot 2 \cdot 2 \cdot 2 = 16$$

$$10^6 = 10 \cdot 10 \cdot 10 \cdot 10 \cdot 10 \cdot 10 = 1{,}000{,}000$$

Negative exponents signify multiplying the *inverse* of the number the indicated number of times. For example,

$$a^{-5} = \frac{1}{a} \cdot \frac{1}{a} \cdot \frac{1}{a} \cdot \frac{1}{a} \cdot \frac{1}{a}$$

$$2^{-4} = \frac{1}{2} \cdot \frac{1}{2} \cdot \frac{1}{2} \cdot \frac{1}{2} = \frac{1}{16}$$

$$10^{-6} = \frac{1}{10} \cdot \frac{1}{10} \cdot \frac{1}{10} \cdot \frac{1}{10} \cdot \frac{1}{10} \cdot \frac{1}{10} = \frac{1}{1{,}000{,}000}$$

This exponential notation in powers of ten is very useful in chemistry in representing very large and very small numbers. For example, one often encounters very large equilibrium constants or very small equilibrium constants, like 1.6×10^{23} and 6.1×10^{-19}. Without the exponential notation, these numbers would look ridiculously long:

$$160000000000000000000000 \text{ and } 0.00000000000000000061$$

Furthermore, one would have incompatibilities with the concept of significant figures in many cases. The number that precedes the exponential factor (power of ten) in a number (e.g., 3.6 in 3.6×10^6) can be called the premultiplier. Often one tries to express numbers such that the premultiplier falls between 1 and 10.

There are very simple rules, which are easily verified, for working with numbers in exponential notation.

I. *In multiplication of numbers in exponential notation, multiply the premultiplier and add the exponents.*

For example,

$$(1.3 \times 10^6)(7.0 \times 10^2) = (1.3 \times 7.0) \times 10^{(6+2)} = 9.1 \times 10^8$$

$$(4.4 \times 10^6)(1.8 \times 10^{-4}) = (4.4 \times 1.8) \times 10^{(6-4)} = 7.9 \times 10^2$$

II. *In division of numbers in exponential notation, divide the premultiplier and subtract the exponents.*

For example,

$$\frac{6.2 \times 10^{11}}{4.8 \times 10^4} = \left(\frac{6.2}{4.8}\right) \times 10^{(11-4)} = 1.3 \times 10^7$$

$$\frac{1.9 \times 10^3}{4.6 \times 10^{-4}} = \left(\frac{1.9}{4.6}\right) \times 10^{(3-[-4])} = 0.41 \times 10^7$$

$$= 4.1 \times 10^6$$

In taking the last step, we have taken advantage of the following property of exponential numbers:

III. *If one multiplies the premultiplier by a specific power of ten (i.e., moves the decimal point a certain number of places to the right), then one compensates by reducing the exponential factor by the same number of powers of ten; the value of the number is thus unchanged.*

IV. Similarly, *if one divides the premultiplier by a specific power of ten (i.e., moves the decimal point a certain number of places to the left), then one compensates by increasing the exponential factor by the same number of powers of ten; the value of the resulting number is thus unchanged.*

For example,

$$4.3 \times 10^3 = (4.3 \times 10)\left(\frac{10^3}{10}\right) = 43 \times 10^2$$

$$= \left(\frac{4.3}{10}\right)(10^3 \times 10) = 0.43 \times 10^4$$

All five of these expressions have the same value.

V. *In addition or subtraction of numbers written in exponential notation, one first converts the numbers to the same power of ten, then adds or subtracts the premultiplier.*

For example,

$$6.3 \times 10^{-6} + 2.4 \times 10^{-5} - 0.1 \times 10^{-4}$$
$$= 0.63 \times 10^{-5} + 2.4 \times 10^{-5} - 1 \times 10^{-5} = 2 \times 10^{-5}$$

3. Logarithms

The **common logarithm** of a number n (referred to as $\log n$) is defined as the exponent (power) to which 10 must be raised to give the number n.

$$10^{\log n} = n$$

Clearly, the logarithm of 10 is 1. That is,

$$10^1 = 10$$

Similarly, the logarithm of 1 is 0; the log of 100 is 2; the log of 1000 is 3; and so on.

$$10^0 = 1 \qquad (\log 10^0 = 0)$$
$$10^2 = 100 \qquad (\log 10^2 = 2)$$
$$10^3 = 1000 \qquad (\log 10^3 = 3)$$

For other numbers that are not simple powers of ten, but range from 1 to 10, one can use a table of logarithms (e.g., Appendix D) or an electronic calculator. For example, from Appendix D we find that the logarithm of 2.61 is 0.4166 (we look for the 2.6 in the left column of the table, and then read the corresponding number in that row in the "1" column, placing the decimal point to the left column line). According to the above definition of logarithms, this means that

$$10^{0.4166} = 2.61$$

A few readily verifiable rules are commonly used to work with logarithms. These are:

I. *The logarithm of a product is the sum of the logarithms.*

$$\log (a \cdot b) = \log a + \log b$$

For example,

$$\log (2 \cdot 4) = \log 8 = \log 2 + \log 4$$
$$= 0.9031 = 0.3010 + 0.6021$$

II. *The logarithm of a ratio is the difference of the logarithms.*

$$\log \frac{a}{b} = \log a - \log b$$

For example,

$$\log \left(\frac{9}{3}\right) = \log 3 = \log 9 - \log 3$$
$$= 0.9542 - 0.4771 = 0.4771$$

These rules can be used to permit one to compute the logarithm of any positive number, even if it is smaller than 1 or larger than 10.

$$\log(3.9 \times 10^{-7}) = \log(3.9) + \log(10^{-7}) = 0.5911 - 7$$
(sometimes written 7.5911)
$$= -6.4089$$

$$\log(7.72 \times 10^{4}) = \log(7.72) + \log(10^{4}) = 0.8876 + 4$$
$$= 4.8876$$

Hence, to find the logarithm of any positive number, convert it to exponential notation with the premultiplier between 1 and 10, and derive the logarithm as shown above.

$$\begin{aligned}\log(0.0426) &= \log(4.26 \times 10^{-2})\\ &= \log 4.26 + \log(10^{-2})\\ &= 0.6294 - 2(= \bar{2}.6294) = -1.3706\end{aligned}$$

III. *An exponent in a number becomes a coefficient in its logarithm.*

$$\log(a^n) = n \log a$$

For example,

$$\begin{aligned}\log(2.14)^3 &= 3 \log 2.14 = (3)(0.3304)\\ &= 0.9912\end{aligned}$$

Often in chemical calculations one must determine, for a given number, what number it is the logarithm of. That is, if we have a number m, we want to find the number p such that

$$\log p = m$$

The number p is called the **antilogarithm** of m.

$$\text{antilog } m = p$$

For numbers (logarithms) between 0 and 1, it is easy to find the antilog by simply looking in a table of logarithms. For example, if one wishes to find the antilog of 0.9460, using Appendix D, one finds that the logarithm of 8.83 is 0.9460. Hence,

$$8.83 = \text{antilog } 0.9460$$

$$0.9460 = \log(8.83)$$

If the number for which one wishes to find the antilog does not fall between 0 and 1, then it should be expressed as the sum of two numbers, one of which falls between 0 and 1 and the other an integer. Then, using the rules stated above, one can find the antilog.

$$\begin{aligned}\text{antilog}(18.5038) &= \text{antilog}(18 + 0.5038)\\ &= (\text{antilog } 18)(\text{antilog } 0.5038)\\ &= (10^{18})(3.19) = 3.19 \times 10^{18}\end{aligned}$$

As a check, we note

$$\log(3.18 \times 10^{18}) = \log 3.18 + \log 10^{18}$$
$$= 0.5038 + 18 = 18.5038$$

Similarly,

$$\text{antilog}(-11.0645) = \text{antilog}(-12 + 0.9355)$$
$$= [\text{antilog}(-12)][\text{antilog } 0.9355]$$
$$= 8.62 \times 10^{-12}$$

Another type of logarithm which occurs in chemistry, especially in thermodynamics and electrochemistry, is the **natural logarithm.** The natural logarithm of a number l (denoted $\ln l$) is defined in terms of the *base* of natural logarithms, $e = 2.71828 \ldots$.

$$e^{\ln l} = l = (2.71828\ldots)^{\ln l}$$

Of course,

$$e^0 = 1 \qquad \text{so } \ln 1 = 0$$
$$e^1 = e \qquad \text{so } \ln 2.71828\ldots = 1$$

Tables of natural logarithms and electronic calculators can be used in a manner analogous to what was described above for common logarithms. However, it is often more convenient to convert a problem expressed in terms of natural logarithms into a problem expressed in terms of common logarithms. This can be done by using the following relationship:

$$\ln a = 2.303 \log a$$

For example,

$$\ln 6.62 = 2.303 \log 6.62 = (2.303)(0.8209) = 1.890$$

INDEX

Page numbers followed by *n.* refer to footnotes.
Page numbers followed by *t.* refer to tables.
Page numbers in italics refer to figures.

1 2 3 4 5 6 7 8 9 0

Physical Constants

Avogadro's number	N_A	6.02209×10^{23} mol^{-1}
Planck's constant	h	6.6252×10^{-27} erg sec
		$9.537 \times 10^{-14} \left(\frac{\text{kcal}}{\text{mol}}\right)$ sec
Speed of light (vacuum)	c	$2.998 \times 10^{10} \frac{\text{cm}}{\text{sec}}$
Faraday constant	F	$96{,}487 \frac{\text{coulombs}}{\text{mol of electrons}}$
		$23{,}061 \frac{\text{calories per volt}}{\text{mol of electrons}}$
Standard atmosphere	atm	760 mm of Hg = 760 torr
		$1.013 \times 10^{6} \frac{\text{dyne}}{\text{cm}^2}$
Elementary charge	e	1.602×10^{-19} coulomb
		4.803×10^{-10} abs esu
Bohr radius	a_o	0.5292×10^{-8} cm
		0.5292 Å
Gas constant	R	$1.987 \frac{\text{cal}}{\text{mol °K}}$
		$8.205 \times 10^{-2} \frac{\text{liter atm}}{\text{mol °K}}$
		$8.314 \frac{\text{joule}}{\text{mol °K}}$
Electron rest mass	m_e	9.1091×10^{-28} g
		5.4876×10^{-4} amu
Proton rest mass	m_p	1.6734×10^{-24} g
		1.00797 amu
Neutron rest mass	m_n	1.6746×10^{-24} g
		1.00867 amu

Conversion Factors

	To convert → *From*	*To*	*Multiply By* ↓
2.206×10^{-3}	lb	g	453.29
1.6606×10^{-24}	mol (gram molecular wt)	molecule	6.0221×10^{23}
1.32×10^{-3}	atm	torr	760
4.184×10^{3}	joule	kcal	2.390×10^{-4}
4.184×10^{10}	erg	kcal	2.390×10^{-11}
6.947×10^{-14}	$\frac{\text{erg}}{\text{molecule}}$	$\frac{\text{kcal}}{\text{mol}}$	1.439×10^{13}
3.498×10^{2}	$\frac{\text{cm}^{-1}}{\text{molecule}}$	$\frac{\text{kcal}}{\text{mol}}$	2.859×10^{-3}
3.338×10^{-10}	coulomb	abs esu (statcoulomb)	2.996×10^{9}
Multiply by ↑	← *To convert* *To*	*From*	